GENETICS

ANALYSIS & PRINCIPLES

second edition

Test Yourself

Take a quiz at the Genetics Online Learning Center to gauge your mastery of chapter content. Each chapter quiz is specifically constructed to test your comprehension of key concepts. Immediate feedback on your responses explains why an answer is correct or incorrect. You can even e-mail your quiz results to your professor!

Tweaking the Experiment

For the in-depth experiments that are found in each chapter, a self-help quiz is found. These questions ask you to predict how changes in the experimental parameters may affect the outcome of the experiment.

Multiple Choice Quiz
(See related pages)

Results Reporter

Out of 25 questions, you answered 3 correctly, for a final grade of 12%.

3 correct (12%)
4 incorrect (16%)
18 unanswered (72%)

Your Results:

The correct answer for each question is indicated by a ☺.

1 CORRECT The nucleotide change _____ is an example of a transversion, while the nucleotide change _____ is an example of a transition.

- A) A→G; C→G
- ☺ B) C→G; A→G
- C) T→C; A→G
- D) C→T; G→A
- E) G→C; C→G

2 CORRECT Which type of mutation is most likely to revert?

- A) deletion
- B) translocation
- C) inversion
- ☺ D) transposition
- E) transition

3 CORRECT Which type of mutation is least likely to revert?

- ☺ A) deletion
- B) translocation
- C) inversion
- D) transposition
- E) transition

4 INCORRECT The hydrolysis of an -NH2 group from a base is called _____; causing _____ by
- A) deamination; transversions

Routing Information

Date: Tue Sep 09 09:13:24 CDT 2003
My name: [   ]

Email these results to:

	Email address:	Format:
Me:	[ ]	Text
My Instructor:	[ ]	Text
My TA:	[ ]	Text
Other:	[ ]	Text

[E-Mail The Results]

Interactive Activities

Fun and exciting learning experiences await you at the Genetics Online Learning Center! Each chapter offers a series of interactive crossword puzzles, vocabulary flashcards, and other engaging activities designed to reinforce learning.

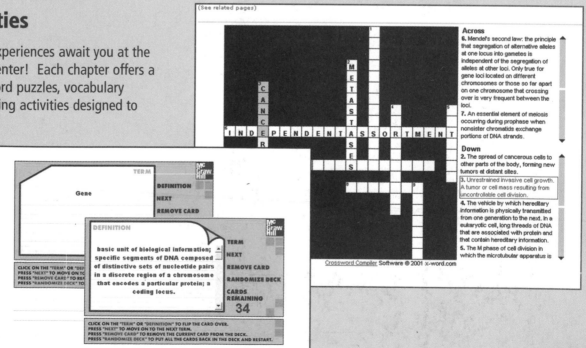

Gene

DEFINITION

basic unit of biological information; specific segments of DNA composed of distinctive sets of nucleotide pairs in a discrete region of a chromosome that encodes a particular protein; a coding locus.

TERM
NEXT
REMOVE CARD
RANDOMIZE DECK
CARDS REMAINING
34

CLICK ON THE "TERM" OR "DEFINITION" TO FLIP THE CARD OVER.
PRESS "NEXT" TO MOVE ON TO THE NEXT TERM.
PRESS "REMOVE CARD" TO REMOVE THE CURRENT CARD FROM THE DECK.
PRESS "RANDOMIZE DECK" TO PUT ALL THE CARDS BACK IN THE DECK AND RESTART.

(See related pages)

Across

6. Mendel's second law: the principle that segregation of alternative alleles at one locus into gametes is independent of the segregation of alleles at other loci. Only true for gene loci located on different chromosomes or those so far apart on one chromosome that crossing over is very frequent between the loci.

7. An essential element of meiosis occurring during prophase when nonsister chromatids exchange portions of DNA strands.

Down

2. The spread of cancerous cells to other parts of the body, forming new tumors at distant sites.

3. Unrestrained invasive cell growth. A tumor or cell mass resulting from uncontrollable cell division.

4. The vehicle by which hereditary information is physically transmitted from one generation to the next. In a eukaryotic cell, long threads of DNA that are associated with protein and that contain hereditary information.

5. The M phase of cell division in which the microtubular apparatus is

Crossword Compiler Software © 2001 x-word.com

IMPORTANT:

HERE IS YOUR REGISTRATION CODE TO ACCESS
YOUR PREMIUM McGRAW-HILL ONLINE RESOURCES.

For key premium online resources you need THIS CODE to gain access. Once the code is entered, you will be able to use the Web resources for the length of your course.

If your course is using **WebCT** or **Blackboard**, you'll be able to use this code to access the McGraw Hill content within your instructor's online course.

Access is provided if you have purchased a new book. If the registration code is missing from this book, the registration screen on our Website, and within your WebCT or Blackboard course, will tell you how to obtain your new code.

Registering for McGraw-Hill Online Resources

TO gain access to your McGraw-Hill web resources simply follow the steps below:

1. USE YOUR WEB BROWSER TO GO TO: **www.mhhe.com/brooker**

2. CLICK ON **FIRST TIME USER**.

3. ENTER THE REGISTRATION CODE* PRINTED ON THE TEAR-OFF BOOKMARK ON THE RIGHT.

4. AFTER YOU HAVE ENTERED YOUR REGISTRATION CODE, CLICK **REGISTER**.

5. FOLLOW THE INSTRUCTIONS TO SET-UP YOUR PERSONAL UserID AND PASSWORD.

6. WRITE YOUR UserID AND PASSWORD DOWN FOR FUTURE REFERENCE.
   KEEP IT IN A SAFE PLACE.

TO GAIN ACCESS to the McGraw-Hill content in your instructor's **WebCT** or **Blackboard** course simply log in to the course with the UserID and Password provided by your instructor. Enter the registration code exactly as it appears in the box to the right when prompted by the system. You will only need to use the code the first time you click on McGraw-Hill content.

Thank you, and welcome to your McGraw-Hill online Resources!

* YOUR REGISTRATION CODE CAN BE USED ONLY ONCE TO ESTABLISH ACCESS. IT IS NOT TRANSFERABLE.

0-07-295589-9 T/A BROOKER: GENETICS, 2/E

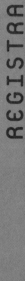

REGISTRATION CODE

MCGRAW-HILL
ONLINE RESOURCES

G33X-7FOY-AJG5-MYVO-U1GS

:: ROBERT J. BROOKER

UNIVERSITY OF MINNESOTA, TWIN CITIES

second edition

GENETICS

ANALYSIS & PRINCIPLES

Boston   Burr Ridge, IL   Dubuque, IA   Madison, WI   New York   San Francisco   St. Louis
Bangkok   Bogotá   Caracas   Kuala Lumpur   Lisbon   London   Madrid   Mexico City
Milan   Montreal   New Delhi   Santiago   Seoul   Singapore   Sydney   Taipei   Toronto

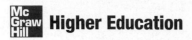

Higher Education

GENETICS: ANALYSIS AND PRINCIPLES, SECOND EDITION

Published by McGraw-Hill, a business unit of The McGraw-Hill Companies, Inc., 1221 Avenue of the Americas, New York, NY 10020. Copyright © 2005 by The McGraw-Hill Companies, Inc. All rights reserved. Previous edition copyright © 1999 by Benjamin/Cummings, an imprint of Addison Wesley Longman, Inc. All rights reserved. No part of this publication may be reproduced or distributed in any form or by any means, or stored in a database or retrieval system, without the prior written consent of The McGraw-Hill Companies, Inc., including, but not limited to, in any network or other electronic storage or transmission, or broadcast for distance learning.

Some ancillaries, including electronic and print components, may not be available to customers outside the United States.

This book is printed on acid-free paper.

International 1 2 3 4 5 6 7 8 9 0 QPV/QPV 0 9 8 7 6 5 4 3
Domestic   1 2 3 4 5 6 7 8 9 0 QPV/QPV 0 9 8 7 6 5 4 3

ISBN 0–07–283512–5
ISBN 0–07–111098–4 (ISE)

Publisher: *Martin J. Lange*
Senior sponsoring editor: *Patrick E. Reidy*
Director of development: *Kristine Tibbetts*
Marketing manager: *Tami Petsche*
Lead project manager: *Peggy J. Selle*
Senior production supervisor: *Laura Fuller*
Senior media project manager: *Jodi K. Banowetz*
Media technology producer: *Renee Russian*
Design manager: *K. Wayne Harms*
Cover/interior designer: *Kaye Farmer*
Cover images: *David Phillips/Photo Researchers Inc.*
Senior photo research coordinator: *John C. Leland*
Photo research: *Chris Hammond/PhotoFind, LLC*
Supplement producer: *Brenda A. Ernzen*
Compositor: *Lachina Publishing Services*
Typeface: *10/12 Minion*
Printer: *Quebecor World Versailles Inc.*

The credits section for this book begins on page 823 and is considered an extension of the copyright page.

Library of Congress Cataloging-in-Publication Data

Brooker, Robert J.
    Genetics : analysis and principles / Robert J. Brooker.—2nd ed.
      p.   cm.
    Includes index.
    ISBN 0–07–283512–5—ISBN 0–07–111098–4
    1. Genetics. I. Title.

QH430.B766   2005
576.5—dc22   2003066612
                                                    CIP

INTERNATIONAL EDITION ISBN 0–07–111098–4
Copyright © 2005. Exclusive rights by The McGraw-Hill Companies, Inc., for manufacture and export. This book cannot be re-exported from the country to which it is sold by McGraw-Hill. The International Edition is not available in North America.

www.mhhe.com

ABOUT THE AUTHOR

Robert J. Brooker (Ph.D., Yale University) is a Professor in the Department of Genetics, Cell Biology, and Development at the University of Minnesota, Twin Cities. He received his B.A. in Biology at Wittenberg University in 1978. At Harvard, he studied the lactose permease, the product of the *lacY* gene of the *lac* operon. He continues this work at the University of Minnesota, where he has an active research lab composed of undergraduates, Ph.D. students, post-doctoral fellows, and laboratory technicians (see photo). Dr. Brooker's laboratory also investigates the structure, function, and regulation of iron transporters found in bacteria, *C. elegans,* and mammals.

At the University of Minnesota, Dr. Brooker teaches undergraduate courses in biology, genetics, and cell biology. He also teaches students at the graduate level and greatly enjoys the more informal method of discussing research papers in class.

Rob Brooker and the Minneapolis skyline as seen from the University of Minnesota, Minneapolis campus.

Brooker lab: (1st row) Rob Brooker, Liz Matzke; (2nd row) Heather Haemig, Brian Wiczer, Peter Franco, Kristopher Carver, and Jerry Johnson.

DEDICATION

To D.S.B.

BRIEF CONTENTS

::

TABLE OF CONTENTS

::

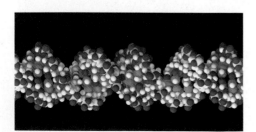

PART V

GENETIC TECHNOLOGIES   489

PREFACE

::

My desire to write this genetics textbook actually stemmed from a cell biology course that I introduced at the University of Minnesota many years ago. Our department decided to offer an intermediate-level cell biology course to bridge the gap between our basic undergraduate cell biology course and a more advanced graduate course, and I was asked to create it. I developed and taught this course for many years. It was a particularly fun experience, but I must admit that it got off to a rocky start. Let me explain.

The goal was to develop an intermediate cell biology course that used an undergraduate level of explanation but incorporated experiments from actual research papers. For each class session, the students were expected to read one or two research papers and understand them enough to discuss them in class. As a young, naive yet enthusiastic assistant professor, I thought this was a good plan. During the first few weeks of this course, however, I would walk into the classroom each day and say to the class, "So, tell me about the first experiment in the paper you were supposed to read for today." I was hoping that lots of volunteers would go to the blackboard and explain the experiment. But instead, I received the "deer in the headlights" response. Obviously something was wrong.

I knew that the problem wasn't the students in the course, because many of our brightest and hardest working students were enrolled. Hoping that I wasn't the problem, I sought another cause. Over time, the students and I uncovered the problem. Their previous coursework had not trained them to analyze experiments according to the scientific method. Sure, they had learned many "facts" and bits and pieces of experiments, but they hadn't been exposed, in a systematic and rigorous way, to the scientific process. We needed to fix that.

To transform the intermediate cell biology course into a success, the students and I had to develop a systematic way to dissect the parts of each research experiment. Looking back on it, the solution seems obvious: we needed to follow the scientific method. To integrate this process into our class discussions, I would ask the class the same series of questions for each experiment:

1. **Why was this experiment done?** The students' answers would have two parts. First, they would need to consider the background work that led to the experiment. And second, the students would have to explicitly state the hypothesis that the researchers were trying to test.
2. **How was the experiment done?** As a class, we would generate a flow diagram that described the experimental methods.
3. **What are the results?** We would look at the raw data and quantitatively analyze it.
4. **What do the results mean?** We would discuss our interpretations of the data. The goal of such interpretations was to derive conceptual ideas that underlie the experimental data. Sometimes, our interpretations were quite different from the researchers who had actually done the experiments. The occasional conflict between our interpretations and those of the researchers provided the most meaningful moments in the classroom. It caused many students to realize that the scientific concepts and principles they had learned about in previous courses are the product of scientific interpretations of research data. And maybe, they aren't always right!

I have to say that the intermediate cell biology course has been my most inspiring teaching experience. What I mean is that the students were inspiring to me. (I hope that I was inspiring to them as well.) On the positive side, it was impressive to see how the students' abilities to analyze experiments improved as the course progressed. The course turned out to be really fun for the students because it was the first time that many of them had taken a college course that allowed them to unleash their curiosity and use their critical thinking skills. It was also fun for me because I could witness this happening. On the negative side, however, it left me with a deeply disappointing sense that most college-level science courses do not provide students with an adequate exposure to the scientific process. To some extent, I felt that textbooks were to blame. In an effort to include too much information, science textbooks had largely deleted the essence of science, namely, the scientific method. At that point, I decided to write a textbook that would incorporate the scientific method into each chapter. Since my primary background is in genetics rather than cell biology, it was logical for me to write a genetics textbook, although I strongly feel that all scientific disciplines would benefit from this approach.

As you will see when you use this textbook, each chapter (beginning with chapter 2) incorporates one or two experiments that are presented according to the scientific method. These experiments are not "boxed off" from the rest of the chapter for you to read in your spare time. Rather, they are integrated within the chapters and flow with the rest of the text. As you are reading the experiments, you will simultaneously explore the scientific method and the genetic principles that have been learned from this approach. For students, I hope this textbook will help you to see the fundamental connection between scientific analysis and principles. For both students and instructors, I expect that this strategy will make genetics much more fun to explore.

HOW WE EVALUATED YOUR NEEDS

ORGANIZATION

In surveying many genetics instructors, it became apparent that most people fall into two camps: **Mendel first** versus **Molecular first.** I have taught genetics both ways on many occasions. As a teaching tool, this textbook has been written with these different teaching strategies in mind. The organization and content lend themselves to such different formats of teaching.

Chapters 2 through 8 are largely inheritance chapters, while chapters 24 to 26 examine quantitative and population genetics. The bulk of the molecular genetics is found in chapters 9 through 23, although I have tried to weave a fair amount of molecular genetics into chapters 2 through 8 as well. The information in chapters 9 to 23 *does not assume* that a student has already covered chapters 2 through 8. Actually, each chapter is written with the perspective that instructors may want to vary the order of their chapters to fit their students' needs.

For those who like to discuss inheritance patterns first, a common strategy would be to cover chapters 1 to 8 first, and then possibly 24 to 26. (However, many instructors like to cover quantitative and population genetics at the end. Either way works fine.) The more molecular and technical aspects of genetics would then be covered in chapters 9 through 23. Alternatively, if you like the "Molecular first" approach, you would probably cover chapter 1, then skip to chapters 9 to 23, then return to chapters 2 to 8, and then cover chapters 24 to 26 at the end of the course. This textbook was written in such a way that either strategy works just fine.

ACCURACY

Both the publisher and I acknowledge the fact that inaccuracies can be a source of frustration for both the instructor and students. Therefore, throughout the writing and production of this textbook we have worked diligently to catch and correct errors during each phase of development and production.

A team of seven instructors worked as an accuracy consultant panel and checked every chapter, illustration, and problem in the textbook. I also had a team of students work through all the problems sets.

The page proofs were double-proofread against the manuscript to ensure the correction of any errors introduced when the manuscript was typeset. The textual examples, practice problems and solutions, end-of-chapter questions and problems, and problem answers were accuracy checked by reviewers and students again at page proof stage after the manuscript was typeset. This last round of corrections was then cross-checked against the solution manuals.

PEDAGOGY

Based on our discussions with instructors from many institutions, some common goals have emerged. Instructors want a broad textbook that clearly explains concepts in a way that is interesting, accurate, concise, and up-to-date. Likewise, most instructors want students to understand the experimentation that revealed these genetic concepts. In this textbook, concepts and experimentation are woven together to provide a story that enables students to learn the important genetic concepts that they will need in their future careers, and also to be able to explain the types of experiments that allowed researchers to derive such concepts. The end-of-chapter problem sets are categorized according to their main focus, either conceptual or experimental, although some problems contain a little of both. The problems are meant to strengthen students' abilities in a wide variety of ways.

- By bolstering their understanding of genetic principles.
- By enabling students to apply genetic concepts to new situations.
- By analyzing scientific data.
- By organizing their thoughts regarding a genetic topic.
- By improving their writing skills.

Finally, since genetics is such a broad discipline ranging from the molecular to the populational levels, many instructors have told us that it is a challenge for students to see both "the forest and the trees." It is commonly mentioned that students often have trouble connecting the concepts they have learned in molecular genetics with the traits that occur at the level of a whole organism. For example, What does transcription have to do with blue eyes? To try to make this connection more meaningful, certain figure legends in each chapter, designated **Genes → Traits,** remind students that molecular and cellular phenomena ultimately lead to the traits that are observed in each species (see fig. 4.13).

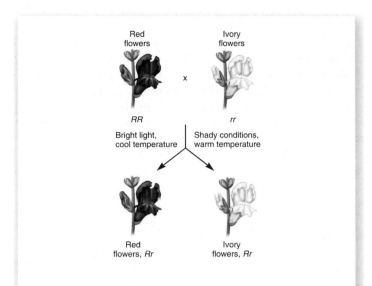

FIGURE 4.13   Variation in the expression of a trait due to environmental effects.

GENES→TRAITS   In this example, snapdragon flower color in the heterozygote, *Rr*, is affected by the amount of light and temperature in the plant's environment. Presumably, the protein encoded by this gene is sensitive to changes in temperature and light intensity.

ILLUSTRATIONS

In surveying students whom I teach, I often hear it said that most of their learning comes from studying the figures. Likewise, instructors frequently use the illustrations from a textbook as a central teaching tool. For these reasons, the greatest amount of effort in improving the second edition has gone into the illustrations. Every illustration of the second edition has been redrawn with four goals in mind:

1. **Completeness** For most figures, it should be possible to understand an experiment or genetic concept by looking at the illustration alone. Students have complained that it is difficult to understand the content of an illustration if they have to keep switching back and forth between the figure and text. In cases where an illustration shows the steps in a scientific process, the steps are described in brief statements that allow the students to understand the whole process (e.g., see fig. 12.9). Likewise, such illustrations should make it easier for instructors to explain these processes in the classroom.

2. **Clarity** The figures have been extensively reviewed by students and instructors. This has helped us to avoid drawing things that may be confusing or unclear. I hope that no one looks at an element in any figure and wonders, "What is that thing?" Aside from being unmistakably drawn, all new elements within each figure are clearly labeled.

3. **Consistency** Before we began to draw the figures for the second edition, we generated a style sheet that contained recurring elements that are found in many places in the

textbook. Examples include the DNA double helix, DNA polymerase, and fruit flies. We agreed upon the best way(s) to draw these elements and also what colors they should be. Therefore, as students and instructors progress through this textbook, they become accustomed to the way things should look.

4. **Realism** An important, albeit tiring, goal of this second edition was to make each figure as realistic as possible. When drawing macroscopic elements (e.g., fruit flies, pea plants), the illustrations were based on real images, not on cartoon-like simplifications. Our most challenging goal, and one that we feel has been the most successful in this second edition, was the realism of our molecular drawings. Whenever possible, we tried to draw molecular elements according to their actual structures, if such structures are known. For example, the ways we have drawn RNA polymerase, DNA polymerase, DNA helicase, and ribosomes are based on their crystal structures. When a student sees a figure in this textbook that illustrates an event in transcription, RNA polymerase is depicted in a way that is as realistic as possible (e.g., see fig. 12.9).

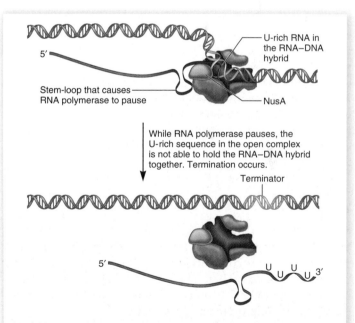

FIGURE 12.9   *ρ*-independent or intrinsic termination. When RNA polymerase reaches the end of the gene, it transcribes a uracil-rich sequence. Soon after this uracil-rich sequence is transcribed, a stem-loop forms just upstream from the open complex. The formation of this stem-loop causes RNA polymerase to pause in its synthesis of the transcript. This pausing is stabilized by NusA, which binds near the region where RNA exits the open complex. While it is pausing, the RNA-DNA hybrid region is a uracil-rich sequence. Since UA hydrogen bonds are relatively weak interactions, the transcript and RNA polymerase dissociate from the DNA.

WRITING STYLE

Motivation in learning often stems from enjoyment. If someone enjoys what they're reading, they are more likely to spend longer amounts of time with it and focus their attention more crisply. The writing style of this book is meant to be interesting, down to earth, and easy to follow. Each section of every chapter begins with an overview of the contents of that section, usually with a table or figure that summarizes the broad points. The section then unfolds as a story that examines how those broad points were discovered experimentally, as well as explaining many of the finer scientific details. Important terms are introduced in a bold-face font. These terms are also found in the glossary.

There are various ways to make a genetics book interesting and inspiring. The subject matter itself is pretty amazing, so it's not difficult to build on that. In addition to describing the concepts and experiments in ways that motivate students, it is important to draw upon examples that bring the concepts to life. In a genetics book, many of these examples come from the medical realm. This textbook contains lots of examples of human diseases that exemplify some of the underlying principles of genetics. Students often say that they remember certain genetic concepts because they remember how defects in certain genes can cause disease. For example, defects in DNA repair genes cause a higher predisposition to develop cancer. In addition, I have tried to be evenhanded in providing examples from the microbial and plant world. Finally, students are often interested in applications of genetics that impact their everyday lives. Since we hear about genetics in the news on a daily basis, it's inspiring for students to learn the underlying basis for such technologies. Chapters 18 to 21 are devoted to genetic technologies, and applications of other technologies are found throughout this textbook. By the end of their genetics course, students should come away with a greater appreciation for the impact of genetics in their lives.

WHAT'S NEW IN THE SECOND EDITION?

Since genetics is a rapidly changing discipline, a major goal of the second edition was to update the entire textbook. In some cases, this involved minor content changes, while in other cases entire sections were added. These major changes are described under the following heading, "Content Changes." In addition, the following general improvements were made:

- **Illustrations** As mentioned earlier in the Preface, all of the illustrations in the second edition were newly drawn using state-of-the-art technology. These illustrations were drawn with an eye toward completeness, clarity, consistency, and realism.
- **Problem sets** A large number of new and challenging problems were added to the problem sets of every chapter.
- **Animations** The student and instructor CD-ROM contains over 40 animations.
- **Website (Online Learning Center)** As described later in this Preface, the website for the instructor and students contains a much wider variety of resources and activities.

SUGGESTIONS WELCOME!

It seems very appropriate to use the word *evolution* to describe the continued development of this textbook. I welcome any and all comments. The refinement of any science textbook requires input from instructors and their students. These include comments regarding writing, illustrations, supplements, factual content, and topics that may need greater or less emphasis. You are invited to contact me at:

Rob Brooker
Dept. of Genetics, Cell Biology, and Development
University of Minnesota
6-160 Jackson Hall
321 Church St.
Minneapolis, MN 55455
brook005@umn.edu

TEACHING AND LEARNING SUPPLEMENTS

McGraw-Hill offers various tools and technology products to support the second edition of *Genetics: Analysis and Principles*. Students can order supplemental study materials via their bookstore. Instructors can obtain teaching supplements by calling the Customer Service Department at 800-338-3987, by visiting our textbook website at www.mhhe.com/brooker, or by contacting their local McGraw-Hill sales representative.

FOR THE INSTRUCTOR:

Digital Content Manager CD-ROM

This multimedia collection of visual resources allows instructors to use artwork from the textbook in multiple formats to create customized classroom presentations, visually based tests and quizzes, dynamic course website content, or attractive printed support material. The digital assets on this cross-platform CD-ROM are grouped by chapter within the following easy-to-use folders:

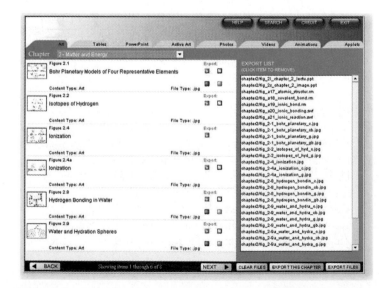

MAJOR CONTENT CHANGES

- **Chapter 4 (Extensions of Mendelian Inheritance)** A greater emphasis is placed on the molecular mechanisms that underlie the various patterns of inheritance.
- **Chapter 6 (Genetic Transfer and Mapping in Bacteria and Bacteriophages)** The process of bacterial conjugation is updated and more realistically illustrated.
- **Chapter 7 (Non-Mendelian Inheritance)** The molecular mechanisms that underlie X inactivation and genomic imprinting are substantially updated.
- **Chapter 10 (Chromosome Organization and Molecular Structure)** The structure of eukaryotic chromosomes is greatly updated including descriptions of the nuclear matrix, chromatin remodeling proteins, and two multiprotein complexes called condensin and cohesin, which play a critical role in chromosomal condensation and sister chromatid alignment.
- **Chapter 11 (DNA Replication)** Descriptions are presented of newly discovered DNA polymerases, such as lesion-replicating polymerases, along with updates regarding the molecular mechanisms of the DNA polymerases that play a major role in DNA replication.
- **Chapter 12 (Gene Transcription and RNA Modification)** This chapter contains updated information regarding the structure of RNA polymerase, the functions of general transcription factors, and the role of chromatin remodeling during transcription.
- **Chapter 15 (Gene Regulation in Eukaryotes)** This area has changed considerably since the first edition. New information regarding the roles of TFIID and mediator, which communicate the effects of regulatory transcription factors, was added, along with the importance of chromatin remodeling for transcriptional activation and silencing. A new subsection on RNA interference also was added.
- **Chapter 16 (Gene Mutation and DNA Repair)** A new subsection was added that explains some of the chemical changes that underlie spontaneous mutations. Also, a subsection was added that concerns the roles of translesion DNA polymerases in the replication of damaged DNA.
- **Chapter 19 (Biotechnology)** The topic of stem cells has received considerable attention in the news lately. In a new subsection, which contains three figures and one table, the genetic issues that underlie the proliferative capacities of stem cells are discussed, along with their potential applications and societal concerns.
- **Chapter 20 (Structural Genomics)** Since the first edition, the area of genomics has expanded to the point where it needs to be divided into two chapters. Chapter 20 contains the updated methods that are used to dissect genomes and sequence them.
- **Chapter 21 (Functional Genomics, Proteomics, and Bioinformatics)** The first two sections of this chapter are new to the second edition. Functional genomics includes updates to topics such as EST libraries, subtractive libraries, and the applications of DNA microarrays. The proteomics section explains the techniques of two-dimensional gel electrophoresis, mass spectrometry, and protein microarrays. The bioinformatics section is supplemented with tools at our Online Learning Center, where students can actually run programs to analyze sequence files. These include programs that can translate a DNA sequence, generate a multiple sequence alignment, or search databases for homologous sequences (i.e., a BLAST search).
- **Chapter 23 (Developmental Genetics)** An entire new section regarding sex determination was added. This section describes the genetic pathways that govern sex determination in *Drosophila, C. elegans,* mammals, and dioecious plants.
- **Chapter 26 (Evolutionary Genetics)** The UPGMA method for constructing phylogenetic trees is described. In addition, a new subsection regarding the role of horizontal gene transfer in the evolution of species was added.

Art Library   Full-color digital files of all illustrations in the textbook, plus the same art saved in gray-scale versions, can be readily incorporated into lecture presentations, exams, or custom-made classroom materials.

TextEdit Art Library   Every art piece is placed into a PowerPoint presentation that allows the user to revise, move, or delete labels as desired for creation of customized presentations and/or for testing purposes.

Active Art Library   Active Art consists of art files that have been converted to a format that allows the artwork to be edited inside PowerPoint. Each piece can be broken down to its core elements, grouped or ungrouped, and edited to create customized illustrations.

Animations Library   Full-color presentations involving key figures in the textbook have been brought to life via animation. Because these animations were created using the art from the textbook, the characters in the animation match those found in the textbook.

Photo Library   Like the Art Library, digital files of all photographs from the textbook are available.

Table Library   Every table that appears in the textbook is provided in electronic form.

PowerPoint Lecture Outlines   Based on the information in the Instructor's Manual (described next), it is possible to create ready-made presentations that combine art and lecture notes for each of the 26 chapters of the textbook. Written by Johnny El-Rady, University of South Florida, these lectures can be used as they are or customized to reflect your preferred lecture topics and sequences.

Instructor's Testing and Resource CD-ROM This cross-platform CD-ROM contains the Instructor's Manual, Test Item File, and Solutions Manual/Study Guide, all available in both Word and PDF formats. The manual follows the order of sections and subsections in the textbook and summarizes the main points in the text, figures, and tables. The Instructor's Manual also includes links to relevant websites, answers to the problem sets at the end of each chapter, and a Test Bank of additional questions that can be used for homework assignments and/or the preparation of exams. These additional questions are found in a computerized test bank using Brownstone Diploma testing software to quickly create customized exams. This user-friendly program allows instructors to search questions by topic, format, or difficulty level; edit existing questions or add new ones; and scramble questions and answer keys for multiple versions of the same test.

Transparencies A set of 600 transparency overheads includes nearly every piece of line art and all the tables from the textbook. The images are printed with better visibility and contrast than ever before, and labels are large and bold for clear projection.

eInstruction This Classroom Performance System (CPS) brings interactivity into the classroom/lecture hall. It is a wireless response system that gives the instructor and students immediate feedback from the entire class. The wireless response pads are essentially remotes that are easy to use and engage students. CPS allows you to motivate student preparation, interactivity, and active learning so you can receive immediate feedback and know what students understand.

Course Delivery Systems With help from our partners, WebCT, Blackboard, TopClass, eCollege, and other course management systems, instructors can take complete control over their course content. These course cartridges also provide online testing and powerful student tracking features. The Brooker Online Learning Center is available within all of these platforms.

FOR THE STUDENT:

Solutions Manual/Study Guide The solutions to the end-of-chapter problems and questions will aid the students in developing their problem-solving skills by providing the steps for each solution. The Study Guide follows the order of sections and subsections in the textbook and summarizes the main points in the text, figures, and tables. It also contains concept-building exercises, self-help quizzes, and practice exams.

Brooker Interactive CD-ROM Set up in an easy to use tabular format, this dual-platform CD-ROM is a fully interactive learning tool. The CD is organized chapter-by-chapter and provides a link directly to the text's Online Learning Center. Standard features include chapter-based quizzes, animations of complex processes, and PowerPoints of all the images found in the textbook. Brooker Interactive CD-ROM offers an indispensable resource for enhancing topics covered within the text.

Online Learning Center Our website, which is organized according to the chapters in the textbook, has a variety of activities and resources that include the following:

Interactive Genetic Problems For chapters involving inheritance patterns and other quantitative problems (chapters 2, 3, 4, 5, 7, 24, and 25), students can set up crosses and attempt to predict the outcome according to genetic principles. Students have the freedom to alter parameters to see how such changes affect the outcome.

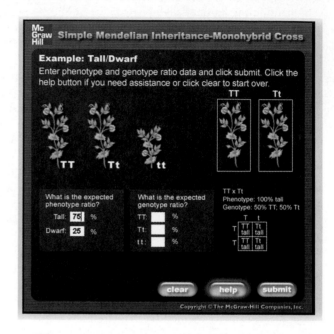

Computer Programs to Analyze Genetic Data For chapter 21, which involves issues of genomics and bioinformatics, students can access files that contain genetic data and run various types of computer programs. For example, students can use a DNA sequence file and run a program that translates it to an amino acid sequence according to the genetic code. This hands-on learning should help students gain a greater appreciation for the tools of bioinformatics.

Tweaking the Experiment For the in-depth experiments that are found in each chapter, a self-help quiz is provided. The questions ask students to predict how changes in the experimental parameters may affect the outcome of the experiment. These quizzes should help students to improve their skills of designing experiments and predicting their outcome according to genetic hypotheses.

Answers to Problem Sets The answers to all of the problems in the textbook are given.

References Classic and modern references to the literature are provided for each chapter.

Links Internet links to a variety of websites related to the course material are available according to the relevant chapter.

ACKNOWLEDGMENTS

The production of a textbook is truly a collaborative effort. I am greatly indebted to a variety of people. The first and second editions went through multiple rounds of rigorous revisions that involved the input of scientists, students, and editors. The collective contributions of all these people are reflected in the final outcome.

There are many people at McGraw-Hill whom I deeply respect and whose efforts are amazing. Since the McGraw-Hill people work as a well-greased and integrated team, it's sometimes difficult to sort out their relative contributions, but I'll do my best. Patrick Reidy, Senior Sponsoring Editor, and Marty Lange, Publisher, have been committed to the second edition since the very beginning. They provided important pedagogical input and maintained a level of financial commitment that was vital to the outcome of this project. Of course, Kris Tibbetts gets my highest amount of praise for her dedicated role of Director of Development. Her ability to organize the various components of this project is pretty incredible. Grinding through the process of writing a second edition can be unnerving at times, and it's really great to have someone you know you can count on. Other people at McGraw-Hill have played key roles in producing an actual book and the supplements that go along with it. In particular, Peggy Selle, Project Manager, was an important interface between me and the art and editorial staff. The input of Designer Wayne Harms was important with regard to the layout and design of the chapters and cover.

Outside of McGraw-Hill, several people/groups were involved with the production of this textbook. My copy editor, Laura Beaudoin, did an exceptional job of strengthening the writing with regard to clarity and consistency. I am deeply grateful. Chris Hammond should be acknowledged for his tireless work at locating photographs for the textbook. I would also like to extend a special thanks to the folks at Imagineering, the art house that produced the illustrations in this second edition. Though we initially considered several art houses, when I saw how Imagineering could draw RNA polymerase, it was clear who the winner was. Their ability to draw three-dimensional molecular structures is just plain outstanding. I particularly want to thank Kierstan Hong at Imagineering, who played a key role in communicating my vision to the art team.

I am also deeply indebted to my wife, Deborah, and kids, Dan, Nate, and Sarah, for keeping me grounded and for being excited about the first edition when it came out. I would also like to acknowledge that my wife, Deb, and my colleague Peter Franco have been my "super proofreaders." They tenaciously read the page proofs for all 26 chapters and provided critical input during the final stages of the second edition.

Finally, I want to thank the many scientists who reviewed the chapters of this textbook. Their broad insights and constructive suggestions were an important factor that shaped its final content and organization. I am truly grateful for their time and effort.

REVIEWERS FROM THE UNITED STATES

Arnold, Robert M., *Colgate University*
Bean, Barry S., *Lehigh University*
Bergeson, Susan E., *University of Texas*
Brownell, Anne G., *Chapman University*
Butler, David, *Montana State University*
Calie, Pat, *Eastern Kentucky University*
Campbell, Arden, *Iowa State University*
Caruso, Steven M., *University of Maryland*
Carver, Kristopher, *University of Minnesota*
Cassill, J. Aaron, *University of Texas, San Antonio*
Coates, Craig J., *Texas A&M University*
Crnekovic, Victoria, *Parkland College*
Davis, Sandra L., *University of Louisiana, Monroe*
D'Surney, Stephen J., *University of Mississippi*
Duhrkopf, Richard E., *Baylor University*
Durica, David S., *University of Oklahoma*
El-Rady, Johnny, *University of South Florida*
Ely, Bert, *University of South Carolina*
Farrell, Robert, *Penn State University, York*
Fore, Stephanie, *Truman State University*
Forrester, Wayne, *Indiana University*

Fowler, Robert G., *San Jose State University*
Franco, Peter, *University of Minnesota*
Frugoli, Julie, *Clemson University*
Galbraith, Anne, *University of Wisconsin, LaCrosse*
Goldstein, Elliott S., *Arizona State University*
Guilfoile, Patrick, *Bemidji State University*
Guralnick, Lonnie J., *Western Oregon University*
Hamilton, Robert, *Mississippi College*
Harris, Randall, *William Carey College*
Higginbotham, Jeri, *Transylvania University*
Hinrichsen, Robert, *Indiana University of Pennsylvania*
Hollenbeck, James E., *Indiana Unversity, Southeast*
Imberski, Richard B., *University of Maryland*
Ivarie, Robert, *University of Georgia*
Jansky, Shelley, *University of Wisconsin, Steven's Point*
Jensen, Phillip, *University of Minnesota*
Kageyama, Glenn H., *California State Polytechnic University, Pomona*
Kass, David, *Eastern Michigan University*
Katz, Alan J., *Illinois State University*
Kelson, Todd, *Brigham Young University*

Krause, Eliot, *Seton Hall University*
Krieger, Kari Beth, *University of Wisconsin, Green Bay*
Kushner, Sidney R., *University of Georgia*
Liao, Min-Ken, *Furman University*
Lorson, Christian L., *Arizona State University*
Mackay, William James, *Edinboro University of Pennsylvania*
MacKrell, Albert, *Bradley University*
Magill, Clint, *Texas A&M University*
Marshall, Pamela A., *SUNY College at Fredonia*
Martinez, Robert M., *Quinnipiac University*
Matthews, Terry C., *Millikin University*
Matzke, Elizabeth, *University of Minnesota*
Mauricio, Rodney, *University of Georgia*
McCabe, Norah R., *Washington State University*
McDonough, Virginia, *Hope College*
McGuire, Terry, *Rutgers University*
Meade, Mark E., *Jacksonville State University*
Meneely, Philip, *Haverford College*
Morvillo, Nancy, *Florida Southern College*
Moss, Robert, *Wofford College*
Owen, Henry R., *Eastern Illinois University*
Palladino, Michael, *Monmouth University*
Perlin, Mike, *University of Louisville*

Perreault, William, *Lawrence University*
Pfaffle, Patrick, *Carthage College*
Ream, Walt, *Oregon State University*
Robinson, James V., *University of Texas, Arlington*
Schug, Malcolm, *University of North Carolina, Greensboro*
Scott, Rodney J., *Wheaton College*
Sears, Barbara B., *Michigan State University*
Shull, J. Kenneth, Jr., *Appalachian State University*
Spitze, Ken, *University of Miami*
Sternick, John, *Mansfield University*
Thompson, Jeffrey S., *Georgian Court College*
Tobin-Janzen, Tammy, *Susquehanna University*
Triman, Kathleen, *Franklin and Marshall College*
Wood, Andrew J., *Southern Illinois University*
Yampolsky, Len, *Eastern Tennessee State University*

REVIEWERS FROM CANADA

Bonham-Smith, Peta, *University of Saskatchewan*
Duncker, Bernard C., *University of Waterloo*
O'Connor, Tracy J., *University of Calgary*
Schoen, Daniel J., *McGill University*

REVIEWERS FROM OUTSIDE NORTH AMERICA

Allison, Heather E., *University of Liverpool*
Bouthwaite, Stephen, *University of Southern Denmark*
Egel, Richard, *University of Copenhagen*
Fletcher, H. L., *Queens University of Belfast*
Jensen, Per, *Linkoping University*
Kuipers, Oscar, *University of Groningen*
Tata, Fred, *University of Leicester*
Wheals, Alan, *Bath University*

ART REVIEW BOARD

Birky, C. William, *University of Arizona*
El Rady, Johnny, *University of South Florida*
Ely, Bert, *University of South Carolina, Columbia*
Forrester, Wayne, *Indiana University*
Freking, Fredrick, *University of California, Los Angeles*
Girton, Jack, *Iowa State University*
Kliebenstein, Dan, *University of California, Davis*
Laten, Howard, *Loyola University of Chicago*
Paquin, Charlotte, *University of Cincinnati*
Pickett, Bryan, *Loyola University of Chicago*

Richmond, Janet, *University of Illinois, Chicago*
Saxena, Inder, *University of Texas, Austin*
Sears, Barbara, *Michigan State University*
Snider, Phil, *University of Houston*
Thrower, Doug, *University of California, Santa Barbara*
Urnov, Fyodor, *University of California, Berkeley*
Weil, Clifford, *Purdue University*
Wooten, Leon, *Xavier University*

ACCURACY REVIEW PANEL

Arnold, Robert, *Colgate University*
Krieger, Kari Beth, *University of Wisconsin, Green Bay*
Marshall, Pam A., *SUNY College at Fredonia*
Meneely, Philip, *Haverford College*
Moss, Robert, *Wofford College*
Scott, Rodney, *Wheaton College*
Shull, J. Kenneth, *Appalachian State University*

MEDIA REVIEW PANEL

Bowling, Scott, *Auburn University*
Cassill, J. Aaron, *University of Texas, San Antonio*
Kageyama, Glenn H., *California State Polytechnic University, Pomona*
Locy, Bob, *Auburn University*
Pendarvis, M. Pat, *Southeast Louisiana University*
Pickett, F. Bryan, *Loyola University of Chicago*
Sears, Barbara B., *Michigan State University*
Skinner, Monica, *Oregon State University*

QUESTIONNAIRE RESPONDENTS

Achenbach, Laurie A., *Southern Illinois University*
Adamkewicz, Laura, *George Mason University*
Allen, Jim, *Florida International University*
Arnold, Robert, *Colgate University*
Bean, Barry, *Lehigh University*
Beck, Barbara N., *Rochester Community & Technical College*
Belk, Colleen, *University of Minnesota, Duluth*
Belote, John, *Syracuse University*
Berger, Edward, *Dartmouth College*
Bhardwaj, Harbans L., *Virginia State University*
Bishr, Mahmoud, *Labette Community College*
Bloom, Kerry, *University of North Carolina, Chapel Hill*
Bouma, Hessel, III, *Calvin College*

Bradley, Jane, *Des Moines Area Community College*
Brooks, George M., *Ohio University*
Brown, Susan J., *Kansas State University*
Brownell, Anne, *Chapman University*
Butler, David, *Montana State University*
Calcamuggio, James, *Pillsbury Baptist Bible College*
Caldwell, Guy A., *University of Alabama*
Calie, Pat, *Eastern Kentucky University*
Campbell, Arden, *Iowa State University*
Caprioglio, Daniel, *University of Southern Colorado*
Carlson, Steve, *University of Wisconsin, River Falls*
Caruso, Steven M., *University of Maryland*
Celenza, John, *Boston University*
Chalfie, Martin, *Columbia University*
Chernin, Mitchell, *Bucknell University*
Choi, Jung, *Georgia Institute of Technology*
Coates, Craig, *Texas A&M University*
Coffin, Candice J., *Briar Cliff University*
Collier, Glen, *University of Tulsa*
Collins, John, *University of New Hampshire*
Collins, John L., *University of Tennessee*
Colson, Carl M., *West Virginia Wesleyan College*
Cronmiller, Claire, *University of Virginia*
Crnekovic, Victoria, *Parkland College*
Darville, Roy, *East Texas Baptist University*
David, John, *University of Missouri*
Davis, Sandra, *University of Louisiana, Monroe*
Demarais, Alyce, *University of Puget Sound*
Dendy, Leslie, *University of New Mexico, Los Alamos*
Dhar, Sujit, *Ellsworth Community College*
Dixon, Linda, *University of Colorado, Denver*
Dorn, Patricia, *Loyola University*
Dorset, Michael A., *Cleveland State Community College*
D'Surney, Stephen J., *University of Mississippi*
Duhrkopf, Richard E., *Baylor University*
Dunbar, Maureen, *Penn State University, Berks*
Durica, David, *University of Oklahoma*
Duronio, Robert, *University of North Carolina, Chapel Hill*
Ehrman, Lee, *SUNY Purchase College*
Eliason, James L., *St. Thomas Aquinas College*
Ellis, Ron, *University of Michigan*
Ellison, John R., *Texas A&M University*
Ely, Bert, *University of South Carolina*
Emmel, Thomas C., *University of Florida*
Esen, Asim, *Virginia Polytechnic Institute of Technology*
Falkinham, Joseph O., III, *Virginia Polytechnic Institute of Technology*
Farrell, Robert, *Penn State University, York*
Fernandes, Pearl Ramola, *University of South Carolina, Sumter*
Fore, Stephanie, *Truman State University*

Forrester, Wayne, *Indiana University*
Fowler, Robert, *San Jose State University*
Frank, Ronald L., *University of Missouri, Rolla*
Fromson, David, *California State University, Fullerton*
Frugoli, Julie, *Clemson University*
Gabriel, Edward G., *Lycoming College*
Galbraith, Anne, *University of Wisconsin, LaCrosse*
Gaydos, Beth, *San Jose City College*
Gerber, Anne S., *University of North Dakota*
Giblin, Tara, *Stephens College*
Given, Mac F., *Neumann College*
Glaubensklee, Carolyn, *University of Southern Colorado*
Goldstein, Elliott S., *Arizona State University*
Guilfoile, Patrick, *Bemidji State University*
Guralnick, Lonnie J., *Western Oregon University*
Hamilton, Meredith, *Oklahoma State University*
Hamilton, Robert, *Mississippi College*
Hanratty, Pamela, *Indiana University*
Hardig, T. M., *University of Montevallo*
Harms, Robert, *St. Louis Community College*
Harris, Randall K., *William Carey College*
Haws, Janice L., *Delaware Valley College*
Hellack, Jenna, *University of Central Oklahoma*
Henderson, Curtis, *MacMurray College*
Hicks, Karen A., *Kenyon College*
Higginbotham, Jeri W., *Transylvania University*
Hinrichsen, Robert, *Indiana University of Pennsylvania*
Hobbie, Lawrence, *Adelphi University*
Hollenbeck, James E., *Iowa Wesleyan College*
Huser, Carl, *Southwest Baptist University*
Hutter, Dottie, *Villanova University*
Imberski, Richard, *University of Maryland*
Ivarie, Robert, *University of Georgia*
Jansky, Shelley, *University of Wisconsin, Steven's Point*
Jerry, Joseph, *University of Massachusetts*
Kaczmarczyk, Walter, *West Virginia University*
Kageyama, Glenn H., *California State Polytechnic University, Pomona*
Kananen, Mary King, *Penn State University, Altoona*
Kass, David, *Eastern Michigan University*
Katz, Alan J., *Illinois State University*
Keil, Clifford, *University of Delaware*
Kelson, Todd, *Brigham Young University*
Kirk, Karen, *Lake Forest College*
Klein, Anita, *University of New Hampshire*
Koehler, Richard, *St. Ambrose University*
Kowles, Richard, *Saint Mary's University of Minnesota*

Krause, Eliot, *Seton Hall University*
Krieger, Kari Beth, *University of Wisconsin, Green Bay*
Kroening, Dubear, *University of Wisconsin, Fox Valley*
LaBar, Martin, *Southern Wesleyan University*
Levinthal, Mark, *Purdue University*
Liao, Min-Ken, *Furman University*
Linhart, Yan, *University of Colorado, Boulder*
Liu, Zhiming, *Eastern New Mexico University*
Lorson, Christian L., *Arizona State University*
Lutze, Margaret, *Loyola University*
Mabry, Michelle, *Davis & Elkins College*
Mackay, William James, *Edinburo University of Pennsylvania*
MacKrell, Albert, *Bradley University*
Magill, Clint, *Texas A&M University*
Marcovitz, Michael Howard, *Midland Lutheran College*
Marshall, Pamela A., *SUNY College, Fredonia*
Martin, Linda C., *Kansas State University*
Martinez, Robert M., *Quinnipiac University*
Matthews, Terry C., *Millikin University*
Mauricio, Rodney, *University of Georgia*
McCabe, Norah R., *Washington State University*
McCommas, Steven, *Southern Illinois University*
McDonough, Virginia, *Hope College*
McMillin, David E., *Brewton Parker College*
Meade, Mark E., *Jacksonville State University*
Meagher, Shawn, *Western Illinois University*
Meneely, Philip, *Haverford College*
Menees, Thomas M., *University of Missouri, Kansas City*
Messley, Karen, *Rock Valley College*
Miller, Gail L., *York College*
Morrow, Glenn, *Medaille College*
Morvillo, Nancy, *Florida Southern College*
Moser, Dan W., *Kansas State University*
Moss, Robert, *Wofford College*
Murnik, Mary, *Ferris State University*
Murphy, Ed, *Columbia State Community College*
Myka, Jennifer, *Brescia University*
Nesbitt, Muriel, *University of California, San Diego*
Norris, William R., *Western New Mexico University*
Nuttall, Ted R., *Lock Haven University*
Oppenheimer, David G., *University of Alabama*
Orive, Maria E., *University of Kansas*
Osterman, John, *University of Nebraska, Lincoln*
Ostrowski, Ronald S., *University of North Carolina, Charlotte*
Owen, Henry R., *Eastern Illinois University*
Paterson, Ann V., *Williams Baptist College*

Perlin, Michael, *University of Louisville*
Perreault, William, *Lawrence University*
Pfaffle, Patrick K., *Carthage College*
Poole, Theresa M., *Georgia State University*
Ray, Dennis, *University of Arizona*
Ream, Walt, *Oregon State University*
Rector, Regina, *William Rainey Harper College*
Redman, Renee, *University of North Carolina, Greensboro*
Robinson, James V., *University of Texas, Arlington*
Savage, Thomas, *Oregon State University*
Schwaner, Terry, *Southern Utah University*
Scott, Rodney J., *Wheaton College*
Sears, Barbara B., *Michigan State University*
Shull, J. Kenneth, Jr., *Appalachian State University*
Slieman, Tony A., *Morningside College*
Spitze, Ken, *University of Miami*
Stark, Ben, *Illinois Institute of Technology*
Steck, Todd, *University of North Carolina, Charlotte*
Sternick, John, *Mansfield University*
Tepper, Craig, *Cornell College*
Theroux, Steven, *Assumption College*
Thompson, Jeffrey S., *Georgian Court College*
Tobin-Janzen, Tammy, *Susquehanna University*
Tolsma, Sara S., *Northwestern College*
Triman, Kathleen, *Franklin and Marshall College*
Vanderbilt, Callie A., *San Juan College*
Visick, Jonathan, *North Central College*
Wales, Melinda, *Texas A&M University*
Weber, David, *Illinois State University*
Weber, Kenneth, *University of Southern Maine*
Wells, Harrington, *University of Tulsa*
West, Robert, *University of Colorado, Boulder*
Whiteman, Leslie, *Virginia Union University*
Wiemerslage, Leslie J., *Southwestern Illinois College*
Williams, Dan, *Fort Peck Community College*
Wilt, Steven D., *Kentucky Wesleyan College*
Wolf, Paul, *Utah State University*
Wood, Andrew J., *Southern Illinois University*
Yampolsky, Len, *Eastern Tennessee State University*
Zeng, Chaoyang, *University of Wisconsin, Milwaukee*
Zimmerer, Edmund J., *Murray State University*
Zink, Brenda, *Northeastern Junior College*

A Visual Guide to:
GENETICS: ANALYSIS & PRINCIPLES

::

ROBERT J. BROOKER
UNIVERSITY OF MINNESOTA, TWIN CITIES

Brooker's Genetics *brings key concepts to life with its unique style of illustration.*

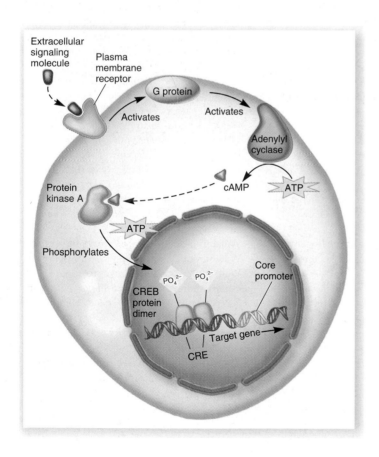

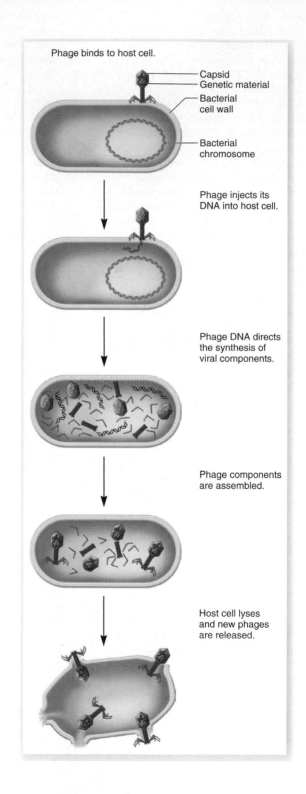

Phage binds to host cell.

Capsid
Genetic material
Bacterial cell wall
Bacterial chromosome

Phage injects its DNA into host cell.

Phage DNA directs the synthesis of viral components.

Phage components are assembled.

Host cell lyses and new phages are released.

The digitally rendered images have a vivid three-dimensional look that will stimulate a student's interest and enthusiasm.

Each figure is carefully designed to follow closely with the text material.

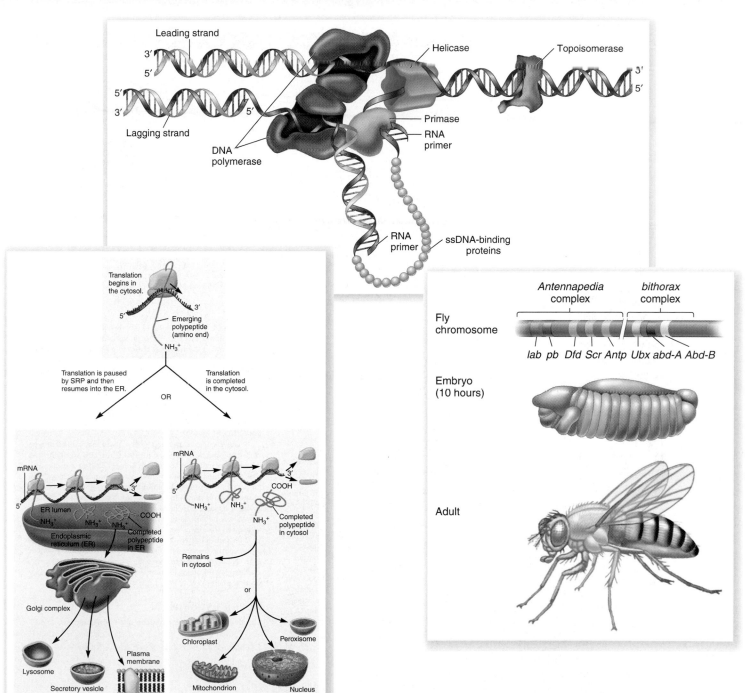

Every illustration was drawn with four goals in mind: completeness, clarity, consistency, and realism.

Each chapter (beginning with chapter 2) incorporates one or two experiments that are presented according to the scientific method. These experiments are integrated within the chapters and flow with the rest of the textbook. As you read the experiments, you will simultaneously explore the scientific method and the genetic principles learned from this approach.

EXPERIMENT 12A

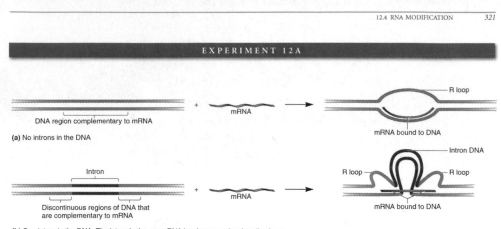

(a) No introns in the DNA

(b) One intron in the DNA. The intron in the pre-mRNA has been previously spliced out.

FIGURE 12.16   Hybridization of mRNA to double-stranded DNA. In this experiment, the DNA is denatured and then allowed to renature under conditions that favor an RNA-DNA hybrid. **(a)** If the DNA does not contain an intron, the binding of the mRNA to the anticoding strand of DNA prevents the two strands of DNA from forming a double helix. The single-stranded DNA will form an R loop. **(b)** When mRNA hybridizes to a gene containing one intron, two single-stranded R loops will form that are separated by a double-stranded DNA region. The intervening double-stranded region occurs because an intron has been spliced out of the mRNA, so that the mRNA cannot hybridize to this segment of the gene.

STEP 1: BACKGROUND OBSERVATIONS

Each experiment begins with a description of the information that led researchers to study an experimental problem. Detailed information about the researchers and the experimental challenges they faced help students to understand actual research.

Introns Were Experimentally Identified Via Microscopy

Although the discovery of ribozymes was surprising, the observation that tRNA and rRNA transcripts are processed to a smaller form did not seem unusual to geneticists and biochemists, because the enzymatic cleavage of RNA was similar to the cleavage that can occur for other macromolecules such as DNA and proteins. In sharp contrast, when splicing was detected in the 1970s, it was a novel concept. Splicing involves cleavage at two sites. An intron is removed, and—in a unique step—the remaining fragments are hooked back together again.

Eukaryotic introns were first detected by comparing the base sequence of genes and mRNAs during viral infection of mammalian cells by adenovirus. This research was carried out in 1977 by two groups headed by Philip Sharp and Richard Roberts. This pioneering observation led to the next question: Are introns a peculiar phenomenon that occurs only in viral genes, or are they found in eukaryotic genes as well?

In the late 1970s, several research groups, including those of Pierre Chambon, Bert O'Malley, and Phillip Leder, investigated the presence of introns in eukaryotic structural genes. The experiments of Leder used electron microscopy to identify introns in the β-globin gene. β-globin is a polypeptide that is a subunit of hemoglobin, the protein that carries oxygen in red blood cells. Prior to Leder's work, the β-globin gene had been cloned. To detect introns within the cloned gene, Leder used a strategy involving hybridization. In this approach, shown in figure 12.16, a segment of double-stranded, chromosomal DNA containing the

β-globin gene was first denatured and mixed with mature mRNA for β-globin. Because this mRNA is complementary to the anticoding strand of the DNA, the anticoding strand and the mRNA will specifically bind to each other. This event is called hybridization. Later, when the DNA is allowed to renature, the binding of the mRNA to the anticoding strand of DNA prevents the two strands of DNA from forming a double helix. In the absence of any introns, the single-stranded DNA will form a loop. Since the RNA has displaced one of the DNA strands, this structure is known as an RNA displacement loop or **R loop,** as shown in figure 12.16a.

However, Leder and colleagues realized that a different type of R loop structure would be formed if the DNA contained an intron. For example, when mRNA is hybridized to a region of a gene containing one intron, two single-stranded R loops will form that are separated by a double-stranded DNA region. This phenomenon is shown in figure 12.16b. The intervening double-stranded region occurs because an intron has been spliced out of the mature mRNA, so that the mRNA cannot hybridize to this segment of the gene.

As shown in steps 1 through 4 of figure 12.17, this hybridization approach was used to identify introns within the β-globin gene. Following hybridization, the samples were placed on a microscopy grid, stained with heavy metal, and then observed by electron microscopy.

■ THE HYPOTHESIS

The mouse β-globin gene contains one or more introns.

STEP 2: HYPOTHESIS

The student is given a statement describing the possible explanation for the observed phenomenon that will be tested. The hypothesis section reinforces the scientific method and allows students to experience the process for themselves.

■ TESTING THE HYPOTHESIS — FIGURE 4.8 Identification of an allele that shows a gene dosage effect.

Starting Material: True-breeding strains of flies that have either the red-eye (normal) allele, white-eye allele, or eosin-eye allele.

Experimental level	Conceptual level

1. Make the following crosses:

A. White-eyed females x red-eyed males

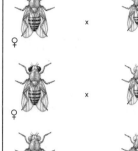

A. $X^w X^w$ x $X^{w^+} Y$

B. Red-eyed females x light eosin-eyed males

B. $X^{w^+} X^{w^+}$ x $X^{w\text{-}e} Y$

C. Eosin-eyed females x white-eyed males

C. $X^{w\text{-}e} X^{w\text{-}e}$ x $X^w Y$

2. Observe the outcome of the three types of crosses. Compare the eye colors of the true-breeding parents with their offspring.

Predicted ratio of offspring

A. 1 red-eyed female ($X^{w^+} X^w$) : 1 white-eyed male ($X^w Y$)

B. 1 red-eyed female ($X^{w^+} X^{w\text{-}e}$) : 1 red-eyed male ($X^{w^+} Y$)

C. 1 light eosin-eyed female ($X^{w\text{-}e} X^w$) : 1 light eosin-eyed male ($X^{w\text{-}e} Y$)

■ THE DATA

Cross	Outcome	Genotype	Phenotype	
White-eyed females ($X^w X^w$) × red-eyed males ($X^{w^+} Y$)	225 red-eyed females ($X^{w^+} X^w$) : 208 white-eyed males ($X^w Y$)	$X^w X^{w^+}$	Red eyes	No difference was observed in the redness of the eyes among these 4 groups.
Red-eyed females ($X^{w^+} X^{w^+}$) × light eosin-eyed males ($X^{w\text{-}e} Y$)	679 red-eyed females ($X^{w^+} X^{w\text{-}e}$) : 747 red-eyed males ($X^{w^+} Y$)	$X^{w^+} X^{w\text{-}e}$	Red eyes	
		$X^{w^+} X^w$	Red eyes	
		$X^{w^+} Y$	Red eyes	
Eosin-eyed females ($X^{w\text{-}e} X^{w\text{-}e}$) × white-eyed males ($X^w Y$)	694 light eosin-eyed females ($X^{w\text{-}e} X^w$) : 579 light eosin-eyed males ($X^{w\text{-}e} Y$)	$X^{w\text{-}e} X^{w\text{-}e}$	Eosin eyes	The heterozygous females had similar eye color to the hemizygous males.
		$X^{w\text{-}e} X^w$	Light-eosin eyes	
		$X^{w\text{-}e} Y$	Light-eosin eyes	
		$X^w X^w$	White eyes	
		$X^w Y$	White eyes	

■ INTERPRETING THE DATA

When Morgan and Bridges crossed females that were homozygous for an eye color allele to males that were hemizygous for a different allele, they always observed a 1:1 ratio among the offspring. This is consistent with the idea that these genes for eye color are X linked and allelic to each other. Let's first consider flies that carried an X^{w^+} allele. All of these flies exhibited red eye color. Furthermore, the eye color was bright red whether the fly had one or two copies of the red allele. As seen in the data, the $X^{w^+} X^{w^+}$ females had red eyes that were indistinguishable from the eyes of flies that had only one copy of the X^{w^+} allele (namely, $X^{w^+} X^{w\text{-}e}$ females, $X^{w^+} X^w$ females, and $X^{w^+} Y$ males). In this case, the red allele appears to be dominant to the eosin and white alleles, and no gene dosage effect occurs.

In contrast, the eosin phenotype is affected by the number of copies of the eosin allele. Homozygous females containing two copies of the $X^{w\text{-}e}$ allele had eosin eyes, while flies with only one copy of this allele (namely, $X^{w\text{-}e} X^w$ females and $X^{w\text{-}e} Y$ males) had light-eosin eyes (see data table in fig. 4.8). In heterozygous $X^{w\text{-}e} X^w$ females, Morgan and Bridges proposed that the white allele (which does not produce color) was having a "dilution effect" on the phenotype. Another way of considering this phenomenon is that the $X^{w\text{-}e}$ allele shows a gene dosage effect. An increase in the dosage of this gene from one copy to two copies yields an increase in the degree of the eosin color in the eyes of the flies. This result supports the idea that the amount of a gene product can influence the outcome of a trait.

A self-help quiz involving this experiment can be found at the Online Learning Center.

STEP 3: TESTING THE HYPOTHESIS

This section illustrates the experimental process, including the actual steps followed by scientists to test their hypothesis. Science comes alive for students with this detailed look at experimentation.

STEP 4: THE DATA

Actual data from the original research paper help students understand how actual research results are reported. Each experiment's results are discussed in the context of the larger genetic principle to help students understand the implications and importance of the research.

STEP 5: INTERPRETING THE DATA

This discussion, which examines whether the experimental data supported or disproved the hypothesis, gives students an appreciation for scientific interpretation.

A large number of new and challenging problems were added to the problem sets of every chapter. These problems are crafted to aid students in developing a wide range of skills. They also develop a student's cognitive, writing, analytical, computational, and collaborative abilities.

CONCEPTUAL QUESTIONS

Test the understanding of basic genetic principles. The student is given many questions with a wide range of difficulty. Some require critical thinking skills, and some require the student to write coherent essay questions.

EXPERIMENTAL QUESTIONS

Test the ability to analyze data, design experiments, or appreciate the relevance of experimental techniques.

STUDENT DISCUSSION/ COLLABORATION QUESTIONS

Encourage students to consider broad concepts and practical problems. Some questions require a substantial amount of computational activities, which can be worked on as a group.

Conceptual Questions

C1. Define the term *epigenetic*, and describe two examples.

C2. Describe the pattern of inheritance that maternal effect genes follow. Explain how the maternal effect occurs at the cellular level. What are the expected functional roles of the proteins that are encoded by maternal effect genes?

C3. A maternal effect gene exists in a dominant *N* (normal) allele and a recessive *n* (abnormal) allele. What would be the ratios of genotypes and phenotypes for the offspring of the following crosses?

  A. *nn* female × *NN* male

  B. *NN* female × *nn* male

  C. *Nn* female × *Nn* male

C4. A *Drosophila* embryo dies during early embryogenesis due to a recessive maternal effect allele called *bicoid*. The wild-type allele is designated *bicoid⁺*. What are the genotypes and phenotypes of the embryo's mother and maternal grandparents?

C5. For Mendelian traits, the on chromosomes in the spring's traits. In contras it is often not enough to mosomes in the nucleus type. For the following t the phenotypic outcome

  A. Dwarfism due to a m

  B. Snail coiling direction

  C. Leber's hereditary opt

C6. Let's suppose a maternal allele and an abnormal r cally abnormal produces of the mother.

C7. A gene that affects the ar ited as a maternal effect and a recessive allele, *h,* with a normal head is m head. All of the offspring of the mother and offspr

C8. Explain why maternal ef early stages of developm

C9. As described in chapter mammals by fusing a cel cell) with an egg that ha

  A. With regard to materi such a cloned animal the oocyte or by the a Explain.

  B. Would the cloned ani animal that donated t the somatic cell? Expl

  C. In what ways would y to or different from th it fair to call such an i donated the nucleus?

C10. With regard to the numbers of sex chromosomes, explain why dosage compensation is necessary.

C11. What is a Barr body? How is its structure different from that of other chromosomes in the cell? How does the structure of a Barr body affect the level of X-linked gene expression?

C12. Among different species, describe three distinct strategies for accomplishing dosage compensation.

C13. Describe when X inactivation occurs and how this leads to phenotypic results at the organismal level. In your answer, you should explain why X inactivation causes results such as variegated coat patterns in mammals. Why do two different calico cats have their patches of orange and black fur in different places? Explain whether or not a variegated coat pattern could occur in marsupials due to X inactivation.

C14. Describe the molecular process of X inactivation. This description should include the three phases of inactivation and the role of the

---

C22. On rare occasions, people are born with a condition known as uniparental disomy. It happens when an individual inherits both copies of a chromosome from one parent and no copies from the other parent. This occurs when two abnormal gametes happen to complement each other to produce a diploid zygote. For example, an abnormal sperm that lacks chromosome 15 could fertilize an egg that contains two copies of chromosome 15. In this situation, the individual would be said to have maternal uniparental disomy 15, because both copies of chromosome 15 were inherited from the mother. Alternatively, an abnormal sperm with two copies of chromosome 15 could fertilize an egg with no copies. This is known as paternal uniparental disomy 15. If a female is born with paternal uniparental disomy 15, would you expect her to be phenotypically normal, have Angelman syndrome, or have Prader-Willi syndrome? Explain. Would you expect her to produce normal offspring or offspring affected with AS or PWS?

Experimental Questions

E1. Figure 7.1 describes an example of a maternal effect gene. Explain how Sturtevant deduced a maternal effect gene based on the F_2 and F_3 generations.

E2. Discuss the types of experimental observations that Mary Lyon had to consider in proposing her hypothesis concerning X inactivation. In your own words, explain how these observations were consistent with her hypothesis.

E3. Chapter 18 describes three blotting methods (i.e., Southern blotting, Northern blotting, and Western blotting) that are used to detect specific genes and gene products. Southern blotting detects DNA, Northern blotting detects RNA, and Western blotting detects proteins. Let's suppose that a female fruit fly is heterozygous for a maternal effect gene, which we will call gene X. The female is *Xx*. The normal allele, *X*, encodes a functional mRNA that is 550 nucleotides long. A recessive allele, *x*, encodes a shorter mRNA that is 375 nucleotides long. (Allele *x* is due to a deletion within this gene.) How could you use one or more of these techniques to show that nurse cells transfer gene products (from gene *X*) to developing oocytes? You may assume that you can dissect the ovaries of fruit flies and isolate oocytes separately from nurse cells. In your answer, describe your expected results.

E4. As a hypothetical example, a trait in mice results in mice with very long tails. You initially have a true-breeding strain with normal tails and a true-breeding strain with long tails. You then make the following types of crosses:

  Cross 1. When true-breeding females with normal tails are crossed to true-breeding males with long tails, all the F_1 offspring have long tails.

  Cross 2. When true-breeding females with long tails are crossed to true-breeding males with normal tails, all the F_1 offspring have normal tails.

  Cross 3. When F_1 females from cross 1 are crossed to true-breeding males with normal tails, all the offspring have normal tails.

  Cross 4. When F_1 males from cross 1 are crossed to true-breeding females with long tails, half the offspring have normal tails and half have long tails.

  Explain the pattern of inheritance of this trait.

E5. You have a female snail that coils to the right, but you do not know its genotype. You may assume that right coiling (*D*) is dominant to left coiling (*d*). You also have male snails at your disposal of known genotype. How would you determine the genotype of this female snail? In your answer, describe your expected results depending on whether the female is *DD*, *Dd*, or *dd*.

E6. On a recent camping trip, you find one male snail on a deserted island that coils to the right. However, in this same area, you find several shells (not containing living snails) that coil to the left. Therefore, you conclude that you are not certain of the genotype of this male snail. On a different island, you find a large colony of snails of the same species. All of these snails coil to the right, and every snail shell that you find on this second island coils to the right. With regard to the maternal effect gene that determines coiling pattern, how would you determine the genotype of the male snail that you found on the deserted island? In your answer, describe your expected results.

E7. Figure 7.6 describes the results of X inactivation in mammals. If fast and slow alleles of glucose-6-phosphate dehydrogenase (*G-6-PD*) exist in other species, what would be the expected results for a heterozygous female of the following species?

  A. Marsupial

  B. *Drosophila melanogaster*

  C. *Caenorhabditis elegans* (Note: We are considering the hermaphrodite in *C. elegans* to be equivalent to a female.)

Questions for Student Discussion/Collaboration

1. Recessive maternal effect genes are identified in flies (for example) when a phenotypically normal mother cannot produce any normal offspring. Since all the offspring are dead, this female fly cannot be used to produce a strain of heterozygous flies that could be used in future studies. How would you identify heterozygous individuals that are carrying a recessive maternal effect allele? How would you maintain this strain of flies in a laboratory over many generations?

2. What is an infective particle? Discuss the similarities and differences between infective particles and organelles such as mitochondria and chloroplasts. Do you think the existence of infective particles supports the endosymbiotic theory of the origin of mitochondria and chloroplasts?

Note: All answers appear at the website for this textbook; the answers to even-numbered questions are in the back of the textbook.

OVERVIEW OF GENETICS

::

1

Hardly a week goes by without a major news story involving a genetic breakthrough. The rising pace of genetic discoveries has become staggering. The Human Genome Project is a case in point. This project began in the United States in 1990, when the National Institutes of Health and the Department of Energy joined forces with international partners to decipher the massive amount of information contained in our genome—the DNA found within all of our chromosomes (fig. 1.1). Working collectively, a large group of scientists from around the world has produced a detailed series of maps that help geneticists navigate through human DNA. Remarkably, in only a decade, they determined the DNA sequence (read in the letters of A, T, G, and C) covering 90% of the human genome. This sequence, published in 2000, is nearly 3 billion nucleotide base pairs in length. The completed sequence has an accuracy greater than 99.99%; fewer than one mistake was made in every 10,000 letters!

Studying the human genome allows us to explore fundamental details about ourselves at the molecular level. It is expected to shed considerable light regarding basic questions, like how many genes we have, how genes direct the activities of living cells, how species evolved, how single cells develop into complex tissues, and how defective genes cause disease. Furthermore, such understanding may lend itself to improvements in modern medicine by leading to better diagnoses of diseases and the development of new medicines to treat them.

As scientists have attempted to unravel the mysteries within our genes, this journey has involved the invention of many new technologies. This textbook emphasizes a large number of these

PART I
Introduction

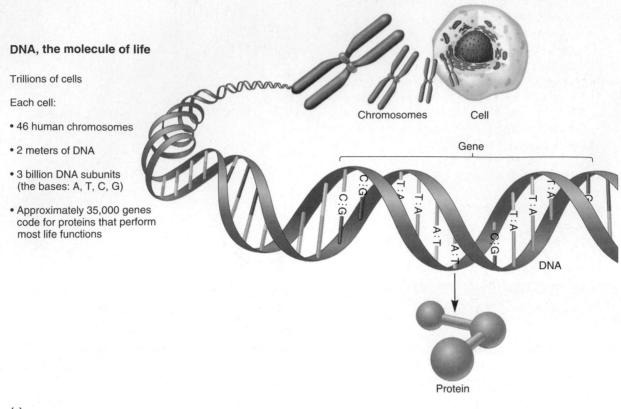

DNA, the molecule of life

Trillions of cells

Each cell:

• 46 human chromosomes

• 2 meters of DNA

• 3 billion DNA subunits
 (the bases: A, T, C, G)

• Approximately 35,000 genes
 code for proteins that perform
 most life functions

Chromosomes   Cell

Gene

C:G
C:G
T:A
T:A
A:T
T:A
A:T
C:G
T:A

DNA

Protein

(a)

Chromosome 4

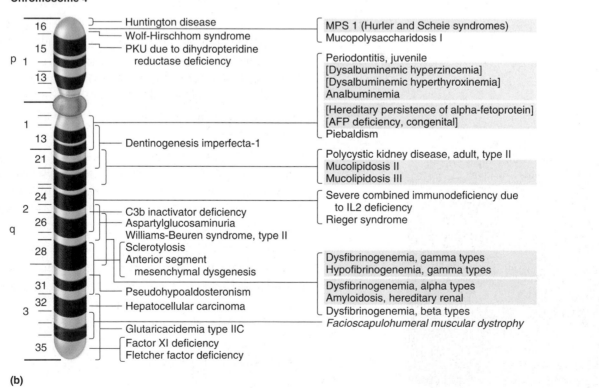

Huntington disease
Wolf-Hirschhom syndrome
PKU due to dihydropteridine
  reductase deficiency

Dentinogenesis imperfecta-1

C3b inactivator deficiency
Aspartylglucosaminuria
Williams-Beuren syndrome, type II
Sclerotylosis
Anterior segment
  mesenchymal dysgenesis
Pseudohypoaldosteronism
Hepatocellular carcinoma
Glutaricacidemia type IIC
Factor XI deficiency
Fletcher factor deficiency

MPS 1 (Hurler and Scheie syndromes)
Mucopolysaccharidosis I

Periodontitis, juvenile
[Dysalbuminemic hyperzincemia]
[Dysalbuminemic hyperthyroxinemia]
Analbuminemia

[Hereditary persistence of alpha-fetoprotein]
[AFP deficiency, congenital]
Piebaldism

Polycystic kidney disease, adult, type II
Mucolipidosis II
Mucolipidosis III

Severe combined immunodeficiency due
  to IL2 deficiency
Rieger syndrome

Dysfibrinogenemia, gamma types
Hypofibrinogenemia, gamma types

Dysfibrinogenemia, alpha types
Amyloidosis, hereditary renal
Dysfibrinogenemia, beta types
Facioscapulohumeral muscular dystrophy

(b)

FIGURE 1.1   The Human Genome Project. (a) The human genome is a complete set of human chromosomes. Collectively, these chromosomes are composed of a DNA sequence that is 3 billion nucleotide base pairs long. Estimates suggest that this sequence contains about 35,000 different genes. **(b)** An important outcome of this work is the identification of genes that contribute to human diseases. This illustration depicts a map of a few genes that are located on human chromosome 4. When these genes carry certain rare mutations, they can cause the diseases designated in this figure.

modern approaches. Paradoxically, one could argue that these advancements may have a greater impact on our lives than the original discoveries upon which they were based. For example, the technology of gene cloning originally arose from discoveries concerning bacterial enzymes that "cut and paste" DNA fragments together. These scientific discoveries have made it possible to create medicines that would otherwise be difficult or impossible to make. A case in point is human recombinant insulin, termed *Humulin,* which is synthesized in a laboratory strain of *Escherichia coli* bacteria that has been genetically altered by the addition of the gene for human insulin. The bacteria are grown in a laboratory and make large amounts of human insulin. As discussed in chapter 19, this insulin is purified and administered to many people with insulin-dependent diabetes.

It is common for new genetic technologies to be met with skepticism and even disdain. An example would be DNA fingerprinting, a molecular method to identify an individual based on a DNA sample. Though this technology is now relatively common in the area of forensic science, it was not always universally accepted. High-profile crime cases in the news cause us to realize that not everyone believes in DNA fingerprinting, in spite of its extraordinary ability to uniquely identify individuals. A second controversial example would be mammalian cloning. In 1997, Ian Wilmut and his colleagues created clones of sheep using mammary cells from an adult animal (fig. 1.2). The first cloned sheep, named Dolly, came as a great surprise to many in the scientific

FIGURE 1.3   The introduction of a jellyfish gene into laboratory mice. A gene that naturally occurs in the jellyfish encodes a protein called green fluorescent protein (GFP). The *GFP* gene was cloned and introduced into mice. When these mice are exposed to UV light, GFP emits a bright green color. These mice glow green, just like jellyfish!

community. This event quickly led to concerns that such technology might be applied to humans. Within a short period of time, legislative bills were introduced that involved bans on human cloning.

Finally, genetic technologies provide the means to modify the traits of animals and plants in ways that would have been unimaginable just a few decades ago. Figure 1.3 illustrates a bizarre example in which scientists introduced a gene from jellyfish into mice. As you may know, certain species of jellyfish appear to emit a "green glow." This is because jellyfish have a gene that encodes a bioluminescent protein called green fluorescent protein (GFP). When exposed to blue or ultraviolet (UV) light, the protein emits a striking green-colored light. Scientists were able to clone the *GFP* gene from a sample of jellyfish cells and then introduce this gene into laboratory mice. The GFP protein is made throughout the cells of their bodies. As a result, their skin, eyes, and organs give off an eerie green glow when exposed to UV light. Only their fur does not glow.

Overall, as we move forward in the twenty-first century, the excitement level in the field of genetics is high, perhaps higher than it has ever been. Nevertheless, the excitement generated by new genetic knowledge and technologies will also create many ethical and societal challenges.

FIGURE 1.2   The cloning of a mammal. The lamb on the *left* is Dolly, a normal looking Finn Dorset. She is the first mammal to be cloned. A description of how Dolly was created is presented in chapter 19. The sheep on the *right* is Dolly's surrogate mother, a Blackface ewe.

1.1 THE RELATIONSHIP BETWEEN GENES AND TRAITS

Genetics is the branch of biology that deals with heredity and variation. It stands as the unifying discipline in biology by allowing us to understand how life can exist at all levels of complexity, ranging from the molecular to the populational level. Genetic variation is the root of the natural diversity that we observe among members of the same species as well as among different species.

Genetics is centered on the study of genes. A **gene** is classically defined as a unit of heredity, but such a vague definition does not do justice to the exciting characteristics of genes as intricate molecular units that manifest themselves as potent contributors to cell structure and function. It is common to describe genes according to the way they affect the **traits** or characteristics of an organism. In humans, for example, we speak of traits such as eye color, hair texture, and height. The ongoing theme of this textbook is the relationship between genes and traits. As an organism grows and develops, its collection of genes provides a blueprint to determine its characteristics.

In this section of chapter 1, we will examine the general features of life, beginning with the molecular level and ending with populations of organisms. As will become apparent, genetics is the common thread that explains the existence of life and its continuity from generation to generation. For most students, this chapter should serve as a cohesive review of topics they learned in other introductory courses such as General Biology. Even so, it is usually helpful to see the "big picture" of genetics before delving into the finer details that are covered in chapters 2 through 26.

Living Cells Are Composed of Biochemicals

To fully understand the relationship between genes and traits, we need to begin with an examination of the composition of living organisms. Every cell is constructed from intricately organized chemical substances. Small organic molecules such as glucose and amino acids are produced from the linkage of atoms via chemical bonds. The chemical properties of organic molecules are essential for cell vitality in two key ways. First, the breakage of chemical bonds during the degradation of small molecules provides energy to drive cellular processes. A second important function of these small organic molecules is their role as the building blocks for the synthesis of larger molecules. Four important categories of larger cellular molecules are **nucleic acids** (i.e., DNA and RNA), **proteins, carbohydrates,** and **lipids.** Three of these, nucleic acids, proteins, and carbohydrates, are considered **macromolecules** because they are composed of many repeating units of smaller building blocks. Proteins, RNA, and carbohydrates are made from hundreds or even thousands of repeating building blocks. DNA is the largest macromolecule found in living cells. A single DNA molecule can be composed of a linear sequence of hundreds of millions of nucleotides!

The formation of cellular structures relies on the interactions of molecules and macromolecules. For example, nucleotides are the building blocks of DNA, which is a constituent of cellular chromosomes (fig. 1.4). In addition, the DNA is associated with a myriad of proteins that provide organization to the structure of chromosomes. Within a cell, the chromosomes are contained in a compartment called the cell nucleus. The nucleus is bounded by a membrane composed of lipids and proteins that shields the chromosomes from the rest of the cell. It is thought that the organization of chromosomes within a cell nucleus protects the chromosomes from mechanical breakage and provides a single

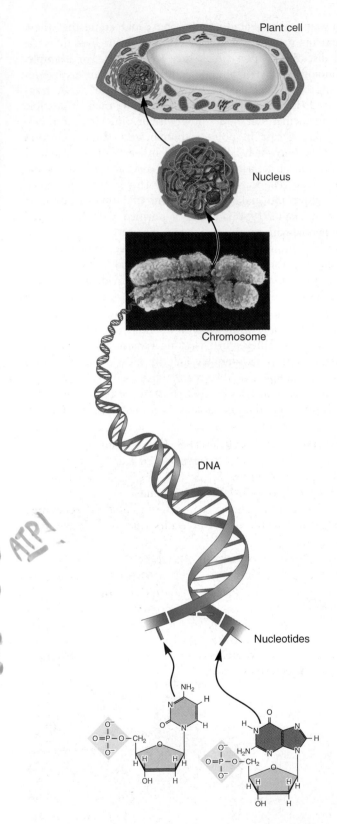

FIGURE 1.4   Chemical composition of living cells. Cellular structures are constructed from smaller building blocks. In this example, DNA is formed from the linkage of nucleotides to produce a very long macromolecule. The DNA associates with proteins to form a chromosome. The chromosomes are located within a membrane-bounded organelle called the nucleus. Many different types of organelles are found within a complete cell.

compartment for genetic activities such as gene transcription. As a general theme, the formation of large cellular structures arises from interactions among different molecules and macromolecules. These cellular structures, in turn, are organized to make a complete living cell.

Each Cell Contains Many Different Proteins That Determine Cellular Structure and Function

To a great extent, the characteristics of a cell depend on the types of proteins that it makes. As we will learn throughout this textbook, proteins are the "workhorses" of all living cells. The range of functions among different types of proteins is truly remarkable. Some proteins help determine the shape and structure of a given cell. For example, the protein known as tubulin can assemble into large structures known as microtubules, which provide the cell with internal structure and organization. Other proteins are inserted into cell membranes and aid in the transport of ions and small molecules across the membrane. Proteins may also function as biological motors. An interesting case is the protein known as myosin, which is involved in the contractile properties of muscle cells. Within multicellular organisms, certain proteins also function in cell-to-cell recognition and signaling. For example, hormones such as insulin are secreted by endocrine cells and bind to the insulin receptor protein found within the plasma membrane of target cells.

Enzymes, which accelerate chemical reactions, are a particularly important category of proteins. Some enzymes play a role in the breakdown of molecules or macromolecules into smaller units. These are known as catabolic enzymes and are important in generating cellular energy. Alternatively, anabolic enzymes function in the synthesis of molecules and macromolecules. Throughout the cell, the synthesis of molecules and macromolecules relies on enzymes and accessory proteins. Thus, the construction of a cell greatly depends on its proteins involved in anabolism, since these are required to synthesize all cellular macromolecules.

Molecular biologists have come to realize that the functions of proteins underlie the cellular characteristics of every organism. At the molecular level, proteins can be viewed as the "active participants" in the enterprise of life.

DNA Stores the Information for Protein Synthesis

The genetic material is composed of a substance called deoxyribonucleic acid, abbreviated DNA. The DNA stores the information needed for the synthesis of all cellular proteins. In other words, the main function of the genetic blueprint is to code for the production of cellular proteins in the correct cell, at the proper time, and in suitable amounts. This is an extremely complicated task, since living cells can make thousands of different proteins. Genetic analyses have shown that a typical bacterium can make a few thousand different proteins, and estimates among higher eukaryotes range in the tens of thousands.

DNA's ability to store information is based on its molecular structure. DNA is composed of a linear sequence of nucleotides.

Each nucleotide contains one nitrogen-containing base, either adenine (A), thymine (T), guanine (G), or cytosine (C). The linear order of these bases along a DNA molecule stores information similar to the way that groups of letters of the alphabet represent words. For example, the "meaning" of the sequence of bases ATGGGCCTTAGC differs from that of TTTAAGCTTGCC. DNA sequences within most genes contain the information to direct the order of amino acids within proteins according to the genetic code. In the code, a three-base sequence specifies one particular amino acid among the 20 possible choices. In this way, the DNA can store the information to specify the sequence of amino acids within a protein.

DNA Sequence	Amino Acid Sequence
ATG GGC CTT AGC	methionine glycine leucine serine
TTT AAG CTT GCC	phenylalanine lysine leucine alanine

In living cells, the DNA is found within large structures known as chromosomes. Figure 1.5 is a photograph of the 46 chromosomes contained in a cell from a human male. The DNA of an average human chromosome is an extraordinarily long, linear, double-stranded structure that contains well over a hundred million nucleotides. Along the immense length of a chromosome, the genetic information is parceled into shorter segments known

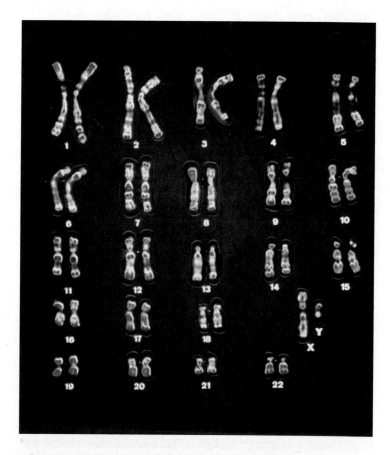

FIGURE 1.5   A micrograph of the 46 chromosomes found in a cell from a human male.

as genes. An average-sized human chromosome is expected to contain between 1,000 and 2,000 different genes.

The Information Within the DNA Is Accessed During the Process of Gene Expression

To synthesize its proteins, a cell must be able to access the information that is stored within its DNA. The process of using a gene sequence to synthesize a cellular protein is referred to as **gene expression.** At the molecular level, the information within genes is accessed in a stepwise process. In the first step, known as **transcription,** the DNA sequence within a gene is copied into a nucleotide sequence of **ribonucleic acid (RNA).** The information within most genes codes for the synthesis of a particular protein. For the synthesis to occur, the sequence of nucleotides transcribed in the RNA must be **translated** (using the genetic code) into the amino acid sequence of a protein (fig. 1.6).

The expression of a gene results in the production of a protein with a specific structure and function. The unique relationship between a gene's sequence and a protein's structure is of paramount importance, because the distinctive structure of each protein determines its function within a living cell or organism. Mediated by the process of gene expression, therefore, the sequence of nucleotides in DNA stores the information required for synthesizing proteins with specific structures and functions.

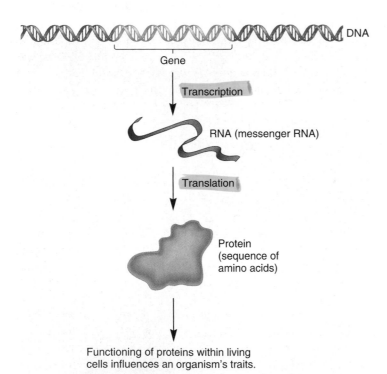

DNA

Gene

Transcription

RNA (messenger RNA)

Translation

Protein
(sequence of
amino acids)

Functioning of proteins within living
cells influences an organism's traits.

FIGURE 1.6   Gene expression. The expression of a gene is a multistep process. As discussed in chapter 12, additional steps such as RNA splicing may occur in eukaryotic species. During transcription, one of the DNA strands is used as a template to make an RNA strand. During translation, the RNA strand is used to specify the sequence of amino acids within a protein. This protein then functions within the cell, thereby influencing an organism's traits.

The Molecular Expression of Genes Within Cells Leads to an Organism's Traits

A **trait** is any characteristic that an organism displays. In genetics, we often focus our attention on **morphological traits** that affect the appearance of an organism. The color of a flower or the height of a pea plant are morphological traits. Geneticists frequently study these types of traits because they are easy to evaluate. For example, an experimenter can simply look at a plant and tell if it has red or white flowers. However, not all traits are morphological. **Physiological traits** affect the ability of an organism to function. For example, the rate at which a bacterium metabolizes a sugar such as lactose is a physiological trait. Like morphological traits, physiological traits are controlled, in part, by the expression of genes.

A difficult, yet very exciting, aspect of genetics is that our observations and theories span four levels of biological organization: molecular, cellular, organismal, and populational. This can make it difficult to appreciate the relationship between genes and traits. To understand this connection, we need to relate the following phenomena:

1. Genes are expressed at the **molecular level.** In other words, gene transcription and translation lead to the production of a particular protein. This is a molecular process.
2. Proteins function at the **cellular level.** The function of a protein within a cell will affect the structure and workings of that cell.
3. An organism's traits are determined by the characteristics of its cells. We do not have microscopic vision, yet when we view morphological traits, we are really observing the properties of an individual's cells. For example, a red flower has its color because the flower cells make a red pigment. The trait of red flower color is an observation at the **organismal level.** Yet it is rooted in the molecular characteristics of the organism's cells.
4. The occurrence of a trait within a species is an observation at the **populational level.** Along with learning how a trait occurs, we also want to understand why a trait becomes prevalent in a particular species. In many cases, we discover that a trait predominates within a population because it promotes the survival or reproduction of the members of the population. This leads to the evolution of beneficial traits.

As a schematic example to illustrate the four levels of genetics, figure 1.7 shows the trait of pigmentation in butterflies. One is light colored and the other is very dark. Now let's consider how we can explain this trait at the molecular, cellular, organismal, and populational level.

At the molecular level, we need to understand the nature of the gene or genes that govern this trait. As shown in figure 1.7a, a gene, which we will call the pigmentation gene, is responsible for the amount of pigment that is produced. The pigmentation gene can exist in two different forms, called **alleles.** In this example, there are two alleles, one that confers a light pigmentation and one that causes a dark pigmentation. Each of these alleles

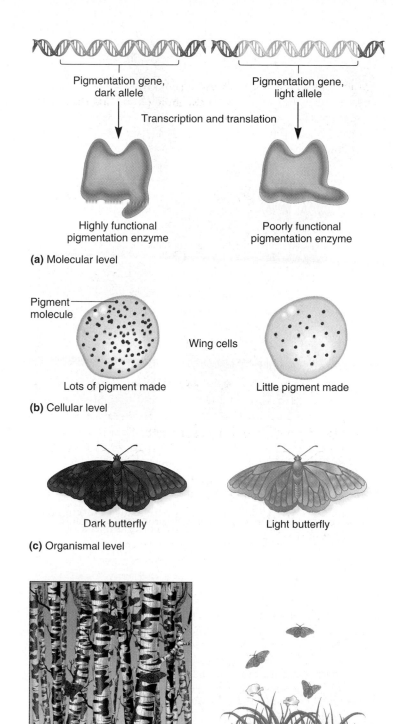

Pigmentation gene, dark allele

Pigmentation gene, light allele

Transcription and translation

Highly functional pigmentation enzyme

Poorly functional pigmentation enzyme

(a) Molecular level

Pigment molecule

Wing cells

Lots of pigment made

Little pigment made

(b) Cellular level

Dark butterfly

Light butterfly

(c) Organismal level

Dark butterflies are usually in forested regions.

Light butterflies are usually in unforested regions.

(d) Populational level

FIGURE 1.7 **The relationship between genes and traits at the (a) molecular, (b) cellular, (c) organismal, and (d) populational levels.**

encodes a protein that functions as a pigment-synthesizing enzyme. However, the DNA sequences of the two alleles differ slightly from each other. This difference in the DNA sequence leads to a variation in the structure and function of the respective pigmentation enzymes.

At the cellular level (fig. 1.7b), the functional differences between the pigmentation enzymes affect the amount of pigment that is produced. The allele causing dark pigmentation, which is shown on the *left,* encodes a protein that functions very well. Therefore, when this gene is expressed in the cells of the wings, a large amount of pigment is made. By comparison, the allele causing light pigmentation encodes an enzyme that functions poorly. Therefore, when this allele is the only pigmentation gene that is expressed, little pigment is made.

At the organismal level (fig. 1.7c), the amount of pigment in the wing cells governs the color of the wings. If the pigment cells produce high amounts of pigment, the wings are dark colored; if the pigment cells produce little pigment, the wings are light.

Finally, at the populational level (fig. 1.7d), geneticists would like to know why a species of butterfly would contain some members with dark wings and other members with light wings. One possible explanation is differential predation. The butterflies with dark wings might avoid being eaten by birds if they live within the dim light of a forest. The dark wings would help to camouflage the butterfly if it were perched on a dark surface such as a tree trunk. In contrast, the lightly colored wings would be an advantage if the butterfly inhabited a brightly lit meadow. Under these conditions, a bird may be less likely to notice a light-colored butterfly that is perched on a sunlit surface. A population geneticist might study this species of butterfly and find that the dark-colored members usually live in forested areas and the light-colored members in unforested regions.

Inherited Differences in Traits Are Due to Genetic Variation

In figure 1.7, we considered how gene expression could lead to variation in a trait of the organism (e.g., dark- versus light-colored butterflies). Variation in traits among members of the same species is very common. For example, some people have brown hair, while others have blond hair; some petunias have white flowers, while others have purple flowers. These are examples of **genetic variation.** This term describes the differences in inherited traits among individuals within a population.

In large populations that occupy a wide geographic range, genetic variation can be quite striking. In fact, morphological differences have often led geneticists to misidentify two members of the same species as belonging to separate species. As an example, figure 1.8 compares four garter snakes that are members of the same species, *Thamnophis ordinoides.* They display dramatic differences in their markings. Such contrasting forms within a single species are termed **morphs.** It is easy to imagine how someone might erroneously conclude that these four snakes were not members of the same species.

FIGURE 1.8   **Four garter snakes showing different morphs within a single species.**

Changes in the nucleotide sequence of DNA underlie the genetic variation that we see among individuals. Throughout this textbook, we will routinely examine how variation in the genetic material results in changes in the outcome of traits. At the molecular level, genetic variation can be attributed to different types of modifications.

1. Small differences can occur within gene sequences. These are called **gene mutations.** This type of variation, which produces two or more alleles of the same gene, was previously described in figure 1.7. In many cases, gene mutations will alter the function of the protein that the gene specifies.
2. Major alterations can also occur in the structure of a chromosome. A large segment of a chromosome can be lost or reattached to another chromosome. These types of changes can greatly affect the characteristics of an organism.
3. There can be variations in the total number of chromosomes. In some cases, there may be one too many or one too few chromosomes. In other cases, an offspring may inherit an extra set of chromosomes.

Variations within the sequences of genes are a common source of genetic variation among members of the same species. In humans, familiar examples of variation involve genes for eye color, hair texture, and skin pigmentation. Chromosome variation (i.e., a change in chromosome structure and/or number) is also found, but this type of change is often detrimental. Many human genetic disorders are the result of chromosomal alterations. The most common example is Down syndrome, which is due to the presence of an extra chromosome (fig. 1.9a). By comparison, chromosome variation in plants is common and often can lead to strains of plants with superior characteristics, such as

(a)   **(b)**

FIGURE 1.9   **Examples of chromosome variation.** (**a**) A person with Down syndrome competing in the Special Olympics. This person has 47 chromosomes rather than the normal number of 46. (**b**) A wheat plant. Modern wheat is derived from the contributions of three related species with two sets of chromosomes each, producing an organism with six sets of chromosomes.

increased resistance to disease. Plant breeders have frequently exploited this observation. The cultivated variety of wheat, for example, has many more chromosomes than the wild species (fig. 1.9b).

Traits Are Governed by Genes and by the Environment

In our discussion thus far, we have considered the role that genes play in the outcome of traits. There is another critical factor—the environment. A variety of factors in an organism's environment profoundly affect its morphological and physiological features. For example, a person's diet greatly influences many traits such as height, weight, and even intelligence. Likewise, the amount of sunlight that a plant receives affects its growth rate and the color of its flowers.

External influences may dictate the way genetic variation is manifested in an individual. An interesting example of this phenomenon is the human genetic disease **phenylketonuria (PKU).**

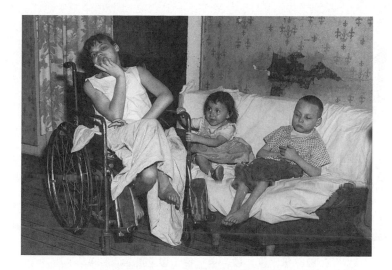

FIGURE 1.10 **Environmental influences on the outcome of PKU within a single family.** All three children pictured here have inherited the alleles that cause PKU. The child in the middle was raised on a phenylalanine-free diet and developed normally. The other two children were born before the benefits of a phenylalanine-free diet were known and were raised on diets that contained phenylalanine. Therefore, they manifest a variety of symptoms, including mental retardation. (Photo from the March of Dimes Birth Defects Foundation.)

Humans possess a gene that encodes an enzyme known as phenylalanine hydroxylase. Most people have two functional copies of this gene, which encodes a fully active phenylalanine hydroxylase protein. People with one or two functional copies of the gene can eat foods containing the amino acid phenylalanine and metabolize it correctly.

A rare variation in the sequence of the phenylalanine hydroxylase gene results in a nonfunctional version of this protein. Individuals with two copies of this rare, inactive allele cannot metabolize phenylalanine. This occurs in about 1 in 8,000 births among Caucasians in the United States. When given a standard diet containing phenylalanine, individuals with this disorder are unable to break down this amino acid and it becomes highly toxic as it accumulates within their bodies. Under these conditions, PKU individuals manifest a variety of detrimental traits including mental retardation, underdeveloped teeth, and foul-smelling urine. In contrast, when PKU individuals are identified at birth and raised on a restricted diet that is free of phenylalanine, they develop normally (fig. 1.10). This is a dramatic example of how the environment and an individual's genes can interact to influence the traits of the organism.

During Sexual Reproduction, Genes Are Passed from Parent to Offspring

Now that we have considered how genes and the environment govern the outcome of traits, we can turn to the issue of inheri-

tance. A centrally important matter in genetics is the manner in which traits are passed from parents to offspring. The foundation for our understanding of inheritance came from the studies of Gregor Mendel. His work revealed that genetic determinants, which we now call genes, are passed from parent to offspring as discrete units. We can now predict the outcome of genetic crosses based on Mendel's laws of inheritance.

The inheritance patterns identified by Mendel can be explained by the existence of chromosomes and their behavior during cell division. As in Mendel's pea plants, it is common for sexually reproducing species to be **diploid.** This means they contain two copies of each chromosome, one from each parent. The two copies are called **homologues** of each other. Since genes are located on the chromosomes, diploid organisms have two copies of most genes. In humans, for example, there are 46 chromosomes, which are found in 23 homologous pairs (fig. 1.11*a*). With the exception of the sex chromosomes (namely, X and Y), each homologous pair contains the same kinds of genes. For example, both copies of human chromosome 12 carry the gene that encodes phenylalanine hydroxylase, which was discussed previously. Therefore, an individual has two copies of this gene; these two copies may or may not be identical alleles.

Most cells of the human body contain 46 chromosomes. The exceptions are the gametes (i.e., sperm and egg cells), which contain half that number (fig. 1.11*b*). The gametes are needed for sexual reproduction. The gametes are considered **haploid.** The union of gametes during fertilization restores the diploid number of chromosomes. The primary advantage of sexual reproduction is that it enhances genetic variation. For example, a tall person with blue eyes and a short person with brown eyes may have short offspring with blue eyes or tall offspring with brown eyes. Therefore, sexual reproduction can result in new combinations of two or more traits that differ from those of either parent.

The Genetic Composition of a Species Evolves over the Course of Many Generations

As we have just seen, sexual reproduction has the potential to enhance genetic variation. This can be an advantage for a population of individuals as they struggle to survive and compete within their natural environment. The term **biological evolution** refers to the phenomenon that the genetic makeup of a population can change over the course of many generations.

As suggested by Charles Darwin, the members of a species are in competition with each other for essential resources. Random genetic changes (i.e., mutations) occasionally occur within an individual's genes, and sometimes these changes lead to a modification of traits that promote survival. For example, over the course of many generations, random gene mutations have lengthened the neck of the giraffe, enabling it to feed on leaves that are high in the trees. When a mutation creates a new allele that is beneficial, the allele may become prevalent in future generations because the individuals carrying the allele are more likely

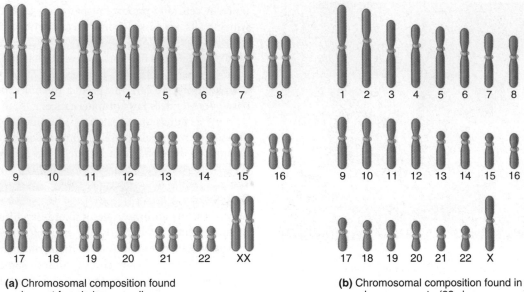

(a) Chromosomal composition found
in most female human cells
(46 chromosomes)

(b) Chromosomal composition found in
a human gamete (23 chromosomes)

FIGURE 1.11   **The complement of human chromosomes in body cells and gametes.** (a) A schematic drawing of the 46 chromosomes of a human. With the exception of the sex chromosomes, these are always found in homologous pairs. (b) The chromosomal composition of a gamete, which contains only 23 chromosomes, one from each pair. This gamete contains an X chromosome. Half of the gametes from human males would contain a Y chromosome instead of the X chromosome.

to reproduce and pass the beneficial allele to their offspring. This process is known as **natural selection.** In this way, a species becomes better adapted to its environment. In addition, neutral mutations, which have little or no impact on survival, can also accumulate in natural populations.

Over a long period of time, the accumulation of many genetic changes leads to rather striking modifications in a species' characteristics. As an example, figure 1.12 depicts the evolutionary changes that led to the development of the modern-day horse. A variety of morphological changes are apparent, including an increase in size, fewer toes, and modified jaw structure.

1.2 FIELDS OF GENETICS

Genetics is a broad discipline encompassing molecular, cellular, organismal, and population biology. Many scientists who are interested in genetics have been trained in supporting disciplines such as biochemistry, biophysics, cell biology, mathematics, microbiology, population biology, ecology, agriculture, and medicine. The study of genetics has been traditionally divided into three areas: transmission, molecular, and population genetics. In this section, we will examine the general questions that scientists in these overlapping areas are attempting to answer.

Transmission Genetics Explores the Inheritance Patterns of Traits As They Are Passed from Parents to Offspring

Transmission genetics is the oldest field of genetics. A scientist working in this field examines the relationship between the transmission of genes from parent to offspring and the outcome of the offspring's traits. For example, how can two brown-eyed parents produce a blue-eyed child? Or why do tall parents tend to produce tall children, but not always? Our modern understanding of transmission genetics began with the studies of Gregor Mendel. His work provided the conceptual framework for transmission genetics. In particular, he originated the idea that genetic determinants, which we now call genes, are passed as discrete units from parent to offspring via sperm and egg cells. Since these pioneering studies of the 1860s, our knowledge of genetic transmission has greatly increased. Many patterns of genetic transmission are more complex than the simple Mendelian patterns that are described in chapter 2. These additional complexities of transmission genetics are examined in chapters 3 through 8.

Experimentally, the fundamental approach of a transmission geneticist is the **genetic cross.** A genetic cross involves a mating between two selected individuals and the subsequent analysis of their offspring in an attempt to understand how traits are passed from parent to offspring. In the case of experimental

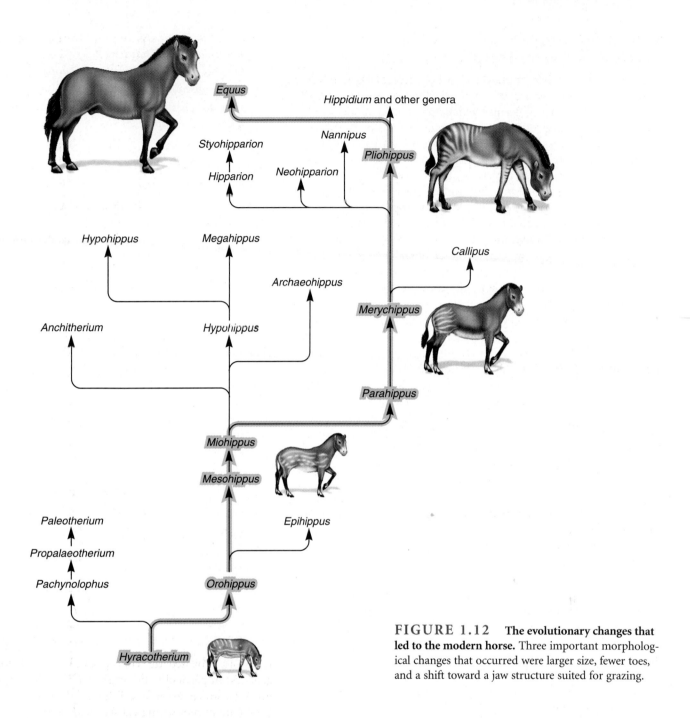

FIGURE 1.12 **The evolutionary changes that led to the modern horse.** Three important morphological changes that occurred were larger size, fewer toes, and a shift toward a jaw structure suited for grazing.

organisms, the experimenter chooses two parents with particular traits and then categorizes the offspring according to the traits they possess. In many cases, this analysis is quantitative in nature. For example, an experimenter may cross two tall pea plants and obtain 100 offspring that fall into two categories: 75 tall and 25 dwarf. As we will see in chapter 2, the ratio of tall and dwarf offspring provides important information concerning the inheritance pattern of this trait.

Throughout chapters 2 to 8, we will learn how transmission genetics seeks to answer many fundamental questions concerning the passage of traits from parents to offspring. Some of these questions are as follows:

What are the common patterns of inheritance for genes? **Chapters 2–4**

When two or more genes are located on the same chromosome, how does this affect the pattern of inheritance? **Chapters 5, 6**

Are there unusual patterns of inheritance that cannot be explained by the simple transmission of genes located on chromosomes in the cell nucleus? **Chapter 7**

How do variations in chromosome structure or chromosome number occur, and how are they transmitted from parents to offspring? **Chapter 8**

Molecular Genetics Seeks a Biochemical Understanding of the Hereditary Material

As the name of the field implies, the goal of molecular genetics is to understand how the genetic material works at the molecular level. In other words, molecular geneticists want to understand the molecular features of DNA and how these molecular features underlie the expression of genes. Experimentally, molecular geneticists often study organisms such as the bacterium *Escherichia coli*, *Saccharomyces cerevisiae* (baker's yeast), *Drosophila*, mice, or *Arabidopsis thaliana*. By limiting their work to a few such "model organisms," researchers can more easily compare their results and begin to unravel the genetic composition of a given species. Furthermore, the genes found in model organisms often function in a similar way to those found in humans. The experiments of molecular geneticists are usually conducted within the confines of a laboratory. Their efforts frequently progress to a detailed analysis of DNA, RNA, and/or protein using a variety of techniques that are described throughout parts III, IV, and V of this textbook.

Molecular geneticists often study mutant genes that have abnormal function. This is called a **genetic approach** to the study of a research question. In many cases, researchers analyze the effects of gene mutations that eliminate the function of a gene. This type of mutation is called a **loss-of-function mutation** or a **loss-of-function allele.** By studying the effects of such mutations, it is often possible to understand the role of the functional, non-mutant gene. For example, let's suppose that a particular plant species produces purple flowers. If a loss-of-function mutation within a given gene causes a plant to produce white flowers, one would suspect the role of the functional gene involves the production of purple pigmentation.

Studies within molecular genetics interface with other disciplines such as biochemistry, biophysics, and cell biology. In addition, advances within molecular genetics have shed considerable light on the areas of transmission and population genetics. We will see that our quest to understand molecular genetics has spawned a variety of modern molecular technologies and computer-based approaches to understand genetics. It is particularly exciting that discoveries within molecular genetics have had widespread applications in agriculture, medicine, and biotechnology.

Some general questions within the field of molecular genetics are the following:

What are the molecular structures of DNA and RNA? **Chapters 9, 18**

What is the composition and conformation of chromosomes? **Chapters 10, 20**

How is the genetic material copied? **Chapter 11**

How are genes expressed at the molecular level? **Chapters 12, 13, 18, 19, 21**

How is gene expression regulated so that it occurs under the appropriate conditions and in the appropriate cell type? **Chapters 14, 15, 18, 23**

What is the molecular nature of mutations? How are mutations repaired? **Chapter 16**

How does the genetic material become rearranged at the molecular level? **Chapter 17**

What is the underlying relationship between genes and genetic diseases? **Chapter 22**

How do genes govern the development of multicellular organisms? **Chapter 23**

Population Genetics Is Concerned with Genetic Variation and Its Role in Evolution

The foundations of population genetics arose during the first few decades of the twentieth century. Although many scientists of this era did not accept the findings of Mendel and/or Darwin, the theories of population genetics provided a compelling way to connect the two viewpoints. Mendel's work and that of many succeeding geneticists gave insight into the nature of genes and how they are transmitted from parents to offspring. The work of Darwin provided a natural explanation for the various types of characteristics observed among the members of a species. To relate these two phenomena, population geneticists have developed mathematical theories to explain the prevalence of certain forms of genes within populations of individuals. The work of population geneticists helps us understand how the forces of nature have produced and favored the existence of individuals that carry particular genes.

Population geneticists are particularly interested in genetic variation and how that variation is related to an organism's environment. In this field, the prevalence of alleles within a population is of central importance. Some general questions in population genetics are the following:

What are the contributions of genetics and environment in the outcome of a trait? **Chapter 24**

How do genetics and the environment influence quantitative traits, such as size and weight? **Chapter 24**

Why are two or more different alleles of a gene maintained in a population? **Chapter 25**

What factors alter the prevalence of alleles within a population? **Chapter 25**

What factors have the most impact on the process of evolution? **Chapter 26**

How does evolution occur at the molecular level? **Chapter 26**

Genetics Is an Experimental Science

Science is a way of knowing about our natural world. The science of genetics allows us to understand how the expression of our genes produces the traits that we possess. The **scientific method** is the basis for conducting science. It is a standard process that scientists follow to reach verifiable conclusions about the world in which we live. Although scientists arrive at their theories in different ways, the scientific method provides a way to validate (or invalidate) a particular hypothesis.

In traditional science textbooks, the emphasis often lies on the product of science. Namely, many textbooks are aimed primarily at teaching the student about the observations that scientists have made and the theories that they have proposed to explain these observations. Along the way, the student is provided with many bits and pieces of experimental techniques and data. Likewise, this textbook also provides you with many observations and theories; however, it attempts to go one step further. Each of the following chapters contains one or two experiments that have been "dissected" into five individual components. As you will see in chapters 2 through 26, these experiments are presented in a way that helps you to understand the entire scientific process.

1. Background information is provided so that you may appreciate what previous observations were known prior to conducting the experiment.

2. The figure states the hypothesis that the scientist (or scientists) was trying to test. In other words, what scientific question was the researcher trying to answer?

3. Next, the figure follows the experimental steps that the scientist took to test the hypothesis. The steps necessary to carry out the experiment are listed in the order they were conducted. In other words, how the experiment was done is shown. The figure contains two parallel illustrations labeled *Experimental Level* and *Conceptual Level*. The illustration shown under the Experimental Level helps you to understand the techniques that were followed. The Conceptual Level helps you to understand what is actually happening at each step in the procedure.

4. The raw data for each experiment are then presented.

5. And finally, an interpretation of the data is offered within the text.

The rationale behind this approach is that it enables you to see the experimental process from beginning to end. Hopefully, you will find this a more interesting and rewarding way to learn about genetics. As you read through the chapters, the experiments will help you to see the relationship between science and scientific theories.

As a student of genetics, you will be given the opportunity to involve your mind in the experimental process. As you are reading an experiment, you may find yourself thinking about different approaches and alternative hypotheses. Different people can view the same data and arrive at very different conclusions. As you progress through the experiments in this textbook, you will enjoy genetics far more if you try to develop your own skills at formulating hypotheses, designing experiments, and interpreting data. Also, some of the questions in the problem sets are aimed at refining these skills.

Finally, it is worthwhile to point out that science is a social discipline. As you develop your skills at scrutinizing experiments, it is fun to discuss your ideas with other people, including fellow students and faculty members. Importantly, you do not need to "know all the answers" before you enter into a scientific discussion. Instead, it is more rewarding to view science as an ongoing and never-ending dialogue.

PROBLEM SETS & INSIGHTS

Solved Problems

S1. A human gene called the *CF* gene (for cystic fibrosis) encodes a protein that functions in the transport of chloride ions across the cell membrane. Most people have two copies of a functional *CF* gene and do not have cystic fibrosis. However, there is a mutant version of the cystic fibrosis gene. If a person has two mutant copies of the gene, he or she develops the disease known as cystic fibrosis. Are the following examples a description of genetics at the molecular, cellular, organismal, or populational level?

A. People with cystic fibrosis have lung problems due to a buildup of mucus in their lungs.

B. The mutant *CF* gene encodes a chloride transporter that doesn't transport chloride ions very well.

C. A defect in the chloride transporter causes a salt imbalance in lung cells.

D. Scientists have wondered why the mutant cystic fibrosis gene is relatively common. It is the most common mutant gene that causes a severe disease in Caucasians. Usually, mutant genes that cause severe diseases are relatively rare. One explanation why the *CF* gene is so common is that people who have one copy of the functional *CF* gene and one copy of the mutant gene are resistant to diarrheal diseases. Therefore, even though individuals with two mutant copies are very sick, people with one mutant copy and one functional copy are more disease resistant than people with two functional copies of the gene.

Answer:

A. Organismal. This is a description of a trait at the level of an entire individual.

B. Molecular. This is a description of a gene and the protein it encodes.

C. Cellular. This is a description of how protein function affects the cell.

D. Populational. This is a possible explanation why two versions of the gene occur within a population.

S2. Explain the relationship between the following pairs of terms:

A. RNA and DNA

B. RNA and transcription

C. Gene expression and trait

D. Mutation and allele

Answer:

A. DNA is the genetic material. DNA is used to make RNA. RNA is then used to specify a sequence of amino acids within a protein.

B. Transcription is a process in which RNA is made using DNA as a template.

C. Genes are expressed at the molecular level to produce functional proteins. The functioning of proteins within living cells ultimately affects an organism's traits.

D. Alleles are alternative forms of the same gene. For example, there is a human gene that affects eye color. The gene can exist as a blue allele and a brown allele. The difference between these two alleles is caused by a mutation. Perhaps the brown allele was the first eye color allele in the human population. Within some ancestral person, however, a mutation may have occurred in the eye color gene that converted the brown allele to the blue allele. Now the human population has both the brown allele and the blue allele.

S3.   How are genes "passed" from generation to generation?

Answer: When a diploid individual makes haploid cells for sexual reproduction, the cells contain half the number of chromosomes. When two haploid cells (e.g., sperm and egg) combine with each other, a zygote is formed that begins the life of a new individual. This zygote has inherited half of its chromosomes (and therefore half of its genes) from each parent. This is how genes are passed from parent to offspring.

Conceptual Questions

C1.   Pick any example of a genetic technology and describe how it has directly impacted your life.

C2.   At the molecular level, what is a gene? Where are genes located?

C3.   Most genes encode proteins. Explain how the structure and function of proteins produce an organism's traits.

C4.   Briefly explain how gene expression occurs at the molecular level.

C5.   A human gene called the β-globin gene encodes a polypeptide that functions as a subunit of the protein known as hemoglobin. Hemoglobin is found within red blood cells; it carries oxygen. In human populations, the β-globin gene can be found as the predominant allele called the Hb^A allele, but it can also be found as the rare Hb^S allele. Individuals who have two copies of the Hb^S allele have the disease called sickle-cell anemia. Are the following examples a description of genetics at the molecular, cellular, organismal, or populational level?

   A. The Hb^S allele encodes a polypeptide that functions slightly differently from the polypeptide encoded by the Hb^A allele.

   B. If an individual has two copies of the Hb^S allele, that person's red blood cells form a sickle shape.

   C. Individuals who have two normal copies of the Hb^A allele are normal, but they are not resistant to malaria. People who have one Hb^A allele and one Hb^S allele are also normal and they are resistant to malaria. People who have two copies of the Hb^S allele have sickle-cell anemia, and this disease may significantly shorten their lives.

   D. Individuals with sickle-cell anemia have anemia because their red blood cells are easily destroyed by the body.

C6.   What is meant by the term *genetic variation?* Give two examples of genetic variation not discussed in chapter 1. What causes genetic variation at the molecular level?

C7.   What is the cause of Down syndrome?

C8.   Your textbook describes how the trait of phenylketonuria is greatly influenced by the environment. Pick a trait in your favorite plant and explain how genetics and environment may play important roles.

C9.   What is meant by the term *diploid?* Which cells of the human body are diploid and which cells are not?

C10. What is a DNA sequence?

C11. What is the genetic code?

C12. Explain the relationships between the following pairs of genetic terms:

   A. Gene and trait

   B. Gene and chromosome

   C. Allele and gene

   D. DNA sequence and amino acid sequence

C13. With regard to biological evolution, which of the following statements is *not* correct? Explain why.

   A. During its lifetime, an animal evolves to become better adapted to its environment.

   B. The process of biological evolution has produced species that are better adapted to their environments.

   C. When an animal is better adapted to its environment, the process of natural selection makes it more likely for that animal to reproduce.

C14. What are the primary interests of researchers working in the following fields of genetics?

   A. Transmission genetics

   B. Molecular genetics

   C. Population genetics

Experimental Questions

E1.   What is a genetic cross?

E2.   The technique known as DNA sequencing (described in chapter 18) enables researchers to determine the DNA sequence of genes. Would this technique be used primarily by transmission geneticists, molecular geneticists, or population geneticists?

E3.   Figure 1.5 shows a micrograph of chromosomes from a normal human cell. If you performed this type of experiment using cells from a person with Down syndrome, what would you expect to see?

E4.   Many organisms are studied by geneticists. Of the following species, do you think it would be more likely for them to be studied by a transmission geneticist, a molecular geneticist, or a population geneticist? Explain your answer. Note: More than one answer may be possible.

   A. Dogs

   B. *E. coli*

   C. Fruit flies

D. Leopards

E. Corn

E5. Pick any trait you like in any species of wild plant or animal. The trait must somehow vary among different members of the species. For example, some moths have dark wings and others have light wings (see fig. 25.13).

　　A. Discuss all of the background information that you already have (from personal observations) regarding this trait.

　　B. Propose a hypothesis that would explain the genetic variation within the species. For example, in the case of the moths, your hypothesis might be that the dark moths survive better in places where the trees are dark, while the light moths survive better in places where the trees are covered with lightly colored lichens.

　　C. Describe the experimental steps you would follow to test your hypothesis.

　　D. Describe the possible data you might collect.

　　E. Interpret your data.

Note: When picking a trait to answer this question, do not pick the trait of wing color in moths.

Note: All answers appear at the website for this textbook; the answers to even-numbered questions are in the back of the textbook.

---

www.mhhe.com/brooker

Visit the Online Learning Center for practice tests, answer keys, and other learning aids for this chapter. Enhance your understanding of genetics with our interactive exercises, web links, news feeds, tutorial service, and much more.

MENDELIAN INHERITANCE

::

2

An appreciation for the concept of heredity can be traced far back in human history. Hippocrates, a famous Greek physician, was the first person to provide an explanation for hereditary traits (ca. 400 B.C.). He suggested that "seeds" are produced by all parts of the body, which are then collected and transmitted to the offspring at the time of conception. Furthermore, he hypothesized that these seeds cause certain traits of the offspring to resemble those of the parents. This theory, known as **pangenesis,** was the first attempt to explain the transmission of hereditary traits from generation to generation.

For the next 2,000 years, the ideas of Hippocrates were accepted by some and rejected by many. After the invention of the microscope in the late seventeenth century, some people observed sperm and thought they could see a tiny creature inside, which they termed a homunculus (little man). This homunculus was hypothesized to be a miniature human waiting to develop within the womb of its mother. These spermists, therefore, suggested that only the father was responsible for creating future generations and that any resemblance between mother and offspring was due to influences "within the womb." During the same time, an opposite school of thought also developed. According to the ovists, it was the egg that was solely responsible for human characteristics. The only role of the sperm was to stimulate the egg onto its path of development. Of course, neither of these ideas was correct.

The first systematic studies of genetic crosses were carried out by Joseph Kölreuter from 1761–66. In crosses between different strains of tobacco plants, he found that the offspring were

PART II
Patterns of Inheritance

usually intermediate in appearance between the two parents. This led Kölreuter to conclude that both parents make equal genetic contributions to their offspring. Furthermore, his observations were consistent with the **blending theory of inheritance.** According to this view, the factors that dictate hereditary traits can blend together from generation to generation. The blended traits would then be passed to the next generation. By combining the notions of pangenesis and the blending theory of inheritance, the popular view before the 1860s was that hereditary traits were rather malleable and could change and blend over the course of one or two generations. However, the pioneering work of Gregor Mendel would prove instrumental in refuting this viewpoint.

In chapter 2, we will first examine the outcome of crosses in pea plants. We begin our inquiry into genetics here because the inheritance patterns observed in peas are fundamentally related to inheritance patterns found in other eukaryotic species such as humans, mice, fruit flies, and corn. Mendel's insights into the patterns of inheritance in pea plants revealed some simple rules that govern the process of inheritance. In chapters 3 through 8, we will explore more complex patterns of inheritance and also consider the role that the chromosomes play as the carriers of the genetic material. In the second part of this chapter, we will become familiar with general concepts in probability and statistics. This is useful in two ways. First, probability calculations allow us to predict the outcomes of genetic crosses described in chapter 2, as well as the outcomes of more complicated crosses described in later chapters. In addition, we will learn how to use statistics to test the validity of genetic hypotheses that attempt to explain the inheritance patterns of traits.

FIGURE 2.1   **Gregor Johann Mendel, the father of genetics.**

2.1 MENDEL'S LAWS OF INHERITANCE

Gregor Johann Mendel, born in 1822, is now remembered as the father of genetics (fig. 2.1). He grew up on a small farm in Heinzendorf in northern Moravia, which was then a part of Austria and is now a part of the Czech Republic. As a young boy, he worked with his father grafting trees to improve the family orchard. Undoubtedly, his success at grafting taught him that precision and attention to detail are important elements of success. These qualities would later be important in his experiments as an adult scientist. Instead of farming, however, Mendel was accepted into the Augustinian monastery of St. Thomas, completed his studies for the priesthood, and was ordained in 1847. Soon after becoming a priest, Mendel worked for a short time as a substitute teacher. To continue that role, he needed to obtain a teaching license from the government. Surprisingly, he failed the licensing exam due to poor answers in the areas of physics and natural history. Therefore, Mendel then enrolled at the University of Vienna to expand his knowledge in these two areas. Mendel's training in physics and mathematics taught him to perceive the world as an orderly place, governed by natural laws. In his studies, Mendel learned that these natural laws could be stated as simple mathematical relationships.

In 1856, Mendel began his historic studies on pea plants. For eight years, he grew and crossed thousands of pea plants on a small 115- by 23-foot plot. He kept meticulously accurate records that included quantitative data concerning the outcome of his crosses. He published his work, entitled "Experiments on Plant Hybrids," in 1866. This paper was largely ignored by scientists at that time, possibly because of its title or because it was published in a rather obscure journal (*The Proceedings of the Brünn Society of Natural History*). Also, another reason his work went unrecognized could be tied to a lack of understanding of chromosome transmission; this topic is discussed in chapter 3. Nevertheless, Mendel's groundbreaking work allowed him to propose the natural laws that now provide a framework for our understanding of genetics.

Prior to his death in 1884, Mendel reflected, "My scientific work has brought me a great deal of satisfaction and I am convinced that it will be appreciated before long by the whole world." Sixteen years later, in 1900, the work of Mendel was independently rediscovered by three biologists with an interest in plant genetics: Hugh de Vries of Holland, Carl Correns of Germany, and Erich von Tschermak of Austria. Within a few years, the impact of Mendel's studies was felt around the world. In chapter 2, we will examine Mendel's experiments and consider their broad significance in the field of genetics.

Mendel Chose Pea Plants as His Experimental Organism

Mendel's study of genetics grew out of his interest in ornamental flowers. Prior to his work with pea plants, many plant breeders had conducted experiments aimed at obtaining flowers with new varieties of colors. When two distinct individuals with different characteristics are mated (or **crossed**) to each other, this is called a **hybridization** experiment and the offspring are referred to as **hybrids.** For example, a hybridization experiment could involve a cross between a purple-flowered plant and a white-flowered plant. From such an experiment, Mendel was particularly intrigued by the consistency with which offspring of subsequent generations showed characteristics of one or the other parent. His intellectual foundation in physics and the natural sciences led him to consider that this regularity might be rooted in natural laws that could be expressed mathematically. To uncover these laws, he realized that he would need to carry out quantitative experiments in which the numbers of offspring carrying certain traits were carefully recorded and analyzed.

Mendel chose the garden pea, *Pisum sativum,* to investigate the natural laws that govern plant hybrids. The morphological features of this plant are shown in figure 2.2a and b. Several properties of this species were particularly advantageous for studying

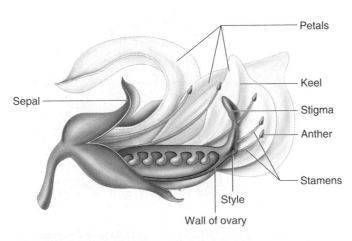

(a) Structure of a pea flower

(b) A flowering pea plant

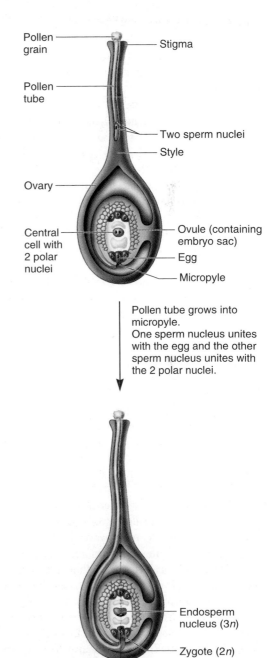

(c) Pollination and fertilization in angiosperms

FIGURE 2.2 **Flower structure and pollination in pea plants. (a)** The pea flower can produce both pollen and egg cells. The pollen grains are produced within the anthers and the egg cells are produced within the ovary. A modified petal called a keel encloses the anthers and ovaries. **(b)** Photograph of a flowering pea plant. **(c)** A pollen grain must first land on the stigma. After this occurs, the pollen sends out a long tube through which two sperm nuclei travel toward the ovules to reach the egg cells. The fusion between a sperm nucleus and an egg cell results in fertilization to create a zygote. A second sperm nucleus fuses with two polar nuclei to create endosperm. The endosperm provides a storage material for the developing embryo.

plant hybridization. First, the species was available in several varieties that had decisively different physical characteristics. As we will discuss, many strains of the garden pea were available that varied in the appearance of their seeds, pods, flowers, and stems.

A second important issue is the ease of making crosses. In flowering plants, reproduction occurs by a pollination event (fig. 2.2c). Male **gametes** (**sperm** nuclei) are produced within **pollen grains** formed in the **stamens,** while the female gametes (**eggs**) are contained within **ovules** that form in the **ovaries.** For fertilization to occur, a pollen grain lands on the **stigma,** which stimulates the growth of a pollen tube. This enables a sperm nucleus to enter the stigma and migrate toward an egg cell. Fertilization occurs when a sperm nucleus fuses with an egg cell. In this textbook, the term *gamete* is used to describe haploid reproductive cells (or nuclei) that can unite to form a zygote. It should be emphasized, however, that the process that produces gametes in animals is quite different from the way that gametes are produced in plants and fungi. These processes are described in greater detail in chapter 3.

In some experiments, Mendel wanted to carry out **self-fertilization.** This means that the pollen and egg are derived from the same plant. In peas, the structure of the flowers greatly favors self-fertilization. The stamens (which produce pollen) and the ovaries (which produce eggs) are covered by a modified petal known as the keel. Because of this covering, pea plants naturally reproduce by self-fertilization. In fact, pollination occurs even before the flower opens.

In other experiments, however, Mendel wanted to make crosses between different plants. Pea plants contain relatively large flowers that are easy to manipulate, making it possible to make crosses between two particular plants and study their outcomes. This process, known as **cross-fertilization,** requires that the pollen from one plant be placed on the stigma of another plant. This procedure is shown in figure 2.3. Mendel was able to pry open immature flowers and remove the stamens before they produced pollen. Therefore, these flowers could not self-fertilize. He would then obtain pollen from another plant by gently touching its mature stamens with a paintbrush. Mendel applied this pollen to the stigma of the flower that already had its stamens removed. In this way, he was able to cross-fertilize his pea plants and thereby obtain any type of hybrid he wanted.

Mendel Studied Seven Traits That Bred True

When he initiated his studies, Mendel obtained several varieties of peas that were considered to be distinct. These plants were different with regard to many morphological characteristics. Such characteristics of an organism are called **characters** or **traits.** Over the course of two years, Mendel tested the strains to determine if their characteristics bred true. This means that a trait did not vary in appearance from generation to generation. For example, if the seeds from a pea plant were yellow, the next generation would also produce yellow seeds. Likewise, if these offspring were allowed to self-fertilize, all of their offspring would also produce

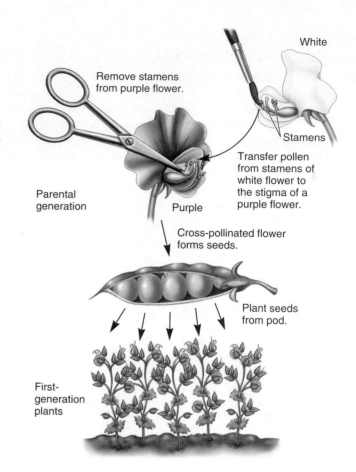

FIGURE 2.3   **How Mendel cross-fertilized two different pea plants.** This illustration depicts a cross between a plant with white flowers and another plant with purple flowers. The offspring that are obtained from this cross are the result of pollination of the purple flower using pollen from a white flower.

yellow seeds, and so on. A variety that continues to produce the same characteristic after several generations of self-fertilization is called a **true-breeding line.**

Mendel next concentrated his efforts on the analysis of characteristics that were clearly distinguishable between different true-breeding lines. Figure 2.4 illustrates the seven characteristics that Mendel eventually chose to follow in his breeding experiments. All seven of Mendel's traits were found in two **variants.** For example, one trait he followed was height. He had variants for this trait, tall and dwarf plants, which he crossed; he then followed the outcomes of these crosses. A cross in which an experimenter is observing only one trait is called a **single-factor cross.** When the two parents are different variants for a given trait, this type of cross produces single-trait hybrids, also known as **monohybrids.** When two monohybrids are crossed to each other, this is referred to as a **monohybrid cross.**

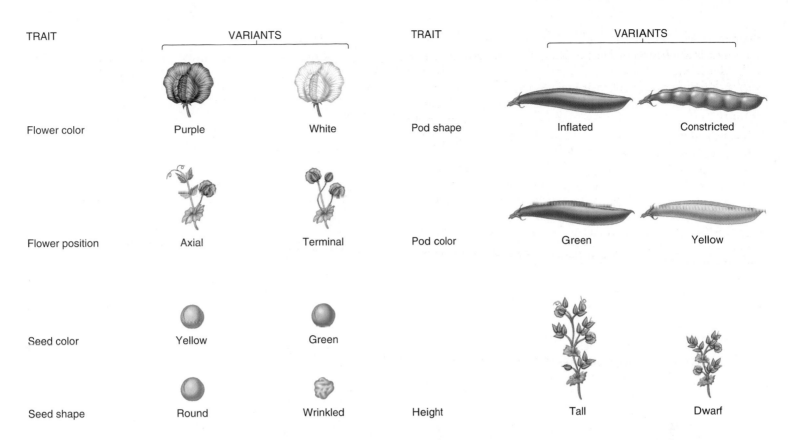

FIGURE 2.4   An illustration of the seven traits that Mendel studied. Each trait was found as two variants that were decisively different from each other.

Mendel Followed the Outcome of a Single Trait for Two Generations

Prior to conducting his experiments, Mendel did not already have a hypothesis to explain the formation of hybrids. However, his educational background caused him to realize that a quantitative analysis of crosses may uncover mathematical relationships that would otherwise be mysterious. His experiments were designed to determine such relationships that govern hereditary traits. This rationale is known as an **empirical approach.** Laws that are deduced from an empirical approach are known as **empirical laws.**

Mendel's experimental procedure is shown in figure 2.5. He began with true-breeding plants that differed with regard to a single trait. These are termed the **parental generation,** or **P gen-** eration. When the true-breeding parents were crossed to each other, this is called a P cross, and the offspring comprise the **F₁ generation.** As seen in the data, all the plants of the F₁ generation showed the phenotype of one parent but not the other. This prompted Mendel to follow the transmission of this trait for one additional generation. To do so, the plants of the F₁ generation were allowed to self-fertilize to produce a second generation called the **F₂ generation.**

THE HYPOTHESIS

The inheritance pattern for a single trait follows quantitative natural "laws."

■ TESTING THE HYPOTHESIS — FIGURE 2.5 Mendel's analysis of single-factor crosses.

Starting material: Mendel began his experiments with true-breeding pea plants that varied with regard to only one of seven different traits (fig. 2.4).

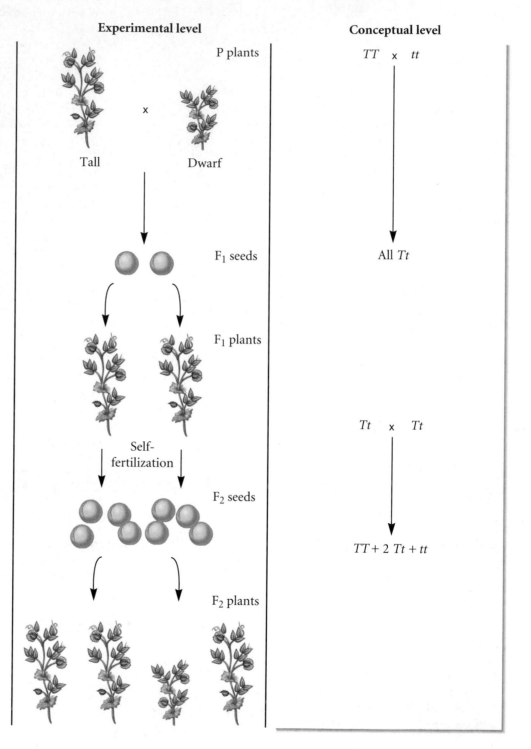

Experimental level

Conceptual level

1. For each of seven traits, Mendel cross-fertilized two different true-breeding lines. Keep in mind that each cross involves two plants that differ only with regard to one trait. The illustration at the right shows one cross between a tall and dwarf plant. This is called a P (parental) cross.

P plants

x

Tall   Dwarf

TT x tt

Note: The P cross produces seeds that are part of the F_1 generation.

F_1 seeds

All Tt

2. Collect many seeds. The following spring, plant the seeds and allow the plants to grow. These are the plants of the F_1 generation.

F_1 plants

3. Allow the F_1 generation plants to self-fertilize. This produces seeds that are part of the F_2 generation.

Self-fertilization

Tt x Tt

F_2 seeds

4. Collect the seeds and plant them the following spring to obtain the F_2 generation plants.

$TT + 2\ Tt + tt$

F_2 plants

5. Analyze the characteristics found in each generation.

THE DATA

P Cross	F₁ Generation	F₂ Generation	Ratio
Round × wrinkled seeds	All round	5,474 round, 1,850 wrinkled	2.96:1
Yellow × green seeds	All yellow	6,022 yellow, 2,001 green	3.01:1
Purple × white flowers	All purple	705 purple, 224 white	3.15:1
Smooth × constricted pods	All smooth	882 smooth, 299 constricted	2.95:1
Green × yellow pods	All green	428 green, 152 yellow	2.82:1
Axial × terminal flowers	All axial	651 axial, 207 terminal	3.14:1
Tall × dwarf stem	All tall	787 tall, 277 dwarf	2.84:1
Total	All dominant	14,949 dom., 5,010 rec.	2.98:1

INTERPRETING THE DATA

The data shown in figure 2.5 are the results of producing an F_1 generation via cross-fertilization, and an F_2 generation via self-fertilization of the F_1 monohybrids. A quantitative analysis of these data allowed Mendel to postulate three important ideas:

1. Mendel's data argued strongly against a blending mechanism of heredity. In all seven cases, the F_1 generation displayed characteristics distinctly like one of the two parents rather than traits intermediate in character. Using genetic terms that Mendel originated and are still used today, we would say that one trait is **dominant** over another trait. For exam-

ple, the trait of green pods is dominant to that of yellow pods. The term **recessive** is used to describe a trait that is masked by the presence of a dominant trait. Yellow pods and dwarf stems are examples of recessive traits.

2. When a true-breeding plant containing a dominant trait was crossed to a true-breeding plant with a recessive trait, the dominant trait was always observed in the F_1 generation. In the F_2 generation, some offspring displayed the dominant characteristic while a lower proportion showed the recessive trait. However, none of the offspring exhibited intermediate traits. This observation caused Mendel to form a second postulate. Since the recessive trait turned up in the second generation, Mendel concluded that the genetic determinants of traits are passed along as "unit factors" from generation to generation. His data were consistent with a **particulate theory of inheritance,** in which the genes that govern traits are inherited as discrete units that remain unchanged as they are passed from parent to offspring. Mendel called them unit factors, but we now call them **genes.**

3. A third important interpretation of Mendel's data is related to the proportions of offspring. When Mendel compared the numbers of dominant and recessive offspring in the F_2 generation, he noticed a recurring pattern. Although there was some experimental variation, he always observed approximately a 3:1 ratio between the dominant trait and the recessive trait. Mendel was the first scientist to apply this type of quantitative analysis in a biological experiment. As described next, this quantitative approach allowed him to conclude that genetic determinants **segregate** from each other during the process that gives rise to gametes.

A self-help quiz involving this experiment can be found at the Online Learning Center.

The 3:1 Phenotypic Ratio That Mendel Observed Is Consistent with the Segregation of Alleles, Now Known as Mendel's Law of Segregation

Mendel's research was aimed at understanding the laws that govern the inheritance of traits. At that time, scientists did not understand the molecular composition of the genetic material or its mode of transmission during gamete formation and fertilization. We now know that the genetic material is composed of deoxyribonucleic acid (DNA), a component of chromosomes. Each chromosome contains hundreds or thousands of shorter segments that function as *genes*—a term that was originally coined by the Danish botanist Wilhelm Johannsen. A gene is defined as a "unit of heredity" that may influence the outcome of an organism's traits. Each of the seven traits that Mendel studied is influenced by a different gene.

Most eukaryotic species, such as pea plants and humans, have their genetic material organized into pairs of chromosomes. For this reason, there are two copies of most eukaryotic genes.

These copies may be the same, or they may differ. The term **allele** refers to different versions of the same gene. With this modern knowledge, the results shown in figure 2.5 are consistent with the idea that each parent transmits only one copy of each gene (i.e., one allele) to each offspring. This is **Mendel's law of segregation:** *The two copies of a gene segregate (or separate) from each other during transmission from parent to offspring.* For this to occur, the two copies of a gene segregate from each other so that only one copy is found in each gamete. At fertilization, two gametes combine randomly, potentially producing different allelic combinations.

Let's use Mendel's cross of tall and dwarf pea plants as an example. The letters *T* and *t* are used to represent the alleles of the gene that determine plant height. By convention, the upper-case letter represents the dominant allele (*T* for tall height, in this case), and the recessive allele is represented by the same letter in lowercase. For the P cross, both parents are true-breeding plants; therefore, we know each has identical copies of the height gene. When an individual possesses two identical copies of a gene, the

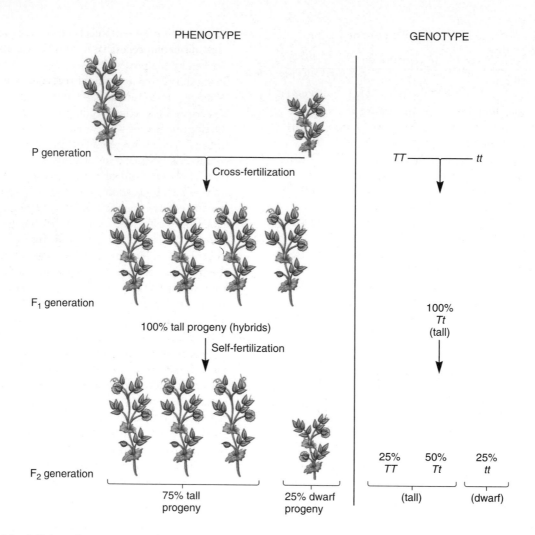

PHENOTYPE   GENOTYPE

P generation

Cross-fertilization

TT ——————— tt

F$_1$ generation

100% tall progeny (hybrids)

100%
Tt
(tall)

Self-fertilization

F$_2$ generation

75% tall
progeny

25% dwarf
progeny

25%   50%   25%
TT Tt tt

(tall)   (dwarf)

FIGURE 2.6   Mendel's law of segregation. This illustration shows a cross between a tall plant and dwarf plant and the subsequent segregation of the tall (T) and dwarf (t) alleles in the F$_1$ and F$_2$ generations.

individual is said to be **homozygous** with respect to that gene. In the P cross, the tall plant is homozygous for the tall allele T, while the dwarf plant is homozygous for the dwarf allele t. The term **genotype** refers to the genetic composition of an individual; TT and tt are the genotypes of the P generation in this experiment. The term **phenotype** refers to an observable characteristic of an organism. In the P generation, half of the plants are phenotypically tall and half are dwarf (fig. 2.6).

In contrast, the F$_1$ generation is **heterozygous,** with the genotype Tt, because every individual carries one copy of the tall allele and one copy of the dwarf allele. A heterozygous individual carries different alleles of a gene. Although these plants are heterozygous, their phenotypes are tall, because they have a copy of the dominant tall allele.

The law of segregation predicts that the phenotypes of the F$_2$ generation will be tall and dwarf in a ratio of 3:1 (fig. 2.6). Since the parents of the F$_2$ generation are heterozygous, each parent can transmit either a T allele or a t allele to a particular offspring, but not both, because each gamete carries only one of the two alleles. Therefore, TT, Tt, and tt are the possible genotypes of

the F$_2$ generation (note that the genotype Tt is the same as tT). By randomly combining these alleles, the genotypes are produced in a 1:2:1 ratio. Because TT and Tt both produce tall phenotypes, a 3:1 phenotypic ratio is observed in the F$_2$ generation.

A Punnett Square Can Be Used to Predict the Outcome of Crosses

An easy way to predict the outcome of simple genetic crosses is to use a **Punnett square,** a method originally proposed by Reginald Punnett. To construct a Punnett square, you must know the genotypes of the parents. With this information, the Punnett square enables you to predict the types of offspring the parents will produce and in what proportions. This section provides a step-by-step description of the Punnett square approach using a cross of heterozygous tall plants as an example.

Step 1.   *Write down the genotypes of both parents.* In this example, a heterozygous tall plant is crossed to another heterozygous tall plant. The plant providing the pollen is considered the male parent and the plant providing the eggs, the female parent.

> Male parent: *Tt*
>
> Female parent: *Tt*

Step 2.   *Write down the possible gametes that each parent can make.* Remember that the law of segregation tells us that a gamete can contain only one copy of each gene.

> Male gametes: *T* or *t*
>
> Female gametes: *T* or *t*

Step 3.   *First, you must create an empty Punnett square.* The number of columns equals the number of male gametes, and the number of rows equals the number of female gametes. In our example, there are two rows and two columns. Place the male gametes across the top of the Punnett square and the female gametes along the side.

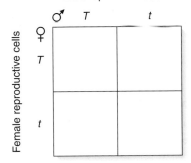

Male reproductive cells

Step 4.   *Fill in the possible genotypes of the offspring by combining the alleles of the gametes in the empty boxes.*

Male reproductive cells

	♂ *T*	*t*
♀ *T*	*TT*	*Tt*
t	*Tt*	*tt*

Female reproductive cells

Step 5.   *Determine the relative proportions of genotypes and phenotypes of the offspring.* The genotypes are obtained directly from the Punnett square. They are contained within the boxes that have been filled in. In this example, the genotypes are *TT*, *Tt*, and *tt* in a 1:2:1 ratio. To determine the phenotypes, you must know the dominant/recessive relationship between the alleles. For plant height, we know that *T* (tall) is dominant to *t* (dwarf). The genotypes *TT* and *Tt* are tall, whereas the genotype *tt* is dwarf. Therefore, our Punnett square shows us that the ratio of phenotypes is 3:1, or 3 tall plants : 1 dwarf plant. Additional problems of this type are provided in the Solved Problems at the end of this chapter.

EXPERIMENT 2B

Mendel Also Analyzed Crosses Involving Two Different Traits

Though his experiments described in figure 2.5 revealed important ideas regarding hereditary laws, Mendel realized that additional insights might be uncovered if he conducted more complicated experiments. In particular, he conducted crosses in which he simultaneously investigated the pattern of inheritance for two different traits. In other words, he carried out **two-factor crosses,** in which he followed the inheritance of two different traits within the same groups of individuals. One of the traits was seed texture, found in round or wrinkled variants. The second trait was seed color, which existed as yellow and green variants. In this two-factor cross, he followed the inheritance pattern for both traits simultaneously. Before we discuss Mendel's results, let's consider different possible patterns of inheritance for two traits (fig. 2.7). One possibility is that the genetic determinants for two different traits are (always) linked to each other and are inherited

as a single unit. A second possibility is that they are not linked and can assort themselves independently into haploid reproductive cells. Keep in mind that the results of figure 2.5 have already shown us that a gamete carries only one allele for each gene.

The experimental protocol of one of Mendel's two-factor crosses is shown in figure 2.8. He began with two different strains of true-breeding pea plants that were different with regard to two traits. In this example, one plant had round, yellow seeds; the other plant had wrinkled, green seeds. As expected, the data revealed that the F_1 seeds displayed a phenotype of round and yellow. This was observed because round and yellow are dominant traits. It is the F_2 generation that supports the independent assortment model and refutes the linkage model.

■ THE HYPOTHESIS

The inheritance pattern for two different traits follows one or more quantitative natural laws.

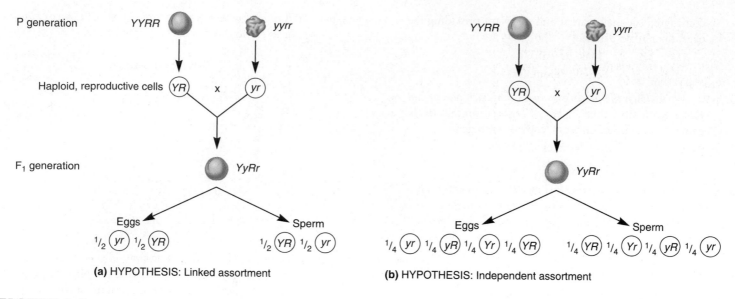

P generation

Haploid, reproductive cells

F$_1$ generation

Eggs   Sperm

(a) HYPOTHESIS: Linked assortment

(b) HYPOTHESIS: Independent assortment

FIGURE 2.7   **Two hypotheses to explain how two different genes assort during gamete formation. (a)** According to the linked hypothesis, the two genes always stay associated with each other. **(b)** In contrast, the independent assortment hypothesis proposes that the two different genes randomly enter into haploid reproductive cells.

■ **TESTING THE HYPOTHESIS** — **FIGURE 2.8**   **Mendel's analysis of two-factor crosses.**

Starting material: In this experiment Mendel began with two types of true-breeding pea plants that were different with regard to two traits. One plant had round, yellow seeds (*RRYY*); the other plant had wrinkled, green seeds (*rryy*).

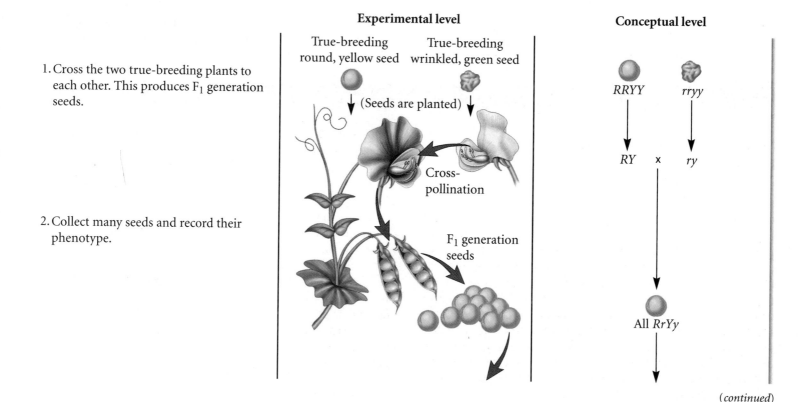

Experimental level

Conceptual level

1. Cross the two true-breeding plants to each other. This produces F$_1$ generation seeds.

2. Collect many seeds and record their phenotype.

True-breeding round, yellow seed

True-breeding wrinkled, green seed

(Seeds are planted)

Cross-pollination

F$_1$ generation seeds

RRYY   *rryy*

RY   x   *ry*

All *RrYy*

(continued)

3. F$_1$ seeds are planted and grown, and the F$_1$ plants are allowed to self-fertilize. This produces seeds that are part of the F$_2$ generation.

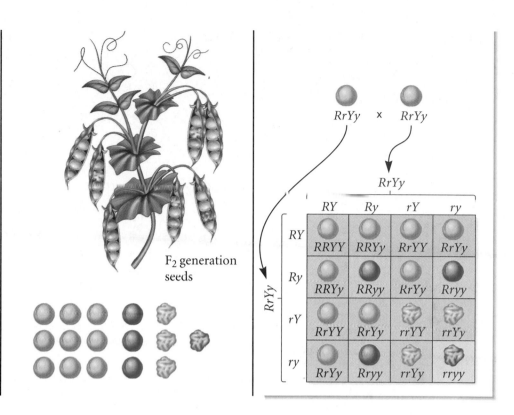

F$_2$ generation seeds

4. Analyze the characteristics found in the F$_2$ generation seeds.

THE DATA

P Cross	F$_1$ Generation	F$_2$ Generation
Round, yellow seeds × wrinkled, green seeds	All round, yellow	315 round, yellow seeds 101 wrinkled, yellow seeds 108 round, green seeds 32 wrinkled, green seeds

INTERPRETING THE DATA

In the F$_2$ generation, there were seeds that are round and green and also seeds that are wrinkled and yellow. These two categories of F$_2$ seeds are called **nonparentals,** because these combinations of traits were not found in the true-breeding plants of the parental generation. The occurrence of nonparental offspring contradicts the linkage model. According to the linkage model, the *R* and *Y* variants should be linked together and so should the *r* and *y* variants. If this were the case, the F$_1$ plants could only produce gametes that are *RY* or *ry*. These would combine to produce *RRYY* (round, yellow), *RrYy* (round, yellow), or *rryy* (wrinkled, green) in a 1:2:1 ratio. There would not be any nonparental seeds produced. However, Mendel did not obtain this result. Instead, he observed a phenotypic ratio of 9:3:3:1 in the F$_2$ generation.

Mendel's results from many dihybrid experiments supported the idea that different traits assort themselves independently during reproduction. Using the modern notion of genes,

the **law of independent assortment** states that *two different genes will randomly assort their alleles during the formation of haploid reproductive cells*. In other words, the allele for one gene will be found within a resulting gamete independently of whether the allele for a different gene is found in the same gamete. Using the example given in figure 2.8, the round and wrinkled alleles will be assorted into haploid reproductive cells independently of the yellow and green alleles. Therefore, a heterozygous *RrYy* parent can produce four different gametes—*RY, Ry, rY,* and *ry*—in equal proportions.

In an F$_1$ self-fertilization experiment, any two gametes can combine randomly during fertilization. This allows for 4^2, or 16, possible offspring, although some offspring will be genetically identical to each other. As shown in figure 2.9, these 16 possible combinations result in seeds with the following phenotypes: 9 round, yellow; 3 wrinkled, yellow; 3 round, green; and 1 wrinkled, green. This 9:3:3:1 ratio is the expected outcome when a dihybrid is allowed to self-fertilize. Mendel was clever enough to realize that the data for his dihybrid experiments were close to a 9:3:3:1 ratio. In figure 2.8, for example, his F$_1$ generation produced F$_2$ seeds with the following characteristics: 315 round, yellow seeds; 101 wrinkled, yellow seeds; 108 round, green seeds; and 32 wrinkled, green seeds. If we divide each of these numbers by 32 (the number of plants with wrinkled, green seeds), the phenotypic ratio of the F$_2$ generation is 9.8:3.2:3.4:1.0. Within experimental error, Mendel's data approximated the predicted 9:3:3:1 ratio for the F$_2$ generation.

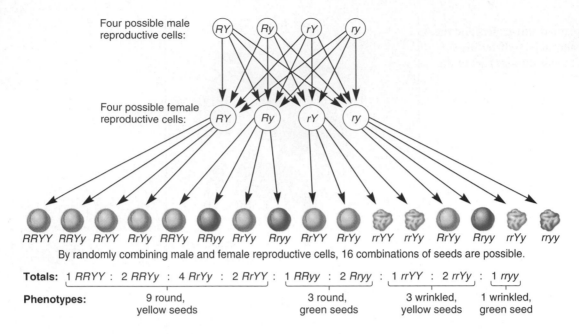

Four possible male reproductive cells: RY   Ry   rY   ry

Four possible female reproductive cells: RY   Ry   rY   ry

RRYY RRYy RrYY RrYy RRYy RRyy RrYy Rryy RrYY RrYY rrYY rrYy RrYY Rryy rrYy rryy

By randomly combining male and female reproductive cells, 16 combinations of seeds are possible.

Totals: 1 *RRYY* : 2 *RRYy* : 4 *RrYy* : 2 *RrYY* : 1 *RRyy* : 2 *Rryy* : 1 *rrYY* : 2 *rrYy* : 1 *rryy*

Phenotypes: 9 round, yellow seeds   3 round, green seeds   3 wrinkled, yellow seeds   1 wrinkled, green seed

FIGURE 2.9   **Mendel's law of independent assortment.**

GENES→TRAITS   The cross is between two parents that are heterozygous for seed shape and seed color (*RrYy* × *RrYy*). Four types of male gametes are possible: *RY, Ry, rY,* and *ry.* Likewise, four types of female gametes are possible: *RY, Ry, rY,* and *ry.* These four types of gametes are the result of the independent assortment of the seed shape and seed color alleles relative to each other. During fertilization, any one of the four male gametes can combine with any one of the four types of female gametes. This results in 16 types of offspring, each one containing two copies of the seed shape gene and two copies of the seed color gene.

The law of independent assortment held true for all seven traits that Mendel studied in pea plants. However, in other cases, the inheritance pattern of two different genes is consistent with the linkage model described earlier in figure 2.7. In chapter 5, we will examine the inheritance of genes that are linked to each other because they are physically within the same chromosome. As we will see, linked genes do not assort independently. Also, we will learn in chapter 5 that the phenomenon of crossing over during meiosis can occasionally lead to the separation of linked genes.

An important consequence of independent assortment is that a single individual can produce a vast array of genetically different gametes. As mentioned in chapter 1, diploid species have pairs of homologous chromosomes, which may differ with respect to the alleles they carry. The phenomenon of independent assortment is rooted in the random pattern by which these homologues assort themselves during the process of meiosis. This topic is addressed in chapter 3. If a species contains a large number of homologous chromosomes, this creates the potential for an enormous amount of genetic diversity. For example, human cells contain 23 pairs of homologous chromosomes. When a person creates gametes via meiosis, these pairs can randomly assort into gametes. Therefore, there are 2^{23}, or over 8 million, possible gametes that an individual can make. The capacity to make so many genetically different gametes enables a species to produce individuals with many different combinations of traits. As discussed in chapter 25, this allows environmental forces to select for those traits that favor reproductive success.

A self-help quiz involving this experiment can be found at the Online Learning Center.

The Punnett Square Can Also Be Used to Solve Independent Assortment Problems

We can use a Punnett square to predict the outcome of crosses involving two or more genes that assort independently. Using the steps described previously, we can consider a cross between two plants that are heterozygous for height and seed color. This cross is *TtYy* × *TtYy*. When we construct a Punnett square for this cross, we must keep in mind that each gamete has a single allele for each of two genes. In this example, the four possible gametes from each parent are

TY, Ty, tY, and *ty*

In this dihybrid experiment, we need to make a larger Punnett square containing 16 boxes. The phenotypes of the resulting offspring are predicted to occur in a ratio of 9:3:3:1.

Cross: *TtYy* x *TtYy*

♀ \ ♂	*TY*	*Ty*	*tY*	*ty*
TY	*TTYY* Tall, yellow	*TTYy* Tall, yellow	*TtYY* Tall, yellow	*TtYy* Tall, yellow
Ty	*TTYy* Tall, yellow	*TTyy* Tall, green	*TtYy* Tall, yellow	*Ttyy* Tall, green
tY	*TtYY* Tall, yellow	*TtYy* Tall, yellow	*ttYY* Dwarf, yellow	*ttYy* Dwarf, yellow
ty	*TtYy* Tall, yellow	*Ttyy* Tall, green	*ttYy* Dwarf, yellow	*ttyy* Dwarf, green

Genotypes: 1 *TTYY* : 2 *TTYy* : 4 *TyYy* : 2 *TtYY* : 1 *TTyy* : 2 *Ttyy* : 1 *ttYY* : 2 *ttYy* : 1 *ttyy*

Phenotypes:

9 tall plants with yellow seeds	3 tall plants with green seeds	3 dwarf plants with yellow seeds	1 dwarf plant with green seeds

In crosses involving three or more genes, it becomes rather unmanageable to construct a single large Punnett square to predict the outcome of crosses. For example, in a trihybrid cross between two pea plants that are *Tt Rr Yy*, each parent can make 2^3, or 8, possible gametes. Therefore, the Punnett square must contain $8 \times 8 = 64$ boxes. As a more reasonable alternative, it is possible to consider each gene separately and then algebraically combine them by multiplying together the expected outcomes for each gene. Two such methods termed the **multiplication method** and the **forked-line method** are shown in solved problem S3 at the end of this chapter.

Independent assortment is also revealed by a **dihybrid testcross.** In this type of experiment, dihybrid individuals are mated to individuals that are doubly homozygous recessive for the two traits. For example, individuals with a *TtYy* genotype could be crossed to *ttyy* plants. As shown here, independent assortment would predict a 1:1:1:1 ratio among the resulting offspring.

	TY	*Ty*	*tY*	*ty*
ty	*TtYy* Tall, yellow	*Ttyy* Tall, green	*ttYy* Dwarf, yellow	*ttyy* Dwarf, green

Modern Geneticists Are Often Interested in the Relationship Between the Molecular Expression of Genes and the Outcome of Traits

Mendel's work with pea plants was critically important, because his laws of inheritance pertain to all eukaryotic organisms such as fruit flies, corn, roundworms, mice, and humans that pass their genes through sexual reproduction. During the past several decades, there has been an increasing interest in the relationship between the phenotypic appearance of traits and the molecular expression of genes. This is a recurring theme throughout this textbook. As mentioned in chapter 1, most genes encode proteins that function within living cells. The specific function of individual proteins affects the outcome of an individual's traits. A genetic approach can help us understand the relationship between a protein's function and its effect on phenotype. Most commonly, a geneticist will try to identify an individual that has a defective copy of a gene to see how that will affect the phenotype of the organism. These defective genes are called **loss-of-function alleles,** and they provide geneticists with a great amount of information.

Unknowingly, Gregor Mendel had studied seven loss-of-function alleles among his strains of plants. The recessive characteristics in his pea plants were due to genes that had been rendered defective by a mutation. It is fairly common for loss-of-function alleles to be inherited in a recessive manner, though this is not always the case. Loss-of-function alleles provide critical clues concerning the purpose of the protein's function within the organism.

For example, we expect the gene affecting flower color (purple versus white) to encode a protein that is necessary for pigment production. This protein may function as an enzyme that is necessary for the synthesis of purple pigment. Furthermore, a reasonable guess is that the white allele is a loss-of-function allele that is unable to express this protein and, therefore, cannot make the purple pigment. To confirm this idea, a biochemist could analyze the petals from purple and white flowers and try to identify the protein that is defective or missing in the white petals but functionally active in the purple ones. The identification and characterization of this protein would provide a molecular explanation for this phenotypic characteristic.

In chapter 5, we will also see that the identification of two alleles for a particular gene makes it possible to determine the gene's chromosomal location by genetic mapping techniques. In chapter 18, we will learn how researchers can make many copies of a particular gene by cloning techniques. This allows a geneticist to better understand the relationship between the molecular characteristics of a gene and its effect on the phenotype of the organism.

Pedigree Analysis Can Be Used to Follow the Mendelian Inheritance of Traits in Humans

Before we end our discussion of simple Mendelian traits, it is interesting to consider how we can analyze inheritance patterns among humans. In Mendel's experiments, he selectively made crosses and then analyzed a large number of offspring. When studying human traits, however, it is not ethical to control parental crosses. Instead, we must rely on the information that is contained within family trees. This type of approach, known as a **pedigree analysis,** is aimed at determining the type of inheritance pattern that a gene will follow. Although this method may be less definitive than the results described in Mendel's experiments, a pedigree analysis can often provide important clues concerning the pattern of inheritance of traits within human families. An expanded discussion of human pedigrees is provided in chapter 22, which concerns the inheritance patterns of many different human diseases, including cancer.

Before discussing the applications of pedigree analyses, we need to understand the symbols and organization of a pedigree (fig. 2.10). The oldest generation is at the top of the pedigree, and the most recent generation is at the bottom. Vertical lines connect each succeeding generation. A man (*square*) and woman (*circle*) who produce one or more offspring are directly connected by a horizontal line. A vertical line connects parents with their offspring. If parents produce two or more offspring, the group of siblings (brothers and sisters) is denoted by two or more individuals projecting from the same horizontal line.

When a pedigree involves the transmission of a human trait or disease, affected individuals are depicted by filled symbols (in this case, black) that distinguish them from unaffected individuals. Each generation is given a Roman numeral designation, and individuals within the same generation are numbered from left to right. Examples of the genetic relationships in figure 2.10 are described here.

Individuals I-1 and I-2 are the grandparents of individual III-2

Individuals III-1, III-2, and III-3 are brother and sisters

Individual III-4 is affected by a genetic disease

Pedigree analysis is commonly used to determine the inheritance pattern of human genetic diseases. Human geneticists are routinely interested in knowing whether a genetic disease is inherited as a recessive or dominant trait. One way to discern the dominant/recessive relationship between two alleles is by a pedigree analysis. Genes that play a role in disease may exist as a normal allele or a mutant allele that causes disease symptoms. If the disease follows a simple Mendelian pattern of inheritance and is caused by a recessive allele, an individual must inherit two copies of the mutant allele to exhibit the disease. Therefore, a recessive

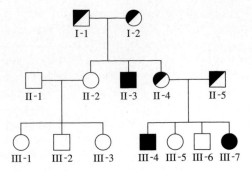

(a) Human pedigree showing cystic fibrosis

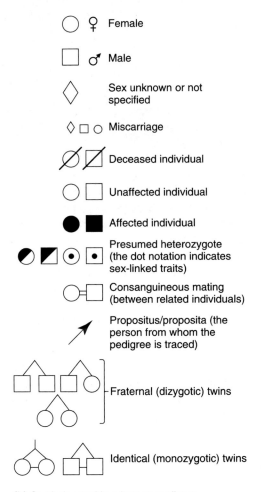

(b) Symbols used in a human pedigree

FIGURE 2.10   Pedigree analysis. (a) A family pedigree in which some of the members are affected with cystic fibrosis. Individuals I-1, I-2, II-4, and II-5 are depicted as presumed heterozygotes because they produce affected offspring. **(b)** The symbols used in a pedigree analysis.

pattern of inheritance makes two important predictions. First, two heterozygous normal individuals will, on average, have 1/4 of their offspring affected. Second, all the offspring of two affected individuals will be affected. Alternatively, a dominant trait predicts that affected individuals will have inherited the gene from at least one affected parent (unless a new mutation has occurred during gamete formation).

The pedigree in figure 2.10a concerns a human genetic disease known as cystic fibrosis (CF). Among Caucasians, approximately 3% of the population are heterozygous carriers of this recessive allele. In homozygotes, the disease symptoms include abnormalities of the pancreas, intestine, sweat glands, and lungs. These abnormalities are caused by an imbalance of ions across the cell membrane. In the lungs, this leads to a buildup of thick, dry mucus. Respiratory problems may lead to early death, although modern treatments have greatly increased the life span of CF patients. In the late 1980s, the gene for CF was identified. It encodes a protein called the cystic fibrosis transmembrane conductance regulator (CFTR). This protein regulates the ion balance across the cell membrane in tissues such as the pancreas, intestine, sweat glands, and lungs. The mutant allele causing CF alters the encoded CFTR protein. The altered CFTR protein functions incorrectly and thereby causes the ionic imbalance. As seen in the pedigree, the pattern of affected and unaffected individuals is consistent with a recessive mode of inheritance. Two unaffected individuals can produce an affected offspring. Although not shown in this pedigree, a recessive mode of inheritance is also characterized by the observation that two affected individuals will produce 100% affected offspring. However, for human genetic diseases that limit survival and/or fertility, there may never be cases where two affected individuals produce offspring.

2.2 PROBABILITY AND STATISTICS

A powerful application of Mendel's work is that the laws of inheritance can be used to predict the outcome of genetic crosses. This is useful in many ways. In agriculture, for example, plant and animal breeders are concerned with the types of offspring that their crosses will produce. This information is used to produce commercially important crops and livestock. In addition, people are often interested in predicting the characteristics of children they may have. This is particularly important to individuals who may carry alleles that cause inherited diseases. Of course, it is not possible to see into the future and definitively predict what will happen. Nevertheless, genetic counselors may help couples to predict the likelihood of having an affected child. This probability is one factor that may influence a couple's decision whether to have children.

In this section, we will see how probability calculations are used in genetic problems to predict the outcome of crosses. To compute probability, we will use three mathematical operations

known as the sum rule, the product rule, and the binomial expansion equation. These methods allow us to determine the probability that a cross between two individuals will produce a particular outcome. To apply these operations, we must have some knowledge regarding the genotypes of the parents and the laws that govern their pattern of inheritance.

Probability calculations can also be used in hypothesis testing. In many situations, a researcher would like to discern the genotypes and patterns of inheritance for traits that are not yet understood. A traditional approach to this problem is to conduct crosses and then analyze their outcomes. The proportions of offspring may provide important clues that allow the experimenter to propose a hypothesis explaining the transmission of the trait from parent to offspring. Statistical methods, such as the chi square test, can then be used to evaluate how well a genetic hypothesis fits the observed data from crosses. We will end this chapter by considering a problem that uses the chi square test as a way to evaluate the validity of a genetic hypothesis.

Probability Is the Likelihood That an Event Will Occur

The chance that an event will occur in the future is called the event's **probability.** For example, if you flip a coin, the probability is 0.50, or 50%, that the head side will be showing when the coin lands. Probability depends on the number of possible outcomes. In this case, there are two possible outcomes (heads and tails), which are equally likely. This allows us to predict there is a 50% chance that a coin flip will produce heads. The general formula for probability is

$$\text{Probability} = \frac{\text{Number of times an event occurs}}{\text{Total number of events}}$$

$$P_{\text{heads}} = 1 \text{ heads}/(1 \text{ heads} + 1 \text{ tails}) = 1/2 = 50\%$$

In genetic problems, we are often interested in the probability that a particular type of offspring will be produced. For example, when two heterozygous tall pea plants (Tt) are crossed, the phenotypic ratio of the offspring is 3 tall:1 dwarf. This information can be used to calculate the probability for either type of offspring.

$$\text{Probability} = \frac{\text{Number of individuals with a given phenotype}}{\text{Total number of individuals}}$$

$$P_{\text{tall}} = 3 \text{ tall}/(3 \text{ tall} + 1 \text{ dwarf}) = 3/4 = 0.75 = 75\%$$

$$P_{\text{dwarf}} = 1 \text{ dwarf}/(3 \text{ tall} + 1 \text{ dwarf}) = 1/4 = 0.25 = 25\%$$

The probability of obtaining a tall plant is 75%, a dwarf plant 25%. When we add together the probabilities of all the possible outcomes (tall and dwarf), we should get a sum of 100% (here, 75% + 25% = 100%).

A probability calculation allows us to predict the likelihood that an event will occur in the future. The accuracy of this prediction, however, depends to a great extent on the size of the sample. For example, if we toss a coin six times, our probability prediction would suggest that 50% of the time we should get heads (i.e., three heads and three tails). In this small sample size, however, we would not be too surprised if we came up with four heads and two tails. Each time we toss a coin, there is a random chance that it will be heads or tails. The deviation between the observed and expected outcomes is due to **random sampling error.** In a small sample, the error between the predicted percentage of heads and the actual percentage observed may be quite large. By comparison, if we flipped a coin 1,000 times, the percentage of heads would be fairly close to the predicted 50% value. In a larger sample, we expect the random sampling error to be a much smaller percentage.

The Sum Rule Can Be Used to Predict the Occurrence of Mutually Exclusive Events

Now that we have an understanding of probability, we can see how mathematical operations using probability values can allow us to predict the outcome of genetic crosses. Our first genetic problem involves the use of the **sum rule,** which states that *the probability that one of two or more mutually exclusive events will occur is equal to the sum of the individual probabilities of the events.*

As an example, we can consider a cross between two mice that are both heterozygous for genes affecting the ears and tail. One gene can be found as an allele designated *de,* which is a recessive allele that causes droopy ears; the normal allele is *De.* An allele of a second gene causes a crinkly tail. This crinkly tail allele *(ct)* is recessive to the normal allele *(Ct).* If a cross is made between two heterozygous mice *(Dede Ctct),* the predicted ratio of offspring is nine with normal ears and normal tails, three with normal ears and crinkly tails, three with droopy ears and normal tails, and one with droopy ears and a crinkly tail. These four phenotypes are mutually exclusive. For example, a mouse with droopy ears and a normal tail cannot have normal ears and a crinkly tail.

The sum rule allows us to determine the probability that we will obtain any one of two or more different types of offspring. For example, in a cross between two heterozygotes *(Dede Ctct × Dede Ctct),* we can ask the following question: What is the probability that an offspring will have normal ears and a normal tail *or* have droopy ears and a crinkly tail? In other words, if we closed our eyes and picked an offspring out of a litter from this cross, what are the chances that we would be holding a mouse that has normal ears and a normal tail or a mouse with droopy ears and a crinkly tail? In this case, the investigator wants to predict whether one of two mutually exclusive events will occur. A strategy for solving such genetic problems using the sum rule is described here.

The Cross: *Dede Ctct × Dede Ctct*

The Question: What is the probability that an offspring will have normal ears and a normal tail *or* have droopy ears and a crinkly tail?

Step 1.   *Calculate the individual probabilities of each phenotype.* This can be accomplished using a Punnett square.

The probability of normal ears and a normal tail is $9/(9 + 3 + 3 + 1) = 9/16$

The probability of droopy ears and a crinkly tail is $1/(9 + 3 + 3 + 1) = 1/16$

Step 2.   *Add together the individual probabilities.*

$$9/16 + 1/16 = 10/16$$

This means that 10/16 is the probability that an offspring will have either normal ears and a normal tail or droopy ears and a crinkly tail. We can convert 10/16 to 0.625, which means that 62.5% of the offspring are predicted to have normal ears and a normal tail *or* droopy ears and a crinkly tail.

The Product Rule Can Be Used to Predict the Probability of Independent Events

We can use probability to make predictions regarding the likelihood of two or more independent outcomes from a genetic cross. When we say that events are independent, we mean that the occurrence of one event does not affect the probability of another event. As an example, let's consider a rare, recessive human trait known as congenital analgesia. Persons with this trait can distinguish between sharp and dull, and hot and cold, but extremes of sensation are not perceived as being painful. The first case of congenital analgesia, described in 1932, was a man who made his living entertaining the public as a "human pincushion."

For a phenotypically normal couple, each being heterozygous for the recessive allele causing congenital analgesia, we can ask the question, What is the probability that the couple's first three offspring will have congenital analgesia? To answer this question, the **product rule** is used. According to this rule, *the probability that two or more independent events will occur is equal to the product of their individual probabilities.* A strategy for solving this type of problem is shown here.

The Cross: *Pp × Pp* (where *P* is the normal allele and *p* is the recessive congenital analgesia allele)

The Question: What is the probability that the couple's first three offspring will have congenital analgesia?

Step 1.   *Calculate the individual probability of this phenotype.* As described previously, this is accomplished using a Punnett square.

The probability of an affected offspring is 1/4.

Step 2. *Multiply the individual probabilities.* In this case, we are asking about the first three offspring, and so we multiply 1/4 three times.

$$1/4 \times 1/4 \times 1/4 = 1/64 = 0.016$$

Thus the probability that the first three offspring will have this trait is 0.016. In other words, we predict that 1.6% of the time the first three offspring of a couple, each heterozygous for the recessive allele, will all have congenital analgesia. In this example, the phenotypes of the first, second, and third offspring are independent events. In other words, the phenotype of the first offspring does not have an effect on the phenotype of the second or third offspring.

In the problem described here, we have used the product rule to determine the probability that the first three offspring will all have the same phenotype (congenital analgesia). We can also apply the rule to predict the probability of a sequence of events that involves combinations of different offspring. For example, consider the question, What is the probability that the first offspring will be normal, the second offspring will have congenital analgesia, and the third offspring will be normal? Again, to solve this problem, begin by calculating the individual probability of each phenotype.

Normal = 3/4

Congenital analgesia = 1/4

The probability that these three phenotypes will occur in this specified order is

$$3/4 \times 1/4 \times 3/4 = 9/64 = 0.14$$

In other words, this sequence of events is expected to occur only 14% of the time.

The product rule can also be used to predict the outcome of a cross involving two or more genes. For example, we can calculate the probability of a given genotype in an offspring if we have information regarding the genotypes of the parents. Let's suppose an individual with the genotype *Aa Bb CC* was crossed to an individual with the genotype *Aa bb Cc*. We could ask the question, What is the probability that an offspring will have the genotype *AA bb Cc*? If the three genes independently assort, the probability of inheriting alleles for each gene is independent of the other two genes. Therefore, we can separately calculate the probability of the desired outcome for each gene.

Cross: *Aa Bb CC* × *Aa bb Cc*

Probability that an offspring will be *AA* = 1/4, or 0.25

Probability that an offspring will be *bb* = 1/2, or 0.5

Probability that an offspring will be *Cc* = 1/2, or 0.5

We can use the product rule to determine the probability that an offspring will be *AA bb Cc*.

$$P = (0.25)(0.5)(0.5) = 0.0625, \text{ or } 6.25\%$$

The Binomial Expansion Equation Can Be Used to Predict the Probability of an Unordered Combination of Events

A third predictive problem in genetics is to determine the probability that a certain proportion of offspring will be produced with specific characteristics; here they can be produced in an unspecified order. For example, we can consider a group of children produced by two heterozygous brown-eyed (*Bb*) individuals. We can ask the question, What is the probability that two out of five children will have blue eyes?

In this case, we are not concerned with the order in which the offspring are born. Instead, we are only concerned with the final numbers of blue-eyed and brown-eyed offspring. One desired outcome would be the following: firstborn child with blue eyes, second child with blue eyes, and then the next three with brown eyes. Another desired outcome could be: firstborn child with brown eyes, second with blue eyes, third with brown eyes, fourth with blue eyes, and fifth with brown eyes. Both of these two situations would satisfy the desire to obtain two offspring with blue eyes and three with brown eyes. In fact, there are many other possible ways to have such a family.

To solve this type of question, the **binomial expansion equation** can be used. This equation represents all of the possibilities for a given set of unordered events.

$$P = \frac{n!}{x! \, (n - x)!} \, p^x q^{n-x}$$

where

P = the probability that the unordered number of events will occur

n = total number of events

x = number of events in one category (e.g., blue eyes)

p = individual probability of *x*

q = individual probability of the other category (e.g., brown eyes)

Note: In this case, *p* + *q* = 1.

The symbol ! denotes a factorial. *n*! is the product of all integers from *n* down to 1. For example, 4! = 4 × 3 × 2 × 1 = 24. An exception is 0!, which equals 1.

The use of the binomial expansion equation is illustrated next.

The Cross: *Bb* × *Bb*

The Question: What is the probability that two out of five offspring will have blue eyes?

Step 1. *Calculate the individual probabilities of the blue-eye and brown-eye phenotypes.* If we constructed a Punnett square, we would find that the probability of blue eyes is 1/4 and the probability of brown eyes is 3/4:

$$p = 1/4$$

$$q = 3/4$$

Step 2.   *Determine the number of events in category* x *(in this case, blue eyes) versus the total number of events.* In this example, the number of events in category *x* is two blue-eyed children among a total number of five.

$$x = 2$$

$$n = 5$$

Step 3.   *Substitute the values for* p, q, x, *and* n *in the binomial expansion equation.*

$$P = \frac{n!}{x!(n-x)!}\,p^x q^{n-x}$$

$$P = \frac{5!}{2!(5-2)!}\,(1/4)^2(3/4)^{5-2}$$

$$P = \frac{5 \times 4 \times 3 \times 2 \times 1}{(2 \times 1)(3 \times 2 \times 1)}\,(1/16)(27/64)$$

$$P = 0.26 = 26\%$$

Thus the probability is 0.26 that two out of five offspring will have blue eyes. In other words, 26% of the time we expect a *Bb* × *Bb* cross yielding five offspring to contain two blue-eyed children and three brown-eyed children.

In solved problem S7 at the end of this chapter, we consider an expanded version of this approach that uses a **multinomial expansion equation.** This method makes it possible to solve unordered genetic problems that involve three or more phenotypic categories.

The Chi Square Test Can Be Used to Test the Validity of a Genetic Hypothesis

We now look at a different issue in genetic problems, namely **hypothesis testing.** Our goal here is to determine if the data from genetic crosses are consistent with a particular pattern of inheritance. For example, a geneticist may study the inheritance of body color and wing shape in fruit flies over the course of two generations. The following question may be asked about the F_2 generation: Do the observed numbers of offspring agree with the predicted numbers based on Mendel's laws of segregation and independent assortment? As we will see in chapters 3 through 8, not all traits follow a simple Mendelian pattern of inheritance. We will see that some genes do not segregate and independently assort themselves the same way that Mendel's seven traits did in pea plants.

To distinguish between inheritance patterns that obey Mendel's laws versus those that do not, a conventional strategy is to make crosses and then quantitatively analyze the offspring. Based on the observed outcome, an experimenter may make a tentative hypothesis. For example, it may seem that the data are obeying Mendel's laws. Hypothesis testing provides an objective, statistical method to evaluate whether or not the observed data really agree with the hypothesis. In other words, we use statistical methods to determine whether the data that have been gathered from crosses are consistent with predictions based on quantitative laws of inheritance.

The rationale behind a statistical approach is to evaluate the **goodness of fit** between the observed data and the data that are predicted from a hypothesis. If the observed and predicted data are very similar, we can conclude that the hypothesis is consistent with the observed outcome. In this case, it is reasonable to accept the hypothesis. However, it should be emphasized that this does not prove that a hypothesis is correct. Statistical methods can never prove that a hypothesis is correct. They can provide insight as to whether or not the observed data seem reasonably consistent with the hypothesis. Alternative hypotheses, perhaps even ones that the experimenter has failed to realize, may also be consistent with the data. In some cases, statistical methods may reveal a poor fit between hypothesis and data. In other words, there would be a high deviation between the observed and expected values. If this occurs, the hypothesis is rejected. Hopefully, the experimenter can subsequently propose an alternative hypothesis that has a reasonable fit with the data.

One commonly used statistical method to determine goodness of fit is the **chi square test** (often written χ^2). We can use the chi square test to analyze population data in which the members of the population fall into different categories. This is the kind of data we have when we evaluate the outcome of genetic crosses, because these usually produce a population of offspring that differ with regard to phenotypes. The general formula for the chi square test is

$$\chi^2 = \sum \frac{(O-E)^2}{E}$$

where

 O = observed data in each category
 E = expected data in each category based on the experimenter's hypothesis
 Σ means to sum this calculation for each category. For example, if the population data fell into two categories, the chi square calculation would be

$$\chi^2 = \frac{(O_1 - E_1)^2}{E_1} + \frac{(O_2 - E_2)^2}{E_2}$$

We can use the chi square test to determine if a genetic hypothesis is consistent with the observed outcome of a genetic cross. The strategy described here provides a step-by-step outline for applying the chi square testing method. In this problem, the experimenter wants to determine if a dihybrid cross is obeying the laws of Mendel. The experimental organism is *Drosophila melanogaster* (the common fruit fly), and the two traits affect wing shape and body color. Normal wing shape and curved wing shape are designated by c^+ and c, respectively; normal (gray) body color and ebony body color are designated by e^+ and e, respectively. Note: In certain species, such as *Drosophila melanogaster*, the convention is to designate the wild-type allele with a plus sign. Recessive mutant alleles are designated with lowercase letters and dominant mutant alleles with capital letters.

The Cross: A true-breeding fly with straight wings and a gray body ($c^+c^+e^+e^+$) is crossed to a true-breeding fly with curved wings and an ebony body (*ccee*). The flies of the F_1 generation are then allowed to mate with each other to produce an F_2 generation.

The Outcome:

F_1 generation: all offspring have straight wings and gray bodies

F_2 generation:
193 straight wings, gray bodies
69 straight wings, ebony bodies
64 curved wings, gray bodies
26 curved wings, ebony bodies

Total: 352

Step 1. *Propose a hypothesis that allows us to calculate the expected values based on Mendel's laws.* The F_1 generation suggests that the trait of straight wings is dominant to curved wings and gray body coloration is dominant to ebony. Looking at the F_2 generation, it appears that offspring are following a 9:3:3:1 ratio. If so, this is consistent with an independent assortment of the two traits.

Based on these observations, the hypothesis is:

Straight (c^+) is dominant to curved (c), and gray (e^+) is dominant to ebony (e). The two traits assort independently from generation to generation.

Step 2. *Based on the hypothesis, calculate the expected values of the four phenotypes.* We first need to calculate the individual probabilities of the four phenotypes. According to our hypothesis, there should be a 9:3:3:1 ratio in the F_2 generation. Therefore, the expected probabilities are

9/16 = straight wings, gray bodies

3/16 = straight wings, ebony bodies

3/16 = curved wings, gray bodies

1/16 = curved wings, ebony bodies

The observed F_2 generation contained a total of 352 individuals. Our next step is to calculate the expected numbers of each type of offspring when the total equals 352. This can be accomplished by multiplying each individual probability by 352.

9/16 × 352 = 198 (expected number with straight wings, gray bodies)

3/16 × 352 = 66 (expected number with straight wings, ebony bodies)

3/16 × 352 = 66 (expected number with curved wings, gray bodies)

1/16 × 352 = 22 (expected number with curved wings, ebony bodies)

Step 3. *Apply the chi square formula, using the data for the expected values that have been calculated in step 2.* In this case, there are four categories within the population, and thus the sum has four terms.

$$\chi^2 = \frac{(O_1 - E_1)^2}{E_1} + \frac{(O_2 - E_2)^2}{E_2} + \frac{(O_3 - E_3)^2}{E_3} + \frac{(O_4 - E_4)^2}{E_4}$$

$$\chi^2 = \frac{(193 - 198)^2}{198} + \frac{(69 - 66)^2}{66} + \frac{(64 - 66)^2}{66} + \frac{(26 - 22)^2}{22}$$

$$\chi^2 = 0.13 + 0.14 + 0.06 + 0.73 = 1.06$$

Step 4. *Interpret the calculated chi square value. This is done using a chi square table.*

Before interpreting the chi square value we have obtained, we must understand how to use table 2.1. The probabilities, called **P values,** listed in the chi square table allow us to determine the likelihood that the amount of variation indicated by a given chi square value is due to random chance alone. For example, let's consider the values listed in row 1. (The meaning of the rows will be explained presently.) Chi square values that are equal to or greater than 0.00393 are expected to occur 95% of the time when a hypothesis is correct. In other words, 95 out of 100 times we would expect that random chance alone would produce a deviation between the experimental data and hypothesized model that is equal to or greater than 0.00393. A low chi square value indicates a high probability that the observed deviations could be due to random chance alone. By comparison, chi square values that are equal to or greater than 3.841 are only expected to occur less than 5% of the time due to random sampling error. If a high chi square value is obtained, an experimenter becomes suspicious that the high deviations have occurred because the hypothesis is incorrect. It is a common convention to reject a hypothesis if the chi square value results in a probability that is less than 0.05 (i.e., less than 5%).

In our problem involving flies with straight or curved wings and gray or ebony bodies, we have calculated a chi square value of 1.06. Before we can determine the probability that this deviation would have occurred as a matter of random chance, we must first determine the degrees of freedom (*df*) in this experiment. The **degrees of freedom** is a measure of the number of categories that are independent of each other. It is typically $n - 1$, where *n* equals the total number of categories. In the preceding problem, $n = 4$ (the categories are the phenotypes: straight wings and gray body; straight wings and ebony body; curved wings and gray body; and curved wings and ebony body); thus, the degrees

TABLE 2.1
Chi Square Values and Probability

Degrees of Freedom	P = 0.99	0.95	0.80	0.50	0.20	0.05	0.01
1	0.000157	0.00393	0.0642	0.455	1.642	3.841	6.635
2	0.020	0.103	0.446	1.386	3.219	5.991	9.210
3	0.115	0.352	1.005	2.366	4.642	7.815	11.345
4	0.297	0.711	1.649	3.357	5.989	9.488	13.277
5	0.554	1.145	2.343	4.351	7.289	11.070	15.086
6	0.872	1.635	3.070	5.348	8.558	12.592	16.812
7	1.239	2.167	3.822	6.346	9.803	14.067	18.475
8	1.646	2.733	4.594	7.344	11.030	15.507	20.090
9	2.088	3.325	5.380	8.343	12.242	16.919	21.666
10	2.558	3.940	6.179	9.342	13.442	18.307	23.209
15	5.229	7.261	10.307	14.339	19.311	24.996	30.578
20	8.260	10.851	14.578	19.337	25.038	31.410	37.566
25	11.524	14.611	18.940	24.337	30.675	37.652	44.314
30	14.953	18.493	23.364	29.336	36.250	43.773	50.892

From Fisher, R. A., and Yates, F. (1943) *Statistical Tables for Biological, Agricultural, and Medical Research.* Oliver and Boyd, London.

of freedom equals 3. We now have sufficient information to interpret our chi square value of 1.06. With *df* = 3, the chi square value of 1.06 is slightly greater than 1.005 (*P* = 0.80). Values equal to or greater than 1.005 are expected to occur 80% of the time based on random chance alone. In other words, 80% of the time, we expect values between 1.005 and infinity, just as a matter of random sampling error. Therefore, in this experiment, it is quite probable that deviations between the observed and expected values can be explained by random sampling error. To reject the hypothesis, the chi square would have to be greater than 7.815. Because it was actually far less than this value, we are inclined to accept that the hypothesis is correct.

We must keep in mind that the chi square test does not *prove* that a hypothesis is correct. It is a statistical method for evaluating whether or not the data and hypothesis have a good fit.

CONCEPTUAL SUMMARY

The work of Gregor Mendel has led to great insight into the transmission of **traits** from parent to offspring. First, his work showed us that the determinants of traits are discrete units that are passed via **gametes** from generation to generation. This **particulate theory of inheritance** refuted previous theories such as the **blending theory of inheritance** and **pangenesis.** In addition, Mendel deduced two fundamental laws that pertain to the transmission of many traits in eukaryotic organisms. The **law of segregation** tells us that two variants for a single trait will segregate during their passage to offspring. This means that the two **alleles** for a particular **gene** will segregate from each other during the formation of haploid reproductive cells, so that a gamete will contain only one copy of an allele. Mendel's **law of independent assortment** says that two different genes assort independently of each other during the formation of haploid reproductive cells.

An understanding of genetic inheritance allows us to make predictions about future offspring and also to test the validity of genetic hypotheses. **Probability** is the likelihood that an event will occur in the future. By knowing the **genotypes** of parents, we can determine the probability of obtaining an offspring with a particular trait. Furthermore, mathematical operations can be used to predict combinations of offspring that may, or may not, occur in a specified order. The **sum rule** is used in genetic problems that involve mutually exclusive events; the **product rule** is used when independent events occur; and the **binomial expansion equation** is used when unordered events are involved. In other genetic problems, the goal is to deduce the pattern of inheritance from parents to offspring. This requires an analysis of genetic **crosses** in order to propose a genetic hypothesis that is consistent with the observed data. A **chi square test** can be used to evaluate whether or not the hypothesis predicts values that have a reasonable agreement with the observed data. In this approach, the amount of deviation between the observed and expected values is used to determine the **goodness of fit** between the data and hypothesis. If the deviation is so large that it would be expected to occur less than 5% of the time as a matter of random sampling error, then it is appropriate to reject the hypothesis. If, on the other hand, the chi square value is low, the hypothesis is accepted.

In chapter 2 we have begun our foundation in the inheritance pattern of simple traits. The two laws that we examined, segregation and independent assortment, provide us with insight into the transmission of traits from parent to offspring. While they form a cornerstone for our understanding of genetics, we will see in later chapters that not all genes obey these laws. In chapter 3, we will take a close look at the composition and behavior of chromosomes, particularly during meiosis. As we will see, the segregation and assortment of chromosomes provides a cellular explanation for Mendel's laws.

EXPERIMENTAL SUMMARY

In chapter 2 we have examined the research studies of Gregor Mendel. By making single-factor crosses and examining the traits of offspring for two generations, Mendel found that the determinants of traits do not blend; rather, they are discrete units that are passed along unaltered from generation to generation. He quantitatively analyzed the outcome of his self-fertilization experiments among the monohybrids and showed that there was a 3:1 phenotypic ratio for the F_2 generation. This observation allowed him to deduce the law of segregation.

In dihybrid experiments, Mendel followed the inheritance pattern of two genes relative to each other. His data supported the idea that two different genes assort independently of each other during the formation of haploid reproductive cells, producing a 9:3:3:1 phenotypic ratio in the F_2 generation. The law of independent assortment held true for all seven traits that Mendel studied. However, as we will see in chapter 5, some genes follow a linked pattern of inheritance.

PROBLEM SETS & INSIGHTS

Solved Problems

S1. A heterozygous pea plant that is tall with yellow seeds, *TtYy*, is allowed to self-fertilize. What is the probability that an offspring will be either tall with yellow seeds, tall with green seeds, or dwarf with yellow seeds?

Answer: This problem involves three mutually exclusive events, and so we use the sum rule to solve it. First, we must calculate the individual probabilities for the three phenotypes. The outcome of the cross can be determined using a Punnett square.

$$P_{\text{Tall with yellow seeds}} = 9/(9 + 3 + 3 + 1) = 9/16$$

$$P_{\text{Tall with green seeds}} = 3/(9 + 3 + 3 + 1) = 3/16$$

$$P_{\text{Dwarf with yellow seeds}} = 3/(9 + 3 + 3 + 1) = 3/16$$

Sum rule: $9/16 + 3/16 + 3/16 = 15/16 = 0.94 = 94\%$

We expect to get one of these three phenotypes 15/16, or 94%, of the time.

Cross: *TtYy* x *TtYy*

♂ \ ♀	TY	Ty	tY	ty
TY	*TTYY* Tall, yellow	*TTYy* Tall, yellow	*TtYY* Tall, yellow	*TtYy* Tall, yellow
Ty	*TTYy* Tall, yellow	*TTyy* Tall, green	*TtYy* Tall, yellow	*Ttyy* Tall, green
tY	*TtYY* Tall, yellow	*TtYy* Tall, yellow	*ttYY* Dwarf, yellow	*ttYy* Dwarf, yellow
ty	*TtYy* Tall, yellow	*Ttyy* Tall, green	*ttYy* Dwarf, yellow	*ttyy* Dwarf, green

S2. As described in chapter 2, a human disease known as cystic fibrosis is inherited as a recessive trait. A normal couple's first child has the disease. What is the probability that their next two children will not have the disease?

Answer: A phenotypically normal couple has already produced an affected child. To be affected, the child must be homozygous for the disease allele and, thus, has inherited one copy from each parent. Therefore, since the parents are unaffected with the disease, we know that both of them must be heterozygous carriers for the recessive disease-causing allele. With this information, we can calculate the probability that they will produce an unaffected offspring. Using a Punnett square, this couple should produce a ratio of 3 unaffected : 1 affected offspring.

N = normal allele
n = cystic fibrosis allele

The probability of a single unaffected offspring is

$$P_{\text{Unaffected}} = 3/(3 + 1) = 3/4$$

To obtain the probability of getting two unaffected offspring in a row (i.e., in a specified order), we must apply the product rule.

$$3/4 \times 3/4 = 9/16 = 0.56 = 56\%$$

There is a 56% chance that their next two children will be unaffected.

S3. A pea plant is heterozygous for three genes ($Tt\ Rr\ Yy$), where $T =$ tall, $t =$ dwarf, $R =$ round seeds, $r =$ wrinkled seeds, $Y =$ yellow seeds, and $y =$ green seeds. If this plant is self-fertilized, what are the predicted phenotypes of the offspring and what fraction of the offspring will occur in each category?

Answer: One could solve this problem by constructing a large Punnett square and filling in the boxes. However, in this case, there are eight possible male gametes and eight possible female gametes: *TRY, TRy, TrY, tRY, trY, Try, tRy,* and *try.* It would become rather tiresome to construct and fill in this Punnett square, which would contain 64 boxes. As an alternative, we can consider each gene separately and then algebraically combine them by multiplying together the expected phenotypic outcomes for each gene. In the cross $Tt\ Rr\ Yy \times Tt\ Rr\ Yy$, the following Punnett squares can be made for each gene:

3 tall : 1 dwarf

3 round : 1 wrinkled

3 yellow : 1 green

Instead of constructing a large, 64-box Punnett square, there are two similar ways to determine the phenotypic outcome of this trihybrid cross. In the **multiplication method,** we can simply multiply these three combinations together:

$$(3 \text{ tall} + 1 \text{ dwarf})(3 \text{ round} + 1 \text{ wrinkled})(3 \text{ yellow} + 1 \text{ green})$$

This multiplication operation can be done in a stepwise manner. First, multiply (3 tall + 1 dwarf) by (3 round + 1 wrinkled).

(3 tall + 1 dwarf)(3 round + 1 wrinkled) = 9 tall, round + 3 dwarf, round + 3 tall, wrinkled + 1 dwarf, wrinkled

Next, multiply this product by (3 yellow + 1 green).

(9 tall, round + 3 dwarf, round + 3 tall, wrinkled + 1 dwarf, wrinkled)(3 yellow + 1 green) = 27 tall, round, yellow + 9 tall, round, green + 9 dwarf, round, yellow + 3 dwarf, round, green + 9 tall, wrinkled, yellow + 3 tall, wrinkled, green + 3 dwarf, wrinkled, yellow + 1 dwarf, wrinkled, green

Even though the multiplication steps are also somewhat tedious, this approach is much easier than making a Punnett square with 64 boxes, filling them in, deducing each phenotype, and then adding them up!

A second approach that is analogous to the multiplication method is the **forked-line method.** In this case, the genetic proportions are determined by multiplying together the probabilities of each phenotype.

Tall or dwarf	Round or wrinkled	Yellow or green	Observed product	Phenotype

$3/4$ tall
- $3/4$ round
  - $3/4$ yellow → $(3/4)(3/4)(3/4) = 27/64$ tall, round, yellow
  - $1/4$ green → $(3/4)(3/4)(1/4) = 9/64$ tall, round, green
- $1/4$ wrinkled
  - $3/4$ yellow → $(3/4)(1/4)(3/4) = 9/64$ tall, wrinkled, yellow
  - $1/4$ green → $(3/4)(1/4)(1/4) = 3/64$ tall, wrinkled, green

$1/4$ dwarf
- $3/4$ round
  - $3/4$ yellow → $(1/4)(3/4)(3/4) = 9/64$ dwarf, round, yellow
  - $1/4$ green → $(1/4)(3/4)(1/4) = 3/64$ dwarf, round, green
- $1/4$ wrinkled
  - $3/4$ yellow → $(1/4)(1/4)(3/4) = 3/64$ dwarf, wrinkled, yellow
  - $1/4$ green → $(1/4)(1/4)(1/4) = 1/64$ dwarf, wrinkled, green

S4. A cross was made between two heterozygous pea plants, $TtYy \times TtYy$. The following Punnett square was constructed:

♂ / ♀	TT	Tt	Tt	tt
YY	TTYY	TtYY	TtYY	ttYY
Yy	TTYy	TtYy	TtYy	ttYy
Yy	TTYy	TtYy	TtYy	ttYy
yy	TTyy	Ttyy	Ttyy	ttyy

Phenotypic ratio:

9 tall, yellow seeds : 3 tall, green seeds : 3 dwarf, yellow seeds : 1 dwarf, green seed

What is wrong with this Punnett square?

Answer: The outside of the Punnett square is supposed to contain the possible types of gametes. A gamete should contain one copy of each type of gene. Instead, the outside of this Punnett square contains two copies of one gene and zero copies of the other gene. The outcome happens to be correct (i.e., it yields a 9:3:3:1 ratio), but this is only a coincidence. The outside of the Punnett square must contain one copy of each type of gene. In this example, the correct possible types of gametes are TY, Ty, tY, and ty for each parent.

S5. For an individual expressing a dominant trait, how can you tell if it is a heterozygote or a homozygote?

Answer: One way is to conduct a cross with an individual that expresses the recessive version of the same trait. If the individual is heterozygous, half of the offspring will show the recessive trait, whereas if the individual is homozygous, none of the offspring will express the recessive trait.

$Dd \times dd$ or $DD \times dd$

1Dd (dominant trait)   All Dd (dominant trait)

1 dd (recessive trait)

Another way to determine heterozygosity involves a more careful examination of the individual at the cellular or molecular level. At the cellular level, the heterozygote may not look exactly like the homozygote. This phenomenon is described in chapter 4. Also, gene cloning methods described in chapter 18 can be used to distinguish between heterozygotes and homozygotes.

S6. In dogs, black fur color is dominant to white. Two heterozygous black dogs are mated. What would be the probability of the following combinations of offspring?

A. A litter of six pups, four with black fur and two with white fur.

B. A litter of six pups, the firstborn with white fur, and among the remaining five pups, two with white fur and three with black fur.

C. A first litter of six pups, four with black fur and two with white fur, and then a second litter of seven pups, five with black fur and two with white fur.

D. A first litter of five pups, four with black fur and one with white fur, and then a second litter of seven pups in which the firstborn is homozygous, the second born is black, and the remaining five pups are three black and two white.

Answer:

A. This is an unordered combination of events, so we use the binomial expansion equation where: $n = 6$, $x = 4$, $p = 0.75$ (probability of black), and $q = 0.25$ (probability of white).

The answer is 0.297, or 29.7%, of the time.

B. We use the product rule because there is a specific order. The first pup is white and then the remaining five are born later. We also need to use the binomial expansion equation to determine the probability of the remaining five pups.

(probability of a white pup)(binomial expansion for the remaining five pups)

The probability of the white pup is 0.25. In the binomial expansion equation, $n = 5$, $x = 2$, $p = 0.25$, and $q = 0.75$.

The answer is 0.066, or 6.6%, of the time.

C. The order of the two litters is specified, so we need to use the product rule. We multiply the probability of the first litter times the probability of the second litter. We need to use the binomial expansion equation for each litter.

(binomial expansion of the first litter)(binomial expansion of the second litter)

For the first litter, $n = 6$, $x = 4$, $p = 0.75$, $q = 0.25$. For the second litter, $n = 7$, $x = 5$, $p = 0.75$, $q = 0.25$.

The answer is 0.092, or 9.2%, of the time.

D. The order of the litters is specified, so we need to use the product rule to multiply the probability of the first litter times the probability of the second litter. We use the binomial expansion equation to determine the probability of the first litter. The probability of the second litter is a little more complicated. The firstborn is homozygous. There are two mutually exclusive ways to be homozygous, *BB* and *bb*. We use the sum rule to determine the probability of the first pup, which equals $0.25 + 0.25 = 0.5$. The probability of the second pup is 0.75, and we use the binomial expansion equation to determine the probability of the remaining pups.

(binomial expansion of first litter)([0.5][0.75][binomial expansion of second litter])

For the first litter, $n = 5$, $x = 4$, $p = 0.75$, $q = 0.25$. For the last five pups in the second litter, $n = 5$, $x = 3$, $p = 0.75$, $q = 0.25$.

The answer is 0.039, or 3.9%, of the time.

S7. In chapter 2 the binomial expansion equation was used in situations where there are only two possible phenotypic outcomes. When there are more than two possible outcomes, it is necessary to use a **multinomial expansion equation** to solve a problem involving an unordered number of events. A general expression for this equation is:

$$P = \frac{n!}{a!\,b!\,c! \, \ldots} \, p^a q^b r^c \ldots$$

where $P =$ the probability that the unordered number of events will occur.

$$n = \text{ total number of events}$$
$$a + b + c + \ldots = n$$
$$p + q + r + \ldots = 1$$

(p is the likelihood of a, q is the likelihood of b, r is the likelihood of c, and so on)

The multinomial expansion equation can be useful in many genetic problems where there are more than two possible combinations of offspring. For example, this formula can be used to solve problems associated with an unordered sequence of events in a dihybrid experiment. This approach is illustrated next.

A cross is made between two heterozygous tall plants with axial flowers (*TtAa*), where tall is dominant to dwarf and axial is dominant to terminal flowers. What is the probability that a group of five offspring will be composed of two tall plants with axial flowers, one tall plant with terminal flowers, one dwarf plant with axial flowers, and one dwarf plant with terminal flowers?

Answer:

Step 1. *Calculate the individual probabilities of each phenotype.* This can be accomplished using a Punnett square.

The phenotypic ratios are 9 tall with axial flowers, 3 tall with terminal flowers, 3 dwarf with axial flowers, and 1 dwarf with terminal flowers.

The probability of a tall plant with axial flowers is $9/(9 + 3 + 3 + 1) = 9/16$.

The probability of a tall plant with terminal flowers is $3/(9 + 3 + 3 + 1) = 3/16$.

The probability of a dwarf plant with axial flowers is $3/(9 + 3 + 3 + 1) = 3/16$.

The probability of a dwarf plant with terminal flowers is $1/(9 + 3 + 3 + 1) = 1/16$.

$$p = 9/16$$
$$q = 3/16$$
$$r = 3/16$$
$$s = 1/16$$

Step 2. *Determine the number of each type of event versus the total number of events.*

$$n = 5$$
$$a = 2$$
$$b = 1$$
$$c = 1$$
$$d = 1$$

Step 3. *Substitute the values in the multinomial expansion equation.*

$$P = \frac{n!}{a!\,b!\,c!\,d!} \, p^a q^b r^c s^d$$

$$P = \frac{5!}{2!\,1!\,1!\,1!} \, (9/16)^2 (3/16)^1 (3/16)^1 (1/16)^1$$

$$P = 0.04 = 4\%$$

This means that 4% of the time we would expect to obtain five offspring with the phenotypes described in the question.

Conceptual Questions

C1. Why did Mendel's work refute the blending theory of inheritance?

C2. What is the difference between cross-fertilization versus self-fertilization?

C3. Describe the difference between genotype and phenotype. Give three examples. Is it possible for two individuals to have the same phenotype but different genotypes?

C4. With regard to genotypes, what is a true-breeding organism?

C5. How can you determine whether an organism is heterozygous or homozygous for a dominant trait?

C6. In your own words, describe what Mendel's law of segregation means. Do not use the word *segregation* in your answer.

C7. Based on genes in pea plants that we have considered in this chapter, which statement(s) is *not* correct?

A. The gene causing tall plants is an allele of the gene causing dwarf plants.

B. The gene causing tall plants is an allele of the gene causing purple flowers.

C. The alleles causing tall plants and purple flowers are dominant.

C8. In a cross between a heterozygous tall pea plant and a dwarf plant, predict the ratios of the offspring's genotypes and phenotypes.

C9. Do you know the genotype of an individual with a recessive trait and/or a dominant trait? Explain your answer.

C10. A cross is made between a pea plant that has constricted pods (a recessive trait; smooth is dominant) and is heterozygous for seed color (yellow is dominant to green) and a plant that is heterozygous for both pod texture and seed color. Construct a Punnett square that depicts this cross. What are the predicted outcomes of genotypes and phenotypes of the offspring?

C11. A pea plant that is heterozygous with regard to seed color (yellow is dominant to green) is allowed to self-fertilize. What are the predicted outcomes of genotypes and phenotypes of the offspring?

C12. Describe the significance of nonparentals with regard to the law of independent assortment. In other words, explain how the appearance of nonparentals refutes a linkage hypothesis.

C13. For the following pedigrees, describe what you think is the most likely inheritance pattern (dominant versus recessive). Explain your reasoning. Filled (*black*) symbols indicate affected individuals.

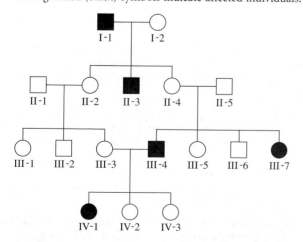

(a)

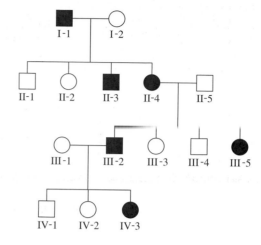

(b)

C14. Ectrodactyly, also known as "lobster claw," is a recessive disorder in humans. If a phenotypically normal couple produces an affected offspring, what are the following probabilities?

A. Both parents are heterozygotes.

B. An offspring is a heterozygote.

C. The next three offspring will be phenotypically normal.

D. Any two out of the next three offspring will be phenotypically normal.

C15. Identical twins are produced from the same sperm and egg (which splits after the first mitotic division), whereas fraternal twins are produced from separate sperm and separate egg cells. If two parents with brown eyes (a dominant trait) produce one twin boy with blue eyes, what are the following probabilities?

A. If the other twin is identical, he will have blue eyes.

B. If the other twin is fraternal, he/she will have blue eyes.

C. If the other twin is fraternal, he/she will transmit the blue eye allele to his/her offspring.

D. The parents are both heterozygotes.

C16. In cocker spaniels, solid coat color is dominant over spotted coat color. If two heterozygous dogs were crossed to each other, what would be the probability of the following combinations of offspring?

A. A litter of five pups, four with solid fur and one with spotted fur.

B. A first litter of six pups, four with solid fur and two with spotted fur, and then a second litter of five pups, all with solid fur.

C. A first litter of five pups, the firstborn with solid fur, and then among the next four, three with solid fur and one with spotted fur, and then a second litter of seven pups in which the firstborn is spotted, the second born is spotted, and the remaining five are composed of four solid and one spotted animal.

D. A litter of six pups, the firstborn with solid fur, the second born spotted, and among the remaining four pups, two with spotted fur and two with solid fur.

C17. A cross was made between a white male dog and two different black females. The first female gave birth to eight black pups, while the second female gave birth to four white and three black pups. What are the likely genotypes of the male parent and the two female parents? Explain whether you are uncertain about any of the genotypes.

C18. In humans, the allele for brown eye color (*B*) is dominant to blue eye color (*b*). If two heterozygous parents produce children, what are the following probabilities?

A. The first two children have blue eyes.

B. A total of four children, two with blue eyes and the other two with brown eyes.

C. The first child has blue eyes, and the next two have brown eyes.

C19. Albinism is a recessive human trait. If a phenotypically normal couple produced an albino child, what is the probability that their next child will be albino?

C20. A true-breeding tall plant was crossed to a dwarf plant. Tallness is a dominant trait. The F_1 individuals were allowed to self-fertilize. What are the following probabilities for the F_2 generation?

A. The first plant is dwarf.

B. The first plant is dwarf or tall.

C. The first three plants are tall.

D. For any seven plants, three are tall and four are dwarf.

E. The first plant is tall, and then among the next four, two are tall and the other two are dwarf.

C21. For pea plants with following genotypes, list the possible gametes that the plant can make:

A. *TT Yy Rr*

B. *Tt YY rr*

C. *Tt Yy Rr*

D. *tt Yy rr*

C22. An individual has the genotype *Aa Bb Cc* and makes an abnormal gamete with the genotype *Aa B c*. Does this gamete violate the law of independent assortment and/or the law of segregation? Explain your answer.

C23. Maple syrup urine disease is a disease found in humans in which the urine smells like maple syrup. An unaffected couple produced six children in the following order: unaffected daughter, affected daughter, unaffected son, unaffected son, affected son, and unaffected son. The youngest unaffected son marries an unaffected woman and has three children in the following order: affected daughter, unaffected daughter, and unaffected son. Draw a pedigree that describes this family. What type of inheritance (dominant or recessive) would you propose to explain maple syrup urine disease?

C24. Marfan syndrome is a rare inherited human disorder characterized by unusually long limbs and digits plus defects in the heart (especially the aorta) and the eyes. Following is the pedigree for this disorder. Affected individuals are shown with filled (*black*) sym-

bols. What type of inheritance pattern do you think is the most likely?

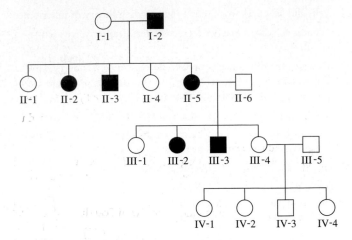

C25. A true-breeding pea plant with round and green seeds was crossed to a true-breeding plant with wrinkled and yellow seeds. Round and yellow seeds are the dominant traits. The F_1 plants were allowed to self-fertilize. What are the following probabilities for the F_2 generation?

A. An F_2 plant with wrinkled, yellow seeds.

B. Three out of three F_2 plants with round, yellow seeds.

C. Five F_2 plants in the following order: two have round, yellow seeds; one has round, green seeds; and two have wrinkled, green seeds.

D. An F_2 plant will not have round, yellow seeds.

C26. A true-breeding tall pea plant was crossed to a true-breeding dwarf plant. What is the probability that an F_1 individual will be true-breeding? What is the probability that an F_1 individual will be a true-breeding tall plant?

C27. What are the expected phenotypic ratios from the following cross: *Tt Rr yy Aa* × *Tt rr YY Aa*, where *T* = tall, *t* = dwarf, *R* = round, *r* = wrinkled, *Y* = yellow, *y* = green, *A* = axial, *a* = terminal; *T, R, Y,* and *A* are dominant alleles. Note: See solved problem S3 for help in answering this problem.

C28. When an abnormal organism contains three copies of a gene (instead of the normal number of two copies), the alleles for the gene usually segregate so that a gamete will contain one or two copies of the gene. Let's suppose that an abnormal pea plant has three copies of the height gene. Its genotype is *T T t*. The plant is also heterozygous for the seed color gene, *Yy*. How many types of gametes can this plant make, and in what proportions? (Assume that it is equally likely that a gamete will contain one or two copies of the height gene.)

C29. Honeybees are unusual in that male bees (i.e., drones) have only one copy of each gene, while female bees have two copies of their genes. That is because male bees develop from eggs that have not been fertilized by sperm cells. In bees, the trait of normal wings is dominant over short wings, and the trait of black eyes is dominant over white eyes. If a drone with short wings and black eyes was mated to a queen bee that is heterozygous for both genes, what are the predicted genotypes and phenotypes of male and female offspring? What are the phenotypic ratios if we assume an equal number of male and female offspring?

C30. A pea plant that is dwarf with green, wrinkled seeds was crossed to a true-breeding plant that is tall with yellow, round seeds. The F_1 generation was allowed to self-fertilize. What types of gametes, and in what proportions, would the F_1 generation make? What would be the ratios of genotypes and phenotypes of the F_2 generation?

C31. A true-breeding plant with round and green seeds was crossed to a true-breeding plant with wrinkled and yellow seeds. The F_1 plants were allowed to self-fertilize. What is the probability of obtaining the following plants in the F_2 generation: two that have round, yellow seeds; one with round, green seeds; and two with wrinkled, green seeds. (Note: See solved problem S7 for help.)

C32. Wooly hair is a rare trait found in people of Scandinavian descent in which the hair resembles the wool of a sheep. It is a dominant trait. A male with wooly hair, who has a mother with normal hair, moves to an island that is inhabited by people who are not of Scandinavian descent. Assuming that he never leaves the island, and that no other Scandinavians immigrate to the island, what is the probability that a great-grandchild of this male will have wooly hair? (Hint: You may want to draw a pedigree to help you figure this out.) If this wooly-haired male has eight great-grandchildren, what is the probability that one out of eight will have wooly hair?

C33. Huntington disease is a rare dominant trait that causes neurodegeneration later in life. A man in his thirties, who already has three children, discovers that his mother has Huntington disease. What are the following probabilities?

A. That the man in his thirties will develop Huntington disease.

B. That his first child will develop Huntington disease.

C. That one out of three of his children will develop Huntington disease.

C34. A woman with achondroplasia (a dominant form of dwarfism) and a phenotypically normal man have seven children, all of whom have achondroplasia. What is the probability of producing such a family if this woman is a heterozygote? What is the probability that the woman is a heterozygote if her eighth child does not have this disorder?

Experimental Questions

E1. Describe three advantages of using pea plants as an experimental organism.

E2. Explain the technical differences between a cross-fertilization experiment versus a self-fertilization experiment.

E3. How long did it take Mendel to complete the experiment in figure 2.5?

E4. For all seven traits described in the data of figure 2.5, Mendel allowed the F_2 plants to self-fertilize. He found that when F_2 plants with recessive traits were crossed to each other, they always bred true. However, when F_2 plants with dominant traits were crossed, some bred true while others did not. A summary of Mendel's results is shown here.

The Ratio of True-Breeding and Non-True-Breeding Parents of the F_2 Generation

F_2 Parents	True-Breeding	Non-True-Breeding	Ratio
Round	193	372	1:1.93
Yellow	166	353	1:2.13
Gray	36	64	1:1.78
Smooth	29	71	1:2.45
Green	40	60	1:1.5
Axial	33	67	1:2.08
Tall	28	72	1:2.57
TOTAL:	525	1,059	1:2.02

When considering the data in this table, keep in mind that it describes the characteristics of the F_2 generation parents that had displayed a dominant phenotype. These data were deduced by analyzing the outcome of the F_3 generation. Based on Mendel's laws, explain the 1:2 ratio obtained in these data.

E5. From the point of view of crosses and data collection, what are the experimental differences between a monohybrid and a dihybrid experiment?

E6. As in many animals, albino is a recessive trait in guinea pigs. Researchers removed the ovaries from an albino female guinea pig and then transplanted ovaries from a true-breeding black guinea pig. They then mated this albino female (with the transplanted ovaries) to an albino male. The albino female produced three offspring. What were their coat colors? Explain the results.

E7. The fungus *Melampsora lini* causes a disease known as "flax rust." Different strains of *M. lini* cause varying degrees of the rust disease. Conversely, different strains of flax are resistant or sensitive to the various varieties of rust. The Bombay variety of flax is resistant to *M. lini*-strain 22 but sensitive to *M. lini*-strain 24. A strain of flax called 770B was just the opposite; it is resistant to strain 24 but sensitive to strain 22. When 770B was crossed to Bombay, all the F_1 individuals were resistant to both *M. lini*-strain 22 and -strain 24. When F_1 individuals were self-fertilized, the following data were obtained:

43 resistant to strain 22 but sensitive to strain 24

9 sensitive to strain 22 and strain 24

32 sensitive to strain 22 but resistant to strain 24

110 resistant to strain 22 and strain 24

Explain the inheritance pattern for flax resistance and sensitivity to *M. lini* strains.

E8. For Mendel's data shown in figure 2.8, conduct a chi square analysis to determine if the data agree with Mendel's law of independent assortment.

E9. Would it be possible to deduce the law of independent assortment from a monohybrid experiment? Explain your answer.

E10. In fruit flies, curved wings are recessive to normal wings, and ebony body is recessive to gray body. A cross was made between true-breeding flies with curved wings and gray bodies to flies with normal wings and ebony bodies. The F_1 offspring that were obtained were then mated to flies with curved wings and ebony bodies to produce an F_2 generation.

A. Diagram the genotypes of this cross, starting with the parental generation and ending with the F_2 generation.

B. What are the predicted phenotypic ratios of the F_2 generation?

C. Let's suppose the following data were obtained for the F_2 generation:

    114 curved wings, ebony body

    105 curved wings, gray body

    111 normal wings, gray body

    114 normal wings, ebony body

Conduct a chi square analysis to determine if the experimental data are consistent with the expected outcome based on Mendel's laws.

E11. A recessive allele in mice results in an abnormally long neck. Sometimes, during early embryonic development, the abnormal neck causes the embryo to die. An experimenter began with a population of true-breeding normal mice and true-breeding mice with long necks. Crosses were made between these two populations to produce an F_1 generation of mice with normal necks. The F_1 mice were then mated to each other to obtain an F_2 generation. For the mice that were born alive, the following data were obtained:

    522 mice with normal necks

    62 mice with long necks

What percentage of homozygous mice (that would have had long necks if they had survived) died during embryonic development?

E12. The data in figure 2.5 show the results of the F_2 generation for seven of Mendel's crosses. Conduct a chi square analysis to determine if these data are consistent with the law of segregation.

E13. Let's suppose you conducted an experiment involving genetic crosses and calculated a chi square value of 1.005. There were four categories of offspring (i.e., the degrees of freedom equaled 3). Explain what the 1.0005 value means. Your answer should include the phrase "80% of the time."

E14. A tall pea plant with axial flowers was crossed to a dwarf plant with terminal flowers. Tall plants and axial flowers are dominant traits. The following offspring were obtained: 27 tall, axial flowers; 23 tall,

terminal flowers; 28 dwarf, axial flowers; and 25 dwarf, terminal flowers. What are the genotypes of the parents?

E15. A cross was made between two strains of plants that are agriculturally important. One strain was disease resistant but herbicide sensitive; the other strain was disease sensitive but herbicide resistant. A plant breeder crossed the two plants and then allowed the F_1 generation to self-fertilize. The following data were obtained:

F_1 generation: all offspring are disease sensitive and herbicide resistant

F_2 generation: 157 disease sensitive, herbicide resistant

              57 disease sensitive, herbicide sensitive

              54 disease resistant, herbicide resistant

              <u>20</u> disease resistant, herbicide sensitive

    Total: 288

Formulate a hypothesis that you think is consistent with the observed data. Test the goodness of fit between the data and your hypothesis using a chi square test. Explain what the chi square results mean.

E16. A cross was made between a plant that has blue flowers and purple seeds to a plant with white flowers and green seeds. The following data were obtained:

F_1 generation: all offspring have blue flowers with purple seeds

F_2 generation: 103 blue flowers, purple seeds

              49 blue flowers, green seeds

              44 white flowers, purple seeds

              <u>104</u> white flowers, green seeds

    Total: 300

Start with the hypothesis that blue flowers and purple seeds are dominant traits and that the two genes are assorting independently. Calculate a chi square value. What does this value mean with regard to your hypothesis? If you decide to reject your hypothesis, which aspect of the hypothesis do you think is incorrect (i.e., blue flowers and purple seeds are dominant traits, or the idea that the two genes are assorting independently)?

Questions for Student Discussion/Collaboration

1. Consider the following cross in pea plants: *Tt Rr yy Aa* $\times$ *Tt rr Yy Aa*, where *T* = tall, *t* = dwarf, *R* = round, *r* = wrinkled, *Y* = yellow, *y* = green, *A* = axial, *a* = terminal. What is the expected phenotypic outcome of this cross? Have one group of students solve this problem by making one big Punnett square, and have another group solve it by making four single-gene Punnett squares and using multiplication. Time each other to see who gets done first.

2. A cross was made between two pea plants, *TtAa* and *Ttaa*, where *T* = tall, *t* = dwarf, *A* = axial, and *a* = terminal. What is the probability that the first three offspring will be tall with axial flowers or dwarf with terminal flowers and the fourth offspring will be tall

with axial flowers. Discuss what operation(s) (e.g., sum rule, product rule, and/or binomial expansion) you used to solve them and in what order they were used.

3. Consider the following tetrahybrid cross: *Tt Rr yy Aa* $\times$ *Tt RR Yy aa*, where *T* = tall, *t* = dwarf, *R* = round, *r* = wrinkled, *Y* = yellow, *y* = green, *A* = axial, *a* = terminal. What is the probability that the first three plants will have round seeds? What is the easiest way to solve this problem?

Note: All answers appear at the website for this textbook; the answers to even-numbered questions are in the back of the textbook.

Visit the Online Learning Center for practice tests, answer keys, and other learning aids for this chapter. Enhance your understanding of genetics with our interactive exercises, web links, news feeds, tutorial service, and much more.

REPRODUCTION AND
CHROMOSOME TRANSMISSION

3

In chapter 2, we considered some patterns of inheritance that explain the passage of traits from parent to offspring. In chapter 3, we will survey reproduction at the cellular level, and we will pay close attention to the inheritance of chromosomes. An examination of chromosomes at the microscopic level provides us with insights to understand the inheritance patterns of traits. To appreciate this relationship, we will first consider how cells distribute their chromosomes during the process of cell division. We will see that in bacteria and most unicellular eukaryotes, simple cell division provides a way to reproduce asexually. Then we will explore a form of cell division called meiosis that produces cells with half the number of chromosomes. By closely examining this process, we will see how the transmission of chromosomes accounts for the inheritance patterns that were observed by Mendel.

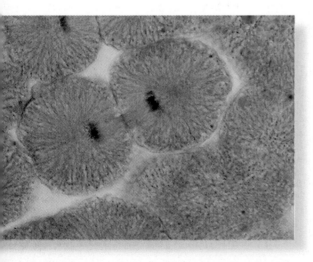

3.1 GENERAL FEATURES OF CHROMOSOMES

The **chromosomes** are structures within living cells that contain the genetic material. Genes are physically located within the chromosomes. Biochemically, chromosomes contain a very long segment of DNA, which is the genetic material, and proteins, which are bound to the DNA and provide it with an organized structure. In chapter 3, we will focus on the cellular mechanics of chromosome transmission to better understand the patterns of gene transmission that we considered in chapter 2. In particular, we will examine how chromosomes are copied and sorted into newly made cells. In later chapters, particularly Chapters 8, 10, and 11, we will examine the molecular features of chromosomes in greater detail.

Before we begin a description of chromosome transmission, we need to consider the distinctive cellular differences between bacteria and eukaryotic species. Bacteria and archaea are referred to as **prokaryotes,** from the Greek meaning prenucleus, because their chromosomes are not contained within a separate nucleus of the cell. Prokaryotes usually contain a single type of circular chromosome in a region of the cytoplasm called the **nucleoid** (fig. 3.1*a*). The cytoplasm is enclosed by a plasma membrane that regulates the uptake of nutrients and the excretion of waste products. Outside the plasma membrane is found a rigid cell wall that protects the cell from breakage. Certain species of bacteria also have an outer membrane that is located beyond the cell wall.

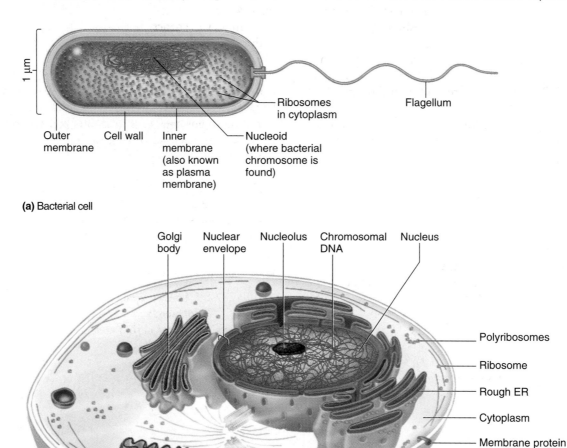

(a) Bacterial cell

(b) Animal cell

FIGURE 3.1 **The basic organization of cells.** (a) A bacterial cell. The example shown here is typical of bacterium such as *Escherichia coli*, which has an outer membrane. (b) A eukaryotic cell. The example shown here would be typical of an animal cell.

Eukaryotes, from the Greek meaning true nucleus, include some simple species, such as single-celled protists and fungi (such as yeast), and more complex multicellular species, such as plants, animals, and other fungi. The cells of eukaryotic species are compartmentalized, which means there are internal membranes that enclose highly specialized compartments (fig. 3.1*b*). These compartments form membrane-bounded **organelles** with specific functions. For example, the lysosome plays a role in the degradation of macromolecules. The endoplasmic reticulum and Golgi body play a role in protein modification and trafficking. A particularly conspicuous organelle is the **nucleus,** which is bounded by two membranes. Most of the genetic material is found within chromosomes that are located in the nucleus. In addition to the nucleus, certain organelles in eukaryotic cells contain a small amount of their own DNA. These include the mitochondrion, which plays a role in ATP synthesis, and the chloroplast, which plays a role in photosynthesis. The DNA found in these organelles is referred to as extranuclear or extrachromosomal DNA to distinguish it from the DNA that is found in the cell nucleus. We will examine the role of mitochondrial and chloroplast DNA in chapter 7.

This section of chapter 3 is focused on the composition of chromosomes found in the nucleus of eukaryotic cells. As we will learn, eukaryotic species contain genetic material that comes in sets of linear chromosomes.

Eukaryotic Chromosomes Are Examined Cytologically to Yield a Karyotype

Insights into inheritance patterns have been gained by observing the behavior of chromosomes under the microscope. **Cytogenetics** is the field of genetics that involves the microscopic examination of chromosomes. The most basic observation that a **cytogeneticist** can make is to examine the chromosomal composition of a particular cell. For eukaryotic species, this is usually accomplished by observing the chromosomes as they are found in actively dividing cells. When a cell is preparing to divide, the chromosomes become more tightly coiled, which shortens them and thereby increases their diameter. The consequence of this shortening is that distinctive shapes and numbers of chromosomes become visible with a light microscope. There is constancy in the chromosome composition within a given species. For example, most human cells contain 23 pairs of chromosomes for a total of 46. However, some individuals may inherit an abnormal number of chromosomes or a chromosome with an abnormal structure. Such abnormalities can often be detected by a microscopic examination of the chromosomes within actively dividing cells. In addition, a cytogeneticist may examine chromosomes as a way to distinguish two closely related species.

Figure 3.2 shows the general procedure for preparing human chromosomes to be viewed by microscopy. In this example, the cells were obtained from a sample of human blood; more specifically, the chromosomes within lymphocytes (a type of white blood cell) were examined. Blood cells are a type of **somatic cell.** This term refers to any cell of the body that is not a gamete. The **gametes** (i.e., sperm and egg cells) are types of **germ cells.**

After the blood cells have been removed from the body, they are treated with drugs that stimulate them to divide. As shown in figure 3.2, these actively dividing cells are centrifuged to concentrate them. The concentrated preparation is then mixed with a hypotonic solution that makes the cells swell. This swelling causes the chromosomes to spread out within the cell and thereby makes it easier to see each individual chromosome. Next, the cells are treated with a fixative that chemically freezes them so that the chromosomes will no longer move around. The cells are then treated with a chemical dye that binds to the chromosomes and stains them. As discussed in greater detail in chapter 8, this gives chromosomes a distinctive banding pattern that greatly enhances their visualization and ability to be uniquely identified. The cells are then placed on a slide and viewed with a light microscope.

In a cytogenetics laboratory, the microscopes are equipped with a camera that can photograph the chromosomes. In recent years, advances in technology have allowed cytogeneticists to scan microscopic images into a computer. On a computer screen, the chromosomes can be organized in a standard way, usually from largest to smallest. As seen at the bottom right of figure 3.2, the human chromosomes have been lined up, and a number is given to designate each type of chromosome. An exception would be the sex chromosomes, which are designated with the letters X and Y. We will consider the sex chromosomes later in this chapter. A photographic representation of the chromosomes within a cell, as in figure 3.2, is called a **karyotype.** A karyotype reveals how many chromosomes are found within an actively dividing somatic cell.

Eukaryotic Chromosomes Are Inherited in Sets

Most eukaryotic species are **diploid,** which means that each type of chromosome is a member of a pair. In diploid species, the somatic cells have two sets of chromosomes. As mentioned, the somatic cells of humans carry two sets of chromosomes with 23 chromosomes per set (46 total). Other diploid species, however, can have different numbers of chromosomes. For example, the dog has 78 chromosomes (39 per set), the fruit fly has 8 chromosomes (4 per set), and the tomato has 24 chromosomes (12 per set).

When a species is diploid, the members of a pair of chromosomes are called **homologues;** each type of chromosome is found in a homologous pair. As shown in figure 3.2, for example, there are two copies of chromosome 1, two copies of chromosome 2, and so forth. Within each pair, the chromosome on the left is a homologue to the one on the right (and vice versa). Homologous chromosomes are very similar to each other. The two chromosomes in a homologous pair are nearly identical in size, have the same banding pattern and centromere location, and contain a similar composition of genetic material. If a particular gene is found on one copy of a chromosome, it is also found on the other homologue. However, the two homologues may carry different alleles of a given gene. As an example, let's consider the eye color gene in humans. One chromosome might carry the dominant brown allele, while its homologue could carry the blue allele.

A sample of blood is collected and subjected to centrifugation.

- Supernatant

Blood cells — - Pellet

The supernatant is discarded, and the cell pellet is suspended in a hypotonic solution. This causes the cells to swell.

— Hypotonic solution

The sample is subjected to centrifugation a second time to concentrate the cells. The cells are suspended in a fixative, stained, and placed on a slide.

Fix
Stain

Blood cells —

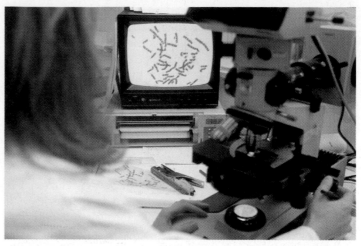

The slide is viewed by a light microscope; the sample is seen on a video screen. The chromosomes can be arranged electronically on the screen.

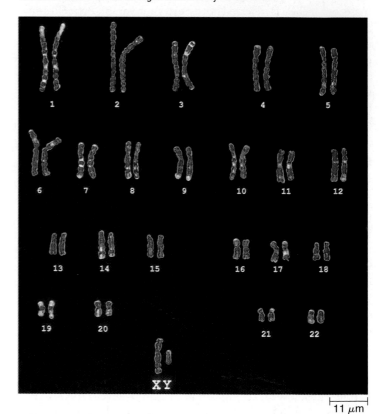

11 μm

FIGURE 3.2   The procedure for making a human karyotype.

For a diploid human cell, two complete sets of chromosomes from a single cell constitute a karyotype of that cell.

The DNA sequences of homologous chromosomes are also very similar. In most cases, the sequence of bases of one homologue would differ by less than 1% compared to the sequence of the other homologue. For example, the DNA sequence of chromosome 1 that you inherited from your mother would be greater than 99% identical to the sequence of chromosome 1 that you inherited from your father. Nevertheless, it should be emphasized that the sequences are not identical. The slight differences in DNA sequence provide the allelic differences in genes. Again, if we use the eye color gene as an example, it is a slight difference in DNA sequence

that distinguishes the brown allele from the blue allele. It should also be noted that the striking similarities between homologous chromosomes do not apply to the pair of sex chromosomes (e.g., X and Y). These chromosomes differ in size and genetic composition. Certain genes that are found on the X chromosome are not found on the Y chromosome, and vice versa. The X and Y chromosomes are not considered homologous chromosomes though they do have short regions of homology. A comparison of the genetic composition of the X and Y chromosomes will be discussed at the end of chapter 3.

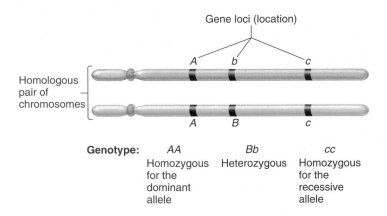

FIGURE 3.3 A comparison of homologous chromosomes. Each pair of homologous chromosomes carries the same types of genes, but, as shown here, the alleles may be different.

Figure 3.3 considers two homologous chromosomes that are labeled with three different genes. An individual carrying these two chromosomes would be homozygous for the dominant allele of gene *A*. The individual would be heterozygous, *Bb,* for the second gene. For the third gene, the individual is homozygous for a recessive allele, *c*. The physical location of a gene is called its **locus** (plural: **loci**). As seen in figure 3.3, for example, the locus of gene *C* is toward one end of this chromosome while the locus of gene *B* is more in the middle.

3.2 CELLULAR DIVISION

Now that we have an appreciation for the chromosomal composition of living cells, we can consider how chromosomes are copied and transmitted when cells divide. One purpose of cell division is **asexual reproduction.** This is the way that some unicellular organisms produce new individuals. In this process, a preexisting cell divides to produce two new cells. By convention, the original cell is usually called the mother cell, and the new cells are the two daughter cells. When species are unicellular, the mother cell is judged to be one individual and the two daughter cells are two new separate organisms. Asexual reproduction is how bacterial cells proliferate. In addition, certain unicellular eukaryotes, such as the amoeba and baker's yeast (*Saccharomyces cerevisiae*), can reproduce asexually.

A second important reason for cell division is multicellularity. Species such as plants, animals, and certain fungi are derived from a single cell (e.g., a fertilized egg) that has undergone repeated cellular divisions. Humans, for example, begin as a single fertilized egg; repeated cellular divisions produce an adult with several trillion cells. As you might imagine, it is critical to have the precise transmission of chromosomes during every cell division, so that all the cells of the body receive the correct amount of genetic material.

In this section, we will consider how the process of cell division requires the duplication, organization, and sorting of the chromosomes. In bacteria, which have a single circular chromo-

some, the sorting process is relatively simple. Prior to cell division, bacteria duplicate their circular chromosome; they then distribute a copy into each of the two daughter cells. This process, known as binary fission, is described next. By comparison, eukaryotes have multiple numbers of chromosomes that occur as sets. Compared to bacteria, this added complexity requires a more complicated sorting process to ensure that each newly made cell receives the correct number and types of chromosomes. As described later in this section, a mechanism known as mitosis entails the organization and sorting of eukaryotic chromosomes during cell division.

Bacteria Reproduce Asexually by Binary Fission

As discussed earlier (see fig. 3.1*a*), bacterial species are unicellular, although individual bacteria may associate with each other to form pairs, chains, or clumps. Unlike eukaryotes, which have their chromosomes in a separate nucleus, the circular chromosomes of bacteria are in direct contact with the cytoplasm. In chapter 10, we will consider the molecular structure of bacterial chromosomes in greater detail.

The capacity of bacteria to divide is really quite astounding. Some species, such as *Escherichia coli* (a common bacterium of the intestine), can divide every 20 to 30 minutes. Prior to cell division, bacterial cells copy, or replicate, their chromosomal DNA. This produces two identical copies of the genetic material as shown at the top of figure 3.4. Following DNA replication, a bacterial cell divides into two daughter cells by a process known

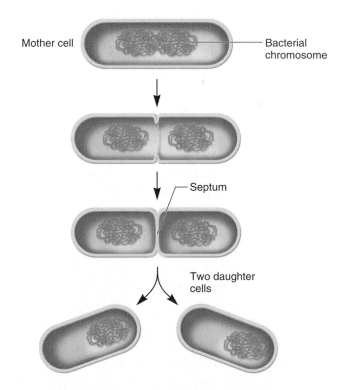

FIGURE 3.4 Binary fission produces two bacterial cells. Prior to division, the chromosome replicates to produce two identical copies. These two copies segregate from each other, with one copy going to each daughter cell.

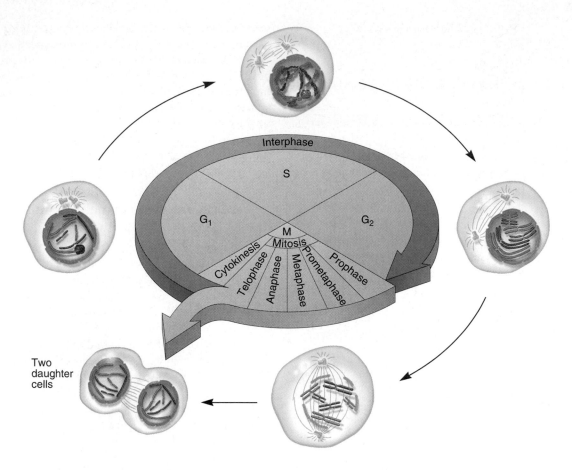

FIGURE 3.5   The eukaryotic cell cycle. Dividing cells progress through a series of phases, denoted G_1, S, G_2, and M phases (mitosis). This diagram shows the progression of a cell through mitosis to produce two daughter cells. The original diploid cell had three pairs of chromosomes, for a total of six individual chromosomes. During S phase, these have replicated to yield 12 chromatids. After mitosis is complete, there are two daughter cells each containing six chromosomes.

as **binary fission.** During this event, the two daughter cells become separated from each other by the formation of a septum. As seen in the figure, each cell receives a copy of the chromosomal genetic material. Except when rare mutations occur, the daughter cells are usually genetically identical, because they contain exact copies of the genetic material from the mother cell.

Binary fission is an asexual form of reproduction, since it does not involve genetic contributions from two different gametes. On occasion, bacteria can exchange small pieces of genetic material with each other. We will consider some interesting mechanisms of genetic exchange in chapter 6.

The Transmission of Chromosomes During the Division of Eukaryotic Cells Requires a Sorting Process Known as Mitosis

The goal of eukaryotic cell division is to produce two daughter cells that have the same number and types of chromosomes as the original mother cell. This requires a replication and sorting process that is more complicated than simple binary fission.

Eukaryotic cells that are destined to divide progress through a series of phases known as the **cell cycle** (fig. 3.5). These are G for gap, S for synthesis (of the genetic material), and M for mitosis. There are two G phases, G_1 and G_2. The G_1 and G_2 phases were originally described as gap phases because it was not microscopically apparent that significant changes were occurring in G_1 and G_2. There seemed to be a gap in any microscopically observable, cellular stages. However, we now know that these are critical stages of the cell cycle. In actively dividing cells, the G_1, S, and G_2 phases are collectively known as **interphase.** In addition, cells may remain for long periods of time in a phase of the cell cycle called G_0. This stage would occur before the S phase. A cell in the G_0 phase has postponed making a decision to divide or, in the case of terminally differentiated cells (e.g., nerve cells in an adult animal), it has made a decision to never divide again.

During the G_1 phase, a cell may prepare to divide. Depending on the cell type and the conditions that a cell encounters, a cell in the G_1 phase may accumulate molecular changes that cause it to progress through the rest of the cell cycle. When this occurs, cell biologists say that a cell has reached a **restriction point** and is

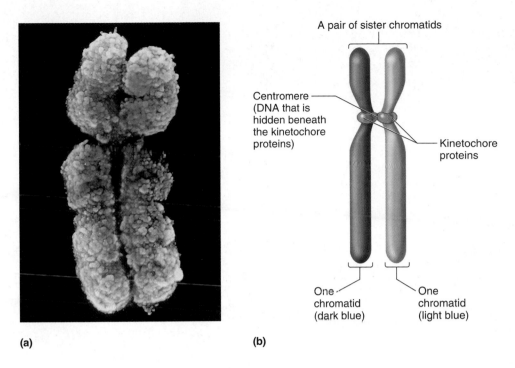

(a)   (b)

FIGURE 3.6   **Metaphase chromosomes.** **(a)** A photomicrograph of a metaphase chromosome that exists in a form called a pair of sister chromatids. **(b)** A schematic drawing of sister chromatids. This structure has two chromatids that lie side by side. As seen here, each chromatid is a distinct unit. The two chromatids are held together by proteins that bind to each other and to the centromeres of each chromatid.

committed on a pathway that leads to cell division. Once past the restriction point, the cell will then advance to the S phase, during which the chromosomes are replicated. After replication, the two copies are called **chromatids.** They are joined to each other to form a unit known as a pair of **sister chromatids** (fig. 3.6). The cohesion between chromatids is discussed in chapter 10 (see fig. 10.24). When S phase is completed, a cell actually has twice as many chromatids compared to the number of chromosomes in the G_1 phase. For example, a human cell in the G_1 phase has 46 distinct chromosomes whereas in G_2 it would have 46 pairs of sister chromatids for a total of 92 chromatids. The term *chromosome* is also used to describe a pair of sister chromatids. This is appropriate since the term originally meant a distinct structure that is observable with the microscope. However, it can be a bit confusing because the chromosomes, which are seen microscopically during the early stages of M phase, actually contain two chromatids. By comparison, the chromosomes at the end of M phase and during G_1 contain the equivalent of one chromatid (refer back to fig. 3.5).

During the G_2 phase, the cell accumulates the materials that are necessary for nuclear and cell division. It then progresses into the M phase of the cell cycle when **mitosis** occurs. The primary purpose of mitosis is to distribute the replicated chromosomes, dividing one cell nucleus into two nuclei, so that each daughter cell receives the same complement of chromosomes. For example, a human cell in the G_2 phase has 92 chromatids, which are found in 46 pairs. It is during mitosis that these pairs of chromatids are separated and sorted so that each daughter cell receives 46 chromosomes. Mitosis is the name given to this sorting process.

Mitosis was first observed microscopically in the 1870s by the German biologist Walter Flemming, who coined the term *mitosis* (from the Greek *mitos,* meaning thread). He studied the large, transparent epithelial cells of salamander larvae and noticed that chromosomes are constructed of two "threads" that are double in appearance along their length. These double threads are divided and move apart, one going to each of the two daughter nuclei. By this mechanism, Flemming pointed out that the two daughter cells receive an identical group of threads, of quantity comparable to the number of threads in the parent cell.

In Figure 3.7, the process of mitosis is shown for a diploid animal cell. In the simplified diagrams shown along the bottom of this figure, the original mother cell contains six chromosomes; it is diploid (2n) and contains three chromosomes per set ($n = 3$). One set is shown in *blue,* and the homologous set is shown in *red.* Mitosis is subdivided into phases known as prophase, prometaphase, metaphase, anaphase, and telophase. Prior to mitosis, the cells are in interphase, during which the chromosomes are decondensed and found in the nucleus (fig. 3.7a). At the start of mitosis, in **prophase,** the chromosomes have already replicated to produce 12 chromatids, joined as six pairs of sister chromatids (fig. 3.7b). As prophase proceeds, the nuclear membrane dissociates into small vesicles. At the same time, the chromatids condense into more compact structures that are readily visible by light microscopy.

In conjunction with these early events of M phase, the **mitotic spindle apparatus** forms (also known simply as the **mitotic spindle**). The organization and structure of the mitotic

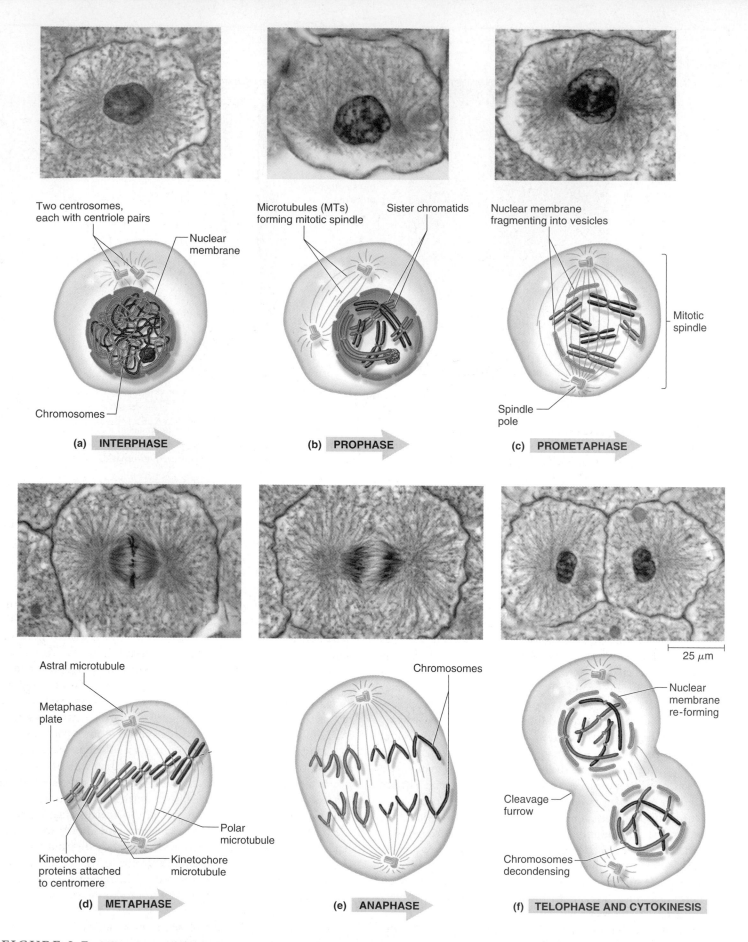

FIGURE 3.7 **The process of mitosis in an animal cell.** The *top* panels illustrate the cells of a fish embryo progressing through mitosis. The *bottom* panels are schematic drawings that emphasize the sorting and separation of the chromosomes. In this case, the original diploid cell had six chromosomes (three in each set). At the start of mitosis, these have already replicated into 12 chromatids. The final result is two daughter cells each containing six chromosomes.

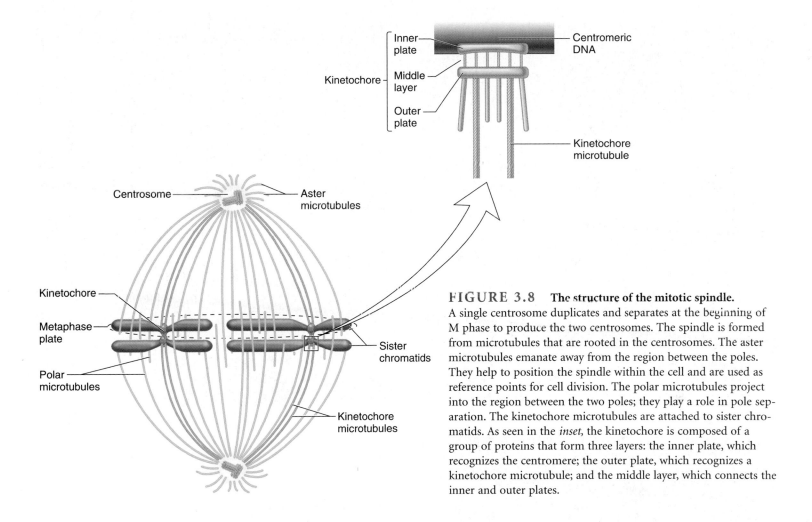

FIGURE 3.8   The structure of the mitotic spindle.
A single centrosome duplicates and separates at the beginning of M phase to produce the two centrosomes. The spindle is formed from microtubules that are rooted in the centrosomes. The aster microtubules emanate away from the region between the poles. They help to position the spindle within the cell and are used as reference points for cell division. The polar microtubules project into the region between the two poles; they play a role in pole separation. The kinetochore microtubules are attached to sister chromatids. As seen in the *inset,* the kinetochore is composed of a group of proteins that form three layers: the inner plate, which recognizes the centromere; the outer plate, which recognizes a kinetochore microtubule; and the middle layer, which connects the inner and outer plates.

spindle is shown in figure 3.8. The mitotic spindle is formed from two structures called the **centrosomes.** Each centrosome organizes the construction of protein fibers called microtubules (MTs). These fibers are produced from the rapid polymerization of tubulin proteins. There are three types of spindle microtubules. The aster microtubules emanate away from the chromosomes. They are important for the positioning of the spindle apparatus within the cell and later in the process of cell division. The polar microtubules project toward the region where the chromosomes will be found during metaphase (i.e., the region between the two poles). Polar microtubules that overlap with each other play a role in the separation of the two poles. They help to "push" the poles away from each other. Finally, the kinetochore microtubules have attachments to the **kinetochore,** which is bound to the **centromere** of individual chromosomes. As seen in the inset to figure 3.8, the kinetochore is composed of a group of proteins that form three layers. The proteins of the inner plate make direct contact with the centromeric DNA, while the outer plate contacts the kinetochore microtubules. The role of the middle layer is to connect these two regions.

With this understanding of the mitotic spindle, we can return our attention to figure 3.7 to examine the dynamic events that occur during mitosis. As mitosis progresses, the centrosomes

move apart and demarcate two poles, one within each of the future daughter cells. In animal cells, centrioles are found at the centrosome. However, centrioles are not found in many other eukaryotic species and are not required for spindle formation. Once the nuclear membrane is gone, the spindle fibers are able to interact with the sister chromatids. This interaction occurs in a phase of mitosis called **prometaphase** (fig. 3.7c). Initially, kinetochore microtubules are rapidly formed and can be seen growing out from the two poles. As it grows, if the end of a kinetochore microtubule happens to make contact with a kinetochore, its end is said to be "captured" and remains firmly attached to the kinetochore. This random process is how sister chromatids become attached to kinetochore microtubules. Alternatively, if the end of a kinetochore microtubule does not collide with a kinetochore, the microtubule eventually depolymerizes and retracts to the centrosome. As the end of prometaphase nears, the two kinetochores on a pair of sister chromatids are attached to kinetochore microtubules from opposite poles. As these events are occurring, the sister chromatids are seen to undergo jerky movements as they are tugged, back and forth, between the two poles.

Eventually, the pairs of sister chromatids align themselves along a plane called the **metaphase plate.** As shown in figure 3.7d, when this alignment is complete, the cell is in **metaphase** of

(a) Prophase

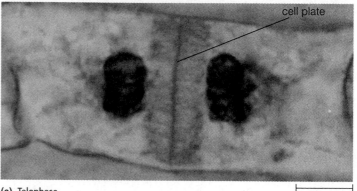

(d) Anaphase

(b) Prometaphase

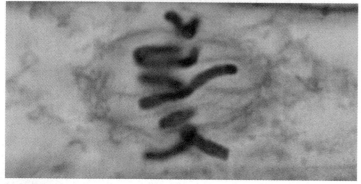

cell plate

(e) Telophase

42 μm

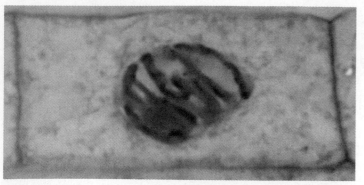

(c) Metaphase

FIGURE 3.9   Mitosis in a plant cell. These photomicrographs show mitosis in cells of an onion root. Cytokinesis occurs via the formation of a cell plate between the two daughter cells.

only one of the two poles. As anaphase proceeds, the chromosomes move toward the pole to which they are attached. This involves a shortening of the kinetochore microtubules. In addition, the two poles themselves move farther away from each other due to the actions of overlapping polar microtubules from opposite poles, which lengthen at the end of anaphase and push against each other.

During **telophase,** the chromosomes have reached their respective poles and decondense. The nuclear membrane now reforms to produce two separate nuclei. In Figure 3.7f, this has produced two nuclei that contain six chromosomes each. In most cases, mitosis is quickly followed by **cytokinesis,** in which the two nuclei are segregated into separate daughter cells. In animal cells, cytokinesis involves the formation of a **cleavage furrow,** which constricts to separate the cells. In plants, the two daughter cells are separated by the formation of a **cell plate** (fig. 3.9).

Mitosis and cytokinesis ultimately produce two daughter cells having the same number of chromosomes as the mother cell. Barring rare mutations, the two daughter cells are genetically identical to each other and to the mother cell from which they were derived. Thus, the critical consequence of this sorting process is to ensure genetic consistency from one cell to the next. The development of multicellularity relies on the repeated process of mitosis and cytokinesis. For diploid organisms that are multicellular, most of the somatic cells are diploid and genetically identical to each other.

mitosis. At this point, each pair of chromatids is attached to both poles by kinetochore microtubules; the pairs of sister chromatids have become organized into a single row along the metaphase plate. When this organizational process is finished, the chromatids can be equally distributed into two daughter cells.

The next step in the sorting process occurs during **anaphase** (fig. 3.7e). At this stage, the connection that is responsible for holding the pairs of chromatids together is broken. A molecular description of sister chromatid cohesion and the process of sister chromatid separation is found in chapter 10 (fig. 10.24). Each chromatid, now an individual chromosome, is linked to

3.3 SEXUAL REPRODUCTION

In the previous section, we considered how a cell could divide to produce two new cells with identical complements of genetic material. Now we will turn our attention to sexual reproduction, a common way for eukaryotic organisms to produce offspring. During sexual reproduction, gametes are made that contain half the amount of genetic material. These gametes fuse with each other in the process of **fertilization** to begin the life of a new organism. Gametes are highly specialized cells designed to locate each other and provide nutrients to the developing embryo.

Some simple eukaryotic species are **isogamous,** which means that the gametes are morphologically similar. Examples of isogamous organisms include many species of fungi and algae. Most eukaryotic species, however, are **heterogamous**—they produce two morphologically different types of gametes. Male gametes, **sperm cells,** are relatively small and usually travel far distances to reach the female gamete. The mobility of the male gamete is an important characteristic making it likely to come in close proximity to the female gamete. The sperm of animals often contain flagella that enable them to swim. This is also true of ferns and nonvascular plants such as bryophytes. In higher plants (e.g., angiosperms and gymnosperms), however, the sperm nuclei are contained within pollen grains. Pollen is a small mobile structure that can be carried by the wind or on the feet or hairs of insects. By comparison, the female gamete, known as the **egg cell** or **ovum,** is usually very large and nonmotile. In animal species, the egg stores a large amount of nutrients that will be available to the growing embryo.

Gametes are typically **haploid,** which means they contain half the number of chromosomes as diploid cells. Compared to a diploid cell, a haploid gamete contains a single set of chromosomes. In this case, the haploid gametes are $1n$ while the diploid cells are $2n$. For example, a diploid human cell contains 46 chromosomes, but a gamete (sperm or egg cell) contains only 23 chromosomes, one from each of the 23 pairs.

During the process known as **meiosis,** haploid cells are produced from cells that were originally diploid. For this to occur, the chromosomes must be correctly sorted and distributed in a way that reduces the chromosome number to half its original value. In the case of human gametes, for example, each gamete must receive half the total number of chromosomes, but not just any 23 chromosomes will do. A gamete must receive one chromosome from each of the 23 pairs. In this section, we will examine the cellular events of gamete development and how the stages of meiosis lead to the formation of cells with a haploid complement of chromosomes.

Meiosis Produces Cells That Are Haploid

The process of meiosis bears striking similarities to mitosis. Like mitosis, meiosis begins after a cell has progressed through the G_1, S, and G_2 phases of the cell cycle. However, meiosis involves two successive divisions rather than one (as in mitosis). In this section, we will emphasize the distinctive features of meiosis that reduce the amount of genetic material.

The stages of meiosis are described in figures 3.10 and 3.12. These simplified diagrams depict a diploid cell $(2n)$ that contains a total of six chromosomes (as in our look at mitosis). Prior to meiosis, the chromosomes are replicated in S phase to produce pairs of sister chromatids. This single replication event is then followed by two sequential cell divisions called meiosis I and II. Like mitosis, each of these is subdivided into prophase, prometaphase, metaphase, anaphase, and telophase.

Figure 3.10 emphasizes some of the important events that occur during prophase I, which is further subdivided into periods known as leptotena, zygotena, pachytena, diplotena, and diakinesis. During **leptotena,** the chromosomes begin to condense and become visible with a light microscope. At first, they appear as threadlike structures, and the individual sister chromatids are not readily discernible. In figure 3.10 (leptotene and zygotene stages), each structure is actually a pair of chromatids. Unlike mitosis, the **zygotene** stage of prophase I involves a recognition process known as **synapsis.** The homologous chromosomes recognize each other and then align themselves along their entire lengths. The associated chromatids are known as **bivalents.** Each bivalent contains two pairs of sister chromatids, or a total of four chromatids.

In most eukaryotic species, a **synaptonemal complex** is formed between the homologous chromosomes. As shown in figure 3.11, this complex is composed of parallel lateral elements, which are bound to the chromosomal DNA, and a central element, which promotes the binding of the lateral elements to each other via transverse filaments. The synaptonemal complex is not required for the pairing of homologous chromosomes, since some species such as *Aspergillus nidulans* and *Schizosaccharomyces pombe* completely lack such a complex, yet their chromosomes synapse correctly. At present, the precise role of the synaptonemal complex is not clearly understood and is the subject of intense research. It may play more than one role. First, although not required for synapsis, the synaptonemal complex could help to maintain homologous pairing in situations where the normal process has failed. Second, the complex may play a role in meiotic chromosome structure. And third, the synaptonemal complex may serve to regulate the process of crossing over, which is described next.

Prior to **pachytena,** when synapsis is complete, an event known as **crossing over** occurs. Crossing over involves a physical exchange of chromosome pieces. Depending on the size of the chromosome and the species, an average eukaryotic chromosome would incur a couple to a couple dozen crossovers. During spermatogenesis in humans, for example, an average chromosome undergoes slightly more than 2 crossovers, while chromosomes in certain plant species may undergo 20 or more crossovers. Recent research has shown that crossing over is critical for the proper segregation of chromosomes. In fact, abnormalities in

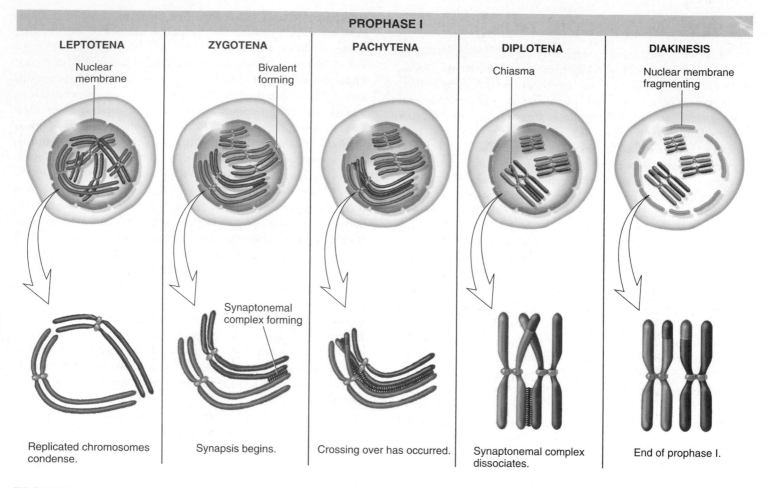

PROPHASE I

LEPTOTENA	ZYGOTENA	PACHYTENA	DIPLOTENA	DIAKINESIS
Nuclear membrane	Bivalent forming		Chiasma	Nuclear membrane fragmenting
	Synaptonemal complex forming			
Replicated chromosomes condense.	Synapsis begins.	Crossing over has occurred.	Synaptonemal complex dissociates.	End of prophase I.

FIGURE 3.10   The events that occur during prophase of meiosis I.

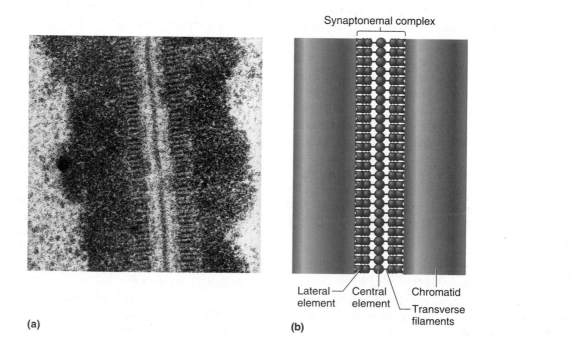

Synaptonemal complex

Lateral element   Central element   Chromatid   Transverse filaments

(a)   (b)

FIGURE 3.11   **The synaptonemal complex formed during prophase I. (a)** Micrograph of a synaptonemal complex. **(b)** Lateral elements are bound to the chromosomal DNA of homologous chromatids. A central element provides a link between the lateral elements via transverse tubules.

chromosome segregation may be related to a defect in crossing over. In a high percentage of people with Down syndrome, in which an individual has three copies of chromosome 21 instead of two, it has been found that the presence of the extra chromosome is associated with a lack of crossing over between homologous chromosomes.

In figure 3.10, crossing over has occurred at a single site between two of the (larger) chromatids. The connection that results from crossing over is called a **chiasma** (plural: **chiasmata**), because it physically resembles the Greek letter chi, χ. We will consider the genetic consequences of crossing over in chapter 5 and the molecular process of crossing over in chapter 18. At the **diplotene** stage, the synaptonemal complex has disappeared. Microscopically, it becomes easier to see that a bivalent is actually composed of four chromatids. This association between four chromatids is also called a **tetrad** (from the prefix *tetra-*, meaning four). In the last stage of prophase I, **diakinesis,** the synaptonemal complex completely disappears.

Figure 3.10 is designed to emphasize the pairing and crossing over that occurs during prophase I. In figure 3.12, we turn our attention to the interactions between the chromatids and spindle apparatus and how these interactions promote chromatid sorting. In prometaphase I, the spindle apparatus is complete, and the chromatids are attached via kinetochore microtubules. At metaphase I, the bivalents are organized along the metaphase plate. However, their pattern of alignment is strikingly different from that observed during mitosis (refer back to fig. 3.7d). Before we consider the rest of meiosis I, a particularly critical feature for you to appreciate is how the bivalents are aligned along the metaphase plate. In particular, the alignment of pairs of sister chromatids is in a double row rather than a single row (as in mitosis). Furthermore, the arrangement of sister chromatids within this double row is random with regard to the (*blue* and *red*) homologues. In figure 3.12, one of the blue homologues is above the metaphase plate and the other two are below, while one of the red homologues is below the metaphase plate and other two are above.

In an organism that is producing many gametes, meiosis in other cells could produce a different arrangement of homologues (e.g., three blues above and none below, or none above and three below, etc.). As discussed later in this chapter, the random arrangement of homologues is consistent with Mendel's law of independent assortment. Since most eukaryotic species have several chromosomes per set, there are many possible ways that the homologues can randomly align themselves along the metaphase plate. For example, consider humans that have 23 chromosomes per set. The possible number of different, random alignments equals 2^n, where n equals the number of chromosomes per set. Thus, in humans, this would equal 2^{23}, or over 8 million, possibilities. Since the homologues are genetically similar but not identical, we see from this calculation that the random alignment of homologous chromosomes provides a mechanism to promote a vast amount of genetic diversity.

In addition to the random arrangement of homologues within a double row, a second distinctive feature of metaphase I is the attachment of kinetochore microtubules to the sister chromatids (fig. 3.13). One pair of sister chromatids is linked to one of the poles, and the homologous pair is linked to the opposite pole. This arrangement is quite different from the kinetochore attachment sites during mitosis (compare figs. 3.8 and 3.13). During anaphase I, the two pairs of sister chromatids within a bivalent separate from each other (return to fig. 3.12). However, the connection that holds sister chromatids together does not break. Instead, each joined pair of chromatids migrates to one pole, and the homologous pair of chromatids moves to the opposite pole. Finally, at telophase I, the sister chromatids have reached their respective poles, and decondensation occurs. The nuclear membrane now re-forms to produce two separate nuclei. If we consider the end result of meiosis I, we see that it has produced two cells, each with three pairs of sister chromatids. It is a reduction division. The original diploid cell had its chromosomes in homologous pairs, while the two cells produced at the end of meiosis I are considered to be haploid; they do not have pairs of homologous chromosomes.

Meiosis I is followed by cytokinesis and then meiosis II. The sorting events that occur during meiosis II are similar to those that occur during mitosis, but the starting point is different. For a diploid organism with six chromosomes, mitosis begins with 12 chromatids that are joined as six pairs of sister chromatids (refer back to fig. 3.7). By comparison, the two cells that begin meiosis II each have six chromatids that are joined as three pairs of sister chromatids. Otherwise, the steps that occur during prophase, prometaphase, metaphase, anaphase, and telophase of meiosis II are analogous to a mitotic division.

If we compare the outcome of meiosis (see fig. 3.12) to that of mitosis (see fig. 3.7), the results are quite different. (A comparison is also made in solved problem S3 at the end of chapter 3.) In these examples, mitosis produced two diploid daughter cells with six chromosomes each, whereas meiosis produced four haploid daughter cells with three chromosomes each. In other words, meiosis has halved the number of chromosomes per cell. With regard to alleles, the results of mitosis and meiosis are also different. The daughter cells produced by mitosis are genetically identical. However, the haploid cells produced by meiosis are not genetically identical to each other, because they contain only one homologous chromosome from each pair. Later in chapter 3, we will consider how the gametes may differ in the alleles that they carry on their homologous chromosomes.

In Animals, Spermatogenesis Produces Four Haploid Sperm Cells and Oogenesis Produces a Single Haploid Egg Cell

In male animals, **spermatogenesis,** the production of sperm, occurs within glands known as the testes. In the testes are spermatogonial cells that divide by mitosis to produce two cells. One

MEIOSIS I

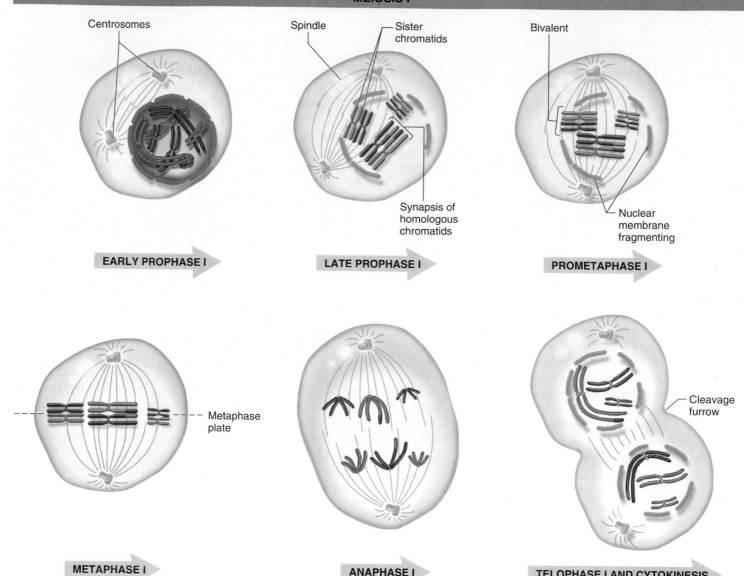

Centrosomes

EARLY PROPHASE I

Spindle · Sister chromatids

Synapsis of homologous chromatids

LATE PROPHASE I

Bivalent

Nuclear membrane fragmenting

PROMETAPHASE I

Metaphase plate

METAPHASE I

ANAPHASE I

Cleavage furrow

TELOPHASE I AND CYTOKINESIS

MEIOSIS II

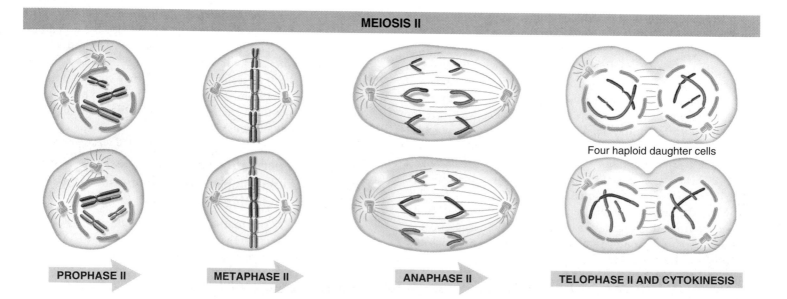

Four haploid daughter cells

PROPHASE II

METAPHASE II

ANAPHASE II

TELOPHASE II AND CYTOKINESIS

FIGURE 3.12   The stages of meiosis in an animal cell. See text for details.

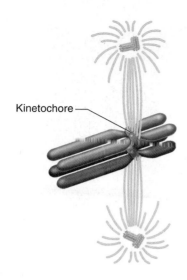

Kinetochore

FIGURE 3.13   Attachment of the kinetochore microtubules to replicated chromosomes during meiosis. The kinetochore microtubules from a given pole are attached to one pair of chromatids in a bivalent, but not both. Therefore, each pair of sister chromatids is attached to only one pole.

of these remains a spermatogonial cell, while the other cell becomes a primary spermatocyte. As shown in figure 3.14a, the spermatocyte progresses through meiosis I and meiosis II to produce four haploid cells, which are known as spermatids. These cells then mature into sperm cells. The structure of a sperm cell includes the long flagellum and a head. The head of the sperm contains little more than a haploid nucleus and an organelle at its tip, known as an acrosome. The acrosome contains digestive enzymes that are released when a sperm meets an egg cell. These enzymes enable the sperm to penetrate the outer protective layers of the egg and gain entry into the egg cell's cytoplasm. In animal species without a mating season, sperm production is a continuous process in mature males. A mature human male, for example, produces several hundred million sperm each day.

In female animals, **oogenesis,** the production of egg cells, occurs within specialized diploid cells of the ovary known as oogonia. Quite early in the development of the ovary, the oogonia initiate meiosis to produce primary oocytes (fig. 3.14b). For example, in humans, approximately 1 million primary oocytes per ovary are produced before birth. These primary oocytes are arrested (i.e., enter a dormant phase) at prophase I of meiosis, remaining at this stage until the female becomes sexually mature.

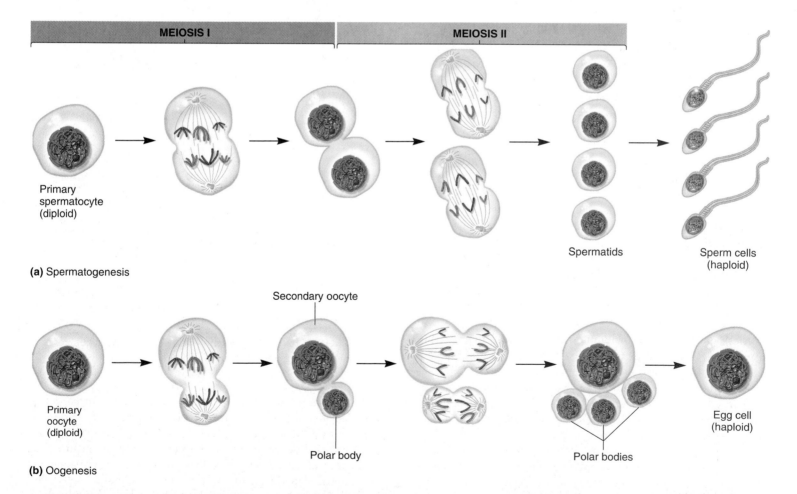

MEIOSIS I	MEIOSIS II

(a) Spermatogenesis

Primary spermatocyte (diploid)

Spermatids

Sperm cells (haploid)

(b) Oogenesis

Primary oocyte (diploid)

Secondary oocyte

Polar body

Polar bodies

Egg cell (haploid)

FIGURE 3.14   Gametogenesis in animals. (a) Spermatogenesis. A diploid spermatocyte undergoes meiosis to produce four haploid (n) spermatids. These differentiate during spermatogenesis to become mature sperm. **(b)** Oogenesis. A diploid oocyte undergoes meiosis to produce one haploid egg cell and three polar bodies. This maximizes the amount of cytoplasm that the egg receives. The polar bodies degenerate.

Beginning at this stage, primary oocytes are periodically activated to progress through the remaining stages of oocyte development. In humans, approximately one primary oocyte per month is activated to undergo this process.

During oocyte maturation, meiosis produces only one cell (rather than four) that is destined to become an egg (fig. 3.14*b*). With each cycle, one or more primary oocytes within the ovary progress through the first meiotic division. This division is very asymmetric and produces a secondary oocyte and a much smaller cell, known as a polar body. Most of the cytoplasm is retained by the secondary oocyte and very little by the polar body. The secondary oocyte then begins meiosis II. While this process is under way, the secondary oocyte is released from the ovary (an event called ovulation) and travels down the oviduct toward the uterus. In humans, typically one secondary oocyte is released per month. During this journey, if a sperm cell penetrates the secondary oocyte, it is stimulated to complete meiosis II to produce a haploid egg and an additional polar body. The haploid egg and sperm nuclei then unite to create the diploid nucleus of a new individual.

Plant Species Alternate Between Haploid (Gametophyte) and Diploid (Sporophyte) Generations

Most species of animals are diploid, and their haploid gametes are considered to be a specialized type of cell. By comparison, the life cycles of plant species alternate between haploid and diploid generations. The haploid generation is called the **gametophyte,** whereas the diploid generation is called the **sporophyte.** Meiosis produces haploid cells called spores, which divide by mitosis to produce the gametophyte. In simpler plants, a haploid spore can produce a large multicellular gametophyte by repeated mitoses and cellular divisions. In higher plants, however, spores develop into gametophytes that contain only a few cells. In this case, the organism that we think of as a "plant" is the sporophyte, while the gametophyte is very inconspicuous. In fact, the gametophytes of many plant species are small structures produced within the much larger sporophyte. Certain cells within the haploid gametophytes then become specialized as haploid gametes.

Figure 3.15 provides an overview of gametophyte development and **gametogenesis** in higher plants. Meiosis occurs within two different structures of the sporophyte: the anthers and the ovaries, which produce male and female gametophytes, respectively. This diagram depicts a flower from an angiosperm, which is a plant that produces seeds within an ovary.

In the anther, diploid cells called microsporocytes undergo meiosis to produce four haploid microspores. These separate into individual microspores. Each microspore undergoes mitosis to produce a two-celled structure containing one tube cell and one generative cell (both of which are haploid). In higher plants, this structure differentiates into a **pollen grain,** which is considered the male gametophyte. Later, the generative cell undergoes mitosis to produce two haploid sperm nuclei. In most plant species, this mitosis occurs only if the pollen grain germinates (refer back to fig. 2.2*c*).

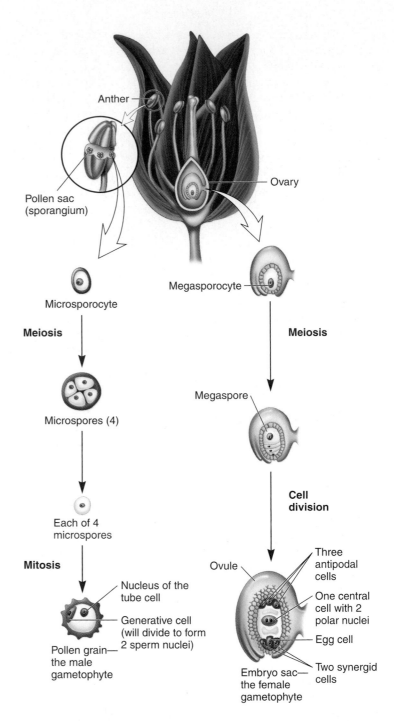

FIGURE 3.15   The formation of male and female gametes by the gametophyte of angiosperms.

By comparison, female gametophytes are produced within the plant ovaries. A cell known as the megasporocyte undergoes meiosis to produce four haploid megaspores. Three of the four megaspores degenerate. The remaining haploid megaspore then undergoes three successive mitotic divisions accompanied by asymmetric cytokinesis to produce seven individual cells. This seven-celled structure, also known as the **embryo sac,** is considered the mature female gametophyte. Each embryo sac is contained within an ovule.

For fertilization to occur, specialized cells within the male and female gametophytes must meet. The steps of plant fertilization were described in chapter 2. To begin this process, a pollen grain lands on a stigma (refer back to fig. 2.2c). This stimulates the tube cell to sprout a tube that grows through the style and ovary, eventually making contact with an ovule. As this is occurring, the generative cell undergoes mitosis to produce two haploid sperm nuclei. The sperm nuclei migrate through the pollen tube and eventually reach the ovule. One of the sperm nuclei enters the central cell, which contains the two polar nuclei. This creates a cell that is triploid (3n). This cell divides mitotically to produce **endosperm,** which acts as a food-storing tissue. The other sperm nucleus enters the egg cell. The egg and sperm nuclei fuse to create a diploid cell, the zygote, which becomes a plant embryo. Therefore, fertilization in higher plants is actually a double fertilization. It is intended to ensure that the endosperm (which uses a large amount of plant resources) will develop only when an egg cell has been fertilized. After fertilization is complete, the ovule develops into a seed, and the surrounding ovary develops into the fruit, which encloses one or more seeds.

3.4 THE CHROMOSOME THEORY OF INHERITANCE AND SEX CHROMOSOMES

Thus far, we have considered how chromosomes are transmitted during cell division and gamete formation. In this section, we will first examine how chromosomal transmission is related to the patterns of inheritance observed by Mendel. This relationship, known as the **chromosome theory of inheritance,** was a major breakthrough in our understanding of genetics, because it established the framework for understanding how chromosomes carry and transmit the genetic determinants that govern the outcome of traits. This theory dramatically unfolded as a result of three lines of scientific inquiry (table 3.1). As discussed in chapter 2, one avenue concerned Mendel's breeding studies, in which he analyzed the transmission of traits from parent to offspring. A second line of inquiry involved the material basis for heredity. A German biologist, August Weismann, and a Swiss botanist, Carl Nägeli, championed the idea that there is a substance in living cells that is responsible for the transmission of traits from parents to offspring. Nägeli also suggested that both parents contribute equal amounts of this substance to their offspring. Several scientists, including Oscar Hertwig, Eduard Strasburger, and Walter Flemming, conducted studies suggesting that the chromosomes are the carriers of the genetic material. We now know that the DNA within the chromosomes is the genetic material.

Finally, the third line of evidence involved the microscopic examination of the processes of fertilization, mitosis, and meiosis. It became increasingly clear that the characteristics of organisms are rooted in the continuity of cells during the life of an organism and from one generation to the next. When the work of Mendel was rediscovered, several scientists noted striking parallels between the segregation and assortment of traits noted by

TABLE 3.1

Chronology for the Development and Proof of the Chromosome Theory of Inheritance

1866	Gregor Mendel: analyzed the transmission of traits from parents to offspring and showed that it follows a pattern of segregation and independent assortment.
1876–77	Oscar Hertwig and Hermann Fol: observed that the nucleus of the sperm enters the egg during animal cell fertilization.
1877	Eduard Strasburger: observed that the sperm nucleus of plants (and no detectable cytoplasm) enters the egg during plant fertilization.
1878	Walter Flemming: described mitosis in careful detail.
1883	Carl Nägeli and August Weismann: proposed the existence of a genetic material, which Nägeli called idioplasm and Weismann called germ plasm.
1883	Wilhelm Roux: proposed that the most important event of mitosis is the equal partitioning of "nuclear qualities" to the daughter cells.
1883	Edouard van Beneden: showed that gametes contain half the number of chromosomes and that fertilization restores the normal diploid number.
1884–85	Hertwig, Strasburger, and Weismann: proposed that chromosomes are carriers of the genetic material.
1889	Theodore Boveri: showed that enucleated sea urchin eggs that are fertilized by sperm from a different species develop into larva that have characteristics that coincide with the sperm's species.
1900	Hugh de Vries, Carl Correns, and Erich von Tschermak: rediscovered Mendel's work.
1901	Thomas Montgomery: maternal and paternal chromosomes pair with each other during meiosis.
1901	C. E. McClung: sex determination in insects is related to differences in chromosome composition.
1902	Boveri: showed that when sea urchin eggs were fertilized by two sperm, the abnormal development of the embryo was related to an abnormal number of chromosomes.
1903	Walter Sutton: showed that even though the chromosomes seem to disappear during interphase, they do not actually disintegrate. Instead, he argued that chromosomes must retain their continuity and individuality from one cell division to the next.
1902–3	Boveri and Sutton: independently proposed tenets of the chromosome theory of inheritance. Some historians primarily credit this theory to Sutton.
1910	Thomas Hunt Morgan: showed that a genetic trait (i.e., white-eyed phenotype in *Drosophila*) was linked to a particular chromosome.
1913	E. Eleanor Carothers: demonstrated that homologous pairs of chromosomes show independent assortment.
1916	Calvin Bridges: studied nondisjunction as a way to prove the chromosome theory of inheritance.

For a description of these experiments, the student is encouraged to read: Voeller, B. R. (1968) *The Chromosome Theory of Inheritance. Classic Papers in Development and Heredity.* New York: Appleton-Century-Crofts.

Mendel and the behavior of chromosomes during meiosis. Among them were Theodore Boveri, a German biologist, and Walter Sutton at Columbia University. They independently proposed the chromosome theory of inheritance, which was a milestone in our understanding of genetics. The tenets of this theory are described at the beginning of this section.

The remainder of this section focuses on **sex chromosomes.** The experimental connection between the chromosomal theory of inheritance and sex chromosomes is profound. Even though an examination of meiosis provided compelling evidence that Mendel's laws could be explained by chromosome sorting, it was still essential to correlate chromosome behavior with the inheritance of particular traits. Since sex chromosomes, such as the X and Y chromosome, are cytologically unique, it was possible experimentally to correlate the inheritance of particular traits with the transmission of specific sex chromosomes. In particular, early studies were able to identify genes on the X chromosome that govern eye color in fruit flies. This phenomenon, called **X-linked inheritance,** is a great connection between chapter 2 and future chapters. On the one hand, X-linked inheritance confirmed that Mendel's laws of inheritance could be explained by the chromosomal transmission of traits. On the other hand, X-linked inheritance showed us that not all traits follow simple Mendelian rules. In later chapters, we will examine a variety of traits that are governed by chromosomal genes yet follow inheritance patterns that are more complex than those observed by Mendel.

The Chromosome Theory of Inheritance Relates the Behavior of Chromosomes with the Mendelian Inheritance of Traits

According to the chromosome theory of inheritance, the inheritance patterns of traits can be explained by the transmission patterns of chromosomes during gametogenesis and fertilization. This theory is based on a few fundamental principles.

1. Chromosomes contain the genetic material, which is transmitted from parent to offspring (and from cell to cell).
2. Chromosomes are replicated and passed along generation after generation from parent to offspring. They are also passed from cell to cell during the multicellular development of an organism. Each type of chromosome retains its individuality during cell division and gamete formation.
3. The nuclei of most eukaryotic cells contain chromosomes that are found in homologous pairs (i.e., they are diploid). One member of each pair is inherited from the mother, the other from the father. At meiosis, one of the two members of each pair segregates into one daughter nucleus, and the homologue segregates into the other daughter nucleus. Gametes contain one set of chromosomes (i.e., they are haploid).
4. During the formation of haploid cells, different types of (nonhomologous) chromosomes segregate independently of each other.

5. Each parent contributes one set of chromosomes to its offspring. The maternal and paternal sets of homologous chromosomes are functionally equivalent; each set carries a full complement of genetic determinants (genes).

The chromosome theory of inheritance allows us to see the relationship between Mendel's laws and chromosomal transmission. As shown in figure 3.16, Mendel's law of segregation can be

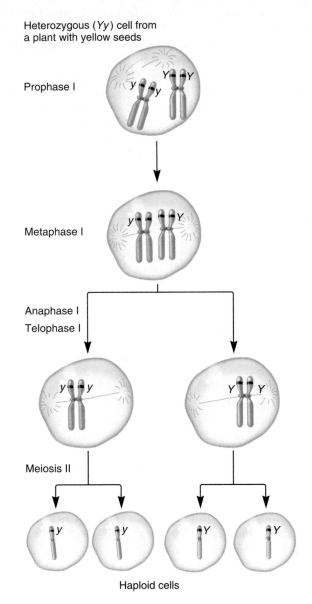

Heterozygous (*Y y*) cell from a plant with yellow seeds

Prophase I

Metaphase I

Anaphase I
Telophase I

Meiosis II

Haploid cells

FIGURE 3.16 **Mendel's law of segregation can be explained by the segregation of homologues during meiosis.** The two copies of a gene are contained on homologous chromosomes. In this example using pea seed color, the two alleles are *Y* (yellow) and *y* (green). During meiosis, the homologous chromosomes segregate from each other, leading to segregation of the two alleles into separate gametes.

GENES→TRAITS The gene for seed color exists in two alleles, *Y* (yellow) and *y* (green). During meiosis, the homologous chromosomes that carry these alleles segregate from each other. The resulting cells receive the *Y* or *y* allele but not both. When two gametes unite during fertilization, the alleles that they carry determine the traits of the resulting offspring.

explained by the homologous pairing and segregation of chromosomes during meiosis. This figure depicts the behavior of a pair of homologous chromosomes that carry a gene for seed color. One of the chromosomes carries a dominant allele that confers yellow seed color, while the homologous chromosome carries a recessive allele that confers green color. As discussed in chapter 2, the heterozygous individual would pass only one of these alleles to each offspring. In other words, a gamete may contain the yellow allele or the green allele but not both. Since homologous chromosomes segregate from each other, a gamete will contain only one copy of each type of chromosome.

The law of independent assortment can also be explained by the relative behavior of different (nonhomologous) chromosomes. Figure 3.17 considers the segregation of two types of chromosomes, each carrying a different gene. One pair of chromosomes carries the gene for seed color: the yellow (*Y*) allele is on one chromosome, and the green (*y*) allele is on the homologue. The other pair of (smaller) chromosomes carries the gene for seed texture: one copy has the round (*R*) allele, and the homologue carries the wrinkled (*r*) allele. At metaphase I of meiosis, the different types of chromosomes have randomly aligned themselves along the metaphase plate. As shown in figure 3.17, this can occur in more than one way. On the *left*, the *R* allele has sorted with the *y* allele, while the *r* allele has sorted with the *Y* allele. On the *right*, the opposite situation has occurred. Therefore, the random alignment of chromatid pairs during meiosis I can lead to an independent assortment of genes that are found on nonhomologous chromosomes. As we will see in chapter 5, this law is violated if two different types of genes are located close to one another on the same chromosome.

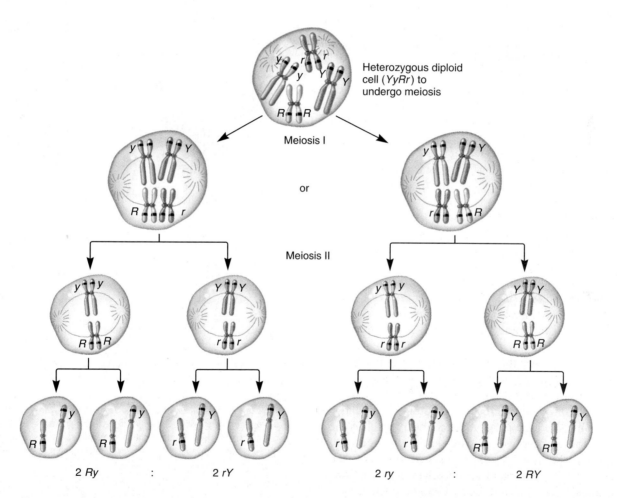

FIGURE 3.17 **Mendel's law of independent assortment can be explained by the random alignment of bivalents during metaphase of meiosis I.** This figure shows the assortment of two genes located on two different chromosomes, using pea seed color and shape as an example (*YyRr*). During metaphase of meiosis I, different possible arrangements of the homologues within bivalents can lead to different combinations of the alleles in the resulting gametes. For example, on the *left*, the dominant *R* allele has sorted with the recessive *y* allele; on the *right*, the dominant *R* allele has sorted with the dominant *Y* allele.

GENES→TRAITS Most species have several different chromosomes that carry many different genes. In this example, the gene for seed color exists in two alleles, *Y* (yellow) and *y* (green), and the gene for seed shape is found as *R* (round) and *r* (wrinkled) alleles. The two genes are found on different (nonhomologous) chromosomes. During meiosis, the homologous chromosomes that carry these alleles segregate from each other. In addition, the chromosomes carrying the *Y* or *y* alleles will independently assort from the chromosomes carrying the *R* or *r* alleles. As shown here, this provides a reassortment of alleles, perhaps creating combinations of alleles that are different from the parental combinations. When two gametes unite during fertilization, the alleles they carry affect the traits of the resulting offspring.

Sex Differences Correlate with the Presence of Sex Chromosomes

According to the chromosome theory of inheritance, the chromosomes carry the genes that determine an organism's traits. Some early evidence supporting this theory involved a consideration of sex. Many species are divided into male and female sexes. In 1901, C. E. McClung, who studied fruit flies, was the first to suggest that male and female sexes are due to the inheritance of particular chromosomes. Since McClung's initial observations, we now know that a pair of chromosomes, called the sex **chromosomes,** determines sex in many different species. Some examples are described in figure 3.18.

In the X-Y system of sex determination, which operates in mammals, the male contains one X chromosome and one Y chromosome, while the female contains two X chromosomes (fig. 3.18*a*). In this case, the male is called the **heterogametic sex,** because he produces two types of sperm: one that carries only the X chromosome, and another type that carries the Y. In contrast, the female is the **homogametic sex,** because all of her eggs carry an X chromosome. The 46 chromosomes carried by humans consist of 1 pair of sex chromosomes and 22 pairs of **autosomes,** chromosomes that are not sex chromosomes. In the human male, each of the four sperm produced during gametogenesis contains 23 chromosomes; two sperm contain an X chromosome, the other two a Y chromosome. The sex of the offspring is determined by whether the sperm that fertilizes the egg carries an X or a Y chromosome. It is the presence of the Y chromosome that causes maleness in mammals. This is known from the analysis of rare individuals who carry chromosomal abnormalities. For example, mistakes that occasionally occur during meiosis may produce an individual who carries two X chromosomes and one Y chromosome. Such an individual develops into a male.

Other mechanisms of sex determination include the X-0, Z-W, and haplo-diploid systems. The X-0 system of sex determination operates in many insects (fig. 3.18*b*). In such species, the male has one sex chromosome (the X) and is designated X0, while the female has a pair (two Xs). In other insect species, such as *Drosophila melanogaster,* the male is XY. For both types of insect species (i.e., X0 or XY males, and XX females), the ratio between X chromosomes and the number of autosomal sets determines sex. If a fly is diploid for the autosomes (2*n*), and has one X chromosome, the ratio is 1/2, or 0.5. This fly will become a male even if it does not receive a Y chromosome. In contrast to mammals, the Y chromosome in the X-0 system does not determine maleness. If a fly is diploid and receives two X chromosomes, the ratio is 2/2, or 1.0, and the fly becomes a female.

For the Z-W system, which determines sex in birds and some fish, the male is ZZ and the female is ZW (fig. 3.18*c*). The letters Z and W are used to distinguish these types of sex chromosomes from those found in the X-Y pattern of sex determination of other species. In the Z-W system, the male is the homogametic sex and the female is heterogametic.

Another interesting mechanism of sex determination, known as the haplo-diploid system, is found in bees (fig. 3.18*d*). The male bee, also known as the drone, is produced from unfer-

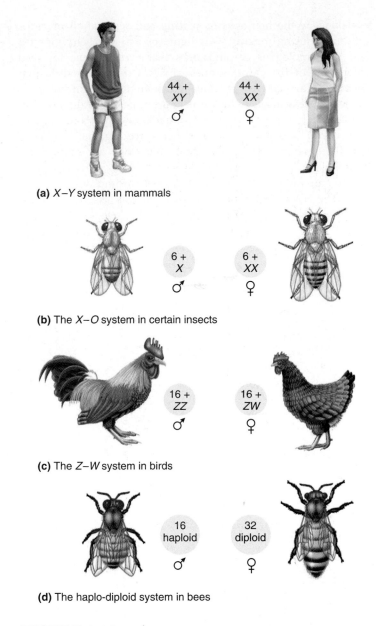

(a) *X–Y* system in mammals

(b) The *X–O* system in certain insects

(c) The *Z–W* system in birds

(d) The haplo-diploid system in bees

FIGURE 3.18   Different mechanisms of sex determination in animals. See text for a description.

GENES→TRAITS   Certain genes that are found on the sex chromosomes play a key role in the development of sex (male versus female). For example, in mammals, genes on the Y chromosome initiate male development. In the X-0 system, the balance between the genes on the X chromosome and the genes on the autosomes plays a key role in governing the pathway of development toward male or female.

tilized haploid eggs. Female bees, both worker bees and queen bees, are produced from fertilized eggs and therefore are diploid.

The chromosomal basis for sex determination is rooted in the location of particular genes on the sex chromosomes. For example, a gene called the *Sry* gene is located on the Y chromosome found in mammals. This gene plays a key role in the developmental pathway that causes a mammal to develop into a male. We will examine the molecular basis for sex determination in chapter 23.

EXPERIMENT 3A

Morgan's Experiments Showed a Connection Between a Genetic Trait and the Inheritance of a Sex Chromosome in *Drosophila*

Thomas Hunt Morgan carried out a particularly influential study that confirmed the chromosome theory of inheritance. Morgan was trained as an embryologist, and much of his early research involved descriptive and experimental work in that field. He was particularly interested in ways that organisms change. He wrote, "The most distinctive problem of zoological work is the change in form that animals undergo, both in the course of their development from the egg (embryology) and in their development in time (evolution)." Throughout his life, he usually had dozens of different experiments going on simultaneously, many of them unrelated to each other. He jokingly said that there are three kinds of experiments—those that are foolish, those that are damn foolish, and those that are worse than that!

In one of his most famous studies, Morgan engaged one of his graduate students to rear the fruit fly *Drosophila melanogaster* in the dark, hoping to produce flies whose eyes would atrophy from disuse and disappear in future generations. Even after many consecutive generations, however, the flies appeared to have no noticeable changes in spite of repeated attempts at inducing mutations by treatments with agents such as X rays and radium. After two years, they finally obtained an interesting result when a true-breeding line of *Drosophila* produced a male fruit fly with white eyes rather than the normal red eyes. Since this had been a true-breeding line of flies, this white-eyed male must have arisen from a new mutation that converted a red-eye allele into a white eye allele. Morgan is said to have carried this fly home with him in a jar, put it by his bedside at night while he slept, and then taken it back to the laboratory during the day.

Much like Mendel, Morgan studied the inheritance of this white-eye trait by making crosses and quantitatively analyzing their outcome. In the experiment described in figure 3.19, he began with his white-eyed male and crossed it to a red-eyed female. All of the F_1 offspring had red eyes, indicating that red is dominant to white. The F_1 offspring were then mated to each other to obtain an F_2 generation.

■ THE HYPOTHESIS

A quantitative analysis of genetic crosses may reveal the inheritance pattern for the white-eye allele.

■ TESTING THE HYPOTHESIS — FIGURE 3.19 Inheritance pattern of an X-linked trait in fruit flies.

Starting material: A true-breeding line of red-eyed fruit flies plus one white-eyed male fly that was discovered in the culture.

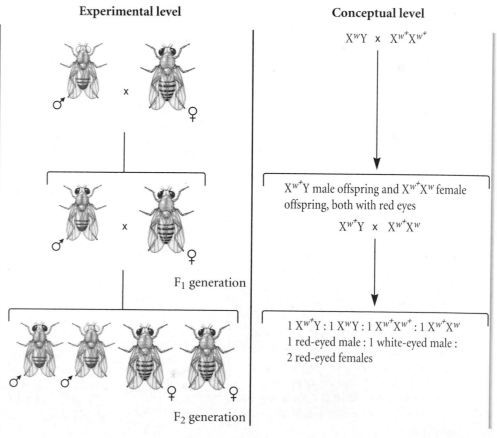

	Experimental level	Conceptual level
1. Cross the white-eyed male to red-eyed females.		$X^wY \times X^{w^+}X^{w^+}$
2. Record the results of the F_1 generation. This involves noting the eye color and sexes of several thousand flies.	F_1 generation	$X^{w^+}Y$ male offspring and $X^{w^+}X^w$ female offspring, both with red eyes $X^{w^+}Y \times X^{w^+}X^w$
3. Cross F_1 offspring with each other to obtain F_2 offspring. Also record the eye color and sex of the F_2 offspring.	F_2 generation	$1\,X^{w^+}Y : 1\,X^wY : 1\,X^{w^+}X^{w^+} : 1\,X^{w^+}X^w$ 1 red-eyed male : 1 white-eyed male : 2 red-eyed females

(continued)

4. In a separate experiment, perform a testcross between a white-eyed male and a red-eyed female from the F_1 generation. Record the results.

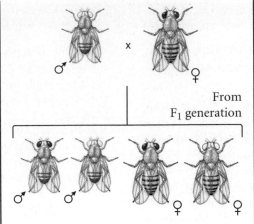

From F_1 generation

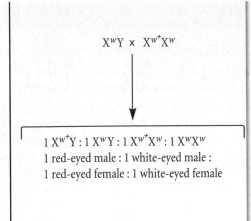

$X^w Y \times X^{w^+} X^w$

$1\ X^{w^+}Y : 1\ X^w Y : 1\ X^{w^+}X^w : 1\ X^w X^w$
1 red-eyed male : 1 white-eyed male :
1 red-eyed female : 1 white-eyed female

■ THE DATA

Cross	Results	
Original white-eyed male to red-eyed females	F_1 generation:	All red-eyed flies
F_1 male to F_1 female	F_2 generation:	2,459 red-eyed females 1,011 red-eyed males 0 white-eyed females 782 white-eyed males
White-eyed male to F_1 female	Testcross:	129 red-eyed females 132 red-eyed males 88 white-eyed females 86 white-eyed males

■ INTERPRETING THE DATA

As seen in the data table, the F_2 generation consisted of 2,459 red-eyed females, 1,011 red-eyed males, and 782 white-eyed males. Most notably, no white-eyed female offspring were observed in the F_2 generation. These results suggested that the pattern of transmission from parent to offspring depends on the sex of the offspring and on the alleles that they carry. As shown in the Punnett square here, it is consistent with the idea that the eye color alleles are located on the X chromosome:

F_1 male is $X^{w^+}Y$
F_1 female is $X^{w^+}X^w$

Male gametes

♀ \ ♂	X^{w^+}	Y
X^{w^+}	$X^{w^+}X^{w^+}$ Red, female	$X^{w^+}Y$ Red, male
X^w	$X^{w^+}X^w$ Red, female	$X^w Y$ White, male

Female gametes

The Punnett square predicts that the F_2 generation will not have any white-eyed females. This prediction was confirmed experimentally. These results indicated that the eye color alleles are located on the X chromosome. Genes that are physically located within the X chromosome are called **X-linked genes** or **X-linked alleles.** However, it should also be pointed out that the experimental ratio in the F_2 generation of red eyes to white eyes is (2,459 + 1,011):782, which equals 4.4:1. This ratio deviates significantly from the predicted ratio of 3:1 shown in the previous Punnett square. The lower-than-expected number of white-eyed flies can be explained by a decreased survival of white-eyed flies.

Morgan also conducted a **testcross** (step 4, fig. 3.19), which is a cross between an individual with a recessive phenotype and an individual whose genotype the experimenter wishes to test. In this case, he mated an F_1 female to a white-eyed male. This cross produced red-eyed males and females, and white-eyed males and females, in approximately equal numbers. The testcross data are also consistent with an X-linked pattern of inheritance. As shown in the following Punnett square, the testcross predicts a 1:1:1:1 ratio:

Testcross:
male is $X^w Y$
F_1 female is $X^{w^+}X^w$

Male gametes

♀ \ ♂	X^w	Y
X^{w^+}	$X^{w^+}X^w$ Red, female	$X^{w^+}Y$ Red, male
X^w	$X^w X^w$ White, female	$X^w Y$ White, male

Female gametes

The observed data are 129:132:88:86, which is a ratio of 1.5:1.5:1:1. Again, the lower-than-expected numbers of white-eyed males and females can be explained by a lower survival rate for white-eyed flies. In his own interpretation, Morgan concluded that R (red eye color) and X (a sex factor that is present in two copies in the female) are combined and have never existed apart. In other words, this gene for eye color is on the X chromosome. Morgan was the first geneticist to receive the Nobel Prize.

Calvin Bridges, a graduate student in the laboratory of Morgan, also examined the transmission of X-linked traits. Bridges conducted hundreds of crosses involving several different types of X-linked alleles. In his crosses, he occasionally obtained offspring that had unexpected phenotypes and abnormalities in sex chromosome composition. In all cases, the parallel between the cytological presence of sex chromosome abnormalities and the occurrence of unexpected traits confirmed the idea that the sex chromosomes carry X-linked genes. Altogether, the work of

Morgan and Bridges provided an impressive body of evidence confirming the idea that traits following an X-linked pattern of inheritance are governed by genes that are physically located on the X chromosome. Bridges wrote, "There can be no doubt that the complete parallelism between the unique behavior of chromosomes and the behavior of sex-linked genes and sex in this case means that the sex-linked genes are located in and borne by the X chromosomes." An example of Bridges's work is described in solved problem S6 at the end of chapter 3.

A self-help quiz involving this experiment can be found at the Online Learning Center.

The Inheritance Pattern of X-Linked Genes Is Not the Same in Reciprocal Crosses

As we have seen, the traits of an organism may be governed by genes that are located on a sex chromosome. We have also learned that many species have males and females that differ in their sex chromosome composition. In these species, the transmission pattern of certain traits may depend on the sex of the parent. In experimental organisms, a **reciprocal cross** can be conducted to discern if a trait is carried on a sex chromosome or on an autosome; X-linked traits do not behave identically in reciprocal crosses. To explain what a reciprocal cross is, let's first consider a cross between a homozygous white-eyed female fly and a red-eyed male. As shown here, the outcome is all males with white eyes and all females with red eyes.

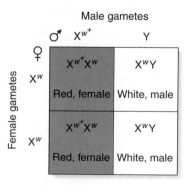

In the reciprocal cross, a homozygous red-eyed female is crossed to a white-eyed male. The result is 100% offspring with red eyes.

When comparing the two Punnett squares, the outcome of the reciprocal cross did not yield the same results. This is expected of X-linked genes, because the male transmits the gene only to his female offspring while the female transmits an X chromosome to both male and female offspring. Since the male parent does not transmit the X chromosome to his sons, he does not contribute to the appearance of their X-linked phenotypes. This explains why X-linked traits do not behave equally in reciprocal crosses. Experimentally, the observation that reciprocal crosses do not yield the same results is an important clue that a trait may be X linked.

When setting up a Punnett square involving X-linked traits, it is worth emphasizing that we must consider the alleles on the X chromosome as well as the fact that males may transmit a Y chromosome instead of the X chromosome. The male makes two types of gametes, one that carries the X chromosome and one that carries the Y. The Punnett square must also include the Y chromosome even though this chromosome does not carry any X-linked genes. The X chromosomes from the female and male are designated with their corresponding alleles. When the Punnett square is filled in, it predicts the X-linked genotypes and sexes of the offspring.

Genes Located on Human Sex Chromosomes Can Be Transmitted in an X-Linked, a Y-Linked, or a Pseudoautosomal Pattern

Our previous discussion of the sex chromosomes has concentrated on X-linked genes that are located on the X chromosome but not on the Y chromosome. The term **hemizygous** is used to describe the single copy of an X-linked gene in the male. A male mammal is said to be hemizygous for his X-linked genes. Since males of certain species such as humans have a single copy of the X chromosome, they are more likely to be affected by rare, recessive X-linked disorders. Chapter 22 considers several examples of such diseases.

The general term **sex linkage** is used to describe genes that are found on one of the two types of sex chromosomes but not on both. Hundreds of X-linked genes have been identified in humans and other mammals; several examples are given in chapter 22. Therefore, the terms *sex-linked genes* and *X-linked genes* tend to be used synonymously. By comparison, there are relatively few genes that are only located on the Y chromosome. These are usually called **holandric genes.** An example of a holandric gene is the *Sry* gene found in mammals. Its expression is necessary for proper male development.

Besides sex-linked genes, the X and Y chromosomes also contain short regions of homology where the X and Y chromo-

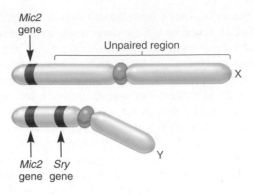

FIGURE 3.20   **A comparison of the homologous and nonhomologous regions of the X and Y chromosome in humans.** A few pseudoautosomal genes, such as *Mic2*, are found on both the X and Y chromosomes in the region of homology. Many X-linked genes are found in the unpaired region of the X chromosome. A few Y-linked genes, such as *Sry*, are found on the Y chromosome, in the unpaired (nonhomologous) region.

somes carry the same genes. As shown in figure 3.20, the human sex chromosomes contain homologous regions at one end of the X and Y chromosomes. These regions, which are evolutionarily related, promote the necessary pairing of the X and Y chromosomes that occurs during meiosis I of spermatogenesis. Only a few genes are located in this homologous region. One example is a human gene called *Mic2*, which is involved in antibody production. The *Mic2* gene is found on both the X and Y chromosomes. It follows a pattern of inheritance called **pseudoautosomal inheritance.** The term *pseudoautosomal* refers to the idea that the inheritance pattern of the *Mic2* gene is the same as the inheritance pattern of a gene located on an autosome even though the *Mic2* gene is actually located on the sex chromosomes. As in autosomal inheritance, males contain two copies of pseudoautosomally inherited genes, and they can transmit the genes to both daughters and sons.

CONCEPTUAL SUMMARY

In chapter 3 we have considered the process of reproduction at the cellular level. During **binary fission** in bacteria and **mitosis** in eukaryotic cells, a mother cell divides to produce two daughter cells that are genetically identical to each other. In bacteria, which contain a single circular chromosome, binary fission provides a way to reproduce asexually. Eukaryotic species, which contain many linear chromosomes, must sort their chromosomes during cell division. This ensures that the chromosomes are correctly distributed to the daughter cells. In simple eukaryotes, mitotic divisions can produce new unicellular offspring; it is a form of **asexual reproduction.** Otherwise, the primary reason that eukaryotic species undergo mitotic cellular divisions is to produce organisms that are multicellular. Diploid eukaryotes develop from a single fertilized egg cell that undergoes repeated mitotic and cellular

divisions to produce an organism containing many cells. Most of the cells of the body are genetically identical to each other.

Sexual reproduction requires the formation of **gametes** (i.e., sperm cells or egg cells) that contain half the genetic material. For **diploid** organisms (those with two sets of chromosomes, or 2n), the process of **meiosis** results in cells that are **haploid** (containing one set, or 1n). Meiosis requires two successive divisions. Meiosis I involves a pairing between homologous chromosomes that does not occur during mitosis. This pairing is necessary to ensure that the resultant cells will contain one complete set of chromosomes. Meiosis II is very similar to a mitotic division, except that the cells entering meiosis II begin with half the number of chromatids as compared to cells entering mitosis.

In animals, meiosis occurs in conjunction with the cellular specialization of gametes. During **spermatogenesis,** meiosis coincides with the production of four motile sperm cells. By comparison, **oogenesis** is aimed at producing a single haploid egg that is rich in nutrients. Fertilization in animals involves the union between a single sperm and egg cell. In plants, sexual reproduction involves an alternation between diploid organisms (**sporophytes**) and haploid organisms (**gametophytes**). However, the gametophytes of more complex plants contain only a few cells, which are primarily involved in producing gametes. Plant reproduction is a double fertilization event in which one sperm fertilizes the egg and a second sperm unites with the central cell to produce the endosperm. The proliferation of endosperm provides nutrients to the plant embryo.

A cellular understanding of mitosis, meiosis, and fertilization allows us to appreciate how the transmission of traits is due to the passage of chromosomes. Genes, which govern the outcome of traits, are physically located within the structure of chromosomes. The **chromosome theory of inheritance,** proposed by Boveri and Sutton, describes how the transmission of chromosomes accounts for Mendel's laws of segregation and independent assortment. Later, it was experimentally possible to correlate the inheritance of particular traits with the transmission of specific sex chromosomes. In particular, early studies were able to identify genes on the X chromosome that govern eye color in fruit flies. This phenomenon, called **X-linked inheritance,** confirmed the notion that chromosomes carry genes that govern the outcome of traits. In some species, chromosomes also play a role in sex determination. Particular chromosomes, termed **sex chromosomes,** are different between males and females.

EXPERIMENTAL SUMMARY

A central method that has been used to understand chromosomal transmission is light microscopy. **Cytogeneticists** can view the chromosomes in actively dividing cells under a microscope. This has allowed researchers to study the behavior of chromosomes during mitosis and meiosis. In the late nineteenth and early twentieth century, microscopic examinations of chromosomes yielded observations that were consistent with the idea that the chromosomes are the carriers of genes. This idea, along with Mendel's experiments in pea plants, led to the formulation of the chromosome theory of inheritance. Many subsequent experiments have confirmed the validity of this theory. In particular, investigations involving the inheritance of the X chromosome were instrumental in establishing that the inheritance of certain traits paralleled the inheritance of the X chromosome. Morgan was the first scientist to show that the inheritance of a recessive X-linked trait was consistent with the respective gene's location on the X chromosome. Since that time, the chromosomal locations of many genes have been firmly established.

PROBLEM SETS & INSIGHTS

Solved Problems

S1. A diploid cell has eight chromosomes, four per set. In the following diagram, what phase of mitosis, meiosis I or meiosis II, is this cell in?

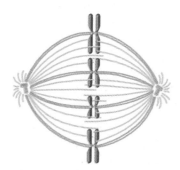

Answer: The cell is in metaphase of meiosis II. You can tell because the chromosomes are lined up along the metaphase plate, and it has only four pairs of sister chromatids. If it were mitosis, the cell would have eight pairs of sister chromatids.

S2. An unaffected woman (i.e., without disease symptoms) who is heterozygous for the X-linked allele causing Duchenne muscular dystrophy has children with a normal man. What are the probabilities of the following combinations of offspring?

A. An unaffected son

B. An unaffected son or daughter

C. A family of three children, all of whom are affected

Answer: The first thing we must do is construct a Punnett square to determine the outcome of the cross. N represents the normal allele, n the recessive allele causing Duchenne muscular dystrophy. The mother is heterozygous and the father is hemizygous for the normal allele.

Male gametes

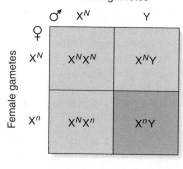

Phenotype ratio is

2 normal daughters :
1 normal son :
1 affected son

A. There are four possible children, one of whom is an unaffected son. Therefore, the probability of an unaffected son is 1/4.

B. Use the sum rule: 1/4 + 1/2 = 3/4.

C. You could use the product rule, since there would be three offspring, in a row, with the disorder: $(1/4)(1/4)(1/4) = 1/64 = 0.016 = 1.6\%$.

S3. What are the major differences between prophase, metaphase, and anaphase when comparing mitosis, meiosis I, and meiosis II?

Answer: The table summarizes key differences.

A Comparison of Mitosis, Meiosis I, and Meiosis II

Phase	Event	Mitosis	Meiosis I	Meiosis II
Prophase	Synapsis:	No	Yes	No
Prophase	Crossing over:	Rarely	Commonly	Rarely
Metaphase	Alignment along the metaphase plate:	Sister chro-matids	Bivalents	Sister chro-matids
Anaphase	Separation of:	Sister chro-matids	Bivalents	Sister chro-matids

S4. Among different plant species, gametophytes can be produced by single individuals or by separate sexes. In some species, such as the garden pea, a single individual can produce both male and female gametophytes. Such a species is called monoecious (or monecious). In monoecious species, fertilization takes place via self-fertilization or cross-fertilization. A monoecious plant species that has a single type of flower producing both pollen and eggs is a monoclinous plant. In other plant species, two different types of flowers produce pollen and eggs. When this occurs, it is most common for the "male flowers" to be produced near the top of the plant and the "female flowers" toward the bottom. There are also a few species of plants that are dioecious. For dioecious species, one individual makes either male flowers or female flowers, but not both.

Give examples of monoecious plants that are monoclinous and some that are not. Give examples of dioecious plants. What would be the advantages and disadvantages of being monoecious compared to dioecious?

Answer:

Monoecious, monoclinous plants—pea plant, tulip, and roses. The same flower produces pollen on the anthers and egg cells within the ovary.

Monoecious, nonmonoclinous plant—corn. In corn, the tassels are the male flowers and the ears result from fertilization within the female flowers.

Dioecious plants—ginkgo and fig trees. Certain individuals produce only pollen while others produce only eggs.

An advantage of being monoecious is that fertilization is relatively easy because the pollen and egg cells are produced on the same individual. This is particularly true for monoclinous plants. The proximity of the pollen to the egg cells makes it more likely for self-fertilization to occur. This is advantageous if the plant population is relatively sparse. On the other hand, a dioecious species can reproduce only via cross-fertilization. The advantage of cross-fertilization is that it enhances genetic variation. Over the long run, this can be an advantage because cross-fertilization is more likely to produce a varied population of individuals, some of which may possess combinations of traits that promote survival.

S5. In humans, why are X-linked recessive traits more likely to occur in males compared to females?

Answer: Because a male is hemizygous for X-linked traits, the phenotypic expression of X-linked traits depends on only a single copy of the gene. When a male inherits a recessive X-linked allele, he will automatically exhibit the trait because he does not have another copy of the gene on the corresponding Y chromosome. This phenomenon is particularly relevant to the inheritance of recessive X-linked alleles that cause human disease. Some examples are described in chapter 22.

S6. To test the chromosome theory of inheritance, Calvin Bridges made crosses involving the inheritance of X-linked traits. One of his experiments involved two different X-linked genes affecting eye color and wing shape. For the eye color gene, the red-eye allele is dominant to the white-eye allele. A second X-linked trait is wing size; the allele called miniature is recessive to the normal allele. In this case, m represents the miniature allele and m^+ the normal allele. A hemizygous male fly containing a miniature allele has small (miniature) wings. A female must be homozygous, mm, in order to have miniature wings.

Bridges made a cross between $X^{w,m^+} X^{w,m^+}$ female flies (white eyes and normal wings) to $X^{w^+,m} Y$ male flies (red eyes and miniature wings). He then examined the eyes, wings, and sexes of thousands of offspring. Most of the offspring were males with white eyes and normal wings, and females with red eyes and normal wings. On rare occasions (i.e., approximately 1 out of 1,700 flies), however, he also obtained female offspring with white eyes or males with red eyes. He also noted the wing shape in these flies and then cytologically examined their chromosome composition using a microscope. The following results were obtained:

Offspring	Eye color	Wing Size	Sex Chromosomes
Expected females	Red	Normal	XX
Expected males	White	Normal	XY
Unexpected females (rare)	White	Normal	XXY
Unexpected males (rare)	Red	Miniature	X0

Data from: Bridges, C. B. "Non-Disjunction as Proof of the Chromosome Theory of Heredity," *Genetics 1*, 1–52, 107–63.

Explain these data.

Answer: Remember that in fruit flies, it is the number of X chromosomes (not the presence of the Y chromosome) that determines sex. As seen in the data, the flies with unexpected phenotypes were abnormal in their sex chromosome composition. The white-eyed female flies were due to the union between an abnormal XX female gamete and a normal Y male gamete. Likewise, the unexpected male offspring contained only one X chromosome and no Y. These male offspring were due to the union between an abnormal egg without any X chromosome and a normal sperm containing one X chromosome. The wing size of the unexpected males was a particularly significant result. The red-eyed males showed a miniature wing size. As noted by Bridges, this means that they inherited their X chromosome from their father rather than their mother. This observation provided compelling evidence that the inheritance of the X chromosome correlates with the inheritance of particular traits.

At the time of his work, Bridges's results were particularly striking since chromosomal abnormalities had been rarely observed in *Drosophila*. Nevertheless, Bridges first predicted how chromosomal abnormalities would cause certain unexpected phenotypes and then he actually observed the abnormal number of chromosomes using a microscope. Together, his work provided evidence confirming the idea that traits that follow an X-linked pattern of inheritance are governed by genes that are physically located on the X chromosome.

Conceptual Questions

C1. The process of binary fission begins with a single mother cell and ends with two daughter cells. Would you expect the mother and daughter cells to be genetically identical? Explain why or why not.

C2. What is a homologue? With regard to genes and alleles, how are homologues similar to and different from each other?

C3. What is a sister chromatid? Are sister chromatids genetically similar or identical? Explain.

C4. With regard to the chromosomes, which phase of mitosis is the organization phase and which is the separation phase?

C5. A species is diploid containing three chromosomes per set. Draw what the chromosomes would look like in the G_1 and G_2 phases of the cell cycle.

C6. How does the attachment of kinetochore microtubules to the kinetochore differ in metaphase I of meiosis compared to metaphase of mitosis? Discuss what you think would happen if a sister chromatid was not attached to a kinetochore microtubule.

C7. For the following events, specify whether they occur during mitosis, meiosis I, and/or meiosis II:

A. Separation of conjoined chromatids within a pair of sister chromatids

B. Pairing of homologous chromosomes

C. Alignment of chromatids along the metaphase plate

D. Attachment of sister chromatids to both poles

C8. Describe the key events during meiosis that result in a 50% reduction in the amount of genetic material per cell.

C9. A cell is diploid and contains three chromosomes per set. Draw the arrangement of chromosomes during metaphase of mitosis and metaphase I and II of meiosis. In your drawing, make one set dark and the other lighter.

C10. The arrangement of homologues during metaphase I of meiosis is a random process. In your own words, explain what this means.

C11. A eukaryotic cell is diploid containing 10 chromosomes (5 in each set). For mitosis and meiosis, how many daughter cells would be produced and how many chromosomes would each one contain?

C12. If a diploid cell contains six chromosomes (i.e., three per set), how many possible random arrangements of homologues could occur during metaphase I?

C13. A cell has four pairs of chromosomes. Assuming that crossing over does not occur, what is the probability that a gamete will contain all of the paternal chromosomes? If n equals the number of chromosomes in a set, which of the following expressions can be used to calculate the probability that a gamete will receive all of the paternal chromosomes: $(1/2)^n$, $(1/2)^{n-1}$, or $n^{1/2}$?

C14. With regard to question C13, how would the phenomenon of crossing over affect the results? In other words, would there be a higher or lower probability of a gamete inheriting only paternal chromosomes? Explain your answer.

C15. Eukaryotic cells must sort their chromosomes during mitosis so that each daughter cell receives the correct number of chromosomes. Why don't bacteria need to sort their chromosomes?

C16. Why is it necessary to condense the chromosomes during mitosis and meiosis? What do you think might happen if the chromosomes were not condensed?

C17. Nine-banded armadillos almost always give birth to four offspring that are genetically identical quadruplets. Explain how you think this happens.

C18. A diploid species contains four chromosomes per set for a total of eight chromosomes in its somatic cells. Draw the cell as it would look in late prophase II of meiosis and prophase of mitosis. Discuss how prophase II of meiosis and prophase of mitosis differ from each other, and explain how the difference originates.

C19. Explain why the products of meiosis may not be genetically identical while the products of mitosis are.

C20. The cellular period between meiosis I and meiosis II is called interphase II. Does DNA replication take place during interphase II? Explain your answer.

C21. Describe several ways in which telophase appears to be the reverse of prophase and prometaphase.

C22. In corn, there are 10 chromosomes per set and the species is diploid. If you performed a karyotype, what is the total number of chromosomes that you would expect to see in the following types of cells?

A. A leaf cell

B. The sperm nucleus of a pollen grain

C. An endosperm cell after fertilization

D. A root cell

C23. The arctic fox has 50 chromosomes (25 per set) and the common red fox has 38 chromosomes (19 per set). These species can interbreed to produce viable but infertile offspring. How many chromosomes would the offspring contain? What problems do you think may occur during meiosis that would explain the offspring's infertility?

C24. Let's suppose that a gene affecting pigmentation is found on the X chromosome (in mammals or insects) or the Z chromosome (in birds) but not on the Y or W chromosome. It is found on an autosome in bees. This gene is found in two alleles, D (dark), which is dominant to d (light). What would be the phenotypic results of crosses between a true-breeding dark female and true-breeding light male, and the reciprocal crosses involving a true-breeding light female and true-breeding dark male, in the following species?

Refer back to figure 3.18 for the mechanism of sex determination in these species.

A. Birds

B. *Drosophila*

C. Bees

D. Humans

C25. Describe the cellular differences between male and female gametes.

C26. At puberty, the testes contain a finite number of cells and produce an enormous number of sperm cells during the life span of a male. Explain why testes do not run out of spermatogonial cells.

C27. Describe the timing of meiosis I and II during human oogenesis.

C28. Three genes (*A, B,* and *C*) are found on three different chromosomes. For the following diploid genotypes, describe all the possible gamete combinations.

A. *Aa Bb Cc*

B. *AA Bb CC*

C. *Aa BB Cc*

D. *Aa bb cc*

C29. Propose the most likely mode of inheritance (autosomal dominant, autosomal recessive, or X-linked recessive) for the following pedigrees. Affected individuals are shown with filled (black) symbols.

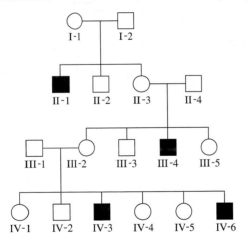

(a)

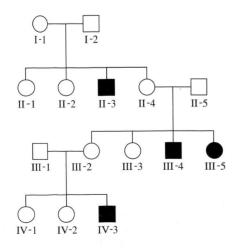

(b)

C30. A human disease known as vitamin D resistant rickets is inherited as an X-linked dominant trait. If a male with the disease produces children with a female who does not have the disease, what is the expected ratio of affected and unaffected offspring?

C31. Duchenne muscular dystrophy is an inherited degenerative disorder that affects the muscles. Based on the following pedigree, propose an inheritance pattern for this disease.

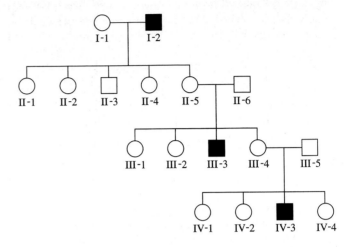

C32. Hemophilia is a recessive X-linked trait in humans. If a heterozygous woman has children with a normal man, what are the odds of the following combinations of children?

A. An affected son

B. Four unaffected offspring in a row

C. An unaffected daughter or son

D. Two out of five offspring that are affected

C33. A form of color blindness is inherited as a recessive X-linked trait. A couple has a color-blind daughter and a normal son. What are the genotypes and phenotypes of the parents? What is the probability that their next two children will be color-blind?

C34. Explain whether the following events could or could not happen.

A. A male has the same X chromosome as his paternal grandfather.

B. A male has the same X chromosome as his maternal grandfather.

C. A female has the same X chromosome as her maternal grandmother.

D. A female has the same X chromosome as her maternal great-great-grandmother.

C35. Incontinentia pigmenti is a rare, dominant X-linked disorder in humans characterized by swirls of pigment in the skin. If an affected female, who had an unaffected father, has children with an unaffected male, what would be the predicted ratios of affected and unaffected sons and daughters?

C36. Red-green color blindness is inherited as a recessive X-linked trait. What are the following probabilities?

A. A woman with phenotypically normal parents and a color-blind brother will have a color-blind son. Assume that she has no previous children.

B. The next child of a phenotypically normal woman, who has already had one color-blind son, will be a color-blind son.

C. The next child of a phenotypically normal woman, who has already had one color-blind son, and who is married to a color-blind man, will have a color-blind daughter.

C37. A phenotypically normal woman with an abnormally long chromosome 13 (and a normal homologue of chromosome 13) marries a phenotypically normal man with an abnormally short chromosome 11 (and a normal homologue of chromosome 11). What is the probability of producing an offspring that will have both a long chromosome 13 and a short chromosome 11? If such a child is produced, what is the probability that this child would eventually pass both abnormal chromosomes to one of his/her offspring?

C38. Assuming that such a fly would be viable, what would be the sex of a fruit fly with the following chromosomal composition?

A. One X chromosome and two sets of autosomes

B. Two X chromosomes, one Y chromosome, and two sets of autosomes

C. Two X chromosomes and four sets of autosomes

D. Four X chromosomes, two Y chromosomes, and four sets of autosomes

C39. What would be the sex of a human with the following numbers of sex chromosomes?

A. XXX

B. X (also described as X0)

C. XYY

D. XXY

Experimental Questions

E1. When studying living cells in a laboratory, researchers sometimes use drugs as a way to make cells remain at a particular stage of the cell cycle. For example, aphidicolin inhibits DNA synthesis in eukaryotic cells and causes them to remain in the G_1 phase because they cannot replicate their DNA. In what phase of the cell cycle—G_1, S, G_2, prophase, metaphase, anaphase, or telophase—would you expect somatic cells to stay if the following types of drug were added?

A. A drug that inhibits microtubule formation

B. A drug that allows microtubules to form but prevents them from shortening

C. A drug that inhibits cytokinesis

D. A drug that prevents chromosomal condensation

E2. In Morgan's experiments, which result do you think is the most convincing piece of evidence pointing to X-linkage of the eye color gene? Explain your answer.

E3. In his original studies of figure 3.19, Morgan first suggested that the original white-eyed male had two copies of the white-eye allele. In this problem, let's assume that he meant the fly was X^wY^w instead of X^wY. Are his data in figure 3.19 consistent with this hypothesis? What crosses would need to be made to rule out the possibility that the Y chromosome carries a copy of the eye color gene?

E4. How would you set up crosses to determine if a gene was Y linked versus X linked?

E5. Occasionally during meiosis, a mistake can happen whereby a gamete may receive zero or two sex chromosomes rather than one. Calvin Bridges made a cross between white-eyed female flies and red-eyed males. As you would expect, most of the offspring were red-eyed females and white-eyed males. On rare occasions, however, he found a white-eyed female or a red-eyed male. These rare flies were not due to new mutations but instead were due to mistakes during meiosis in the parent flies. Consider the mechanism of sex determination in fruit flies and propose how this could happen. In your answer, describe the sex chromosome composition of these rare flies.

E6. Let's suppose that you have karyotyped a female fruit fly with red eyes and found that it has three X chromosomes instead of the normal two. Although you do not know its parents, you do know that this fly came from a mixed culture of flies in which some had red eyes, some had white eyes, and some had eosin eyes. Eosin is an allele of the same gene that causes white or red eyes. The red allele is dominant and the white allele is recessive. The expression of the eosin allele, however, depends on copy number. When females have two copies of this allele they have eosin eyes. When females are heterozygous for the eosin allele and white allele, they have light-eosin eyes. When females are heterozygous for the red allele and the eosin allele, they have red eyes. Males that are hemizygous for the eosin allele have light-eosin eyes.

You cross this female with a white-eyed male and count the number of offspring. You may assume that this unusual female makes half of its gametes with one X chromosome and half of its gametes with two X chromosomes. The following results were obtained:

	Females*	Males
Red eyes	50	11
White eyes	0	0
Eosin	20	0
Light-eosin	21	20

*A female offspring can be XXX, XX, or XXY.

Explain the 3:1 ratio between female and male offspring. What was the genotype of the original mother, which had red eyes and three X chromosomes? Construct a Punnett square that is consistent with these data.

E7. With regard to thickness and length, what do you think the chromosomes would look like if you microscopically examined them during interphase? How would that compare to their appearance during metaphase?

E8. White-eyed flies have a lower survival rate compared to red-eyed flies. Based on the data in figure 3.19, what percentage of white-eyed flies survived compared to red-eyed flies assuming 100% survival of red-eyed flies?

E9. The *Mic2* gene in humans is present on both the X and Y chromosome. Let's suppose the *Mic2* gene exists in a dominant *Mic2* allele, which results in normal antibody production, and a recessive *mic2* allele, which results in defective antibody production. Using molecular techniques, it is possible to distinguish homozygous and heterozygous individuals. By following the transmission of the *Mic2* and *mic2* alleles in a large human pedigree, would it be possible to distinguish pseudoautosomal inheritance from autosomal inheritance? Explain your answer.

E10. A rare form of dwarfism that also included hearing loss was found to run in a particular family. It is inherited in a dominant manner. It was discovered that an affected individual had one normal copy of chromosome 15 and one abnormal copy of chromosome 15 that was unusually long. How would you determine if the unusually long chromosome 15 was causing this disorder?

E11. Discuss why crosses (i.e., the experiments of Mendel) and the microscopic observations of chromosomes during mitosis and meiosis were both needed to deduce the chromosome theory of inheritance.

E12. A cross was made between female flies with white eyes and miniature wings (both X-linked recessive traits) to male flies with red eyes and normal wings. On rare occasions, female offspring were produced with white eyes. If we assume that these females are due to errors in meiosis, what would be the most likely chromosomal composition of such flies? What would be their wing shape?

E13. Experimentally, how do you think researchers were able to determine that the Y chromosome causes maleness in mammals, whereas the ratio of X chromosomes to the sets of autosomes causes sex determination in fruit flies?

E14. When examining a human pedigree, what features do you look for to distinguish between X-linked recessive inheritance versus autosomal recessive inheritance? How would you distinguish X-linked dominant inheritance from autosomal dominant inheritance in a human pedigree?

Questions for Student Discussion/Collaboration

1. In figure 3.19, Morgan obtained a white-eyed male fly in a population containing many red-eyed flies that he thought were true-breeding. As mentioned in the experiment, he crossed this fly with several red-eyed sisters, and all the offspring had red eyes. But actually this is not quite true. Morgan observed 1,237 red-eyed flies and 3 white-eyed males. Provide two or more explanations why he obtained 3 white-eyed males in the F_1 generation.

2. A diploid eukaryotic cell has 10 chromosomes (5 per set). As a group, take turns having one student draw the cell as it would look

during a phase of mitosis, meiosis I, or meiosis II; then have the other students guess which phase it is.

3. Discuss the principles of the chromosome theory of inheritance. Which principles were deduced via light microscopy, and which were deduced from crosses? What modern techniques could be used to support the chromosome theory of inheritance?

Note: All answers appear at the website for this textbook; the answers to even-numbered questions are in the back of the textbook.

www.mhhe.com/brooker

Visit the Online Learning Center for practice tests, answer keys, and other learning aids for this chapter. Enhance your understanding of genetics with our interactive exercises, web links, news feeds, tutorial service, and much more.

EXTENSIONS OF
MENDELIAN INHERITANCE

::

4

The term **Mendelian inheritance** describes inheritance patterns that obey two laws: the law of segregation and the law of independent assortment. Until now, we have mainly considered traits that are affected by a single gene that is found in two different alleles. In these cases, one allele was dominant over the other. This type of inheritance is sometimes called **simple Mendelian inheritance,** because the observed ratios in the offspring readily obey Mendel's laws. For example, when two different true-breeding pea plants are crossed (e.g., tall and dwarf) and the F_1 generation is allowed to self-fertilize, the F_2 generation shows a 3:1 phenotypic ratio of tall and dwarf offspring. Similarly, in chapter 3, we examined the phenomenon that some genes are located on the sex chromosomes. The pattern of transmission of these traits also depended on the pattern of transmission of the X chromosome, yielding differing but predictable ratios among male and female offspring. X-linked inheritance follows a Mendelian pattern because the sex chromosomes obey the laws of segregation and independent assortment.

In chapter 4, we will extend our understanding of Mendelian inheritance by examining the transmission patterns for several traits that do not display a simple dominant/recessive relationship. Such patterns usually don't produce the ratios of offspring that are expected from a simple Mendelian relationship. This does not mean that Mendel was wrong. Rather, the inheritance patterns of many traits are more complex and interesting than he had realized. Later, in chapters 5 and 7, we will examine eukaryotic inheritance patterns that actually violate the laws of segregation and independent assortment.

4.1 INHERITANCE PATTERNS OF SINGLE GENES

We begin chapter 4 with the further exploration of traits that are influenced by a single gene. Table 4.1 describes the general features of several types of inheritance patterns that have been observed by modern geneticists. These various patterns occur because there are many different ways that two alleles may govern the outcome of a trait. In this section we will examine these patterns with two goals in mind. First, we want to understand how the molecular expression of genes can account for an individual's phenotype. In other words, we will study the underlying relationship between molecular genetics (i.e., the expression of a gene to produce a functional protein) and the trait of an individual. Our second goal concerns the outcome of crosses. Many of the inheritance patterns described in table 4.1 do not produce a 3:1 phenotypic ratio when two heterozygotes produce offspring. In this section, we consider how allelic interactions produce ratios that

differ from a simple Mendelian pattern. However, as our starting point, we will begin by reconsidering a simple dominant/recessive relationship from a molecular perspective.

Recessive Alleles Often Cause a Reduction in the Amount or Function of the Encoded Protein

For any given gene, geneticists refer to prevalent alleles in a population as **wild-type alleles.** In most instances, a wild-type allele encodes a protein that is made in the proper amount and functions normally. By comparison, alleles that have been altered by mutation are called **mutant alleles;** these tend to be rare in natural populations. Among Mendel's seven traits discussed in chapter 2, the wild-type alleles are purple flowers, axial flowers, yellow seeds, round seeds, smooth pods, green pods, and tall plants. The mutant alleles are white flowers, terminal flowers, green seeds, wrinkled seeds, constricted pods, yellow pods, and dwarf plants. You may have already noticed that the seven wild-type alleles are dominant over the seven mutant alleles. Likewise, red eyes and

TABLE 4.1
Different Types of Mendelian Inheritance Patterns

Type	Description
Simple Mendelian	**Inheritance:** This term is commonly applied to the inheritance of alleles that obey Mendel's laws and follow a strict dominant/recessive relationship. In chapter 4, we will see that some genes can be found in three or more alleles, making the relationship more complex. **Molecular:** 50% of the protein normally encoded by two copies of the dominant allele is sufficient to produce the dominant trait.
X linked	**Inheritance:** It involves the inheritance of genes that are located on the X chromosome. In mammals and fruit flies, males are hemizygous for X-linked genes while females have two copies. **Molecular:** If a pair of X-linked alleles shows a simple dominant/recessive relationship, 50% of the protein encoded by two copies of the dominant allele is sufficient to produce the dominant trait (in the female).
Lethal alleles	**Inheritance:** An allele that has the potential of causing the death of an organism. **Molecular:** Lethal alleles are most commonly loss-of-function alleles that encode proteins that are necessary for survival. In rare cases, the alleles may be in nonessential genes that change a protein to function with abnormal and detrimental consequences.
Incomplete dominance	**Inheritance:** This pattern occurs when the heterozygote has a phenotype that is intermediate between either corresponding homozygote. For example, a plant produced from a cross between red-flowered and white-flowered parents will have pink flowers. **Molecular:** 50% of the protein encoded by two copies of the normal (i.e., wild-type) allele is not sufficient to produce the normal trait.
Codominance	**Inheritance:** This pattern occurs when the heterozygote expresses both alleles simultaneously. For example, in blood typing, an individual carrying the A and B alleles will have an AB blood type. **Molecular:** The codominant alleles encode proteins that function slightly differently from each other, and the function of each protein, in the heterozygote, affects the phenotype uniquely.
Overdominance	**Inheritance:** This pattern occurs when the heterozygote has a trait that is more beneficial than either homozygote. **Molecular:** Three common ways that heterozygotes may have benefits include: their cells may be resistant to infection by microorganisms, they may produce protein dimers with enhanced function, or they may produce proteins that function under a wider range of conditions.
Incomplete penetrance	**Inheritance:** This pattern occurs when a dominant phenotype is not expressed even though an individual carries a dominant allele. An example is an individual who carries the polydactyly allele but has a normal number of fingers and toes. **Molecular:** Even though a dominant gene may be present, the protein encoded by the gene may not exert its effects. This can be due to environmental influences or due to other genes that may encode proteins that counteract the effects of the protein encoded by the (seemingly) dominant allele.
Sex-influenced inheritance	**Inheritance:** This pattern refers to the impact of sex on the phenotype of the individual. Some alleles are recessive in one sex and dominant in the opposite sex. An example would be baldness in humans. **Molecular:** Sex hormones may regulate the molecular expression of genes. This can have an impact on the phenotypic effects of alleles.
Sex-limited inheritance	**Inheritance:** This refers to traits that occur in only one of the two sexes. An example would be breast development in mammals. **Molecular:** Sex hormones may regulate the molecular expression of genes. This can have an impact on the phenotypic effects of alleles. In this case, sex hormones that are primarily produced in only one sex are essential to produce a particular phenotype.

TABLE 4.2

Examples of Recessive Human Diseases

Disease	Protein That Is Produced by the Normal Gene*	Description
Phenylketonuria	Phenylalanine hydroxylase	Inability to metabolize phenylalanine. The disease can be prevented by a phenylalanine-free diet. If the diet is not followed early in life, symptoms can develop, including severe mental retardation and physical degeneration.
Albinism	Tyrosinase	Lack of pigmentation in the skin, eyes, and hair.
Tay-Sachs disease	Hexosaminidase A	Defect in lipid metabolism. Leads to paralysis, blindness, and early death.
Sandhoff disease	Hexosaminidase B	Defect in lipid metabolism. Muscle weakness in infancy, early blindness, and progressive mental and motor deterioration.
Cystic fibrosis	Chloride transporter	Inability to regulate ion balance across epithelial cells. Leads to production of thick lung mucus and chronic lung infections.
Lesch-Nyhan syndrome	Hypoxanthine-guanine phosphoribosyl transferase	Inability to metabolize purines, which are bases found in DNA and RNA. Leads to self-mutilation behavior, poor motor skills, and usually mental retardation and kidney failure.

*Individuals who exhibit the disease are homozygous (or hemizygous) for a recessive allele that results in a defect in the amount or function of the normal protein.

normal wings are examples of wild-type alleles in *Drosophila*, and white eyes and miniature wings are recessive mutant alleles. Because random mutations are more likely to disrupt gene function, it is common for a mutant allele to be defective in its ability to express a functional protein. In other words, mutations that produce mutant alleles are likely to decrease or eliminate the synthesis or functional activity of a protein. Such mutations are often inherited in a recessive fashion.

The idea that recessive alleles usually cause a substantial decrease in the expression of a functional protein is supported by analyses of many human genetic diseases. Keep in mind that a genetic disease is caused by a mutant allele. Table 4.2 lists several examples of human genetic diseases in which the recessive allele fails to produce a specific cellular protein in its active form. In many cases, molecular techniques have enabled researchers to clone these genes and determine the differences between the normal and recessive alleles. They have found that the recessive allele usually contains a mutation that causes a defect in the synthesis of a fully functional protein.

To understand why many defective mutant alleles are inherited recessively, we need to take a quantitative look at protein function. With the exception of sex-linked traits, diploid individuals have two copies of every gene. In a simple dominant/recessive relationship, the recessive allele does not affect the phenotype of the heterozygote. In other words, a single copy of the dominant (normal) allele is sufficient to mask the effects of the recessive allele. If the recessive allele cannot produce a functional protein, how do we explain the wild-type phenotype of the heterozygote? As described in figure 4.1, the usual explanation is that 50% of the functional protein encoded by two copies of the normal gene is adequate to provide a normal phenotype. In this example, the *PP* homozygote and *Pp* heterozygote each make sufficient functional protein to yield purple flowers. This means that the normal homozygous individual is making much more of the

wild-type protein than it really needs. Therefore, if the normal amount is reduced to 50%, as in the heterozygote, the individual still has plenty of this protein to accomplish whatever cellular function it performs. The phenomenon that "50% of the normal protein is enough" is fairly common among many genes. A second possible explanation is that the heterozygote actually produces more than 50% of the functional protein. Due to the phenomenon of gene regulation, the expression of the normal gene may be increased in the heterozygote. In this case, the normal gene is "up regulated" to compensate for the lack of function of the defective allele. The topic of gene regulation is discussed in chapters 14 and 15.

Normal allele: *P* (purple)
Recessive (defective) allele: *p* (white)

Genotype	*PP*	*Pp*	*pp*
Amount of functional protein P	100%	50%	0%

Phenotype	Purple	Purple	White
Simple dominant/ recessive relationship			

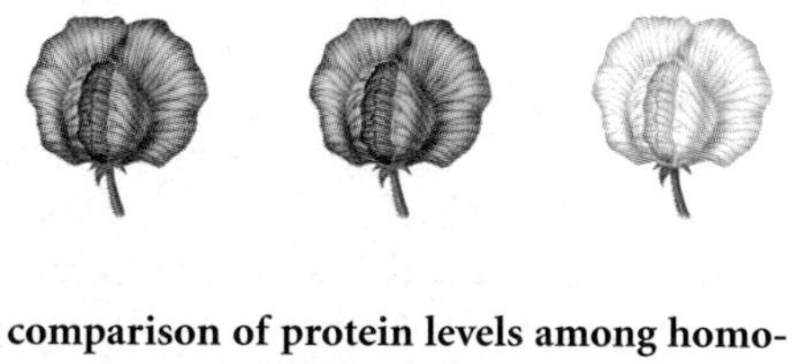

FIGURE 4.1   **A comparison of protein levels among homozygous and heterozygous genotypes *PP*, *Pp*, and *pp*.**
GENES→TRAITS   In a simple dominant/recessive relationship, 50% of the protein normally encoded by two copies of the dominant allele is sufficient to produce the normal phenotype, in this case purple flowers. A complete lack of the normal protein results in white flowers.

Mutations That Cause a Loss of Function in an Essential Gene Result in a Lethal Phenotype

When the absence of a specific protein results in a lethal pheno-type, the gene that encodes the protein is considered an **essential gene** for survival. An allele that has the potential to cause the death of an organism is called a **lethal allele.** These are usually inherited in a recessive manner. Though it varies according to species, it is estimated that approximately 1/3 of all genes are essential for survival. By comparison, **nonessential genes** are not absolutely required for survival, although they are likely to be beneficial to the organism. A loss-of-function mutation in a nonessential gene will not cause death. On rare occasions, how-ever, a nonessential gene may acquire a mutation that causes the gene product to be abnormally expressed in a way that leads to a lethal phenotype. A mutation that causes the overexpression of a nonessential gene may interfere with normal cell function and thereby cause lethality. Therefore, not all lethal mutations are in essential genes, although the great majority are.

Many lethal alleles prevent cell division and thereby kill an organism at a very early stage. Others, however, may only exert their effects later in life, or under certain environmental condi-tions. For example, a human genetic disease known as Hunting-ton disease is characterized by a progressive degeneration of the nervous system, dementia, and early death. The age when these symptoms appear, or the **age of onset,** is usually between 30 and 50. Therefore, this lethal allele acts relatively late in life.

Other lethal alleles may kill an organism only when certain environmental conditions prevail. These are called **conditional lethal alleles.** They have been extensively studied in experimental organisms. For example, some conditional lethals will kill an organism only in a particular temperature range. These alleles, called **temperature-sensitive (*ts*) lethals,** have been observed in many organisms such as *Drosophila*. A *ts* allele may kill a devel-oping larva at a high temperature (30°C), but it will survive if grown at a lower temperature (22°C). Temperature-sensitive alleles are typically mutations that alter the structure of the encoded protein so it does not function correctly at the nonper-missive temperature, or the protein becomes unfolded and is rap-idly degraded. Conditional lethal alleles may also be identified when an individual is exposed to a particular agent in the envi-ronment. For example, people with a defect in the gene that encodes the enzyme glucose-6-phosphate dehydrogenase (G-6-PD) have a negative reaction to the ingestion of fava beans. This can lead to an acute hemolytic syndrome with 10% mortality if not treated properly.

Finally, it is surprising that certain lethal alleles act only in some individuals. These are called **semilethal alleles.** Of course, any particular individual cannot be semidead. However, within a population, a semilethal allele will kill some individuals but not all of them. The reasons for semilethality are not always under-stood, but environmental conditions and the actions of other genes within the organism may help to prevent the detrimental effects of certain semilethal alleles.

In some cases, a lethal allele may produce ratios that seem-ingly deviate from Mendelian ratios. An example is an allele in chickens that causes a phenotype known as "creeper" in which the chicken has shortened legs and must creep along rather than walk normally. Such birds also have shortened wings. Chickens with a creeper phenotype are heterozygous for the creeper allele. When a creeper is mated to a normal bird, half the offspring are normal and half are creepers.

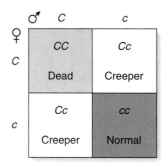

This result suggests that creeper is a dominant allele. However, when crosses are made between two creepers, the offspring are either normal or creepers, with a ratio of 1 normal : 2 creepers. About 1/4 of the eggs from this type of cross fail to hatch. This can be explained because the creeper allele is lethal in the homozygous condition.

Incomplete Dominance Occurs When Two Alleles Produce an Intermediate Phenotype

In some cases, a heterozygote that carries two different alleles may exhibit a phenotype that is intermediate between the corre-sponding homozygous individuals. This phenomenon is known as **incomplete dominance.** In flowering plants such as the four-o'clock (*Mirabilis jalapa*), this is observed for flower color alleles. This was first discovered by the German botanist Carl Correns in 1905. Figure 4.2 describes Correns's experiment in which a homo-zygous red-flowered four-o'clock plant was crossed to a homo-zygous white-flowered plant. The wild-type allele for red flower color is designated C^R, and the white allele C^W. As shown here, the offspring had pink flowers. If these F_1 offspring were allowed to self-fertilize, the F_2 generation showed a ratio of 1/4 red-flowered plants, 1/2 pink-flowered plants, and 1/4 white-flowered plants. The pink plants in the F_2 generation are heterozygotes with an intermediate phenotype. As noted in the Punnett square in figure 4.2, the F_2 generation displays a 1:2:1 ratio, which is different from the 3:1 ratio observed for simple Mendelian inheritance.

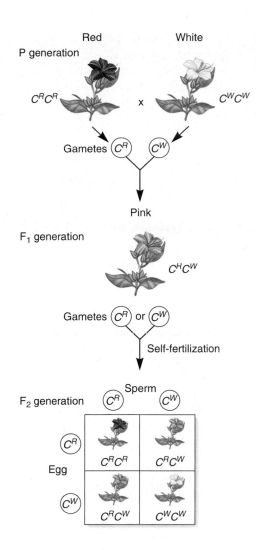

FIGURE 4.2   **Incomplete dominance in the four-o'clock plant.**

GENES→TRAITS   When two different homozygotes ($C^R C^R$ and $C^W C^W$) are crossed, the resulting heterozygote, $C^R C^W$, has an intermediate phenotype of pink flowers. In this case, 50% of the protein normally encoded by two copies of the C^R gene is not sufficient to produce a red phenotype.

In figure 4.2, an incompletely dominant relationship results in a heterozygote with an intermediate phenotype. At the molecular level, the allele that causes a white phenotype is expected to result in a lack of a functional protein required for pigmentation. The heterozygotes are expected to contain 50% of the normal protein, but this is not sufficient to produce the same phenotype as the homozygote that makes twice as much of this protein. In this example, 50% of the normal protein cannot accomplish the same level of pigment synthesis that 100% of the protein can.

Finally, our opinion of whether a trait is dominant or incompletely dominant may depend on how closely we examine the trait in the individual. The more closely we look, the more likely we are to discover that the heterozygote is not quite the same as the wild-type homozygote. For example, Mendel studied the characteristic of pea seed shape and visually concluded that

the *RR* and *Rr* genotypes produced round seeds and the *rr* genotype produced wrinkled seeds. The peculiar morphology of the wrinkled seed is caused by a great decrease in the amount of starch deposition in the seed due to a defective *r* allele. Since the time of Mendel's work, other scientists have dissected round and wrinkled seeds and examined their contents under the microscope. They have found that round seeds from heterozygotes actually contain an intermediate number of starch grains compared to seeds from the corresponding homozygotes (fig. 4.3). Within the seed, an intermediate amount of the functional protein is not enough to produce as many starch grains as in the homozygote carrying two copies of the *R* allele. Nevertheless, at the level of our naked eyes, heterozygotes produce seeds that appear to be round. With regard to phenotypes, the *R* allele is dominant to the *r* allele at the level of visual examination, but the *R* and *r* alleles are incompletely dominant at the level of starch biosynthesis.

Some Genes Exist as Three or More Different Alleles

Within a population of organisms, some genes are found in three or more alleles. In other words, a gene can exist in **multiple alleles** that are different from each other. An interesting example is coat color in rabbits. Figure 4.4 illustrates the relationship between genotype and phenotype for a combination of four different alleles, which are designated *C* (full coat color), c^{ch} (chinchilla pattern of coat color), c^h (himalayan pattern of coat color), and *c* (albino). Any particular rabbit can inherit only two copies of these alleles. A rabbit's phenotype, therefore, depends on the

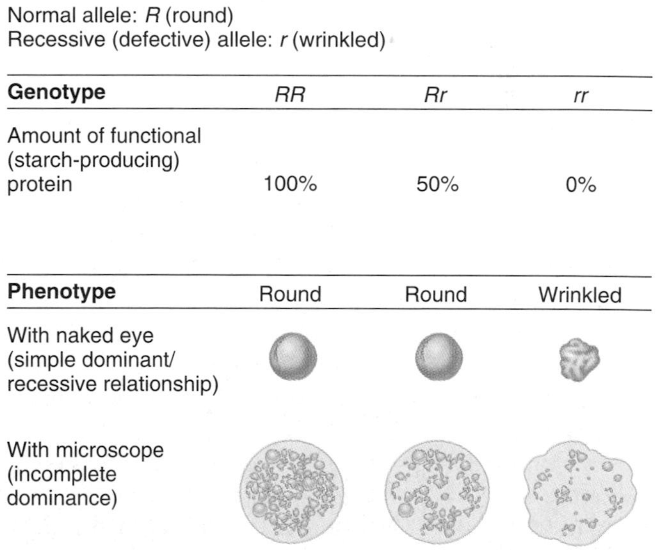

Normal allele: *R* (round)
Recessive (defective) allele: *r* (wrinkled)

Genotype	*RR*	*Rr*	*rr*
Amount of functional (starch-producing) protein	100%	50%	0%

Phenotype	Round	Round	Wrinkled
With naked eye (simple dominant/recessive relationship)			
With microscope (incomplete dominance)			

FIGURE 4.3   **A comparison of phenotype at the macroscopic and microscopic levels.**

GENES→TRAITS   This illustration shows the effects of the heterozygote having only 50% of the functional protein needed for starch production. This seed appears to be as round as those of the homozygote carrying the *R* allele, but when examined microscopically, it has produced only half the amount of starch.

(a)   (b)

(c)   (d)

FIGURE 4.4   **The relationship between genotype and phenotype in rabbit coat color.** (a) Full coat color *CC*, *Cc^{ch}*, *Cc^h*, or *Cc*. (b) Chinchilla coat color *c^{ch}c^{ch}*, *c^{ch}c^h*, or *c^{ch}c*. (c) Himalayan coat color *c^hc^h* or *c^hc*. (d) Albino coat color *cc*.

dominant/recessive relationships among combinations of alleles. These are as follows:

> *C* is dominant to *c^{ch}*, *c^h*, and *c*
>
> *c^{ch}* is recessive to *C* but dominant to *c^h* and *c*
>
> *c^h* is recessive to *C* and *c^{ch}* but dominant to *c*
>
> *c* is recessive to *C*, *c^{ch}*, and *c^h*

These alleles are also interesting at the molecular level. The wild-type allele, *C*, provides full coat color. The albino allele *c* is a defective allele that cannot produce a protein necessary for pigment synthesis. The chinchilla allele produces a partial defect in pigmentation. The himalayan allele is particularly unusual, because it results in pigmentation in certain parts of the body. We can understand how this works by considering gene expression at the molecular level.

The himalayan pattern of coat color is an example of a **temperature-sensitive conditional allele.** The mutation in this gene has caused a slight change in the structure of the encoded protein, so that it only works enzymatically at low temperature. Because of this property, the enzyme functions only in cooler regions of the body, primarily the ends of the extremities, the tail, the paws, and the tips of the nose and ears. As shown in figures 4.4c and figure 4.5, similar types of temperature-sensitive alleles have been found in other species of domestic animals, such as the Siamese cat.

FIGURE 4.5   **Photographs of different species showing the expression of a temperature-sensitive conditional allele that produces a Siamese or Himalayan pattern of coat color.**

GENES→TRAITS   The allele shown here encodes a pigment-producing protein that functions only at lower temperatures. For this reason, the dark fur is produced only in the cooler parts of the animal, including the tips of the nose, ears, and paws.

It is interesting to note that a breed of dairy cattle called Brown Swiss has an opposite phenotype to the himalayan rabbit and Siamese cat. As shown in figure 4.6, the coat in the cooler parts of the body (e.g., around the nose and along the legs) is light-colored while the coat on the trunk is dark. This allele is likely to be a cold-sensitive allele. In other words, it is likely to encode a pigment-producing protein that does not work well at lower body temperatures.

Alleles of the ABO Blood Group Can Be Dominant, Recessive, or Codominant

The ABO group of antigens, which determine blood type in humans, provides another example of multiple alleles. As shown in table 4.3, the red blood cells of humans contain structures on their plasma membranes known as surface antigens. Antigens are substances that are recognized by antibodies produced by the immune system. On red blood cells, three different types of surface antigens, known as A, B, and O, may be found. The synthesis of these surface antigens is controlled by three alleles designated I^A, I^B, and i, respectively. The i allele is recessive to both I^A and I^B. A person who is ii homozygous will have type O blood. A homozygous I^AI^A or heterozygous I^Ai individual will have type A blood. The red blood cells of this individual will contain the surface antigen known as A. Similarly, a homozygous I^BI^B or heterozygous I^Bi individual will produce the surface antigen B. As suggested by the drawing in table 4.3, surface antigens A and B have significantly different molecular structures. A person who is

FIGURE 4.6 **Photograph of a Brown Swiss breed of cattle.** The color pattern of the coat appears opposite to that of the Siamese cat and Himalayan rabbit (shown in fig. 4.5).

I^AI^B will have the blood type AB and express both surface antigens A and B. The phenomenon in which two alleles are both expressed in the heterozygous individual is called **codominance.** In this case, the I^A and I^B alleles are codominant to each other.

Biochemists have analyzed the carbohydrate tree produced in people of differing blood types. In type O people, the tree is smaller than type A or type B individuals because a sugar has not been attached to a specific site on the tree. This idea is schematically shown in table 4.3. Type O people contain a mutation in a gene that encodes an enzyme that attaches a sugar at this site. This type of enzyme, called a glycosyl transferase, is defective in type O people. By comparison, type A and type B antigens have different sugars attached to this site on the carbohydrate tree. This occurs because the glycosyl transferase encoded by the I^A allele has a different substrate specificity compared to the one encoded by the I^B allele. Because of differences in the part of the gene that encodes the amino acid sequence of the glycosyl transferase, the two allelic enzymes have different structures in their active sites. The active site is the part of the protein that recognizes the sugar molecule that will be attached to the carbohydrate tree. The glycosyl transferase encoded by the I^A allele recognizes UDP-GalNAc and attaches GalNAc (i.e., N-acetylgalactosamine) to the carbohydrate tree. N-acetylgalactosamine is symbolized as a *green hexagon* in table 4.3. This creates the structure of surface antigen A. In contrast, the glycosyl transferase encoded by the I^B allele recognizes UDP-galactose and attaches galactose to the carbohydrate tree. Galactose is symbolized as an *orange triangle* in table 4.3. This creates the molecular structure of surface antigen B.

A small difference in the structure of the carbohydrate tree, namely, an N-acetylgalactosamine in antigen A versus galactose in antigen B, is why the two antigens are different from each other at the molecular level. These differences enable them to be recognized by different antibodies. A person who is blood type A will make antibodies to blood type B. The antibodies against

TABLE 4.3
The ABO Blood Group

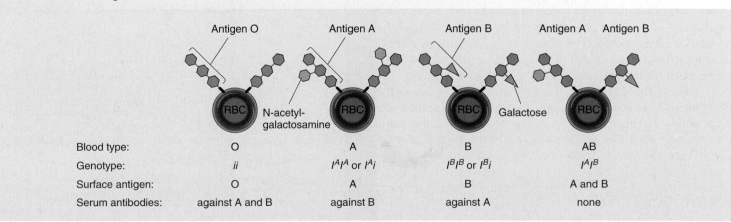

	O	A	B	AB
Blood type:	O	A	B	AB
Genotype:	ii	I^AI^A or I^Ai	I^BI^B or I^Bi	I^AI^B
Surface antigen:	O	A	B	A and B
Serum antibodies:	against A and B	against B	against A	none

blood type B must require a galactose in the carbohydrate tree for their proper recognition. Their antibodies will not recognize and destroy their own blood cells, but they will recognize and destroy the blood cells from a type B person.

Blood typing is essential for safe blood transfusions. When a person receives a blood transfusion, as for medical reasons, the donor's blood must be an appropriate match with the recipient's blood. Such a match is critical due to the occurrence of antibodies within the blood of the recipient. Antibodies bind very specif-ically to certain antigens. A person with type O blood naturally produces antibodies that recognize the A antigen and those that recognize the B antigen. If a person with type O blood is given type A, type B, or type AB blood, the antibodies in the recipient will react with the donated blood cells and cause them to agglutinate (clump together). This is a life-threatening situation, since it causes the blood vessels to clog. Other incompatible combinations include a type A person receiving type B or type AB blood and a type B person receiving type A or type AB blood.

EXPERIMENT 4A

Certain Genes Exhibit a Gene Dosage Effect

As we have seen, dominant and recessive relationships often depend on the amount of protein necessary to produce a phenotype. In a simple dominant/recessive relationship, 50% of the functional protein allows the heterozygote to have the dominant phenotype. By comparison, incomplete dominance occurs when 50% will not produce the dominant phenotype. In the experiment we are about to consider, the inheritance pattern for a single gene depends on the particular alleles of the gene. In this example, we again turn to the X-linked eye color gene in *Drosophila*. As discussed in chapter 3, the red allele is dominant to the white allele. Besides these two alleles, Thomas Hunt Morgan and Calvin Bridges found another allele of this gene that they called eosin (fig. 4.7). This allele had the curious phenotypic effect of making the intensity of the eye color different in males and females. In female flies homozygous for this allele, the eye color is eosin, but in males, the eyes are much paler. The eyes of males are termed *light eosin*.

Morgan and Bridges hypothesized that this observed variation might be caused by the difference in the number of X chromosomes between the female and male. They suggested that the eyes of the female contain more color because it is diploid with regard to the "eosin color producer" allele, whereas the paler eyes of the male are due to it having a single copy of this allele. This is an example of a **gene dosage effect.** In this case, two copies (and, thus, doses) of the allele provide more color than one copy of the allele. This phenomenon is analogous to incomplete dominance. By comparison, a gene dosage effect is not seen with the red allele. One copy of the red allele is sufficient to exhibit full red color.

To test this hypothesis, Morgan and Bridges conducted experiments to look at gene dosage effects among these three eye color alleles. In this experiment shown in figure 4.8, they conducted crosses involving combinations of the red, white, and eosin alleles; these alleles are designated X^{w^+}, X^w, and $X^{w\text{-}e}$, respectively.

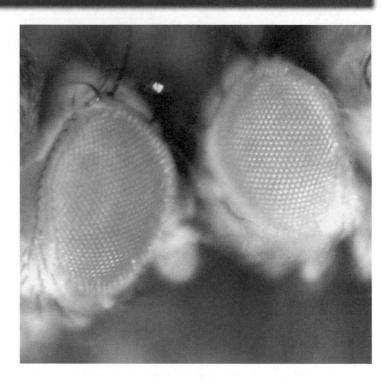

FIGURE 4.7   **Eosin eye color in *Drosophila*.** The eye on the left is that of a female that is homozygous for the eosin allele. The eye on the right is that of a male that is hemizygous for the eosin allele.

■ THE HYPOTHESIS

The phenotypic effects of the eosin eye color allele are related to the number of copies of the allele.

■ TESTING THE HYPOTHESIS — FIGURE 4.8 Identification of an allele that shows a gene dosage effect.

Starting Material: True-breeding strains of flies that have either the red-eye (normal) allele, white-eye allele, or eosin-eye allele.

	Experimental level	Conceptual level

1. Make the following crosses:

   A. White-eyed females x red-eyed males

   x ♀ ♂

   A. X^wX^w x $X^{w^+}Y$

   B. Red-eyed females x light eosin-eyed males

   x ♀ ♂

   B. $X^{w^+}X^{w^+}$ x $X^{w\text{-}e}Y$

   C. Eosin-eyed females x white-eyed males

   x ♀ ♂

   C. $X^{w\text{-}e}X^{w\text{-}e}$ x X^wY

2. Observe the outcome of the three types of crosses. Compare the eye colors of the true-breeding parents with their offspring.

Predicted ratio of offspring

A. 1 red-eyed female ($X^{w^+}X^w$) : 1 white-eyed male (X^wY)

B. 1 red-eyed female ($X^{w^+}X^{w\text{-}e}$) : 1 red-eyed male ($X^{w^+}Y$)

C. 1 light eosin-eyed female ($X^{w\text{-}e}X^w$) : 1 light eosin-eyed male ($X^{w\text{-}e}Y$)

■ THE DATA

Cross	Outcome	Genotype	Phenotype	
White-eyed females (X^wX^w) × red-eyed males ($X^{w^+}Y$)	225 red-eyed females ($X^{w^+}X^w$) : 208 white-eyed males (X^wY)	$X^{w^+}X^{w^+}$	Red eyes	No difference was observed in the redness of the eyes among these 4 groups.
		$X^{w^+}X^{w\text{-}e}$	Red eyes	
Red-eyed females ($X^{w^+}X^{w^+}$) × light eosin-eyed males ($X^{w\text{-}e}Y$)	679 red-eyed females ($X^{w^+}X^{w\text{-}e}$) : 747 red-eyed males ($X^{w^+}Y$)	$X^{w^+}X^w$	Red eyes	
		$X^{w^+}Y$	Red eyes	
Eosin-eyed females ($X^{w\text{-}e}X^{w\text{-}e}$) × white-eyed males ($X^wY$)	694 light eosin-eyed females ($X^{w\text{-}e}X^w$) : 579 light eosin-eyed males ($X^{w\text{-}e}Y$)	$X^{w\text{-}e}X^{w\text{-}e}$	Eosin eyes	The heterozygous females had similar eye color to the hemizygous males.
		$X^{w\text{-}e}X^w$	Light-eosin eyes	
		$X^{w\text{-}e}Y$	Light-eosin eyes	
		X^wX^w	White eyes	
		X^wY	White eyes	

■ INTERPRETING THE DATA

When Morgan and Bridges crossed females that were homozygous for an eye color allele to males that were hemizygous for a different allele, they always observed a 1:1 ratio among the offspring. This is consistent with the idea that these genes for eye color are X linked and allelic to each other. Let's first consider flies that carried an X^{w+} allele. All of these flies exhibited red eye color. Furthermore, the eye color was bright red whether the fly had one or two copies of the red allele. As seen in the data, the $X^{w+}X^{w+}$ females had red eyes that were indistinguishable from the eyes of flies that had only one copy of the X^{w+} allele (namely, $X^{w+}X^{w-e}$ females, $X^{w+}X^{w}$ females, and $X^{w+}Y$ males). In this case, the red allele appears to be dominant to the eosin and white alleles, and no gene dosage effect occurs.

In contrast, the eosin phenotype is affected by the number of copies of the eosin allele. Homozygous females containing two copies of the X^{w-e} allele had eosin eyes, while flies with only one copy of this allele (namely, $X^{w-e}X^{w}$ females and $X^{w-e}Y$ males) had light-eosin eyes (see data table in fig. 4.8). In heterozygous $X^{w-e}X^{w}$ females, Morgan and Bridges proposed that the white allele (which does not produce color) was having a "dilution effect" on the phenotype. Another way of considering this phenomenon is that the X^{w-e} allele shows a gene dosage effect. An increase in the dosage of this gene from one copy to two copies yields an increase in the degree of the eosin color in the eyes of the flies. This result supports the idea that the amount of a gene product can influence the outcome of a trait.

A self-help quiz involving this experiment can be found at the Online Learning Center.

Overdominance Occurs When Heterozygotes Have Superior Traits

Let's now turn our attention to another way that two alleles may interact to affect the phenotype of the individual. For certain genes, heterozygotes may display characteristics that are more beneficial to their survival and reproduction than either corresponding homozygote. For example, a heterozygote may be larger, disease resistant, or better able to withstand harsh environmental conditions. The phenomenon in which a heterozygote is more vigorous than both of the corresponding homozygotes is called **overdominance** or **heterozygote advantage.**

A well-documented example involves a human allele that causes sickle-cell anemia in homozygous individuals. This disease is an autosomal recessive disorder in which the affected individual produces an abnormal form of the protein hemoglobin, which carries oxygen within red blood cells. Normal individuals carry the Hb^A allele and make hemoglobin A. Individuals affected with this disease are homozygous for the Hb^S allele and produce only hemoglobin S. This causes their red blood cells to deform into a sickle shape under conditions of low oxygen concentration (fig. 4.9). The sickling phenomenon causes the life span of these cells to be greatly shortened, to only a few weeks compared with a normal span of four months. Therefore, anemia results. In addition, abnormal sickled cells become clogged in the capillaries throughout the body, leading to localized areas of oxygen depletion, tissue damage, and painful crises. For these reasons, the homozygous Hb^SHb^S individual usually has a shortened life span compared to an individual producing hemoglobin A.

In spite of the harmful consequences to homozygotes, the sickle-cell allele has been found at a fairly high frequency among human populations that are exposed to malaria. The protozoan that causes malaria, *Plasmodium* sp., spends part of its life cycle within the *Anopheles* mosquito and another part within the red blood cells of humans who have been bitten by an infected mosquito. However, red blood cells of heterozygotes, Hb^AHb^S, are likely to rupture when infected by this parasite, thereby preventing it from propagating. This provides people who are heterozygous, Hb^AHb^S, with better resistance to malaria compared to normal, Hb^AHb^A, homozygotes. Therefore, even though the homozygous Hb^SHb^S condition is detrimental, the greater survival

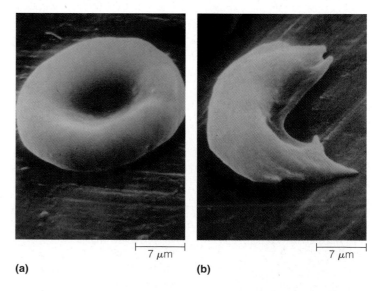

(a)   (b)

FIGURE 4.9 **A comparison of (a) normal red blood cells and (b) those from a person with sickle-cell anemia.** The cells in (b) exhibit an abnormal sickle-cell morphology.

of the heterozygote has selected for the presence of the *Hb^S* allele within populations where malaria is prevalent. Therefore, when viewing survival in such a region, the sickle-cell allele exhibits overdominance. In chapter 25, we will consider the role that natural selection plays in maintaining alleles that are beneficial to the heterozygote but harmful to the homozygote.

It is worth noting that one could also view the sickle-cell allele as exhibiting incomplete dominance. Although most heterozygotes do not manifest a clinical case of sickle-cell anemia, they can show some symptoms under certain conditions such as high levels of physical exertion. With regard to sickle-cell anemia, the heterozygote shows symptoms that are intermediate between the *Hb^AHb^A* and *Hb^SHb^S* homozygotes.

At the molecular level and cellular level, overdominance is usually due to two alleles that produce proteins with slightly different amino acid sequences. There are several ways to explain how two protein variants, rather than a single protein as in the homozygote, can produce a favorable phenotype. In the case of sickle-cell anemia, the phenotype is related to the infectivity of *Plasmodium* (fig. 4.10*a*). When hemoglobin function is optimal, the infectious agent is better able to propagate within red blood cells. The heterozygote has a slightly reduced hemoglobin function. Though this reduced function is not enough to cause serious side effects, it is enough to prevent the propagation of *Plasmodium*. Interestingly, several other examples of alleles in humans confer disease resistance in the heterozygous condition but are detrimental in the homozygous state. These include cystic fibrosis, in which the heterozygote is resistant to diarrheal disease; PKU, in which the heterozygous fetus is resistant to miscarriage caused by a fungal toxin; and Tay-Sachs disease, in which the heterozygote is resistant to tuberculosis.

A second way to explain heterozygote advantage is related to the subunit composition of proteins. In some cases, proteins function as a complex of multiple subunits, each composed of a separate polypeptide. When a protein is composed of two subunits, it is known as a dimer. When both subunits are encoded by the same gene, the protein is a homodimer. Figure 4.10*b* considers a situation in which a gene exists in two alleles that encode polypeptides designated A1 and A2. Homozygous individuals can produce only A1A1 or A2A2 homodimers, whereas a heterozygote can also produce an A1A2 homodimer. Thus, heterozygotes produce three forms of the homodimer, homozygotes only one. For some proteins, A1A2 homodimers may have better functional activity because they are more stable or able to function under a wider range of conditions. The greater activity of the homodimer protein may be the underlying reason why a heterozygote has characteristics superior to either homozygote.

A third molecular explanation of overdominance is that the protein encoded by each allele functions under different conditions. For example, suppose that a gene encodes a metabolic enzyme that can be found in two forms (corresponding to the two alleles), one that functions better at a lower temperature and the other that functions optimally at a higher temperature (fig. 4.10*c*). The heterozygote, which makes a mixture of both enzymes,

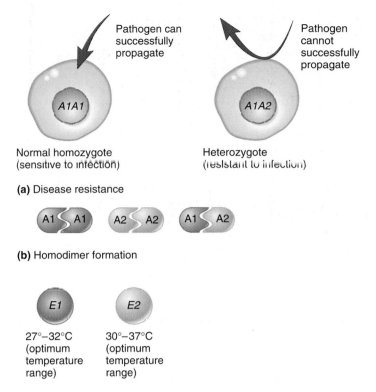

(a) Disease resistance

(b) Homodimer formation

(c) Variation in functional activity

FIGURE 4.10 **Three possible explanations for overdominance at the molecular level. (a)** The successful infection of cells by certain microorganisms depends on the optimal function of particular cellular proteins. Heterozygotes have one altered copy of the gene that encodes the protein, which diminishes its function. Though the heterozygote may be phenotypically normal, it becomes resistant to infection by the microorganism. **(b)** Some proteins function as homodimers. In this example, a gene exists in two alleles designated *A1* and *A2,* which encode polypeptides also designated A1 and A2. The homozygotes that are *A1A1* or *A2A2* will make homodimers that are A1A1 and A2A2, respectively. The *A1A2* heterozygote can make A1A1 and A2A2 and can also make A1A2 homodimers. **(c)** In this example, a gene exists in two alleles designated *E1* and *E2.* The *E1* allele encodes an enzyme that functions well in the temperature range of 27° to 32°C. *E2* encodes an enzyme that functions in the range of 30° to 37°C. A heterozygote, *E1E2,* would produce both enzymes and have a broader temperature range (i.e., 27° to 37°C) in which the enzyme would function.

may be at an advantage under a wider temperature range than either of the corresponding homozygotes.

Before ending this section, it is interesting to compare overdominance with a related phenomenon. Among plant and animal breeders, a common mating strategy is to begin with two different highly inbred strains and cross them together to produce hybrids. When the hybrids display traits that are superior to both corresponding parental strains, the outcome is known as **heterosis** or **hybrid vigor.** Within the field of agriculture, heterosis has been particularly valuable in improving quantitative traits such as size, weight, growth rate, and disease resistance. It was first described by George Shull in crosses involving different strains of corn. Heterosis is different from overdominance, because the hybrid may be heterozygous for many genes, not just a single gene. Nevertheless, some of the beneficial effects of heterosis may be caused by the occurrence of overdominance in one or more heterozygous genes. However, as we will see in chapter 24, heterosis can also result from the masking of deleterious recessive alleles that tend to accumulate in highly inbred domesticated strains.

Dominant Traits Can Skip a Generation Due to Incomplete Penetrance

Dominant alleles are expected to influence the outcome of a trait when they are present in heterozygotes. Occasionally, however, this may not occur. Figure 4.11 illustrates a human pedigree for a dominant trait known as polydactyly. This trait causes the affected individual to have additional fingers and/or toes (fig. 4.12). A single copy of this gene is sufficient to cause this condition. Sometimes, however, individuals carry the dominant allele but do not exhibit the trait. In figure 4.11, individual III-2 has inherited the polydactyly allele from his mother and passed the trait to a daughter and son. However, individual III-2 does not actually exhibit the trait himself, even though he is a heterozygote. This phenomenon is called **incomplete penetrance.** This term indicates that a dominant allele does not always "penetrate" into the phenotype of the individual. The measure of penetrance is described at the populational level. For example, if 60% of the heterozygotes carrying a dominant allele exhibit the trait, we would say that this trait is 60% penetrant. In any particular individual, the trait is either present or not.

FIGURE 4.12   **Antonio Alfonseca, a baseball player with polydactyly.**

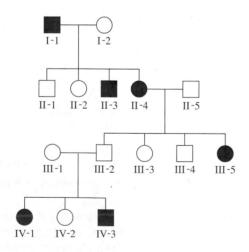

FIGURE 4.11   **A family pedigree for polydactyly, a dominant trait showing incomplete penetrance.** Affected individuals are shown in *black*. Notice that offspring IV-1 and IV-3 have inherited the trait from a parent, III-2, who is heterozygous and phenotypically normal.

Another issue in evaluating the outcome of traits is the degree to which the trait is expressed. In the case of polydactyly, the number of extra digits can vary. For example, one individual may have an extra toe on only one foot, whereas a second individual may have extra digits on both the hands and feet. This variation is known as the **expressivity** of the trait. Using genetic terminology, a person with several extra digits would have high expressivity of this trait, while a person with a single extra digit would have a low expressivity. The extreme case, when a heterozygote does not express the trait at all, is incomplete penetrance.

The molecular basis for expressivity and incomplete penetrance may not always be understood. In most cases, the range of phenotypes is thought to be due to environmental influences and/or due to other genes that encode proteins that counteract the effects of the protein encoded by the (seemingly) dominant allele. We will consider the issue of the environment next. The effects of modifier genes are discussed later in chapter 4.

The Expression of a Trait Can Be Influenced by the Environment

Throughout this textbook, our study of genetics tends to focus on the roles of genes in the outcome of traits. In addition to genetics, environmental conditions may have a great impact on the phenotype of the individual. For example, the temperature and degree of sunlight have been shown to affect the flower color of certain snapdragons (fig. 4.13). When a true-breeding red-flowered plant is crossed to a true-breeding ivory-flowered plant, the outcome of the F_1 generation depends on light intensity and temperature. When grown in bright light at relatively low temperatures, the heterozygotes develop red flowers, but under shady conditions and warmer temperatures, the flowers are ivory.

A dramatic example of the relationship between environment and phenotype can be seen in the human genetic disease known as phenylketonuria (PKU). This autosomal recessive disease is caused by a defect in a gene that encodes the enzyme phenylalanine hydroxylase. Homozygous individuals with this defective allele are unable to metabolize the amino acid phenylalanine. When given a normal diet containing phenylalanine,

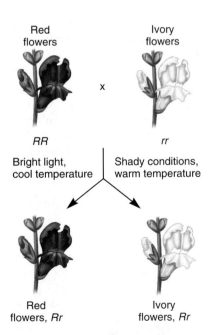

Red flowers   Ivory flowers

x

RR   *rr*

Bright light, cool temperature | Shady conditions, warm temperature

Red flowers, *Rr*   Ivory flowers, *Rr*

FIGURE 4.13   **Variation in the expression of a trait due to environmental effects.**

GENES→TRAITS   In this example, snapdragon flower color in the heterozygote, *Rr*, is affected by the amount of light and temperature in the plant's environment. Presumably, the protein encoded by this gene is sensitive to changes in temperature and light intensity.

PKU individuals manifest a variety of detrimental traits including mental retardation, underdeveloped teeth, and foul-smelling urine. In contrast, when PKU individuals are diagnosed early and given a restricted diet that is free of phenylalanine, they develop normally (refer back to fig. 1.10). Since the 1960s, testing methods have been developed that can determine if an individual is lacking the phenylalanine hydroxylase enzyme. These tests permit the identification of infants who have PKU; their diets can then be modified before the harmful effects of phenylalanine ingestion have occurred. As a result of government legislation, more than 90% of infants born in the United States are now tested for PKU. This test prevents a great deal of human suffering. Furthermore, it is cost-effective. In the United States, the annual cost of PKU testing is estimated to be a few million dollars, whereas the cost of treating severely retarded individuals with the disease would be hundreds of millions of dollars.

The Outcome of Certain Traits Can Be Influenced by the Sex of the Individual

The inheritance pattern for certain traits is governed by the sex of the individual. The term **sex-influenced inheritance** refers to the phenomenon in which an allele is dominant in one sex but recessive in the opposite sex. Therefore, sex influence is a phenomenon of heterozygotes. In humans, pattern baldness provides an example of a sex-influenced trait. As shown in figure 4.14, it is characterized by a balding pattern on the male's head in which hair loss occurs on the front and top but not on the sides. The gene that causes pattern baldness is inherited as an autosomal trait. (A common misconception is that this gene is X linked.) When a male is heterozygous for the baldness allele, he will become bald.

Genotype	Phenotype	
	Males	**Females**
BB	bald	bald
Bb	bald	nonbald
bb	nonbald	nonbald

In contrast, a heterozygous female will not be bald. Women who are homozygous for the baldness allele will develop the trait, but in women it is usually characterized by a significant thinning of the hair that occurs relatively late in life. The sex-influenced nature of pattern baldness appears to be related to levels of male sex hormones. In females, a rare tumor of the adrenal gland can cause the secretion of large amounts of male sex hormones. If this occurs in a heterozygous *Bb* female, she will become bald. When the tumor is removed surgically, her hair returns to its normal condition.

The autosomal nature of pattern baldness has been revealed by the analysis of many human pedigrees. An example is shown in figure 4.15. A bald male may inherit the bald allele from either parent, and thus a striking observation is that bald

(a) John Adams

(b) John Quincy Adams

(c) Charles Francis Adams

(d) Henry Adams

FIGURE 4.14   Pattern baldness in the Adams family line.

spring if they are homozygous for the bald allele and/or the mother also carries one or two copies of the bald allele. For example, a heterozygous bald male and heterozygous (nonbald) female will produce 75% bald sons, while a homozygous bald male or homozygous bald female will produce all bald sons.

Another example in which sex affects an organism's phenotype is provided by **sex-limited traits,** which are those that occur in only one of the two sexes. In humans, for example, breast development is a trait that is normally limited to females, whereas beard growth is limited to males. Among many types of birds, the male of the species has more ornate plumage than the female. As shown in figure 4.16, roosters have a larger comb and wattles, and longer neck, tail, and sickle feathers than do hens. These are sex-limited features that may be found in roosters but never in normal hens. However, some varieties of chickens have males that are hen-feathered. Hen-feathering is controlled by a dominant allele that is expressed both in males and females, whereas cock-feathering is controlled by a recessive allele that is expressed only in males.

Genotype	Phenotype	
	Females	**Males**
hh	hen-feathered	cock-feathered
Hh	hen-feathered	hen-feathered
HH	hen-feathered	hen-feathered

Like pattern baldness in humans, the pattern of hen-feathering depends on the production of sex hormones. If a newly hatched female with an *hh* genotype has her single ovary removed surgically, she will develop cock-feathering and look indistinguishable from a male.

Sex-influenced and sex-limited inheritance are often confused with the different phenomenon of sex linkage. As discussed in chapter 3, some genes are located on the sex chromosomes, such as the X chromosome in mammals. In the examples of sex-influenced and sex-limited traits described in chapter 4, however, the genes are not located on the sex chromosomes. Baldness in

fathers can pass this trait to their sons. This could not occur if the trait was X linked, since fathers transmit Y chromosomes to their sons. The analyses of many human pedigrees have shown that bald fathers, on average, have at least 50% bald sons. They are expected to produce an even higher percentage of bald male off-

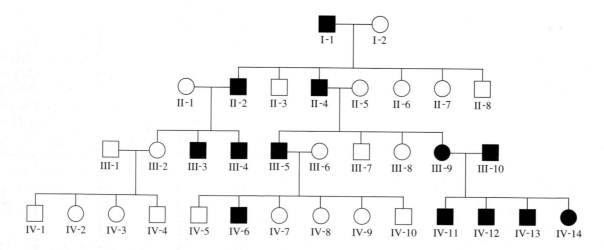

FIGURE 4.15   A pedigree of human pattern baldness, an example of the sex-influenced expression of an autosomal gene. Bald individuals are shown in *black.*

FIGURE 4.16 **Differences in the feathering pattern between male and female chickens, an example of sex-limited inheritance.**

humans and feather plumage in chickens are cases where the genes are located on autosomes. Nevertheless, the expression of these genes is affected by the sex of the individual (i.e., male versus female).

4.2 GENE INTERACTIONS

In the preceding section, we considered the effects of a single gene on the outcome of a trait. Most traits, however, are affected by the contributions of two or more genes. Morphological features such as height, weight, growth rate, and pigmentation are all affected by the expression of many different genes in combination with environmental factors. We often discuss the effects of a single gene on the outcome of a single trait as a way to simplify the genetic analysis. Even so, the trait may not be completely determined by a single gene. For example, Mendel studied one gene that affected the height of pea plants (tall versus dwarf alleles). Actually, there are many other genes in pea plants that can also affect height, but Mendel did not happen to study variants in other height genes. In general, most traits of an organism are determined by complex contributions of many different genes that collectively influence the individual's phenotype. When two or more different genes influence the outcome of a single trait, this is known as a **gene interaction.**

In this section, we will consider the diverse ways that two genes may interact to affect a single trait. We will examine three different examples, all involving two genes that exist in two alleles. To compare these examples, we will focus our attention on the outcome of a cross between two individuals that are heterozygous for both genes. In a general way, we can illustrate this cross as

$AaBb \times AaBb$ where A is dominant to a, and B is dominant to b

If these two genes govern two different traits, Mendel's laws of segregation and independent assortment predict a 9:3:3:1 ratio among the offspring. However, in this section, the two genes will affect the same trait. As we will see, there are many different ways that the alleles of two genes may interact to affect a trait. We will appreciate their interactions by considering how they affect the 9:3:3:1 ratio. Additional examples are found in the solved problems at the end of chapter 4.

A Cross Involving a Two-Gene Interaction Can Produce a 9:3:3:1 Ratio in Offspring When Four Distinct Phenotypes Are Produced

The first case of two different genes interacting to affect a single trait was discovered by William Bateson and Reginald Punnett in 1906 while they were investigating the inheritance of comb morphology in chickens. Several common varieties of chicken possess combs with different morphologies, as illustrated in figure 4.17a. In their studies, Bateson and Punnett crossed a Wyandotte breed with a rose comb to a Brahma having a pea comb. All the F_1 offspring had a walnut comb.

When these F_1 offspring were mated to each other, the F_2 generation consisted of chickens with four types of combs in the following phenotypic ratio: 9 walnut : 3 rose : 3 pea : 1 single. As we have seen in chapter 2, a 9:3:3:1 ratio is obtained in the F_2 generation when the F_1 generation is heterozygous for two different genes and these genes assort independently. However, an important difference here is that we have four distinct categories of a single trait. Based on the 9:3:3:1 ratio, Bateson and Punnett reasoned that a single trait (comb morphology) was determined by two different genes.

> R (rose comb) is dominant to r
>
> P (pea comb) is dominant to p
>
> R and P are codominant (walnut comb)
>
> $rrpp$ produces single comb

As shown in the Punnett square of figure 4.17b, each of the genes can exist in two alleles, and the two genes show independent assortment.

A Cross Involving a Two-Gene Interaction Can Produce a 9:7 Ratio When Both Genes Are Epistatic to Each Other

Bateson and Punnett also discovered an unexpected gene interaction when studying crosses involving the sweet pea, *Lathyrus odoratus*. The wild sweet pea normally has purple flowers. However, they obtained several true-breeding varieties with white flowers. Not surprisingly, when they crossed a true-breeding purple-flowered plant to a true-breeding white-flowered plant, the F_1 generation contained all purple-flowered plants and the F_2 generation (produced by self-fertilization of the F_1 generation) consisted of purple- and white-flowered plants in a 3:1 ratio. The surprising result came in an experiment where they crossed two different varieties of white-flowered plants (fig. 4.18). All the F_1 generation plants had purple flowers! When these were allowed to self-fertilize, the F_2 generation contained purple and white flowers in a ratio of 9 purple : 7 white. From this result, Bateson and Punnett deduced that two different genes were involved.

> C (one purple-color-producing) allele is dominant to c (white)
>
> P (another purple-color-producing) allele is dominant to p (white)
>
> cc or pp masks the P or C alleles, producing white color

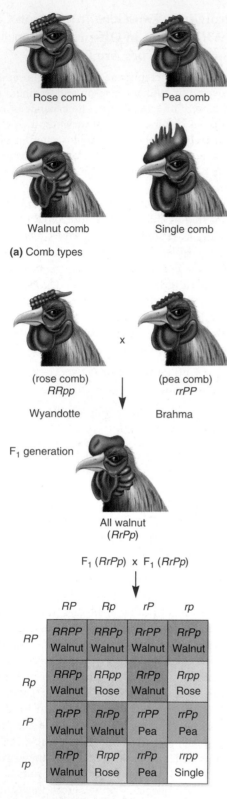

(a) Comb types

(b) The crosses of Bateson and Punnett

FIGURE 4.17   **Inheritance of comb morphology in chickens.**
This trait is influenced by two different genes, which can each exist in
two alleles. **(a)** Four phenotypic outcomes are possible. **(b)** The crosses
of Bateson and Punnett to examine the interaction of the two genes.

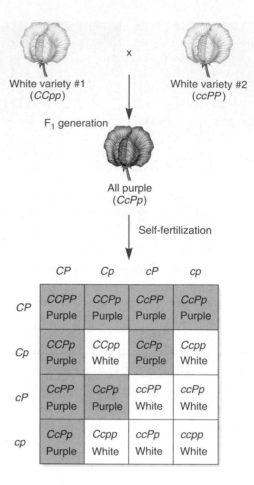

FIGURE 4.18   **A cross between two different white varieties of
the sweet pea.**
GENES→TRAITS   The color of the sweet pea flower is controlled by two genes,
which are epistatic to each other. Each gene is necessary for the production of an
enzyme required for pigment synthesis. The recessive allele of either gene encodes
a defective enzyme. If an individual is homozygous recessive for either of the two
genes, the purple pigment cannot be synthesized. This results in a white phenotype.

If a plant was homozygous for either recessive white allele,
it would develop a white flower even though the other gene may
contain a purple-producing allele. The term **epistasis** is used to
describe the situation in which one gene can mask the phenotypic
effects of a different gene. In this case, homozygosity for the
white allele of one gene masks the purple-producing allele of
another gene.

Epistatic interactions often arise because two (or more) different proteins participate in a common cellular function. For
example, two or more proteins may be part of an enzymatic
pathway leading to the formation of a single product. To illustrate
this idea, let's consider the formation of a purple pigment in the
sweet pea.

Colorless precursor $\xrightarrow{\text{Enzyme C}}$ Colorless intermediate $\xrightarrow{\text{Enzyme P}}$ Purple pigment

In this example, a colorless precursor molecule must be acted on by two different enzymes to produce the purple pigment. Gene *C* encodes a functional protein called enzyme C, which converts the colorless precursor into a colorless intermediate. Two copies of the recessive allele (*cc*) result in a lack of production of this enzyme in the homozygote. Gene *P* encodes a functional enzyme P, which converts the colorless intermediate into the purple pig-ment. Like the *c* allele, the recessive *p* allele encodes a defective enzyme P. If an individual is homozygous for either recessive allele (*cc* or *pp*), it will not make any functional enzyme C or enzyme P, respectively. When one of these enzymes is missing, it is impossible to make the purple pigment. Therefore, the flowers remain white.

EXPERIMENT 4B

Bridges Observed an 8:4:3:1 Ratio Because the Cream-Eye Allele Can Modify the Eosin Allele but Not the Red or White Alleles

Bridges also discovered an example of two different genes interacting to affect the phenotype of a single trait. Within true-breeding cultures of flies with eosin eyes, he occasionally found a fly that had a noticeably different eye color. In particular, he identified a rare fly with cream-colored eyes. He reasoned that this new eye color could be explained in two different ways. One possibility is that the cream-colored phenotype could be the result of a new mutation that changed the eosin allele into a cream allele. A second possibility is that a different gene may have incurred a mutation that modified the expression of the eosin allele. This second possibility is an example of a gene interaction. To distinguish between these two possibilities, he carried out the crosses described in figure 4.19. He crossed males with cream-colored eyes to wild-type females and then allowed the F$_1$ generation flies to mate with each other. As shown in the data, all the F$_2$ females had red eyes, while the males had red eyes, light-eosin eyes, or cream eyes.

■ THE HYPOTHESIS

Cream-colored eyes in fruit flies are due to the effect of a second gene that modifies the expression of the eosin allele.

■ TESTING THE HYPOTHESIS — FIGURE 4.19 A gene interaction between the cream allele and eosin allele.

Starting material: From a culture of flies with eosin eyes, Bridges obtained a fly with cream-colored eyes. The allele was called *cream a*. This fly was used to produce a true-breeding culture of flies with cream-colored eyes.

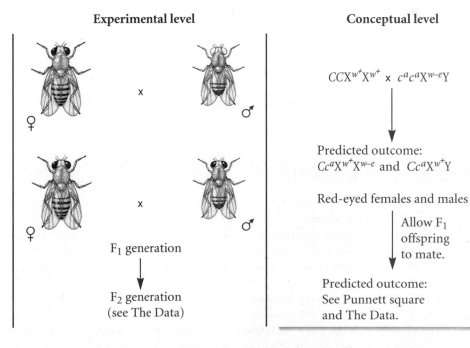

	Experimental level	Conceptual level
1. Cross males with cream-colored eyes to wild-type females.	♀ x ♂	$CCX^{w^+}X^{w^+}$ x $c^ac^aX^{w-e}Y$
2. Observe the F$_1$ offspring and then allow the offspring to mate with each other.	♀ x ♂ F$_1$ generation	Predicted outcome: $Cc^aX^{w^+}X^{w-e}$ and $Cc^aX^{w^+}Y$ Red-eyed females and males Allow F$_1$ offspring to mate.
3. Observe and record the eye color and sex of the F$_2$ generation.	F$_2$ generation (see The Data)	Predicted outcome: See Punnett square and The Data.

■ THE DATA

Cross	Outcome
P cross:	
Cream-eyed male × wild-type female	F_1: all red eyes
F_1 cross:	
F_1 brother × F_1 sister	F_2: 104 females with red eyes
	47 males with red eyes
	44 males with light-eosin eyes
	14 males with cream eyes

■ INTERPRETING THE DATA

To interpret these data, keep in mind that Bridges already knew that the eosin allele is X linked. However, he did not know whether the cream allele was in the same gene as the eosin allele, in a different gene on the X chromosome, or on an autosome. The F_2 generation indicates that the cream allele is not in the same gene as the eosin allele. If the cream allele was in the same gene as the eosin allele, none of the F_2 males would have had light-eosin eyes; there would have been a 1:1 ratio of red-eyed males and cream-eyed males in the F_2 generation. This result was not obtained. Instead, the actual results are consistent with the idea that the male flies of the parental generation possessed both the eosin and cream alleles. Therefore, Bridges concluded that the cream allele was an allele of a different gene.

One possibility is that the cream allele is an autosomal recessive allele. If so, we could let C represent the normal allele (which does not modify the eosin phenotype) and c^a represent the cream allele that modifies the eosin color to cream. We already know that the eosin allele is X linked and recessive to the red allele. The parental cross is expected to produce all red-eyed F_1 flies in which the males are $Cc^aX^{w^+}Y$ and the females are $Cc^aX^{w^+}X^{w-e}$. According to these ideas, when these F_1 offspring are allowed to mate with each other, the Punnett square shown here would predict the following outcome:

Outcome:

$1\ CCX^{w^+}X^{w^+}$: $1\ CCX^{w^+}X^{w-e}$: $2\ Cc^aX^{w^+}X^{w^+}$: $2\ Cc^aX^{w^+}X^{w-e}$: $1\ c^ac^aX^{w^+}X^{w^+}$: $1\ c^ac^aX^{w^+}X^{w-e}$ = **8 red-eyed females**

$1\ CCX^{w^+}Y$: $2\ Cc^aX^{w^+}Y$: $1\ c^ac^aX^{w^+}Y$ = **4 red-eyed males**

$1\ CCX^{w-e}Y$: $2\ Cc^aX^{w-e}Y$ = **3 light eosin-eyed males**

$1\ c^ac^aX^{w-e}Y$ = **1 cream-eyed male**

This phenotypic outcome proposes that the specific modifier allele, c^a, can modify the phenotype of the eosin allele but not the red-eye allele. The eosin allele can be modified only when the c^a allele is homozygous. The predicted 8:4:3:1 ratio agrees reasonably well with Bridges's data. Even so, these results would be difficult to distinguish from the classic 9:3:3:1 ratio that we have seen in previous experiments.

A self-help quiz involving this experiment can be found at the Online Learning Center.

CONCEPTUAL SUMMARY

In chapter 4, we have considered many examples that extend our understanding of Mendelian inheritance. For single genes, the relationship between dominance and recessiveness has been broadened to include situations that deviate from the simple dominant/recessive relationship observed by Mendel. For example, many genes have **multiple alleles.** This creates a more complex relationship where one allele can be dominant to a second allele but recessive to a third allele. Alternatively, two alleles in the heterozygote may follow a pattern that is different from clear-cut dominance. In **codominance,** two alleles in the heterozygote both contribute to the phenotype. **Incomplete dominance** produces a heterozygote with an intermediate phenotype, whereas **overdominance** results in a heterozygote with a superior phenotype. Certain alleles can also exhibit a **gene dosage effect,** in which the degree of the phenotype is correlated with the number of copies of the allele. In addition, the outcome of a trait may exhibit a range of phenotypes. The **expressivity** of a trait may be very high, very low, or in the case of **incomplete penetrance,** even zero. Both environmental and genetic factors contribute to the expressivity of inherited traits.

We have also explored cases in which two different genes affect the outcome of a single trait. This is known as a **gene interaction.** Even though geneticists tend to study single genes that have a great impact on a single trait, it is more realistic to expect that traits are due to complex contributions from many genes. As in the case of chicken combs, two alleles may interact in several different ways to alter an organism's morphology. In addition, alleles of one gene may act to mask the phenotypic effects of the alleles of a second gene. This phenomenon is known as **epistasis.** For example, two genes affect flower color in the sweet pea; homozygosity for a white-flower allele of one gene masks the phenotypic expression of a purple-flower allele of a second gene.

Many particularly gratifying and exciting advances in genetics have involved an explanation of Mendelian inheritance at the molecular level. As described in parts III and IV of this textbook, there continues to be a great amount of effort toward understanding molecular genetics. Recessive alleles often are the result of a defective gene that is unable to produce a functionally active protein. In many cases, the normal (dominant) gene can compensate for this defect and still produce a normal phenotype. Recessive lethal alleles are usually loss-of-function alleles of essential genes. In an incomplete dominant relationship, 50% of the normal protein is not sufficient to yield the normal phenotype. Overdominance can sometimes be explained by disease resistance, by protein subunit associations, or by enzymes that can function under different conditions. Codominant alleles may alter the function of cellular proteins in specific ways. Also, we have seen that conditional alleles may alter protein structure so that the protein will function only under certain circumstances. Other factors, such as sex and the environment, may also have a dramatic impact on the phenotypic expression of certain genes.

EXPERIMENTAL SUMMARY

To understand how inheritance may deviate from a simple Mendelian pattern, either a genetic or molecular approach can be taken. In a genetic approach, a researcher makes crosses and then analyzes the phenotypes of the offspring. In chapter 4, we have seen that the experimenter must weigh many factors when designing a cross and analyzing its outcome. These include the dominant/recessive relationships of alleles (namely, dominant, recessive, codominant, incompletely dominant, or overdominant), the presence of multiple alleles in a population, the relationship between the sex of the offspring and their phenotype, and the influence of the environment. In addition, complicating factors such as incomplete penetrance and gene interactions must be considered. Based on these factors, a researcher can construct a Punnett square to see if a pattern of inheritance fits the available data obtained from crosses.

Another level of understanding in Mendelian inheritance is to appreciate how the molecular expression of genes underlies their impact on the phenotype of an organism. Using modern molecular tools that are described later in this textbook, researchers can quantitatively compare the level of gene expression between wild-type and mutant alleles. This helps us to understand how the amount of gene expression is correlated with the phenotype of the organism. Also, when gene interactions occur, an investigator may want to understand, at the molecular level, how the gene products participate in a common cellular function such as an enzymatic pathway. Therefore, the molecular investigation of gene interactions frequently involves research collaborations between geneticists, cell biologists, and biochemists.

PROBLEM SETS & INSIGHTS

Solved Problems

S1. Coat color in rodents is determined by a gene interaction between two genes. If a true-breeding black rat is crossed to a true-breeding albino rat, the result is a rat with agouti (brownish/dark gray) coat color. If two agouti animals of the F_1 generation are crossed to each other, they produce agouti, black, and albino animals in a 9:3:4 ratio. Explain the pattern of inheritance for this trait.

Answer: Since the parental generation is true-breeding, we know that the F_1 offspring are heterozygous for two genes. One strategy for solving

this problem is to consider how a 9:3:4 ratio deviates from the 9:3:3:1 ratio we have already encountered in (two-gene) independent assortment problems. A 9:3:3:1 ratio has four categories of phenotypes. If we consider that the last two categories are combined via a gene interaction, this would yield a 9:3:4 ratio. With this idea in mind, we can then proceed to fill in a Punnett square to deduce the pattern of inheritance.

The F$_1$ offspring are heterozygous for two genes. Let's call them *A* (for albino) and *C* (for colored).

CcAa x CcAa

♂ CA	CA	Ca	cA	ca
♀				
CA	CCAA Agouti	CCAa Agouti	CcAA Agouti	CcAa Agouti
Ca	CCAa Agouti	CCaa Albino	CcAa Agouti	Ccaa Albino
cA	CcAA Agouti	CcAa Agouti	ccAA Black	ccAa Black
ca	CcAa Agouti	Ccaa Albino	ccAa Black	ccaa Albino

In this case, *C* is dominant to *c*, and *A* is dominant to *a*. If an animal has at least one copy of both dominant alleles, it will have the agouti coat color. If an animal has a dominant *A* allele but is *cc* homozygous, it will develop a black coat. The four cases of albino animals all are *aa* homozygous. This occurs even when an animal carries the dominant *C* allele. Therefore, *a* is epistatic to *C*. The converse, however, does not yield the same result. Animals with *ccAA* or *ccAa* genotypes are black.

S2. In Ayrshire cattle, the spotting pattern of the animals can be either red and white or mahogany and white. The mahogany and white pattern is caused by the allele *M*. The red and white phenotype is controlled by the allele *m*. When mahogany and white animals are mated to red and white animals, the following results are obtained:

Genotype	Phenotype	
	Females	Males
MM	mahogany and white	mahogany and white
Mm	red and white	mahogany and white
mm	red and white	red and white

Explain the pattern of inheritance.

Answer: The inheritance pattern for this trait is sex-influenced inheritance. The *M* allele is dominant in males but recessive in females, while the *m* allele is dominant in females but recessive in males.

S3. George Shull, a botanist at Princeton University, conducted a genetic study of a weed known as shepherd's purse, a member of the mustard family. One trait he followed was the shape of the seed capsule, which can be triangular or a small ovate shape. When he crossed a true-breeding plant with triangular capsules to a plant having ovate capsules, the F$_1$ generation all had triangular capsules. When the F$_1$ plants were self-fertilized, the surprising result came in the F$_2$ generation. Shull observed a 15:1 ratio of plants having triangular capsules to ovate capsules. Explain this ratio based on a two-gene interaction.

Answer: This result can be explained by a two-gene interaction in which each gene is found as a dominant or recessive allele. The presence of

one dominant allele in either gene results in a triangular capsule. If we let *T* and *t* represent the dominant and recessive alleles of one gene and let *V* and *v* represent the dominant and recessive alleles of a second gene, the results of Shull's crosses are outlined in the following Punnett square:

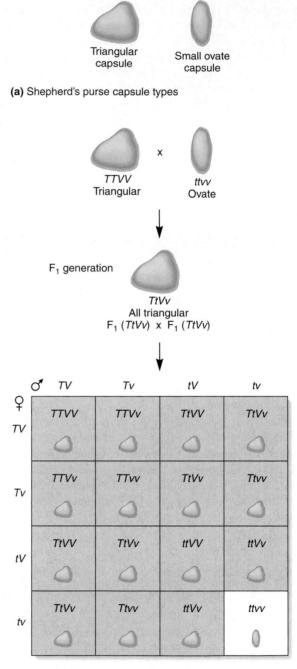

(a) Shepherd's purse capsule types

Triangular capsule   Small ovate capsule

TTVV Triangular   ×   *ttvv* Ovate

F$_1$ generation

TtVv All triangular F$_1$ (*TtVv*) × F$_1$ (*TtVv*)

♂ TV	TV	Tv	tV	tv
♀				
TV	TTVV	TTVv	TtVV	TtVv
Tv	TTVv	TTvv	TtVv	Ttvv
tV	TtVV	TtVv	ttVV	ttVv
tv	TtVv	Ttvv	ttVv	ttvv

(b) The crosses of Shull

This experiment of George Shull demonstrates an important phenomenon, confirmed in subsequent genetic studies. Many eukaryotic genes appear to be redundant. In other words, there are two (or more) genes that can play analogous roles in the life of the organism. In this example, either of the dominant alleles, *T* or *V*, which are alleles of different genes, is sufficient to cause triangular capsules. If the expression of one of the two genes is abolished by a recessive mutation, a dominant allele of the other gene can still produce the triangular phenotype. At the molecular level, gene redundancy can sometimes be explained by the formation of extra copies of a gene during evolution.

S4. For the following pedigree involving a single gene causing an inherited disease, indicate which modes of inheritance are *not* possible. (Affected individuals are shown as filled symbols.)

   A. Recessive

   B. Dominant

   C. X linked, recessive

   D. Sex influenced, dominant in females

   E. Sex limited, recessive in females

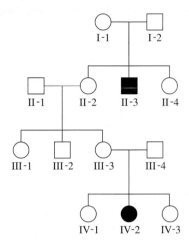

Answer:

   A. It could be recessive.

   B. It is probably not dominant unless it is incompletely penetrant.

   C. It could not be X-linked recessive because individual IV-2 does not have an affected father.

   D. It could not be sex influenced, dominant in females because individual II-3 (who would have to be homozygous) has an unaffected mother (who would have to be heterozygous and affected).

   E. It is not sex limited because individual II-3 is an affected male and IV-2 is an affected female.

S5. Pattern baldness is an example of a sex-influenced trait that is dominant in males and recessive in females. A nonbald couple produced a bald son. What are the genotypes of the parents?

Answer: Since the father is not bald, we know he must be homozygous, *bb*. Otherwise, he would be bald. A nonbald female can be either *Bb* or *bb*. Since she has produced a bald son, we know that she must be *Bb* in order to pass the *B* allele to her son.

S6. Two pink-flowered four-o'clocks were crossed to each other. What are the following probabilities?

   A. A red-flowered plant.

   B. The first three plants examined will be white.

   C. A plant will be either white or pink.

   D. A group of six plants contain one pink, two whites, and three reds.

Answer: The first thing we need to do is construct a Punnett square to determine the individual probabilities for each type of offspring.

   Since flower color is incompletely dominant, the cross is *Rr* × *Rr*.

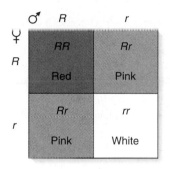

Phenotypic ratio is 1 red : 2 pink : 1 white. In other words, 1/4 are expected to be red, 1/2 pink, and 1/4 white.

   A. The probability of a red-flowered plant is 1/4, which equals 25%.

   B. Use the product rule.

$$1/4 \times 1/4 \times 1/4 = 1/64 = 1.6\%$$

   C. Use the sum rule because these are mutually exclusive events. A given plant cannot be both white and pink.

$$1/4 + 1/2 = 3/4 = 75\%$$

   D. Use the multinomial expansion. See solved problem S7 in chapter 2 for an explanation of the multinomial expansion. In this case, there are three possible phenotypes.

$$P = \frac{n!}{a!b!c!}\, p^a q^b r^c$$

where n = total number of offspring = 6
   a = number of reds = 3
   p = probability of reds = (1/4)
   b = number of pinks = 1
   q = probability of pink = (1/2)
   c = number of whites = 2
   r = probability of whites = (1/4)

If we substitute these values into the equation:

$$P = \frac{6!}{3!1!2!}\, (1/4)^3 (1/2)^1 (1/4)^2$$

$$P = 0.029 = 2.9\%$$

This means that 2.9% of the time we would expect to obtain six plants, three with red flowers, one with pink flowers, and two with white flowers.

Conceptual Questions

C1. Describe the differences among dominance, incomplete dominance, codominance, and overdominance.

C2. Discuss the differences among sex-influenced, sex-limited, and sex-linked inheritance. Describe examples.

C3. What is meant by gene interaction? How can gene interaction be explained at the molecular level?

C4. Let's suppose a recessive allele encodes a completely defective protein. If the normal allele is dominant, what does that tell us about the amount of the normal protein that is sufficient to cause the phenotype? What if the allele is incompletely dominant?

C5. A nectarine is a peach without the fuzz. The difference is controlled by a single gene that is found in two alleles, D and d. At the molecular level, would it make more sense to you that the nectarine is homozygous for a recessive allele or that the peach is homozygous for the recessive allele? Explain your reasoning.

C6. An allele in *Drosophila* produces a "star-eye" trait in the heterozygous individual. However, the star-eye allele is lethal in homozygotes. What would be the ratio and phenotypes of surviving flies if star-eyed flies were crossed to each other?

C7. A seed dealer wants to sell four-o'clock seeds that will produce only red, white, or pink flowers. Explain how this should be done.

C8. At the molecular level, how would you explain a gene dosage effect?

C9. The serum from one individual (let's call this person individual 1) is known to agglutinate the red blood cells from a second individual (individual 2). List the pairwise combinations of possible genotypes that individuals 1 and 2 could be. If individual 1 is the parent of individual 2, what are his/her possible genotypes?

C10. Which blood phenotypes (A, B, AB, and/or O) provide an unambiguous genotype? Is it possible for a couple to produce a family of children with all four blood types? If so, what would be the genotypes of the parents have to be?

C11. A woman with type B blood has a child with type O blood. What are the possible genotypes and blood types of the father?

C12. A type A woman is the daughter of a type O father and type A mother. If she has children with a type AB man, what are the following probabilities?

A. A type AB child

B. A type O child

C. The first three children with type AB blood

D. A family containing two children with type B blood and one child with type AB

C13. In Shorthorn cattle, coat color is controlled by a single gene that can exist as a red allele (R) or white allele (r). The heterozygotes (Rr) have a color called roan that looks less red than the RR homozygotes. However, when examined carefully, the roan phenotype in cattle is actually due to a mixture of completely red hairs and completely white hairs. Should this trait be called incompletely dominant, codominant, or something else? Explain your reasoning.

C14. With regard to genes and the functions of the proteins they encode, suggest a molecular explanation for solved problem S1.

C15. In chickens, the Leghorn variety has white feathers due to a dominant allele. Silkies have white feathers due to a recessive allele in a second (different) gene. If a true-breeding white Leghorn is crossed to a true-breeding white Silkie, what is the expected phenotype of the F_1 generation? If members of the F_1 generation are mated to each other, what is the expected phenotypic outcome of the F_2 generation? (Assume that the chickens in the parental generation are homozygous for the white allele at one gene and homozygous for the brown allele at the other gene. In subsequent generations, nonwhite birds will be brown.)

C16. With regard to pattern baldness in humans (a sex-influenced trait), a woman who is not bald and whose mother is bald has children with a bald man whose father is not bald. What are their probabilities of having the following types of families?

A. Their first child will not become bald.

B. Their first child will be a male who will not become bald.

C. Their first three children will be nonbald females.

C17. In rabbits, the color of body fat is controlled by a single gene with two alleles, designated Y and y. The outcome of this trait is affected by the diet of the rabbit. When raised on a standard vegetarian diet, the dominant Y allele confers white body fat and the y allele confers yellow body fat. However, when raised on a xanthophyll-free diet, the homozygote yy animal has white body fat. If a heterozygous animal is crossed to a rabbit with yellow body fat, what are the proportions of offspring with white and yellow body fat when raised on a standard vegetarian diet? How do the proportions change if the offspring are raised on a xanthophyll-free diet?

C18. The X-linked eye color gene discussed in figure 4.7 also exists in apricot and coral alleles, in addition to red, eosin, and white. Compared to red (the wild-type allele), homozygous females have the following amounts of pigment: white eyes = 0%, eosin eyes = 3%, apricot eyes = 6%, and coral eyes = 8%. Assuming that the red-eye allele is dominant, and that the other alleles exhibit a gene dosage effect, what amount of pigment would you expect in the male and female offspring of the following crosses?

A. Homozygous apricot-eyed female × white-eyed male

B. Heterozygous apricot/coral female × red-eyed male

C. Heterozygous red/apricot female × white-eyed male

D. Homozygous coral-eyed female × apricot-eyed male

C19. A Siamese cat that spends most of its time outside was accidentally injured in a trap and required several stitches in its right front paw. The veterinarian had to shave the fur from the paw and leg, which originally had rather dark fur. Later, when the fur grew back, it was much lighter than the fur on the other three legs. Do you think this injury occurred in the summer or winter? Explain your answer.

C20. A true-breeding male fly with eosin eyes is crossed to a white-eyed female that is heterozygous for the normal (C) and cream alleles (c^a). What are the expected proportions of their offspring?

C21. In chapter 4, we considered the trait of hen- versus cock-feathering. Starting with two heterozygous fowl that are hen-feathered, explain how you would obtain a true-breeding line that always produced cock-feathered males.

C22. In the pedigree shown here for a trait determined by a single gene (affected individuals are shown in *black*), state whether it would be possible for the trait to be inherited in each of the following ways:

A. Recessive

B. X-linked recessive

C. Dominant, complete penetrance

D. Sex influenced, dominant in males

E. Sex limited

F. Dominant, incomplete penetrance

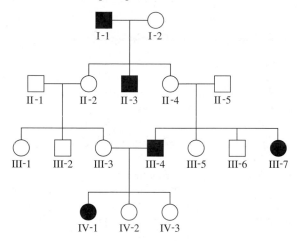

C23. The pedigree shown here also concerns a trait determined by a single gene (affected individuals are shown in *black*). Which of the following patterns of inheritance are possible?

A. Recessive

B. X-linked recessive

C. Dominant

D. Sex influenced, recessive in males

E. Sex limited

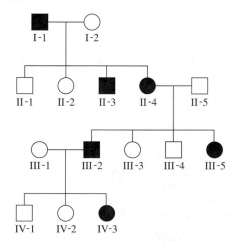

C24. Let's suppose that you have pedigree data from thousands of different families. For a particular genetic disease, how would you decide whether the disease is inherited as a recessive trait as opposed to one that is dominant with incomplete penetrance?

C25. Compare phenotypes at the molecular, cellular, and organismal levels for individuals who are homozygous for the normal hemoglobin allele ($Hb^A Hb^A$) and the sickle-cell allele ($Hb^S Hb^S$).

C26. A very rare dominant allele that causes the little finger to be crooked has a penetrance value of 80%. In other words, 80% of heterozygotes carrying the allele will have a crooked little finger. What is the probability that a heterozygote will produce offspring with crooked little fingers?

C27. A sex-influenced trait in humans is one that affects the length of the index finger. A "short" allele is dominant in males and recessive in females. Heterozygous males have an index finger that is significantly shorter than the fourth finger. The gene affecting index finger length is located on an autosome. A woman with short index fingers has children with a man who has normal index fingers. They produce five children in the following order: female, male, male, female, male. The oldest female offspring marries a man with normal fingers and then has one daughter. The youngest male among the five children marries a woman with short index fingers, and then they have two sons. Draw the pedigree for this family. Indicate the phenotypes of every individual (filled symbols for individuals with short index fingers and open symbols for individuals with normal index fingers).

C28. In horses, there are three coat color patterns termed *cremello* (beige), *chestnut* (brown), and *palomino* (golden with light mane and tail). If two palomino horses are mated, they produce about 1/4 cremello, 1/4 chestnut, and 1/2 palomino offspring. In contrast, cremello horses and chestnut horses breed true. (In other words, two cremello horses will produce only cremello offspring and two chestnut horses will produce only chestnut offspring.) Explain this pattern of inheritance.

Experimental Questions

E1. Mexican hairless dogs have little hair and few teeth. When a Mexican hairless is mated to another breed of dog, about half the puppies are hairless. When two Mexican hairless dogs are mated to each other, about 1/3 of the surviving puppies have hair and about 2/3 of the surviving puppies are hairless. However, about two out of eight puppies from this type of cross are born grossly deformed and do not survive. Explain this pattern of inheritance.

E2. An animal breeder has purchased five chinchilla rabbits of unknown genotypes. We will call them chinchilla 1, 2, 3, 4, and 5. To determine their genotypes, he made the following crosses:

A. Chinchilla 1 (male) mated to chinchilla 3 produced six chinchilla offspring and one albino offspring.

B. Chinchilla 1 mated to chinchilla 4 produced seven chinchilla offspring.

C. Chinchilla 2 (male) mated to chinchilla 3 produced five chinchilla offspring and two Himalayan offspring.

D. Chinchilla 2 mated to chinchilla 5 produced four chinchilla offspring and two Himalayan offspring.

E. Chinchilla 2 mated to chinchilla 4 produced eight chinchilla offspring.

What are the possible genotypes of the five rabbits? Are there any genotypes that you are not sure about?

E3. In chickens, some varieties have feathered shanks (legs) while others do not. In a cross between a Black Langhans (feathered shanks) and Buff Rocks (unfeathered shanks), the shanks of the F_1 generation are all feathered. When the F_1 generation is crossed, the F_2 generation contains chickens with feathered shanks to unfeathered shanks in a ratio of 15:1. Suggest an explanation for this result.

E4. In sheep, the formation of horns is a sex-influenced trait; the allele that results in horns is dominant in males and recessive in females. Females must be homozygous for the horned allele to have horns. A horned ram was crossed to a polled (unhorned) ewe, and the first offspring they produced was a horned ewe. What are the genotypes of the parents?

E5. A particular breed of dog can have long hair or short hair. When true-breeding long-haired animals were crossed to true-breeding short-haired animals, the offspring all had long hair. The F_2 generation produced a 3:1 ratio of long- to short-haired offspring. A second trait involves the texture of the hair. The two variants are wiry hair and straight hair. F_1 offspring from a cross of these two varieties all had wiry hair, and F_2 offspring showed a 3:1 ratio of wiry- to straight-haired puppies. Recently, a breeder of the short-, wiry-haired dogs found a female puppy that was albino. Similarly, another breeder of the long-, straight-haired dogs found a male puppy that was albino. Since the albino trait is always due to a recessive allele, the two breeders got together and mated the two dogs. Surprisingly, all the puppies in the litter had black hair. How would you explain this result?

E6. In the clover butterfly, males are always yellow but females can be yellow or white. In females, white is a dominant allele. Two yellow butterflies were crossed to yield an F_1 generation consisting of 50% yellow males, 25% yellow females, and 25% white females. Describe how this trait is inherited and the genotypes of the parents.

E7. Duroc Jersey pigs are typically red, but a sandy variation is also seen. When two different varieties of true-breeding sandy pigs were crossed to each other, they produced F_1 offspring that were red. When these F_1 offspring were crossed to each other, they produced red, sandy, and white pigs in a 9:6:1 ratio. Explain this pattern of inheritance.

E8. As discussed in chapter 4, comb morphology in chickens is governed by a gene interaction. Two walnut comb chickens were crossed to each other. They produced only walnut comb and rose comb offspring, in a ratio of 3:1. What are the genotypes of the parents?

E9. In certain species of summer squash, fruit color is determined by two interacting genes. A dominant allele, W, determines white color, while a recessive allele allows the fruit to be colored. In a homozygous ww individual, a second gene determines fruit color: G (green) is dominant to g (yellow). A white squash and a yellow squash were crossed, and the F_1 generation yielded approximately 50% white fruit and 50% green fruit. What are the genotypes of the parents?

E10. Certain species of summer squash can exist in long, spherical, or disk shapes. When a true-breeding long-shaped strain was crossed to a true-breeding disk-shaped strain, all the F_1 offspring were disk-shaped. When the F_1 offspring were allowed to self-fertilize, the F_2 generation consisted of a ratio of 9 disk-shaped : 6 round-shaped : 1 long-shaped. Assuming that the shape of summer squash is governed by two different genes, with each gene existing in two alleles, propose a mechanism to account for this 9:6:1 ratio.

E11. In the experiment in figure 4.8, explain why the heterozygous $X^w X^{w-e}$ female fly had the same eye color as the $X^{w-e} Y$ male fly.

E12. In *Drosophila*, many other eye colors have been found to be allelic to white. In addition to eosin, there are ivory, pearl, coral, cherry, apricot, and blood, to name a few. Based on the intensity of eye color alone, which alleles might you expect to have a gene dosage effect? Describe the types of observations and experiments you would carry out to determine if a gene dosage effect occurs.

E13. For the data in figure 4.8, conduct a chi square analysis to see if the experimental data has a good fit with the hypothesis.

E14. In a species of plant, two genes control flower color. The red allele (R) is dominant to the white allele (r); the color-producing allele (C) is dominant to the non-color-producing allele (c). You suspect that either an rr homozygote or a cc homozygote will produce white flowers. In other words, rr is epistatic to C, and cc is epistatic to R. To test your hypothesis, you allowed heterozygous plants ($RrCc$) to self-fertilize and counted the offspring. You obtained the following data: 201 plants with red flowers and 144 with white flowers. Conduct a chi square analysis to see if your observed data are consistent with your hypothesis.

E15. In *Drosophila*, red eyes is the wild-type phenotype. There are several different genes (with each gene existing in two or more alleles) that affect eye color. One allele causes purple eyes, and a different allele causes sepia eyes. Both of these alleles are recessive compared to red eye color. When flies with purple eyes were crossed to flies with sepia eyes, all the F_1 offspring had red eyes. When the F_1 offspring were allowed to mate with each other, the following data were obtained:

146 purple eyes
151 sepia eyes
 50 purplish sepia eyes
444 red eyes

Explain this pattern of inheritance. Conduct a chi square analysis to see if the experimental data fit with your hypothesis.

E16. As mentioned in experimental question E15, red eyes is the wild-type phenotype in *Drosophila* and there are several different genes (with each gene existing in two or more alleles) that affect eye color. One allele causes purple eyes and a different allele causes vermilion eyes. The purple and vermilion alleles are recessive compared to red eye color. The following crosses were made and the following data were obtained:

Cross 1: Males with vermilion eyes × females with purple eyes

354 offspring, all with red eyes

Cross 2: Males with purple eyes × females with vermilion eyes

212 male offspring with vermilion eyes
221 female offspring with red eyes

Explain the pattern of inheritance based on these results. What additional crosses might you make to confirm your hypothesis?

E17. Let's suppose that you were looking through a vial of fruit flies in your laboratory and noticed a male fly that has pink eyes. What crosses would you make to determine if the pink allele is an X-linked gene? What crosses would you make to determine if the pink allele is an allele of the same X-linked gene that has white and eosin alleles? Note: The white and eosin alleles are discussed in figure 4.8.

Questions for Student Discussion/Collaboration

1. Let's suppose a gene exists as a functional wild-type allele and a nonfunctional mutant allele. At the organismal level, the wild-type allele is dominant. In a heterozygote, discuss whether dominance occurs at the cellular or molecular levels. Discuss examples in which the issue of dominance depends on the level of examination.

2. A true-breeding rooster with a rose comb, feathered shanks, and cock-feathering was crossed to a hen that is true-breeding for pea comb and unfeathered shanks but is heterozygous for hen-feathering. If you assume that the four genes involved can assort independently, what is the expected outcome of the F_1 generation?

3. In oats, the color of the chaff is determined by a two-gene interaction. When a true-breeding black plant was crossed to a true-breeding white plant, the F_1 generation was composed of all black plants. When the F_1 offspring were crossed to each other, the ratio produced was 12 black : 3 gray : 1 white. First, construct a Punnett square that accounts for this pattern of inheritance. Which genotypes produce the gray phenotype? Second, at the level of protein function, how would you explain this type of inheritance?

Note: All answers appear at the website for this textbook; the answers to even-numbered questions are in the back of the textbook.

www.mhhe.com/brooker

Visit the Online Learning Center for practice tests, answer keys, and other learning aids for this chapter. Enhance your understanding of genetics with our interactive exercises, web links, news feeds, tutorial service, and much more.

LINKAGE AND GENETIC MAPPING IN EUKARYOTES

5

::

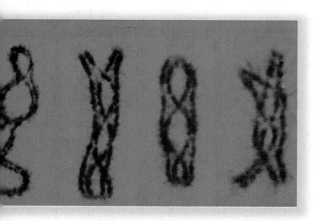

In chapter 2, we were introduced to the law of independent assortment. According to Mendel, we expect that two different genes will segregate and independently assort themselves during the process that creates gametes. After Mendel's work was rediscovered at the turn of the twentieth century, chromosomes were identified as the cellular structures that carry genes. The chromosome theory of inheritance explained how the transmission of chromosomes is responsible for the passage of genes from parents to offspring.

When geneticists first realized that chromosomes contain the genetic material, they began to suspect that a conflict might sometimes occur between Mendel's law of independent assortment of genes and the behavior of chromosomes during meiosis. In particular, geneticists assumed that each species of organism must contain thousands of different genes, yet cytological studies revealed that most species have at most a few dozen chromosomes. Therefore, it seemed likely (and turned out to be true) that each chromosome would carry many hundreds or even thousands of different genes. The transmission of genes located close to each other on the same chromosome violates the law of independent assortment.

In chapter 5, we will consider the pattern of inheritance that occurs when different genes are situated on the same chromosome. In addition, we will briefly explore how the data from genetic crosses are used to construct genetic maps, which describe the order of genes along a chromosome. However, newer strategies for gene mapping, which are described in chapter 20, have largely superseded the traditional genetic mapping, which is described in chapter 5. Nevertheless, an understanding of traditional mapping studies strengthens our appreciation for newer molecular approaches. Moreover, traditional mapping studies further illustrate how the location of two or more genes on the same chromosome can affect the transmission patterns from parents to offspring.

5.1 LINKAGE AND CROSSING OVER

In eukaryotic species, each linear chromosome contains a very long piece of DNA. As mentioned previously, a chromosome contains many individual functional units—called genes—that influence an organism's traits. A typical chromosome is expected to contain many hundreds or perhaps a few thousand different genes. The term **linkage** has two related meanings. It refers to the phenomenon that two or more genes can be located on the same chromosome. The genes are physically linked to each other, because each eukaryotic chromosome contains a single, continuous, linear piece of DNA. Secondly, genes that are close together on the same chromosome tend to be transmitted as a unit. This second meaning indicates that linkage has an influence on inheritance patterns.

As an example of linkage, let's consider a human karyotype that contains 46 chromosomes. All the genes on a particular chromosome are physically linked to each other. Chromosomes thus are sometimes called **linkage groups,** because a chromosome contains a group of genes that are linked together. However, as discussed later, genes that are very far apart on the same chromosome may independently assort from each other due to the phenomenon of **crossing over,** even though they are in a single linkage group. In a particular species, there are as many linkage groups as there are types of chromosomes. For example, human somatic cells have 46 chromosomes, which are composed of 22 types of autosomes that come in pairs plus one pair of sex chromosomes. There are two types of sex chromosomes, the X and Y. Therefore, humans have 22 autosomal linkage groups, an X chromosome linkage group, and a Y chromosome linkage group.

Geneticists are often interested in the transmission of two or more traits in a genetic cross. When a geneticist follows two traits in a cross, this is called a **dihybrid cross;** when three traits are followed, it is a **trihybrid cross;** and so forth. The outcome of a dihybrid or trihybrid cross depends on whether or not the genes are linked to each other along the same chromosome. In this section, we will examine how linkage affects the transmission patterns of two or more traits.

Crossing Over May Produce Recombinant Phenotypes

Even though the alleles for different genes may be linked along the same chromosome, the linkage can be altered during meiosis. In diploid eukaryotic species, homologous chromosomes can exchange pieces with each other by crossing over. This event occurs during prophase I of meiosis. As discussed in chapter 3, the replicated chromosomes, known as sister chromatids, associate

with the homologous sister chromatids to form a structure known as a **bivalent.** A bivalent is composed of two pairs of sister chromatids. In prophase I, a sister chromatid of one pair commonly crosses over with a sister chromatid from the homologous pair.

Figure 5.1 considers meiosis when two genes are linked on the same chromosome. One of the parental chromosomes carries the A and B alleles, while the homologue carries the a and b alleles. In figure 5.1a, no crossing over has occurred. Therefore, the haploid cells contain the same combination of alleles as the original chromosomes. In this case, two haploid cells carry the A and B alleles, and the other two carry the recessive a and b alleles. The arrangement of linked alleles has not been altered.

In contrast, figure 5.1b illustrates what can happen when crossing over occurs. Two of the haploid cells contain combinations of alleles (namely, A and b, a and B) that differ from those in the original chromosomes. In these two cells, the grouping of linked alleles has been changed. An event such as this, leading to a new combination of alleles, is known as **genetic recombination.** The haploid cells carrying the A and b, or the a and B,

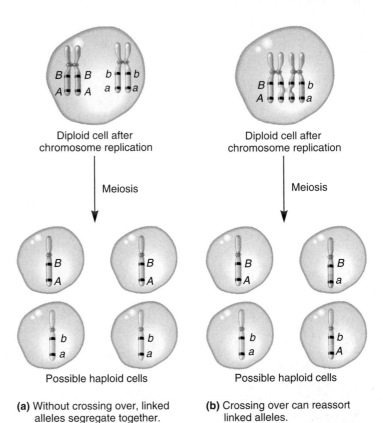

(a) Without crossing over, linked alleles segregate together.

(b) Crossing over can reassort linked alleles.

FIGURE 5.1   **Consequences of crossing over during meiosis.** (a) In the absence of crossing over, the A and B alleles and the a and b alleles are maintained in the same arrangement found in the parental chromosomes. (b) Crossing over has occurred in the region between the two genes, creating two nonparental haploid cells with a new combination of alleles.

alleles are called **nonparental** or **recombinant cells.** Likewise, if such haploid cells were gametes that participated in fertilization, the resulting offspring are called nonparental or recombinant offspring. These offspring can display combinations of traits that are different from those of either parent. In contrast, offspring that have inherited the same combination of alleles that are found in the chromosomes of their parents are known as **parental** or **nonrecombinant** offspring.

In chapter 5, we will consider how crossing over affects the pattern of inheritance for genes linked on the same chromosome. In chapter 17, we will consider the molecular events that cause crossing over to occur.

Bateson and Punnett Discovered Two Traits That Did Not Assort Independently

An early study indicating that some traits may not assort independently was carried out by William Bateson and Reginald Punnett in 1905. As mentioned in chapter 4, they were interested in the pattern of inheritance of genes in several organisms, including the chicken and the sweet pea. According to Mendel's law of indepen-

dent assortment, a dihybrid cross between two individuals, heterozygous for two genes, should yield a 9:3:3:1 phenotypic ratio among the offspring. However, a surprising result occurred when Bateson and Punnett conducted a cross in the sweet pea involving two different traits, flower color and pollen shape (fig. 5.2).

As seen in figure 5.2, they began by crossing a true-breeding strain with purple flowers (PP) and long pollen (LL) to a strain with red flowers (pp) and round pollen (ll). This yielded an F_1 generation of plants that all had purple flowers and long pollen ($PpLl$). The unexpected result came from the F_2 generation. Even though there were four different phenotypic categories among the F_2 generation, the observed numbers of offspring did not conform to a 9:3:3:1 ratio. Bateson and Punnett found that the F_2 generation had a much greater proportion of the two phenotypes found in the parental generation: purple flowers with long pollen, and red flowers with round pollen. Therefore, they suggested that the transmission of these two traits from the parental generation to the F_2 generation was somehow coupled and not easily assorted in an independent manner. However, Bateson and Punnett did not realize that this coupling was due to the linkage of the flower color gene and the pollen shape gene on the same chromosome.

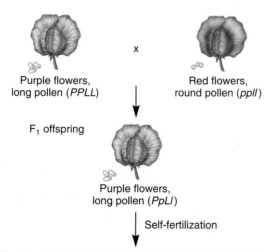

F_2 offspring	Observed number	Ratio	Expected number	Ratio
Purple flowers, long pollen	296	15.6	240	9
Purple flowers, round pollen	19	1.0	80	3
Red flowers, long pollen	27	1.4	80	3
Red flowers, round pollen	85	4.5	27	1

FIGURE 5.2 **An experiment of Bateson and Punnett with sweet peas showing that independent assortment does not always occur.**

GENES→TRAITS Two genes that govern flower color and pollen shape are found on the same chromosome. Therefore, the offspring tend to inherit the parental combinations of alleles (*PL* or *pl*). Due to occasional crossing over, a lower percentage of offspring inherit nonparental combinations of alleles (*Pl* or *pL*).

Morgan Provided Evidence for the Linkage of Several X-Linked Genes and Proposed That Crossing Over Between X Chromosomes Can Occur

The first direct evidence that different genes are physically located on the same chromosome came from the studies of Thomas Hunt Morgan. In chapters 3 and 4, we considered some of Morgan's studies involving X-linked traits. These earlier studies provided the groundwork to demonstrate that the genes governing X-linked traits are physically linked on the X chromosome.

Morgan investigated the inheritance pattern of many different traits that had been shown to follow an X-linked pattern of inheritance. Figure 5.3 illustrates an experiment involving three traits that Morgan studied. His parental crosses were wild-type male fruit flies to females that had yellow bodies (yy), white eyes (ww), and miniature wings (mm). The wild-type alleles for these three genes are designated y^+ (gray body), w^+ (red eyes), and m^+ (normal wings). As expected, the phenotypes of the F_1 generation were wild-type females and males with yellow bodies, white eyes, and miniature wings. The linkage of these genes was revealed when the F_1 flies were mated to each other and the F_2 generation examined.

Instead of equal proportions of the eight possible phenotypes, Morgan observed a much higher proportion of the combinations of traits found in the parental generation. There were 758 flies with gray bodies, red eyes, and normal wings, and 700 flies with yellow bodies, white eyes, and miniature wings. The former combination (gray body, red eyes, and normal wings) was found in the males of the parental generation, the latter combination (yellow body, white eyes, and miniature wings) in the females of the parental generation. Morgan's explanation for this higher

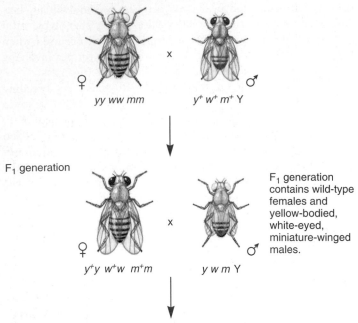

F₁ generation

F₁ generation contains wild-type females and yellow-bodied, white-eyed, miniature-winged males.

F₂ generation	Females	Males	Total
Gray body, red eyes, normal wings	439	319	758
Gray body, red eyes, miniature wings	208	193	401
Gray body, white eyes, normal wings	1	0	1
Gray body, white eyes, miniature wings	5	11	16
Yellow body, red eyes, normal wings	7	5	12
Yellow body, red eyes, miniature wings	0	0	0
Yellow body, white eyes, normal wings	178	139	317
Yellow body, white eyes, miniature wings	365	335	700

FIGURE 5.3   A trihybrid cross of Morgan's involving three X-linked traits in *Drosophila*.

GENES→TRAITS   Three genes that govern body color, eye color, and wing length are all found on the X chromosome. Therefore, the offspring tend to inherit the parental combinations of alleles ($y^+ w^+ m^+$ or $y\ w\ m$). Figure 5.4 explains how single and double crossovers can create nonparental combinations of alleles.

proportion of parental combinations was that all three genes are located on the X chromosome and, therefore, tend to be transmitted together as a unit.

However, to fully account for the data shown in figure 5.3, Morgan needed to interpret two other key observations. First, he needed to explain why a significant proportion of the F₂ generation had nonparental combinations of alleles. Along with the two parental phenotypes, there were five other phenotypic combinations that were not found in the parental generation. Second, he needed to explain why there was a quantitative difference between nonparental combinations involving body color and eye color versus eye color and wing length. This quantitative difference is revealed by reorganizing the data from Morgan's cross by pairs of genes.

Gray body, red eyes	1,159	
Yellow body, white eyes	1,017	
Gray body, white eyes	17	Nonparental
Yellow body, red eyes	12	offspring
Total	2,205	

Red eyes, normal wings	770	
White eyes, miniature wings	716	
Red eyes, miniature wings	401	Nonparental
White eyes, normal wings	318	offspring
Total	2,205	

We see that there were substantial differences between the numbers of nonparental offspring when pairs of genes were considered separately. It was fairly common for nonparental combinations to occur when just eye color and wing length were examined (401 + 318 nonparental offspring). In sharp contrast, it was rare to obtain nonparental combinations when looking at body color and eye color (17 + 12 nonparental offspring).

To explain these data, Morgan considered the previous studies of the French cytologist F. A. Janssens, who proposed that crossing over involves a physical exchange between homologous chromosomes that occurs prior to the microscopic visualization of chiasmata. Morgan shrewdly realized that crossing over between homologous X chromosomes was consistent with his data. He assumed that crossing over did not occur between the X and Y chromosome and that these three genes are not found on the Y chromosome. With these ideas in mind, he made three important hypotheses to explain his results:

1. The genes for body color, eye color, and wing length are all located on the same chromosome (the X chromosome). Therefore, it is most likely for all three traits to be inherited together.
2. Due to crossing over, the homologous X chromosomes (in the female) can exchange pieces of chromosomes and create new (nonparental) combinations of alleles.
3. The likelihood of crossing over depends on the distance between two genes. If two genes are far apart from each other, it is more likely that crossing over will occur between them.

With these ideas in mind, figure 5.4 illustrates the possible events that occurred in the F₁ female flies of Morgan's experiment. One of the X chromosomes contained all three dominant alleles, the other all three recessive alleles. During oogenesis in the F₁ female flies, crossing over may or may not have occurred in this region of the X chromosome. If no crossing over occurred, the parental phenotypes were produced in the F₂ offspring. Alternatively, a crossover sometimes occurred between the eye color gene and the wing length gene to create nonparental offspring (namely, gray body, red eyes, and miniature wings; or yellow body, white eyes, and normal wings). According to Morgan's proposal, this is a fairly likely event, because these two genes are far

Sex chromosomes in:

X chromosomes in F₁ female before oogenesis — F₁ female gametes — F₁ male gametes — F₂ offspring phenotypes

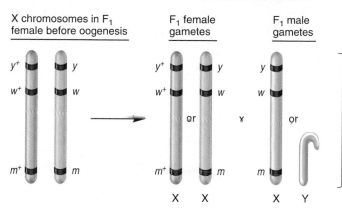

Gray body, red eyes, normal wings; total = 758

Yellow body, white eyes, miniature wings; total = 700

(a) No crossing over between y^+ and m^+ (most likely)

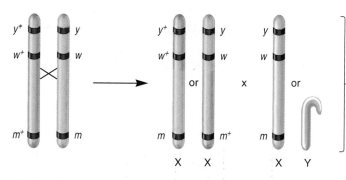

Gray body, red eyes, miniature wings; total = 401

Yellow body, white eyes, normal wings; total = 317

(b) Crossing over between w^+ and m^+ (fairly likely)

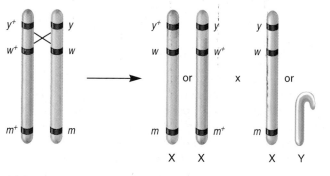

Gray body, white eyes, miniature wings; total = 16

Yellow body, red eyes, normal wings; total = 12

(c) Crossing over between y^+ and w^+ (unlikely)

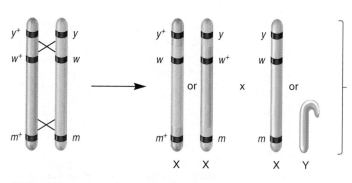

Gray body, white eyes, normal wings; total = 1

Yellow body, red eyes, miniature wings; total = 0

(d) Double crossing over (very unlikely)

FIGURE 5.4   **The likelihood of crossing over provides an explanation of Morgan's trihybrid cross. (a)** In the absence of crossing over, the parental chromosomes carry the $y^+w^+m^+$ or $y\,w\,m$ alleles. Alternatively, a crossover may occur between homologous X chromosomes in the female parent during oogenesis. As described in chapter 3, crossing over actually occurs at the bivalent stage, but for simplicity, this figure shows only two X chromosomes (one of each homologue) rather than four chromatids, which would occur during the bivalent stage of meiosis. In part **(b)**, a crossover has occurred between the eye color and wing length genes to produce recombinant chromosomes that carry the y^+w^+m or $y\,w\,m^+$ alleles. Since the eye color and wing length genes are relatively far apart, such recombinant offspring are common. **(c)** A crossover occurs between the body color and eye color genes to produce recombinant chromosomes that contain the $y^+w\,m$ or $y\,w^+m^+$ alleles. Since these two genes are very close together, these two types of recombinant offspring are fairly uncommon. **(d)** Finally, a double crossover could produce chromosomes carrying the $y^+w\,m^+$ or $y\,w^+m$ alleles. This is a very rare event. Note: In this experiment, crossing over occurred between homologous X chromosomes in the female during oogenesis. In *Drosophila*, crossing over does not occur during spermatogenesis. This is a peculiar characteristic of *Drosophila*. In many other animal species, homologous recombination commonly occurs during spermatogenesis. Also note that this figure shows only a portion of the X chromosome. A map of the entire X chromosome is shown in figure 5.8.

apart from each other on the X chromosome. Because of the long distance, it was fairly likely for a crossover to occur in this region. In contrast, he proposed that the body color and eye color genes are very close together, which makes crossing over between them an unlikely event. Nevertheless, it occasionally occurred, yielding offspring with gray bodies, white eyes, and miniature wings, or with yellow bodies, red eyes, and normal wings. Finally, it was also possible for two homologous chromosomes to cross over twice. This double crossover is very unlikely. Among the 2,205 offspring Morgan examined, he found only one fly (gray body, white eyes, normal wings) that could be explained by this phenomenon.

A Chi Square Analysis Can Be Used to Distinguish Between Linkage and Independent Assortment

Now that we have an appreciation for linkage and the production of recombinant offspring, let's consider how an experimenter can objectively decide whether two genes are linked or assort independently. In chapter 2, chi square analysis was introduced to evaluate the goodness of fit between a genetic hypothesis and observed experimental data. This method can be used to determine if the outcome of a dihybrid cross is consistent with linkage or independent assortment.

To conduct a chi square analysis, we must first propose a hypothesis. In a dihybrid cross, the standard hypothesis is that the two genes are not linked. This hypothesis is chosen even if the observed data suggest linkage, because an independent assortment hypothesis allows us to calculate the expected number of offspring based on the genotypes of the parents and the law of independent assortment. In contrast, for two linked genes that have not been previously mapped, we cannot calculate the expected number of offspring from a genetic cross, because we do not know how likely it is for a crossover to occur between the two genes. Without expected numbers of recombinant and parental offspring, we cannot use a chi square test. Therefore, we begin with the hypothesis that the genes are not linked; then, we determine whether or not our data fit this hypothesis. If the chi square value is low and we cannot reject our hypothesis, we infer that the genes assort independently. On the other hand, if the chi square value is so high that our hypothesis is rejected, we will accept the alternative hypothesis. Namely, we will conclude that a linkage hypothesis is correct. Of course, it is important to remember that a statistical analysis cannot prove that a hypothesis is true. If the chi square value is high, we accept the linkage hypothesis because we are assuming that there are only two possible explanations for a genetic outcome: the genes are either linked or not linked. If other factors affect the outcome of the cross (e.g., viability of particular phenotypes, etc.), this can cause deviations between the observed and expected values, and this assumption may not be valid.

With these ideas in mind, let's consider the data concerning body color and eye color. This cross produced the following offspring: 1,159 gray body, red eyes; 1,017 yellow body, white eyes; 17 gray body, white eyes; and 12 yellow body, red eyes. However, when a heterozygous female is crossed to a hemizygous male

$(X^{yw}Y)$, an independent assortment hypothesis predicts the following outcome:

The independent assortment hypothesis predicts a 1:1:1:1 ratio among the four phenotypes. The observed data mentioned obviously seem to conflict with this hypothesis. Nevertheless, we stick to the strategy just discussed. First, we propose that the two genes are not linked, and then we use a chi square analysis to see if the data fit this hypothesis. If the data do not fit, we will reject the idea that the genes assort independently and conclude the genes are linked.

An example of a chi square approach to determine linkage is shown here.

Step 1. *Propose a hypothesis.* Even though the observed data appear inconsistent with this hypothesis, we propose that the two genes for eye color and body color are X linked but somehow are able to obey Mendel's law of independent assortment. This hypothesis allows us to calculate expected values. We actually anticipate that the chi square analysis will allow us to reject the independent assortment hypothesis in favor of a linkage hypothesis.

Step 2. *Based on the hypothesis, calculate the expected values of each of the four phenotypes.* Each phenotype has an equal probability of occurring (see the Punnett square given previously). Therefore, the probability of each phenotype is 1/4. The observed F_2 generation contained a total of 2,205 individuals. Our next step is to calculate the expected numbers of offspring with each phenotype when the total equals 2,205; 1/4 of the offspring should be each of the four phenotypes:

$1/4 \times 2,205 = 551$ (expected number of each phenotype)

Step 3. *Apply the chi square formula, using the data for the observed values (O) and the expected values (E) that have*

been calculated in step 2. In this case, there are four categories within the population.

$$\chi^2 = \frac{(O_1 - E_1)^2}{E_1} + \frac{(O_2 - E_2)^2}{E_2} + \frac{(O_3 - E_3)^2}{E_3} + \frac{(O_4 - E_4)^2}{E_4}$$

$$\chi^2 = \frac{(1,159 - 551)^2}{551} + \frac{(17 - 551)^2}{551}$$

$$+ \frac{(12 - 551)^2}{551} + \frac{(1,017 - 551)^2}{551}$$

$$\chi^2 = 670.9 + 517.5 + 527.3 + 394.1 = 2,109.8$$

Step 4. *Interpret the calculated chi square value.* This is done with a chi square table, as discussed in chapter 2. Since there are four experimental categories ($n = 4$), the degrees of freedom is $n - 1 = 3$.

The calculated chi square value is enormous! Thus, the deviation between observed and expected values is very large. According to table 2.1, such a large deviation is expected to occur by chance alone less than 1% of the time. Therefore, we reject the hypothesis that the two genes assort independently. In other words, we conclude that the genes are linked.

EXPERIMENT 5A

Creighton and McClintock Correlated Crossing Over That Produced New Combinations of Alleles with the Exchange of Homologous Chromosomes

As we have seen, Morgan's studies were consistent with the hypothesis that crossing over occurs between homologous chromosomes to produce new combinations of alleles. To obtain direct evidence that genetic recombination is due to crossing over, Harriet Creighton and Barbara McClintock used an interesting strategy involving parallel observations. First, they made crosses involving two linked genes to produce parental and recombinant offspring. Second, they used a microscope to view the structures of the chromosomes in the parents and in the offspring. Because the parental chromosomes had some unusual structural features, they could microscopically distinguish the two homologous chromosomes within a pair. As we will see, this enabled them to correlate the occurrence of recombinant offspring with microscopically observable exchanges in segments of homologous chromosomes.

Creighton and McClintock focused much of their attention on the pattern of inheritance of traits in corn. In previous cytological examinations of corn chromosomes, some strains were found to have an unusual chromosome (9) with a darkly staining knob at one end. McClintock also identified an abnormal version of this chromosome that had an extra piece of a different chromosome (8) at the other end (called a **translocation**). As shown in figure 5.5a, this unusual version of chromosome 9 had changes at both ends that could be distinguished under the microscope.

Creighton and McClintock insightfully realized that this abnormal chromosome could be used to demonstrate that two homologous chromosomes physically exchange segments as a result of crossing over. They knew that a gene was located near the knobbed end of chromosome 9 that provided color to corn kernels. It existed in two alleles, the dominant allele *C* (colored) and the recessive allele *c* (colorless). Toward the other end of the

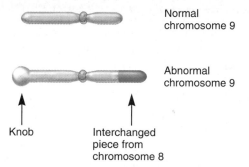

(a) Normal and abnormal chromosome 9

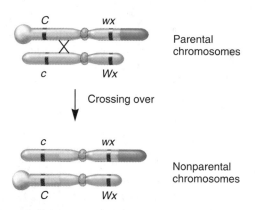

(b) Crossing over between normal and abnormal chromosome 9

FIGURE 5.5 **(a) Normal and abnormal chromosome 9 in corn used by Creighton and McClintock.** A normal chromosome 9 is compared to an abnormal chromosome 9 that contains a knob at one end and a translocation (i.e., an interchange) at the opposite end. **(b) Crossing over between normal and abnormal chromosomes in corn.** A crossover produces a chromosome that contains only a knob at one end and another chromosome that contains only a translocation at the other end.

chromosome was located a second gene that affected the texture of the kernel endosperm. The dominant allele *Wx* caused starchy endosperm, while the recessive *wx* allele caused waxy endosperm. Creighton and McClintock reasoned that a crossover involving a normal chromosome 9 and a knobbed/translocated chromosome 9 would produce a chromosome that had either a knob or a translocation but not both. These two types of chromosomes would be distinctly different from either of the parental chromosomes (fig. 5.5*b*).

As shown in the experiment of figure 5.6, Creighton and McClintock began with a corn strain that carried an abnormal chromosome that had a knob at one end and a translocation at the other. Genotypically, this chromosome was *C wx*. The cytologically normal chromosome in this strain was *c Wx*. This corn plant, which was termed parent A, was crossed to a strain (termed parent B) that carried two cytologically normal chromosomes and was genotypically *cc Wxwx*. They then observed the kernels in two ways. First, they examined the phenotypes of the kernels to see if they were colored or colorless, and waxy or starchy. Second, the chromosomes in each kernel were examined under a microscope. Altogether, they observed a total of 25 kernels (see data of fig. 5.6).

■ THE HYPOTHESIS

Offspring with nonparental phenotypes are the product of a crossover. This crossover should create nonparental chromosomes via an exchange of chromosomal segments between homologous chromosomes.

■ TESTING THE HYPOTHESIS — FIGURE 5.6 Experimental correlation between genetic recombination and crossing over.

Starting materials: Two different strains of corn. One strain (referred to as parent A) has an abnormal chromosome 9 (knobbed/translocation) with a dominant *C* allele and a recessive *wx* allele. It also contains a cytologically normal copy of chromosome 9 that carries the recessive *c* allele and the dominant *Wx* allele. Its genotype is *Cc Wxwx*. The other strain (referred to as parent B) has two normal versions of chromosome 9. The genotype of this strain is *cc Wxwx*.

Experimental level	Conceptual level

1. Cross the two strains described. The tassel is the pollen-bearing structure and the silk (equivalent to the stigma and style) is connected to the ovary. After fertilization, the ovary will develop into an ear of corn.

2. Observe the kernels from this cross.

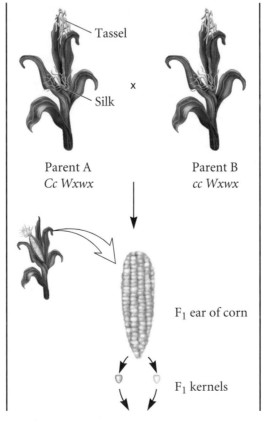

Tassel

Silk

Parent A
Cc Wxwx

Parent B
cc Wxwx

F₁ ear of corn

F₁ kernels

Each kernel is a separate seed that has inherited a set of chromosomes from each parent.

(continued)

3. Microscopically examine chromosome 9
   in the kernels.

Microscope

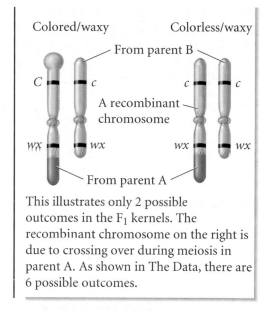

This illustrates only 2 possible
outcomes in the F₁ kernels. The
recombinant chromosome on the right is
due to crossing over during meiosis in
parent A. As shown in The Data, there are
6 possible outcomes.

■ THE DATA

Phenotype of F₁ Kernel	Number of Kernels Analyzed	Cytological Appearance of a Chromosome in F₁ Offspring*		Did a Crossover Occur During Gamete Formation in Parent A?
Colored/waxy	3	Knobbed/translocation C wx	Normal c wx	No
Colorless/starchy	11	Knobless/normal c Wx	Normal c or Wx c wx	No
Colorless/starchy	4	Knobless/translocation c wx	Normal c Wx	Yes
Colorless/waxy	2	Knobless/translocation c wx	Normal c wx	Yes
Colored/starchy	5	Knobbed/normal C Wx	Normal c Wx	Yes

*In this table, the chromosome on the *left* was inherited from parent A and the chromosome on the *right* was inherited from parent B.

■ INTERPRETING THE DATA

In this experiment, the researchers were interested in whether or not crossing over had occurred in parent A, which is heterozygous for both genes. This parent can produce four types of reproductive cells, while parent B can produce only two types.

Parent A	Parent B
C wx (nonrecombinant)	*c Wx*
c Wx (nonrecombinant)	*c wx*
C Wx (recombinant)	
c wx (recombinant)	

By combining these reproductive cells in a Punnett square, the following types of offspring can be produced:

Parent B

As seen in this Punnett square, two of the phenotypic categories, colored, starchy (*Cc Wxwx* or *Cc WxWx*) and colorless, starchy (*cc WxWx* or *cc Wxwx*), are ambiguous because they could arise from a nonrecombinant and from a recombinant gamete. In other words, these phenotypes can be produced whether or not recombination occurs in parent A. Therefore, let's begin by considering the two unambiguous phenotypic categories: colored, waxy (*Cc wxwx*) and colorless, waxy (*cc wxwx*). The colored, waxy phenotype can occur only if recombination did not occur in parent A and if parent A passed the knobbed, translocated chromosome to its offspring. As shown in the data table, three kernels were obtained with this phenotype, and all of them contained the knobbed, translocated chromosome. By comparison, the colorless, waxy phenotype can be obtained only if genetic recombination did occur in parent A and this parent passed a chromosome 9 that had a translocation but was knobless. Two kernels were obtained with this phenotype, and both of them contained the expected chromosome that had a translocation but was knobless. Taken together, these results show a perfect correlation between genetic recombination of alleles and the cytological presence of a chromosome displaying a genetic exchange of chromosomal pieces from parent A.

Overall, the observations described in this experiment were consistent with the idea that a crossover occurred in the region between the *C* and *wx* genes that involved an exchange of segments between two homologous chromosomes. As stated by Creighton and McClintock, "Pairing chromosomes, heteromorphic in two regions, have been shown to exchange parts at the same time they exchange genes assigned to these regions." These results supported the view that genetic recombination involves a physical exchange between homologous chromosomes. This microscopic evidence helped to convince geneticists that recombinant offspring arise from the physical exchange of segments of homologous chromosomes. As shown in solved problem S4 at the end of chapter 5, another experiment by Curt Stern was consistent with the idea that genetic recombination is due to the physical exchange of chromosomal segments.

A self-help quiz involving this experiment can be found at the Online Learning Center.

Crossing Over Occasionally Occurs During Mitosis

In multicellular organisms, the union of egg and sperm is followed by many cellular divisions, which occur in conjunction with mitotic divisions of the cell nuclei. As discussed in chapter 3, mitosis normally does not involve the homologous pairing of chromosomes to form a bivalent. Therefore, crossing over during mitosis is expected to occur much less frequently than during meiosis. Nevertheless, it does occur on rare occasions. When it happens, mitotic crossing over may produce a pair of recombinant chromosomes that have a new combination of alleles. This is known as **mitotic recombination.** If it occurs during an early stage of embryonic development, the daughter cells containing the recombinant chromosomes continue to divide many times to produce a patch of tissue in the adult. This may result in a portion of tissue with characteristics different from those of the rest of the organism.

Curt Stern proposed that unusual patches on the bodies of certain *Drosophila* strains were due to mitotic recombination. As shown in figure 5.7, he was working with strains carrying X-linked alleles affecting body color and bristle morphology. The recessive *y* allele confers yellow body color, and the recessive *sn*

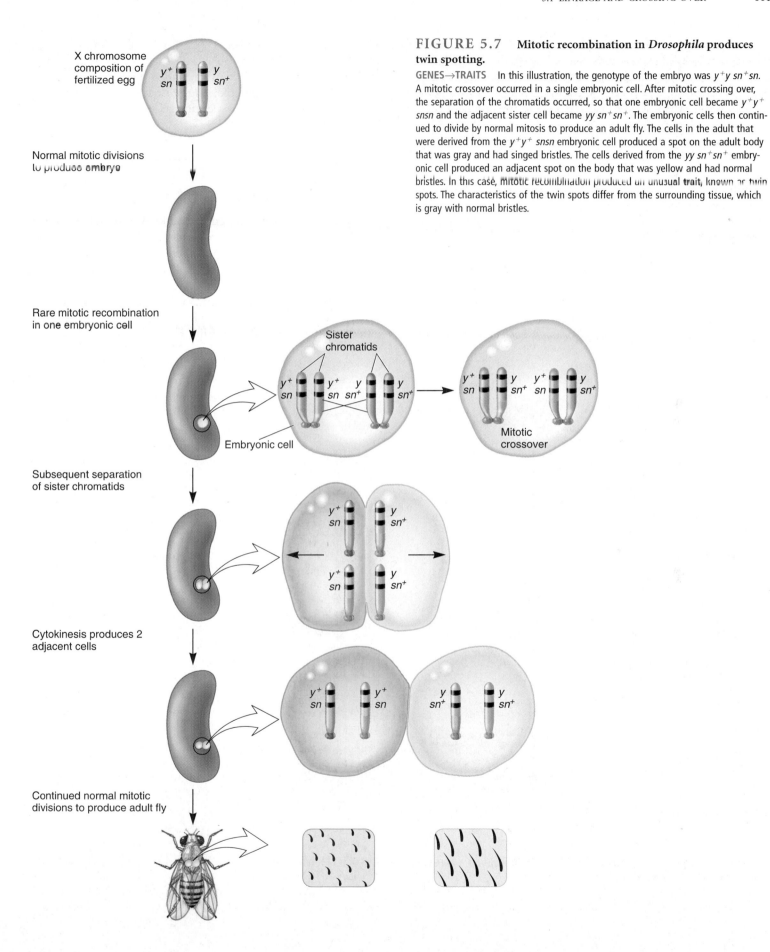

FIGURE 5.7   **Mitotic recombination in *Drosophila* produces twin spotting.**

GENES→TRAITS   In this illustration, the genotype of the embryo was y^+y sn^+sn. A mitotic crossover occurred in a single embryonic cell. After mitotic crossing over, the separation of the chromatids occurred, so that one embryonic cell became y^+y^+ $snsn$ and the adjacent sister cell became yy sn^+sn^+. The embryonic cells then continued to divide by normal mitosis to produce an adult fly. The cells in the adult that were derived from the y^+y^+ $snsn$ embryonic cell produced a spot on the adult body that was gray and had singed bristles. The cells derived from the yy sn^+sn^+ embryonic cell produced an adjacent spot on the body that was yellow and had normal bristles. In this case, mitotic recombination produced an unusual trait, known as twin spots. The characteristics of the twin spots differ from the surrounding tissue, which is gray with normal bristles.

X chromosome composition of fertilized egg

y^+ sn y sn^+

Normal mitotic divisions to produce embryo

Rare mitotic recombination in one embryonic cell

Sister chromatids

y^+ sn y^+ sn y sn^+ y sn^+

Embryonic cell

y^+ sn y sn^+ y^+ sn y sn^+

Mitotic crossover

Subsequent separation of sister chromatids

y^+ sn y sn^+

y^+ sn y sn^+

Cytokinesis produces 2 adjacent cells

y^+ sn y^+ sn y sn^+ y sn^+

Continued normal mitotic divisions to produce adult fly

allele confers shorter body bristles that look singed. The corresponding wild-type alleles confer gray body color (y^+) and normal bristles (sn^+). Females that are $y^+y\ sn^+sn$ are expected to have gray body color and normal bristles. This was generally the case. However, when Stern carefully observed the bodies of these female flies under a low-power microscope, he occasionally noticed places in which two adjacent regions were different from the rest of the body. This is called a twin spot. He concluded that twin spotting was too frequent to be explained by the random positioning of two independent single spots that happened to occur close together. Stern proposed that twin spots are due to a single mitotic recombination within one cell during embryonic development.

As shown in figure 5.7, the X chromosomes of the female fly are y^+sn and $y\ sn^+$. Rarely, though, a crossover can occur during mitosis to produce two adjacent daughter cells that are y^+y^+ $snsn$ and $yy\ sn^+sn^+$. As embryonic development proceeds, the cell on the left will continue to divide to produce many cells, eventually producing a patch on the body that has gray color with singed bristles. The daughter cell next to it will produce a patch of yellow body color with normal bristles. These two adjacent patches (a twin spot) will be surrounded by cells that are y^+y sn^+sn and, thus, have gray color and normal bristles. These infrequent twin spots provide evidence that mitotic recombination occasionally occurs.

5.2 GENETIC MAPPING IN PLANTS AND ANIMALS

The purpose of **genetic mapping** (also known as gene mapping or chromosome mapping) is to determine the linear order and distance of separation among genes that are linked to each other along the same chromosome. Figure 5.8 illustrates a simplified genetic map of *Drosophila melanogaster* depicting the locations of many different genes along the individual chromosomes. As shown, each gene has its own unique **locus** at a particular site within a chromosome. For example, the gene designated *vestigial*

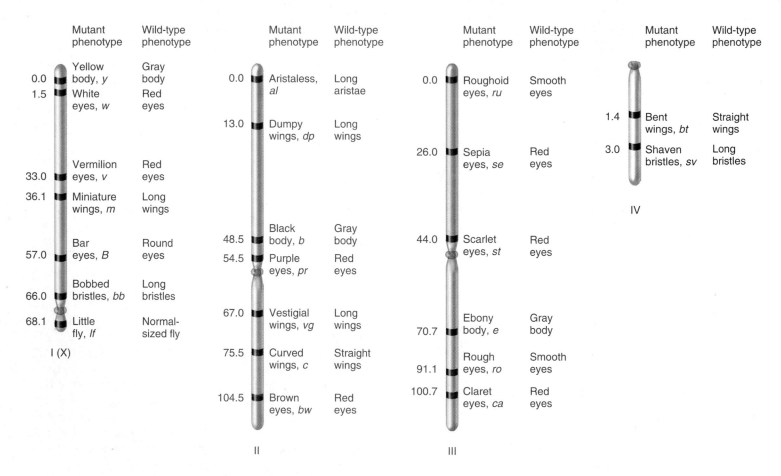

FIGURE 5.8   A simplified genetic map of *Drosophila melanogaster*. This simplified map illustrates a few of the many hundreds of genes that have been identified in this organism.

wing (*vg*), which affects wing length, is located on chromosome 2. The gene designated *black body* (*b*), which affects body color, is found a moderate distance away on the same chromosome.

Even though it is an enormous amount of work, the construction of a genetic map is useful in many ways. First, it allows geneticists to understand the overall complexity and genetic organization of a particular species. The genetic map of a species portrays the underlying basis for the inherited traits that an organism displays. In some cases, the known locus of a gene within a genetic map can help molecular geneticists to clone that gene and thereby obtain greater information about its molecular features. In addition, genetic maps are useful from an evolutionary point of view. A comparison of the genetic maps among different species can improve our understanding of the evolutionary relationships among these species.

Along with these scientific uses, genetic maps have many practical benefits. For example, many human genes that play a role in human disease have been genetically mapped. This information can be used to diagnose and perhaps someday to treat inherited human diseases. It can also help genetic counselors predict the likelihood that a couple will produce children with certain inherited diseases. In addition, genetic maps are gaining increasing importance in agriculture. A genetic map can provide plant and animal breeders with helpful information for improving agriculturally important strains through selective breeding programs.

In this section, we will examine traditional genetic mapping techniques that involve an analysis of crosses of individuals that are heterozygous for two or more genes. The frequency of nonparental offspring provides a way to deduce the linear order of genes along a chromosome. As depicted in figure 5.8, this linear arrangement of genes is shown in an illustration known as a **genetic linkage map.** This approach has been useful for analyzing organisms that are easily crossed and produce a large number of offspring in a short period of time. It has been used to successfully map the genes of several plant species and certain species of animals, such as *Drosophila.* For many organisms, however, traditional mapping approaches are difficult due to long generation times or the inability to carry out crosses (e.g., humans). Fortunately, many alternative methods of gene mapping have been developed to replace the need to carry out crosses. As described in chapter 20, cytological and molecular approaches are now used to map genes.

The Frequency of Recombination Between Two Genes Can Be Correlated with Their Map Distance Along a Chromosome

Genetic mapping allows us to estimate the relative distances between linked genes, based on the likelihood that a crossover will occur between them. If two genes are very close together on the same chromosome, a crossover is unlikely to begin in the region between them. However, if two genes are very far apart, a crossover is more likely to be initiated in this region and thereby recombine the alleles of the two genes. Experimentally, the basis for genetic mapping is that the percentage of recombinant offspring is correlated with the distance between two genes. If two genes are far apart, many recombinant offspring will be produced. However, if two genes are close together, very few recombinant offspring will be observed.

To interpret a genetic mapping experiment, the experimenter must know if the characteristics of an offspring are due to crossing over during meiosis in a parent. This is accomplished by conducting a **testcross.** Most testcrosses are between an individual that is heterozygous for two or more genes and an individual that is recessive and homozygous for these same genes. The goal of the testcross is to determine if recombination has occurred during meiosis in the heterozygous parent. New combinations of alleles cannot occur in the other parent, which is homozygous for these genes.

Figure 5.9 illustrates how a testcross provides an experimental strategy to distinguish between recombinant and nonrecombinant offspring. This cross concerns two linked genes affecting bristle length and body color in fruit flies. The recessive alleles are *s* (short bristles) and *e* (ebony body), and the dominant (wild-type) alleles are s^+ (normal bristles) and e^+ (gray body). One parent displays both recessive traits. Therefore, we know this parent is homozygous for the recessive alleles of the two genes (i.e., *ssee*). The other parent is heterozygous for the linked genes affecting bristle length and body color. This parent was produced from a cross involving a true-breeding wild-type fly and a true-breeding fly with short bristles and an ebony body. Therefore, in this heterozygous parent, we know that the *s* and *e* alleles are located on one chromosome and the corresponding s^+ and e^+ alleles are located on the homologous chromosome.

Now let's take a look at the four possible types of offspring these parents can produce. The offspring's phenotypes are normal bristles, gray body; short bristles, ebony body; short bristles, gray body; and normal bristles, ebony body. All four types of offspring have inherited a chromosome carrying the *s* and *e* alleles from their homozygous parent (shown on the *right* in each pair). Focus your attention on the other chromosome. The offspring with short bristles and ebony bodies have also inherited a second chromosome carrying the *s* and *e* alleles from their other parent. This chromosome is not the product of a crossover in the heterozygous parent. The offspring with normal bristles and gray bodies have inherited a chromosome carrying the s^+ and e^+ alleles from the heterozygous parent. Again, this chromosome is not the product of a crossover.

The other two types of offspring, however, can be produced only if crossing over has occurred in the region between these two genes. Those with normal bristles and ebony bodies or short bristles and gray bodies have inherited a chromosome that is the product of a crossover during meiosis in the heterozygous parent. As noted in figure 5.9, the recombinant offspring are fewer in number than are the nonrecombinant offspring.

The data shown at the bottom of figure 5.9 can be used to estimate the distance between the two genes. The **map distance** is

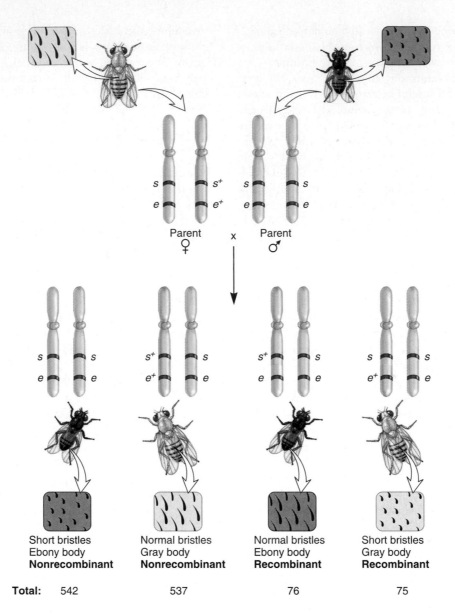

FIGURE 5.9   An example of a testcross. The cross involves one *Drosophila* parent that is homozygous recessive for short bristles (*ss*) and ebony body (*ee*), and one parent heterozygous for these recessive traits (*s⁺s e⁺e*). (Note: *Drosophila* geneticists normally designate the short allele as *ss* and a homozygous fly with short bristles as *ssss*. In the text, the allele causing short bristles is designated with a single *s* to avoid confusion between the allele designation and the genotype of the fly.)

defined as the number of recombinant offspring divided by the total number of offspring, multiplied by 100. We can calculate the map distance between the *s* and *e* alleles using this formula:

$$\text{Map distance} = \frac{\text{Number of recombinant offspring}}{\text{Total number of offspring}} \times 100$$

$$= \frac{76 + 75}{542 + 537 + 76 + 75} \times 100$$

$$= 12.3 \text{ map units}$$

The units of distance are called **map units** (mu) or sometimes **centiMorgans (cM)** in honor of Thomas Hunt Morgan. One map unit is equivalent to a 1% recombination frequency. In this example, we would say that the *s* and *e* alleles are 12.3 map units apart from each other along the same chromosome.

Alfred Sturtevant Used the Frequency of Crossing Over in Dihybrid Crosses to Produce the First Genetic Map

In 1911, the first individual to construct a (very small) genetic map was Alfred Sturtevant, an undergraduate who spent time in the laboratory of Thomas Hunt Morgan. Sturtevant wrote: "In conversation with Morgan . . . I suddenly realized that the variations in the strength of linkage, already attributed by Morgan to differences in the spatial separation of the genes, offered the possibility of determining sequences [of different genes] in the linear dimension of a chromosome. I went home and spent most of the night (to the neglect of my undergraduate homework) in producing the first chromosome map, which included the sex-linked genes, *y, w, v, m,* and *r,* in the order and approximately the relative spacing that they still appear on the standard maps."

In the experiment of figure 5.10, Sturtevant considered the outcome of crosses involving six different mutant alleles that altered the phenotype of normal flies. All of these alleles were known to be recessive and X linked. They are *y* (yellow body color), *w* (white eye color), *w-e* (eosin eye color), *v* (vermilion eye color), *m* (miniature wings), and *r* (rudimentary wings). The *w* and *w-e* alleles are alleles of the same gene. In contrast, the *v* allele (vermilion eye color) is an allele of a different gene that also affects eye color. The two alleles that affect wing length, *m* and *r*, are also in different genes. Therefore, Sturtevant studied the inheritance of six recessive alleles, but since *w* and *w-e* are alleles of the same gene, his genetic map contained only five genes. The corresponding wild-type alleles are y^+ (gray body), w^+ (red eyes), v^+ (red eyes), m^+ (normal wings), and r^+ (normal wings).

■ THE HYPOTHESIS

When genes are located on the same chromosome, the distance between the genes can be estimated from the proportion of recombinant offspring. This provides a way to map the order of genes along a chromosome.

■ TESTING THE HYPOTHESIS — FIGURE 5.10   **The first genetic mapping experiment.**

Starting materials: Sturtevant began with several different strains of *Drosophilia* that contained the six alleles already described.

Experimental level	Conceptual level

1. Cross a female that is heterozygous for two different genes to a male that is hemizygous recessive for the same two genes. In this example, cross a female that is $X^{y^+w^+}X^{yw}$ to a male that is $X^{yw}Y$.

   This strategy was employed for many dihybrid combinations of the six alleles already described.

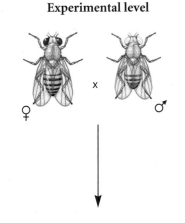

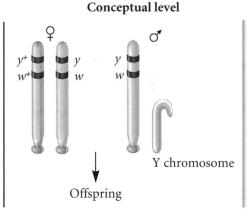

Offspring

Y chromosome

(continued)

2. Observe the outcome of the crosses.

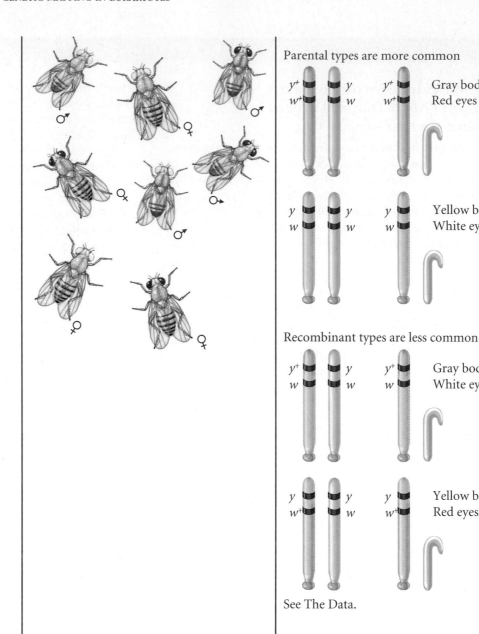

Parental types are more common

y^+ y y^+ Gray bodies
w^+ w w^+ Red eyes

y y y Yellow bodies
w w w White eyes

Recombinant types are less common

y^+ y y^+ Gray bodies
w w w White eyes

y y y Yellow bodies
w^+ w w^+ Red eyes

See The Data.

3. Calculate the percentages of offspring that are the result of crossing over (number of nonparental/total).

■ THE DATA

Alleles Concerned	Number Recombinant/Total Number	Percent Recombinant Offspring
y and w/w-e	214/21,736	1.0
y and v	1,464/4,551	32.2
y and r	115/324	35.5
y and m	260/693	37.5
w/w-e and v	471/1,584	29.7
w/w-e and r	2,062/6,116	33.7
w/w-e and m	406/898	45.2
v and r	17/573	3.0
v and m	109/405	26.9

■ INTERPRETING THE DATA

As shown in figure 5.10, Sturtevant made pairwise testcrosses and then counted the number of offspring in the four phenotypic categories. Two of the categories would be recombinant (i.e., requiring a crossover between the X chromosomes in the female heterozygote) and two would be nonrecombinant. Let's begin by contrasting the results between particular pairs of genes. In some dihybrid crosses, the percentage of nonparental offspring was rather low. For example, dihybrid crosses involving the y allele and the w or w-e allele yielded 1% recombinant offspring. This result suggested that these two genes are very close together. By comparison, other dihybrid crosses showed a higher percentage of nonparental offspring. For example, crosses involving the v

and m alleles produced 26.9% recombinant offspring. These two genes are expected to be farther apart.

To construct his map, Sturtevant began with the assumption that the map distances would be more accurate between genes that are closely linked. Therefore, his map is based on the distance between y and w (1.0), w and v (29.7), v and r (3.0), and v and m (26.9). He also considered other features of the data to deduce the order of the genes. For example, the percentage of crossovers between w and r was 33.7. The percentage of crossovers between w and v was only 29.7, suggesting that v is between w and r, but closer to r. The proximity of v and r is confirmed by the low percentage of crossovers between v and r (3.0). Sturtevant collectively considered all these data and proposed the genetic map shown here.

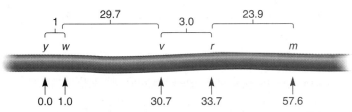

In this genetic map, Sturtevant began at the y allele and mapped the genes from left to right. For example, the y and v alleles are 30.7 mu apart, and the v and r alleles are 3.0 mu apart. This study by Sturtevant was a major breakthrough, since it showed how to map the locations of genes along chromosomes by making the appropriate crosses.

If you look carefully at Sturtevant's data, you will notice that there were two observations that do not agree very well with his genetic map. The percentage of recombinant offspring for the y and m dihybrid cross was 37.5 (but the map distance is 57.6), and the crossover percentage between w and m was 45.2 (but the map distance is 56.6).

As the percentage of recombinant offspring approaches a value of 50%, this value becomes a progressively more inaccurate measure of map distance (fig. 5.11). When the distance between two genes is large, the likelihood of multiple crossovers in the region between them causes the observed number of recombinant offspring to underestimate this distance. In addition, multiple crossovers set a quantitative limit on the relationship between map distance and the percentage of recombinant offspring. Even though two different genes can be on the same chromosome and more than 50 mu apart, a testcross is expected to yield a maximum of only 50% recombinant offspring. This idea is also discussed in solved problem S5 at the end of chapter 5.

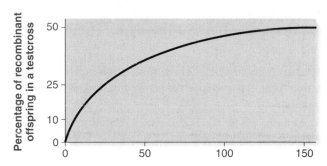

FIGURE 5.11   **Relationship between the percentage of recombinant offspring in a testcross and the actual map distance between genes.** The y-axis depicts the percentage of recombinant offspring that would be observed in a dihybrid testcross. The actual map distance shown on the x-axis is calculated by analyzing the percentages of recombinant offspring from a series of many dihybrid crosses involving closely linked genes.

A self-help quiz involving this experiment can be found at the Online Learning Center.

Trihybrid Crosses Can Be Used to Determine the Order and Distance Between Linked Genes

Until now, we have been considering the construction of genetic maps using dihybrid testcrosses to compute map distance. The data from trihybrid crosses can also yield information about map distance and gene order. In a trihybrid cross, the experimenter crosses two individuals that differ in three traits. The following experiment outlines a common strategy for using trihybrid crosses to map genes. In this experiment, the parental generation consists of fruit flies that differ in body color, eye color, and wing shape. We must begin with true-breeding lines so that we know which alleles are initially linked to each other on the same chromosome. In this example, all the dominant alleles are linked on the same chromosome.

Step 1. *Cross two true-breeding strains that differ with regard to three alleles.* In this example, we will cross a fly that has a black body (bb), purple eyes ($prpr$), and vestigial wings

($vgvg$) to a homozygous wild-type fly (b^+b^+ pr^+pr^+ vg^+vg^+):

Parental flies

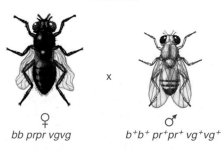

♀ *bb prpr vgvg*   ♂ *b⁺b⁺ pr⁺pr⁺ vg⁺vg⁺*

The goal in this step is to obtain F_1 individuals that are heterozygous for all three alleles. In the F_1 heterozygotes, all the dominant alleles are located on one chromosome and all the recessive alleles on the other homologous chromosome.

Step 2. *Perform a testcross by mating F_1 female heterozygotes to male flies that are homozygous recessive for all three alleles* (bb prpr vgvg).

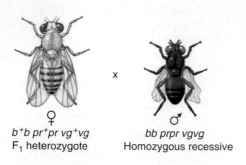

x

$b^+b\ pr^+pr\ vg^+vg$
F_1 heterozygote

$bb\ prpr\ vgvg$
Homozygous recessive

During gametogenesis in the heterozygous female F_1 flies, crossovers may produce new combinations of the three alleles.

Step 3. *Collect data for the F_2 generation.* There are eight possible phenotypic combinations.

Phenotype	Number of Observed Offspring (males and females)
Gray body, red eyes, normal wings	411
Gray body, red eyes, vestigial wings	61
Gray body, purple eyes, normal wings	2
Gray body, purple eyes, vestigial wings	30
Black body, red eyes, normal wings	28
Black body, red eyes, vestigial wings	1
Black body, purple eyes, normal wings	60
Black body, purple eyes, vestigial wings	412

Analysis of the F_2 generation flies will allow us to map these three genes. Since the three genes exist as two alleles each, there are $2^3 = 8$ possible combinations of offspring. If these alleles assorted independently, all eight combinations would occur in equal proportions. However, we see that the proportions of the eight phenotypes are far from equal.

The genotypes of the parental generation correspond to the phenotypes gray body, red eyes, and normal wings and black body, purple eyes, and vestigial wings. In crosses involving linked genes, the parental phenotypes occur most frequently in the offspring. The remaining six phenotypes are due to crossing over.

The double crossover is always expected to be the least frequent category of offspring. Two of the phenotypes (namely, gray body, purple eyes, normal wings and black body, red eyes, vestigial wings) arise from a double crossover between two pairs of genes. Also, the combination of traits in the double crossover tells us which gene is in the middle. When a chromatid undergoes a double crossover, the gene in the middle becomes separated from the other two genes at either end.

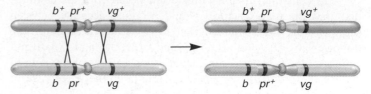

In the double crossover categories, the recessive purple-eye allele is separated from the other two recessive alleles. When mated to a homozygous recessive fly in the testcross, this yields flies with gray bodies, purple eyes, and normal wings, or with black bodies, red eyes, and vestigial wings. This observation indicates that the gene for eye color lies between the genes for body color and wing shape.

Step 4. *Calculate the map distance between pairs of genes.* To do this, one strategy is to regroup the data according to pairs of genes. From the parental generation, we know that the dominant alleles are initially linked to each other, as are the recessive alleles. This allows us to group pairs of genes into parental and nonparental combinations. The parental combinations are composed of a pair of dominant or a pair of recessive genes, whereas nonparental combinations have one dominant and one recessive gene. After we have regrouped the data in this way, the map distance between two genes can be calculated.

Parental Offspring	Total	Nonparental Offspring	Total
Gray body, red eyes (411 + 61)	472	Gray body, purple eyes (30 + 2)	32
Black body, purple eyes (412 + 60)	472	Black body, red eyes (28 + 1)	29
	944		61
Gray body, normal wings (411 + 2)	413	Gray body, vestigial wings (30 + 61)	91
Black body, vestigial wings (412 + 1)	413	Black body, normal wings (28 + 60)	88
	826		179
Red eyes, normal wings (411 + 28)	439	Red eyes, vestigial wings (61 + 1)	62
Purple eyes, vestigial wings (412 + 30)	442	Purple eyes, normal wings (60 + 2)	62
	881		124

The map distance between body color and eye color is

$$\text{Map distance} = \frac{61}{944 + 61} \times 100 = 6.1 \text{ mu}$$

The map distance between body color and wing shape is

$$\text{Map distance} = \frac{179}{826 + 179} \times 100 = 17.8 \text{ mu}$$

The map distance between eye color and wing shape is

$$\text{Map distance} = \frac{124}{881 + 124} \times 100 = 12.3 \text{ mu}$$

Step 5. *Construct the map.* Based on the map unit calculation, the body color and wing shape genes are farthest apart. The eye color gene must lie in the middle. As mentioned earlier, this order of genes is also confirmed by the pattern of traits found in the double crossovers. To construct the map, we use the distances between the genes that are closest together.

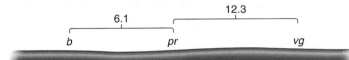

In our example, we have placed the body color gene first and the wing shape gene last. The data also are consistent with a map in which the wing shape gene comes first and the body color gene comes last. In detailed genetic maps, the locations of genes are mapped relative to the centromere.

The method just described for calculating map distance between three genes involved the separation of the data into pairs of genes (see step 4). An alternative way to calculate map distance does not require this manipulation. As shown here, we can first calculate the percentage of single and double crossovers using the trihybrid data directly.

Phenotype	Number of Observed Offspring		
Gray body, purple eyes, vestigial wings	30	Single crossover between *b* and *pr*	$\frac{30 + 28}{1,005} = 0.058$
Black body, red eyes, normal wings	28		
Gray body, red eyes, vestigial wings	61	Single crossover between *pr* and *vg*	$\frac{61 + 60}{1,005} = 0.120$
Black body, purple eyes, normal wings	60		
Gray body, purple eyes, normal wings	2	Double crossover, between *b* and *pr*, and between *pr* and *vg*	$\frac{2 + 1}{1,005} = 0.003$
Black body, red eyes, vestigial wings	1		

To determine the map distance between the genes, we need to consider both single and double crossovers.

To calculate the distance between *b* and *pr*:

$$\text{Map distance} = (0.058 + 0.003) \times 100 = 6.1 \text{ mu}$$

To calculate the distance between *pr* and *vg*:

$$\text{Map distance} = (0.120 + 0.003) \times 100 = 12.3 \text{ mu}$$

As you can see, we get the same answer as we did when we separated the data into gene pairs. However, the method shown here is a bit simpler, because we did not have to regroup the data. We could also calculate the distance between *b* and *vg*. In this case, however, we would need to multiply the double crossover frequency by two, because both crossovers can occur between *b* and *vg*, and we also need to include the single crossovers between *b* and *pr*, and between *pr* and *vg*. To calculate the map distance between *b* and *vg*:

$$\text{Map distance} = (0.058 + 0.120 + 2[0.003]) \times 100 = 18.4 \text{ mu}$$

We would get the same value if we simply added together the map distances between *b* and *pr*, and between *pr* and *vg* (i.e., 6.1 + 12.3). In our previous method of grouping genes into pairs, the distance between *b* and *vg* was found to be a slightly lower value of 17.8 mu. This value was a small underestimate because the first method did not consider the double crossovers in the calculation.

Interference Can Influence the Number of Double Crossovers That Occur in a Short Region

In chapter 2, we considered the product rule to determine the probability that two independent events will both occur. The product rule allows us to predict the expected likelihood of a double crossover provided we know the individual probabilities of each single crossover. Let's reconsider the data of the trihybrid testcross just described to see if the frequency of double crossovers is what we would expect based on the product rule. If we multiply the likelihood of a single crossover between *b* and *pr* (0.061) times the likelihood of a single crossover between *pr* and *vg* (0.123), then the product rule predicts

$$\text{Expected likelihood of a double crossover} = 0.061 \times 0.123$$
$$= 0.0075 = 0.75\%$$

Based on a total of 1,005 offspring produced,

Expected number of offspring due to a double crossover
$$= 1,005 \times 0.0075 = 7.5$$

In other words, we would expect about seven or eight offspring to be produced as a result of a double crossover. The observed number of offspring was only three (namely, two with gray bodies, purple eyes, and normal wings, and one with a black body, red eyes, and vestigial wings). This lower-than-expected value is probably not due to random sampling error. Instead, the likely cause is a common genetic phenomenon known as **positive interference.** When a crossover occurs in one region of a chromosome, it often decreases the probability that another crossover will occur nearby. In other words, the first crossover interferes with the ability to form a second crossover in the immediate vicinity. To provide interference with a quantitative value, we first calculate the coefficient of coincidence (C).

$$C = \frac{\text{Observed number of double crossovers}}{\text{Expected number of double crossovers}}$$

Interference (I) is expressed as

$$I = 1 - C$$

For the data of the trihybrid testcross, the observed number of crossovers is 3 and the expected number is 7.5, and so the coefficient of coincidence equals 3/7.5 = 0.40. In other words, only 40% of the expected number of double crossovers were actually observed. The value for interference equals $1 - 0.4 = 0.60$, or 60%. This means that 60% of the expected number of crossovers did not occur. Since I has a positive value, this is positive interference. Rarely, the outcome of a testcross yields a negative value for interference. A negative interference value suggests that a first crossover enhances the rate of a second crossover in a nearby region. Although the molecular mechanisms that cause interference are not entirely understood, most organisms regulate the number of crossovers so that very few occur per chromosome. The reasons for positive and negative interference will require further research.

5.3 GENETIC MAPPING IN HAPLOID EUKARYOTES

Before ending our discussion of genetic mapping, it is interesting to consider some pioneering studies that involved the genetic mapping of haploid organisms. It may be surprising to you that certain species of lower eukaryotes, particularly unicellular algae and fungi, which spend part of their life cycle in the haploid state, have also been used in mapping studies. The sac fungi (ascomycetes) have been particularly useful to geneticists because of their unique style of sexual reproduction. In fact, much of our earliest understanding of genetic recombination came from the genetic analyses of fungi.

Fungi may be unicellular or multicellular organisms. Fungal cells are typically haploid ($1n$) and can reproduce asexually. In addition, fungi can also reproduce sexually by the fusion of two haploid cells to create a diploid zygote ($2n$) (fig. 5.12). The diploid zygote can then proceed through meiosis to produce four haploid cells, which are called **spores.** This group of four spores is known as a **tetrad** (not to be confused with a tetrad of four sister chromatids). In some species, meiosis is followed by a mitotic division to produce eight cells, known as an **octad.** In *Ascomycete*

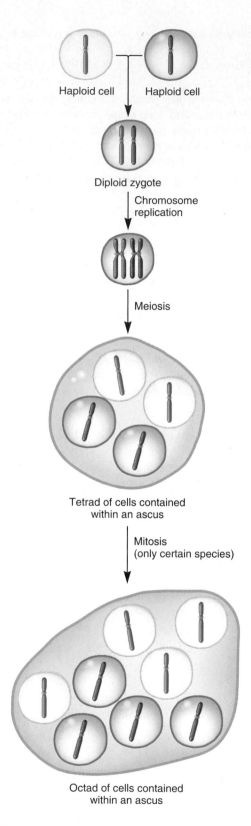

FIGURE 5.12   **Sexual reproduction in ascomycetes.** For simplicity, this diagram shows each haploid cell as having only one chromosome per haploid set. However, fungal species actually contain several chromosomes per haploid set.

fungi, the cells of a tetrad or octad are contained within a sac known as an **ascus** (plural: *asci*). In other words, the products of a single meiotic division are contained within one sac. This is a key feature that is useful to geneticists, and it dramatically differs from sexual reproduction in animals and plants. For example, in animals, oogenesis produces a single functional egg, and spermatogenesis occurs in the testes, where the resulting sperm become mixed with millions of other sperm.

Using a microscope, researchers can dissect asci and study the traits of each haploid spore. In this way, these organisms offer a unique opportunity for geneticists to identify and study all the cells that are derived from a single meiotic division. In this section, we will consider how the analysis of asci can be used to map genes in fungi.

Ordered Tetrad Analysis Can Be Used to Map the Distance Between a Gene and the Centromere

The arrangement of spores within an ascus varies from species to species (fig. 5.13a). In some cases, the ascus provides enough

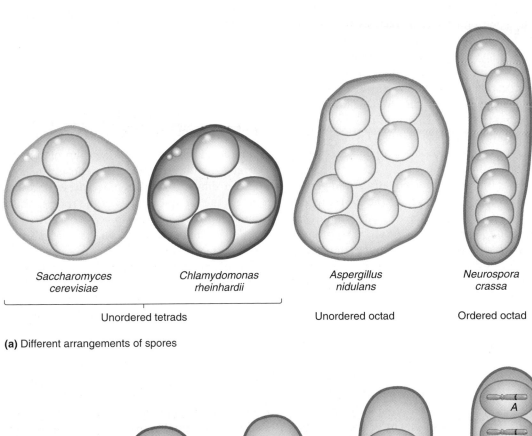

(a) Different arrangements of spores

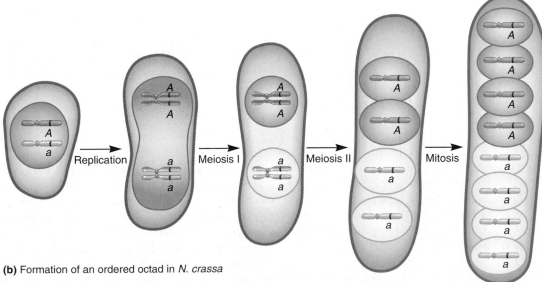

(b) Formation of an ordered octad in *N. crassa*

FIGURE 5.13 **Arrangement of spores within asci of different species. (a)** *Saccharomyces cerevisiae* and *Chlamydomonas rheinhardii* (an alga) produce unordered tetrads, *Aspergillus nidulans* produces an unordered octad, and *Neurospora crassa* produces an ordered octad. **(b)** Ordered octads are produced in *Neurospora* by meiosis and mitosis in such a way that the eight resulting cells are arranged linearly.

space for the tetrads or octads of spores to randomly mix together. This is known as an **unordered tetrad** or **octad.** These occur in fungal species such as *Saccharomyces cerevisiae* and *Aspergillus nidulans* and also in certain unicellular algae (*Chlamydomonas rheinhardii*). By comparison, other species of fungi produce a very tight ascus that prevents spores from randomly moving around. This creates an **ordered tetrad** or **octad.** Figure 5.13*b* illustrates how an ordered octad is formed in *Neurospora crassa*. In this example, spores that carry the *A* allele have orange pigmentation, while spores having the *a* (albino) allele are white.

A key feature of ordered tetrads or octads is that the position and order of spores within the ascus reflects their relationship to each other as they were produced by meiosis and mitosis. This idea is schematically shown in figure 5.13*b*. After the original diploid cell has undergone chromosome replication, the first meiotic division produces two cells that are arranged next to each other within the sac. The second meiotic division then produces four cells that are also arranged in a straight row. Due to the tight enclosure of the sac around the cells, each pair of daughter cells is forced to lie next to each other in a linear fashion. Likewise, when each of these four cells divides by mitosis, each pair of daughter cells is located next to each other.

In the case of species that make ordered tetrads and octads, experimenters can determine the genotypes of the spores within the asci and map the distance between a single gene and the centromere. Since the location of the centromere can be seen under the microscope, the mapping of a gene relative to the centromere provides a way to correlate a gene's location with the cytological characteristics of a chromosome. This approach has been extensively exploited in *N. crassa*.

Figure 5.14 compares the arrangement of cells within a *Neurospora* ascus depending on whether or not a crossover has occurred between two homologues that differ at a gene with alleles *A* (orange pigmentation) and *a* (albino, which results in a white phenotype). In figure 5.14*a*, a crossover has not occurred, and so the octad contains a linear arrangement of four haploid cells carrying the *A* allele, which are adjacent to four haploid cells that contain the *a* allele. This 4:4 arrangement of spores within the ascus is called a first-division segregation (FDS) pattern or an M1 pattern. It is called a first-division segregation pattern because the *A* and *a* alleles have segregated from each other after the first meiotic division.

In contrast, as shown in figure 5.14*b*, if a crossover occurs between the centromere and the gene of interest, the ordered octad will deviate from the 4:4 pattern. Depending on the relative locations of the two chromatids that participated in the crossover, the ascus will contain a 2:2:2:2 or 2:4:2 pattern. These are called second-division segregation (SDS) patterns or M2 patterns. In this case, the *A* and *a* alleles do not segregate until the second meiotic division is completed.

Since a pattern of second-division segregation is a result of crossing over, the percentage of M2 asci can be used to calculate the map distance between the centromere and the gene of inter-

est. To understand why this is possible, let's consider the relationship between a crossover site and the centromere. As shown in figure 5.15, a crossover will separate a gene from its original centromere only if it begins in the region between the centromere and that gene. Therefore, the chances of getting a 2:2:2:2 or 2:4:2 pattern depend on the distance between the gene of interest and the centromere.

To determine the map distance between the centromere and a gene, the experimenter must count the number of SDS asci and the total number of asci. In SDS asci, only half of the spores are actually the product of a crossover. Therefore, the map distance is calculated as

$$\text{Map distance} = \frac{(1/2)(\text{Number of SDS asci})}{\text{Total number of asci}} \times 100$$

Unordered Tetrad Analysis Can Be Used to Map Genes in Dihybrid Crosses

Unordered tetrads contain a group of spores that are randomly arranged and the product of meiosis. An experimenter can conduct a dihybrid cross, remove the spores from each ascus, and determine the phenotypes of the spores. This analysis can determine if two genes are linked or assort independently. If two genes are linked, a tetrad analysis can also be used to compute map distance.

Figure 5.16 illustrates the possible outcomes starting with two haploid cells that create a diploid yeast zygote that has the genotype *ura⁺ura-2 arg⁺arg-3*. *Ura⁺* and *arg⁺* are normal alleles required for uracil and arginine biosynthesis, respectively. *Ura-2* and *arg-3* are defective alleles that result in yeast strains that require uracil and arginine in the growth medium. This diploid cell was produced from the fusion of two haploid cells that were *ura⁺arg⁺* and *ura-2 arg-3*. After the diploid cell has completed meiosis, there are three distinct possible combinations of four haploid cells. One possibility is that the tetrad will contain four spores with the parental combinations of alleles. This ascus is said to have the parental ditype (PD). Alternatively, an ascus with a nonparental ditype (NPD) contains four cells with nonparental genotypes. Finally, it is possible to have an ascus that has two parental cells and two nonparental cells. This is called a tetratype (T).

When two genes assort independently, the number of asci having a parental ditype is expected to equal the number having a nonparental ditype, thus yielding 50% recombinant spores. For linked genes, figure 5.17 illustrates the relationship between crossing over and the type of ascus that will result. If no crossing over occurs in the region between the two genes, the parental ditype will be created. A single crossover event will produce a tetratype. Double crossovers can yield a parental ditype, tetratype, or nonparental ditype depending on the combination of chromatids that are involved. A nonparental ditype is produced when a double crossover involves all four chromatids. A tetratype

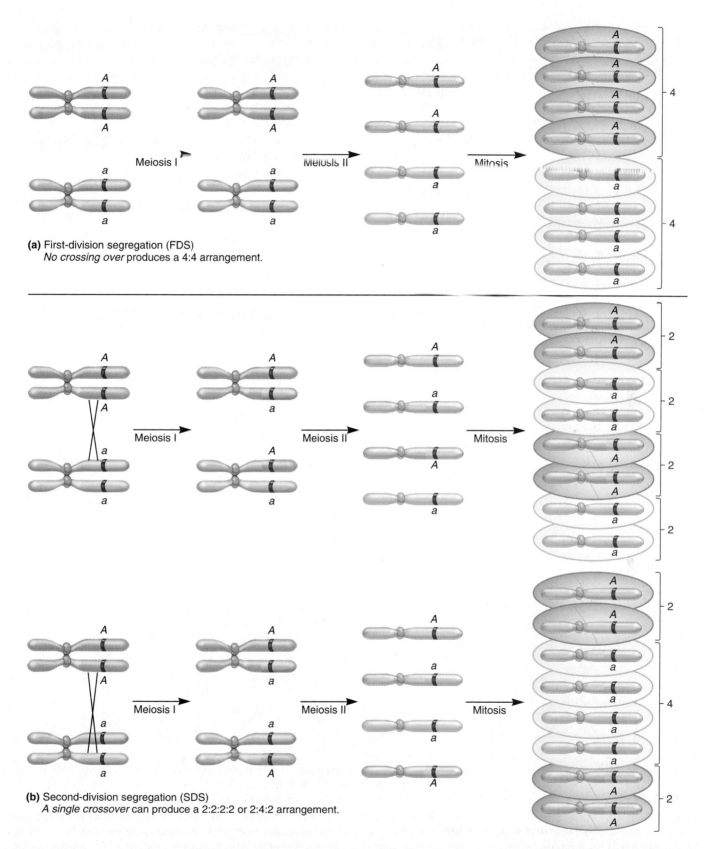

(a) First-division segregation (FDS)
   No crossing over produces a 4:4 arrangement.

(b) Second-division segregation (SDS)
   A single crossover can produce a 2:2:2:2 or 2:4:2 arrangement.

FIGURE 5.14   **A comparison of the arrangement of cells within an ordered octad, depending on whether or not crossing over has occurred.**
(a) If no crossing over has occurred, the octad will have a 4:4 arrangement of spores known as an FDS or M1 pattern. (b) If a crossover has occurred between the centromere and the gene of interest, a 2:2:2:2 or 2:4:2 pattern, known as an SDS or M2 pattern, is observed.

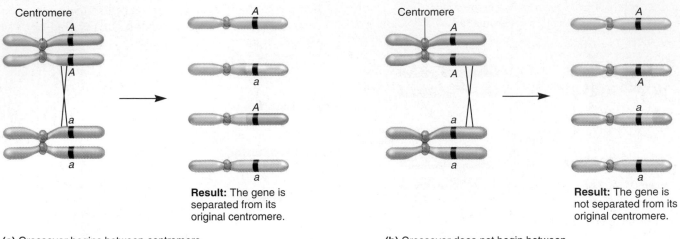

Result: The gene is separated from its original centromere.

(a) Crossover begins between centromere and gene of interest.

Result: The gene is not separated from its original centromere.

(b) Crossover does not begin between centromere and gene of interest.

FIGURE 5.15   The relationship between a crossover site and the separation of an allele from its original centromere. (a) If a crossover initially forms between the centromere and the gene of interest, the gene will be separated from its original centromere. **(b)** If a crossover initiates outside this region, the gene remains attached to its original centromere.

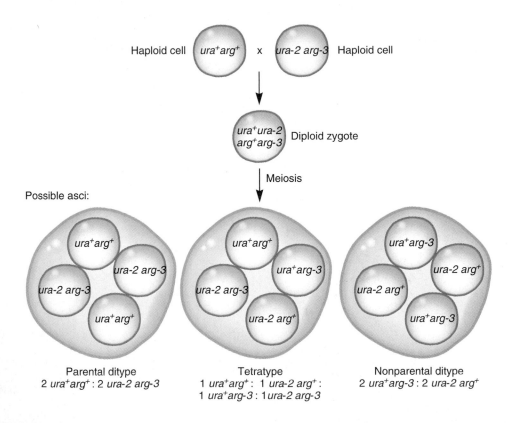

Parental ditype
2 ura⁺arg⁺ : 2 ura-2 arg-3

Tetratype
1 ura⁺arg⁺ : 1 ura-2 arg⁺ :
1 ura⁺arg-3 : 1 ura-2 arg-3

Nonparental ditype
2 ura⁺arg-3 : 2 ura-2 arg⁺

FIGURE 5.16   The assortment of two genes in an unordered tetrad. If the tetrad contains 100% parental cells, this ascus has the parental ditype. If it contains 50% parental and 50% recombinant cells, it is a tetratype. Finally, an ascus with 100% recombinant cells is called a nonparental ditype. This figure does not illustrate the chromosomal locations of the alleles. In this type of experiment, the goal is to determine whether the two genes are linked on the same chromosome.

will result from a three-chromatid crossover. Finally, a double crossover between the same two chromatids will produce the parental ditype.

The data from a tetrad analysis can be used to calculate the map distance between two linked genes. As in conventional mapping, the map distance is calculated as the percentage of offspring that carry recombinant chromosomes. As mentioned, a tetratype contains 50% recombinant chromosomes, a nonparental ditype 100%. Therefore, the map distance is computed as

$$\text{Map distance} = \frac{\text{NPD} + (1/2)(\text{T})}{\text{Total number of asci}} \times 100$$

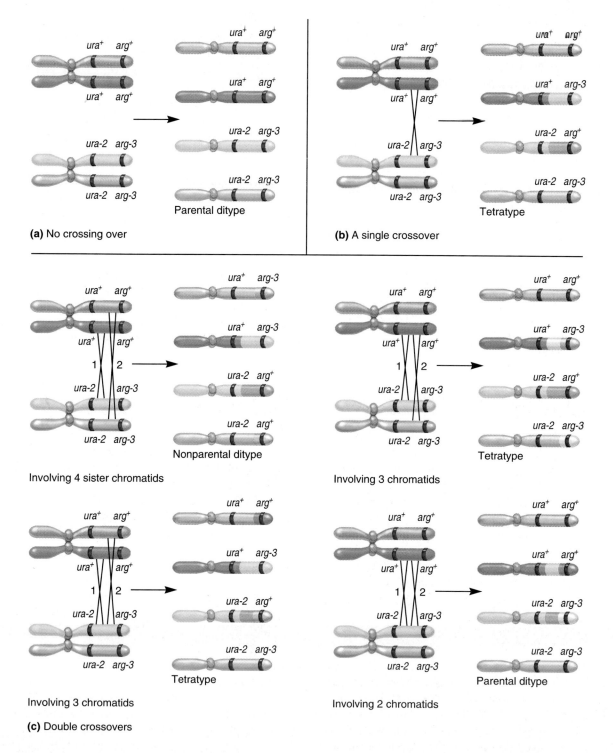

FIGURE 5.17 **Relationship between crossing over and the production of the parental ditype, tetratype, and nonparental ditype for two linked genes.** In the case of double crossovers, this figure shows the outcome in which the crossover on the left occurs first.

Over short map distances, this calculation provides a fairly reliable measure of distance. However, it does not adequately account for double crossovers. When two genes are far apart on the same chromosome, the calculated map distance using this equation underestimates the actual map distance due to double crossovers. Fortunately, a particular strength of tetrad analysis is that we can derive another equation that accounts for double crossovers and thereby provides a more accurate value for map distance. To begin this derivation, let's consider a more precise way to calculate map distance.

$$\text{Map distance} = \frac{\text{Single crossover tetrads} + (2)(\text{Double crossover tetrads})}{\text{Total number of asci}} \times 0.5 \times 100$$

This equation includes the number of single and double crossovers in the computation of map distance. The total number of crossovers equals the number of single crossovers plus two times the number of double crossovers. Overall, the tetrads that contain single and double crossovers also contain 50% nonrecombinant chromosomes. To calculate map distance, therefore, we divide the total number of crossovers by the total number of asci and multiply by 0.5.

To be useful, we need to relate this equation to the number of parental ditypes, nonparental ditypes, and tetratypes that are obtained by experimentation. To derive this relationship, we must consider the types of tetrads that are produced from no crossing over, a single crossover, and double crossovers. To do so, let's take another look at figure 5.17. As shown there, the parental ditype and tetratype are ambiguous. The parental ditype can be derived from no crossovers or a double crossover; the tetratype can be derived from a single crossover or a double crossover. However, the nonparental ditype is unambiguous, since it can only be produced from a double crossover. We can use this observation as a way to determine the actual number of single and double crossovers. As seen in figure 5.17, 1/4 of all the double crossovers are nonparental ditypes. Therefore, the total number of double crossovers equals four times the number of nonparental ditypes.

Next, we need to know the number of single crossovers. A single crossover will yield a tetratype, but double crossovers can also yield a tetratype. Therefore, the total number of tetratypes overestimates the true number of single crossovers. Fortunately, we can compensate for this overestimation. Since there are two types of tetratypes that are due to a double crossover, the actual number of tetratypes arising from a double crossover should equal 2NPD. Therefore, the true number of single crossovers is calculated as T − 2NPD.

Now we have accurate measures of both single and double crossovers. The number of single crossovers equals T − 2NPD, and the number of double crossovers equals 4NPD. We can substitute these values into our previous equation.

$$\text{Map distance} = \frac{(T - 2NPD) + (2)(4NPD)}{\text{Total number of asci}} \times 0.5 \times 100$$

$$= \frac{T + 6NPD}{\text{Total number of asci}} \times 0.5 \times 100$$

This equation provides a more accurate measure of map distance, since it considers both single and double crossovers.

CONCEPTUAL SUMMARY

Linkage refers to the phenomenon that many different genes may be located on the same chromosome. Chromosomes are sometimes called **linkage groups** because they contain a group of linked genes. Linkage affects the pattern of inheritance, because closely linked genes do not assort independently during meiosis. This produces a greater percentage of offspring that display parental phenotypes. Nevertheless, nonparental offspring can be produced as a result of **crossing over.**

The likelihood of crossing over depends on the distance between two genes. If two genes are far apart from each other on the same chromosome, it is more likely that crossing over will occur between them. Therefore, when two genes are widely separated, a substantial percentage of recombinant offspring will be obtained from a **testcross.** However, the percentage of recombinant offspring cannot exceed a value of 50%, even when two genes are more than 50 **map units** apart on the same chromosome. The relationship between the percentage of recombinant offspring and the linear distance between genes is the basis for **genetic mapping.**

EXPERIMENTAL SUMMARY

Experimentally, the phenomenon of linkage was deduced from genetic crosses. Bateson and Punnett were the first scientists to notice that certain genes do not assort independently. Morgan conducted crosses involving X-linked traits in fruit flies and correctly proposed that linkage is due to the location of particular genes on the same chromosome. He also hypothesized that recombinant phenotypes occur because of crossing over during meiosis. Morgan realized that the likelihood of crossing over depends on the distance between two genes. The proposal that genetic recombination is due to crossing over was confirmed cytologically by the studies of Creighton and McClintock, which showed that the production of recombinant offspring correlates with the physical exchange of material between chromosomes.

Genetic mapping is the determination of gene order and distance along chromosomes. In this chapter, we have considered how testcrosses are conducted as a method to map genes. Sturtevant was the first person to understand that the percentage of recombinant offspring in a testcross could be used as a measure of the relative distance between two genes. **Map distance** is computed as the number of recombinant offspring divided by the total number of offspring times 100. This approach can be readily applied to map genes using dihybrid and trihybrid testcrosses. Genetic mapping is most accurate when map distances are calculated between closely linked genes. As the map distance approaches 50 map units (mu) and above, the percentage of recombinant offspring is not a reliable measure of map distance. In chapter 20, several molecular methods of genetic mapping are described.

Chapter 5 ended with a discussion of gene mapping methods in fungi. A group of fungi known as the ascomycetes have been extensively used in genetic studies, because all the products of a single meiosis are within an ascus. For fungi such as *Neurospora* that make an **ordered octad,** the spores are arranged in a manner that reflects their relationship to each other during meiosis (and mitosis). Ordered octads can be analyzed to map the location of a single gene relative to the centromere. Fungal species that produce unordered asci have also been used in mapping studies. In this case, dihybrid crosses are made, and the distance between the two genes can be computed by determining the proportions of parental ditypes, tetratypes, and nonparental ditypes.

In chapter 6, we will consider the linkage of genes within bacterial chromosomes and bacteriophages. Although bacteria normally reproduce asexually, they still can transfer genetic material by various mechanisms. As we will see, these mechanisms also provide a way to map genes along the bacterial chromosome.

PROBLEM SETS & INSIGHTS

Solved Problems

S1. In the garden pea, orange pods (*orp*) are recessive to green pods (*Orp*), and sensitivity to pea mosaic virus (*mo*) is recessive to resistance to the virus (*Mo*). A plant with orange pods and sensitivity to the virus was crossed to a true-breeding plant with green pods and resistance to the virus. The F_1 plants were then testcrossed to plants with orange pods and sensitivity to the virus. The following results were obtained:

160	orange pods, virus sensitive
165	green pods, virus resistant
36	orange pods, virus resistant
39	green pods, virus sensitive
400	total

A. Conduct a chi square analysis to see if these genes are linked.

B. If they are linked, calculate the map distance between the two genes.

Answer:

A. Chi square analysis.

1. Our hypothesis is that the genes are not linked.

2. Calculate the predicted number of offspring based on the hypothesis. The testcross is

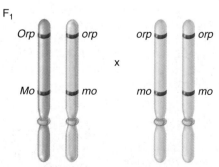

The predicted outcome of this cross under our hypothesis is a 1:1:1:1 ratio of plants with the four possible phenotypes. In other words, 1/4 should have the phenotype orange pods, virus sensitive; 1/4 should have green pods, virus resistant; 1/4 should have orange pods, virus resistant; and 1/4 should have green pods, virus sensitive. Since there were a total of 400 offspring, our hypothesis predicts 100 offspring in each category.

3. Calculate the chi square.

$$\chi^2 = \frac{(O_1 - E_1)^2}{E_1} + \frac{(O_2 - E_2)^2}{E_2} + \frac{(O_3 - E_3)^2}{E_3} + \frac{(O_4 - E_4)^2}{E_4}$$

$$\chi^2 = \frac{(160 - 100)^2}{100} + \frac{(165 - 100)^2}{100} + \frac{(36 - 100)^2}{100} + \frac{(39 - 100)^2}{100}$$

$$\chi^2 = 36 + 42.3 + 41 + 37.2 = 156.5$$

4. Interpret the chi square value. The calculated chi square value is quite large. This indicates that the deviation between observed and expected values is very high. For three degrees of freedom in table 2.1, such a large deviation is expected to occur by chance alone less than 1% of the time. Therefore, we reject the hypothesis that the genes assort independently. As an alternative, we may infer that the two genes are linked.

B. Calculate the map distance.

$$\text{Map distance} = \frac{\text{Number of nonparental offspring}}{\text{Total number of offspring}} \times 100$$

$$= \frac{36 + 39}{36 + 39 + 160 + 165} \times 100$$

$$= 18.8 \text{ mu}$$

The genes are approximately 18.8 mu apart.

S2. Two recessive disorders in mice, droopy ears and flaky tail, are caused by genes that are located 6 mu apart on chromosome 3. A true-breeding mouse with normal ears (*De*) and a flaky tail (*ft*) was crossed to a true-breeding mouse with droopy ears (*de*) and a normal tail (*Ft*). The F_1 offspring were then crossed to mice with droopy ears and flaky tails. If this testcross produced 100 offspring, what is the expected outcome?

Answer: The testcross is

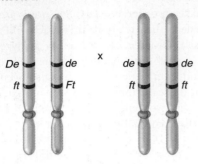

The parental offspring are

Dede   *ftft*   normal ears, flaky tail
dede   *Ftft*   droopy ears, normal tail

The recombinant offspring are

dede   *ftft*   droopy ears, flaky tail
Dede   *Ftft*   normal ears, normal tail

Since the two genes are located 6 mu apart on the same chromosome, 6% of the offspring will be recombinants. Therefore, the expected outcome for 100 offspring is

   3 droopy ears, flaky tail
   3 normal ears, normal tail
47 normal ears, flaky tail
47 droopy ears, normal tail

S3. The following X-linked recessive traits are found in fruit flies: vermilion eyes are recessive to normal (red) eyes, miniature wings are recessive to normal (long) wings, and sable body is recessive to normal (gray) body. A cross was made between wild-type males and females with vermilion eyes, miniature wings, and sable bodies. The heterozygous females from this cross, which had a wild-type phenotype, were then crossed to males with vermilion eyes, miniature wings, and sable bodies. The following outcome was obtained:

Males and Females

1,320 vermilion eyes, miniature wings, sable body
1,346 red eyes, long wings, gray body
  102 vermilion eyes, miniature wings, gray body
   90 red eyes, long wings, sable body
   42 vermilion eyes, long wings, gray body
   48 red eyes, miniature wings, sable body
    2 vermilion eyes, long wings, sable body
    1 red eyes, miniature wings, gray body

A. Calculate the map distance between the three genes.

B. Is positive interference occurring?

Answer:

A. The first step is to determine the order of the three genes. We can do this by evaluating the pattern of inheritance in the double crossovers. The double crossover group occurs with the lowest frequency. Thus, the double crossovers are vermilion eyes, long wings, sable body and red eyes, miniature wings, gray

body. Compared to the parental combinations of alleles (vermilion eyes, miniature wings, sable body and red eyes, long wings, gray body), the gene for wing length has been reassorted. Two flies have long wings associated with vermilion eyes and sable body, and one fly has miniature wings associated with red eyes and gray body. Taken together, these results indicate that the wing length gene is found in between the eye color and body color genes.

Eye color——wing length——body color

We now calculate the distance between eye color and wing length, and between wing length and body color. To do this, one strategy is to rearrange the data according to gene pairs:

vermilion eyes, miniature wings = 1,320 + 102 = 1,422
red eyes, long wings = 1,346 + 90 = 1,436
vermilion eyes, long wings = 42 + 2 = 44
red eyes, miniature wings = 48 + 1 = 49

The recombinants are vermilion eyes, long wings and red eyes, miniature wings. The map distance between these two genes is

$$(44 + 49)/(1422 + 1436 + 44 + 49) \times 100 = 3.2 \text{ mu}$$

Likewise, the other gene pair is wing length and body color.

miniature wings, sable body = 1,320 + 48 = 1,368
long wings, gray body = 1,346 + 42 = 1,388
miniature wings, gray body = 102 + 1 = 103
long wings, sable body = 90 + 2 = 92

The recombinants are miniature wings, gray body and long wings, sable body. The map distance between these two genes is

$$(103 + 92)/(1,368 + 1,388 + 103 + 92) \times 100 = 6.6 \text{ mu}$$

With these data, we can produce the following genetic map:

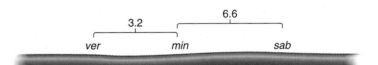

B. To calculate the interference value, we must first calculate the coefficient of coincidence.

$$C = \frac{\text{Observed number of double crossovers}}{\text{Expected number of double crossovers}}$$

Based on our mapping in part A, the percentage of single crossovers equals 3.2% (0.032) and 6.6% (0.066). The expected number of double crossovers equals 0.032 times 0.066, which is 0.002, or 0.2%. There were a total of 2,951 offspring produced. If we multiply 2,951 times 0.002, we get 5.9, which is the expected number of double crossovers. The observed number was 3. Therefore:

$$C = 3/5.9 = 0.51$$

$$I = 1 - C = 1 - 0.51 = 0.49$$

In other words, approximately 49% of the expected double crossovers did not occur due to interference.

S4. Around the same time as the study of Creighton and McClintock described in figure 5.6, Curt Stern conducted similar experiments with *Drosophila*. He had strains of flies with microscopically detectable abnormalities in the X chromosome. In one case, the X chromosome was shorter than normal due to a deletion at one end. In another case, the X chromosome was longer than normal because an extra piece of the Y chromosome was attached at the other end of the X chromosome where the centromere is located. He had female flies that had both abnormal chromosomes. On the short X chromosome, a recessive allele (*car*) was located that results in carnation colored eyes, and a dominant allele (*B*) that causes bar-shaped eyes was also found on this chromosome. On the long X chromosome in this female fly were located the wild-type alleles for these two genes (designated *car*$^+$ and *B*$^+$), which confer red eyes and round eyes. Stern realized that a crossover between the two X chromosomes in the female fly would result in recombinant chromosomes that would be cytologically distinguishable from the parental chromosomes. If a crossover occurred between the *B* and *car* genes on the X chromosome, this is expected to produce a normal-sized X chromosome and an abnormal chromosome with a deletion at one end and an extra piece of the Y chromosome at the other end.

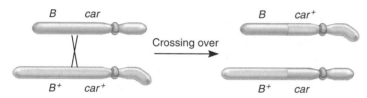

Stern crossed these female flies to male flies that contained a normal X chromosome with the *car* allele and the allele for normal-shaped (round) eyes (*car B*$^+$). Using a microscope, it was possible for him to discriminate between the morphologies of parental chromosomes (like those contained within the original parental flies) and recombinant chromosomes that may be found in the offspring. What would be the predicted phenotypes and chromosome characteristics in the offspring if crossing over did or did not occur between the X chromosomes in the female flies of this cross?

Answer: To demonstrate that genetic recombination is due to crossing over, it was necessary for Stern to correlate recombinant phenotypes (due to genetic recombination) with the inheritance of recombinant chromosomes (due to crossing over). Since he knew the arrangement of alleles in the female fly, it was possible for him to predict the phenotypes of parental and nonparental offspring. The male fly could contribute the *car* and *B*$^+$ alleles (on a cytologically normal X chromosome) or contribute a Y chromosome. In the absence of crossing over, the female fly could contribute a short X chromosome with the *car* and *B* alleles or a long X chromosome with the wild-type alleles. If crossing over occurred in the region between the *car* and *B* alleles, the female fly could also contribute recombinant X chromosomes. One possible recombinant X chromosome would be normal-sized and carry the *car* and *B*$^+$ alleles while the other recombinant X chromosome would be

deleted at one end with a piece of the Y chromosome at the other end and carry the *car*$^+$ and *B* alleles. When combined with an X or Y chromosome from the male, the parental offspring would have carnation, bar eyes or wild-type eyes; the nonparental offspring would have carnation, round eyes or red, bar eyes.

Male gametes

	♂ *carB*$^+$	Y	Phenotype	X chromosome from female
♀ *carB*	*carB* / *carB*$^+$	*carB* / Y	Carnation, bar eyes	Short X chromosome
car$^+$*B*$^+$	*car*$^+$*B*$^+$ / *carB*$^+$	*car*$^+$*B*$^+$ / Y	Red, round eyes	Long X chromosome with a piece of Y
carB$^+$	*carB*$^+$ / *carB*$^+$	*carB*$^+$ / Y	Carnation, round eyes	Normal-sized X chromosome
car$^+$*B*	*car*$^+$*B* / *carB*$^+$	*car*$^+$*B* / Y	Red, bar eyes	Short X chromosome with a piece of Y

(left axis label: Female gametes)

The results shown in the Punnett square are the actual results that Stern observed. His interpretation was that crossing over between homologous chromosomes (in this case, the X chromosome) accounts for the formation of offspring with recombinant phenotypes.

S5. There is a limit to the relationship between map distance and the percentage of recombinant offspring. Even though it is possible for two genes on the same chromosome to be much more than 50 mu apart, we do not expect to obtain greater than 50% recombinant offspring in a testcross. You may be wondering why this is so. The answer lies in the pattern of multiple crossovers. At the pachytene stage of meiosis, a single crossover in the region between two genes will produce only 50% recombinant chromosomes (see fig. 5.1*b*). Therefore, to exceed a 50% recombinant level, it would seem like it would be necessary to have multiple crossovers within the tetrad.

Let's suppose that two genes are far apart on the same chromosome. A testcross is made between a heterozygous individual, *AaBb*, and a homozygous individual, *aabb*. In the heterozygous individual, the dominant alleles (*A* and *B*) are linked on the same chromosome and the recessive alleles (*a* and *b*) are linked on the same chromosome. Draw out all of the possible double crossovers (between two, three, or four chromatids) and determine the average number of recombinant offspring, assuming an equal probability of all of the double crossover possibilities.

Answer: A double crossover occurring between the two genes could involve two chromatids, three chromatids, or four chromatids. There are two different ways that crossover could involve three chromatids. The possibilities for all types of double crossovers are shown here:

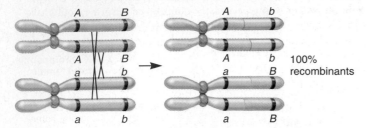

100% recombinants

Double crossover (involving 4 chromatids)

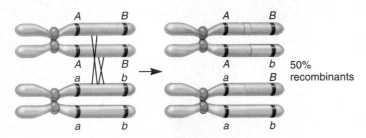

50% recombinants

Double crossover (involving 3 chromatids)

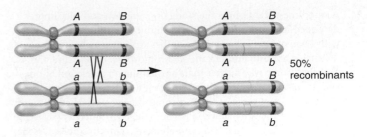

50% recombinants

Double crossover (involving 3 chromatids)

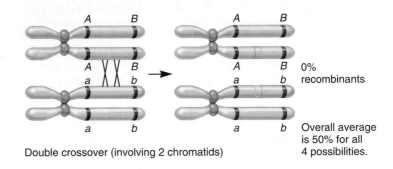

0% recombinants

Overall average is 50% for all 4 possibilities.

Double crossover (involving 2 chromatids)

This drawing considers the situation where two crossovers are expected to occur in the region between the two genes. Since the tetrad is composed of two pairs of homologues, there are several possible ways that a double crossover could occur between homologues. Since all of these double crossing over events are equally probable, we take the average of them to determine the maximum recombination frequency. This average equals 50%.

Conceptual Questions

C1. What is the difference in meaning between the terms *genetic recombination* and *crossing over*?

C2. When applying a chi square approach in a linkage problem, explain why an independent assortment hypothesis is used.

C3. What is mitotic recombination? A heterozygous individual (*Bb*) with brown eyes has one eye with a small patch of blue. Provide two or more explanations for how the blue patch may have occurred.

C4. Mitotic recombination can occasionally produce a twin spot. Let's suppose an animal species can be heterozygous for two genes that govern fur color and length: one gene affects pigmentation, with dark pigmentation (*A*) dominant to albino (*a*); long hair (*L*) is dominant to short hair (*l*). The two genes are linked on the same chromosome. Let's assume an animal is *AaLl*; *A* is linked to *l*, and *a* is linked to *L*. Draw the chromosomes labeled with these alleles, and explain how mitotic recombination could produce a twin spot with one spot having albino pigmentation and long fur, the other having dark, short fur.

C5. A crossover has occurred in the bivalent shown here.

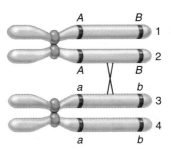

If a second crossover occurs in same the region between these two genes, which two chromatids would be involved in order to produce the following outcomes?

A. 100% recombinants

B. 0% recombinants

C. 50% recombinants

C6. A crossover has occurred in the bivalent shown here.

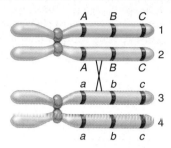

What would be the outcome of this single crossover event? If a second crossover occurs somewhere between *A* and *C*, explain which two chromatids it would involve and where it would occur (i.e., between which two genes) to produce the types of chromosomes shown here.

A. *A B C, A b C, a B c,* and *a b c*

B. *A b c, A b c, a B C,* and *a B C*

C. *A B c, A b c, a B C,* and *a b C*

D. *A B C, A B C, a b c,* and *a b c*

C7. A diploid organism has a total of 14 chromosomes and about 20,000 genes per haploid genome. Approximately how many genes are there in each linkage group?

C8. If you try to throw a basketball into a basket, the likelihood of succeeding depends on the size of the basket. It is more likely that you will get the ball into the basket if the basket is bigger. In your own words, explain how this analogy also applies to the idea that the likelihood of crossing over is greater when two genes are far apart compared to when they are close together.

C9. By conducting testcrosses, researchers have found that the sweet pea has seven linkage groups. How many chromosomes would you expect to find in leaf cells?

C10. In humans, a rare dominant disorder known as nail-patella syndrome causes abnormalities in the fingernails, toenails, and kneecaps. Researchers have examined family pedigrees with regard to this disorder and, within the same pedigree, also examined the individuals with regard to their blood types. (A description of blood genotypes is found in chapter 4.) In the following pedigree, individuals affected with nail-patella disorder are shown with filled symbols. The genotype of each individual with regard to their ABO blood type is also shown. Does this pedigree suggest any linkage between the gene that causes nail-patella syndrome and the gene that causes blood type?

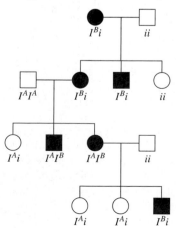

C11. When true-breeding mice with brown fur and short tails (*BBtt*) were crossed to true-breeding mice with white fur and long tails (*bbTT*), all the F_1 offspring had brown fur and long tails. The F_1 offspring were crossed to mice with white fur and short tails. What are the possible phenotypes of the F_2 offspring? Which F_2 offspring are recombinant and which are nonrecombinant? What will be the ratios of the F_2 offspring if independent assortment is taking place? How will the ratios be affected by linkage?

C12. Though we often think of genes in terms of the phenotypes they produce (e.g., curly leaves, flaky tail, brown eyes), the molecular function of most genes is to encode proteins. Many cellular proteins function as enzymes. The table that follows describes the map distances between six different genes that encode six different enzymes: *Ada*, adenosine deaminase; *Hao-1*, hydroxyacid oxidase-1; *Hdc*, histidine decarboxylase; *Odc-2*, ornithine decarboxylase-2; *Sdh-1*, sorbitol dehydrogenase-1; and *Ass-1*, arginosuccinate synthetase-1.

Map distances between two genes:

	Ada	*Hao-1*	*Hdc*	*Odc-2*	*Sdh-1*	*Ass-1*
Ada		14		8	28	
Hao-1	14		9		14	
Hdc		9		15	5	
Odc-2	8		15			63
Sdh-1	28	14	5			43
Ass-1				63	43	

Construct a genetic map that describes the locations of all six genes.

C13. If the likelihood of a single crossover in a particular chromosomal region is 10%, what is the theoretical likelihood of a double or triple crossover in that same region? How does positive interference affect these theoretical values?

C14. Except for fungi that form asci, in most dihybrid crosses involving linked genes, we cannot tell if a double crossover between the two genes has occurred because the offspring will inherit the parental combination of alleles. How does the inability to detect double crossovers affect the calculation of map distance? Is map distance underestimated or overestimated because of our inability to detect double crossovers? Explain your answer.

C15. Researchers have discovered that certain regions of chromosomes are much more likely to cross over compared to other regions. We might call such a region a "hot spot" for crossing over. Let's suppose that two genes, gene *A* and gene *B*, are 5,000,000 nucleotides apart on the same chromosome. Gene *A* and *B* are in a hot spot for crossing over. Two other genes, let's call them gene *C* and gene *D*, are also 5,000,000 nucleotides apart but are not in a hot spot for recombination. If we conducted dihybrid crosses to compute the map distance between genes *A* and *B*, and other dihybrid crosses to compute the map distance between genes *C* and *D*, would the map distances be the same between *A* and *B* compared to *C* and *D*? Explain.

C16. Describe the unique features of ascomycetes that lend themselves to genetic analysis.

C17. In fungi, what is the difference between a tetrad and an octad? What cellular process occurs in an octad that does not occur in a tetrad?

C18. Explain the difference between an unordered versus an ordered octad.

C19. In *Neurospora*, a cross is made between a wild-type and an albino mutant strain (which produce orange and white spores, respectively). Draw two different ways that an octad might look if it was displaying second-division segregation.

C20. One gene in *Neurospora* (let's call it gene *A*) is located close to a centromere, while a second gene (let's call it gene *B*) is located more toward the end of the chromosome. Would the percentage of octads exhibiting first-division segregation be higher with respect to gene *A* or gene *B*? Explain your answer.

Experimental Questions (Includes Most Mapping Questions)

E1. Figure 5.2 shows the first experimental results that indicated linkage between two different genes. Conduct a chi square analysis to confirm that the genes are really linked and the data could not be explained by independent assortment.

E2. In the experiment of figure 5.6, the researchers followed the inheritance pattern of chromosomes that were abnormal at both ends to correlate genetic recombination with the physical exchange of chromosome pieces. Is it necessary to use a chromosome that is abnormal at both ends, or could they have used a strain with two abnormal chromosome 9s, one with a knob at one end and its homologue with a translocation at the other end?

E3. The experiment of figure 5.6 is not like a standard testcross, because neither parent is homozygous recessive for both genes. If you were going to carry out this same kind of experiment to verify that crossing over can explain the recombination of alleles of different genes, how would you modify this experiment to make it a standard testcross? For both parents, you should designate which alleles are found on an abnormal chromosome (i.e., knobbed, translocation chromosome 9) and which alleles are found on normal chromosomes.

E4. How would you determine that genes in mammals are located on the Y chromosome linkage group? Is it possible to conduct crosses (let's say in mice) to map the distances between genes along the Y chromosome? Explain.

E5. Explain the rationale behind a testcross. Is it necessary for one of the parents to be homozygous recessive for the genes of interest? In the heterozygous parent of a testcross, must all the dominant alleles be linked on the same chromosome and all the recessive alleles be linked on the homologue?

E6. In your own words, explain why a testcross cannot produce more than 50% recombinant offspring. When a testcross does produce 50% recombinant offspring, what do these results mean?

E7. Explain why the percentage of recombinant offspring in a testcross is a more accurate measure of map distance when two genes are close together. When two genes are far apart, is the percentage of recombinant offspring an underestimate or overestimate of the actual map distance?

E8. If two genes are more than 50 mu apart, how would you ever be able to show experimentally that they are located on the same chromosome?

E9. In Morgan's trihybrid testcross of figure 5.3, he realized that crossing over was more frequent between the eye color and wing length genes than between the body color and eye color genes. Explain how he determined this.

E10. In the experiment of figure 5.10, list the gene pairs from the particular dihybrid crosses that Sturtevant used to construct his genetic map.

E11. In the tomato, red fruit (*R*) is dominant over yellow fruit (*r*), and yellow flowers (*Wf*) are dominant over white flowers (*wf*). A cross was made between true-breeding plants with red fruit and yellow flowers and plants with yellow fruit and white flowers. The F_1 generation plants were then crossed to plants with yellow fruit and white flowers. The following results were obtained:

> 333 red fruit, yellow flowers
>
> 64 red fruit, white flowers
>
> 58 yellow fruit, yellow flowers
>
> 350 yellow fruit, white flowers

Calculate the map distance between the two genes.

E12. Two genes are located on the same chromosome and are known to be 12 mu apart. An *AABB* individual was crossed to an *aabb* individual to produce *AaBb* offspring. The *AaBb* offspring were then crossed to *aabb* individuals.

A. If this cross produces 1,000 offspring, what are the predicted numbers of offspring with each of the four genotypes: *AaBb*, *Aabb*, *aaBb*, and *aabb*?

B. What would be the predicted numbers of offspring with these four genotypes if the parental generation had been *AAbb* and *aaBB* instead of *AABB* and *aabb*?

E13. Two genes, designated *A* and *B*, are located 10 mu from each other. A third gene, designated *C*, is located 15 mu from *B* and 5 mu from *A*. A parental generation consisting of *AA bb CC* and *aa BB cc* individuals were crossed to each other. The F_1 heterozygotes were then testcrossed to *aa bb cc* individuals. If we assume that there are no double crossovers in this region, what percentage of offspring would you expect with the following genotypes?

A. *Aa Bb Cc*

B. *aa Bb Cc*

C. *Aa bb cc*

E14. Two genes in tomatoes are 61 mu apart; normal fruit (*F*) is dominant to fasciated fruit, and normal numbers of leaves (*Lf*) is dominant to leafy (*lf*). A true-breeding plant with normal leaves and fruit was crossed to a leafy plant with fasciated fruit. The F_1 offspring were then crossed to leafy plants with fasciated fruit. If this cross produced 600 offspring, what are the expected numbers of plants in each of the four possible categories: normal leaves, normal fruit; normal leaves, fasciated fruit; leafy, normal fruit; and leafy, fasciated fruit?

E15. In the tomato, three genes are linked on the same chromosome. Tall is dominant to dwarf, fruit skin that is smooth is dominant to skin that is peachy, and fruit with a normal tomato shape is dominant to oblate shape. A plant that is true-breeding for the dominant traits was crossed to a dwarf plant with peachy, oblate fruit. The F_1 plants were then testcrossed to dwarf plants with peachy, oblate fruit. The following results were obtained:

151 tall, smooth, normal

33 tall, smooth, oblate

11 tall, peach, oblate

2 tall, peach, normal

155 dwarf, peach, oblate

29 dwarf, peach, normal

12 dwarf, smooth, normal

0 dwarf, smooth, oblate

Construct a genetic map that describes the order of these three genes and the distances between them.

E16. A trait in garden peas involves the curling of leaves. A dihybrid cross was made involving a plant with yellow pods and curling leaves to a wild-type plant with green pods and normal leaves. All the F_1 offspring had green pods and normal leaves. The F_1 plants were then crossed to plants with yellow pods and curling leaves. The following results were obtained:

117 green pods, normal leaves

115 yellow pods, curling leaves

78 green pods, curling leaves

80 yellow pods, normal leaves

A. Conduct a chi square analysis to determine if these two genes are linked.

B. If they are linked, calculate the map distance between the two genes. How accurate do you think this distance is?

E17. In mice, the gene that encodes the enzyme inosine triphosphatase is 12 mu from the gene that encodes the enzyme ornithine decarboxylase. Let's suppose you have identified a strain of mice homozygous for a defective inosine triphosphatase gene that does not produce any of this enzyme and is also homozygous for a defective ornithine decarboxylase gene. In other words, this strain of mice cannot make either enzyme. You crossed this homozygous recessive strain to a normal strain of mice to produce heterozygotes. The heterozygotes were then backcrossed to the strain that cannot produce either enzyme. What is the probability of obtaining a mouse that cannot make either enzyme?

E18. In the garden pea, several different genes affect pod characteristics. A gene affecting pod color (green is dominant to yellow) is approximately 7 mu away from a gene affecting pod width (wide is dominant to narrow). Both genes are located on chromosome 5. A third gene, located on chromosome 4, affects pod length (long is dominant to short). A true-breeding wild-type plant (green, wide, long pods) was crossed to a plant with yellow, narrow, short pods. The F_1 offspring were then testcrossed to plants with yellow, narrow, short pods. If the testcross produced 800 offspring, what are the expected numbers of the eight possible phenotypic combinations?

E19. A sex-influenced trait is dominant in males and causes bushy tails. The same trait is recessive in females. Fur color is not sex influenced. Yellow fur is dominant to white fur. A true-breeding female with a bushy tail and yellow fur was crossed to a white male without a bushy tail. The F_1 females were then crossed to white males without bushy tails. The following results were obtained:

Males	Females
28 normal tails, yellow	102 normal tails, yellow
72 normal tails, white	96 normal tails, white
68 bushy tails, yellow	0 bushy tails, yellow
29 bushy tails, white	0 bushy tails, white

A. Conduct a chi square analysis to determine if these two genes are linked.

B. If the genes are linked, calculate the map distance between them. Explain which data you used in your calculation.

E20. Three recessive traits in garden pea plants are as follows: yellow pods are recessive to green pods, bluish green seedlings are recessive to green seedlings, creeper (a plant that cannot stand up) is recessive to normal. A true-breeding normal plant with green pods and green seedlings was crossed to a creeper with yellow pods and bluish green seedlings. The F_1 plants were then crossed to creepers with yellow pods and bluish green seedlings. The following results were obtained:

2,059 green pods, green seedlings, normal

151 green pods, green seedlings, creeper

281 green pods, bluish green seedlings, normal

15 green pods, bluish green seedlings, creeper

2,041 yellow pods, bluish green seedlings, creeper

157 yellow pods, bluish green seedlings, normal

282 yellow pods, green seedlings, creeper

11 yellow pods, green seedlings, normal

Construct a genetic map that describes the map distance between these three genes.

E21. In mice, a trait called snubnose is recessive to a wild-type nose, a trait called pintail is dominant to a normal tail, and a trait called jerker (a defect in motor skills) is recessive to normal movement. Jerker mice with a snubnose and pintail were crossed to normal mice, and then the F_1 mice were crossed to jerker mice that have a snubnose and normal tail. The outcome of this cross was as follows:

560 jerker, snubnose, pintail

548 normal movement, normal nose, normal tail

102 jerker, snubnose, normal tail

104 normal movement, normal nose, pintail

77 jerker, normal nose, normal tail

71 normal movement, snubnose, pintail

11 jerker, normal nose, pintail

9 normal movement, snubnose, normal tail

Make a genetic map that describes the order and distance between these genes.

E22. In chapter 4, multiple alleles that affect the pattern of coat color in rabbits are described. This gene affects the amount of pigment distributed throughout the hairs of the rabbit's body. However, this gene does not determine the color of the pigment. A second gene affects pigment color. It can exist in two alleles, B and b, where B is dominant and confers black coat color, while b is recessive and confers brown coat color. (Incidentally, all the rabbits shown in chapter 4 are *bb*.) When true-breeding black chinchilla rabbits were crossed to true-breeding brown Himalayan rabbits, all the F_1 offspring were black chinchillas. When the F_1 offspring were mated to brown Himalayan rabbits, the following results were obtained:

Black chinchilla	368
Black Himalayan	160
Brown chinchilla	194
Brown Himalayan	352

Conduct a chi square analysis to determine if the color gene and coat pattern gene are independently assorting. If so, calculate the map distance between the genes.

E23. In *Drosophila*, an allele causing vestigial wings is 12.5 mu away from another gene that causes purple eyes. A third gene that affects body color has an allele that can cause black body color. This third gene is 18.5 mu away from the vestigial wings gene and 6 mu away from the gene causing purple eyes. The alleles causing vestigial wings, purple eyes, and black body are all recessive. The dominant (wild-type) traits are long wings, red eyes, and gray body. A researcher crossed wild-type flies to flies with vestigial wings, purple eyes, and black bodies. All the F_1 flies were wild type. F_1 female flies were then crossed to male flies with vestigial wings, purple eyes, and black bodies. If 1,000 offspring were observed, what are the expected numbers of the following types of flies?

Long wings, red eyes, gray body

Long wings, purple eyes, gray body

Long wings, red eyes, black body

Long wings, purple eyes, black body

Short wings, red eyes, gray body

Short wings, purple eyes, gray body

Short wings, red eyes, black body

Short wings, purple eyes, black body

Which kinds of flies can be produced only by a double crossover event?

E24. Three autosomal genes are linked along the same chromosome. The distance between gene *A* and *B* is 7 mu, the distance between *B* and *C* is 11 mu, and the distance between *A* and *C* is 4 mu. An individual who is *AA bb CC* was crossed to an individual who is *aa BB cc* to produce heterozygous F_1 offspring. The F_1 offspring were then crossed to homozygous *aa bb cc* individuals to produce F_2 offspring.

A. Draw the arrangement of alleles on the chromosomes in the parents and in the F_1 offspring.

B. Where would a crossover have to occur to produce an F_2 offspring that was heterozygous for all three genes?

C. If we assume that no double crossovers occur in this region, what percentage of F_2 offspring are likely to be homozygous for all three genes?

E25. Let's suppose that two different X-linked genes exist in mice, designated with the letters *N* and *L*. Gene *N* exists in a dominant, nor-

mal allele and a recessive allele, *n*, that is lethal. Similarly, gene *L* exists in a dominant, normal allele and in a recessive allele, *l*, that is lethal. Heterozygous females are normal, while males that carry either recessive allele are born dead. Would it be possible to map the distance between these two genes by making crosses and analyzing the number of living and dead offspring? You may assume that you have strains of mice in which females are heterozygous for one or both genes.

E26. *His-5* and *lys-1* are alleles found in baker's yeast that require histidine and lysine for growth, respectively. A cross was made between two haploid yeast that are *his-5 lys-1* and *his⁺ lys⁺*. From the analysis of 818 individual tetrads, the following numbers of tetrads were obtained:

2 spores: *his-5 lys⁺* + 2 spores: *his⁺ lys-1* = 4

2 spores: *his-5 lys-1* + 2 spores: *his⁺ lys⁺* = 502

1 spore: *his-5 lys-1* + 1 spore: *his-5 lys⁺* + 1 spore: *his⁺ lys-1* + 1 spore: *his⁺ lys⁺* = 312

A. Compute the map distance between these two genes using the method of calculation that considers double crossovers and the one that does not. Which method gives a higher value? Explain why.

B. What is the frequency of single crossovers between these two genes?

C. Based on your answer to part B, how many NPDs are expected from this cross? Explain your answer. Is positive interference occurring?

E27. On chromosome 4 in *Neurospora*, the allele *pyr-1* results in a pyrimidine requirement for growth. A cross was made between a *pyr-1* and a *pyr⁺* (wild-type) strain, and the following results were obtained:

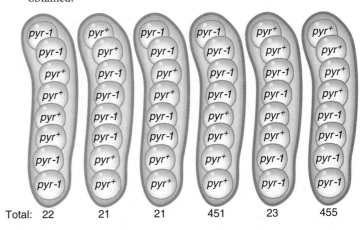

| Total: | 22 | 21 | 21 | 451 | 23 | 455 |

What is the distance between the *pyr-1* gene and the centromere?

E28. On chromosome 3 in *Neurospora*, the *pro-1* gene is located approximately 9.8 mu from the centromere. Let's suppose a cross was made between a *pro-1* and a *pro⁺* strain and 1,000 asci were analyzed.

A. What are the six types of asci that can be produced?

B. What are the expected numbers of each type of asci?

Questions for Student Discussion/Collaboration

1. In mice, a dominant gene that causes a short tail is located on chromosome 2. On chromosome 3, a recessive gene causing droopy ears is 6 mu away from another recessive gene that causes a flaky tail. A recessive gene that causes a jerker (uncoordinated) phenotype is located on chromosome 4. A jerker mouse with droopy ears and a short, flaky tail was crossed to a normal mouse. All the F_1 generation mice were phenotypically normal, except they have short tails. These F_1 mice were then testcrossed to jerker mice with droopy ears and long, flaky tails. If this cross produced 400 offspring, what would be the proportions of the 16 possible phenotypic categories?

2. In chapter 3, we discussed the idea that the X and Y chromosomes have a few genes in common. These genes are inherited in a pseudoautosomal pattern. With this phenomenon in mind, discuss whether or not the X and Y chromosomes are really distinct linkage groups.

3. Mendel studied seven traits in pea plants, and the garden pea happens to have seven different chromosomes. It has been pointed out that Mendel was very lucky not to have conducted crosses involving two traits governed by genes that are closely linked on the same chromosome, since the results would have confounded his theory of independent assortment. It has even been suggested that, perhaps, Mendel did not publish data involving traits that were linked! An article by Blixt, S. (1975) *Nature 256*, p. 206, considers this issue. Look up this article and discuss why Mendel did not find linkage.

Note: All answers appear at the website for this textbook; the answers to even-numbered questions are in the back of the textbook.

www.mhhe.com/brookcr

Visit the Online Learning Center for practice tests, answer keys, and other learning aids for this chapter. Enhance your understanding of genetics with our interactive exercises, web links, news feeds, tutorial service, and much more.

GENETIC TRANSFER AND MAPPING IN BACTERIA AND BACTERIOPHAGES

::

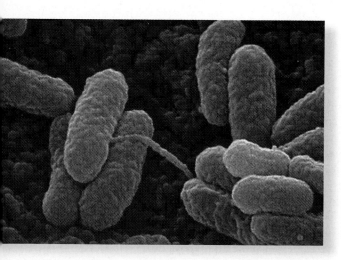

T hus far, our attention in part II of this textbook has focused on genetic analyses of eukaryotic species such as fungi, plants, and animals. As we have seen, these organisms are amenable to genetic studies for two reasons. First, allelic differences, such as white versus red eyes in *Drosophila* and tall versus dwarf pea plants, provide readily discernible traits among different individuals. Second, since most eukaryotic species reproduce sexually, crosses can be made, and the pattern of transmission of traits from parent to offspring can be analyzed. The ability to follow allelic differences in a genetic cross is a basic tool in the genetic examination of eukaryotic species.

In chapter 6, we turn our attention to the genetic analysis of bacteria. Like their eukaryotic counterparts, bacteria often possess allelic differences that affect their cellular traits. Common allelic variations among bacteria that are readily discernible involve traits such as sensitivity to antibiotics and differences in their nutrient requirements for growth. In these cases, the allelic differences are between different strains of bacteria because any given bacterium is usually haploid for a particular gene. In fact, the haploid nature of bacteria is one advantage that makes it easier to identify mutations that produce phenotypes such as altered nutritional requirements. Loss-of-function mutations, which are often recessive in diploid eukaryotes, are not masked by dominant genes in haploid species. Throughout chapter 6, we will consider interesting experiments that examine bacterial strains with allelic differences.

Compared to eukaryotes, another striking difference in prokaryotic species is their mode of reproduction. Since bacteria reproduce asexually, crosses are not used in the genetic analysis of bacterial species. Instead, researchers rely on a similar phenomenon called **genetic transfer.** In this process, a segment of bacterial DNA is transferred from one bacterium to another. Researchers have used genetic transfer to map the locations of genes along the chromosome of many bacterial species.

In the second part of chapter 6, we will examine **bacteriophages** (also known as **phages**), which are viruses that infect bacteria. Bacteriophages contain their own genetic material that governs the traits of the phage. As we will see, the genetic analysis of phages can yield a highly detailed genetic map of a short region of DNA. These types of analyses have provided researchers with insights regarding the structure and function of genes.

6.1 GENETIC TRANSFER AND MAPPING IN BACTERIA

Genetic transfer is a process whereby genetic material from one bacterium is transferred to another bacterium. Like sexual reproduction in eukaryotes, genetic transfer in bacteria is thought to enhance the genetic diversity of bacterial species. For example, a bacterial cell carrying a gene that provides antibiotic resistance may transfer this gene to another bacterial cell, allowing that bacterial cell to survive exposure to the antibiotic.

There are three naturally occurring ways that bacteria can transfer genetic material. The first route, known as **conjugation,** involves direct physical interaction between two bacterial cells. One bacterium acts as a donor and transfers genetic material to a recipient cell. A second means of transfer is called **transduction.** This occurs when a virus infects a bacterium and then transfers bacterial genetic material from one bacterium to another. The last mode of genetic transfer is **transformation.** In this case, genetic material is released into the environment when a bacterial cell dies. This material then binds to a living bacterial cell, which can take it up. These three mechanisms of genetic transfer have been extensively investigated in research laboratories, and their molecular mechanisms continue to be studied. In this section, we will examine these three systems of genetic transfer in greater detail.

We will also learn how genetic transfer between bacterial cells has provided unique ways to accurately map bacterial genes. The methods described in chapter 6 have been largely replaced by molecular approaches, which are described in chapter 20. Even so, the mapping of bacterial genes serves to illustrate the mechanisms by which genes are transferred between bacterial cells.

Bacteria Can Transfer Genetic Material During Conjugation

The natural ability of some bacteria to transfer genetic material between each other was first recognized by Joshua Lederberg and Edward Tatum in 1946. They were studying strains of *Escherichia coli* (*E. coli*) that had different nutritional requirements for growth. (The general methods for growing bacteria in a laboratory are described in the appendix.) As shown in the experiment

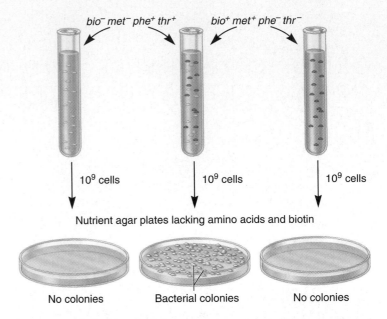

FIGURE 6.1   **Experiment of Lederberg and Tatum demonstrating genetic transfer during conjugation in *E. coli*.** When the *bio⁻ met⁻ phe⁺ thr⁺* or *bio⁺ met⁺ phe⁻ thr⁻* strains were plated on a growth medium lacking amino acids and biotin, they were unable to grow. However, if they were mixed together and then plated, some colonies were observed. These colonies are due to the transfer of genetic material between these two strains by conjugation. Note: In bacteria, it is common to give genes a three-letter name (shown in italics) that is related to the function of the gene. A plus (+) superscript indicates a functional gene while a minus (−) superscript indicates a mutation that has caused the gene or gene product to be inactive. In some cases, several genes have related functions. These may have the same three-letter name followed by different capital letters. For example, different genes involved with phenylalanine biosynthesis may be called *pheA, pheB, pheC,* etc. In the experiment described in figure 6.1, the genes involved in phenylalanine biosynthesis had not been distinguished, so the gene involved is simply referred to as *phe⁺* (for a functional gene) and *phe⁻* (for a nonfunctional gene).

in figure 6.1, they studied one strain, designated *bio⁻ met⁻ phe⁺ thr⁺*, which required one vitamin, biotin (bio), and one amino acid, methionine (met), in order to grow. This strain did not require the amino acids phenylalanine (phe) or threonine (thr) for growth. Another strain, designated *bio⁺ met⁺ phe⁻ thr⁻*, had just the opposite requirements. It needed phenylalanine and threonine for growth, but not biotin and methionine. These differences in nutritional requirements correspond to variations in the genetic material of the two strains. The first strain had two defective genes encoding enzymes necessary for biotin and methionine synthesis. The second strain contained two defective genes required to make phenylalanine and threonine.

Figure 6.1 compares the results when the two strains were mixed together and when they were not mixed. Without mixing, about 1 billion (10^9) *bio⁻ met⁻ phe⁺ thr⁺* cells were applied to plates on a growth medium lacking amino acids and biotin; no colonies were observed to grow. This result is expected, since the plates do not contain biotin and methionine. Likewise, when 10^9 bio^+ met^+ phe^- thr^- cells were plated, no colonies were observed, because phenylalanine and threonine were also missing from this growth medium. However, when bio^- met^- phe^+ thr^+ and bio^+ met^+ phe^- thr^- cells were mixed together and then 10^9 cells plated, approximately 100 cells were observed to multiply and form visible bacterial colonies. Since growth occurred, the genotype of the cells within these colonies must have been bio^+ met^+ phe^+ thr^+. To explain these results, Lederberg and Tatum hypothesized that some genetic material was transferred between the two strains. One possibility is that the genetic material providing the ability to synthesize phenylalanine and threonine (phe^+ thr^+) was transferred to the bio^+ met^+ phe^- thr^- strain. Alternatively, the ability to synthesize biotin and methionine (bio^+ met^+) may have been transferred to the bio^- met^- phe^+ thr^+ cells. The results of this experiment cannot distinguish between these two possibilities.

Bernard Davis subsequently conducted experiments showing that the two strains of bacteria must make physical contact with each other to transfer genetic material. The apparatus he used, known as a U-tube, is shown in figure 6.2. At the bottom of the U-tube is a filter with pores small enough to allow the passage of genetic material (i.e., DNA molecules) but too small to permit the passage of bacterial cells. On one side of the filter, Davis added a bacterial strain with a certain combination of nutritional requirements (the bio^- met^- phe^+ thr^+ strain), on the other side he added a different bacterial strain (the bio^+ met^+ phe^- thr^- strain). The application of alternating pressure and suction promoted the movement of liquid through the filter. Since the bacteria were too large to fit through the pores, the movement of liquid did not allow the two types of bacterial strains to mix with each other. However, genetic material (which could have been released from a bacterium) could pass through the filter.

After incubation in a U-tube, bacteria from either side of the tube were placed on plates that could select for the growth of strains that were bio^+ met^+ phe^+ thr^+. These selective plates lacked biotin, methionine, phenylalanine, and threonine but contained all other nutrients essential for growth. In this case, no bacterial colonies grew on the plates. Thus, without physical contact, the two bacterial strains did not transfer genetic material to one another.

The term *conjugation* is now used to describe the natural process of genetic transfer between bacterial cells that requires direct cell-to-cell contact. Many, but not all, species of bacteria can conjugate. Working independently, Joshua and Esther Lederberg, William Hayes, and Luca Cavalli-Sforza discovered in the early 1950s that only certain bacterial strains can act as donors of genetic material. For example, only about 5% of natural isolates of *E. coli* can act as donor strains. Research studies showed that a strain incapable of acting as a donor could subsequently be converted to a donor strain after being mixed with another donor strain. Hayes correctly proposed that donor strains contain a fertility factor that can be transferred to conjugation-defective strains to make them conjugation proficient.

We now know that certain donor strains of *E. coli* contain a small circular segment of genetic material known as an **F factor** (for Fertility factor) in addition to their circular chromosome. Strains of *E. coli* that contain an F factor are designated F^+, strains without F factors are F^-. The more general term **plasmid** is used to describe circular pieces of DNA that exist independently of the chromosome. Plasmids are also considered in chapter 18.

In recent years, the molecular details of the conjugation process have been extensively studied. Though the mechanisms vary somewhat from one bacterial species to another, some general themes have emerged. Plasmids, such as F factors, that are transmitted via conjugation are termed **conjugative plasmids.** Such plasmids carry several genes that are required for conjugation to occur. For example, figure 6.3 shows the arrangement of genes on the F factor found in certain strains of *E. coli*. The functions of the proteins encoded by these genes are needed to transfer a strand of DNA from the conjugative plasmid to a recipient cell.

Figure 6.4*a* describes the molecular events that occur during conjugation in *E. coli*. Contact between donor and recipient cells is a key step that initiates the conjugation process. **Sex pili** (singular: **pilus**) are made by F^+ strains (fig. 6.4*b*). The gene encoding the pilin protein is located on the F factor. The pili act as attachment sites that promote the binding of bacteria to each other. In this way, an F^+ strain makes physical contact with an F^- strain. In certain species, such as *E. coli,* long pili are made that project from the F^+ cells and attempt to make contact with nearby F^- cells. Once contact is made, the pili shorten and thereby draw the donor and recipient cells closer together.

The successful contact between a donor and recipient cell stimulates the donor cells to begin the transfer process. Genes

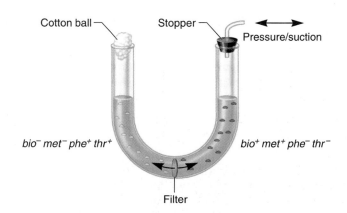

Cotton ball — Stopper — Pressure/suction

bio^- met^- phe^+ thr^+ bio^+ met^+ phe^- thr^-

Filter

FIGURE 6.2   **A U-tube apparatus like that used by Bernard Davis.** The fluid in the tube is forced through the filter by alternating suction and pressure. However, the pores in the filter are too small for the passage of bacteria.

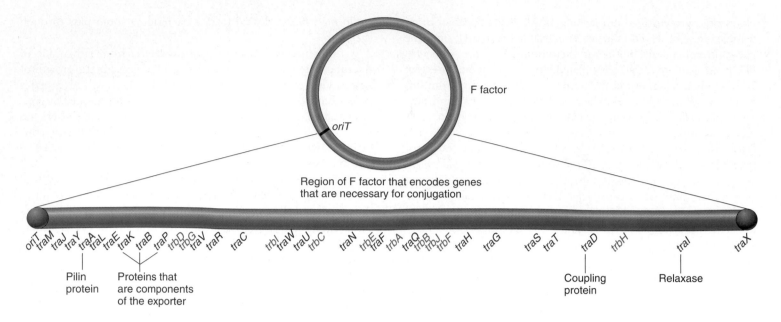

FIGURE 6.3   **Genes on the F factor that play a role during conjugation.** The transfer region of the F factor contains a few dozen genes that play a role in the conjugative process. Since they play a role in the *tr*ansfer of DNA from donor to recipient cell, the genes are designated with the three-letter names of *tra* or *trb* followed by a capital letter. The *tra* genes are shown in red and the *trb* genes are shown in blue. The functions of a few examples are indicated. The origin of transfer is designated *oriT*.

within the F factor encode a protein complex called the **relaxosome.** This complex first recognizes a DNA sequence in the F factor known as the **origin of transfer.** Upon recognition, the site is cut in one DNA strand. This cut strand of DNA is called **T DNA** because it is the strand that will be transferred to the recipient cell. The relaxosome also catalyzes the separation of the T-DNA strand from its complementary strand. As the DNA strands separate, most of the proteins within the relaxosome are released, but one protein called *relaxase* remains bound to the end of the T DNA. The complex between the T DNA and relaxase is called a **nucleoprotein** since it contains both nucleic acid (i.e., DNA) and protein (i.e., relaxase).

The next phase of conjugation involves the export of the nucleoprotein complex from the donor cell. To begin this process, the T-DNA/relaxase complex is recognized by a coupling factor that promotes the entry of the nucleoprotein into the exporter. As shown in figure 6.4*a*, the export process in Gram-negative bacteria such as *E. coli* requires the movement of the DNA across the inner and outer membrane. This export is facilitated by a large complex of proteins that span both membranes. In such bacterial species, this complex is formed from 10 to 15 different proteins that are encoded by genes within the conjugative plasmid. Once the T-DNA/relaxase complex is pumped out of the donor cell, it is thought to travel through the pilus and then into the recipient cell. As noted in figure 6.4*a*, the other strand of the F factor DNA remains in the donor cell. DNA replication in the donor cell restores the F factor DNA to its original

double-stranded condition. After the recipient cell receives a single strand of the F factor DNA, relaxase catalyzes the joining of the ends of the linear DNA molecule to form a circular plasmid. This single-stranded plasmid DNA is replicated in the recipient cell to become double stranded.

As seen in figure 6.4*a*, the result of conjugation is that the recipient cell has acquired an F factor, converting it from an F⁻ to an F⁺ cell. The genetic composition of the donor strain has not been changed. In some cases, the F⁺ factor that is transferred to the recipient strain may carry genes that were once found in the bacterial chromosome. These types of F factors are called **F′ factors.** For example, in the previous experiments of Lederberg and Tatum described in figure 6.1, an F′ factor carrying the *bio⁺* and *met⁺* genes or an F′ factor carrying the *phe⁺* and *thr⁺* genes was transferred to the recipient strain. Therefore, conjugation may introduce new genes into the recipient strain and thereby alter its genotype.

Hfr Strains Contain an F Factor Integrated into the Bacterial Chromosome

An important advance in our understanding of bacterial genetics occurred when Luca Cavalli-Sforza discovered a strain of *E. coli* that was very efficient at transferring many chromosomal genes to recipient *F⁻* strains. Cavalli-Sforza designated this bacterial strain an ***Hfr* strain** (for High frequency of recombination). It was originally derived from an *F⁺* strain. In an *Hfr* strain, an

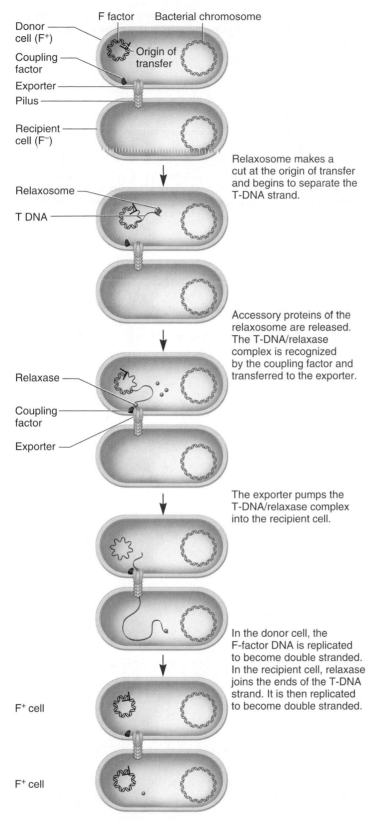

(a) Transfer of an F factor via conjugation

Donor cell (F⁺)
F factor
Bacterial chromosome
Origin of transfer
Coupling factor
Exporter
Pilus
Recipient cell (F⁻)

Relaxosome
T DNA

Relaxosome makes a cut at the origin of transfer and begins to separate the T-DNA strand.

Accessory proteins of the relaxosome are released. The T-DNA/relaxase complex is recognized by the coupling factor and transferred to the exporter.

Relaxase
Coupling factor
Exporter

The exporter pumps the T-DNA/relaxase complex into the recipient cell.

In the donor cell, the F-factor DNA is replicated to become double stranded. In the recipient cell, relaxase joins the ends of the T-DNA strand. It is then replicated to become double stranded.

F⁺ cell

F⁺ cell

FIGURE 6.4   The transfer of a fertility factor during bacterial conjugation. (a) The relaxosome complex recognizes the origin of transfer and cuts one of the DNA strands called the T DNA. This strand is separated from its complementary strand. The proteins of the relaxosome are released, except for relaxase, which remains bound to the end of the T DNA. The T-DNA/relaxase complex is recognized by a coupling factor that promotes the entry of the nucleoprotein into the exporter. This export is facilitated by a large complex of proteins that span both membranes. The T-DNA/relaxase complex is pumped out of the donor cell and travels through the sex pilus into the recipient cell. The other strand of the F factor DNA remains in the donor cell and is replicated to restore its original double-stranded condition. In the recipient cell, relaxase catalyzes the joining of the ends of the linear T-DNA molecule, and then the circular molecule is replicated to a double-stranded F factor. (b) Two *E. coli* cells in the act of conjugation. This is a micrograph showing two conjugating *E. coli* cells. The cell on the *left* is F⁺, and the one on the *right* is F⁻. The two cells make contact with each other via sex pili that are made by the F⁺ cell.

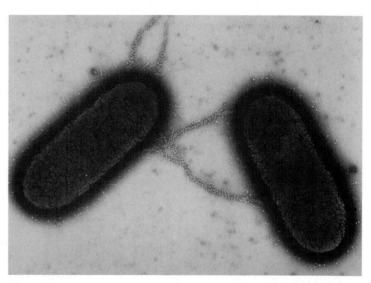

(b)

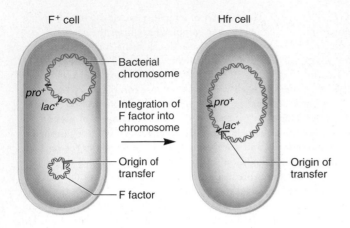

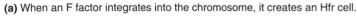

(a) When an F factor integrates into the chromosome, it creates an Hfr cell.

FIGURE 6.5   **The transfer of the *E. coli* chromosome occurs in a linear manner during an Hfr-mediated conjugation. (a)** An Hfr cell is created when an F factor integrates into the bacterial chromosome. **(b)** The transfer of the bacterial chromosome begins at the origin of transfer and then proceeds around the circular chromosome. After a segment of chromosome has been transferred to the F⁻ recipient cell, it recombines with the recipient cell's chromosome. If mating occurs for a brief period, only a short segment of the chromosome is transferred. If mating is prolonged, a longer segment of the bacterial chromosome is transferred.

GENES→TRAITS   The F⁻ recipient cell was originally *lac*⁻ (unable to metabolize lactose) and *pro*⁻ (unable to synthesize proline). If mating occurred for a short period of time, the recipient cell acquired *lac*⁺, allowing it to metabolize lactose. If mating occurred for a longer period of time, the recipient cell also acquired *pro*⁺, enabling it to synthesize proline.

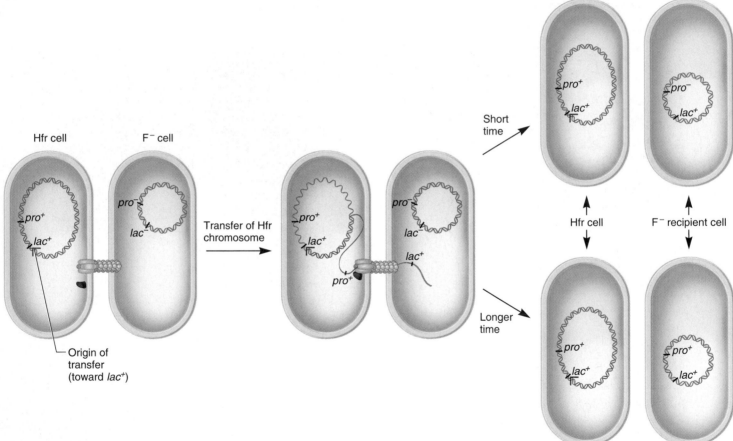

(b) An Hfr donor cell can pass a portion of its chromosome to an F⁻ recipient cell.

F factor has become integrated into the bacterial chromosome (fig. 6.5*a*). The term **episome** is used to describe a segment of bacterial DNA that can exist as a plasmid and also integrate into the chromosome.

Hayes, who independently isolated another *Hfr* strain, demonstrated that conjugation between an *Hfr* strain and an F⁻ strain involves the transfer of a portion of the bacterial chromosome from the *Hfr* strain to the F⁻ cell (fig. 6.5*b*). The origin of

transfer within the integrated F factor determines the starting point and direction of this transfer process. One of the DNA strands is cut at the origin of transfer. This cut, or nicked, site is the starting point that will enter the F⁻ recipient cell. Behind this starting point, a strand of the DNA from the bacterial chromosome begins to enter the F⁻ cell in a linear manner. This transfer process occurs in conjunction with chromosomal replication, so that the Hfr cell retains its original chromosomal composition. It generally takes about 1.5 to 2 hours for the entire Hfr chromosome to be passed into the F⁻ cell. Since most matings do not last that long, usually only a portion of the Hfr chromosome is transmitted to the F⁻ cell.

Once inside the F⁻ cell, the chromosomal material from the Hfr cell can swap, or recombine, with the homologous region of the recipient cell's chromosome. The molecular steps of homologous recombination are considered in chapter 17. The net result is that the recipient cell has received a new segment of chromosomal DNA from the Hfr cell. As illustrated in figure 6.5b, this recombination may provide the recipient cell with a new combination of alleles. In this example, the recipient strain was originally *lac⁻* (unable to metabolize lactose) and *pro⁻* (unable to synthesize proline). If mating occurred for a short time, the recipient cell received a short segment of chromosomal DNA from the donor. In this case, the recipient cell has become

lac⁺ but remains *pro⁻*. If the mating is prolonged, the recipient cell will receive a longer segment of chromosomal DNA from the donor. After a longer mating, the recipient becomes *lac⁺* and *pro⁺*. As noted in figure 6.5b, an important feature of Hfr mating is that the bacterial chromosome is transferred linearly to the recipient strain. In this example, *lac⁺* is always transferred first and *pro⁺* is transferred later.

In any particular *Hfr* strain, the origin of transfer has a specific orientation that promotes either a counterclockwise or clockwise transfer of genes. Also, among different *Hfr* strains, the origin of transfer may be located in different regions of the chromosome. Therefore, the order of gene transfer depends on the location and orientation of the origin of transfer. For example, another *Hfr* strain could have its origin of transfer next to *pro⁺* and transfer *pro⁺* first and then *lac⁺*.

EXPERIMENT 6A

Conjugation Experiments Can Be Used to Map Genes Along the *E. coli* Chromosome

The first genetic mapping experiments in bacteria were carried out by Elie Wollman and François Jacob in the 1950s. At the time of their studies, not much information was known about the organization of bacterial genes along the chromosome. A few key advances made Wollman and Jacob realize that the process of genetic transfer could be used to map the order of genes in *E. coli*. Most importantly, the discovery of conjugation by Lederberg and Tatum, and the identification of *Hfr* strains by Cavalli-Sforza and Hayes, made it clear that bacteria can transfer genes from donor to recipient cells in a linear fashion. In addition, Wollman and Jacob were aware of previous microbiological studies concerning bacteriophages that bind to *E. coli* cells and subsequently infect them. These studies showed that bacteriophages can be sheared from the surface of *E. coli* cells if they are spun in a blender. In this treatment, the bacteriophages are detached from the surface of the bacterial cells, but the bacteria themselves remain healthy and viable. Wollman and Jacob reasoned that a blender treatment could also be used to separate bacterial cells that were in the act of conjugation without killing them. This technique is known as an **interrupted mating.**

The rationale behind Wollman and Jacob's mapping strategy is that the time it takes for genes to enter a donor cell is directly related to their order along the bacterial chromosome. They hypothesized that the chromosome of the donor strain in an Hfr mating is transferred in a linear manner to the recipient strain. If so, the order of genes along the chromosome can be deduced by determining the time it takes various genes to enter the recipient strain. Assuming that the Hfr chromosome is transferred linearly, they realized that interruptions of mating at different times would lead to various lengths of the Hfr chromosome being transferred to the F⁻ recipient cell. If two bacterial cells had mated for a short period of time, only a small segment of the Hfr chromosome

would be transferred to the recipient bacterium. However, if the bacterial cells were allowed to mate for a longer period of time before being interrupted, a longer segment of the Hfr chromosome could be transferred. By determining which genes were transferred during short matings and which required longer mating times, Wollman and Jacob were able to deduce the order of particular genes along the *E. coli* chromosome.

As shown in the experiment of figure 6.6, Wollman and Jacob began with two *E. coli* strains. The donor (*Hfr*) strain had the following genetic composition:

thr⁺: able to synthesize threonine, an essential amino acid for growth

leu⁺: able to synthesize leucine, an essential amino acid for growth

aziˢ: sensitive to killing by azide (a toxic chemical)

tonˢ: sensitive to infection (i.e., a plaque formation) by bacteriophage T1. As discussed later in chapter 6, bacteriophages are viruses that infect bacteria and may cause lysis, resulting in plaque formation.

lac⁺: able to metabolize lactose and use it for growth

gal⁺: able to metabolize galactose and use it for growth

strˢ: sensitive to killing by streptomycin (an antibiotic)

The recipient (*F⁻*) strain had the opposite genotype: *thr⁻ leu⁻ aziʳ tonʳ lac⁻ gal⁻ strʳ* (r = resistant). Before the experiment, Wollman and Jacob already knew that the *thr⁺* gene was transferred first, followed by the *leu⁺* gene, and both were transferred relatively soon (5–10 minutes) after mating. Their main goal in this experiment was to determine the times at which the other genes (*aziˢ tonˢ lac⁺ gal⁺*) were transferred to the recipient strain. The transfer of the *strˢ* gene was not examined because streptomycin was used to kill the donor strain following conjugation.

Before discussing the conclusions of this experiment, it is helpful to describe how Wollman and Jacob monitored gene transfer. To determine if particular genes had been transferred after mating, they took the mated cells and first plated them on growth media that lacked threonine and leucine but contained streptomycin. On these plates, the original donor and recipient strains could not grow, because the donor strain was streptomycin sensitive and the recipient strain required threonine and leucine. However, mated cells in which the donor had transferred chromosomal DNA carrying the *thr⁺* and *leu⁺* genes to the recipient cell would be able to grow.

To determine the order of gene transfer of the *aziˢ*, *tonˢ*, *lac⁺*, and *gal⁺* genes, Wollman and Jacob picked colonies from the first plates and restreaked them on plates that contained azide or bacteriophage T1, or on plates that contained lactose or galactose as the sole source of energy for growth. The plates were incubated overnight to observe the formation of visible bacterial growth. Whether or not the bacteria could grow depended on their genotypes. For example, a cell that is *aziˢ* cannot grow on plates containing azide, and a cell that is *lac⁻* cannot grow on plates containing lactose as the carbon source for growth. By comparison, a cell that is *aziʳ* and *lac⁺* can grow on both types of plates.

■ THE HYPOTHESIS

The chromosome of the donor strain in an Hfr mating is transferred in a linear manner to the recipient strain. The order of genes along the chromosome can be deduced by determining the time various genes take to enter the recipient strain.

■ TESTING THE HYPOTHESIS — FIGURE 6.6 The use of conjugation to map the order of genes along the *E. coli* chromosome.

Starting materials: The two *E. coli* strains already described, one *Hfr* strain (*thr⁺ leu⁺ aziˢ tonˢ lac⁺ gal⁺ strˢ*) and one *F⁻* (*thr⁻ leu⁻ aziʳ tonʳ lac⁻ gal⁻ strʳ*).

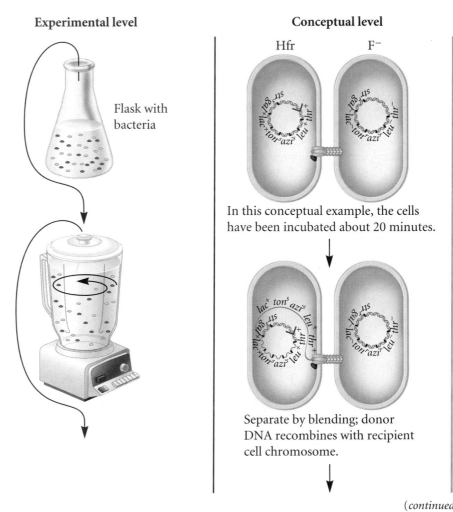

Experimental level

1. Mix together a large number of Hfr donor and F⁻ recipient cells.

Flask with bacteria

2. After different periods of time, take a sample of cells and interrupt conjugation in a blender.

Conceptual level

Hfr   F⁻

In this conceptual example, the cells have been incubated about 20 minutes.

Separate by blending; donor DNA recombines with recipient cell chromosome.

(*continued*)

3. Plate the cells on growth media lacking threonine and leucine but containing streptomycin. Note: The general methods for growing bacteria in a laboratory are described in the appendix.

4. Pick each surviving colony, which would have to be *thr+ leu+ str^r*, and test to see if it is sensitive to killing by azide, sensitive to infection by T1 bacteriophage, and able to metabolize lactose or galactose.

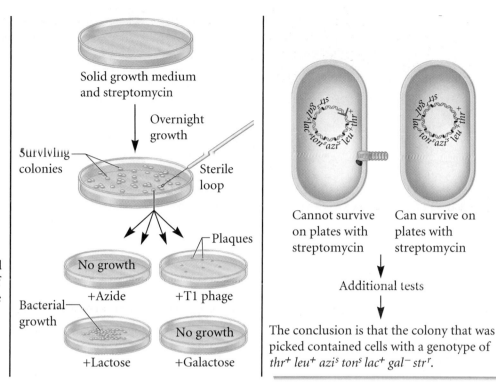

Solid growth medium and streptomycin

Overnight growth

Surviving colonies

Sterile loop

No growth   Plaques

+Azide   +T1 phage

Bacterial growth

No growth

+Lactose   +Galactose

Cannot survive on plates with streptomycin

Can survive on plates with streptomycin

Additional tests

The conclusion is that the colony that was picked contained cells with a genotype of *thr+ leu+ azi^s ton^s lac+ gal− str^r*.

THE DATA

Minutes that Bacterial Cells Were Allowed to Mate Before Blender Treatment	Percent of Surviving Bacterial Colonies with the Following Genotypes:				
	thr+ leu+	*azi^s*	*ton^s*	*lac+*	*gal+*
5	—*	—	—	—	—
10	100	12	3	0	0
15	100	70	31	0	0
20	100	88	71	12	0
25	100	92	80	28	0.6
30	100	90	75	36	5
40	100	90	75	38	20
50	100	91	78	42	27
60	100	91	78	42	27

*There were no surviving colonies within the first 5 minutes of mating.

INTERPRETING THE DATA

Now let's discuss the data shown in figure 6.6. After the first plating, all survivors would be cells in which the *thr+* and *leu+* alleles had been transferred to the *F−* recipient strain (which is already streptomycin resistant). As seen in the data, 5 minutes was not sufficient time to transfer the *thr+* and *leu+* alleles, since no surviving colonies were observed. After 10 minutes or longer, how-ever, surviving bacterial colonies with the *thr+ leu+* genotype were obtained. To determine the order of the remaining genes (*azi^s*, *ton^s*, *lac+*, and *gal+*) each surviving colony was tested to see if it was sensitive to killing by azide, sensitive to infection by T1 bacteriophage, able to use lactose for growth, or able to use galactose for growth. A consistent pattern emerged from the data. The gene that conferred sensitivity to azide (*azi^s*) was transferred first, followed by *ton^s*, *lac+*, and finally *gal+*. From these data, as well as those from other experiments, Wollman and Jacob constructed a genetic map that described the order of these genes along the *E. coli* chromosome.

| thr | leu | azi | ton | lac | gal |

This work provided the first method for bacterial geneticists to map the order of genes along the bacterial chromosome. Throughout the course of their studies, Wollman and Jacob identified several different *Hfr* strains in which the origin of transfer had been integrated at different places along the bacterial chromosome. When they compared the order of genes among different *Hfr* strains, their results were consistent with the idea that the *E. coli* chromosome is circular (see solved problem S2).

A self-help quiz involving this experiment can be found at the Online Learning Center.

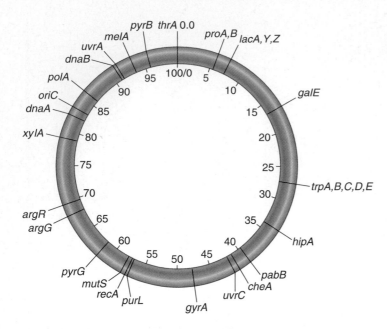

FIGURE 6.7   A simplified genetic map of the *E. coli* chromosome indicating the positions of several genes. *E. coli* has a circular chromosome with about 4,200 different genes. This map shows the locations of several of them. The map is scaled in units of minutes, and it proceeds in a clockwise direction. The starting point on the map is the gene *thrA*.

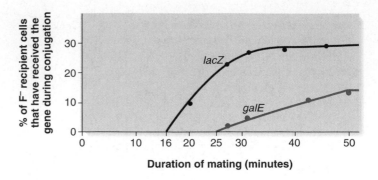

FIGURE 6.8   Time course of an interrupted *E. coli* conjugation experiment. By extrapolating the data back to the origin, the approximate time of entry of the *lacZ* gene is found to be 16 minutes, that of the *galE* gene 25 minutes. Therefore, the distance between these two genes is 9 minutes.

A Detailed Genetic Map of the *E. coli* Chromosome Has Been Obtained from Many Conjugation Studies

Conjugation experiments have been used to map more than 1,000 genes along the circular *E. coli* chromosome. A map of the *E. coli* chromosome is shown in figure 6.7. This simplified map shows the locations of only a few dozen genes. Since the chromosome is circular, we must arbitrarily assign a starting point on the map. The gene *thrA* is at the location of 0 minutes. The *E. coli* genetic map is 100 minutes long, which is approximately the time that it takes to transfer the complete chromosome during an Hfr mating.

As shown in figure 6.7, we scale genetic maps from bacterial conjugation studies in units of **minutes.** This unit refers to the relative time it takes for genes to first enter an F^- recipient strain during a conjugation experiment. The distance between two genes is determined by comparing their times of entry during a conjugation experiment. As shown in figure 6.8, the time of entry is found by conducting mating experiments at different time intervals before interruption. We compute the time of entry by extrapolating the time back to the origin. In this experiment, the time of entry of the *lacZ* gene was approximately 16 minutes, and that of the *galE* gene was 25 minutes. Therefore, these two genes are approximately 9 minutes apart from each other along the *E. coli* chromosome.

Bacteriophages Can Also Transfer Genetic Material from One Bacterial Cell to Another in the Process of Transduction

We now turn to a second method of genetic transfer involving bacteriophages. Before we discuss the ability of bacteriophages to transfer genetic material between bacterial cells, let's consider some general features of a phage's life cycle. Bacteriophages are composed of genetic material that is surrounded by a protein coat. As shown in figure 6.9, bacteriophages can bind to the surface of a bacterium and inject their genetic material into the bacterial cytoplasm. Depending on the specific type of virus and its growth conditions, a phage may follow a **lytic cycle** or a **lysogenic cycle. Virulent phages** only follow a lytic cycle. During the lytic cycle, the bacteriophage directs the synthesis of many copies of the phage genetic material and coat proteins (see *left* side of figure 6.9). These components then assemble to make new phages. When synthesis and assembly are completed, the bacterial host cell is lysed (broken apart), releasing the newly made phages into the environment.

In other cases, a bacteriophage infects a bacterium and follow the lysogenic cycle (see *right* side of figure 6.9). A bacteriophage that usually exists in the lysogenic life cycle is called a **temperate phage.** Under most conditions, temperate phages do not produce new phages and will not kill the host bacterial cell. During the lysogenic cycle, a phage integrates its genetic material into the chromosome of the bacterium. This integrated phage DNA is known as a **prophage.** A prophage can exist in a dormant state for a long time, during which no new bacteriophages are made. When a bacterium containing a lysogenic prophage divides to produce two daughter cells, it will copy the prophage's genetic material along with its own chromosome. Therefore, both daughter cells will inherit the prophage. At some later time, a prophage may become activated to excise itself from the bacterial chromosome and enter the lytic cycle. Then, once again, it will promote the synthesis of new phages and eventually lyse the host cell.

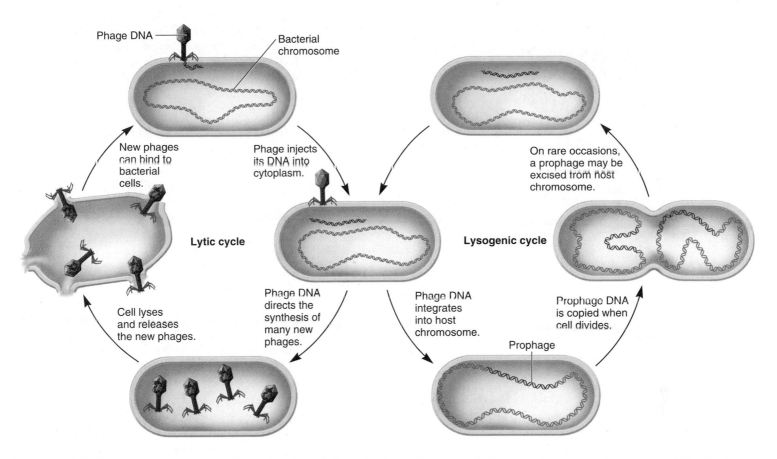

FIGURE 6.9   **The lytic and lysogenic life cycles of certain bacteriophages.** Some bacteriophages, such as temperate phages, can follow both cycles. Other phages, known as virulent phages, can only follow a lytic cycle.

With a general understanding of bacteriophage life cycles, we may now examine the ability of phages to transfer genetic material between bacteria. This process is called transduction. Examples of phages that can transfer bacterial chromosomal DNA from one bacterium to another are the P22 and P1 phages, which infect the bacterial species *Salmonella typhimurium* and *E. coli,* respectively. The P22 and P1 phages can follow either the lytic or lysogenic cycle. While in the lytic cycle, the phage directs the synthesis of phage DNA and coat proteins, which then assemble to make new phages. During this process, the bacterial chromosome becomes fragmented into small pieces of DNA (fig. 6.10).

Occasionally, a mistake can happen in which a piece of bacterial DNA assembles with the coat proteins. This creates a phage that contains bacterial chromosomal DNA. When phage synthesis is completed, the bacterial cell is lysed and releases the newly made phage into the environment. Following release, this abnormal phage can still bind to a living bacterial cell and inject its genetic material into the bacterium. This DNA fragment (which was derived from the bacterial chromosomal DNA) can then recombine with the bacterial chromosome inside the infected cell. In the example of figure 6.10, any piece of the bacterial chromosomal DNA can be incorporated into the phage. This type of transduction is called **generalized transduction.**

Transduction was first discovered in 1952 by Joshua Lederberg and Norton Zinder using an experimental strategy similar to that of figure 6.1. They mixed together two strains of the bacterium *S. typhimurium.* One strain, designated LA-22, was *phe⁻ trp⁻ met⁺ his⁺* (unable to synthesize phenylalanine or tryptophan but able to synthesize methionine and histidine); the other strain, LA-2, was *phe⁺ trp⁺ met⁻ his⁻* (able to synthesize phenylalanine and tryptophan, but unable to synthesize methionine or histidine). When a mixture of these cells was placed on plates with growth media lacking these four amino acids, approximately 1 cell in 100,000 was observed to grow. The genotype of the surviving bacterial cells must have been *phe⁺ trp⁺ met⁺ his⁺.* Therefore, Lederberg and Zinder concluded that genetic material had been transferred between the two strains.

A novel result occurred when Lederberg and Zinder repeated this experiment using a U-tube apparatus as previously described in figure 6.2. They placed the LA-22 strain (*phe⁻ trp⁻ met⁺ his⁺*) on one side of the filter and LA-2 (*phe⁺ trp⁺ met⁻ his⁻*) on the other. They removed samples from either side of the tube and plated the cells on plates lacking the four amino acids. Surprisingly, they obtained colonies from the side of the tube that contained LA-22 but not from the side that contained LA-2. From these results, they concluded that some filterable agent was being transferred from LA-2 to LA-22 that converted LA-22 to a *phe⁺ trp⁺ met⁺ his⁺* genotype. In other words, some filterable agent carrying the *phe⁺* and *trp⁺* genes was produced on the side of the tube containing LA-2, traversed the filter, and then was

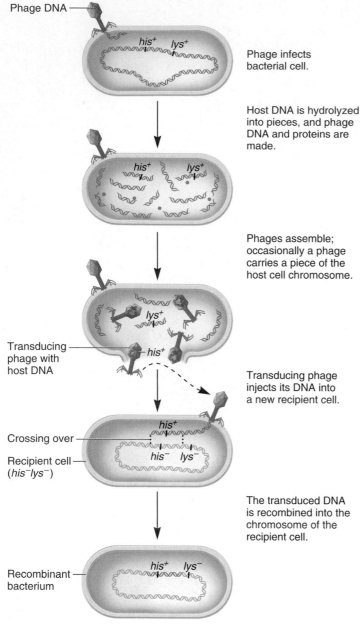

Phage DNA

Phage infects bacterial cell.

Host DNA is hydrolyzed into pieces, and phage DNA and proteins are made.

Phages assemble; occasionally a phage carries a piece of the host cell chromosome.

Transducing phage with host DNA

Transducing phage injects its DNA into a new recipient cell.

Crossing over

Recipient cell (*his⁻lys⁻*)

The transduced DNA is recombined into the chromosome of the recipient cell.

Recombinant bacterium

The recombinant bacterium has a genotype (*his⁺lys⁻*) that is different from recipient bacterial cell (*his⁻lys⁻*).

FIGURE 6.10   Transduction in bacteria.

GENES→TRAITS   The transducing phage introduced DNA into a new host cell that was originally *his⁻ lys⁻* (unable to synthesize histidine and lysine). During transduction, it received a segment of bacterial chromosomal DNA that carried *his⁺*. Following recombination, the host cell's genotype was changed to *his⁺*, so that it now could synthesize histidine.

taken up by the LA-22 strain. By conducting this type of experiment using filters with different pore sizes, Lederberg and Zinder found that the filterable agent was slightly less than 0.1 μm in diameter, a size much smaller than a bacterium. They correctly concluded that the filterable agent in these experiments was a bacteriophage. In this case, the LA-2 strain contained a prophage, such as P22. On occasion, the prophage switched to the lytic cycle and packaged a segment of DNA carrying the *phe⁺* and *trp⁺* genes. This phage then traversed the filter and injected the genes into LA-22.

Cotransduction Can Be Used to Map Genes That Are Within 2 Minutes of Each Other

During transduction, there is a maximum size of the DNA segment that can be packaged by bacteriophages. P1 cannot package pieces that are greater than 2 to 2.5% of the entire *E. coli* chromosome, and P22 cannot package pieces that are greater than 1% of the length of the *S. typhimurium* chromosome. If two genes are close together along the chromosome, a bacteriophage may package a single piece of the chromosome that carries both genes and transfer that piece to another bacterium. This phenomenon is called **cotransduction.** The likelihood that two genes will be cotransduced depends on how close together they lie. If two genes are far apart along the bacterial chromosome, they will never be cotransduced, because the bacteriophage cannot physically package a DNA fragment that is larger than 1 to 2.5% of the bacterial chromosome. In genetic mapping studies, cotransduction is used to determine the order and distance between genes that lie fairly close to each other.

To map genes using cotransduction, a researcher selects for the transduction of one gene and then monitors whether or not a second gene is cotransduced along with it. As an example, let's consider a donor strain of *E. coli* that is *arg⁺ met⁺ strˢ* (able to synthesize arginine and methionine but sensitive to killing by streptomycin) and a recipient strain that is *arg⁻ met⁻ strʳ* (fig. 6.11). The donor strain is infected with phage P1. Some of the *E. coli* cells are lysed by P1, and this P1 lysate is mixed with the recipient cells. After allowing sufficient time for transduction, the recipient cells are plated on a growth medium that contains arginine and streptomycin but not methionine. Therefore, these plates select for the growth of cells in which the *met⁺* gene has been transferred to the recipient strain, but they do not select for the growth of cells in which the *arg⁺* gene has been transferred, since the plates are supplied with arginine. Nevertheless, the *arg⁺* gene may have been cotransduced with the *met⁺* gene if the two genes are close together. To determine this, a sample of cells from each bacterial colony can be picked up with a wire loop and restreaked on plates that lack both amino acids. If the colony can grow, it must have also obtained the *arg⁺* gene during transduction. In other words, cotransduction of both the *arg⁺* and *met⁺* genes has occurred. Alternatively, if the restreaked colony does not grow, it must have received only the *met⁺* gene during transduction. Data from this type of experiment are shown at the *bottom* of figure 6.11. These data indicate a cotransduction frequency of 21/50 = 0.42.

T. T. Wu derived a mathematical expression that relates cotransduction frequency with map distance obtained from conjugation experiments. This equation is

$$\text{Cotransduction frequency} = (1 - d/L)^3$$

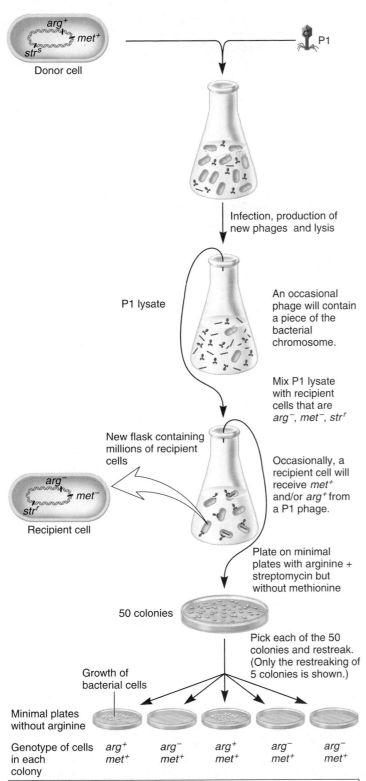

Genotype of cells in each colony:

| arg^+ | arg^- | arg^+ | arg^- | arg^- |
| met^+ | met^+ | met^+ | met^+ | met^+ |

Results				
Selected gene	Nonselected gene	Number of colonies that grew on		Cotransduction frequency
		minimal + arginine	minimal − arginine	
met^+	arg^+	50	21	0.42

where,

d = distance between two genes in minutes

L = the size of the chromosomal pieces (in minutes) that the phage carries during transduction
(For P1 transduction, this size is approximately 2% of the bacterial chromosome, which equals about 2 minutes.)

This equation assumes that the bacteriophage randomly packages pieces of the bacterial chromosome that are similar in size. Depending on the type of phage used in a transduction experiment, this assumption may not always be valid. Nevertheless, this equation has been fairly reliable in computing map distance for P1 transduction experiments in *E. coli*. We can use this equation to estimate the distance between the two genes described in figure 6.11.

$$0.42 = (1 - d/2)^3$$

$$(1 - d/2) = \sqrt[3]{0.42}$$

$$1 - d/2 = 0.75$$

$$d/2 = 0.25$$

$$d = 0.5 \text{ minutes}$$

This equation tells us that the distance between the met^+ and arg^+ genes is approximately 0.5 minutes.

Historically, genetic mapping strategies in bacteria often involved data from both conjugation and transduction experiments. Conjugation has been used to determine the relative order and distance of genes, particularly those that are far apart along the chromosome. By comparison, transduction experiments can provide very accurate mapping data for genes that are fairly close together.

FIGURE 6.11   The steps in a cotransduction experiment. The donor strain is arg^+ met^+ str^s (able to synthesize arginine and methionine but sensitive to streptomycin), and the recipient strain is arg^- met^- str^r. The donor strain is infected with phage P1. Some of the *E. coli* cells are lysed by P1, and this P1 lysate is mixed with the recipient cells. P1 phages in this lysate may carry fragments of the donor cell's chromosome, and the P1 phage may inject that DNA into the recipient cells. To identify recipient cells that have received the met^+ gene from the donor strain, the recipient cells are plated on a growth medium that contains arginine and streptomycin but not methionine. To determine if the arg^+ gene has also been cotransduced, cells from each bacterial colony are lifted with a sterile wire loop and streaked on plates that lack both amino acids. If the cells can grow, cotransduction of the arg^+ and met^+ genes has occurred. Alternatively, if the restreaked colony does not grow, it must have received only the met^+ gene during transduction. Note: P1 can lyse cells in two ways. One possibility is that P1 can exist as a prophage that switches from the lysogenic to the lytic cycle. Alternatively, as in this experiment, the uptake of P1 DNA into the donor cell leads immediately to the lytic cycle.

Bacteria Can Also Transfer Genetic Material by Transformation

A third mechanism for the transfer of genetic material from one bacterium to another is known as **natural transformation.** This process was first discovered by Frederick Griffith in 1928 while working with strains of *Streptococcus pneumoniae.* During natural transformation, a living bacterial cell will take up a strand of DNA released from a dead bacterium. This DNA strand may then recombine with the bacterial chromosome, producing a bacterium with genetic material that it has received from the dead bacterium. Alternatively, scientists have devised ways to force DNA into cells. This is termed **artificial transformation.** For example, a technique known as electroporation, in which an electric current causes the uptake of DNA, can be used by researchers to promote the transport of DNA into a bacterial cell. These types of methods are described in chapter 19.

Since the initial studies of Griffith, we have learned a great deal about the events that occur in natural transformation. This form of genetic transfer has been reported in a wide variety of Gram-negative and Gram-positive bacteria. Bacterial cells that are able to take up DNA are known as **competent cells.** Those that can take up DNA naturally carry genes that encode proteins called **competence factors.** These proteins facilitate the binding of DNA fragments to the cell surface, the uptake of DNA into the cytoplasm, and its subsequent incorporation into the bacterial chromosome. Temperature, ionic conditions, and nutrient availability can affect whether or not a bacterium will be competent to take up genetic material from its environment. These conditions influence the expression of competence genes.

In recent years, geneticists have unraveled some of the steps that occur when competent bacterial cells are transformed by genetic material in their environment. Figure 6.12 describes these steps. First, a large fragment of genetic material binds to the surface of the bacterial cell. Competent cells express DNA receptors that promote such binding. Before entering the cell, however, this large piece of chromosomal DNA must be cut into smaller fragments. This cutting is accomplished by a bacterial enzyme known as an endonuclease, which makes occasional random cuts in the long piece of chromosomal DNA. At this stage, the DNA fragments are composed of double-stranded DNA.

The next step is for the DNA fragment to begin its entry into the bacterial cytoplasm. For this to occur, the double-stranded DNA interacts with proteins in the bacterial membrane. One of the DNA strands is degraded and the other strand enters the bacterial cytoplasm via an uptake system, which is similar to the one described for conjugation.

To be stably inherited, the DNA strand must be incorporated into the bacterial chromosome. If the DNA strand has a sequence that is similar to a region of DNA in the bacterial chromosome, the DNA may be incorporated into the chromosome by a process known as **homologous recombination.** The details of homologous recombination are discussed in chapter 17. For this to occur, the single-stranded DNA aligns itself with the homologous location on the bacterial chromosome. In the example shown in figure 6.12, the foreign DNA carries a functional *lac*⁺

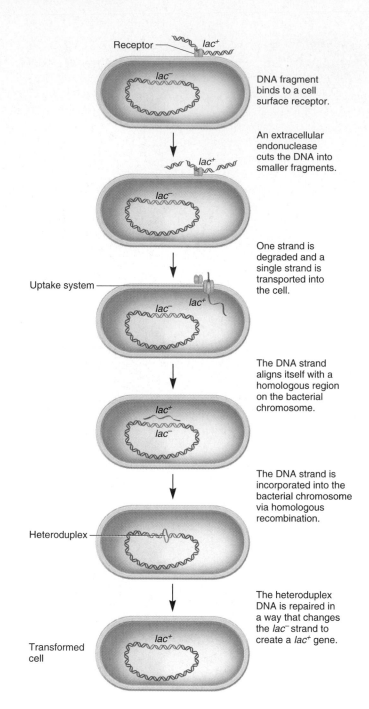

FIGURE 6.12   The steps of bacterial transformation. A large piece of genetic material containing the *lac*⁺ gene binds to cell surface receptors of a competent bacterium. The genetic material is digested into smaller fragments by an extracellular endonuclease. One of the DNA strands is degraded, and the other, which contains the *lac*⁺ gene, enters the bacterial cell. The newly imported DNA strand aligns with the matching section of the bacterial chromosome, in this case where the *lac*⁻ gene is located. The imported strand is substituted for one of the chromosomal strands. The resulting temporary mismatch creates a heteroduplex. The heteroduplex is then repaired.

GENES→TRAITS   Bacterial transformation can also lead to new genetic traits for the recipient cell. The original cell was *lac*⁻ (unable to metabolize lactose). Following transformation, it became *lac*⁺. This result would transform the original bacterial cell into a cell that could grow on plates that contain lactose. Before transformation, the original *lac*⁻ cell would not have been able to grow on plates containing lactose.

gene. This strand of DNA aligns itself with a nonfunctional (mutant) *lac⁻* gene already present within the bacterial chromosome. The foreign DNA then recombines with one of the strands in the bacterial chromosome of the competent cell. In other words, the foreign DNA replaces one of the chromosomal strands of DNA, which is subsequently degraded. During recombination, alignment of the *lac⁻* and the *lac⁺* alleles results in a region of mismatch called a **heteroduplex.** However, the heteroduplex is a temporary situation. DNA repair enzymes in the recipient cell recognize the heteroduplex and repair it. In this case, the heteroduplex has been repaired by eliminating the mutation that caused the *lac⁻* genotype, thereby creating a *lac⁺* gene. In this example, the recipient cell has been transformed from a *lac⁻* strain to a *lac⁺* strain. Alternatively, a DNA fragment that has entered a cell may not be homologous to any genes that are already found in the bacterial chromosome. In this case, the DNA strand may be incorporated at a random site in the chromosome. This process, known as **nonhomologous** or **illegitimate recombination,** is less efficient compared to homologous recombination.

Transformation has also been used to map many bacterial genes using methods similar to the cotransduction experiments described earlier. If two genes are close together, the **cotransformation** frequency is expected to be high, whereas genes that are far apart will have a cotransformation frequency that is very low or even zero. Like cotransduction, genetic mapping via cotransformation is used only to map genes that are relatively close together.

Horizontal Gene Transfer Is the Transfer of Genes Among Different Species

Gene transfer via conjugation, transformation, and transduction is relatively widespread among bacterial species. When a gene is transferred between two different species, this is termed **horizontal gene transfer.** By comparison, the transmission of genes from mother cell to daughter cell or from parents to offspring is called **vertical gene transfer.** When analyzing the genomes of bacterial species, it has become increasingly clear that a sizable fraction of their genes are derived from horizontal gene transfer. For example, in *E. coli* and *Salmonella typhimurium,* roughly 17% of their genes have been acquired via horizontal gene transfer during the past 100 million years.

The types of genes that have been acquired via horizontal gene transfer are quite varied, though they commonly involve functions that are readily acted upon by natural selection. These include genes that confer antibiotic resistance, the ability to degrade toxic compounds, and pathogenicity. Geneticists have suggested that much of the speciation that has occurred in prokaryotic species is the result of horizontal gene transfer. In many cases, the acquisition of new genes allows a novel survival strategy that has led to the formation of a new species. These phenomena are also considered in chapters 25 and 26.

The medical relevance of horizontal gene transfer is quite profound. Antibiotics are commonly prescribed to treat many bacterial illnesses. These include infections of the respiratory tract, urinary tract, skin, ear, eye, etc. In addition, antibiotics are used in agriculture as a supplement in animal feed and to control certain bacterial diseases of high-value fruits and vegetables. Unfortunately, however, the widespread and uncontrolled use of antibiotics has promoted the prevalence of antibiotic-resistant strains of bacteria. This phenomenon, termed **acquired antibiotic resistance,** may occur via genetic alterations in the bacteria's own genome or by the horizontal transfer of resistance genes from a resistant to a sensitive strain. Resistant strains carry genes that counteract the effects of antibiotics. Such resistance genes encode proteins that either break down the drug, pump the drug out of the cell, or prevent the drug from inhibiting cellular processes.

Bacterial resistance to antibiotics in community-acquired respiratory tract infections as well as other medical illnesses is a serious problem, and it is increasing in prevalence worldwide at an alarming rate. As often mentioned in the news media, antibiotic resistance has increased dramatically over the past 10 years. Resistance has been reported in almost all species of bacteria. In many countries, for example, penicillin resistance in *Streptococcus pneumoniae* is found in nearly 50% of all strains, with resistance to other drugs rising as well. Likewise, the antibiotic-resistance problem in hospitals continues to worsen. Resistant strains of *Klebsiella pneumoniae* and *Enterococci* sp. are significant causes of morbidity and mortality among critically ill patients in intensive care units. Treating infections caused by these pathogens presents difficult therapeutic dilemmas.

6.2 INTRAGENIC MAPPING IN BACTERIOPHAGES

We now turn our attention to viruses. Biologists do not consider viruses to be living, because they rely on a host cell for their existence and proliferation. Nevertheless, we can think of viruses as having traits, because they have unique biological structures and functions. Each type of virus has its own genetic material, which contains many genes. In this section, we will focus our attention on a bacteriophage called T4. Its genetic material contains several dozen different genes encoding proteins that carry out a variety of functions. For example, some of the genes encode proteins needed for the synthesis of new viruses and the lysis of the host cell. Other genes encode the viral coat proteins that are found in the head, shaft, base plate, and tail fibers. Figure 6.13 illustrates the locations of several coat proteins that make up the structure of T4. As seen here, five different proteins bind to each other to form a tail fiber (see inset to fig. 6.13). The expression of T4 genes to make these proteins provides the bacteriophage with the trait of having tail fibers, enabling it to attach to the surface of a bacterium.

The study of viral genes has been instrumental in our basic understanding of how the genetic material works. During the 1950s, Seymour Benzer embarked on a 10-year study that focused on the function of viral genes in the T4 bacteriophage. In this section, we will examine some of his pivotal results, which

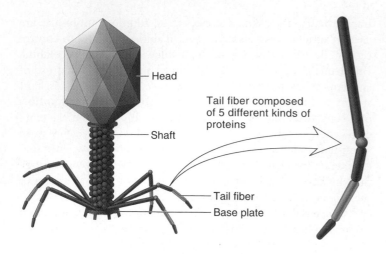

Head

Tail fiber composed
of 5 different kinds of
proteins

Shaft

Tail fiber

Base plate

FIGURE 6.13   Structure of the T4 virus.

GENES→TRAITS   The inset to this figure shows the five different proteins, encoded by five different genes, which make up the tail fiber. The expression of these genes provides the bacteriophage with the trait of having tail fibers, enabling it to attach itself to the surface of a bacterium.

advanced our knowledge of gene function. We will also explore how he conducted a detailed type of genetic mapping known as **intragenic** or **fine structure mapping.** The difference between intragenic and intergenic mapping is shown here:

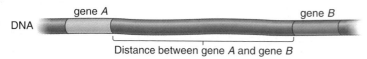

gene *A*   gene *B*

DNA

Distance between gene *A* and gene *B*

Intergenic mapping

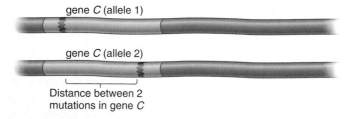

gene *C* (allele 1)

gene *C* (allele 2)

Distance between 2
mutations in gene *C*

Intragenic mapping (also known as fine structure mapping)

We considered intergenic mapping earlier in this chapter and in chapter 5. In that type of mapping, the goal is to determine the distance between two different genes. As shown here, the determination of the distance between genes *A* and *B* is an example of intergenic mapping. By comparison, intragenic mapping seeks to ascertain distances within the same gene. For example, in a population, gene *C* may exist as two different mutant alleles. One allele may be due to a mutation near the beginning of the gene, the second to a mutation near the end. The goal of intragenic mapping is to determine the distance between the two mutations that occur in the same gene. In this section, we will explore the pioneering studies that showed the feasibility of intragenic mapping.

Mutations in Viral Genes Can Alter Plaque Morphology

As they progress through the lytic cycle, bacteriophages ultimately produce new phages, which are released when the bacterial cell lyses (refer back to fig. 6.9). In the laboratory, researchers can visually observe the consequences of bacterial cell lysis in the following way (fig. 6.14). A sample of bacterial cells and lytic bacteriophages are mixed together. This mixture of cells and phages is then poured onto petri plates that contain nutrient agar for bacterial cell growth. Bacterial cells, which are not infected by a bacteriophage, will rapidly grow and divide to produce a lawn of bacteria. This lawn of bacteria is opaque (i.e., you cannot see through it to the underlying agar). In the experiment shown in figure 6.14, eleven bacterial cells have been infected by bacteriophages, and these infected cells are found at random locations in the lawn of uninfected bacteria. The infected cells will lyse and release newly made bacteriophages. These bacteriophages will then infect the nearby bacteria within the lawn. These cells eventually lyse and also release newly made phages. Over time, these repeated cycles of infection and lysis will produce, around each site of an original phage infection, an observable clear area, or **plaque,** where the bacteria have been lysed.

Now that we have an appreciation for the composition of a viral plaque, let's consider how the characteristics of a plaque can be viewed as traits of a bacteriophage. As we have seen in chapter 6 and previous chapters, the genetic analysis of any organism requires strains with allelic differences. Since bacteriophages can be visualized only with an electron microscope, it would be rather difficult for geneticists to analyze mutations that affect phage morphology. However, some mutations in the bacteriophage's genetic material can alter the ability of the phage to cause plaque formation. Therefore, we can view the morphology of plaques as a trait of the bacteriophage. Since plaques are visible with the naked eye, mutations affecting this trait lend themselves to a much easier genetic analysis. An example is a rapid-lysis mutant of bacteriophage T4, which tends to form unusually large plaques (fig. 6.15). The plaques are large because the mutant phages lyse the bacterial cells more rapidly than do the wild-type phages. Rapid-lysis mutants form large, clearly defined plaques, as opposed to wild-type bacteriophages that produce smaller, fuzzy-edged plaques. Mutations in several different bacteriophage genes can produce a rapid-lysis phenotype.

Benzer studied one category of T4 phage mutants, designated *rII* (*r* stands for rapid lysis). In the bacterial strain called *E. coli B*, *rII* phages produce abnormally large plaques. Nevertheless, *E. coli B* strains produce low yields of *rII* phages, because the *rII* phages lyse the bacterial cells so quickly they do not have sufficient time to produce many new phages. To help study this phage, Benzer wanted to obtain large quantities of it. Therefore, to improve his yield of *rII* phage, he decided to test its yield in other bacterial strains.

On the day Benzer decided to do this, he happened to be teaching a phage genetics class. For that class, he was growing two *E. coli* strains designated *E. coli K12S* and *E. coli K12(λ)*. He was growing these two strains to teach his class about the lysogenic

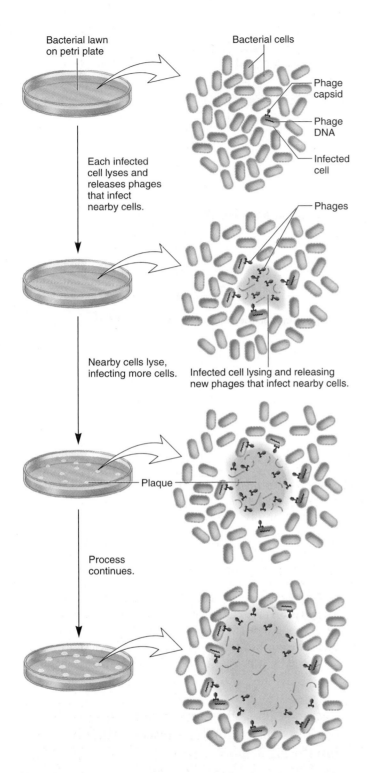

FIGURE 6.14   The formation of phage plaques on a lawn of bacteria in the laboratory. In this experiment, bacterial cells were mixed with a small number of lytic bacteriophages. In the figure, eleven bacterial cells were initially infected by phages. The cells were poured onto petri plates containing nutrient agar for bacterial cell growth. Bacterial cells will rapidly grow and divide to produce an opaque lawn of densely packed bacteria. The eleven infected cells will lyse and release newly made bacteriophages. These bacteriophages will then infect the nearby bacteria within the lawn. Likewise, these newly infected cells will lyse and release new phages. By this repeated process, the area of cell lysis creates a clear zone known as a viral plaque.

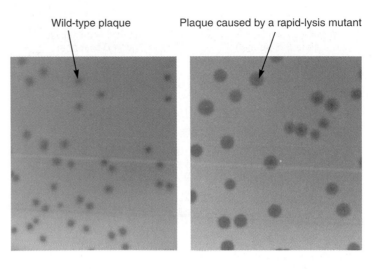

Plaques caused by wild-type bacteriophages

Plaques caused by rapid-lysis bacteriophage strains

FIGURE 6.15   A comparison of plaques produced by the wild-type T4 bacteriophage and a rapid-lysis mutant.
GENES→TRAITS   The plaques on the *left* side were caused by the infection and lysis of *E. coli* cells by the wild-type T4 phage. On the *right,* a mutation in a phage gene, called a rapid-lysis mutation, caused the phage to lyse *E. coli* cells more quickly. A phage carrying a rapid-lysis mutant allele yields much larger plaques.

cycle. *E. coli K12(λ)* has DNA from another phage, called lambda, integrated into its chromosome, whereas *E. coli K12S* does not. To see if the use of these strains might improve phage yield, *E. coli B*, *E. coli K12S*, and *E. coli K12(λ)* were infected with the *rII* and wild-type T4 phage strains. As expected, the wild-type phage could infect all three bacterial strains. However, the *rII* mutant strains behaved quite differently. In *E. coli B*, the *rII* strains produced large plaques that had poor yields of bacteriophage. In *E. coli K12S*, the *rII* mutants produced normal plaques that gave good yields of phage. Surprisingly, in *E. coli K12(λ)*, the *rII* mutants were unable to produce plaques at all, for reasons that were not understood. Nevertheless, as we will see later, this fortuitous observation was a critical feature that allowed intragenic mapping in this bacteriophage.

A Complementation Test Can Reveal if Mutations Are in the Same Gene or in Different Genes

In his experiments, Benzer was interested in a single trait, namely, the ability to form plaques. He had isolated many *rII* mutant strains that could form large plaques in *E. coli B* but could not form plaques in *E. coli K12(λ)*. Prior to his intragenic mapping studies, he needed to know if all the *rII* mutations were in the same gene or if they involved mutations in different genes. To accomplish this, he conducted **complementation** experiments. In this type of approach, the goal is to determine if two different mutations (which affect the same trait) are in the same gene or in two different genes. The possible outcomes of complementation experiments involving mutations that affect plaque formation are shown in figure 6.16.

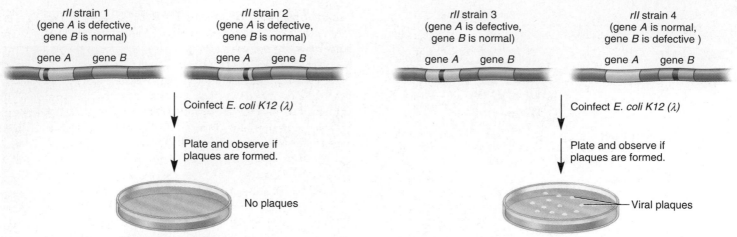

(a) **Noncomplementation:** The phage mutations are in the same gene.

(b) **Complementation:** The phage mutations are in different genes.

FIGURE 6.16   **A comparison of complementation and noncomplementation.** Four different T4 phage strains (designated 1 through 4) that carry *rII* mutations were coinfected into *E. coli K12(λ)*. (**a**) If the two *rII* phage strains possess mutations in the same gene, noncomplementation will occur. (**b**) If the *rII* mutations are in different genes (say, gene *A* and gene *B*), a coinfected cell will have two mutant genes but also two wild-type genes. Doubly infected cells with a wild-type copy of each gene can produce new phages and form plaques. This result is called complementation, because the defective genes in each *rII* strain are complemented by the corresponding wild-type genes.

Four different *rII* mutations in T4 bacteriophage, designated strains 1 through 4, prevented plaque formation in *E. coli K12(λ)*. To conduct this complementation experiment, bacterial cells were coinfected with two different strains of phage. Two distinct outcomes are possible. If the two *rII* phage strains possess deleterious mutations in the same gene, they will not be able to make new phages when coinfected into an *E. coli K12(λ)* cell. For example, if both *rII* phages contain deleterious mutations in gene *A*, they cannot make plaques because the coinfected cell cannot make a wild-type gene *A* product. This phenomenon is called **noncomplementation.**

Alternatively, if each *rII* mutation was in a different phage gene (e.g., gene *A* and gene *B*), a bacterial cell that is coinfected by both types of phages will have two mutant genes but also two wild-type genes. If the mutant phage genes behave in a recessive fashion, the doubly infected cell will have a wild-type phenotype. This is because the coinfected cells will produce normal proteins that are encoded by the wild-type versions of genes *A* and *B*. For this reason, coinfected cells will be lysed in the same manner as if infected by the wild-type strain. Therefore, this coinfection should be able to produce plaques in *E. coli K12(λ)*. This is called *complementation*, because the defective genes in each *rII* strain are complemented by the corresponding wild-type genes. It should be noted, however, that intergenic complementation may not always work for a variety of reasons. One possibility is that a mutation may behave in a dominant fashion. In addition, mutations that affect regulatory genetic regions rather than the protein-coding region may not show complementation.

By carefully considering the pattern of complementation and noncomplementation, Benzer found that the *rII* mutations occurred in two different genes, which were termed *rIIA* and *rIIB*. The identification of two distinct genes affecting plaque formation was a necessary step that preceded his intragenic mapping analysis, which is described next. Benzer coined the term **cistron** to refer to the smallest genetic unit that gives a negative complementation test. In other words, if two mutations occur within the same cistron, they cannot complement each other. Since these studies, it has become clear that a cistron is equivalent to a gene. In recent decades, the term *gene* has gained wide popularity while the term *cistron* is not commonly used. Nevertheless, it would be correct to refer to gene *A* as cistron *A* or gene *B* as cistron *B*.

Intragenic Maps Were Constructed Using Data from a Recombinational Analysis of Mutants Within the rII Region

As we have seen in figure 6.16, the ability of a coinfection to produce many viral plaques is due to complementation. Noncomplementation occurs when two different strains have mutations in the same gene. However, at an extremely low rate, two noncomplementing strains of viruses can produce an occasional viral plaque if intragenic recombination has taken place. For example, figure 6.17 shows a coinfection experiment between two phage strains that both contain *rII* mutations in gene *A*. These mutations are located at different places within the same gene. Rarely,

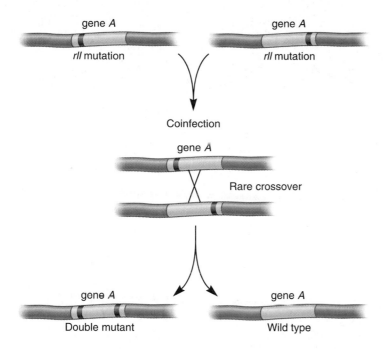

gene A   gene A

rII mutation   *rII* mutation

Coinfection

gene A

Rare crossover

gene A   gene A

Double mutant   Wild type

FIGURE 6.17 **Intragenic recombination.** Following coinfection, a rare crossover has occurred between the sites of the two mutations. This produces a wild-type phage with no mutations and a doubly mutant phage with both mutations.

a crossover occurs in the very short region between each mutation. This produces a gene A with two mutations and also a wild-type gene A. Since this event has produced a wild-type gene A, the function of the protein encoded by gene A will be restored. Therefore, new phages can be made in *E. coli K12(λ)*, resulting in the formation of viral plaques.

Figure 6.18 describes the general strategy for intragenic mapping of *rII* phage mutations. Bacteriophages from two different noncomplementing *rII* phage mutants (here, *r103* and *r104*) were mixed together in equal numbers and then infected into *E. coli B*. In this strain, the *rII* mutants could grow and propagate. When these two different mutants coinfected the same cell, intragenic recombination may occur, producing wild-type phages and double mutant phages. However, these intragenic recombinants were produced at a very low rate. Following coinfection and lysis of *E. coli B*, a new population of phages was isolated. This population was expected to mostly contain nonrecombinant phages. However, due to intragenic recombination, it would also contain a very low percentage of wild-type phages and double mutant phages (refer back to fig. 6.17).

To determine the relative numbers of parental and recombinant phages, Benzer took advantage of the observation that *rII* phages cannot grow in *E. coli K12(λ)*. He took his new population of phages and used some of them to infect *E. coli B* and some to infect *E. coli K12(λ)*. After plating, the *E. coli B* infection was used to determine the total number of phages, since *rII* mutants as well as wild-type phages can produce plaques in this strain. The overwhelming majority of these phages were expected to be

nonrecombinant phages. The *E. coli K12(λ)* infection was used to determine the number of rare intragenic recombinants that produce wild-type phages.

Figure 6.18 illustrates the great advantage of this experimental system in detecting a low percentage of recombinants. In the laboratory, phage preparations containing several billion phages per milliliter are readily made. Among billions of phages, a low percentage (e.g., 1 in every 1,000) may be wild-type phages arising from intragenic recombination. The wild-type recombinants can produce plaques in *E. coli K12(λ)*, whereas the *rII* mutant strains cannot. In other words, only the tiny fraction of wild-type recombinants would produce plaques in *E. coli K12(λ)*.

It is possible to determine the frequency of recombinant phages by comparing the number of recombinant (wild-type) phages and the total number of phages. As shown in figure 6.18, the total number of phages can be deduced from the number of plaques obtained from the infection of *E. coli B*. In this experiment, the phage preparation was diluted by 10^8 (1:100,000,000), and 1 ml was used to infect *E. coli B*. Since this plate produced 66 plaques, the total number of phages in the original preparation was $66 \times 10^8 = 6.6 \times 10^9$, or 6.6 billion phages per milliliter. By comparison, the phage preparation used to infect *E. coli K12(λ)* was diluted by only 10^6 (1 in 1,000,000). This plate produced 11 plaques. Therefore, the number of wild-type phages was 11×10^6, which equals 11 million wild-type phages per milliliter.

As we have already seen in chapter 5, genetic mapping distance is computed by dividing the number of recombinants by the total population (nonrecombinants and recombinants) times 100. In this experiment, intragenic recombination produces an equal number of two types of recombinants: wild-type phages and double mutant phages. Only the wild-type phages are detected in the infection of *E. coli K12(λ)*. Therefore, to obtain the total number of recombinants, the number of wild-type phages must be multiplied by two. With all this information, we can compute the frequency of recombinants using the experimental approach described in figure 6.18.

$$\text{Frequency of recombinants} = \frac{2[\text{Wild-type plaques obtained in } E.\ coli\ K12(\lambda)]}{\text{Total number of plaques obtained in } E.\ coli\ B}$$

We can use this equation to calculate the frequency of recombinants obtained in the experiment described in figure 6.18.

$$\text{Frequency of recombinants} = \frac{2(11 \times 10^6)}{6.6 \times 10^9}$$

$$= 3.3 \times 10^{-3} = 0.0033$$

In this example, there were approximately 3.3 recombinants per 1,000 phages.

The frequency of recombinants provides a measure of map distance. In eukaryotic mapping studies, we compute the map distance by multiplying the frequency of recombinants by 100 to give a value in map units (also known as centiMorgans). Similarly, in

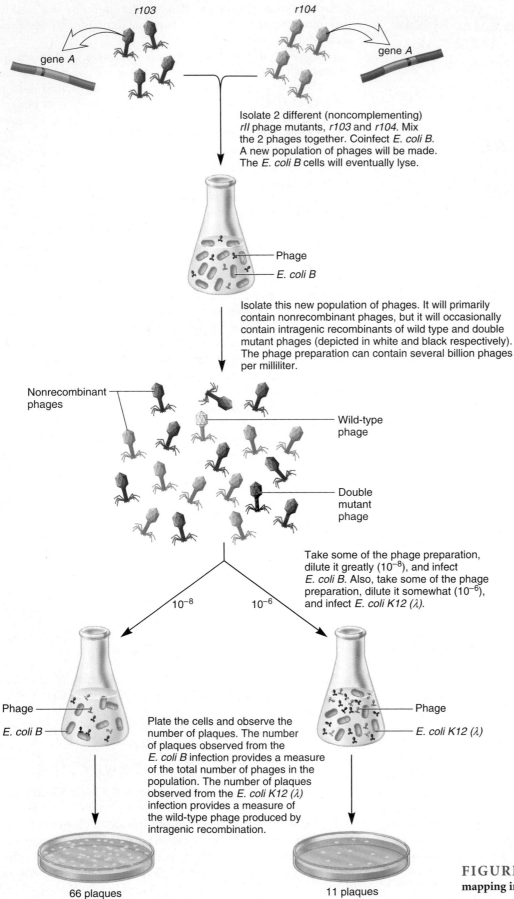

r103

gene *A*

r104

gene *A*

Isolate 2 different (noncomplementing) *rII* phage mutants, *r103* and *r104*. Mix the 2 phages together. Coinfect *E. coli* B. A new population of phages will be made. The *E. coli* B cells will eventually lyse.

Phage

E. coli B

Isolate this new population of phages. It will primarily contain nonrecombinant phages, but it will occasionally contain intragenic recombinants of wild type and double mutant phages (depicted in white and black respectively). The phage preparation can contain several billion phages per milliliter.

Nonrecombinant phages

Wild-type phage

Double mutant phage

Take some of the phage preparation, dilute it greatly (10^{-8}), and infect *E. coli* B. Also, take some of the phage preparation, dilute it somewhat (10^{-6}), and infect *E. coli* K12 (λ).

10^{-8} 10^{-6}

Phage

E. coli B

Plate the cells and observe the number of plaques. The number of plaques observed from the *E. coli* B infection provides a measure of the total number of phages in the population. The number of plaques observed from the *E. coli* K12 (λ) infection provides a measure of the wild-type phage produced by intragenic recombination.

Phage

E. coli K12 (λ)

66 plaques

11 plaques

FIGURE 6.18   **Benzer's method of intragenic mapping in the *rII* region.**

these experiments, the frequency of recombinants can provide a measure of map distance along the bacteriophage chromosome. In this case, the map distance is between two mutations within the same gene. Like intergenic mapping, the frequency of intragenic recombinants is correlated with the distance between the two mutations; the farther apart they are, the higher the frequency of recombinants. If two mutations happen to be located at exactly the same site within a gene, they would not be able to produce any wild-type recombinants, and so the map distance would be zero. These are known as **homoallelic** mutations.

Deletion Mapping Can Be Used to Localize Many *rII* Mutations to Specific Regions in the *rIIA* or *rIIB* Genes

Now that we have seen the general approach to intragenic mapping, let's consider a method to efficiently map hundreds of *rII* mutations within the two genes designated *rIIA* and *rIIB*. As you may have realized, the coinfection experiments described in figure 6.18 are quite similar to Sturtevant's strategy of making dihybrid crosses to map genes along the X chromosome of *Drosophila* (refer back to chapter 5). Similarly, Benzer wanted to coinfect different *rII* mutants in order to map the sites of the mutations within the *rIIA* and *rIIB* genes. During the course of his work, he obtained hundreds of different *rII* mutant strains that he wanted to map. However, making all the pairwise combinations would have been an overwhelming task. Instead, Benzer used an approach known as **deletion mapping** as a first step in localizing his *rII* mutations to a fairly short region within gene *A* or gene *B*.

Figure 6.19 describes the general strategy used in deletion mapping. This approach is easier to understand if we use an example. Let's suppose that the goal is to know the approximate location of an *rII* mutation, such as *r103*. To do so, *E. coli* K12(λ) would be coinfected with *r103* and a deletion strain. Each deletion strain is a T4 bacteriophage that is missing a known segment of the *rIIA* and/or *rIIB* gene. If the deleted region includes the same region that contains the *r103* mutation, it will be impossible for a coinfection to produce intragenic wild-type recombinants. Therefore, plaques will not be formed. However, if a deletion strain recombines with *r103* to produce a wild-type phage, the deleted region does not overlap with the *r103* mutation. In the example shown in figure 6.19, the *r103* strain produced wild-type recombinants when coinfected with deletion strains *PB242, A105,* and *638*. However, coinfection of *r103* with *PT1, J3, 1241,* and *1272* did not produce intragenic wild-type recombinants. Since coinfection with *PB242* produced recombinants and *PT1* did not, the *r103* mutation must be located in the region that is missing in *PT1* but not missing in *PB242*. This region is called A4 (A refers to the *rIIA* gene). In other words, the *r103* mutation is located somewhere within the A4 region, but not in the other six regions (A1, A2, A3, A5, A6, and B).

As described in figure 6.19, this first step in the deletion mapping strategy localized an *rII* mutation to one of seven regions; six of these were in *rIIA* and one was in *rIIB*. Other dele-

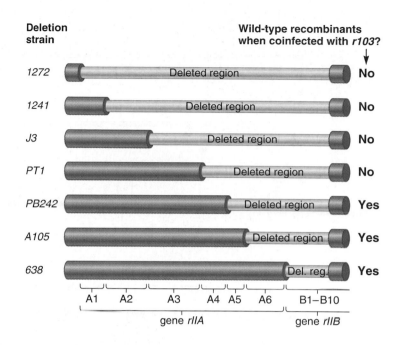

FIGURE 6.19 **The use of deletion strains to localize *rII* mutants to short regions within the *rIIA* or *rIIB* gene.** The deleted regions are shown in gray.

tion strains were used to eventually localize each *rII* mutation to one of 47 short regions; 36 were in *rIIA*, 11 in *rIIB*. At this point, pairwise coinfections were made between mutant strains that had been localized to the same region by deletion mapping. For example, 24 mutations were deletion mapped to a region called A5d. Pairwise coinfection experiments were conducted among this group of 24 mutants to precisely map their locations relative to each other in the A5d region. Similarly, all the mutants in each of the 46 other groups were mapped by pairwise coinfections. In this way, a fine structure map was constructed depicting the locations of hundreds of different *rII* mutations (fig. 6.20). As seen here, certain locations contained a relatively high number of mutations compared to other sites. These were termed **hot spots** for mutation.

Intragenic Mapping Experiments Provided Insight into the Relationship Between Traits and Molecular Genetics

Intragenic mapping studies were a pivotal achievement in our early understanding of gene structure. Since the time of Mendel, geneticists had considered a gene to be a unit of heredity that provided an organism with its inherited traits. In the late 1950s, however, the molecular nature of the gene was not understood. Since it was a unit of heredity, some scientists envisioned a gene as being a particle-like entity that could not be further subdivided into additional parts. However, intragenic mapping studies revealed, convincingly, that this is not the case. These studies

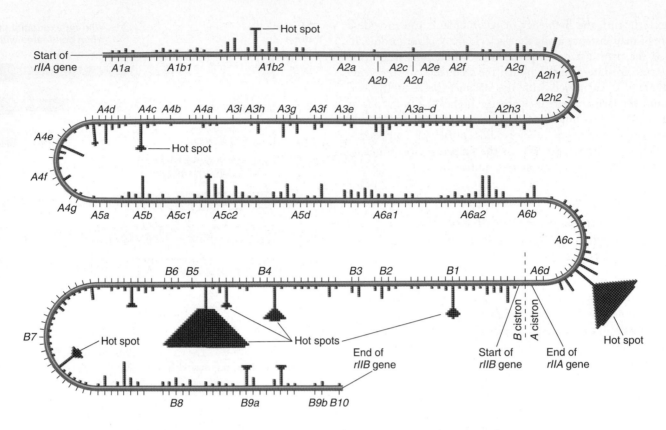

FIGURE 6.20   **The outcome of intragenic mapping of many *rII* mutations.** The *blue line* represents the linear sequence of the *rIIA* and *rIIB* genes, which are found within the T4 phage's genetic material. Each small *purple box* attached to the blue line symbolizes a mutation that was mapped by intragenic mapping. Among hundreds of independent mutant phages, several mutations sometimes mapped to the same site. In this figure, mutations at the same site form columns of boxes. Hot spots contain a large number of mutations and are represented as a triangular group of boxes attached to a column of boxes. A hot spot contains many mutations at the same site within the *rIIA* or *rIIB* gene.

showed that mutations can occur at many different sites within a single gene. Furthermore, intragenic crossing over can recombine these mutations, resulting in wild-type genes. Therefore, rather than being an indivisible particle, a gene must be composed of a large structure, which can be subdivided during crossing over.

Benzer's results were published in the late 1950s and early 1960s, not long after the structure of DNA had been elucidated by Watson and Crick. We now know that a gene is a segment of DNA that is composed of smaller building blocks called nucleotides. A typical gene is a linear sequence of several hundred to several thousand nucleotides. As the genetic map of Figure 6.20 indicates, mutations can occur at many sites along the linear structure of a gene; intragenic crossing over can recombine mutations that are located at different sites within the same gene.

CONCEPTUAL SUMMARY

This chapter has been concerned with genetic transfer and mapping in bacteria. There are three mechanisms for the transfer of genetic material from one bacterial cell to another. During **conjugation,** bacteria make direct physical contact with each other. Donor cells can be either **F⁺** cells or **Hfr** cells. An F⁺ cell contains a circular piece of genetic material, a **plasmid** called an **F factor** that is transferred to the recipient (F⁻) cell during mating. By comparison, an Hfr (for High frequency of recombination) bacterium transfers chromosomal DNA to the recipient cell. A second route of genetic transfer is **transduction.** In this case, a bacteriophage accidentally packages a portion of the bacterial chromosome, which is then transferred to another bacterium

upon infection. Finally, a third mechanism of genetic transfer is **transformation.** For this to occur, a dead bacterial cell must release its genetic material into the environment. A living bacterial cell in a **competent** state subsequently imports the DNA, and then genetic recombination causes the imported DNA to replace segments of genetic material along the bacterial chromosome. Overall, these three mechanisms have promoted gene transfer within bacterial species and also allowed **horizontal gene transfer** among different bacterial species. From a medical perspective, this phenomenon has led to a dramatic rise in the prevalence of strains that are resistant to antibiotics.

Chapter 6 concluded with a description of **intragenic mapping** studies in T4 bacteriophages. Benzer constructed a **fine structure** genetic map of two bacteriophage genes, *rIIA* and *rIIB*. In the early 1960s, his fine structure map provided important insights into the molecular nature of the gene. It revealed that a gene is not an indivisible particle. Rather, his results indicated that mutations can occur at many sites along the linear structure of a gene and that intragenic crossing over can recombine mutations that are located at different sites within the same gene. As we will learn in chapters 9 and 12, these results are consistent with the molecular structure of genes.

EXPERIMENTAL SUMMARY

Research studies of genetic transfer in bacteria have provided a unique experimental strategy for mapping the linear order and relative locations of bacterial genes. Conjugation has been used to map the locations of many genes along the bacterial chromosome. In this approach, map distances are determined by the number of **minutes** it takes for a gene to enter a recipient cell during conjugation. In addition, **cotransduction** and **cotransformation** experiments have been commonly used to accurately map bacterial genes that are relatively close to each other on the chromosome.

As mentioned in the Conceptual Summary, intragenic mapping studies of T4 bacteriophages have provided great insight into the molecular nature of genes. A fine structure genetic map was constructed of two bacteriophage genes, *rIIA* and *rIIB*. This was accomplished using a coinfection strategy, in which the very low percentage of intragenic recombinants yielding wild-type phages could be identified by their unique ability to infect a strain of bacteria known as *E. coli K12(λ)*. This approach illustrates the power of phage genetics compared to the analysis of offspring in eukaryotic species. It is technically easy to isolate huge numbers of phages. The selective inability of *rII* mutants to lyse *E. coli K12(λ)* allowed detection of the very few intragenic recombinants which produced wild-type phages that could lyse *E. coli K12(λ)*. In this way, the very short distance that occurs between mutations within a single gene could be determined.

PROBLEM SETS & INSIGHTS

Solved Problems

S1. In *E. coli*, the gene *bioD* encodes an enzyme involved in biotin synthesis, and *galK* encodes an enzyme involved in galactose utilization. An *E. coli* strain that contained wild-type versions of both genes was infected with P1, and then a P1 lysate was obtained. This lysate was used to transduce a strain that was *bioD⁻* and *galK⁻*. The cells were plated on media containing galactose as the sole carbon source for growth to select for transduction of the *galK* gene. These plates also were supplemented with biotin. The colonies were then restreaked on plates that lacked biotin to see if the *bioD* gene had been cotransduced. The following results were obtained:

Selected Gene	Non-selected Gene	Number of Colonies That Grew On:		Cotransduction Frequency
		Galactose + Biotin	Galactose − Biotin	
galK	*bioD*	80	10	0.125

How far apart are these two genes?

Answer: We can use the cotransduction frequency to calculate the distance between the two genes (in minutes) using the equation:

$$\text{Cotransduction frequency} = (1 - d/2)^3$$

$$0.125 = (1 - d/2)^3$$

$$1 - d/2 = \sqrt[3]{0.125}$$

$$1 - d/2 = 0.5$$

$$d/2 = 1 - 0.5$$

$$d = 1.0 \text{ minute}$$

The two genes are approximately 1 minute apart on the *E. coli* chromosome.

S2. By conducting mating experiments between a single *Hfr* strain and a recipient strain, Wollman and Jacob mapped the order of many bacterial genes. Throughout the course of their studies, they identified several different *Hfr* strains in which the F factor DNA had been integrated at different places along the bacterial chromosome. A sample of their experimental results is shown in the following table:

Hfr strain	Origin	Order of Transfer of Several Different Bacterial Genes								
		First								Last
H	O	*thr*	*leu*	*azi*	*ton*	*pro*	*lac*	*gal*	*str*	*met*
1	O	*leu*	*thr*	*met*	*str*	*gal*	*lac*	*pro*	*ton*	*azi*
2	O	*pro*	*ton*	*azi*	*leu*	*thr*	*met*	*str*	*gal*	*lac*
3	O	*lac*	*pro*	*ton*	*azi*	*leu*	*thr*	*met*	*str*	*gal*
4	O	*met*	*str*	*gal*	*lac*	*pro*	*ton*	*azi*	*leu*	*thr*
5	O	*met*	*thr*	*leu*	*azi*	*ton*	*pro*	*lac*	*gal*	*str*
6	O	*met*	*thr*	*leu*	*azi*	*ton*	*pro*	*lac*	*gal*	*str*
7	O	*ton*	*azi*	*leu*	*thr*	*met*	*str*	*gal*	*lac*	*pro*

A. Explain how these results are consistent with the idea that the bacterial chromosome is circular.

B. Draw a map that shows the order of genes and the locations of the origins of transfer among these different *Hfr* strains.

Answer:

A. In comparing the data among different *Hfr* strains, the order of the nine genes was always the same or the reverse of the same order. For example, *HfrH* and *Hfr1* have the same order of genes but are reversed relative to each other. In addition, the *Hfr* strains showed an overlapping pattern of transfer with regard to the origin. For example, *Hfr1* and *Hfr2* had the same order of genes, but *Hfr1* began with *leu* and ended with *azi* while *Hfr2* began with *pro* and ended with *lac*. From these findings, Wollman and Jacob concluded that the segment of DNA that was the origin of transfer had been inserted at different points within a circular *E. coli* chromosome in different *Hfr* strains. They also concluded that the origin can be inserted in either orientation, so that the direction of gene transfer can be clockwise or counterclockwise around the circular bacterial chromosome.

B. A genetic map that is consistent with these results is shown here.

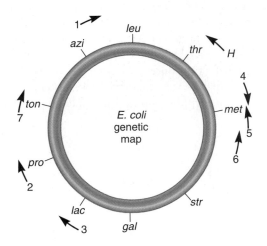

S3. An *Hfr* strain that is *leuA⁺* and *thiL⁺* was mated to a strain that is *leuA⁻* and *thiL⁻*. In the data points shown here, the mating was interrupted and the percentage of recombinants for each gene was determined by streaking on plates that lacked either leucine or thiamine. The results are shown.

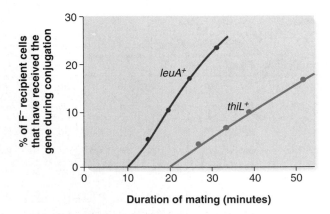

What is the map distance (in minutes) between these two genes?

Answer: This problem is solved by extrapolating the data points to the x-axis to determine the time of entry. For *leuA⁺*, they extrapolate back to 10 minutes. For *thiL⁺*, they extrapolate back to 20 minutes. Therefore, the distance between the two genes is approximately 10 minutes.

S4. Genetic transfer via transformation can also be used to map genes along the bacterial chromosome. In this approach, fragments of chromosomal DNA are isolated from one bacterial strain and used to transform another strain. The experimenter examines the transformed bacteria to see if they have incorporated two or more different genes. For example, the DNA may be isolated from a donor *E. coli* bacterium that has functional copies of the *araB* and *leuD* genes. Let's call these genes *araB⁺* and *leuD⁺* to indicate that the genes are functional. These two genes are required for arabinose metabolism and leucine synthesis, respectively. To map the distance between these two genes via transformation, a recipient bacterium would be used that is *araB⁻* and *leuD⁻*. Following transformation, the recipient bacterium may become *araB⁺* and *leuD⁺*. This phenomenon is called cotransformation because two genes from the donor bacterium have been transferred to the recipient via transformation. In this type of experiment, the recipient cell is exposed to a fairly low concentration of donor DNA, making it unlikely that the recipient bacterium will take up more than one fragment of DNA. Therefore, under these conditions, cotransformation is likely only when two genes are fairly close together and are found on one fragment of DNA.

In a cotransformation experiment, a researcher has isolated DNA from an *araB⁺* and *leuD⁺* donor strain. This DNA was transformed into a recipient strain that was *araB⁻* and *leuD⁻*. Following transformation, the cells were plated on media containing arabinose and leucine. On these plates, only bacteria that are *araB⁺* can grow. The bacteria can be either *leuD⁺* or *leuD⁻* because leucine is provided in the plates. Colonies, which grew on these plates, were then restreaked on plates that contained arabinose but lacked leucine. Only *araB⁺* and *leuD⁺* cells could grow on these secondary plates. Following this protocol, a researcher obtained the following results:

Number of colonies growing on arabinose plus leucine plates: 57

Number of colonies that grew when restreaked on arabinose plates without leucine: 42

What is the map distance between these two genes? Note: This problem can be solved using the strategy of a cotransduction experiment except that the researcher must determine the average size of DNA fragments that are taken up by the bacterial cells. This would correspond to the value of *L* in a cotransduction experiment.

Answer: As mentioned, the basic principle of gene mapping via cotransformation is identical to the method of gene mapping via cotransduction described in chapter 6. One way to calculate the map distance is to use the same equation that we used for cotransduction data, except that we substitute cotransformation frequency for cotransduction frequency.

$$\text{Cotransformation frequency} = (1 - d/L)^3$$

(Note: Cotransformation is not quite as accurate as cotransduction because the sizes of chromosomal pieces tend to vary significantly from experiment to experiment, so that the value of *L* is not quite as reliable. Nevertheless, cotransformation has been used extensively to map the order and distance between closely linked genes along the bacterial chromosome.)

The researcher needs to experimentally determine the value of *L* by running the DNA on a gel and estimating the average size of the DNA fragments. Let's assume that they are about 2% of the bacterial chromosome, which, for *E. coli*, would be about 80,000 base pairs in length. So *L* equals 2 minutes, which is the same as 2%.

$$\text{Cotransformation frequency} = (1 - d/L)^3$$

$$42/57 = (1 - d/2)^3$$

$$d = 0.2 \text{ minutes}$$

The distance between *araB* and *leuD* is approximately 0.2 minutes.

S5. In our discussion of transduction via P1 or P22, the life cycle of the bacteriophage sometimes resulted in the packaging of many different pieces of the bacterial chromosome. For other bacteriophages, however, transduction may only involve the transfer of a few specific genes from the donor cell to the recipient. This phenomenon is known as specialized transduction. The key event that causes specialized transduction to occur is that the lysogenic phase of the phage life cycle involves the integration of the viral DNA at a single specific site within the bacterial chromosome. The transduction of particular bacterial genes involves an abnormal excision of the phage DNA from this site within the chromosome that would carry adjacent bacterial genes. For example, a bacteriophage called lambda (λ) that infects *E. coli* specifically integrates between two genes designated *gal*+ and *bio*+ (required for galactose utilization and biotin synthesis, respectively). Either of these genes could be packaged into the phage if an abnormal excision event occurred. How would specialized transduction be different from generalized transduction?

Answer: Generalized transduction can involve the transfer of any bacterial gene, while specialized transduction can transfer only genes that are adjacent to the site where the phage integrates. As mentioned, a bacteriophage that infects *E. coli* cells, known as lambda (λ), provides a well-studied example of specialized transduction. In the case of phage lambda, the lysogenic life cycle results in the integration of the phage DNA at a site that is called the attachment site. This is described in chapter 17. The attachment site is located between two bacterial genes, *gal*+ and *bio*+. An *E. coli* strain that is lysogenic for phage lambda will have the lambda DNA integrated between these two bacterial genes. On occasion, the phage may enter the lytic cycle and excise its DNA from the bacterial chromosome. When this occurs normally, the phage excises its entire viral DNA from the bacterial chromosome. The excised phage DNA is then replicated and becomes packaged into newly made phages. However, an abnormal excision does occur at a low rate (i.e., about one in a million). In this abnormal event, the phage DNA is excised in such a way that an adjacent bacterial gene is included and some of the phage DNA is not included in the final product. For example, the abnormal excision may yield a fragment of DNA that includes the *gal*+ gene and some of the lambda DNA but is missing part of the lambda DNA. If this DNA fragment is packaged into a virus, it is called a defective phage because it is missing some of the phage DNA. If it carries the *gal*+ gene, it is designated λ*dgal* (the letter *d* designates a defective phage). Alternatively, an abnormal excision may carry the *bio*+ gene. This phage is designated λ*dbio*. Defective lambda phages can then transduce the *gal*+ or *bio*+ genes to other *E. coli* cells.

Conceptual Questions

C1. The terms *conjugation, transduction,* and *transformation* are used to describe three different natural forms of genetic transfer between bacterial cells. Briefly discuss the similarities and differences between these processes.

C2. Conjugation is sometimes called "bacterial mating." Is it a form of sexual reproduction? Explain.

C3. If you mix together an equal number of F+ and F− cells, how would you expect the proportions to change over time? In other words, do you expect an increase in the relative proportions of F+ or of F− cells? Explain your answer.

C4. What is the difference between an F+ and an *Hfr* strain? Which type of strain do you expect to be able to transfer many bacterial genes to recipient cells?

C5. What is the role of the origin of transfer during F+- and Hfr-mediated conjugation? What is the significance of the direction of transfer in Hfr-mediated conjugation?

C6. What is the role of sex pili during conjugation?

C7. Think about the structure and transmission of F factors and discuss how you think F factors may have originated.

C8. Each species of bacteria has its own distinctive cell surface. The characteristics of the cell surface play an important role in processes such as conjugation and transduction. For example, certain strains of *E. coli* have pili on their cell surface; these pili enable *E. coli* to mate with other *E. coli,* and the pili also enable certain bacteriophages (such as M13) to bind to the surface of *E. coli* and gain entry into the cytoplasm. With these ideas in mind, explain which forms of genetic transfer (i.e., conjugation, transduction, and transformation) are more likely to occur between different species of bacteria. Discuss some of the potential consequences of interspecies genetic transfer.

C9. Briefly describe the lytic and lysogenic cycles of bacteriophages. In your answer, explain what a prophage is.

C10. What is cotransduction? What determines the likelihood that two genes will be cotransduced?

C11. When bacteriophage P1 causes *E. coli* to lyse, the resulting material is called a P1 lysate. What type of genetic material would be found in most of the P1 phages in the lysate? What kind of genetic material would occasionally be found within a P1 phage?

C12. As described in figure 6.10 of your textbook, host DNA is hydrolyzed into small pieces, which are occasionally packaged into a phage coat. If the breakage of the chromosomal DNA is not random (i.e., it is more likely to break at certain spots compared to other spots), how might nonrandom breakage affect cotransduction frequency?

C13. Describe the steps that occur during bacterial transformation. What is a competent cell? What factors may determine whether a cell will be competent?

C14. Which bacterial genetic transfer process does not require recombination with the bacterial chromosome?

C15. Researchers who study the molecular mechanism of transformation have identified many proteins in bacteria that function in the uptake of DNA from the environment and its recombination into the host cell's chromosome. This means that bacteria have evolved molecular mechanisms for the purpose of transformation by extracellular DNA. Of what advantage(s) would it be for a bacterium to import DNA from the environment and/or incorporate it into its chromosome?

C16. Antibiotics such as tetracycline, streptomycin, and bacitracin are small organic molecules that are synthesized by particular species of bacteria. Microbiologists have hypothesized that the reason why certain bacteria make antibiotics is to kill other species that occupy the same environment. Bacteria that produce an antibiotic may be able to kill competing species. This provides more resources for the antibiotic-producing bacteria. In addition, bacteria that have the genes necessary for antibiotic biosynthesis contain genes that confer resistance to the same antibiotic. For example, tetracycline is made by the soil bacterium *Streptomyces aureofaciens*. Besides the genes that are needed to make tetracycline, *S. aureofaciens* also contains genes that confer tetracycline resistance; otherwise it would kill itself when it makes tetracycline. In recent years, however, many other species of bacteria that do not synthesize tetracycline have acquired the genes that confer tetracycline resistance. For example, certain strains of *E. coli* carry tetracycline-resistance genes, even though *E. coli* does not synthesize tetracycline. When these genes are analyzed at the molecular level, it has been found that they are evolutionarily related to the genes in *S. aureofaciens*. This observation indicates that the genes from *S. aureofaciens* have been transferred to *E. coli*.

A. What form of genetic transfer (i.e., conjugation, transduction, or transformation) would be most likely in interspecies gene transfer?

B. Since *S. aureofaciens* is a nonpathogenic soil bacterium, and *E. coli* is an enteric bacterium, do you think it was direct gene transfer, or do you think it may have occurred in multiple steps (i.e., from *S. aureofaciens* to other bacterial species, and then to *E. coli*)?

C. How could the widespread use of antibiotics to treat diseases have contributed to the proliferation of many bacterial species that are resistant to antibiotics?

C17. What does the term *complementation* mean? If two different mutations that produce the same phenotype can complement each other, what can you conclude about the locations of each mutation?

C18. Intragenic mapping is sometimes called interallelic mapping. Explain why the two terms mean the same thing. In your own words, explain what an intragenic map is.

C19. As discussed in chapter 12, genes are composed of a sequence of nucleotides. A typical gene in a bacteriophage is a few hundred or a few thousand nucleotides in length. If two different strains of bacteriophage T4 have a mutation in the *rIIA* gene that gives a rapid-lysis phenotype, yet they never produce wild-type phages by intragenic recombination when they are coinfected into *E.coli B*, what would you conclude about the locations of the mutations in the two different T4 strains?

Experimental Questions

E1. In the experiment of figure 6.1, a *bio⁻ met⁻ phe⁺ thr⁺* cell could become *bio⁺ met⁺ phe⁺ thr⁺* by a (rare) double mutation that converts the *bio⁻ met⁻* genetic material into *bio⁺ met⁺*. Likewise, a *bio⁺ met⁺ phe⁻ thr⁻* cell could become *bio⁺ met⁺ phe⁺ thr⁺* by double mutations that convert the *phe⁻ thr⁻* genetic material into *phe⁺ thr⁺*. From the results of figure 6.1, how do you know that the occurrence of 100 *bio⁺ met⁺ phe⁺ thr⁺* colonies is not due to these types of rare double mutations?

E2. In the experiment of figure 6.1, Lederberg and Tatum could not discern whether *bio⁺met⁺* genetic material was transferred to the *bio⁻ met⁻ phe⁺ thr⁺* strain or if *phe⁺ thr⁺* genetic material was transferred to the *bio⁺ met⁺ phe⁻ thr⁻* strain. Let's suppose that one strain is streptomycin resistant (say, *bio⁺ met⁺ phe⁻ thr⁻*) while the other strain is sensitive to streptomycin. Describe an experiment that could determine whether the *bio⁺ met⁺* genetic material was transferred to the *bio⁻ met⁻ phe⁺ thr⁺* strain or if *phe⁺ thr⁺* genetic material was transferred to the *bio⁺ met⁺ phe⁻ thr⁻* strain.

E3. Explain how a U-tube apparatus can distinguish between genetic transfer involving conjugation versus transduction. Do you think a U-tube could be used to distinguish between transduction and transformation?

E4. What is an interrupted mating experiment? What type of experimental information can be obtained from this type of study? Why is it necessary to interrupt mating?

E5. In a conjugation experiment, what is meant by the term *time of entry*? How is it determined experimentally?

E6. In your laboratory, you have an *F⁻* strain of *E. coli* that is resistant to streptomycin and is unable to metabolize lactose, but it can metabolize glucose. Therefore, this strain can grow on plates that contain glucose and streptomycin, but it cannot grow on plates containing lactose. A researcher has sent you two *E. coli* strains in two separate tubes. One strain, let's call it strain *A*, has an F factor that carries the genes that are required for lactose metabolism. On its chromosome, it also has the genes that are required for glucose metabolism. However, it does not have the genes that confer streptomycin resistance. This strain can grow on plates containing lactose or glucose, but it cannot grow if streptomycin is added to the plates. The second strain, let's call it strain *B*, is an *F⁻* strain. On its chromosome, it has the genes that are required for lactose and glucose metabolism. Strain *B* is also sensitive to streptomycin. Unfortunately, when strains *A* and *B* were sent to you, the labels had fallen off the tubes. Describe how you could determine which tubes contain strain *A* and strain *B*.

E7. As mentioned in solved problem S2, origins of transfer can be located in many different locations, and their direction of transfer can be clockwise or counterclockwise. Let's suppose a researcher mated six different *Hfr* strains that were *thr⁺ leu⁺ tonˢ strʳ aziˢ lac⁺*

gal^+ pro^+ met^+ to an F^- strain that was thr^- leu^- ton^r str^s azi^r lac^- gal^- pro^- met^-, and obtained the following results:

Strain	Order of Gene Transfer
1	ton^s azi^s leu^+ thr^+ met^+ str^r gal^+ lac^+ pro^+
2	leu^+ azi^s ton^s pro^+ lac^+ gal^+ str^r met^+ thr^+
3	lac^+ gal^+ str^r met^+ thr^+ leu^+ azi^s ton^s pro^+
4	leu^+ thr^+ met^+ str^r gal^+ lac^+ pro^+ ton^s azi^s
5	ton^s pro^+ lac^+ gal^+ str^r met^+ thr^+ leu^+ azi^s
6	met^+ str^r gal^+ lac^+ pro^+ ton^s azi^s leu^+ thr^+

Draw a circular map of the *E. coli* chromosome and describe the locations and orientations of the origins of transfer in these six *Hfr* strains.

E8. An *Hfr* strain that is $hisE^+$ and $pheA^+$ was mated to a strain that is $hisE^-$ and $pheA^-$. The mating was interrupted and the percentage of recombinants for each gene was determined by streaking on plates that lacked either histidine or phenylalanine. The following results were obtained:

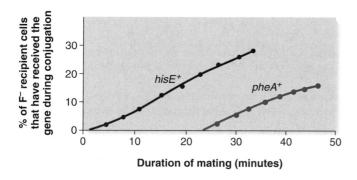

A. Determine the map distance (in minutes) between these two genes.

B. In a previous experiment, it was found that *hisE* is 4 minutes away from the gene *pabB*. *PheA* was shown to be 17 minutes from this gene. Draw a genetic map describing the locations of all three genes.

E9. Acridine orange is a chemical that inhibits the replication of F factor DNA but it does not affect the replication of chromosomal DNA, even if the chromosomal DNA contains an Hfr. Let's suppose that you have an *E. coli* strain that is unable to metabolize lactose and has an F factor that carries a streptomycin-resistant gene. You also have an F^- strain of *E. coli* that is sensitive to streptomycin and has the genes that allow the bacterium to metabolize lactose. This second strain can grow on lactose plates. How would you generate an *Hfr* strain that is resistant to streptomycin and can metabolize lactose? (Hint: F factors occasionally integrate into the chromosome to become *Hfr* strains, and occasionally *Hfr* strains excise their DNA from the chromosome to become F^+ strains that carry an F′ factor.)

E10. In a P1 transduction experiment, the P1 lysate contains phages that carry pieces of the host chromosomal DNA, but the lysate also contains broken pieces of chromosomal DNA (see fig. 6.10). If a P1 lysate is used to transfer chromosomal DNA to another bacterium, how could you show experimentally that the recombinant bacterium has been transduced (i.e., taken up a P1 phage with a piece

of chromosomal DNA inside) versus transformed (i.e., taken up a piece of chromosomal DNA that is not within a P1 phage coat)?

E11. Could you devise an experimental strategy to get P1 phage to transduce the entire lambda genome from one strain of bacterium to another strain? (Note: The general features of phage lambda's life cycle are described in chapter 14.) Phage lambda has a genome size of 48,502 nucleotides (about 1% of the size of the *E. coli* chromosome) and can follow the lytic or lysogenic life cycle. Growth of *E. coli* on minimal growth media favors the lysogenic life cycle, whereas growth on rich media and/or UV light promotes the lytic cycle.

E12. Let's suppose a new strain of P1 has been identified that packages larger pieces of the *E. coli* chromosome. This new P1 strain packages pieces of the *E. coli* chromosome that are 5 minutes long. If two genes are 0.7 minutes apart along the *E. coli* chromosome, what would be the cotransduction frequency using a normal strain of P1 and using this new strain of P1 that packages large pieces? What would be the experimental advantage of using this new P1 strain?

E13. If two bacterial genes are 0.6 minutes apart on the bacterial chromosome, what frequency of cotransductants would you expect to observe in a P1 transduction experiment?

E14. In an experiment involving P1 transduction, the cotransduction frequency was 0.53. How far apart are the two genes?

E15. In a cotransduction experiment, the transfer of one gene is selected for and the presence of the second gene is then determined. If 0 out of 1,000 P1 transductants that carry the first gene also carry the second gene, what would you conclude about the minimum distance between the two genes?

E16. In a cotransformation experiment (see solved problem S4), DNA was isolated from a donor strain that was $proA^+$ and $strC^+$ and sensitive to tetracycline. The *proA* and *strC* genes confer the ability to synthesize proline and confer streptomycin resistance, respectively. A recipient strain is $proA^-$ $strC^-$ and resistant to tetracycline. After transformation, the bacteria were first streaked on plates containing proline, streptomycin, and tetracycline. Colonies were then restreaked on plates containing streptomycin and tetracycline. (Note: Both types of plates had carbon and nitrogen sources for growth.) The following results were obtained:

70 colonies grew on plates containing proline, streptomycin, and tetracycline while only 2 of these 70 colonies grew when restreaked on plates containing streptomycin and tetracycline but lacking proline.

A. If we assume that the average size of the DNA fragments is 2 minutes, how far apart are these two genes?

B. What would you expect the cotransformation frequency to be if the average size of the DNA fragments was 4 minutes and the two genes are 1.4 minutes apart?

E17. If you took a pipette tip and removed a phage plaque from a petri plate, what would it contain?

E18. Phages with *rII* mutations cannot produce plaques in *E. coli* K12(λ), but wild-type phages can. From an experimental point of view, explain why this observation is so significant.

E19. In the experimental strategy described in figure 6.18, explain why it was necessary to dilute the *E. coli* B infection much more than the *E. coli* K12(λ) infection.

E20. Here are data from several complementation experiments, involving rapid-lysis mutations in genes *rIIA* and *rIIB*. The strain designated *L51* is known to have a mutation in *rIIB*.

Phage Mixture	Complementation
L91 and L65	No
L65 and L62	No
L33 and L47	Yes
L40 and L51	No
L47 and L92	No
L51 and L47	Yes
L51 and L92	Yes
L33 and L40	No
L91 and L92	Yes
L91 and L33	No

List which groups of mutations are in the *rIIA* gene and which groups are in the *rIIB* gene.

E21. A researcher has several different strains of T4 phage with single mutations in the same gene. In these strains, the mutations render the phage temperature sensitive. In this case, this means that temperature-sensitive phages can propagate when the bacterium (i.e., *E. coli*) is grown at 32°C but cannot propagate themselves when *E. coli* is grown at 37°C. Think about Benzer's strategy for intragenic mapping, and propose an experimental strategy to map the temperature-sensitive mutations.

E22. Explain how Benzer's results indicated that a gene is not an indivisible unit.

E23. Explain why deletion mapping was used as a step in the intragenic mapping of *rII* mutations.

Questions for Student Discussion/Collaboration

1. Discuss the advantages of the genetic analysis of bacteria and bacteriophages. Make a list of the types of allelic differences among bacteria and phages that are suitable for genetic analyses.

2. Complementation occurs when two defective alleles in two different genes are found within the same organism and produce a normal phenotype. What other examples of complementation have we encountered in previous chapters of this textbook?

Note: All answers appear at the website for this textbook; the answers to even-numbered questions are in the back of the textbook.

www.mhhe.com/brooker

Visit the Online Learning Center for practice tests, answer keys, and other learning aids for this chapter. Enhance your understanding of genetics with our interactive exercises, web links, news feeds, tutorial service, and much more.

NON-MENDELIAN INHERITANCE

7

::

Mendelian inheritance patterns involve genes that directly influence the outcome of an offspring's traits and obey Mendel's laws. To predict phenotype, we must consider several factors. These include the dominant/recessive relationships of alleles, gene interactions that may affect the expression of a single trait, and the roles that sex and the environment play in influencing the individual's phenotype. Once these factors are understood, the phenotypes of offspring can be predicted from their genotypes.

Most genes in eukaryotic species follow a Mendelian pattern of inheritance. However, there are many that do not. In chapter 5, for example, genes that are closely linked did not obey Mendel's law of independent assortment. One could argue that such inheritance patterns are also non-Mendelian. In chapter 7, we will examine several additional and even bizarre types of inheritance patterns that deviate from a Mendelian pattern. In the first two sections of this chapter, we will consider two important examples of non-Mendelian inheritance called the maternal effect and epigenetic inheritance. Even though these inheritance patterns involve genes on chromosomes within the cell nucleus, the genotype of the offspring does not directly govern their phenotype in ways predicted by Mendel. We will see how the timing of gene expression and gene inactivation can cause a non-Mendelian pattern of inheritance.

In the third section, we will examine a deviation from Mendelian inheritance that arises because some genetic material is not located in the cell nucleus. There are certain cellular organelles, such as mitochondria and chloroplasts, that contain their own genetic material. We will survey the inheritance of organellar genes and a few other examples in which traits are influenced by genetic material outside the cell nucleus.

7.1 MATERNAL EFFECT

We will begin by considering genes that have a **maternal effect.** This term refers to an inheritance pattern for certain **nuclear genes** (i.e., genes located on chromosomes that are found in the cell nucleus) in which the genotype of the mother directly determines the phenotypic traits of her offspring. Surprisingly, for maternal effect genes, the genotypes of the father and offspring themselves do not affect the phenotype of the offspring. We will see that this phenomenon is explained by the accumulation of gene products that the mother provides to her developing eggs.

The Genotype of the Mother Determines the Phenotype of the Offspring for Maternal Effect Genes

The first example of a maternal effect gene was studied in the 1920s by A. E. Boycott and involved morphological features of the water snail, *Limnea peregra.* In this species, the shell and internal organs can be arranged in either a right-handed (dextral) or left-handed (sinistral) direction. The dextral orientation is more common and is dominant to the sinistral orientation. Whether a snail's body plan curves in a dextral or sinistral direction depends on the cleavage pattern of the egg immediately following fertilization.

Figure 7.1 describes the results of a genetic analysis carried out by Boycott. In this experiment, he began with two different true-breeding strains of snails with either a dextral or sinistral morphology. Many combinations of crosses produced results that could not be explained by a Mendelian pattern of inheritance. When a dextral female was crossed to a sinistral male, all the offspring were dextral. However, in the **reciprocal cross,** where a sinistral female was crossed to a dextral male, the opposite result was obtained. Taken together, these results contradict a Mendelian pattern of inheritance.

The idea that snail coiling is due to a maternal effect gene that exists as dextral (*D*) and sinistral (*d*) alleles was later proposed by Alfred Sturtevant. His conclusions were drawn from the inheritance patterns of the F_2 and F_3 generations (fig. 7.1). In this experiment, the genotype of the F_1 generation is expected to be heterozygous (*Dd*). When these F_1 individuals were crossed to each other, a genotypic ratio of 1 *DD* : 2 *Dd* : 1 *dd* is predicted for the F_2 generation. Since the *D* allele is dominant to the *d* allele, a 3:1 phenotypic ratio of dextral to sinistral snails would be produced according to a Mendelian pattern of inheritance. Instead of this predicted phenotypic ratio, however, the F_2 generation was composed of all dextral snails. This incongruity with Mendelian inheritance is due to the maternal effect. The phenotype of the offspring depended solely on the genotype of the mother. The F_1 mothers were *Dd*; the *D* allele in the mothers is dominant to the *d* allele and caused the offspring to be dextral, even if the offspring's genotype was *dd*. When the members of the F_2 generation were crossed, the F_3 generation exhibited a 3:1 ratio of dextral to sinistral snails. This is due to the genotypes of the F_2 females, which were the mothers of the F_3 generation. The ratio

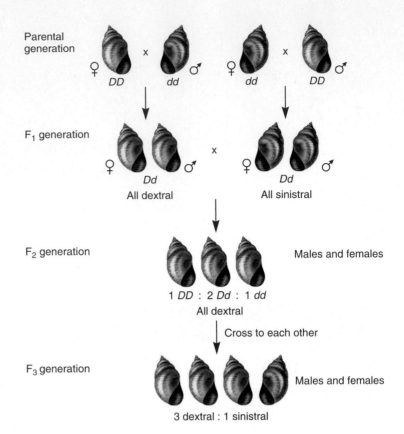

FIGURE 7.1 **Experiment showing the inheritance pattern of snail coiling.** In this experiment, *D* (dextral) is dominant to *d* (sinistral). The genotype of the mother determines the phenotype of the offspring. This phenomenon is known as the maternal effect. In this case, a *DD* or *Dd* mother produces dextral offspring and a *dd* mother produces sinistral offspring. The genotypes of the father and offspring themselves do not affect the offspring's phenotype.

of genotypes for the F_2 females was 1 *DD* : 2 *Dd* : 1 *dd*. The *DD* and *Dd* females produced dextral offspring, while the *dd* females produced sinistral offspring. This explains the 3:1 ratio of dextral and sinistral offspring in the F_3 generation.

Female Gametes Receive Gene Products from the Mother That Affect Early Developmental Stages of the Embryo

At the molecular and cellular level, the non-Mendelian inheritance pattern of maternal effect genes can be explained by the process of oogenesis in female animals (fig. 7.2a). As an animal oocyte matures, surrounding maternal cells called nurse cells provide it with nutrients. The nurse cells are diploid cells as opposed to the oocyte, which will become haploid (refer back to fig. 3.14). In the example of figure 7.2a, a female is heterozygous for the snail-coiling maternal effect gene, with the alleles designated *D* and *d*. Depending on the outcome of meiosis, the haploid egg may receive the *D* allele or the *d* allele, but not both. Within the diploid nurse cells, however, both genes are activated to produce

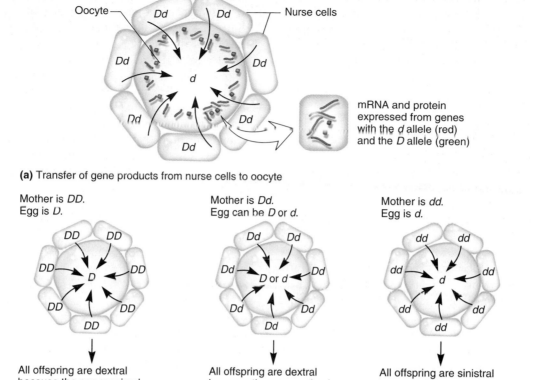

(a) Transfer of gene products from nurse cells to oocyte

Mother is *DD*.
Egg is *D*.

All offspring are dextral
because the egg received
the gene products of the *D* allele.

Mother is *Dd*.
Egg can be *D* or *d*.

All offspring are dextral
because the egg received
the gene products of the
dominant *D* allele.

Mother is *dd*.
Egg is *d*.

All offspring are sinistral
because the egg only
received the gene products
of the *d* allele.

(b) Maternal effect in snail coiling

FIGURE 7.2 The mechanism of maternal effect in snail coiling. (a) Transfer of gene products from nurse cells to an oocyte. The nurse cells are heterozygous (*Dd*). Both the *D* and *d* alleles are activated in the nurse cells to produce *D* and *d* gene products (mRNA and/or proteins). These products are transported into the cytoplasm of the oocyte, where they accumulate to significant amounts. **(b)** Explanation of the maternal effect in snail coiling. **GENES→TRAITS** If the nurse cells are *DD* or *Dd*, they will transfer the *D* gene product to the oocyte and thereby cause the resulting offspring to be dextral. If the nurse cells are *dd,* only the *d* gene product will be transferred to the oocyte, so the resulting offspring will be sinistral.

their gene products (mRNA and proteins). These gene products are then transported into the oocyte. As shown here, the egg has received both the *D* allele gene product and the *d* allele gene product, though the egg itself only carries the *d* allele. These gene products persist for a significant time after the egg has been fertilized and begins its embryonic development. In this way, the gene products of the nurse cells, which reflect the genotype of the mother, influence the early developmental stages of the embryo.

Now that we have an understanding of the relationship between oogenesis and maternal effect genes, let's reconsider the topic of snail coiling. As shown in figure 7.2*b*, a female snail that is *DD* will transmit only the *D* gene products to the eggs. During the early stages of embryonic development, these gene products will cause the egg cleavage to occur in a way that promotes a right-handed body plan. A heterozygous female will transmit both *D* and *d* gene products. Since the *D* allele is dominant, the maternal effect will also be a right-handed body plan. Finally, a *dd* mother will contribute only *d* gene products that promote a left-handed body plan, even if the egg is fertilized by a sperm car-

rying a *D* allele. The sperm's genotype is irrelevant, because the expression of the sperm's gene would occur too late.

Since these initial studies, researchers have found that maternal effect genes encode RNA and proteins that play important roles in the early steps of embryogenesis. Maternal effect genes often play a role in cell division, cleavage pattern, and body axis orientation. Therefore, defective alleles in maternal effect genes tend to have a dramatic effect on the phenotype of the individual. Abnormal maternal effect alleles usually alter major features of morphology, often with dire consequences.

Our understanding of maternal effect genes has been greatly aided by their identification in experimental organisms such as *Drosophila melanogaster*. In such organisms with a short generation time, geneticists have successfully searched for mutant alleles that prevent the normal process of embryonic development. In *Drosophila,* for example, geneticists have identified several dozen maternal effect genes with profound effects on the early stages of development. The pattern of development of a *Drosophila* embryo occurs along axes, such as the antero-posterior axis and the dorso-ventral axis. The proper development of each axis requires a distinct set of maternal gene products. Mutant alleles of maternal effect genes often lead to abnormalities in the antero-posterior or the dorso-ventral pattern of development. Chapter 23 examines the relationships among the actions of several maternal effect genes during embryonic development.

7.2 EPIGENETIC INHERITANCE

As we have just seen, events during oogenesis can cause the inheritance pattern of traits to deviate from a Mendelian pattern. Likewise, **epigenetic inheritance** is a pattern in which a modification occurs to a nuclear gene or chromosome that alters gene expression, but the expression is not permanently changed over the course of many generations. As we will see, epigenetic inheritance patterns are caused by DNA and chromosomal modifications that occur during oogenesis, spermatogenesis, or early stages of embryogenesis. Once they are initiated during these early stages, epigenetic changes alter the expression of particular genes in a way that may be fixed during an individual's lifetime. Therefore, epigenetic changes can permanently affect the phenotype of the individual. However, epigenetic modifications are not permanent over the course of many generations and they do not change the actual DNA sequence. For example, a gene may undergo an epigenetic change that inactivates it for the lifetime of an individual. However, when this individual makes gametes, the gene may become activated and remain operative during the lifetime of an offspring who inherits the active gene.

In this section, we will examine two examples of epigenetic inheritance called **dosage compensation** and **genomic imprinting.** The effect of dosage compensation is to offset differences in the number of active sex chromosomes. One of the sex chromosomes is altered with the result that males and females have similar levels of gene expression, even though they do not possess the same complement of sex chromosomes. In mammals, dosage compensation is initiated during the early stages of embryonic development. By comparison, genomic imprinting happens prior to fertilization; it involves a change in a single gene or chromosome during gamete formation. Depending on whether the modification occurs during spermatogenesis or oogenesis, imprinting governs whether an offspring expresses a gene that has been inherited from its mother or father.

Dosage Compensation Is Necessary to Ensure Genetic Equality Between the Sexes

Dosage compensation refers to the phenomenon that the level of expression of many genes on the sex chromosomes (such as the X chromosome) is similar in both sexes even though males and females have a different complement of sex chromosomes. This term was coined in 1932 by Hermann Muller to explain the effects of eye color mutations in *Drosophila*. Muller observed that female flies homozygous for certain X-linked eye color alleles had a similar phenotype to hemizygous males. For example, an X-linked gene conferring an apricot eye color produces a very similar phenotype in homozygous females and hemizygous males. In contrast, a female that has one copy of the apricot allele and a deletion of the apricot gene on the other X chromosome has eyes of paler color. Therefore, one copy of the allele in the female is not equivalent to one copy of the allele in the male. Instead, two copies of the allele in the female produce a phenotype that is similar to one copy in the male. In other words, the difference in gene dosage (two copies in females versus one copy in males) is being compensated at the level of gene expression. In *Drosophila*, however, dosage compensation does not occur for all eye color alleles. As described in chapter 4, the eosin eye color allele exhibits a gene dosage effect. In this case, two copies of the eosin allele cause a deeper color compared to one copy. The reasons why most X-linked genes show dosage compensation, and a few do not, are not understood.

Since these initial studies, dosage compensation has been studied extensively in mammals, *Drosophila*, and *Caenorhabditis elegans*. Depending on the species, dosage compensation occurs via different mechanisms. Table 7.1 describes how sex chromosome compensation is accomplished in different species. Mammals equalize the expression of X-linked genes by turning off one X chromosome in the somatic cells of females. This process is known as **X inactivation.** In the nematode worm, *C. elegans,* the XX animal is a hermaphrodite that produces both sperm and egg cells, while an animal carrying a single X chromosome is a male that produces only sperm. The XX hermaphrodite diminishes the

---

TABLE 7.1

Mechanisms of Dosage Compensation Among Different Species

| Species | Sex Chromosomes in: | | Mechanism of Compensation |
	Females	Males	
Placental mammals	XX	XY	One of the X chromosomes in the somatic cells of females is inactivated. In certain species, the paternal X chromosome is inactivated while in other species, such as humans, either the maternal or paternal X chromosome is randomly inactivated throughout the female's body.
Marsupial mammals	XX	XY	The paternally derived X chromosome is inactivated in the somatic cells of females.
Drosophila melanogaster	XX	XY	The level of expression of genes on the X chromosome in males is increased 2-fold.
Caenorhabditis elegans	XX*	X0	The level of expression of genes on both X chromosomes in hermaphrodites is decreased to 50% levels compared to males.

*In *C. elegans*, an XX individual is a hermaphrodite, not a female.

(a)   (b)

FIGURE 7.3 **X chromosome inactivation in the cells of female mammals.** (**a**) Nucleus from a human female cell. The *arrow* denotes the Barr body. (**b**) A calico cat.

GENES ›TRAITS   The pattern of black and orange fur on this cat is due to random X inactivation during embryonic development. The orange patches of fur are due to the inactivation of the X chromosome that carries a black allele; the black patches are due to the inactivation of the X chromosome that carries the orange allele. In general, only heterozygous female cats can be calico; a rare exception would be a male cat (XXY) that has an abnormal composition of sex chromosomes.

expression of X-linked genes in a different manner. The level of expression of genes on both X chromosomes is decreased to approximately 50% of that in the male. In *Drosophila*, the male accomplishes dosage compensation by doubling the expression of most X-linked genes.

In certain species, such as birds and fish, the phenomenon of dosage compensation is not well understood. In birds, the Z chromosome is a large chromosome, usually the fourth or fifth largest, and it contains almost all the known sex-linked genes. The W chromosome is generally a much smaller microchromosome, containing a high proportion of repeat sequence DNA that does not encode genes. Males are ZZ while females are ZW. Several years ago, researchers studied the level of expression of a Z-linked gene that encodes an enzyme called aconitase. It was discovered that males express twice as much aconitase as females do. These results suggested that dosage compensation does not occur in birds. However, more recently the expression of nine Z-linked genes was examined in chickens and it was found that at least six of them showed expression levels that were similar in males and females. The interpretation of these recent results is that dosage compensation usually occurs in birds, but perhaps not for every gene. (Also, as we will see later in this section, not all genes on the female X chromosome are inactivated in mammals.) The molecular mechanism of dosage compensation is not understood in birds. Cytologically, a highly compacted chromosome is not observed in the somatic cells of male birds, so it appears that the Z chromosome in males does not undergo condensation like one of the X chromosomes in female mammals. Alternatively, genes on the Z chromosome in males could be down regulated by 50% or the corresponding genes in the female could be up regulated twofold.

Dosage Compensation Occurs in Female Mammals by the Random Inactivation of One X Chromosome

In 1961, Mary Lyon proposed that dosage compensation in mammals occurs by the inactivation of a single X chromosome in females. Liane Russell also proposed the same theory around the same time. This proposal brought together two lines of study. The first type of evidence came from cytological studies. In 1949, Murray Barr and Ewart Bertram identified a highly condensed structure in the interphase nuclei of somatic cells in female cats that was not found in male cats. This structure became known as the **Barr body** (fig. 7.3*a*). In 1960, Susumu Ohno correctly proposed that the Barr body is a highly condensed X chromosome.

In addition to this cytological evidence, Lyon was also familiar with mammalian mutations in which the coat color had a variegated pattern. An example is shown in figure 7.3*b*, which illustrates a calico cat. Calico cats are females that are heterozygous for an X-linked gene that can occur as an orange or a black allele. (The white underside is due to a dominant allele in a different gene.) The orange and black patches are randomly distributed in different female individuals. Similar kinds of mutations have also been identified in the mouse. Based on these genetic observations and the cytological data, Lyon suggested that these results can be explained by X inactivation in the cells of female mammals.

The mechanism of X inactivation, also known as the **Lyon hypothesis,** is schematically illustrated in figure 7.4. This example involves a white and black variegated coat color found in certain strains of mice. As shown here, a female mouse has inherited an X chromosome from its mother that carries an allele conferring white coat color (X^b); the X chromosome from its father carries a

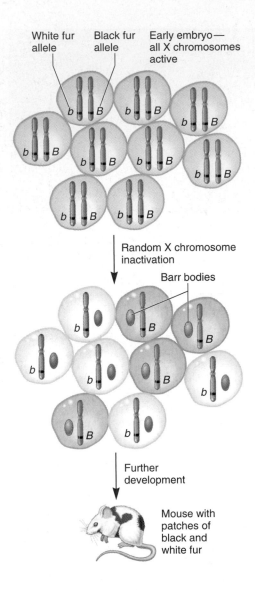

White fur allele   Black fur allele   Early embryo— all X chromosomes active

Random X chromosome inactivation

Barr bodies

Further development

Mouse with patches of black and white fur

FIGURE 7.4   The mechanism of X chromosome inactivation.

GENES→TRAITS   The *top* of this figure represents a mass of several cells that compose the early embryo. Initially, both X chromosomes are active. At an early stage of embryonic development, random inactivation of one X chromosome occurs in each cell. This inactivation pattern is maintained as the embryo matures into an adult. As noted in this schematic diagram, the pattern of X inactivation in the embryo parallels the pattern of white and black fur found in the adult mouse.

black coat color allele (X^B). Initially, both X chromosomes are active. However, at an early stage of embryonic development, one of the two X chromosomes is randomly inactivated in each somatic cell. For example, one embryonic cell may have the X^B chromosome inactivated. As the embryo continues to grow and mature, this embryonic cell will divide and may eventually give rise to billions of cells in the adult animal. The epithelial (skin) cells that are derived from this embryonic cell will produce a patch of white fur, because the X^B chromosome has been permanently inactivated. Alternatively, another embryonic cell may have the other X chromosome inactivated (i.e., X^b). The epithelial cells derived from this embryonic cell will produce a patch of black fur. Since the primary event of X inactivation is a random process that occurs at an early stage of development, the result is an animal with patches of white fur and other patches of black fur. This is the basis for the variegated phenotype.

During inactivation, the chromosomal DNA becomes highly compacted so that most of the genes on the inactivated X chromosome cannot be expressed. When cell division occurs and the inactivated X chromosome is replicated, both copies remain highly compacted and inactive. Likewise, during subsequent cell divisions, X inactivation is passed along to all future somatic cells.

EXPERIMENT 7A

In Adult Female Mammals, One X Chromosome Has Been Permanently Inactivated

According to the Lyon hypothesis, each somatic cell of female mammals will express the genes on one of the X chromosomes, but not both. If an adult female is heterozygous for an X-linked gene, only one of two alleles will be expressed in any given cell. In 1963, Ronald Davidson, Harold Nitowsky, and Barton Childs set out to test the Lyon hypothesis at the cellular level. To do so, they analyzed the expression of a human X-linked gene that encodes an enzyme involved with sugar metabolism known as glucose-6-phosphate dehydrogenase (*G-6-PD*).

Prior to the Lyon hypothesis, biochemists had found that individuals vary with regard to the G-6-PD enzyme. This variation can be detected when the enzyme is subjected to gel electrophoresis (see the appendix for a description of gel electrophoresis). One *G-6-PD* allele encodes a G-6-PD enzyme that migrates very quickly during gel electrophoresis (the "fast" enzyme), whereas another *G-6-PD* allele produces an enzyme that migrates more slowly (the "slow" enzyme). As shown in figure 7.5, a sample of cells from heterozygous adult females produces both types of enzymes, whereas hemizygous males produce either the fast or slow type. The difference in migration between the fast and slow G-6-PD enzymes is due to minor differences in the structures of

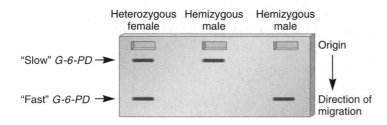

FIGURE 7.5 Mobility of G-6-PD on gels. G-6-PD can exist as a fast allele that encodes a protein that migrates more quickly to the bottom of the gel and a slow allele that migrates more slowly. The protein encoded by the fast allele is seen more toward the bottom of the gel.

these enzymes. These minor differences do not significantly affect G-6-PD function, but they do enable geneticists to distinguish the proteins encoded by the two X-linked alleles.

As shown in figure 7.6, Davidson, Nitowsky, and Childs tested the Lyon hypothesis using cell culturing techniques. They removed small samples of epithelial cells from a heterozygous female and grew them in the laboratory. When combined together, these samples contained a mixture of both types of enzymes because the adult cells were derived from many different embryonic cells, some that had the slow allele inactivated and some that had the fast allele inactivated. In the experiment of figure 7.6, these cells were sparsely plated on solid growth media. After several days, a single cell grows and divides to produce a colony of cells. This is called a **clone** of cells, because all the cells within the colony have been derived from a single cell. Davidson, Nitowsky, and Childs reasoned that all the cells within a clone would express only one of the two *G-6-PD* alleles if the Lyon hypothesis was correct.

■ THE HYPOTHESIS

According to the Lyon hypothesis, an adult female who is heterozygous for the fast and slow *G-6-PD* alleles should express only one of the two alleles in any particular somatic cell and its descendants, but not both.

■ TESTING THE HYPOTHESIS   —   **FIGURE 7.6 Evidence that adult female mammals contain one X chromosome that has been permanently inactivated.**

Starting material: Small skin samples taken from a woman who was heterozygous for the fast and slow alleles of *G-6-PD*.

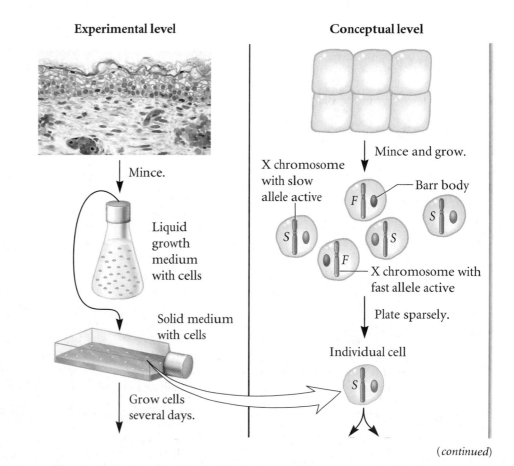

Experimental level

1. Mince the tissue to separate the individual cells.

   Mince.

   Liquid growth medium with cells

   Solid medium with cells

   Grow cells several days.

2. Grow the cells in a liquid growth medium and then plate (sparsely) onto solid growth medium. The cells then divide to form a clone of many cells.

Conceptual level

Mince and grow.

X chromosome with slow allele active

Barr body

X chromosome with fast allele active

Plate sparsely.

Individual cell

(continued)

3. Take nine isolated clones and grow in liquid cultures. (Only three are shown here.)

4. Take cells from the liquid cultures, prepare protein lysates, and subject to gel electrophoresis. (This technique is described in the appendix.)

Note: As a control, prepare protein lysates from cells in step 1, and subject the lysate to gel electrophoresis. This control sample is not from a clone. It is a mixture of cells derived from a woman's skin sample.

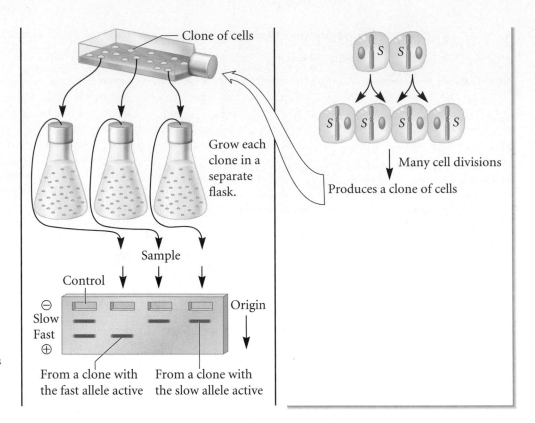

THE DATA

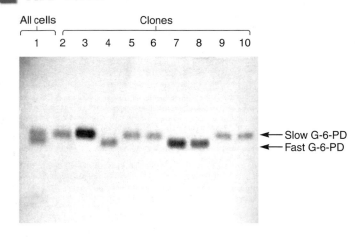

INTERPRETING THE DATA

In the data shown in figure 7.6, lane 1 contains a mixture of epithelial cells from a heterozygous woman who produced both types of G-6-PD enzymes. Bands corresponding to the fast and slow enzymes were observed in this lane. As described in steps 2 to 4, this mixture of epithelial cells was also used to generate nine clones, which are shown in lanes 2 to 10. Each clone was a population of cells independently derived from a single epithelial cell. Since the epithelial cells were obtained from an adult female, the Lyon hypothesis predicts that each epithelial cell would already have one of its X chromosomes permanently inactivated and would pass this trait to its progeny cells. For example, suppose that an epithelial cell had inactivated the X chromosome that encoded the fast G-6-PD. If this cell was allowed to form a clone of cells on a plate, all the cells in this clonal population would be expected to have the same X chromosome inactivated (namely, the X chromosome encoding the fast G-6-PD). Therefore, this clone of cells should express only the slow G-6-PD. As shown in the data, all nine clones expressed either the fast or slow G-6-PD, but not both. These results are consistent with the hypothesis that X inactivation has already occurred in any given epithelial cell and that this pattern of inactivation is passed to all of its progeny cells.

A self-help quiz involving this experiment can be found at the Online Learning Center.

X Inactivation in Mammals Depends on the *Xic* Locus and the *Xist* Gene

Since the Lyon hypothesis was confirmed, researchers have become interested in the genetic control of X inactivation. They have found that human cells possess the ability to count their X chromosomes and allow only one of them to remain active. In normal females, two X chromosomes are counted and one is inactivated; in males, one X chromosome is counted and none inactivated. On occasion, however, people are born with abnormalities in the number of sex chromosomes.

Phenotype	Chromosome Composition	Number of Barr Bodies
Normal female	XX	1
Normal male	XY	0
Turner syndrome (female)	X0	0
Triple X syndrome (female)	XXX	2
Klinefelter syndrome (male)	XXY	1

As seen here, the cells of humans (as well as other mammals) have the ability to count their X chromosomes. If the number of X chromosomes exceeds one, additional X chromosomes are converted to Barr bodies.

Although the genetic control of inactivation is not entirely understood at the molecular level, a short region on the X chromosome called the **X-inactivation center (Xic)** is known to play a critical role (fig. 7.7). Eeva Therman and Klaus Patau identified Xic from its key role in X inactivation. The counting of human X chromosomes is accomplished by counting the number of Xics. There is an absolute requirement for a Xic on each X chromosome for inactivation to occur. Therman and Patau found that if

one of the X chromosomes is missing its Xic due to a chromosome mutation, neither X chromosome will be inactivated. This is a lethal condition for a human female embryo.

Though the topic of gene expression is discussed in chapters 12 to 15, it is interesting to consider how the molecular expression of certain genes controls X inactivation. The expression of a specific gene within the X-inactivation center is required for the compaction of the X chromosome into a Barr body. This gene, discovered in 1991, is named *Xist* (for X-inactive specific transcript). The *Xist* gene on the inactivated X chromosome is expressed, which is unusual since most other genes on the inactivated X chromosome are silenced. The *Xist* gene product is a long RNA molecule that does not encode a protein. Instead, the role of the *Xist* RNA is to coat the inactive X chromosome. After coating, other proteins associate with the *Xist* RNA and promote chromosomal compaction into a Barr body.

A second region termed the **X chromosomal controlling element (Xce)** affects the choice of the X chromosome to be inactivated. This choice occurs during embryonic development and is maintained in all subsequent cell divisions. In females heterozygous for different *Xce* alleles, an X chromosome that carries a strong *Xce* allele is more likely to remain active than one that carries a weak *Xce* allele, thereby leading to skewed X inactivation. However, the degree of skewing is rarely more than 70 to 30%. As shown in figure 7.7, the Xce region is very close to the end of the Xic region. A gene designated *TsiX*, which is expressed only during early embryonic development, also plays a role in chromosome choice. The *TsiX* gene encodes an RNA that is complementary to the *Xist* RNA. For this reason, it is termed *antisense RNA*. In this case, the *Xist* RNA would be the *sense RNA*, since it has a defined function of promoting X inactivation. *TsiX* antisense RNA is believed to bind to *Xist* RNA and inhibit its function. In other words, the binding of *TsiX* RNA to *Xist* RNA may prevent inactivation during embryonic development. Therefore, the expression of *TsiX* during early embryonic development may play a role in choosing the active X chromosome.

The process of X inactivation can be divided into three phases: initiation, spreading, and maintenance (fig. 7.8). During initiation, one of the X chromosomes is targeted to remain active and the other is chosen to be inactivated. During the spreading phase, the chosen X chromosome is inactivated. This spreading requires the expression of the *Xist* gene. The *Xist* RNA coats the inactivated X chromosome and promotes condensation. It is called the spreading phase because inactivation begins near the X-inactivation center and spreads in both directions along the X chromosome. Both initiation and spreading occur during embryonic development. Subsequently, the inactivated X chromosome is maintained as such during future cell divisions. When a cell divides, the Barr body is replicated and both copies remain compacted.

A few genes on the inactivated X chromosome are expressed in the somatic cells of adult female mammals. These genes are said to escape the effects of X inactivation. As mentioned, *Xist* is an example of a gene that is expressed from the highly condensed Barr body. In addition, researchers have identified a few other

Portion of the X chromosome

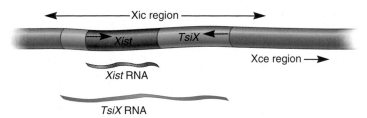

FIGURE 7.7 **The Xic region of the X chromosome.** The *Xist* gene is transcribed into RNA from the inactive X chromosome but not from the active X chromosome. This *Xist* RNA binds to the inactive X chromosome, which promotes compaction. The *TsiX* gene is transcribed in the opposite direction and encodes an antisense RNA that is complementary to the *Xist* RNA. During embryonic development, the *TsiX* RNA is thought to control the choice of the active X chromosome by inhibiting the ability of the *Xist* RNA to promote compaction of one X chromosome. The Xce region is adjacent to one end of the Xic region. It may regulate the transcription of the *TsiX* gene and thereby influence the choice of the X chromosome that remains active.

Initiation: Occurs during embryonic development. The number of X-inactivation centers (*Xics*) are counted and one of the X chromosomes remains active and the other is targeted for inactivation.

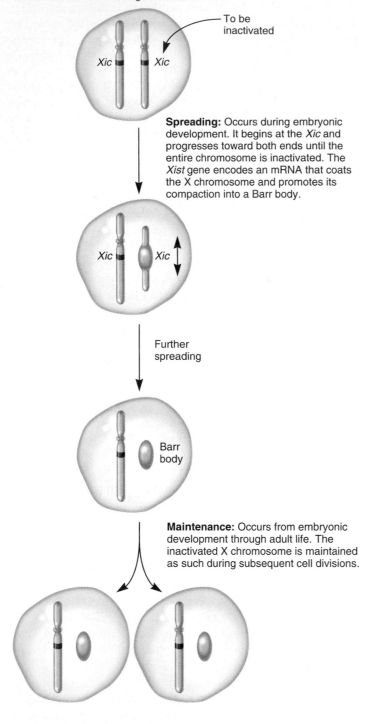

To be inactivated

Xic   *Xic*

Spreading: Occurs during embryonic development. It begins at the *Xic* and progresses toward both ends until the entire chromosome is inactivated. The *Xist* gene encodes an mRNA that coats the X chromosome and promotes its compaction into a Barr body.

Xic   *Xic*

Further spreading

Barr body

Maintenance: Occurs from embryonic development through adult life. The inactivated X chromosome is maintained as such during subsequent cell divisions.

FIGURE 7.8   The function of the *Xic* locus during X chromosome inactivation.

human X-linked genes that escape X inactivation. Among these are the pseudoautosomal genes that are also found on the Y chromosome (described in chapter 3). Dosage compensation is not necessary for X-linked pseudoautosomal genes, since they are located on both the X and Y chromosomes.

The Expression of an Imprinted Gene Depends on the Sex of the Parent from Which the Gene Was Inherited

As we have just seen, dosage compensation changes the level of expression of many genes located on the X chromosome. We now turn to another epigenetic phenomenon, known as imprinting. The term *imprinting* implies a type of marking process that has a memory. For example, newly hatched birds identify marks on their parents, allowing them to distinguish their parents from other individuals. The term **genomic imprinting** refers to an analogous situation in which a segment of DNA is marked, and that mark is retained and recognized throughout the life of the organism inheriting the marked DNA. The phenotypes caused by imprinted genes follow a non-Mendelian pattern of inheritance, because the marking process causes the offspring to distinguish between maternally and paternally inherited alleles. Depending on how the genes are marked, the offspring expresses one of the two alleles, but not both. This is termed **monoallelic expression.**

To understand genomic imprinting, it is helpful to consider a specific example. In the mouse, a gene designated *Igf-2* encodes a growth hormone called insulin-like growth factor 2. Imprinting occurs in a way that results in the expression of the paternal *Igf-2* allele but not the maternal allele. The paternal allele is transcribed into RNA while the maternal allele is transcriptionally silent. With regard to phenotype, a functional *Igf-2* gene is necessary for normal size. A mutant allele of this gene, designated *Igf-2m,* yields a defective Igf-2 protein. This may cause a mouse to be a dwarf, but the dwarfism depends on whether the mutant allele is inherited from the male or female parent, as shown in figure 7.9. On the *left* side, an offspring has inherited the *Igf-2* allele from its father and the *Igf-2m* allele from its mother. Due to imprinting, only the *Igf-2* allele is expressed in the offspring. Therefore, this mouse grows to a normal size. Alternatively, in the reciprocal cross on the *right* side, an individual has inherited the *Igf-2m* allele from its father and the *Igf-2* allele from its mother. In this case, the *Igf-2* allele is not expressed. In this mouse, the *Igf-2m* allele would be transcribed into mRNA, but the mutation renders the Igf-2 protein defective. Therefore, the offspring on the *right* has a dwarf phenotype. As shown here, both offspring have the same genotype; they are heterozygous for the *Igf-2* alleles (i.e., *Igf-2 Igf-2m*). They are phenotypically different, however, because the allele that is expressed in their somatic cells depends on the parent from which the allele was inherited.

At the cellular level, imprinting is an epigenetic process that can be divided into three stages: the establishment of the imprint during gametogenesis, the maintenance of the imprint during embryogenesis and in adult somatic cells, and the erasure and reestablishment of the imprint in the germ cells. These stages are described in figure 7.10. This example also considers the imprinting of the *Igf-2* gene. The two mice shown here have inherited the *Igf-2* allele from their father and the *Igf-2m* allele from their mother. Due to imprinting, both mice express the *Igf-2* allele in their somatic cells. However, this pattern of imprinting is maintained only in the somatic cells. In the germ cells (i.e., sperm and eggs), the imprint is erased; it will be reestablished according to

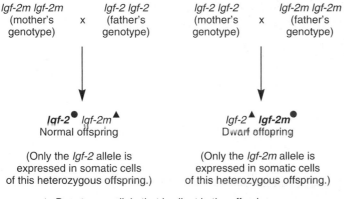

Igf-2m Igf-2m   *Igf-2 Igf-2*   *Igf-2 Igf-2*   *Igf-2m Igf-2m*
(mother's   x   (father's   (mother's   x   (father's
genotype)   genotype)   genotype)   genotype)

Igf-2 ● *Igf-2m* ▲   *Igf-2* ▲ *Igf-2m* ●
Normal offspring   Dwarf offspring

(Only the *Igf-2* allele is   (Only the *Igf-2m* allele is
expressed in somatic cells   expressed in somatic cells
of this heterozygous offspring.)   of this heterozygous offspring.)

▲ Denotes an allele that is silent in the offspring
● Denotes an allele that is expressed in the offspring

FIGURE 7.9   **An example of genomic imprinting in the mouse.**
In the cross on the *left,* a homozygous male with the normal *Igf-2* allele
is crossed to a homozygous female carrying a defective allele, designated
Igf-2m. An offspring is heterozygous and normal, because the paternal
allele is active. In the reciprocal cross on the *right,* a homozygous male
carrying the defective allele is crossed to a homozygous normal female.
In this case, the offspring is dwarf. This is because the paternal allele is
defective due to mutation and the maternal allele is not expressed.

the sex of the animal. The female mouse on the *left* will transmit
transcriptionally inactive alleles to her offspring. The male mouse
on the *right* will transmit transcriptionally active alleles. However,
since this male is a heterozygote, it will transmit either a functionally active *Igf-2* allele or a functionally inactive mutant allele
(*Igf-2m*). An *Igf-2m* allele, which is inherited from a male mouse,
can be expressed into mRNA (i.e., it is transcriptionally active),
but it will not produce a functional Igf-2 protein due to the deleterious mutation that created the *Igf-2m* allele.

As seen in figure 7.10, genomic imprinting is permanent in
the somatic cells of an animal, but the marking of alleles can be
altered from generation to generation. For example, the female
mouse on the *left* has an active copy of the *Igf-2* allele but may
transmit a transcriptionally inactive copy of this allele to its offspring.

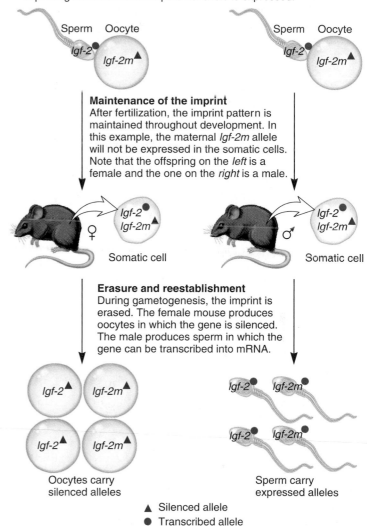

Establishment of the imprint
In this example, imprinting occurs in the *Igf-2* gene, which exists in the
Igf-2 allele from the male and the *Igf-2m* allele from the female. This
imprinting occurs so that the paternal allele is expressed.

Sperm   Oocyte   Sperm   Oocyte
Igf-2 ●   *Igf-2* ●
        Igf-2m ▲   *Igf-2m* ▲

Maintenance of the imprint
After fertilization, the imprint pattern is
maintained throughout development. In
this example, the maternal *Igf-2m* allele
will not be expressed in the somatic cells.
Note that the offspring on the *left* is a
female and the one on the *right* is a male.

Igf-2 ●   *Igf-2* ●
Igf-2m ▲   *Igf-2m* ▲
♀ Somatic cell   ♂ Somatic cell

Erasure and reestablishment
During gametogenesis, the imprint is
erased. The female mouse produces
oocytes in which the gene is silenced.
The male produces sperm in which the
gene can be transcribed into mRNA.

Igf-2 ▲ *Igf-2m* ▲   *Igf-2* ● *Igf-2m* ●

Igf-2 ▲ *Igf-2m* ▲   *Igf-2* ● *Igf-2m* ●

Oocytes carry   Sperm carry
silenced alleles   expressed alleles

▲ Silenced allele
● Transcribed allele

FIGURE 7.10   **Genomic imprinting during gametogenesis.**
This example involves a mouse gene *Igf-2,* which is found in two alleles
designated *Igf-2* and *Igf-2m.* The *left* side shows a female mouse that
was produced from a sperm carrying the *Igf-2* allele and an egg carrying
the *Igf-2m* allele. In the somatic cells of this female animal, the *Igf-2*
allele is active. However, when this female produces eggs, both alleles are
transcriptionally inactive when they are transmitted to offspring. The
right side of this figure shows a male mouse that was also produced
from a sperm carrying the *Igf-2* allele and an egg carrying the *Igf-2m*
allele. In the somatic cells of this male animal, the *Igf-2* allele is active.
However, the sperm from this male will contain either an active *Igf-2*
allele or a defective *Igf-2m* allele. Note: The defective *Igf-2m* allele transmitted by the sperm can be expressed into mRNA in the somatic cells of
offspring, but the expression will not produce a functional Igf-2 protein
due to the mutation, which inhibits the protein's function.

Genomic imprinting occurs in several species including numerous insects, plants, and mammals. Imprinting may involve a single gene, a part of a chromosome, an entire chromosome, or even all the chromosomes from one parent. Helen Crouse discovered the first example of imprinting, which involved an entire chromosome in the housefly, *Sciara coprophilia*. In this species, the fly normally inherits three sex chromosomes (rather than two as in most other species); one X chromosome is inherited from the female, two from the male. In male flies, both paternal X chromosomes are lost from somatic cells during embryogenesis. In female flies, only one of the paternal X chromosomes is lost. In both sexes, the maternally inherited X chromosome is never lost. These results indicate that the maternal X chromosome is marked to promote its retention or paternal X chromosomes are marked to promote their loss.

Genomic imprinting can also be correlated with the process of X inactivation described previously. In certain species, imprinting is involved in the choice of the X chromosome that will be inactivated. In marsupials, the paternal X chromosome is always inactivated in the somatic cells of the individual. In placental mammals, X inactivation of the paternal X chromosome occurs in the extraembryonic tissue (e.g., the placenta), while X inactivation occurs randomly between the maternal and paternal X chromosomes in the embryo itself.

The Imprinting of Genes and Chromosomes Is a Molecular Marking Process That Involves DNA Methylation

As we have seen, genomic imprinting must involve a marking process. A particular gene or chromosome must be marked dissimilarly during spermatogenesis versus oogenesis. After fertilization takes place, this differential marking affects the expression of particular genes. At the molecular level, the imprinting of several genes is known to involve **differentially methylated regions (DMRs)** that are located near the imprinted genes. Depending on the particular gene, the DMRs are methylated in the oocyte or the sperm, but not both. The DMRs contain binding sites for one or more proteins that regulate the transcription of nearby genes.

For most imprinted genes, methylation at a DMR results in an inhibition of gene expression. Methylation could enhance the binding of proteins that inhibit transcription and/or inhibit the binding of proteins that enhance transcription. (The relationship between methylation and gene expression is also considered in chapter 15.) For this reason, imprinting is usually described as a marking process that silences gene expression by preventing transcription. However, this is not always the case. Two imprinted human genes, *H19* and *Igf-2*, provide an interesting example. These two genes lie close to each other on human chromosome 11 and appear to be controlled by the same DMR, which is a 2,000 base pair region that lies upstream of the *H19* gene (fig. 7.11*a*). It contains binding sites for proteins that regulate the transcription of the *H19* or *Igf-2* genes. This DMR is highly methylated on the paternally inherited chromosome. Methylation of the DMR silences the *H19* gene because a specific protein or

group of proteins bind to the DMR only after it has been methylated; protein binding inhibits the transcription of the *H19* gene. In contrast, when this same DMR is methylated, the *Igf-2* gene is activated. Though the activation mechanism is not fully understood, it is possible that a protein that inhibits *Igf-2* transcription cannot bind to the methylated DMR. The inability of the inhibitory protein to bind has the effect of activating the *Igf-2* gene. Since this DMR in the maternally inherited chromosome is not methylated, this explains why the maternal copy of *Igf-2* is not expressed. It may seem odd that proteins binding at the DMR would affect the transcription of the *Igf-2*, because the DMR and *Igf-2* gene are not immediately adjacent. As discussed in chapter 15, transcription factors sometimes influence the transcription of genes by a looping mechanism that brings two separated regions close together. This looping mechanism may bring the DMR very close to the promoter of the *Igf-2* gene and thereby influence its transcription.

Now that we have an understanding of how methylation may affect gene transcription, let's consider the methylation process during the life of an animal. A key issue is that certain imprinted genes are differentially methylated during oogenesis compared to spermatogenesis (fig. 7.11*b*). In this example, a male and female offspring have inherited a methylated DMR from their father and a nonmethylated DMR from their mother. If this was the DMR next to the *H19* gene, the somatic cells of the offspring would express only the maternal *H19* allele. This pattern of imprinting is maintained in the somatic cells of both offspring. However, when the female offspring makes gametes, the imprinting is erased during early oogenesis so that the female will pass an unmethylated DMR to its offspring. In the male, erasure of the imprint also occurs during early spermatogenesis, but then *de novo* (new) methylation occurs in both DMRs. Therefore, if this DMR is next to the *H19* gene, the male transmits inactive alleles of this gene.

Genomic imprinting is a fairly new and exciting area of research. To date, imprinting has been identified in several mammalian genes (table 7.2). In some cases, the female alleles are

TABLE 7.2

Examples of Mammalian Genes and Inherited Human Diseases That Involve Imprinted Genes

Gene	Allele Expressed	Function
WT1	Maternal	Wilms tumor-suppressor gene—suppresses uncontrollable cell growth
INS	Paternal	Insulin—hormone involved in cell growth and metabolism
Igf-2	Paternal	Insulin-like growth factor II—similar to insulin
Igf-2R	Maternal	Receptor for insulin-like growth factor II
H19	Maternal	Unknown
SNRPN	Paternal	Splicing factor
Gabrb	Maternal	Neurotransmitter receptor

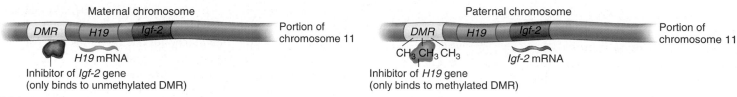

(a) Actions of regulatory proteins that bind to a methylated DMR or to an unmethylated DMR

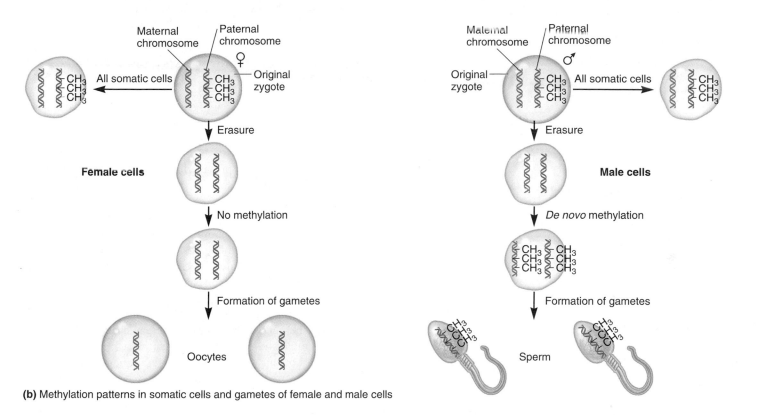

(b) Methylation patterns in somatic cells and gametes of female and male cells

FIGURE 7.11   **DNA methylation in the imprinting process.** (a) The effects of methylation on gene transcription. A DMR is located next to the two genes called *H19* and *Igf-2*. This DMR is methylated on the paternally inherited chromosome. When it is methylated, it binds a protein that inhibits the *H19* gene. Therefore, the paternally inherited allele is silenced. In contrast, when the DMR is not methylated, it binds a different protein that inhibits the *Igf-2* gene. In this way, the maternally inherited *Igf-2* allele is silenced. (b) The pattern of DMR methylation during the life of an animal. In this example, a male and a female offspring have inherited a methylated DMR and nonmethylated DMR from their father and mother, respectively. Maintenance methylation retains the imprinting in somatic cells during embryogenesis and in adulthood. Demethylation occurs in cells that are destined to become gametes. *De novo* methylation occurs only in cells that are destined to become sperm. Haploid male gametes transmit a methylated DMR, whereas haploid female gametes transmit an unmethylated DMR.

transcriptionally active in the somatic cells of offspring, whereas in other cases the male alleles are active. The biological significance of imprinting is still a matter of speculation.

Imprinting plays a role in the inheritance of certain human diseases such as Prader-Willi syndrome (PWS) and Angelman syndrome (AS). PWS is characterized by reduced motor function, obesity, and mental deficiencies. AS patients are hyperactive, have unusual seizures and repetitive symmetrical muscle movements, and show mental deficiencies. Most commonly, PWS and AS involve a small deletion in human chromosome 15. If this deletion is inherited from the maternal parent, it leads to Angelman syndrome; paternal inheritance leads to Prader-Willi syndrome (fig. 7.12).

Researchers have discovered that this region contains closely linked but distinct genes that are maternally or paternally imprinted. Current evidence suggests that AS results from the lack of expression of a single gene, *UBE3A*, which codes for a protein called E6-AP that functions to transfer small ubiquitin molecules to certain proteins to target their degradation. The paternal allele of *UBE3A* is silenced. Therefore, if the maternal allele is deleted, as in the *left* side of figure 7.12, the individual will develop AS because he or she will not have an active copy of the *UBE3A* gene.

The gene(s) responsible for PWS has not been definitively determined, although several imprinted genes in this region of chromosome 15 are known. The most likely candidate involved

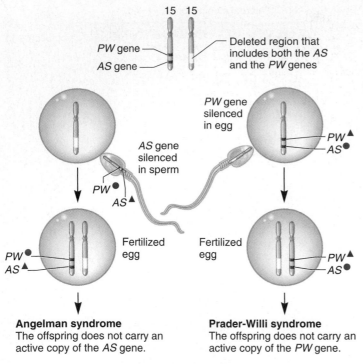

FIGURE 7.12   The role of imprinting in the development of Angelman and Prader-Willi syndromes.

GENES→TRAITS   A small region on chromosome 15 contains two different genes designated the *AS* gene and *PW* gene in this figure. If a chromosome 15 deletion is inherited from the maternal parent, it leads to Angelman syndrome. This phenotype occurs because the offspring does not inherit an active copy of the *AS* gene. Alternatively, the chromosome 15 deletion may be inherited from the male parent, leading to Prader-Willi syndrome. The phenotype of this syndrome occurs because the offspring does not inherit an active copy of the *PW* gene.

in PWS is a gene designated *SNRPN*. The gene product is part of a small nuclear ribonucleoprotein polypeptide N, which is a complex that controls gene splicing and is necessary for the synthesis of critical proteins in the brain. The maternal allele of *SNRPN* is silenced.

7.3 EXTRANUCLEAR INHERITANCE

Thus far, we have considered several types of non-Mendelian inheritance patterns. These include maternal effect genes, dosage compensation, and genomic imprinting. All of these inheritance patterns involve genes found on chromosomes in the cell nucleus. Another cause of non-Mendelian inheritance patterns concerns genes that are not located in the cell nucleus. In eukaryotic species, the most biologically important example of extranuclear inheritance involves genetic material in cellular organelles. Since these organelles are found within the cytoplasm of the cells, the inheritance of organellar genetic material is called **extranuclear**

inheritance (the prefix *extra-* means outside of) or **cytoplasmic inheritance.** In addition to the cell nucleus, where the linear chromosomes reside, the mitochondria and plastids (e.g., chloroplasts) contain their own genetic material.

In this section, we will examine the genetic composition of mitochondria and plastids and explore the pattern of transmission of these organelles from parent to offspring. We will also consider a few other examples of inheritance patterns that cannot be explained by the transmission of nuclear genes.

Mitochondria and Chloroplasts Contain Circular Chromosomes with Many Genes

In 1951, Y. Chiba was the first to suggest that chloroplasts contain their own DNA. He based his conclusion on the staining properties of a DNA-specific dye known as Feulgen. Researchers later developed techniques to purify organellar DNA. In addition, electron microscopy studies provided interesting insights into the organization and composition of mitochondrial and chloroplast chromosomes. More recently, the advent of molecular genetic techniques in the 1970s and 1980s has allowed researchers to determine the DNA sequences of organellar DNAs. From these types of studies, the chromosomes of mitochondria and chloroplasts were found to resemble smaller versions of bacterial chromosomes.

The genetic material of mitochondria and chloroplasts is located inside the organelle in a region known as the **nucleoid** (fig. 7.13). The genome is a single circular chromosome (composed of double-stranded DNA), although a nucleoid contains several copies of this chromosome. In addition, a mitochondrion or chloroplast often contains more than one nucleoid. In mice, for example, each mitochondrion has one to three nucleoids, with each nucleoid containing two to six copies of the circular mitochondrial genome. On average, each mouse mitochondrion contains five to six copies of the mitochondrial genome. However, this number is variable and depends on the type of cell and the stage of development. In comparison, the plastids of algae and higher plants tend to have more nucleoids per organelle. Table 7.3 describes the genetic composition of mitochondria and chloroplasts for a few selected species.

Besides variation in copy number, the sizes of mitochondrial and plastid genomes also vary greatly among different species. For example, there is a 400-fold variation in the size of the mitochondrial chromosomes. In general, the mitochondrial genomes of animal species tend to be fairly small; those of fungi, algae, and protists are intermediate in size; and those of plant cells tend to be fairly large. Among algae and more complex plants, there is substantial variation in the size of plastid chromosomes.

Figure 7.14 illustrates the genome of a human mitochondrion; this genetic material is called **mtDNA.** Each copy of the mitochondrial chromosome consists of a circular DNA molecule that is only 17,000 base pairs (bp) in length. This size is less than 1% of a typical bacterial chromosome. The mtDNA carries relatively few genes. There are genes that encode ribosomal RNA and

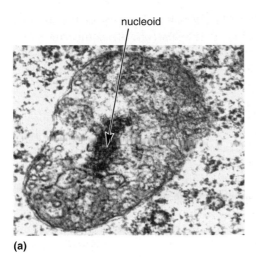

nucleoid

(a)

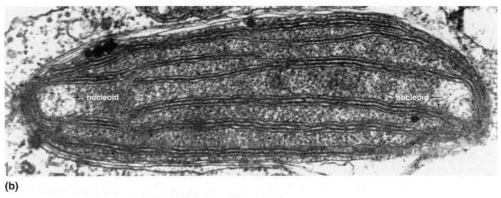

nucleoid   nucleoid

(b)

FIGURE 7.13 **Nucleoids within (a) mitochondria and (b) chloroplasts.** The mitochondrial and chloroplast chromosomes are found within the nucleoid region of the organelle.

TABLE 7.3

Genetic Composition of Mitochondria and Chloroplasts

Species	Organelle	Nucleoids per Organelle	Total Number of Chromosomes per Organelle
Tetrahymena	Mitochondrion	1	6–8
Mouse	Mitochondrion	1–3	5–6
Chlamydomonas	Chloroplast	5–6	~80
Euglena	Chloroplast	20–34	100–300
Higher plants	Chloroplast	12–25	~60

Data from: Gillham, N. W. (1994). *Organelle Genes and Genomes.* Oxford University Press, New York.

transfer RNA. These RNAs are necessary for the synthesis of proteins inside the mitochondrion. In addition, there are 13 genes encoding proteins that function within the mitochondrion. The primary functional role of mitochondria is to provide cells with the bulk of their ATP (adenosine triphosphate), which is used as an energy source to drive cellular reactions. These 13 polypeptides function in a process known as oxidative phosphorylation, which enables the mitochondrion to synthesize ATP. However, mitochondria require many additional proteins to carry out oxidative phosphorylation and other mitochondrial functions. Most mitochondrial proteins are encoded by genes within the cell nucleus. When these nuclear genes are expressed, the mitochondrial proteins are first synthesized outside the mitochondrion in the cytosol of the cell. After synthesis, nuclearly encoded mitochondrial proteins are then transported into the mitochondria.

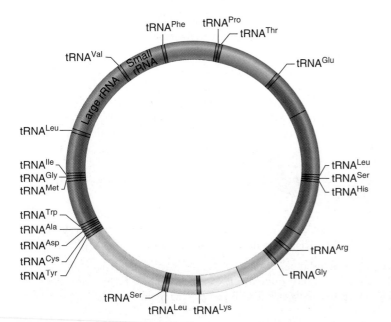

■ Ribosomal RNA genes
■ Transfer RNA genes
■ NADH dehydrogenase genes
■ Cytochrome *b* gene
■ Coenzyme Q–cytochrome *c* reductase genes
▨ Cytochrome *c* oxidase genes
□ ATP synthase genes
■ Noncoding DNA

FIGURE 7.14 **A genetic map of the human mitochondrial genome.** This diagram illustrates the locations of many genes along the circular mitochondrial chromosome. The genes shown in *red* are genes that encode tRNAs. For example, tRNA^arg encodes the tRNA that carries arginine. The genes that encode ribosomal RNA are shown in *pale orange*. The remaining genes encode proteins that function within the mitochondrion. The mitochondrial genomes from numerous species have been determined.

Chloroplast genomes tend to be larger than mitochondrial genomes, and they have a correspondingly greater number of genes. A typical chloroplast genome is approximately 100,000 to 200,000 bp in length, which is about 10 times larger than the mitochondrial genome of animal cells. Figure 7.15 shows the chloroplast genome, called **cpDNA,** of the tobacco plant. It is a circular DNA molecule that contains 156,000 bp. It carries between 110 and 120 different genes. These include genes that encode ribosomal RNAs, transfer RNAs, and many proteins required for photosynthesis. As with mitochondria, many chloroplast proteins are encoded by genes found in the plant cell nucleus. These proteins contain chloroplast-targeting signals that direct them into the chloroplasts.

Extranuclear Inheritance Produces Non-Mendelian Results in Reciprocal Crosses

In a diploid eukaryotic species, most genes within the nucleus obey a Mendelian pattern of inheritance because the homologous pairs of chromosomes segregate during gamete formation. Except for sex-linked traits, offspring inherit one copy of each gene from both the maternal and paternal parents. By comparison, the inheritance pattern of extranuclear genetic material does not display a Mendelian pattern. This is because mitochondria and plastids do not segregate into the gametes in the same way as nuclear chromosomes.

Carl Correns discovered a trait that showed a non-Mendelian pattern of inheritance involving a pigmentation trait in *Mirabilis jalapa* (the four-o'clock plant). Leaves could be green, white, or variegated (with both green and white sectors). Correns discovered that the pigmentation of the offspring depended solely on the maternal parent and not at all on the paternal parent (fig. 7.16). If the female parent had white pigmentation, all the offspring had white leaves. Similarly, if the female was green, all the offspring were green. When the female was variegated, the offspring could be green, white, or variegated.

The pattern of inheritance observed by Correns is a type of extrachromosomal inheritance called **maternal inheritance** (not to be confused with maternal effect). In this example, maternal inheritance occurs because the chloroplasts (a type of plastid),

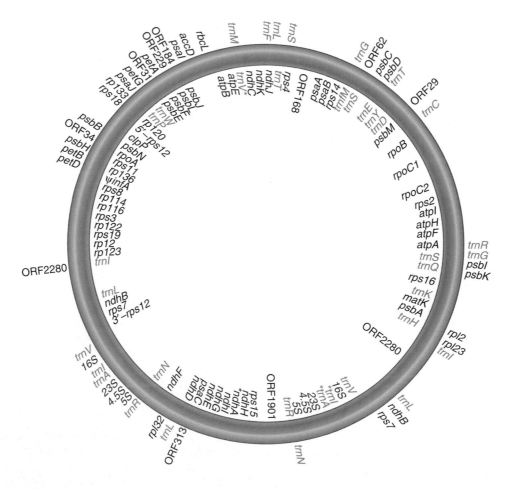

FIGURE 7.15   A genetic map of the tobacco chloroplast genome. This diagram illustrates the locations of many genes along the circular chloroplast chromosome. The genes shown in *blue* encode transfer RNAs. The genes that encode ribosomal RNA are shown in *red*. The remaining genes shown in *black* encode polypeptides that function within the chloroplast. The genes designated ORF (open reading frame) encode polypeptides with unknown functions.

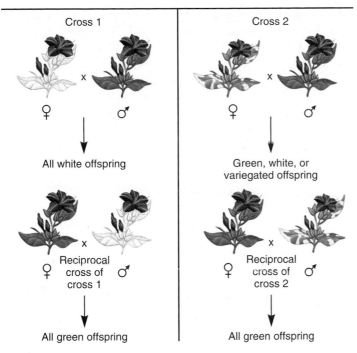

Correns' crosses

FIGURE 7.16   Maternal inheritance in the four-o'clock plant, *Mirabilis jalapa.* The reciprocal crosses of four-o'clock plants by Carl Correns. A pair of crosses between white-leaved and green-leaved plants. A pair of crosses between variegated- and green-leaved plants.

GENES→TRAITS   In this example, the white phenotype is due to chloroplasts that carry a mutant allele that diminishes green pigmentation. The variegated phenotype is due to a mixture of chloroplasts, some of which carry the normal (green) allele and some of which carry the white allele. In the crosses shown here, the parent providing the eggs determines the phenotypes of the offspring. This is due to maternal inheritance. The egg contains the chloroplasts that are inherited by the offspring. If the plant providing the eggs has a white phenotype, all the offspring will inherit chloroplasts that carry the mutant allele and will have white leaves. Similarly, if the plant providing the eggs is green, all the offspring will inherit normal chloroplasts and will be green. If the plant providing the eggs is variegated, it can transmit normal chloroplasts, mutant chloroplasts, or a mixture of these to the offspring. This will produce green, white, or variegated offspring, respectively. Note: the defective chloroplasts that give rise to white sectors are not completely defective in chlorophyll synthesis. Therefore, entirely white plants can survive though they are smaller than green or variegated plants.

which are the site of green pigment synthesis, are inherited only through the cytoplasm of the egg. The pollen grains of *M. jalapa* do not transmit plastids to the offspring.

The phenotypes of leaves can be explained by the types of chloroplasts within the leaf cells. The green phenotype is the wild-type condition. It is due to the presence of normal chloroplasts that can make green pigment. By comparison, the white phenotype is due to a mutation in a gene within the chloroplast DNA that diminishes the synthesis of green pigment. It is possible for a cell to contain both types of chloroplasts, a condition known as **heteroplasmy.** If a leaf cell contained both types of chloroplasts, it would be green because the normal chloroplasts would produce green pigment.

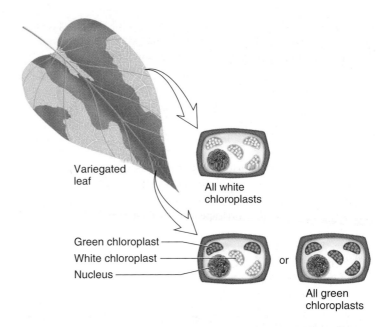

FIGURE 7.17   A cellular explanation of the variegated phenotype in *Mirabilis jalapa.* The plant shown originally inherited two types of chloroplasts: those that can produce green pigment and those that are defective. As this plant grows, the two types of chloroplasts are irregularly distributed to daughter cells. On occasion, a cell may receive only the chloroplasts that are defective at making green pigment. Such a cell continues to divide and produces a sector of the plant that is entirely white. Cells that contain both types of chloroplasts, or cells that contain only green chloroplasts, would produce green tissue, which may be adjacent to a sector of white tissue. This is the basis for the variegated phenotype of the leaves.

Figure 7.17 considers the leaf of a plant that began from a fertilized egg that contained both types of chloroplasts (i.e., a heteroplasmic cell). As a plant grows, the two types of chloroplasts are irregularly distributed to daughter cells. On occasion, a cell may receive only the chloroplasts that have a defect in making green pigment. Such a cell continues to divide and produce a sector of the plant that is entirely white. This is how the variegated phenotype is created. Similarly, if we consider the results of figure 7.16, a female parent that is variegated may transmit green, white, or a mixture of these types of chloroplasts to the egg cell and thereby produce green, white, or variegated offspring, respectively.

Studies in Yeast and *Chlamydomonas* Provided Genetic Evidence for Extranuclear Inheritance of Mitochondria and Chloroplasts

The studies of Correns and others indicated that some traits (e.g., leaf pigmentation) are inherited in a non-Mendelian manner. However, such studies did not definitively determine that maternal inheritance was due to genetic material within organelles. Further progress in the investigation of extranuclear inheritance was provided by detailed genetic analyses of eukaryotic microorganisms such as yeast and algae by isolating and characterizing mutant phenotypes that specifically affected the chloroplasts or mitochondria.

During the 1940s and 1950s, yeasts and molds became model eukaryotic organisms for investigating the inheritance of mitochondria. Because mitochondria produce energy for cells in the form of ATP, mutations that yield defective mitochondria are expected to make cells grow much more slowly. Boris Ephrussi and his colleagues identified mutations in *Saccharomyces cerevisiae* that had such a phenotype. These mutants were called **petites** to describe their formation of small colonies on agar plates (as opposed to wild-type strains that formed larger colonies). Biochemical and physiological evidence indicated that petite mutants had defective mitochondria. The researchers found that petite mutants could not grow when the cells only had an energy source requiring the metabolic activity of mitochondria, but they could form small colonies when grown on sugars that are metabolized by the glycolytic pathway, which occurs outside the mitochondria.

Genetic analyses showed that petite mutants are inherited in different ways. Segregational petite mutants were shown to have mutations in genes located in the nucleus. These mutations affect genes encoding proteins necessary for mitochondrial function. Many nuclear genes encode proteins that are synthesized in the cytosol and are then taken up by the mitochondria, where they perform their functions. Segregational petites get their name because they segregate in a Mendelian manner during meiosis (fig. 7.18a). By comparison, the second category of petite mutants, known as vegetative petite mutants, did not segregate in a Mendelian manner. This is because vegetative petites involve mutations in the mitochondrial genome itself.

Ephrussi identified two types of vegetative petites, called neutral petites and suppressive petites. Since yeast exist in two mating types designated a and α, he was able to cross vegetative petites to a wild-type strain. This revealed a non-Mendelian pattern of inheritance (fig. 7.18b). In a cross between a wild-type strain and a neutral petite, all four haploid progeny were wild type. This contradicts the normal 2:2 ratio expected for the segregation of Mendelian traits (see the discussion of tetrad analysis in chapter 5). By comparison, a cross between a wild-type strain and a suppressive petite yielded all petite colonies. Thus, vegetative petites are defective in mitochondrial function and show a non-Mendelian pattern of inheritance. Taken together, these results provided compelling evidence that the mitochondrion has its own genetic material.

Since these initial studies, researchers have found that neutral petites lack most of their mitochondrial DNA, whereas suppressive petites usually lack small segments of the mitochondrial genetic material. When two yeast cells are mated, the progeny inherit mitochondria from both parents. For example, in a cross between a wild-type and a neutral petite strain, the progeny inherit both types of mitochondria. Since wild-type mitochondria are inherited, the cells display a normal phenotype. The inheritance pattern of suppressive petites is more difficult to explain since the offspring inherit both normal and suppressive petite mitochondria. One possibility is that the suppressive petite mitochondria replicate more rapidly so that the wild-type mitochondria are not maintained in the cytoplasm for many doublings. Alternatively, some experimental evidence suggests that genetic exchanges between the genomes of wild-type and suppressive petites may ultimately produce a defective population of mitochondria.

Let's now turn our attention to the inheritance of chloroplasts that are found in eukaryotic species capable of photosynthesis (namely, algae and higher plants). The unicellular alga *Chlamydomonas reinhardtii* has been used as a model organism to investigate the inheritance of chloroplasts. This organism contains a single chloroplast that occupies approximately 40% of the cell volume. Genetic studies of chloroplast inheritance began when Ruth Sager identified a mutant strain of *Chlamydomonas*

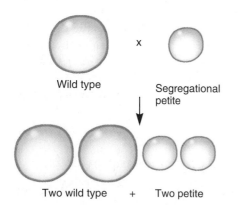

(a) A cross between a wild type and a segregational petite

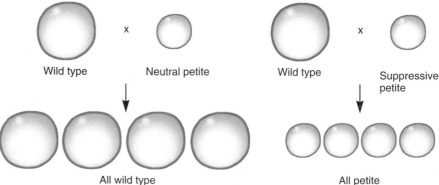

(b) A cross between a wild type and a neutral or suppressive vegetative petite

FIGURE 7.18   **Transmission of the petite trait in *Saccharomyces cerevisiae*.** **(a)** A wild-type strain crossed to a segregational petite. **(b)** A wild-type strain crossed to a neutral and to a suppressive vegetative petite.

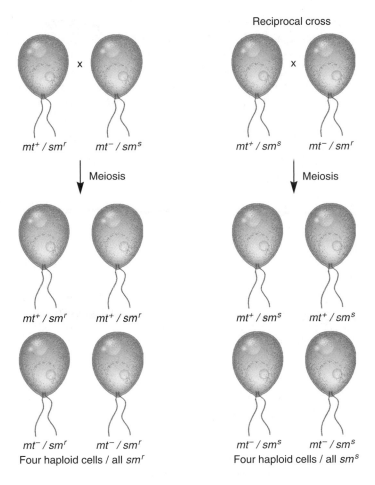

Reciprocal cross

mt^+ / sm^r × mt^- / sm^s mt^+ / sm^s × mt^- / sm^r

Meiosis   Meiosis

mt^+ / sm^r mt^+ / sm^r mt^+ / sm^s mt^+ / sm^s

mt^- / sm^r mt^- / sm^r mt^- / sm^s mt^- / sm^s

Four haploid cells / all sm^r Four haploid cells / all sm^s

FIGURE 7.19   **Chloroplast inheritance in *Chlamydomonas*.**
Mt^+ and mt^- indicate the two mating types of the organism. Sm^r
indicates streptomycin resistance, whereas sm^s indicates sensitivity
to this antibiotic.

TABLE 7.4

Transmission of Organelles Among Different Species

Species	Organelle	Transmission
Mammals	Mitochondria	Maternal inheritance
Mussels	Mitochondria	Biparental
S. cerevisiae	Mitochondria	Biparental
Molds	Mitochondria	Usually maternal inheritance; paternal inheritance has been found in the genus *Allomyces*
Chlamydomonas	Mitochondria	Inherited from the parent with the mt^- mating type
Chlamydomonas	Chloroplasts	Inherited from the parent with the mt^+ mating type
Plants Angiosperms	Mitochondria and plastids	Often times maternally inherited although biparental inheritance is not uncommon among many species
Gymnosperms	Mitochondria and plastids	Usually paternal inheritance

Data from: Gillham, N. W. (1994). *Organelle Genes and Genomes.* Oxford University Press, New York.

with resistance to the antibiotic streptomycin (sm^r). By comparison, most strains are sensitive to killing by streptomycin (sm^s). In
1954, Sager isolated an sm^r mutant that was not inherited in a
Mendelian manner (fig. 7.19). Like yeast, *Chlamydomonas* is an
organism that can be found in two mating types, in this case, designated mt^+ and mt^-. Mating type is due to nuclear inheritance
and segregates in a 1:1 manner. By comparison, Sager and her
colleagues discovered that the sm^r trait was inherited from the
mt^+ parent but not from the mt^- parent. In subsequent studies,
they mapped several genes, including the sm^r gene, to a single
chromosome that was not inherited in Mendelian manner. This
linkage group is the chloroplast chromosome.

The Pattern of Inheritance of Mitochondria and Plastids Varies Among Different Species

The inheritance of traits via genetic material within mitochondria and plastids is now a well-established phenomenon that
geneticists have investigated in many different species. In **heterogamous** species, two kinds of gametes are made. The female
gamete tends to be large and provides most of the cytoplasm to

the zygote, while the male gamete is small and often provides little more than a nucleus. Therefore, mitochondria and plastids are
most often inherited from the maternal parent. However, this is
not always the case. Table 7.4 describes the inheritance patterns
of mitochondria and plastids in several selected species.

In species where maternal inheritance is generally observed,
the paternal parent may, on rare occasions, provide mitochondria
via the sperm. This is called **paternal leakage.** It occurs in many
species that primarily exhibit maternal inheritance of their organelles. In the mouse, for example, approximately one to four paternal mitochondria are inherited for every 100,000 maternal mitochondria per generation of offspring. This means that most
offspring do not inherit any paternal mitochondria, but a rare
individual may inherit a mitochondrion from the sperm.

A Few Rare Human Diseases Are Caused by Mitochondrial Mutations

The human mitochondrial genome has 13 genes that encode
polypeptides necessary for the synthesis of ATP. In addition, the
mtDNA has genes that encode ribosomal RNA and transfer RNA
molecules. Human mtDNA is maternally inherited, since it is
transmitted from mother to offspring via the cytoplasm of the
egg. Therefore, the transmission of human mitochondrial diseases follows a strict maternal inheritance pattern.

TABLE 7.5

Examples of Human Mitochondrial Diseases

Disease	Mitochondrial Gene Mutated
Leber's hereditary optic neuropathy	A mutation in one of several mitochondrial genes that encode respiratory proteins: *ND1, ND2, CO1, ND4, ND5, ND6,* and *cytb*
Neurogenic muscle weakness	A mutation in the *ATPase6* gene that encodes a subunit of the mitochondrial ATP-synthetase, which is required for ATP synthesis
Mitochondrial encephalomyopathy, lactic acidosis, and strokelike episodes	A mutation in genes that encode tRNAs for leucine and lysine
Mitochondrial myopathy	A mutation in a gene that encodes a tRNA for leucine
Maternal myopathy and cardiomyopathy	A mutation in a gene that encodes a tRNA for leucine
Myoclonic epilepsy and ragged-red muscle fibers	A mutation in a gene that encodes a tRNA for lysine

Data from: Wallace, D. C. (1993). Mitochondrial diseases: genotype versus phenotype. *Trends Genet. 9,* 128–33.

Table 7.5 describes several mitochondrial diseases that have been discovered in humans and are caused by mutations in mitochondrial genes. These are usually chronic degenerative disorders that affect the brain, heart, muscles, kidneys, and endocrine glands. For example, Leber's hereditary optic neuropathy (LHON) affects the optic nerve. It may lead to the progressive loss of vision in one or both eyes. LHON can be caused by a defective mutation in one of several different mitochondrial genes. However, it is still unclear how a defect in these mitochondrial genes produces the symptoms of this disease.

Extranuclear Genomes of Mitochondria and Chloroplasts Evolved from an Endosymbiotic Relationship

The idea that the nucleus, mitochondria, and chloroplasts all contain their own separate genetic material may at first seem puzzling. It would appear simpler to have all the genetic material in one place in the cell. The underlying reason for distinct genomes of mitochondria and chloroplasts can be traced back to their evolutionary origin, which is thought to involve a symbiotic association.

A symbiotic relationship occurs when two different species live together in direct contact with each other in a mutually beneficial relationship. The symbiont is the smaller of the two species. The term **endosymbiosis** describes a symbiotic relationship in which the symbiont actually lives inside (*endo-*, inside) the larger of the two species. In 1883, Andreas Schimper proposed that plas-

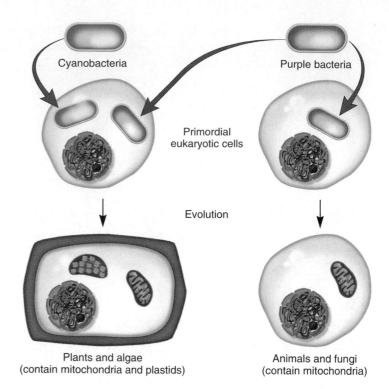

FIGURE 7.20 **The endosymbiotic origin of mitochondria and chloroplasts.** According to the endosymbiotic theory, plastids descended from an endosymbiotic relationship between cyanobacteria and eukaryotic cells. This arose when a bacterium took up residence within a primordial eukaryotic cell. Over the course of evolution, the intracellular bacterial cell gradually changed its characteristics, eventually becoming a plastid. Similarly, mitochondria were derived from an endosymbiotic relationship between purple bacteria and eukaryotic cells.

tids were descended from an endosymbiotic relationship between cyanobacteria and eukaryotic cells. This idea, now known as the **endosymbiosis theory,** suggested that the ancient origin of plastids was initiated when a cyanobacterium took up residence within a primordial eukaryotic cell (fig. 7.20). Over the course of evolution, the characteristics of the intracellular bacterial cell gradually changed to those of a plastid. In 1922, Ivan Wallin also proposed an endosymbiotic origin for mitochondria.

In spite of these hypotheses, the question of endosymbiosis was largely ignored until researchers in the 1950s discovered that chloroplasts and mitochondria contain their own genetic material. The issue of endosymbiosis was hotly debated after Lynn Margulis published a book entitled *Origin of Eukaryotic Cells.* During the 1970s and 1980s, the advent of molecular genetic techniques allowed researchers to analyze genes from chloroplasts, mitochondria, bacteria, and eukaryotic nuclear genomes. They found that genes in chloroplasts and mitochondria are very similar to bacterial genes but not as similar to those found within the nucleus of eukaryotic cells. This observation provided strong support for the endosymbiotic origin of mitochondria and chloroplasts.

Symbiosis occurs because the relationship is beneficial to one or both species. In the case of the endosymbiosis theory, the relationship provided eukaryotic cells with useful cellular characteristics. Plastids were derived from cyanobacteria, a bacterial species that is capable of photosynthesis. The ability to carry out photosynthesis is beneficial to plant cells, providing them with the ability to use the energy from sunlight. It is less clear how the relationship would have been beneficial to a cyanobacterium. By comparison, mitochondria are thought to have been derived from a different type of bacteria known as Gram-negative nonsulfur purple bacteria. In this case, the endosymbiotic relationship enabled eukaryotic cells to synthesize greater amounts of ATP.

During the evolution of eukaryotic species, most genes that were originally found in the genome of the primordial cyanobacteria and purple bacteria have been lost or transferred from the organelles to the nucleus. The sequences of certain genes within the nucleus are consistent with their origin within an organelle. Such genes are more similar in DNA sequence to known bacterial genes than to their eukaryotic counterparts. Therefore, researchers have concluded that these genes have been removed from the mitochondrial and chloroplast chromosomes and relocated to the nuclear chromosomes. This has occurred many times throughout evolution, so that modern mitochondria and chloroplasts have lost most of the genes that are still found in present-day purple bacteria and cyanobacteria.

Most of this gene transfer occurred early in mitochondrial and chloroplast evolution. It appears that the functional transfer of mitochondrial genes has ceased in animals, but gene transfer from mitochondria and chloroplasts to the nucleus continues to occur in plants at a low rate. The molecular mechanism of gene transfer is not entirely understood, but the direction of transfer is well established. During evolution, gene transfer has occurred primarily from the organelles to the nucleus. Transfer of genes from the nucleus to the organelles has almost never occurred, although there is one example of a nuclear gene in plants that has been transferred to the mitochondrial genome. This unidirectional gene transfer from organelles to the nucleus partly explains why the organellar genomes now contain relatively few genes. In addition, gene transfer can occur between organelles. It can happen between two mitochondria, between two chloroplasts, and between a chloroplast and mitochondrion. Overall, the transfer of genetic material between the nucleus, chloroplasts, and mitochondria is an established phenomenon, although its biological benefits remain unclear.

Eukaryotic Cells Occasionally Contain Symbiotic Infective Particles

Other unusual endosymbiotic relationships have been identified in eukaryotic organisms. There are a few rare cases where infectious particles establish a symbiotic relationship with their host. In some cases, research indicates that symbiotic infectious particles are bacteria that exist within the cytoplasm of eukaryotic cells. While examples of symbiotic infectious particles are relatively rare, they have provided interesting and even bizarre examples of the extranuclear inheritance of traits.

In the 1940s, Tracy Sonneborn studied a trait in the protozoan *Paramecia aurelia* known as the killer trait. Killer paramecia secrete a substance called paramecin, which kills some but not all strains of paramecia. The killer strains themselves are resistant to paramecin. Sonneborn found that killer strains contain particles in their cytoplasm known as kappa particles. Each kappa particle is 0.4 μm long and has its own DNA. Genes within the kappa particle encode the paramecin toxin. In addition, other kappa particle genes provide the killer paramecia with resistance to paramecin. When nonkiller strains are mixed with a cell extract derived from killer paramecia, the kappa particles within the extract can infect the nonkiller strains and convert them into killer strains. In other words, the extranuclear particle that determines the killer trait is infectious.

Infectious particles have also been identified in fruit flies. P. l'Heritier identified strains of *Drosophila melanogaster* that are highly sensitive to killing by CO_2. Reciprocal crosses between CO_2-sensitive and normal flies revealed that the trait is inherited in a non-Mendelian manner. Furthermore, cell extracts from a sensitive fly can infect a normal fly and make it sensitive to CO_2.

Another example of an infectious particle in fruit flies involves a trait, known as sex ratio, that was discovered by Chana Malogolowkin and D. Poulson. This trait was found in one strain of *Drosophila willistoni* where most of the offspring of female flies were daughters; nearly all the male offspring died. The sex ratio trait is transmitted from mother to offspring. The rare surviving males do not transmit this trait to their male or female offspring. This result indicates a maternal inheritance pattern for the sex ratio trait. The agent in the cytoplasm of female flies was found to be a symbiotic microorganism. Its presence is usually lethal to males but not to females. This infective agent can be extracted from the tissues of adult females and used to infect the females of a normal strain of flies.

CONCEPTUAL SUMMARY

In chapter 7, we have considered inheritance patterns that differ from Mendelian inheritance. In some cases, genes that are physically located on nuclear chromosomes fail to follow a Mendelian pattern of inheritance. For example, genes that display a **maternal effect** are expressed in the mother's nurse cells during oogenesis and affect the phenotype of the offspring. **Epigenetic inheritance** describes another class of examples in which **nuclear genes** do not follow a Mendelian pattern of inheritance. During **dosage compensation,** the level of expression of genes on the sex chromosomes is altered. In mammals, dosage compensation occurs via **X inactivation.** According to the **Lyon hypothesis,** one of the two X chromosomes in females is randomly inactivated during embryonic development to produce a transcriptionally inactive **Barr body.** The active and inactive X chromosomes are then inherited during later stages of development to create a mosaic animal that is a random patchwork in which a single type of X chromosome is active in each cell. A second example of epigenetic inheritance is known as **genomic imprinting.** In this case, a gene or even an entire chromosome is marked in such as way as to be inherited in an inactive form. DNA methylation is the key event in this marking process.

Other examples of non-Mendelian inheritance are due to the existence of genetic material outside the nucleus. This is known as **extranuclear** or **cytoplasmic inheritance.** The two most important examples of extranuclear inheritance are due to genetic material within mitochondria and chloroplasts. In many cases, extranuclear genetic material follows a **maternal inheritance** pattern. This is because the oocyte is large and, in most species, is more likely than the sperm to transmit organelles to the offspring. However, some species display biparental and even paternal patterns of organelle inheritance. Mitochondria and plastids contain genetic material due to their **endosymbiotic** origin. Mitochondria were derived from purple bacteria, plastids from cyanobacteria. During evolution, many genes have been transferred from the organellar genomes to the nuclear genomes, so that modern mitochondria and chloroplast genomes are small and contain relatively few genes. Other examples of extranuclear inheritance include symbiotic infective particles.

EXPERIMENTAL SUMMARY

The primary way that researchers have identified non-Mendelian patterns of inheritance is by making a series of genetic crosses and then analyzing the phenotypes of the offspring. In many cases, **reciprocal crosses** have yielded results that deviate from Mendelian inheritance. For genes with a maternal effect, reciprocal crosses reveal that the genotype of the mother governs the phenotype of the offspring. In genomic imprinting, reciprocal crosses indicate that one parent transmits an active form of a gene while the other parent transmits an allele that is inactive. Genetic crosses are also consistent with the phenomenon of X inactivation. For example, the trait of variegated coat pattern occurs only in heterozygous female animals.

Similarly, the outcomes of reciprocal crosses are usually different for extrachromosomal inheritance. In most species, organelles are inherited from the maternal parent, although many exceptions are known. The inheritance of the genetic material within mitochondria and chloroplasts, first identified in studies of yeast and algae, is thus non-Mendelian.

Researchers have also investigated non-Mendelian inheritance at the cellular and molecular levels. The cellular explanation of maternal effect genes is that the nurse cells transfer the gene products to the developing oocyte. In chapter 23, we will examine methods that demonstrate the accumulation of maternal effect gene products within the oocyte. Genomic imprinting is also being investigated at the molecular level. Researchers exploring the biochemistry of imprinted genes have identified DNA methylation as a key to marking. Further experimentation is needed to elucidate the mechanisms of erasure and *de novo* methylation that occur during gametogenesis.

Likewise, researchers have explored the cellular and molecular mechanisms of X inactivation. Cytological examination of female mammalian cells revealed the presence of a highly condensed X chromosome, called the Barr body. A study of G-6-PD expression at the cellular level confirmed that one X chromosome is permanently inactivated in the somatic cell lineages of females. A site on the X chromosome called *Xic* is required for the proper counting and inactivation of X chromosomes. Expression of the *Xist* gene is required for the X-inactivation mechanism. Another site termed *Xce* is important in the choice of the inactivated chromosome. The embryonic expression of a gene termed *TsiX* may play a role in choosing the active X chromosome.

Finally, molecular tools have been used to dissect the genetic composition of organellar chromosomes. Some of these methods, which include DNA cloning and sequencing, are described in chapter 18. These techniques have enabled researchers to identify the genes located in mtDNA and cpDNA.

PROBLEM SETS & INSIGHTS

Solved Problems

S1. Our understanding of maternal effect genes has been greatly aided by their identification in experimental organisms such as *Drosophila melanogaster* and *Caenorhabditis elegans.* In experimental organisms with a short generation time, geneticists have successfully searched for mutant alleles that prevent the normal process of embryonic development. In many cases, the offspring will die at early embryonic or larval stages. These are called **maternal** **effect lethal alleles.** How would a researcher identify a mutation that produced a recessive, maternal effect lethal allele?

Answer: A maternal effect lethal allele can be identified when a phenotypically normal mother produces only offspring with gross developmental abnormalities. For example, let's call the normal allele *N* and the maternal effect lethal allele *n.* A parental cross between two flies that are heterozygous for a maternal effect lethal allele would produce 1/4 of the offspring with a homozygous genotype, *nn*. These flies are viable because of the maternal effect. Their mother would be *Nn* and provide the *n* oocyte with a sufficient amount of *N* gene product so that the *nn* flies would develop properly. However, homozygous *nn* females cannot provide their oocytes with any normal gene product. Therefore, all their offspring are abnormal and die during early stages.

32. A maternal effect gene in *Drosophila*, called *torso*, is found as a recessive allele that prevents the correct development of anterior- and posterior-most structures. A wild-type male is crossed to a female of unknown genotype. This mating produces 100% larva that are missing their anterior- and posterior-most structures and therefore die during early development. What is the genotype and phenotype of the female fly in this cross? What are the genotypes and phenotypes of the female fly's parents?

Answer: Since this cross produces 100% abnormal offspring, the female fly must be homozygous for the abnormal *torso* allele. Even so, the female fly must be phenotypically normal in order to reproduce. This female fly would have had a mother that was heterozygous for a normal and abnormal *torso* allele and a father that was either heterozygous or homozygous for the abnormal *torso* allele.

$$torso^+ \ torso^- \quad \times \quad torso^+ \ torso^- \ \text{or} \ torso^- \ torso^-$$
(grandmother)   (grandfather)
$$\downarrow$$
$$torso^- \ torso^- \quad \text{(female that is mother of}$$
100% abnormal offspring)

This female fly is phenotypically normal because its mother was heterozygous and provided the gene products of the *torso*+ allele from the nurse cells. However, this homozygous female will produce only abnormal offspring, because it cannot provide them the normal *torso*+ gene products.

S3. An individual with Angelman syndrome produced an offspring with Prader-Willi syndrome. Why does this occur? What are the sexes of the parent with Angelman syndrome and the offspring with Prader-Willi syndrome?

Answer: These two different syndromes are most commonly caused by a small deletion in chromosome 15. In addition, genomic imprinting plays a role, since genes in this deleted region are differentially imprinted depending on sex. If this deletion is inherited from the paternal parent, the offspring develops Prader-Willi syndrome. Therefore, in this problem, the individual with Angelman syndrome must have been a

male, since he produced a child with Prader-Willi syndrome. The child could be either a male or female.

S4. In yeast, a haploid petite mutant also carries a mutant gene that requires the amino acid histidine for growth. The *petite his⁻* strain is crossed to a wild-type strain to yield the following tetrad:

2 cells: *petite his⁻*
2 cells: *petite his⁺*

Explain the inheritance of the *petite* and *his⁻* mutations.

Answer: The *his⁻* and *his⁺* alleles are segregating in a 2:2 ratio. This indicates a nuclear pattern of inheritance. By comparison, all four cells in this tetrad have a *petite* phenotype. This is a suppressive petite that arises from a mitochondrial mutation.

S5. Let's suppose that you are a horticulturist who has recently identified an interesting plant with variegated leaves. How would you determine if this trait is nuclearly or cytoplasmically inherited?

Answer: Make reciprocal crosses involving normal and variegated strains. In many species, plastid genomes are inherited maternally, although this is not always the case. In addition, a significant percentage of paternal leakage may occur. Nevertheless, when reciprocal crosses yield different outcomes, an organellar mode of inheritance is possibly at work.

S6. A phenotype that is similar to a yeast suppressive petite was also identified in the mold *Neurospora crassa*. Mary and Hershel Mitchell identified a slow-growing mutant that they called *poky*. Unlike yeast, which are isogamous (i.e., produce one type of gamete), *Neurospora* is sexually dimorphic and produces male and female reproductive structures. The "female" reproductive structure is the ascogonium and the "male" reproductive structure is the antherium. When a *poky* strain of *Neurospora* was crossed to a wild-type strain, the results were different between reciprocal crosses. If a *poky* mutant was the female parent, all the spores exhibited the *poky* phenotype. By comparison, if the wild-type strain was the female parent, all the spores were wild type. Explain these results.

Answer: These genetic studies indicate that the *poky* mutation is maternally inherited. This is because the cytoplasm of the female reproductive cells provides the offspring with their mitochondria. Besides these genetic studies, the Mitchells and their collaborators showed that *poky* mutants are defective in certain cytochromes, which are iron-containing proteins that are known to be located in the mitochondria.

Conceptual Questions

C1. Define the term *epigenetic,* and describe two examples.

C2. Describe the pattern of inheritance that maternal effect genes follow. Explain how the maternal effect occurs at the cellular level. What are the expected functional roles of the proteins that are encoded by maternal effect genes?

C3. A maternal effect gene exists in a dominant *N* (normal) allele and a recessive *n* (abnormal) allele. What would be the ratios of genotypes and phenotypes for the offspring of the following crosses?

A. *nn* female × *NN* male

B. *NN* female × *nn* male

C. *Nn* female × *Nn* male

C4. A *Drosophila* embryo dies during early embryogenesis due to a recessive maternal effect allele called *bicoid*. The wild-type allele is designated *bicoid*+. What are the genotypes and phenotypes of the embryo's mother and maternal grandparents?

C5. For Mendelian traits, the nuclear genotype (i.e., the alleles found on chromosomes in the cell nucleus) directly influences an offspring's traits. In contrast, for non-Mendelian inheritance patterns, it is often not enough to know the combination of alleles on chromosomes in the nucleus in order to predict the offspring's phenotype. For the following traits, what do you need to know to predict the phenotypic outcome?

A. Dwarfism due to a mutant *Igf-2* allele

B. Snail coiling direction

C. Leber's hereditary optic neuropathy

C6. Let's suppose a maternal effect gene exists as a normal dominant allele and an abnormal recessive allele. A mother who is phenotypically abnormal produces all normal offspring. Explain the genotype of the mother.

C7. A gene that affects the anterior morphology in fruit flies is inherited as a maternal effect gene. The gene exists in a normal allele, *H,* and a recessive allele, *h,* which causes a small head. A female fly with a normal head is mated to a true-breeding male with a small head. All of the offspring have small heads. What are the genotypes of the mother and offspring? Explain your answer.

C8. Explain why maternal effect genes exert their effects during the early stages of development.

C9. As described in chapter 19, researchers have been able to "clone" mammals by fusing a cell with a diploid nucleus (i.e., a somatic cell) with an egg that has had its (haploid) nucleus removed.

A. With regard to maternal effect genes, would the phenotype of such a cloned animal be determined by the animal that donated the oocyte or by the animal that donated the somatic cell? Explain.

B. Would the cloned animal inherit extranuclear traits from the animal that donated the oocyte or by the animal that donated the somatic cell? Explain.

C. In what ways would you expect this cloned animal to be similar to or different from the animal that donated the somatic cell? Is it fair to call such an animal a "clone" of the animal that donated the nucleus?

C10. With regard to the numbers of sex chromosomes, explain why dosage compensation is necessary.

C11. What is a Barr body? How is its structure different from that of other chromosomes in the cell? How does the structure of a Barr body affect the level of X-linked gene expression?

C12. Among different species, describe three distinct strategies for accomplishing dosage compensation.

C13. Describe when X inactivation occurs and how this leads to phenotypic results at the organismal level. In your answer, you should explain why X inactivation causes results such as variegated coat patterns in mammals. Why do two different calico cats have their patches of orange and black fur in different places? Explain whether or not a variegated coat pattern could occur in marsupials due to X inactivation.

C14. Describe the molecular process of X inactivation. This description should include the three phases of inactivation and the role of the *Xic* locus. Explain what happens to X chromosomes during embryogenesis, in adult somatic cells, and during oogenesis.

C15. On rare occasions, an abnormal human male is born who is somewhat feminized compared to normal males. Microscopic examination of the cells of one such individual revealed that he has a single Barr body in each cell. What is the chromosomal composition of this individual?

C16. How many Barr bodies would you expect to find in humans with the following abnormal compositions of sex chromosomes?

A. XXY

B. XYY

C. XXX

D. X0 (a person with just a single X chromosome)

C17. Certain forms of human color blindness are inherited as X-linked recessive traits. Hemizygous males are color-blind while heterozygous females are not. However, heterozygous females sometimes have partial color blindness.

A. Discuss why heterozygous females may have partial color blindness.

B. Doctors identified an unusual case in which a heterozygous female was color-blind in her right eye but had normal color vision in her left eye. Explain how this might have occurred.

C18. A black female cat (X^BX^B) and an orange male cat (X^OY) were mated to each other and produced a male cat that was calico. Which sex chromosomes did this male offspring inherit from its mother and father? Remember that the presence of the Y chromosome determines maleness in mammals.

C19. What is the spreading stage of X inactivation? Why do you think it is called a "spreading" stage? Discuss the role of the *Xist* gene in the spreading stage of X inactivation.

C20. When does the erasure and reestablishment phase of imprinting occur? Explain why it is necessary to erase an imprint and then reestablish it in order to always maintain imprinting from the same sex of parent.

C21. In what types of cells would you expect *de novo* methylation to occur? In what cell types would it not occur?

C22. On rare occasions, people are born with a condition known as uniparental disomy. It happens when an individual inherits both copies of a chromosome from one parent and no copies from the other parent. This occurs when two abnormal gametes happen to complement each other to produce a diploid zygote. For example, an abnormal sperm that lacks chromosome 15 could fertilize an egg that contains two copies of chromosome 15. In this situation, the individual would be said to have maternal uniparental disomy 15, because both copies of chromosome 15 were inherited from the mother. Alternatively, an abnormal sperm with two copies of chromosome 15 could fertilize an egg with no copies. This is known as paternal uniparental disomy 15. If a female is born with paternal uniparental disomy 15, would you expect her to be phenotypically normal, have Angelman syndrome, or have Prader-Willi syndrome? Explain. Would you expect her to produce normal offspring or offspring affected with AS or PWS?

C23. Genes that cause Prader-Willi syndrome and Angelman syndrome are closely linked along the same chromosome (i.e., chromosome 15). Although people with these syndromes do not usually reproduce, let's suppose that a couple produces two children with

Angelman syndrome. The oldest child (named Pat) grows up and has two children with Prader-Willi syndrome. The second child (named Lynn) grows up and has one child with Angelman syndrome.

    A. Are Pat and Lynn's parents both normal or does one of them have Angelman or Prader-Willi syndrome? If one of them has a disorder, explain why it is the mother or the father.

    B. What are the sexes of Pat and Lynn? Explain.

C24. How is the process of X inactivation similar to genomic imprinting? How do they differ?

C25. What is extranuclear inheritance? Describe three examples.

C26. What is a reciprocal cross? Let's suppose that a gene is found as a wild-type allele and a recessive mutant allele. What would be the expected outcomes of reciprocal crosses if a true-breeding normal individual were crossed to a true-breeding individual carrying the mutant allele? What would be the results if the gene were maternally inherited?

C27. Among different species, does extranuclear inheritance always follow a maternal inheritance pattern? Why or why not?

C28. What is the phenotype of a petite mutant? Where can a petite mutation occur—in nuclear genes, extranuclear genetic material, or both? What is the difference between a neutral and suppressive petite?

C29. Extranuclear inheritance often correlates with maternal inheritance. Even so, paternal leakage is not uncommon. Explain what paternal leakage is. If a cross produced 200 offspring and the rate of mitochondrial paternal leakage was 3%, how many offspring would be expected to contain paternal mitochondria?

C30. Discuss the structure and organization of the mitochondrial and chloroplast genomes. How large are they, how many genes do they contain, and how many copies of the genome are there per organelle?

C31. Explain the likely evolutionary origin of mitochondrial and chloroplast genomes. How have the sizes of the mitochondrial and chloroplast genomes changed since their origin? How has this occurred?

C32. Which of the following traits or diseases are determined by nuclear genes?

    A. Snail coiling pattern

    B. Prader-Willi syndrome

    C. Streptomycin resistance in *Chlamydomonas*

    D. Leber's hereditary optic neuropathy

C33. Acute murine leukemia virus (AMLV) causes leukemia in mice. This virus is easily passed from mother to offspring through the mother's milk. (Note: Even though newborn offspring acquire the virus, they may not develop leukemia until much later in life. Testing can determine if an animal carries the virus.) Describe how the formation of leukemia via AMLV resembles a maternal inheritance pattern. How could you prove that this form of leukemia is not caused by extranuclear inheritance?

C34. Describe how a biparental pattern of extranuclear inheritance would resemble a Mendelian pattern of inheritance for a particular gene. How would they differ?

C35. According to the endosymbiosis theory, mitochondria and chloroplasts are derived from bacteria that took up residence within eukaryotic cells. At one time, therefore, these bacteria were free living (prior to being taken up by eukaryotic cells). However, we cannot take a mitochondrion or chloroplast out of a living eukaryotic cell and get it to survive and replicate on its own. Why not?

Experimental Questions

E1. Figure 7.1 describes an example of a maternal effect gene. Explain how Sturtevant deduced a maternal effect gene based on the F_2 and F_3 generations.

E2. Discuss the types of experimental observations that Mary Lyon had to consider in proposing her hypothesis concerning X inactivation. In your own words, explain how these observations were consistent with her hypothesis.

E3. Chapter 18 describes three blotting methods (i.e., Southern blotting, Northern blotting, and Western blotting) that are used to detect specific genes and gene products. Southern blotting detects DNA, Northern blotting detects RNA, and Western blotting detects proteins. Let's suppose that a female fruit fly is heterozygous for a maternal effect gene, which we will call gene X. The female is Xx. The normal allele, X, encodes a functional mRNA that is 550 nucleotides long. A recessive allele, x, encodes a shorter mRNA that is 375 nucleotides long. (Allele x is due to a deletion within this gene.) How could you use one or more of these techniques to show that nurse cells transfer gene products (from gene X) to developing oocytes? You may assume that you can dissect the ovaries of fruit flies and isolate oocytes separately from nurse cells. In your answer, describe your expected results.

E4. As a hypothetical example, a trait in mice results in mice with very long tails. You initially have a true-breeding strain with normal tails and a true-breeding strain with long tails. You then make the following types of crosses:

Cross 1. When true-breeding females with normal tails are crossed to true-breeding males with long tails, all the F_1 offspring have long tails.

Cross 2. When true-breeding females with long tails are crossed to true-breeding males with normal tails, all the F_1 offspring have normal tails.

Cross 3. When F_1 females from cross 1 are crossed to true-breeding males with normal tails, all the offspring have normal tails.

Cross 4. When F_1 males from cross 1 are crossed to true-breeding females with long tails, half the offspring have normal tails and half have long tails.

Explain the pattern of inheritance of this trait.

E5. You have a female snail that coils to the right, but you do not know its genotype. You may assume that right coiling (D) is dominant to left coiling (d). You also have male snails at your disposal of known genotype. How would you determine the genotype of this female snail? In your answer, describe your expected results depending on whether the female is *DD, Dd,* or *dd.*

E6. On a recent camping trip, you find one male snail on a deserted island that coils to the right. However, in this same area, you find several shells (not containing living snails) that coil to the left. Therefore, you conclude that you are not certain of the genotype of this male snail. On a different island, you find a large colony of snails of the same species. All of these snails coil to the right, and every snail shell that you find on this second island coils to the right. With regard to the maternal effect gene that determines coiling pattern, how would you determine the genotype of the male snail that you found on the deserted island? In your answer, describe your expected results.

E7. Figure 7.6 describes the results of X inactivation in mammals. If fast and slow alleles of glucose-6-phosphate dehydrogenase (*G-6-PD*) exist in other species, what would be the expected results for a heterozygous female of the following species?

A. Marsupial

B. *Drosophila melanogaster*

C. *Caenorhabditis elegans* (Note: We are considering the hermaphrodite in *C. elegans* to be equivalent to a female.)

E8. Two male mice, which we will call male A and male B, are both phenotypically normal. Male A was from a litter that contained half phenotypically normal mice and half dwarf mice. The mother of male A was known to be homozygous for the normal *Igf-2* allele. Male B was from a litter of eight mice that were all phenotypically normal. The parents of male B were a phenotypically normal male and a dwarf female. Male A and male B were put into a cage with two female mice that we will call female A and female B. Female A is dwarf and female B is phenotypically normal. The parents of these two females were unknown, although it was known that they were from the same litter. The mice were allowed to mate with each other and the following data were obtained:

Female A gave birth to three dwarf babies and four normal babies.

Female B gave birth to four normal babies and two dwarf babies.

Which male(s) mated with female A and female B? Explain.

E9. In the experiment of figure 7.6, why does a clone of cells produce only one type of G-6-PD enzyme? What would you expect to happen if a clone was derived from an early embryonic cell? Why does the initial sample of tissue produce both forms of G-6-PD?

E10. Chapter 18 describes a blotting method known as Northern blotting that can be used to determine the amount of mRNA produced by a particular gene. In this method, the amount of a specific mRNA produced by cells is detected as a band on a gel. If one type of cell produces twice as much of a particular mRNA compared to another cell, the band will appear twice as intense. Also, sometimes mutations affect the length of mRNA that is transcribed from a gene. For example, a small deletion within a gene may shorten an mRNA. Northern blotting also can discern the sizes of mRNAs.

Northern blot

Lane 1 is a Northern blot of mRNA from cell type A that is 800 nucleotides long.

Lane 2 is a Northern blot of the same mRNA from cell type B.

(Cell type B produces twice as much of this RNA compared to cell type A.)

Lane 3 shows a heterozygote in which one of the two genes has a deletion, which shortens the mRNA by 200 nucleotides.

Now here is the question. Let's suppose an X-linked gene exists as two alleles, which we will call *B* and *b*. Allele *B* encodes an mRNA that is 750 nucleotides long, while allele *b* encodes a shorter mRNA that is 675 nucleotides long. Draw the expected results of a Northern blot using mRNA isolated from the same type of somatic cells taken from the following individuals:

A. First lane is mRNA from an X^{b}Y male fruit fly.

Second lane is mRNA from an X^{b}X^b female fruit fly.

Third lane is mRNA from an X^{B}X^b female fruit fly.

B. First lane is mRNA from an X^{B}Y male mouse.

Second lane is mRNA from an X^{B}X^b female mouse.*

Third lane is mRNA from an X^{B}X^B female mouse.*

*The sample is taken from an adult female mouse. It is not a clone of cells. It is a tissue sample, like the one described in the experiment of figure 7.6.

C. First lane is mRNA from an X^{B}Y male *C. elegans*.

Second lane is mRNA from an X^{B}X^b hermaphrodite *C. elegans*.

Third lane is mRNA from an X^{B}X^B hermaphrodite *C. elegans*.

E11. A variegated trait in plants is analyzed using reciprocal crosses. The following results are obtained:

Variegated female × Normal male   Normal female × Variegated male
↓   ↓
1,024 variegated + 52 normal   1,113 normal + 61 variegated

Explain this pattern of inheritance.

E12. Sager and her colleagues discovered that the mode of inheritance of streptomycin resistance in *Chlamydomonas* could be altered if the *mt*$^+$ cells were exposed to UV irradiation prior to mating. This exposure dramatically increased the frequency of biparental inheritance. What would be the expected outcome of a cross between an *mt*$^+$ *sm*^{r} and an *mt*$^-$ *sm*^{s} strain in the absence of UV irradiation? How would the result differ if the *mt*$^+$ strain was exposed to UV light?

E13. Take a look at figure 7.18 and describe how you could experimentally distinguish between yeast strains that are neutral petites versus suppressive petites.

Questions for Student Discussion/Collaboration

1. Recessive maternal effect genes are identified in flies (for example) when a phenotypically normal mother cannot produce any normal offspring. Since all the offspring are dead, this female fly cannot be used to produce a strain of heterozygous flies that could be used in future studies. How would you identify heterozygous individuals that are carrying a recessive maternal effect allele? How would you maintain this strain of flies in a laboratory over many generations?

2. What is an infective particle? Discuss the similarities and differences between infective particles and organelles such as mitochondria and chloroplasts. Do you think the existence of infective particles supports the endosymbiotic theory of the origin of mitochondria and chloroplasts?

Note: All answers appear at the website for this textbook; the answers to even-numbered questions are in the back of the textbook.

www.mhhe.com/brooker

Visit the Online Learning Center for practice tests, answer keys, and other learning aids for this chapter. Enhance your understanding of genetics with our interactive exercises, web links, news feeds, tutorial service, and much more.

VARIATION IN CHROMOSOME STRUCTURE AND NUMBER

8

∷

The term **genetic variation** refers to genetic differences between members of the same species, or those between different species. Throughout chapters 2 to 7, we have focused primarily on variation in specific genes. This is called **allelic variation.** Allelic differences are due to small changes (i.e., mutations) that occur within a particular gene. In chapter 8, our emphasis will shift to larger types of genetic changes that affect the chromosomal composition of eukaryotic organisms. As we will see, these alterations may affect the expression of many genes and lead to changes in phenotypes.

In the first section of chapter 8, we will examine how the structure of a eukaryotic chromosome can be modified, either by altering the total amount of genetic material or by rearranging the order of genes along a chromosome. A substantial change in chromosome structure, possibly affecting more than a single gene, is referred to as a **chromosome mutation** or **chromosomal aberration.** In most cases, a mutant chromosome can be detected microscopically; its structure will differ visually from a normal chromosome. In chapter 8, we will examine how chromosome mutations occur, how they are transmitted from parent to offspring, and how they affect the phenotype of the individual who inherits them.

The rest of chapter 8 is concerned with changes in the total number of chromosomes. A change in chromosome number is called a **genome mutation.** This type of mutation comprises two subtypes: changes in the number of sets of chromosomes, and changes in the numbers of individual chromosomes within a set.

Natural variation in the number of sets of chromosomes is relatively common among different species, particularly in the plant kingdom. Within the same species, changes in chromosome number within a set can also occur, but such alterations are usually detrimental. In chapter 8, we will explore how variation in chromosome number occurs and consider examples where it has significant phenotypic consequences.

8.1 VARIATION IN CHROMOSOME STRUCTURE

We will begin our discussion of chromosome variation by examining several ways in which the structures of eukaryotic chromosomes can be altered. As discussed in chapters 3 and 5, chromosomes in the nuclei of eukaryotic cells contain long, linear DNA molecules that carry hundreds or even thousands of genes. In this section, we will explore how the composition of a chromosome can be changed. As you will see, segments of a chromosome can be lost, duplicated, or rearranged in a new way.

The study of chromosomal variation is important for several reasons. First, geneticists have discovered that variations in chromosome structure can have major effects on the phenotype of an organism. For example, we now know that several human genetic diseases are caused by changes in chromosome structure. In other cases, an individual possessing a chromosomal rearrangement may have a normal phenotype yet have a high probability of producing offspring with genetic abnormalities. Finally, changes in chromosome structure have been an important force in the evolution of new species. This topic will be considered in greater detail in chapter 26.

In this section, we will also examine the cellular mechanisms that underlie changes in chromosome structure. We will explore unusual events during meiosis that affect how altered chromosomes are transmitted from parents to offspring. Also, we will consider many examples where chromosomal alterations affect an organism's phenotypic characteristics.

Natural Variation Exists in Chromosome Structure

Before we begin to discuss how chromosome structure can be altered, we need to have a reference point for a normal set of chromosomes. To determine what the normal chromosomes of a species look like, a **cytogeneticist** microscopically examines the chromosomes from several members of the species. In most cases, two phenotypically normal individuals of the same species will have the same number and types of chromosomes.

To determine the chromosomal composition of a species, the chromosomes in actively dividing cells are examined microscopically. Figure 8.1a shows micrographs of chromosomes from three species: humans, fruit flies, and corn. As seen here, humans have 46 chromosomes (23 pairs), fruit flies have 8 chromosomes (4 pairs), and corn has 20 chromosomes (10 pairs). Except for the sex chromosomes, which differ between males and females, most members of the same species have very similar chromosomes. For

example, the overwhelming majority of people have 46 chromosomes in their somatic cells. When comparing different species, it is common for distantly related species, such as humans and fruit flies, to have very different chromosomal compositions. Nevertheless, a total of 46 chromosomes is normal for humans, while 8 chromosomes is the norm for fruit flies.

Chromosomes from any given species can vary considerably in size and shape. Cytogeneticists have various ways to classify and identify chromosomes. The three most commonly used features are size, location of the centromere, and banding patterns that are revealed when the chromosomes are treated with stains. As shown in figure 8.1b, chromosomes are classified as **metacentric, submetacentric, acrocentric,** and **telocentric.** Because the centromere is never exactly in the center of a chromosome, each chromosome has a short arm and a long arm. The short arm is designated with the letter *p* (for the French, petite), while the long arm is designated with the letter *q*. In the case of telocentric chromosomes, the short arm may be nearly nonexistent.

Figure 8.1c shows a human **karyotype;** the procedure for making a karyotype was described in chapter 3 (see fig. 3.2). A karyotype is a micrograph in which all the chromosomes within a single cell have been arranged in a standard fashion. When preparing a karyotype, the chromosomes are aligned with the short arms on top and the long arms on the bottom. By convention, the chromosomes are numbered according to their size, with the largest chromosomes having the smallest numbers. For example, human chromosomes 1, 2, and 3 are relatively large, whereas 21 and 22 are the two smallest. An exception to the numbering system involves the sex chromosomes, which are designated with letters (for humans, X and Y).

Since different chromosomes often have similar sizes and centromeric locations (e.g., compare human chromosomes 8, 9, and 10), geneticists must use additional methods to accurately identify each type of chromosome within a karyotype. For detailed identification, chromosomes are treated with stains to produce characteristic banding patterns. Several different staining procedures are used by cytogeneticists to identify specific chromosomes. An example is **G banding,** which is shown in figure 8.1c. In this procedure, chromosomes are treated with mild heat or with proteolytic enzymes that partially digest chromosomal proteins. When exposed to the dye called Giemsa, some chromosomal regions bind the dye heavily and produce a dark band; in other regions, the stain hardly binds at all and a light band results. Though the mechanism of staining is not completely understood, it is thought that the dark bands represent regions that are more tightly compacted. As shown in figure 8.1c and d, the alternating pattern of G bands is a unique feature for each chromosome.

In the case of human chromosomes, approximately 300 G bands can be distinguished during metaphase, while 2,000 G bands are seen in prophase. Figure 8.1d shows the conventional numbering system that is used to designate G bands along a set of human chromosomes. The left chromatid in each pair of sister chromatids shows the expected banding pattern during metaphase, while the right chromatid shows the banding pattern as it would appear during prophase.

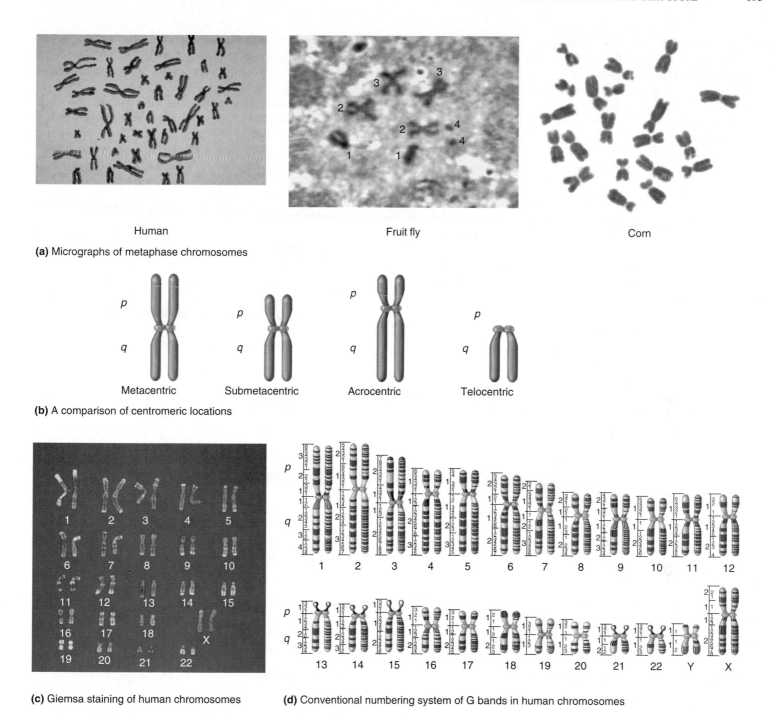

Human

Fruit fly

Corn

(a) Micrographs of metaphase chromosomes

Metacentric   Submetacentric   Acrocentric   Telocentric

(b) A comparison of centromeric locations

(c) Giemsa staining of human chromosomes

(d) Conventional numbering system of G bands in human chromosomes

FIGURE 8.1 **Features of normal chromosomes. (a)** Micrographs of chromosomes from humans, fruit flies, and corn. **(b)** A comparison of centromeric locations. Centromeres can be metacentric (roughly in the middle), submetacentric (slightly off center), acrocentric (near one end), or telocentric (at the end). **(c)** Human chromosomes that have been stained with Giemsa. **(d)** The conventional numbering of bands in Giemsa-stained human chromosomes of midmetaphase chromosomes. The numbering is divided into broad regions, which then are subdivided into smaller regions. The numbers increase as the region gets farther away from the centromere. For example, if you take a look at the *left* chromatid of chromosome 1, the uppermost dark band would be at a location designated p35. The banding patterns of chromatids change as the chromatids condense. The *left* chromatid of each pair of sister chromatids shows the banding pattern of a chromatid in midmetaphase, and the right side shows a chromatid as it would appear in prophase.

The banding pattern of eukaryotic chromosomes is useful in several ways. First, individual chromosomes can be distinguished from each other, even if they have similar sizes and centromeric locations. For example, compare the differences in banding patterns between human chromosomes 8 and 9 (fig. 8.1c and

d). These differences permit us to distinguish these two chromosomes even though their sizes and centromeric locations are very similar. Banding patterns are also used to detect changes in chromosome structure. As discussed next, chromosomal rearrangements or changes in the total amount of genetic material are more easily detected when viewing banded chromosomes. Also, chromosome banding may reveal evolutionary relationships among the chromosomes of closely related species.

Mutations Can Alter Chromosome Structure

With an understanding that normal chromosomes can come in a variety of shapes and sizes, we now can consider how the structures of normal chromosomes can be modified. As mentioned earlier, a chromosome mutation is a substantial change in the structure of a chromosome. There are two primary ways that the structure of chromosomes can be altered. First, the total amount of genetic material within a single chromosome can be increased or decreased significantly. Second, the genetic material in one or more chromosomes may be rearranged without affecting the total amount of material. As shown in figure 8.2, these alterations are categorized as deficiencies, duplications, inversions, and translocations.

Deficiencies and duplications are changes in the total amount of genetic material within a single chromosome. In figure 8.2, human chromosomes are labeled according to their normal G banding patterns. When a **deficiency** occurs, a segment of chromosomal material is missing. In other words, the affected chromosome is deficient in a significant amount of genetic material. The term **deletion** is also used to describe a missing region of a chromosome. In chapter 8, the terms *deficiency* and *deletion* will be used interchangeably. In contrast, a **duplication** occurs when a section of a chromosome is repeated compared to the normal parent chromosome.

Inversions and translocations are chromosomal rearrangements. An **inversion** involves a change in the direction of the genetic material along a single chromosome. For example, in figure 8.2c, a segment of one chromosome has been inverted, so that the order of four G bands is opposite to that of the parent chromosomes. A **translocation** occurs when one segment of a chromosome becomes attached to a different chromosome or to a

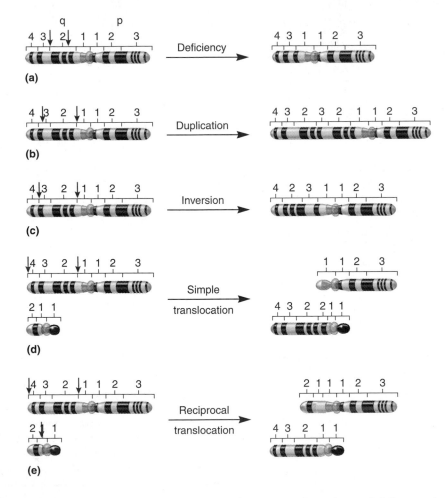

FIGURE 8.2   Types of changes in chromosome structure. The larger chromosome shown in *purple* is human chromosome 1. The smaller chromosome is human chromosome 21. (**a**) A deficiency occurs that removes a large portion of the q2 region. (**b**) A duplication occurs that doubles the q2-q3 region. (**c**) An inversion occurs that inverts the q2-q3 region. (**d**) The q2-q4 region of chromosome 1 is translocated to chromosome 21. A region of a chromosome cannot be translocated directly to the tip of another chromosome, because telomeres at the tips of chromosomes would prevent such an event. In this example, a small piece at the end of chromosome 21 would have to be removed in order for the q2-q4 region of chromosome 1 to be attached to chromosome 21. (**e**) The q2-q4 region of chromosome 1 is exchanged with most of the q1-q2 region of chromosome 21.

different part of the same chromosome. When a single piece of chromosome is attached to another chromosome, this is called a **simple translocation.** In other cases, two different types of chromosomes can exchange pieces, thereby producing two abnormal chromosomes carrying translocations. This is a **reciprocal translocation.**

Figure 8.2 illustrates the common ways that the structure of chromosomes can be changed. Throughout the rest of this section, we will consider how these changes occur, how the changes are detected experimentally, and how they affect the phenotypes of the individuals who inherit them.

The Loss of Genetic Material in a Deficiency Tends to Be Detrimental to an Organism

A chromosomal deficiency occurs when a chromosome breaks in one or more places and a fragment of the chromosome is lost. In figure 8.3a, a normal chromosome has broken into two separate pieces. The piece without the centromere will be lost and degraded. Therefore, this event produces a chromosome with a **terminal deficiency.** In figure 8.3b, a chromosome has broken in two places to produce three chromosomal fragments. The central fragment is lost, and the two outer pieces reattach to each other. This process has created a chromosome with an **interstitial defi-**

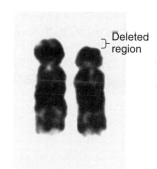
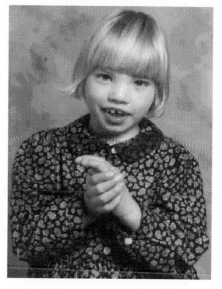

(a) Chromosome 5   **(b)** A child with cri-du-chat syndrome

FIGURE 8.4   Cri-du-chat syndrome. (a) Chromosome 5 from the karyotype of an individual with this disorder. A section of the short arm of the chromosome is missing. **(b)** An affected individual.
GENES→TRAITS   Compared to a normal individual who has two copies of each gene on chromosome 5, an individual with cri-du-chat syndrome has only one copy of the genes that are located within the missing segment. This genetic imbalance (one versus two copies of many genes on chromosome 5) causes the phenotypic characteristics of this disorder. These include a cat-like cry, short stature, characteristic facial anomalies (e.g., a triangular face, almond-shaped eyes, broad nasal bridge, and low-set ears), and microencephaly.

ciency. Deficiencies can also be created when recombination takes place at incorrect locations between two homologous chromosomes. The products of this type of aberrant recombination event are one chromosome with a deficiency and another chromosome with a duplication. This process is examined later in this chapter.

The phenotypic consequences of a chromosomal deficiency depend on the size of the deletion and whether it includes genes or portions of genes that are vital to the development of the organism. When deletions have a phenotypic effect, they are usually detrimental. Larger deletions tend to be more harmful because more genes are missing. Many examples are known in which deficiencies have significant phenotypic influences. In humans, for example, a genetic disease known as cri-du-chat syndrome is caused by a deficiency in a segment of the short arm of human chromosome 5 (fig. 8.4a). Individuals who carry a single copy of this abnormal chromosome along with a normal chromosome 5 display an array of abnormalities including severe mental retardation, unique facial anomalies, and an unusual cat-like cry, which is the meaning of the French name for the syndrome (fig. 8.4b). Some other human genetic diseases, such as Angelman syndrome and Prader-Willi syndrome, described in chapter 7, are due to a deficiency in chromosome 15. Many types of chromosomal deletions have also been associated with human cancers. This topic is discussed in chapter 22.

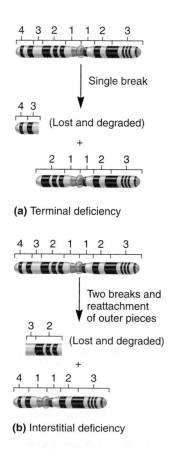

FIGURE 8.3   Production of terminal and interstitial deficiencies. This illustration shows the production of deletions in human chromosome 1.

Deficiencies Can Be Detected Using Cytological, Genetic, and Molecular Techniques

A variety of experimental techniques are used to detect a deletion within a chromosome. These include cytological (i.e., microscopic), genetic, and molecular methods. Microscopic techniques, which are relatively quick and easy, can detect chromosomal deficiencies large enough to be seen with a light microscope. Cri-du-chat and Prader-Willi syndromes are chromosomal deletions easily detectable via light microscopy. Fairly small deletions, which may involve only one or a few genes, are often difficult to detect with simple microscopy. However, more complicated microscopy techniques, such as *in situ* hybridization, have enhanced our ability to detect small deletions. This technique is described in chapter 20.

When a mutation causes a phenotypic effect, genetic analysis can sometimes show that it is due to a deletion. When a deletion occurs in a particular gene, it cannot revert back to the wild-type gene via a spontaneous mutation. At the phenotypic level, a mutant phenotype cannot revert back to a wild-type phenotype. Among a large population of experimental organisms harboring the mutant phenotype, the wild-type phenotype cannot arise by a spontaneous mutation if the original mutation involves a deletion. In addition, deletions can sometimes be revealed by a phenomenon known as **pseudodominance.** This occurs when one copy of a gene has been deleted from a chromosome and the remaining copy of a recessive allele on the homologous chromosome is phenotypically expressed. Under these conditions, the individual is hemizygous for the recessive allele (see solved problem S5).

Duplications Tend to Be Less Harmful Than Deletions

Duplications result in extra genetic material. They are usually caused by abnormal events during recombination. Under normal circumstances, crossing over occurs at analogous sites between homologous chromosomes. On rare occasions, a crossover may occur at misaligned sites on the homologues (fig. 8.5). This results in one chromatid with an internal duplication and another chromatid with a deletion. In figure 8.5, the chromosome with the extra genetic material carries a **gene duplication,** because the number of copies of gene *C* has been increased from one to two. In most cases, gene duplications happen as rare, spo-

radic events during the evolution of species. Later in this section, we will consider how multiple copies of genes can evolve into a family of genes with specialized functions.

Like deletions, the phenotypic consequences of duplications tend to be correlated with size. Duplications are more likely to have phenotypic effects if they involve a large piece of the chromosome. In general, small duplications are less likely to have harmful effects than are deletions of comparable size. This observation suggests that having one copy of a gene is more harmful than having three copies. In humans, relatively few well-defined syndromes are caused by small chromosomal duplications. An example is Charcot-Marie-Tooth disease (type 1A), which is a relatively common peripheral neuropathy characterized by numbness in the hands and feet. It is caused by a small duplication on the short arm of chromosome 17.

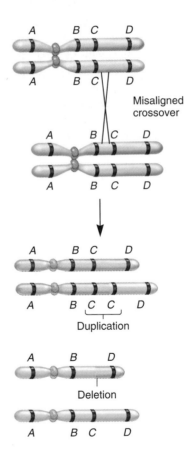

FIGURE 8.5 **Abnormal crossing over, leading to a duplication and a deficiency.** A crossover has occurred at sites between genes *C* and *D* in one chromatid and between genes *B* and *C* in another chromatid. After crossing over is completed, one chromatid contains a duplication and the other contains a deletion.

EXPERIMENT 8A

A Gene Duplication Produced the Bar-Eye Phenotype in *Drosophila*

Insight into the causes of gene duplications came from studies in *Drosophila* involving a trait that affects the number of facets in the eye. In 1914, Sabra Colby Tice discovered a fly that had a reduced number of facets. This trait was called bar eyes. A genetic analysis of the trait revealed that it is an X-linked trait that shows

incomplete dominance. Females homozygous for the *bar* allele have a more severe phenotype than heterozygous females that have one *bar* allele and one wild-type allele (fig. 8.6).

In 1921, Charles Zeleny identified some rare mutants in a stock of flies that was originally homozygous for the *bar* allele. These flies had eyes with even fewer facets than the *bar* homozygote. Zeleny called these flies ultra-bar (also known as double-bar). This trait also shows incomplete dominance, since a female

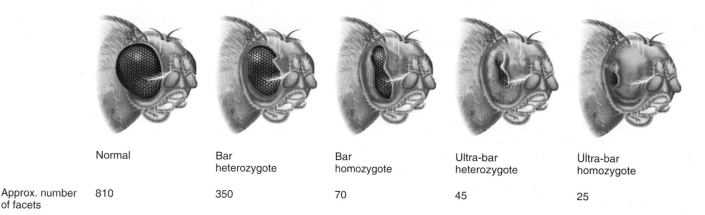

	Normal	Bar heterozygote	Bar homozygote	Ultra-bar heterozygote	Ultra-bar homozygote
Approx. number of facets	810	350	70	45	25

FIGURE 8.6   **A comparison of facet number in the eyes of normal female flies versus those that carry the *bar* or *ultra-bar* mutation in the homozygous or heterozygous condition.**

fly that is homozygous for the *ultra-bar* allele has fewer facets than a heterozygous female fly carrying one *ultra-bar* allele and one normal allele.

To gain further insight into the cause of the bar and ultra-bar phenotypes, Calvin Bridges investigated the bar/ultra-bar phenomenon at the cytological level in the early 1930s. In cells of the *Drosophila* salivary gland, the chromosomes are easy to study under the microscope because they replicate many times to form gigantic **polytene chromosomes.** Since the formation of polytene chromosomes involves a change in chromosome number, the structure of polytene chromosomes will be described later in chapter 8. (If you are not familiar with polytene chromosomes, you may wish to study figure 8.21 before continuing with this experiment.) Because polytene chromosomes are so large, their banding pattern is very easy to see in great detail. It is thus possible to detect very small changes in chromosome structure. In fact, it is possible to detect the duplication or deletion of a single gene.

In the experiment described in figure 8.7, Bridges investigated changes in chromosome structure associated with the bar and ultra-bar phenotypes. He began with a stock of flies that were true-breeding for the bar phenotype. Within this strain, two kinds of rare variants arose in approximately equal numbers. These were the ultra-bar phenotype already described and bar revertants, which had wild-type eyes. Bridges dissected the salivary glands from the three phenotypic classes (i.e., bar, ultra-bar, and bar revertants) and examined them under the microscope.

■ THE HYPOTHESIS

Information concerning the nature of the bar and ultra-bar phenotypes may be revealed by a cytological examination of polytene chromosomes.

■ TESTING THE HYPOTHESIS — FIGURE 8.7 A cytological examination of *Drosophila* chromosomes to understand the nature of the bar and ultra-bar phenotype.

Starting materials: A strain of flies homozygous for the *bar* allele and a homozygous normal strain.

Experimental level	**Conceptual level**

1. Within the strain of homozygous bar-eyed flies, identify rare flies that have normal eyes (bar revertants) or ultra-bar eyes.

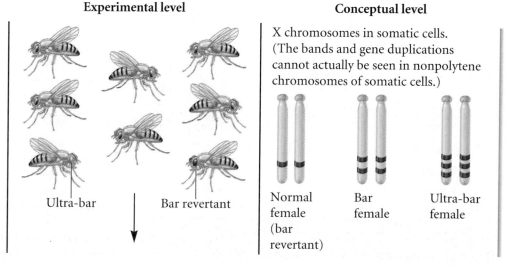

Ultra-bar   Bar revertant

X chromosomes in somatic cells. (The bands and gene duplications cannot actually be seen in nonpolytene chromosomes of somatic cells.)

Normal female (bar revertant)   Bar female   Ultra-bar female

(*continued*)

2. Dissect the salivary glands from the larva of a normal strain, a homozygous bar strain, a bar-revertant strain, and an ultra-bar strain.

3. Prepare the salivary cells for the microscopic examination of the polytene chromosomes. Note: The microscopic examination of chromosomes is described in chapter 3. It involves gently breaking open the cells, staining the chromosomes with dyes, and squashing the preparation on a microscope slide underneath a coverslip.

4. View the banding patterns of the polytene chromosomes under the microscope.

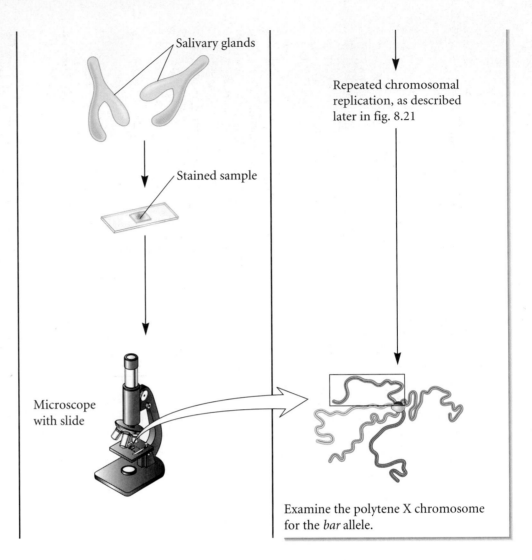

Salivary glands

Stained sample

Microscope with slide

Repeated chromosomal replication, as described later in fig. 8.21

Examine the polytene X chromosome for the *bar* allele.

■ THE DATA

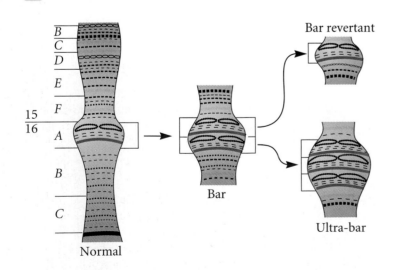

Note: This is a drawing of a short segment of a polytene chromosome that corresponds to the region of the X chromosome where the *bar* allele is located. The *bar* allele is found within the region designated 16A.

■ INTERPRETING THE DATA

The cytological examination of region 16A, which carries the X-linked gene that influences facet number, provided direct insight into the nature of the bar and ultra-bar phenotypes. As shown schematically in the data, Bridges found that the bar phenotype itself is caused by a duplication of the genetic material in region 16A of the X chromosome. Furthermore, homozygous *bar* stocks can occasionally produce phenotypically normal flies (bar revertants) and ultra-bar flies by altering gene number. The polytene chromosomes of bar revertants showed that the 16A region had returned to the wild-type banding pattern. In other words, the duplicated region was returned to a single copy. By comparison, the opposite situation occurred for the ultra-bar phenotype. In this case, an additional duplication occurred, so there were three copies of the 16A region. As stated by Bridges, "The bar-eye reduction [in the number of facets] is thus seen to be interpretable as the effect of increasing the action of certain genes by doubling or triplicating their number."

The mechanism for the formation of the *bar* allele can be explained by a misaligned crossover as described previously in figure 8.5. The formation of *ultra-bar* and *bar-revertant* alleles can likewise be explained by unequal crossing over. As shown in figure 8.8, a misaligned crossover between two X chromosomes carrying the duplicated *bar* allele can result in one chromosome carrying one copy of the gene (*bar revertant*) and the other chromosome carrying three copies (*ultra-bar*).

The gene duplication and triplication seen with the *bar* and *ultra-bar* alleles are also associated with a phenomenon known as a **position effect.** A female that is homozygous for the *bar* allele has four copies of this gene, and a female that is heterozygous for the *ultra-bar* allele (three copies) and the wild-type allele (one copy) also has four copies. However, a female that is heterozygous for the *ultra-bar* and normal alleles has fewer facets (45) than does a *bar* homozygote (70) (see fig. 8.6). This is a position effect—the positioning of three copies next to each other on an X chromosome increases the severity of the defect. As discussed in chapter 16, position effects are caused by the influences of chromosome structure or genetic regulatory regions on the level of gene expression.

A self-help quiz involving this experiment can be found at the Online Learning Center.

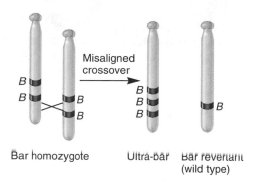

FIGURE 8.8 **A misaligned crossover produces an *ultra-bar* and *bar revertant*.** During oogenesis in a homozygous female, the two X chromosomes are misaligned, and a crossover occurs at the misaligned region. Following meiosis, this creates a chromosome with three copies of the gene (*ultra-bar*) and one copy of the gene (*bar revertant*).

Duplications Provide Additional Material for Gene Evolution, Sometimes Leading to the Formation of Gene Families

In contrast to the gene duplication that causes the bar phenotype in *Drosophila,* the majority of small chromosomal duplications have no phenotypic effect. Nevertheless, they are vitally important because they provide raw material for the addition of more genes into a species' chromosomes. Over the course of many generations, this can lead to the formation of a **gene family** consisting of two or more genes that are similar to each other. As shown in figure 8.9, the members of a gene family are derived from the same ancestral gene. Over time, two copies of an ancestral gene can accumulate different mutations. Therefore, after many generations, the two genes will be similar but not identical. During evolution, this type of event can occur several times, creating a family of many similar genes.

When two or more genes are derived from a single ancestral gene, the genes are said to be **homologous.** Homologous genes within a single species are called **paralogues** and constitute a gene family. A well-studied example of a gene family is shown in figure 8.10, which illustrates the evolution of the globin gene family found in humans. The globin genes encode polypeptides, which are subunits of proteins that function in oxygen binding. For example, hemoglobin is found in red blood cells; its function is to carry oxygen throughout the body. The globin gene family is composed of 14 homologous genes that were originally derived from a single ancestral globin gene. According to an evolutionary analysis, the ancestral globin gene first duplicated about 800 million years ago. Since that time, additional duplication events and chromosomal rearrangements have occurred to produce the current number of 14 genes on three different human chromosomes.

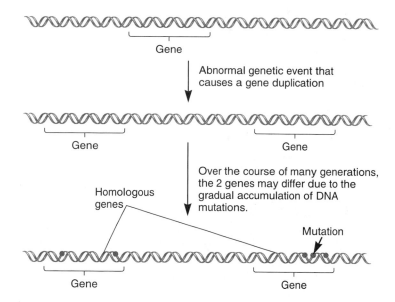

FIGURE 8.9 **Gene duplication and the evolution of homologous genes.** An abnormal crossover event like the one described in figure 8.5 leads to a gene duplication. Over time, each gene accumulates different mutations.

Gene families have been important in the evolution of traits. Even though all the globin polypeptides are subunits of proteins that play a role in oxygen binding, the accumulation of different mutations in the various family members has created globins that are more specialized in their function. For example, myoglobin is better at binding and storing oxygen in muscle cells, whereas the hemoglobins are better at binding and transporting oxygen via the red blood cells. Also, different globin genes are

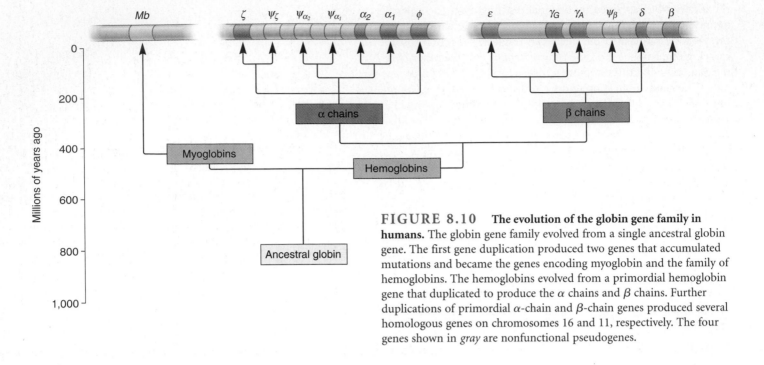

FIGURE 8.10   The evolution of the globin gene family in humans. The globin gene family evolved from a single ancestral globin gene. The first gene duplication produced two genes that accumulated mutations and became the genes encoding myoglobin and the family of hemoglobins. The hemoglobins evolved from a primordial hemoglobin gene that duplicated to produce the α chains and β chains. Further duplications of primordial α-chain and β-chain genes produced several homologous genes on chromosomes 16 and 11, respectively. The four genes shown in *gray* are nonfunctional pseudogenes.

expressed during different stages of human development. The ε- and ζ-globin genes are expressed very early in embryonic life, while the γ-globin genes exhibit maximal expression during the second and third trimesters of gestation. Following birth, the γ-globin genes are turned off and the β-globin gene is turned on. These differences in the expression of the globin genes reflect the differences in the oxygen transport needs of humans during the embryonic, fetal, and postpartum stages of life.

Inversions Often Occur Without Phenotypic Consequences

We now turn our attention to changes in chromosome structure that involve a rearrangement in the genetic material. A chromosome with an inversion contains a segment that has been flipped to the opposite orientation. Geneticists classify inversions according to the location of the centromere. If the centromere lies within the inverted region of the chromosome, the inverted region is known as a **pericentric inversion.** Alternatively, if the centromere is found outside the inverted region, the inverted region is called a **paracentric inversion.** The two types of inversions are illustrated in figure 8.11*b* and *c*.

When a chromosome contains an inversion, the total amount of genetic material remains the same as in a normal chromosome. As long as a gene's promoter and coding sequence are maintained in the same direction, the genes within the inverted region would be transcribed correctly. Therefore, the great majority of inversions do not have any phenotypic consequences. In rare cases, however, an inversion can alter the phenotype of an individual. Whether or not this occurs is related to the boundaries of the inverted segment. When an inversion occurs, the chromo-

some is broken in two places and the center piece flips around to produce the inversion. If either breakpoint occurs within a vital gene, the function of the gene is expected to be disrupted, possibly producing a phenotypic effect. For example, some people with hemophilia (type A) have inherited an X-linked inversion that has inactivated a gene for factor VIII, which is a blood clotting protein. In other cases, an inversion or translocation may reposition a gene on a chromosome in a way that alters its normal level of expression. This is another type of position effect.

Since inversions seem like an unusual genetic phenomenon, it is perhaps surprising that they are found in human populations in quite significant numbers. About 2% of the human population

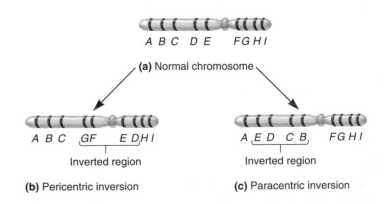

FIGURE 8.11   Types of inversions. (a) Depicts a normal chromosome with the genes ordered from *A* through *I*. A pericentric inversion (**b**) includes the centromere, whereas a paracentric inversion (**c**) does not.

carries inversions that are detectable with a light microscope. In most cases, these individuals are phenotypically normal and live their lives without knowing they contain this abnormality. In a few cases, however, an individual with an inversion chromosome may produce offspring with genetic abnormalities. This event may prompt a physician to request a microscopic examination of the individual's chromosomes. In this way, phenotypically normal individuals may discover they have a chromosome with an inversion. Next, we will examine why an individual carrying an inversion may produce offspring with phenotypic abnormalities.

Inversion Heterozygotes May Produce Abnormal Chromosomes Due to Crossing Over

An individual who carries one copy of a normal chromosome and one copy of an inverted chromosome is known as an **inver-sion heterozygote.** Such an individual, though possibly phenotypically normal, may have a high probability of producing haploid cells that are abnormal in their total genetic content. This likelihood depends on the size of the inverted segment. During meiosis, an inversion heterozygote with a fairly large inverted segment may produce a sizable fraction of abnormal haploid cells (perhaps 1/3 or even higher).

The underlying cause of abnormality is the phenomenon of crossing over within the inverted region. During meiosis I, pairs of homologous sister chromatids synapse with each other. Figure 8.12 illustrates how this occurs in an inversion heterozygote. For the normal chromosome and inversion chromosome to synapse properly, an **inversion loop** must form to permit the homologous genes on both chromosomes to align next to each other despite the inverted sequence. If a crossover occurs within the inversion loop, highly abnormal chromosomes are produced.

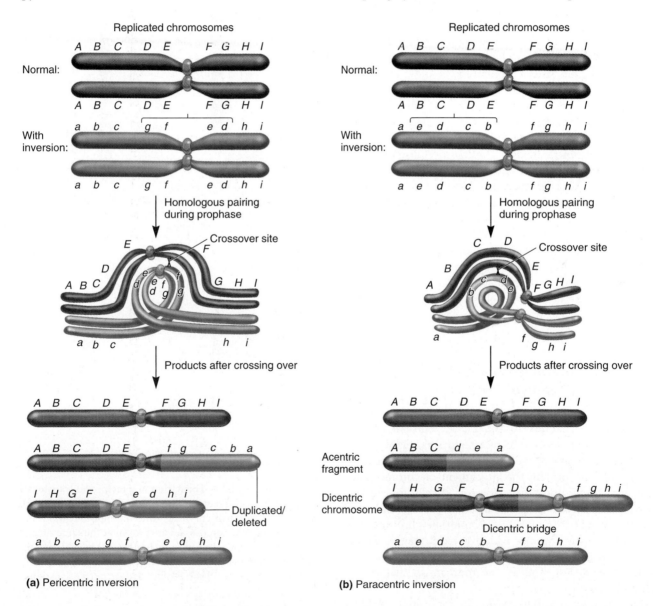

(a) Pericentric inversion

(b) Paracentric inversion

FIGURE 8.12   **The consequences of crossing over in the inversion loop. (a)** Crossover within a pericentric inversion. **(b)** Crossover within a paracentric inversion.

The consequences of this type of crossover depend on whether the inversion is pericentric or paracentric. Figure 8.12*a* describes a crossover in the inversion loop when one of the homologues has a pericentric inversion. This is a single crossover that involves only two of the four sister chromatids. Following the completion of meiosis, this single crossover yields two abnormal chromosomes. Both of these abnormal chromosomes have a segment that is deleted and a different segment that is duplicated. In this example, one of the abnormal chromosomes is missing genes *A*, *B*, and *C* and has extra copies of genes *H* and *I*. The other abnormal chromosome has the opposite situation; it is missing genes *H* and *I* and has an extra copy of genes *A*, *B*, and *C*. If these abnormal chromosomes are passed to offspring, they are likely to produce phenotypic abnormalities depending on the amount and nature of the duplicated/deleted genetic material. A large deficiency is likely to be lethal. As discussed in chapter 26, these types of inversions can be important in the formation of new species.

The outcome of a crossover involving a paracentric inversion is shown in figure 8.12*b*. A single crossover has occurred between two homologous sister chromatids; the other two sister chromatids have not participated in a crossover. This single crossover event produces a very strange outcome. One chromosome contains two centromeres. This is called a **dicentric chromosome;** the region of chromosome connecting the two centromeres is called a **dicentric bridge.** The crossover also produces a piece of chromosome without any centromere. This **acentric fragment** will be lost and degraded in subsequent cell divisions. The dicentric chromosome is a temporary condition. If the two centromeres try to move toward opposite poles during anaphase, the dicentric bridge will be forced to break at some random location. Therefore, the net result of this crossover is to produce two normal chromosomes (namely, the chromosomes that did not cross over) and two chromosomes that contain deletions. These two chromosomes result from the breakage of the dicentric chromosome. They are missing the genes that were located on the acentric fragment.

Unbalanced Translocations Usually Have Detrimental Phenotypic Effects

Another type of chromosomal rearrangement is a translocation. As described in chapters 10 and 11, the ends of normal chromosomes have **telomeres,** which contain specialized repeat sequences of DNA. Telomeres allow cells to distinguish natural chromosome ends and prevent the attachment of chromosomal DNA to the ends of chromosomes. However, if cells are exposed to agents that cause chromosomes to break, the broken ends lack telomeres and are said to be reactive. If multiple chromosomes are broken, DNA repair enzymes within the cell may recognize these reactive ends and join them together to produce abnormal chromosomes (fig. 8.13*a*). This is one mechanism that causes translocations to occur.

A second mechanism may involve an abnormal crossover. As shown in figure 8.13*b*, a reciprocal translocation can be produced when two nonhomologous chromosomes cross over. This

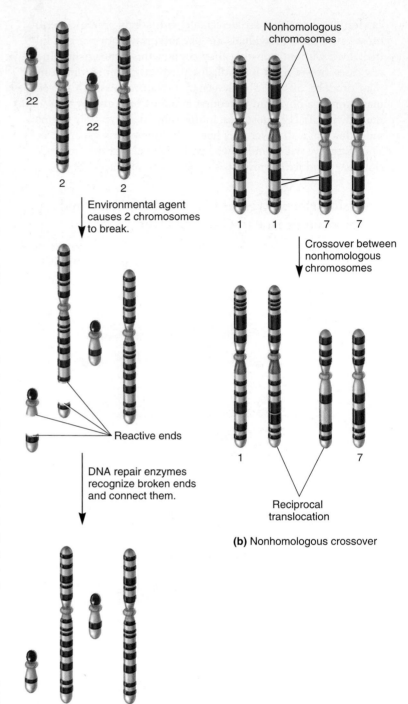

(a) Chromosomal breakage and DNA repair

(b) Nonhomologous crossover

FIGURE 8.13   Formation of a reciprocal translocation. **(a)** When two different chromosomes break, the broken ends are recognized by DNA repair enzymes, which attempt to reattach broken ends. If two different chromosomes are broken at the same time, the incorrect ends may become attached to each other, as shown here. **(b)** A crossover has occurred between chromosome 1 and chromosome 7. This crossover yields two chromosomes that carry translocations.

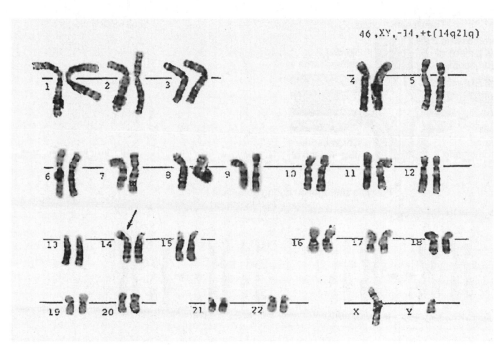

46,XY,-14,+t(14q21q)

(b)

(a)

FIGURE 8.14   **An individual affected with familial Down syndrome.** (a) This individual carries a simple translocation in which a large segment of chromosome 21 has been translocated to chromosome 14 (see arrow). In addition, the individual also carries two normal copies of chromosome 21. (b) A different person with this disorder.

type of rare aberrant event results in a rearrangement of the genetic material, though not a change in the total amount of genetic material. For this reason, a reciprocal translocation is also called a **balanced translocation.** Like inversions, balanced translocations are usually without any phenotypic consequences, because the individual has a normal amount of genetic material. In a few cases, balanced translocations can result in position effects in a fashion similar to inversions.

The inheritance of a simple translocation, in which one fragment of a chromosome is attached to another chromosome, can result in an **unbalanced translocation.** Unbalanced translocations are generally associated with phenotypic abnormalities or even lethality. An inherited human syndrome known as familial Down syndrome provides an example of this phenomenon. In this condition, the majority of chromosome 21 is attached to chromosome 14 (fig. 8.14a). In addition, this individual already has two normal copies of chromosome 21. This results in an imbalance in genetic material, since the individual has three copies of the genes that are found on a large segment of chromosome 21. In familial Down syndrome (fig. 8.14b), the person exhibits characteristics like those of an individual with the more common form of Down syndrome, which is due to three entire copies of chromosome 21. This common form of Down syndrome is described later in this chapter.

Familial Down syndrome is an example of a **Robertsonian translocation.** This term refers to a translocation in which the centromeric region of two nonhomologous acrocentric chromosomes become fused to form a single chromosome. For this to occur, there are breaks at the extreme ends of the short arms of two nonhomologous acrocentric chromosomes. The small acentric fragments are lost and then the larger chromosomal segments fuse at their centromeric regions. This creates a chromosome that is metacentric or submetacentric (see fig. 8.14a). These types of fusions between two nonhomologous acrocentric chromosomes are the most common types of chromosome rearrangements in humans.

Individuals with Balanced Translocations May Produce Abnormal Gametes Due to the Segregation of Chromosomes

Individuals who carry balanced translocations have a greater risk of producing gametes with unbalanced combinations of chromosomes. Whether or not this occurs depends on the segregation pattern during meiosis I. This idea is illustrated in figure 8.15. In this example, the parent carries a balanced, reciprocal translocation and is likely to be phenotypically normal. During meiosis, the homologous chromosomes attempt to synapse with each other. Because of the translocations, the pairing of homologous regions leads to the formation of an unusual structure that contains four pairs of sister chromatids (i.e., eight chromatids), termed a **translocation cross.**

To understand the segregation of translocated chromosomes, pay close attention to the centromeres, which are numbered in figure 8.15. For these translocated chromosomes, the expected segregation is due to the segregation of the centromeres; each haploid gamete should receive a centromere located on chromosome 1 and a centromere located on chromosome 2. This

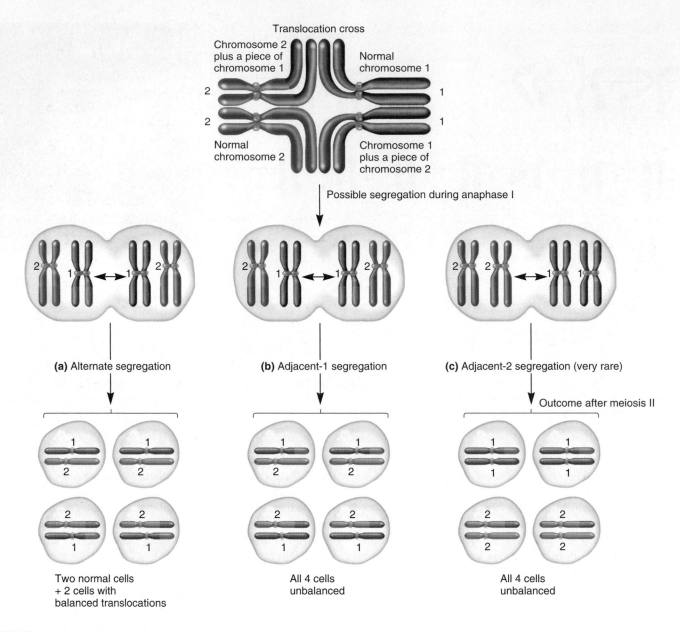

FIGURE 8.15 **Meiotic segregation of a reciprocal translocation.** Follow the numbered centromeres through each process.

can occur in two ways. One possibility is alternate segregation. As shown in figure 8.15a, this occurs when the chromosomes on opposite sides of the translocation cross sort into the same cell. One daughter cell receives two normal chromosomes, and the other cell gets two translocated chromosomes. Following meiosis II, four haploid cells are produced: two have normal chromosomes and two have reciprocal (balanced) translocations.

Another possible segregation pattern is called adjacent-1 segregation (fig. 8.15b). This occurs when adjacent chromosomes (one of each type of centromere) segregate into the same cell. Following anaphase I, each daughter cell receives one normal chromosome and one translocated chromosome. After meiosis II is completed, this produces four haploid cells, all of which are genetically unbalanced because part of one chromosome has

been deleted and part of another has been duplicated. If these haploid cells give rise to gametes that unite with a normal gamete, the zygote is expected to be abnormal genetically and possibly phenotypically.

Alternate and adjacent-1 segregation patterns are the likely outcomes when an individual carries a reciprocal translocation. Depending on the sizes of the translocated segments, both types may be equally likely to occur. In many cases, the haploid cells from adjacent-1 segregation are not viable, thereby lowering the fertility of the parent. This condition is called **semisterility.** On very rare occasions, adjacent-2 segregation can occur (fig. 8.15c). In this case, the centromeres do not segregate as they should. One daughter cell has received both copies of the centromere on chromosome 1, the other both copies of the centromere on chromosome 2. This rare segregation pattern also yields four abnormal haploid cells that contain an unbalanced combination of chromosomes.

8.2 VARIATION IN CHROMOSOME NUMBER

As we have seen in the preceding section, chromosome structure can be altered in a variety of ways. The total number of chromosomes, too, can vary. Eukaryotic species typically contain several chromosomes that are inherited as one or more sets. Variations in chromosome number can be categorized in two ways: variation in the number of sets of chromosomes, and variation in the number of particular chromosomes within a set.

Organisms that are **euploid** have a chromosome number that is an exact multiple of a chromosome set. In *Drosophila melanogaster,* for example, a normal individual has 8 chromosomes. The species is diploid, having two sets of 4 chromosomes each (fig. 8.16a). We can also recognize that a normal fruit fly is euploid because 8 chromosomes divided by 4 chromosomes per set equals two exact sets. On rare occasions, an abnormal fruit fly can be produced with 12 chromosomes, containing three sets of 4

chromosomes each. This alteration in euploidy produces a **triploid** fruit fly. Such a fly is also euploid, because it contains exactly three sets of chromosomes. Organisms with three or more sets of chromosomes are also called **polyploid** (fig. 8.16b). Geneticists use the letter *n* to represent a set of chromosomes. A diploid organism is referred to as $2n$, a triploid organism as $3n$, a **tetraploid** organism as $4n$, and so forth.

A second way in which chromosome number can vary is by **aneuploidy.** This refers to a variation that involves an alteration in the number of particular chromosomes, so that the total number of chromosomes is not an exact multiple of a set; it is not an exact multiple of *n*. For example, an abnormal fruit fly could contain nine chromosomes instead of eight because it had three copies of chromosome 2 instead of the normal two copies (fig. 8.16c). Such an animal is said to have trisomy 2 or to be **trisomic.** Instead of being perfectly diploid (i.e., $2n$), a trisomic animal is $2n + 1$. By comparison, a fruit fly could be lacking a single chromosome, such as chromosome 1, and contain a total of seven chromosomes ($2n-1$). This animal would be **monosomic** and be described as having monosomy 1.

In this section, we will examine euploid and aneuploid variation in many eukaryotic species. We will learn that euploid variation among different species occurs occasionally in animals and quite frequently in plants. We will see how natural variation

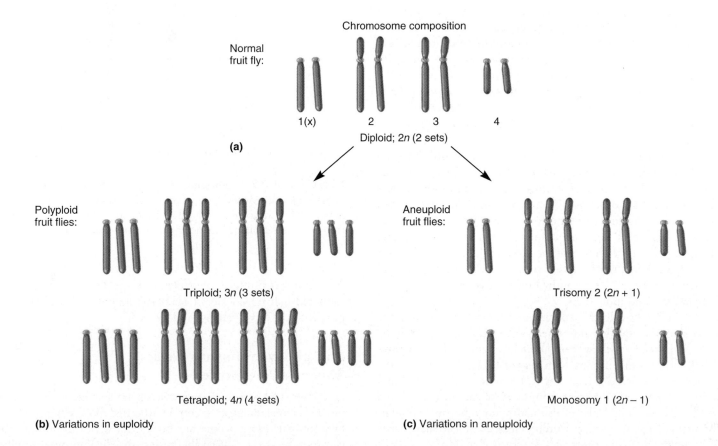

FIGURE 8.16 **Types of variation in chromosome number.** (a) Depicts the normal diploid number of chromosomes in *Drosophila*. (b) Examples of polyploidy. (c) Examples of aneuploidy.

in the number of sets of chromosomes affects phenotypic variation. By comparison, aneuploidy is generally regarded as an abnormal condition. We will begin by considering several examples in which aneuploidy has a negative impact on an organism's phenotype.

Aneuploidy Causes an Imbalance in Gene Expression That Is Often Detrimental to the Phenotype of the Individual

The phenotype of every eukaryotic species is influenced by thousands of different genes. In humans, for example, it is estimated that a single set of chromosomes contains approximately 35,000 different genes. Most genes are expressed only in certain cell types or during particular stages of development. To produce a phenotypically normal individual, intricate coordination has to occur in the expression of thousands of genes. In the case of humans and many other diploid species, evolution has resulted in a developmental process that works correctly when there are two copies of each chromosome per cell. In other words, when a human is diploid, the balance of gene expression among many different genes usually produces a person with a normal phenotype.

Aneuploidy commonly causes an abnormal phenotype. To understand why, consider the relationship between gene expression and chromosome number (fig. 8.17). For many but not all genes, the level of expression is correlated with the number of genes per cell. Compared to a diploid cell, if a gene is carried on a chromosome that is present in three copies instead of two, approximately 150% of the normal amount of gene product will be made. Alternatively, if only one copy of that gene is present due to a missing chromosome, it is common for only 50% of the gene product to be made. Therefore, in trisomic and monosomic individuals, there is an imbalance in the level of gene expression between the majority of chromosomes found in pairs versus the one type that is not.

At first glance, the difference in gene expression between euploid and aneuploid individuals may not seem terribly drastic. Keep in mind, however, that a eukaryotic chromosome carries hundreds or even thousands of different genes. Therefore, when an organism is trisomic or monosomic, many gene products will occur in excessive or deficient amounts. This imbalance among many genes appears to underlie the abnormal phenotypic effects that aneuploidy frequently causes. In most cases, these effects are detrimental and produce an individual that is less likely to survive than a euploid individual.

In the 1920s, Albert Blakeslee and his colleagues were the first to recognize the harmful effects of aneuploidy by studying the Jimson weed (*Datura stramonium*). Figure 8.18 compares normal and trisomic strains with regard to morphology of the capsule, which is the dry fruit. All 12 trisomies had capsules that were morphologically different from those of the normal diploid strain. These aneuploid plants also had many other morphologically distinguishable traits, including changes in leaf shape, size, and so forth. In many cases, the observed changes in chromo-

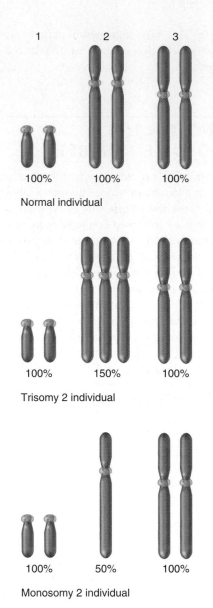

FIGURE 8.17 **Imbalance of gene products in trisomic and monosomic individuals.** Aneuploidy of chromosome 2 (i.e., trisomy and monosomy) leads to an imbalance in the amount of gene products from chromosome 2 compared to the amounts from chromosomes 1 and 3.

some number produced detrimental traits. For example, Blakeslee noted that the cocklebur plant (trisomy 6) is "weak and lopping with the leaves narrow and twisted."

Aneuploidy in Humans Causes Abnormal Phenotypes

One important reason that geneticists are so interested in aneuploidy is its relationship to certain inherited disorders in humans. Even though most people are born with a normal number of chromosomes (i.e., 46), alterations in chromosome number

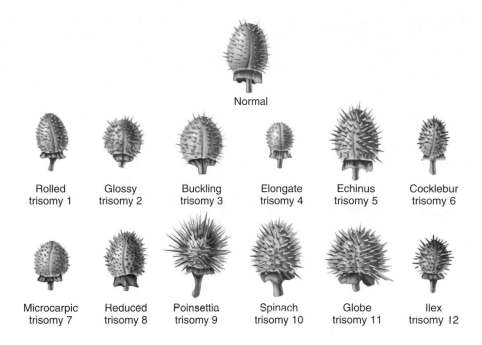

Normal

Rolled
trisomy 1

Glossy
trisomy 2

Buckling
trisomy 3

Elongate
trisomy 4

Echinus
trisomy 5

Cocklebur
trisomy 6

Microcarpic
trisomy 7

Reduced
trisomy 8

Poinsettia
trisomy 9

Spinach
trisomy 10

Globe
trisomy 11

Ilex
trisomy 12

FIGURE 8.18   **The effects of trisomy on the phenotype of the capsules of the *Datura* plant.**

GENES→TRAITS   As described in figure 8.17, trisomy leads to an imbalance in the copy number of genes on the trisomic chromosomes (here, 150% of normal) versus those on the other chromosomes (100%). This drawing compares the morphology of the capsule of the *Datura* plant in normal individuals and those carrying a trisomy of each of the 12 types of chromosomes in this species. As seen here, trisomy of each chromosome causes changes in the phenotype of the capsule. This is due to changes in the balance of gene expression.

occur fairly frequently during gamete formation. About 5 to 10% of all fertilized human eggs result in an embryo with an abnormality in chromosome number. In most cases, these abnormal embryos do not develop properly and result in a spontaneous abortion very early in pregnancy. Approximately 50% of all spontaneous abortions are due to alterations in chromosome number.

In some cases, an abnormality in chromosome number produces an offspring that can survive. Several human disorders involve abnormalities in chromosome number. The most common are trisomies of chromosomes 21, 18, or 13 and abnormalities in the number of the sex chromosomes (table 8.1). Most of the known trisomies involve chromosomes that are relatively small (i.e., chromosomes 21, 18, and 13; refer back to fig. 8.1). Trisomies of the other human autosomes and monosomies of all autosomes are presumed to produce a lethal phenotype, and many have been found in spontaneously aborted embryos and fetuses. For example, all possible human trisomies have been found in spontaneously aborted embryos except trisomy 1. It is likely that trisomy 1 is lethal at such an early stage that it prevents the successful implantation of the embryo and/or causes an abortion before a woman knows that she is pregnant.

Variation in the number of X chromosomes, unlike the case for most other large chromosomes, is often nonlethal. The survival of trisomy X individuals is explained by X inactivation, which is described in chapter 7. In an individual with more than

TABLE 8.1
Aneuploid Conditions in Humans

Condition	Frequency	Syndrome	Characteristics
Autosomal			
Trisomy 21	1/800	Down	Mental retardation, abnormal pattern of palm creases, slanted eyes, flattened face, short stature
Trisomy 18	1/6,000	Edward	Mental and physical retardation, facial abnormalities, extreme muscle tone, early death
Trisomy 13	1/15,000	Patau	Mental and physical retardation, wide variety of defects in organs, large triangular nose, early death
Sex Chromosomal			
XXY	1/1,000 (males)	Klinefelter	Sexual immaturity (no sperm), breast swelling
XYY	1/1,000 (males)	Jacobs	Tall
XXX	1/1,500 (females)	Triple X	Tall and thin, menstrual irregularity
X0	1/5,000 (females)	Turner	Short stature, webbed neck, sexually undeveloped

one X chromosome, all additional X chromosomes are converted to Barr bodies in the somatic cells of adult tissues. In an individual with trisomy X, for example, two out of three X chromosomes are converted to inactive Barr bodies. Unlike the autosomes, the normal level of expression for X-linked genes is from a single X chromosome. In other words, the correct level of mammalian gene expression results from two copies of each autosomal gene and one copy of each X-linked gene. This explains how the expression of X-linked genes in males (XY) can be maintained at the same levels as in females (XX). It may also explain why monosomy X and trisomy X are not lethal conditions. The phenotypic effects noted in table 8.1, involving sex chromosomal abnormalities, may be due to the expression of X-linked genes prior to embryonic X inactivation. In addition, sex chromosomal abnormalities could result from an imbalance in the expression of pseudoautosomal genes, which are expressed from both the X and Y chromosome. Having one or three copies of the sex chromosomes would result in an underexpression or overexpression of these sex-linked genes, respectively.

Some human abnormalities in chromosome number are influenced by the age of the parents. Older parents are more likely to produce children with abnormalities in chromosome number. Down syndrome provides an example. The common form of this disorder is caused by the inheritance of three copies of chromosome 21. The incidence of Down syndrome rises with the age of either parent. In males, however, the rise occurs relatively late in life, usually past the age when most men have children. By comparison, the likelihood of having a child with Down syndrome rises dramatically during the reproductive ages of women (fig. 8.19). This syndrome was first described by the English physician John Langdon Down in 1866. The association between maternal age and Down syndrome was later discovered by L. S. Penrose in 1933, even before the chromosomal basis for the disorder was

identified by the French scientist Jerome Lejeune in 1959. Down syndrome is most commonly caused by **nondisjunction,** which means that the chromosomes do not segregate properly. In this case, it has been found that nondisjunction of chromosome 21 most commonly occurs during meiosis I in the oocyte.

Different theories have been proposed to explain the relationship between maternal age and Down syndrome. One popular idea suggests that it may be due to the age of the oocytes. Human primary oocytes are produced within the ovary of the female fetus prior to birth and are arrested at prophase I of meiosis I until the time of ovulation. Therefore, as a woman ages, her primary oocytes have been in prophase I for a progressively longer period of time. This added length of time may contribute to an increased frequency of nondisjunction. About 5% of the time, Down syndrome is due to an extra paternal chromosome. Prenatal tests can determine if a fetus has Down syndrome and some other genetic abnormalities. The topic of fetal testing is discussed in chapter 22.

Variations in Euploidy Occur Naturally in a Few Animal Species

We now turn our attention to changes in the number of sets of chromosomes, referred to as variations in euploidy. Most species of animals are diploid. In many cases, changes in euploidy are not well tolerated. For example, polyploidy in mammals is generally a lethal condition. However, a few examples of naturally occurring variations in euploidy occur. Male bees, which are called drones, contain a single set of chromosomes; they are produced from unfertilized eggs. By comparison, female bees are diploid. Geneticists often refer to male bees as being haploid, although in a strict genetic sense, the term *haploid* describes cells that contain half the genetic material of the parent's somatic cells. An organism with a single set of chromosomes within its somatic cells is more accurately called **monoploid.**

A few examples of vertebrate polyploid animals have been discovered. Interestingly, on rare occasions, animals that are morphologically very similar to each other can be found as a diploid species as well as a separate polyploid species. This situation occurs among certain amphibians and reptiles. Figure 8.20 shows photographs of a diploid and a tetraploid frog. As you can see, they look fairly similar to each other. Their differences in euploidy can be revealed only by an examination of the chromosome number in the somatic cells of the animals.

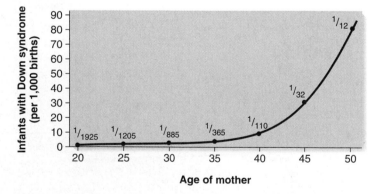

FIGURE 8.19   **The incidence of Down syndrome births according to the age of the mother.** The y-axis shows the number of children born with Down syndrome per 1,000 live births, while the x-axis plots the age of the mother at the time of birth. The fractions next to the data points indicate the fraction of live offspring born with Down syndrome.

Variations in Euploidy Can Occur in Certain Tissues Within an Animal

Thus far, we have considered variations in chromosome number that occur at fertilization, so that all the somatic cells of an individual contain this variation. In many animals, certain tissues of the body display normal variations in the number of sets of chromosomes. Diploid animals sometimes produce tissues that are

(a) *Hyla chrysocelis*

(b) *Hyla versicolor*

FIGURE 8.20 **Differences in euploidy in two closely related frog species.** The frog in **(a)** is diploid, whereas the frog in **(b)** is tetraploid. Both species grow to the same size.

GENES→TRAITS A doubling of the chromosomal composition from diploid to tetraploid has only a minor effect on the phenotypes of these two species. At the level of gene expression, this observation suggests that the copy number of each gene (two versus four) does not critically affect the phenotype of these two species.

polyploid. The cells of the human liver, for example, can vary to a great degree in their ploidy. Liver cells contain nuclei that can be triploid, tetraploid, and even octaploid ($8n$). This phenomenon is known as **endopolyploidy.** The biological significance of endopolyploidy is not completely understood. One possibility is that the increase in chromosome number in certain cells may enhance their ability to produce specific gene products that are needed in great abundance.

An unusual example of natural variation in the ploidy of somatic cells occurs in *Drosophila* and a few other insects. Within certain tissues, such as the salivary glands, the chromosomes undergo repeated rounds of chromosome replication without cellular division. For example, in the salivary gland cells of *Drosophila*, the pairs of chromosomes double approximately nine times ($2^9 = 512$). Figure 8.21*a* illustrates how repeated rounds of chromosomal replication produce a bundle of chromosomes that lie together in a parallel fashion. This bundle, termed a **polytene chromosome,** was first observed by E. G. Balbiani in 1881. Later, in the 1930s, Theophilus Painter and his colleagues recognized that the size and morphology of polytene chromosomes provided geneticists with unique opportunities to study chromosome structure and gene organization.

Figure 8.21*b* shows a micrograph of a polytene chromosome. Prior to the formation of polytene chromosomes, *Drosophila* cells contain eight chromosomes (two sets of four chromosomes each). In the salivary gland cells, the homologous chromosomes synapse with each other and replicate to form a polytene structure. During this process, the four types of chromosomes aggregate to form a single structure with several polytene arms. The central point where the chromosomes aggregate is known as the **chromocenter.** Each of the four types of chromosome is attached to the chromocenter near its centromere. The X and Y and chromosome 4 are telocentric, and chromosomes 2 and 3 are metacentric. Therefore, chromosomes 2 and 3 have two arms that radiate from the chromocenter while the X and Y and chromosome 4 have a single arm projecting from the chromocenter (fig. 8.21*c*).

Because of their considerable size, polytene chromosomes lend themselves to an easy microscopic examination. Ordinarily, we use light microscopy to visualize the highly condensed metaphase chromosomes seen during mitosis or meiosis, as in figure 8.1. Since polytene chromosomes are so large, we can see them during interphase, when normal chromosomes are not readily visible. Remarkably, a polytene chromosome during interphase is actually 100 to 200 times larger than the average metaphase chromosome. As shown in figure 8.21*b*, polytene chromosomes exhibit a characteristic banding pattern. Each dark band is known as a **chromomere.** The structure of the genetic material within a dark band is more compact than the interband region. More than 95% of the DNA is found within these dark bands. The banding patterns of polytene chromosomes are much more detailed than those observed in metaphase chromosomes. Cytogeneticists have identified approximately 5,000 bands along polytene chromosomes. At one time, each chromomere was thought to correspond to one gene. However, this idea was found to be incorrect since the sequencing of the entire *Drosophila* genome has revealed an approximate gene number of 14,000.

Polytene chromosomes have allowed geneticists to study the organization and functioning of interphase chromosomes in great detail. When a gene is deleted or duplicated via mutation, researchers can map the change by observing abnormalities in the structure of polytene chromosomes. The experiment of figure 8.7 presented an example of this approach. In addition, since polytene chromosomes can be observed during interphase, the expression of particular genes can be correlated with changes in the compaction of certain bands in the polytene chromosome.

(a) Repeated chromosome replication produces polytene chromosome.

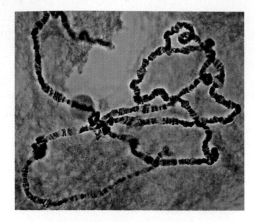

(b) A polytene chromosome.

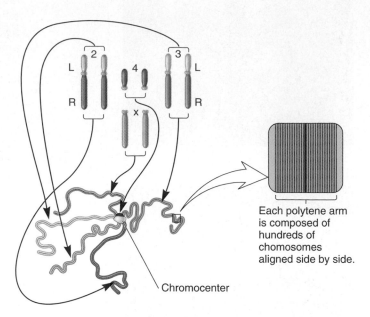

Each polytene arm is composed of hundreds of chomosomes aligned side by side.

Chromocenter

(c) Composition of polytene chromosome from regular *Drosophila* chromosomes.

FIGURE 8.21   **Polytene chromosomes in *Drosophila*.** (a) A schematic illustration of the formation of polytene chromosomes. Several rounds of repeated replication result in a bundle of sister chromatids that lie side by side. Both homologues also lie parallel to each other. This replication does not occur in highly condensed, heterochromatic DNA near the centromere. (b) A photograph of a polytene chromosome. (c) This drawing shows the relationship between the four pairs of chromosomes and the formation of a polytene chromosome in the salivary gland. The heterochromatic regions of the chromosomes aggregate at the chromocenter, and the arms of the chromosomes project outward.

Variations in Euploidy Are Common in Plants

We now turn our attention to variations of euploidy that occur in plants. In contrast to animals, plants commonly exhibit polyploidy. Among ferns and flowering plants, about 30 to 35% of species are polyploid. Polyploidy is also important in agriculture. Many of the fruits and grains we eat are produced from polyploid plants. For example, the species of wheat that we use to make bread, *Triticum aestivum,* is a hexaploid that arose from the union of diploid genomes from three closely related species (fig. 8.22*a*).

In many instances, polyploid strains of plants display outstanding agricultural characteristics. They are often larger in size and more robust. These traits are clearly advantageous in the production of food. In addition, polyploids tend to exhibit a greater adaptability, which allows them to withstand harsher environmental conditions. Also, polyploid ornamental plants often produce larger flowers than their diploid counterparts (fig. 8.22*b*).

Polyploids having an odd number of chromosome sets, such as triploids (3*n*) or pentaploids (5*n*), are usually sterile. The sterility arises because they produce highly aneuploid gametes. During prophase I of meiosis, homologous pairs of sister chromatids will form bivalents. However, if there is an odd number of chromosomes, such as three, there will be an unequal separation of homologous chromosomes during anaphase I in a triploid organism (fig. 8.23). Because there are three copies of each repli-

cated chromosome, an odd number cannot be divided equally between two daughter cells. For each type of chromosome, a daughter cell randomly gets one or two copies. For example, one daughter cell might receive one copy of chromosome 1, two copies of chromosome 2, two copies of chromosome 3, one copy of chromosome 4, and so forth. For a triploid species containing many different chromosomes in a set, it is very unlikely that meiosis will produce a cell that is euploid. If we assume that a daughter cell will receive either one copy or two copies of each kind of chromosome, the probability that meiosis will produce a cell that will be perfectly haploid or diploid is $(1/2)^{n-1}$, where *n* is the number of chromosomes in a set. As an example, in a triploid organism containing 20 chromosomes per set, the probability of producing a haploid or diploid cell is 1 in 524,288. Thus, meiosis is almost certain to produce cells that contain one copy of some chromosomes and two copies of the other chromosomes. This high probability of aneuploidy underlies the reason for triploid sterility.

Though sterility is generally a detrimental trait, it can be desirable agriculturally since it may result in a seedless fruit. For example, seedless watermelons and bananas are triploid varieties. The domestic banana, which is triploid, was originally derived from a normal diploid species and has been asexually propagated by humans via cuttings. The small black spots in the center of a domestic banana are degenerate seeds. In the case of flowers, the

Tetraploid

Diploid

(a) A hexaploid species

(b) A comparison of diploid and tetraploid petunias

FIGURE 8.22   Examples of polyploid plants. (a) Cultivated wheat, *Triticum aestivum,* is an allohexaploid. It was derived from three different diploid species of grasses that originally were found in the Middle East and were cultivated by ancient farmers in this region. During the course of cultivation, two of these species must have interbred to create an allotetraploid, and then a third species interbred with the allotetraploid to create an allohexaploid. Modern varieties of wheat have been produced from this allohexaploid species. **(b)** Differences in euploidy may exist in two closely related petunia species. The flower on the *bottom* is diploid, whereas the large one on the *top* is tetraploid.

GENES→TRAITS   An increase in chromosome number from diploid to tetraploid or hexaploid affects the phenotype of the individual. In the case of many plant species, a polyploid individual is larger and more robust than its diploid counterpart. This suggests that having additional copies of each gene is somewhat better than having two copies of each gene. This phenomenon in plants is rather different from the situation in animals. Tetraploidy in animals may have little effect (as in fig. 8.20*b*), but it is more common for polyploidy in animals to be detrimental.

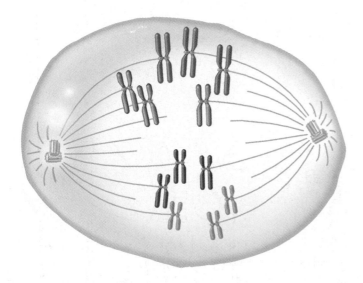

FIGURE 8.23   Schematic representation of anaphase I in a triploid organism containing three sets of four chromosomes each. In this example, the homologous chromosomes (three each) do not evenly separate during anaphase. Each cell receives one copy of some chromosomes and two copies of other chromosomes. This produces aneuploid gametes.

seedless phenotype can also be beneficial. Seed producers such as Burpee have developed triploid varieties of flowering plants. The triploid marigold is sterile and unable to set seed. According to Burpee, "They bloom and bloom, unweakened by seed bearing."

8.3 NATURAL AND EXPERIMENTAL WAYS TO PRODUCE VARIATIONS IN CHROMOSOME NUMBER

As we have seen, variations in chromosome number are fairly widespread and usually have a significant impact on the phenotypes of plants and animals. For these reasons, researchers have wanted to understand the cellular mechanisms that cause variations in chromosome number. In some cases, a change in chromosome number is the result of nondisjunction. The term *nondisjunction* refers to the event in which the chromosomes do not separate properly during anaphase. As we will see, it may be caused by an improper separation of homologous pairs in a bivalent, or a failure of the centromeres to disconnect during meiosis II or mitosis.

Meiotic nondisjunction can produce haploid cells that have too many or too few chromosomes. If such a cell gives rise to a gamete that fuses with a normal gamete during fertilization, the resulting individual will have an abnormal chromosomal composition in all the cells of the organism. An abnormal nondisjunction event also may occur after fertilization in one of the somatic cells of the body. This second mechanism is known as **mitotic nondisjunction.** When this occurs during embryonic stages of development, it may lead to a patch of tissue in the organism that has an altered chromosomal composition.

Finally, a third common way in which the chromosome composition of an organism can vary is by interspecies crosses. An **alloploid** organism contains sets of chromosomes from two or more different species. This term refers to the occurrence of chromosome sets (ploidy) from the genomes of different (*allo-*) species.

In this section, we will examine these three mechanisms in greater detail. Also, in the past few decades, researchers have devised several methods to manipulate chromosome number in experimentally and agriculturally important species. As we will learn, the experimental manipulation of chromosome number has had an important impact on genetic research and agriculture.

Meiotic Nondisjunction Can Produce Aneuploidy or Polyploidy

Figure 8.24 illustrates nondisjunction during meiosis. This type of abnormality can occur during meiosis I or II. If it happens during anaphase of meiosis I, an entire bivalent migrates to one pole (fig. 8.24*a*). Following the completion of meiosis, the four resulting haploid cells produced from this event are abnormal. A second possibility is that nondisjunction can occur during anaphase of meiosis II (fig. 8.24*b*). In this case, the net result is two normal and two abnormal haploid cells. When a gamete that is missing a chromosome participates in fertilization, the resulting offspring is monosomic for the missing chromosomes. Alternatively, if a gamete carrying an extra chromosome unites with a normal gamete, the offspring is trisomic.

In rare cases, all the chromosomes can undergo nondisjunction and migrate to one of the daughter cells. The net result of **complete nondisjunction** is a diploid cell and a cell without any chromosomes. While the cell without chromosomes is nonviable, the diploid cell might participate in fertilization with a normal haploid gamete to produce a triploid individual. Therefore, complete nondisjunction can produce individuals that are polyploid.

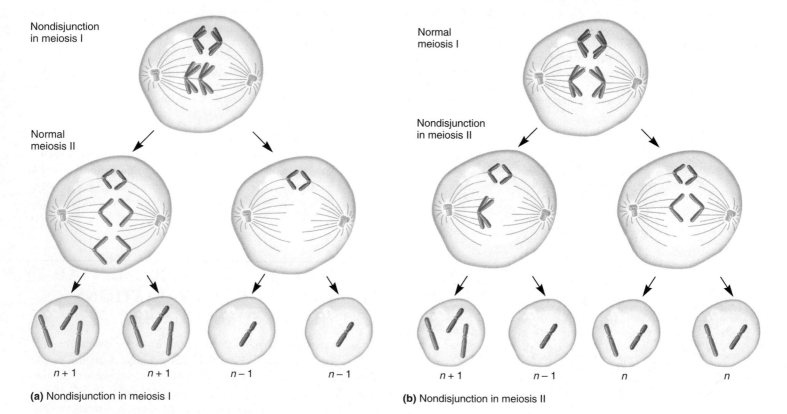

(a) Nondisjunction in meiosis I

(b) Nondisjunction in meiosis II

FIGURE 8.24 **Nondisjunction during meiosis I and II.** The chromosomes shown in *purple* are behaving properly during meiosis I and II, so that each cell receives one copy of this chromosome. The chromosomes shown in *blue* are not disjoining correctly. In (**a**), nondisjunction occurred in meiosis I, so that the resulting four cells receive either two copies of the blue chromosome or zero copies. In (**b**), nondisjunction occurred during meiosis II, so that one cell has two blue chromosomes and another cell has zero. The remaining two cells are normal.

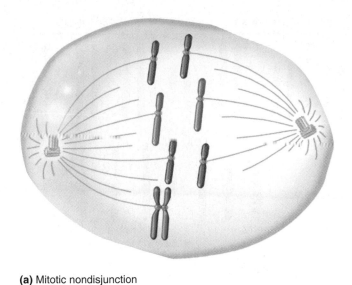

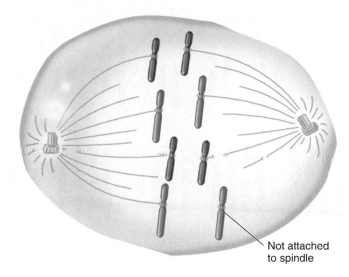

(a) Mitotic nondisjunction

(b) Chromosome loss

Not attached
to spindle

FIGURE 8.25   **Nondisjunction and chromosome loss during mitosis in somatic cells.** (**a**) Mitotic nondisjunction produces a trisomic and a monosomic daughter cell. (**b**) Chromosome loss produces a normal and a monosomic daughter cell.

Mitotic Nondisjunction or Chromosome Loss Can Produce a Patch of Tissue with an Altered Chromosome Number

Abnormalities in chromosome number occasionally occur after fertilization takes place. In this case, the abnormal event happens during mitosis rather than meiosis. One possibility is that the sister chromatids could separate improperly, so that one daughter cell would have three copies of that chromosome while the other daughter cell would have only one (fig. 8.25a). Alternatively, the sister chromatids could separate during anaphase of mitosis but one of the chromosomes could be improperly attached to the spindle, so that it would not migrate to a pole (fig. 8.25b). If this happens, a chromosome will be degraded if it is left outside the nucleus when the nuclear membrane re-forms. In this case, one of the daughter cells would have two copies of that chromosome, while the other would have only one.

When genetic abnormalities occur after fertilization, part of the organism will contain cells that are genetically different from the rest of the organism. This condition is referred to as **mosaicism.** The size and location of the mosaic region depend on the timing and location of the original abnormal event. If a genetic alteration happens very early in the embryonic development of an organism, the abnormal cell will be the precursor for a large section of the organism. In the most extreme case, an abnormality could take place at the first mitotic division. As a bizarre example, consider a fertilized *Drosophila* egg that is XX. One of the X chromosomes may be lost during the first mitotic division, producing one daughter cell that is X0 and one that is XX. Flies that are XX develop into females, while X0 flies develop into males. Therefore, in this example, one-half of the organism will become male and one-half will become female. This peculiar

and rare individual is referred to as a **bilateral gynandromorph** (fig. 8.26).

Changes in Euploidy Can Occur by Autopolyploidy, Alloploidy, and Allopolyploidy

Different mechanisms account for changes in the number of chromosome sets among natural populations of plants and animals (fig. 8.27). As mentioned earlier, complete nondisjunction, due to a general defect in the spindle apparatus, can produce an individual with one or more extra sets of chromosomes. This individual is known as an **autopolyploid** (fig. 8.27a). The terms *auto-* (meaning self) and *polyploid* (meaning many sets of chromosomes) refer to an increase in the number of chromosome sets within a single species.

A much more common mechanism for change in chromosome number, called **alloploidy,** is a result of interspecies crosses

FIGURE 8.26   **A bilateral gynandromorph of *Drosophila melanogaster.***

GENES→TRAITS   In *Drosophila,* the ratio between genes on the X chromosome and genes on the autosomes determines sex. This fly began as an XX female. One X chromosome carried the recessive white-eye and miniature wing alleles, while the other X chromosome carried the wild-type alleles. The X chromosome carrying the wild-type alleles was lost from one of the cells during the first mitotic division, producing one XX cell and one X0 cell. The XX cell became the precursor for one side of the fly, which developed as female. The X0 cell became the precursor for the other side of the fly, which developed as male with a white eye and a miniature wing.

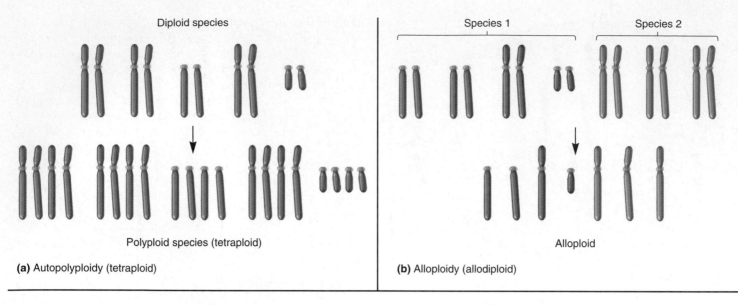

(a) Autopolyploidy (tetraploid)

(b) Alloploidy (allodiploid)

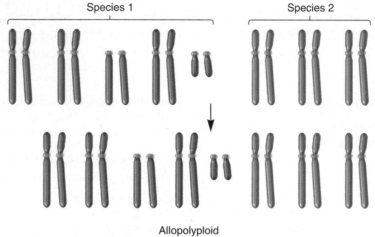

(c) Allopolyploidy (allotetraploid)

FIGURE 8.27   A comparison of autopolyploidy, alloploidy, and allopolyploidy.

(fig. 8.27*b*). An alloploid that has one set of chromosomes from two different species is called an **allodiploid.** This event is most likely to occur between species that are close evolutionary relatives. For example, closely related species of grasses may interbreed to produce allodiploids. As shown in figure 8.27*c*, an **allopolyploid** contains a combination of both autopolyploidy and alloploidy. In this case, the **allotetraploid** contains two complete sets of chromosomes from two different species for a total of four sets. In nature, allotetraploids usually arise from allodiploids. This can occur when a somatic cell in an allodiploid undergoes complete nondisjunction to create an allotetraploid cell. In plants, such a cell can continue to grow and produce a section of the plant that is allotetraploid. This part of the plant could produce seeds. Such seeds would give rise to offspring that are allotetraploids.

Allodiploids Are Often Sterile, but Allotetraploids Are More Likely to Be Fertile

Geneticists are interested in the production of allodiploids and allopolyploids as ways to generate interspecies hybrids with desirable traits. For example, if one species of grass can withstand hot temperatures and a closely related species is adapted to survive cold winters, a plant breeder may attempt to produce an interspecies hybrid that combines both qualities (i.e., good growth in the heat and survival through the winter). Such an alloploid may be desirable in climates with both hot summers and cold winters.

An important determinant of success in producing a fertile allodiploid is the degree of similarity of the different species' chromosomes. In two very closely related species, the number and types of chromosomes might be very similar. Figure 8.28 shows a

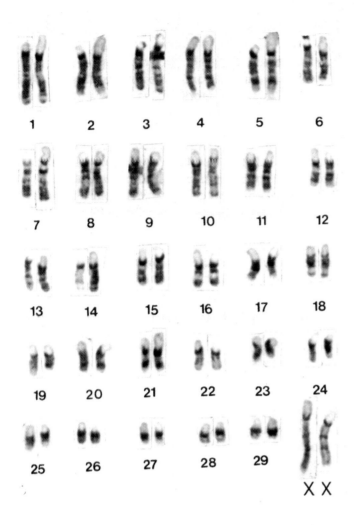

FIGURE 8.28   The karyotype of a hybrid animal produced from two closely related antelope species. In each chromosome pair in this karyotype, one chromosome was inherited from the roan antelope and the other from the sable antelope. For chromosomes with slightly different banding patterns between the two species, the roan chromosomes are shown on the left side of each pair.

karyotype of an interspecies hybrid between the roan antelope (*Hippotragus equinus*) and the sable antelope (*Hippotragus niger*). As seen here, these two closely related species have the same number of chromosomes. Moreover, the sizes and banding patterns of the chromosomes show that they correspond to one another. For example, chromosome 1 from both species is fairly large and has very similar banding patterns. This chromosome from both species carries many of the same genes. Evolutionarily related chromosomes from two different species are called **homeologous** chromosomes (not to be confused with homologous). This allodiploid is fertile because the homeologous chromosomes can properly synapse during meiosis to create haploid gametes.

The critical relationship between chromosome pairing and fertility was first recognized by the Russian cytogeneticist Georgi Karpechenko in 1928. He crossed a radish (*Raphanus*) and a cabbage (*Brassica*), both of which are diploid and contain 18 chro-

mosomes. Each of these organisms produces haploid cells containing 9 chromosomes. Therefore, the allodiploid produced from this interspecies cross contains 18 chromosomes. However, since the radish and cabbage are not closely related species, the nine *Raphanus* chromosomes are distinctly different from the nine *Brassica* chromosomes. During meiosis I, the radish and cabbage chromosomes cannot synapse with each other. This prevents the proper chromosome pairing and results in a high degree of aneuploidy (fig. 8.29*a*). Therefore, the radish/cabbage hybrid is sterile.

Among his strains of sterile alloploids, Karpechenko discovered that on rare occasions a plant would produce a viable seed. When such seeds were planted and subjected to karyotyping, it was observed that the plants were allotetraploids with two sets of chromosomes from each of the two species. In the example here, a radish/cabbage allotetraploid would contain 36 chromosomes instead of 18. The homologous chromosomes from each of the two species can synapse properly (fig. 8.29*b*). When anaphase I occurs, the pairs of synapsed chromosomes can disjoin equally to produce cells with 18 chromosomes each (a haploid set from the radish plus a haploid set from the cabbage). These cells can give rise to gametes with 18 chromosomes that can combine with each other to produce an allotetraploid containing 36 chromosomes. In this way, the allotetraploid is a fertile organism. Unfortunately, it is not agriculturally useful, because its leaves are like the radish and its roots are like the cabbage! Nevertheless, Karpechenko's scientific contribution was still important because he showed that it is possible to artificially produce a new self-perpetuating species of plant by creating an allotetraploid.

Modern plant breeders employ several different strategies to produce allotetraploids. One method is to start with two different tetraploid species. Since these make diploid gametes, the hybrid from this cross would contain a diploid set from each species. Alternatively, if an allodiploid containing one set of chromosomes from each species already exists, a second approach is to create an allotetraploid by using agents such as colchicine that alter chromosome number. The effects of colchicine are described next.

Experimental Treatments Can Promote Polyploidy

Because polyploid and allopolyploid plants often exhibit desirable traits, the development of polyploids is of considerable interest among plant breeders. Experimental studies on the ability of environmental agents to promote polyploidy began in the early 1900s. Since that time, various agents have been shown to promote nondisjunction and thereby lead to polyploidy. These include abrupt temperature changes during the initial stages of seedling growth and the treatment of plants with chemical agents that interfere with the formation of the spindle apparatus.

The drug colchicine is commonly used to promote polyploidy. Once inside the cell, colchicine binds to tubulin (a protein found in the spindle apparatus) and thereby interferes with normal chromosome segregation during mitosis or meiosis. In 1937, Alfred Blakeslee and Amos Avery were the first to apply colchicine to plant tissue and, at high doses, were able to cause complete

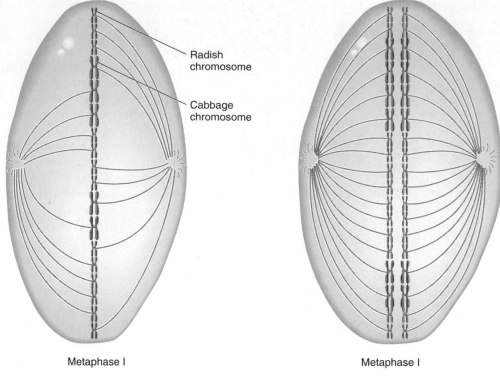

Radish
chromosome

Cabbage
chromosome

Metaphase I   Metaphase I

(a) Alloploid with a monoploid set from each species   **(b)** Allopolyploid with a diploid set from each species

FIGURE 8.29   **A comparison of metaphase I in an allodiploid and an allotetraploid.** The *purple* chromosomes are from the radish, and the *blue* are from the cabbage. (**a**) In the allodiploid, the radish and cabbage chromosomes have not synapsed and are randomly aligned during metaphase. (**b**) In the allotetraploid, the homologous radish and homologous cabbage chromosomes are properly aligned along the metaphase plate.

mitotic nondisjunction. As shown in figure 8.30, this strategy can be used to produce new polyploid strains of plants. Colchicine can be applied to seeds, young embryos, or rapidly growing regions of a plant. This application may produce aneuploidy, which is usually an undesirable outcome, but it often produces polyploid cells, which may grow faster than the surrounding diploid tissue. In a diploid plant, colchicine may cause complete mitotic nondisjunction, yielding tetraploid (4*n*) cells. As the tetraploid cells continue to divide, they generate a portion of the plant that is often morphologically distinguishable from the remainder. For example, a tetraploid stem may have a larger diameter and produce larger leaves and flowers. Since individual plants can be propagated asexually from pieces of plant tissue (i.e., cuttings), the polyploid portion of the plant can be removed, treated with the proper growth hormones, and grown as a separate plant. Alternatively, the tetraploid region of a plant may have flowers that produce seeds; a tetraploid flower will produce diploid pollen and eggs, which combine to produce tetraploid offspring. In this way, the use of colchicine provides a straightforward method to produce polyploid strains of plants.

Cell Fusion Techniques Can Be Used to Make Hybrid Plants

Thus far, we have examined several mechanisms that produce variations in chromosome number. Some of these processes can occur

naturally and have figured prominently in speciation and evolution. In addition, agricultural geneticists can administer treatments such as colchicine to promote nondisjunction and thereby obtain useful strains of organisms. More recently, researchers have developed cellular approaches to produce hybrids with altered chromosomal composition. As described here, these cellular approaches have important applications in research and agriculture.

In the technique known as **cell fusion,** individual cells are mixed together and made to fuse. In agriculture, cell fusion can create new strains of plants. It is particularly useful since it enables the crossing of two species that cannot interbreed naturally. As an example, figure 8.31 illustrates the use of cell fusion to produce a hybrid grass. The parent cells were derived from tall fescue grass (*Festuca arundinacea*) and Italian ryegrass (*Lolium multiflorum*). Prior to fusion, the cells from these two species were treated with agents that gently digest the cell wall without rupturing the plasma membrane. A plant cell without a cell wall is called a **protoplast.** The protoplasts were mixed together and treated with agents that promote fusion. Immediately after this takes place, a cell containing two separate nuclei is formed. This cell, known as a **heterokaryon,** will spontaneously go through a nuclear fusion process to create a **hybrid cell** with a single nucleus. After nuclear fusion, the hybrid cells can be grown on laboratory media and eventually regenerate an entire plant. The allotetraploid shown in figure 8.31 has phenotypic characteristics that are intermediate between tall fescue grass and Italian ryegrass.

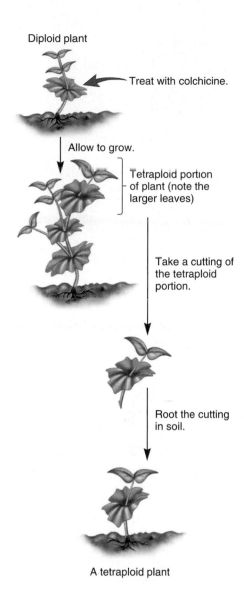

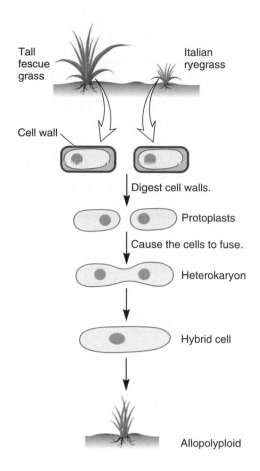

FIGURE 8.31   **The technique of cell fusion.** This technique is shown here with cells from tall fescue grass and Italian ryegrass. The resulting strain is an allotetraploid.

GENES→TRAITS   The allotetraploid contains two copies of genes from each parent. In this case, the allotetraploid displays characteristics that are intermediate between the tall fescue grass and the Italian ryegrass.

FIGURE 8.30   **Use of colchicine to promote polyploidy in plants.** Colchicine interferes with the mitotic spindle apparatus and promotes nondisjunction. If complete nondisjunction occurs in a diploid cell, a daughter cell will be formed that is tetraploid. Such a tetraploid cell may continue to divide and produce a segment of the plant with more robust characteristics. This segment may be cut from the rest of the plant and rooted. In this way, a tetraploid plant can be propagated.

Monoploids Produced in Agricultural and Genetic Research Can Be Used to Create Homozygous and Hybrid Strains

A goal of some plant breeders is to have diploid strains of crop plants that are homozygous for all of their genes. One true-breeding strain can then be crossed to a different true-breeding strain to produce an F_1 hybrid that is heterozygous for many genes. Such hybrids are often more vigorous than the corresponding homozygous strains. This phenomenon, known as **hybrid vigor** or **heterosis,** is described in greater detail in chapter 24.

Seed companies often use this strategy to produce hybrid seed for many crops, such as corn and alfalfa. To achieve this goal, the companies must have homozygous parental strains that can be crossed to each other to produce the hybrid seed. One way to obtain these homozygous strains involves inbreeding over many generations. This may be accomplished after several rounds of self-fertilization. As you might imagine, this can be a rather time-consuming endeavor.

As an alternative, the production of monoploids can be used as part of an experimental strategy to develop homozygous diploid strains of plants. Monoploids have been used to improve many agricultural crops such as wheat, rice, corn, barley, and potato. In 1964, Sipra Guha and Satish Maheshwari developed a method to produce monoploid plants directly from pollen grains.

Figure 8.32 describes the experimental technique called **anther culture,** which has been extensively used to produce diploid strains of crop plants that are homozygous for all of their genes. It involves alternation between monoploid and diploid generations. The parental plant is diploid but not homozygous for all of its genes. The anthers from this diploid plant are collected, and the haploid pollen grains are induced to begin development by a cold shock treatment. After several weeks, monoploid plantlets emerge, and these can be grown on agar media in a laboratory. However, due to the presence of deleterious alleles that are recessive, many of the pollen grains may fail to produce viable plantlets. Therefore, anther culture has been described as a "monoploid sieve" that weeds out individuals that carry deleterious recessive alleles.

Eventually, plantlets that are healthy can be transferred to small pots. After the plants grow to a reasonable size, a section of the monoploid plant can be treated with colchicine to convert it to diploid tissue. A cutting from this diploid section can then be used to generate a separate plant. This diploid plant is homozygous for all of its genes because it was produced by the chromosomal doubling of a monoploid strain.

In certain animal species, monoploids can be produced by experimental treatments that induce the eggs to begin development without fertilization by sperm. This process is known as **parthenogenesis.** In many cases, however, the haploid zygote will develop only for a short period of time before it dies. Nevertheless, a short phase of development can be useful to research scientists. For example, the zebrafish (*Brachydanio rerio*), a common aquarium fish, has recently gained popularity among researchers interested in vertebrate development. The haploid egg can be induced to begin development by exposure to sperm rendered biologically inactive by UV irradiation.

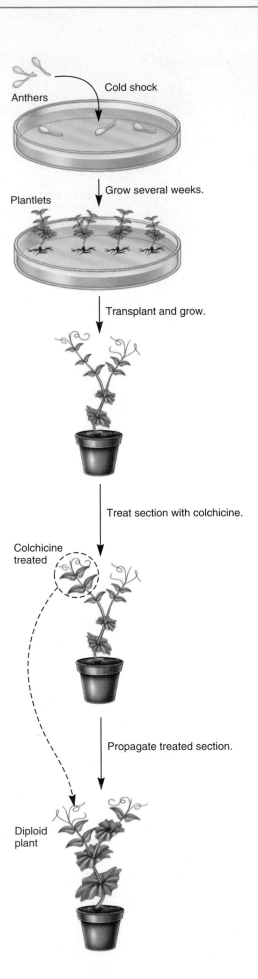

FIGURE 8.32   The experimental production of monoploids. In the technique of anther culture, the anthers are collected, and the pollen within them is induced to begin development by a cold shock treatment. After several weeks, monoploid plantlets emerge, and these can be grown on an agar medium. Eventually, the plantlets can be transferred to small pots. After the plantlets have grown to a reasonable size, a section of the monoploid plantlet can be treated with colchicine to convert it to diploid tissue. A cutting from this diploid section can then be used to generate a diploid plant.

CONCEPTUAL SUMMARY

We classify variations in chromosome structure as **deletions** (or **deficiencies**), **duplications, inversions,** and **translocations.** Deletions and duplications involve changes in the total amount of genetic material; inversions and translocations are genetic rearrangements. Deletions tend to be detrimental to the phenotype of the individual, although this depends on the size and location of the deletion. By comparison, duplications tend to be less harmful. Gene duplications provide the raw material for the evolution of **gene families** in which a group of genes encodes proteins that carry out similar yet specialized functions.

Inversions and translocations are chromosomal rearrangements; they often do not affect the phenotype of the individual that carries them. **Reciprocal translocations** are **balanced translocations,** because the individual has a normal amount of genetic material. In a few cases, inversions and translocations can affect the phenotype because a breakpoint is within a vital gene or the rearrangement results in a **position effect** that alters gene expression or regulation. Inversions and translocations are often associated with fertility problems and a higher probability of producing abnormal offspring. Even though an individual carrying an inversion or balanced translocation may be phenotypically normal, crossing over and chromosomal segregation may lead to the production of gametes that do not have a balanced amount of genetic material. Such gametes are likely to produce phenotypic abnormality or even lethality.

Chromosome number is another critical factor that determines the phenotype of an organism. In the condition known as **aneuploidy,** an individual may have an extra chromosome (**trisomy**) or may be missing a chromosome (**monosomy**). This is usually detrimental to the individual's phenotype. For example, various human genetic diseases, such as Down syndrome (trisomy 21), are due to irregularities in chromosome number. Likewise, in plants, aneuploidy also significantly affects the phenotype of an individual.

Variations in the number of sets of chromosomes are also common, particularly in the plant kingdom. This has dramatically influenced agriculture. Many of our modern crops are polyploid plants. These frequently exhibit characteristics superior to those of their diploid counterparts. Polyploids with an odd number of chromosome sets are usually sterile, and these are useful in producing seedless varieties of plants. In animals, some cells of the body may be **endopolyploid**—that is, they have more sets of chromosomes than other somatic cells. Extreme examples are the **polytene chromosomes** found in *Drosophila.*

Changes in the chromosome number can arise via several different mechanisms. In natural populations, the most important of these are **nondisjunction** and interspecies crosses. Meiotic nondisjunction produces cells with alterations in chromosome number, which can lead to aneuploidy or even polyploidy. Either nondisjunction or chromosome loss can occur after fertilization to produce an individual that is a **genetic mosaic.** Another way to produce new combinations of chromosomes involves interspecies crosses. These produce **allodiploids,** which have one set of chromosomes from each species, or **allopolyploids,** which have two or more sets from each species. **Allotetraploids** are more likely to be fertile because the chromosomes are able to form homologous pairs during meiosis.

EXPERIMENTAL SUMMARY

The primary experimental strategy for studying variation in chromosome structure and number is the cytological (i.e., microscopic) examination of chromosomes. A **karyotype** is a photographic representation of the chromosomes from actively dividing eukaryotic cells. When preparing a karyotype, the chromosomes are usually stained with a dye such as Giemsa, which gives individual chromosomes their own characteristic **G banding** pattern. By observing the banding patterns and number of chromosomes under a microscope, a cytogeneticist can determine if an individual carries a change in chromosome structure or number. In chapter 20, we will learn about methods such as *in situ* hybridization, which can detect relatively small changes in chromosome structure (e.g., small deletions), which are difficult to visualize by standard light microscopy.

Experimental methods are also available to cause changes in chromosome number. For example, drugs such as colchicine can alter chromosome number by promoting nondisjunction. In addition, laboratory methods can produce hybrids by cell fusion techniques. This can be useful in making fertile interspecies hybrids. Also, monoploid plants and animals can be produced by the experimental activation of haploid gametes.

PROBLEM SETS & INSIGHTS

Solved Problems

S1. Describe how a gene family is produced. Discuss the common and unique features of the family members in the globin gene family.

Answer: A gene family is produced when a single gene is copied one or more times by a gene duplication event. This duplication occurs by an abnormal (misaligned) crossover, which produces a chromosome with a deficiency and another chromosome with a gene duplication.

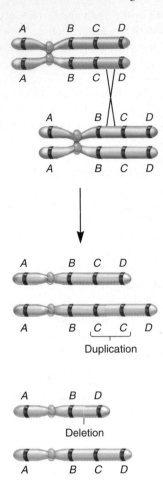

Duplication

Deletion

Over time, this type of duplication may occur several times to produce many copies of a particular gene. In addition, translocations may move the duplicated genes to other chromosomes, so that the members of the gene family may be dispersed among several different chromosomes. Eventually, each member of a gene family will accumulate mutations, which may subtly alter their function.

All the members of the globin gene family bind oxygen. Myoglobin tends to bind it more tightly; therefore, it is good at storing oxygen. Hemoglobin binds it more loosely, so it can transport oxygen throughout the body (via red blood cells) and release it to the tissues that need oxygen. The polypeptides that form hemoglobins are predominantly expressed in red blood cells, whereas myoglobin genes are expressed in many different cell types. The expression pattern of the globin genes changes during different stages of development. The ε- and ζ-globin genes are expressed in the early embryo, they are turned off near the end of the first trimester, and then the γ-globin genes exhibit their maximal expression during the second and third trimesters of gestation. Following birth, the γ-globin genes are silenced and the β-globin gene is expressed for the rest of a person's life. These differences in the expression of the globin genes reflect the differences in the oxygen transport needs of humans during the different stages of life. Overall, the evolution of gene families has resulted in gene products that are better suited to a particular tissue or stage of development. This has allowed a better "fine-tuning" of human traits.

S2. An inversion heterozygote has the following inverted chromosome:

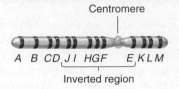

What is the result if a crossover occurs between genes *F* and *G* on one inversion and one normal chromosome?

Answer: The resulting product is four chromosomes. One chromosome is normal, one is an inversion chromosome, and two chromosomes have duplications and deficiencies. The two duplicated/deficient chromosomes are shown here:

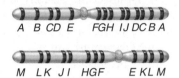

S3. In humans, the number of chromosomes per set equals 23. Even though the following conditions are lethal, what would be the total number of chromosomes for the following individuals?

A. Trisomy 22

B. Monosomy 11

C. Triploid individual

Answer:

A. 47 (the diploid number, 46, plus 1)

B. 45 (the diploid number, 46, minus 1)

C. 69 (3 times 23)

S4. A diploid species with 44 chromosomes (i.e., 22/set) is crossed to another diploid species with 38 chromosomes (i.e., 19/set). What would be the number of chromosomes in an allodiploid or allotetraploid produced from this cross? Would you expect the offspring to be sterile or fertile?

Answer: An allodiploid would have $22 + 19 = 41$ chromosomes. This individual would likely be sterile, because all the chromosomes would not have homeologous partners to pair with during meiosis. This yields aneuploidy, which usually causes sterility. An allotetraploid would have $44 + 38 = 82$ chromosomes. Since each chromosome would have a homologous partner, the allotetraploid would likely be fertile.

S5. Pseudodominance occurs when a single copy of a recessive allele is phenotypically expressed because the second copy of the gene has been deleted from the homologous chromosome; the individual is hemizygous for the recessive allele. As an example, we can consider the "notch" phenotype in *Drosophila,* which is an X-linked trait. Fruit flies with this condition have wings with a notched appearance at their edges. Female flies that are heterozygous for this mutation have notched wings; homozygous females and hemizygous males are unable to survive. The notched phenotype is due to a defect in a single gene called *notch* (*N*). When geneticists have studied fruit flies with this phenotype, some of the mutant flies are due to a small deletion that includes the *notch* gene as well as a few genes on either side of it. Other notch mutations are due to small mutations confined within the *notch* gene itself. A genetic analysis can distinguish between "notched" fruit flies carrying a deletion versus those that carry a single-gene mutation. This is possible because the *notch* gene happens to be located next to the red/white eye color gene on the X chromosome. How would you distinguish between a notched phenotype due to a deletion that included the *notch* gene and the adjacent eye color gene versus a notch phenotype due to a small mutation only within the *notch* gene itself?

Answer: To determine if the notch mutation is due to a deletion, red-eyed females with the notched phenotype can be crossed to white-eyed males. Only the daughters with notched wings need to be analyzed. If they have red eyes, this means that the notch mutation has not deleted the red-eye allele from the X chromosome. Alternatively, if they have white eyes, this indicates that the red-eye allele has been deleted from the X chromosome, which carries the notch mutation. In this case, the white-eye allele is expressed because it is present in a single copy in a female fly with two X chromosomes. This phenomenon is pseudodominance.

S6. Albert Blakeslee began using the Jimson weed (*Datura stramonium*) as an experimental organism to teach his students the laws of Mendelian inheritance. Although this plant has not gained widespread use in genetic studies, Blakeslee's work provided a convincing demonstration that changes in chromosome number have an impact on the phenotype of organisms (also see fig. 8.18). Blakeslee's assistant, B. T. Avery, identified a Jimson weed mutant, which he called "globe" because the capsule is more rounded than normal. In genetic crosses, he found that the globe mutant had a peculiar pattern of inheritance. The globe trait was passed to about 25%

of the offspring when the globe plants were allowed to self-fertilize. Unexpectedly, about 25% of the offspring also had the globe phenotype when globe plants were pollinated by a normal plant. In contrast, when pollen from a globe plant was used to pollinate a normal plant, less than 2% of the offspring had the globe phenotype. This non-Mendelian pattern of inheritance caused Blakeslee and his colleagues to investigate the nature of this trait further. We now know that the globe phenotype is due to trisomy 11. Can you explain this unusual pattern of inheritance knowing that it is due to trisomy 11?

Answer: This unusual pattern of inheritance of these aneuploid strains can be explained by the viability of euploid versus aneuploid gametes and/or gametophytes. When an individual is trisomic, there is a 50% chance that an egg or sperm nucleus will inherit an extra chromosome and a 50% chance that a gamete will be normal. Blakeslee's results indicate that when a pollen grain inherited an extra copy of chromosome 11, it was almost always nonviable and unable to produce an aneuploid offspring. However, a significant percentage of aneuploid eggs were viable so that some (25%) of the offspring were aneuploid. Since an aneuploid plant should produce a 1:1 ratio between euploid and aneuploid eggs, the observation that only 25% of the offspring were aneuploid also indicates that about half of the female gametophytes or aneuploid eggs from such gametophytes were also nonviable. Overall, these results provide compelling evidence that imbalances in chromosome number can alter reproductive viability and also cause significant phenotypic consequences.

Conceptual Questions

C1. Which changes in chromosome structure cause a change in the total amount of genetic material, and which do not?

C2. Explain why small deletions and duplications are less likely to have a detrimental effect on an individual's phenotype than large ones. If a small deletion on a single chromosome happens to have a phenotypic effect, what would you conclude about the genes in this region?

C3. How does a chromosomal duplication occur?

C4. What is a gene family? How are gene families produced over time? With regard to gene function, what is the biological significance of a gene family?

C5. Following a gene duplication, two genes will accumulate different mutations, causing them to have slightly different sequences. In figure 8.10, which two genes would you expect to have more similar sequences, α_1 and α_2 or ψ_{α_1} and α_2? Explain your answer.

C6. Two chromosomes have the following order of genes:

Normal: *A B C* centromere *D E F G H I*

Abnormal: *A B G F E D* centromere *C H I*

Does the abnormal chromosome have a pericentric or paracentric inversion? Draw a sketch showing how these two chromosomes would pair during prophase I of meiosis.

C7. An inversion heterozygote has the following inverted chromosome:

Centromere

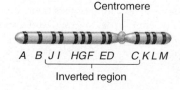

A B JI HGF ED   C KLM

Inverted region

What would be the products if a crossover occurred between genes *H* and *I* on one inverted and one normal chromosome?

C8. An inversion heterozygote has the following inverted chromosome:

Centromere

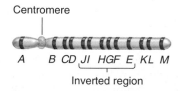

A   B CD JI HGF E KL M

Inverted region

What would be the products if a crossover occurred between genes *H* and *I* on one inverted and one normal chromosome?

C9. Explain why inversions and reciprocal translocations do not usually cause a phenotypic effect. In a few cases, however, they do. Explain how.

C10. An individual has the following reciprocal translocation:

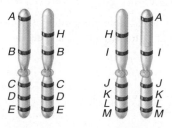

What would be the outcome of alternate and adjacent-1 segregation?

C11. A phenotypically normal individual has the following combinations of abnormal chromosomes:

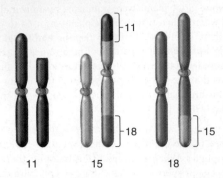

The normal chromosomes are shown on the *left* of each pair. Suggest a series of events (breaks, translocations, crossovers, etc.) that may have produced this combination of chromosomes.

C12. Two phenotypically normal parents produce an abnormal child in which chromosome 5 is missing part of its long arm but has a piece of chromosome 7 attached to it. The child also has one normal copy of chromosome 5 and two normal copies of chromosome 7. With regard to chromosomes 5 and 7, what do you think are the chromosomal compositions of the parents?

C13. In the segregation of centromeres, why is adjacent-2 segregation less frequent than alternate or adjacent-1 segregation?

C14. Which of the following types of chromosomal changes would you expect to have phenotypic consequences? Explain your choices.

A. Pericentric inversion

B. Reciprocal translocation

C. Deficiency

D. Unbalanced translocation

C15. Explain why a translocation cross occurs during metaphase I when a cell contains a reciprocal translocation.

C16. A phenotypically abnormal individual has a phenotypically normal father with an inversion on one copy of chromosome 7 and a normal mother without any changes in chromosome structure. The order of genes along chromosome 7 in the father is as follows:

R T D M centromere *P U X Z C* (normal chromosome 7)

R T D U P centromere *M X Z C* (inverted chromosome 7)

The phenotypically abnormal offspring has a chromosome 7 with the following order of genes:

R T D M centromere *P U D T R*

How was this chromosome formed? In your answer, explain where the crossover occurred (i.e., between which two genes).

C17. A diploid fruit fly has eight chromosomes. How many total chromosomes would be found in the following flies?

A. Tetraploid

B. Trisomy 2

C. Monosomy 3

D. $3n$

E. $4n + 1$

C18. A person is born with one X chromosome, zero Y chromosomes, trisomy 21, and two copies of the other chromosomes. How many chromosomes does this person have altogether? Explain whether this person is euploid or aneuploid.

C19. Two phenotypically normal parents produce two children with familial Down syndrome. With regard to chromosomes 14 and 21, what are the chromosomal compositions of the parents?

C20. Aneuploidy is typically detrimental, whereas polyploidy is sometimes beneficial, particularly in plants. Discuss why you think this is the case.

C21. Explain how aneuploidy, deletions, and duplications cause genetic imbalances. Why do you think that deletions and monosomies are more detrimental than duplications and trisomies?

C22. Female fruit flies homozygous for the X-linked white-eye allele are crossed to males with red eyes. On very rare occasions, an offspring is a male with red eyes. Assuming that these rare offspring are not due to a new mutation in one of the mother's X chromosomes that converted the white-eye allele into a red-eye allele, explain how this red-eyed male arose.

C23. A cytogeneticist has collected tissue samples from members of the same butterfly species. Some of the butterflies were located in Canada, while others were found in Mexico. Upon karyotyping, the cytogeneticist discovered that chromosome 5 of the Canadian butterflies had a large inversion compared to the Mexican butterflies. The Canadian butterflies were inversion homozygotes, whereas the Mexican butterflies had two normal copies of chromosome 5.

A. Explain whether a mating between the Canadian and Mexican butterflies would produce phenotypically normal offspring.

B. Explain whether the offspring of a cross between Canadian and Mexican butterflies would be fertile.

C24. Why do you think that human trisomies 13, 18, and 21 can survive but the other trisomies are lethal? Even though X chromosomes are large, aneuploidies of this chromosome are also tolerated. Explain why.

C25. A zookeeper has collected a male and female lizard that look like they belong to the same species. They mate with each other and produce phenotypically normal offspring. However, the offspring are sterile. Suggest one or more explanations for their sterility.

C26. What is endopolyploidy? What is its biological significance?

C27. What is a genetic mosaic? How is it produced?

C28. Explain how polytene chromosomes of *Drosophila melanogaster* are produced and how they form a six-armed structure.

C29. Describe some of the advantages of polyploid plants. What are the consequences of having an odd number of chromosome sets?

C30. While conducting field studies on a chain of islands, you decide to karyotype two phenotypically identical groups of turtles, which are found on different islands. The turtles on one island have 24 chromosomes while the turtles on another island have 48 chromosomes. How would you explain this observation? How do you think the turtles with 48 chromosomes came into being? If you mated the two types of turtles together, would you expect their offspring to be phenotypically normal? Would you expect them to be fertile? Explain.

C31. A diploid fruit fly has eight chromosomes. Which of the following terms should *not* be used to describe a fruit fly with four sets of chromosomes?

A. Polyploid

B. Aneuploid

C. Euploid

D. Tetraploid

E. 4n

C32. Which of the following terms should *not* be used to describe a human with three copies of chromosome 12?

A. Polyploid

B. Triploid

C. Aneuploid

D. Euploid

E. 2n + 1

F. Trisomy 12

C33. The kidney bean, *Phaseolus vulgaris,* is a diploid species containing a total of 22 chromosomes in somatic cells. How many possible types of trisomic individuals could be produced in this species?

C34. The karyotype of a young girl with Down syndrome revealed a total of 46 chromosomes. Her older brother, however, who is phenotypically normal, actually had 45 chromosomes. Explain how this could happen. What would you expect to be the chromosomal number in the parents of these two children?

C35. A triploid plant has 18 chromosomes (i.e., 6 chromosomes per set). If we assume there is an equal probability of a gamete receiving one or two copies of each of the six types of chromosome, what are the odds of this plant producing a monoploid or a diploid gamete? What are the odds of producing an aneuploid gamete? If the plant is allowed to self-fertilize, what are the odds of producing a euploid offspring?

C36. Describe three naturally occurring ways that the chromosome number can change.

C37. Meiotic nondisjunction is much more likely than mitotic nondisjunction. Based on this observation, would you conclude that meiotic nondisjunction is usually due to nondisjunction during meiosis I or meiosis II? Explain your reasoning

C38. A woman who is heterozygous, *Bb,* has brown eyes. *B* (brown) is a dominant allele, while *b* (blue) is recessive. In one of her eyes, however, there is a patch of blue color. Give three different explanations for how this might have occurred.

C39. What is an allodiploid? What factor determines the fertility of an allodiploid? Why are allotetraploids more likely to be fertile?

C40. What are homeologous chromosomes?

C41. Meiotic nondisjunction usually occurs during meiosis I. What is not separating properly: bivalents or sister chromatids? What is not separating properly during mitotic nondisjunction?

C42. Table 8.1 shows that Turner syndrome occurs when an individual inherits one X chromosome but lacks a second sex chromosome. Can Turner syndrome be due to nondisjunction during oogenesis, spermatogenesis, or both? If a phenotypically normal couple has a color-blind child (due to a recessive X-linked allele) with Turner syndrome, did nondisjunction occur during oogenesis or spermatogenesis in this child's parents? Explain your answer.

C43. Male honeybees, which are monoploid, produce sperm by meiosis. Explain what unusual event (compared to other animals) must occur during spermatogenesis in honeybees to produce sperm? Does this unusual event occur during meiosis I or meiosis II?

Experimental Questions

E1. With regard to the analysis of chromosome structure, explain the experimental advantage that polytene chromosomes offer. Discuss why changes in chromosome structure are more easily detected in polytene chromosomes compared to ordinary (nonpolytene) chromosomes.

E2. Describe how colchicine can be used to alter chromosome number.

E3. Describe the steps you would take to produce a tetraploid plant that is homozygous for all of its genes.

E4. In agriculture, what is the primary purpose of anther culture?

E5. What are some experimental advantages of cell fusion techniques as opposed to interbreeding approaches?

E6. It is an exciting time to be a plant breeder because so many options are available for the development of new types of agriculturally useful plants. Let's suppose you wish to develop a seedless tomato that could grow in a very hot climate and is resistant to a viral pathogen that commonly infects tomato plants. At your disposal, you have a seed-bearing tomato strain that is heat resistant and produces great-tasting tomatoes. You also have a wild strain of tomato plants (which have lousy-tasting tomatoes) that is resistant to the viral pathogen. Suggest a series of steps that you might follow to produce a great-tasting, seedless tomato that is resistant to heat and the viral pathogen.

E7. What is a G band? Discuss how G bands are useful in the analysis of chromosome structure.

E8. A female fruit fly contains one normal X chromosome and one X chromosome with an internal deficiency. The deficiency is in the middle of the X chromosome and is about 10% of the entire length of the X chromosome. If you stained and observed the chromosomes of this female fly in salivary gland cells, draw what the polytene arm of the X chromosome would look like. Explain your drawing.

E9. In the experiment of figure 8.7, researchers began with a true-breeding strain of bar-eyed flies and identified rare flies whose eyes appeared phenotypically different. If this same kind of experiment were conducted on a true-breeding culture of ultra-bar-eyed flies, what kinds of rare female flies would be produced? In other words, estimate how many facets would be contained within the eyes of the rare female flies. Note: These rare female flies have inherited an X chromosome from their mother that is the product of an unequal crossover event. They also would inherit a chromosome carrying three copies of the *bar* gene from their fathers. In your answer, also explain how many copies of the *bar* gene would be contained within these rare female flies.

E10. As described in chapter 18, the technique of Northern blotting can be used to determine the amount of mRNA that is transcribed from a gene. In this technique, the amount of mRNA is visualized as a band on a gel. If the level of transcription is increased, the intensity of the band is greater (i.e., the band on the gel is wider and darker). Let's assume that a decrease in the number of eye facets in bar-eyed flies is due to an increase in the transcription of the *bar* gene. If so, rank the intensity of the mRNA band from the *bar* gene (from most intense to least intense) that we would expect to see if we conducted a Northern blot on tissues from the female flies with the following genotypes:

Genotype 1. 1 X chromosome-normal and 1 X chromosome-*bar*
Genotype 2. 1 X chromosome-normal and 1 X chromosome-*ultra-bar*
Genotype 3. 2 X chromosomes-*bar*
Genotype 4. 1 X chromosome-*bar*, 1 X chromosome-*ultra-bar*

Note: An X chromosome-normal has one copy of the gene, an X chromosome-*bar* has two adjacent copies, and an X chromosome-*ultra-bar* has three copies.

E11. Describe two different experimental strategies to create an allotetraploid from two different diploid species of plants.

E12. In the procedure of anther culture (see fig. 8.32), an experimenter may begin with a diploid plant and then cold shock pollen to get them to grow as haploid plantlets. In some cases, the pollen may come from a (phenotypically) vigorous plant that is heterozygous for many genes. Even so, many of the haploid plantlets appear rather weak and nonvigorous. In fact, many of them fail to grow at all. In contrast, some of the plantlets are fairly healthy. Explain why some plantlets would be weak, while others could be quite healthy.

Questions for Student Discussion/Collaboration

1. A chromosome involved in a reciprocal translocation also has an inversion. The cell also contains two normal chromosomes.

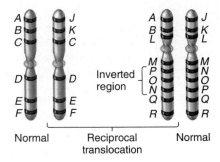

Make a drawing that shows how these chromosomes will pair during metaphase I.

2. Besides the ones mentioned in this textbook, look for other examples of variations in euploidy. Perhaps you might look in more advanced textbooks concerning population genetics, ecology, or the like. Discuss the phenotypic consequences of these changes.

3. Cell biology textbooks often discuss cellular proteins encoded by genes that are members of a gene family. Examples of such proteins include myosins and glucose transporters. Take a look through a cell biology textbook and identify some proteins encoded by members of gene families. Discuss the importance of gene families at the cellular level.

4. Discuss how variation in chromosome number has been useful in agriculture.

Note: All answers appear at the website for this textbook; the answers to even-numbered questions are in the back of the textbook.

www.mhhe.com/brooker

Visit the Online Learning Center for practice tests, answer keys, and other learning aids for this chapter. Enhance your understanding of genetics with our interactive exercises, web links, news feeds, tutorial service, and much more.

MOLECULAR STRUCTURE OF DNA AND RNA

::

9

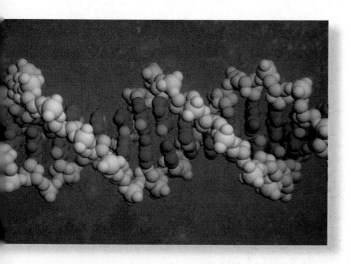

In chapters 2 through 8, we focused on the relationship between the inheritance of genes and chromosomes and the outcome of an organism's traits. In chapter 9, we will shift our attention to **molecular genetics,** the study of DNA structure and function at the molecular level. An exciting goal of molecular genetics is to use our knowledge of DNA structure to understand how DNA functions as the genetic material. Using molecular techniques, researchers have determined the organization of many genes. This information, in turn, has helped us understand how the expression of such genes governs the outcome of an individual's inherited traits.

The past several decades have seen dramatic advances in techniques and approaches to investigate and even to alter the genetic material. Not only have these advances greatly expanded our understanding of molecular genetics, they have also provided key insights into the mechanisms underlying transmission and population genetics. Molecular genetic technology is also widely used in supporting disciplines such as biochemistry, cell biology, and microbiology.

To a large extent, our understanding of genetics comes from our knowledge of the molecular structure of DNA (deoxyribonucleic acid) and RNA (ribonucleic acid). In chapter 9, we will begin by considering classic experiments that showed that DNA is the genetic material. We will then survey the molecular features of DNA and RNA that underlie their function.

PART III
Molecular Structure and Replication of the Genetic Material

9.1 IDENTIFICATION OF DNA AS THE GENETIC MATERIAL

In his classic experiments, Gregor Mendel studied several different traits in pea plants. By conducting the appropriate crosses, he showed that traits are inherited as discrete units as they pass from parent to offspring. This implies that living organisms contain a genetic material that governs an individual's traits, a substance that is transferred during the process of reproduction.

To fulfill its role, the genetic material must meet several criteria.

1. **Information:** The genetic material must contain the information necessary to construct an entire organism. In other words, it must provide the blueprint that determines all the inherited traits that an organism will have.
2. **Transmission:** During reproduction, the genetic material must be passed from parent to offspring.
3. **Replication:** Since the genetic material is passed from parents to offspring, and from mother cell to daughter cells during cell division, it must be copied.
4. **Variation:** Within any species, a significant amount of phenotypic variability occurs. For example, Mendel studied several traits in pea plants that were variable among different plants. These included height (i.e., tall vs. dwarf) and seed texture (i.e., round vs. wrinkled). Therefore, the genetic material must also have variation that can account for the known phenotypic differences within each species.

Along with Mendel's work, the data of many other geneticists in the early 1900s were consistent with these four properties: information, transmission, replication, and variation. However, the experimental study of genetic crosses cannot, by itself, identify the chemical nature of the genetic material.

As discussed in chapter 3, August Weismann and Karl Nägeli championed the idea that a chemical substance within living cells is responsible for the transmission of traits from parents to offspring. The chromosome theory of inheritance was developed, and experimentation demonstrated that the chromosomes are the carriers of the genetic material. Nevertheless, the story was not complete because chromosomes contain three classes of macromolecules, namely, DNA, RNA, and proteins. Therefore, further research was needed to precisely identify the genetic material. In this section, we will examine the first experimental approaches to achieve this goal.

Experiments with Pneumococcus Suggested That DNA Is the Genetic Material

Some early work in microbiology was important in developing an experimental strategy to identify the genetic material. Frederick Griffith studied a type of bacterium known then as pneumococci and now classified as *Streptococcus pneumoniae*. Certain strains of *S. pneumoniae* secrete a polysaccharide capsule, whereas other strains do not. When streaked on petri plates containing solid growth media, capsule-secreting strains have a smooth colony

morphology, whereas those strains unable to secrete a capsule have a rough appearance.

When comparing two different smooth strains of *S. pneumoniae*, researchers found that the chemical composition of their capsules can differ significantly. Using biochemical techniques, these different types of smooth strains were classified. For example, a type II smooth strain and a type III smooth strain both make a capsule, but their capsules differ from each other biochemically. This idea is schematically shown in figure 9.1. Rare mutations can occasionally convert a smooth bacterium into a rough bacterium, and vice versa. These infrequent interconversions are type-specific. For example, a type III smooth strain can mutate to become a rough strain. On rare occasions, a bacterium from this rough strain may mutate back to become a smooth strain. In this case, it can mutate to become only a type III smooth strain, never a type II smooth strain.

In 1928, Griffith conducted experiments that involved the injection of live and/or heat-killed bacteria into mice; he then observed whether or not the bacteria caused a lethal infection. He was working with two strains of *S. pneumoniae*, a type IIIS (*S* for smooth) and a type IIR (*R* for rough). When injected into a live mouse, the type IIIS bacteria proliferated within the mouse's bloodstream and ultimately killed the mouse (fig. 9.2a). In this process, the capsule is very important in preventing the mouse's immune system from inhibiting the proliferation of the bacteria. Following the death of the mouse, Griffith found many type IIIS bacteria within the mouse's blood. In contrast, when type IIR

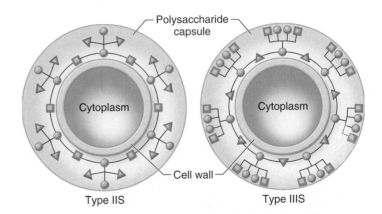

FIGURE 9.1 Schematic illustration of biochemical differences between the capsules of two strains of pneumococci. The capsule is composed of sugar units linked to form a polysaccharide material. This figure illustrates how the types of sugars (represented by *circles, squares,* and *triangles*) and their arrangement to form a polysaccharide capsule can differ among different strains. This figure is not meant to represent the actual polysaccharide capsule, which is much more complex than shown here. Instead, it emphasizes that the structures of the polysaccharide capsules are different in type IIS and type IIIS bacteria.

GENES→TRAITS The chemical composition of the polysaccharide capsule is governed by genes within the bacterial cell. These genes encode enzymes that synthesize the polysaccharide capsule. The genes within type IIS bacteria encode enzymes that synthesize the polysaccharide capsule depicted on the *left*. The genes within type IIIS bacteria encode enzymes that function somewhat differently to synthesize the capsule shown on the *right*.

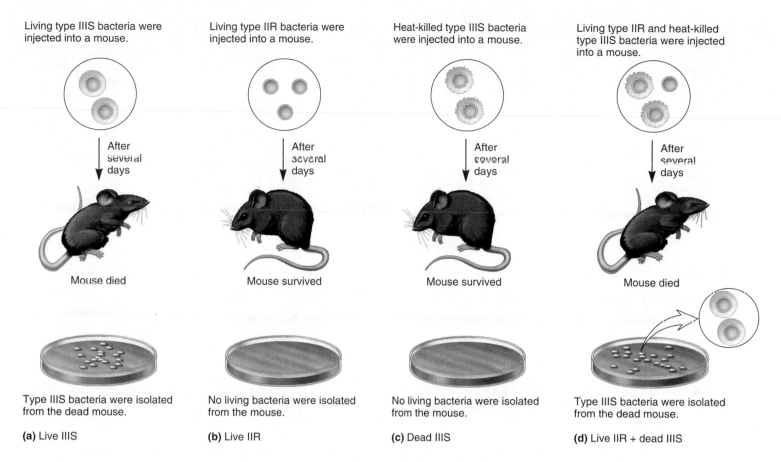

Living type IIIS bacteria were injected into a mouse.

After several days

Mouse died

Type IIIS bacteria were isolated from the dead mouse.

(a) Live IIIS

Living type IIR bacteria were injected into a mouse.

After several days

Mouse survived

No living bacteria were isolated from the mouse.

(b) Live IIR

Heat-killed type IIIS bacteria were injected into a mouse.

After several days

Mouse survived

No living bacteria were isolated from the mouse.

(c) Dead IIIS

Living type IIR and heat-killed type IIIS bacteria were injected into a mouse.

After several days

Mouse died

Type IIIS bacteria were isolated from the dead mouse.

(d) Live IIR + dead IIIS

FIGURE 9.2   **Griffith's experiments on genetic transformation in pneumococcus.**

bacteria were injected into a mouse, they did not cause a lethal infection (fig. 9.2*b*). To verify that the proliferation of the smooth bacteria was causing the death of the mouse, Griffith killed the smooth bacteria with heat treatment before injecting them into the mouse. In this case, the mouse survived (fig. 9.2*c*).

The critical and unexpected result was obtained in the experiment outlined in figure 9.2*d*. In this experiment, live type IIR bacteria were mixed with heat-killed type IIIS bacteria. As shown here, the mouse died. Furthermore, extracts from tissues of the dead mouse were found to contain living type IIIS bacteria! Since type IIR bacteria cannot revert to type IIIS, the interpretation of these data is that something from the dead type IIIS bacteria was transforming the type IIR bacteria into type IIIS. He called this process **transformation,** and the unidentified substance that was causing this to occur was termed the *transformation principle.* The steps of bacterial transformation are described in chapter 6 (see fig. 6.12).

At this point, let's look at what Griffith's observations mean in genetic terms. The transformed bacteria acquired the *information* to make a type III capsule. Among different strains, *variation* exists both in the ability to make a capsule and the type that is made. The genetic material for the type III trait must be *replicated* so that it can be *transmitted* from mother to daughter cells during cell division. Taken together, these observations are consistent with the idea that the formation of a capsule is governed by the

bacteria's genetic material, meeting the four criteria described previously. Significantly, Griffith's experiments showed that some genetic material from the dead bacteria was being transferred to the living bacteria and providing them with a new trait. However, Griffith did not know what the transforming substance was.

Important scientific discoveries often take place when researchers recognize that someone else's experimental observations can be used to address a particular scientific question. Oswald Avery, Colin MacLeod, and Maclyn McCarty realized that Griffith's observations could be used as part of an experimental strategy to identify the genetic material. They asked the question, What substance is being transferred from the dead type IIIS bacteria to the live type IIR? To answer this question, they incorporated additional biochemical techniques into their experimental methods as shown in figure 9.3.

At the time of these experiments in the 1940s, researchers already knew that DNA, RNA, proteins, and carbohydrates are major constituents of living cells. To separate these components, and to determine if any of them was the genetic material, Avery, MacLeod, and McCarty used established biochemical purification procedures and prepared bacterial extracts from type IIIS strains containing each type of molecule. Only one of the extracts, namely the one that contained purified DNA, was able to convert the type IIR bacteria into type IIIS. As shown in figure 9.3, when this extract was mixed with type IIR bacteria, some of

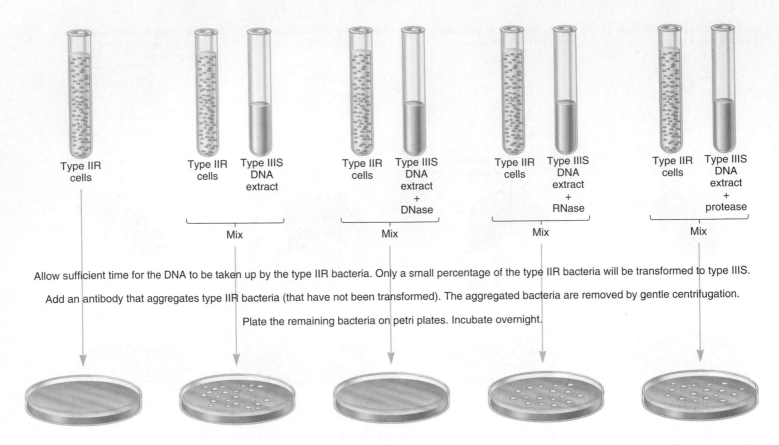

FIGURE 9.3   Experimental protocol used by Avery, MacLeod, and McCarty to identify the transformation principle. Antibodies are molecules that can specifically recognize the molecular structure of macromolecules. In this experiment, the antibodies recognized the cell surface of type IIR bacteria. When the antibodies recognized the type IIR bacteria, they cause them to clump together. The clumped bacteria were removed by a gentle centrifugation step. Only the bacteria that were not recognized by the antibody (namely, the type IIIS bacteria) remained in the supernatant. The cells in the supernatant were plated on solid growth media. After overnight incubation, visible colonies may be observed.

the bacteria were converted to type IIIS. However, if no DNA extract was added, no type IIIS bacterial colonies were observed on the petri plates.

In the experiment outlined in figure 9.3, a biochemist might point out that the DNA extract may not be 100% pure. In fact, any purified extract might contain small traces of some other substances. Therefore, one can argue that a small amount of contaminating material in the DNA extract might actually be the genetic material. The most likely contaminating substances in this case would be protein or RNA. To further verify that the DNA in the extract was indeed responsible for the transformation, Avery, MacLeod, and McCarty treated samples of the DNA extract with enzymes that digest DNA (called **DNase**), RNA

(**RNase**), or protein (**protease**). When the DNA extracts were treated with RNase or protease, they still converted type IIR bacteria into type IIIS. These results indicated that contaminating RNA or protein in the extract was not acting as the genetic material. However, when the extract was treated with DNase, its ability to convert type IIR into type IIIS was lost. These results indicated that the degradation of the DNA in the extract by DNase prevented conversion of type IIR to type IIIS. This interpretation is consistent with the hypothesis that DNA is the genetic material. A more elegant way of saying this is that the transforming principle is DNA.

EXPERIMENT 9A

Hershey and Chase Provided Evidence That the Genetic Material Injected into the Bacterial Cytoplasm Is T2 Phage DNA

A second experimental approach indicating that DNA is the genetic material came from the studies of Alfred Hershey and

Martha Chase in 1952. Their research centered on the study of a virus known as T2. This virus infects *Escherichia coli* bacterial cells and is therefore known as a **bacteriophage** or simply a **phage.** As shown in figure 9.4, the external structure of the T2 phage, known as the phage coat, contains a capsid, sheath, base plate, and tail fibers. Its structure is very similar to T4 phage,

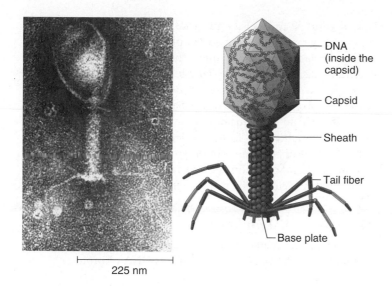

FIGURE 9.4 **Structure of the T2 bacteriophage.** The T2 bacteriophage is composed of a phage coat and genetic material inside the phage capsid. The phage coat is divided into regions called the capsid, sheath, base plate, and tail fibers. These components are composed of proteins. The genetic material is composed of DNA.

GENES→TRAITS   The genetic material of a bacteriophage contains many genes, which provide the blueprint for making new viruses. When the bacteriophage injects its genetic material into a bacterium, these genes are activated and direct the host cell to make new bacteriophages, as described in figure 9.5.

which was discussed in chapter 6. Biochemically, the phage coat is composed entirely of protein, which includes several different polypeptides. DNA is found inside the T2 capsid. From a molecular point of view, this virus is rather simple, since it is composed of only two types of macromolecules: DNA and proteins.

Although the viral genetic material contains the blueprint to make new viruses, a virus itself cannot synthesize new viruses. Instead, a virus must introduce its genetic material into the cytoplasm of a living cell. In the case of T2, this first involves the attachment of its tail fibers to the bacterial cell wall and the subsequent injection of its genetic material into the cytoplasm of the cell (fig. 9.5). The phage coat remains attached on the outside of the bacterium and does not enter the cell. After the entry of the viral genetic material, the bacterial cytoplasm provides all the synthetic machinery necessary to make viral proteins and DNA. The viral proteins and DNA assemble to make new viruses that are subsequently released from the cell by **lysis** (i.e., cell breakage).

To verify that DNA is the genetic material of T2, Hershey and Chase devised a method to separate the phage coat, which is attached to the outside of the bacterium, from the DNA, which is injected into the cytoplasm. They were aware of microscopy experiments by T. F. Anderson showing that the T2 phage attaches itself to the outside of a bacterium by its tail fibers. Hershey and Chase reasoned that this is a fairly precarious attachment that could be disrupted by subjecting the bacteria to high shear forces such as those produced in a blender. Their method was to expose bacteria to T2 phage, allowing sufficient time for the viruses to attach

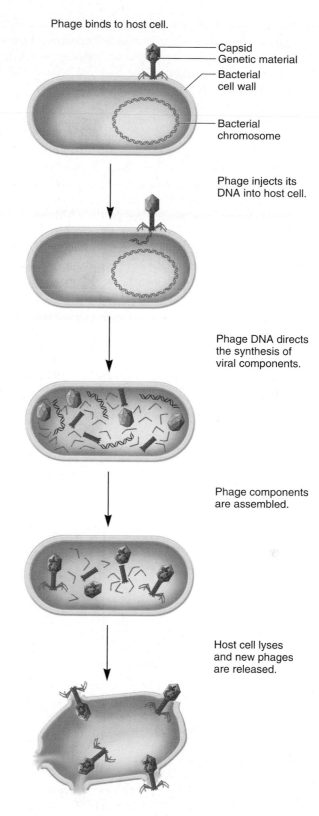

FIGURE 9.5 **Life cycle of the T2 bacteriophage.**

to bacteria and inject their genetic material. They then sheared the phage coats from the surface of the bacteria by a blender treatment. In this way, the phages' genetic material, which had been injected into the cytoplasm of the bacterial cells, could be separated from the phage coats that were sheared away.

Hershey and Chase used radioisotopes to distinguish DNA from proteins. Sulfur atoms are found in proteins but not in DNA, whereas phosphorus atoms are found in DNA but not in phage proteins. Therefore, ^{35}S (a radioisotope of sulfur) and ^{32}P (a radioisotope of phosphorus) were used to specifically label proteins and DNA, respectively. Researchers can grow *E. coli* cells in media that contain ^{35}S or ^{32}P and then infect the *E. coli* cells with T2 phages. When new phages are produced, these will be labeled with ^{35}S or ^{32}P. In the experiment described in figure 9.6, they began with *E. coli* cells and two preparations of T2 phage that were obtained in this manner. One preparation was labeled with ^{35}S to label the phage proteins; the other preparation was

labeled with ^{32}P to label the phage DNA. In separate flasks, each type of phage was mixed with a sample of *E. coli* cells. The phages were given sufficient time to inject their genetic material into the bacterial cells, and then the sample was subjected to shearing force using a blender. This treatment is expected to remove the phage coat from the surface of the bacterial cell. The sample was then subjected to centrifugation at a speed that would cause the heavier bacterial cells to form a pellet at the bottom of the tube, while the light phage coats would remain in the supernatant. The amount of radioactivity in the supernatant (emitted from either ^{35}S or ^{32}P) was determined using a scintillation counter.

■ THE HYPOTHESIS

Only the genetic material of the phage is injected into the bacterium. Isotope labeling will reveal if it is DNA or protein.

■ TESTING THE HYPOTHESIS — FIGURE 9.6 Evidence that DNA is the genetic material of T2 bacteriophage.

Starting materials: The starting materials are *E. coli* cells and two preparations of T2 phage. One preparation is labeled with ^{35}S to label the phage proteins; the other preparation is labeled with ^{32}P to label the phage DNA.

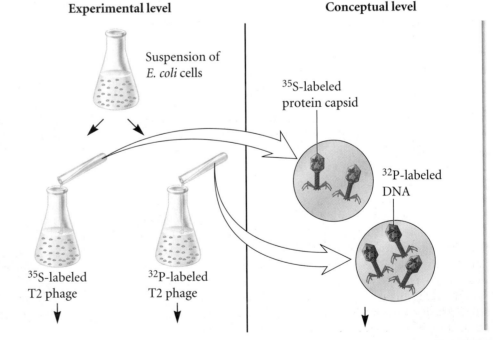

Experimental level   Conceptual level

1. Grow bacterial cells. Divide into two flasks.

Suspension of *E. coli* cells

^{35}S-labeled protein capsid

^{32}P-labeled DNA

2. Into one flask, add ^{35}S-labeled phage; in the second flask, add ^{32}P-labeled phage.

^{35}S-labeled T2 phage

^{32}P-labeled T2 phage

3. Allow infection to occur.

(continued)

4. Agitate solutions in blenders for different lengths of time to shear the empty phages off the bacterial cells.

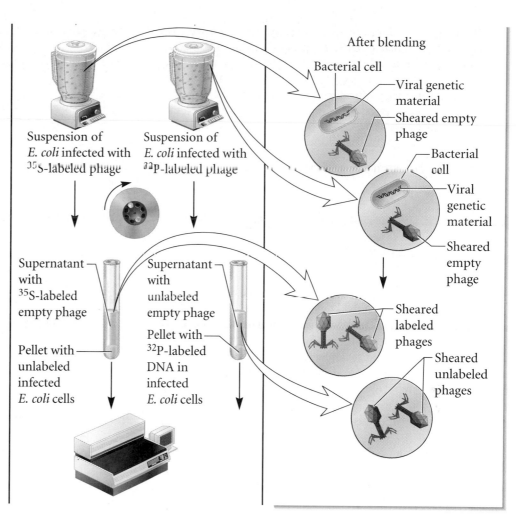

5. Centrifuge at 10,000 rpm.

6. The heavy bacterial cells sediment to the pellet, while the lighter phages remain in the supernatant. (See appendix for explanation of centrifugation.)

7. Count the amount of radioisotope in the supernatant with a scintillation counter (see appendix). Compare it with the starting amount.

■ THE DATA

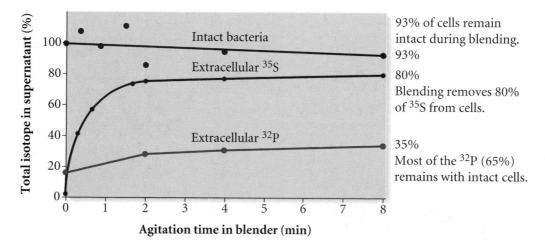

■ INTERPRETING THE DATA

As seen in the data, most of the ^{35}S isotope was found in the supernatant. Since the shearing force was expected to remove the phage coat, this result indicates that the empty phages contain primarily protein. By comparison, only about 35% of the ^{32}P was found in the supernatant following shearing. Therefore, most of the DNA was located within the bacterial cells in the pellet. These results are consistent with the idea that the DNA is injected into the bacterial cytoplasm during infection. This is the expected result if DNA is the genetic material.

By themselves, the results described in figure 9.6 were not conclusive evidence that DNA is the genetic material. For example, you may have noticed that less than 100% of the phage protein was found in the supernatant. Therefore, some of the phage protein could have been introduced into the bacterial cells (and could function as the genetic material). Nevertheless, the results of Hershey and Chase were consistent with the conclusion that the genetic material is DNA rather than protein. Overall, their studies of the T2 phage were quite influential in promoting the belief that DNA is the genetic material.

A self-help quiz involving this experiment can be found at the Online Learning Center.

RNA Functions as the Genetic Material in Some Viruses

We now know that bacteria, protozoa, fungi, algae, plants, and animals all use DNA as their genetic material. As mentioned, viruses also have their own genetic material. Hershey and Chase concluded from their experiments that this genetic material is DNA. In the case of T2 bacteriophage, that is the correct conclusion. However, many viruses use RNA, rather than DNA, as their genetic material. In 1956, A. Gierer and G. Schramm isolated RNA from the tobacco mosaic virus (TMV), which infects plant cells. When this purified RNA was applied to plant tissue, the plants developed the same types of lesions that occurred when they were exposed to intact tobacco mosaic viruses. Gierer and Schramm correctly concluded that the viral genome of tobacco mosaic virus is composed of RNA. Since that time, many other viruses have been found to contain RNA as their genetic material. Table 9.1 compares the genetic compositions of several different types of viruses.

TABLE 9.1
Examples of DNA- and RNA-Containing Viruses

Virus	Host	Nucleic Acid
Tomato bushy stunt virus	Tomato	RNA
Tobacco mosaic virus	Tobacco	RNA
Influenza virus	Humans	RNA
HIV	Humans	RNA
f2	*E. coli*	RNA
Qβ	*E. coli*	RNA
Cauliflower mosaic virus	Cauliflower	DNA
Herpes virus	Humans	DNA
SV40	Primates	DNA
Epstein-Barr virus	Humans	DNA
T2	*E. coli*	DNA
M13	*E. coli*	DNA

9.2 NUCLEIC ACID STRUCTURE

Geneticists, biochemists, and biophysicists have been interested in the molecular structure of nucleic acids for many decades. Both DNA and RNA are macromolecules composed of smaller building blocks. To fully appreciate their structures, we need to consider several levels of complexity (fig. 9.7):

1. **Nucleotides** form the repeating structural unit of nucleic acids.
2. Nucleotides are linked together in a linear manner to form a **strand** of DNA or RNA.
3. Two strands of DNA (and sometimes RNA) can interact with each other to form a **double helix.**
4. The three-dimensional structure of DNA results from the folding and bending of the double helix. Within living

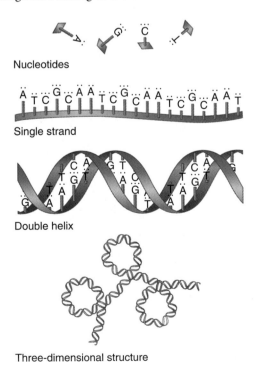

Nucleotides

Single strand

Double helix

Three-dimensional structure

FIGURE 9.7   Levels of nucleic acid structure.

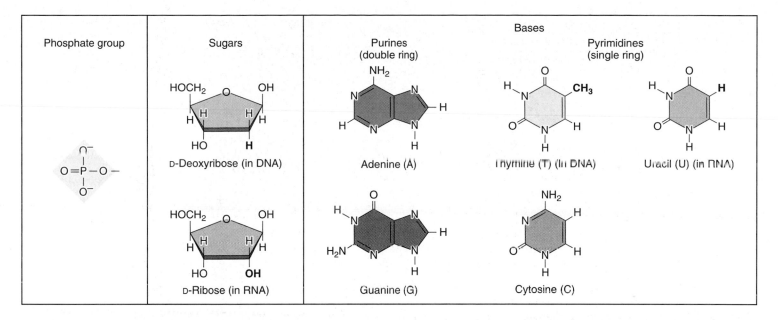

Phosphate group	Sugars	Bases

FIGURE 9.8 **The components of nucleotides.** The three building blocks of a nucleotide are one or more phosphate groups, a sugar, and a base. The bases are categorized as purines (adenine and guanine) and pyrimidines (thymine, uracil, and cytosine).

cells, DNA is associated with a wide variety of proteins that influence its structure. Chapter 10 is devoted to the roles of these proteins and the three-dimensional structure of DNA found within chromosomes.

In this section, we will first examine the structure of individual nucleotides and then progress through the structural features of DNA and RNA. Along the way, we will also review some of the pivotal experiments that led to the discovery of the double helix.

Nucleotides Are the Building Blocks of Nucleic Acids

The nucleotide is the repeating structural unit of DNA and RNA. A nucleotide has three components: a phosphate group, a pentose sugar, and a nitrogenous base. As shown in figure 9.8, nucleotides can vary with regard to the sugar and the nitrogenous base. There are two types of sugars, ribose and deoxyribose. The five different bases are subdivided into two categories, the **purines** and the **pyrimidines.** The purine bases, **adenine (A)** and **guanine (G),** contain a double-ring structure; the pyrimidine bases, **cytosine (C), thymine (T),** and **uracil (U),** contain a single-ring structure. The chemical differences between nucleotides are important in distinguishing DNA and RNA (fig. 9.9). For example, the sugar in DNA is always **deoxyribose;** in RNA, it is **ribose.** Also, the base thymine is not found in RNA. Rather, uracil is found in RNA instead of thymine. Adenine, cytosine, and guanine occur in both DNA and RNA.

The terminology used to describe nucleic acid units is based on three structural features: the type of base, the type of sugar, and the number of phosphate groups. When a base is attached to only a sugar, we call this pair a **nucleoside.** For example, if adenine is attached to ribose, this is called adenosine (fig. 9.10). Like-

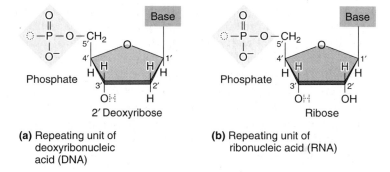

(a) Repeating unit of deoxyribonucleic acid (DNA)

(b) Repeating unit of ribonucleic acid (RNA)

FIGURE 9.9 **The structure of nucleotides found in (a) DNA and (b) RNA.** DNA contains deoxyribose as its sugar and the bases A, T, G, and C. RNA contains ribose as its sugar and the bases A, U, G, and C. In a DNA or RNA strand, the oxygen on the 3′ carbon is linked to the phosphorus atom of phosphate in the adjacent nucleotide. The two atoms (O and H) shown with *dotted lines* would be found within individual nucleotides but would be removed when nucleotides are joined together to make strands of DNA and RNA.

wise, nucleosides containing guanine, thymine, cytosine, or uracil are called guanosine, thymidine, cytidine, and uridine, respectively. If the five bases are attached to deoxyribose, they would be called deoxyguanosine, deoxythymidine, deoxycytidine, deoxyadenosine, and deoxyuridine. The covalent attachment of one or more phosphate molecules to a nucleoside creates a nucleotide. If a nucleotide contains adenine, ribose, and three phosphate groups, it is called adenosine triphosphate, abbreviated ATP. Or if it contains guanine, ribose, and three phosphate groups, it is guanosine triphosphate, or GTP. If a nucleotide contains adenine, ribose, and one phosphate, it is adenosine monophosphate (AMP). A

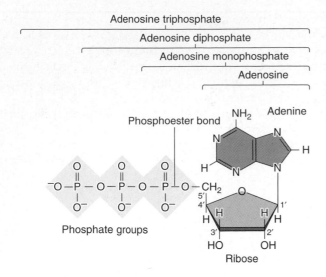

FIGURE 9.10 **A comparison between the structures of an adenine-containing nucleoside and nucleotides.**

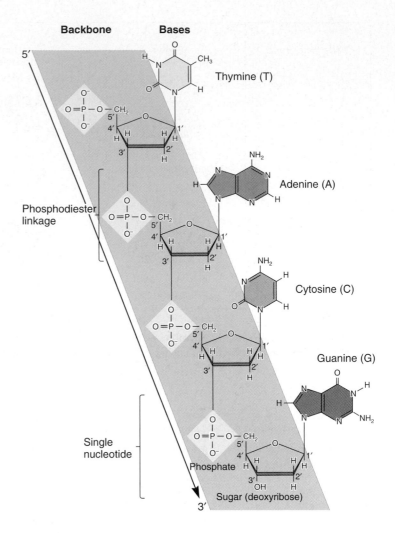

FIGURE 9.11 **A short strand of DNA containing four nucleotides.** Nucleotides are covalently linked together to form a strand of DNA.

nucleotide can also be composed of adenine, deoxyribose, and three phosphate groups. This nucleotide is deoxyadenosine triphosphate.

The locations of the attachment sites of the base and phosphate to the sugar molecule are important to the nucleotide's function. Figures 9.9 and 9.10 illustrate the conventional numbering of the carbon atoms in a pentose sugar. In the sugar ring, carbon atoms are numbered in a clockwise direction, beginning with a carbon atom adjacent to the ring oxygen atom. The fifth carbon is outside the ring structure. These numbers are designated 1′ (i.e., one prime), 2′, 3′, 4′, and 5′. In a single nucleotide, the base is always attached to the 1′ carbon atom and one or more phosphate groups are attached at the 5′ position. As discussed in the next section, the −OH group attached to the 3′ carbon is important in allowing nucleotides to form covalent linkages with each other.

Nucleotides Are Linked Together to Form a Strand

A strand of DNA or RNA contains nucleotides that are covalently attached to each other in a linear fashion. Figure 9.11 depicts a short strand of DNA with four nucleotides. A few structural features here are worth noting. First, the linkage involves an ester bond between a phosphate group on one nucleotide and the sugar molecule on the adjacent nucleotide. Another way of viewing this linkage is to notice that a phosphate group is connecting two sugar molecules. For this reason, the linkage in DNA or RNA strands is called a **phosphodiester linkage.** The phosphates and sugar molecules form the **backbone** of a DNA or RNA strand. The bases project from the backbone. The backbone is negatively charged due to a negative charge on each phosphate.

A second important structural feature is the orientation of the nucleotides. As mentioned, the carbon atoms in a sugar molecule are numbered in a particular way. A phosphodiester linkage

involves a phosphate attachment to the 5′ carbon in one nucleotide and to the 3′ carbon in the other. In a strand, all sugar molecules are oriented in the same direction. As shown in figure 9.11, all the 5′ carbons in every sugar molecule are above the 3′ carbons. Therefore, a strand has a **directionality** based on the orientation of the sugar molecules within that strand. In figure 9.11, the direction of the strand is 5′ to 3′ when going from top to bottom.

A critical aspect regarding DNA and RNA structure is that a strand contains a specific sequence of bases. In figure 9.11, the sequence of bases is thymine–adenine–cytosine–guanine. This sequence is abbreviated TACG. Furthermore, to show the directionality, the strand should be abbreviated 5′–TACG–3′. The nucleotides within a strand are covalently attached to each other so that the sequence of bases cannot shuffle around and become rearranged. Therefore, the sequence of bases in a DNA strand will remain the same over time, except in rare cases when mutations occur. As we will see throughout this textbook, the sequence of bases within DNA and RNA is the defining feature that allows them to carry information.

A Few Key Events Led to the Discovery of the Double-Helix Structure

A discovery of paramount importance in molecular genetics was made in 1953 by James Watson and Francis Crick. At that time, DNA was already known to be composed of nucleotides. However, it was not understood how the nucleotides are bonded together to form the structure of DNA. Watson and Crick committed themselves to determine the structure of DNA because they felt this knowledge was needed to understand the functioning of genes. Others, like Rosalind Franklin and Maurice Wilkins, shared this view. Before we examine the characteristics of the double helix, let's consider the events that provided the scientific framework for Watson and Crick's breakthrough.

In the early 1950s, Linus Pauling proposed that regions of proteins can fold into a secondary structure known as an α helix (fig. 9.12a). To elucidate this structure, Pauling built large models by linking together simple ball-and-stick units (fig. 9.12b). By carefully scaling the objects in his models, he could visualize if atoms fit together properly in a complicated three-dimensional structure. This approach is still widely used today except that

researchers construct their putative three-dimensional models on computers. As we will see, Watson and Crick used a ball-and-stick approach to solve the structure of the DNA double helix. Interestingly, they were well aware that Pauling might figure out the structure of DNA before they could. This provided a stimulating rivalry between the researchers.

A second important development that led to the elucidation of the double helix was X-ray diffraction data. When a purified substance, such as DNA, is subjected to X rays, it will produce a well-defined diffraction pattern if the molecule is organized into a regular structural pattern. An interpretation of the diffraction pattern (using mathematical theory) can ultimately provide information concerning the structure of the molecule. Rosalind Franklin (fig. 9.13a), working in the same laboratory as Maurice Wilkins, used X-ray diffraction to study wet DNA

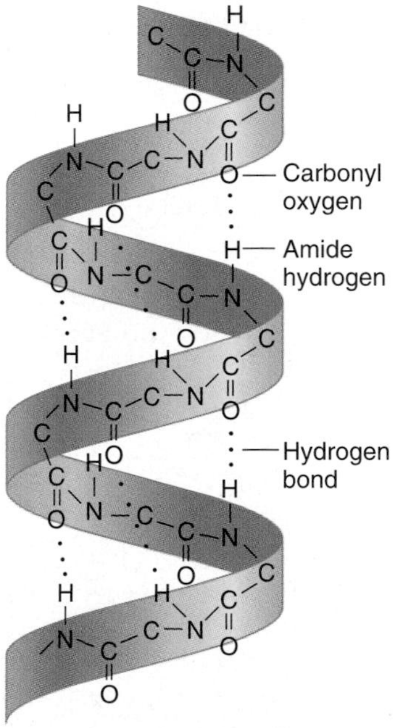

Carbonyl oxygen

Amide hydrogen

Hydrogen bond

(a) An α helix in a protein    **(b)** Linus Pauling

FIGURE 9.12 **Linus Pauling and the α-helix protein structure.** (a) An α helix is a secondary structure found in proteins. This structure emphasizes the polypeptide backbone (shown as a *tan ribbon*), which is composed of amino acids linked together in a linear fashion. Hydrogen bonding between hydrogen and oxygen atoms stabilizes the helical conformation. (b) Linus Pauling with a ball-and-stick model.

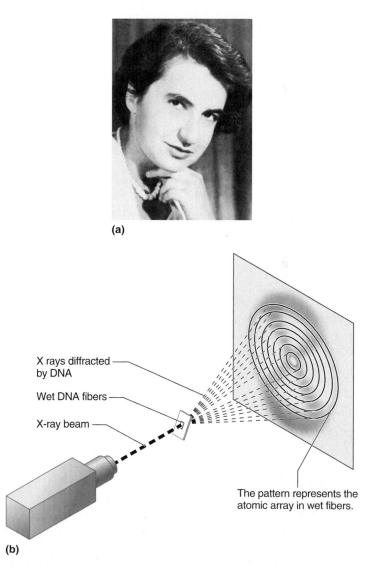

(a)

X rays diffracted by DNA

Wet DNA fibers

X-ray beam

The pattern represents the atomic array in wet fibers.

(b)

FIGURE 9.13 **X-ray diffraction of DNA.** (a) Rosalind Franklin. (b) A diffraction pattern of wet DNA fibers.

fibers. Franklin made marked advances in X-ray diffraction techniques while working with DNA. She adjusted her equipment to produce an extremely fine beam of X rays. She extracted finer DNA fibers than ever before and arranged them in parallel bundles. And she studied the fibers' reactions to humid conditions. The diffraction pattern of Franklin's DNA fibers is shown in figure 9.13*b*. This pattern suggested several structural features of DNA. First, it was consistent with a helical structure. Second, the diameter of the helical structure was too wide to be only a single-stranded helix. Finally, the diffraction pattern indicated that the helix contains about 10 base pairs per complete turn. These observations were instrumental in solving the structure of DNA.

EXPERIMENT 9B

Chargaff Found That DNA Has a Biochemical Composition in Which the Amount of A Equals T and the Amount of G Equals C

Another piece of information that led to the discovery of the double-helix structure came from the studies of Erwin Chargaff. He pioneered many of the biochemical techniques for the isolation, purification, and measurement of nucleic acids from living cells. This is not a trivial undertaking because the biochemical composition of living cells is complex. At the time of Chargaff's work, it was already known that the building blocks of DNA are nucleotides containing the bases adenine, thymine, cytosine, or guanine. Chargaff analyzed the base composition of DNA, which was isolated from many different species. He expected that the results might provide important clues concerning the structure of DNA.

The experimental protocol of Chargaff is described in figure 9.14. He began with various types of cells as starting material. The chromosomes were extracted from cells and then treated with protease to separate the DNA from chromosomal proteins. The DNA was then subjected to a strong acid treatment that cleaved the bonds between the sugars and bases. Therefore, the strong acid treatment would release the individual bases from the DNA strands. This mixture of bases was then subjected to paper chromatography to separate the four types. The amounts of the four bases were determined spectroscopically.

■ THE HYPOTHESIS

An analysis of the base composition of DNA in different organisms may reveal important features about the structure of DNA.

■ TESTING THE HYPOTHESIS — FIGURE 9.14 An analysis of base composition among different DNA samples.

Starting material: The following types of cells were obtained: *Escherichia coli*, *Streptococcus pneumoniae* (type III), yeast, turtle red blood cells, salmon sperm cells, chicken red blood cells, and human liver cells.

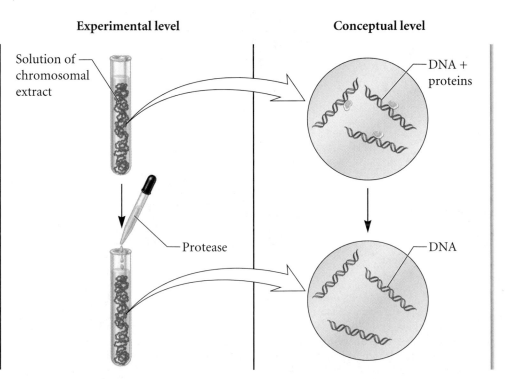

1. For each type of cell, extract the chromosomal material. This can be done in a variety of ways including the use of high salt, detergent, or mild alkali treatment. Note: The chromosomes contain both DNA and protein.

2. Remove the protein. This can be done in several ways including chloroform extraction or treament with protease.

(continued)

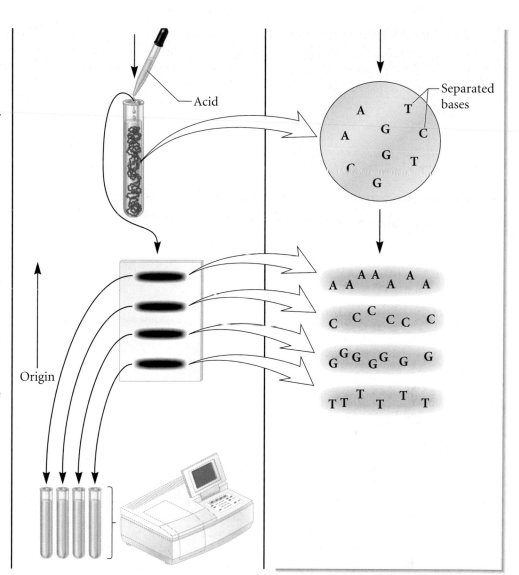

3. Hydrolize the DNA to remove the bases. A common way to do this is by strong acid treatment.

4. Separate the bases by chromatography. Paper chromatography provides an easy way to separate the four types of bases. (The technique of chromatography is described in the appendix.)

5. Extract bands from paper into solutions and determine the amounts of each base by spectroscopy. Each base will absorb light at a particular wavelength. By examining the absorption profile of a sample of base, it is then possible to calculate the amount of the base. (Spectroscopy is described in the appendix.)

6. Compare the results of DNA samples from different organisms.

■ THE DATA

Base Content in the DNA from a Variety of Organisms*

Organism	% of Bases (based on molarity)			
	Adenine	Thymine	Guanine	Cytosine
Escherichia coli	26.0	23.9	24.9	25.2
Streptococcus pneumoniae (type III)	29.8	31.6	20.5	18.0
Yeast	31.7	32.6	18.3	17.4

(continued)

Organism	% of Bases (based on molarity)			
	Adenine	Thymine	Guanine	Cytosine
Turtle red blood cells	28.7	27.9	22.0	21.3
Salmon sperm	29.7	29.1	20.8	20.4
Chicken red blood cells	28.0	28.4	22.0	21.6
Human liver cells	30.3	30.3	19.5	19.9

*When the base contents from different tissues within the same species were measured, similar results were obtained.

INTERPRETING THE DATA

The data shown in figure 9.14 are only a small sampling of Chargaff's results. Chargaff published many papers concerned with the chemical composition of DNA from biological sources. Hundreds of measurements were made. The compelling observation was that the amount of adenine was similar to thymine and the amount of cytosine was similar to guanine. These results were not sufficient to propose a model for the structure of DNA. However, they provided the important clue that DNA is structured so

that each molecule of adenine interacts with thymine, and each molecule of guanine interacts with cytosine. A DNA structure in which A binds to T, and G to C, would explain the equal amounts of A and T, and G and C observed in Chargaff's experiments. As we will see, this observation became crucial evidence that Watson and Crick used to elucidate the structure of the double helix.

A self-help quiz involving this experiment can be found at the Online Learning Center.

Watson and Crick Deduced the Double-Helical Structure of DNA

Thus far, we have examined key pieces of information used to determine the structure of DNA. In particular, the X-ray diffraction work of Franklin suggested a helical structure composed of two or more strands with 10 bases per turn. In addition, the work of Chargaff indicated that the amount of A equals T and the amount of G equals C. Furthermore, Watson and Crick were familiar with Pauling's success in using ball-and-stick models to deduce the secondary structure of proteins. With these key observations, they set out to solve the secondary structure of DNA.

Watson and Crick assumed that DNA is composed of nucleotides that are linked together in a linear fashion. They also assumed that the chemical linkage between two nucleotides is always the same. With these ideas in mind, they tried to build ball-and-stick models that incorporated all the known experimental observations. Since the diffraction pattern suggested that there must be two (or more) strands within the helix, a critical question was how two strands could interact. As discussed in the book *The Double Helix* (by James Watson), in an early attempt at model building, Watson and Crick considered the possibility that the negatively charged phosphate groups together with magnesium ions were promoting an interaction between the backbones of DNA strands (fig. 9.15). However, more detailed diffraction data were not consistent with this model.

Because the magnesium hypothesis for DNA structure appeared to be incorrect, it was back to the drawing board (or back to the ball-and-stick units) for Watson and Crick. During

this time, Rosalind Franklin had produced even clearer X-ray diffraction patterns, which provided greater detail concerning the relative locations of the bases and backbone of DNA. This was a major breakthrough that suggested a two-strand interaction that was helical. In their model building, the emphasis shifted to models containing the two backbones on the outside of the model, with the bases projecting toward each other. At first, a structure was considered in which the bases form hydrogen bonds with the identical base in the opposite strand (i.e., A to A, T to T, G to G, and C to C). However, the model building revealed that the bases could not fit together this way. The final hurdle was

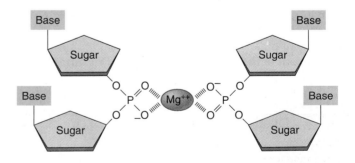

FIGURE 9.15   An incorrect hypothesis for the structure of the DNA double helix. This illustration shows an early hypothesis of Watson and Crick's suggesting that two DNA strands interact by a cross-link between the negatively charged phosphate groups in the backbone and divalent Mg^{++} cations.

(a) Watson and Crick

FIGURE 9.16   **Watson and Crick and their model of the DNA double helix.** (a) James Watson is shown here on the *left* and Francis Crick on the *right*. (b) The molecular model they originally proposed for the double helix in which A hydrogen bonds to T and G hydrogen bonds with C.

(b) Original model of the DNA double helix

overcome when it was realized that the hydrogen bonding of adenine to thymine was structurally similar to that of cytosine to guanine. With an interaction between A and T and between G and C, the ball-and-stick models showed that the two strands would fit together properly. This ball-and-stick model, shown in figure 9.16, was consistent with all the known data regarding DNA structure.

For their work, Watson, Crick, and Maurice Wilkins (discussed earlier) were awarded the Nobel Prize in 1962. The contribution of Rosalind Franklin to the discovery of the double helix was also critical and has been discussed in several books. Franklin was independently trying to solve the structure of DNA. However, Wilkins, who worked in the same laboratory, shared the X-ray data of Franklin with Watson and Crick, presumably without her knowledge. Working with their ball-and-stick models, this provided important information that helped them solve the structure of DNA, which was published in the journal *Nature*. Though she was not given credit in the original publication of

the double-helix structure, her key contribution became known in later years. Unfortunately, however, Rosalind Franklin died before 1962, and the Nobel Prize is not awarded posthumously.

The Molecular Structure of the DNA Double Helix Has Several Key Features

The general structural features of the double helix are shown in figure 9.17. In a DNA double helix, two DNA strands are twisted together around a common axis to form a structure that resembles a circular staircase. There are 10 **base pairs** within a complete twist. This double-stranded structure is stabilized by hydrogen bonding between the base pairs (shown as *dotted lines*). A distinguishing feature of this hydrogen bonding is its specificity. An adenine base in one strand hydrogen bonds with a thymine base in the opposite strand, or a guanine base hydrogen bonds with a cytosine. This **AT/GC** rule (also known as Chargaff's rule)

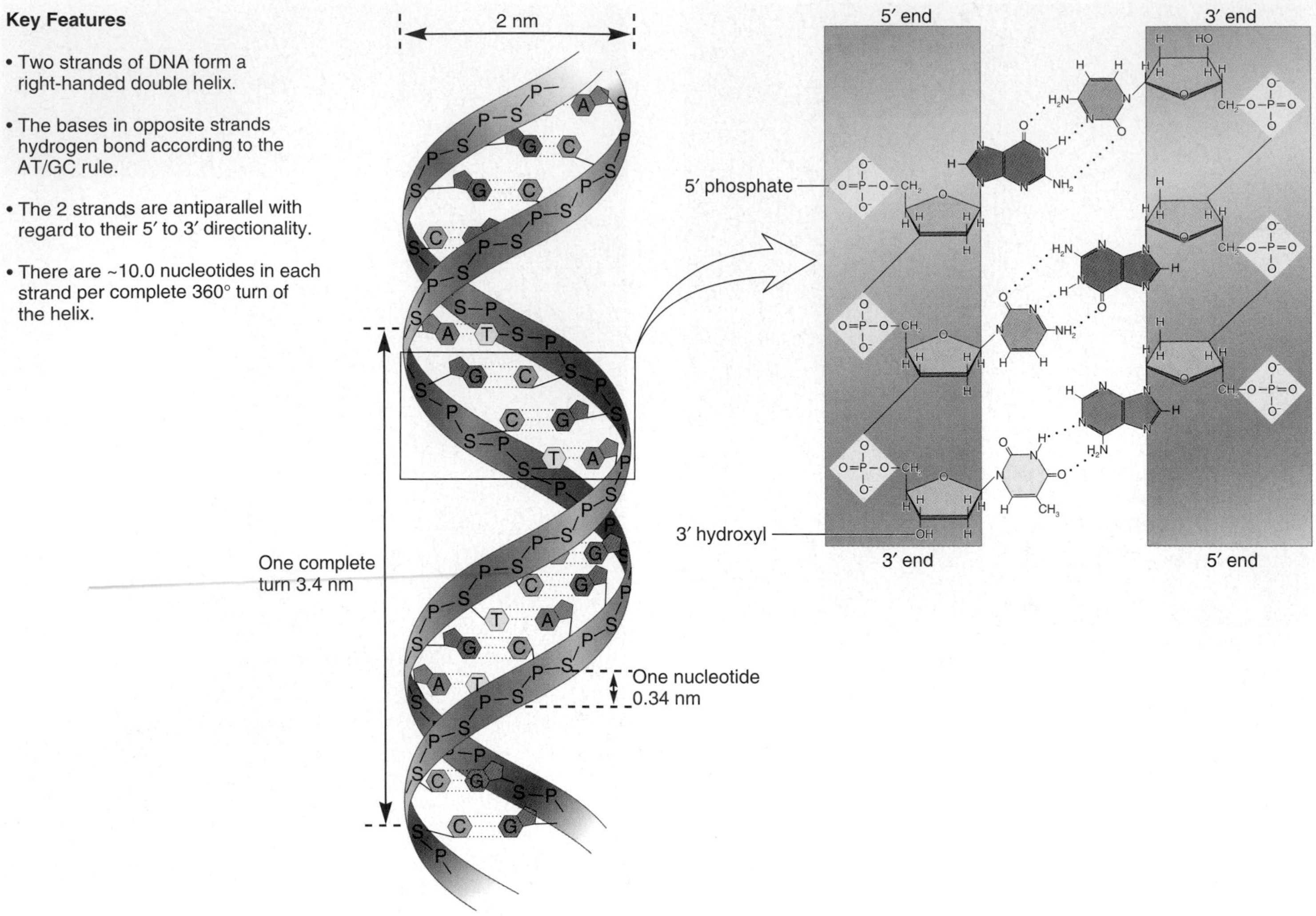

Key Features

- Two strands of DNA form a right-handed double helix.

- The bases in opposite strands hydrogen bond according to the AT/GC rule.

- The 2 strands are antiparallel with regard to their 5′ to 3′ directionality.

- There are ~10.0 nucleotides in each strand per complete 360° turn of the helix.

FIGURE 9.17   Key features of the structure of the double helix.

explained the earlier data of Chargaff showing that the DNA from many organisms contains equal amounts of A and T, and equal amounts of G and C (see fig. 9.14). The AT/GC rule indicates that purines (A and G) always bond with pyrimidines (T and C). This keeps the width of the double helix relatively constant. As noted in figure 9.17, there are three hydrogen bonds between G and C but only two between A and T. For this reason, DNA sequences that have a high proportion of G and C tend to form more stable double-stranded structures.

The AT/GC rule implies that base sequences within two DNA strands are **complementary** to each other. We can predict the sequence in one DNA strand if the sequence in the opposite strand is known. For example, if one strand has the sequence of 5′–ATGGCGGATTT–3′, then the opposite strand must be 3′–TACCGCCTAAA–5′. In genetic terms, we would say that these two sequences are complementary to each other, or that the

two sequences exhibit complementarity. In addition, you may have noticed that the sequences are labeled with 5′ and 3′ ends. These numbers designate the direction of the DNA backbones. The direction of DNA strands is depicted in the inset to figure 9.17. When going from the top of this figure to the bottom, one strand is running in the 5′ to 3′ direction while the other strand is 3′ to 5′. This opposite orientation of the two DNA strands is referred to as an **antiparallel** arrangement. An antiparallel structure was initially proposed in the models of Watson and Crick.

Figure 9.18*a* is a schematic model that emphasizes certain molecular features of DNA structure. The DNA backbone is what forms a helical structure. Counting the bases, if you move past 10 base pairs, you have gone 360° around the backbone. In addition, the bases in this model are depicted as flat rectangular structures that are hydrogen bonding in pairs. (The hydrogen bonds are the *dotted lines.*) Although the bases are not actually rectangular, they

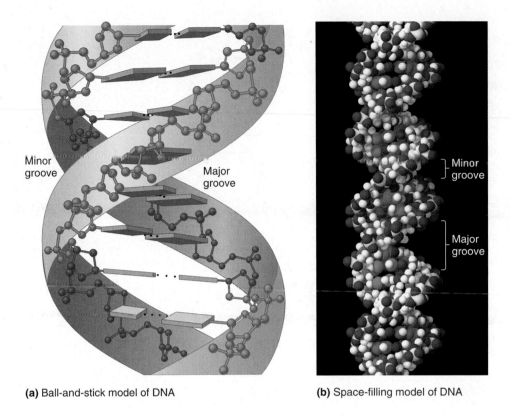

(a) Ball-and-stick model of DNA **(b)** Space-filling model of DNA

FIGURE 9.18 **Two models of the double helix.** (a) Ball-and-stick model of the double helix. The ribose–phosphate backbone is shown in detail, while the bases are depicted as *flattened rectangles.* (b) Space-filling model of the double helix.

do form flattened planar structures. Within DNA, the bases are oriented so that the flattened regions are facing each other. This is referred to as base stacking. In other words, if you think of the bases as flat plates, these plates are stacked on top of each other in the double-stranded DNA structure. Along with hydrogen bonding, base stacking is a structural feature that stabilizes the double helix.

By convention, the direction of the DNA double helix shown in figure 9.18a spirals in a direction that is called "right-handed." To understand this terminology, imagine that a double helix is laid on your desk; one end of the helix is close to you and the other end is at the opposite side of the desk. As it spirals *away* from you, a right-handed helix turns in a clockwise direction. By comparison, a left-handed helix would spiral in a counterclockwise manner. Both strands in figure 9.18a spiral in a right-handed direction.

Figure 9.18b is a space-filling model for DNA in which the atoms are represented by spheres. This model emphasizes what the surface of DNA is like. Note that the backbone (composed of sugar and phosphate groups, respectively) is on the outermost surface. In a living cell, the backbone has the most direct contact with water. In contrast, the bases are more internally located within the double-stranded structure. Biochemists use the term

grooves to describe the indentations where the atoms of the bases are in contact with the surrounding water. In the helical structure of DNA, there are actually two grooves as you go around the structure. These are called the **major groove** and the **minor groove.** As will be discussed in later chapters, certain proteins can bind within these grooves and interact with a particular sequence of bases, thereby regulating gene expression.

DNA Can Form Alternative Types of Double Helices

The DNA double helix can form different types of secondary structures. Figure 9.19 compares the structures of **A DNA, B DNA,** and **Z DNA.** The highly detailed structures shown here were deduced by X-ray crystallography on short segments of DNA. B DNA is the predominant form of DNA found in living cells. However, under certain in vitro conditions, the two strands of DNA can twist into A DNA or Z DNA, which differ significantly from B DNA. A and B DNA are right-handed helices; Z DNA has the opposite, left-handed, orientation. In addition, the helical backbone in Z DNA appears to slightly zigzag as it winds itself around the double-helical structure. The numbers of base pairs per 360° turn are 11.0, 10.0, and 12.0 in A, B, and Z DNA,

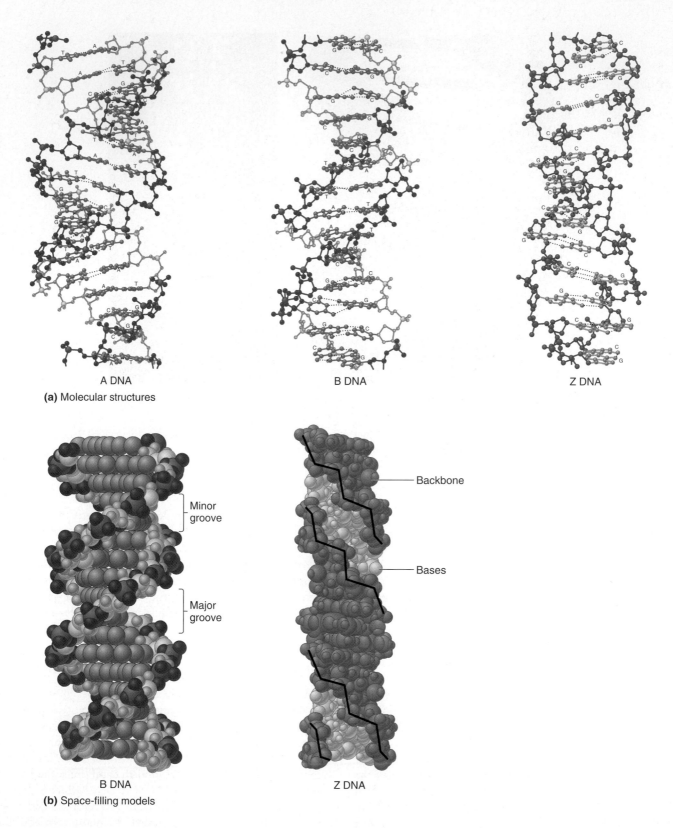

FIGURE 9.19   Comparison of the structures of A DNA, B DNA, and Z DNA. (a) The highly detailed structures shown here were deduced by X-ray crystallography performed on short segments of DNA, rather than the less detailed structures obtained from DNA wet fibers. The diffraction pattern obtained from the crystallization of short segments of DNA provides much greater detail concerning the exact placement of atoms within a double-helical structure. Alexander Rich, Richard Dickerson, and their colleagues were the first researchers to crystallize a short piece of DNA. (b) Space-filling models of the B-DNA and Z-DNA structures. In the case of Z DNA, the *bold line* connects the phosphate groups in the DNA backbone. As seen here, they travel along the backbone in a zigzag pattern.

respectively. In B DNA, the bases tend to be centrally located and the hydrogen bonds between base pairs occur relatively perpendicular to the central axis. In contrast, the bases in A DNA and Z DNA are substantially tilted relative to the central axis.

The ability of the predominant B DNA to adopt A DNA and Z DNA conformations depends on certain conditions. In X-ray diffraction studies, A DNA occurs under conditions of low humidity. The ability of a double helix to adopt a Z DNA conformation depends on various factors. At high ionic strength (i.e., high salt concentration), formation of a Z DNA conformation is favored by a sequence of bases that alternates between purines and pyrimidines. One such sequence is

```
5'-GCGCGCGCG-3'
3'-CGCGCGCGC-5'
```

At lower ionic strength, the methylation of cytosine bases can favor Z-DNA formation. Cytosine **methylation** occurs when a cellular enzyme attaches a methyl group (–CH$_3$) to the cytosine base. In addition, negative supercoiling (a topic to be discussed in chapter 10) favors the Z-DNA conformation. An important issue is the functional significance of A and Z DNA. There is little evidence to indicate that A DNA is biologically important, but several studies in recent years have suggested a possible role for Z DNA in the processes of transcription and recombination.

DNA Can Form a Triple Helix, Called Triplex DNA

A surprising discovery made by Alexander Rich, David Davies, and Gary Felsenfeld during the late 1950s was that DNA can form a triple-helical structure called **triplex DNA.** This triplex was formed in vitro using pieces of DNA that were made synthetically. Although this result was interesting, it seemed to have little, if any, biological relevance.

About 30 years later, interest in triplex DNA was renewed by the observation that triplex DNA can form in vitro by mixing natural double-stranded DNA and a third short strand that is synthetically made. The synthetic strand binds into the major groove of the naturally occurring double-stranded DNA (fig. 9.20). As shown here, an interesting feature of triplex DNA formation is that it is sequence specific. In other words, the synthetic third strand incorporates itself into a triple helix only due to specific interactions between the synthetic DNA and the biological DNA. The pairing rules are that a thymine in the synthetic DNA will hydrogen bond at an AT pair in the biological DNA, and that a cytosine in the synthetic DNA will hydrogen bond at a GC pair.

The formation of triple-helical DNA has been implicated in several cellular processes, including transcription, replication, and recombination. While there is no direct evidence for triplexes in vivo, cellular proteins that specifically recognize triplex DNA have been discovered recently. This suggests that triplex DNA may have biological functions, though further research is needed to unravel its potential roles.

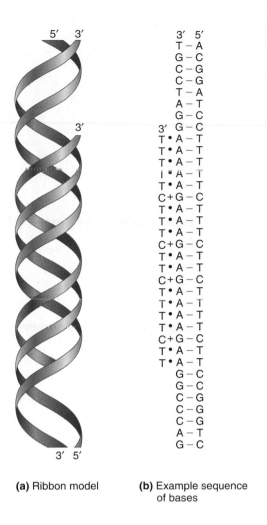

(a) Ribbon model   **(b)** Example sequence of bases

FIGURE 9.20 **The structure of triplex DNA. (a)** As seen in the ribbon model, the third, synthetic strand binds within the major groove of the double-stranded structure. **(b)** Within triplex DNA, the third strand hydrogen bonds according to the rule T to AT, and C to GC. The cytosine bases in the third strand are protonated (i.e, positively charged).

The Three-Dimensional Structure of DNA Within Chromosomes Requires Additional Folding and the Association with Proteins

To fit within a living cell, the long double-helical structure of chromosomal DNA must be extensively compacted into a three-dimensional conformation. The double helix becomes greatly twisted and folded with the aid of DNA-binding proteins. Figure 9.21 depicts the relationship between the DNA double helix and the compaction that occurs within a eukaryotic chromosome. Chapter 10 is devoted to the topic of chromosome structure and organization.

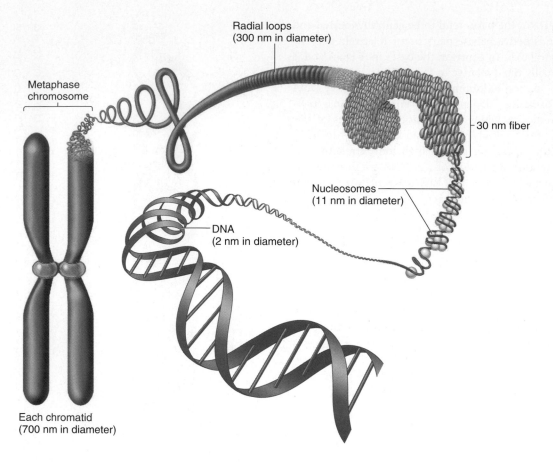

FIGURE 9.21 **The steps in eukaryotic chromosomal compaction leading to the metaphase chromosome.** The DNA double helix is wound around histone proteins and then further compacted to form a highly condensed metaphase chromosome. The levels of DNA compaction will be described in greater detail in chapter 10.

RNA Molecules Are Composed of Strands That Can Fold into Secondary and Tertiary Structures

Let's now turn our attention to RNA structure, which bears many similarities to DNA structure. The primary structure of an RNA strand is much like a DNA strand (fig. 9.22). Strands of RNA are typically several hundred or several thousand nucleotides in length, which is much shorter than chromosomal DNA. When RNA is made during transcription, the DNA is used as a template to make a copy of single-stranded RNA. In most cases, only one of the two DNA strands is used as a template for RNA synthesis; only one complementary strand of RNA is usually made. Nevertheless, relatively short sequences within one RNA molecule or between two separate RNA molecules can be complementary to each other. These sequences can form short double-stranded regions.

The secondary structure of RNA molecules is due to the ability of complementary regions to form base pairs between A and U and between G and C. This base pairing allows short regions to form a double helix. RNA double helices are antiparallel and typically of the right-handed A form with 11 to 12 base pairs per turn. As shown in figure 9.23, different types of secondary structural patterns are possible. These include bulges, internal loops, multibranched junctions, and stem-loops (also called hairpins). Note that these structures contain regions of complementarity punctuated by regions of noncomplementarity. In figure 9.23, the complementary regions are held together by connecting hydrogen bonds, while the noncomplementary regions have their bases projecting away from the double-stranded region.

Many factors contribute to the tertiary structure of RNA molecules. Within the RNA molecule itself, these include the base-paired double-stranded helices, stacking between bases, and hydrogen bonding between bases and backbone regions. In addition, interactions with ions, small molecules, and large proteins may influence RNA tertiary structure. Figure 9.24 depicts the tertiary structure of a transfer RNA molecule known as tRNA[phe] (i.e., a tRNA molecule that carries the amino acid phenylalanine). It was the first naturally occurring RNA to have its structure solved. Note that this RNA molecule contains several double-stranded and single-stranded regions. In a living cell, these regions spontaneously interact with each other to produce the three-dimensional structure.

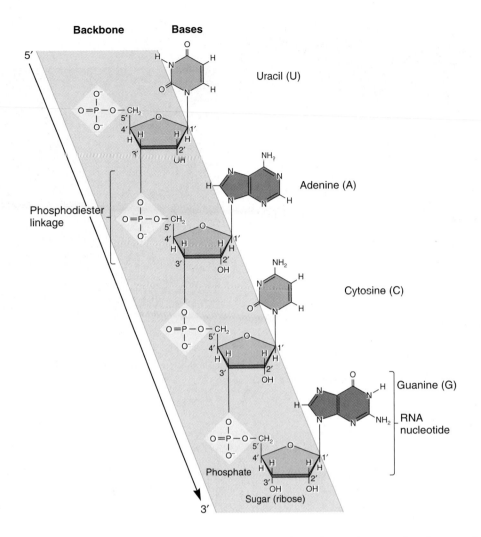

FIGURE 9.22   A strand of RNA. This structure is very similar to a DNA strand (see fig. 9.11), except that the sugar is ribose instead of deoxyribose and uracil is substituted for thymine.

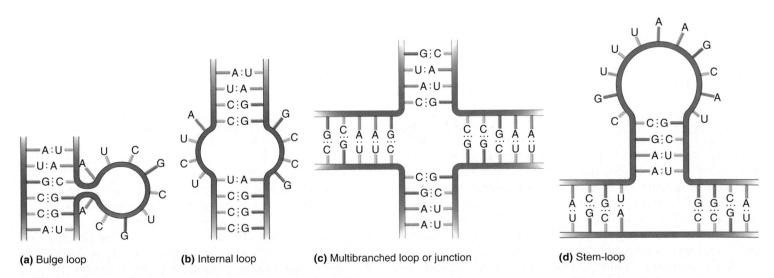

(a) Bulge loop   **(b)** Internal loop   **(c)** Multibranched loop or junction   **(d)** Stem-loop

FIGURE 9.23   Possible secondary structures of RNA molecules. The double-stranded regions are depicted by connecting hydrogen bonds. Loops are noncomplementary regions that are not hydrogen bonded with complementary bases. Double-stranded RNA structures can form within a single RNA molecule or between two separate RNA molecules.

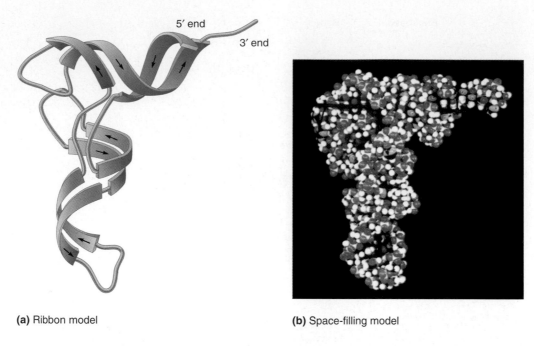

5′ end

3′ end

(a) Ribbon model   **(b)** Space-filling model

FIGURE 9.24   **The tertiary structure of tRNA^{phe}, the transfer RNA molecule that carries phenylalanine. (a)** Here the double-stranded regions of the molecule are shown as antiparallel ribbons. **(b)** A space-filling model of tRNA.

CONCEPTUAL SUMMARY

The molecular structure of nucleic acids underlies their function. **Nucleotides,** which are composed of a sugar, phosphate, and nitrogenous base, form the repeating structural unit of nucleic acids. A **strand** of DNA or RNA contains a linear sequence of nucleotides. Strands of DNA (and occasionally RNA) can form a **double helix** where **complementary strands** interact due to hydrogen bonding between **adenine** and **thymine** (or **uracil**) and between **guanine** and **cytosine.** Base stacking also stabilizes a double-stranded structure. The most common form of DNA is a right-handed double helix; the **backbones** in the double helix are composed of sugar–phosphate linkages, with the bases projecting inward from the backbones and hydrogen bonding with each other. The two strands are **antiparallel.** In DNA, different helical conformations have been identified, including **A DNA, B DNA,** and **Z DNA.** B DNA is the predominant form found in living cells, but short regions of Z DNA may play an important functional role. Within chromosomes, DNA is folded into a three-dimensional conformation with the aid of proteins. The structure of chromosomes will be discussed in chapter 10. It is also common for short segments within RNA to form double-helical structures such as hairpins, bulges, loops, and junctions. As we will see in later chapters, RNA secondary structures can also play many important functional roles. The final tertiary structure of RNA is dictated by several factors including double-helical regions, base stacking, hydrogen bonding between bases and backbone, and interactions with other molecules.

EXPERIMENTAL SUMMARY

Two different experimental approaches were used to show that DNA is the genetic material. Avery, MacLeod, and McCarty took advantage of Griffith's observations regarding **transformation** in pneumococci. They purified DNA from type IIIS strains and used that purified DNA to transform type IIR strains into type IIIS. Furthermore, DNase prevented the transformation, whereas RNase and protease did not. These observations indicated that DNA is the transforming substance that provides type IIR bacteria with a new trait. Hershey and Chase showed that T2 bacteriophage injects DNA into bacterial cells, indicating that DNA is the genetic material of this virus. Taken together, these two studies were instrumental in convincing scientists that the genetic material is composed of DNA.

Several experimental observations enabled Watson and Crick to solve the structure of DNA. The X-ray diffraction work of Franklin indicated a helical structure with two strands and 10 base pairs per turn. Furthermore, the work of Chargaff indicated that the amount of A equals T and the amount of G equals C. With these observations, Watson and Crick followed Pauling's strategy of using ball-and-stick models to visualize the structure of macromolecules. In this way, they deduced the structure of DNA, which is a double helix.

PROBLEM SETS & INSIGHTS

Solved Problems

S1. A hypothetical sequence at the beginning of an mRNA molecule is

5′-AUUUGCCCUAGCAAACGUAGCAAACG...rest of the coding sequence

Using two out of the three underlined sequences, draw two possible models for potential stem-loop structures at the 5′ end of this mRNA.

Answer:

S2. Describe the previous experimental evidence that led Watson and Crick to the discovery of the DNA double helix.

Answer:

1. The chemical structure of single nucleotides was understood by the 1950s.

2. Watson and Crick assumed that DNA is composed of nucleotides that are linked together in a linear fashion. They also assumed that the chemical linkage between two nucleotides is always the same.

3. Franklin's diffraction patterns suggested several structural features. First, it was consistent with a helical structure. Second, the diameter of the helical structure was too wide to be only a single-stranded helix. Finally, the line spacing on the diffraction pattern indicated that the helix contains about 10 base pairs per complete turn.

4. In the chemical analysis of the DNA from different species, the work of Chargaff indicated that the amount of adenine equaled the amount of thymine and that the amount of cytosine equaled the amount of guanine.

5. In the early 1950s, Linus Pauling proposed that regions of proteins can fold into a secondary structure known as an α helix. To discover this, Pauling built large models by linking together simple ball-and-stick units. In this way, it becomes possible to determine if atoms fit together properly in a complicated three-dimensional structure. A similar approach was used by Watson and Crick to solve the structure of the DNA double helix.

S3. Within living cells, a myriad of different proteins play important functional roles by binding to DNA and RNA. As described throughout your textbook, the dynamic interactions between nucleic acids and proteins lie at the heart of molecular genetics. Some proteins bind to DNA (or RNA) but not in a sequence-specific manner. For example, histones are proteins that are important in the formation of chromosome structure. In this case, the positively charged histone proteins actually bind to the negatively charged phosphate groups in DNA. In addition, several other proteins interact with DNA but do not require a specific nucleotide sequence to carry out their function. For example, DNA polymerase, which catalyzes the synthesis of new DNA strands, does not bind to DNA in a sequence-dependent manner. By comparison, many other proteins do interact with nucleic acids in a sequence-dependent fashion. This means that a specific sequence of bases can provide a structure that will be recognized by a particular protein. Throughout the textbook, the functions of many of these proteins will be described. Some examples include transcription factors that affect the rate of transcription, proteins that bind to centromeres, and proteins that bind to origins of replication. With regard to the three-dimensional structure of DNA, where would you expect DNA-binding proteins to bind if they recognize a specific base sequence? What about DNA-binding proteins that do not recognize a base sequence?

Answer: DNA-binding proteins that recognize a base sequence must bind into the major or minor groove of the DNA. This is where the bases would be accessible to a DNA-binding protein. Most DNA-binding proteins, which recognize a base sequence, actually fit into the major groove. By comparison, other DNA-binding proteins, such as histones, which do not recognize a base sequence, bind to the DNA backbone.

S4. The formation of a double-stranded structure must obey the rule that adenine hydrogen bonds to thymine (or uracil) and cytosine hydrogen bonds to guanine. Based on your previous understanding of genetics (from this course or general biology courses), discuss reasons why complementarity is an important feature of DNA and RNA structure and function.

Answer: One way that complementarity underlies function is that it provides the basis for the synthesis of new strands of DNA and RNA. During replication, the synthesis of the new DNA strands occurs in such a way that adenine hydrogen bonds to thymine and cytosine hydrogen bonds to guanine. In other words, it is the molecular feature of a complementary double-stranded structure that makes it possible to make exact copies of DNA. Likewise, the ability to transcribe DNA into RNA is based on complementarity. During transcription, one strand of DNA is used as a template to make a complementary strand of RNA.

In addition to the synthesis of new strands of DNA and RNA, complementarity is important in other ways. As mentioned in chapter 9, the folding of RNA into a particular secondary structure is driven by the hydrogen bonding of complementary regions. This event is necessary to produce functionally active tRNA molecules. Likewise, stem-loop structures also occur in other types of RNA. For example, the rapid formation of stem-loop structures is known to occur as RNA is being transcribed and to affect the termination of transcription.

A third way that complementarity can be functionally important is that it can promote the interaction of two separate RNA molecules. During translation, codons in mRNA bind to the anticodons in tRNA (see chapter 13). This binding is due to complementarity. For example, if a codon is 5′-AGG-3′, the anticodon is 3′-UCC-5′. This type of specific interaction between codons and anticodons is an important step that enables the nucleotide sequence in mRNA to code for an amino acid sequence within a protein. In addition, many other examples of RNA-RNA interactions are known and will be described throughout this textbook.

Eukaryotic Chromosomes Contain Many Functionally Important Sequences Including Genes, Origins of Replication, Centromeres, and Telomeres

Each eukaryotic chromosome contains a long, linear DNA molecule (fig. 10.11). Three types of DNA sequences are required for chromosomal replication and segregation: origins of replication, centromeres, and telomeres. Origins of replication are nucleotide sequences that are necessary to initiate DNA replication. Unlike most bacterial chromosomes, which contain only one, eukaryotic chromosomes contain many origins, interspersed approximately every 100,000 bp apart. The function of origins is discussed in greater detail in chapter 11. As described in chapter 3, centro-

meres are DNA sequences required for the proper segregation of chromosomes during mitosis and meiosis. For most species, each eukaryotic chromosome contains a single centromere. The centromere serves as an attachment site for a group of proteins that form the **kinetochore.** During mitosis and meiosis, the kinetochore is linked to the spindle apparatus and thereby ensures the proper segregation of the chromosomes to each daughter cell. Finally, at the ends of linear chromosomes are found specialized regions known as **telomeres.** Telomeres serve several important functions. As discussed in chapter 8, telomeres prevent chromosomes from having "sticky ends" and thereby inhibit chromosomal rearrangements such as translocations. In addition, they prevent chromosome shortening in two ways. First, the telomeres protect chromosomes from digestion via enzymes that recognize the ends of DNA (i.e., exonucleases). Secondly, as we will learn in chapter 11, an unusual form of DNA replication occurs at the telomere to ensure that eukaryotic chromosomes do not become shortened with each round of DNA replication.

Genes are located between the centromeric and telomeric regions along the entire eukaryotic chromosome. A single chromosome usually has a few hundred to several thousand different genes. The sequence of a typical eukaryotic gene is a few thousand to tens of thousands of nucleotides in length. In less complex eukaryotes such as yeast, genes are relatively small and contain primarily the nucleotide sequences that encode the amino acid sequences within proteins (i.e., structural genes). In more complex eukaryotes such as mammals and higher plants, structural genes tend to be much longer due to the presence of introns (noncoding intervening sequences). Introns range in size from less than 100 bp to more than 10,000 bp. Therefore, the presence of large introns can greatly increase the lengths of eukaryotic genes. Species such as mammals and higher plants tend to have more introns within their genes compared to lower eukaryotes (e.g., yeast).

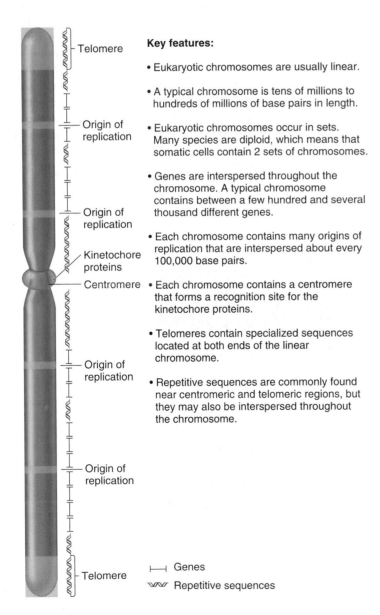

Key features:

• Eukaryotic chromosomes are usually linear.

• A typical chromosome is tens of millions to hundreds of millions of base pairs in length.

• Eukaryotic chromosomes occur in sets. Many species are diploid, which means that somatic cells contain 2 sets of chromosomes.

• Genes are interspersed throughout the chromosome. A typical chromosome contains between a few hundred and several thousand different genes.

• Each chromosome contains many origins of replication that are interspersed about every 100,000 base pairs.

• Each chromosome contains a centromere that forms a recognition site for the kinetochore proteins.

• Telomeres contain specialized sequences located at both ends of the linear chromosome.

• Repetitive sequences are commonly found near centromeric and telomeric regions, but they may also be interspersed throughout the chromosome.

⊢—⊣ Genes

〜〜 Repetitive sequences

FIGURE 10.11   **Organization of DNA sequences within eukaryotic chromosomes.**

The Genome of Eukaryotes Contains Sequences That Are Unique, Moderately Repetitive, and Highly Repetitive

The term **sequence complexity** refers to the number of times a particular base sequence appears throughout the genome. **Unique** or **nonrepetitive sequences** are those found once or a few times within the genome. Unique sequences include DNA segments that encode structural genes as well as intergenic regions. The vast majority of proteins in eukaryotic cells are encoded by genes that are present in one or a few copies. In the case of humans, unique sequences make up roughly 60 to 70% of the entire genome.

Moderately repetitive sequences are found a few hundred to several thousand times in the genome. In a few cases, moderately repetitive sequences are multiple copies of the same gene. For example, the genes that encode ribosomal RNA (rRNA) are found in many copies. Ribosomal RNA is necessary for the functioning of ribosomes. Cells need a large amount of rRNA for making ribosomes. This is accomplished by having multiple copies of the genes that encode rRNA. Likewise, the histone genes are also found in multiple copies because a large number of histone proteins are needed for the structure of chromatin. In addition, other types of functionally important sequences can be moderately repetitive. For example, multiple copies of origins of replication are found within eukaryotic chromosomes. Other moderately repetitive sequences may play a role in the regulation of gene transcription and translation. By comparison, some moderately repetitive sequences do not play a functional role and are derived from **transposable elements,** which are segments of DNA that have the ability to move within a genome. This category of repetitive sequence is discussed in greater detail in chapter 17.

Highly repetitive sequences are those that are found tens of thousands or even millions of times throughout the genome. Each copy of a highly repetitive sequence is relatively short, ranging from a few nucleotides to several hundred in length. A widely studied example is the *Alu* family of sequences found in humans and other primates. The *Alu* sequence is approximately 300 bp long. This sequence derives its name from the observation that it contains a site for cleavage by a restriction enzyme known as *Alu*I. (The function of restriction enzymes is described in chapter 18.) The *Alu* sequence is present in 500,000 to 1 million copies in the human genome. It represents about 5 to 6% of the total human DNA and occurs approximately every 5,000 to 6,000 base pairs. Evolutionary studies suggest that the *Alu* sequence arose 65 million years ago from a section of a single ancestral gene known as the 7SL RNA gene. Since that time, this gene has become a type of transposable element called a **retroelement,** which can be transcribed into RNA, copied into DNA, and then inserted into

the genome. Remarkably, over the course of 65 million years, the *Alu* sequence has been copied and inserted into the human genome to achieve the modern number of more than 500,000 copies. The mechanism for the proliferation of *Alu* sequences will be described in chapter 17.

Some highly repetitive sequences, like the *Alu* family, are interspersed throughout the genome. However, other highly repetitive sequences are clustered together in a tandem array, also known as tandem repeats. In a tandem array, a very short nucleotide sequence is repeated many times in a row. In *Drosophila,* for example, 19% of the chromosomal DNA is highly repetitive DNA found in tandem arrays. An example is shown here.

```
AATATAATATAATATAATATAATATATAATAT
TTATATTATATTATATTATATTATATATTATA
```

In this particular tandem array, two related sequences, AATAT and AATATAT, are repeated many times. Highly repetitive sequences, which contain tandem arrays of short sequences, are commonly found in centromeric regions of chromosomes and can be quite long, sometimes more than 1 million bp in length!

Whether highly repetitive sequences play any significant functional role is controversial. Some experiments in *Drosophila* indicate that highly repetitive sequences may be important in the proper segregation of chromosomes during meiosis. It is not yet clear if highly repetitive DNA plays the same role in other species. The sequences within repetitive DNA vary greatly from species to species. In fact, as noted earlier, the amount of highly repetitive DNA can vary a great deal even among closely related species (as shown earlier in fig. 10.10).

EXPERIMENT 10A

Highly Repetitive DNA Can Be Separated from the Rest of the Chromosomal DNA by Equilibrium Density Centrifugation

As we will examine later in this chapter, centromeric and telomeric regions tend to be highly compacted, or **heterochromatic.** The folding of chromosomes into compact structures is called **condensation.** Heterochromatic regions are highly condensed. During the 1960s and 1970s, many scientists were interested in the relationship between heterochromatic regions and repetitive sequences. At that time, cytological evidence suggested that highly repetitive sequences are localized to heterochromatic regions. To gain further insight into this question, biochemical methods were developed to detect and isolate highly repetitive DNA. *Drosophila* was considered an ideal organism to study because a large proportion (19%) of its genome is heterochromatic. Therefore, if it is correct that heterochromatic DNA is composed of highly repetitive sequences, experiments should reveal that *Drosophila* has a large amount of highly repetitive DNA.

When repetitive DNA is in a tandem array, its base composition may be quite different from the rest of the chromosomal DNA. For example, the *Drosophila* tandem array discussed previously is 100% AT, whereas the average base composition of nonrepetitive chromosomal DNA in *Drosophila* is approximately 60% AT and 40% GC. One way to distinguish an AT pair from a GC pair is their relative densities. An AT pair is slightly less dense than a GC pair. Therefore, a DNA fragment with a high proportion of AT pairs will have a lighter density than a fragment with a higher percentage of GC pairs.

In the experiment described in figure 10.12, W. Peacock, D. Brutlag, E. Goldring, R. Appels, C. Hinton, and D. Lindsley took advantage of the experimental observation that molecules with different densities can be separated from each other via equilibrium density centrifugation (see the appendix for a description of density centrifugation). They began with a sample of nuclei isolated from *Drosophila* cells. The DNA was extracted into a phenol phase, resulting in a mixture of many different DNA fragments. Some of these fragments would be derived from repetitive sequences while others would be derived from nonrepetitive sequences. The mixture of DNA fragments was then subjected to centrifugation. In this experiment, the centrifuge tube contained a gradient of a dissolved salt called cesium chloride (CsCl). The CsCl concentration was progressively higher, going from the top to the bottom of the tube. Such a gradient facilitates the accurate separation of molecules with slightly different densities. Following centrifugation, the liquid within the tube was separated into fractions; early fractions contained liquid from the bottom of the tube, while later fractions contained liquid nearer the top. The samples were analyzed spectrophotometrically to determine the amount of DNA in each fraction.

■ THE HYPOTHESIS

Highly repetitive DNA may have a base composition that is significantly different from the rest of the chromosomal DNA. If so, it may be possible to separate repetitive DNA from the rest of the chromosomal DNA by equilibrium density centrifugation.

■ TESTING THE HYPOTHESIS — FIGURE 10.12 Identification of highly repetitive DNA.

Starting material: Nuclei isolated from *Drosophila melanogaster*.

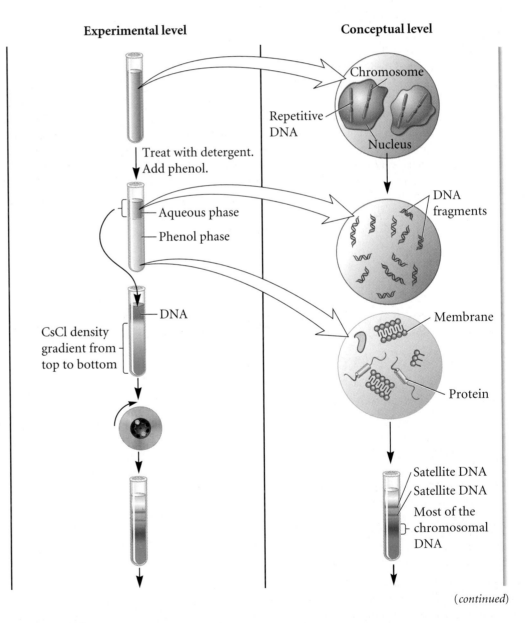

Experimental level

Conceptual level

1. Extract the DNA from the nuclei. This involves treatment with detergent (to dissolve the nuclear membrane) and then addition of phenol (to remove the protein). The DNA remains in the aqueous phase, while most of the protein goes into the phenol phase. During this procedure, the chromosomal DNA is broken into small fragments.

Treat with detergent.
Add phenol.

Chromosome

Repetitive DNA

Nucleus

Aqueous phase

Phenol phase

DNA fragments

2. Load the aqueous phase, which contains the DNA fragments, onto a CsCl density gradient. (Note: Density gradient centrifugation is described in the appendix.)

DNA

CsCl density gradient from top to bottom

Membrane

Protein

3. Centrifuge for 18 hours until the DNA fragments reach their equilibrium density.

Satellite DNA
Satellite DNA
Most of the chromosomal DNA

(continued)

4. Collect fractions along the gradient.

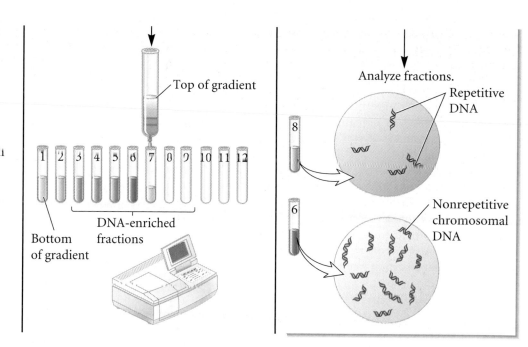

Top of gradient

Analyze fractions.

Repetitive DNA

Nonrepetitive chromosomal DNA

5. Determine the amount of DNA in each fraction using a spectrophotometer. DNA absorbs light in the UV range. (The use of a spectrophotometer is described in the appendix.)

Bottom of gradient

DNA-enriched fractions

6. Plot the amount of DNA in each fraction.

THE DATA

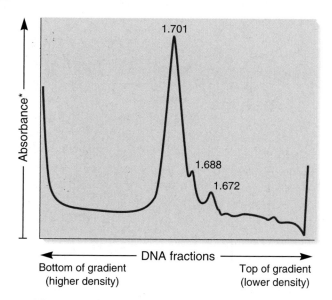

*Higher absorbance means a higher amount of DNA in the fractions.

INTERPRETING THE DATA

As shown in the data of figure 10.12, most of the DNA was found in a fraction of density 1.701 g/cm³. This peak contained about 80% of the *Drosophila* DNA. Most of this peak corresponds to the chromosomal DNA that is not heterochromatic. We now know that this peak primarily contains the DNA pieces derived from the nonrepetitive chromosomal DNA, though a small amount could correspond to repetitive DNA that happens to have a similar density. Its average base composition is approximately 60% AT and 40% GC. In addition, the data show two other notable peaks of DNA found in fractions that have a lighter density (namely, 1.672 and 1.688 g/cm³). These peaks of DNA are referred to as **satellite DNA,** since they are separated from the main peak. In this experiment, the two peaks of satellite DNA have a base composition that is significantly lighter than the rest of the chromosomal DNA. Other experiments have been carried out to identify satellite DNA in *Drosophila*. By changing the DNA preparation techniques and the centrifugation conditions, several more satellite DNAs have been identified in this organism.

The experiment described in figure 10.12 demonstrated that some of the DNA in *Drosophila* has a base composition that is significantly different from rest of the chromosomal DNA. Furthermore, density centrifugation provided a way to biochemically separate the satellite DNA from the rest of the DNA. This enabled researchers to clone the satellite DNA and subject it to DNA sequencing. The sequencing of satellite DNA confirmed that it is composed of highly repetitive sequences. The satellite with a density of 1.672 g/cm³ has been sequenced; it is the highly repetitive DNA discussed earlier (i.e., the tandem repeat of AATAT and AATATAT). Since it is 100% AT, its density is lighter than the main peak of chromosomal DNA.

A self-help quiz involving this experiment can be found at the Online Learning Center.

Sequence Complexity Can Be Evaluated in a Renaturation Experiment

A second approach that has proven useful in understanding genome complexity has come from renaturation studies. These kinds of experiments were first carried out by Roy Britten and David Kohne in 1968. In a renaturation study, the DNA is broken up into pieces containing several hundred base pairs. The double-stranded DNA is then "melted" (separated) into single-stranded pieces by heat treatment (fig. 10.13a). When the temperature is lowered, the pieces of DNA that are complementary can renature with each other to form double-stranded molecules.

The rate of renaturation of complementary DNA strands provides a way to distinguish unique, moderately repetitive, and highly repetitive sequences. For a given category of DNA sequences, the renaturation rate will depend on the concentration of its complementary partner. Highly repetitive DNA sequences renature much faster, because there are many copies of the complementary sequences. In contrast, unique sequences, such as those found within most genes, take longer to renature because of the added time it takes for the unique sequences to find each other.

The renaturation of two DNA strands is a bimolecular reaction that involves the collision of two complementary DNA strands. Its rate is proportional to the product of the concentrations of both strands. If C is the concentration of a single-stranded DNA, then for any DNA derived from a double-stranded fragment, the concentration of one DNA strand (denoted C_1) equals the concentration of its complementary partner (denoted C_2). Letting $C = C_1 = C_2$, we see the rate of renaturation is represented by the second-order equation

$$-dC/dt = kC^2$$

(This is called a second-order equation because the rate depends on the concentration of both reactants—i.e., C_1 and C_2. In this case, this product is simplified to C^2 because $C_1 = C_2$.)

This equation says that a change in concentration of a single strand ($-dC$) with respect to time (dt) equals a rate constant (k) times the concentration of the single-stranded molecule squared (C^2). This equation can then be integrated to determine how the concentration of the single-stranded DNA will change from time zero to a later time.

$$\frac{C}{C_0} = \frac{1}{1 + k_2 C_0 t}$$

Where
- C = the concentration of single-stranded DNA at a later time, t
- C_0 = the concentration of single-stranded DNA at time zero
- k_2 = the second-order rate constant for renaturation

In this equation, C/C_0 is the fraction of DNA still in single-stranded form after a given length of time. For example, if C/C_0 equals 0.4 after a certain period of time, 40% of the DNA is still in the single-stranded form while 60% has renatured into the double-stranded form. A renaturation experiment can provide quantita-

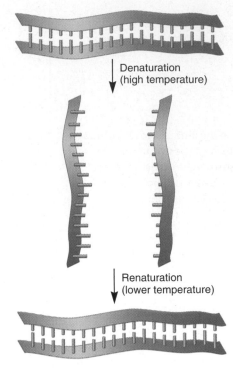

(a) Renaturation of DNA strands

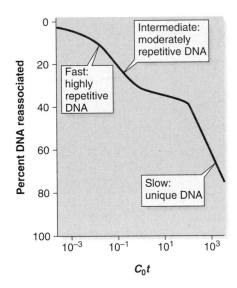

(b) Human chromosomal DNA $C_0 t$ curve

FIGURE 10.13   Renaturation and DNA sequence complexity. **(a)** Denaturation and renaturation (or reassociation) of DNA strands. **(b)** A $C_0 t$ curve for human chromosomal DNA.

tive information about the complexity of DNA sequences within chromosomal DNA. In the experiment shown in figure 10.13b, human DNA was sheared into small pieces (each about 600 bp in length), subjected to heat, and then allowed to renature at a lower temperature. The rates of renaturation for the DNA pieces can be represented in a plot of C/C_0 versus $C_0 t$. This is referred to as a C_0t **curve** (called a "cot" curve). A small amount of the DNA renatures very rapidly. This is the highly repetitive DNA. Some DNA renatures at a moderate rate, but most of the DNA renatures fairly slowly. From these data, the relative amounts of highly repetitive, moderately repetitive, and unique DNA sequences can be approximated. As seen in figure 10.13*b*, 60 to 70% of human DNA fragments are unique DNA sequences that renature slowly.

Eukaryotic Chromatin Must Be Compacted to Fit Within the Cell

We now turn our attention to ways that eukaryotic chromosomes are folded to fit within a living cell. A typical eukaryotic chromosome contains a single, linear double-stranded DNA molecule that may be hundreds of millions of base pairs in length. If a single set of human chromosomes were stretched from end to end, the length would be over 1 meter! By comparison, most eukaryotic cells are only 10 to 100 μm in diameter, and the cell nucleus is only about 2 to 4 μm in diameter. Therefore, the DNA in a eukaryotic cell must be folded and packaged by a staggering amount to fit inside the nucleus.

The compaction of linear DNA within eukaryotic chromosomes is accomplished through mechanisms that involve interactions between DNA and several different proteins. In recent years, it has become increasingly clear that the proteins bound to chromosomal DNA are subject to change during the life of the cell. These changes in protein composition, in turn, affect the degree of compaction of the chromatin. Chromosomes are very dynamic structures that alternate between tight and loose compaction states in response to changes in protein composition. In the remaining parts of chapter 10, we will focus our attention on two issues of chromosome structure. First, we will consider how chromosomes are compacted and organized during interphase (i.e.,

the period of the cell cycle that includes the G_1, S, and G_2 phases). Next, we will examine the additional compaction that is necessary to produce the highly condensed chromosomes found in M phase. In later chapters, we will learn that gene transcription is greatly influenced by a process known as **chromatin remodeling.** This term refers to changes in chromatin structure that regulate the ability of transcription factors to gain access to genes so they may be transcribed into RNA.

Linear DNA Wraps Around Histone Proteins to Form Nucleosomes, the Repeating Structural Unit of Chromatin

The repeating structural unit within eukaryotic chromatin is the **nucleosome,** which is composed of double-stranded DNA wrapped around an octamer of **histone proteins** (fig. 10.14*a*). Each octamer contains eight histone subunits (two copies each of four different histone proteins). The DNA lies on the surface and makes 1.65 negative superhelical turns around the histone octamer. The amount of DNA that is required to wrap around the histone octamer is 146 bp. At its widest point, a single nucleosome is about 11 nm in diameter.

The chromatin of eukaryotic cells contains a repeating pattern in which the nucleosomes are connected by linker regions of DNA that vary in length from 20 to 100 bp, depending on the species and cell type. It has been suggested that the overall structure of connected nucleosomes resembles beads on a string. This structure shortens the length of the DNA molecule about sevenfold.

Histone proteins are very basic proteins because they contain a large number of positively charged lysine and arginine amino acids. The arginine residues, in particular, play a major role in binding to the DNA. Arginine residues within the histone proteins form electrostatic and hydrogen-bonding interactions

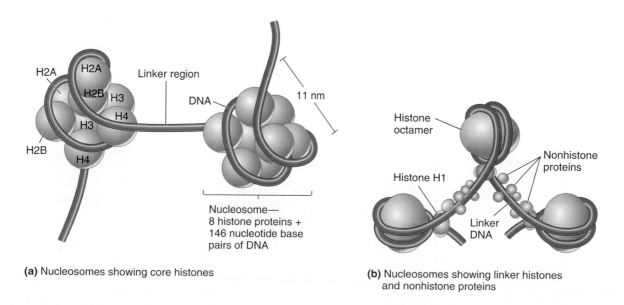

(a) Nucleosomes showing core histones

(b) Nucleosomes showing linker histones and nonhistone proteins

FIGURE 10.14 **Nucleosome structure.** (a) 146 bp of DNA are wrapped around an octamer of core histones. (b) The linker DNA connects adjacent nucleosomes. The linker histone H1 and nonhistone proteins also bind to this linker region.

with the phosphate groups along the DNA backbone. The octamer of histones contains two molecules each of four different histone proteins: H2A, H2B, H3, and H4. These are called the core histones. Another histone, H1, is found in most eukaryotic cells and is called the linker histone. It binds to the DNA on one side of a nucleosome and to the linker region of DNA (fig. 10.14*b*). The linker histones are less tightly bound to the DNA than are the core histones. In addition, nonhistone proteins are bound to this linker region. These proteins also play a role in the organization and compaction of chromosomes, and their presence may affect the expression of nearby genes.

EXPERIMENT 10B

The Repeating Nucleosome Structure Is Revealed by Digestion of the Linker Region

The model of nucleosome structure was originally proposed by Roger Kornberg in 1974. He based his proposal on several observations. Biochemical experiments had shown that chromatin contains a ratio of one molecule of each of the four core histones (namely, H2A, H2B, H3, and H4) per 100 bp of DNA. Approximately one H1 protein was found per 200 bp of DNA. In addition, purified core histone proteins were observed to bind to each other via specific pairwise interactions. X-ray diffraction studies showed that chromatin is composed of a repeating pattern of smaller units. And finally, electron microscopy of chromatin fibers revealed a diameter of approximately 11 nm. Taken together, these observations led Kornberg to propose a model in which the DNA double helix is wrapped around an octamer of core histone proteins. Including the linker region, this would involve about 200 bp of DNA.

Markus Noll decided to test Kornberg's model by digesting chromatin with DNase I, an enzyme that cuts the DNA backbone, and then accurately measuring the molecular mass of the DNA fragments by gel electrophoresis. The rationale behind this experiment is that the linker region of DNA will be more accessible to DNase I. In other words, DNase I is more likely to make cuts in the linker region than in the 146-bp region that is tightly bound to the core histones. If this is correct, incubation with DNase I is expected to make cuts in the linker region and thereby produce DNA pieces that are approximately 200 bp in length. (Note: The size of the DNA fragments may vary somewhat, since the linker region is not of constant length and because the cut within the linker region may occur at different sites.)

Figure 10.15 describes the experimental protocol of Noll. He began with nuclei from rat liver cells and incubated them with low, medium, or high concentrations of DNase I. The DNA was extracted into an aqueous phase, and then loaded onto an agarose gel that separated the fragments according to their molecular mass. The DNA fragments within the gel were stained with a UV-sensitive dye, ethidium bromide, which made it possible to view the DNA fragments under UV illumination.

■ THE HYPOTHESIS

This experiment seeks to test the beads-on-a-string model for chromatin structure. According to this model, DNase I should preferentially cut the DNA in the linker region and thereby produce DNA pieces that are about 200 bp in length.

■ **TESTING THE HYPOTHESIS** — **FIGURE 10.15** DNase I cuts chromatin into repeating units containing 200 bp of DNA.

Starting material: Nuclei from rat liver cells.

1. Incubate the nuclei with low, medium, and high concentrations of DNase I. The conceptual level illustrates a low DNase I concentration.

Experimental level

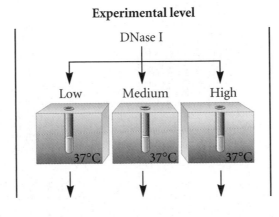

Conceptual level

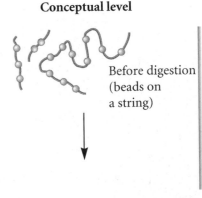

Before digestion (beads on a string)

(*continued*)

2. Extract the DNA (see step 1 of fig. 10.12).

3. Load the DNA into a well of an agarose gel and run the gel to separate the DNA pieces according to size. On this gel, also load DNA fragments of known molecular mass (marker lane).

4. Visualize the DNA fragments by staining the DNA with ethidium bromide. This is a dye that binds to DNA and is fluorescent when excited by UV light.

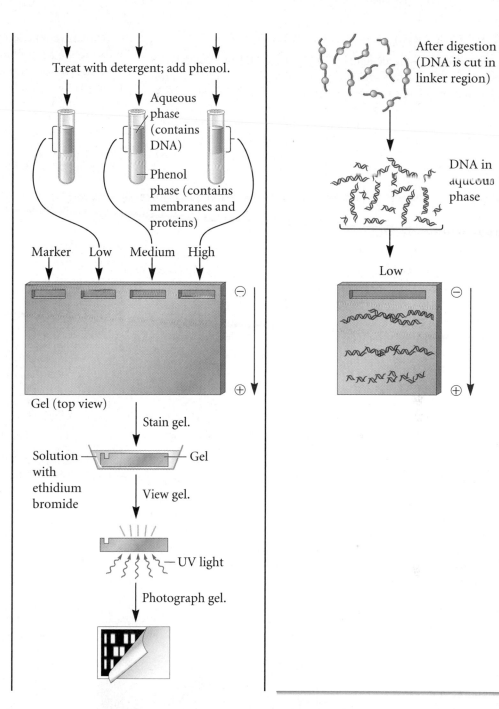

Treat with detergent; add phenol.

Aqueous phase (contains DNA)

Phenol phase (contains membranes and proteins)

Marker   Low   Medium   High

Gel (top view)

Stain gel.

Solution with ethidium bromide — Gel

View gel.

— UV light

Photograph gel.

After digestion (DNA is cut in linker region)

DNA in aqueous phase

Low

■ **THE DATA**

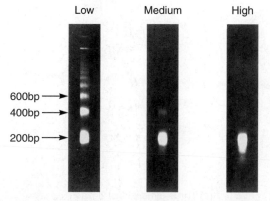

Low   Medium   High

600bp →
400bp →
200bp →

DNase concentration: 30 units ml⁻¹   150 units ml⁻¹   600 units ml⁻¹

INTERPRETING THE DATA

As shown in the data of figure 10.15, at high DNase I concentrations, the entire sample of chromosomal DNA was digested into fragments of approximately 200 bp in length. This is the result predicted by the beads-on-a-string model. Furthermore, at lower DNase I concentrations, longer pieces were observed, and these were in multiples of 200 bp (400, 600, etc.). These longer pieces may be explained by occasional uncut linker regions at lower DNase I concentrations. For example, a DNA piece might contain two nucleosomes and be 400 bp in length. If two consecutive linker regions were not cut, this would produce a piece with three nucleosomes containing about 600 bp of DNA. And so on. Taken together, these results strongly supported the nucleosome model for chromatin structure.

A self-help quiz involving this experiment can be found at the Online Learning Center.

Nucleosomes Become Closely Associated to Form a 30 nm Fiber

In eukaryotic chromatin, nucleosomes associate with each other to form a more compact structure that is 30 nm in diameter. Evidence for the packaging of nucleosomes was obtained in the microscopy studies of F. Thoma in 1977. As shown in figure 10.16, chromatin samples were treated with or without solutions of moderate salt concentration (100 mM NaCl) and then observed with an electron microscope. Moderate salt concentrations are expected to remove the H1 histone but not the core histones, because the H1 histone is more loosely attached to the DNA. At moderate salt concentrations (fig. 10.16a), the chromatin exhibited the classic beads-on-a-string morphology. At low salt concentrations (when H1 is expected to remain bound to the DNA), these "beads" associated with each other into a more compact conformation (fig. 10.16b). These results suggest that the nucleosomes are packaged to create a more compact unit. Furthermore, the results are consistent with a role for H1 in the packaging of nucleosomes into more compact structures. However, the precise role of H1 in chromatin compaction remains unclear. Recent data suggest that the core histones also play a key role in the compaction and relaxation of chromatin.

The experiment of figure 10.16 and other experiments have established that nucleosome units are organized into a more compact structure that is 30 nm in diameter, known as the **30 nm fiber** (fig. 10.17a). The 30 nm fiber shortens the total length of DNA another sevenfold. The structure of the 30 nm fiber has proven difficult to determine, because the conformation of the

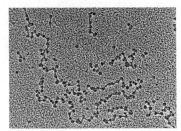
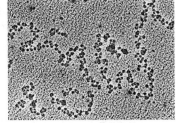

(a) At moderate salt concentration   **(b)** At low salt concentration

FIGURE 10.16 **The nucleosome structure of eukaryotic chromatin as viewed by electron microscopy.** The chromatin in (**a**) has been treated with moderate salt concentrations to remove the linker histone H1. It exhibits the classic beads-on-a-string morphology. The chromatin in (**b**) has been incubated at lower salt concentrations and shows a more compact morphology.

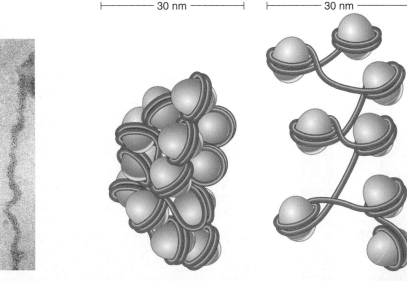

(a) Micrograph of a 30 nm fiber   **(b)** Solenoid model (not correct)   **(c)** Three-dimensional zigzag model

FIGURE 10.17 **The 30 nm fiber.** (**a**) A photomicrograph of the 30 nm fiber. (**b**) In the solenoid model, the nucleosomes are packed in a spiral configuration containing six nucleosomes per turn. (**c**) In the three-dimensional zigzag model, the linker DNA forms a more irregular structure and there is less contact between adjacent nucleosomes. The three-dimensional zigzag model is consistent with more recent data regarding chromatin conformation.

DNA may be substantially altered when it is extracted from living cells. An early model by Thoma, known as the solenoid model, suggested a helical structure in which nucleosome contacts produce a symmetrically compact structure within the 30 nm fiber (fig. 10.17b). However, more recent data suggest that the 30 nm fiber does not form such a regular structure. Instead, a newer model based on cryoelectron microscopy (i.e., electron microscopy at low temperature) was proposed by Rachel Horowitz and Christopher Woodcock in the 1990s (fig. 10.17c). According to their model, linker regions within the 30 nm structure are variably bent and twisted, and there is little face-to-face contact between nucleosomes. The 30 nm fiber forms an asymmetric, three-dimensional zigzag of nucleosomes. At this level of compaction, the overall picture of chromatin that emerges is an irregular, fluctuating, three-dimensional zigzag structure with stable nucleosome units connected by deformable linker regions.

Chromosomes Are Further Compacted by Anchoring the 30 nm Fiber into Radial Loop Domains Along the Nuclear Matrix

Thus far, we have examined two mechanisms that compact eukaryotic DNA. These involve the wrapping of DNA within nucleosomes and the arrangement of nucleosomes to form a 30 nm fiber. Taken together, these two events shorten the DNA about 50-fold. A third level of compaction involves interactions between the 30 nm fibers and a filamentous network of proteins in the nucleus called the **nuclear matrix.** As shown in figure 10.18a, the nuclear matrix consists of two parts: the nuclear lamina, which is a collection of filamentous proteins that line the inner nuclear membrane, and an internal nuclear matrix, which is connected to the lamina and fills the nucleus interior. This internal matrix is an intricate fine network of irregular 10 nm

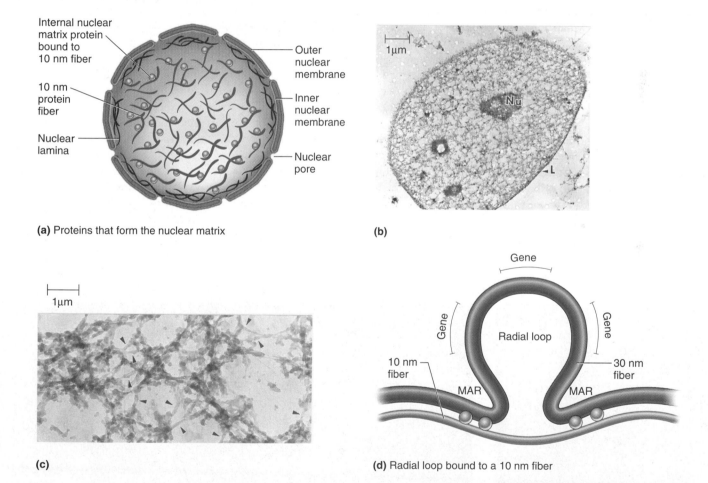

(a) Proteins that form the nuclear matrix

(b)

(c)

(d) Radial loop bound to a 10 nm fiber

FIGURE 10.18   **Structure of the nuclear matrix. (a)** This schematic drawing shows the arrangement of the matrix within a cell nucleus. The nuclear lamina (depicted in *red*) is a collection of fibrous proteins that line the inner nuclear membrane. The internal nuclear matrix is composed of 10 nm filaments (depicted in *green*) that are interconnected. These fibers also have many other proteins associated with them (depicted in *orange*). **(b)** An electron micrograph of the nuclear matrix after the chromatin has been removed. The nucleolus is labeled *Nu,* and the lamina is labeled *L.* **(c)** At higher magnification, the 10 nm filaments are more easily seen. **(d)** The matrix-attachment regions (MARs), which contain a high percentage of A and T residues, are anchored to the nuclear matrix to create radial loops. This causes a greater compaction of eukaryotic chromosomal DNA.

fibers plus many other proteins that bind to these fibers. Even when the chromatin is extracted from the nucleus, the internal nuclear matrix may remain intact (fig. 10.18b and c). The overall composition of the nuclear matrix is very complex, consisting of dozens or perhaps hundreds of different proteins. This complexity has made it difficult to propose models regarding its overall organization. Further research will be necessary to understand how the proteins fit together to form the interwoven structure of the nuclear matrix.

The third mechanism that compacts the DNA involves the formation of **radial loop domains,** similar to those described for the bacterial chromosome. During interphase, chromatin is organized into loops, often 25,000 to 200,000 bp in size, which are anchored to the nuclear matrix. The chromosomal DNA of eukaryotic species contains sequences called **matrix-attachment regions (MARs)** or **scaffold-attachment regions (SARs),** which are interspersed at regular intervals throughout the genome. The MARs bind to specific proteins in the nuclear matrix, thus forming chromosomal loops (fig. 10.18d).

The attachment of radial loops to the nuclear matrix is important in two ways. First, as discussed in chapter 15, it is thought to be important in gene regulation. Certain proteins can influence the compaction level of an entire radial loop and thereby affect the expression of many genes within the loop. Second, the nuclear matrix serves to organize the chromosomes within the nucleus. Each chromosome in the cell nucleus is located in a discrete and nonoverlapping **chromosome territory.** These can be viewed when interphase cells are exposed to multiple fluorescent molecules that recognize specific sequences on each chromosome. Figure 10.19 shows an experiment in which chicken chromosomes were exposed to a mixture of probes that recognize multiple sites along each of the chromosomes. Figure 10.19a shows the chromosomes in metaphase. The probes label

each type of metaphase chromosome with a different color. Figure 10.19b shows the use of the same probes during interphase, when the chromosomes are less condensed and found in the cell nucleus. As seen here, each chromosome occupies its own distinct territory.

Before ending the topic of interphase chromosome compaction, it should be mentioned that the compaction level of interphase chromosomes is not completely uniform. This variability can be seen with a light microscope and was first observed by the German cytologist E. Heitz in 1928. He used the term **heterochromatin** to describe the tightly compacted regions of chromosomes. In general, these regions of the chromosome are transcriptionally inactive. By comparison, the less condensed regions, known as **euchromatin,** reflect areas that are capable of gene transcription. Euchromatin is the form of chromatin in which the 30 nm fiber forms radial loop domains. In heterochromatin, these radial loop domains are compacted even further.

Figure 10.20 illustrates the distribution of euchromatin and heterochromatin in a typical eukaryotic chromosome during interphase. The chromosome contains regions of both heterochromatin and euchromatin. Heterochromatin is most abundant in the centromeric regions of the chromosomes and, to a lesser extent, in the telomeric regions. The term **constitutive heterochromatin** refers to chromosomal regions that are always heterochromatic and permanently inactive with regard to transcription. Constitutive heterochromatin usually contains highly repetitive DNA sequences, such as tandem repeats, rather than gene sequences. **Facultative heterochromatin** refers to chromatin that can occasionally interconvert between euchromatin and heterochromatin. An example of facultative heterochromatin occurs in female mammals in which one of the two X chromosomes is converted to a heterochromatic **Barr body.** As was discussed in chapter 7, most of the genes on the Barr body are transcriptionally inactive,

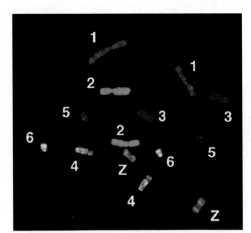

(a) Metaphase chromosomes

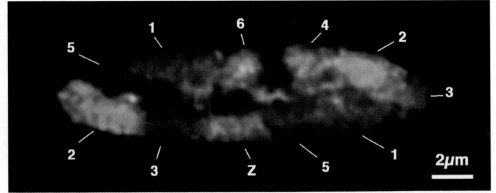

(b) Chromosomes in the cell nucleus during interphase

FIGURE 10.19   Chromosome territories in the cell nucleus. (a) Metaphase chromosomes from the chicken were labeled with chromosome-specific probes. Each of the seven types of chicken chromosomes (i.e., 1, 2, 3, 4, 5, 6, and Z) is labeled a different color. **(b)** The same probes were used to label interphase chromosomes in the cell nucleus. Each of these chromosomes occupies its own distinct, nonoverlapping territory within the cell nucleus.

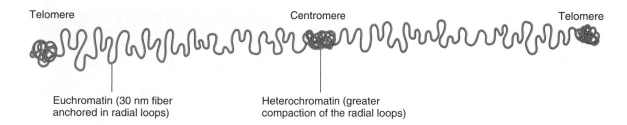

Telomere   Centromere   Telomere

Euchromatin (30 nm fiber   Heterochromatin (greater
anchored in radial loops)   compaction of the radial loops)

FIGURE 10.20 **Chromatin structure during interphase.** Heterochromatic regions are more highly condensed and tend to be localized in centromeric and telomeric regions.

and so only these genes on the other (euchromatic) X chromosome can function. The conversion of one X chromosome to heterochromatin occurs during embryonic development in the somatic cells of the body.

Condensin and Cohesin Promote the Formation of Metaphase Chromosomes

As we have learned, there are several ways to alter the level of chromosomal compaction. Furthermore, the degree of compaction can vary along a single eukaryotic chromosome. When cells prepare to divide, the chromosomes become even more compacted or condensed. This aids in their proper sorting during metaphase. Figure 10.21 illustrates the levels of compaction that lead to a metaphase chromosome. During interphase, most of the chromosomal DNA is found in the 300 nm configuration. This corresponds to euchromatin, in which the 30 nm fibers form radial loop domains. The 300 nm configuration can be further compacted to a diameter of approximately 700 nm. This is the compaction level found in heterochromatin. During interphase, most chromosomal regions are euchromatic and some localized regions (e.g., near centromeres) are heterochromatic.

As cells enter M phase, the level of compaction changes dramatically. By the end of prophase, sister chromatids are entirely heterochromatic. Two parallel chromatids would have a larger diameter of approximately 1,400 nm but a much shorter length compared to interphase chromosomes. These highly condensed metaphase chromosomes undergo little gene transcription because it is difficult for transcription proteins to gain access to the compacted DNA. Therefore, most transcriptional activity ceases during M phase, although a few specific genes may be transcribed. M phase is usually a short period of the cell cycle.

In highly condensed chromosomes, such as those found in metaphase, the radial loops are highly compacted and remain anchored to a **scaffold,** which is formed from the nuclear matrix. Experimentally, it is possible to delineate the proteins of the scaffold that hold the loops in place. This point is emphasized in the micrographs shown in figure 10.22. At the *left* is a human metaphase chromosome. In this condition, the radial loops of DNA are in a very compact configuration. If this chromosome is treated with a high concentration of salt to remove both the core

and linker histones, the highly compact configuration is lost, but the bottoms of the elongated loops remain attached to the scaffold. An *arrow* points to an elongated DNA strand emanating from the darkly staining scaffold. It is remarkable that the scaffold retains the shape of the original metaphase chromosome even though the DNA strands have become greatly elongated. These results illustrate the importance of both the nuclear matrix proteins (which form the scaffold) and the histones (which are needed to compact the radial loops) for the structure of metaphase chromosomes.

Researchers are trying to understand the steps that lead to the formation and organization of metaphase chromosomes. During the past several years, studies in yeast and frog oocytes have been aimed at the identification of proteins that promote the conversion of interphase chromosomes into metaphase chromosomes. It has been possible to isolate mutants in yeast that have alterations in the condensation or the segregation of chromosomes. Similarly, biochemical studies using frog oocytes resulted in the purification of protein complexes that promote chromosomal condensation or sister chromatid alignment. These two lines of independent research produced the same results. It was found that cells contain two multiprotein complexes called **condensin** and **cohesin,** which play a critical role in chromosomal condensation and sister chromatid alignment, respectively.

Condensin and cohesin are two completely distinct complexes, but both contain a category of proteins called **SMC proteins.** The acronym SMC stands for <u>s</u>tructural <u>m</u>aintenance of <u>c</u>hromosomes. SMC proteins use energy from ATP to catalyze changes in chromosome structure. Together with topoisomerases, SMC proteins have been shown to promote major changes in DNA structure. These include effects on DNA supercoiling and the formation of **catenanes,** which are intertwined DNA molecules. The topic of catenanes is discussed in chapter 11.

As their names suggest, cohesin and condensin play different roles in metaphase chromosome structure. Prior to M phase, condensin is found outside the nucleus (fig. 10.23). However, as M phase begins, condensin is observed to coat the individual chromatids and bring about the conversion of euchromatin into heterochromatin. Though the molecular mechanism of compaction is not fully understood, the function of condensin is to promote the compaction of the radial loops. This idea is shown in the insets to figure 10.23.

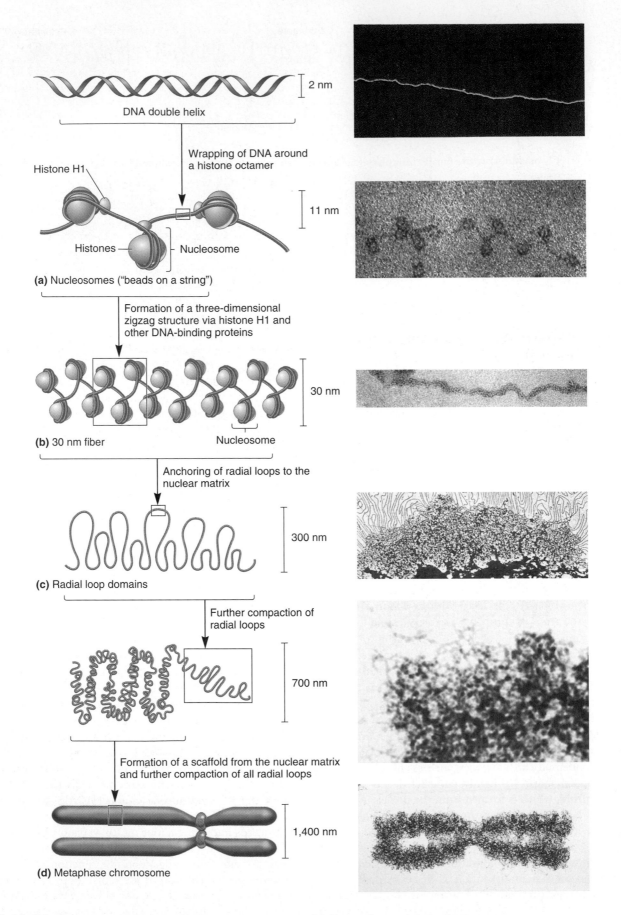

2 nm

DNA double helix

Wrapping of DNA around
a histone octamer

Histone H1

11 nm

Histones   Nucleosome

(a) Nucleosomes ("beads on a string")

Formation of a three-dimensional
zigzag structure via histone H1 and
other DNA-binding proteins

30 nm

(b) 30 nm fiber   Nucleosome

Anchoring of radial loops to the
nuclear matrix

300 nm

(c) Radial loop domains

Further compaction of
radial loops

700 nm

Formation of a scaffold from the nuclear matrix
and further compaction of all radial loops

1,400 nm

(d) Metaphase chromosome

FIGURE 10.21   The steps in eukaryotic chromosomal compaction leading to the metaphase chromosome.

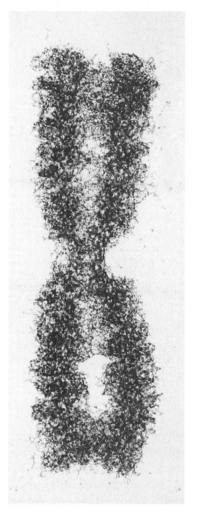

(a) Metaphase chromosome

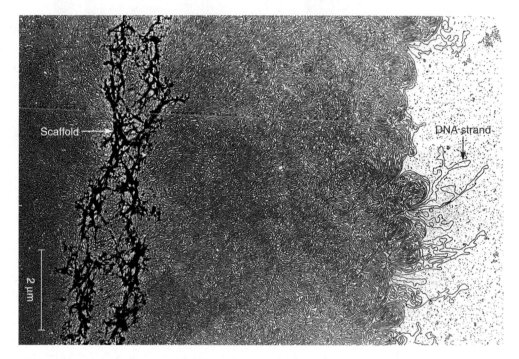

Scaffold

DNA strand

2 μm

(b) Metaphase chromosome treated with high salt to remove histone proteins

FIGURE 10.22   **The importance of histones and scaffolding proteins in the compaction of eukaryotic chromosomes.** (a) A metaphase chromosome. (b) A metaphase chromosome following treatment with high salt concentration to remove the histone proteins. The *arrow* on the *left* points to the scaffold, which anchors the bases of the radial loops. The *arrow* on the *right* points to an elongated strand of DNA.

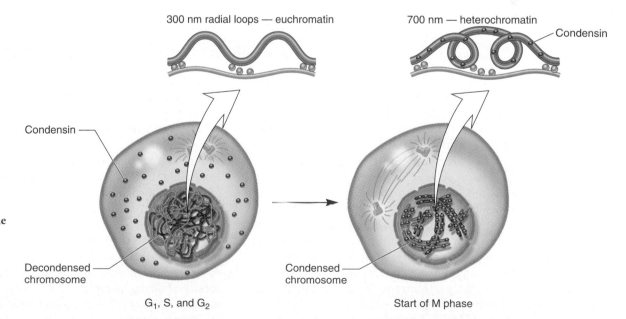

300 nm radial loops — euchromatin

700 nm — heterochromatin

Condensin

Condensin

Decondensed chromosome

Condensed chromosome

G_1, S, and G_2

Start of M phase

FIGURE 10.23   **The condensation of a metaphase chromosome by condensin.** During interphase (i.e., G_1, S, and G_2) most of the condensin protein is found outside the nucleus. The interphase chromosomes are largely euchromatic. At the start of M phase, condensin travels into the nucleus and binds to the chromosomes, causing them to become heterochromatic. As depicted in the insets, the effect of condensin is to cause a greater compaction of the radial loop domains.

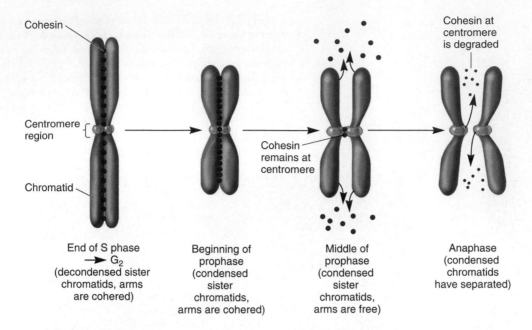

End of S phase → G₂ (decondensed sister chromatids, arms are cohered)

Beginning of prophase (condensed sister chromatids, arms are cohered)

Middle of prophase (condensed sister chromatids, arms are free)

Anaphase (condensed chromatids have separated)

FIGURE 10.24   The alignment of sister chromatids via cohesin. After S phase is completed, many cohesin complexes bind along each chromatid, thereby facilitating their attachment to each other. During the middle of prophase, cohesin is released from the chromosome arms, but some cohesin remains in the centromeric regions. At anaphase, the remaining cohesin complexes are rapidly degraded, which promotes sister chromatid separation.

By comparison, the function of cohesin is to promote the binding (i.e., cohesion) between sister chromatids. After S phase and until the middle of prophase, sister chromatids remain attached to each other along their length. As shown in figure 10.24, the attachment is promoted by cohesin, which is found along the entire length of each chromatid. During prophase, cohesins that are located along the chromosome arms are released. This allows the chromosome arms to separate. However, some cohesins remain attached to the centromeric regions, leaving the centromeric region as the only linkage before anaphase. At anaphase, the cohesins bound to the centromere are rapidly degraded, thereby allowing sister chromatid separation.

CONCEPTUAL SUMMARY

The chromosomal location and function of DNA sequences are central to our understanding of genetics. **Viruses** contain relatively small genomes that are composed of DNA or RNA. The **viral genome** is packaged into mature virus particles in an *assembly* process. In bacteria, several thousand different genes are interspersed throughout a circular chromosome; a single **origin of replication** is also present. Highly repetitive sequences, which may play a role in chromosome structure and gene function, are found throughout the chromosome. By comparison, eukaryotes have many linear chromosomes. A single **centromere** sequence is found on each chromosome; this functions as a recognition site for a **kinetochore** complex. At the ends of the linear chromosomes are **telomeres.** Hundreds or even thousands of different genes are located between the centromeres and telomeres. An unusual feature of many eukaryotic species is that their DNA contains an abundance of **highly repetitive sequences.** The functional importance of these sequences remains unclear and is discussed further in chapter 17.

A second important issue that underlies chromosomal structure and function is the amount of compaction that occurs within chromosomal DNA. In bacteria, two main events, chromosomal looping and **DNA supercoiling,** are responsible for the folding of the circular chromosome into a compact structure within a **nucleoid.** Certain DNA-binding proteins are likely involved in this process. In eukaryotic **chromatin,** the DNA is first folded into **nucleosomes,** which contain the DNA double helix wrapped around **histone proteins.** The nucleosomes are then compacted into a 30 nm structure. The **30 nm fiber** is further organized into **radial loop domains,** which are anchored to the **nuclear matrix.** This is termed **euchromatin.** During interphase, eukaryotic chromosomes are found within the cell nucleus. They consist of short, highly compacted regions that are **heterochromatic** and less condensed **euchromatic** regions. During M phase, the chromosomes become highly condensed and entirely heterochromatic. **Condensin** plays a key role in this heterochromatic conversion. **Cohesin** promotes the alignment of sister chromatids.

Our understanding of genome complexity and organization has been aided by various cytological, genetic, biochemical, and molecular techniques. In chapter 10, we have considered equilibrium density centrifugation and renaturation experiments as techniques to probe the complexity of genomes. In chapters 5 and 6, we examined ways to genetically map the locations of genes along eukaryotic and bacterial chromosomes. In chapter 20 we will explore the use of cytological techniques such as *in situ* hybridization to identify particular sequences within an intact chromosome. In addition, chapter 20 will describe molecular methods in which segments of a genome are cloned and analyzed molecularly.

The second topic in chapter 10 was the compaction of chromosomes to fit within living cells. This phenomenon can also be examined via several different approaches. As seen throughout chapter 10, the microscopic analysis of chromosomes provides key insight into their structure. In addition, chromatin can be analyzed biochemically by many different methods. For example, the repeating nucleosome structure was verified by DNase I digestion (as described in fig. 10.15). More recently, the detailed structure of nucleosomes has been solved by crystallization studies. Currently, much research is aimed at identifying the proteins that are bound to chromosomal DNA and elucidating the roles that these proteins play in determining chromatin structure.

PROBLEM SETS & INSIGHTS

Solved Problems

S1. Here is a C_0t curve for a hypothetical eukaryotic species:

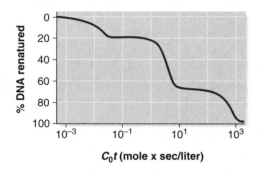

Estimate the amount of highly repetitive DNA, moderately repetitive DNA, and unique DNA.

Answer: About 20% is highly repetitive and renatures quickly, about 50% is moderately repetitive, and about 30% is unique and renatures very slowly.

S2. Let's suppose that a bacterial DNA molecule is given a left-handed twist. How does this affect the structure and function of the DNA?

Answer: A left-handed twist is negative supercoiling. Negative supercoiling makes the bacterial chromosome more compact. It also promotes DNA functions that involve strand separation. These include gene transcription and DNA replication.

S3. To hold bacterial DNA in a more compact configuration, specific proteins must bind to the DNA and stabilize its conformation (as shown in fig. 10.5). Several different proteins are probably involved in this process. These proteins have been collectively referred to as "histonelike" due to their possible functional similarity to the histone proteins found in eukaryotes. Based on your knowledge of eukaryotic histone proteins, what biochemical properties would you expect from bacterial histonelike proteins?

Answer: The histonelike proteins have the properties that are expected for proteins involved in DNA folding. They are all small proteins that are found in relative abundance within the bacterial cell. In some cases, the histonelike proteins are biochemically similar to eukaryotic histones. For example, they tend to be basic (i.e., positively charged) and bind to DNA in a non-sequence-dependent fashion. However, other proteins appear to bind to bacterial DNA at specific sites in order to promote DNA bending.

S4. If an organism is shown to have two different satellite DNAs by equilibrium density gradient centrifugation, what does this tell you about the types of sequences within the satellites? In other words, do the two satellites have the same or different repeated sequences? Explain your answer.

Answer: The two satellites must have different sequences. The density of the DNA depends on its base sequence.

Conceptual Questions

C1. In viral replication, what is the difference between self-assembly and directed assembly?

C2. Bacterial chromosomes have one origin of replication, while eukaryotic chromosomes have several. Would you expect viral chromosomes to have an origin of replication? Why or why not?

C3. What is a bacterial nucleoid? With regard to cellular membranes, what is the difference between a bacterial nucleoid and a eukaryotic nucleus?

C4. In part II of this textbook, we have considered inheritance patterns for diploid eukaryotic species. Bacteria frequently contain two or more nucleoids. With regard to genes and alleles, how is a bacterium that contains two nucleoids similar to a diploid eukaryotic cell, and how is it different?

C5. Describe the two main mechanisms by which the bacterial DNA becomes compacted.

C6. As described in chapter 9, 1 bp of DNA is approximately 0.34 nm in length. A bacterial chromosome is about 4 million bp in length and is organized into about 100 loops that are about 40,000 bp in length.

A. How long (in micrometers) is one loop, if it was stretched out linearly?

B. If a bacterial chromosomal loop is circular, what would be its diameter? (Note: Circumference = πD, where D is the diameter of the circle.)

C. Is the diameter of the circular loop calculated in part B small enough to fit inside a bacterium? The dimensions of a bacterium, such as *E. coli*, are roughly 0.5 μm wide and 1.0 μm long.

C7. Why is DNA supercoiling called supercoiling rather than just coiling? Why is positive supercoiling called overwinding and negative supercoiling called underwinding? How would you define the terms *positive* and *negative supercoiling* for Z DNA (described in chapter 9)?

C8. Coumarins and quinolones are two classes of drugs that inhibit bacterial growth by directly inhibiting DNA gyrase. Discuss two reasons why inhibiting DNA gyrase might inhibit bacterial growth.

C9. Take two pieces of string that are approximately 10 inches each, and create a double helix by wrapping the two strings around each other to make 10 complete turns. Tape one end of the strings to a table, and now twist the strings three times (360° each time) in a right-handed direction. Note: As you are looking down at the strings from above, a right-handed twist is in the clockwise direction.

A. Did the three turns create more or fewer turns in your double helix? How many turns are now in your double helix?

B. Is your double helix right-handed or left-handed? Explain your answer.

C. Did the three turns create any supercoils?

D. If you had coated your double helix with rubber cement and allowed the cement to dry before making the three additional right-handed turns, would the rubber cement make it more or less likely for the three turns to create supercoiling? Would a pair of cemented strings be more or less like a real DNA double helix compared to an uncemented pair of strings? Explain your answer.

C10. Try to explain the function of DNA gyrase with a drawing.

C11. How are two topoisomers different from each other? How are they the same?

C12. On rare occasions, a chromosome can suffer a small deletion that removes the centromere. When this occurs, the chromosome usually is not found within subsequent daughter cells. Explain why a chromosome without a centromere is not transmitted very efficiently from mother to daughter cells. (Note: If a chromosome is located outside the nucleus after telophase, it is degraded.)

C13. What is the function of a centromere? At what stage of the cell cycle would you expect the centromere to be the most important?

C14. Describe the characteristics of highly repetitive DNA.

C15. Describe the structures of a nucleosome and a 30 nm fiber.

C16. Beginning with the G_1 phase of the cell cycle, describe the level of compaction of the eukaryotic chromosome. How does the level of compaction change as the cell progresses through the cell cycle? Why is it necessary to further compact the chromatin during mitosis?

C17. If you assume that the average length of linker DNA is 50 bp, approximately how many nucleosomes are there in the haploid human genome, which contains 3 billion bp?

C18. Draw the binding between the nuclear matrix and MARs.

C19. Compare heterochromatin and euchromatin. What are the differences between them?

C20. Compare the structure and cell localization of chromosomes during interphase and M phase.

C21. What types of genetic activities occur during interphase? Explain why these activities cannot occur during M phase.

C22. Let's assume that the linker DNA averages 54 bp in length. How many molecules of H2A would you expect to find in a DNA sample that is 46,000 bp in length?

C23. In figure 10.16, what are we looking at in part b? Is this an 11 nm fiber, a 30 nm fiber, or a 300 nm fiber? Does this DNA come from a cell during M phase or interphase?

C24. What are the roles of the core histone proteins compared to histone H1 in the compaction of eukaryotic DNA?

C25. A typical eukaryotic chromosome found in humans contains about 100 million bp of DNA. As described in chapter 9, 1 bp of DNA has a linear length of 0.34 nm.

A. What is the linear length of the DNA for a typical human chromosome in micrometers?

B. What is the linear length of a 30 nm fiber of a typical human chromosome?

C. Based on your calculation of part B, would a typical human chromosome fit inside the nucleus (with a diameter of 5 μm) if the 30 nm fiber were stretched out in a linear manner? If not, explain how a typical human chromosome fits inside the nucleus during interphase.

C26. Which of the following terms should not be used to describe a Barr body?

A. Chromatin

B. Euchromatin

C. Heterochromatin

D. Chromosome

E. Genome

C27. Discuss the differences in the compaction levels of metaphase chromosomes compared to interphase chromosomes. When would you expect gene transcription and DNA replication to take place, during M phase or interphase? Explain why.

Experimental Questions

E1. Two circular DNA molecules, which we will call molecule A and molecule B, are topoisomers of each other. When viewed under the electron microscope, molecule A appears more compact compared to molecule B. The level of gene transcription is much lower for molecule A. Which of the following three possibilities could account for these observations?

First possibility: molecule A has +3 supercoils and molecule B has −3.

Second possibility: molecule A has +4 supercoils and molecule B has −1.

Third possibility: molecule A has 0 supercoils and molecule B has −3.

E2. Explain how a renaturation experiment can provide quantitative information about genome sequence complexity.

E3. In a renaturation experiment, does the copy number affect only the rate of renaturation or does it also affect the rate of denaturation? Explain your answer.

E4. Here are the results of an equilibrium density centrifugation experiment carried out on two different species (a mammal and an amphibian).

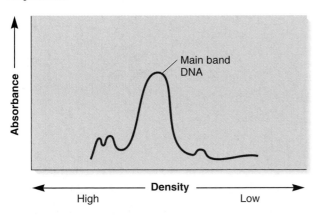

(a) Mammal

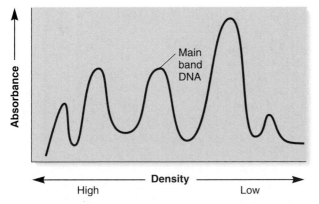

(b) Amphibian

The amphibian has 11 times as much DNA as the mammal. Discuss why the amphibian has so much more DNA.

E5. Let's suppose that you have isolated DNA from a cell and have viewed it under a microscope. It looks supercoiled. What experiment would you perform to determine if it is positively or nega-

tively supercoiled? In your answer, describe your expected results. You may assume that you have purified topoisomerases at your disposal.

E6. We seem to know more about the structure of eukaryotic chromosomal DNA than bacterial DNA. Discuss several experimental procedures that have yielded important information concerning the compaction of eukaryotic chromatin.

E7. Is it possible for repetitive DNA to sediment at the same density as the main band DNA? Explain.

E8. An organism contains 20% highly repetitive DNA, 10% moderately repetitive DNA, and 70% unique sequences. Draw the expected C_0t curve that would be obtained from this organism.

E9. In the experiment of figure 10.12, a sample of DNA is layered on top of a CsCl gradient. The bottom of the gradient contained approximately 70% CsCl and the top contained approximately 30% CsCl.

A. How would you make a CsCl gradient?

B. Would the gradient last forever?

C. Is it necessary to place the DNA on the top of the gradient or could you put it in the middle of the gradient and get the same results?

E10. Figure 10.12 describes the use of CsCl density gradient centrifugation to separate satellite DNA from the rest of the chromosomal DNA. This technique can be used to purify satellite DNA. For example, the fraction containing the 1.672 g/cm³ DNA can be separated away from the rest of the chromosomal DNA, including other satellite bands. If a purified fraction of satellite DNA was obtained by this method, and then subjected to a renaturation experiment, what results would you expect?

E11. When chromatin is treated with a moderate salt concentration, the linker histone H1 is removed (see fig. 10.16a). Higher salt concentration removes the rest of the histone proteins (see fig. 10.22b). If the experiment of fig. 10.15 were carried out after the DNA was treated with moderate or high salt, what would be the expected results?

E12. Let's suppose you have isolated chromatin from some bizarre eukaryote with a linker region that is usually 300 to 350 bp in length. The nucleosome structure is the same as in other eukaryotes. If you digested this eukaryotic organism's chromatin with a high concentration of DNase I, what would be your expected results?

E13. If you were given a sample of chromosomal DNA and asked to determine if it is bacterial or eukaryotic, what experiment would you perform and what would be your expected results?

E14. Consider how histone proteins bind to DNA and then explain why a high salt concentration can remove histones from DNA (as shown in fig. 10.22b).

E15. In chapter 20, the technique of fluorescence *in situ* hybridization (FISH) is described. This is another method that can be used to examine sequence complexity within a genome. In this method, a particular DNA sequence, such as a particular gene sequence, can be detected within an intact chromosome by using a DNA probe that is complementary to the sequence. For example, let's consider the β-globin gene, which is found on human chromosome 11. A probe that is complementary to the β-globin gene will bind to the β-globin gene and show up as a brightly colored spot on human chromosome 11. In this way, it is possible to detect where the β-globin gene is located within a set of chromosomes. Since the β-globin gene is unique, and since human cells are diploid (i.e., have two copies of each chromosome), a FISH experiment would show two bright spots per cell; the probe would bind to each copy of

chromosome 11. What would you expect to see if you used the following types of probes?

A. A probe that is complementary to the *Alu*I sequence

B. A probe that is complementary to a tandemly repeated sequence near the centromere of the X chromosome

Questions for Student Discussion/Collaboration

1. Bacterial and eukaryotic chromosomes are very compact. Discuss the advantages and disadvantages of having a compact structure.

2. The prevalence of highly repetitive sequences seems rather strange to many geneticists. Do they seem strange to you? Why or why not? Discuss whether or not you think they have an important function.

3. Discuss and make a list of the similarities and differences between bacterial and eukaryotic chromosomes.

Note: All answers appear at the website for this textbook; the answers to even-numbered questions are in the back of the textbook.

www.mhhe.com/brooker

Visit the Online Learning Center for practice tests, answer keys, and other learning aids for this chapter. Enhance your understanding of genetics with our interactive exercises, web links, news feeds, tutorial service, and much more.

DNA REPLICATION

::

11

As we have seen throughout chapters 2 to 8, genetic material is transmitted from parent to offspring and from cell to cell. For transmission to occur, the genetic material must be copied. During this process, known as **DNA replication,** the original DNA strands are used as templates for the synthesis of new DNA strands. We will begin chapter 11 with a consideration of the structural features of the double helix that pertain to the replication process. Then we will examine how chromosomes are replicated within living cells: where does DNA replication begin, how does it proceed, and where does it end? At the molecular level, it is rather remarkable that the replication of chromosomal DNA occurs very quickly, very accurately, and at the appropriate time in the life of the cell. For this to happen, many cellular proteins play vital roles. In chapter 11, we will examine the mechanism of DNA replication and consider the functions of several proteins involved in the process.

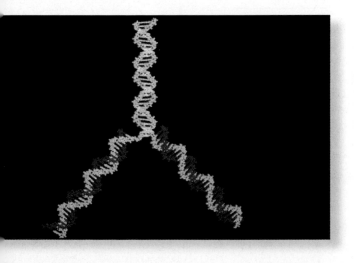

11.1 STRUCTURAL OVERVIEW OF DNA REPLICATION

We begin by recalling a few important structural features of the double helix from chapter 9, since they bear directly on the replication process. The double helix is composed of two DNA strands, and the individual building blocks of each strand are nucleotides. The nucleotides contain one of four bases: adenine, thymine, guanine, or cytosine. The double-stranded structure is held together by hydrogen bonding between the bases in opposite strands. A critical feature of the double helix structure is that it will fit together only if adenine hydrogen bonds with thymine and guanine hydrogen bonds with cytosine. This rule, known as the AT/GC rule or Chargaff's rule, is the basis for the complementarity of the base sequences in double-stranded DNA.

Another feature worth noting is that the strands within a double helix have an antiparallel alignment. This directionality is determined by the orientation of sugar residues within the sugar-phosphate backbone. If one strand is running in the 5′ to 3′ direction, the complementary strand is running in the 3′ to 5′ direction. The issue of directionality will be important later in chapter 11, when we consider the function of the enzymes that synthesize new DNA strands. In this section, we will consider how the structure of the DNA double helix provides a basis for DNA replication.

Existing DNA Strands Act as Templates for the Synthesis of New Strands

As shown in figure 11.1a, DNA replication relies on the complementarity of DNA strands according to the AT/GC rule. During the replication process, the two complementary strands of DNA come apart and serve as **template strands** for the synthesis of two new strands of DNA. After the double helix has separated,

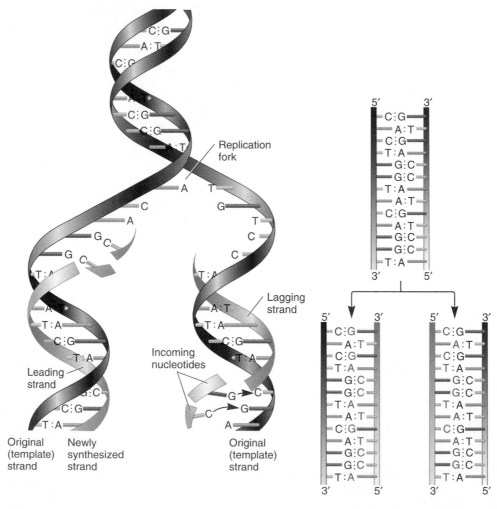

(a) The mechanism of DNA replication

(b) The products of replication

FIGURE 11.1   **The structural basis for DNA replication. (a)** The mechanism of DNA replication as originally proposed by Watson and Crick. As discussed later in chapter 11, the synthesis in the leading strand occurs in the direction toward the replication fork, while the synthesis of the lagging strand occurs in small segments away from the replication fork. **(b)** DNA replication produces two copies of DNA with the same sequence as the original DNA molecule.

individual nucleotides have access to the template strands. Hydrogen bonding between individual nucleotides and the template strands must obey the AT/GC rule. To complete the replication process, a covalent bond is formed between the phosphate on one nucleotide and the sugar on the previous nucleotide. The two newly made strands are referred to as the **daughter strands,** while the original strands are called the **template strands** or **parental strands.** Note that the base sequences are identical in both double-stranded molecules after replication (fig. 11.1*b*). Therefore, DNA can be replicated so that both copies retain the same information (i.e., the same base sequence) as the original molecule.

EXPERIMENT 11A

Three Different Models Were Proposed That Described the Net Result of DNA Replication

Scientists in the late 1950s had considered three different mechanisms to produce the net result of DNA replication. These mechanisms are shown in figure 11.2. The first is referred to as a **conservative** model. According to this hypothesis, both strands of parental DNA remain together following DNA replication. In this model, the original arrangement of parental strands is completely conserved, while the two newly made daughter strands are also together following replication. The second is called a **semiconservative** model. As seen in figure 11.2*b*, the double-stranded DNA is half conserved following the replication process. In other words, the double-stranded DNA contains one parental strand and one daughter strand. The third, **dispersive,** model proposes that segments of parental DNA and newly made DNA are interspersed in both strands following the replication process. As we will see, only the semiconservative model shown in figure 11.2*b* is actually correct.

In 1958, Matthew Meselson and Franklin Stahl devised a method to investigate this question in which they could experimentally distinguish newly made daughter strands from the original parental strands. The technique they used involved heavy isotope labeling. Nitrogen, which is found within the bases of DNA, occurs in a light (^{14}N) form and a heavy (^{15}N) form. Prior to their experiment, they grew *E. coli* cells in the presence of ^{15}N for many generations. This produced a population of cells in which all the DNA was heavy labeled. At the start of their experiment, shown in figure 11.3 (generation 0), they switched the bacteria to a medium that contained only ^{14}N and then collected samples of cells after various time points. Under the growth conditions they employed, 30 minutes is the time required for one doubling, or one generation time. Since the bacteria were doubling in a medium that contained only ^{14}N, all of the newly made DNA strands would be labeled with light nitrogen, while the original strands would remain heavy.

Meselson and Stahl then analyzed the density of the DNA by centrifugation using a CsCl gradient. (The procedure of gradient

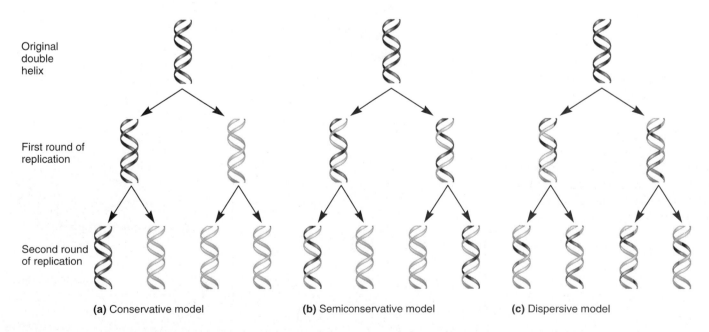

Original double helix

First round of replication

Second round of replication

(a) Conservative model   **(b)** Semiconservative model   **(c)** Dispersive model

FIGURE 11.2   **Three possible models for DNA replication.** The two original parental DNA strands are shown in *purple,* and the newly made strands after one or two generations are shown in *light blue.*

centrifugation is described in the appendix and was also explained in the experiment of fig. 10.12.) If both DNA strands contained ^{14}N, the DNA would have a light density and sediment near the top of the tube. If one strand contained ^{14}N and the other strand contained ^{15}N, the DNA would be half-heavy and have an intermediate density. Finally, if both strands contained ^{15}N, the DNA would be heavy and sediment closer to the bottom of the centrifuge tube.

■ **THE HYPOTHESIS**

One of the three models shown in figure 11.2 accurately describes the net result of DNA replication. The purpose of this experiment is to determine which of the three models is correct.

■ **TESTING THE HYPOTHESIS** — **FIGURE 11.3** Evidence that DNA replication is semiconservative.

Starting material: A strain of *E. coli* that has been grown for many generations in the presence of ^{15}N. All of the nitrogen in the DNA is labeled with ^{15}N.

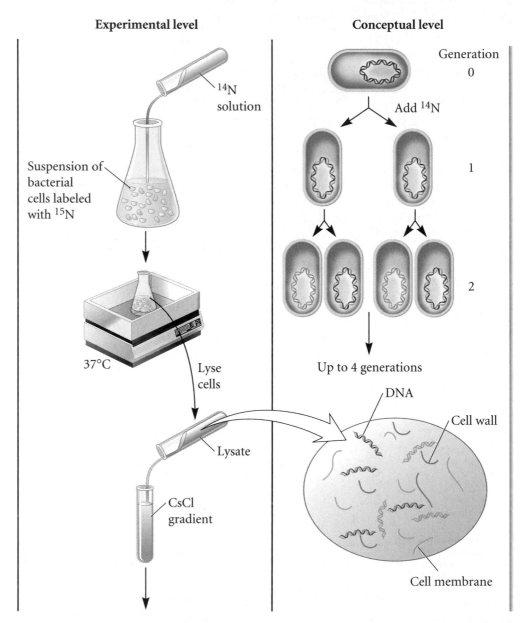

Experimental level

Conceptual level

1. Add an excess of ^{14}N-containing compounds to the bacterial cells so that all the newly made DNA will contain ^{14}N.

   ^{14}N solution

   Suspension of bacterial cells labeled with ^{15}N

   Generation 0

   Add ^{14}N

   1

2. Incubate the cells for various lengths of time. Note: The ^{15}N-labeled DNA is shown in *purple* and the ^{14}N-labeled DNA is shown in *blue*.

   2

3. Lyse the cells by the addition of lysozyme and detergent, which disrupt the bacterial cell wall and cell membrane respectively.

   37°C   Lyse cells

   Up to 4 generations

4. Load a sample of the lysate onto a CsCl gradient that contains ethidium bromide. (Note: The average density of DNA is around 1.7 gm/cm^3, which is well isolated from other cellular macromolecules.)

   Lysate

   CsCl gradient

   DNA

   Cell wall

   Cell membrane

(continued)

5. Centrifuge the gradients until the DNA molecules reach their equilibrium densities.

6. DNA within the gradient can be observed with a UV light since the DNA is now stained with ethidium bromide, a UV-sensitive dye.

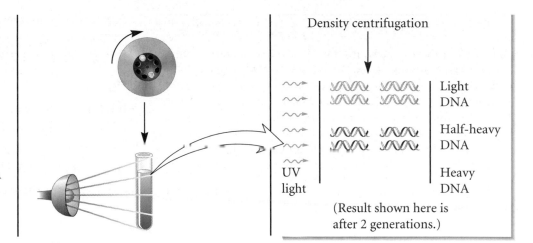

Density centrifugation

Light DNA

Half-heavy DNA

Heavy DNA

UV light

(Result shown here is after 2 generations.)

■ THE DATA

Generations After ^{14}N Addition

4.1   3.0   2.5   1.9   1.5   1.1   1.0   0.7   0.3

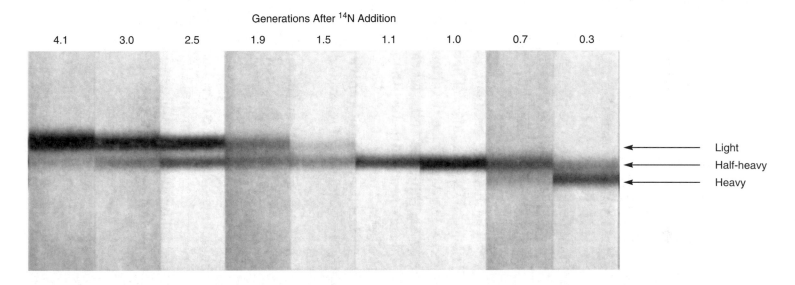

Light
Half-heavy
Heavy

■ INTERPRETING THE DATA

As seen in the data of figure 11.3, after one complete generation (i.e., one round of DNA replication), all of the DNA sedimented at a density that was half-heavy. These results are consistent with both the semiconservative and dispersive models. The conservative model would predict two separate DNA types: a light type and a heavy type. Since all the DNA had sedimented as a single band, this model was disproved. According to the semiconservative model, the replicated DNA would contain one original strand (a heavy strand) and a newly made daughter strand (a light strand). Likewise, in a dispersive model, there should have been all half-heavy DNA after one generation as well. To determine which of these two remaining models is correct, therefore, Meselson and Stahl had to investigate future generations.

After approximately two generations (i.e., 1.9 generations), there was a mixture of light DNA and half-heavy DNA. This observation was also consistent with the semiconservative mode of DNA replication, because some DNA molecules should contain all light DNA while other molecules should be half-heavy (see fig. 11.2b). In a dispersive model, however, the DNA after two generations would have been 1/4 heavy. This is because a dispersive model predicts that the heavy nitrogen would be evenly dispersed among four strands, each strand containing 1/4 heavy nitrogen and 3/4 light nitrogen (see fig. 11.2c). However, this result was not obtained. Instead, the results of the Meselson and Stahl experiment provided compelling evidence in favor of the semiconservative model for DNA replication.

A self-help quiz involving this experiment can be found at the Online Learning Center.

11.2 BACTERIAL DNA REPLICATION

Thus far, we have considered how a complementary, double-stranded structure underlies the ability of DNA to be copied. In addition, the experiments of Meselson and Stahl showed that DNA replication results in two double helices, each one containing an original parental strand and a newly made daughter strand. We will now turn our attention to how DNA replication actually occurs within living cells. Much research has focused on the bacterium *E. coli.* The results of these studies have provided

the foundation for our current molecular understanding of DNA replication. The replication of the bacterial chromosome is a stepwise process in which many cellular proteins participate. In this section, we will follow this process from beginning to end.

Bacterial Chromosomes Contain a Single Origin of Replication

Figure 11.4 presents an overview of the process of bacterial chromosomal replication. The site on the bacterial chromosome where

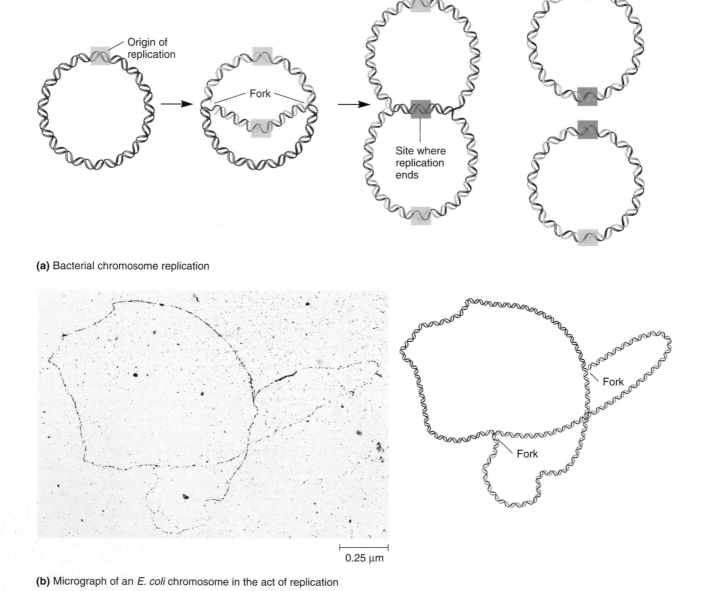

(a) Bacterial chromosome replication

(b) Micrograph of an *E. coli* chromosome in the act of replication

FIGURE 11.4 **The process of bacterial chromosome replication.** (a) An overview of the process of bacterial chromosomal replication. **(b)** A replicating *E. coli* chromosome visualized by autoradiography. This chromosome was radiolabeled by growing bacterial cells in media containing radiolabeled thymidine. The diagram at the *right* shows the locations of the two replication forks. The chromosome is about one-third replicated. New strands are shown in *blue.*

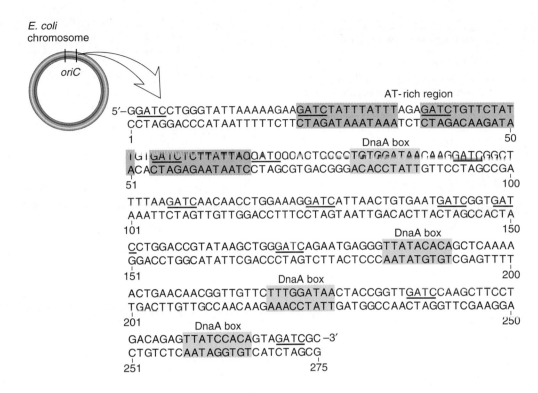

FIGURE 11.5 **The sequence of *oriC* in *E. coli*.** The GATC methylation sites are *underlined*. The DnaA boxes are highlighted in *orange*. The AT-rich region is composed of three tandem repeats that are 13 base pairs long and highlighted in *blue*. The repeats have similar sequences and are AT rich. The region immediately upstream of the first tandem repeat is also AT rich.

DNA synthesis begins is known as the **origin of replication.** Bacterial chromosomes have a single origin of replication. The synthesis of new daughter strands is initiated within the origin and proceeds **bidirectionally** (in both directions) around the bacterial chromosome. This means that two replication forks move in opposite directions outward from the origin. Eventually, these replication forks meet each other on the opposite side of the bacterial chromosome to complete the replication process.

Replication Is Initiated by the Binding of DnaA Protein to the Origin of Replication

Considerable research has focused on the origin of replication in *E. coli*. This origin is named *oriC* (for *ori*gin of *C*hromosomal replication; fig. 11.5). There are three types of DNA sequences within *oriC* that are functionally important: an AT-rich region, DnaA box sequences, and GATC methylation sites. The third functional sequence, GATC methylation sites, will be discussed later when we consider the regulation of replication.

DNA replication is initiated by the binding of **DnaA proteins** to sequences within the origin known as **DnaA box sequences.** The DnaA box serves as a recognition site for the binding of the DnaA protein. As shown in figure 11.6, DnaA proteins bind to the four DnaA boxes in *oriC* to initiate DNA repli-

cation. After a DnaA protein is bound to each DnaA box, this stimulates the cooperative binding of an additional 20 to 40 DnaA proteins to form a large complex. With the aid of other DNA-binding proteins such as HU and IHF, this causes the DNA to bend around the complex of DnaA proteins and results in the separation of the AT-rich region. Since AT base pairs form only two hydrogen bonds while GC base pairs form three, it is easier to separate two DNA strands at an AT-rich region.

Following denaturation of the AT-rich region, the DnaA proteins, with the help of the DnaC protein, recruit **DNA helicase** enzymes (also known as DnaB helicase, or simply as helicase) to bind to this site. Two DNA helicases begin strand separation within the *oriC* region and continue to separate the DNA strands beyond the origin. These enzymes use the energy from ATP hydrolysis to catalyze the separation of the double-stranded parental DNA (fig. 11.6). DNA helicases bind to single-stranded DNA and travel along the DNA in a 5′ to 3′ direction to keep the fork moving. When a helicase encounters a double-stranded region, it breaks the hydrogen bonds between the two strands, thereby generating two single strands. As shown in figure 11.6, the action of DNA helicases promotes the movement of two forks outward from *oriC* in opposite directions. This initiates the replication of the bacterial chromosome in both directions, an event termed **bidirectional replication.**

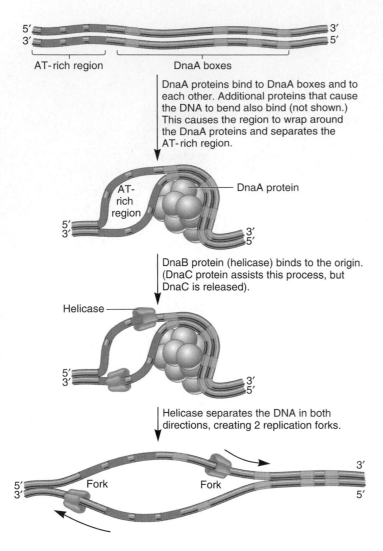

FIGURE 11.6   The events that occur at *oriC* to initiate the DNA replication process. To initiate DNA replication, DnaA proteins bind to the four DnaA boxes, which causes the DNA strands to separate. DnaA and DnaC proteins then recruit helicase into this region. Each helicase is composed of six subunits, which form a ring around one DNA strand and migrates in the 5′ to 3′ direction. As shown here, the movement of two helicase proteins serves to separate the DNA strands beyond the *oriC* region.

DNA Strand Separation and the Synthesis of RNA Primers Are Necessary Before Daughter Strands of DNA Can Be Made

Figure 11.7 provides an overview of the molecular events that occur as one of the two replication forks moves around the bacterial chromosome. Table 11.1 gives a summary of the functions of the major proteins involved in *E. coli* DNA replication. Throughout the next several pages, we will focus on the steps that occur at the replication fork. This will require that you occasionally refer back to figure 11.7 and table 11.1 to get an overview of the entire process.

Let's begin with strand separation. To act as a template for DNA replication, the strands of a double helix must separate. As

TABLE 11.1
Proteins Involved in *E. coli* DNA Replication

Common Name	Function
DnaA protein	Binds to DnaA boxes within the origin to initiate DNA replication
DNA helicase (DnaB)	Separates double-stranded DNA
DnaC protein	Aids DnaA in the recruitment of helicase to the origin
Topoisomerase	Removes supercoils ahead of the replication fork
DNA primase	Synthesizes short RNA primers
DNA polymerase (polIII)	Synthesizes DNA in the leading and lagging strands
DNA polymerase (polI)	Removes RNA primers, fills in gaps
DNA ligase	Covalently attaches adjacent Okazaki fragments
Tus	Binds to *ter* sequences and prevents the advancement of the replication fork

mentioned previously, the function of DNA helicase is to break the hydrogen bonds between base pairs and thereby separate the strands. The action of DNA helicase is believed to generate positive supercoiling ahead of each replication fork. As shown in figure 11.7, an enzyme known as **DNA gyrase** (a **topoisomerase type II** enzyme described in chapter 10) travels ahead of the helicase enzyme and alleviates this supercoiling. Its mechanism of action is also described in chapter 10.

After the two parental DNA strands have been separated and the supercoiling relaxed, they must be kept that way until the complementary daughter strands have been made. The function of the **single-strand binding protein** is to bind to both of the single strands of parental DNA and prevent them from re-forming a double helix. In this way, the bases within the parental strands are kept in an exposed condition that enables them to hydrogen bond with single nucleotides.

The next event in DNA replication involves the synthesis of a short strand of RNA (rather than DNA) called an **RNA primer.** This strand of RNA is synthesized by the linkage of ribonucleotides via an enzyme known as **DNA primase** or simply **primase.** This enzyme synthesizes short strands of RNA, typically 10 to 12 nucleotides in length. These short RNA strands start, or prime, the process of DNA replication (fig. 11.7). At a later stage in DNA replication, the RNA primers are removed.

DNA Polymerases Link Nucleotides to Synthesize the Daughter Strands

The enzymes known as **DNA polymerases** are responsible for covalently attaching nucleotides together to make new daughter strands. In *E. coli,* there are five distinct proteins that function as DNA polymerases, designated polI, II, III, IV, and V. PolI and polIII are involved in normal DNA replication, while polII, polIV, and polV play a role in DNA repair and the replication of damaged DNA. DNA polymerase I is composed of a single subunit;

Functions of key proteins involved with DNA replication

- DNA helicase breaks the hydrogen bonds between the DNA strands.

- Topoisomerase alleviates positive supercoiling.

- Single-strand binding proteins keep the parental strands apart.

- Primase synthesizes an RNA primer.

- DNA polymerase III synthesizes a daughter strand of DNA.

- DNA polymerase I excises the RNA primers and fills in with DNA (not shown).

- DNA ligase covalently links the DNA fragments together.

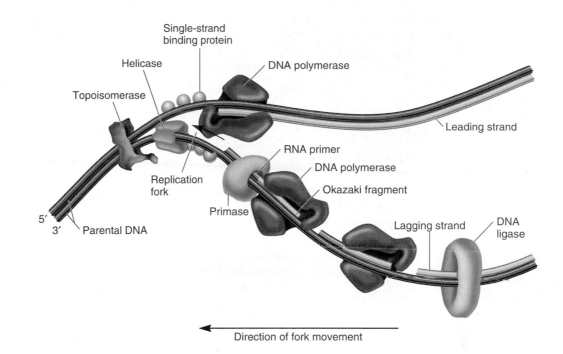

FIGURE 11.7   Enzymology of DNA replication.

its role during DNA replication is to fill in small regions where RNA primers were located. By comparison, DNA polymerase III, which is responsible for most of the DNA replication, is a large enzyme composed of multiple subunits. Table 11.2 summarizes the subunit composition of polIII. As seen here, polIII consists of 10 different subunits that play various roles in the DNA replication process. The α subunit actually catalyzes the bond formation between adjacent nucleotides, while the remaining 9 subunits fulfill other functions. The complex of all 10 subunits together is called DNA polymerase III holoenzyme.

Though the various DNA polymerases in *E. coli* and other bacterial species vary in their subunit composition, several common structural features have emerged. The catalytic subunit of all DNA polymerases has a structure that resembles a human right hand. As shown in figure 11.8, the template DNA is threaded through the palm of the hand; the thumb and fingers are wrapped around the DNA.

As researchers began to unravel the function of DNA polymerase, there were two features that seemed unusual. These oddities are depicted in figure 11.9. As seen here, DNA polymerase cannot begin DNA synthesis by linking together two individual nucleotides. Rather, this enzyme can only elongate a strand starting with an RNA primer or existing DNA strand (fig. 11.9*a*). A second unusual feature is the directionality of strand synthesis. DNA polymerases can attach nucleotides only in the 5′ to 3′ direction, not in the 3′ to 5′ direction (fig. 11.9*b*).

At first glance, these two characteristics may seem to pose problems during the synthesis of DNA. Let's now turn back to figure 11.7 and see how these problems are overcome. Additional

TABLE 11.2

Subunit Composition of DNA Polymerase (polIII) Holoenzyme from *E. coli*

Subunit	Function
α	Synthesizes DNA
ε	3′ to 5′ proofreading (removes mismatched nucleotides)
θ	Accessory protein that stimulates the proofreading function
τ	Promotes the dimerization of two polIII proteins together at the replication fork; also, stimulates DNA helicase
β	Clamp protein, which allows DNA polymerase to slide along the DNA without falling off
γ	Clamp loader protein; initially helps the clamp protein to bind to the DNA
δ	Accessory protein that binds to β
δ'	Accessory protein that stimulates γ
ψ	Accessory protein that stimulates γ
χ	Accessory protein that binds to single-strand binding protein

enzymes within the replication fork compensate for these two unusual features of DNA polymerases. Also, the two new DNA strands (namely, the leading and lagging strands) are synthesized in different ways.

The synthesis of an RNA primer by DNA primase allows DNA polymerase to begin the synthesis of complementary

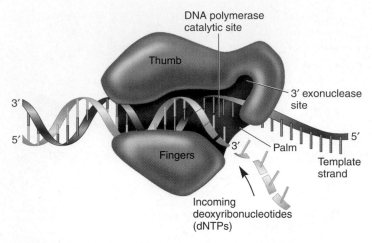

(a) Schematic side view of DNA polymerase

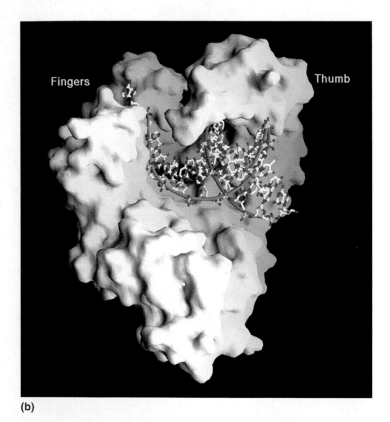

(b)

FIGURE 11.8 **The action of DNA polymerase.** (a) DNA polymerase slides along the template strand as it synthesizes a new strand by connecting deoxyribonucleoside triphosphates (dNTPs) in a 5′ to 3′ direction. The structure of DNA polymerase resembles a hand that is wrapped around the template strand. In this regard, the movement of DNA polymerase along the template strand is similar to a hand that is sliding along a rope. (b) The molecular structure of DNA polymerase I from the bacterium *Thermus aquaticus.* This model shows a portion of DNA polymerase I that is bound to DNA. This molecular structure depicts a front view of DNA polymerase, while part (a) is a schematic side view.

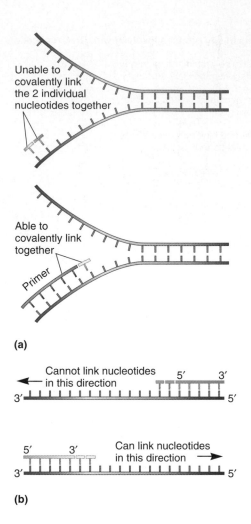

(a)

(b)

FIGURE 11.9 **Unusual features of DNA polymerase function.** (**a**) DNA polymerase can elongate a strand only from an RNA primer or existing DNA strand. (**b**) DNA polymerase can attach nucleotides only in a 5′ to 3′ direction. Note the template strand is in the opposite, 3′ to 5′, direction.

daughter strands of DNA. DNA polymerase catalyzes the attachment of nucleotides to the 3′ end of the primer, in a 5′ to 3′ direction. In the **leading strand,** one RNA primer is made at the origin and then DNA polymerase III can attach nucleotides in a 5′ to 3′ direction as it slides toward the opening of the replication fork (fig. 11.7).

In the **lagging strand,** the synthesis of DNA is also in a 5′ to 3′ manner, but it occurs in the direction away from the replication fork. In the lagging strand, short segments of DNA are made. The length of these fragments in bacteria is approximately 1,000 to 2,000 nucleotides. Each fragment contains a short RNA primer at the 5′ end, which is made by primase; the remainder of the fragment is a strand of DNA made by DNA polymerase III. To make the next fragment, primase is released from the DNA and then binds closer to the replication fork. The DNA fragments made in this manner are known as **Okazaki fragments,** after Reiji and Tuneko Okazaki, who initially discovered them in the late 1960s. In their studies, they incubated *E. coli* cells with radiolabeled thymidine for 15 seconds and then added an excess of nonlabeled thymidine. This is termed a pulse/chase experiment because the cells were given the radiolabeled compound for a brief period of time (i.e., a pulse) followed by an excess amount of unlabeled compound (i.e., a chase). The Okazakis then isolated DNA from samples of cells at timed intervals after the pulse/chase. The DNA was denatured into single-stranded molecules and the sizes of the radiolabeled DNA strands were determined by centrifugation. At short time intervals (i.e., only a few seconds following the thymidine incubation), the fragments were found to be small, in the range of 1,000 to 2,000 nucleotides in length. At longer time intervals, the radiolabeled strands became much longer. At these later time points, the adjacent Okazaki fragments had enough time to link together.

To complete the synthesis of Okazaki fragments within the lagging strand, three additional events must occur: removal of the RNA primers, synthesis of DNA in the area where the primers

have been removed, and the covalent attachment of adjacent fragments of DNA (refer back to fig. 11.7). In *E. coli*, the RNA primers are removed by the action of DNA polymerase I. This enzyme has a 5′ to 3′ exonuclease activity, which means that polI digests away the RNA primers in a 5′ to 3′ direction leaving a vacant area. PolI can then fill in this region by synthesizing DNA there. After the gap has been completely filled in, a covalent bond is still missing between the last nucleotide added by polI and the first nucleotide in the next DNA fragment. An enzyme known as **DNA ligase** catalyzes a covalent bond between these adjacent DNA fragments to complete the process in the lagging strand.

DNA Polymerase III Is a Processive Enzyme That Uses Deoxynucleoside Triphosphates

Let's now turn our attention to the enzymatic features of DNA polymerase. As shown in figure 11.10, DNA polymerases catalyze the covalent attachment between the phosphate in one nucleotide

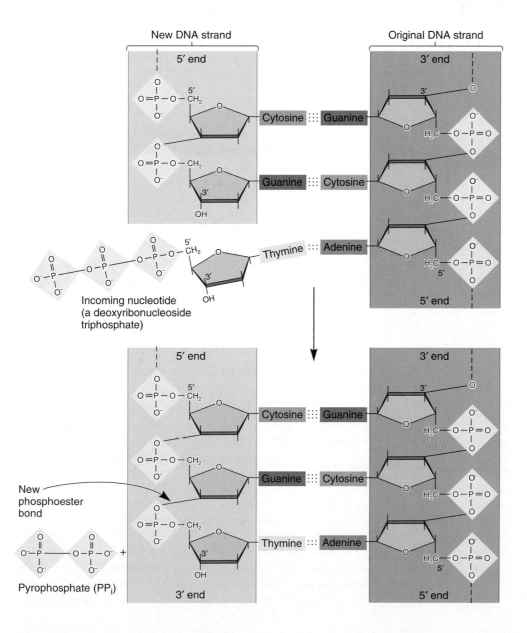

FIGURE 11.10 The enzymatic action of DNA polymerase. DNA polymerase catalyzes the formation of a covalent bond between the 3′ —OH group on the previous nucleotide and the innermost 5′ phosphate group on the incoming nucleotide. Pyrophosphate (PPᵢ) is released.

and the sugar in the previous nucleotide. Prior to this event, the nucleotide that is about to be attached to the growing strand is a deoxyribonucleoside triphosphate. It contains three phosphate groups attached at the $5'$ –OH group of deoxyribose. The deoxyribonucleoside triphosphate first hydrogen bonds to the template strand according to the AT/GC rule. Next, the $3'$ –OH group on the previous nucleotide reacts with the phosphate group adjacent to the sugar on the incoming nucleotide to form a covalent bond. The formation of this covalent bond causes the newly made strand to grow in the $5'$ to $3'$ direction. As shown in figure 11.10, pyrophosphate (PO_4–PO_3) is released.

DNA polymerase catalyzes the covalent attachment of nucleotides with great speed. In *E. coli,* DNA polymerase III attaches approximately 750 nucleotides per second! DNA polymerase III can catalyze the synthesis of the daughter strands so quickly because it is a **processive enzyme.** This means that it does not dissociate from the growing strand after it has catalyzed the covalent joining of two nucleotides. Rather, as depicted earlier in figure 11.8, it remains clamped to the DNA template strand and slides along the template as it catalyzes the synthesis of the daughter strand. The β subunit of the holoenzyme, also known as the clamp protein, promotes the association of the holoenzyme to DNA as it glides along the template strand (refer back to table 11.2). As you might expect, the β subunit is the shape of a ring; the hole of the ring is large enough to accommodate a double-stranded DNA molecule, and its width is about one turn of DNA. The γ subunit is needed for the β subunit to initially clamp onto the DNA. For this reason, the γ subunit is sometimes called the clamp-loader protein. It initially loads the holoenzyme onto the template strand. Other accessory subunits known as δ, δ', and ψ are needed for the optimal function of the β and γ subunits.

The effects of processivity are really quite remarkable. In the absence of the β subunit, DNA polymerase can synthesize DNA only at a rate of approximately 20 nucleotides per second; on average, it falls off the DNA template after a few dozen nucleotides have been linked together. By comparison, when the β subunit is present, as in the holoenzyme, the synthesis rate is approximately 750 nucleotides per second and (in the leading strand) it typically synthesizes a segment of DNA that is over 50,000 nucleotides in length before it inadvertently falls off.

Replication Is Terminated When the Replication Forks Meet at the Terminus Sequences

On the opposite side of the *E. coli* chromosome from *oriC* is a pair of **termination sequences** called *ter* sequences. A protein known as the termination utilization substance (Tus) binds to the *ter* sequences and stops the movement of the replication forks. As shown in figure 11.11, one of the *ter* sequences designated *T1* allows the advancement of clockwise-moving forks but prevents the movement of counterclockwise-moving forks (see the inset to fig. 11.11). Alternatively, *T2* permits the advancement of counterclockwise-moving forks but prevents the advancement of clockwise-moving forks. In any given cell, only one *ter* sequence is required to stop the advancement of one replication fork, and then the other fork ends its synthesis of DNA when it reaches the

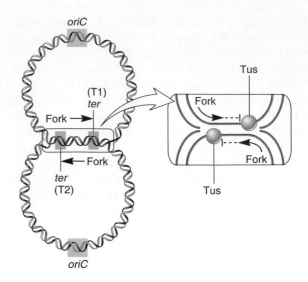

FIGURE 11.11   **The termination of DNA replication.** Two sites in the bacterial chromosome, shown with *rectangles,* are *ter* sequences designated *T1* and *T2.* The *T1* site prevents the further advancement of counterclockwise-moving forks while *T2* prevents the advancement of clockwise-moving forks. As shown in the *inset,* the binding of Tus prevents the replication forks from proceeding past the *ter* sequences in a particular direction.

halted replication fork. In other words, DNA replication ends when oppositely advancing forks meet, usually at *T1* or at *T2.* Finally, DNA ligase covalently links the two daughter strands, creating two circular, double-stranded molecules.

After DNA replication is completed, one last problem may exist. DNA replication often results in two intertwined DNA molecules (fig. 11.12). Interlocked circular molecules are known as **catenanes.** Fortunately, catenanes are only transient structures in DNA replication. Topoisomerases introduce a temporary break into the DNA strands and then rejoin them after the strands have become unlocked. This allows the catenanes to be separated into individual circular molecules.

Certain Enzymes of DNA Replication Bind to Each Other to Form a Complex

Figure 11.13 provides a more three-dimensional view of the DNA replication process. The replication of double-stranded DNA involves an assembly of proteins called the **replisome.** It is an amazing example of a finely crafted biological machine. Its greatest challenge is to use two strands of DNA that are running in opposite directions as templates for the synthesis of new strands. Since DNA polymerase can make new strands only in the $5'$ to $3'$ direction, and since it requires a primer, the synthesis of the lagging strand is discontinuous, while the leading strand can be made continuously.

DNA helicase and primase are physically bound to each other to form a complex known as a **primosome.** This complex leads the way at the replication fork. The primosome tracks along the DNA, separating the parental strands and synthesizing RNA

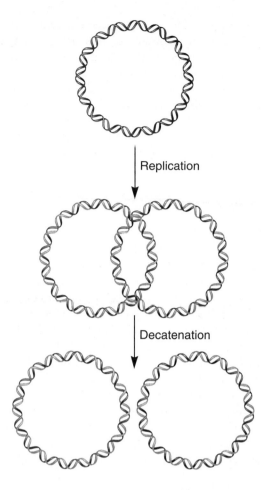

FIGURE 11.12 **Separation of catenanes.** When DNA replication is complete, the two circular chromosomes may be interlocked. These catenanes can be separated by the action of topoisomerases.

primers at regular intervals along the lagging strand. By acting within a primosome, the actions of helicase and primase can be better coordinated.

The primosome is physically associated with two DNA polymerase holoenzymes to form a replisome. As shown in figure 11.13, two DNA polymerase III proteins act in concert to replicate the leading and lagging strands. The term **dimeric DNA polymerase** is used to describe two DNA polymerase holoenzymes that move as a unit toward the replication fork. For this to occur, the lagging strand is looped out with respect to the DNA polymerase that synthesizes the lagging strand. This loop allows the lagging-strand polymerase to make DNA in a 5′ to 3′ direction yet move toward the opening of the replication fork. Interestingly, when this DNA polymerase reaches the end of an Okazaki fragment, it must be released from the template DNA and "hop" to the RNA primer that is closest to the fork. Though the mechanism of this hopping is not fully understood, the association of DNA polymerase with primase is likely to be important in the ability of the lagging-strand polymerase to relocate to the newest RNA primer.

The Fidelity of DNA Replication Is Ensured by Proofreading Mechanisms

With replication occurring so rapidly, one might imagine that mistakes could happen in which the wrong nucleotide could be incorporated into the growing daughter strand. Although mistakes can happen during DNA replication, they are extraordinarily rare. In the case of DNA synthesis via polIII, only one mistake per 100 million nucleotides is made. Therefore, DNA synthesis occurs with relatively few errors. Another way of saying this is that DNA replication exhibits a high degree of **fidelity.**

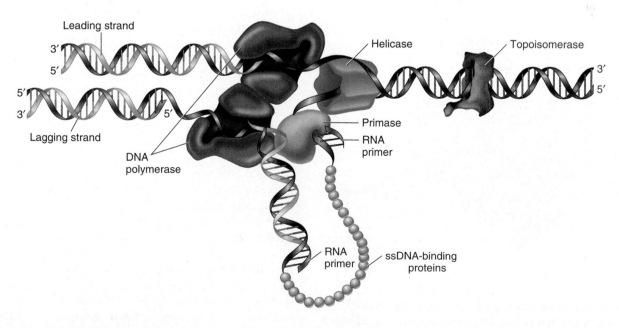

FIGURE 11.13   **A three-dimensional view of DNA replication.** DNA helicase and primase associate together to form a primosome. The primosome associates with two DNA polymerase enzymes to form a replisome. Note: After primase synthesizes an RNA primer in the 5′ to 3′ direction, it must hop over the primer and synthesize the next primer closer to the replication fork.

There are several reasons why fidelity is high. First, the hydrogen bonding between G and C or A and T is much more stable than between mismatched pairs. However, this accounts for only part of the fidelity, since mismatching due to stability considerations would account for one mistake per 1,000 nucleotides. Two additional characteristics of DNA polymerases contribute to their fidelity of DNA replication. The active site of DNA polymerase is such that it will preferentially catalyze the attachment of nucleotides when the correct bases are hydrogen bonding in opposite strands. In other words, DNA polymerase is unlikely to catalyze bond formation between adjacent nucleotides if a mismatched base pair is formed. This induced-fit phenomenon decreases the error rate to a range of 1 in 100,000 to 1 million.

Another way that DNA polymerase decreases the error rate is by the enzymatic removal of mismatched nucleotides. As shown in figure 11.14, DNA polymerase can identify a mismatched nucleotide and remove it from the daughter strand. This

Mismatch causes DNA polymerase to pause, leaving mismatched nucleotide at the 3′ end.

3′ exonuclease site

Template strand

Base pair mismatch at the 3′ end

The 3′ end enters the exonuclease site, which cuts the DNA backbone to release the mismatched nucleotide.

Site where DNA backbone is cut

A schematic drawing of proofreading

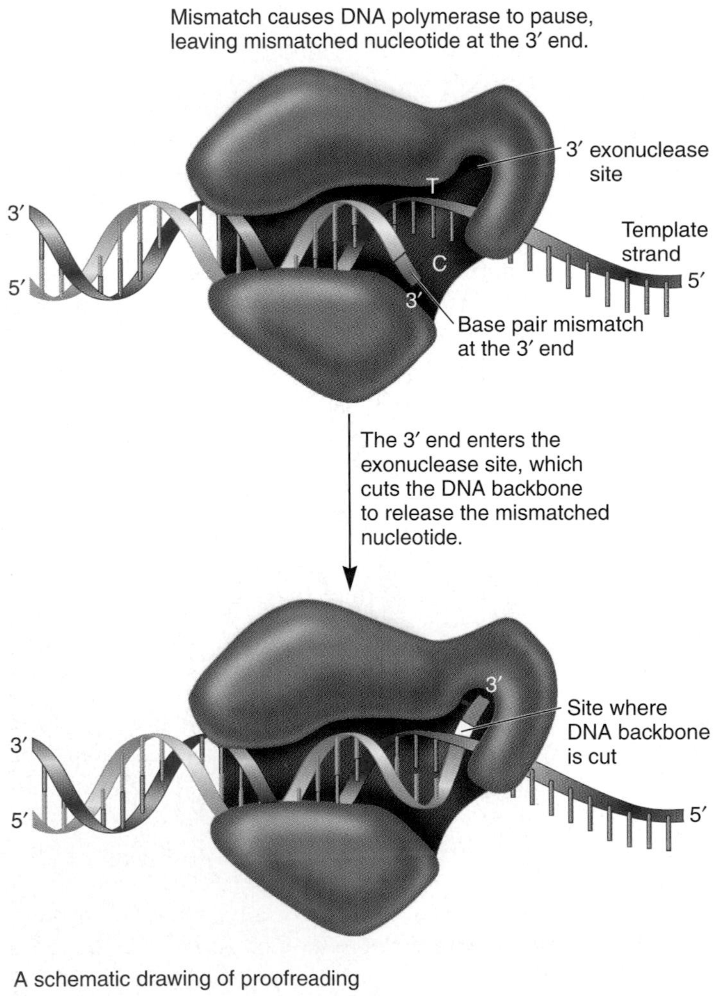

FIGURE 11.14   The proofreading function of DNA polymerase. When a mismatch is formed between the template strand and newly made strand, the end of the newly made strand is shifted into the 3′ exonuclease site. The DNA backbone is then cleaved to release the incorrect nucleotide.

occurs by cleavage of the bond between adjacent nucleotides at the 3′ end of the newly made strand. This process is referred to as **exonuclease** cleavage (the prefix *exo-* indicates that DNA polymerase digests away nucleotides at the end of a DNA strand). The ability to remove mismatched bases by this mechanism has been called the **proofreading function** of DNA polymerase. Proofreading occurs in the 3′ to 5′ direction. After the mismatched nucleotide is removed, DNA polymerase changes direction and resumes DNA synthesis in the 5′ to 3′ direction.

Bacterial DNA Replication Is Coordinated with Cell Division

Bacterial cells can divide into two daughter cells at an amazing rate. Under optimal conditions, certain bacteria such as *E. coli* can divide every 20 to 30 minutes. It is important that DNA replication take place only when a cell is about to divide. If DNA replication occurs too frequently, there will be too many copies of the bacterial chromosome per cell. Alternatively, if DNA replication does not occur frequently enough, a daughter cell will be left without a chromosome. Therefore, cell division in bacterial cells must be coordinated with DNA replication.

Bacterial cells regulate the DNA replication process by controlling the initiation of replication at the origin. This control has been extensively studied in *E. coli*. In this bacterium, several different mechanisms may control DNA replication. In general, the regulation prevents the premature initiation of DNA replication at *oriC*. Two different mechanisms are described here.

First, the initiation of replication is coupled to cell division by the action of the DnaA protein (fig. 11.15). As discussed previously in this chapter, the DnaA protein binds to DnaA boxes within the origin of replication and opens the DNA helix at the AT-rich region within the origin. To initiate DNA replication, the concentration of the DnaA protein must be high enough to bind all the DnaA boxes and form a complex. Immediately following DNA replication, there are twice as many DnaA boxes, and so there is insufficient DnaA protein to initiate a second round of replication. Also, some of the DnaA protein may be rapidly degraded and some of it may be unavailable for DNA binding because it becomes attached to the cell membrane during cell division. Since it takes time to accumulate newly made DnaA protein, DNA replication cannot occur until the daughter cells have had time to grow.

Another way to regulate DNA replication involves the GATC sites within *oriC*. These sites can be methylated by an enzyme known as **Dam methylase.** The acronym stands for <u>D</u>NA <u>a</u>denine <u>m</u>ethyltransferase. The Dam methylase recognizes the 5′–GATC–3′ sequence, binds there, and attaches a methyl group onto the adenine base (fig. 11.16a). DNA methylation within *oriC* helps regulate the replication process. Prior to DNA replication, these sites are methylated in both strands. This full methylation of the 5′–GATC–3′ sites facilitates the initiation of DNA replication at the origin. Following DNA replication, the newly made strands are not methylated, since adenine rather than methyladenine is found in the daughter strands (fig. 11.16b). The initiation

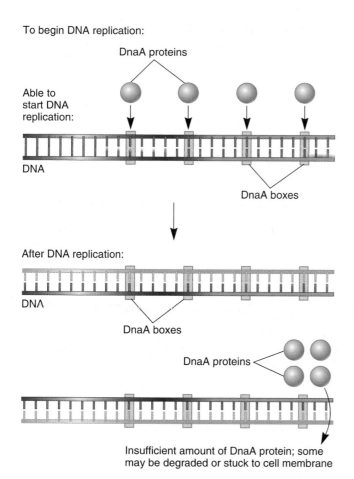

To begin DNA replication:

DnaA proteins

Able to start DNA replication:

DNA

DnaA boxes

After DNA replication:

DNA

DnaA boxes

DnaA proteins

Insufficient amount of DnaA protein; some may be degraded or stuck to cell membrane

FIGURE 11.15   **The amount of DnaA protein provides a way to regulate DNA replication.** To begin replication, there must be enough DnaA protein to bind to all the DnaA boxes. Immediately after DNA replication, there is insufficient DnaA protein to reinitiate a second (premature) round of DNA replication. This is because there are twice as many DnaA boxes after DNA replication and because some DnaA proteins may be degraded or stuck to the cell membrane after cell division.

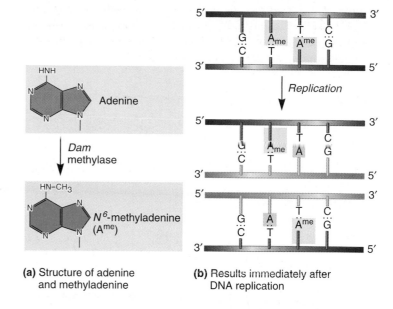

(a) Structure of adenine and methyladenine

(b) Results immediately after DNA replication

FIGURE 11.16   **Methylation of GATC sites in *oriC*.** (a) The action of Dam methylase, which covalently attaches a methyl group to adenine. (b) Prior to DNA replication, the action of Dam methylase causes both adenines within the GATC sites to be methylated. After DNA replication, only the adenines in the original strands are methylated. These are called hemimethylated double helices. Several minutes will pass before Dam methylase will methylate these unmethylated adenines.

of DNA replication at the origin does not readily occur until after it has become fully methylated. Since it takes several minutes for Dam methylase to methylate all the 5′–GATC–3′ sequences within this region, DNA replication will not occur again too quickly.

DNA Replication Can Be Studied In Vitro

Much of our understanding of bacterial DNA replication has come from thousands of experiments in which DNA replication has been studied in vitro. A common experimental strategy is to purify proteins from cell extracts and to determine their roles in the replication process. In other words, purified proteins, such as those described previously in table 11.1, can be mixed with nucleotides, template DNA, and other substances in a test tube, and the synthesis of new DNA strands may occur. This approach was pioneered by Arthur Kornberg in the 1950s, who received a Nobel Prize for his efforts in 1959.

Although we will not consider the procedures for purifying replication proteins, the experiment in figure 11.17 describes a standard way to monitor DNA replication in vitro. In this experi-

ment, an extract of proteins from *E. coli* was used. Alternatively, it would be possible to purify individual proteins from the extract and study their functions individually. In either case, the proteins would be mixed with template DNA and radiolabeled nucleotides. Kornberg correctly hypothesized that deoxyribonucleoside triphosphates are the precursors for DNA synthesis. Also, he knew that deoxyribonucleoside triphosphates are soluble in an acidic solution, whereas long strands of DNA are not. Rather, DNA strands will precipitate out of solution at an acidic pH. This precipitation event provides a method to separate nucleotides (e.g., deoxyribonucleoside triphosphates) from strands of DNA. Therefore, after the proteins, nucleotides, and template DNA were incubated for a sufficient time to allow the synthesis of new strands, step 3 of this procedure involved the addition of perchloric acid. This would precipitate strands of DNA, which were then separated from the radiolabeled nucleotides via centrifugation. Newly made strands of DNA (which would be radiolabeled) sediment to the pellet, while radiolabeled nucleotides that had not been incorporated into new strands would remain in the supernatant.

■ THE HYPOTHESIS

DNA synthesis can occur in vitro if all the necessary components are present.

■ TESTING THE HYPOTHESIS — FIGURE 11.17 In vitro synthesis of DNA strands.

Starting material: An extract of proteins from *E. coli*.

Experimental level   **Conceptual level**

1. Mix together the extract of *E. coli* proteins, template DNA that is not radiolabeled, and ^{32}P-radiolabeled deoxyribonucleoside triphosphates. This is expected to be a complete system that contains everything necessary for DNA synthesis. As a control, a second sample is made in which the template DNA was omitted from the mixture.

2. Incubate the mixture for 30 minutes at 37°C.

3. Add perchloric acid to precipitate DNA. It does not precipitate free nucleotides.

4. Centrifuge the tube. Note: The radiolabeled deoxyribonucleoside triphosphates that have not been incorporated into DNA will remain in the supernatant.

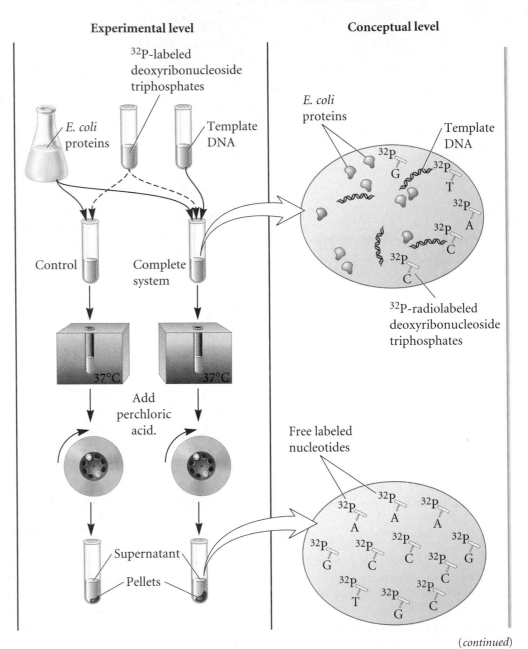

(continued)

5. Collect the pellet, which contains precipitated DNA and proteins. (The control pellet is not expected to contain DNA.)

6. Count the amount of radioactivity in the pellet using a scintillation counter. (See the appendix.)

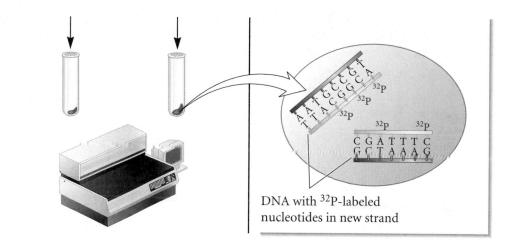

DNA with ^{32}P-labeled nucleotides in new strand

THE DATA

Conditions	Amount of Radiolabeled DNA*
Complete system	3,300
Template DNA omitted	0.0

*Calculated in picomoles of ^{32}P-labeled DNA.

INTERPRETING THE DATA

As shown in the data of figure 11.17, when the *E. coli* proteins were mixed with nonlabeled template DNA and radiolabeled deoxyribonucleoside triphosphates, an acid-precipitable, radiolabeled product was formed. This product was newly synthesized DNA strands. As a control, if nonlabeled template DNA was omitted from the assay, no radiolabeled DNA was made. This is the expected result, since the template DNA is necessary to make new daughter strands. Taken together, these results indicate that this technique can be used to measure the synthesis of DNA in vitro.

The in vitro approach has provided the foundation to study the replication process at the molecular level. Researchers have been able to purify the individual proteins described in table 11.1 and determine their specific roles. This approach still continues, particularly as we try to understand the added complexities of eukaryotic DNA replication.

A self-help quiz involving this experiment can be found at the Online Learning Center.

The Isolation of Mutants Has Been Instrumental to Our Understanding of DNA Replication

In the previous experiment, we considered an experimental strategy to study DNA synthesis in vitro. In his early experiments, Arthur Kornberg used crude extracts containing *E. coli* proteins and monitored their ability to synthesize DNA. In such extracts, the predominant polymerase enzyme is DNA polymerase I. Surprisingly, its activity is so high that it is nearly impossible to detect the activities of the other DNA polymerases. For this reason, researchers in the 1950s and 1960s thought that DNA polymerase I was the only enzyme responsible for DNA replication. This situation dramatically changed as a result of mutant isolation.

In 1969, Paula DeLucia and John Cairns isolated a mutant in which DNA polymerase I lacked its 5′ to 3′ polymerase function but retained its 5′ to 3′ exonuclease function. This mutant was identified by screening thousands of bacterial colonies that had been subjected to agents that cause mutations. This result indicated that the DNA-synthesizing function of DNA polymerase I is not absolutely required for bacteria to replicate their DNA, since the strain harboring this mutation was able to grow normally. Therefore, researchers began to search for other DNA polymerases in *E. coli*.

The isolation of mutants was one way that helped researchers identify additional DNA polymerase enzymes, namely DNA polymerase II and III. In addition, mutant isolation played a key role in the identification of other proteins that are needed to replicate the leading and lagging strands, as well as proteins that recognize the origin of replication and the *ter* sites. Since DNA replication is vital for cell division, most mutations that block DNA replication would be lethal to a growing population of bacterial cells. For this reason, if researchers want to identify loss-of-function mutations in vital genes, they must screen for **conditional mutants.** A type of conditional mutant is a **temperature-sensitive (ts) mutant.** In the case of a vital gene, an organism harboring a temperature-sensitive mutation can survive at the permissive temperature but not at the nonpermissive temperature. For example, a *ts* mutant might survive and grow at 30°C (the permissive temperature) but fail to grow at 42°C (the nonpermissive temperature). The higher temperature inactivates the function of the protein encoded by the mutant gene.

Figure 11.18 shows a general strategy for the isolation of *ts* mutants. Researchers can expose bacterial cells to a mutagen that increases the likelihood of mutations. Such mutagenized cells can be plated on growth media and incubated at the permissive temperature. The colonies can then be replica plated onto two plates,

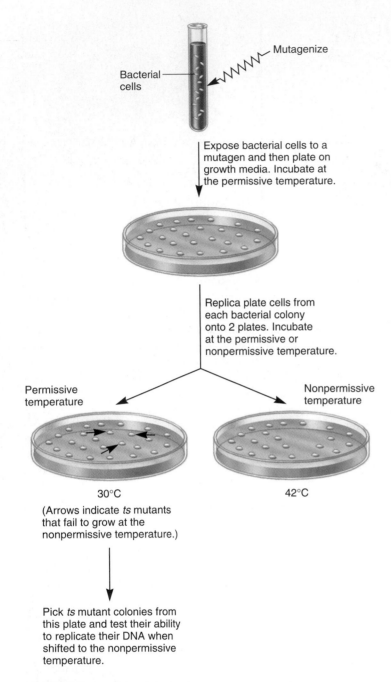

Bacterial cells

Mutagenize

Expose bacterial cells to a mutagen and then plate on growth media. Incubate at the permissive temperature.

Replica plate cells from each bacterial colony onto 2 plates. Incubate at the permissive or nonpermissive temperature.

Permissive temperature

Nonpermissive temperature

30°C

42°C

(Arrows indicate *ts* mutants that fail to grow at the nonpermissive temperature.)

Pick *ts* mutant colonies from this plate and test their ability to replicate their DNA when shifted to the nonpermissive temperature.

FIGURE 11.18   A strategy to identify *ts* mutations in vital genes. In this approach, bacteria are mutagenized, which increases the likelihood of mutation, and then grown at the permissive temperature. Colonies are then replica plated and grown at the permissive and non-permissive temperatures. (Note: The procedure of replica plating is shown in chapter 16, fig. 16.7). *Ts* mutants fail to grow at the nonpermissive temperature. The appropriate colonies can be picked from the plates grown at the permissive temperature and analyzed to see if DNA replication is altered at the nonpermissive temperature.

one that is incubated at the permissive temperature and one at the nonpermissive temperature. As seen here, this enables researchers to identify *ts* mutations that are lethal at the nonpermissive temperature.

With regard to the study of DNA replication, researchers analyzed a large number of *ts* mutants to discover if any of them

had a defect in DNA replication. For example, one could expose a *ts* mutant to radiolabeled thymidine (a nucleotide that is incorporated into DNA), shift to the nonpermissive temperature, and determine if a mutant strain could make radiolabeled DNA, using procedures that are similar to those previously described in figure 11.17. Since *E. coli* has many vital genes that are not involved with DNA replication, only a small subset of *ts* mutants would be expected to have mutations in genes that encode proteins that are vital for the replication process. Therefore, researchers in the 1960s had to screen many thousands of *ts* mutants to identify the few that were involved in DNA replication. This is sometimes called a "brute force" genetic screen.

Table 11.3 summarizes some of the genes that were identified using this type of strategy. The genes were originally designated with the acronym *dna*, followed by a capital letter that generally refers to the order in which they were discovered. When shifted to the nonpermissive temperature, certain mutants showed a rapid defect in DNA synthesis. These rapid-stop mutations inactivated genes that encode enzymes that are needed for DNA replication. By comparison, other mutants were able to complete their current round of replication but could not start another round of replication. As shown in table 11.3, these slow-stop mutants involved genes that encode proteins that are needed for the initiation of replication at the origin.

The isolation of *dna* mutants was important in several ways. First, a biochemical analysis comparing mutant and non-mutant strains made it possible to identify the proteins that were defective in the mutant strains. Researchers could purify proteins from mutant and nonmutant strains, follow the experimental strategy described previously in figure 11.17, and determine which ones were unable to function at the nonpermissive temperature. Secondly, the locations of the mutations could be mapped along the *E. coli* chromosome using methods that are

TABLE 11.3

Examples of *ts* Mutants Involved in DNA Replication in *E. coli*

Gene Name	Protein Function
Rapid-Stop Mutants	
dnaE	α subunit of DNA polymerase III, synthesizes DNA
dnaX	τ subunit of DNA polymerase III, promotes the dimerization of two polIII proteins together at the replication fork and stimulates DNA helicase
dnaN	β subunit of DNA polymerase III, functions as a clamp protein that makes DNA polymerase a processive enzyme
dnaZ	γ subunit of DNA polymerase III, helps the β subunit bind to the DNA
dnaG	Primase, needed to make RNA primers
dnaB	Helicase, needed to unwind the DNA strands during replication
Slow-Stop Mutants	
dnaA	DnaA protein that recognizes the DnaA boxes at the origin
dnaC	DnaC protein that recruits DNA helicase to the origin

described in chapter 6. Finally, as DNA cloning and sequencing methods emerged during the 1970s, the *ts* mutants provided an important starting point for the subsequent cloning of these genes. As described in chapter 18, fragments of DNA can be inserted into plasmids and introduced into bacterial cells. Researchers used this approach to insert nonmutant genes into plasmids and then introduced those genes into *ts* mutant strains (i.e., *dna* mutants). If the wild-type gene on the plasmid was the same gene as the *ts* mutant gene on the bacterial chromosome, the bacteria carrying this plasmid would be able to grow at the nonpermissive temperature, since they would now have a copy of the normal gene. This strategy provided a way to clone and sequence many *dna* genes.

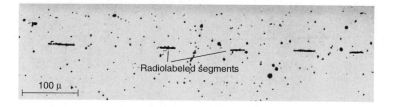

FIGURE 11.19 **Evidence for multiple origins of replication in eukaryotic chromosomes.** In this experiment, cells were given a pulse/chase of ^{3}H-thymidine and unlabeled thymidine. The chromosomes were isolated and subjected to autoradiography. In this micrograph, radiolabeled segments were interspersed among nonlabeled segments, indicating that eukaryotic chromosomes contain multiple origins of replication.

11.3 EUKARYOTIC DNA REPLICATION

Eukaryotic DNA replication is not as well understood as bacterial replication. Much research has been carried out on a variety of experimental organisms, particularly yeast and mammalian cells. Many of these studies have found extensive similarities between the general features of DNA replication in prokaryotes and eukaryotes. For example, the types of bacterial enzymes described earlier in table 11.1 have also been identified in eukaryotes. DNA helicases, primases, DNA polymerases, single-strand binding proteins, DNA ligases, and topoisomerases have all been found in eukaryotic species. Nevertheless, at the molecular level, eukaryotic DNA replication appears to be substantially more complex. These additional intricacies of eukaryotic DNA replication are related to several features of eukaryotic cells. In particular, eukaryotic cells have larger, linear chromosomes, the chromatin is tightly packed within nucleosomes, and cell cycle regulation is much more complicated. This section will emphasize some of the unique features of eukaryotic DNA replication that are a consequence of these complexities.

Initiation Occurs at Multiple Origins of Replication on Linear Eukaryotic Chromosomes

Since eukaryotes have long linear chromosomes, they require multiple origins of replication so that the DNA can be replicated in a reasonable length of time. In 1968, Joel Huberman and Arthur Riggs provided evidence for multiple origins by adding a radiolabeled nucleotide (i.e., ^{3}H-thymidine) to a culture of actively dividing cells. The radiolabeled thymidine was taken up by the cells and incorporated into their newly made DNA strands. After a few minutes, the cells were given an excess amount of unlabeled thymidine. After this pulse/chase procedure, the chromosomes were isolated from the cells and subjected to autoradiography. As seen in the micrograph of figure 11.19, radiolabeled segments were interspersed among nonlabeled segments. This result is consistent with the idea that eukaryotic chromosomes contain multiple origins of replication.

As shown schematically in figure 11.20*a*, DNA replication proceeds bidirectionally from many origins of replication. The multiple replication forks eventually make contact with each other to complete the replication process. Figure 11.20*b* shows a region of eukaryotic DNA in the process of replication. As DNA replication radiates bidirectionally from each origin, regions are formed that contain two double helices. These regions are punctuated by other regions that have not yet replicated and consist of one double helix.

The molecular features of eukaryotic origins of replication may have some similarities to the origins found in bacteria. At the molecular level, eukaryotic origins have been extensively studied in the yeast *Saccharomyces cerevisiae*. In this organism, several replication origins have been identified and sequenced. They have been named **ARS elements** (Autonomously Replicating Sequence). ARS elements, which are 100 to 150 bp in length, are necessary to initiate chromosome replication in vivo. There are two common sequence features among different ARS elements. First, they have a higher percentage of A and T bases than the rest of the chromosomal DNA. In addition, three or four copies of the sequence (A or T)TTTAT(A or G)TTT(A or T) are interspersed within all ARS elements. This arrangement is similar to the DnaA boxes found in bacterial origins. Thus far, the identification of DNA sequences that function as origins of replication in plants and animals has been more difficult to determine.

The **origin recognition complex (ORC)** is a six-subunit protein complex that acts as the initiator of eukaryotic DNA replication. It appears to be found in all eukaryotic species. ORC was originally identified in yeast as a protein complex that binds directly to ARS elements. ORC requires ATP to bind to ARS elements. Single-stranded DNA stimulates ORC to hydrolyze ATP. This suggests that DNA unwinding at the origin stimulates ATP hydrolysis by ORC. However, further research will be needed to elucidate the steps that initiate the formation of two replication forks that emanate from eukaryotic origins of replication.

Eukaryotes Contain Several Different DNA Polymerases

In mammalian cells, there are well over a dozen different DNA polymerases (table 11.4). Four of these, designated α (alpha), δ (delta), ε (epsilon), and γ (gamma), have the primary function of replicating DNA. DNA polymerase γ functions in the mitochondria to replicate mitochondrial DNA, whereas α, δ, and ε are

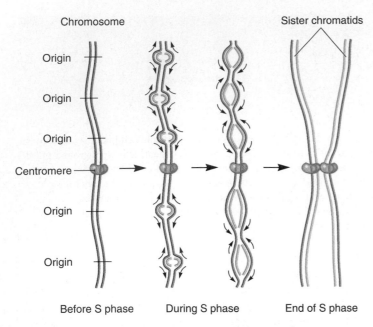

(a) DNA replication from multiple origins of replication

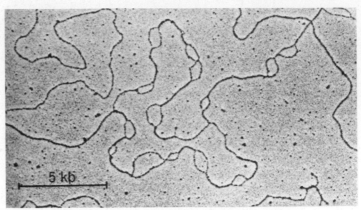

(b) A micrograph of a replicating, eukaryotic chromosome

FIGURE 11.20   The replication of eukaryotic chromosomes. **(a)** At the beginning of the S phase of the cell cycle, eukaryotic chromosome replication begins from multiple origins of replication. As the S phase continues, the replication forks move bidirectionally to replicate the DNA. By the end of the S phase, all the replication forks have merged. The net result is two sister chromatids that are attached to each other at the centromere. **(b)** A micrograph of a replicating chromosome in a eukaryotic species. At different locations, DNA replication radiates bidirectionally from each origin creating regions that contain two double helices. These regions are connected by sites that have not yet replicated and consist of one double helix.

involved with DNA replication in the cell nucleus during S phase. In the nucleus, DNA polymerase α is the only eukaryotic polymerase that associates with primase. The functional role of the DNA polymerase α/primase complex is to synthesize a short RNA-DNA hybrid of approximately 10 RNA nucleotides followed by 20 to 30 DNA nucleotides. This short RNA-DNA strand is then used by DNA polymerase δ or ε for the processive elongation of the leading and lagging strands (fig. 11.21). The exchange of DNA polymerase α for δ or ε is called a **polymerase switch.** It occurs only after the RNA-DNA primer has been made by the complex of DNA polymerase α and primase. The relative importance of DNA polymerase δ versus ε is not yet understood. Current evidence suggests that DNA polymerase δ may have a greater role in DNA replication.

TABLE 11.4
Eukaryotic DNA Polymerases

Polymerase Types*	Function
$\alpha, \delta, \varepsilon$	Replication of nondamaged DNA in the cell nucleus during S phase
γ	Replication of mitochondrial DNA
$\eta, \kappa, \iota, \zeta$ (lesion-replicating polymerases)	Replication of damaged DNA
$\alpha, \beta, \delta, \varepsilon, \sigma, \lambda, \mu, \phi, \theta$	DNA repair or other functions†

*The designations are those of mammalian enzymes.
†Many DNA polymerases have dual functions. For example, DNA polymerase δ is involved in the replication of normal DNA, and it also plays a role in DNA repair. In cells of the immune system, certain genes that encode antibodies (i.e., immunoglobulin genes) undergo a phenomenon known as hypermutation. This increases the variation in the kinds of antibodies the cells can make. Certain polymerases in this list, such as μ, may play a role in hypermutation of immunoglobulin genes. DNA polymerase σ may play a role in sister chromatid cohesion, discussed in chapter 10.

DNA polymerases also play an important role in DNA repair, a topic that will be examined in chapter 16. DNA polymerase β, which is not involved in the replication of normal DNA, does plays an important role in removing incorrect bases from damaged DNA. This phenomenon, called **base excision repair (BER),** is described in chapter 16.

DNA polymerases $\alpha, \beta, \delta, \varepsilon$, and γ have been studied for several decades, and their roles in DNA replication and repair are becoming increasingly clear. More recently, several additional DNA polymerases have been identified. While their precise roles have not been elucidated, an interesting category are the **lesion-replicating polymerases.** When DNA polymerase α, δ, or ε encounter abnormalities in DNA structure (e.g., abnormal bases, cross-links, etc.) they may be unable to replicate over the aberration. When this occurs, lesion-replicating polymerases are attracted to the damaged DNA and have special properties that enable them to synthesize a complementary strand over the abnormal region. Each type of lesion-replicating polymerase may be able to replicate over a different kind of DNA damage.

Nucleosomes Containing New Histone Proteins Are Quickly Formed After DNA Replication

The DNA within eukaryotic cells is wrapped around histone proteins to form a nucleosome structure. Since replication doubles the amount of DNA, the cell must synthesize more histone proteins to accommodate this increase. Like DNA replication, the synthesis of histones occurs during the S phase of the cell cycle. These histones are assembled into octamer structures

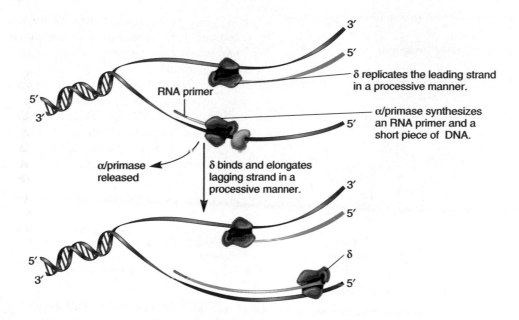

FIGURE 11.21   Polymerase switching of α for δ. After DNA polymerase α/primase have made an RNA primer and short segment of DNA, this complex is released and a processive polymerase, such as DNA polymerase δ, attaches to the template strand and rapidly synthesizes DNA. A similar mechanism could switch α for ε.

and associate with the newly made DNA very near the replication fork. Following DNA replication, each daughter strand contains a random mixture of original histone octamers and newly assembled histone octamers (fig. 11.22).

The Ends of Eukaryotic Chromosomes Are Replicated by Telomerase

Linear eukaryotic chromosomes contain telomeres at both ends. The term **telomere** refers to the complex of telomeric sequences within the DNA and the special proteins that are bound to these sequences. As shown in figure 11.23, telomeric sequences consist of a moderately repetitive tandem array and a 3′ overhang region that is 12 to 16 nucleotides in length.

The tandem array that occurs within the telomere has been studied in a wide variety of eukaryotic organisms. A common

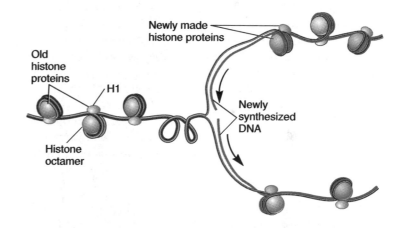

FIGURE 11.22   Newly assembled histone octamers form nucleosomes near the replication fork.

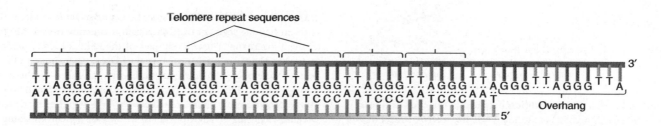

FIGURE 11.23   General structure of telomeric sequences. The sequence consists of a tandemly repeated sequence and a 12- to 16-nucleotide overhang.

TABLE 11.5
Telomeric Sequences Within Selected Organisms

Group	Example	Telomeric Repeat Sequence
Mammals	Humans	TTAGGG
Slime molds	*Physarum, Didymium*	TTAGGG
	Dictyostelium	AG$_{(1-8)}$
Filamentous fungi	*Neurospora*	TTAGGG
Budding yeast	*Saccharomyces cerevisiae*	TG$_{(1-3)}$
Ciliates	*Tetrahymena*	TTGGGG
	Paramecium	TTGGG(T/G)
	Euplotes	TTTTGGGG
Higher plants	*Arabidopsis*	TTTAGGG

feature is that the telomeric sequence contains several guanine nucleotides and often many thymine nucleotides (table 11.5). Depending on the species and the cell type, this sequence can be tandemly repeated up to several hundred times in the telomere region.

One reason why telomeric sequences are needed is because DNA polymerase is unable to replicate the 3′ ends of DNA strands. As discussed previously, DNA polymerases synthesize DNA only in a 5′ to 3′ direction, and they cannot link together two individual nucleotides. DNA polymerases can only elongate preexisting strands. In other words, they need an RNA primer or an existing DNA strand. These two features of DNA polymerase function pose a problem at the 3′ ends of linear chromosomes. As shown in figure 11.24, the strand with the 3′ end cannot be replicated by DNA polymerase. Upstream from this point, there

is no place on the parental DNA strand for a primer to be made. Therefore, if this problem were not solved, the linear chromosome would become progressively shorter with each round of DNA replication.

To prevent the loss of genetic information due to chromosome shortening, additional DNA sequences are attached to the ends of telomeres. This specialized mechanism requires an enzyme known as **telomerase,** which recognizes telomeric sequences at the ends of eukaryotic chromosomes and synthesizes additional repeats of telomeric sequences. Figure 11.25 shows the rather

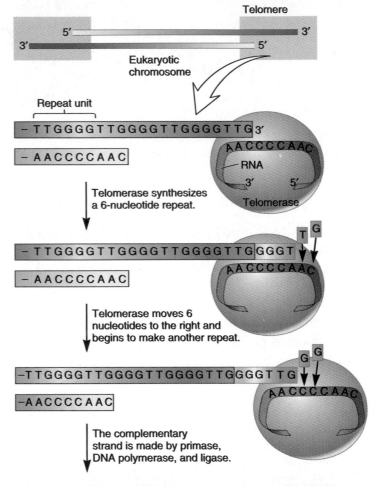

FIGURE 11.25   The enzymatic action of telomerase. A short, three nucleotide segment of RNA within telomerase causes it to bind to the 3′ overhang. The adjacent part of the RNA is used as a template to make a short, six-nucleotide repeat of DNA. After the repeat is made, telomerase moves six nucleotides to the right and then synthesizes another repeat. This process is repeated many times to lengthen the *top* strand shown in this figure. The *bottom* strand would be made by DNA polymerase, using an RNA primer at the end of the chromosome that would be complementary to the telomeric repeat sequence in the top strand.

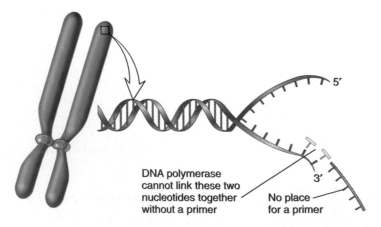

FIGURE 11.24   The replication problem at the ends of linear chromosomes. DNA polymerase cannot synthesize a DNA strand that is complementary to the 3′ end, because there is no primer upstream from this site.

interesting mechanism by which telomerase works. The telomerase enzyme contains both protein and RNA. The RNA part of telomerase contains a sequence that is complementary to the DNA sequence found in the telomeric repeat. This allows telomerase to bind to the 3′ overhang region of the telomere. Following binding, the RNA sequence beyond the binding site functions as a template allowing the attachment of a six-nucleotide sequence to the end of the DNA strand. This is called polymerization, because it is analogous to the function of DNA polymerase. The telomerase can then move (i.e., translocate) to the new end of this DNA strand and attach another six nucleotides to the end. And so on. This binding-polymerization-translocation cycle can occur many times and thereby greatly lengthen one of the DNA strands in the telomeric region. The complementary strand would then be synthesized by DNA polymerase. In this way, the progressive shortening of eukaryotic chromosomes is prevented.

CONCEPTUAL SUMMARY

The structural basis for **DNA replication** is the double-stranded helix of DNA, in which the AT/GC rule is obeyed. This complementarity between strands allows DNA to be replicated in a **semiconservative** fashion. DNA replication begins at a specific site within the DNA, known as the origin of replication. Circular bacterial chromosomes have a single origin of replication. Specific proteins, such as the **DnaA protein** found in *E. coli*, recognize the origin and initiate the DNA replication process. The synthesis of DNA strands proceeds **bidirectionally** from the origin. This synthesis requires various proteins. **Helicase** breaks the hydrogen bonding between the **parental strands, topoisomerase** removes positive supercoils, and **single-strand binding proteins** hold the parental strands in a single-stranded state. **Primase** synthesizes RNA primers, which are necessary for **DNA polymerase** to elongate new **daughter strands** in a 5′ to 3′ direction. In the **leading strand,** synthesis of the daughter strand is continuous. In the **lagging strand,** the synthesis of DNA occurs in short **Okazaki fragments.** The RNA primers are removed, DNA polymerase fills in the gaps, and **DNA ligase** covalently attaches the fragments together. DNA replication is terminated when both replication forks reach the *ter* sequences, and a possible **catenane** structure is resolved.

Coordination of cell division and DNA replication is accomplished via the cellular regulation of replication. In bacteria, the control of DNA replication occurs at the origin. Several mechanisms can prevent premature DNA replication. Two examples are the availability of the DnaA protein and the methylation of GATC sites.

DNA replication in eukaryotes has several unique features. For example, eukaryotic chromosomes contain multiple origins of replication. Also, since eukaryotic chromosomes are linear, a specialized mechanism exists for the replication at the ends of the chromosomes within the **telomere.** An enzyme known as **telomerase** attaches telomeric repeat sequences. This prevents chromosome shortening with each round of DNA replication. Also, since eukaryotic chromosomes exist in a nucleosome structure, it is necessary to assemble the DNA and histone octamers following replication. This assembly occurs in the vicinity of the replication fork, and replicated chromosomes contain a random mixture of new and old histone octamers.

EXPERIMENTAL SUMMARY

The semiconservative mechanism of DNA replication was initially proposed based on the double helix structure determined by Watson and Crick. Nevertheless, researchers needed to experimentally demonstrate that this mechanism was correct. The isotope labeling experiments of Meselson and Stahl were consistent with a semiconservative mode of replication and were inconsistent with conservative and dispersive models.

Once the semiconservative model was established, researchers focused their attention on the details of DNA replication within living cells. Arthur Kornberg and his colleagues developed methods to synthesize bacterial DNA in vitro. In conjunction with the isolation of *ts* mutants (e.g., *dna* mutants), this allowed researchers to identify the components required for DNA replication. Many of these components were found to be enzymes, such as DNA polymerase, primase, helicase, topoisomerase, and ligase. Also, an analysis of DNA revealed two sequences, *oriC* and *ter,* that are necessary for the initiation and termination of DNA replication. In eukaryotes, similar findings have been obtained, but there are more enzymes and the sequences of the origins of replication are more complex. These added complexities have made it more difficult to analyze DNA replication in eukaryotes, although much progress has been made. As we will see in chapter 22, the regulation of DNA replication is an important topic and an area of intense research, because it may underlie the proliferation of cancer cells.

PROBLEM SETS & INSIGHTS

Solved Problems

S1. Describe three ways to account for the high fidelity of DNA replication. Discuss the quantitative contributions of each of the three ways.

Answer:

First: AT and GC pairs are preferred in the double helix structure. This provides fidelity to around one mistake per 1,000.

Second: Induced fit by DNA polymerase prevents covalent bond formation unless the proper nucleotides are in place. This increases fidelity another 100- to 1,000-fold, to about one error in 100,000 to 1 million.

Third: Exonuclease proofreading increases fidelity another 100- to 1,000-fold, to about one error per 100 million nucleotides added.

S2. What do you think would happen if the *ter* sequences were deleted from the bacterial DNA?

Answer: Instead of meeting at the *ter* sequences, the two replication forks would meet somewhere else. This would depend on how fast they were moving. For example, if the counterclockwise-moving fork was advancing faster than the clockwise-moving fork, they would meet closer to where the clockwise-moving fork started. In fact, researchers have actually conducted this experiment. Interestingly, *E. coli* without the *ter* sequences seemed to survive just fine.

S3. Summarize the steps that occur in the process of chromosomal DNA replication in *E. coli*.

Answer:

Step 1. DnaA proteins bind to the origin of replication. The AT-rich region is denatured after binding.

Step 2. DNA helicase breaks the hydrogen bonds between the DNA strands, topoisomerases alleviate positive supercoiling, and single-strand binding proteins hold the parental strands apart.

Step 3. Primase synthesizes one RNA primer in the leading strand and many primers in the lagging strand. DNA polymerase III then synthesizes the daughter strands of DNA. DNA polymerase I excises the RNA primers and fills in with DNA. DNA ligase covalently links the DNA fragments together.

Step 4. The processes described in steps 2 and 3 continue until the two forks reach the *ter* sequences on the other side of the circular bacterial chromosome.

Step 5. Topoisomerases unravel the intertwined chromosomes, if necessary.

S4. If a strain of *E. coli* overproduced the methylase enzyme, how would that affect the DNA replication process? Would you expect such a strain to have more or less chromosomes per cell, compared to a normal strain of *E. coli*? Explain why.

Answer: If a strain overproduced the methylase enzyme, it would be easier to replicate the DNA. The GATC sites in the origin of replication have to be fully methylated for DNA replication to occur. Immediately after DNA replication, there is a delay in the next round of DNA replication because the two copies of newly replicated DNA are hemimethylated. A strain that overproduces methylase would rapidly convert the hemimethylated DNA into fully methylated DNA and more quickly allow the next round of DNA replication to occur. For this reason, the overproducing strain might have more copies of the *E. coli* chromosome because it would not have a long delay in DNA replication.

Conceptual Questions

C1. What are the key structural features of the DNA molecule that underlie its ability to be faithfully replicated?

C2. With regard to DNA replication, define the term *bidirectionality*.

C3. Which of the following statements is *not* true? Explain why.

A. A DNA strand can serve as a template strand on many occasions.

B. Following semiconservative DNA replication, one strand is a newly made daughter strand and the other strand is a parental strand.

C. A DNA double helix may contain two strands of DNA that were made at the same time.

D. A DNA double helix obeys the AT/GC rule.

E. A DNA double helix could contain one strand that is 10 generations older than its complementary strand.

C4. The compound known as nitrous acid is a reactive chemical that replaces amino groups ($-NH_2$) with keto groups ($=O$). When nitrous acid reacts with the bases in DNA, it can change cytosine to uracil and change adenine to hypoxanthine. A DNA double helix has the following sequence:

```
TTGGATGCTGG
AACCTACGACC
```

A. What would be the sequence of this double helix immediately after reaction with nitrous acid? Let the letter *H* represent hypoxanthine and *U* represent uracil.

B. Let's suppose this DNA was reacted with nitrous acid. The nitrous acid was then removed and the DNA was replicated for two generations. What would be the sequences of the DNA products after the DNA had replicated two times? Your answer should contain the sequences of four double helices. Note: During DNA replication, uracil hydrogen bonds with adenine, and hypoxanthine hydrogen bonds with cytosine.

C5. One way that bacterial cells regulate DNA replication is by GATC methylation sites within the origin of replication. Would this mechanism work if the DNA was conservatively (rather than semiconservatively) replicated?

C6. The chromosome of *E. coli* contains 4.6 million bp. How long will it take to replicate its DNA? Assuming that polIII is the primary enzyme involved, and this enzyme can actively proofread during DNA synthesis, how many base pair mistakes will be made in one round of DNA replication in a bacterial population containing 1,000 bacteria?

C7. Here are two strands of DNA.

————————————————DNA polymerase→

The one on the bottom is a template strand, and the one on the top is being synthesized by DNA polymerase in the direction shown by the arrow. Label the 5′ and 3′ ends of the top and bottom strands.

C8. A DNA strand has the following sequence:

5′-GATCCCGATCCGCATACATTTACCAGATCACCACC-3′

In which direction would DNA polymerase slide along this strand (from left to right or from right to left)? If this strand was used as a template by DNA polymerase, what would be the sequence of the newly made strand? Indicate the 5′ and 3′ ends of the newly made strand.

C9. List and briefly describe the three types of sequences within bacterial origins of replication that are functionally important.

C10. As shown in figure 11.5, there are four DnaA boxes that are found within the origin of replication in *E. coli*. Take a look at these four sequences carefully.

A. Are the sequences of the four DnaA boxes very similar to each other? (Hint: Remember that DNA is double stranded; think about these sequences in the forward and reverse direction.)

B. What is the most common sequence for the DnaA box? In other words, what is the most common base in the first position, second position, and so on until the ninth position? The most common sequence is called the consensus sequence.

C. The *E. coli* chromosome is about 4.6 million bp long. Based on random chance, is it likely that the consensus sequence for a DnaA box occurs elsewhere in the *E. coli* chromosome? If so, why aren't there multiple origins of replication in *E. coli*?

C11. Obtain two strings of different colors (e.g., black and white) that are the same length. A length of 20 inches is sufficient. Tie a knot at one end of the black string and tie a knot at one end of the white string. Each knot designates the 5′ end of your strings. Make a double helix with your two strings. Now tape one end of the double helix to a table so that the tape is covering the knot on the black string.

A. Pretend your hand is DNA helicase and use your hand to unravel the double helix, beginning at the end that is not taped to the table. Should your hand be sliding along the white string or the black string?

B. As in figure 11.13, imagine that your two hands together form a dimeric replicative DNA polymerase. Unravel your two strings halfway to create a replication fork. Grasp the black string with your left hand and the white string with your right hand. Your thumbs should point toward the 5′ end of each string. You need to loop one of the strings so that one of the DNA polymerases can synthesize the lagging strand. With such a loop, it is possible for the dimeric replicative DNA polymerase to move toward the replication fork and synthesize both DNA strands in the 5′ to 3′

direction. In other words, with such a loop, your two hands can touch each other with both of your thumbs pointing toward the fork. Should the black string be looped or should the white string be looped?

C12. Sometimes DNA polymerase makes a mistake and the wrong nucleotide is added to the growing DNA strand. With regard to pyrimidines and purines, two general types of mistakes are possible. The addition of an incorrect pyrimidine instead of the correct pyrimidine (e.g., adding cytosine where thymine should be added) is called a transition. If a pyrimidine is incorrectly added to the growing strand instead of purine (e.g., adding cytosine where an adenine should be added), this type of mistake is called a transversion. If a transition or transversion is not detected by DNA polymerase, this will create a mutation that permanently changes the DNA sequence. Though both types of mutations are rare, it has been found that transition mutations are more frequent than transversion mutations. Based on your understanding of DNA replication and DNA polymerase, offer three explanations why transition mutations are more common.

C13. Researchers have found that a short genetic sequence, which may be recognized by DNA primase, is repeated many times throughout the *E.coli* chromosome. It has been hypothesized that DNA primase may recognize this sequence as a site to begin the synthesis of an RNA primer for DNA replication. The *E. coli* chromosome is roughly 4.6 million bp in length. How many copies of the DNA primase recognition sequence would be necessary to replicate the entire *E. coli* chromosome?

C14. Single-strand binding proteins keep the two parental strands of DNA separated from each other until DNA polymerase has an opportunity to replicate the strands. Suggest how single-strand binding proteins keep the strands separated and yet do not impede the ability of DNA polymerase to replicate the strands.

C15. The ability of DNA polymerase to digest a DNA strand from one end is called its exonuclease activity. Exonuclease activity is used to digest RNA primers and also to proofread a newly made DNA strand. Note: DNA polymerase I does not change direction while it is removing an RNA primer and synthesizing new DNA. It does change direction during proofreading.

A. In which direction, 5′ to 3′ or 3′ to 5′, is the exonuclease activity occurring during the removal of RNA primers and during the proofreading and removal of mistakes following DNA replication?

B. Figure 11.14b shows a drawing of the 3′ exonuclease site. Do you think this site would be used by DNA polymerase I to remove RNA primers? Why or why not?

C16. In the following drawing, the top strand is the template DNA and the bottom strand shows the lagging strand prior to the action of DNA polymerase I. The lagging strand contains three Okazaki fragments. The RNA primers have not yet been removed.

The top strand is the template DNA

3′————————————————————————————5′
5′************_____*************_____***********_____3′

RNA primer   ↑ RNA primer ↑   RNA primer

|————————————||———————————————||———————————|
Left Okazaki   Middle Okazaki   Right Okazaki
fragment   fragment   fragment

A. Which Okazaki fragment was made first, the one on the left or the one on the right?

B. Which RNA primer would be the first one to be removed by DNA polymerase I, the primer on the left or the primer on the right? For this primer to be removed by DNA polymerase I and for the gap to be filled in, is it necessary for the Okazaki fragment in the middle to have already been synthesized? Explain why.

C. Let's consider how DNA ligase connects the left Okazaki fragment with the middle Okazaki fragment. After DNA polymerase I removes the middle RNA primer and fills in the gap with DNA, where does DNA ligase function? See the arrows on either side of the middle RNA primer. Is ligase needed at the left arrow, at the right arrow, or both?

D. When connecting two Okazaki fragments, DNA ligase needs to use ATP as a source of energy to catalyze this reaction. The ATP nucleotide is not incorporated into the DNA strand. Explain why DNA ligase needs another source of energy to connect two nucleotides, but DNA polymerase needs nothing more than the incoming nucleotide and the existing DNA strand. Note: You may want to refer to figure 11.10 to answer this question.

C17. What is DNA methylation? Why is DNA in a hemimethylated condition immediately after DNA replication? What are the functional consequences of methylation in the regulation of DNA replication?

C18. Describe the three important functions of the DnaA protein.

C19. If a strain of bacteria was making too much DnaA protein, how would you expect this to affect its ability to regulate DNA replication? With regard to the number of chromosomes per cell, how might this strain differ from a normal bacterial strain?

C20. Draw a picture that describes how helicase works.

C21. What is an Okazaki fragment? In which strand of DNA are Okazaki fragments found? Why are they necessary?

C22. Discuss the similarities and differences in the synthesis of DNA in the lagging and leading strands. What is the advantage of a primosome and a replisome as opposed to having all replication enzymes functioning independently of each other?

C23. Explain the proofreading function of DNA polymerase.

C24. What is a processive enzyme? Explain why this is an important feature of DNA polymerase.

C25. How and when are nucleosomes assembled during DNA replication?

C26. Why is it important for living organisms to regulate DNA replication?

C27. What enzymatic features of DNA polymerase prevent it from replicating one of the DNA strands at the ends of linear chromosomes? Compared to DNA polymerase, how is telomerase different in its ability to synthesize a DNA strand? What does telomerase use as its template for the synthesis of a DNA strand? How does the use of this template result in a telomere sequence that is tandemly repetitive?

C28. As shown in figure 11.25, telomerase attaches additional DNA, six nucleotides at a time, to the ends of eukaryotic chromosomes. However, it works in only one DNA strand. Describe how the opposite strand is replicated.

C29. If a eukaryotic chromosome has 25 origins of replication, how many replication forks does it have at the beginning of DNA replication?

C30. A diagram of a linear chromosome is shown here. The end of each strand is labeled with an A, B, C, or D. Which ends could not be replicated by DNA polymerase? Why not?

$$5'-A \rule{6cm}{0.4pt} B-3'$$
$$3'-C \rule{6cm}{0.4pt} D-5'$$

C31. As discussed in chapter 10, some viruses contain RNA as their genetic material. Certain RNA viruses can exist as a provirus in which the viral genetic material has been inserted into the chromosomal DNA of the host cell. For this to happen, the viral RNA must be copied into a strand of DNA. An enzyme called reverse transcriptase, encoded by the viral genome, copies the viral RNA into a complementary strand of DNA. The strand of DNA is then used as a template by DNA polymerase to make a double-stranded DNA molecule. This double-stranded DNA molecule is then inserted into the chromosomal DNA, where it may exist as a provirus for a long period of time.

A. How is the function of reverse transcriptase similar to the function of telomerase?

B. Unlike DNA polymerase, reverse transcriptase does not have a proofreading function. How might this affect the proliferation of the virus?

C32. Telomeres contain a 3' overhang region as shown in figure 11.23. Does telomerase require a 3' overhang to replicate the telomere region? Explain.

Experimental Questions

E1. Answer the following questions that pertain to the experiment of figure 11.3.

A. What would be the expected results if the Meselson and Stahl experiment were carried out for four or five generations?

B. What would be the expected results of the Meselson and Stahl experiment after three generations if the mechanism of DNA replication were dispersive?

C. As shown in the data, explain why three different bands (i.e., light, half-heavy, and heavy) can be observed in the CsCl gradient.

E2. An absentminded researcher follows the steps of figure 11.3 and when the gradient is viewed under UV light, the researcher does not see any bands at all. Which of the following mistakes could account for this observation? Explain how.

A. The researcher forgot to add ^{14}N-containing compounds.

B. The researcher forgot to add lysozyme.

C. The researcher forgot to add ethidium bromide.

E3. Figure 11.4 shows an autoradiograph of a replicating bacterial chromosome. If you analyzed many replicating chromosomes, what types of information could you learn about the mechanism of DNA replication?

E4. The experiment of Figure 11.17 described a method for determining the amount of DNA that is made during replication. Let's suppose that you can purify all the proteins required for DNA replication. You then want to "reconstitute" DNA synthesis by mixing together all the purified components that are necessary to synthesize a complementary strand of DNA. If you started with single-stranded DNA as a template, what additional proteins and molecules would you have to add for DNA polymerization to occur? What additional proteins would be necessary if you started with a double-stranded DNA molecule?

E5. Using the reconstitution strategy described in experimental question E4, what components would you have to add to measure the ability of telomerase to synthesize DNA? Be specific about the type of template DNA that you would add to your mixture.

E6. As described in figure 11.17, perchloric acid precipitates strands of DNA but it does not precipitate free nucleotides. (Note: The term *free nucleotide* means nucleotides that are not connected covalently to other nucleotides.) Explain why this is a critical step in the experimental procedure. If a researcher used a different acid that precipitated DNA strands and free nucleotides instead of using perchloric acid (which precipitates only DNA strands), how would that affect the results?

E7. Would the experiment of figure 11.17 work if the ^{32}P-labeled nucleotides were deoxyribonucleoside monophosphates instead of deoxyribonucleoside triphosphates? Explain why or why not.

E8. To synthesize DNA in vitro, single-stranded DNA can be used as a template. As described in figure 11.17, you also need to add DNA polymerase, deoxyribonucleotides, and a primer in order to synthesize a complementary strand of DNA. The primer can be a short sequence of DNA or RNA. The primer must be complementary to the template DNA. Let's suppose a single-stranded DNA molecule is 46 nucleotides long and has the following sequence:

GCCCCGGTACCCCGTAATATACGGGACTAGGCCGGAGGTCCGGGCG

This template DNA is mixed with a primer with the sequence 5'–CGCCCGGACC–3', DNA polymerase, and deoxyribonucleotides. In this case, a double-stranded DNA molecule is made. However, if the researcher substitutes a primer with the sequence 5'-CCAGGCCCGC-3', a double-stranded DNA molecule is not made.

A. Which is the 5' end of the DNA molecule shown, the left end or the right end?

B. If you added a primer that was 10 nucleotides long and complementary to the left end of the single-stranded DNA, what would be the sequence of the primer? You should designate the 5' and 3' ends of the primer. Could this primer be used to replicate the single-stranded DNA?

E9. The technique of dideoxy DNA sequencing is described in chapter 18. The technique relies on the use of dideoxyribonucleotides (shown in figures 18.14 and 18.15). A dideoxyribonucleotide has a hydrogen atom attached to the 3' carbon atom instead of an –OH group. When a dideoxyribonucleotide is incorporated into a newly made strand, the strand cannot grow any longer. Explain why.

E10. Another technique described in chapter 18 is the polymerase chain reaction (PCR) (see fig. 18.6). This method is based on our understanding of DNA replication. In this method, a small of amount of double-stranded template DNA is mixed with a high concentration of primers. Nucleotides and DNA polymerase are also added. The template DNA strands are separated by heat treatment, and when the temperature is lowered, the primers can bind to the single-stranded DNA, and then DNA polymerase replicates the DNA. This increases the amount of DNA that is made from the primers. This cycle of steps (i.e., 1. heat treatment, 2. lower temperature, 3. allow DNA replication to occur) is repeated again and again and again. Since the cycles are repeated many times, this method is called a chain reaction. It is called a polymerase chain reaction because DNA polymerase is the enzyme that is needed to increase the amount of DNA with each cycle. In a PCR experiment, the template DNA is placed in a tube and the primers, nucleotides, and DNA polymerase are added to the tube. The tube is then placed in a machine called a thermocycler, which raises and lowers the temperature. During one cycle, the temperature is raised (e.g., to 95°C) for a brief period and then lowered (e.g., to 60°C) to allow the primers to bind. The sample is incubated at the lower temperature for a few minutes to allow DNA replication to proceed. In a typical PCR experiment, the tube may be left in the thermocycler for 25 to 30 cycles. The total time for a PCR experiment is a few hours.

A. Why is helicase not needed in a PCR experiment?

B. How is the sequence of each primer important in a PCR experiment? Do the two primers recognize the same strand or opposite strands? In figure 18.6, are the arrowheads on the primers at the 3' end of the primer or at the 5' end?

C. The DNA polymerase used in PCR experiments is a DNA polymerase isolated from thermophilic bacteria. Why is this kind of polymerase used?

D. If a tube initially contained 10 copies of double-stranded DNA, how many copies of double-stranded DNA (in the region flanked by the two primers) would be obtained after 27 cycles?

Questions for Student Discussion/Collaboration

1. The complementarity of double-stranded DNA is the underlying reason that DNA can be faithfully copied. Propose alternative chemical structures that could be faithfully copied.

2. The technique described in figure 11.17 makes it possible to measure DNA synthesis in vitro. Let's suppose that you have purified the following enzymes: DNA polymerase, helicase, ligase, primase, single-strand binding protein, and topoisomerase. You also have the following reagents available:

A. Radiolabeled nucleotides (labeled with ^{32}P, a radioisotope of phosphorus)

B. Nonlabeled double-stranded DNA

C. Nonlabeled single-stranded DNA

D. An RNA primer that binds to one end of the nonlabeled single-stranded DNA

With these reagents, how could you show that helicase is necessary for strand separation and primase is necessary for the synthesis of an RNA primer? Note: In this question, think about conditions where helicase or primase would be necessary to allow DNA replication and other conditions where they would be unnecessary.

3. DNA replication is fast, virtually error-free, and coordinated with cell division. Discuss which of these three features you think is the most important.

Note: All answers appear at the website for this textbook; the answers to even-numbered questions are in the back of the textbook.

www.mhhe.com/brooker

Visit the Online Learning Center for practice tests, answer keys, and other learning aids for this chapter. Enhance your understanding of genetics with our interactive exercises, web links, news feeds, tutorial service, and much more.

GENE TRANSCRIPTION AND RNA MODIFICATION

::

12

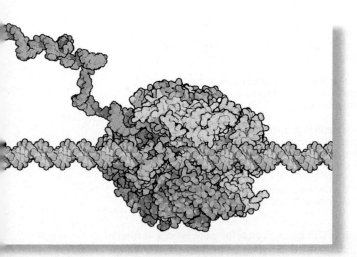

The chromosomal DNA of living organisms contains thousands of different genes. In chapter 12, we will consider the first step in gene expression, **transcription,** and the modifications that may occur after an RNA molecule is made. The process of transcription involves two fundamental concepts. First, DNA sequences provide the underlying information for gene transcription. We will examine the sequences that determine the starting site for transcription and its ending site. In chapters 14 and 15, we will consider DNA sequences involved in the regulation of transcription. These regulatory sites influence whether a gene is turned on or off. The second important concept that underlies transcription is the function of proteins. DNA sequences, in and of themselves, just exist. For genes to be actively transcribed, proteins must recognize particular DNA sequences and act upon them in a way that affects the transcription process. In addition, other proteins bind to the RNA transcript and modify it in ways that make it functionally active. Much of chapter 12 is devoted to an understanding of proteins that play important roles in the synthesis of RNA or its subsequent modification.

Gene transcription is one of the hottest topics in molecular biology. The transcription of genes is of central importance in a variety of research areas including developmental biology, cancer biology, and biotechnology. In chapter 12, we will focus on the basic components that are required to transcribe genes, in both bacterial and eukaryotic cells. In chapters 14 and 15, we will examine the ability to turn genes on and off by transcriptional regulation.

PART IV
Molecular Properties of Genes

12.1 OVERVIEW OF TRANSCRIPTION

In chapters 9 through 11, we focused our attention on the molecular structure of DNA and chromosomes, and the process by which chromosomes are replicated. The underlying function of the genetic material is that of a blueprint. It stores the information that is necessary to create an organism. In chapter 12, we begin to explore the molecular mechanisms that explain how this information is accessed. The first step in this process is transcription, which literally means the act or process of making a copy. In genetics, this term refers to the copying of a DNA sequence into an RNA sequence. The structure of DNA is not altered as a result of transcription. Rather, the DNA nucleotide sequence has only been accessed to make a copy in the form of RNA. Therefore, the same DNA can continue to store information. In this introductory section, we will consider the general steps in the transcription process and the types of RNA transcripts that can be made.

Gene Expression Requires Base Sequences That Perform Different Functional Roles

At the molecular level, a gene is a **transcriptional unit;** it can be transcribed into RNA. Several different types of base sequences perform different roles during the process of gene expression. Figure 12.1 shows a common organization of sequences that are needed to create a structural gene that functions in a bacterium such as *E. coli*. Each type of base sequence performs its role during a specific stage of gene expression. For example, the promoter and terminator are nucleotide base sequences that are used during gene transcription. Specifically, the **promoter** sequence provides a site to begin transcription, and the **terminator** specifies the end of transcription. Therefore, these two sequences cause RNA synthesis to occur within a defined location. As shown in figure 12.1, the DNA is transcribed into RNA in a region that spans the end of the promoter to the terminator. The base sequence in the RNA transcript is complementary to the **template strand.** For structural genes that encode proteins, the non-template DNA strand is called the **coding strand** or the **sense strand.**

A second category of sequences are those involved in the regulation of gene expression. These are termed **regulatory sequences,** which act as binding sites for genetic regulatory proteins. The binding of regulatory proteins to a regulatory sequence affects gene expression. For example, certain transcription factors bind to regulatory sequences and thereby influence the rate of transcription in a positive or negative way. A variety of gene regulation mechanisms are considered in chapters 14 and 15.

A third category of sequences ensures that the nucleotide sequence within an mRNA will be translated into the amino acid sequence of a polypeptide. These sequences are used within the

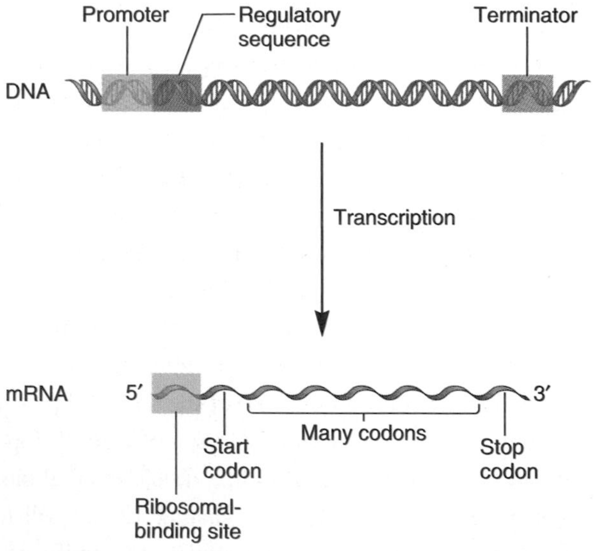

DNA:

• Promoter: site for RNA polymerase binding; signals the beginning of transcription.

• Terminator: signals the end of transcription.

• Regulatory sequences: site for the binding of regulatory proteins; the role of regulatory proteins is to influence the rate of transcription. In eukaryotes, regulatory sequences can be found in a variety of locations.

mRNA:

• Ribosomal-binding site: site for ribosome binding; translation begins near this site in the mRNA. In eukaryotes, a ribosome scans the mRNA for a start codon.

• Start codon: specifies the first amino acid in a protein sequence, usually a formylmethionine (in bacteria) or a methionine (in eukaryotes).

• Codons: a 3-nucleotide sequence within the mRNA that specifies a particular amino acid. The sequence of codons within mRNA determines the sequence of amino acids within a polypeptide.

• Bacterial mRNA may be polycistronic, which means it encodes two or more polypeptides.

FIGURE 12.1   Organization of sequences of a bacterial gene and its mRNA transcript. This figure depicts the general organization of sequences that are needed to create a functional gene that encodes an mRNA. Within a chromosome that contains many genes, the direction of transcription of any given gene, and the use of a particular DNA strand as a template strand, are variable. For example, the gene shown here is transcribed from left to right and uses only one DNA strand as the template strand. Another gene in the same chromosome could have a promoter that directs transcription from right to left and uses the other DNA strand as a template. Within any chromosome, some genes will be transcribed in one direction and use a particular strand as a template, while the remaining genes will be transcribed in the opposite direction and use the complementary strand as a template.

mRNA during the translation process. We will consider them in greater detail in chapter 13. In bacteria, a short sequence within the mRNA, the **ribosomal-binding site,** provides a location for the ribosomes to recognize and begin translation. The ribosome recognizes this site because it is complementary to a sequence in ribosomal RNA. During polypeptide synthesis, the sequence of nucleotides within the mRNA is read as groups of three nucleotides, known as **codons.** The first codon, which is very close to the ribosomal-binding site, is known as the **start codon.** This is followed by many more codons that dictate the sequence of amino acids within the synthesized polypeptide. The relationship between the codons and the genetic code is discussed in chapter 13. Finally, a **stop codon** signals the end of translation.

The Three Stages of Transcription Are Initiation, Elongation, and Termination

Transcription occurs in three stages: **initiation, synthesis** of the RNA transcript (also called **elongation**), and **termination.** These steps involve protein-DNA interactions in which proteins such as RNA polymerase interact with DNA sequences. As shown in figure 12.2, the initiation stage in the transcription process is a recognition step. The sequence of bases within the promoter region is recognized by proteins known as transcription factors. The specific binding of transcription factors to the promoter sequence identifies the starting site for transcription.

Transcription factors and RNA polymerase first bind to the promoter region when the DNA is in the form of a double helix. This is termed a **closed promoter complex.** For transcription to occur, the DNA strands must be separated. This allows one of the two strands to be used as a template for the synthesis of a complementary strand of RNA. This synthesis occurs as RNA polymerase slides along the DNA, creating a small bubblelike structure known as the **open promoter complex,** or simply as an **open complex.** Eventually, RNA polymerase reaches a termination sequence, which causes it and the newly made RNA transcript to dissociate from the DNA. This event is transcriptional termination.

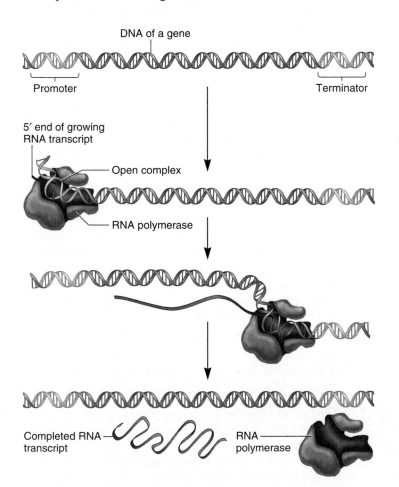

DNA of a gene

Promoter   **Terminator**

Initiation: The promoter functions as a recognition site for transcription factors (not shown). The transcription factor(s) enables RNA polymerase to bind to the promoter. Following binding, the DNA is denatured into a bubble known as the open complex.

5′ end of growing RNA transcript

Open complex

RNA polymerase

Elongation/synthesis of the RNA transcript: RNA polymerase slides along the DNA in an open complex to synthesize RNA.

Termination: A termination signal is reached that causes RNA polymerase and the RNA transcript to dissociate from the DNA.

Completed RNA transcript   RNA polymerase

FIGURE 12.2   **Stages of transcription.**

GENES→TRAITS   The ability of genes to produce an organism's traits relies on the molecular process of gene expression. Transcription is the first step in gene expression. During transcription, the gene's sequence within the DNA is used as a template to make a complementary copy of RNA. In chapter 13, we will examine how the sequence in mRNA is translated into a polypeptide chain. After polypeptides are made within a living cell, they fold into functional proteins that govern an organism's traits. For example, once the gene encoding phenylalanine hydroxylase is transcribed and then translated, the enzyme phenylalanine hydroxylase is made, and the cells can metabolize phenylalanine. As described in chapter 1, individuals who possess two defective copies of this gene cannot synthesize this enzyme and thus cannot metabolize phenylalanine, causing a disease known as phenylketonuria (PKU).

RNA Transcripts Have Different Functions

Once they are made, RNA transcripts play different functional roles (table 12.1). As we have already seen, when a **structural gene** is transcribed, the product is an RNA transcript known as **messenger RNA (mRNA)**. The function of mRNA is to specify the amino acid sequence of a polypeptide. As will be described in chapter 13, the process of translation uses the codon sequence in mRNA to synthesize a polypeptide with a defined amino acid sequence. Well over 90% of all genes are structural genes.

In addition, several other genes encode RNAs that are never translated. As described in table 12.1, the RNA transcripts from nonstructural genes have various important cellular functions. In some cases, the RNA transcript becomes part of a complex that contains both protein subunits and one or more RNA molecules. Examples of protein-RNA complexes include ribosomes, signal recognition particles, spliceosomes, and certain enzymes such as RNaseP. The function of spliceosomes and RNaseP are described later in chapter 12.

TABLE 12.1
Functions of RNA Molecules

Type of RNA	Description
mRNA	Messenger RNA (mRNA) encodes the sequence of amino acids within a polypeptide.
tRNA	Transfer RNA (tRNA) is necessary for the translation of mRNA.
rRNA	Ribosomal RNA (rRNA) is necessary for the translation of mRNA. rRNAs are components of ribosomes that are composed of both rRNAs and protein subunits. The structure and function of ribosomes are examined in chapter 13.
Other Small RNAs	
7S RNA	7S RNA, which is found in eukaryotes, is necessary in the targeting of proteins to the endoplasmic reticulum. 7S RNA is a component of a complex known as signal recognition particle (SRP), which is composed of 7S RNA and six different protein subunits.
scRNA	Small cytoplasmic RNA found in bacteria. Its sequence is similar to 7S RNA found in eukaryotes. ScRNA is needed for protein secretion.
RNA of RNaseP	RNaseP is an enzyme that is necessary in the processing of all bacterial tRNA molecules. The RNA is the catalytic component of this enzyme. RNaseP is composed of a 350- to 410-nucleotide RNA and one protein subunit.
snRNA	Small nuclear RNA (snRNA) is necessary in the splicing of eukaryotic pre-mRNA. SnRNAs are components of a spliceosome, which is composed of both snRNAs and protein subunits. The structure and function of spliceosomes are examined later in chapter 12.
snoRNA	Small nucleolar RNA (snoRNA) is necessary in the processing of eukaryotic rRNA transcripts. SnoRNAs are also associated with protein subunits. In eukaryotes, snoRNAs are found in the nucleolus where rRNA processing and ribosome assembly occur.
Viral RNAs	Some types of viruses use RNA as their genome, which is packaged within the viral capsid.

12.2 TRANSCRIPTION IN BACTERIA

Our molecular understanding of gene transcription initially came from studies involving bacteria and bacteriophages. Several early investigations focused on the production of viral RNA after bacteriophage infection. The first suggestion that RNA is derived from the transcription of DNA was made by Eliot Volkin and Lazarus Astrachan in 1956. When *E. coli* cells were exposed to T7 bacteriophage, it was observed that the RNA made immediately after infection had a base composition that was substantially different from the base composition of RNA prior to infection. Furthermore, the base composition after infection was very similar to the base composition in the T7 DNA (except that the RNA contained uracil instead of thymine). These results were consistent with the idea that the bacteriophage DNA was being used as a template in the synthesis of bacteriophage RNA.

In 1960, Matthew Meselson and Francöis Jacob found that proteins are synthesized on ribosomes. One year later, Jacob and his colleague Jacques Monod proposed that a certain type of RNA acts as a genetic messenger (from the DNA to the ribosome) to provide the information for protein synthesis. They hypothesized that this RNA, which they called messenger RNA, is transcribed from the sequence within DNA and then directs the synthesis of particular polypeptides. In the early 1960s, this was a remarkable proposal, considering that it was made before the actual isolation and characterization of the mRNA molecules in vitro. In 1961, the mRNA hypothesis was confirmed by Sydney Brenner in collaboration with Jacob and Meselson. They found that when a virus infects a bacterial cell, a virus-specific RNA is made that rapidly associates with preexisting ribosomes in the cell.

Since these pioneering studies, a great deal has been learned about the molecular features of bacterial gene transcription. Much of our knowledge comes from studies in *E. coli*. In this section, we will examine the three steps in the gene transcription process as they occur in bacteria.

A Promoter Is a Short Sequence of DNA That Is Necessary to Initiate Transcription

The type of DNA sequence known as the promoter gets its name from the idea that it "promotes" gene expression. More precisely, this sequence of nucleotides directs the exact location for the initiation of RNA transcription. Much of the promoter region is located just upstream from the site where transcription of a gene actually begins. By convention, the bases in a promoter sequence are numbered in relation to the transcriptional start site (fig. 12.3). The first base used as a template for RNA transcription is denoted +1. The bases preceding this are numbered in a negative direction. There is no base numbered 0. Therefore, most of the promoter region is labeled with negative numbers that describe the number of bases preceding the beginning of transcription.

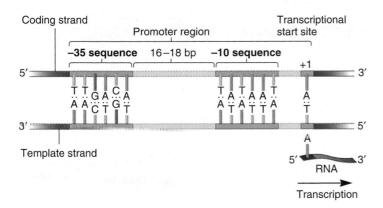

FIGURE 12.3 **The conventional numbering system of promoters.** The first nucleotide that acts as a template for transcription is designated +1. The numbering of nucleotides to the left of this spot is in a negative direction, while the numbering to the right is in a positive direction. For example, the nucleotide that is immediately to the left of the +1 nucleotide is numbered −1, and the nucleotide to the right of the +1 nucleotide is numbered +2. There is no zero nucleotide in this numbering system. In many bacterial promoters, sequence elements at the −35 and −10 regions play a key role in promoting transcription.

Although the promoter may encompass a region that is several dozen nucleotides in length, there are short **sequence elements** that are particularly critical for promoter recognition. By comparing the sequence of DNA bases within many promoters, researchers have learned that certain sequences of bases are necessary to create a functional promoter. In many promoters found in *E. coli* and similar species, two sequence elements are important. These are located at approximately the −35 and −10 sites in the promoter region (fig. 12.3). The sequence at the −35 region is 5′–TTGACA–3′, and the one at the −10 region is 5′–TATAAT–3′. The TATAAT sequence is sometimes called the **Pribnow box** after David Pribnow, who initially discovered it in 1975.

The sequences at the −35 and −10 sites can vary among different genes. For example, figure 12.4 illustrates the sequences found in several different *E. coli* promoters. The most commonly occurring bases within a sequence element form the **consensus sequence.** It is usually the sequence that is most efficiently recognized. For many bacterial genes, there is a good correlation between the rate of RNA transcription and the degree to which the −35 and −10 regions agree with their consensus sequences.

Bacterial Transcription Is Initiated When RNA Polymerase Holoenzyme Binds at a Promoter Sequence

Thus far, we have considered the DNA sequences that constitute a functional promoter. Let's now turn our attention to a consideration of the proteins that recognize those sequences and carry out the transcription process. The enzyme that catalyzes the synthesis of RNA is **RNA polymerase.** In *E. coli,* the **core enzyme** is composed of four subunits, $\alpha_2\beta\beta'$. The association of a fifth subunit, **sigma factor,** with the core enzyme is referred to as RNA polymerase **holoenzyme.** The different subunits within the holoen-

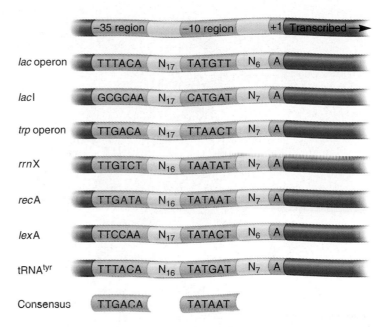

FIGURE 12.4 **Examples of −35 and −10 sequences within a variety of bacterial promoters.** This figure shows the −35 and −10 sequences for seven different bacterial and bacteriophage promoters. The consensus sequence is shown at the *bottom.* The spacer regions contain the designated number of nucleotides between the −35 and −10 region or between the −10 region and the transcriptional start site. For example, N_{17} means that there are 17 nucleotides between the end of the −35 region and the beginning of the −10 region.

zyme play distinct functional roles. The two α subunits are important in the proper assembly of the holoenzyme and in the process of binding to DNA. The β and β' subunits are also needed for binding to the DNA and are critical in the catalytic synthesis of RNA. The holoenzyme is required to initiate transcription; the primary role of sigma factor is to recognize the promoter. Proteins, such as sigma factor, that influence the function of RNA polymerase are known as **transcription factors.** The core enzyme is necessary for RNA synthesis.

After RNA polymerase holoenzyme is assembled into its five subunits, it binds loosely to the DNA. It then scans along the DNA much as a train rolls down the tracks. When it encounters a promoter region, sigma factor recognizes the promoter elements at both the −35 and −10 positions. A region within the sigma factor protein that contains a **helix-turn-helix structure** is involved in a tighter binding to the DNA. Alpha helices within the protein can fit into the major groove of the DNA double helix and form hydrogen bonds with the bases. This phenomenon is shown in figure 12.5. Hydrogen bonding occurs between nucleotides in the −35 and −10 regions of the promoter and amino acid side chains in the helix-turn-helix structure of sigma factor.

As shown in figure 12.6, the process of transcription is initiated when sigma factor within the holoenzyme has bound to the promoter region to form the **closed complex.** For transcription to begin, the double-stranded DNA must then be unwound into an open complex. This unwinding first occurs at the TATAAT box in the −10 region, which contains mostly AT base

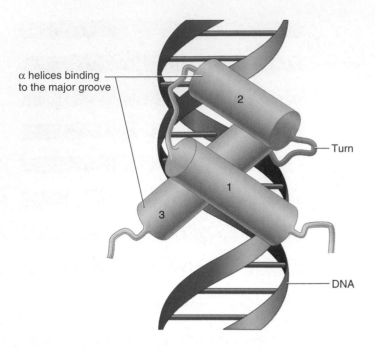

FIGURE 12.5 **The binding of a transcription factor protein to the DNA double helix.** In this example, the protein contains a helix-turn-helix motif. Two α helices of the protein (labeled *2* and *3*) can fit within the major groove of the DNA. Amino acids within the α helices form hydrogen bonds with the bases in the DNA.

pairs, as shown earlier in figure 12.4. As you may recall from chapter 9, AT base pairs form only two hydrogen bonds, whereas GC pairs form three. Therefore, it is easier to separate DNA in an AT-rich region, since fewer hydrogen bonds must be broken. A short strand of RNA is made within the open complex and then sigma factor is released from the core enzyme. The release of sigma factor marks the transition to the elongation phase of transcription. The core enzyme may now slide down the DNA to synthesize a strand of RNA.

The RNA Transcript Is Synthesized During the Elongation Stage

After the initiation stage of transcription is completed, the RNA transcript is made in the elongation stage of transcription. During the synthesis of the RNA transcript, RNA polymerase moves along the DNA, causing it to unwind (fig. 12.7). The DNA strand that is used as a template for RNA synthesis is called the **template** or **noncoding strand.** The opposite DNA strand is called the **coding strand;** it has the same sequence as the RNA transcript except that T in the DNA corresponds to U in the RNA. Within a given gene, only the template strand is used for RNA synthesis while the coding strand is never used. As it moves down the DNA, the open complex formed by the action of RNA polymerase is approximately 17 bp long. On average, the rate of RNA synthesis is about 43 nucleotides per second! Behind the open complex, the DNA rewinds back into a double helix.

As described in figure 12.7, the chemistry of transcription by RNA polymerase is similar to the synthesis of DNA via DNA

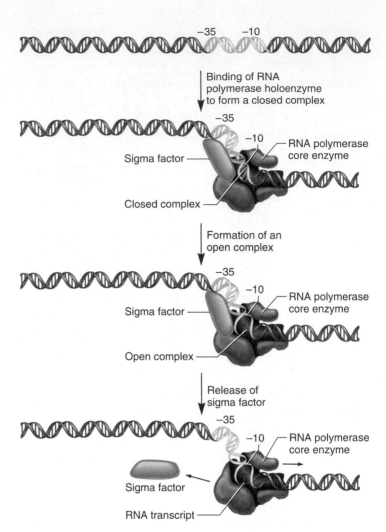

FIGURE 12.6 **The initiation stage of transcription in bacteria.** The sigma factor subunit of the RNA polymerase holoenzyme recognizes the −35 and −10 regions of the promoter. The DNA unwinds in the −10 region to form an open complex. Sigma factor then dissociates from the holoenzyme, and the RNA polymerase core enzyme can proceed down the DNA to transcribe RNA.

polymerase, which was discussed in chapter 11. RNA polymerase always connects nucleotides in the 5′ to 3′ direction. During this process, RNA polymerase catalyzes the formation of a bond between the 5′ phosphate group on one nucleotide and the 3′ −OH group on the previous nucleotide. The complementarity rule is similar to the AT/GC rule, except that uracil substitutes for thymine in the RNA. In other words, RNA synthesis obeys an $A_{RNA}T_{DNA}/U_{RNA}A_{DNA}/G_{RNA}C_{DNA}/C_{RNA}G_{DNA}$ rule.

Transcription Is Terminated by Either an RNA-Binding Protein or an Intrinsic Terminator

The end of RNA synthesis is referred to as termination. Prior to termination, the hydrogen bonding between the DNA and RNA within the open complex is of central importance in preventing dissociation of the RNA polymerase from the template strand.

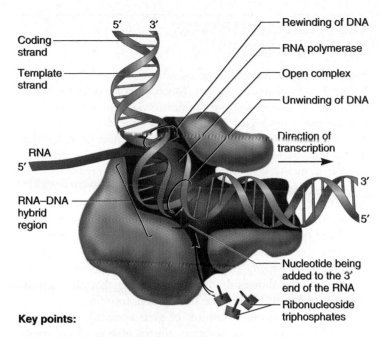

Key points:

- RNA polymerase slides along the DNA, creating an open complex as it moves.

- The DNA strand known as the template strand is used to make a complementary copy of RNA as an RNA–DNA hybrid.

- The RNA is synthesized in a 5′ to 3′ direction using ribonucleoside triphosphates as precursors. Pyrophosphate is released (not shown).

- The complementarity rule is the same as the AT/GC rule except that U is substituted for T in the RNA.

FIGURE 12.7 **Synthesis of the RNA transcript.**

Termination occurs when this short RNA-DNA hybrid region is forced to separate, thereby releasing the newly made RNA transcript as well as RNA polymerase. In *E. coli*, two different mechanisms for termination have been identified. For certain genes, a protein known as **ρ (rho)** is responsible for terminating transcription, a mechanism called **ρ-dependent termination.** For other genes, termination does not require ρ. This is referred to as **ρ-independent termination.** Both mechanisms will be described here.

In ρ-dependent termination, a sequence near the 3′ end of the newly made RNA called the *rut* site (for r̲ho u̲tilization site) acts as a recognition site for the binding of the ρ protein (fig. 12.8). Rho protein functions as a helicase, an enzyme that can separate RNA-DNA hybrid regions. After the *rut* site is synthesized in the RNA, ρ binds to the RNA and moves in the direction of RNA polymerase. In the DNA, a sequence, slightly upstream from the terminator, encodes an RNA sequence that forms a stem-loop structure containing several GC base pairs. As you may recall from chapter 9, a stem-loop structure can form due to complementary sequences within the RNA. These RNA stem-loops can form almost immediately after they are synthesized. The stem-loop causes RNA polymerase to pause in its synthesis of RNA. This allows ρ to catch up to the stem-loop, pass through it, and break the hydrogen bonds between the DNA and RNA

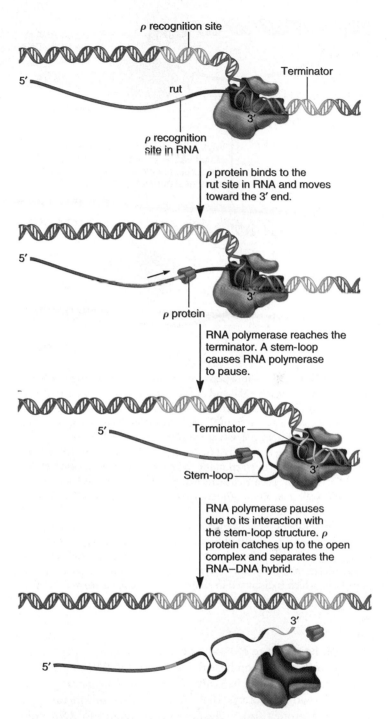

FIGURE 12.8 **ρ-dependent termination.**

within the open complex. When this occurs, the completed RNA strand is separated from the DNA along with RNA polymerase.

We now turn our attention to ρ-independent termination, a process that is facilitated by two sequence elements within the RNA (fig. 12.9). One element is a uracil-rich sequence located at the 3′ end of the RNA. The second sequence is slightly upstream from the uracil-rich sequence, near the 3′ end; it promotes the formation of a stem-loop structure.

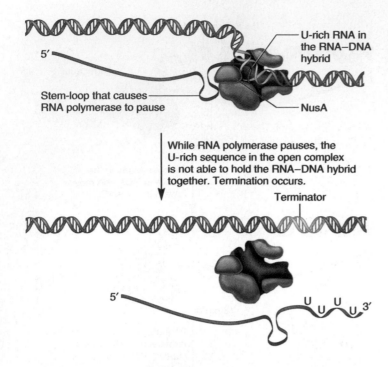

FIGURE 12.9 *ρ*-independent or intrinsic termination. When RNA polymerase reaches the end of the gene, it transcribes a uracil-rich sequence. Soon after this uracil-rich sequence is transcribed, a stem-loop forms just upstream from the open complex. The formation of this stem-loop causes RNA polymerase to pause in its synthesis of the transcript. This pausing is stabilized by NusA, which binds near the region where RNA exits the open complex. While it is pausing, the RNA-DNA hybrid region is a uracil-rich sequence. Since UA hydrogen bonds are relatively weak interactions, the transcript and RNA polymerase dissociate from the DNA.

Interestingly, proteins such as NusA, which enhance the efficiency of transcriptional termination in many bacterial genes, were first discovered because they actually prevent termination in certain viral genes. NusA aids in the prevention of termination via a viral protein called the N protein, which is discussed in chapter 14. Its name is derived from an acronym for N utilization substance.

In the sequence of events described in figure 12.9, the formation of the stem-loop near the 3' end of the RNA causes RNA polymerase to pause in its synthesis of RNA. This pausing is stabilized by other proteins that bind to RNA polymerase. For example, a protein called NusA, which is bound to RNA polymerase, promotes pausing at stem-loop sequences. At the precise time RNA polymerase pauses, the uracil-rich sequence in the RNA transcript happens to be bound to the DNA template strand. As previously mentioned, the hydrogen bonding of RNA to DNA keeps RNA polymerase clamped onto the DNA. However, the binding of this uracil-rich sequence to the DNA template strand is thought to be unstable, causing the RNA transcript to spontaneously dissociate from the DNA and terminate further transcription. Because this process does not require a protein to physically remove the RNA transcript from the DNA, the sequence elements that cause this type of termination are referred to as **intrinsic terminators.**

12.3 TRANSCRIPTION IN EUKARYOTES

Many of the basic features of gene transcription are very similar in bacterial and eukaryotic species. Much of our understanding of transcription has come from studies in *Saccharomyces cerevisiae* (baker's yeast) and higher eukaryotic species such as mammals. In general, gene transcription in eukaryotes is more complex than that of their bacterial counterparts. Eukaryotic cells are larger and contain a variety of compartments known as organelles. This added level of cellular complexity dictates that eukaryotes contain many more genes encoding cellular proteins. In addition, higher eukaryotic species are multicellular, being composed of many different cell types. Multicellularity adds the requirement that genes be transcribed in the correct type of cell and during the proper stage of development. Therefore, in any given species, the transcription of the thousands of different genes that an organism possesses requires appropriate timing and coordination. In this section, we will examine features of gene transcription that are unique to eukaryotes. In addition, the regulation of eukaryotic gene transcription is covered in chapter 15.

Eukaryotes Have Multiple RNA Polymerases That Are Structurally Similar to the Bacterial Enzyme

The genetic material within the nucleus of a eukaryotic cell is transcribed by three different RNA polymerase enzymes, designated I, II, and III. Each of the three RNA polymerases transcribes different categories of genes. RNA polymerase I transcribes all of the genes that encode ribosomal RNA (rRNA) except for the 5S rRNA. RNA polymerase II plays a major role in cellular transcription since it transcribes all of the structural genes. It also transcribes certain snRNA genes, which are needed for pre-mRNA splicing. RNA polymerase II is responsible for the synthesis of all mRNA. RNA polymerase III transcribes all of the tRNA genes and the 5S rRNA gene.

All three RNA polymerases are very similar structurally and are composed of many subunits. They contain two large subunits that are similar to the β and β' subunits of bacterial RNA polymerase. The structures of a few RNA polymerases have been determined by X-ray crystallography. There is a remarkable similarity between the bacterial enzyme and its eukaryotic counterparts. Figure 12.10*a* compares the structures of a bacterial RNA polymerase with RNA polymerase II from yeast. As seen here, both enzymes have a very similar structure. Also, it is very exciting that this structure provides a way to envision how the transcription process works. As seen in figure 12.10*b*, the DNA that is about to be transcribed lies on a surface within RNA polymerase termed the bridge, which is nested between two regions called the jaw and clamp. A location called the wall forces the DNA-RNA hybrid to make a right-angle turn. This bend facilitates the ability of nucleotides to bind to the template strand. Mg^{++} is located at the catalytic site, which is precisely at the 3' end of the growing RNA strand. Nucleoside triphosphates (NTPs) can enter the catalytic site via a funnel and pore region. The correct nucleotide

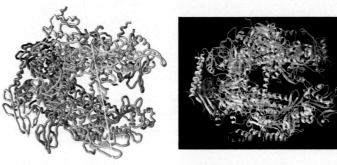

Structure of a bacterial
RNA polymerase

Structure of a eukaryotic
RNA polymerase II

(a)

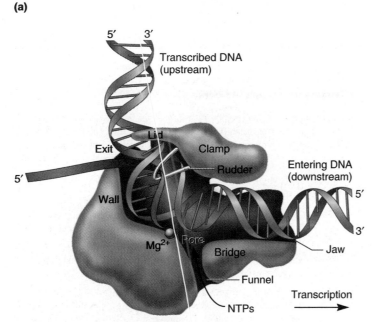

(b) Schematic structure of RNA polymerase

FIGURE 12.10   Structure and molecular function of RNA polymerase. (a) A comparison of the crystal structures of a bacterial RNA polymerase (*left*) to a eukaryotic RNA polymerase II (*right*). The bacterial enzyme is from *Thermus aquaticus.* The eukaryotic enzyme is from *Saccharomyces cerevisiae.* **(b)** A molecular mechanism for transcription based on the crystal structure. In this diagram, the direction of transcription is from left to right. The double-stranded DNA enters the polymerase along a bridge surface that is between the jaw and clamp. At a region termed the wall, the DNA-RNA hybrid is forced to make a right-angle turn, which enables nucleotides to bind to the template strand. Mg^{++} is located at the catalytic site. Nucleoside triphosphates (NTPs) enter the catalytic site via a funnel and pore region and bind to the template DNA. At the catalytic site, the nucleotides are covalently attached to the 3' end of the RNA. As RNA polymerase slides down the template, a small region of the protein termed the rudder separates the RNA-DNA hybrid. The single-stranded RNA then exits under a small lid.

would bind to the template DNA and then be covalently attached to the 3' end. As RNA polymerase slides down the template, a rudder, which is about 9 bp away from the 3' end of the RNA, forces the RNA-DNA hybrid apart. The single-stranded RNA then exits under a small lid.

Eukaryotic Structural Genes Have a Core Promoter and Regulatory Elements

In eukaryotes, the promoter sequence is more variable and often more complex than that found in bacteria. For structural genes, at least three features are found in most promoters: a **transcriptional start site,** a **TATA box,** and **regulatory elements.** Figure 12.11 shows a common pattern of sequences found within the promoters of eukaryotic structural genes. The **core promoter** is relatively short. It consists of a TATAAAA sequence called the TATA box and a transcriptional start site. The TATA box, which is usually about 25 bp upstream from a transcriptional start site, is important in determining the precise starting point for transcription. If it is missing from the core promoter, the transcription starting point becomes undefined, and transcription may start at a variety of different locations. The core promoter, by itself, produces a low level of transcription. This is termed **basal transcription.**

Regulatory elements affect the ability of RNA polymerase to recognize the core promoter and begin the process of transcription. There are two categories of regulatory elements. Activating sequences, known as **enhancers,** are needed to stimulate transcription. In the absence of enhancer sequences, most eukaryotic genes have very low levels of basal transcription. Under certain conditions, it is necessary to prevent transcription of a given gene. This occurs via repressing sequences, known as **silencers,** which inhibit transcription. In chapter 12, we will briefly consider the locations of regulatory elements relative to the core promoter. The functions of eukaryotic regulatory elements, and the **regulatory transcription factors** that bind to them, are examined in greater detail in chapter 15. As seen in figure 12.11, a common location for regulatory elements is the -50 to -100 region. However, the locations of regulatory elements are quite variable among different eukaryotic genes. Such elements can be far away from the core promoter yet exert strong effects on the ability of RNA polymerase to initiate transcription.

With regard to terminology, there are two ways to view the relative locations of factors that control the expression of genes. DNA sequences such as the TATA box, enhancers, and silencers exert their effects only over a nearby gene. They are called *cis-acting elements.* The term *cis* comes from chemistry nomenclature meaning "next to." By comparison, the regulatory proteins that bind to such elements are called *trans-acting factors,* the term *trans* meaning "across from." The proteins that control the expression of a particular gene are also encoded by genes (i.e., regulatory genes), but such regulatory genes may be far away from the genes they control. Nevertheless, when a regulatory gene encoding a *trans*-acting factor is expressed, the protein that is made can diffuse throughout the cell and bind to its appropriate *cis*-acting element. Let's now turn our attention to the function of proteins that bind to *cis*-acting elements.

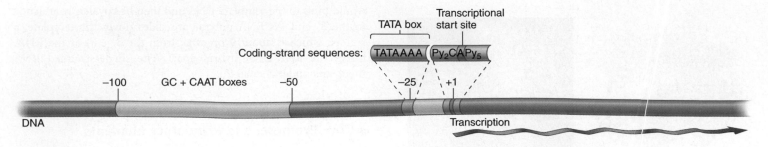

FIGURE 12.11 **A common pattern found within the promoter of structural genes recognized by RNA polymerase II.** The startpoint usually occurs at adenine; there are two pyrimidines and a cytosine that precede this adenine, and five pyrimidines that follow it. A TATA box is approximately 25 bp upstream. Regulatory elements, such as GC or CAAT boxes, are variable in their locations but often are found in the −50 to −100 region.

Transcription of Eukaryotic Structural Genes Is Initiated When RNA Polymerase II and General Transcription Factors Bind to a Promoter Sequence

Thus far, we have considered the DNA sequences that play a role in the promoter region of eukaryotic structural genes. By studying transcription in a variety of eukaryotic species, researchers have discovered that three categories of proteins are needed for basal transcription at the core promoter (table 12.2). These are RNA polymerase II, five different proteins called **general transcription factors (GTFs),** and a protein complex called **mediator.** We will examine their roles next.

Figure 12.12 describes the assembly of general transcription factors and RNA polymerase II at the TATA box. As shown here, a series of interactions leads to the formation of the open complex. Transcription factor IID (TFIID) first binds to the TATA box and thereby plays a critical role in the recognition of the promoter. TFIID is composed of several subunits including TATA-binding protein (TBP), which directly binds to the TATA box, and several other proteins called TBP-associated factors (TAFs). After TFIID binds to the TATA box, it associates with TFIIB. TFIIB promotes the binding of RNA polymerase II and TFIIF to the core promoter. Lastly, TFIIE and TFIIH bind to the complex. This completes the assembly of proteins to form a closed complex, also known as a **preinitiation complex.**

TFIIH plays a major role in the formation of the open complex. TFIIH has several subunits that perform different functions. One subunit hydrolyzes ATP and phosphorylates a domain in RNA polymerase II known as the carboxyl terminal domain (CTD). Phosphorylation of the CTD releases the contact between RNA polymerase II and TFIIB. Other subunits in TFIIH function as helicases, which break the hydrogen bonding between the double-stranded DNA and thereby promote the formation of the open complex. After the open complex has formed, TFIIB, TFIIE, and TFIIH dissociate. RNA polymerase II is then free to proceed to the elongation stage of transcription.

In vitro, when researchers mix together TFIID, TFIIB, TFIIE, TFIIF, TFIIH, RNA polymerase II, and a DNA sequence

TABLE 12.2

Proteins Needed for Transcription via the Core Promoter of Eukaryotic Structural Genes

RNA polymerase II: The enzyme that catalyzes the linkage of ribonucleotides in the 5′ to 3′ direction, using DNA as a template. Essentially all eukaryotic RNA polymerase II proteins are composed of 12 subunits. The two largest subunits are structurally similar to the β and β' subunits found in *E. coli* RNA polymerase.

General transcription factors:

TFIID: Composed of TATA-binding protein (TBP) and other TBP-associated factors (TAFs). Recognizes the TATA-binding sequence of eukaryotic structural gene promoters.

TFIIB: Binds to TFIID and then enables RNA polymerase II to bind to the core promoter.

TFIIF: Binds to RNA polymerase II and plays a role in its ability to bind to TFIIB and the core promoter. Also plays a role in the ability of TFIIE and TFIIH to bind to RNA polymerase II.

TFIIE: Plays a role in the formation and/or the maintenance of the open complex. It may exert its effects by facilitating the binding of TFIIH to RNA polymerase II and regulating the activity of TFIIH.

TFIIH: A multisubunit protein that has multiple roles. First, certain subunits have helicase activity and promote the formation of the open complex. Other subunits phosphorylate the CTD of RNA polymerase II, which releases its interaction with TFIIB and thereby allows RNA polymerase II to proceed to the elongation phase.

Mediator: A multisubunit complex that mediates the effects of regulatory transcription factors on the function of RNA polymerase II. Though mediator typically has certain core subunits, many of its subunits are variable. This variation may depend on the cell type and environmental conditions. The ability of mediator to affect RNA polymerase II function is thought to occur via the CTD of RNA polymerase II. Mediator can influence the ability of TFIIH to phosphorylate CTD, and subunits within mediator itself have the ability to phosphorylate CTD. Since CTD phosphorylation is needed to release RNA polymerase II from TFIIB, mediator plays a key role in the ability of RNA polymerase II to switch from the initiation to the elongation stage of transcription.

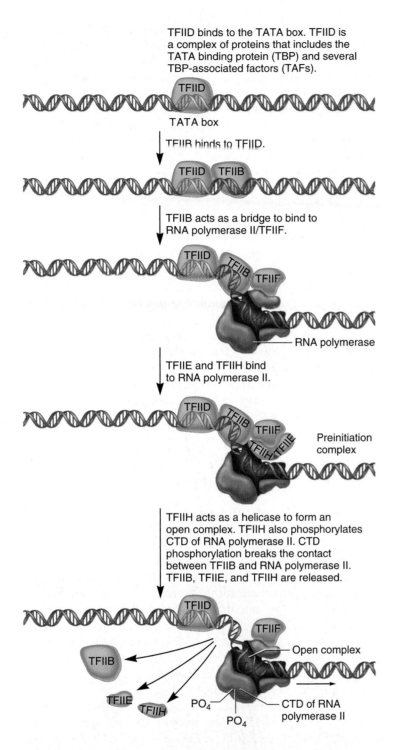

TFIID binds to the TATA box. TFIID is a complex of proteins that includes the TATA binding protein (TBP) and several TBP-associated factors (TAFs).

TATA box

TFIIB binds to TFIID.

TFIIB acts as a bridge to bind to RNA polymerase II/TFIIF.

RNA polymerase

TFIIE and TFIIH bind to RNA polymerase II.

Preinitiation complex

TFIIH acts as a helicase to form an open complex. TFIIH also phosphorylates CTD of RNA polymerase II. CTD phosphorylation breaks the contact between TFIIB and RNA polymerase II. TFIIB, TFIIE, and TFIIH are released.

TFIID

TFIIF

TFIIB

Open complex

TFIIE

TFIIH

PO_4

PO_4

CTD of RNA polymerase II

FIGURE 12.12 **Steps leading to the formation of the open complex.** TFIID first binds to the TATA box. A subunit of TFIID, known as the TATA-binding protein, recognizes the TATA box sequence. TFIIB then binds to TFIID. TFIIB promotes the binding of RNA polymerase II/TFIIF. Transcription factors TFIIE and TFIIH become bound to RNA polymerase to form the closed complex. To form the open complex, TFIIH hydrolyzes ATP and phosphorylates a region in RNA polymerase II known as the carboxyl terminal domain (CTD). RNA polymerase II is released from TFIIB. TFIIH also functions as a helicase, which breaks the hydrogen bonding between the double-stranded DNA and thereby promotes the formation of the open complex. After the open complex has formed, TFIIB, TFIIE, and TFIIH dissociate, and RNA polymerase II proceeds to the elongation stage of transcription.

containing a TATA box and transcriptional start site, the DNA is transcribed into RNA. Therefore, these components are referred to as the **basal transcription apparatus.** In a living cell, however, additional components regulate transcription and allow it to proceed at a reasonable rate.

A third component for transcription is a large protein complex termed *mediator.* The complex derives its name from the observation that it mediates interactions between RNA polymerase II and regulatory transcription factors that bind to enhancers or silencers. It serves as an interface between RNA polymerase II and many, diverse regulatory signals. The subunit composition of mediator is quite complex and variable (see table 12.2). The core subunits form an elliptical-shaped complex that partially wraps around RNA polymerase II. Mediator appears to regulate the ability of TFIIH to phosphorylate the CTD portion of RNA polymerase II. Therefore, it can play a pivotal role in the switch between transcriptional initiation and elongation. The function of mediator during eukaryotic gene regulation is explored in greater detail in chapter 15.

Chromatin Structure Plays a Key Role in Gene Transcription

As we have learned, transcription involves the binding of general transcription factors and RNA polymerase to the promoter region and the subsequent movement of RNA polymerase along the DNA double helix, allowing one strand to function as a template for transcription. The compaction of DNA to form chromatin, described in chapter 10, can be an obstacle to the transcription process. During interphase, when most transcription occurs, the chromatin of eukaryotes is found in 30 nm fibers that are organized into radial loop domains. Within the 30 nm fiber, the DNA is wound around histone octamers to form nucleosomes. The size of a histone octamer is roughly five times smaller than the complex of RNA polymerase II and GTFs. Therefore, since RNA polymerase is a very large enzyme compared to a nucleosome, the tight wrapping of DNA within a nucleosome is expected to inhibit the ability of RNA polymerase to transcribe the DNA. To circumvent this problem, the chromatin structure is significantly loosened during transcription.

Two common mechanisms alter chromatin structure (fig. 12.13). First, the amino terminal ends of histone proteins are covalently modified in a variety of ways including acetylation of lysines, methylation of lysines, and phosphorylation of serines. These covalent modifications play a key role in the compaction of chromatin. For example, positively charged lysine residues within the core histone proteins can be acetylated by enzymes called **histone acetyltransferases.** The attachment of the acetyl group ($-COCH_3$) may have two effects. First, it eliminates the positive charge on the lysine side chain and thereby disrupts the favorable interaction between the histone protein and the negatively charged DNA backbone. This may loosen the structure of nucleosomes and the 30 nm fiber and thereby facilitate the ability of RNA polymerase to transcribe a

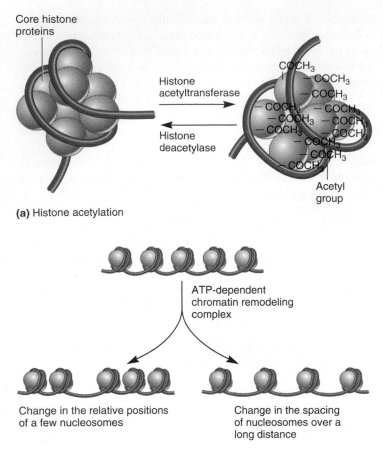

(a) Histone acetylation

(b) Chromatin remodeling

Change in the relative positions of a few nucleosomes

Change in the spacing of nucleosomes over a long distance

ATP-dependent chromatin remodeling complex

Core histone proteins

Histone acetyltransferase

Histone deacetylase

COCH₃

Acetyl group

FIGURE 12.13 **Mechanisms that disrupt the tight binding of DNA and histones within nucleosomes.** **(a)** The covalent modification of histones, such as acetylation, alters their ability to bind to the DNA backbone. Histone acetyltransferase catalyzes the attachment of acetyl groups to lysine residues, thereby loosening the interactions between histones and DNA. Histone deacetylase reverses this process. **(b)** ATP-dependent chromatin remodeling. Proteins of the SWI/SNF family catalyze an ATP-dependent change in the locations of nucleosomes. This may involve a shift in nucleosomes to a new location (*left*) or a change in the spacing of nucleosomes over a long stretch of DNA (*right*). These effects may significantly alter gene transcription.

gene. Some studies suggest that histones are completely displaced, whereas others suggest they are loosened but remain attached to the DNA. Secondly, histones that are covalently modified can be recognized by proteins that alter the compaction level of the chromatin.

Other enzymes, called **histone deacetylases,** can remove the acetyl groups and thereby restore a tighter interaction (fig. 12.13*a*). As discussed further in chapter 15, the competing actions of acetyltransferases and deacetylases play an important role in the transcriptional regulation of genes.

A second way to alter chromatin structure is a process called **ATP-dependent chromatin remodeling.** The energy of ATP hydrolysis is used to drive a change in the structure of nucleosomes and thereby make the DNA more accessible for transcription. This process is carried out by a protein complex that recognizes nucleosomes and uses ATP to alter their configuration. The complexes that catalyze ATP-dependent chromatin remodeling are members of the **SWI/SNF family,** which is a large group of proteins found in all eukaryotic species. They were first identified in yeast as proteins that play a role in the activation of gene transcription. The acronyms SWI and SNF refer to the effects that occur in yeast when these remodeling enzymes are defective. Mutants in SWI were defective in mating type <u>switching</u>, and mutations in SNF created a <u>s</u>ucrose <u>n</u>on-<u>f</u>ermenting phenotype.

As shown in figure 12.13*b,* one result of ATP-dependent chromatin remodeling is a change in the locations of nucleosomes. This may involve a shift in nucleosomes to a new location or a change in the spacing of nucleosomes over a long stretch of DNA. With regard to transcription, these effects on nucleosome location may substantially alter the ability of transcription factors to recognize the promoter and also influence the ability of RNA polymerase to transcribe a given gene. More recently, it has been suggested that chromatin remodeling may also affect the configuration of the 30 nm fiber and/or the radial loop domains found in euchromatin.

12.4 RNA MODIFICATION

During the 1960s and 1970s, studies in bacteria established the physical structure of the gene. The analysis of bacterial genes showed that the sequence of DNA within the coding strand corresponds to the sequence of nucleotides in the mRNA. During translation, the sequence of codons in the mRNA is then read, providing the instructions for the correct amino acid sequence in a polypeptide. The one-to-one correspondence between the sequence of codons in the DNA coding strand and the amino acid sequence of the polypeptide has been termed the **colinearity** of gene expression. Based on these early results in bacteria, researchers concluded that modification of mRNA transcripts is unnecessary for gene expression.

The situation dramatically changed in the late 1970s, when the tools became available to study eukaryotic genes at the molecular level. The scientific community was astonished by the discovery that eukaryotic structural genes are not always colinear with their functional mRNAs. Instead, the coding sequences within many eukaryotic genes are separated by DNA sequences that are not translated into protein. The coding sequences are called **exons,** and the sequences that interrupt them are called **intervening sequences** or **introns.** During transcription, an RNA is made corresponding to the entire gene sequence. Subsequently, the sequences in the RNA that correspond to the introns are removed, while the RNA sequences derived from the exons are connected, or spliced together. This process is called **RNA splicing.** Since the 1970s, it has become apparent that splicing is a common genetic phenomenon in eukaryotic species. Splicing occurs occasionally in bacteria as well.

Aside from splicing, research has also shown that RNA transcripts can be modified in several other ways. Table 12.3 describes the general types of RNA modifications. For example, rRNAs and tRNAs are synthesized as long transcripts that are cleaved into smaller functional pieces. In addition, most eukaryotic mRNAs have a cap attached to their 5′ end and a tail attached at their 3′ end. In this section, we will examine the molecular mechanisms that account for several types of RNA modifications and consider why they are functionally important.

TABLE 12.3
Modifications That May Occur to RNAs

Modification	Description
Trimming	The cleavage of a large RNA transcript into smaller pieces. One or more of the smaller pieces becomes a functional RNA molecule. Trimming occurs for rRNA and tRNA transcripts.
5′ capping	The attachment of a 7-methylguanosine cap to the 5′ end of mRNA. This occurs in eukaryotic mRNAs. As discussed in chapter 13, the cap plays a role in translational initiation.
3′ polyA tailing	The attachment of a string of adenine-containing nucleotides to the 3′ end of mRNA at a site where the mRNA is cleaved (see upward *arrow*). This occurs in eukaryotic mRNAs. It is important for RNA stability and translation.
Splicing	Splicing involves both cleavage and joining of RNA molecules. The RNA is cleaved at two sites, which allows an internal segment of RNA, known as an intron, to be removed. After the intron is removed, the two ends of the RNA molecules are joined together. Splicing is common among eukaryotic structural genes, but it also occurs in the genes that encode rRNA, tRNA, and a few bacterial genes.
RNA editing	The change of the base sequence of an RNA after it has been transcribed. Described in chapter 15.
Base modification	The covalent modification of a base within an RNA molecule. As described in chapter 13, base modification commonly occurs in tRNA molecules.

Some Large RNA Transcripts Are Processed to Smaller Functional Transcripts by Enzymatic Cleavage

For many nonstructural genes, the RNA transcript initially made during gene transcription is processed into smaller functional RNA molecules. This involves cleavage of the RNA to make smaller pieces. As an example, figure 12.14 shows the processing of mammalian ribosomal RNA. The ribosomal RNA gene is transcribed by RNA polymerase I to make a long primary transcript, known as 45S rRNA. (The term *45S* refers to the sedimentation characteristics of this transcript in Svedberg units.) Following the synthesis of the 45S rRNA, cleavage occurs at several points. Three of the fragments, termed *5.8S, 18S,* and *28S* rRNA, are functional rRNA molecules that play a key role in creating the structure of the ribosome. In eukaryotes, the processing of 45S rRNA and the assembly of ribosomal subunits occur in a structure within the cell nucleus known as the **nucleolus.** The structure and biosynthesis of ribosomes are described further in chapter 13.

The production of tRNA molecules also requires processing. Like ribosomal RNA, tRNAs are synthesized as large precursor RNAs that must be cleaved at both the 5′ and 3′ ends to produce mature, functional tRNAs (i.e., ones that bind to amino acids). This processing event has been studied extensively in *E. coli.* Figure 12.15 shows the trimming of a precursor tRNA that carries tyrosine (tRNA^tyr). Interestingly, the cleavage occurs differently at the 5′ end and the 3′ end. At the 5′ end, the precursor tRNA is recognized by an enzyme known as **RNaseP.** This enzyme is an endonuclease that cuts the precursor tRNA. The action of RNaseP produces the correct 5′ end of the mature tRNA. At the 3′ end,

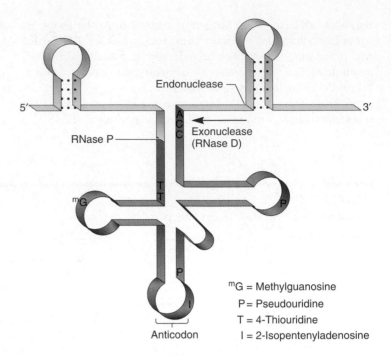

FIGURE 12.15   **The processing of precursor tRNA molecules.** Near the 3′ end of the precursor tRNA, an endonuclease makes a cut and then RNaseD removes nine nucleotides at the 3′ end. RNaseP makes a cut near the 5′ end. In addition to these cleavage steps, several bases within the tRNA molecule are modified to other bases. This phenomenon is discussed in chapter 13.

mG = Methylguanosine
P = Pseudouridine
T = 4-Thiouridine
I = 2-Isopentenyladenosine

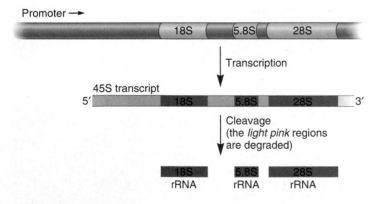

FIGURE 12.14   **The processing of ribosomal RNA in eukaryotes.** The large ribosomal RNA gene is transcribed into a long 45S rRNA primary transcript. This transcript is cleaved to produce 5.8S, 18S, and 28S rRNA molecules, which become associated with protein subunits in the ribosome. This processing occurs within the nucleolus of the cell.

two different enzymes trim the tRNA precursor. First, an endonuclease cleaves the precursor RNA to remove a 170-nucleotide segment. Next, an exonuclease, designated RNaseD, binds to the 3′ end and digests the RNA in the 3′ to 5′ direction. When it reaches a CCA sequence, the exonuclease stops digesting the precursor RNA molecule. Therefore, all tRNAs in *E. coli* have a CCA sequence at their 3′ ends. Finally, certain bases in tRNA molecules may be covalently modified to alter their structure. The modification of bases in tRNAs is described further in chapter 13.

As researchers studied tRNA processing, they found certain enzymatic features that were very unusual and exciting, changing the way biologists view the actions of enzymes. RNaseP has been found to be an enzyme that contains both RNA and protein subunits. In 1983, Sidney Altman and colleagues made the surprising discovery that the RNA portion of this enzyme contains the catalytic ability to cleave the precursor tRNA. RNaseP is an example of a **ribozyme,** which means that its catalytic ability is due to the action of RNA, not protein. Prior to the study of RNaseP and the identification of self-splicing RNAs, biochemists had staunchly believed that only proteins could function as enzymes.

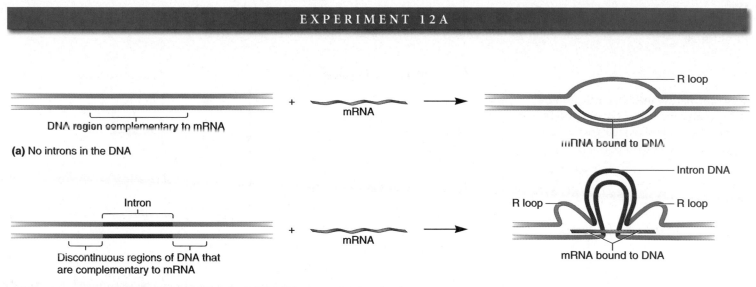

(a) No introns in the DNA

(b) One intron in the DNA. The intron in the pre-mRNA has been previously spliced out.

FIGURE 12.16   **Hybridization of mRNA to double-stranded DNA.** In this experiment, the DNA is denatured and then allowed to renature under conditions that favor an RNA-DNA hybrid. **(a)** If the DNA does not contain an intron, the binding of the mRNA to the anticoding strand of DNA prevents the two strands of DNA from forming a double helix. The single-stranded DNA will form an R loop. **(b)** When mRNA hybridizes to a gene containing one intron, two single-stranded R loops will form that are separated by a double-stranded DNA region. The intervening double-stranded region occurs because an intron has been spliced out of the mRNA, so that the mRNA cannot hybridize to this segment of the gene.

Introns Were Experimentally Identified Via Microscopy

Although the discovery of ribozymes was surprising, the observation that tRNA and rRNA transcripts are processed to a smaller form did not seem unusual to geneticists and biochemists, because the enzymatic cleavage of RNA was similar to the cleavage that can occur for other macromolecules such as DNA and proteins. In sharp contrast, when splicing was detected in the 1970s, it was a novel concept. Splicing involves cleavage at two sites. An intron is removed, and—in a unique step—the remaining fragments are hooked back together again.

Eukaryotic introns were first detected by comparing the base sequence of genes and mRNAs during viral infection of mammalian cells by adenovirus. This research was carried out in 1977 by two groups headed by Philip Sharp and Richard Roberts. This pioneering observation led to the next question: Are introns a peculiar phenomenon that occurs only in viral genes, or are they found in eukaryotic genes as well?

In the late 1970s, several research groups, including those of Pierre Chambon, Bert O'Malley, and Phillip Leder, investigated the presence of introns in eukaryotic structural genes. The experiments of Leder used electron microscopy to identify introns in the β-globin gene. β-globin is a polypeptide that is a subunit of hemoglobin, the protein that carries oxygen in red blood cells. Prior to Leder's work, the β-globin gene had been cloned. To detect introns within the cloned gene, Leder used a strategy involving hybridization. In this approach, shown in figure 12.16, a segment of double-stranded, chromosomal DNA containing the

β-globin gene was first denatured and mixed with mature mRNA for β-globin. Because this mRNA is complementary to the anticoding strand of the DNA, the anticoding strand and the mRNA will specifically bind to each other. This event is called hybridization. Later, when the DNA is allowed to renature, the binding of the mRNA to the anticoding strand of DNA prevents the two strands of DNA from forming a double helix. In the absence of any introns, the single-stranded DNA will form a loop. Since the RNA has displaced one of the DNA strands, this structure is known as an RNA displacement loop or **R loop,** as shown in figure 12.16*a*.

However, Leder and colleagues realized that a different type of R loop structure would be formed if the DNA contained an intron. For example, when mRNA is hybridized to a region of a gene containing one intron, two single-stranded R loops will form that are separated by a double-stranded DNA region. This phenomenon is shown in figure 12.16*b*. The intervening double-stranded region occurs because an intron has been spliced out of the mature mRNA, so that the mRNA cannot hybridize to this segment of the gene.

As shown in steps 1 through 4 of figure 12.17, this hybridization approach was used to identify introns within the β-globin gene. Following hybridization, the samples were placed on a microscopy grid, stained with heavy metal, and then observed by electron microscopy.

▆ THE HYPOTHESIS

The mouse β-globin gene contains one or more introns.

■ TESTING THE HYPOTHESIS — FIGURE 12.17 RNA hybridization to the β-globin gene reveals an intron.

Starting material: A cloned fragment of chromosomal DNA that contains the mouse β-globin gene.

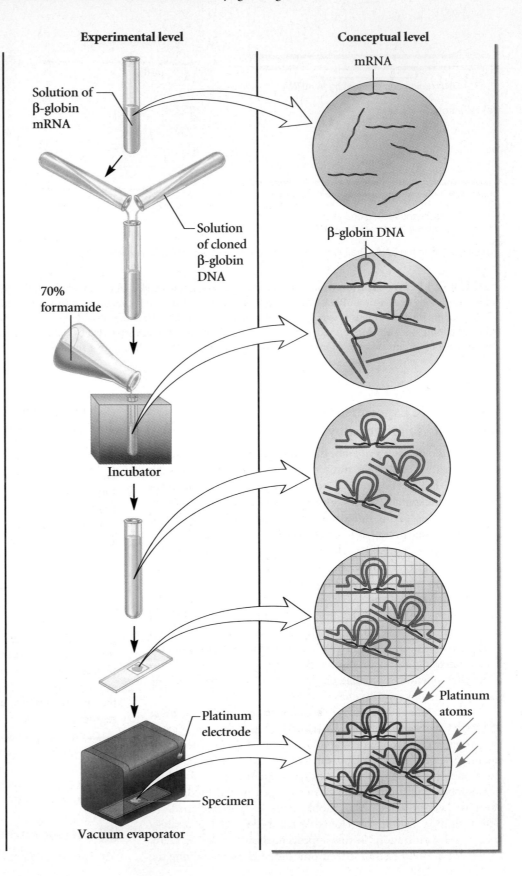

Experimental level

Conceptual level

1. Isolate mature mRNA for the mouse β-globin gene. Note: Globin mRNA is abundant in reticulocytes, which are immature red blood cells.

Solution of β-globin mRNA

Solution of cloned β-globin DNA

mRNA

2. Mix the mRNA and cloned DNA together.

70% formamide

β-globin DNA

3. Separate the double-stranded DNA and allow the mRNA to hybridize. This is done using 70% formamide, at 52°C, for 16 hours.

Incubator

4. Dilute the sample to decrease the formamide concentration. This allows the DNA to re-form a double-stranded structure. Note: The DNA cannot form a double-stranded structure in regions where the mRNA has already hybridized.

5. Spread the sample onto a microscopy grid.

6. Stain with uranyl acetate and shadow with heavy metal. Note: The technique of electron microscopy is described in the appendix.

Platinum electrode

Platinum atoms

7. View the sample under the electron microscope.

Specimen

Vacuum evaporator

THE DATA

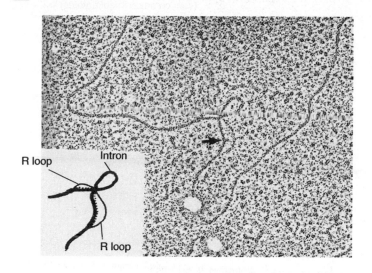

R loop

Intron

R loop

INTERPRETING THE DATA

As seen in the electron micrograph, the mRNA hybridized to the DNA, which caused the formation of two R loops separated by a double-stranded DNA region. These data were consistent with the idea that the DNA of the β-globin gene contains an intron. Similar results were obtained by Chambon and O'Malley for other structural genes. Since these initial discoveries, introns have been found in many eukaryotic genes. The occurrence and biological significance of introns are discussed later in this chapter.

DNA sequencing methods developed in the late 1970s have permitted an easier and more precise way of detecting introns. As described in chapter 18, researchers can clone a fragment of chromosomal DNA that contains a particular gene. This is called a **genomic clone.** In addition, mRNA can be used as a starting material to make a copy of DNA known as **complementary DNA (cDNA).** The cDNA will *not* contain introns, because the introns have been previously removed during RNA splicing. In contrast, if a gene contains introns, a genomic clone for a eukaryotic gene will also contain introns. Therefore, a comparison of the DNA sequences from genomic and cDNA clones can provide direct evidence that a particular gene contains introns.

A self-help quiz involving this experiment can be found at the Online Learning Center.

Three Different Splicing Mechanisms Can Remove Introns

Since the original discovery of introns, the investigations of many research groups have shown that most structural genes among higher eukaryotes contain one or more introns. Less commonly, introns can occur within tRNA and rRNA genes. At the molecular level, three different RNA splicing mechanisms have been identified (fig. 12.18). In all three cases, splicing leads to the removal of the intron RNA and the covalent connection of the exon RNA by a phosphodiester linkage.

The chemistry of splicing among **group I and II introns** is called **self-splicing.** It is given this name because the splicing event does not require the aid of enzymes. Instead, the RNA functions as its own ribozyme. Group I and II differ in the ways that the introns are removed and the exons reconnected. Group I introns that occur within the rRNA of *Tetrahymena* (a protozoan) have been studied extensively by Thomas Cech and colleagues. In this organism, the splicing process involves the binding of a single guanosine to a site within the intron (fig. 12.18*a*). This leads to the cleavage of RNA at the 3′ end of exon 1. Next, the bond between a different guanine nucleotide (that is found within the intron sequence) and the 5′ end of exon 2 is cleaved. This event allows the 3′ end of exon 1 to then form a phosphoester bond with the 5′ end of exon 2. The intron RNA is subsequently degraded. In this example, the RNA molecule functions as its own ribozyme, because it splices itself without the aid of a catalytic protein.

In group II introns, a similar splicing mechanism occurs, except the 2′ –OH group on ribose found in an adenine nucleotide (already within the intron strand) begins the catalytic process (fig. 12.18*b*). Experimentally, group I and II self-splicing can occur in vitro without the addition of any proteins. However, in a living cell (i.e., in vivo), proteins known as **maturases** often enhance the rate of splicing of group I and II introns.

In eukaryotes, the transcription of structural genes produces a long transcript known as **pre-mRNA,** which is located within the nucleus. These large RNA transcripts are also known as **heterogenous nuclear RNA (hnRNA).** This pre-mRNA is usually altered by splicing and other modifications before it exits the nucleus. Unlike the group I and II introns, which may undergo self-splicing, pre-mRNA splicing requires the aid of a multicomponent structure known as the **spliceosome.** As discussed later, this is needed to recognize the intron boundaries and to properly remove it.

Table 12.4 describes the occurrence of introns among the genes of different species. The biological significance of group I and II introns is not understood. By comparison, pre-mRNA splicing is a widespread phenomenon among higher eukaryotes. In mammals and higher plants, most structural genes have at least one intron that can be located anywhere within the gene. In some cases, a single gene can have many introns. As an extreme example, the human dystrophin gene that causes Duchenne muscular dystrophy (when mutant) has 79 exons punctuated by 78 introns. The functional significance of eukaryotic introns is discussed later in this chapter and in chapter 15.

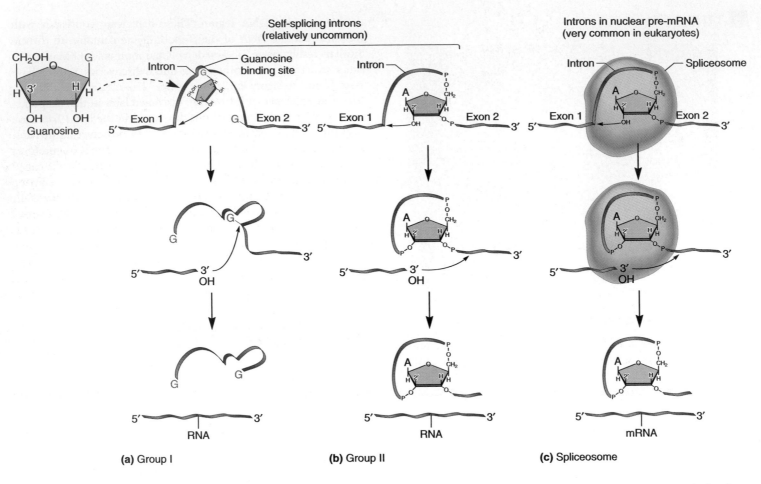

FIGURE 12.18   Mechanisms of RNA splicing. Group I and II introns are self-splicing. **(a)** The splicing of group I introns involves the binding of a free guanosine to a site within the intron, leading to the cleavage of RNA at the 3′ end of exon 1. The bond between a different guanine nucleotide (in the intron strand) and the 5′ end of exon 2 is cleaved. The 3′ end of exon 1 then forms a phosphoester bond with the 5′ end of exon 2. **(b)** In group II introns, a similar splicing mechanism occurs, except that the 2′ –OH group on an adenine nucleotide (already within the intron) begins the catalytic process. **(c)** Pre-mRNA splicing requires the aid of a multicomponent structure known as the spliceosome. The mechanism of pre-mRNA splicing is described later in figure 12.22.

TABLE 12.4
Occurrence of Introns

Type of Intron	Mechanism of Removal	Occurrence
Pre-mRNA	Spliceosome	Very commonly found in structural genes within the nucleus of eukaryotes.
Group I	Self-splicing	Found in rRNA genes within the nucleus of *Tetrahymena* and other lower eukaryotes. Found in a few structural, tRNA, and rRNA genes within the mitochondrial DNA (fungi and plants) and chloroplast DNA. Found very rarely in tRNA genes within bacteria.
Group II	Self-splicing	Found in a few structural, tRNA, and rRNA genes within the mitochondrial DNA (fungi and plants) and in chloroplast DNA. Also found rarely in bacterial genes.

The Ends of Eukaryotic Pre-mRNAs Have a 5′ Cap and a 3′ Tail

At their 5′ end, most mature mRNAs have a 7-methylguanosine covalently attached, an event known as **capping.** Capping occurs while the pre-mRNA is being made by RNA polymerase II, usually when the transcript is only 20 to 25 nucleotides in length. As shown in figure 12.19, it is a three-step process. The nucleotide at the 5′ end of the transcript has three phosphate groups. An enzyme called RNA 5′-triphosphatase removes one of the phosphates, and then a second enzyme, guanylyltransferase, uses GTP and attaches a guanosine monophosphate (GMP) to the 5′ end. Finally, a methyltransferase attaches a methyl group to the guanine base.

The 7-methylguanosine cap structure is recognized by cap-binding proteins, which perform various functions. For example, cap-binding proteins are needed for the proper exit of certain RNAs from the nucleus. Also, as discussed in chapter 13, the cap

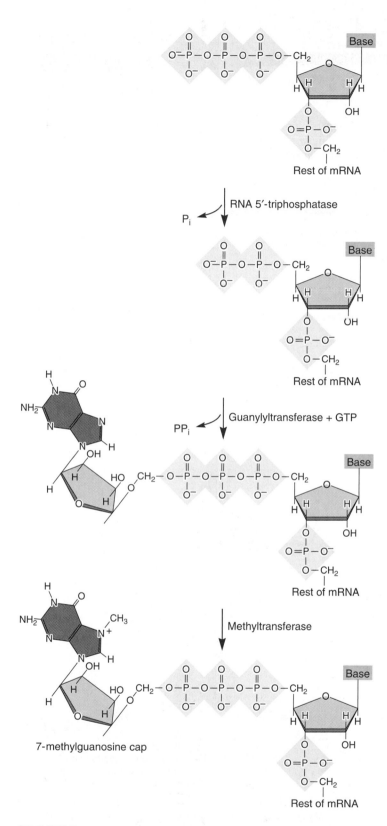

FIGURE 12.19 **Attachment of a 7-methylguanosine cap to the 5′ end of mRNA.** When the transcript is about 20 to 25 nucleotides in length, RNA 5′-triphosphatase removes one of the three phosphates, and then a second enzyme, guanylyltransferase, attaches GMP to the 5′ end. Finally, a methyltransferase attaches a methyl group to the guanine base.

structure is recognized by cap-binding proteins that play an essential role during the early stages of translation. Finally, the cap structure may be important in the efficient splicing of introns, particularly the first intron located nearest the 5′ end.

Let's now turn our attention to the other end of the RNA molecule, namely the 3′ end. Most mature mRNAs have a string of adenine nucleotides, referred to as a **polyA tail,** which appears to be important in the stability of mRNA and in the translation of polypeptides. The polyA tail is not encoded in the gene sequence. Instead, it is added enzymatically after the pre-mRNA has been completely transcribed.

The steps required to synthesize a polyA tail are shown in figure 12.20. To acquire a polyA tail, the pre-mRNA contains a **polyadenylation sequence** near its 3′ end. In higher eukaryotes, the consensus sequence is AAUAAA. This sequence is downstream (toward the 3′ end) from the stop codon in the pre-mRNA. An endonuclease recognizes the polyadenylation sequence and cuts the pre-mRNA at a location that is several nucleotides beyond the 3′ end of the AAUAAA site. The fragment beyond the 3′ cut is degraded. Next, an enzyme known as polyA-polymerase attaches many adenine-containing nucleotides. The length of the polyA tail varies among different mRNAs from a few dozen to several hundred adenine nucleotides.

Besides capping and tailing, a pre-mRNA may also be spliced before it is ready to exit the nucleus. The mechanism of pre-mRNA splicing will be described next.

Pre-mRNA Splicing Occurs by the Action of a Spliceosome

As noted previously, the spliceosome is a large complex that splices pre-mRNA. It is composed of several subunits known as **snRNPs** (pronounced "snurps"). Each snRNP contains small nuclear RNA and a set of proteins. During splicing, the subunits of a spliceosome carry out several functions. First, spliceosome

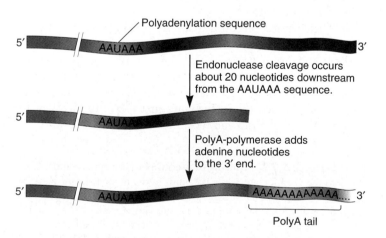

FIGURE 12.20 **Attachment of a polyA tail.** First, an endonuclease cuts the RNA at a location that is 11 to 30 nucleotides after the AAUAAA polyadenylation sequence, making the RNA shorter at its 3′ end. Adenine-containing nucleotides are then attached, one at a time, to the 3′ end by the enzyme polyA-polymerase.

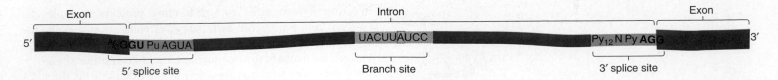

FIGURE 12.21   **Consensus sequences for pre-mRNA splicing in higher eukaryotes.** Consensus sequences exist at the intron-exon boundaries and at a branch site that is found within the intron itself. The adenine nucleotide shown in *blue* in this figure corresponds to the adenine nucleotide at the branch site in figure 12.22. The nucleotides shown in **bold** are highly conserved. Designations: *A/C* = A or C, *Pu* = purine, *Py* = pyrimidine, *N* = any of the four bases.

subunits bind to an intron sequence and precisely recognize the intron-exon boundaries. In addition, the spliceosome must hold the pre-mRNA in the correct configuration so that splicing together of the exons can occur. And finally, the spliceosome catalyzes the chemical reactions that cause the introns to be removed and the exons to be covalently linked.

Intron RNA is defined by particular sequences within the intron and at the intron-exon boundaries. The consensus sequences for the splicing of mammalian pre-mRNA are shown in figure 12.21. These sequences serve as recognition sites for the binding of the spliceosome. The most highly conserved bases are shown in bold.

The molecular mechanism of pre-mRNA splicing is depicted in figure 12.22. The snRNP designated U1 binds to the 5′ splice site, while U2 binds to the branch site. This is followed by the binding of a trimer of three snRNPs, a U4/U6 dimer plus U5. The intron loops outward, and the two exons are brought closer together. The 5′ splice site is then cut, and the 5′ end of the intron becomes covalently attached to the 2′ –OH group of a specific adenine nucleotide in the branch site. U1 and U4 are released, and U5 and U6 shift their positions. In the final step, the 3′ splice site is cut, and then the exons are covalently attached to each other. The three snRNPs, U2, U5, and U6, remain attached to the intron, which is found in a lariat configuration. Eventually, the intron will be degraded, and the snRNPs will be used again to splice other pre-mRNAs.

The chemical reactions that occur during pre-mRNA splicing are not completely understood. Though further research is needed, evidence is accumulating that certain snRNA molecules within the spliceosome may play an enzymatic role in the removal of introns and the connection of exons. In other words, snRNAs may function as ribozymes that cleave the RNA at the exon-intron boundaries and reconnect the remaining exons.

FIGURE 12.22   **Splicing of pre-mRNA via a spliceosome.** U1 binds to the 5′ splice site, while U2 binds to the branch site. A trimer of three snRNPs, a U4/U6 dimer plus U5, binds to the intron region, causing it to form a loop. This brings the exons close together. The 5′ splice site is cut, and the 5′ end of the intron is attached to the adenine nucleotide in the branch site. U1 and U4 are released, and then U5 and U6 shift their positions. In the final step, the 3′ splice site is cut, and then the exons are connected together. The three snRNPs, U2, U5, and U6, remain bound to the intron, which is found in a lariat configuration. Later, the intron will be degraded, and the snRNPs will be used over again.

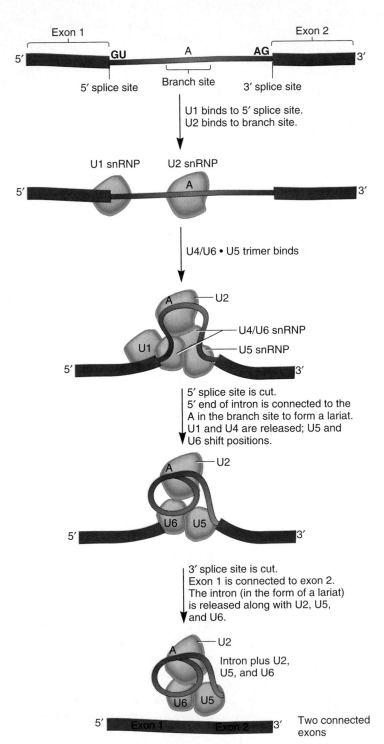

Before ending our discussion of pre-mRNA splicing, it is important to consider why it may be an advantage for a species to have genes that contain introns. One benefit is a phenomenon called **alternative splicing.** When a pre-mRNA has multiple introns, there may be variation in the pattern of splicing so that a mature mRNA may contain different combinations of exons. The variation in splicing may occur in different cell types or during different stages of development. Such alternative splicing allows mature mRNAs to contain different patterns of exons.

As an example, let's suppose a mammalian pre-mRNA contained seven exons. In one cell type (e.g., muscle cells), it may be spliced to produce a mature mRNA with the following pattern of exons: 1-2-4-5-6-7. In a different cell type (e.g., nerve cells), it might be spliced in an alternative pattern: 1-2-3-5-6-7. In this example, the mRNA from muscle cells contains exon 4 while the mRNA from nerve cells contains exon 3. When alternative splicing occurs, proteins with significant differences in their amino acid sequences are produced. In most cases, the alternative versions of a protein will have similar functions, because much of their amino acid sequences will be identical to each other. Nevertheless, alternative splicing produces differences in amino acid sequences that will provide each protein with its own unique characteristics. The biological advantage of alternative splicing is that two (or more) different proteins can be derived from a single gene. This allows an organism to carry fewer genes in its genome. The molecular mechanism of alternative splicing is examined in chapter 15. It involves the actions of proteins (not shown in fig. 12.22) that influence whether or not U1 and U2 can begin the splicing process.

CONCEPTUAL SUMMARY

Transcription is the process of RNA synthesis. Different types of RNA transcripts can be made. These include **mRNA,** which specifies the sequence of amino acids within a polypeptide, and tRNA and rRNA, which are necessary to translate mRNA. In addition, several other small RNA molecules, such as snRNA, snoRNA, 7S RNA, and the RNA component of **RNaseP,** have various functions in the cell.

The synthesis of RNA during gene transcription is a three-step process: **initiation, elongation,** and **termination.** During the initiation stage, proteins that bind to DNA recognize specific sequences within the promoter region. In the case of many bacterial promoters, the −35 and −10 sequences are recognized by **sigma factor,** which is part of the **RNA polymerase** holoenzyme. To begin the synthesis of a transcript, the DNA must be converted to an **open complex** by denaturing a small double-stranded region. At this point, RNA polymerase can slide down the DNA, synthesizing a complementary RNA molecule. At the end of the gene, a termination signal is found. In bacteria, there are two different ways that termination signals can promote transcriptional termination. In **ρ-dependent termination,** a protein known as ρ is responsible for dissociating the RNA transcript and RNA polymerase. In **ρ-independent,** or **intrinsic, termination,** the formation of a stem-loop within the RNA causes RNA polymerase to stall and eventually fall off the DNA.

Transcription in eukaryotes is similar but more complex compared to bacteria. In eukaryotes, there are three RNA polymerases designated I, II, and III that transcribe different types of genes. Transcriptional initiation of **structural genes** requires five **general transcription factors** that enable RNA polymerase II to recognize a **TATA box** in the −25 region. Transcription is initiated by a complicated assembly process that also involves interactions with mediator. During elongation, the histone proteins must be completely or partially removed so that RNA polymerase can transcribe the template DNA. This involves covalent modifications of histone proteins and the actions of chromatin remodeling complexes.

Following transcription, a newly made RNA transcript may be modified in various ways. For example, rRNA and tRNA transcripts are processed to smaller forms by endo- and exonucleases. Some genes contain **introns** that must be removed from the RNA via splicing. **Group I** and **Group II introns** are **self-splicing,** which means that they catalyze their own removal. These types of introns are found in the rRNA genes of *Tetrahymena* and other simpler eukaryotes, in organelle DNA, and occasionally in bacterial genes. In contrast, introns are commonly found in eukaryotic **pre-mRNA,** where they are removed by a multimeric complex known as a **spliceosome.** Besides splicing, eukaryotic pre-mRNA is also **capped** at its 5′ end, and a **polyA tail** may be attached to its 3′ end. It is particularly interesting that many eukaryotic pre-mRNAs can be **alternatively spliced.** The net result of alternative splicing is that a single gene can produce more than one type of polypeptide.

EXPERIMENTAL SUMMARY

Experimentally, there are several methods for studying the transcription of genes at the molecular level. Most of these are described in chapter 18. Several of these techniques are considered in the experimental questions at the end of chapter 12. For example, the quantity of an RNA transcript can be determined by Northern blotting, in which RNA transcripts are identified through a DNA-RNA hybridization technique. This method can also be used to detect alternative splicing of mRNAs.

Our understanding of the sequences required for transcription, such as promoter sequences, has come from the isolation of mutants in which the promoter sequence was altered. More recently, the technique of site-directed mutagenesis (also described in chapter 18) has been developed to alter gene sequences. Researchers now can examine how an alteration in a gene sequence affects genetic processes such as transcription.

The binding and assembly of proteins during transcription can be examined in a variety of ways. In chapter 11, we considered how Arthur Kornberg developed an in vitro system to study DNA replication. Similarly, researchers have succeeded in developing in vitro systems to study transcription. In this way, they have identified the basal transcriptional apparatus needed for transcription to occur.

In chapter 12, we have also considered how hybridization and electron microscopy can be used to detect introns. In this approach, the hybridization of mRNA to a DNA sequence blocks the DNA's ability to re-form double-stranded regions. Using an electron microscope to examine the pattern of single-stranded DNA and double-stranded DNA regions, the hybridization of RNA to DNA allowed researchers to determine if a gene has introns. DNA sequencing methods that are described in chapter 18 provide an easier and more precise way of detecting introns. A comparison of the DNA sequences of genomic and cDNA clones can determine if a particular gene contains introns.

PROBLEM SETS & INSIGHTS

Solved Problems

S1. Describe the important events that occur during the three stages of gene transcription in bacteria. What proteins play critical roles in the three stages?

Answer: The three stages are initiation, elongation, and termination.

Initiation: RNA polymerase holoenzyme scans along the DNA until sigma factor recognizes the sequence of a promoter. Sigma factor binds tightly to this sequence, forming a closed complex. The DNA is then denatured to form a bubblelike structure known as the open complex.

Elongation: RNA polymerase core enzyme slides along the DNA, synthesizing RNA as it goes. The α subunits of RNA polymerase keep the enzyme bound to the DNA, while the β subunits are responsible for the catalytic synthesis of RNA. During elongation, RNA is made according to the AU/GC rule, with nucleotides being added in the $5'$ to $3'$ direction.

Termination: RNA polymerase eventually reaches a sequence at the end of the gene that signals the cessation of transcription. In ρ-independent termination, the properties of the termination sequences in the DNA are sufficient to cause termination. In ρ-dependent termination, the ρ protein recognizes a sequence within the RNA, binds there, and travels toward RNA polymerase. When the formation of a stem-loop structure causes RNA polymerase to pause, ρ catches up and knocks it off the DNA.

S2. What is the difference between a structural gene and a nonstructural gene?

Answer: Structural genes encode mRNA that is translated into a polypeptide sequence. Nonstructural genes encode RNAs that are never translated. Examples of nonstructural genes include tRNA and rRNA, which function during translation; 7S RNA, which is part of SRP; the RNA of RNaseP; snoRNA, which is involved in rRNA trimming; and snRNA, which is part of spliceosomes. In many cases, the RNA from nonstructural genes becomes part of a complex composed of RNA molecules and protein subunits.

S3. When RNA polymerase transcribes DNA, only one of the two DNA strands is used as a template. Take a look at figure 12.3 and explain how RNA polymerase determines which DNA strand is the template strand.

Answer: The binding of sigma factor and RNA polymerase depends on the sequence of the promoter. RNA polymerase binds to the promoter in such a way that the -35 sequence TTGACA and the -10 sequence TATAAT are within the coding strand, while the -35 sequence AACTGT and the -10 sequence ATATTA are within the template strand.

S4. The process of transcriptional termination is not as well understood in eukaryotes as it is in bacteria. Nevertheless, current evidence suggests that there are several different mechanisms for termination. Like bacteria, the termination of certain genes appears to occur via intrinsic terminators (i.e., like ρ-independent termination) while the termination of other genes may involve RNA-binding proteins (i.e., like ρ-dependent termination). A third type of mechanism is also found for the termination of rRNA genes by RNA polymerase I. In this case, a protein known as TTFI (transcription termination factor I) binds to the DNA downstream from the termination site. Discuss how the binding of a protein downstream from the termination site could promote transcriptional termination.

Answer: There are several possibilities. First, the binding of TTFI could act as a roadblock to the movement of RNA polymerase I. Second, TTFI could promote the dissociation of the RNA transcript and RNA polymerase I from the DNA; it may act like a helicase. Or third, it could cause a change in the structure of the DNA that prevents RNA polymerase from moving past the termination site. Though multiple effects are possible, the third effect seems the most likely since TTFI is known to cause a bend in the DNA when it binds to the termination sequence.

Conceptual Questions

C1. Genes may be structural genes that encode polypeptides or they may be nonstructural genes.

  A. Describe three examples of genes that are not structural genes.

  B. For structural genes, one DNA strand is called the template strand and the complementary strand is called the coding strand. Are these two terms appropriate for nonstructural genes? Explain.

  C. Do nonstructural genes have a promoter and terminator?

C2. In bacteria, what event marks the end of the initiation stage of transcription?

C3. What is the meaning of the term *consensus sequence*? Give an example. Describe the locations of consensus sequences within bacterial promoters. What are their functions?

C4. What is the consensus sequence of the following six DNA molecules?

```
GGCATTGACT
GCCATTGTCA
CGCATAGTCA
GGAAATGGGA
GGCTTTGTCA
GGCATAGTCA
```

C5. Mutations in bacterial promoters may increase or decrease the level of gene transcription. Promoter mutations that increase transcription are termed *up promoter* mutations while those that decrease transcription are termed *down promoter* mutations. As shown in figure 12.4, the sequence of the −10 region of the promoter for the *lac* operon is TATGTT. Would you expect the following mutations to be up promoter or down promoter mutations?

  A. TATGTT to TATATT

  B. TATGTT to TTTGTT

  C. TATGTT to TATGAT

C6. According to the examples shown in figure 12.4, which positions of the −35 sequence (i.e., first, second, third, fourth, fifth, and/or sixth) are more tolerant of changes? Do you think that these positions play a more or less important role in the binding of sigma factor? Explain why.

C7. In chapter 9, we considered the dimensions of the double helix (see fig. 9.17). In an α helix of a protein, there are 3.6 amino acids per complete turn. Each amino acid advances the α helix by 0.15 nm; a complete turn of an α helix is 0.54 nm in length. As shown in figure 12.5, two α helices (labeled *2* and *3*) of a transcription factor occupy the major groove of the DNA. According to figure 12.5, estimate the number of amino acids that bind to this region. How many complete turns of the α helices occupy the major groove of DNA?

C8. A mutation within a gene sequence changes the start codon to a stop codon. How will this mutation affect the transcription of this gene?

C9. What is the subunit composition of bacterial RNA polymerase holoenzyme? What are the functional roles of the different subunits?

C10. At the molecular level, describe how sigma factor recognizes bacterial promoters. Be specific about the structure of sigma factor and the type of chemical bonding.

C11. Let's suppose that a DNA mutation changes the consensus sequence at the −35 location so that sigma factor is no longer able to bind there. Explain how a mutation would prevent sigma factor from binding to the DNA. Look at figure 12.4 and describe two specific base substitutions that you think would inhibit the binding of sigma factor. Explain why you think your base substitutions would have this effect.

C12. What is the complementarity rule that governs the synthesis of an RNA molecule during transcription? An RNA transcript has the following sequence:

5′-GGCAUGCAUUACGGCAUCACACUAGGGAUC-3′

What is the sequence of the template and coding strands of the DNA that encodes this RNA? On which side (5′ or 3′) of the template strand is the promoter located?

C13. Describe the movement of the open complex down the RNA.

C14. Describe what happens to the chemical bonding interactions when transcriptional termination occurs. Be specific about the type of chemical bonding.

C15. Discuss the differences between ρ-dependent and ρ-independent termination.

C16. In chapter 11, we discussed the function of DNA helicase, which is involved in DNA replication. The structure and function of DNA helicase and ρ protein are rather similar to each other. Explain how the function of these two proteins is similar and different.

C17. Discuss the similarities and differences between RNA polymerase (described in chapter 12) and DNA polymerase (described in chapter 11).

C18. Mutations, which occur at the end of a gene, may alter the sequence of the gene and prevent transcriptional termination.

  A. What types of mutations would prevent ρ-independent termination?

  B. What types of mutations would prevent ρ-dependent termination?

  C. If a mutation prevented transcriptional termination at the end of a gene, where would gene transcription end? Or would it end?

C19. For each of the following RNA polymerases, if it were missing from a eukaryotic cell, what types of genes would not be transcribed?

  A. RNA polymerase I

  B. RNA polymerase II

  C. RNA polymerase III

C20. What sequence elements are found within the core promoter of structural genes in eukaryotes? Describe their locations and specific functions.

C21. For each of the following transcription factors, how would eukaryotic transcriptional initiation be affected if it were missing?

  A. TFIIB

  B. TFIID

  C. TFIIH

C22. Discuss how the binding of DNA to histones in a nucleosome structure may affect the process of transcription. Do you think that the nucleosome structure would affect the ability of transcription factors to recognize the promoter sequence in eukaryotic genes? Explain.

C23. Which eukaryotic transcription factor(s) shown in figure 12.12 plays an equivalent role to sigma factor found in bacterial cells?

C24. The initiation phase of eukaryotic transcription via RNA polymerase II is considered an assembly and disassembly process. Which types of biochemical interactions (i.e., H bonding, ionic bonding, covalent bonding, and/or hydrophobic interactions) would you expect to drive the assembly and disassembly process? How would temperature and salt concentration affect assembly and disassembly?

C25. A eukaryotic structural gene contains two introns and three exons: exon 1–intron 1–exon 2–intron 2–exon 3. The splice donor site at the boundary between exon 2 and intron 2 has been eliminated by a small deletion in the gene. Describe how the pre-mRNA encoded by this mutant gene would be spliced. Indicate which introns and exons would be found in the mRNA after splicing occurs.

C26. Describe the processing events that occur during the production of tRNA in *E. coli*.

C27. Describe the structure and function of a spliceosome. Speculate on why you think the spliceosome subunits contain snRNA. In other words, what do you think is the functional role(s) of snRNA during splicing?

C28. What is the unique feature of ribozyme function? Give two examples described in chapter 12.

C29. What does it mean to say that gene expression is colinear?

C30. What is meant by the term *self-splicing*? What types of introns are self-splicing?

C31. In eukaryotes, what types of modification occur to pre-mRNA?

C32. What is a transesterification reaction? Draw the chemical bonds that are broken and re-formed.

C33. What is alternative splicing? What is its biological significance?

C34. The processing of ribosomal RNA in eukaryotes is shown in figure 12.14. Why is this called trimming or processing but not splicing?

C35. In the group I category of RNA splicing shown in figure 12.18, does the 5′ end of the intron have a phosphate group? Explain.

C36. According to the mechanism shown in figure 12.22, several snRNPs play different roles in the splicing of pre-mRNA. Identify the snRNP that recognizes the following sites:

A. 5′ splice site

B. 3′ splice site

C. Branch site

C37. After the intron (i.e., the lariat) is released during pre-mRNA splicing, there is a brief moment before the two exons are connected to each other. Which snRNP(s) holds the exons in place so they can be covalently connected to each other?

C38. A lariat contains a closed loop and a linear end. An intron has the following sequence: 5′–GUPuAGUA–60 nucleotides–UACUUAUCC–100 nucleotides–Py_{12}NPyAG–3′. Which sequence would be found within the closed loop of the lariat, the 60-nucleotide sequence or the 100-nucleotide sequence?

Experimental Questions

E1. A research group has sequenced the cDNA and genomic DNA from a particular gene. As described in chapter 18, cDNA is derived from mRNA, so that it does not contain introns. Here are the DNA sequences.

cDNA:

```
5′-ATTGCATCCAGCGTATACTATCTCGGGCCCAATTAATGCCAGCG
GCCAGACTATCACCCAACTCGGTTACCTACTAGTATATCCCATATA
CTAGCATATATTTTACCCATAATTTGTGTGTGGGTATACAGTATAA
TCATATA-3′
```

Genomic DNA (contains one intron):

```
5′-ATTGCATCCAGCGTATACTATCTCGGGCCCAATTAATGCCAGCG
GCCAGACTATCACCCAACTCGGCCCACCCCCCAGGTTTACACAGTC
ATACCATACATACAAAAATCGCAGTTACTTATCCCAAAAAAACCTA
GATACCCCACATACTATTAACTCTTTCTTTCTAGGTTACCTACTAG
TATATCCCATATACTAGCATATATTTTACCCATAATTTGTGTGTGG
GTATACAGTATAATCATATA-3′
```

Indicate where the intron is located. Does the intron contain the normal consensus splice site sequences based on those described in figure 12.21? Underline the splice site sequences, and indicate whether or not they fit the consensus sequence.

E2. What is an R loop? In an R loop experiment, to which strand of DNA does the mRNA bind, the coding strand or template strand?

E3. If a gene contains three introns, draw what it would look like in an R loop experiment.

E4. Chapter 18 describes a blotting method known as Northern blotting that can be used to detect RNA that is transcribed from a par-

ticular gene. In this method, a specific RNA is detected using a short segment of cloned DNA as a probe. The DNA probe, which is radioactive, is complementary to the RNA that the researcher wishes to detect. After the radioactive probe DNA binds to the RNA, the RNA is visualized as a dark (radioactive) band on an X-ray film. As shown here, the method of Northern blotting can be used to determine the amount of a particular RNA transcribed in a given cell type. If one type of cell produces twice as much of a particular mRNA compared to another cell, the band will appear twice as intense. Also, the method can distinguish if alternative RNA splicing has occurred to produce an RNA that has a different molecular mass.

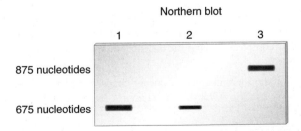

Northern blot

Lane 1 is a sample of RNA isolated from nerve cells.

Lane 2 is a sample of RNA isolated from kidney cells. Nerve cells produce twice as much of this RNA compared to kidney cells.

Lane 3 is a sample of RNA isolated from spleen cells. Spleen cells produce an alternatively spliced version of this RNA that is about

200 nucleotides longer than the RNA produced in nerve and kidney cells.

Now here is the question. Let's suppose a researcher was interested in the effects of mutations on the expression of a particular structural gene in eukaryotes. The gene has one intron that is 450 nucleotides long. After this intron is removed from the pre-mRNA, the mRNA transcript is 1,100 nucleotides in length. There are two copies of this gene in diploid somatic cells. Make a drawing that shows the expected results of a Northern blot using mRNA from the cytosol of somatic cells, which were obtained from the following individuals:

Lane 1:   a normal individual

Lane 2:   an individual homozygous for a deletion that removes the −50 to −100 region of the gene that encodes this mRNA

Lane 3:   an individual heterozygous in which one gene is normal and the other gene had a deletion that removes the −50 to −100 region

Lane 4:   an individual homozygous for a mutation that introduces an early stop codon into the middle of the coding sequence of the gene

Lane 5:   an individual homozygous for a three-nucleotide deletion that removes the AG sequence at the splice acceptor site

E5. A gel retardation assay can be used to study the binding of proteins to a segment of DNA. This method is described in chapter 18. When a protein binds to a segment of DNA, it retards the movement of the DNA through a gel, so the DNA appears at a higher point in the gel (see below).

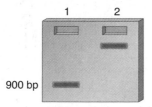

Lane 1: 900 bp fragment alone

Lane 2: 900 bp fragment plus a protein that binds to the 900 bp fragment

In this example, the segment of DNA is 900 bp in length, and the binding of a protein causes the DNA to appear at a higher point in the gel. If this 900 bp fragment of DNA contains a eukaryotic promoter for a structural gene, draw a gel that shows the relative locations of the 900 bp fragment under the following conditions:

Lane 1:   900 bp plus TFIID

Lane 2:   900 bp plus TFIIB

Lane 3:   900 bp plus TFIID and TFIIB

Lane 4:   900 bp plus TFIIB and RNA polymerase II

Lane 5:   900 bp plus TFIID, TFIIB, and RNA polymerase II/TFIIF

E6. As described in chapter 18 and in experimental question E5, a gel retardation assay can be used to determine if a protein binds to DNA. This method can also be used to determine if a protein binds to RNA. In the combinations described here, would you expect the migration of the RNA to be retarded due to the binding of a protein?

A. mRNA from a gene that is terminated in a ρ-independent manner plus ρ-protein

B. mRNA from a gene that is terminated in a ρ-dependent manner plus ρ-protein

C. pre-mRNA from a structural gene that contains two introns plus the snRNP called U1

D. Mature mRNA from a structural gene that contains two introns plus the snRNP called U1

E7. The technique of DNA footprinting is described in chapter 18. If a protein binds over a region of DNA, it will protect that region from digestion by DNase I. To carry out a DNA footprinting experiment, a researcher has a sample of a cloned DNA fragment. The fragments are exposed to DNase I in the presence and absence of a DNA-binding protein. Regions of the DNA fragment that are not covered by the DNA-binding protein will be digested by DNase I, and this will produce a series of bands on a gel. Regions of the DNA fragment that are not digested by DNase I (because a DNA-binding protein is preventing DNase I from gaining access to the DNA) will be revealed, because a region of the gel will not contain any bands.

In the DNA footprinting experiment shown here, a researcher began with a sample of cloned DNA that was 300 bp in length. This DNA contained a eukaryotic promoter for RNA polymerase II. For the sample loaded in lane 1, no proteins were added. For the sample loaded in lane 2, the 300 bp fragment was mixed with RNA polymerase II plus TFIID and TFIIB.

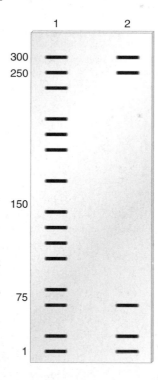

A. How long of a region of DNA is "covered up" by the binding of RNA polymerase II and the transcription factors?

B. Describe how this binding would occur if the DNA was within a nucleosome structure? (Note: The structure of nucleosomes is described in chapter 10.) Do you think that the DNA is in a nucleosome structure when RNA polymerase and transcription factors are bound to the promoter? Explain why or why not.

E8. As described in table 12.1, several different types of RNA are made, especially in eukaryotic cells. Researchers are oftentimes interested in focusing their attention on the transcription of structural genes in eukaryotes. Such researchers want to study mRNA. One method that is widely used to isolate mRNA is column chromatography. (Note: See the appendix for a general description of chromatography.) Researchers can covalently attach short pieces of DNA that contain stretches of thymine (i.e., TTTTTTTTTTTT) to the column matrix. This is called a poly-dT column. When a cell extract is poured over the column, mRNA binds to the column, while other types of RNA do not.

A. Explain how you would use a poly-dT column to obtain a purified preparation of mRNA from eukaryotic cells. In your description, explain why mRNA binds to this column and what you would do to release the mRNA from the column.

B. Can you think of ways to purify other types of RNA, such as tRNA or rRNA?

Questions for Student Discussion/Collaboration

1. Based on your knowledge of introns and pre-mRNA splicing, discuss whether or not you think alternative splicing fully explains the existence of introns. Can you think of other possible reasons to explain the existence of introns?

2. Discuss the types of RNA transcripts and the functional roles that they play. Why do you think that some RNAs form complexes with protein subunits?

Note: All answers appear at the website for this textbook; the answers to even-numbered questions are in the back of the textbook.

www.mhhe.com/brooker

Visit the Online Learning Center for practice tests, answer keys, and other learning aids for this chapter. Enhance your understanding of genetics with our interactive exercises, web links, news feeds, tutorial service, and much more.

TRANSLATION OF mRNA

::

13

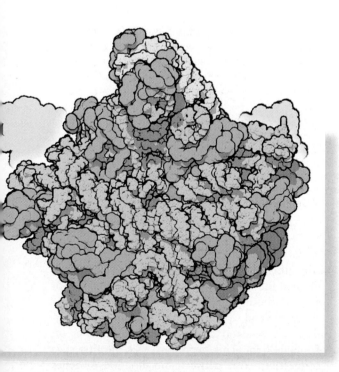

The synthesis of cellular proteins occurs via the **translation** of the codons within mRNA into a sequence of amino acids of a polypeptide. The general steps that occur in this process have already been outlined in chapter 1. In chapter 13 we reexamine translation with an eye toward the specific molecular interactions that are responsible for this process. During the past few decades, the concerted efforts of geneticists, cell biologists, and biochemists have profoundly advanced our understanding of translation. Even so, many questions remain unanswered, and this topic continues to be an exciting area of research investigation.

In chapter 13, we will discuss the current state of knowledge regarding the molecular features of mRNA translation. A variety of cellular components play important roles in translation. These include many different proteins, RNAs, and small molecules. In the first section of chapter 13, we will begin by considering a couple of classic experiments that indicated the purpose of some genes is to encode proteins that function as enzymes. Next, we will examine how the genetic code is used to decipher the information within mRNA to produce a polypeptide with a specific amino acid sequence. The rest of chapter 13 is devoted to a molecular understanding of translation as it occurs in living cells. This will involve an examination of the components that are needed for the translation process. We will consider the structure and function of tRNA molecules, which act as the translators of the genetic information within mRNA, and then examine the composition of ribosomes. Finally, we will explore the stages of translation in bacterial and eukaryotic cells.

13.1 THE GENETIC BASIS FOR PROTEIN SYNTHESIS

Proteins are critically important as active participants in cell structure and function. The primary role of DNA is to store the information needed for the synthesis of all the proteins that an organism can make. Genes that encode an amino acid sequence are known as **structural genes.** As discussed in chapter 12, the RNA that is transcribed from structural genes is called **messenger RNA (mRNA).** The main function of the genetic material is to encode the production of cellular proteins in the correct cell, at the proper time, and in suitable amounts. This is an extremely complicated task, since living cells make thousands of different proteins. Genetic analyses have shown that a typical bacterium can make a few thousand different proteins, and estimates for eukaryotes range from several thousand in lower eukaryotes to tens of thousands in higher eukaryotes.

In this section, we will begin by considering a couple of early experiments that indicated the role of genes is to encode proteins. Then we will examine how the genetic code is used to decipher the nucleotide sequence of mRNA and thereby produce an amino acid sequence within a protein.

Archibald Garrod Proposed That Some Genes Code for the Production of a Single Enzyme

The idea that a relationship exists between genes and the production of proteins was first suggested at the beginning of the twentieth century by Archibald Garrod, a British physician. Prior to Garrod's studies, biochemists had studied many metabolic pathways within living cells. These pathways consist of a series of metabolic conversions of one molecule to another, each step catalyzed by an enzyme. Each enzyme is a distinctly different protein that catalyzes a particular chemical reaction. For example, figure 13.1 illustrates part of the metabolic pathway for the degradation of phenylalanine, an amino acid commonly found in human diets. The enzyme phenylalanine hydroxylase catalyzes the conversion of phenylalanine to tyrosine. A different enzyme, tyrosine aminotransferase, converts tyrosine into *p*-hydroxyphenylpyruvic acid. In all of the steps shown in figure 13.1, a specific enzyme catalyzes a single type of chemical reaction.

Garrod studied patients who had defects in their ability to metabolize certain compounds. He was particularly interested in the inherited disease known as **alkaptonuria.** In this disorder, the patient's body accumulates abnormal levels of homogentisic acid (alkapton), which is excreted in the urine causing it to appear black. In addition, the disease is characterized by bluish black discoloration of cartilage and skin (ochronosis). Garrod proposed that the accumulation of homogentisic acid in these patients is due to a missing enzyme, namely homogentisic acid oxidase (fig. 13.1).

Furthermore, Garrod already knew that alkaptonuria is an inherited trait that follows a recessive pattern of inheritance; thus, an individual with alkaptonuria must have inherited the gene that causes this disorder from each parent. From these observations, Garrod proposed that a relationship exists between the inheritance

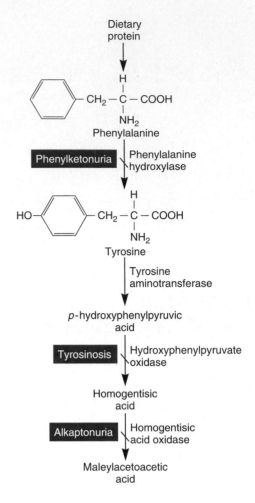

FIGURE 13.1   The metabolic pathway of phenylalanine metabolism and related genetic diseases. This diagram shows part of the metabolic pathway of phenylalanine breakdown. This pathway consists of enzymes that successively convert one molecule to another. Certain human genetic diseases are caused when enzymes in this pathway are missing or defective.

GENES→TRAITS   When a person inherits two defective copies of the gene that encodes homogentisic acid oxidase, he or she cannot convert homogentisic acid into maleylacetoacetic acid. Such a person accumulates large amounts of homogentisic acid in his or her urine and has other symptoms of the disease known as alkaptonuria. Similarly, if a person has two mutant alleles of the gene encoding phenylalanine hydroxylase, he or she is unable to synthesize the enzyme phenylalanine hydroxylase and has the disease called phenylketonuria (PKU).

of the trait and the inheritance of a defective enzyme. Namely, if an individual inherited the defective gene (which caused a loss of enzyme function) from both parents, he or she would not produce any normal enzyme and would thereby be unable to metabolize homogentisic acid. Garrod described alkaptonuria as an **inborn error of metabolism.** This hypothesis was the first suggestion that a connection exists between the function of genes and the production of enzymes. At the turn of the century, this was a particularly insightful idea, since the structure and function of the genetic material were completely unknown.

Beadle and Tatum's Experiments with *Neurospora* Led Them to Propose the One Gene–One Enzyme Theory

In the early 1940s, George Beadle and Edward Tatum were also interested in the relationship among genes, enzymes, and traits. Their important contribution to genetics was their formulation of an experimental system for investigating the relationship between genes and the production of particular enzymes. Consistent with the ideas of Garrod, the underlying assumption behind their approach was that a relationship exists between genes and the production of enzymes. However, the quantitative nature of this relationship was unclear. In particular, they wondered whether one gene was controlling the production of one enzyme or whether one gene was controlling the synthesis of many enzymes involved in a complex biochemical pathway.

At the time of their studies, many geneticists were also trying to understand the nature of the gene by studying morphological traits. However, Beadle and Tatum realized that morphological traits are "likely to be based on systems of biochemical reactions so complex as to make analysis exceedingly difficult." Therefore, they turned their genetic studies to the analysis of simple nutritional requirements in *Neurospora crassa,* a common bread mold. *Neurospora* can be easily grown in the laboratory and has few nutritional requirements. The minimum requirements for growth are a carbon source (namely, sugar), inorganic salts, and one vitamin known as biotin. Otherwise, *Neurospora* has many different cellular enzymes that can synthesize all the small molecules, such as amino acids and many vitamins, that are essential for growth.

Beadle and Tatum wanted to understand how enzymes are controlled by genes. They reasoned that a mutation in a gene, causing a defect in an enzyme needed for the cellular synthesis of an essential molecule, would prevent that mutant strain from growing on minimal medium (which contains only a carbon source, inorganic salts, and biotin). For example, if a *Neurospora* mutant strain could not make the vitamin pantothenic acid, it would be unable to grow on minimal medium. However, it would grow if the growth medium was supplemented with pantothenic acid.

Beadle and Tatum analyzed more than 2,000 strains that had been irradiated to induce mutations. They found three strains that were unable to grow on minimal medium (table 13.1). In each of these cases, the mutant strain required only a single vitamin to restore its growth on minimal medium. One of these strains required pyridoxine for growth, the second strain required thiamine, and the third strain required *p*-aminobenzoic acid. In the normal strain, these vitamins are synthesized by cellular enzymes. In the mutant strains, a genetic defect in one gene prevented the synthesis of a functional enzyme required for the synthesis of a particular vitamin. Therefore, Beadle and Tatum concluded that a single gene controlled the synthesis of a single enzyme. This was referred to as the **one gene–one enzyme theory.**

In later decades, this theory had to be modified in two ways. First, enzymes are only one category of cellular proteins. All proteins are encoded by genes, and many of them do not func-

TABLE 13.1

Identification of Mutant Strains That Were Unable to Synthesize Particular Vitamins

Single Vitamin Added to Minimal Media	Observed Growth:		
	Strain 1	Strain 2	Strain 3
None	No	No	No
Riboflavin	No	No	No
Thiamin	No	Yes	No
Pantothenic acid	No	No	No
Pyridoxine	Yes	No	No
Niacin	No	No	No
Folic acid	No	No	No
p-aminobenzoic acid	No	No	Yes

tion as enzymes. Second, some proteins are composed of two or more different polypeptides. Therefore, it is more accurate to say that a structural gene encodes a polypeptide. The term **polypeptide** denotes structure; it is a linear sequence of amino acids. A single structural gene encodes a single polypeptide. By comparison, the term **protein** denotes function. Some proteins are composed of one polypeptide. In such cases, a single gene would encode a single protein. In other cases, however, a functional protein is composed of two or more different polypeptides. An example would be hemoglobin, which is composed of two α-globin and two β-globin polypeptides. In this example, the expression of two genes (i.e., the α-globin and β-globin genes) is needed to create a functional protein.

During Translation, the Genetic Code Within mRNA Is Used to Make a Polypeptide with a Specific Amino Acid Sequence

Let's now turn to a general description of the translation process. As its name implies, translation involves an interpretation of one language into another. In this case, the language of mRNA, which is a nucleotide sequence, is translated into the language of proteins, which is an amino acid sequence. The ability of mRNA to be translated into a specific sequence of amino acids relies on the **genetic code.** The sequence of bases within an mRNA molecule provides coded information that is read in groups of three nucleotides known as **codons** (fig. 13.2). The sequence of three bases in most codons specifies a particular amino acid. For example, the codon CCC specifies the amino acid proline, while the codon GGC encodes the amino acid glycine. The codon AUG, which specifies methionine, is used as a **start codon;** it is usually the first codon that begins a polypeptide sequence. The AUG codon can also be used to specify additional methionines within the coding sequence. Finally, three codons are used to end the process of translation. These are UAA, UAG, and UGA, which are known as **termination,** or **stop, codons.**

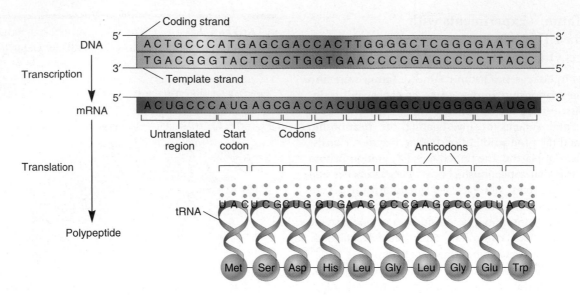

FIGURE 13.2   **The relationships among the DNA coding sequence, mRNA codons, tRNA anticodons, and amino acids in a polypeptide.** The sequence of nucleotides within DNA is transcribed to make a complementary sequence of nucleotides within mRNA. This sequence of nucleotides in mRNA is translated into a sequence of amino acids of a protein. tRNA molecules act as intermediates in this translation process.

The details of the genetic code are shown in table 13.2. Since there are 20 kinds of amino acids, a minimum of 20 codons is needed to specify each type. With four types of bases in mRNA (A, U, G, and C), a genetic code containing two bases in a codon would not be sufficient because it would only have 4^2, or 16, pos-sible types. By comparison, a three-base codon system can specify 4^3, or 64, different codons. Because the number of possible codons exceeds 20 (i.e., the number of different types of amino acids), the genetic code is **degenerate.** This means that more than one codon can specify the same amino acid. For example, the codons GGU, GGC, GGA, and GGG all specify the amino acid glycine. In most instances, the third base in the codon is the degenerate base. The third base is sometimes referred to as the **wobble base.** This term is derived from the idea that the comple-mentary base in the tRNA can "wobble" a bit during the recogni-tion of the third base of the codon in mRNA. The significance of the wobble base is discussed later in this chapter.

From the analysis of many different species, including bac-teria, protozoa, fungi, plants, and animals, researchers have found that the genetic code is nearly **universal.** Only a few rare excep-tions to the genetic code have been noted. Examples are described in table 13.3. As discussed in chapter 7, the eukaryotic

TABLE 13.2
The Genetic Code

		Second position			
	U	**C**	**A**	**G**	
U	UUU } Phe UUC } UUA } Leu UUG }	UCU } UCC } UCA } Ser UCG }	UAU } Tyr UAC } UAA Stop UAG Stop	UGU } Cys UGC } UGA Stop UGG Trp	U C A G
C	CUU } CUC } CUA } Leu CUG }	CCU } CCC } CCA } Pro CCG }	CAU } His CAC } CAA } Gln CAG }	CGU } CGC } CGA } Arg CGG }	U C A G
A	AUU } AUC } Ile AUA } AUG Met/start	ACU } ACC } ACA } Thr ACG }	AAU } Asn AAC } AAA } Lys AAG }	AGU } Ser AGC } AGA } Arg AGG }	U C A G
G	GUU } GUC } GUA } Val GUG }	GCU } GCC } GCA } Ala GCG }	GAU } Asp GAC } GAA } Glu GAG }	GGU } GGC } GGA } Gly GGG }	U C A G

(First position on left axis; Third position on right axis)

TABLE 13.3
Examples of Exceptions to the Genetic Code*

Codon	Universal Meaning	Exception
AUA	Isoleucine	Methionine in yeast and mammalian mitochondria
UGA	Stop	Tryptophan in mammalian mitochondria
CUU, CUA, CUC, CUG	Leucine	Threonine in yeast mitochondria
AGA, AGG	Arginine	Stop codon in ciliated protozoa and in yeast and mammalian mitochondria
UAA, UAG	Stop	Glutamine in ciliated protozoa

*Several other exceptions, sporadically found among various species, are also known.

organelles known as mitochondria have their own DNA, which encodes a few structural genes. In mammals, the mitochondrial genetic code contains differences such as AUA = methionine and UGA = tryptophan. Also, in mitochondria and certain ciliated protozoa, AGA and AGG specify stop codons instead of arginine. Except for rare exceptions such as the ones listed in table 13.3, the genetic code shown in table 13.2 has usually been found to be universal in all species studied thus far.

The first evidence indicating that the genetic code is read in triplets came from studies of Francis Crick and his colleagues in 1961. These experiments involved the isolation of mutants in a bacteriophage called T4. As described in chapter 6, mutations in T4 genes that affect plaque morphology are easily identified. In particular, it was already known that loss-of-function mutations within certain T4 genes, designated *rII*, resulted in plaques that were larger and had a clear boundary (see fig. 6.15). By comparison, wild-type phages, designated r^+, produced smaller plaques with a fuzzy boundary. To obtain *rII* mutations from a wild-type population of phages, Crick and colleagues exposed T4 phages to

a chemical called proflavin that causes single-nucleotide deletions or additions to gene sequences. The mutagenized phages were plated to identify large (*rII*) plaques. Though proflavin can cause either single deletions or additions, the first mutant strain the researchers identified was arbitrarily called a (+) mutation. Many years later, when methods of DNA sequencing became available, it was determined that the (+) mutation was a single-nucleotide addition. As shown in the hypothetical example of table 13.4, a single addition would alter the reading frame and thereby abolish the proper function of the encoded protein.

The (+) mutant strain was then subjected to a second round of mutagenesis via proflavin. Several plaques were identified that now had a wild-type morphology. By analyzing these strains using recombinational methods, described in chapter 6, it was determined that each one contained a second mutation that was close to the original (+) mutation. These second mutations were designated (−) mutations. Three different (−) mutations are considered in table 13.4. Each of these (−) mutations, designated a, b, and c, were single-nucleotide deletions that were close to the

TABLE 13.4

Evidence That the Genetic Code Is Read in Triplets*

Strain	Plaque Phenotype[†]	DNA Sequence/Polypeptide Sequence	Downstream Sequence[‡]
Wild type	r^+	ATG GGG CCC GTC CAT CCG TAC GCC GGA ATT ATA Met Gly Pro Val His Pro Tyr Ala Gly Ile Ile----------	In frame
(+)	rII	↓[§] ATG GGG ACC CGT CCA TCC GTA CGC CGG AAT TAT A Met Gly Thr Arg Pro Ser Val Arg Arg Asn Tyr---------	Out of frame
(+)(−)ₐ	r^+	↓ ↑ ATG GGG ACC GTC CAT CCG TAC GCC GGA ATT ATA Met Gly Thr Val His Pro Tyr Ala Gly Ile Ile----------	In frame
(+)(−)_b	r^+	↓ ↑ ATG GGG ACC CGC CAT CCG TAC GCC GGA ATT ATA Met Gly Thr Arg His Pro Tyr Ala Gly Ile Ile----------	In frame
(+)(−)_c	r^+	↓ ↑ ATG GGG ACC CTC CAT CCG TAC GCC GGA ATT ATA Met Gly Thr Leu His Pro Tyr Ala Gly Ile Ile----------	In frame
(−)ₐ	rII	↑ ATG GGG CCG TCC ATC CGT ACG CCG GAA TTA TA Met Gly Pro Ser Ile Arg Thr Pro Glu Leu---------------	Out of frame
(−)ₐ(−)_b	rII	↑ ↑ ATG GGG CCG CCA TCC GTA CGC CGG AAT TAT A Met Gly Pro Pro Ser Val Arg Arg Asn Tyr----------------	Out of frame
(−)ₐ(−)_b(−)_c	r^+	↑↑↑ ATG GGG CCC CAT CCG TAC GCC GGA ATT ATA Met Gly Pro His Pro Tyr Ala Gly Ile Ile--------------	In frame

Only Val is missing

*This table provides a hypothetical example of how single-nucleotide insertions and deletions could alter the reading frame of a polypeptide. The coding sequence of most genes encodes polypeptides that are a few hundred amino acids in length. This table shows only a small portion of a hypothetical coding sequence.

[†]A loss-of-function mutation within a phage gene causes an rII phenotype that forms large plaques with a clear boundary. In this case, a frameshift mutation causes a loss of function because it substantially alters the amino acid sequence of the encoded protein. The wild-type phages (r^+) produce smaller, fuzzy plaques.

[‡]The term *downstream sequence* refers to the remaining part of the sequence that is not shown in this figure. It could include hundreds of codons. An "in frame" sequence would be wild type whereas an "out of frame" sequence (caused by the addition or deletion of one or two base pairs) would not.

[§]A down arrow (↓) indicates the location of a single nucleotide insertion, while an up arrow (↑) indicates the location of a single nucleotide deletion.

original (+) mutation. Therefore, they restored the reading frame and produced a protein with a nearly normal amino acid sequence.

The critical experiment that suggested that the genetic code is read in triplets came by combining different (−) mutations together. As discussed in chapter 6, double mutations in the same phage can be separated by intragenic recombination. Similarly, mutations in different phages can be brought together into the same phage via the same type of event. Using such an approach, the researchers constructed strains containing one, two, or three (−) mutations. It was found that a wild-type plaque morphology was obtained only when three (−) mutations were combined in the same phage (table 13.4). These results were consistent with the idea that the genetic code is read in multiples of three nucleotides.

EXPERIMENT 13A

Synthetic RNA Helped to Decipher the Genetic Code

During the early 1960s, the genetic code was deciphered. The first steps in this process were taken by several different research groups, headed by Marshall Nirenberg, Severo Ochoa, and H. Gobind Khorana. We will first consider the work of Nirenberg and his colleagues. Prior to their studies, several laboratories had already determined that extracts from bacterial cells are able to synthesize polypeptides. This is termed an in vitro or cell-free translation system. In these earlier studies, the proteins made in a cell-free translation system were specified by the mRNAs that were already present in the extract. It was then found that the addition of DNase to the extract prevented protein synthesis. DNase was expected to digest the DNA found within the extract. Since DNA acts as the template to make mRNA, a DNase-treated extract cannot make new mRNA. Therefore, after a short period of time, a DNase-treated extract becomes depleted of mRNA because bacterial mRNA usually has a very short half-life (a few minutes). In this way, the extract becomes unable to translate new polypeptides.

An important advance occurred when Nirenberg and his colleagues were able to add RNA back to DNase-treated extracts and thereby regain polypeptide synthesis. If radiolabeled amino acids were added to these extracts, the synthesized polypeptides would be radiolabeled and easy to detect. This approach enabled researchers to obtain information that helped to decipher the genetic code. They did this by making RNA molecules of a known base composition, adding them to the DNase-treated cell extract, and then analyzing the amino acid composition of the resultant polypeptides. For example, if an RNA molecule consisted of a string of adenine-containing nucleotides (e.g., 5′–AAAAAAAAAAAAAAAAAAAA–3′), researchers could add this polyA RNA to an extract and ask the question, Which amino acid is specified by a codon that contains only adenine nucleotides?

Before discussing the details of this experiment, let's consider how the synthetic RNA molecules were made. To synthesize RNA, an enzyme known as polynucleotide phosphorylase was used. This enzyme, which was originally discovered by Severo Ochoa in 1955, can catalyze the digestion of RNAs as well their synthesis. In the presence of excess ribonucleoside diphosphates (NDPs), it catalyzes the covalent linkage of nucleotides (more specifically, ribonucleotides) together to make a polymer of RNA. Because it does not use a template, the order of the nucleotides is random. For example, if only uracil-containing diphosphates, UDPs, are added, then a polyU RNA (e.g., 5′–UUUUUUUUUU-

UUUUUU–3′) is made. If nucleotides containing two different bases, such as uracil and guanine, are added, then the phosphorylase makes a random polymer containing both nucleotides (e.g., 5′–UGGGUGUUUUGUGUG–3′). An experimenter can control the amounts of the nucleotides that are added. For example, if 70% G and 30% U are mixed together, the predicted amounts of the codons within the random polymer are as follows:

Codon Possibilities	Percentage in the Random Polymer
GGG	$0.7 \times 0.7 \times 0.7 = 0.34 =$ 34%
GGU	$0.7 \times 0.7 \times 0.3 = 0.15 =$ 15%
GUU	$0.7 \times 0.3 \times 0.3 = 0.06 =$ 6%
UUU	$0.3 \times 0.3 \times 0.3 = 0.03 =$ 3%
UGG	$0.3 \times 0.7 \times 0.7 = 0.15 =$ 15%
UUG	$0.3 \times 0.3 \times 0.7 = 0.06 =$ 6%
UGU	$0.3 \times 0.7 \times 0.3 = 0.06 =$ 6%
GUG	$0.7 \times 0.3 \times 0.7 = 0.15 =$ 15%
	100%

By controlling the amounts of the nucleotides in the phosphorylase reaction, the relative amounts of the possible codons can be predicted. The first experiment, which demonstrated the ability to synthesize polypeptides from synthetic RNA, was performed by Marshall Nirenberg and J. Heinrich Matthaei in 1961. They began with a DNase-treated extract and added it to 20 different tubes. An RNA template made via polynucleotide phosphorylase was also added to each tube. In the example shown in figure 13.3, the RNA was made using 30% U and 70% G in the phosphorylase reaction. Next, the 20 amino acids were added to each tube, but each tube differed with regard to the type of radiolabeled amino acid. For example, radiolabeled glycine would be found in only 1 of the 20 tubes. The tubes were incubated a sufficient length of time to allow translation to occur. The newly made polypeptides were then precipitated onto a filter by treatment with trichloroacetic acid. Finally, the amount of radioactivity trapped on the filter was determined by liquid scintillation counting.

■ THE HYPOTHESIS

The sequence of bases in RNA determines the incorporation of specific amino acids into a polypeptide. (The purpose of this experiment was to provide information that would help to decipher the relationship between base composition and particular amino acids.)

■ TESTING THE HYPOTHESIS — FIGURE 13.3 Elucidation of the genetic code.

Starting material: A bacterial cell extract that has been treated with DNase.

Experimental level **Conceptual level**

1. Add the DNase-treated cell extract to each of 20 tubes.

2. To each tube, add random RNA polymers of G and U made via polynucleotide phosphorylase using 70% G and 30% U.

3. Add a different radiolabeled amino acid to each tube, and add the other 19 non-radiolabeled amino acids.

4. Incubate for 60 minutes to allow translation to occur.

5. Add 15% trichloroacetic acid (TCA). Note: This precipitates polypeptides but not amino acids.

6. Place the precipitate onto a filter and wash.

7. Count the radioactivity on the filter in a scintillation counter (see the appendix for a description).

8. Calculate the amount of radiolabeled amino acids in the precipitated polypeptides.

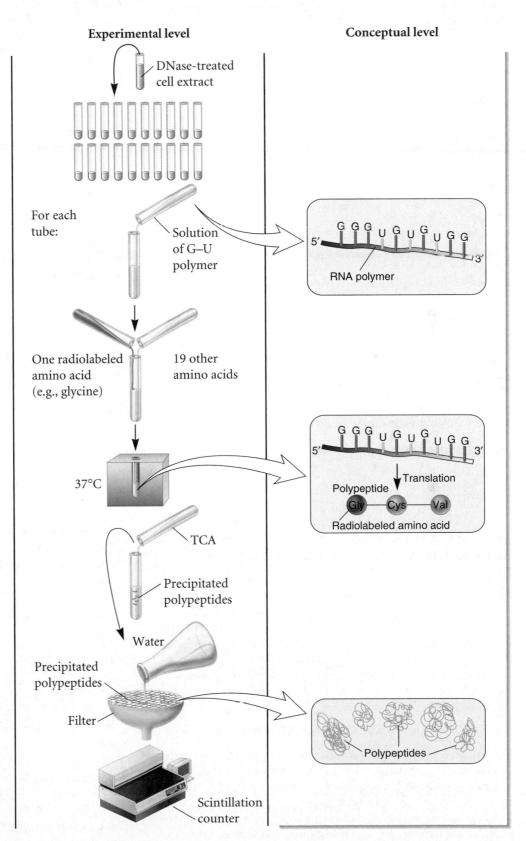

■ THE DATA

Radiolabeled Amino Acid Added	Relative Amount of Radiolabeled Amino Acid Incorporated into Translated Polypeptides (% of total)
Alanine	0
Arginine	0
Asparagine	0
Aspartic acid	0
Cysteine	6
Glutamic acid	0
Glutamine	0
Glycine	49
Histidine	0
Isoleucine	0
Leucine	6
Lysine	0
Methionine	0
Phenylalanine	3
Proline	0
Serine	0
Threonine	0
Tryptophan	15
Tyrosine	0
Valine	21

■ INTERPRETING THE DATA

According to the calculation previously described, there should be the following percentages of codons: 34% GGG, 15% GGU, 6% GUU, 3% UUU, 15% UGG, 6% UUG, 6% UGU, and 15% GUG. In the data shown in figure 13.3, the value of 49% for glycine is due to two codons: GGG (34%) and GGU (15%). The 6% cysteine is due to UGU. And so on. It is important to realize that the genetic code was not deciphered in a single experiment such as the one described here. Furthermore, this kind of experiment yields information regarding only the nucleotide content of codons, not the specific order of nucleotides within a single codon. For example, this experiment indicates that a cysteine codon contains two Us and one G. However, it does not tell us that a cysteine codon is UGU. Based on these data alone, it is possible that a cysteine codon is UUG, GUU, or UGU. However, by comparing many different RNA polymers, it became possible for the laboratories of Nirenberg and Ochoa to establish patterns between the specific sequences of codons and the amino acids they encode. In their first experiments, Nirenberg and Matthaei showed that a random polymer containing only uracil produced a polypeptide containing only phenylalanine. From this result, they inferred that UUU specifies phenylalanine. This idea is consistent with the data shown in the data table. In the random G and U polymer, it is expected that 3% of the codons will be UUU. Likewise, 3% of the amino acids within the polypeptides were found to be phenylalanine.

A self-help quiz involving this experiment can be found at the Online Learning Center.

The Use of RNA Copolymers and the Triplet Binding Assay Also Helped to Crack the Genetic Code

In the 1960s, H. Gobind Khorana and his collaborators developed a novel method to synthesize RNA. They first created short RNA molecules, two to four nucleotides in length, that had a defined sequence. For example, RNA molecules with the sequence 5′–AUC–3′ were synthesized chemically. These short RNAs were then linked together enzymatically, in a 5′ to 3′ manner, to create long copolymers. For example, several RNAs with the sequence 5′–AUC–3′ could be linked together to create a strand of RNA with the sequence.

5′-AUCAUCAUCAUCAUCAUCAUCAUCAUC-3′

This is called a copolymer, because it is made from the linkage of several smaller molecules. Such a copolymer would contain three different codons: AUC (isoleucine), UCA (serine), and CAU (histidine). Using a cell-free translation system like the one described in figure 13.3, it was found that such a copolymer produced polypeptides containing isoleucine, histidine, and serine. Table 13.5 summarizes some of the copolymers that were made using this

TABLE 13.5

Examples of Copolymers That Were Analyzed by Khorana and Colleagues

Synthetic RNA*	Codon Possibilities	Amino Acids Incorporated into Peptides
UC	UCU, CUC	Serine, leucine
AG	AGA, GAG	Arginine, glutamic acid
UG	UGU, GUG	Cysteine, valine
AC	ACA, CAC	Threonine, histidine
UUC	UUC, UCU, CUU	Phenylalanine, serine, leucine
AAG	AAG, AGA, GAA	Lysine, arginine, glutamic acid
UUG	UUG, UGU, GUU	Leucine, cysteine, valine
CAA	CAA, AAC, ACA	Glutamine, asparagine, threonine
UAUC	UAU, AUC, UCU, CUA	Tyrosine, isoleucine, serine, leucine
UUAC	UUA, UAC, ACU, CUU	Leucine, tyrosine, threonine

*The synthetic RNAs were linked together to make copolymers.

approach and the amino acids that were incorporated into polypeptides.

Finally, another method that helped to decipher the genetic code also involved the chemical synthesis of short RNA molecules. In 1964, Marshall Nirenberg and Philip Leder discovered that RNA molecules containing three nucleotides (i.e., a triplet) could stimulate ribosomes to bind a tRNA. In other words, the RNA triplet acted like a codon. Ribosomes were able to bind RNA triplets, and then a tRNA with the appropriate anticodon could subsequently bind to the ribosome. To establish the relationship between triplet sequences and specific amino acids, samples containing ribosomes and a particular triplet were exposed to tRNAs with different radiolabeled amino acids.

As an example, in one experiment the researchers began with a sample of ribosomes that were mixed with 5′–CCA–3′ triplets. Portions of this sample were then added to separate tubes that had tRNAs with different radiolabeled amino acids. For example, one tube would contain radiolabeled histidine, a second tube would contain radiolabeled proline, a third tube would contain radiolabeled glycine, etc. There was only one radiolabeled amino acid in each tube. After allowing sufficient time for tRNAs to bind to the ribosomes, the samples were filtered; only the large ribosomes and anything bound to them were trapped on the filter. Unbound tRNAs would pass through the filter. Finally, the researchers determined the amount of radioactivity that was trapped on each filter. If the filter contained a large amount of radioactivity, it meant the triplet encoded the amino acid that was radiolabeled.

Using this approach, Nirenberg and Leder were able to establish relationships between particular triplet sequences and the binding of tRNAs carrying specific (radiolabeled) amino acids. In the case of the 5′–CCA–3′ triplet, it was determined that tRNAs carrying radiolabeled proline were bound to the ribosomes. Unfortunately, in some cases, a triplet could not promote sufficient tRNA binding to yield unambiguous results. Nevertheless, the triplet binding assay aided in the identification of a majority of codons.

The Amino Acid Sequences of Polypeptides Determine the Structure and Function of Proteins

Now that we understand how the genetic code was deciphered, let's consider the structure and function of the gene product. Following gene transcription and mRNA translation, the net result is a polypeptide with a defined amino acid sequence. This sequence is known as the **primary structure** of a polypeptide. Figure 13.4 shows the primary structure of an enzyme called lysozyme, a relatively small protein containing 129 amino acids. The primary structure of a typical polypeptide may be several hundred or even several thousand amino acids in length. Within a living cell, a newly made polypeptide is not usually found in a long linear state for a significant length of time. Rather, most polypeptides adopt a compact three-dimensional structure. The folding process begins while the polypeptide is still being translated. The progression from the primary sequence of a polypeptide to the three-dimensional structure of a protein is dictated by the amino acid

FIGURE 13.4 **An example of a protein's primary structure.** This is the amino acid sequence of the enzyme lysozyme, which contains 129 amino acids in its primary structure. As you may have noticed, the first amino acid is not methionine; instead, it is lysine. The first methionine residue in this polypeptide sequence is removed after or during translation. The removal of the first methionine occurs in many (but not all) proteins.

sequence within the polypeptide. In particular, the chemical properties of the amino acid side chains play a central role in determining the folding pattern of a protein.

As shown in figure 13.5, there are 20 amino acids that may be found within polypeptides. Each amino acid contains a different **side chain,** or **R group,** that has its own particular chemical properties. For example, aliphatic and aromatic amino acids are nonpolar, which means they are less likely to associate with water. These hydrophobic (meaning water-fearing) amino acids are often buried within the interior of a folded protein. In contrast, the polar amino acids are hydrophilic (meaning water-loving) and are more likely to be on the surface of a protein, where they can favorably interact with the surrounding water. For a given protein, the chemical properties of the amino acids and their

FIGURE 13.5   **The 20 amino acids that are found within proteins.**

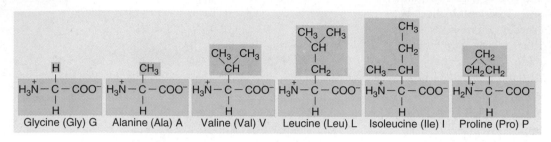

(a) Nonpolar, aliphatic amino acids

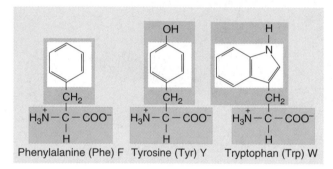

(b) Nonpolar, aromatic amino acids

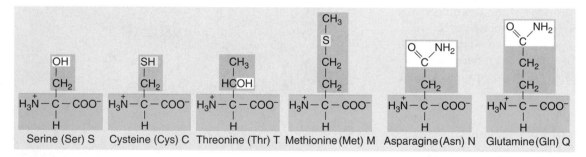

(c) Polar, neutral amino acids

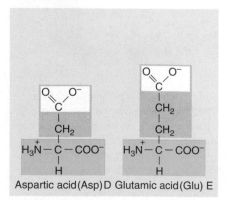

(d) Polar, acidic amino acids

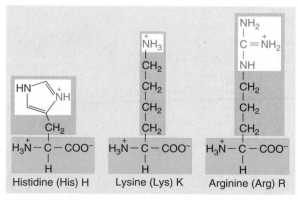

(e) Polar, basic amino acids

sequence in the primary structure are critical factors that determine the unique structure of that protein.

To become a functional unit, the polypeptides that constitute proteins must fold into a three-dimensional shape. This folding process is governed by the amino acid sequence and occurs in multiple stages. The first stage involves the formation of regular, repeating shapes known as **secondary structures.** The two types of secondary structure are the **α helix** and the **β sheet** (fig. 13.6b). A single polypeptide may have some regions that fold into an α helix and other regions that fold into a β sheet. Because of the geometry of secondary structures, certain amino acids (e.g., glutamic acid, alanine, and methionine) are good candidates to form an α helix, whereas others (e.g., valine, isoleucine, and tyrosine) are more likely to be found in a β-sheet conformation. Secondary structures within polypeptides are primarily stabilized by the formation of hydrogen bonds.

The short regions of secondary structure within a polypeptide are folded relative to each other to make the **tertiary structure** of a polypeptide. As shown in figure 13.6c, α-helical regions and β-sheet regions are connected by irregularly shaped segments to determine the three-dimensional or tertiary structure of the polypeptide. The folding of a polypeptide into its secondary and then tertiary conformation can occur spontaneously, because it is a thermodynamically favorable process. Various factors influence the folding process. These include the tendency of hydrophobic amino acids to avoid water, ionic interactions among charged amino acids, hydrogen bonding among amino acids in the folded polypeptide, and weak bonding known as van der Waals interactions.

As mentioned earlier, a protein is a functional unit that can be composed of one or more polypeptides. In cases where a protein is composed of a single polypeptide, the tertiary structure of the polypeptide is identical to the three-dimensional structure of the final protein. Other proteins, however, are composed of two or more polypeptides that associate with each other to make a functional protein with a **quaternary structure** (fig. 13.6d). The individual polypeptides are called **subunits** of the protein. Each subunit has its own tertiary structure. The association of multiple subunits is the quaternary structure of a protein.

Cellular Proteins Are Primarily Responsible for the Characteristics of Living Cells and an Organism's Traits

To a great extent, the characteristics of a cell depend on the types of proteins that it makes. Proteins can perform a variety of functions (table 13.6). Some proteins are important in determining

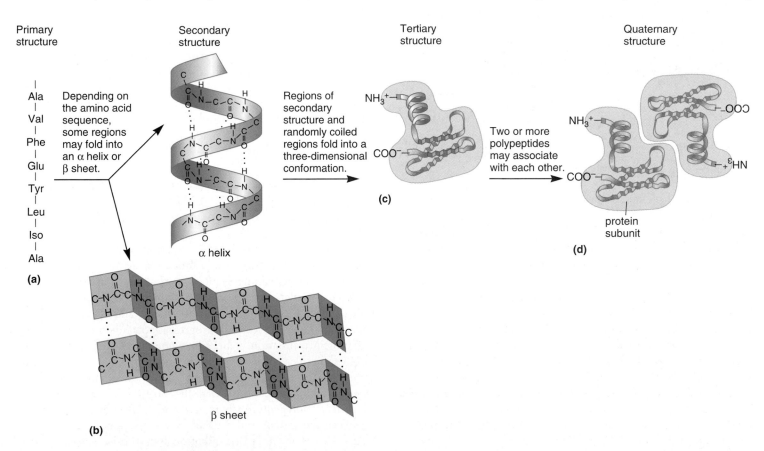

FIGURE 13.6 **Levels of structures formed in proteins.** (a) The primary structure of a polypeptide within a protein is its amino acid sequence. (b) Certain regions of a primary structure will fold into a secondary structure; the two types of secondary structures are called α helices and β sheets. (c) Both of these secondary structures can be found within the tertiary structure of a polypeptide. (d) Some polypeptides associate with each other to form a protein with a quaternary structure.

TABLE 13.6
Functions of Selected Cellular Proteins

Function	Examples
Cell shape and organization	Tubulin: forms cytoskeletal structures known as microtubules
	Ankyrin: anchors cytoskeletal proteins to the plasma membrane
Transport	Lactose permease: transports lactose across the bacterial cell membrane
	Sodium channels: transport sodium ions across the nerve cell membrane
	Hemoglobin: transports oxygen in red blood cells
Movement	Myosin: involved in muscle cell contraction
	Kinesin: involved in the movement of chromosomes during cell division
Cell signaling	Insulin: a hormone that influences target cell metabolism and growth
	Epidermal growth factor: a growth factor that promotes cell division
Cell surface recognition	Insulin receptor: recognizes insulin and initiates a cell response
	Integrins: bind to large extracellular proteins
Enzymes	Hexokinase: phosphorylates glucose during the first step in glycolysis
	β-galactosidase: cleaves lactose into glucose and galactose
	Glycogen synthetase: uses glucose as building blocks to synthesize a large carbohydrate known as glycogen
	Acyl transferase: links together fatty acids and glycerol phosphate during the synthesis of phospholipids
	RNA polymerase: uses ribonucleotides as building blocks to synthesize RNA
	DNA polymerase: uses deoxyribonucleotides as building blocks to synthesize DNA

the shape and structure of a given cell. For example, the protein tubulin assembles into large cytoskeletal structures known as microtubules, which provide eukaryotic cells with internal structure and organization. Some proteins are inserted into membranes and aid in the transport of ions and small molecules across the membrane. An example is a sodium channel that transports sodium ions into nerve cells. Another interesting category of proteins are those that function as biological motors, such as myosin, which is involved in the contractile properties of muscle cells. Within multicellular organisms, certain proteins function in cell-to-cell recognition and signaling. For example, hormones such as insulin are secreted by endocrine cells and bind to the insulin receptor protein found within the plasma membrane of target cells.

A key category of proteins are **enzymes,** which function to accelerate chemical reactions within the cell. Some enzymes assist in the breakdown of molecules or macromolecules into smaller units. These are known as catabolic enzymes and are important in generating cellular energy. In contrast, anabolic enzymes function in the synthesis of molecules and macromolecules. Several anabolic enzymes are listed in table 13.6, including DNA polymerase, which is required for the synthesis of DNA from nucleotide building blocks. Throughout the cell, the synthesis of molecules and macromolecules relies on enzymes and accessory proteins. Ultimately, then, the construction of a cell greatly depends on its anabolic enzymes, since these are required to synthesize all cellular macromolecules.

We can extend our understanding of protein function by considering an example that illustrates how protein function affects an organism's traits. In *Drosophila,* the color of the eye provides an interesting case in point. The eye of the fruit fly is composed of several types of cells. Some of them are pigment cells that are responsible for eye color. In the pigment cells of the red-eyed fly, the X chromosomes contain the red-eye allele (w^+), which codes for a protein that is located in the cell membrane. In the 1990s, the cloning and DNA sequencing of this gene indicated that the encoded protein transports colorless pigment precursor molecules into the cell. In the pigment cells of the red eye, the w^+ genes are transcribed and translated to produce this transport protein (fig. 13.7). This allows the cells to take up the pigment precursor molecules. Once inside, additional enzymes (that are encoded by different genes) convert the colorless precursors into red pigment. Therefore, the eyes of this fly appear red. In contrast, the white allele (w) of this gene is an example of a defective or loss-of-function allele. In the pigment cells of the white-eyed fly, this gene is not expressed properly, so that little or no functional transport protein is made. Without this transport protein, the pigment precursor molecules are not transported so the red pigment cannot be made. Because of this defect, the eyes of the fly remain white.

13.2 STRUCTURE AND FUNCTION OF tRNA

Thus far, we have considered the general features of translation and surveyed the structure and functional significance of cellular proteins. The remaining sections of chapter 13 are devoted to a molecular understanding of translation as it occurs in living cells. Biochemical studies of protein synthesis and tRNA molecules began in the 1950s. As work progressed toward an understanding of protein translation, it became evident that different kinds of RNA molecules are involved in the incorporation of amino acids into growing polypeptides. Francis Crick and Mahlon Hoagland were the first to propose that the position of an amino acid within a polypeptide chain is determined by the bonding between the mRNA and a tRNA carrying a specific amino acid. This idea, known as the **adaptor hypothesis,** suggested that tRNAs play a direct role in the recognition of the codons within the mRNA. In particular, this hypothesis proposed that a tRNA

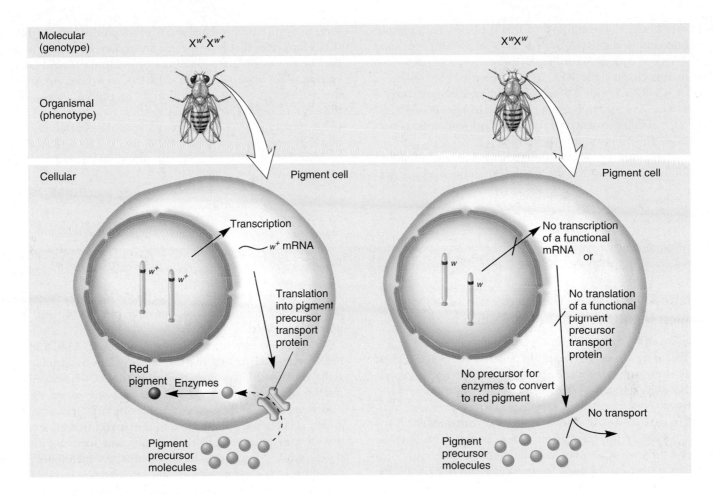

FIGURE 13.7 **The relationship between genes and traits at the molecular, organismal, and cellular levels.**

GENES→TRAITS   The wild-type allele encodes a protein that transports a colorless precursor into the pigment cells of the eye. This precursor is converted to a red pigment, so that the eyes become red. The mutant *w* allele cannot produce a functional transporter, so that little or none of the precursor is transported into the pigment cells. Therefore, the $X^w X^w$ homozygous fly has colorless (white) eyes.

has two functions: recognizing a three-base codon sequence in mRNA, and carrying an amino acid that is specific for that codon. In this section, we will begin by examining the general function of tRNA molecules and describe an experiment that was critical in supporting the adaptor hypothesis. We will then explore some of the important structural features that underlie tRNA function.

The Function of a tRNA Depends on the Specificity Between the Amino Acid It Carries and Its Anticodon

The codons within an mRNA are recognized by tRNA molecules that carry the correct amino acids to the site of polypeptide synthesis. During mRNA-tRNA recognition, the **anticodon** in a tRNA molecule binds to a codon in mRNA due to their complementary sequences (fig. 13.8). Importantly, the anticodon in the

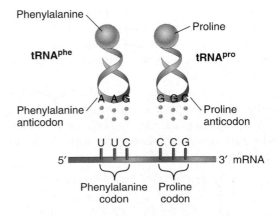

FIGURE 13.8 **Recognition between tRNAs and mRNA.** The anticodon in the tRNA binds to a complementary sequence in the mRNA. At its other end, the tRNA carries the amino acid that corresponds to the codon in the mRNA via the genetic code.

tRNA corresponds to the amino acid that it carries. For example, if the anticodon in the tRNA is 3′–AAG–5′, it is complementary to a 5′–UUC–3′ codon. According to the genetic code, described earlier in this chapter, the UUC codon specifies phenylalanine. Therefore, the tRNA with a 3′–AAG–5′ anticodon must carry a phenylalanine. As another example, if the tRNA has a 3′–GGC–5′ anticodon, it is complementary to a 5′–CCG–3′ codon that specifies proline. This tRNA must carry proline.

As mentioned earlier, the genetic code has 3 stop codons and 61 different codons that specify the 20 amino acids. There-fore, to synthesize proteins, a cell must produce many different tRNA molecules that have specific anticodon sequences. To do so, the chromosomal DNA contains many distinct tRNA genes that encode tRNA molecules with different sequences. According to the adaptor hypothesis, the anticodon in a tRNA always specifies the type of amino acid that it carries. Due to this specificity, tRNA molecules are named according to the type of amino acid that they bear. For example, a tRNA that carries a phenylalanine is described as tRNA^{phe}, while a tRNA that carries proline is tRNA^{pro}.

EXPERIMENT 13B

Translation Depends on the Recognition Between the Codon in mRNA and the Anticodon in tRNA

In 1962, Francöis Chapeville and his colleagues conducted exper-iments that were aimed at testing the adaptor hypothesis. Their technical strategy was similar to that of the Nirenberg experi-ments that helped to decipher the genetic code. In this approach, a translation system was used in which polypeptides were syn-thesized in vitro by isolating cell extracts that contained the com-ponents necessary for translation. These components include ribosomes, tRNAs, and other translation factors. An in vitro translation system can be used to investigate the role of specific factors by adding a particular mRNA template and varying indi-vidual components required for translation.

According to the adaptor hypothesis, the amino acid attached to a tRNA is not directly involved in codon recognition. Chapeville reasoned that if this were true, the alteration of an amino acid already attached to a tRNA should cause that altered amino acid to be incorporated into the polypeptide instead of the normal amino acid. For example, consider a tRNA^{cys} that carries the amino acid cysteine. If the attached cysteine is changed to an alanine, then this tRNA^{cys} will insert an alanine into a polypep-tide where it would normally put a cysteine. Fortunately, Chape-ville could carry out this strategy because he had a reagent, known as Raney nickel, that can chemically convert cysteine to alanine.

An elegant aspect of the experimental design was the choice of the mRNA template. Chapeville and his colleagues syn-thesized an mRNA template that contained only U and G. There-fore, this template could contain only the following codons (refer back to the genetic code in table 13.2):

UUU = phenylalanine	GUU = valine
UUG = leucine	GUG = valine
UGU = cysteine	GGU = glycine
UGG = tryptophan	GGG = glycine

Among the eight possible codons, there is one cysteine codon, but there are no alanine codons that can be formed from a polyUG template.

As shown in the experiment of figure 13.9, Chapeville began with a cell extract that contained tRNA molecules. Amino acids, which would become attached to tRNAs, were added to this mixture. Only cysteine was radiolabeled; the other 19 amino acids were not radiolabeled. After allowing sufficient time for the amino acids to become attached to the correct tRNAs, the sample was divided into two tubes. One tube was treated with Raney nickel, while the control tube was not. As mentioned, Raney nickel converts cysteine into alanine by removing the –SH group. However, it would not remove the radiolabel which was a ^{14}C-label within the cysteine amino acid. The tRNAs were then mixed with other components that are necessary for translation. Finally, the polyUG mRNA was added as a template. At this point, the mRNA should be translated into a polypeptide. In the control tube, we would expect the polypeptide to contain phenylalanine, leucine, cysteine, tryptophan, valine, and glycine, because these are the codons that contain only U and G. However, in the Raney nickel–treated sample, we would also expect to see alanine, if the tRNA^{cys} was using its anticodon region to recognize the mRNA.

Following translation, the polypeptides were isolated and hydrolyzed via a strong acid treatment, and then the individual amino acids were separated by column chromatography. Based on previous experiments, the column would separate cysteine from alanine; alanine would elute in a later fraction. The amount of radioactivity in each fraction was determined by liquid scintil-lation counting.

■ THE HYPOTHESIS

Codon recognition is dictated only by the tRNA antidocon; the chemical structure of the amino acid attached to the tRNA does not play a role.

TESTING THE HYPOTHESIS — **FIGURE 13.9** Evidence that tRNA uses its anticodon sequence to recognize mRNA.

Starting material: A bacterial cell extract containing the components required for translation, such as ribosomes, tRNAs, etc..

Experimental level   **Conceptual level**

1. Isolate an extract containing tRNAs. Note: This drawing only emphasizes tRNA^{cys}, even though the extract contains all types of tRNAs. When the extract is isolated, a substantial proportion of the tRNAs do not have an attached amino acid. The extract also contains enzymes that attach amino acids to tRNAs. (These enzymes will be described later in the chapter.)

2. Add amino acids, including radiolabeled cysteine, to the extract. An enzyme within the extract will specifically attach the radiolabeled cysteine to tRNA^{cys}. The other tRNAs will have unlabeled amino acids attached to them.

3. In one tube, treat the tRNAs with Raney nickel. This removes the –SH group from cysteine, converting it to alanine. In the control tube, do not add Raney nickel.

4. Add the tRNAs to the other components that are necessary for translation: ribosomes, translation factors, and so forth.

5. Add polyUG mRNA as a template. As noted, polyUG contains one cysteine codon but no alanine codons.

6. Allow translation to proceed.

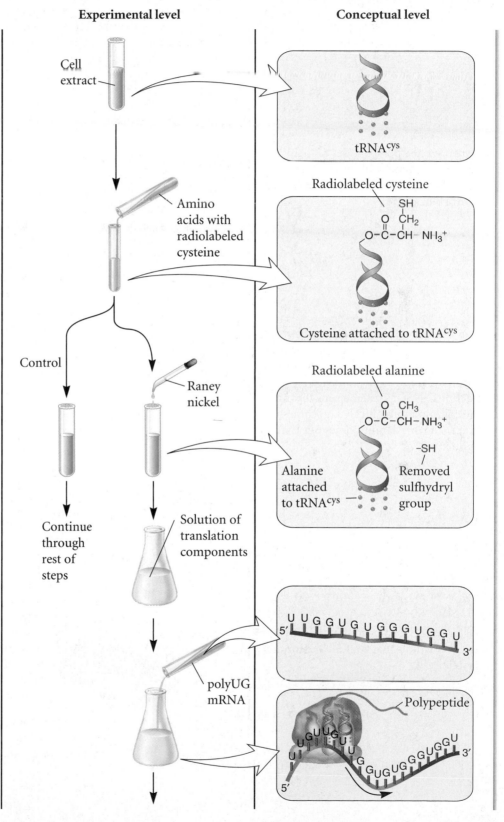

(continued)

7. Isolate the newly made polypeptides by precipitating them with trichloroacetic acid and then isolating the precipitated polypeptides on a filter.

8. Hydrolyze the polypeptides to their individual amino acids by treatment with a solution containing concentrated hydrochloric acid.

9. Run the sample over a column that separates cysteine and alanine. (See the appendix for a description of column chromatography.) Separate into fractions. Note: Cysteine runs through the column more quickly and comes out in fraction 3. Alanine comes out later, in fraction 7.

10. Determine the amount of radioactivity in the fractions that contain alanine and cysteine.

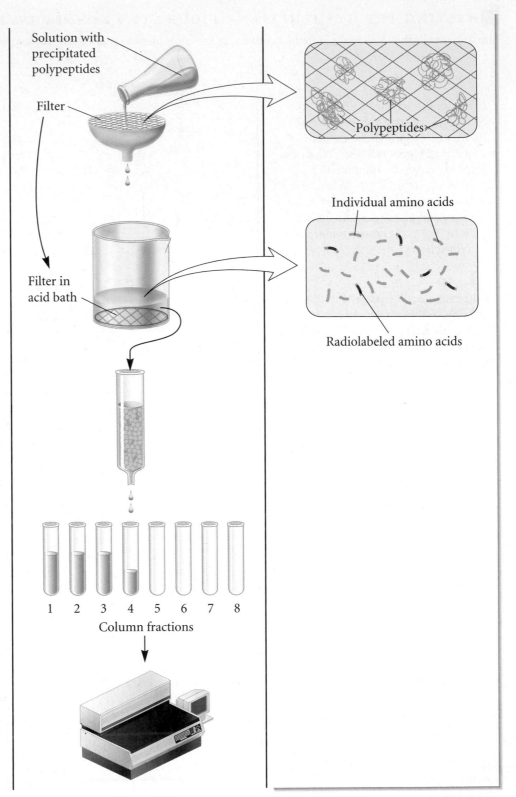

Solution with precipitated polypeptides

Filter

Polypeptides

Individual amino acids

Filter in acid bath

Radiolabeled amino acids

1 2 3 4 5 6 7 8

Column fractions

■ THE DATA

Conditions	Amount of Radiolabeled Amino Acids Incorporated into Polypeptide (cpm)*		
	Cysteine	Alanine	Total
Control, untreated tRNA	2,835	83	2,918
Raney nickel–treated tRNA	990	2,020	3,010

*Cpm is the counts per minute of radioactivity in the sample.
Adapted from Chapeville *et al* (1962) *PNAS 48*, 1086–1092.

■ INTERPRETING THE DATA

In the control sample, nearly all the radioactivity was found in the fraction containing cysteine (see data in fig. 13.9). This was expected, since the only radiolabeled amino acid added to the tRNAs was cysteine. The low radioactivity (83 counts per minute) in the alanine fraction probably represents contamination of this fraction by a small amount of cysteine. By comparison, when the tRNAs were treated with Raney nickel, a substantial amount of radiolabeled alanine became incorporated into polypeptides. This occurred even though the mRNA template did not contain any alanine codons. These results are consistent with the explanation that a tRNA^{cys}, which carried alanine instead of cysteine, incorporated alanine into the synthesized polypeptide. Thus, these observations indicate that the codons in mRNA are identified directly by the tRNA, with the attached amino acid playing no role in codon recognition.

As seen in the data of figure 13.9, the Raney nickel–treated sample still had 990 cpm of cysteine incorporated into polypeptides. This is about one-third of the total amount of radioactivity (namely, 990/3,010). In other experiments conducted in this study, the researchers showed that the Raney nickel did not react with about one-third of the tRNA^{cys}. Therefore, this proportion of the Raney nickel–treated tRNA^{cys} would still carry cysteine. This observation was consistent with the data shown here. Overall, the results of this experiment supported the adaptor hypothesis, indicating that tRNAs act as adaptors to carry the correct amino acid to the ribosome based on their anticodon sequence.

A self-help quiz involving this experiment can be found at the Online Learning Center.

tRNAs Share Common Structural Features

To understand how tRNAs act as carriers of the correct amino acids during translation, researchers have examined the structural characteristics of these molecules in great detail. Though a cell makes many different tRNAs, all tRNAs share common structural features. As originally proposed by Robert Holley in 1965, the secondary structure of tRNAs exhibits a cloverleaf pattern. There are three stem-loop structures, a variable region, and an acceptor stem with a 3′ single-stranded region (fig. 13.10). A conventional numbering system for the nucleotides within a tRNA molecule begins at the 5′ end and proceeds toward the 3′ end. Among different types of tRNA molecules, there are three variable regions that can differ in the number of nucleotides they contain. The anticodon is located in the second stem-loop region.

The actual three-dimensional or tertiary structure of tRNA molecules involves additional folding of the secondary structure. In the tertiary structure of tRNA, the stem-loop regions are folded into a much more compact molecule. The ability of RNA molecules to form stem-loop structures and the tertiary folding of tRNA molecules were described in chapter 9 (fig. 9.24). Interestingly, in addition to the normal A, U, G, and C nucleotides, tRNA molecules commonly contain modified nucleotides within their primary structures. For example, figure 13.10 illustrates a tRNA that contains several modified bases. Among many different species, researchers have found that more than 60 different nucleotide modifications can occur in tRNA molecules. The significance of modified bases in codon recognition will be described later in this chapter.

Aminoacyl-tRNA Synthetases Charge tRNAs by Attaching the Appropriate Amino Acid

To function correctly, each type of tRNA must have the appropriate amino acid attached to its 3′ end. The enzymes that catalyze the attachment of amino acids to tRNA molecules are known as **aminoacyl-tRNA synthetases.** There are 20 types of aminoacyl-tRNA synthetase enzymes, one for each of the 20 distinct amino acids. Each aminoacyl-tRNA synthetase is named for the specific amino acid it attaches to tRNA. For example, alanyl-tRNA synthetase recognizes a tRNA with an alanine anticodon (i.e., tRNA^{ala}) and attaches an alanine to it.

Aminoacyl-tRNA synthetases catalyze a chemical reaction involving three different molecules: an amino acid, a tRNA molecule, and ATP. As shown in figure 13.11, the first step of the reaction involves the activation of an amino acid by the covalent attachment of an AMP molecule. Pyrophosphate is released. During the second step, the activated amino acid is attached to the 3′ end of the tRNA molecule at the acceptor stem; AMP is released. Finally, the tRNA with its attached amino acid is released from the enzyme. At this stage, the tRNA is called a **charged tRNA.** In a charged tRNA molecule, the amino acid is attached to the 3′ end of the tRNA by an ester bond.

The ability of the aminoacyl-tRNA synthetases to recognize tRNAs has sometimes been called the "second genetic code." This recognition process is necessary to maintain the fidelity of genetic information. The frequency of error for aminoacyl-tRNA synthetases is less than 10^{-5}. In other words, the wrong amino acid will be attached to a tRNA less than once in 100,000 times! As

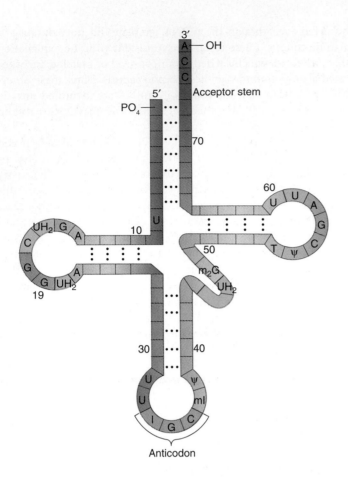

FIGURE 13.10   **Structures of tRNA and modified bases.** This figure depicts the secondary structure of tRNA. The conventional numbering of nucleotides begins at the 5′ end and proceeds toward the 3′ end. In all tRNAs, the nucleotides at the 3′ end contain the sequence CCA. Certain locations can have additional nucleotides that are not found in all tRNA molecules; these variable sites are shown in *blue*. The figure also shows the locations of a few modified bases that are specifically found in a yeast tRNA that carries alanine. The modified bases are: I = inosine, mI = methylinosine, T = ribothymidine, UH_2 = dihydrouridine, m_2G = dimethylguanosine, and ψ = pseudouridine.

you might expect, the anticodon region of the tRNA is usually important for recognition by the correct aminoacyl-tRNA synthetase. In studies of *Escherichia coli* synthetases, 17 of the 20 types of aminoacyl-tRNA synthetases recognize the anticodon region of the tRNA. However, other regions of the tRNA are also important recognition sites. These include the acceptor stem and bases in the stem-loop regions.

As mentioned earlier in this chapter, tRNA molecules frequently contain bases within their structure that have been chemically modified. These modified bases can have important effects on tRNA function. For example, modified bases within tRNA molecules affect the rate of translation and the recognition of tRNAs by aminoacyl-tRNA synthetases. Positions 34 and 37 contain the largest variety of modified nucleotides; position 34 is the first base in the anticodon that matches the third base in the

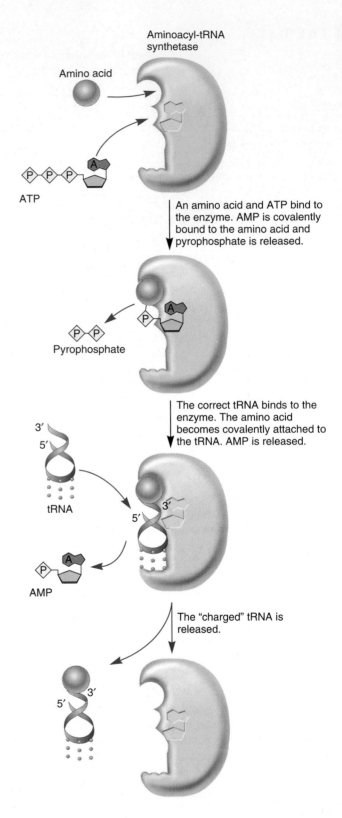

FIGURE 13.11   **Catalytic function of aminoacyl-tRNA synthetase.** Aminoacyl-tRNA synthetase contains binding sites for ATP, a specific amino acid, and a particular tRNA. In the first step, the enzyme catalyzes the covalent attachment of AMP to an amino acid, yielding an activated amino acid. In the second step, the activated amino acid is attached to a tRNA.

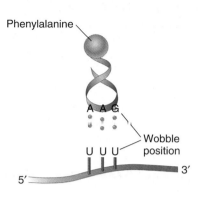

(a) Location of wobble position

Nucleotide of anticodon	Third nucleotide of codon
G	C, U
C	G
A	U, C, (A), G
U	A, U, G, (C)
I	U, C, A
xm^5s^2U	
xm^5Um	
Um	A, (G)
xm^5U	
xo^5U	U, A, G
k^2C	A

(b) Revised wobble rules

FIGURE 13.12 **Wobble position and base pairing rules. (a)** The wobble position occurs between the first base (meaning the first base in the 5′ to 3′ direction) in the anticodon and the third base in the mRNA codon. **(b)** The revised wobble rules are slightly different from those originally proposed by Crick. (The revised rules shown here are from Yokoyama, S., and Nishimura, S. (1995) *tRNA Structure, Biosynthesis, and Function,* Soll, D., and RajBhandary, U. L., eds., p. 209, ASM Press, Washington, D.C.) The standard bases found in RNA are G, C, A, and U. In addition, the structures of bases in tRNAs may be modified. Some modified bases that may occur in the wobble position in tRNA are I = inosine; xm^5s^2U = 5-methyl-2-thiouridine; xm^5Um = 5-methyl- 2′–O-methyluridine; Um = 2′–O-methyluridine; xm^5U = 5-methyluridine; xo^5U = 5-hydroxyuridine; k^2C = lysidine (a cytosine derivative). The mRNA bases in parentheses are recognized very poorly by the tRNA.

codon of mRNA. As discussed next, a modified base at position 34 can have important effects on codon-anticodon recognition.

Mismatches That Follow the Wobble Rule Can Occur at the Third Position in Codon-Anticodon Pairing

After considering the structure and function of tRNA molecules, it is interesting to reexamine some subtle features of the genetic code. As discussed earlier, the genetic code is degenerate. This means that more than one codon can specify the same amino acid. With the exception of serine, arginine, and leucine, this degeneracy always occurs at the third position in the codon. For example, valine is specified by GUU, GUC, GUA, and GUG. In all four cases, the first two bases are G and U. The third base, however, can be U, C, A, or G. To explain this pattern of degeneracy, Francis Crick proposed in 1966 that it is due to "wobble" at the third position in the codon-anticodon recognition process. According to the **wobble hypothesis,** the first two positions pair strictly according to the AU/GC rule. However, the third position can tolerate certain types of mismatches (fig. 13.12). This proposal suggested that the base at the third position can actually move a bit so that hydrogen bonding can occur between the codon and anticodon.

Because of the wobble rules, there is some flexibility in the recognition between a codon and anticodon during the process of translation. When two or more tRNAs that differ at the wobble base are able to recognize the same codon, these are termed **isoacceptor tRNAs.** As an example, tRNAs with an anticodon of 3′–CCA–5′ or 3′–CCG–5′ would be able to recognize a codon with the sequence of 5′–GGU–3′. In addition, the wobble rules enable a single type of tRNA to recognize more than one codon. For example, a tRNA with an anticodon sequence of 3′–AAG–5′ can recognize a 5′–UUC–3′ and a 5′–UUU–3′ codon. The 5′–UUC–3′ codon is a perfect match with this tRNA. The 5′–UUU–3′ codon is mismatched according to the standard RNA-RNA hybridization rules (namely, G in the anticodon is mismatched to U in the codon), but the two can fit according to the wobble rules described in figure 13.12. Likewise, the modifi-

cation of the wobble base to an inosine can allow a tRNA to recognize three different codons. At the cellular level, the ability of a single tRNA to recognize more than one codon makes it unnecessary for a cell to make 61 different tRNA molecules with anticodons that are complementary to the 61 possible codons.

13.3 RIBOSOME STRUCTURE AND ASSEMBLY

In the previous section, we examined how the structure and function of tRNA molecules are important in translation. According to the adaptor hypothesis, tRNAs bind to mRNA due to complementarity between the anticodons and codons. Concurrently, the tRNA molecules have the correct amino acid attached to their 3′ ends.

To synthesize a polypeptide, additional events must occur. In particular, the bond between the 3′ end of the tRNA and the amino acid must be broken, and a peptide bond must be formed between the adjacent amino acids. To facilitate these events, translation occurs on the surface of a large macromolecular complex known as the **ribosome.** The ribosome can be thought of as the macromolecular arena where translation takes place.

In this section, we will begin by examining the biochemical compositions of ribosomes in bacterial and eukaryotic cells. We will then examine the key functional sites on ribosomes for the translation process.

Bacterial and Eukaryotic Ribosomes Are Assembled from rRNA and Proteins

Bacterial cells have one type of ribosome that is found within the cytoplasm. By comparison, eukaryotic cells are compartmentalized into cellular organelles that are bounded by membranes. Eukaryotic cells contain biochemically distinct ribosomes in different cellular locations. The most abundant type of ribosome functions in the cytosol, which is the region of the eukaryotic cell that is inside the plasma membrane but outside the organelles.

Besides the cytosolic ribosomes, all eukaryotic cells have ribosomes within the mitochondria. In addition, plant cells and algae have ribosomes in their chloroplasts. The compositions of mitochondrial and chloroplast ribosomes are quite different from that of the cytosolic ribosomes. Unless otherwise noted, the term *eukaryotic ribosome* refers to ribosomes in the cytosol, not to those found within organelles. Likewise, the description of eukaryotic translation refers to translation via cytosolic ribosomes.

Each ribosome is composed of structures called the large and small subunits (fig. 13.13). This term is perhaps misleading, since each ribosomal subunit itself is formed from the assembly of many different proteins and rRNA molecules. In bacterial ribosomes, the 30S subunit is formed from the assembly of 21 different ribosomal proteins and one rRNA molecule; the 50S subunit contains 34 different proteins and two different rRNA molecules (fig. 13.13a). The designations 30S and 50S refer to the rate that these subunits sediment when subjected to a centrifugal force. This rate is described as a sedimentation coefficient in Svedberg units (S), in honor of Theodor Svedberg, who invented the ultracentrifuge. Together, the 30S and 50S subunits form a 70S ribosome. (Note: Svedberg units do not add up linearly.) In bacteria, the ribosomal proteins and rRNA molecules are synthesized in the cytoplasm, and the ribosomal subunits are assembled there.

The synthesis of eukaryotic rRNA occurs within the nucleus, and the ribosomal proteins are made in the cytosol where translation occurs. The 40S subunit is composed of 33 proteins and an 18S rRNA; the 60S subunit is made up of 49 proteins and 5S, 5.8S, and 28S rRNAs (fig. 13.13b). The assembly of

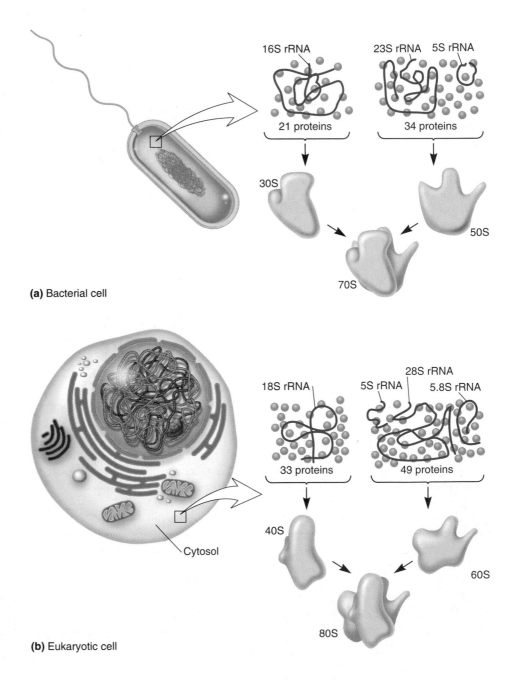

FIGURE 13.13 **Composition of bacterial and eukaryotic ribosomes.** (a) Bacterial ribosomes contain two subunits: 30S and 50S. The 30S subunit contains 21 different proteins and 16S rRNA; the 50S subunit contains 34 proteins and 5S and 23S rRNAs. (b) Eukaryotic ribosomes found in the cytosol contain two subunits: 40S and 60S. The 40S subunit contains 33 proteins and 18S rRNA; the 60S subunit contains 49 proteins and 5S, 5.8S, and 28S rRNAs. The 40S and 60S subunits assemble in the nucleolus and then are exported into the cytosol.

(a) Bacterial cell

(b) Eukaryotic cell

the rRNAs and ribosomal proteins to make the 40S and 60S subunits occurs within the **nucleolus,** a region of the nucleus that is specialized for this purpose. The 40S and 60S subunits are then exported into the cytosol, where they associate to form an 80S ribosome during translation.

Components of Ribosomal Subunits Form Functional Sites for Translation

To understand the structure and function of the ribosome at the molecular level, researchers must determine the locations and functional roles of the individual ribosomal proteins and rRNAs. In recent years, many advances have been made toward a molecular understanding of ribosomes. Microscopic and biophysical methods have been used to study ribosome structure. An electron micrograph of bacterial ribosomes is shown in figure 13.14a. A few research groups have succeeded in crystallizing ribosomal subunits, and even intact ribosomes. Considering the large size of the ribosome, this is an amazing technical feat. Figure 13.14b shows the crystal structure of bacterial ribosomal subunits. The overall shape of each subunit is largely determined by the structure of the rRNAs, which constitute most of the mass of the ribosome. The

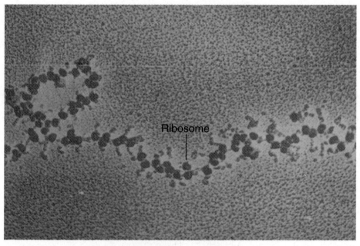

(a) Ribosomes as seen with electron microscopy.

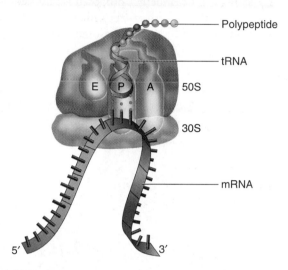

(c) Model for ribosome structure

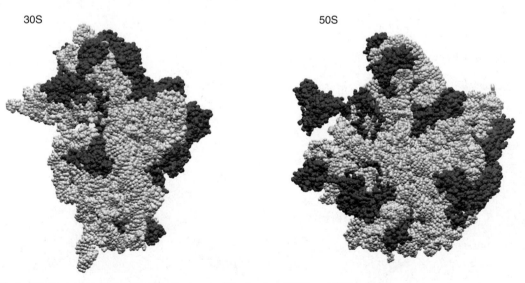

(b) An interface view of the crystal structures of the bacterial 30S and 50S subunits.

FIGURE 13.14 **Ribosomal structure.** (a) Electron micrograph of ribosomes attached to an mRNA molecule. (b) Crystal structure of the 30S and 50S subunits from bacteria. This model shows the interface between the two subunits. rRNA is shown in *light gray* and proteins are shown in *dark gray.* (c) A model depicting the sites where tRNA and mRNA bind to an intact ribosome. The mRNA lies on the surface of the 30S subunit. The A, P, and E sites are formed at the interface between the large and small subunits. The growing polypeptide chain exits through a hole in the 50S subunit. If you want to relate the crystal structures shown in part (b) with the intact structure shown in part (c), you may wish to do the following: Lay both of your hands, palms up, on a desk. The palms of your hands are the interfaces between the ribosomal subunits. These are the structures you are viewing in part (b). Now bring your palms together so they are touching each other. This would be the model shown in part (c) if it were turned 90° clockwise. The A, P, and E sites are in a cavity between your two palms.

interface between the 30S and 50S subunits is primarily composed of ribosomal RNA. Ribosomal proteins cluster on the outer surface of the ribosome and on the periphery of the interface.

During bacterial translation, the mRNA lies on the surface of the 30S subunit within a space between the 30S and 50S subunits. As the polypeptide is being synthesized, it exits through a channel within the 50S subunit (fig. 13.14c). Ribosomes contain discrete sites where tRNAs bind and the polypeptide is synthesized. In 1964, James Watson was the first to propose a two-site model for tRNA binding to the ribosome. These sites are known as the **peptidyl site (P site)** and **aminoacyl site (A site)**. In 1984, Knud Nierhaus and Hans-Jorg Rheinberger proposed a three-site model. This model incorporated the observation that uncharged tRNA molecules can bind to a site on the ribosome that is distinct from the P and A sites. This third site is now known as the **exit site (E site)**. The locations of the A, P, and E sites are shown

in figure 13.14c. In the next section, we will examine the roles of these sites during the three stages of translation.

13.4 STAGES OF TRANSLATION

Like transcription, the process of translation can be viewed as occurring in three stages: **initiation, elongation,** and **termination.** Figure 13.15 presents an overview of these stages. During initiation, the mRNA, initiator tRNA (i.e., the tRNA bound to the start codon), and ribosomal subunits assemble to form a complex. After the initiation complex is formed, the ribosome slides along the mRNA in the 5′ to 3′ direction, moving over the codons. This is the elongation stage of translation. As the ribosome moves, tRNA molecules sequentially bind to the mRNA at the A site in the ribosome, bringing with them the appropriate

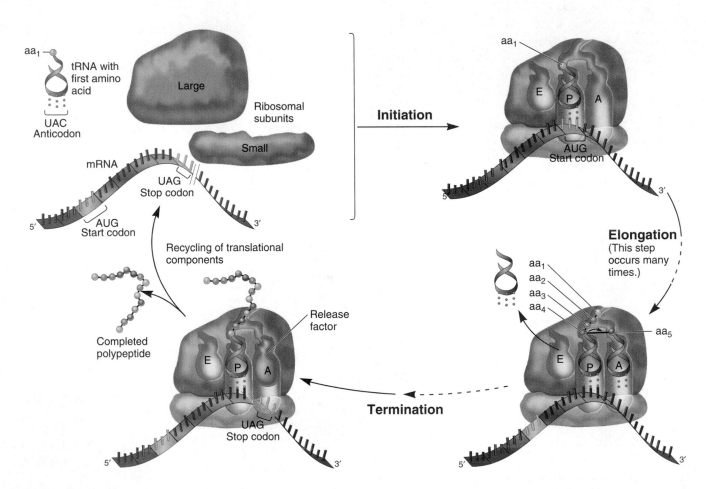

FIGURE 13.15 **Overview of the stages of translation.** The initiation stage involves the assembly of the ribosomal subunits, mRNA, and the tRNA carrying the first amino acid. During elongation, the ribosome slides along the mRNA and synthesizes a polypeptide chain. Translation ends when a stop codon is reached and the polypeptide is released from the ribosome. Note: In this figure, and succeeding figures in this chapter, the ribosomes are drawn schematically to emphasize different aspects of the translation process. The actual structures of ribosomes were described in figures 13.13 and 13.14.

GENES→TRAITS The ability of genes to produce an organism's traits relies on the molecular process of gene expression. During translation, the codon sequence within mRNA (which is derived from a gene sequence during transcription) is translated into a polypeptide sequence. After polypeptides are made within a living cell, they function as proteins to govern an organism's traits. For example, once the globin polypeptide is made, it functions within the hemoglobin protein and provides red blood cells with the ability to carry oxygen, a vital trait for survival. Translation allows functional proteins to be made within living cells.

amino acids. Therefore, amino acids are linked in the order dictated by the codon sequence in the mRNA. Finally, a stop codon is reached, signaling the termination of translation. At this point, disassembly occurs and the newly made polypeptide is released. In this section, we will examine the components that are required for the translation process and consider their functional roles during the three stages of translation.

The Initiation Stage Involves the Binding of mRNA and the Initiator tRNA to the Ribosomal Subunits

During initiation, an mRNA and the first tRNA bind to the ribosomal subunits. A specific tRNA functions as the **initiator tRNA,** which recognizes the start codon in the mRNA. In bacteria, the initiator tRNA, which is also designated tRNA^{fmet}, carries a methionine that has been covalently modified to N-formylmethionine. In this modification, a formyl group (—CHO) is attached to the nitrogen atom in methionine after the methionine has been attached to the tRNA.

Figure 13.16 outlines the steps that occur during the initiation stage of translation in bacteria. During translational initiation, the mRNA, tRNA^{fmet}, and ribosomal subunits associate with each other to form an initiation complex. The formation of this complex requires the participation of three initiation factors: IF1, IF2, and IF3. The binding of the mRNA to the 30S ribosomal subunit is facilitated by IF3, which requires GTP for its function. In addition, complementarity between a short region of the mRNA and rRNA within the 30S subunit plays a key role. A sequence within bacterial mRNAs, known as the **ribosomal-binding site** or **Shine-Dalgarno sequence,** is involved in the binding of the mRNA to the 30S subunit. The location of this sequence is shown in figure 13.16 and in more detail in figure 13.17. The Shine-Dalgarno sequence is complementary to a short sequence within the 16S rRNA. This complementarity promotes the hydrogen bonding of the mRNA to the 30S subunit.

The binding of tRNA^{fmet} to the 30S subunit and mRNA requires the function of IF2, which also requires GTP for its function. The tRNA^{fmet} binds to the first start codon, which is typically a few nucleotides downstream from the Shine-Dalgarno sequence. The start codon is usually AUG, but in some cases it can be GUG or UUG. Even when the start codon is GUG (which normally encodes valine) or UUG (which normally encodes leucine), the first amino acid in the polypeptide is still a formylmethionine because only a tRNA^{fmet} can initiate translation. During or after translation of the entire polypeptide, the formyl group or the entire formylmethionine may be removed. Therefore, it is common for polypeptides to not have formylmethionine or methionine as their first amino acid.

The initiation stage of bacterial translation is completed when the 50S ribosomal subunit associates with the 30S subunit. For this to occur, the initiation factors must be released from the 30S subunit. Much later, after translation is completed, IF1 is necessary to dissociate the 50S and 30S ribosomal subunits so that the 30S subunit can reinitiate with another mRNA molecule. For this reason, it is viewed as an initiation factor. As noted in figure 13.16, the tRNA^{fmet} binds to the P site on the ribosome. During

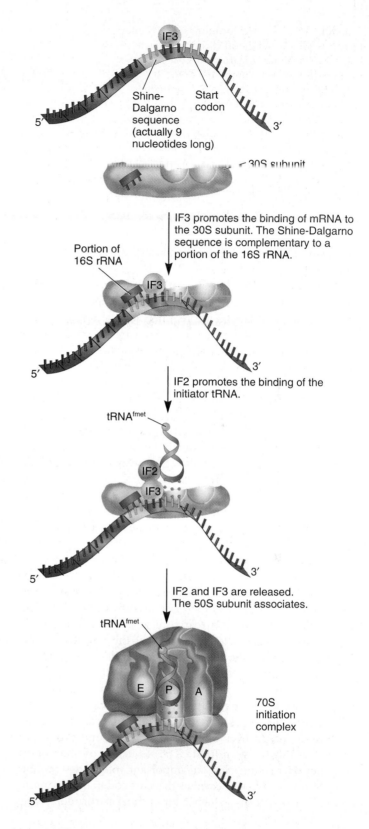

FIGURE 13.16   Translational initiation in bacteria. IF3 promotes the binding of mRNA to the 30S subunit. In addition, the Shine-Dalgarno sequence hydrogen bonds with a portion of the 16S rRNA. IF2 then binds to the initiator tRNA, which binds to the start codon in the P site. IF2 and IF3 are released, and the 50S subunit associates with the 30S subunit to form a 70S initiation complex. This marks the end of the initiation stage.

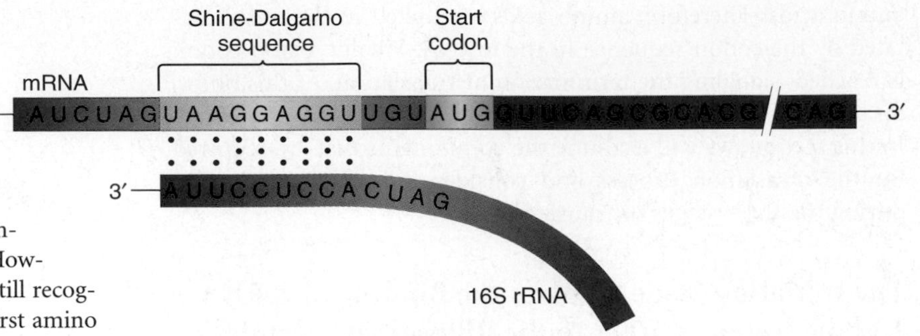

FIGURE 13.17 **The locations of the Shine-Dalgarno sequence and the start codon in bacterial mRNA.** The Shine-Dalgarno sequence is complementary to a sequence in the 16S rRNA. It hydrogen bonds with the 16S rRNA to promote initiation. The start codon is typically a few nucleotides downstream from the Shine–Dalgarno sequence. In bacteria, the start codon is usually AUG. Occasionally, the start codon can be GUG (valine) or UUG (leucine). However, in these rare cases, the GUG and UUG start codons are still recognized by tRNA^fmet, so that a formylmethionine occurs as the first amino acid instead of valine or leucine.

the elongation stage, which is discussed later, all of the other tRNAs initially bind to the A site.

In eukaryotes, the assembly of the initiation complex bears many similarities to that in bacteria. However, as described in table 13.7, additional factors are required for the initiation process. Note that eukaryotic Initiation Factors are designated eIF to distinguish them from bacterial initiation factors. The initiator tRNA in eukaryotes carries methionine rather than formylmethionine (as in bacteria). A eukaryotic initiation factor, eIF2, binds directly to tRNA^met to recruit it to the 40S subunit. Also, several initiation factors (CBPI, eIF4A, eIF4B, eIF4F, and others) bind to the mRNA. These initiation factors recognize the 7-methylguanosine cap structure, unwind any secondary structure in the mRNA so that it can bind to the ribosome, and aid in the identification of a start codon.

The identification of the correct AUG start codon in eukaryotes differs greatly from that in bacteria. As mentioned, eukaryotic mRNAs contain a 7-methylguanosine cap structure at their 5′ end. The recognition of this cap structure is key to the initial binding of mRNA to the ribosome. After this occurs, the next step is to locate an AUG start codon that is somewhere downstream from the 5′ cap structure. Marilyn Kozak proposed that the ribosome begins at the 5′ end and then scans along the mRNA in the 3′ direction in search of an AUG start codon. In many, but not all, cases, the ribosome uses the first AUG codon that it encounters as a start codon. When a start codon is identified, the 60S subunit assembles with the aid of eIF5.

By analyzing the sequences of many eukaryotic mRNAs, researchers have found that not all AUG codons near the 5′ end of mRNA can function as start codons. In some cases, the scanning ribosome passes over the first AUG codon and chooses an AUG farther down the mRNA. The sequence of nucleotides around the AUG codon plays an important role in determining whether or not it will be selected as the start codon by a scanning ribosome. The consensus sequence for optimal start codon recognition is shown here.

Start
Codon

G   C   C   (A/G)   C   C   A   U   G   G

−6   −5   −4   −3   −2   −1   +1   +2   +3   +4

TABLE 13.7

A Comparison of Translational Protein Factors in Bacteria and Eukaryotes

Bacterial Factor	Eukaryotic Factor	Function
Initiation Factors		
IF1	eIF2	Involved in forming the initiation complex
IF2	eIF3	Involved in forming the initiation complex
IF3	eIF4C	Involved in forming the initiation complex
	CBPI	Binds to the 7-methylguanosine cap
	eIF4A, eIF4B, eIF4F	Involved in the search for a start codon
	eIF5	Helps dissociate eIF2, eIF3, and eIF4C
	eIF6	Helps dissociate 60S subunit from inactive ribosomes
Elongation Factors		
EF-Tu	eEF1α	Binding of tRNAs to the A site
EF-Ts	eEF1βγ	Recycling factor for eEF1α
EF-G	eEF2	Required for translocation
Termination Factors		
RF1	eRF	Recognizes stop codon and promotes termination
RF2		
RF3		

Aside from an AUG codon itself, a guanosine at the +4 position and a purine, preferably an adenine, at the −3 position are the most important for start codon selection. These rules for optimal translation initiation are called **Kozak's rules.** This consensus sequence is recognized by a ribosomal component that facilitates the initiation of translation.

Polypeptide Synthesis Occurs During the Elongation Stage

During the elongation stage of translation, amino acids are added, one at a time, to the polypeptide chain. The addition of each amino acid occurs via a series of steps outlined in figure 13.18. These steps occur during the **elongation stage.** Even though this process is rather complex, it occurs at a remarkable rate. Under normal cellular conditions, a polypeptide chain can elongate at a rate of 15 to 18 amino acids per second in bacteria and 6 amino acids per second in eukaryotes!

During the elongation stage, a tRNA brings a new amino acid to the ribosome so that it can be attached to the end of the growing polypeptide chain. At the *top* of figure 13.18, a short polypeptide is attached to the tRNA located at the P site of the ribosome. A new tRNA carrying a single amino acid binds to the A site. This binding occurs because the anticodon in the tRNA is complementary to the codon in the mRNA. The hydrolysis of GTP by EF-Tu provides the energy for the binding of the tRNA to the A site. As discussed earlier, only the first two bases in the mRNA are usually needed for correct recognition of the anticodon in the tRNA. In addition, the 16S rRNA, which is a component of the small ribosomal subunit, plays a key role that ensures the proper recognition between the mRNA and correct tRNA. The 16S rRNA can detect when an incorrect tRNA is bound at the A site and will prevent elongation until the mispaired tRNA is released from the A site. This phenomenon, termed the **decoding function** of the ribosome, is important in maintaining high fidelity of mRNA translation. An incorrect amino acid is incorporated into a growing polypeptide at a rate of approximately 10^{-4} (i.e., one mistake per 10,000).

The formation of a peptide bond between the amino acid at the A site and the growing polypeptide chain occurs via a **peptidyl transfer** reaction. This term means that the polypeptide is removed from the tRNA in the P site and transferred to the amino acid at the A site. This transfer is accompanied by the formation of a peptide bond between the amino acid at the A site and the polypeptide chain, lengthening the chain by one amino acid. The peptidyl transfer reaction is catalyzed by a component of the 50S subunit known as **peptidyltransferase,** which is composed of several proteins and rRNA. Interestingly, based on the crystal structure of the 50S subunit, Thomas Steitz, Peter Moore, and their colleagues concluded that the 23S rRNA is responsible for catalyzing bond formation between adjacent amino acids. In other words, the ribosome contains a ribozyme!

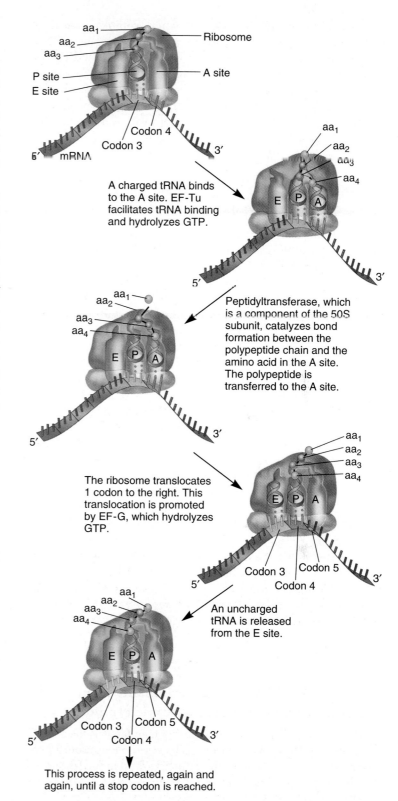

FIGURE 13.18 **The elongation stage of translation in bacteria.** This process begins with the binding of an incoming tRNA. The hydrolysis of GTP by EF-Tu provides the energy for the binding of the tRNA to the A site. A peptide bond is then formed between the incoming amino acid and the last amino acid in the growing polypeptide chain. This moves the polypeptide chain to the A site. The ribosome then translocates in the 3′ direction so that the two tRNAs are moved to the E and P sites. This translocation requires the hydrolysis of GTP via EF-G. The uncharged tRNA in the E site is released from the ribosome. Now the process is ready to begin again. Each cycle of elongation causes the polypeptide chain to grow by one amino acid.

After the peptidyl transfer reaction is complete, the ribosome moves, or translocates, to the next codon in the mRNA. This moves the tRNAs at the P and A sites to the E and P sites, respectively. You should notice that the next codon in the mRNA is now exposed in the unoccupied A site. Finally, the uncharged tRNA exits the E site. At this point, a charged tRNA can enter the empty A site, and the same series of steps can occur to add the next amino acid to the polypeptide chain.

Termination Occurs When a Stop Codon Is Reached in the mRNA

The final stage of translation, known as **termination,** occurs when a stop codon is reached in the mRNA. In most species, the three stop codons are UAG, UAA, and UGA. These codons are sometimes referred to as amber (UAG), opal (UAA), and ocher (UGA). (Note: The term *amber,* or brownstone, is the English translation of the name Bernstein, a graduate student who was involved in the discovery of the UAG codon. In keeping with this tradition, the lighthearted names ocher and opal were given to the other two stop codons.) The stop codons, also known as **nonsense codons,** are not recognized by a tRNA with a complementary sequence. Instead, they are recognized by proteins known as **release factors.** Interestingly, the three-dimensional structure of release factor proteins mimics the structure of tRNAs. In bacteria, RF1 recognizes UAA and UAG, and RF2 recognizes UGA and UAA. A third release factor, RF3, is also required. In eukaryotes, a single release factor, eRF, recognizes all three stop codons.

Figure 13.19 illustrates the termination stage of translation. At the *top* of this figure, the completed polypeptide chain is attached to a tRNA in the P site. A stop codon is located at the A site. In the next step, RF1 or RF2 binds to the stop codon at the A site and RF3 binds at a different location on the ribosome. After RF1 (or RF2) and RF3 have bound, the bond between the polypeptide and the tRNA is hydrolyzed. The polypeptide is then released from the ribosome. The final step in translational termination is the disassembly of mRNA, ribosomal subunits, and the release factors.

A Polypeptide Chain Has Directionality from Its Amino Terminus to Its Carboxyl Terminus

Polypeptide synthesis has a directionality that parallels the 5′ to 3′ orientation of the mRNA. During each cycle of elongation, a **peptide bond** is formed between the carboxyl group in the last amino acid of the polypeptide chain and the amino group in the amino acid being added. As shown in figure 13.20a, this occurs via a condensation reaction that releases a water molecule. The newest amino acid added to a growing polypeptide always has a free carboxyl group. Figure 13.20b compares the sequence of a very short polypeptide with the mRNA that encodes it. The first amino acid is said to be at the **N-terminal** or **amino terminal end** of the polypeptide. The term *N-terminal* refers to the presence of a nitrogen atom (N) at this end. By comparison, the last amino acid

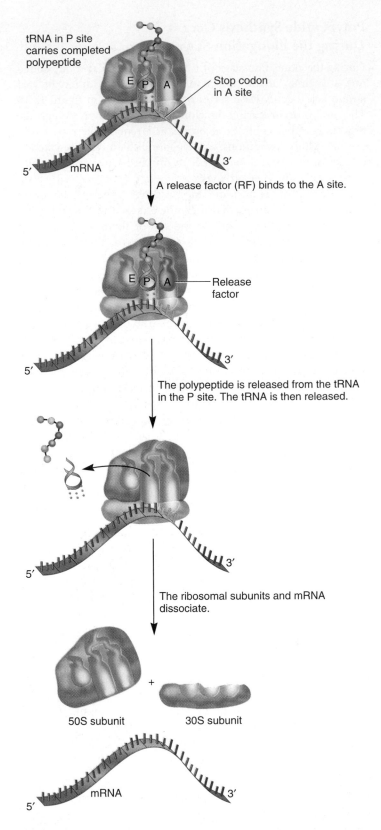

FIGURE 13.19   The termination of translation in bacteria. When a stop codon is reached, RF1 or RF2 binds to the A site. (RF3 binds elsewhere and uses GTP to facilitate the termination process.) The polypeptide is cleaved from the tRNA in the P site and released. The tRNA is released, and the rest of the components disassemble.

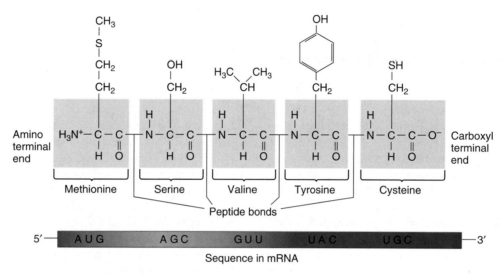

(a) Attachment of an amino acid to a peptide chain

Last peptide bond formed in the growing chain of amino acids

(b) Directionality in a polypeptide and mRNA

FIGURE 13.20   **The directionality of polypeptide synthesis.** (a) An amino acid is connected to a polypeptide chain via a condensation reaction that releases a water molecule. (b) The first amino acid in a polypeptide chain (usually methionine) is located at the amino terminus, and the last amino acid is at the carboxyl terminus. Thus, the directionality of amino acids in a polypeptide chain is from the amino to carboxyl terminus, which corresponds to the 5′ to 3′ orientation of codons in mRNA.

in a completed polypeptide is located at the **C-terminal** or **carboxyl terminal end.** A carboxylic acid group (COO⁻) is always found at this site in the polypeptide chain.

Bacterial Translation Can Begin Before Transcription Is Completed

While most of our knowledge concerning transcription and translation has come from genetic and biochemical studies, electron microscopy has also been an important tool in elucidating the mechanisms of transcription and translation. As described earlier in this chapter, electron microscopy (EM) has been a critical technique in facilitating our understanding of ribosome struc-

ture. In addition, EM can be used to visualize genetic processes such as translation.

The first success in the EM observation of gene expression was achieved by Oscar Miller Jr. and his colleagues in 1967. Figure 13.21 shows an EM photograph of a bacterial gene in the act of gene expression. Prior to this experiment, biochemical and genetic studies had suggested that the translation of a bacterial structural gene begins before the mRNA transcript is completed. In other words, as soon as an mRNA strand is long enough, a ribosome will attach to the 5′ end and begin translation—even before RNA polymerase has reached the transcriptional termination site within the gene. This phenomenon is termed the coupling between transcription and translation in bacterial cells.

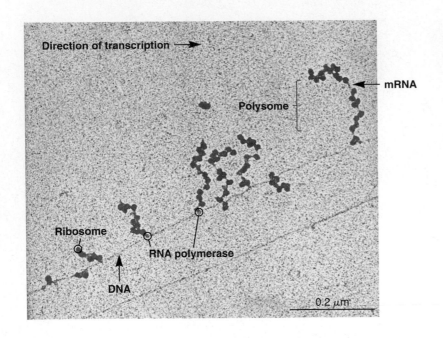

Direction of transcription →

mRNA

Polysome

Ribosome

RNA polymerase

DNA

0.2 μm

FIGURE 13.21 **Coupling between transcription and translation in bacteria.** An electron micrograph showing the simultaneous transcription and translation processes. The DNA is transcribed by many RNA polymerases that move along the DNA from left to right. Note that the RNA transcripts are getting longer as you go from *left* to *right*. Ribosomes attach to the mRNA, even before transcription is completed. The binding of many ribosomes to the mRNA is called a polyribosome or a polysome. Several polyribosomes are seen here.

As shown in figure 13.21, only one of the DNA strands is used as a template for the synthesis of mRNA transcripts. Several RNA polymerase enzymes have recognized this gene and begun to transcribe it. Since the transcripts on the *right* side are longer than those on the *left,* Miller concluded that transcription was proceeding from *left* to *right* in the micrograph. This EM image also shows the process of translation. Relatively small mRNA transcripts, near the *left* side of the figure, had a few ribosomes attached to them. As the transcripts became longer, additional ribosomes were attached to them. The term **polyribosome** or **polysome** is used to describe an mRNA transcript that has many bound ribosomes in the act of translation. In this study, the nascent polypeptide chains were too small for researchers to observe. In later studies, as EM techniques became more refined, the polypeptide chains emerging from the ribosome were also visible.

The Amino Acid Sequences of Proteins Contain Sorting Signals

The amino acid sequence of a protein may contain a sorting signal that will direct the protein to its correct location. In bacteria, for example, some proteins contain a sequence that will direct their secretion from the cytoplasm. In eukaryotic cells, sorting is a more complicated phenomenon because they are compartmentalized into many different membrane-bounded organelles (e.g., mitochondria, lysosomes, nucleus, etc.). In general, any particular protein is meant to function in only a single location. For example, the enzyme F_0F_1-ATP-synthase functions in the mitochondrion and is not found in other locations in the cell. Cytoskeletal proteins such as tubulin and actin are found in the cytosol but not within the lumen of cellular organelles. To fulfill its function,

each type of protein must be sorted into the correct cellular compartment. In eukaryotes, this sorting can occur during translation (termed **cotranslational sorting**) or after translation is completed (termed **posttranslational sorting**).

Most eukaryotic proteins begin their synthesis on ribosomes in the cytosol. Translation is completed in the cytosol for those proteins destined for the cytosol, nucleus, mitochondrion, chloroplast, or peroxisome (fig. 13.22). The uptake of proteins into the nucleus, mitochondrion, chloroplast, and peroxisome then occurs posttranslationally. By comparison, the synthesis of other eukaryotic proteins begins in the cytosol and then is temporarily halted until the ribosome has become bound to the membrane of the endoplasmic reticulum (ER). After the ribosome has bound to the ER membrane, translation resumes, and the polypeptide is synthesized into the ER lumen or ER membrane. This event is termed **cotranslational import.** Proteins that are destined for the ER, Golgi, lysosome, plasma membrane, or secretion are first directed to the ER via this type of cotranslational import mechanism.

It is common to focus attention on the importance of the amino acid sequence in dictating the structure and function of a protein. Even so, another vital aspect of eukaryotic gene sequences is that they encode short amino acid sequences that play a role in protein sorting. In eukaryotic proteins, there are short stretches of amino acid sequences that direct each protein to its correct cellular location. These short amino acid sequences are called **sorting** or **traffic signals.** Table 13.8 provides a general description of the sorting signals contained within eukaryotic proteins. Each traffic signal is recognized by specific cellular components that facilitate the sorting of the protein to its correct compartment, either cotranslationally or posttranslationally.

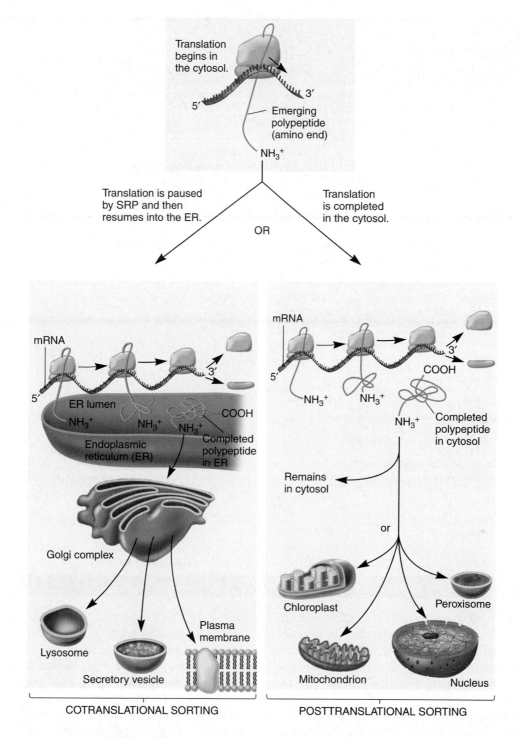

FIGURE 13.22 **Cotranslational and posttranslational protein sorting in eukaryotic cells.** The mRNA encoded by genes in the cell nucleus is exported from the nucleus and begins its translation in the cytosol. Some polypeptides contain an amino acid sequence at their amino terminal end that is recognized by the signal recognition particle (SRP). SRP directs these proteins to the ER, where they are sorted to the ER, Golgi, lysosome, or plasma membrane or they are secreted. Other proteins destined for the nucleus, mitochondrion, chloroplast, or peroxisome are completely synthesized in the cytosol and then are sorted posttranslationally.

TABLE 13.8

Sorting Signals in Eukaryotic Proteins

Type of Signal	Description
Mitochondrial-targeting signal	Usually a short sequence at the amino terminus of a protein that contains several positively charged residues (i.e., lysine and arginine). This signal folds into an α helix in which the positive charges are on one face of the helix.
Nuclear-localization signal	Can be located almost anywhere in the polypeptide sequence. The signal is four to eight amino acids in length and contains several positively charged residues and usually one or more prolines.
Peroxisomal-targeting signal	Either a specific sequence of three amino acids (serine, lysine, leucine) that is located near the carboxyl terminus of the protein or a 26-amino-acid sequence at the amino terminus.
SRP signal	A sequence of ~20 amino acids near the amino terminus that is composed of mostly nonpolar amino acids.
ER retention signal	A sequence of four amino acids (lysine, aspartic acid, glutamic acid, leucine) that is located at the carboxyl terminus of the protein.
Golgi retention signal	A sequence of 20 hydrophobic amino acids that forms a transmembrane domain that is flanked by positively charged residues.
Lysosomal-targeting signal	A patch of amino acids within the polypeptide sequence. Positively charged residues within this patch are thought to play an important role. This patch causes lysosomal proteins to be covalently modified to contain a mannose-6-phosphate that directs the protein to the lysosome.

Destination of a Cellular Protein	Type of Signal the Protein Contains Within Its Amino Acid Sequence
Cytosol	No signal required.
Mitochondrion	Mitochondrial-targeting signal.
Nuclear	Nuclear-localization signal.
Peroxisome	Peroxisomal-targeting signal.
ER	SRP signal and an ER retention signal.
Golgi	SRP signal and a Golgi retention signal.
Lysosome	SRP signal and a lysosomal-targeting signal.
Plasma membrane	SRP signal and the protein contains a hydrophobic transmembrane domain that anchors it in the membrane.
Secretion	SRP signal. No additional signal required.

CONCEPTUAL SUMMARY

An appreciation for the molecular characteristics of gene structure and function is central to our understanding of genetics. Early ideas of Garrod and of Beadle and Tatum suggested that there is a link between genes and the production of **enzymes.** Since their time, the molecular nature of that link has become clear. Segments of DNA within the chromosomes are organized into units called genes. Most genes are **structural genes,** which means that they provide the code for the amino acid sequence of a polypeptide. The function of most genes, therefore, is to store the information for the synthesis of a particular polypeptide—ultimately, for the synthesis of a functional protein. It is the action of proteins within living cells and organisms that directly determines the outcome of an organism's traits.

During gene expression in eukaryotes, DNA is transcribed into a pre-mRNA, which is subsequently modified to form mature mRNA that is **translated** into an amino acid sequence. The translation process employs a **genetic code,** which is nearly **universal** in all species. The use of this code involves the recognition of three-base **codons** within the mRNA by complementary **anti-codons** in tRNA molecules. The tRNAs carry the correct amino acid, allowing the synthesis of a polypeptide with a defined amino acid sequence. This sequence is known as a polypeptide's **primary structure.** The sequence and chemical structures of the amino acid **side chains** within a polypeptide cause it to fold into **secondary** and **tertiary structures.** Some proteins also exhibit a **quaternary structure** if they are composed of two or more **subunits.** The final three-dimensional structure of a protein determines its functional role.

The translation of mRNA into a polypeptide sequence requires many cellular components, including **ribosomes, tRNAs, mRNA,** and small effector molecules. According to the **adaptor hypothesis,** the anticodons in tRNA molecules act as translators of the genetic code within mRNA. Prior to translation, enzymes known as **aminoacyl-tRNA synthetases** recognize tRNAs and attach the correct amino acid to their 3′ end. During translation, the anticodon in a tRNA binds to a codon in the mRNA due to their complementary sequences. Concurrently, the tRNA carries the correct amino acid, which is then added to the growing polypeptide chain. At the **wobble** position, the first base in the anticodon of the tRNA binds to the third base in the codon of the mRNA. In many cases, the wobble position can accommodate more than one type of base. In addition, modified bases in the tRNA can often be recognized by different bases in the mRNA.

The ribosomes are large macromolecular structures that provide a site for translation to occur. Each ribosome is composed of one small and one large subunit; each subunit contains several proteins and rRNAs that assemble together. In eukaryotes, this assembly occurs within the **nucleolus;** the ribosomal sub units are then exported into the cytosol, where translation occurs.

The translation process occurs in three stages, known as **initiation, elongation,** and **termination.** During initiation, the ribosomal subunits, mRNA, and initiator tRNA assemble together. In bacteria, the **ribosomal-binding site** is located a few nucleotides upstream from the start codon. In eukaryotes, the ribosome binds at the 5′ end of the mRNA and then scans along the mRNA until it finds a start codon. After the start codon has been identified, translation can proceed to the elongation phase, in which the polypeptide is made. During elongation, there are three sites on the ribosome that are used for tRNA binding. An appropriate tRNA first binds to the **A site,** where its anticodon

recognizes the codon in the mRNA. A peptide bond then forms between the amino acid attached to the tRNA at the A site and the growing polypeptide chain attached to a tRNA in the **P site.** The tRNA at the P site can then exit the ribosome from the **E site.** Finally, during the termination stage, the stop codon in the mRNA is reached. At this point, termination factors bind to the ribosome and cause the disassembly of the ribosomal subunits, tRNA, mRNA, and completed polypeptide chain.

Some interesting differences are found between bacterial and eukaryotic translation due to cell compartmentalization. Since bacteria contain a single intracellular compartment, translation of mRNA can begin before transcription is completed. This is referred to as coupling between transcription and translation. Compartmentalization in eukaryotic cells, due to the presence of different types of organelles, requires a more complicated system of sorting the proteins into their correct locations. This occurs via the presence of **sorting signals** within the translated polypeptides. The targeting of cytosolic, nuclear, mitochondrial, chloroplast, and peroxisomal proteins occurs **posttranslationally,** whereas proteins destined for the ER, Golgi, lysosomes, plasma membrane, or secretion are initially targeted to the ER in a **cotranslational** manner.

EXPERIMENTAL SUMMARY

The first biochemical insights into the function of genes came from the experimental observations of Garrod. By making the connection between inherited defects in metabolism and the absence of a specific functional enzyme, he proposed that the function of genes is related to the production of enzymes. Later, Beadle and Tatum analyzed mutant genes in *Neurospora* and concluded that a single gene encodes a single enzyme. In chapter 13, we also examined the experimental strategy used by Nirenberg and Ochoa to establish the relationship between the mRNA sequence and the polypeptide sequence. By synthesizing RNAs in vitro with a known nucleotide composition, they were able to correlate the predicted ratio of codons with the observed ratio of amino acids within a polypeptide chain. This helped to decipher the genetic code by determining the relationship between codon composition and the amino acid it specifies. Similarly, the synthesis of RNA copolymers and the triplet binding assay were instrumental in the elucidation of the genetic code.

Our molecular understanding of translation has come from several lines of experimentation, including biochemistry, genetics,

and microscopy. The structure and composition of tRNAs, mRNAs, and ribosomes have been biochemically investigated for many decades. From such studies, we know the structures of tRNAs, mRNAs, and even the ribosomes, which are highly complex. Electron microscopy has aided our elucidation of ribosome structure and has also provided a method to demonstrate that transcription and translation are coupled in bacteria.

The steps of translation have also been investigated at the molecular level. By modifying the amino acid attached to tRNA, Chapeville showed that the anticodon sequence in the tRNA is the determining factor during translation. This experimental observation supported the adaptor hypothesis proposed by Crick. We have also learned a great deal about the translation process by examining genetic sequences. In addition, geneticists have identified the functional roles of sequences such as the Shine-Dalgarno sequence in bacteria, and Kozak's rules in eukaryotes, by analyzing mutations that interfere with translation.

PROBLEM SETS & INSIGHTS

Solved Problems

S1. The first amino acid in a certain bacterial polypeptide chain is methionine. The start codon in the mRNA is GUG, which codes for valine. Why isn't the first amino acid formylmethionine or valine?

Answer: The first amino acid in a polypeptide chain is carried by the initiator tRNA, which always carries formylmethionine. This occurs even when the start codon is GUG (valine) or UUG (leucine). The formyl group can be later removed to yield methionine as the first amino acid.

S2. A tRNA has the anticodon sequence 3′–CAG–5′. What amino acid does it carry?

Answer: Since the anticodon is 3′–CAG–5′, it would be complementary to a codon with the sequence 5′–GUC–3′. According to the genetic code, this codon specifies the amino acid valine. Therefore, this tRNA must carry valine at its acceptor site.

S3. In eukaryotic cells, the assembly of ribosomal subunits occurs in the nucleolus. As discussed in chapter 12, the 5.8S, 18S, and 28S rRNAs are actually synthesized as a single 45S rRNA precursor that is processed by cleavage to produce the three rRNAs. The genes that encode the 45S precursor are found in multiple copies (i.e., they are moderately repetitive). Within the nucleus, the segments of chromosomes that contain the 45S rRNA genes align themselves at the center of the nucleolus. This site is called the **nucleolar-organizing center.** In this region, active transcription of the 45S gene takes place. Briefly explain how the assembly of the ribosomal subunits occurs.

Answer: In the nucleolar-organizing center, the 45S RNA is processed to the 5.8S, 18S, and 28s rRNAs. This processing was described in chapter 12. The other components, 5S rRNA and ribosomal proteins, must also be imported into the nucleolar region. Since proteins are made in the cytosol, they must enter the nucleus through the nuclear pores. When all the components are present, they assemble into 40S and 60S ribosomal subunits. Following assembly, the ribosomal subunits exit the nucleus through the nuclear pores and enter the cytosol.

S4. Throughout chapter 13, we have seen that the general mechanism for bacterial and eukaryotic translation is very similar at the molecular level. In both cases, a polypeptide with a defined amino acid sequence is made on the surface of a ribosome due to the sequential interactions between the anticodons in tRNA molecules and the codons in an mRNA. In addition, we have also seen some interesting distinctions between these processes in bacteria and eukaryotes. Summarize the differences between bacterial and eukaryotic translation.

Answer: See table.

A Comparison of Bacterial and Eukaryotic Translation

	Bacterial	**Eukaryotic**

Ribosome composition:	70S ribosomes:
	30S subunit—
	  21 proteins + 1 rRNA
	50S subunit—
	  34 proteins + 3 rRNAs
		80S ribosomes:
		40S subunit—
		  33 proteins + 1 rRNA
		60S subunit—
		  49 proteins + 3 rRNAs

Initiator tRNA:	tRNA^{fmet}	tRNA^{met}
Formation of the initiation complex:	Requires IF1, IF2, and IF3	Requires more initiation factors compared to bacterial initiation
Initial binding of mRNA to the ribosome:	Requires a Shine-Dalgarno sequence	Requires a 7-methylguanosine cap
Selection of a start codon:	AUG, GUG, or UUG located just downstream from the Shine-Dalgarno sequence	According to Kozak's rules
Termination:	Requires RF1, RF2, and RF3	Requires eRF
Coupled to transcription:	Yes	Not usually
Coordinated with cell compartmentalization:	Less complex, due to a lack of internal organelles	Yes, via many different sorting signals such as the SRP signal

S5. An **antibiotic** is a drug that kills or inhibits the growth of microorganisms. The use of antibiotics has been of paramount importance in the battle against many infectious diseases that are caused by microorganisms. In the case of many antibiotics, their mode of action is to inhibit the translation process within bacterial cells. Certain antibiotics selectively bind to bacterial (70S) ribosomes but do not inhibit eukaryotic (80S) ribosomes. Their ability to inhibit translation can occur at different steps in the translation process. For example, tetracycline prevents the attachment of tRNA to the ribosome while erythromycin inhibits the translocation of the ribosome along the mRNA. Why would an antibiotic bind to a bacterial ribosome but not to a eukaryotic ribosome? How does inhibition of translation by antibiotics such as tetracycline prevent bacterial growth?

Answer: Because bacterial ribosomes have a different protein and rRNA composition compared to eukaryotic ribosomes, certain drugs can recognize these different components and bind specifically to bacterial ribosomes and thereby interfere with the process of translation. In other words, the surface of a bacterial ribosome must be somewhat different compared to the surface of a eukaryotic ribosome so that certain drugs will bind to the surface of only bacterial ribosomes. If a bacterial cell is exposed to tetracycline or other antibiotics, it cannot synthesize new polypeptides. Since polypeptides form functional proteins that are needed for processes such as cell division, the bacterium will be unable to grow and proliferate.

Conceptual Questions

C1. An mRNA has the following sequence:

5′-GGCGAUGGGCAAUAAACCGGGCCAGUAAGC-3′

Identify the start codon and determine the complete amino acid sequence that would be translated from this mRNA.

C2. What does it mean when we say that the genetic code is degenerate? Discuss the universality of the genetic code.

C3. According to the adaptor hypothesis, are the following statements true or false?

A. The sequence of anticodons in tRNA directly recognizes codon sequences in mRNA, with some room for wobble.

B. The amino acid attached to the tRNA directly recognizes codon sequences in mRNA.

C. The amino acid attached to the tRNA affects the binding of the tRNA to codon sequences in mRNA.

C4. In bacteria, researchers have isolated strains that carry mutations within tRNA genes. These mutations can change the sequence of the anticodon. For example, a normal tRNA^{trp} gene would encode a tRNA with the anticodon 3′–ACC–5′. A mutation could change this sequence to 3′–CCC–5′. When this mutation occurs, the tRNA still carries a tryptophan at its 3′ acceptor site, even though the anticodon sequence has been altered.

A. How would this mutation affect the translation of polypeptides within the bacterium?

B. What does this mutation tell you about the recognition between tryptophanyl-tRNA synthetase and tRNA^{trp}? Does the enzyme primarily recognize the anticodon or not?

C5. The covalent attachment of an amino acid to a tRNA is an endergonic reaction. In other words, it requires an input of energy for the reaction to proceed. Where does the energy come from to attach amino acids to tRNA molecules?

C6. The wobble rules for tRNA-mRNA pairing are shown in figure 13.12. If we assume that the tRNAs do not contain modified bases (e.g., inosine), what is the minimum number of tRNAs that are needed to efficiently recognize the codons for the following types of amino acids?

A. Leucine

B. Methionine

C. Serine

C7. How many different sequences of mRNA could encode a peptide with the sequence proline glycine methionine serine?

C8. If a tRNA molecule carries a glutamic acid, what are the two possible anticodon sequences that it could contain? Be specific about the 5′ and 3′ ends.

C9. A tRNA has an anticodon sequence 3′–GGU–5′. What amino acid does it carry?

C10. If a tRNA has an anticodon sequence 3′–CCI–5′, what codon(s) can it recognize?

C11. Describe the anticodon of a single tRNA that could recognize the codons 5′–AAC–3′ and 5′–AAU–3′. How would this tRNA need to be modified for it to also recognize 5′–AAA–3′?

C12. Describe the structural features that all tRNA molecules have in common.

C13. In the tertiary structure of tRNA, where is the anticodon region relative to the attachment site for the amino acid? Are they adjacent to each other?

C14. What is an aminoacyl-tRNA synthetase? Why has the function of this type of enzyme been described as the second genetic code?

C15. What is an activated amino acid?

C16. Discuss the significance of modified bases within tRNA molecules.

C17. How and when does formylmethionine become attached to the initiator tRNA in bacteria?

C18. Is it necessary for a cell to make 61 different tRNA molecules, corresponding to the 61 codons for amino acids? Explain your answer.

C19. List the components required for translation. Describe the relative sizes of these different components. In other words, which components are small molecules, macromolecules, or assemblies of macromolecules?

C20. Describe the components of eukaryotic ribosomal subunits and where the assembly of the subunits occurs within living cells.

C21. The term *subunit* can be used in a variety of ways. Compare the use of the term *subunit in proteins* versus *ribosomal subunit*.

C22. Do the following events during bacterial translation occur primarily within the 30S subunit, within the 50S subunit, or at the interface between these two ribosomal subunits?

A. mRNA-tRNA recognition

B. Peptidyl transferase reaction

C. Exit of the polypeptide chain from the ribosome

D. Binding of initiation factors IF1, IF2, and IF3

C23. What are the three stages of translation? Discuss the main events that occur during these three stages.

C24. Describe the sequence in bacterial mRNA that promotes recognition by the 30S subunit.

C25. For each of the following initiation factors, how would eukaryotic initiation of translation be affected if it were missing?

A. eIF2

B. eIF4A

C. eIF5

C26. How does a eukaryotic ribosome select its start codon? Describe the sequences in eukaryotic mRNA that provide an optimal context for a start codon.

C27. For each of the following sequences, rank them in order (from best to worst) as sequences that could be used to initiate translation according to Kozak's rules.

GACGCCAUGG
GCCUCCAUGC
GCCAUCAAGG
GCCACCAUGG

C28. Explain the functional roles of the A, P, and E sites during translation.

C29. An mRNA has the following sequence: 5′–AUG UAC UAU GGG GCG UAA–3′. Draw the primary structure that would be encoded by this mRNA. Be specific about the amino and carboxyl terminal ends.

C30. For eukaryotic proteins, what is the function of a sorting signal? Describe an example.

C31. What is the difference between posttranslational and cotranslational protein sorting? What cellular components are required for cotranslational sorting?

C32. Which steps during the translation of bacterial mRNA involve an interaction between complementary strands of RNA?

C33. What is the function of the nucleolus?

C34. In which of the ribosomal sites, the A site, P site, and/or E site, could the following be found?

A. A tRNA without an amino acid attached

B. A tRNA with a polypeptide attached

C. A tRNA with a single amino acid attached

C35. What is a polysome?

C36. According to the drawing in figure 13.18, explain why the ribosome translocates along the mRNA in a 5′ to 3′ direction rather than a 3′ to 5′ direction.

C37. The lactose permease of *E. coli* is a protein composed of a single polypeptide that is 417 amino acids in length. By convention, the amino acids within a polypeptide are numbered from the amino terminus to the carboxyl terminus. Are the following questions about the lactose permease true or false?

A. Since the sixty-fourth amino acid is glycine and the sixty-eighth amino acid is aspartic acid, the codon for glycine-64 is closer to the 3′ end of the mRNA compared to the codon for aspartic acid-68.

B. The mRNA that encodes the lactose permease must be greater than 1,241 nucleotides in length.

C38. An mRNA encodes a polypeptide that is 312 amino acids in length. The fifty-third codon in this polypeptide is a tryptophan codon. A mutation in the gene that encodes this polypeptide changes this tryptophan codon into a stop codon. How many amino acids would be in the resulting polypeptide: 52, 53, 259, or 260?

C39. Explain what is meant by the *coupling of transcription and translation* in bacteria. Does coupling occur in bacterial and/or eukaryotic cells? Explain.

Experimental Questions

E1. In the experiment of figure 13.3, what would be the predicted amounts of amino acids incorporated into polypeptides if the RNA was a random polymer containing 50% C and 50% G?

E2. With regard to the experiment described in figure 13.9, answer the following questions:

A. Why was a polyUG mRNA template used?

B. Would you radiolabel the cysteine with the isotope ^{14}C or ^{35}S? Explain your choice.

C. What would be the expected results if the experiment was followed in the same way except that a polyGC template was used? Note: A polyGC template could contain two different alanine codons (GCC and GCG), but it could not contain any cysteine codons.

E3. An experimenter has a chemical reagent that modifies threonine to another amino acid. Following the protocol described in figure 13.9, an mRNA is made that is composed of 50% C and 50% A. The amino acid composition of the resultant polypeptides is 12.5% lysine, 12.5% asparagine, 25% serine, 12.5% glutamine, 12.5% histidine, and 25% proline. One of the amino acids present in this polypeptide is due to the modification of threonine. Which amino acid is it? Based on the structure of the amino acid side chains, explain how the structure of threonine has been modified.

E4. Polypeptides can be translated in vitro. Would a bacterial mRNA be translated in vitro by eukaryotic ribosomes? Would a eukaryotic mRNA be translated in vitro by bacterial ribosomes? Why or why not?

E5. Discuss how the elucidation of the structure of the ribosome can help us to understand its function.

E6. Figure 13.21 shows an electron micrograph of a bacterial gene as it is being transcribed and translated. In this figure, label the 5′ and 3′ ends of the DNA and RNA strands. Place an arrow where you think the start codons are found in the mRNA transcripts.

E7. Chapter 18 describes a blotting method known as Western blotting that can be used to detect the production of a polypeptide that is translated from a particular mRNA. In this method, a protein is detected with an antibody that specifically recognizes and binds to its amino acid sequence. The antibody acts as a probe to detect the presence of the protein. In this method, a mixture of cellular proteins is separated using gel electrophoresis according to their molecular masses. After the antibody has bound to the protein of interest within a blot of a gel, the protein is visualized as a dark band. For example, an antibody that recognizes the β-globin polypeptide could be used to specifically detect the β-globin polypeptide in a blot. As shown here, the method of Western blotting can be used to determine the amount and relative size of a particular protein that is produced in a given cell type.

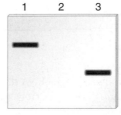

Western blot

Lane 1 is a sample of proteins isolated from normal red blood cells.

Lane 2 is a sample of proteins isolated from kidney cells. Kidney cells do not produce β-globin.

Lane 3 is a sample of proteins isolated from red blood cells from a patient with β-thalassemia. This patient is homozygous for a mutation that results in the shortening of the β-globin polypeptide.

Now here is the question. A protein called troponin contains 334 amino acids. Since each amino acid weighs 120 Daltons (on average), the molecular mass of this protein is about 40,000 Daltons, or 40 kDa. Troponin functions in muscle cells, and it is not expressed in nerve cells. Draw the expected results of a Western blot for the following samples:

Lane 1: proteins isolated from muscle cells

Lane 2: proteins isolated from nerve cells

Lane 3: proteins isolated from the muscle cells of an individual who is homozygous for a mutation that introduces a stop codon at codon 177

E8. A blotting method known as Western blotting can be used to detect proteins that are translated from a particular mRNA. This method is described in chapter 18 and also in experimental question E7. Let's suppose a researcher was interested in the effects of mutations on the expression of a structural gene that encodes a protein that we will call protein X. This protein is expressed in skin cells and contains 572 amino acids. Its molecular mass is approximately 68,600 Daltons, or 68.6 kDa. Make a drawing that shows the expected results of a Western blot using proteins isolated from the skin cells that were obtained from the following individuals:

Lane 1: a normal individual

Lane 2: an individual who is homozygous for a deletion, which removes the promoter for this gene

Lane 3: an individual who is heterozygous in which one gene is normal and the other gene had a mutation that introduces an early stop codon at codon 421

Lane 4: an individual who is homozygous for a mutation that introduces an early stop codon at codon 421

Lane 5: an individual who is homozygous for a mutation that changes codon 198 from a valine codon into a leucine codon

E9. The protein known as tyrosinase is needed to make certain types of pigments. Tyrosinase is composed of a single polypeptide with 511 amino acids. Since each amino acid weighs 120 Daltons (on average), the molecular mass of this protein is approximately 61,300 Daltons, or 61.3 kDa. People who carry two defective copies of the tyrosinase gene have the condition known as albinism. They are unable to make pigment in the skin, eyes, and hair. A blotting method known as Western blotting can be used to detect proteins that are translated from a particular mRNA. This method is described in chapter 18 and also in experimental question E7. Skin samples were collected from a normal individual (lane 1), and from three unrelated albino individuals (lanes 2, 3, and 4), and subjected to a Western blot analysis using an antibody that recognizes tyrosinase. Explain the possible cause of albinism in these three individuals.

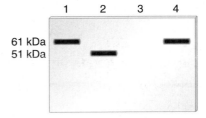

E10. Although there are 61 codons that specify the 20 amino acids, most species display a codon bias. This means that certain codons are used much more frequently than other codons. For example, UUA, UUG, CUU, CUC, CUA, and CUG all specify leucine. In yeast, however, the UUG codon is used to specify leucine approximately 80% of the time.

A. The experiment of figure 13.3 shows the use of an in vitro translation system. In this experiment, the RNA, which was used for translation, was chemically synthesized. Instead of using a chemically synthesized RNA, researchers can isolate mRNA from living cells and then add the mRNA to the in vitro translation system. If a researcher isolated mRNA from kangaroo cells and then added it to an in vitro translation system that came from yeast cells, how might the phenomenon of codon bias affect the production of proteins?

B. Discuss potential advantages and disadvantages of codon bias for translation.

Questions for Student Discussion/Collaboration

1. Discuss why you think the ribosome needs to contain so many proteins and rRNA molecules. Does it seem like a waste of cellular energy to make so large a structure so that translation can occur?

2. Discuss and make a list of the similarities and differences in the events that occur during the initiation, elongation, and termination stages of transcription (see chapter 12) and translation (chapter 13).

3. Which events during translation involve molecular recognition of a nucleotide base sequence within RNA? Which events involve recognition between different protein molecules?

Note: All answers appear at the website for this textbook; the answers to even-numbered questions are in the back of the textbook.

Visit the Online Learning Center for practice tests, answer keys, and other learning aids for this chapter. Enhance your understanding of genetics with our interactive exercises, web links, news feeds, tutorial service, and much more.

GENE REGULATION IN BACTERIA AND BACTERIOPHAGES

14

::

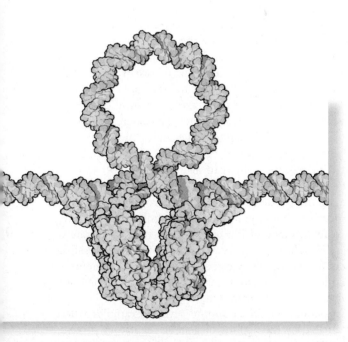

Chromosomes of bacteria, such as *Escherichia coli*, contain a few thousand different genes. Some of these genes are regulated; others are not. The term **gene regulation** means that the level of gene expression can vary under different conditions. By comparison, unregulated genes have essentially constant levels of expression in all conditions over time. These are called **constitutive genes.** Frequently, constitutive genes encode proteins that are always necessary for the survival of the bacterium. In contrast, the majority of genes are **regulated** so that the proteins they encode can be produced at the proper times and in the proper amounts.

The benefit of regulating genes is that the encoded proteins will be produced only when they are required. Therefore, gene regulation allows a cell to avoid wasting its valuable cellular energy making proteins it does not need. From the viewpoint of natural selection, this enables a bacterium to compete as efficiently as possible for a limited amount of available resources. This is particularly important since bacteria find themselves in an environment that is frequently changing with regard to temperature, nutrients, and many other factors. A few common processes that are regulated at the genetic level are listed here.

1. **Metabolism:** Some proteins function in the metabolism of small molecules. For example, certain enzymes are needed for a bacterium to metabolize particular sugars. These enzymes are required only when the bacterium is exposed to such sugars in its environment.

2. **Response to environmental stress:** Certain proteins help a bacterium to survive environmental stress (e.g., osmotic shock, heat shock). These proteins are required only when the bacterium is confronted with the stress.

3. **Cell division:** Some proteins are needed for cell division. These are necessary only when the bacterial cell is getting ready to divide.

The expression of structural genes, which encode distinct polypeptides, ultimately leads to the production of functional cellular proteins. As we have seen in chapters 12 and 13, gene expression is a multistep process that proceeds from transcription to translation, and it may involve posttranslational effects on protein structure and function. As shown in figure 14.1, regulation can occur at any of these steps in the pathway of gene expression. In chapter 14, we will examine the molecular mechanisms that account for these types of gene regulation. We will also consider how gene regulation affects the phenotype of the bacterium.

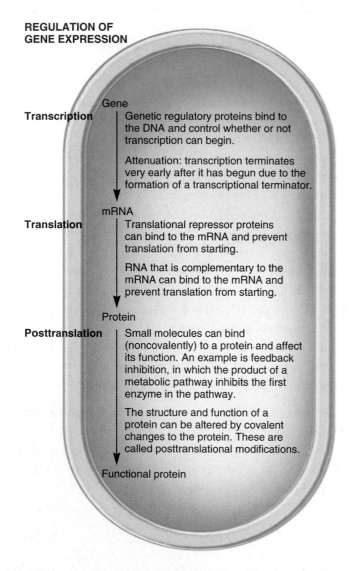

REGULATION OF GENE EXPRESSION

Gene

Transcription Genetic regulatory proteins bind to the DNA and control whether or not transcription can begin.

Attenuation: transcription terminates very early after it has begun due to the formation of a transcriptional terminator.

mRNA

Translation Translational repressor proteins can bind to the mRNA and prevent translation from starting.

RNA that is complementary to the mRNA can bind to the mRNA and prevent translation from starting.

Protein

Posttranslation Small molecules can bind (noncovalently) to a protein and affect its function. An example is feedback inhibition, in which the product of a metabolic pathway inhibits the first enzyme in the pathway.

The structure and function of a protein can be altered by covalent changes to the protein. These are called posttranslational modifications.

Functional protein

FIGURE 14.1   **Common points where regulation of gene expression in bacteria occurs.**

14.1 TRANSCRIPTIONAL REGULATION

In bacteria, the most common way to regulate gene expression is by influencing the rate at which transcription is initiated. Although we frequently refer to genes as being "turned on or off," it is more accurate to say that the level of gene expression is increased or decreased. At the level of transcription, this means that the rate of RNA synthesis can be increased or decreased.

In most cases, transcriptional regulation involves the actions of regulatory proteins that can bind to the DNA and affect the rate of transcription of one or more nearby genes. There are two common types of regulatory proteins. A **repressor** is a regulatory protein that binds to the DNA and inhibits transcription, whereas **activators** are proteins that increase the rate of transcription. The term **negative control** refers to transcriptional regulation by repressor proteins, **positive control** to regulation by activator proteins.

In conjunction with regulatory proteins, small effector molecules often play a critical role in transcriptional regulation. However, small effector molecules do not bind directly to the DNA to alter transcription. Rather, an effector molecule exerts its effects by binding to a regulatory protein such as an activator or repressor. The binding of the effector molecule causes a conformational change in the regulatory protein and thereby influences whether or not the protein can bind to the DNA. As shown in figure 14.2, genetic regulatory proteins that respond to small effector molecules have two functional domains. One domain is a site where the protein binds to the DNA; the other domain is the binding site for the effector molecule.

Small effector molecules are given names that describe how they affect transcription when they are present in the cell at a sufficient concentration to exert their effect. By comparison, regulatory proteins are given names that describe how they affect transcription when they are bound to the DNA (fig. 14.2). An **inducer** is a small effector molecule that causes transcription to increase. An inducer may accomplish this in two ways: it could bind to an activator protein and cause it to bind to the DNA; or it could bind to a repressor protein and prevent it from binding to the DNA. In either case, the transcription rate is increased. Genes that are regulated in this manner are called **inducible.**

Alternatively, the presence of a small effector molecule may inhibit transcription. This can also occur in two ways. A **corepressor** is a small molecule that binds to a repressor protein, thereby causing the protein to bind to the DNA. An **inhibitor** binds to an activator protein and prevents it from binding to the DNA. Both corepressors and inhibitors act to reduce the rate of transcription. Therefore, the genes they regulate are **repressible.** Unfortunately, this terminology can be confusing, since a repressible system could involve an activator protein or an inducible system could involve a repressor protein.

In this section, we will examine several examples where genes are regulated by the actions of genetic regulatory proteins that influence the rate of transcription. We will see how gene regulation provides a way for bacteria to synthesize proteins in an efficient manner.

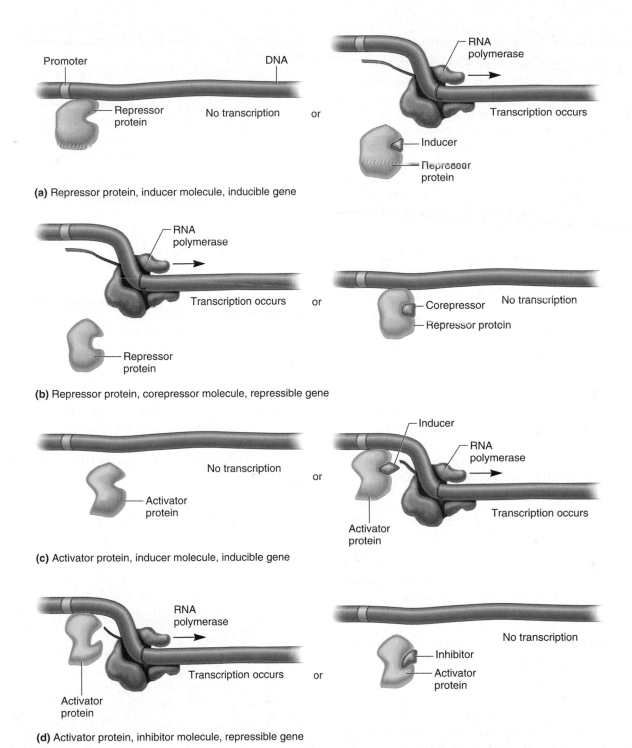

(a) Repressor protein, inducer molecule, inducible gene

(b) Repressor protein, corepressor molecule, repressible gene

(c) Activator protein, inducer molecule, inducible gene

(d) Activator protein, inhibitor molecule, repressible gene

FIGURE 14.2   Binding sites on a genetic regulatory protein. In these examples, a regulatory protein has two binding sites, one for a small effector molecule and one for DNA. The binding of the small effector molecule will change the conformation of the regulatory protein, which alters the DNA-binding-site structure and thereby influences whether the protein can bind to the DNA. **(a)** In the absence of the inducer, a repressor protein shown here blocks transcription. The presence of an inducer causes a conformational change that inhibits the ability of a repressor protein to bind to the DNA. **(b)** In the absence of a corepressor, this repressor protein will not bind to the DNA. Therefore, transcription can occur. When the corepressor is bound to the repressor protein, this causes a conformational change that allows the protein to bind to the DNA and inhibit transcription. **(c)** This activator protein cannot bind to the DNA unless an inducer is present. When the inducer is bound to the activator protein, this enables the activator protein to bind to the DNA and activate transcription. **(d)** Some activator proteins will bind to the DNA without the aid of an effector molecule. The presence of an inhibitor causes a conformational change that releases the activator protein from the DNA. This inhibits transcription.

The Phenomenon of Enzyme Adaptation Is Due to the Synthesis of Cellular Proteins

To a significant extent, our understanding of gene regulation can be traced back to the creative minds of Francöis Jacob and Jacques Monod at the Pasteur Institute in Paris, France. Their research into genes and gene regulation stemmed from an interest in the phenomenon known as **enzyme adaptation,** which had been identified at the turn of the twentieth century. Enzyme adaptation refers to the observation that a particular enzyme appears within a living cell only after the cell has been exposed to the substrate for that enzyme. When a bacterium is not exposed to a particular substance, it does not make the enzymes needed to metabolize that substance.

Jacob and Monod focused their attention on lactose metabolism in *E. coli* to investigate this problem. Key experimental observations that led to an understanding of this genetic system are listed here.

1. The exposure of bacterial cells to lactose increased the levels of lactose-utilizing enzymes by 1,000- to 10,000-fold.
2. Antibody and labeling techniques revealed that the increase in the activity of these enzymes was due to the increased synthesis of the enzymes.
3. The removal of lactose from the environment caused an abrupt termination in the synthesis of the enzymes.
4. Mutations that prevented the synthesis of particular enzymes involved with lactose utilization showed that a separate gene encoded each enzyme.

These critical observations indicated that enzyme adaptation is due to the synthesis of specific cellular proteins in response to lactose in the environment. In the following sections, we will learn how Jacob and Monod discovered that this phenomenon is due to the interactions between genetic regulatory proteins and small effector molecules. In other words, we will see that enzyme adaptation is due to the transcriptional regulation of genes.

The *lac* Operon Encodes Proteins That Are Involved in Lactose Metabolism

In bacteria, it is common for a few structural genes to be arranged together in a regulatory unit that is under the transcriptional control of a single promoter. This arrangement is known as an **operon.** It encodes a **polycistronic mRNA** that contains the coding sequence for two or more **structural genes.** The biological advantage of an operon organization is that it allows a bacterium to coordinately regulate a group of genes that encode proteins with a common functional goal. The key feature of an operon is that the expression of the structural genes occurs as a single unit.

An operon contains several different regions. For transcription to take place, an operon is flanked by a **promoter** to signal the beginning of transcription and a **terminator** to signal the end of transcription. Two or more structural genes are found between these two sequences. To control the ability of RNA polymerase to transcribe an operon, an additional DNA sequence, known as the **operator site,** is present.

Figure 14.3*a* shows the organization of the genes involved with lactose utilization and their transcriptional regulation. There are actually two distinct transcriptional units. The first unit, known as the *lac* operon, contains a promoter (*lacP*), operator (*lacO*), CAP site, and three structural genes, *lacZ, lacY,* and *lacA. LacZ* encodes the enzyme β-galactosidase, which enzymatically cleaves lactose and lactose analogues. As a side reaction, β-galactosidase also converts a small percentage of lactose into allolactose, a structurally similar sugar (fig. 14.3*b*). As we will see later, allolactose acts as a small effector molecule to regulate the *lac* operon. The *lacY* gene encodes lactose permease, a membrane protein required for the active transport of lactose into the cytoplasm of the bacterium. The *lacA* gene encodes a transacetylase enzyme that can covalently modify lactose and lactose analogues. Although the functional necessity of the transacetylase remains unclear, the acetylation of nonmetabolizable lactose analogues may prevent their toxic buildup within the bacterial cytoplasm.

In the vicinity of the *lac* promoter are two regulatory sites, designated the operator and the CAP site. The **operator** is a sequence of nucleotides that provides a binding site for a repressor protein, and the **CAP site** is a DNA sequence recognized by an activator protein called the <u>c</u>atabolite <u>a</u>ctivator <u>p</u>rotein (CAP).

A second transcriptional unit involved in genetic regulation is the *lacI* gene (fig. 14.3*a*). It is not considered a part of the *lac* operon. The *lacI* gene encodes the **lac repressor,** a protein that is important for the regulation of the *lac* operon. The *lac* repressor functions as a tetramer, composed of four identical subunits. The *lacI* gene, which is constitutively expressed at fairly low levels, has its own promoter, the *i* promoter. The amount of *lac* repressor made is approximately 10 tetramer proteins per cell. Only a small amount of the *lac* repressor protein is needed to repress the *lac* operon.

The *lac* Operon Is Regulated by a Repressor Protein

The *lac* operon can be transcriptionally regulated in more than one way. The first method that we will examine is an inducible, negative control mechanism. As shown in figure 14.4, this form of regulation involves the *lac* repressor protein, which binds to the sequence of nucleotides found within the *lac* operator site. Once bound, the *lac* repressor prevents RNA polymerase from recognizing the promoter and transcribing the *lacZ, lacY,* and *lacA* genes (fig. 14.4*a*).

The ability of the *lac* repressor to bind to the operator site depends on whether or not allolactose is bound to it. The repressor protein functions as a tetramer; each of the four subunits has a single binding site for the inducer (namely, allolactose). When allolactose binds to the repressor, a conformational change occurs that prevents the *lac* repressor from binding to the operator site. Under these conditions, RNA polymerase is now free to transcribe the operon (fig. 14.4*b*). In genetic terms, we would say that the operon has been **induced.**

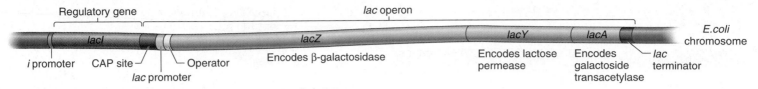

(a) Organization of DNA sequences in the *lac* region of the *E.coli* chromosome

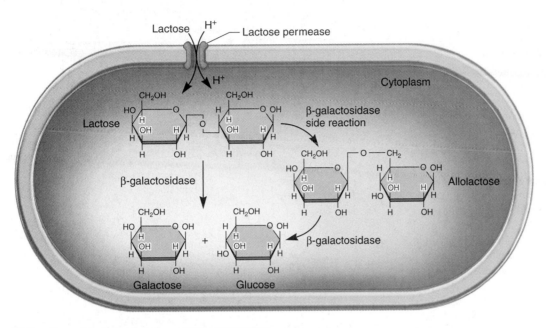

(b) Functions of lactose permease and β-galactosidase

FIGURE 14.3   Organization of the *lac* operon and other genes involved with lactose metabolism in *E. coli*. (a) The CAP site is the binding site for CAP; the operator provides a binding site for the *lac* repressor. The *lac* promoter (*lacP*) is responsible for the transcription of the *lacZ*, *lacY*, and *lacA* genes as a single unit. The *i* promoter is responsible for the transcription of the *lacI* gene. (b) The lactose permease is a membrane protein that allows the uptake of lactose into the bacterial cytoplasm. It cotransports lactose with an H^+. Since bacteria maintain an H^+ gradient across their cytoplasmic membrane, this cotransport permits the active accumulation of lactose against a gradient. β-galactosidase is a cytoplasmic enzyme that cleaves lactose and related compounds into galactose and glucose. As a minor side reaction, β-galactosidase also converts lactose into allolactose. As shown here, allolactose can also be broken down into galactose and glucose.

To better appreciate this form of regulation at the cellular level, let's consider the process as it occurs over time. Figure 14.5 illustrates the effects of external lactose on the regulation of the *lac* operon. In the absence of lactose, no inducer is available to bind to the *lac* repressor. Therefore, the *lac* repressor binds to the operator site and inhibits transcription. In reality, the repressor does not completely inhibit transcription, so that very small amounts of β-galactosidase, lactose permease, and transacetylase are made. However, the levels are far too low for the bacterium to readily use lactose. When the bacterium is exposed to lactose, a small amount can be transported into the cytoplasm via the lactose permease, and β-galactosidase converts some of it to allolactose. As this occurs, the cytoplasmic level of allolactose gradually rises; eventually, allolactose binds to the *lac* repressor. This promotes a conformational change that prevents the repressor from binding to the *lac* operator site and thereby allows transcription of the *lacZ*, *lacY*, and *lacA* genes to occur. Translation of the encoded polypeptides will produce the proteins needed for lactose uptake and metabolism.

To understand how the induction process is shut off in a lactose-depleted environment, let's consider the interaction between allolactose and the *lac* repressor. The *lac* repressor has a measurable affinity for allolactose; it binds reversibly to the repressor protein. The likelihood that allolactose will bind to the repressor depends on the allolactose concentration. During induction of the operon, the concentration of allolactose rises and approaches the affinity for the repressor protein. This makes it likely that allolactose will bind to the *lac* repressor, thereby causing it to be released from the operator site. Later on, however, the bacterial cells metabolize the sugars and thereby lower the concentration of allolactose below its affinity for the repressor. At this point, it is unlikely that the *lac* repressor will have bound allolactose. This enables the *lac* repressor to bind to the operator site; the *lac* operon shuts down when lactose is depleted from the environment. After repression occurs, the proteins encoded by the *lac* operon will be degraded since they have a finite half-life (fig. 14.5).

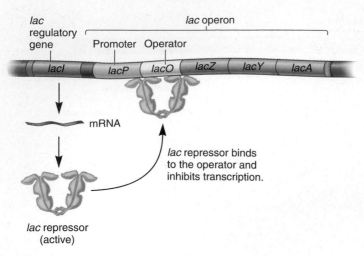

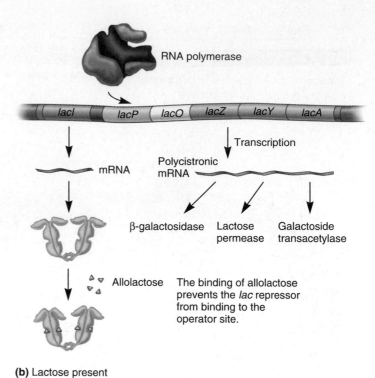

FIGURE 14.4 **Mechanism of induction of the *lac* operon.**
(a) In the absence of the inducer allolactose, the repressor protein is tightly bound to the operator site, thereby inhibiting the ability of RNA polymerase to transcribe the operon. (b) When there is sufficient allolactose, it binds to the repressor. This alters the conformation of the repressor protein in such a way as to prevent it from binding to the operator site. Therefore, RNA polymerase can transcribe the operon. (The CAP site is not shown in this drawing.)

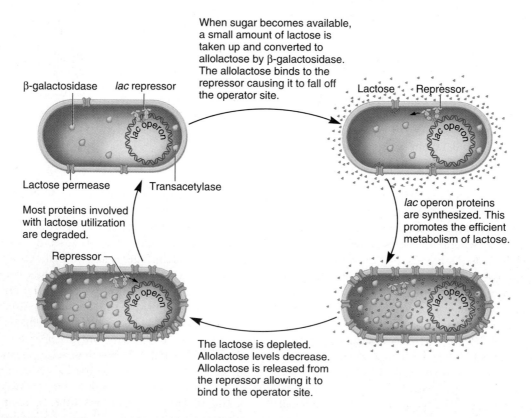

FIGURE 14.5 **The cycle of *lac* operon induction and repression.**
GENES→TRAITS The genes and the genetic regulation of the *lac* operon provide the bacterium with the trait of being able to metabolize lactose when it is present in the environment. When lactose is present, the genes of the *lac* operon are induced, and the bacterial cell can efficiently metabolize this sugar. When lactose is absent, these genes are repressed, so that the bacterium does not waste its energy expressing these genes.

The *lacI* Gene Encodes a Diffusable Repressor Protein

Now that we have an understanding of the *lac* operon, let's consider one of the experimental approaches that was used to elucidate the regulation of the *lac* operon. In the 1950s, Jacob and Monod, and their colleague, Arthur Pardee, had identified a few rare mutant strains of bacteria that had abnormal lactose adaptation. One type of mutant, which involved a defect in the *lacI* gene (designated *lacI⁻*), resulted in the constitutive expression of the *lac* operon even in the absence of lactose. They were able to genetically map *lacI⁻* mutations using conjugation methods described in chapter 6. The *lacI⁻* mutations mapped very close to the *lac* operon.

From these observations, Jacob, Monod, and Pardee proposed two different possible functions of the *lacI* gene as described in figure 14.6. One hypothesis was that the *lacI* gene encodes a repressor protein that binds to the operon and inhibits transcription, which we have already learned is the correct hypothesis. At the time of their work, however, this was not yet known. Alternatively, since the *lacI* gene is physically next to the operon, a second possibility was that the *lacI* gene functions as a site for the binding of a repressor protein produced by some other unknown gene. In other words, a second (incorrect) possibility would be that a *lacI* mutation prevents a repressor protein from binding to a site adjacent to the *lac* operon.

Jacob, Monod, and Pardee applied a genetic approach to distinguish between these two alternative possibilities. Even though bacterial conjugation is described in chapter 6, it is necessary to briefly review this process to understand this experiment. On occasion, bacteria can transfer circular segments of DNA, known as F factors. Sometimes an F factor also carries genes that were originally found within the bacterial chromosome. These types of

F factors are called F′ factors. In their studies, Jacob, Monod, and Pardee identified F′ factors that carried portions of the *lac* operon. These F′ factors can be transferred from one cell to another by bacterial conjugation. A strain of bacteria containing F′ factor genes is called a **merozygote,** or partial diploid.

The production of merozygotes was instrumental in allowing Jacob, Monod, and Pardee to elucidate the function of the *lacI* gene. There are two key points. First, the two *lacI* genes in a merozygote may be different alleles. For example, the *lacI* gene on the chromosome may be a defective *lacI⁻* allele while the *lacI* gene on the F′ factor may be normal. Second, the genes on the F′ factor and the genes on the bacterial chromosome are not physically adjacent to each other. According to the hypothesis shown in figure 14.6 (*left*), the expression of the *lacI* gene on an F′ factor should produce repressor proteins that could diffuse within the cell and eventually bind to the operator site of the *lac* operon located on the chromosome. In contrast, the alternative hypothesis suggests that the *lacI* gene provides a binding site for a repressor protein. In this case, it must be physically adjacent to the *lac* operon to exert its effects.

As shown in the experiment of figure 14.7, the regulation of the *lac* operon was analyzed in a mutant strain that was already known to constitutively express the *lac* operon and compared to that of the corresponding merozygote, which also carried a normal *lacI* gene on an F′ factor. These two strains were grown and then divided into two tubes each. In half of the tubes, the strains were incubated with lactose to determine if lactose was needed to induce the expression of the operon. In the other tubes, lactose was omitted. The cells were lysed by sonication, and then a lactose analogue, β-ONPG, was added. This molecule is colorless, but β-galactosidase cleaves it into a product that has a yellow color. Therefore, the amount of yellow color produced in a given amount of time is a measure of the amount of β-galactosidase that is being expressed from the *lac* operon.

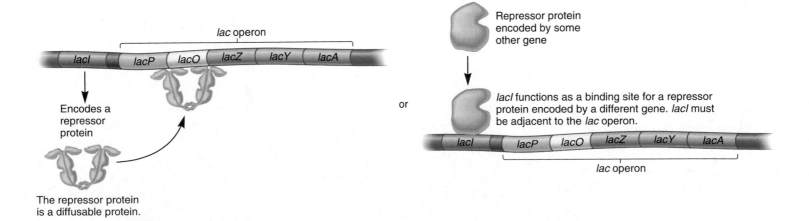

Hypothesis 1

Hypothesis 2

FIGURE 14.6 **Alternative hypotheses to explain how a *lacI* mutation could cause the constitutive expression of the *lac* operon.** The *left* side shows the correct hypothesis. In this case, the *lacI* gene encodes a repressor protein that binds to the operator site to inhibit transcription. The *right* side shows an incorrect hypothesis which suggests that the *lacI* region functions as a binding site for a repressor protein.

■ **THE HYPOTHESIS**

If the *lacI* gene encodes a repressor protein, then the *lacI* gene itself does not have to be physically next to the *lac* operon to repress it; the protein can diffuse throughout the cell and bind to an operator site regardless of the physical location of the *lacI* gene. As an alternative hypothesis, the *lacI* region could act as a binding site for a repressor protein.

■ **TESTING THE HYPOTHESIS** — **FIGURE 14.7** Evidence that the *lacI* gene encodes a diffusable repressor protein.

Starting material: The genotype of the mutant strain was *lacI⁻ lacZ⁺ lacY⁺ lacA⁺*. The merozygote strain had an F′ factor that was *lacI⁺ lacZ⁺ lacY⁺ lacA⁺*, which had been introduced into the mutant strain via conjugation.

Experimental level

Conceptual level

1. Grow mutant strain and merozygote strain separately.

2. Divide each strain into two tubes.

3. In one of the two tubes, add lactose.

4. Incubate the cells long enough to allow *lac* operon induction.

5. Burst the cells with a sonicator. This allows β-galactosidase to escape from the cells.

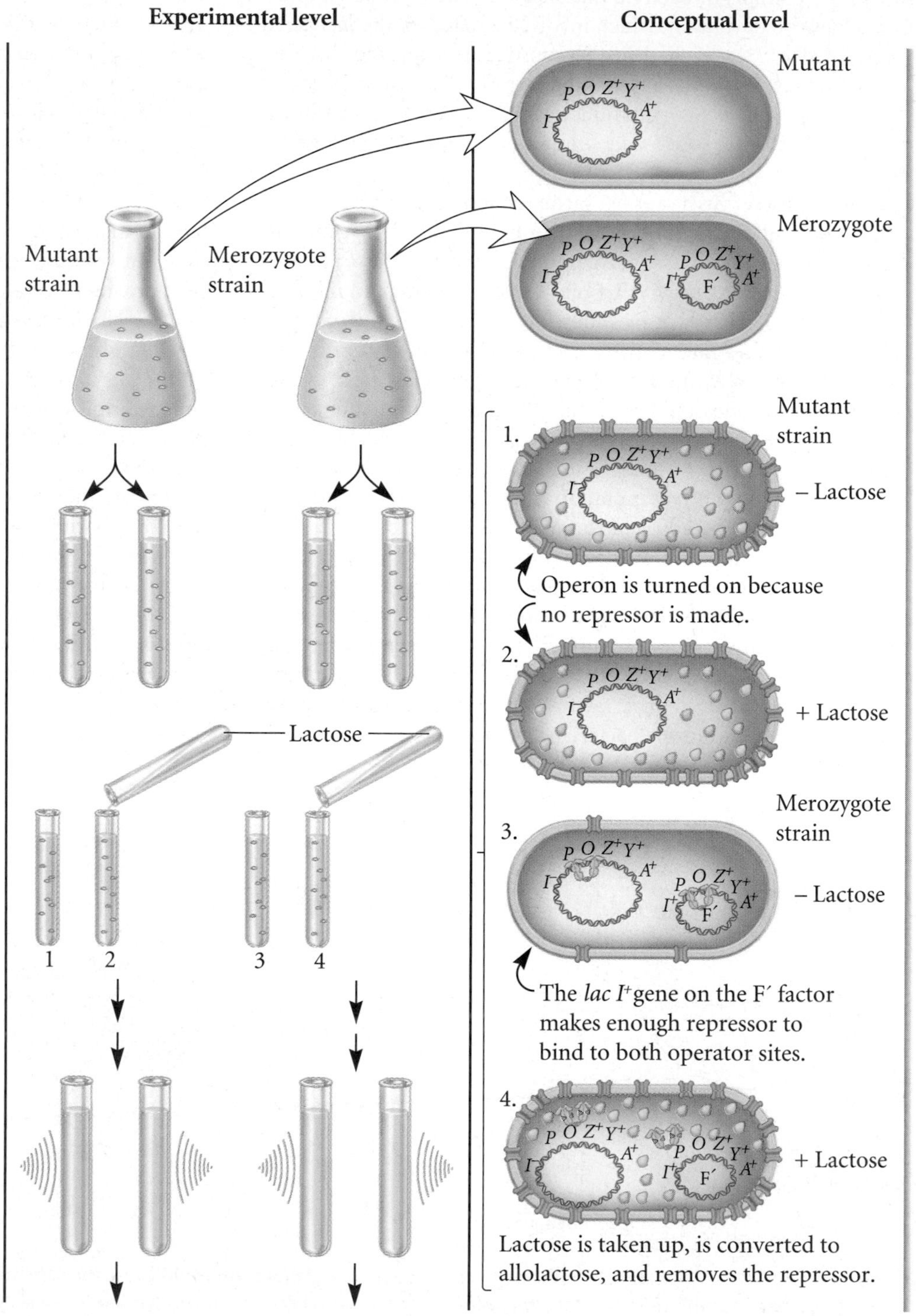

Operon is turned on because no repressor is made.

The *lac I⁺* gene on the F′ factor makes enough repressor to bind to both operator sites.

Lactose is taken up, is converted to allolactose, and removes the repressor.

(continued)

6. Add β-o-nitrophenylgalactoside (β-ONPG). This is a colorless compound. β-galactosidase will cleave the compound to produce galactose and o-nitrophenol (O-NP). O-nitrophenol has a yellow color. The deeper the yellow color, the more β-galactosidase was produced.

β-o-nitrophenyl-galactoside

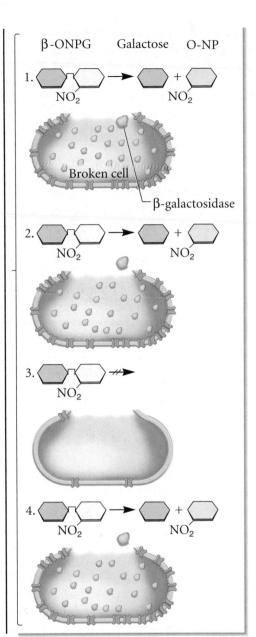

7. Incubate the sonicated cells to allow β-galactosidase time to cleave β-o-nitrophenylgalactoside.

1   2   3   4

8. Measure the yellow color produced with a spectrophotometer. (See the appendix for a description of spectrophotometry.)

THE DATA

Strain	Addition of Lactose	Amount of β-Galactosidase (percentage of parent strain)
Mutant	No	100%
Mutant	Yes	100%
Merozygote	No	<1%
Merozygote	Yes	220%

INTERPRETING THE DATA

As seen in the data, the yellow production in the original mutant strain was the same in the presence or absence of lactose. This is expected, since the expression of β-galactosidase in the *lacI⁻* mutant strain was already known to be constitutive. In other words, the presence of lactose was no longer needed to induce the operon due to a defective *lacI* gene. In the merozygote, however, a different result was obtained. In the absence of lactose, the *lac* operons were repressed—even the operon on the bacterial chromosome. Because the normal *lacI* gene on the F′ factor was not physically located next to the chromosomal *lac* operon, this result is consistent with the hypothesis that the *lacI* gene codes for a repressor protein that can diffuse throughout the cell and bind to any *lac* operon. In the presence of lactose, both operons (i.e, the one on the chromosome and the one on the F′ factor) were induced, yielding a higher level of β-galactosidase activity in the merozygote.

As mentioned in chapter 12, the interactions between regulatory proteins and DNA sequences illustrated in this experiment have led to the definition of two genetic terms. A ***trans*-effect** is a form of genetic regulation that can occur even though two DNA segments are not physically adjacent. The action of the *lac* repressor on the *lac* operon is a *trans*-effect. In contrast, a **cis-effect** or a **cis-acting element** is a DNA segment that must be adjacent to the gene(s) that it regulates. The *lac* operator site is an example of a *cis*-acting element. A *trans*-effect is mediated by genes that encode regulatory proteins, whereas a *cis*-effect is mediated by DNA sequences that are bound by regulatory proteins.

Table 14.1 summarizes the effects of mutations based on their locations in the *lacI* regulatory gene versus *lacO* and their analysis in merozygotes. As seen here, a loss-of-function mutation in a gene encoding a repressor protein has the same effect as a mutation in an operator site that cannot bind a repressor protein. In both cases, the genes of the *lac* operon are constitutively expressed. In a merozygote, however, the results are quite different. When a normal *lacI* gene is introduced into a cell harboring a defective *lacI* gene, the normal gene can regulate both operons. In contrast, when a normal operator site is introduced into a cell with a defective operator site, the operon with the defective operator site continues to be constitutively expressed. Overall, a mutation in a *trans*-acting factor can be complemented by the introduction of a second gene with normal function. However, a

TABLE 14.1

A Comparison of Loss-of-Function Mutations in the *lacI* Gene Versus the Operator Site

Chromosome	F' factor	Expression of the *lac* Operon	
		With lactose	Without lactose
Wild type	None	100%	<1%
lacI⁻	None	100%	100%
lacO⁻	None	100%	100%
lacI⁻	*lacI⁺* and a normal *lac* operon	200%	<1%
lacO⁻	*lacI⁺* and a normal *lac* operon	200%	100%

mutation in a *cis*-acting element is not affected by the introduction of another *cis*-acting element into the cell.

A self-help quiz involving this experiment can be found at the Online Learning Center.

The *lac* Operon Is Also Regulated by an Activator Protein

The *lac* operon can be transcriptionally regulated in a second way, known as **catabolite repression** (as we shall see, a somewhat imprecise term). This form of transcriptional regulation is influenced by the presence of glucose, which is a catabolite (it is broken down—i.e., catabolized—inside the cell). The presence of glucose ultimately leads to repression of the *lac* operon. Catabolite repression enables the bacterium to use two sugars more efficiently. When exposed to both glucose and lactose, *E. coli* cells first use glucose, and catabolite repression prevents the use of lactose. This is advantageous to the bacterium because it does not have to express all the genes that are necessary for both glucose and lactose metabolism. If the glucose is used up, however, catabolite repression is alleviated and the bacterium then expresses the *lac* operon. The sequential use of two sugars by a bacterium is known as **diauxic growth.** It is a common phenomenon among many bacterial species. It is typical that glucose is metabolized first, and then a second sugar is metabolized only after glucose has been depleted from the environment.

Glucose, however, is not itself the small effector molecule that binds directly to a genetic regulatory protein. Instead, this form of regulation involves a small effector molecule, **cyclic AMP (cAMP),** that is produced from ATP via an enzyme known as adenylyl cyclase. cAMP binds to an activator protein called the catabolite activator protein (CAP) and also known as the cAMP receptor protein (CRP).

As shown in figure 14.8, the cAMP-CAP complex is an example of genetic regulation that is inducible and under positive control. When cAMP binds to CAP, the cAMP-CAP complex binds to the CAP site near the *lac* promoter and increases the rate of transcription. The term *catabolite repression* may seem puzzling, since this regulation involves the action of an inducer and activator protein rather than a repressor. The term was coined before the action of the cAMP-CAP complex was understood. At that time, the primary observation was that glucose (a catabolite) inhibited (repressed) lactose metabolism. We now know that when a bacterium is exposed to glucose, the transport of glucose into the cell stimulates a signaling pathway that causes the intracellular concentration of cAMP to decrease because it inhibits adenylyl cyclase, an enzyme needed for cAMP synthesis. When this happens, cAMP is no longer available to bind to the CAP. Therefore, the unoccupied CAP does not bind to the CAP site, causing the transcription rate to decrease.

Many bacterial promoters that transcribe genes involved in the breakdown of other sugars (e.g., maltose, arabinose, and melibiose) also have binding sites for CAP. Therefore, when glucose levels are high, these operons are inhibited, thereby promoting diauxic growth.

Further Studies Have Revealed That the *lac* Operon Has Three Operator Sites for the *lac* Repressor

Our traditional view of the regulation of the *lac* operon has been modified as we have gained a greater molecular understanding of

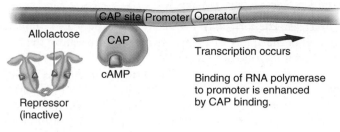

(a) cAMP and lactose

Allolactose

CAP site | Promoter | Operator

CAP

cAMP

Transcription occurs

Binding of RNA polymerase to promoter is enhanced by CAP binding.

Repressor (inactive)

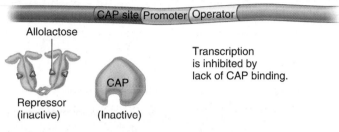

(b) Lactose but no cAMP

Allolactose

CAP site | Promoter | Operator

Transcription is inhibited by lack of CAP binding.

CAP

Repressor (inactive)   (Inactive)

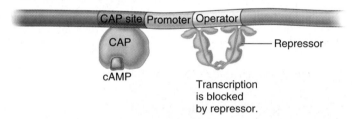

(c) cAMP but no lactose

CAP site | Promoter | Operator

CAP

cAMP

Repressor

Transcription is blocked by repressor.

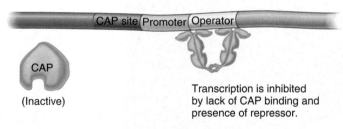

(d) No cAMP and no lactose

CAP site | Promoter | Operator

CAP

(Inactive)

Transcription is inhibited by lack of CAP binding and presence of repressor.

FIGURE 14.8 **The roles of the *lac* repressor and catabolite activator protein (CAP) in the regulation of the *lac* operon.** Under conditions of low extracellular glucose, cAMP is made. This small effector molecule binds to CAP, causing it to change its conformation and bind to the region adjacent to the *lac* promoter. The binding of the CAP-cAMP complex to this region facilitates the binding of RNA polymerase to the promoter. This leads to an enhanced rate of transcription of the *lac* operon. By comparison, the *lac* repressor protein binds to the operator and inhibits the operon when lactose levels are low. This figure illustrates how the *lac* operon will be regulated depending on the levels of cAMP and lactose within the cell.

GENES→TRAITS The mechanism of catabolite repression provides the bacterium with the trait of being able to choose between two sugars. When exposed to both glucose and lactose, the bacterium chooses glucose first. After the glucose is used up, it then expresses the genes that are necessary for lactose metabolism. This trait allows the bacterium to more efficiently use sugars from its environment.

the process. In particular, detailed genetic and crystallographic studies have shown that the binding of the *lac* repressor is more complex than originally realized. The site in the *lac* operon that is commonly called the operator site was first identified by mutations that prevented *lac* repressor binding. These mutations, called *lacO^C* mutants, resulted in the constitutive expression of the *lac* operon even in strains that make a normal *lac* repressor protein. *LacO^C* mutations were localized in the *lac* operator site (which is now known as O_1). This led to the simple view that a single operator site was bound by the *lac* repressor to inhibit transcription (as in figure 14.4).

In the late 1970s and 1980s, two additional operator sites were identified. As shown at the top of figure 14.9, these sites are called O_2 and O_3. O_1 is the operator site that is next to the promoter. O_2 is located downstream in the *lacZ* coding sequence, and O_3 is located slightly upstream from the promoter. The O_2 and O_3 operators were initially called pseudo-operators, because the absence of either one of them only decreased repression by two- to threefold. However, studies by Benno Müller-Hill and his colleagues revealed a surprising result. As shown in figure 14.9, if both O_2 and O_3 are missing, repression is dramatically reduced even when O_1 is present. When O_1 is missing, repression is nearly abolished.

The results of figure 14.9 supported the hypothesis that the *lac* repressor must bind to two out of three operators to cause repression. According to this view, the *lac* repressor can readily

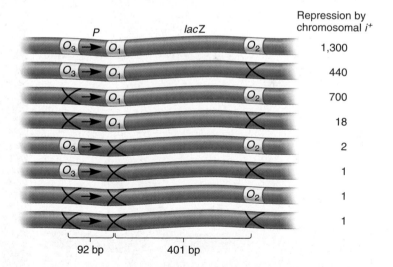

FIGURE 14.9 **The identification of three *lac* operator sites.** The *top* of this figure shows the locations of three *lac* operator sites, designated O_1, O_2, and O_3. O_1 is analogous to the *lac* operator shown in previous figures. The *arrows* depict the starting site for transcription. Defective operator sites are indicated with an X. When all three operators are present, the repression of the *lac* operon is 1,300-fold; this means there is 1/1,300 the level of expression than there is when lactose is present. This figure also shows the amount of repression when one or more operator sites are removed. A repression value of 1.0 indicates that no repression is occurring. In other words, a value of 1.0 indicates constitutive expression.

bind to O_1 and O_2, or to O_1 and O_3, but not to O_2 and O_3. If either O_2 or O_3 were missing, maximal repression is not achieved because it is less likely for the repressor to bind to two operator sites. If you take a look at figure 14.9, you will notice that the operator sites are a fair distance away from each other. For this reason, it was proposed from these studies that the binding of the *lac* repressor to two operator sites requires the DNA to form a loop. A loop in the DNA could bring the operator sites closer together and thereby facilitate the binding of the repressor protein.

In 1996, the proposal that the *lac* repressor binds to two operator sites was confirmed by studies in which the *lac* repressor was crystallized by Mitchell Lewis and his colleagues. The crystal structure of the *lac* repressor has provided exciting insights into its mechanism of action. As mentioned earlier in this chapter, the *lac* repressor is a tetramer of four identical subunits. The crystal structure revealed that each dimer within the tetramer recognizes one operator site. Figure 14.10b is a molecular model illustrating the binding of the *lac* repressor to the O_1 and O_3 sites. The amino acid side chains in the protein interact directly with bases in the major groove of the DNA helix. This is how genetic regulatory proteins recognize specific DNA sequences. Each dimer within the tetramer recognizes a single operator site. The association of two dimers to form a tetramer requires that the two operator

sites are close to each other. For this to occur, a loop must form in the DNA. The formation of this loop is expected to dramatically inhibit the ability of RNA polymerase to recognize the promoter and transcribe this region.

Figure 14.10b also shows the binding of the cAMP-CAP complex to the CAP site (i.e., see the protein within the loop). A particularly striking observation is that the binding of the cAMP-CAP complex to the DNA causes a 90° bend in the DNA structure. When the repressor is active (i.e., not bound to allolactose), the cAMP-CAP complex facilitates the binding of the *lac* repressor to the O_1 and O_3 sites. When the repressor is inactive, this bending also appears to be important in the ability of RNA polymerase to initiate transcription slightly downstream from the bend.

The *ara* Operon Can Be Regulated Positively or Negatively by the Same Regulatory Protein

Now that we have considered the regulation of the *lac* operon, let's compare its regulation to that of other genes in the bacterial chromosome. Another operon in *E. coli* involved in sugar metabolism is the *ara* (arabinose) operon. The sugar arabinose is a constituent of the cell walls of a few types of plants. As shown in fig-

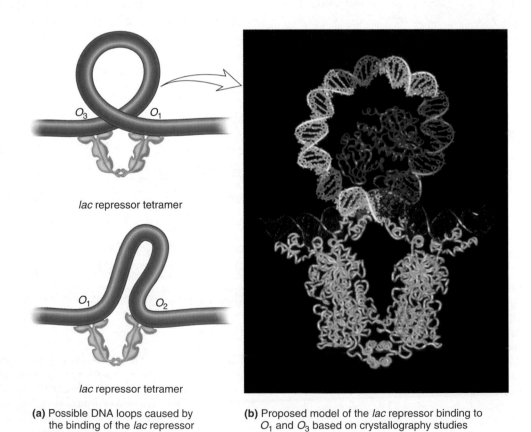

(a) Possible DNA loops caused by the binding of the *lac* repressor

(b) Proposed model of the *lac* repressor binding to O_1 and O_3 based on crystallography studies

FIGURE 14.10   The binding and structure of the *lac* repressor. (a) The binding of the *lac* repressor protein to the O_1 and O_2 or to the O_1 and O_3 operator sites. (b) A molecular model for the binding of the *lac* repressor to O_1 and O_3. Each repressor dimer binds to one operator site, so that the repressor tetramer brings the two operator sites together. This causes the formation of a DNA loop in the intervening region. Note that the DNA loop contains the -35 and -10 regions, which are recognized by the sigma factor of RNA polymerase. This loop also contains the binding site for the CAP-cAMP complex, which is shown here.

ure 14.11, the *ara* operon contains three structural genes, *araB*, *araA*, and *araD*, encoding three enzymes involved in arabinose metabolism. The actions of the three enzymes metabolize arabinose into D-xylulose-5-phosphate.

Like the *lac* operon, the *ara* operon contains a single promoter, designated P_{BAD}. This operon also contains a CAP site for the binding of the catabolite activator protein. The *araC* gene, which has its own promoter (P_C), is adjacent to the *ara* operon. *AraC* encodes a regulatory protein, called the AraC protein, that can bind to operator sites designated *araI*, *araO_1*, and *araO_2*.

As we have seen with the *lac* operon, some regulatory proteins such as the *lac* repressor inhibit transcription whereas others such as CAP turn on transcription. The AraC protein is rather interesting and unusual because it can act as either a negative or positive regulator of transcription, depending on whether or not arabinose is present. Figure 14.12 illustrates the possible actions of the AraC protein. In the absence of arabinose, the AraC protein binds to the *araI*, *araO_1*, and *araO_2* operator sites. An AraC protein dimer is bound to *araO_1*, while monomers are bound at *araO_2* and *araI*. The binding of the AraC proteins to the *araO_1* site inhibits the transcription of the *araC* gene. In other words, the AraC protein is a negative regulator of the *araC* gene. This keeps AraC protein levels fairly low. The AraC proteins bound at *araO_2* and *araI* also repress the *ara* operon, but only in the absence of arabinose. As shown in figure 14.12a, the AraC proteins at *araO_2* and *araI* can bind to each other by causing a loop in the DNA, as originally proposed by Robert Schleif and his colleagues in 1990. This DNA loop prevents RNA polymerase from transcribing the *ara* operon via P_{BAD}. Therefore, in the absence of arabinose, the *ara* operon is turned off.

Figure 14.12b illustrates the activation of the *ara* operon in the presence of arabinose. When arabinose is bound to the AraC protein, the interaction between the AraC proteins at the *araO_2* and *araI* sites is broken. This opens the DNA loop. In addition, a second AraC protein binds at the *araI* site. This AraC dimer at the *araI* operator site activates transcription. This activation can occur in conjunction with the activation of the *ara* operon by CAP and cAMP if glucose levels are low. When the *ara* operon is activated, the bacterial cell can efficiently metabolize arabinose.

The *trp* Operon Is Regulated by a Repressor Protein and Also by Attenuation

The *trp* operon (pronounced "trip") encodes enzymes that are needed for the biosynthesis of the amino acid tryptophan (fig. 14.13). The *trpE*, *trpD*, *trpC*, *trpB*, and *trpA* genes encode enzymes involved in tryptophan biosynthesis. The *trpL* and *trpR* genes are involved in regulating the *trp* operon in two different ways. The *trpR* gene encodes the **trp repressor** protein. When tryptophan levels within the cell are very low, the *trp* repressor cannot bind to the operator site. Under these conditions, RNA polymerase transcribes the operon (fig. 14.13a). In this way, the cell expresses the genes that are required for the synthesis of tryptophan. When the tryptophan levels within the cell become high, tryptophan acts as a corepressor that binds to the *trp* repressor protein. This causes a conformational change in the *trp* repressor that allows it to bind to the *trp* operator site (fig. 14.13b). This inhibits the ability of RNA polymerase to transcribe the operon. Therefore, when there is a high level of tryptophan within the cell (i.e., when the cell does not need to make more tryptophan), the *trp* operon is turned off.

In the 1970s, after the action of the *trp* repressor was elucidated, Charles Yanofsky and coworkers made a few unexpected observations. Two *trp* operon mutations were identified in which a region including the *trpL* gene was missing from the operon. These mutations resulted in higher levels of expression of the other genes in the *trp* operon. In addition, other mutant strains were found that lacked the *trp* repressor protein. Surprisingly, these mutant strains still could inhibit expression of the *trp* operon in the presence of tryptophan. As is often the case,

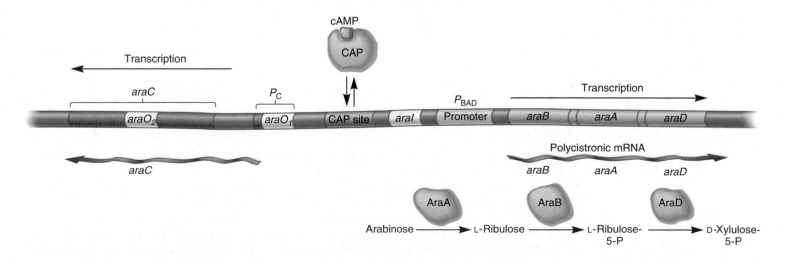

FIGURE 14.11 **Organization of the *ara* operon and other genes involved in arabinose metabolism.** *AraC* encodes a genetic regulatory protein called the AraC protein. This protein can bind to three different regulatory sites, called *araO_1*, *araO_2*, and *araI*. P_C is the promoter for the *araC* gene; P_{BAD} is the promoter for the *ara* operon, which contains three genes (*araB*, *araA*, and *araD*) encoding proteins involved in arabinose metabolism.

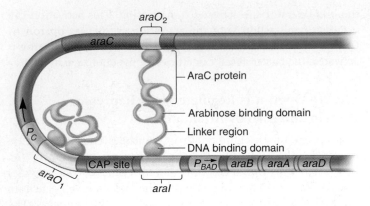

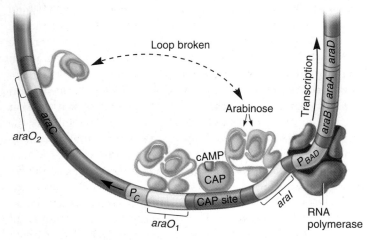

(a) Operon inhibited in the absence of arabinose

(b) Operon activated in the presence of arabinose

FIGURE 14.12   DNA looping and unlooping via the AraC protein. (a) In the absence of arabinose, an AraC protein binds to the *araI* operator site and another to the *araO₂* site. These two AraC proteins interact to promote a loop in the DNA. This loop prevents RNA polymerase from transcribing the *ara* operon from the P_{BAD} promoter. (b) In the presence of arabinose, an AraC dimer is bound to the *araI* operator site. The interaction between the AraC proteins bound at *araI* and *araO₂* no longer occurs. This causes the DNA loop to be broken. Under these conditions, RNA polymerase can transcribe the *ara* operon.

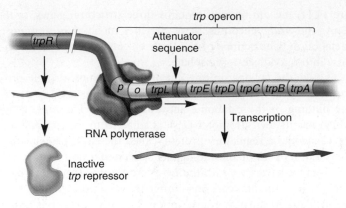

(a) Low tryptophan levels, transcription of the entire *trp* operon occurs

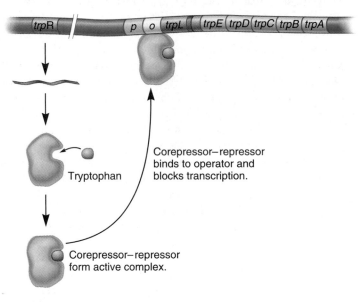

(b) High tryptophan levels, repression occurs

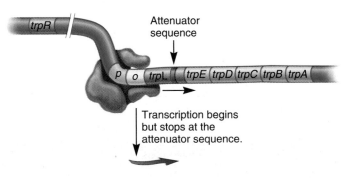

(c) High tryptophan levels, attenuation occurs

FIGURE 14.13   Organization of the *trp* operon and regulation via the *trp* repressor protein. When tryptophan levels are low, tryptophan does not bind to the repressor protein; the latter thus cannot bind to the operator site. Under these conditions, RNA polymerase can transcribe the operon, which leads to the expression of the *trpE, trpD, trpC, trpB,* and *trpA* genes. These genes encode enzymes involved in tryptophan biosynthesis. (b) When tryptophan levels are high, tryptophan acts as a corepressor that binds to the *trp* repressor protein. The tryptophan-*trp* repressor complex then binds to the operator site to inhibit transcription. (c) Another mechanism of regulation is attenuation. When attenuation occurs, the RNA is transcribed only to the attenuator sequence, and then transcription is terminated.

unusual observations can lead people into interesting avenues of study. By pursuing this research further, Yanofsky discovered a second regulatory mechanism in the *trp* operon, called **attenuation.** This mechanism is mediated by the *trpL* gene (fig. 14.13*c*).

Attenuation can occur in bacteria because the processes of transcription and translation are coupled. As described in chapter 13, bacterial ribosomes quickly attach to the 5′ end of mRNA soon after its synthesis begins via RNA polymerase. During attenuation, transcription actually begins, but it is terminated before the entire mRNA is made. A segment of DNA, termed the **attenuator,** is important in facilitating this termination. When attenuation occurs, the mRNA encoding the *trp* operon is made as a short piece that terminates shortly past the *trpL* region (fig. 14.13*c*). Because this short mRNA has been terminated before RNA polymerase has transcribed the *trpE, trpD, trpC, trpB,* and *trpA* genes, it will not encode the proteins required for tryptophan biosynthesis. In this way, attenuation inhibits the further production of tryptophan in the cell.

The segment of the *trp* operon immediately downstream from the operator site plays a critical role during attenuation. The first gene in the *trp* operon is the *trpL* gene, which encodes a short peptide called the *L*eader peptide. As shown in figure 14.14, two features are key in the attenuation mechanism. First, there are two tryptophan codons within the sequence that encodes the *trp* leader peptide. As we will see later, these two codons provide a way to sense whether or not there is sufficient tryptophan for the bacterium to translate its proteins. Second, the RNA that is transcribed from this region can form stem-loop structures. The type of stem-loop structure that forms underlies attenuation.

Different combinations of stem-loop structures are possible due to interactions among four sequences within the RNA transcript (see the *yellow* regions in fig. 14.14). Region 2 is complementary to region 1 and also to region 3. Region 3 is complementary to region 2 as well as to region 4. Therefore, three stem-loop structures are possible: 1–2, 2–3, and 3–4. Even so, keep in mind that a particular segment of RNA can participate in the formation of only one stem-loop structure. For example, if region 2 forms a stem-loop with region 1, it cannot (at the same time) form a stem-loop with region 3. Alternatively, if region 2 forms a stem-loop with region 3, then region 3 cannot form a stem-loop with region 4. Though three stem-loop structures are possible, the 3–4 stem-loop structure is functionally unique; it can act as a transcriptional terminator. The 3–4 stem-loop together with the U-rich attenuator sequence acts as an intrinsic terminator (i.e., ρ-independent terminator, as described in chapter 12). Therefore, the formation of the 3–4 stem-loop causes RNA polymerase to pause and the U-rich sequence dissociates from the DNA. This terminates transcription at the end of the *trpL* gene. By comparison, if region 3 forms a stem-loop with region 2, transcription will not be terminated because a 3–4 stem-loop cannot form.

Conditions that favor the formation of the 3–4 stem-loop ultimately rely on the translation of the *trpL* gene. As shown in figure 14.15, there are three possible scenarios. In figure 14.15*a*, translation is not coupled with transcription. Region 1 rapidly hydrogen bonds to region 2, and region 3 is left to hydrogen bond to region 4. Therefore, the terminator stem-loop forms, and transcription will be terminated just past the *trpL* gene. In figure 14.15*b*, coupled transcription and translation occur under

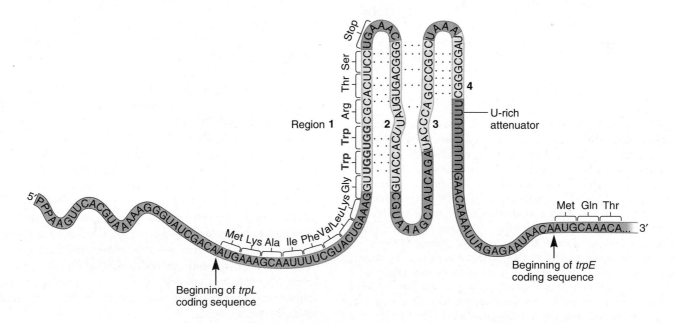

FIGURE 14.14 **Sequence of the *trpL* mRNA produced during attenuation.** A second method of regulation of the *trp* operon is attenuation. At high tryptophan levels, it may occasionally happen that the *trp* repressor protein does not bind to the operator site to inhibit transcription. If so, a short mRNA is made that encodes the *trpL* region. As shown here, this mRNA has several regions that are complementary to each other. The hydrogen bonding between *yellow* regions 1 and 2, 2 and 3, and 3 and 4 is also shown. The last U in the *purple* attenuator sequence is the last nucleotide that would be transcribed during attenuation. At low tryptophan concentrations, however, transcription would occur beyond the end of *trpL* and proceed through the *trpE* gene and the rest of the *trp* operon.

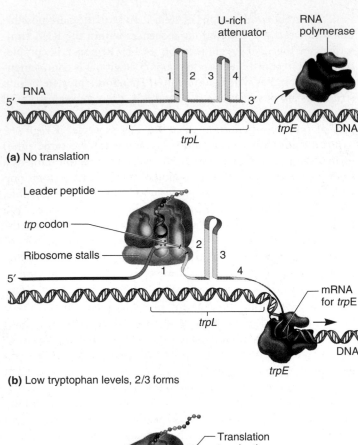

(a) No translation

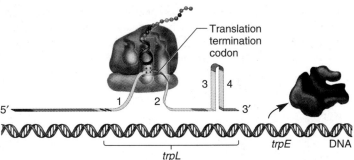

(b) Low tryptophan levels, 2/3 forms

(c) High tryptophan levels, 3/4 forms

FIGURE 14.15 **Possible stem-loop structures formed from** ***trpL*** **mRNA under different conditions of translation.** (a) When translation is not coupled with transcription, the most stable form of the mRNA occurs when region 1 hydrogen bonds to region 2 and region 3 hydrogen bonds to region 4. A terminator stem-loop forms, and transcription will be terminated just past the *trpL* gene. (b) Coupled transcription and translation have occurred under conditions in which the tryptophan concentration is low. The ribosome pauses at the *trp* codons in the *trpL* gene because of insufficient amounts of charged tRNA^{trp}. This pause blocks region 1 of the mRNA, so that region 2 can hydrogen bond only with region 3. Since region 3 is already hydrogen bonded to region 2, the 3–4 stem-loop structure cannot form. Transcriptional termination does not occur, and RNA polymerase transcribes the rest of the operon. (c) Coupled transcription and translation occurs under conditions in which there is a sufficient amount of tryptophan in the cell. Translation of the *trpL* gene progresses to its stop codon, where the ribosome pauses. This blocks region 2 from hydrogen bonding with any region and thereby enables region 3 to hydrogen bond with region 4. This terminates transcription at the end of the *trpL* gene.

conditions where the tryptophan concentration is low. When tryptophan levels are low, the cell cannot make a sufficient amount of charged tRNA^{trp}. As we see in figure 14.15b, the ribosome pauses at the *trp* codons in the *trpL* gene, because it is waiting for charged tRNA^{trp}. This pause occurs in such a way that the ribosome shields region 1 of the mRNA. This sterically prevents region 1 from hydrogen bonding to region 2. As an alternative, region 2 hydrogen bonds to region 3. Therefore, since region 3 is already hydrogen bonded to region 2, the 3-4 stem-loop structure cannot form. Under these conditions, transcriptional termination does not occur, and RNA polymerase transcribes the rest of the operon. This ultimately enables the bacterium to make more tryptophan.

Finally, in figure 14.15c, coupled transcription and translation occur under conditions in which there is sufficient tryptophan in the cell. In this case, translation of the *trpL* gene progresses to its stop codon, where the ribosome pauses. The pausing at the stop codon prevents region 2 from hydrogen bonding with any region and thereby enables region 3 to hydrogen bond with region 4. As in figure 14.15a, this terminates transcription. Of course, keep in mind that the *trpL* gene contains two tryptophan codons. For the ribosome to smoothly progress to the *trpL* stop codon, there must have been enough charged tRNA^{trp} in the cell to translate this mRNA. It follows that the bacterium must have a sufficient amount of tryptophan. Under these conditions, the rest of the transcription of the operon is terminated.

Repressible Operons Usually Encode Anabolic Enzymes, and Inducible Operons Encode Catabolic Enzymes

Thus far, we have seen that bacterial genes can be transcriptionally regulated in a positive or negative way—oftentimes both. The *lac* operon and *ara* operon are inducible systems regulated by sugar molecules that activate transcription of these operons. By comparison, the *trp* operon is a repressible operon regulated by a corepressor, tryptophan, that binds to the repressor and turns the operon off. In addition, an abundance of charged tRNA^{trp} in the cytoplasm can turn the operon off via attenuation.

In studying the genetic regulation of many operons, geneticists have noticed a general trend concerning inducible versus repressible regulation. When the genes in an operon encode proteins that function in the breakdown (i.e., catabolism) of a substance, they are usually regulated in an inducible manner. The substance to be broken down (or a related compound) often acts as the inducer. For example, allolactose and arabinose act as inducers of the *lac* and *ara* operons, respectively. An inducible form of regulation allows the bacterium to phenotypically express the appropriate genes only when they are needed to catabolize these sugars.

In contrast, other enzymes are important for synthesizing small molecules (i.e., anabolism). The genes that encode these anabolic enzymes tend to be regulated by a repressible mechanism. The inhibitor or corepressor is commonly the small molecule that is the product of the enzymes' biosynthetic activities. For example, tryptophan is produced by the sequential action of several enzymes that are encoded by the *trp* operon. Tryptophan itself acts as a corepressor that can bind to the *trp* repressor protein when the intracellular levels of tryptophan become relatively high. This mechanism turns off the genes required for tryptophan biosynthesis when enough of this amino acid has been made. Therefore, genetic regulation via repression provides the bacterium with a way to prevent the overproduction of the product of a biosynthetic pathway.

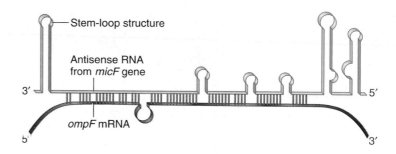

FIGURE 14.16 **The double-stranded RNA structure formed between the *micF* antisense RNA and the *ompF* mRNA.** The *ompF* mRNA within this double-stranded structure cannot be translated.

14.2 TRANSLATIONAL AND POSTTRANSLATIONAL REGULATION

Though genetic regulation in bacteria is exercised predominantly at transcription, there are many examples where regulation is exerted at a later stage in gene expression. In some cases, specialized mechanisms have evolved to regulate the translation of certain mRNAs. As was described in chapter 13, the translation of mRNA occurs in three stages: initiation, elongation, and termination. Genetic regulation of translation is usually aimed at preventing the initiation step. In this section, we will examine a couple of ways that bacteria can regulate the initiation of translation.

The net result of translation is the synthesis of a protein. It is the activities of proteins within living cells and organisms that ultimately determine an individual's traits. Therefore, any modification that alters protein function can be considered to affect gene expression. The term **posttranslational** regulation refers to the functional control of proteins that are already present in the cell rather than regulation of transcription or translation. Posttranslational control can either activate or inhibit the function of a protein. Compared to transcriptional or translational control, posttranslational control can be relatively fast, occurring in a matter of seconds, which is an important advantage. In contrast, transcriptional and translational control typically require several minutes or even hours to take effect because these two mechanisms involve the synthesis and turnover of mRNA and polypeptides. In this section, we will also examine ways that protein function can be regulated posttranslationally.

Repressor Proteins and Antisense RNA Can Inhibit Translation

For some bacterial genes, the translation of mRNA is regulated by the binding of proteins or other RNA molecules that influence the ability of ribosomes to translate the mRNA into a polypeptide. A **translational regulatory protein** recognizes sequences within the mRNA, much as transcription factors recognize DNA sequences. In most cases, translational regulatory proteins act to inhibit translation. These are known as **translational repressors.** When a translational repressor protein binds to the mRNA, it can inhibit translational initiation in one of two ways. One possibility is that it can bind in the vicinity of the Shine-Dalgarno sequence and/or the start codon and thereby sterically block the ribosome's ability to initiate translation in this region. Alternatively, the repressor protein may bind outside the Shine-Dalgarno/start codon

region but stabilize an mRNA secondary structure that prevents initiation. Translational repression is also a form of genetic regulation found in eukaryotic species, and we will consider specific examples in chapter 15.

A second way to regulate translation is via the synthesis of **antisense RNA.** This term refers to an RNA strand that is complementary to a strand of mRNA. The mRNA strand has the same sequence as the DNA sense strand (also known as the coding strand). To understand this form of genetic regulation, let's consider a trait known as osmoregulation, which is essential for the survival of most bacteria. Osmoregulation refers to the ability to control the amount of water inside the cell. Since the solute concentrations in the external environment may rapidly change between hypotonic and hypertonic conditions, bacteria must have an osmoregulation mechanism to maintain their internal cell volume. Otherwise, bacteria would be susceptible to the harmful effects of lysis or shrinking.

In *E. coli,* an <u>o</u>uter <u>m</u>embrane <u>p</u>rotein encoded by the *ompF* gene is important in osmoregulation. At low osmolarity, the ompF protein is preferentially produced, whereas at high osmolarity its synthesis is decreased. The expression of another gene, known as *micF,* is responsible for inhibiting the expression of the *ompF* gene at high osmolarity. As shown in figure 14.16, the inhibition occurs because the *micF* RNA is complementary to the *ompF* mRNA. When the *micF* gene is transcribed, its RNA product binds to the *ompF* mRNA via hydrogen bonding between their complementary regions. The binding of the *micF* RNA to the *ompF* mRNA prevents the *ompF* mRNA from being translated. The RNA transcribed from the *micF* gene is called antisense RNA because it is complementary to the *ompF* mRNA, which is a sense strand of mRNA that encodes a protein. The *micF* RNA does not encode a protein.

Posttranslational Regulation Can Occur Via Feedback Inhibition and Covalent Modifications

A common mechanism to regulate the activity of metabolic enzymes is **feedback inhibition.** The synthesis of many cellular molecules such as amino acids, vitamins, and nucleotides occurs via the action of a series of enzymes that convert precursor molecules to a particular product. The final product in a metabolic

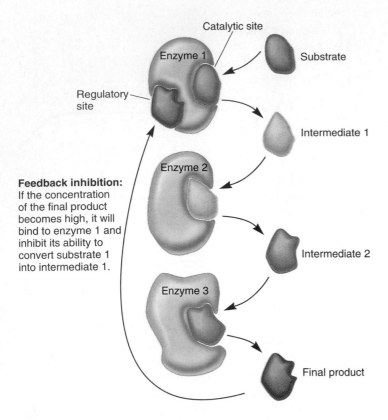

FIGURE 14.17   Feedback inhibition in a metabolic pathway. The substrate is converted to a product by the sequential action of three different enzymes. Enzyme 1 has a catalytic site that recognizes the substrate, and it also has a regulatory site that recognizes the final product. When the final product binds to the regulatory site, it inhibits enzyme 1.

Feedback inhibition: If the concentration of the final product becomes high, it will bind to enzyme 1 and inhibit its ability to convert substrate 1 into intermediate 1.

pathway often can inhibit an enzyme that acts early in the pathway. This idea is schematically shown in figure 14.17, in which the final product inhibits enzyme 1 of the pathway.

In figure 14.17, enzyme 1 is an example of an **allosteric enzyme.** It contains two different binding sites. The catalytic site is responsible for the binding of the substrate and its conversion to intermediate 1. The second site is a regulatory site. This site binds the final product of the metabolic pathway. When bound to the regulatory site, the final product inhibits the catalytic ability of enzyme 1.

To appreciate feedback inhibition at the cellular level, we can consider the relationship between the product concentration and the regulatory site on enzyme 1. As the final product is made within the cell, its concentration will gradually increase. Once the final product concentration has reached a level that is similar to its affinity for enzyme 1 (i.e., its association constant), it becomes likely that the final product will bind to the regulatory site on enzyme 1 and inhibit its function. In this way, the net result is that the final product of a metabolic pathway inhibits the further synthesis of more products. Under these conditions, the concentration of the final product has reached a level that is sufficient for the purpose of the bacterium.

A second strategy to control the function of proteins is by the covalent modification of their structure. Certain types of modifications are involved primarily in the assembly and construction of a functional protein. These alterations include proteolytic processing, disulfide bond formation, and the attachment of prosthetic groups, sugars, or lipids. These are typically irreversible changes required to produce a functional protein. In contrast, other types of modifications, such as phosphorylation ($—PO_4$), acetylation ($—COCH_3$), and methylation ($—CH_3$), are often reversible modifications that transiently affect the function of a protein.

14.3   GENE REGULATION IN THE BACTERIOPHAGE LIFE CYCLE

Viruses are small particles that contain genetic material surrounded by a protein coat. They can infect a living cell and then propagate themselves by using the energy and metabolic machinery of the host cell. For this to occur, the genetic material of viruses orchestrates an intricate series of steps, involving the expression of many viral genes. During the past several decades, the reproduction of viruses has presented an interesting and challenging problem for geneticists to investigate. The study of bacteriophages, which are viruses that infect bacteria, has greatly advanced our basic knowledge of how genetic regulatory proteins work. In addition, the study of viruses has been instrumental in our ability to devise medical strategies aimed at combating viral disease. For example, our knowledge of the life cycles of human viruses has led to the development of drugs that are used to inhibit viral growth. For example, azidothymidine (AZT), which is used to combat HIV (human immunodeficiency virus), suppresses the production of viral DNA by inhibiting a viral gene product called reverse transcriptase that is involved in viral DNA synthesis.

In this section, we will focus on the function of bacteriophage genes that encode genetic regulatory proteins. The structural genes of bacteriophages are often in an operon arrangement. This enables the genes within an operon to be controlled by regulatory proteins that bind to operator sites and influence the function of nearby promoters. Like bacterial operons, phage operons can be controlled by repressor proteins or activator proteins. To understand how this works, we will carefully examine the two life cycles of a virus called phage λ (lambda), which was discovered by Andre Lwoff and his colleagues in the 1940s. Since its discovery, phage λ has been investigated extensively and has provided geneticists with a model on which to base our understanding of viral proliferation.

Phage λ Can Follow a Lytic or Lysogenic Life Cycle

Phage λ can bind to the surface of a bacterium and inject its genetic material into the bacterial cytoplasm. After this occurs, the phage proceeds along only one of two alternative life cycles, known as the **lytic cycle** and the **lysogenic cycle.** This topic was also discussed in chapter 6. You may want to review the material found in figure 6.9. During the lytic cycle, the genetic instructions of the bacteriophage direct the synthesis of many copies of the phage genetic material and coat proteins. These are then assembled to make new phages. When synthesis and assembly are completed, the bacterial host cell is lysed, and the newly made phages are released into the environment.

Alternatively, phage λ can infect a bacterium and not progress to the lytic cycle. Instead, this phage can act as a **temperate phage,** which will usually not produce new phages and will not kill the bacterial cell that acts as its host. Rather, the phage follows a lysogenic cycle. During the lysogenic cycle, phage λ integrates its genetic material into the chromosome of the bacterium. This integrated phage DNA is known as a **prophage.** A prophage can exist in a dormant state for a long time, during which no new bacteriophages are made. When a bacterium containing a lysogenic prophage divides to produce two daughter cells, it will copy the prophage's genetic material along with its own chromosome. Therefore, both daughter cells inherit the prophage. At some later time, a prophage may become activated to excise itself from the bacterial chromosome and enter the lytic cycle. This process is called induction. During induction, the phage promotes the synthesis of new phages and, eventually, lyses the host cell.

Figure 14.18 shows the genome of phage λ. Inside the virion head, phage λ DNA is linear. After injection into the bacterium, the two ends of the DNA become covalently attached to each other to form a circular piece of DNA. The organization of the genes within this circular structure reflects the two alternative life cycles of this virus. The genes in the *top center* of the figure are transcribed very soon after infection. This occurs at the beginning of either life cycle. As examined later, the expression pattern of these early genes determines whether the lytic or lysogenic cycle prevails.

The genes on the *left* side of the viral genome encode proteins that are needed for the lysogenic cycle, while the genes on the *right* side of the genome promote the lytic cycle. If the lysogenic cycle prevails, the integrase (*int*) gene is subsequently turned on. The integrase gene encodes an enzyme that integrates the λ DNA into the bacterial chromosome. The mechanism of phage λ integration is described in chapter 17. If the lysogenic cycle is not chosen, the genes on the *right* side of the figure will be transcribed. This operon contains many genes that are necessary for λ assembly and release. These genes encode replication proteins, coat proteins, proteins involved in coat assembly, proteins involved in packaging the DNA into the phage head, and enzymes that cause the bacterium to lyse.

During the Lysogenic Cycle, the cII-cIII Complex Activates Expression of the λ Repressor

Now that we have an understanding of the phage λ life cycles and genome organization, let's examine how the decision is made

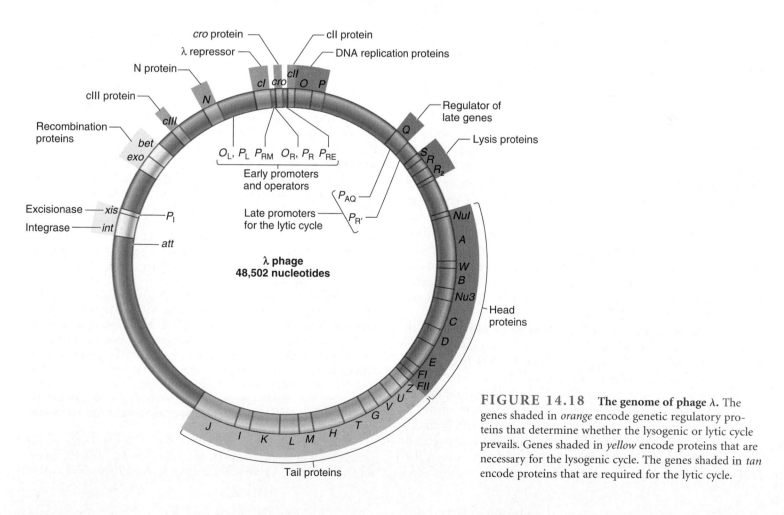

FIGURE 14.18   The genome of phage λ. The genes shaded in *orange* encode genetic regulatory proteins that determine whether the lysogenic or lytic cycle prevails. Genes shaded in *yellow* encode proteins that are necessary for the lysogenic cycle. The genes shaded in *tan* encode proteins that are required for the lytic cycle.

between the lytic and lysogenic cycles. This choice depends on the actions of several genetic regulatory proteins. Our molecular understanding of the phage λ life cycles represents an extraordinary accomplishment in the field of genetic regulation. This process is quite detailed, since it involves a series of intricate steps in which regulatory proteins bind to several different sites in the λ genome. To simplify these events, we will begin by considering the steps that occur when the lysogenic cycle prevails. In the following section, we will examine the steps that occur when the lytic cycle is followed.

Soon after the λ DNA enters the bacterial cell, two promoters, designated P_L and P_R, are used for transcription. This initiates a competition between the lytic and lysogenic cycles (fig. 14.19). Initially, transcription from P_R and P_L results in the synthesis of two short RNA transcripts that encode two proteins called the cro protein and the N protein, both of which are genetic regulatory proteins. We will consider the function of the cro protein later in this chapter. The N protein is a genetic regulatory protein with an interesting function that we have not yet considered. Its function, known as **antitermination,** is to prevent transcriptional termination. The N protein inhibits termination at three sites, designated t_L, t_{R1}, and t_{R2}. The N protein actually binds to RNA polymerase and prevents transcriptional termination when these sites are being transcribed. The transcript from P_R is extended to include cII, O, P, and Q. The cII gene encodes an activator protein, the O and P genes encode enzymes needed for the initiation of λ DNA synthesis, and the Q gene encodes another antiterminator that is required for the lytic cycle. When the N protein prevents termination at t_L, the transcript from P_L is extended to include the int, xis, and cIII genes. The int gene encodes integrase that is involved with integrating λ DNA into the E. coli chromosome. The xis gene encodes excisionase that can excise the λ DNA if a switch is made from the lysogenic to the lytic cycle. The cIII protein helps to stabilize the cII activator protein.

As shown on the left side of figure 14.19, if the cII-cIII complex accumulates to sufficient levels, the lysogenic cycle is favored. Once it is made, the cII protein activates two different promoters in the λ genome. When the cII protein binds to the promoter P_{RE}, it turns on the transcription of cI, a gene that encodes the λ repressor. The cII protein also activates the int gene by binding to the promoter P_I. The λ repressor and integrase proteins play central roles in promoting the lysogenic life cycle. When the λ repressor is made in sufficient quantities, it binds to operator sites that are adjacent to P_R and P_L. When the λ repressor is bound to O_R, it inhibits the expression of the genes that are required for the lytic cycle.

As you are looking at the left side of figure 14.19, you may have noticed that the binding of the λ repressor to O_R will inhibit the expression of cII. This may seem counterintuitive since the cII protein was initially required to activate the cI gene (which encodes the λ repressor). You may be thinking that the inhibition of the cII gene will eventually prevent the expression of the λ repressor protein. However, the cI gene actually has two promoters, P_{RE} (which is activated by the cII protein) and also a second promoter called P_{RM}. Transcription from P_{RE} occurs early in the lysogenic cycle. P_{RE} gets its name because the use of this pro-

moter results in the expression of the λ Repressor during the Establishment of the lysogenic cycle. The transcript made from the use of P_{RE} is very stable and quickly leads to a buildup of the λ repressor protein. This causes an abrupt inhibition of the lytic cycle because the binding of the λ repressor protein to O_R blocks the P_R promoter. Later in the lysogenic cycle, it is no longer necessary to make a large amount of the λ repressor. At this point, the use of the P_{RM} promoter is sufficient to make enough Repressor protein to Maintain the lysogenic cycle. Interestingly, the P_{RM} promoter is activated by the λ repressor protein. The λ repressor was named when it was understood that it repressed the lytic cycle. Later studies revealed that it also activates its own transcription from P_{RM}.

The Lytic Cycle Depends on the Action of the cro Protein

As we have just seen, the λ repressor protein binds to O_R and prevents the expression of the operons needed for the lytic cycle. For the lytic cycle to occur, the λ repressor must be prevented from inhibiting P_R (fig. 14.19). This is the role of the cro protein. If the activity of the cro protein exceeds the activity of the cII protein, the lytic cycle prevails. As was mentioned, an early step in the expression of λ genes is the transcription from P_R to produce the cro protein. If the concentration of the cro protein builds to sufficient levels, it will bind to two operator regions, O_R and O_L. The

FIGURE 14.19   The sequence of events that occur during the lysogenic and lytic cycles of phage λ. The *top* part of this figure shows the region of the phage λ genome that regulates the choice between the lytic and lysogenic cycles. DNA is shown in *blue.* The names of genes that encode proteins are shown *above* the DNA. Promoters and operator sites are shown *below* the DNA. RNA transcripts are shown in *red* or *green.* The key regulatory proteins are indicated as *spheres.* Immediately after infection, P_L and P_R are used to make two short mRNAs. These mRNAs encode two early proteins, designated N and cro. The N protein prevents transcriptional termination at three sites in the RNA (t_L, t_{R1}, and t_{R2}). This allows the transcription of the delayed early genes, which include cIII, cII, O, P, and Q. The expression of the O, P, and Q genes is necessary only for the lytic cycle. If the lysogenic cycle is chosen, transcription of these genes is abruptly inhibited.

The *left* side shows the steps that lead to the lysogenic cycle. The cII-cIII complex binds to P_{RE}. This activates the transcription of the cI gene, which encodes the λ repressor. The cII-cIII complex also binds to P_I to activate the transcription of the int gene. The λ repressor binds to O_R and O_L to inhibit transcription from P_R and P_L. This prevents the lytic cycle. The integrase protein catalyzes the integration of the λ DNA into the E. coli chromosome. Later in the lysogenic cycle, the cI gene is transcribed from the P_{RM} promoter. This results in a low but steady synthesis of the λ repressor, which is necessary to further maintain the lysogenic state.

The *right* side shows the steps that lead to the lytic cycle. The cro protein binds to O_R and blocks transcription from P_{RM}. This prevents the synthesis of the λ repressor. Transcription from P_R is still allowed to

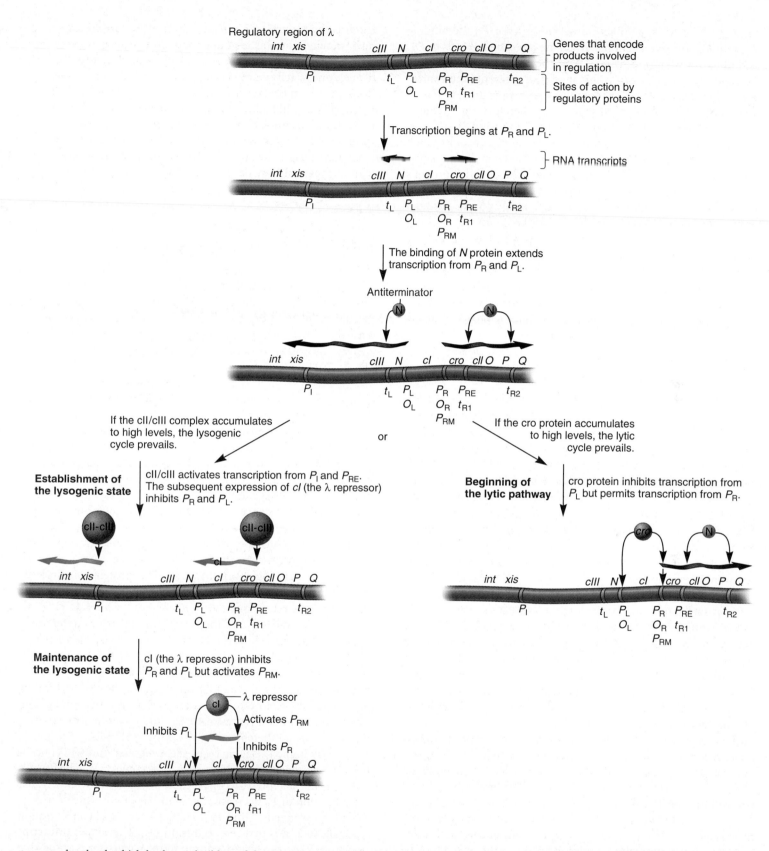

occur at a low level, which leads to a buildup of the O, P, and Q proteins. The O and P proteins catalyze the replication of additional λ DNA. The Q protein stimulates transcription from $P_{R'}$ (not shown in this figure, but see fig. 14.18). This leads to the synthesis of many proteins that are necessary to make new λ phages and to lyse the cell.

GENES→TRAITS   The ability to choose between two alternative life cycles can be viewed as a trait of this bacteriophage. As described here, the choice between the two life cycles depends on the pattern of gene regulation.

binding of cro to O_L inhibits transcription from P_L; the binding of cro to O_R has several effects. When the cro protein binds to O_R, it inhibits transcription from P_{RM} in the leftward direction (see the *right* side of fig. 14.19). This inhibition prevents the expression of the *cI* gene, which encodes the λ repressor; the λ repressor is needed to maintain the lysogenic state. Therefore, the λ repressor cannot successfully shut down transcription from P_R.

The binding of the cro protein to O_R also allows a low level of transcription from P_R in the rightward direction. This enables the transcription of the O, P, and Q genes. The O and P proteins are necessary for the replication of the λ DNA. The Q protein is an antiterminator protein that permits transcription through another promoter, designated $P_{R'}$ (not to be confused with P_R). The $P_{R'}$ promoter controls a very large operon that encodes the proteins necessary for the phage coat, the assembly of the coat proteins, the packaging of the λ DNA, and the lysis of the bacterial cell (refer back to fig. 14.18). These proteins are made toward the end of the lytic cycle. The expression of these late genes leads to the synthesis and assembly of many new λ phages that are released from the bacterial cell when it lyses.

Cellular Proteases Influence the Choice Between the Lytic and Lysogenic Cycle

As you may have noticed from figure 14.19, the first two steps of the lysogenic and lytic cycles are identical. Whether the lysogenic or lytic cycle prevails depends on the steps that occur after the early genes are transcribed. In particular, the activity of the cII protein plays a key role in directing λ to the lysogenic or lytic cycle. A critical physiological issue is that the cII protein is easily degraded by cellular proteases that are produced by *E. coli*. Whether or not these proteases are made depends on the environmental conditions. If the growth conditions are very favorable (e.g., a rich growth medium), the intracellular protease levels are relatively high, and the cII protein tends to be degraded. When cII protein is degraded, P_{RE} cannot be activated and therefore the λ repressor is not made. Instead, the cro protein slowly accumulates to sufficient levels (as in the *right* side of fig. 14.19). The binding of the cro protein to O_R prevents transcription of the λ repressor gene from P_{RM} and, at the same time, allows the lytic cycle to proceed. In this way, environmental conditions that are favorable for growth promote the lytic cycle. This makes sense because a sufficient supply of nutrients is necessary to synthesize new bacteriophages.

Alternatively, starvation conditions favor the lysogenic cycle. When nutrients are limiting, cellular proteases are relatively inactive. Under these conditions, the cII protein will build up much more quickly than the cro protein. Therefore, the cII protein will turn on P_{RE} and lead to the transcription of the λ repressor (as in fig. 14.19). This event favors the lysogenic cycle. From the perspective of the bacteriophage, it is advantageous to favor lysogeny under starvation conditions because there may not be sufficient nutrients available for the production of new λ phages.

After lysogeny is established, certain environmental conditions can also favor induction to the lytic cycle at some later time. For example, exposure to UV light promotes induction. This also is caused by the activation of cellular proteases. In this case, a cellular protein known as recA (a protein ordinarily involved in facilitating recombination between DNA molecules) detects the DNA damage and is activated to become a mediator of protein cleavage. RecA protein mediates cleavage of the λ repressor and thereby inactivates it. This allows transcription from P_R and eventually leads to the accumulation of the cro protein. This favors the lytic cycle. Under these conditions, it may be advantageous for λ to make new phages and lyse the cell, because the exposure to UV light may have already damaged the bacterium to the point where further bacterial growth and division are prevented.

The O_R Region Provides a Genetic Switch Between the Lytic and Lysogenic Cycles

Before we end this section on the λ life cycles, it is interesting to consider how the O_R region acts as a genetic switch between the lytic and lysogenic life cycles. Depending on the binding of genetic regulatory proteins to this region, the switch can be turned to favor the lytic or lysogenic cycle. To understand how this switch works, we need to take a closer look at the O_R region (fig. 14.20).

The O_R region contains three operator sites, designated O_{R1}, O_{R2}, and O_{R3}. These operator sites control two promoters, P_R and P_{RM}, which transcribe in opposite directions. The λ repressor protein or the cro protein can bind to any or all of the three operator sites. The binding of these two proteins at these sites governs the switch between the lysogenic and lytic cycles. Two critical issues influence this binding event. The first is the relative affinities that the regulatory proteins have for these operator sites. The second is the concentrations of the λ repressor protein and the cro protein within the cell.

Let's first consider how an increasing concentration of the λ repressor protein can switch on the lysogenic cycle and switch off the lytic cycle (fig. 14.20, *left* side). This protein was first isolated by Mark Ptashne and his colleagues in 1967. Their studies showed that λ repressor binds with highest affinity to O_{R1}, followed by O_{R2} and O_{R3}. As the concentration of the λ repressor builds within the cell, a dimer of the λ repressor protein first binds to O_{R1}, because it has the highest affinity for this site. Next, a second λ repressor dimer binds to O_{R2}. This occurs very rapidly, because the binding of the first dimer to O_{R1} favors the binding of a second dimer to O_{R2}. This is called a cooperative interaction. The binding of the λ repressor to O_{R1} and O_{R2} inhibits transcription from P_R and thereby switches off the lytic cycle.

Early in the lysogenic cycle, the λ repressor protein concentration may become so high that it occupies O_{R3}. Eventually, however, the λ repressor concentration begins to drop, because the inhibition of P_R decreases the synthesis of cII, which activates the λ repressor gene (from P_{RE}). As the λ repressor concentration gradually falls, it is first removed from O_{R3}. This allows transcription from P_{RM}. As mentioned, the term λ *repressor* is somewhat misleading because the binding of the λ repressor at only O_{R1} and O_{R2} acts as an activator of P_{RM}. The ability of the λ repressor to activate its own transcription allows the switch to the lysogenic cycle to be maintained.

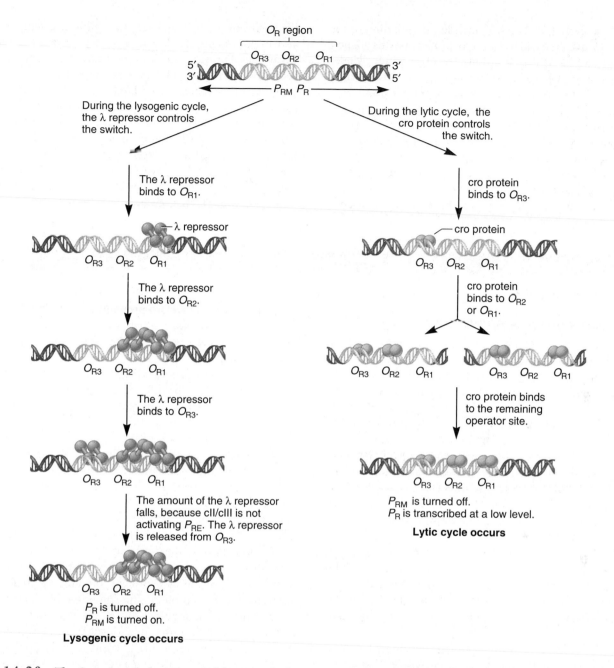

FIGURE 14.20 **The O_R region, the genetic switch between the lysogenic and lytic cycles.** The O_R region contains three operator sites and two promoters, P_{RM} and P_R, which transcribe in opposite directions. The *left* side of the figure depicts the events that promote the lysogenic cycle. The λ repressor protein contains two globular domains connected by a short link. A dimer binds much more tightly to each operator site than a monomer would. The λ repressor protein dimer first binds to O_{R1}, because it has the highest affinity for this site. A second λ repressor dimer binds to O_{R2}. This occurs very rapidly, because the binding of the first dimer to O_{R1} favors the binding of a second dimer to O_{R2}. The binding of the λ repressor to O_{R1} and O_{R2} inhibits transcription from P_R and thereby switches off the lytic cycle. Early in the lysogenic cycle, the λ repressor protein concentration may become so high that it occupies O_{R3}. Later, the λ repressor concentration begins to drop, because the inhibition of P_R decreases the synthesis of cII protein, which activates the λ repressor gene (from P_{RE}). As the λ repressor concentration gradually falls, it is first removed from O_{R3}. This allows transcription from P_{RM} and maintains the lysogenic cycle.

The *right* side depicts events at the O_R region that promote the lytic cycle. The cro repressor protein is a small globular protein that also binds to each operator site as a dimer. The cro protein has its highest affinity for O_{R3}, and so it binds there first. This blocks transcription from P_{RM} and thereby switches off the lysogenic cycle. The cro protein has a similar affinity for O_{R2} and O_{R1}, and so it may occupy either of these sites next. Later in the lytic cycle, the cro protein concentration will continue to rise, so that eventually it binds to both O_{R2} and O_{R1}. This turns down the expression from P_R, which is not needed in the later stages of the lytic cycle.

In the lytic cycle (fig. 14.20, *right* side), the binding of the cro protein controls the switch. The cro protein has its highest affinity for O_{R3} and has a similar affinity for O_{R2} and O_{R1}. Under conditions that favor the lytic cycle, the cro protein accumulates, and a cro dimer first binds to O_{R3}. This blocks transcription from P_{RM} and thereby switches off the lysogenic cycle. Later in the lytic cycle, the cro protein concentration continues to rise, so that eventually it binds to O_{R2} and O_{R1}. This turns down expression from P_R, which is not needed in the later stages of the lytic cycle.

Genetic switches, like the one just described for phage λ, represent an important form of genetic regulation. As we have just seen, a genetic switch can be used to control two alternative life cycles of a bacteriophage. In addition, genetic switches are also important in the developmental pathways for bacteria and eukaryotes. For example, certain species of bacteria can grow in a vegetative state when nutrients are abundant but will sporulate when conditions are unfavorable for growth. The choice between sporulation and vegetative growth involves genetic switches. Likewise, genetic switches operate in the developmental pathways in eukaryotes. As described in chapter 23, they are key events in the initiation of cell differentiation during development. Studies of the phage λ life cycle have provided fundamental information with which to understand how these other switches can operate at the molecular level.

CONCEPTUAL SUMMARY

Genes can be **regulated** in different ways. In bacterial cells, some genes are organized into multigene units called **operons.** An operon typically consists of a **promoter,** an **operator site,** several **structural genes,** and a **terminator.** The organization of an operon allows two or more genes to be coordinately regulated as a single unit. For transcriptional regulation, an important concept is that the binding of regulatory proteins to the operon can greatly influence the rate of transcription. Furthermore, the binding of small effector molecules to regulatory proteins can influence whether or not the regulatory proteins can bind to the DNA. Some regulatory proteins that exert **negative control,** such as the *lac* **repressor** and *trp* **repressor,** inhibit transcription. Others, such as the **catabolite activator protein (CAP),** exert **positive control** by enhancing transcription. The AraC protein is interesting because it can act as both a **repressor** and an **activator** depending on the presence or absence of arabinose. Experiments involving the use of F′ factors showed that regulatory proteins are synthesized and can diffuse within the cell to ultimately bind to a distant operator site and cause repression. This action of a regulatory protein is called a ***trans*-effect.** By comparison, operators exert a ***cis*-effect** by providing a binding site for genetic regulatory proteins.

Our understanding of transcriptional regulation has been greatly aided by studying the life cycles of viruses. In chapter 14, we have considered the ability of phage λ to choose between the lytic and lysogenic cycles. This choice is determined by the actions of several genetic regulatory proteins that can bind to the λ DNA and influence the transcription of nearby genes. The O_R region provides a genetic switch for the lytic or lysogenic cycle. If the λ repressor controls the switch, the lysogenic cycle is favored, whereas binding of the cro protein to this switch favors the lytic cycle. Cellular proteases, which are influenced by environmental conditions, play a key role in the choice between the lytic and lysogenic cycles.

In addition to direct transcriptional regulation, gene expression in bacteria can be affected at later stages in the expression process. During **attenuation** of the *trp* operon, for example, transcription actually begins, but the length of the transcript is determined by the translation of the *trpL* gene. This form of regulation is unique to bacteria because they couple the processes of transcription and translation. The translation of a complete mRNA transcript can also be regulated. One form of translational regulation involves the binding of **translational repressor** proteins that prevent translational initiation. A second way involves the binding of **antisense RNA** to an mRNA to block translation. Finally, the function of proteins that have already been made can be influenced **posttranslationally. Feedback inhibition** involves the noncovalent binding of molecules to an **allosteric** site on a protein. This inhibition blocks the function of an enzyme that is required during an early step in a metabolic pathway. In addition, the function of proteins can be controlled by irreversible and reversible covalent modifications.

Overall, it is clear that genes can be regulated in a variety of different ways. If you are a scientist trying to understand the genetic regulation of a particular gene, your research can be a complicated, frustrating, and challenging task. At the same time, it can be a great deal of fun.

EXPERIMENTAL SUMMARY

Early insights into the molecular mechanisms of gene expression came from experiments involving lactose metabolism in *E. coli.* Jacob and Monod examined the phenomenon of enzyme adaptation and concluded that the proteins involved in lactose metabolism were made after the bacterium was exposed to lactose. This observation led them to investigate the genes involved in this regu-lation process. By constructing **merozygotes,** they discovered that the *lacI* gene encodes a repressor protein that is bound to the operator site in the absence of lactose. More recently, studies by Müller-Hill showed that there are actually three operator sites. The crystal structure of the *lac* repressor tetramer bound at two out of three sites has been determined. Similarly, biophysical studies have determined the structure of CAP, a positive regulator of the *lac* operon.

A different form of genetic regulation, known as attenuation, was discovered by Yanofsky and colleagues. They identified mutant strains of bacteria that were missing the *trp* repressor yet still could repress the *trp* operon when tryptophan levels were high. Additional mutations in the *trpL* region destroyed the ability to attenuate transcription. An analysis of the DNA sequence in this region revealed the ability of the mRNA to form alternative stem-loop structures. The 3–4 stem-loop structure forms a transcriptional terminator and thereby prevents the further synthesis of the *trp* operon.

An analysis of genetic sequences also revealed a posttranscriptional method of gene regulation involving antisense RNA. By comparing the sequence of the *ompF* mRNA and the *micF*

RNA, researchers discovered that the two were complementary to each other. This explains how the synthesis of the *micF* RNA inhibits the translation of the *ompF* mRNA. When the (antisense) *micF* RNA hybridizes to the *ompF* mRNA, the latter cannot be translated by the ribosomes.

We concluded chapter 14 with a description of gene regulation in bacteriophage λ. Many of the same kinds of experimental approaches were used to elucidate the molecular mechanisms that underlie the choice between the lysogenic and lytic cycles. In particular, many researchers have studied how mutations in particular genes in the λ genome favor the lysogenic or lytic cycles. In addition, biochemical and biophysical studies have analyzed the detailed interactions between λ regulatory proteins and the operator sites that they recognize.

PROBLEM SETS & INSIGHTS

Solved Problems

S1. Researchers have identified mutations in the promoter region of the *lacI* gene that make it more difficult for the *lac* operon to be induced. These are called *lacI^Q* mutants, because a greater Quantity of *lac* repressor protein is made. Explain why an increased transcription of the *lacI* gene makes it more difficult to induce the *lac* operon.

Answer: An increase in the amount of *lac* repressor protein makes it easier to repress the *lac* operon. When the cells become exposed to lactose, allolactose levels slowly rise. Some of the allolactose binds to the *lac* repressor protein and causes it to be released from the operator. If there are many more *lac* repressor proteins found within the cell, it takes more allolactose to ensure that there are no unoccupied repressor proteins that can repress the operon.

S2. Explain how the pausing of the ribosome in the presence or absence of tryptophan affects the formation of a terminator stem-loop.

Answer: The key issue is the location where the ribosome stalls. In the absence of tryptophan, it stalls over the *trp* codons in the *trpL* mRNA. Stalling at this site shields region 1 in the attenuator region. Since region 1 is unavailable to hydrogen bond with region 2, region 2 hydrogen bonds with region 3. Therefore, region 3 cannot form a terminator stem-loop with region 4. Alternatively, if there is sufficient tryptophan in the cell, the ribosome pauses over the stop codon in *trpL*. In this case, the ribosome shields region 2. Therefore, region 3 and 4 hydrogen bond with each other to form a terminator stem-loop. This abruptly halts the continued transcription of the *trp* operon.

S3. With regard to the key proteins that affect the choice between the lytic and lysogenic cycles, which one(s) may be degraded by cellular proteases?

Answer: After infection, the key protein affected by cellular proteases is cII. If protease levels are high, as under good growth conditions, cII is degraded. This promotes the lytic cycle. Under starvation conditions, the protease levels are low. This prevents the degradation of cII and thereby promotes the lysogenic cycle. After lysogeny has been established, the key protein affected by cellular proteases is the λ repressor. Agents such as UV light activate cellular proteases that digest the λ repressor. This permits induction of the lytic cycle.

S4. In bacteria, it is common for two or more structural genes to be arranged together in an operon. Discuss the arrangement of genetic sequences within an operon. What is the biological advantage of an operon organization?

Answer: An operon contains several different DNA sequences that play specific roles. For transcription to take place, an operon is flanked by a promoter to signal the beginning of transcription and a terminator to signal the end of transcription. In between these two sequences are found structural genes that encode different proteins. An operon has two or more structural genes. A key feature of an operon is that the expression of the structural genes occurs as a unit. When transcription takes place, a polycistronic mRNA is made that encodes all of the structural genes.

In order to control the ability of RNA polymerase to transcribe an operon, an additional DNA sequence, known as the operator site, is usually present. The base sequence within the operator site can serve as a binding site for genetic regulatory proteins such as activators or repressors. The advantage of an operon organization is that it allows a bacterium to coordinately regulate a group of genes whose encoded proteins have a common function. For example, an operon may contain a group of genes involved in lactose breakdown, or a group of genes involved in tryptophan synthesis, etc. The genes within an operon usually encode proteins within a common metabolic pathway or cellular function.

S5. The sequential use of two sugars by a bacterium is known as diauxic growth. It is a common phenomenon among many bacterial species. When glucose is one of the two sugars available, it is typical that the bacterium metabolizes glucose first, and then a second sugar after the glucose has been used up. Among *E. coli* and related species, diauxic growth is regulated by intracellular cAMP levels and the catabolite activator protein. Summarize the effects of glucose and lactose on the ability of the *lac* repressor and the cAMP-CAP complex to regulate the *lac* operon.

Answer: In the absence of lactose, the *lac* repressor has the dominating effect of shutting off the *lac* operon. Even when glucose is also absent and the cAMP-CAP complex is formed, the presence of the bound *lac* repressor prevents the expression of the *lac* operon. The effects of the cAMP-CAP complex are exerted only in the presence of lactose. When lactose is present and glucose is absent, the cAMP-CAP complex acts to enhance the rate of transcription. However, when both lactose and glucose are present, the inability of CAP to bind to the *lac* operon decreases the rate of transcription. The table shown here summarizes these effects.

lac Operon Regulation

Sugar Present	Transcription of the *lac* Operon
None	The operon is turned off due to the dominating effect of the *lac* repressor protein.
Lactose	The operon is maximally turned on. The repressor protein is removed from the operator site and the cAMP-CAP complex is bound to the CAP site.
Glucose	The operon is turned off due to the dominating effect of the *lac* repressor protein.
Lactose and glucose	The expression of the *lac* operon is greatly decreased. The *lac* repressor is removed from the operator site and the catabolite activator protein is not bound to the CAP site. The absence of CAP at the CAP site makes it difficult for RNA polymerase to begin transcription. However, there is more transcription under these conditions than in the absence of lactose.

S6. The ability of DNA-binding proteins to promote a loop in DNA structure is an interesting phenomenon that is important in the structure and function of DNA. Besides the regulation of genes and operons, DNA looping is required in the compaction of DNA within the nucleoid of a bacterium and the nucleus of a eukaryotic cell (see chapter 10). In addition, DNA looping is frequently involved in the expression of eukaryotic genes (see chapter 15). In this solved problem, we will examine an experimental approach that made it possible for Robert Schleif and his colleagues to determine that the AraC protein causes a loop to form in the DNA. This work relied on the mobility of DNA in an acrylamide gel. A segment of DNA that contains a loop is more compact than the same DNA segment without a loop. Therefore, when these two alternative structures (looped versus unlooped) are run through a gel, the looped structure migrates more quickly to the bottom of the gel, since it can more easily penetrate the gel matrix. As a starting material, Schleif had a sample of DNA that contained a portion of the *ara* operon including both the *araI* and *araO₂* sites. In the gel show here, this segment of DNA was exposed to the following conditions before it was run on the gel:

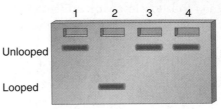

Unlooped / Looped

Lane 1. No further additions
Lane 2. Add araC protein
Lane 3. Add arabinose
Lane 4. Add araC protein and arabinose

Explain these results.

Answer: In lane 2, AraC was added and arabinose was not. As expected from the DNA looping hypothesis, a DNA loop was formed as evidenced from the faster mobility on the acrylamide gel. In lane 1, no AraC protein was added, so that no DNA loop was able to form. These results confirm the idea that it is the AraC protein that is causing the loop to form in the sample loaded into lane 2. In lane 3, only the sugar arabinose is present but not the AraC protein. Since the DNA remains unlooped, the sugar by itself has no effect. In the sample loaded into lane 4, AraC protein was added and arabinose was added as well. In this case, no DNA loop is observed. These results are consistent with the idea that arabinose binds to the AraC protein and breaks the DNA loop that is promoted by the AraC protein.

Conceptual Questions

C1. What is the difference between a constitutive gene and a regulated gene?

C2. In general, why is it important to regulate genes? Discuss examples of situations in which it would be advantageous for a bacterial cell to regulate genes.

C3. If a gene is repressible and under positive control, describe what kind of effector molecule and regulatory protein are involved. Explain how the binding of the effector molecule affects the regulatory protein.

C4. Transcriptional regulation often involves a regulatory protein that binds to a segment of DNA and a small effector molecule that binds to the regulatory protein. Do the following terms apply to a regulatory protein, a segment of DNA, or a small effector molecule?

A. Repressor

B. Inducer

C. Operator

D. Corepressor

E. Activator

F. Attenuator

G. Inhibitor

C5. An operon is repressible (i.e., a small effector molecule turns off transcription). Which combinations of small effector molecules and regulatory proteins could be used to describe it?

A. An inducer plus a repressor

B. A corepressor plus a repressor

C. An inhibitor plus an activator

D. An inducer plus an activator

C6. Some mutations have a *cis*-effect, whereas others have a *trans*-effect. Explain the molecular differences between *cis*- and *trans*-mutations. Which type of mutation (*cis* or *trans*) can be complemented in a merozygote experiment?

C7. What is enzyme adaptation? From a genetic point of view, how does it occur?

C8. In the *lac* operon, how would gene expression be affected if one of the following segments were missing?

A. *lac* operon promoter

B. Operator site

C. *lacA* gene

C9. If an abnormal repressor protein could still bind allolactose but the binding of allolactose did not alter the conformation of the repressor protein, how would this affect the expression of the *lac* operon?

C10. What is diauxic growth? Explain the roles of cAMP and the catabolite activator protein in this process.

C11. Mutations may have an effect on the expression of the *lac* operon, the *ara* operon, and the *trp* operon. Would the following mutations have a *cis-* or *trans-*effect on the expression of the structural genes in the operon?

A. A mutation in the operator site that prevents the *lac* repressor from binding

B. A mutation in the *lacI* gene that prevents the *lac* repressor from binding

C. A mutation in the *araC* gene that prevents two AraC proteins from binding to each other and forming a loop

D. A mutation in *trpL* that prevents attenuation

C12. Would a mutation that inactivated the *lac* repressor and prevented it from binding to the *lac* operator result in the constitutive expression of the *lac* operon under all conditions? Explain. What is the disadvantage to the bacterium of having a constitutive *lac* operon?

C13. Describe the function of the AraC protein. How does it positively and negatively regulate the *ara* operon?

C14. Explain how a mutation would affect the regulation of the *ara* operon if the mutation prevented AraC protein from binding to the following sites:

A. *araO₂*

B. *araO₁*

C. *araI*

D. *araO₂* and *araI*

C15. What is meant by the term *attenuation?* Is it an example of gene regulation at the level of transcription or translation? Explain your answer.

C16. As described in figure 14.14, there are four regions within the *trpL* gene that can form stem-loop structures. Let's suppose that mutations have been previously identified that prevent the ability of a particular region to form a stem-loop structure with a complementary region. For example, a region 1 mutant cannot form a 1–2 stem-loop structure but it can still form a 2–3 or 3–4 structure. Likewise, a region 4 mutant can form a 1–2 or 2–3 stem-loop but not a 3–4 stem-loop. Under the following conditions, would attenuation occur?

A. Region 1 is mutant, tryptophan is high, and translation is not occurring.

B. Region 2 is mutant, tryptophan is low, and translation is occurring.

C. Region 3 is mutant, tryptophan is high, and translation is not occurring.

D. Region 4 is mutant, tryptophan is low, and translation is not occurring.

C17. As described in chapter 13, enzymes known as aminoacyl-tRNA synthetases are responsible for attaching amino acids to tRNAs. Let's suppose that tryptophanyl-tRNA synthetase was partially defective at attaching tryptophan to tRNA; its activity was 10% of that found in a normal bacterium. How would that affect attenuation of the *trp* operon? Would it be more or less likely to be attenuated? Explain your answer.

C18. The 3–4 stem-loop and U-rich attenuator found in the *trp* operon (see fig. 14.14) is an example of ρ-independent termination. The function of ρ-independent terminators is described in chapter 12. Would you expect attenuation to occur if the tryptophan levels were high and the UUUUUUUU sequence at the end of *trpL* was changed to UGGUUGUC? Explain why or why not.

C19. Mutations in tRNA genes can create tRNAs that recognize stop codons. Since stop codons are sometimes called nonsense codons, these types of mutations that affect tRNAs are called nonsense suppressors. For example, a normal tRNAgly has an anticodon sequence CCU that recognizes a glycine codon in mRNA (GGA) and puts in a glycine during translation. However, a mutation in the gene that encodes tRNAgly could change the anticodon to ACU. This mutant tRNAgly would still carry glycine, but it would recognize the stop codon UGA. Would this mutation affect attenuation of the *trp* operon? Explain why or why not. Note: To answer this question, you need to look carefully at figure 14.14 and see if you can identify any stop codons that may exist beyond the UGA stop codon that is found after region 1.

C20. Translational control is usually aimed at preventing the initiation of translation. With regard to cellular efficiency, why do you think this is the case?

C21. What is antisense RNA? How does it affect the translation of a complementary mRNA?

C22. A species of bacteria can synthesize the amino acid histidine so that it does not require histidine in its growth medium. A key enzyme, which we will call histidine synthetase, is necessary for histidine biosynthesis. When these bacteria are given histidine in their growth media, they stop synthesizing histidine intracellularly. Based on this observation alone, propose three different regulatory mechanisms to explain why histidine biosynthesis ceases when histidine is in the growth medium. To explore this phenomenon further, you measure the amount of intracellular histidine synthetase protein when cells are grown in the presence and absence of histidine. In both conditions, the amount of this protein is identical. Which mechanism of regulation would be consistent with this observation?

C23. Using three examples, describe how allosteric sites are important in the function of genetic regulatory proteins.

C24. In what ways are the actions of the *lac* repressor and *trp* repressor similar and different? In other words, discuss the similarities and differences with regard to their binding to operator sites, their effects on transcription, and the influences of small effector molecules.

C25. Transcriptional repressor proteins (e.g., *lac* repressor), antisense RNA, and feedback inhibition are three different mechanisms that can turn off the expression of genes and gene products. Which of these three mechanisms would be most effective in each of the following situations?

A. Shutting down the synthesis of a polypeptide

B. Shutting down the synthesis of mRNA

C. Shutting off the function of a protein

For your answers in parts A–C that have more than one mechanism, which mechanism would be the fastest and most efficient?

C26. What are key features that distinguish the lytic and lysogenic cycles?

C27. With regard to promoting the lytic or lysogenic cycle, what would happen if the following genes were missing from the λ genome?

A. *cro*

B. *cI*

C. *cII*

D. *int*

E. *cII* and *cro*

C28. How do the λ repressor and the cro protein affect the transcription from P_R and P_{RM}? Explain where these proteins are binding to cause their effects.

C29. In your own words, explain why it is necessary for the *cI* gene to have two promoters. What would happen if it only had P_{RE}?

C30. A mutation in P_R causes its transcription rate to be increased 10-fold. Do you think that this mutation would favor the lytic or lysogenic cycle? Explain your answer.

C31. When an *E. coli* bacterium already has a λ prophage integrated into its chromosome, another λ phage cannot usually infect the cell and establish the lysogenic or lytic cycle. Based on your understanding of the genetic regulation of the λ life cycles, why do you think the other phage would be unsuccessful?

C32. If a bacterium were exposed to a drug that inhibited the N protein, what would you expect to happen if the bacterium was later infected by phage λ? Would phage λ follow the lytic cycle, the lysogenic cycle, or neither? Explain your answer.

C33. Figure 14.20 shows a genetic switch that controls the choice between the lytic and lysogenic life cycles of phage λ. What is a genetic switch? Compare the roles of a genetic switch and a simple operator site (like the one found in the *lac* operon) in gene regulation.

C34. This question combines your knowledge of conjugation (described in chapter 6) and the genetic regulation that directs the phage λ life cycles. When donor *Hfr* strains conjugate with recipient F^- bacteria that are lysogenic for phage λ, the conjugated cells survive normally. However, if donor *Hfr* strains that are lysogenic for phage λ conjugate with recipient F^- bacteria that do not contain any phage λ, the conjugated cells often lyse, due to the induction of λ into the lytic cycle. Based on your knowledge of the regulation of the two λ life cycles, explain this observation.

Experimental Questions

E1. Answer the following questions, which pertain to the experiment of figure 14.7.

A. Why was β-ONPG used? Why was no yellow color observed in one of the four tubes? Can you propose alternative methods to measure the level of expression of the *lac* operon?

B. The optical density values were twice as high for the mated strain as for the parent strain. Why was this result obtained?

E2. Chapter 18 describes a blotting method known as Northern blotting, which can be used to detect RNA that is transcribed from a particular gene or a particular operon. In this method, a specific RNA is detected by using a short segment of cloned DNA as a probe. The DNA probe, which is radioactive, is complementary to the RNA that the researcher wishes to detect. After the radioactive probe DNA binds to the RNA within a blot of a gel, the RNA is visualized as a dark (radioactive) band on an X-ray film. For example, a DNA probe that is complementary to the mRNA of the *lac* operon could be used to specifically detect the *lac* operon mRNA on a gel blot. As shown here, the method of Northern blotting can be used to determine the amount of a particular RNA transcribed under different types of growth conditions. In this Northern blot, bacteria containing a normal *lac* operon were grown under different types of conditions, and then the mRNA was isolated from the cells and subjected to a Northern blot, using a probe that is complementary to the mRNA of the *lac* operon.

Lane 1. Growth in media containing glucose
Lane 2. Growth in media containing lactose
Lane 3. Growth in media containing glucose and lactose
Lane 4. Growth in media that doesn't contain glucose or lactose

Based on your understanding of the regulation of the *lac* operon, explain these results. Which is more effective at shutting down the *lac* operon, the binding of the *lac* repressor or the removal of CAP? Explain your answer based on the results shown in the Northern blot.

E3. As described in experimental question E2 and also in chapter 18, the technique of Northern blotting can be used to detect the transcription of RNA. Draw the results you would expect from a Northern blot if bacteria were grown in media containing lactose (and no glucose) but had the following mutations:

Lane 1. Normal strain

Lane 2. Strain with a mutation that inactivates the *lac* repressor

Lane 3. Strain with a mutation that prevents allolactose from binding to the *lac* repressor

Lane 4. Strain with a mutation that inactivates CAP.

How would your results differ if these bacterial strains were grown in media that did not contain lactose or glucose?

E4. An absentminded researcher follows the protocol described in figure 14.7 and (at the end of the experiment) does not observe any yellow color in any of the tubes. Yikes! Which of the following mistakes could account for this observation?

A. Forgot to sonicate the cells

B. Forgot to add lactose to two of the tubes

C. Forgot to add β-ONPG to the four tubes

E5. Explain how the data shown in figure 14.9 indicate that two operator sites are necessary for repression of the *lac* operon. What would the results have been if all three operators were required for the binding of the *lac* repressor?

E6. A mutant strain has a defective *lac* operator site that results in the constitutive expression of the *lac* operon. Outline an experiment you would carry out to demonstrate that the operator site must be physically adjacent to the genes that it influences. Based on your knowledge of the *lac* operon, describe the results you would expect to get.

E7. Let's suppose you have isolated a mutant strain of *E. coli* in which the *lac* operon is constitutively expressed. To understand the nature of this defect, you create a merozygote in which the mutant strain contains an F′ factor with a normal *lac* operon and a normal *lacI* gene. You then compare the mutant strain and the merozygote with regard to their β-galactosidase activities in the presence and absence of lactose. You obtain the following results:

	Addition of Lactose	Amount of β-Galactosidase (percentage of mutant strain in the presence of lactose)
Mutant	No	100
Mutant	Yes	100
Merozygote	No	100
Merozygote	Yes	200

Explain the nature of the defect in the mutant strain.

E8. In the experiment of figure 14.7, a *lacI⁻* mutant was conjugated to a strain that had a functional *lacI* gene on an F′ factor. The results of this experiment were important in determining the action of the *lac* repressor protein. What results would you expect if you used the same approach to investigate the regulation of the *ara* operon via AraC? In other words, what would be the level of expression of the *ara* operon in an *araC⁻* strain in the presence and absence of arabinose, and how would the level of expression change when a functional *araC* gene was introduced into the strain via conjugation?

E9. A segment of DNA that contains a loop is more compact than the same DNA segment without a loop. Therefore, when these two alternative structures (looped versus unlooped) are electrophoresed through a gel, the looped structure migrates more quickly to the bottom of the gel, since it can more easily penetrate the gel matrix. Let's suppose that a mutant *E. coli* strain has been identified in which the AraC protein represses the *ara* operon, even in the presence of arabinose. In the experiment shown here, mutant or normal AraC protein was mixed with a segment of DNA containing the *ara* operon in the absence or presence of arabinose and then run on a gel.

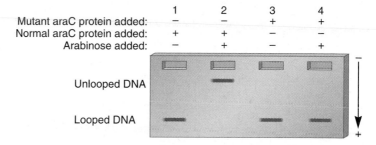

Describe the defect in this mutant AraC protein.

E10. The use of antisense RNA as a research tool is gaining wider popularity. For example, developmental geneticists may want to know if a particular gene product (e.g., a polypeptide) plays a role in early stages of development. To determine this, they can inject antisense RNA (that is complementary to a particular mRNA) into a fertilized frog egg, and then watch the egg develop into an embryo. Explain what antisense RNA will do. What results would you expect if the gene does not play a role in the early stages of development?

Questions for Student Discussion/Collaboration

1. Discuss the advantages and disadvantages of genetic regulation at the different levels described in figure 14.1.

2. As you look at figure 14.10, discuss possible "molecular ways" that the cAMP-CAP complex and *lac* repressor may influence RNA polymerase function. In other words, try to explain how the bending and looping in DNA may affect the ability of RNA polymerase to initiate transcription.

3. Certain environmental conditions such as UV light are known to activate lysogenic λ prophages and cause them to progress into the lytic cycle. UV light initially causes the repressor protein to be proteolytically degraded. Make a flow diagram that describes the subsequent events that would lead to the lytic cycle. Note: The *xis* gene codes for an enzyme that is necessary to excise the λ prophage from the *E. coli* chromosome. The integrase enzyme is also necessary to excise the λ prophage.

Note: All answers appear at the website for this textbook; the answers to even-numbered questions are in the back of the textbook.

Visit the Online Learning Center for practice tests, answer keys, and other learning aids for this chapter. Enhance your understanding of genetics with our interactive exercises, web links, news feeds, tutorial service, and much more.

GENE REGULATION IN EUKARYOTES

::

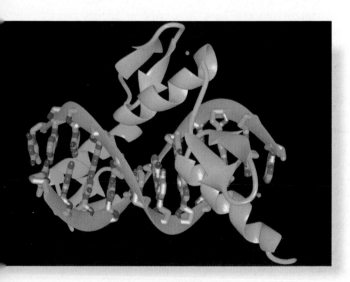

Eukaryotic organisms, a category that includes protozoa, algae, fungi, plants, and animals, have many benefits from regulating their genes. Like their prokaryotic counterparts described in chapter 14, eukaryotic cells need to adapt to changes in their environment. For example, eukaryotic cells can respond to changes in nutrient availability by enzyme adaptation much as prokaryotic cells do. Eukaryotic cells can also respond to environmental stresses such as UV radiation by inducing genes that provide protection against this harmful environmental agent. An example is the ability of sunbathers to develop a tan. The tanning response helps to protect a person's cells against the damaging effects of ultraviolet rays.

Among plants and animals, multicellularity and a more complex cell structure also demand a much greater level of gene regulation. The life cycle of higher eukaryotic organisms involves the progression through several developmental stages to achieve a mature organism. Some genes are expressed only during early stages of development (e.g., the embryonic stage), whereas others are expressed in the adult. In addition, complex eukaryotic species are composed of many different tissues that contain a variety of cell types. Gene regulation is necessary to ensure the differences among distinct cell types. It is amazing that the various cells within a multicellular organism usually contain the same genetic material yet, phenotypically, may look quite different. For example, the appearance of a human nerve cell seems about as similar to a muscle cell as an amoeba is to a paramecium. In spite of these phenotypic differences, a human nerve cell and muscle cell actually contain the same complement of

REGULATION OF GENE EXPRESSION

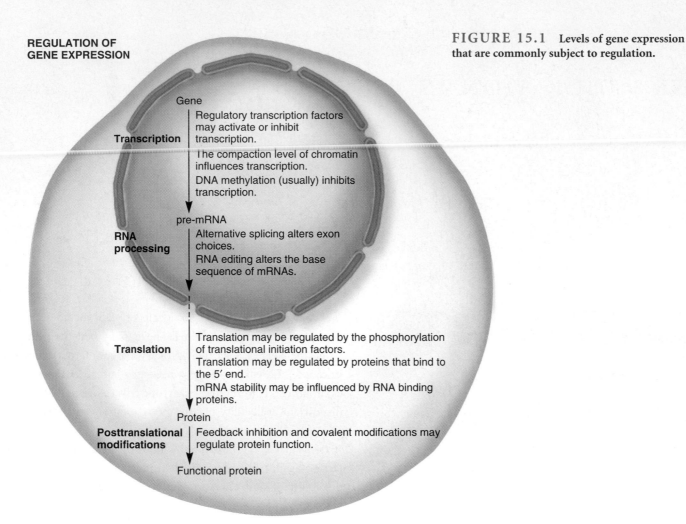

chromosomes of the human genome. Nerve and muscle cells are strikingly different because of gene regulation rather than differences in DNA content. Many genes are expressed in the nerve cell and not the muscle cell, and vice versa.

The molecular mechanisms that underlie gene regulation in eukaryotes bear many similarities to the ways that bacteria regulate their genes. Figure 15.1 describes common points of genetic control where eukaryotic gene expression can be affected. Regulation of gene transcription is an important form of control. In addition, research in the past few decades has revealed that eukaryotic organisms frequently regulate gene expression at points other than transcription. In this chapter, we will discuss some well-studied examples in which genes are regulated at many of these control points.

15.1 REGULATORY TRANSCRIPTION FACTORS

The term **transcription factor** is broadly used to describe proteins that influence the ability of RNA polymerase to transcribe a given

gene. In this section of chapter 15, we will focus our attention on transcription factors that affect the ability of RNA polymerase to begin the transcription process. Such transcription factors can regulate the binding of the transcriptional apparatus to the core promoter and/or control the switch from the initiation to the elongation stage of transcription. In this regard, there are two categories of transcription factors. In chapter 12, we considered **general transcription factors,** which are required for the binding of RNA polymerase to the core promoter and its progression to the elongation stage. General transcription factors are necessary for a basal level of transcription. In addition, eukaryotic cells possess an interesting array of **regulatory transcription factors** that serve to regulate the rate of transcription of target genes. These will be a focus of chapter 15. As we will learn, regulatory transcription factors exert their effects by influencing the ability of RNA polymerase to begin transcription of a particular gene.

Regulatory transcription factors typically recognize *cis* regulatory elements that are located in the vicinity of the core promoter. These DNA sequences are analogous to the operator sites found near bacterial promoters. In eukaryotes, DNA sequences recognized by regulatory transcription factors are known as **response**

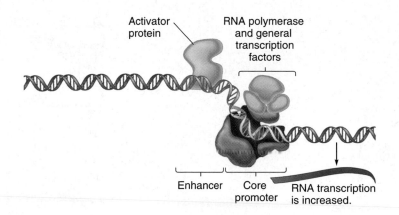

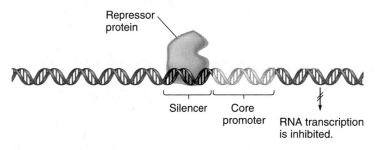

(a) Gene activation

(b) Gene repression

FIGURE 15.2 **Overview of transcriptional regulation by regulatory transcription factors.** These factors can act as either **(a)** an activator to increase the rate of transcription or **(b)** a repressor to decrease that rate.

elements, control elements, or **regulatory elements.** When a regulatory transcription factor binds to a response element, it affects the transcription of an associated gene (fig. 15.2). For example, the binding of regulatory transcription factors may enhance the rate of transcription (fig. 15.2a). Such a transcription factor would be termed an **activator,** and the sequence it binds to would be called an **enhancer.** Alternatively, regulatory transcription factors may act as **repressors** by binding to elements called **silencers** and preventing transcription from occurring (fig. 15.2b). In this section, we will examine several features of regulatory transcription factor function. We will begin by considering the structural features of transcription factor proteins and their response elements. We will then examine two well-studied examples that illustrate how the function of a transcription factor is modulated within a living cell.

Structural Features of Regulatory Transcription Factors Allow Them to Bind to DNA

Genes that encode general and regulatory transcription factor proteins have been identified and sequenced from a wide variety of eukaryotic species including yeast, plants, and animals. Several different families of evolutionarily related transcription factors have been discovered. In recent years, the molecular structures of transcription factor proteins have become an area of intense research. Transcription factor proteins contain regions, called **domains,** that have specific functions. For example, one domain of a transcription factor may have a DNA-binding function, while another may provide a binding site for a small effector molecule. When a domain or portion of a domain has a very similar structure in many different proteins, such a structure is called a **motif.**

Figure 15.3 depicts several different domain structures found in transcription factor proteins. The protein secondary structure known as an α helix is frequently important in the recognition of the DNA double helix. In helix–turn–helix and helix–loop–helix motifs, an α helix called the recognition helix makes contact with and recognizes a base sequence along the major groove of the DNA (fig. 15.3a and b). As discussed in chapter 9, the major groove is a region of the DNA double helix where the bases contact the surrounding water in the cell. Hydrogen bonding between an α helix and nucleotide bases is one way that a transcription factor can bind to a specific DNA sequence. Similarly, a zinc finger motif is composed of one α-helix and two

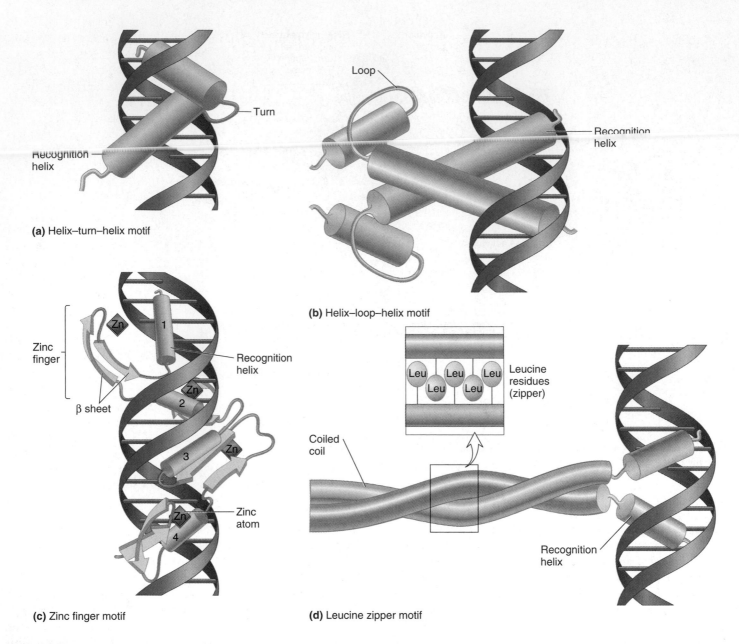

FIGURE 15.3   Structural domains found in transcription factor proteins. Certain types of protein secondary structure are found in many different transcription factors. In this figure, α helices are shown as *cylinders* and β sheets as *flattened arrows*. (a) Helix–turn–helix motif: Two α helices are connected by a turn. The α helices lie in the major groove of the DNA. (b) Helix–loop–helix motif: A short α helix is connected to a longer α helix by a loop. In this illustration, a dimer is formed from the interactions of two helix–loop–helix motifs, and the longer helices are binding to the DNA. (c) Zinc finger motif: Each zinc finger is composed of one α helix and two antiparallel β sheets. A zinc atom, shown in *red,* holds the zinc finger together. This illustration shows four zinc fingers in a row. (d) Leucine zipper motif: The leucine zipper promotes the dimerization of two transcription factors. Two α helices (a coiled coil) are intertwined due to the leucine residues (*see inset*).

β-sheet structures that are held together by a zinc (Zn^{++}) metal ion (see fig. 15.3c). The zinc finger also can recognize DNA sequences within the major groove.

A second interesting feature of certain motifs is that they promote protein dimerization. The leucine zipper (fig. 15.3d) and helix–loop–helix motif (see fig. 15.3b) mediate protein dimerization. For example, figure 15.3d depicts the dimerization and

DNA binding of two proteins that have leucine zippers. Alternating leucine residues in both proteins interact ("zip up"), resulting in protein dimerization. In some cases, two identical transcription factor proteins will come together to form a **homodimer,** or two different transcription factors can form a **heterodimer.** As discussed later in this chapter, the dimerization of transcription factors can be an important way to modulate their function.

Regulatory Transcription Factors Recognize Response Elements That Function as Enhancers or Silencers

As mentioned, when the binding of a regulatory transcription factor to a response element increases transcription, the response element is known as an enhancer. Such elements can stimulate or **up regulate** transcription 10- to 1,000-fold. Alternatively, response elements that serve to inhibit transcription are called **silencers**. This is called **down regulation.**

Many response elements are **orientation independent** or **bidirectional.** This means that the response element can function in the forward or reverse orientation. For example, if the forward orientation of an enhancer is

5'-GATA-3'
3'-CTAT-5'

this enhancer will also be bound by a regulatory transcription factor and enhance transcription even when it is oriented in the reverse direction:

5'-TATC-3'
3'-ATAG-5'

There is also striking variation in the location of response elements relative to a gene's promoter. In general, most response elements are located in a region within a few hundred nucleotides upstream from the promoter site. Nevertheless, response elements are sometimes found several thousand nucleotides away from the actual promoter site. In some cases, response elements are located downstream from the promoter site and may even be found within introns! As you may imagine, the variation in response element orientation and location profoundly complicates the efforts of geneticists to identify the response elements that affect the expression of any given gene.

Regulatory Transcription Factors May Exert Their Effects Through TFIID and Mediator

Different mechanisms have been proposed to explain how a regulatory transcription factor can bind to a response element and thereby affect gene transcription. Indeed, more than one mechanism may be involved. The net effect of a regulatory transcription factor is to influence the ability of RNA polymerase to transcribe a given gene. However, most regulatory transcription factors do not bind directly to RNA polymerase. Instead, they may influence the function of RNA polymerase by interacting with other proteins that bind to RNA polymerase. Two common protein complexes that communicate the effects of regulatory transcription factors are **TFIID** and **mediator.**

Figure 15.4 depicts how regulatory transcription factors may exert their effects. In some cases, regulatory transcription factors bind to a response element and then influence the function of TFIID (fig. 15.4*a*). As discussed in chapter 12, TFIID is a general transcription factor that binds to the TATA box and is needed to recruit RNA polymerase to the core promoter. Activator proteins would be expected to enhance the ability of TFIID to initiate transcription. One possibility is that activator proteins could help recruit TFIID to the TATA box or they could enhance the function of TFIID in a way that facilitates its ability to bind RNA polymerase. In contrast, repressors inhibit the function of TFIID; they could exert their effects by preventing the binding of TFIID to the TATA box, or by inhibiting the ability of TFIID to recruit RNA polymerase to the core promoter.

A second way that regulatory transcription factors control RNA polymerase is via mediator (fig. 15.4*b*). In fact, the term *mediator* refers to the observation that it mediates the interaction between RNA polymerase and regulatory transcription factors. As discussed in chapter 12, mediator controls the ability of RNA polymerase to progress to the elongation stage of transcription. Transcriptional activators stimulate the ability of mediator to facilitate the switch to the elongation stage, while repressors have the opposite effect. When a repressor protein interacts with mediator, RNA polymerase cannot progress to the elongation stage of transcription.

A third way that regulatory transcription factors can influence transcription is by recruiting proteins to the promoter region that affect chromatin compaction. For example, certain transcriptional activators can recruit proteins, such as histone acetyltransferase and ATP-dependent remodeling enzymes, to the promoter region and thereby promote the conversion of chromatin from a closed to an open conformation. This topic will be discussed later in this chapter.

The Function of Regulatory Transcription Factor Proteins Can Be Modulated in Three Ways

Thus far, we have considered the structures of regulatory transcription factors and the molecular mechanisms that account for their abilities to control transcription. It is important to keep in mind that regulatory transcription factors themselves must be regulated. This is necessary so that the genes they control are turned on only at the proper time, in the correct cell type, and under the appropriate environmental conditions. For these reasons, eukaryotes have evolved different ways to modulate the functions of regulatory transcription factors.

There are three common ways that the function of regulatory transcription factor proteins can be affected. These include the binding of an effector molecule, protein–protein interactions, and covalent modification of the transcription factor itself. Figure 15.5 depicts these three means of modulating regulatory transcription factor function. Usually, one or more of these modulating effects are important in determining whether a transcription factor can bind to the DNA and/or influence transcription by

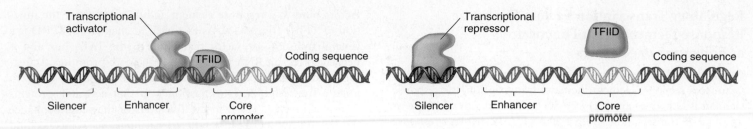

The transcriptional activator recruits TFIID to the core promoter and/or activates its function. Transcription will be activated.

The transcriptional repressor inhibits the binding of TFIID or inhibits its function. Transcription is repressed.

(a) Regulatory transcription factors and TFIID

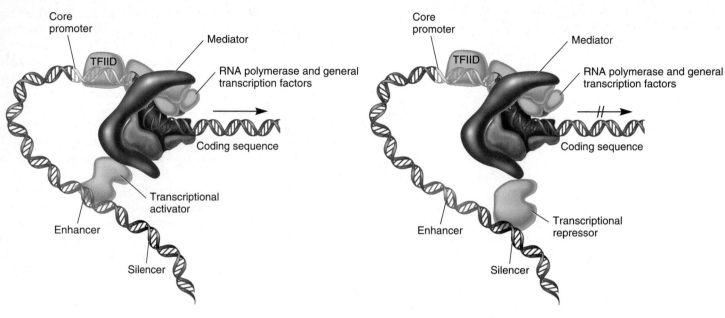

The transcriptional activator interacts with mediator. This enables RNA polymerase to form a preinitiation complex that can proceed to the elongation phase of transcription.

The transcriptional repressor interacts with mediator so that transcription is repressed.

(b) Regulatory transcription factors and mediator

FIGURE 15.4   Ability of regulatory transcription factors to affect transcription. (a) Some regulatory transcription factors exert their effects through TFIID. Transcriptional activators would stimulate the ability of TFIID to recruit RNA polymerase to the core promoter, while repressors would inhibit TFIID binding or its activity. **(b)** A second way that regulatory transcription factors may function is via mediator. By interacting with mediator in different ways, activators would stimulate transcription, while repressors would inhibit transcription. In this example, the enhancer and silencer are relatively far away from the core promoter. Therefore, a loop must form in the DNA so that the regulatory transcription factor can interact with mediator.

RNA polymerase. For example, an effector molecule may bind to a regulatory transcription factor and promote its binding to the DNA (fig. 15.5*a*). Shortly, we will see that steroid hormones function in this manner. Another important way is via protein–protein interactions (fig. 15.5*b*). The formation of homodimers and heterodimers is a fairly common means of controlling transcription. Finally, the function of regulatory transcription factors can be affected by covalent modifications such as the attachment of a phosphate group (fig. 15.5*c*). As discussed later, the phosphorylation of activators can promote their ability to stimulate transcription.

Steroid Hormones Exert Their Effects by Binding to a Regulatory Transcription Factor

Thus far in this section, we have considered the general properties of transcription factor structure and function. We now turn to specific examples that illustrate how regulatory transcription factors function within living cells. Our first example is a category that responds to steroid hormones. This type of regulatory transcription factor is known as a **steroid receptor,** because the steroid hormone binds directly to the protein.

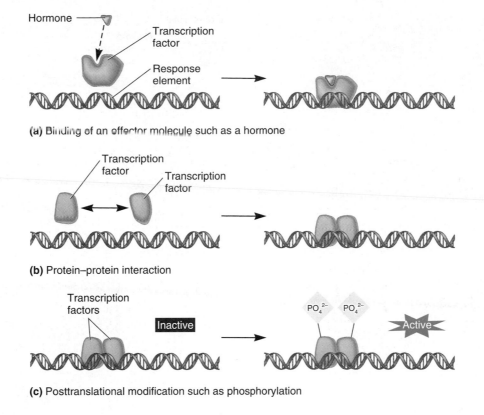

(a) Binding of an effector molecule such as a hormone

(b) Protein–protein interaction

(c) Posttranslational modification such as phosphorylation

FIGURE 15.5 **Common ways to modulate the function of regulatory transcription factors. (a)** The binding of an effector molecule such as a hormone may influence the ability of a transcription factor to bind to the DNA. **(b)** Protein–protein interactions among transcription factor proteins may influence their functions. **(c)** Posttranslational modifications such as phosphorylation may alter transcription factor function.

The ultimate action of a steroid hormone is to affect gene transcription. Steroid hormones act as signaling molecules that are synthesized by endocrine glands of animals and secreted into the bloodstream. The hormones are then taken up by cells that can respond to the hormones in different ways. For example, glucocorticoid hormones influence nutrient metabolism in most body cells by promoting glucose utilization, fat mobilization, and protein breakdown. Other steroid hormones, such as estrogen and testosterone, are called gonadocorticoids because they influence the growth and function of the gonads.

Figure 15.6 shows the stepwise action of a glucocorticoid hormone. In this example, the hormone is transported into the cytosol of a cell. Once inside, **glucocorticoid receptors** can specifically bind the hormone. Prior to hormone binding, the glucocorticoid receptor is complexed with proteins known as heat shock proteins, an example being HSP90. After the hormone binds to the glucocorticoid receptor, HSP90 is released. This exposes a nuclear localization signal (NLS) within the receptor that allows it to travel into the nucleus through a nuclear pore. Two glucocorticoid receptors form a homodimer and then travel through a nuclear pore into the nucleus. The glucocorticoid receptor homodimer binds to glucocorticoid response elements (GREs) that are next to particular genes. The GREs function as

enhancers. The binding of the glucocorticoid receptor homodimer to GREs activates the transcription of the adjacent gene, eventually leading to the synthesis of the encoded protein.

Animal cells usually have a large number of glucocorticoid receptors within the cytoplasm. Since GREs are located near dozens of different genes, the uptake of many hormone molecules can activate many glucocorticoid receptors and thereby up regulate many different genes. For this reason, a cell can respond to the presence of the hormone in a very complex way. Glucocorticoid hormones stimulate many genes that encode proteins involved in several different cellular processes, including the synthesis of glucose, the breakdown of proteins, and the mobilization of fats.

The CREB Protein Is an Example of a Regulatory Transcription Factor Modulated by Protein-Protein Interaction and Covalent Modification

As we have just seen, steroid hormones function as hormonal signaling molecules that bind directly to regulatory transcription factors to alter their function. This enables a cell to respond to a hormone by up regulating a particular set of genes. Most extracellular signaling molecules, however, do not enter the cell and

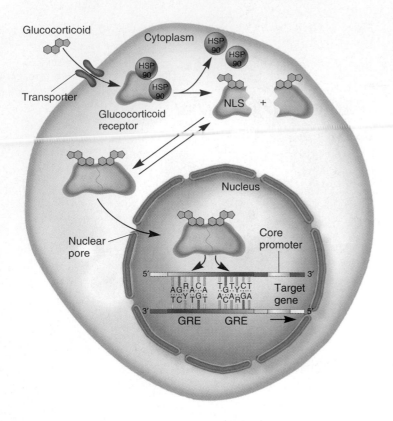

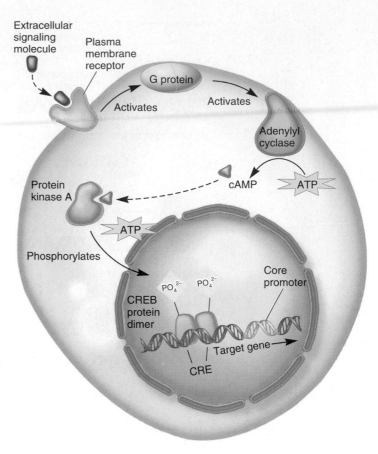

FIGURE 15.6 **The action of glucocorticoid hormones.** Once inside the cell, the hormone binds to the glucocorticoid receptor, releasing it from a protein known as HSP90. This exposes an NLS. Two glucocorticoid receptors then form a dimer and travel into the nucleus, where the dimer binds to glucocorticoid response elements (GREs) that are next to particular genes. The binding of the glucocorticoid receptors to the GREs activates the transcription of the adjacent target gene.

GENES→TRAITS Glucocorticoid hormones are produced by the endocrine glands in response to eating and activity. They enable the body to regulate its metabolism properly. When glucocorticoids are produced, they are taken into cells and bind to the glucocorticoid receptors. This eventually leads to the activation of genes that encode proteins involved in the synthesis of glucose, the breakdown of proteins, and the mobilization of fats.

FIGURE 15.7 **The activity of the CREB protein.** When an extracellular signaling molecule binds to a receptor in the plasma membrane, this activates a G protein, which activates adenylyl cyclase, leading to the synthesis of cAMP. cAMP activates protein kinase A, which travels into the nucleus and phosphorylates the CREB protein. Once phosphorylated, the CREB protein acts as a transcriptional activator.

bind directly to transcription factors. Instead, most signaling molecules must bind to receptors in the plasma membrane. This binding activates the receptor and leads to the synthesis of an intracellular signal that causes a cellular response. One type of cellular response is to affect the transcription of particular genes within the cell.

As our second example of regulatory transcription factor function within living cells, we will examine the **cAMP response element–binding (CREB) protein.** The CREB protein is a regulatory transcription factor that becomes activated in response to cell-signaling molecules that cause an increase in the cytoplasmic concentration of the molecule **cyclic-adenosine monophosphate (cAMP).** This transcription factor recognizes a response element with the consensus sequence 5′–TGACGTCA–3′. This response

element, which is found near many different genes, has been termed a **cAMP response element (CRE).**

Figure 15.7 shows the steps leading to the activation of the CREB protein. A wide variety of hormones, growth factors, neurotransmitters, and other signaling molecules can bind to plasma membrane receptors to initiate an intracellular response. In this case, the response involves the production of a second messenger, cAMP. The extracellular signaling molecule itself is considered the primary messenger. When the signaling molecule binds to the receptor, it activates a G protein that subsequently activates the enzyme adenylyl cyclase. The activated adenylyl cyclase catalyzes the synthesis of cAMP. The cAMP molecule then activates a second enzyme, protein kinase A. This enzyme can phosphorylate several different cellular proteins, including the CREB protein. When phosphorylated, CREB proteins stimulate transcription. In contrast, unphosphorylated CREB proteins can still bind to CREs but do not activate RNA polymerase.

15.2 CHANGES IN CHROMATIN STRUCTURE

Changes in chromatin structure can involve direct alterations in the structure of DNA and/or changes in the level of chromosomal compaction. Since genes are segments of DNA, it is not surprising that alterations in DNA structure can affect gene expression in a variety of ways (table 15.1). One possibility is that the number of copies of a gene can be increased by **gene amplification.** However, this is an uncommon way to regulate gene expression. Another change in DNA that affects gene expression is the rearrangement of the DNA structure. In chapter 17, we will see that **gene rearrangement** occurs in the DNA of immune system cells to generate a diverse array of antibody proteins. A third, more common way to alter DNA structure is **DNA methylation.**

Later in this section, we will examine how DNA methylation occurs and how it can regulate genes in a tissue-specific manner.

As discussed in chapter 12, changes in chromatin compaction are a vital step in gene transcription. For this reason, the regulation of gene transcription is not simply dependent on the activity of regulatory transcription factors that influence RNA polymerase via TFIID and mediator to turn genes on or off. Transcription must also involve changes in chromatin structure that affect the ability of transcription factors to gain access to and bind their target sequences in the promoter region. Since DNA is found within chromosomes, the three-dimensional packing of chromatin is an important parameter affecting gene expression. If the chromatin is very tightly packed or in a **closed conformation,** transcription may be difficult or impossible. By comparison, chromatin that is in an **open conformation** is more easily accessible to transcription factors and RNA polymerase, so that transcription can take place. In this section, we will begin by considering how the conversion between the closed and open conformations is associated with the expression of genes. We will then examine molecular mechanisms that explain how chromosome compaction in the vicinity of specific genes is altered.

TABLE 15.1

Gene Regulation That Occurs via Changes in DNA or Chromatin Structure

Modification	Description
Gene amplification	Though not a common method of gene regulation, gene amplification involves an increase in gene number. For example, in the oocytes of *Xenopus laevis* (a South African frog), the number of rRNA genes is increased through gene amplification. The mechanism involves an unusual DNA replication event in which the rRNA genes are replicated to form many copies of circular DNA molecules called minichromosomes.
Gene rearrangement	Again, this is not a common method of gene regulation. As discussed in chapter 17, gene rearrangement occurs within immunoglobulin genes to generate a diverse array of genes that encode antibody proteins.
DNA methylation	A common DNA modification in vertebrates and plants that involves the attachment of methyl groups to cytosine bases. When DNA methylation occurs in the promoter region, it inhibits transcription. This inhibition may be due to the inability of transcription factors to recognize the promoter region and/or a conversion of the chromatin to the closed conformation.
Chromatin compaction	This is a common way to regulate eukaryotic genes. For a gene to be transcribed, it must be converted from a closed (highly compacted) conformation to an open conformation. Histone acetyltransferases and ATP-dependent remodeling enzymes are needed for this conversion. Transcriptional activators recruit these enzymes to promoter regions to activate gene transcription.

Gene Accessibility Can Be Controlled by Changes in Chromatin Structure

Chromatin is composed of DNA and proteins that are organized into a compact structure that fits inside the nucleus of the cell. The DNA is wound around histone proteins to form nucleosomes that are 11 nm in diameter. This 11 nm fiber is condensed further to a 30 nm fiber, which is the predominant form of euchromatin found within the nucleus during interphase, when gene expression primarily occurs. Nevertheless, chromatin is a very dynamic structure that can alternate between highly condensed and highly extended conformations. The dynamic nature of chromatin is important in regulating gene transcription.

Variations in the degree of chromatin packing occur along the length of eukaryotic chromosomes during interphase. Tightly packed chromatin in a closed conformation cannot be transcribed. During gene activation, such chromatin must be converted to an open conformation that is less tightly packed than a 30 nm fiber. In certain cells, researchers can microscopically observe the decondensation of the 30 nm fiber when transcription is occurring. Figure 15.8 shows photomicrographs of a chromosome from an amphibian oocyte, the genes of which are being actively transcribed. This chromosome does not form a uniform, compact 30 nm fiber. Instead, many decondensed loops radiate from the central axis of the chromosome. These loops are regions of DNA in which the genes are being actively transcribed. These chromosomes have been named *lampbrush chromosomes,* because their feathery appearance resembles the brushes that were once used to clean kerosene lamps.

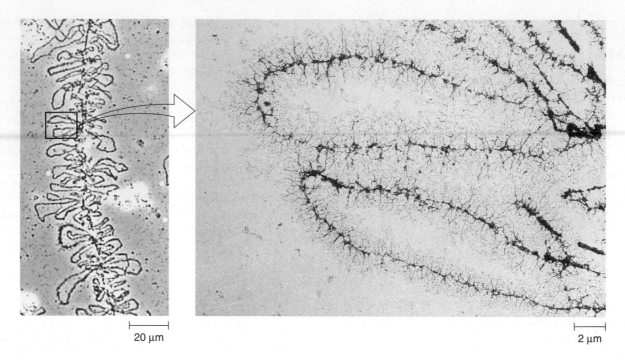

20 μm   2 μm

FIGURE 15.8   Lampbrush chromosomes in amphibian oocytes. The loops that radiate from a chromosome are regions of DNA containing genes that are being actively transcribed.

EXPERIMENT 15A

DNase I Sensitivity Can Be Used to Study Changes in Chromatin Structure

To understand the interconversions between open and closed chromatin conformations, methods are needed to evaluate the degree of chromatin packing as it occurs in living cells. An important technique is the use of DNase I to monitor DNA conformation. As you may recall from chapter 10 (see fig. 10.15), DNase I is an endonuclease that cleaves DNA. DNase I is much more likely to cleave DNA in an open conformation than a closed conformation, because a looser conformation allows greater accessibility of DNase I to the DNA.

In 1976, Harold Weintraub and Mark Groudine used DNase I sensitivity as a tool to evaluate differences in chromatin structure that occur when a gene is actively transcribed. In particular, they focused their attention on the β-globin gene. Humans possess several different genes that encode globin polypeptides. These polypeptides are the subunits of the oxygen-carrying protein hemoglobin. Prior to this work, it had been well established that globin genes are specifically expressed in reticulocytes

(immature red blood cells) but not in other cell types such as brain cells and fibroblasts. Weintraub and Groudine asked the question, "Is there a difference in the chromatin packing of globin genes in cells that can actively transcribe the globin genes compared with that in cells in which the globin genes are turned off?" To answer this question, they used DNase I sensitivity to compare the degree of globin gene packing in reticulocytes versus that in brain cells and fibroblasts.

Before discussing the steps in their experimental protocol, it is helpful if we consider the rationale behind their experimental approach (fig. 15.9). Since the globin genes are but a small part of the total chromosomal DNA, having a way to specifically monitor the digestion of the β-globin gene was a vital aspect of Weintraub and Groudine's experimental protocol. They accomplished this by using a cloned fragment of DNA (i.e., a probe) that was complementary to the β-globin gene. This fragment was hybridized to the chromosomal DNA to determine specifically if the chromosomal β-globin gene was intact. If DNase I had digested the chromosomal β-globin gene, it would not hybridize to the probe DNA because the corresponding chromosomal DNA would have been

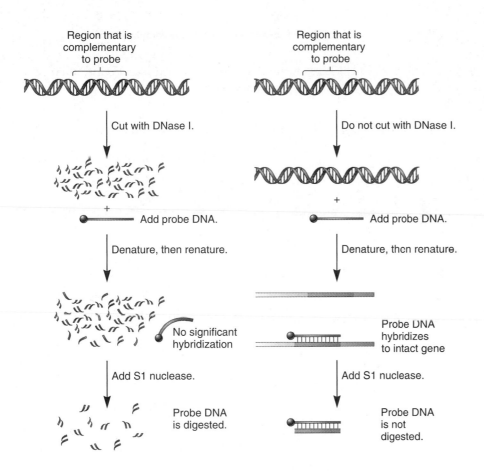

Region that is complementary to probe

Cut with DNase I.

+

Add probe DNA.

Denature, then renature.

No significant hybridization

Add S1 nuclease.

Probe DNA is digested.

Region that is complementary to probe

Do not cut with DNase I.

+

Add probe DNA.

Denature, then renature.

Probe DNA hybridizes to intact gene

Add S1 nuclease.

Probe DNA is not digested.

FIGURE 15.9   The use of DNase I and S1 nuclease to probe the chromatin structure of the β-globin gene. On the *left* side, the DNA has been digested into small pieces via DNase I. These small pieces cannot hybridize with the probe DNA. When S1 nuclease is added, single-stranded probe DNA is digested. On the *right* side, the DNA is not cut with DNase I. The probe DNA can hybridize with a strand of chromosomal DNA. When S1 nuclease is added, it will digest only the single-stranded regions of DNA, not the double-stranded region where the probe DNA and chromosomal DNA are bound to each other.

digested. However, if the chromosomal β-globin gene had not been digested by DNase I, it could hybridize to the probe (fig. 15.9).

In Weintraub and Groudine's experiments, a cloned DNA probe was radiolabeled to allow detection of its presence. Following hybridization, the samples were then exposed to another enzyme, known as S1 nuclease. This enzyme digests DNA strands but, interestingly, cuts DNA only when it is single stranded, not when it is double stranded. As shown in figure 15.9, S1 nuclease would digest the radiolabeled DNA probe if it had not hybridized with the complementary chromosomal strand. In contrast, S1 nuclease would be unable to digest the probe if the chromosomal DNA remained intact and hybridized to the probe. In other words, an intact strand of chromosomal DNA would protect the probe from S1 nuclease digestion.

After S1 incubation, Weintraub and Groudine reasoned that their sample would contain an intact radiolabeled DNA probe if the chromosomal DNA were in a closed conformation. This is because DNase I would not digest the chromosomal gene, and therefore, the β-globin gene would be available to hybridize to the radiolabeled DNA probe. However, if the chromosomal globin gene were in an open conformation, DNase I would digest the chromosomal DNA, preventing it from hybridizing with the radiolabeled DNA probe. In this situation, the radiolabeled DNA strand would be digested by S1 nuclease. Therefore, the susceptibility of the radiolabeled DNA to S1 nuclease digestion allowed

them to evaluate whether the chromosomal globin gene was in an open or closed conformation.

With these ideas in mind, let's examine the steps in their experimental procedure (fig. 15.10). They began with three cell types: reticulocytes, brain cells, and fibroblasts. Only the reticulocytes express the β-globin gene. The nuclei were extracted from these cells and incubated with DNase I. It was expected that if the globin gene was expressed, the DNA would be accessible to digestion. The DNA was then isolated from the three types of nuclei. Following the isolation procedure, the chromosomal DNA was broken into fragments, but the average fragment size would still be larger than the probe DNA. The DNA fragments and probe were mixed together, denatured, and then allowed to hybridize. The cooled samples were then divided into two tubes each. Into one of the two tubes, S1 nuclease was added. A key point is that the probe would not be digested by S1 nuclease if it had hybridized to a complementary strand from the globin gene. The DNA fragments were then precipitated using trichloroacetic acid. The amount of radioactivity in the pellet was then determined by scintillation counting.

THE HYPOTHESIS

A loosening of chromatin structure occurs when β-globin genes are transcriptionally active.

■ **TESTING THE HYPOTHESIS** — **FIGURE 15.10** Using DNase I sensitivity to study the compaction of the β-globin gene.

Starting material: Nuclei were isolated from three different cell types in chicken: reticulocytes (immature red blood cells), brain cells, and fibroblasts. The globin genes are expressed in red blood cells but not in brain cells and fibroblasts.

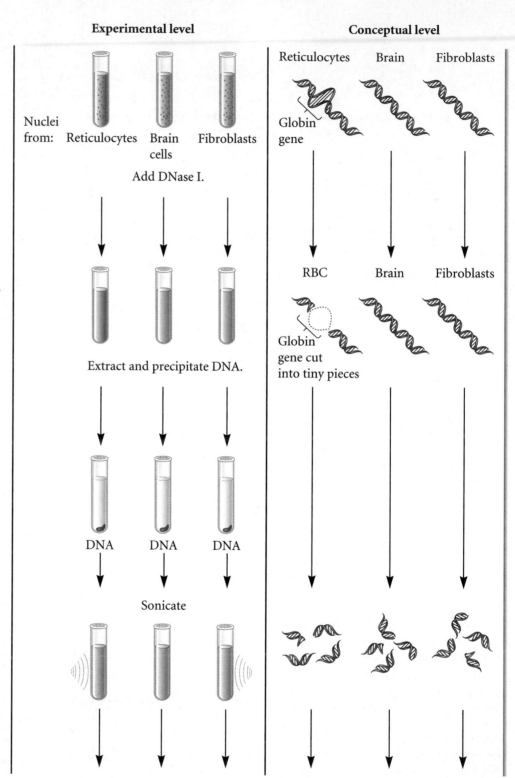

1. Treat all three types of nuclei with the same amount of DNase I.

2. Extract the DNA from the nuclei. This involves lysing the nuclei with detergent, removing the protein by treatment with a phenol–chloroform mixture, and then precipitating the DNA by adding ethanol. The precipitated DNA forms a pellet at the bottom of a test tube following centrifugation. The DNA at the bottom of the tube can then be resuspended in an appropriate solution for the next step.

3. Subject the DNA to sound waves (i.e., sonication) to break the DNA into fragments of an average length of 500 bp.

4. Add a radiolabeled DNA probe that is complementary to the β-globin gene.

5. Denature the DNA into single strands by treatment with high temperature. Then cool it to 65°C to allow the complementary DNA strands to hybridize with each other.

6. Reduce the temperature and divide each sample into two tubes. Into one tube of each sample add S1 nuclease. Omit S1 nuclease from the second tube.

7. Precipitate the DNA with trichloroacetic acid.

8. Count the amount of radioactivity. (The technique of scintillation counting is described in the appendix.) The amount of radioactivity in the presence of S1 nuclease divided by the amount of radioactivity in the absence of S1 nuclease provides a measure of the percentage of radiolabeled DNA that has hybridized to the chromosomal DNA.

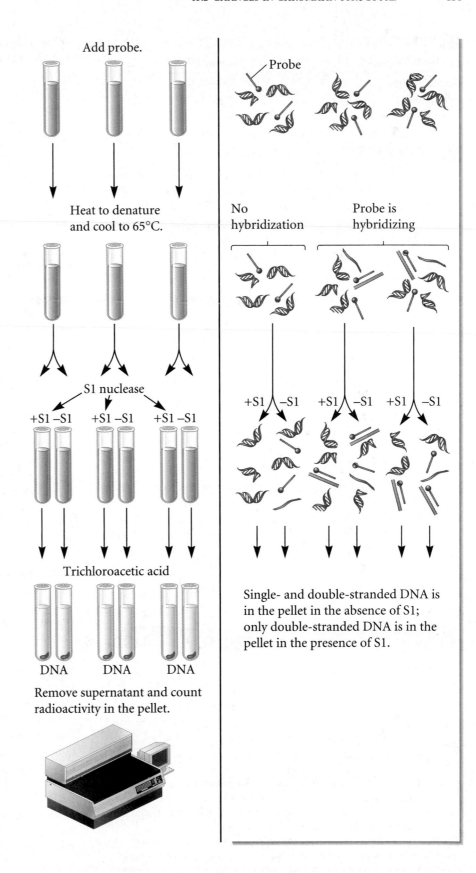

Add probe.

Heat to denature and cool to 65°C.

S1 nuclease

+S1 −S1   +S1 −S1   +S1 −S1

Trichloroacetic acid

DNA   DNA   DNA

Remove supernatant and count radioactivity in the pellet.

Probe

No hybridization   Probe is hybridizing

+S1 −S1   +S1 −S1   +S1 −S1

Single- and double-stranded DNA is in the pellet in the absence of S1; only double-stranded DNA is in the pellet in the presence of S1.

■ THE DATA

Source of Nuclei	% Hybridization of DNA Probe
Reticulocytes	25%
Brain cells	>94%
Fibroblasts	>94%

■ INTERPRETING THE DATA

As shown in the data of figure 15.10, a much smaller percentage of the radiolabeled DNA probe hybridized to the chromosomal DNA in reticulocytes compared to brain cells and fibroblasts. These results indicated that DNase I had digested the globin gene in the chromatin of reticulocytes into small fragments that were too small to hybridize to the radiolabeled DNA probe. In other words, the globin genes in reticulocytes were more sensitive to

DNase I digestion. By comparison, the globin genes in brain cells and fibroblasts were relatively resistant. Since the globin genes are expressed in reticulocytes but not in brain cells and fibroblasts, these results are consistent with the hypothesis that the globin gene is less tightly packed when it is being expressed. In cells where the globin genes should not be expressed (e.g., brain cells and fibroblasts), the chromatin containing the β-globin gene is tightly compacted. However, in reticulocytes where this gene is expressed, the chromatin is more loosely packed so that transcription of the globin genes can occur. This phenomenon provides one way to regulate globin gene expression among different cell types.

> *A self-help quiz involving this experiment can be found at the Online Learning Center.*

Chromatin Packing and Nucleosome Location Are Altered During Globin Gene Expression

Since the studies of Weintraub and Groudine, more detailed information has been gathered from molecular research in globin gene expression. Before discussing these studies, let's briefly consider globin gene organization and expression. Although the family of globin genes is expressed in reticulocytes, individual members are expressed at different stages of development. For example, β-globin is expressed in adult reticulocytes, whereas γ_A-globin is expressed in fetal cells. As shown in figure 15.11a, several of the globin genes are located adjacent to each other on the same chromosome.

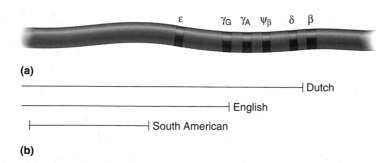

(a)

(b)

FIGURE 15.11 **Globin genes and thalassemias. (a)** The organization of globin genes on human chromosome 11. **(b)** These lines depict the segments of DNA that are deleted in certain thalassemias found in Dutch, English, and South American populations.

GENES→TRAITS When the segment of DNA depicted by the lines is deleted, as has been found in some Dutch, English, and South American populations, the β-globin gene is not expressed even though it is still present. This is because the locus control region (LCR), which is necessary for the expression of the globin genes, is missing. Without the LCR, the β-globin gene is not expressed, and this causes a blood disorder known as thalassemia, which is a type of hemoglobinopathy.

Worldwide studies have been conducted that focus on inherited defects in hemoglobin composition known as hemoglobinopathies. When an individual shows a defect in the expression of one or more globin genes, this type of hemoglobinopathy is called thalassemia. An intriguing observation among certain patients is that they cannot synthesize β-globin even though the DNA sequence of the β-globin gene is perfectly normal. As shown in figure 15.11b, this type of thalassemia has been found in Dutch, English, and South American populations. It involves a DNA deletion that occurs upstream from the β-globin gene, although the β-globin gene itself is intact. Nevertheless, the β-globin gene is turned off in these patients. This unexpected finding prompted further investigations into the regulation of the globin genes.

Since the initial studies of this type of thalassemia, a DNA region upstream from the β-globin gene has been identified as necessary for globin gene expression. This region, known as a **locus control region (LCR),** is involved in the regulation of chromatin opening and closing. It is missing in certain persons with thalassemias. As depicted in figure 15.12, this DNA locus provides recognition sites for proteins that promote the general opening of the entire region. This gives RNA polymerase and transcription factors access to the region. Not only does the locus control region affect the transcription of β-globin, it also influences the other globin genes in this region.

Aside from the degree of chromatin packing, a second structural issue to consider is the position of nucleosomes. In chromatin, the nucleosomes are usually positioned at regular intervals along the DNA. The position of a nucleosome may greatly influence whether or not a gene can be transcribed. For example, if the TATA box is tightly bound to the histone core, it may be inaccessible to general transcription factors and RNA polymerase. Nucleosomes have been shown to change position in cells that normally express a particular gene but not in cells

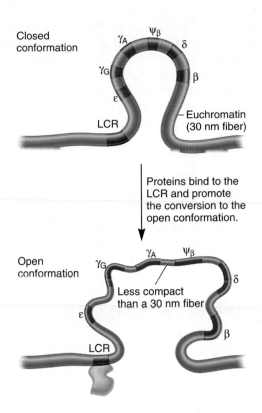

FIGURE 15.12   The function of the globin locus control region. In reticulocytes that express globin genes, proteins may bind to the locus control region (LCR) and promote the general opening of chromatin in the region, so that all the globin genes are accessible to transcription factors and RNA polymerase.

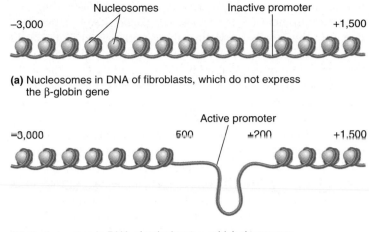

(a) Nucleosomes in DNA of fibroblasts, which do not express the β-globin gene

(b) Nucleosomes in DNA of reticulocytes, which do express the β-globin gene

FIGURE 15.13   Changes in nucleosome position during the activation of the β-globin gene. (a) In fibroblasts, which do not express β-globin, nucleosome positioning is uninterrupted. **(b)** In reticulocytes that express this gene, disruption occurs in the positions of nucleosomes from the −500 to +200 region. (Position +1 is the beginning of transcription.) It is not yet clear whether the histones are removed, the histones are partially displaced, and/or other proteins are bound to this region.

where the gene is inactive. For example, in fibroblasts that do not express the β-globin gene, nucleosomes are positioned at regular intervals from nucleotides −3,000 to +1,500 (fig. 15.13a). However, in reticulocytes that can express the β-globin gene, a disruption in nucleosome positioning occurs in the region from nucleotide −500 to +200 (fig. 15.13b). This disruption may be an important first step in gene activation.

Transcriptional Activators Recruit Chromatin-Remodeling Enzymes to the Promoter Region

In recent years, geneticists have been trying to identify the steps that promote the interconversion between the closed and open conformations of chromatin. As discussed for globin genes, changes in chromatin structure occur in specific regions and are promoted by DNA sequences such as the locus control region. Based on the analysis of many genes, researchers have discovered that a key role of some transcriptional activators is to orchestrate changes in chromatin compaction from the closed to the open conformation. In other words, one of the key roles of certain gene-specific activator proteins is to target promoter regions for the unfolding or remodeling of chromatin structure.

As discussed in chapter 12, there are two common ways that chromatin structure is altered. First, nucleosome arrangement may be altered by a process called **ATP-dependent chromatin remodeling.** The complexes that catalyze ATP-dependent chromatin remodeling are members of the SWI/SNF family, which is a large group of proteins found in all eukaryotic species. The net result of ATP-dependent chromatin remodeling is a change in the locations of nucleosomes. This may involve a shift in nucleosomes to a new location or a change in the spacing of nucleosomes over a long stretch of DNA. As mentioned in chapter 12, the acronyms SWI and SNF refer to the effects that occur in yeast when these remodeling enzymes are defective. Mutants in SWI were defective in mating type switching, and mutations in SNF created a sucrose nonfermenting phenotype.

A second mechanism that leads to an alteration in chromatin structure involves the covalent modification of histones. The amino terminal ends of histone proteins are covalently modified in several ways including the acetylation of lysines, methylation of lysines, and phosphorylation of serines. Because these covalent modifications influence the compaction of chromatin, they play a key role in transcriptional regulation. For example, positively charged lysine residues within the core histone proteins can be acetylated by enzymes called **histone acetyltransferases.** The attachment of the acetyl group (−COCH$_3$) may have two effects. First, it eliminates the positive charge on the lysine side chain and thereby disrupts the favorable interaction between the histone protein and the negatively charged DNA backbone. Secondly, histones that are covalently modified are recognized by DNA-binding proteins that promote changes in the compaction level of the chromatin. In the case of acetylation, the net effect is to diminish the interactions between DNA and nucleosomes and thereby loosen the compaction of the 30 nm fiber. This would facilitate the ability of RNA polymerase to transcribe a gene. Some studies suggest that histones are completely displaced, whereas others suggest they are loosened but remain attached to the DNA.

An important role of transcriptional activators is to recruit histone acetyltransferase and ATP-dependent remodeling enzymes to the promoter region. Though the order of recruitment may differ among specific transcriptional activators, this appears to be a critical step that is necessary to initiate transcription. A well-studied example involves a gene in yeast that is involved in mating. Yeast can exist in two mating types, termed *a* and *α*. The yeast *HO* gene encodes an enzyme that is required for the switch from one mating type to the other.

A proposed mechanism for the transcription of this gene is shown in figure 15.14. A regulatory transcription factor (SWI5P) binds to an enhancer element in the vicinity of the core promoter. This transcription factor then recruits an ATP-dependent remodeling enzyme (SWI/SNF) to the region, which promotes the conversion of the chromatin from the closed to the open conformation. Next, a histone acetyltransferase (SAGA) is recruited to this region, which serves to further open the chromatin conformation. Overall, the actions of the ATP-dependent remodeling enzyme and the histone acetyltransferase remodel a region around the promoter that is approximately 1,000 bp in length. A second regulatory transcription factor (SBP) is then able to bind to an enhancer element near the core promoter and thereby initiate the binding of general transcription factors and RNA polymerase.

DNA Methylation Inhibits Gene Transcription

We now turn our attention to a change in chromatin structure that silences gene expression. As discussed in previous chapters,

DNA structure can be modified by the covalent attachment of methyl groups. **DNA methylation** is common in some eukaryotic species but certainly not all. For example, yeast and *Drosophila* have little or no detectable methylation of their DNA, whereas DNA methylation in vertebrates and plants is relatively abundant. In mammals, approximately 2 to 7% of the DNA is methylated. As shown in figure 15.15, eukaryotic DNA methylation occurs on

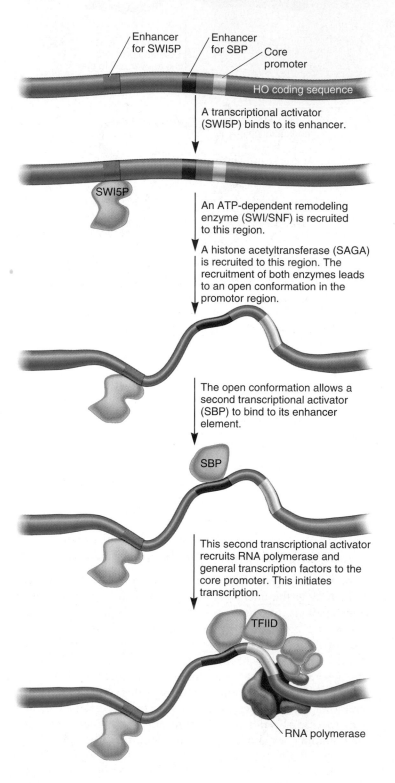

Enhancer for SWI5P   Enhancer for SBP   Core promoter

HO coding sequence

A transcriptional activator (SWI5P) binds to its enhancer.

SWI5P

An ATP-dependent remodeling enzyme (SWI/SNF) is recruited to this region.

A histone acetyltransferase (SAGA) is recruited to this region. The recruitment of both enzymes leads to an open conformation in the promotor region.

The open conformation allows a second transcriptional activator (SBP) to bind to its enhancer element.

SBP

This second transcriptional activator recruits RNA polymerase and general transcription factors to the core promoter. This initiates transcription.

TFIID

RNA polymerase

FIGURE 15.14 **An example of chromatin remodeling of a promoter region that leads to transcriptional activation.** The *HO* gene is involved in yeast mating type switching. For this gene to be transcribed, the promoter region must be converted from a closed to an open conformation. A transcription factor termed SWI5P (SWI refers to mating type <u>swi</u>tching) binds to an enhancer in the promoter region and recruits an ATP-dependent remodeling enzyme to the region. This begins the opening process. A histone acetyltransferase termed *SAGA* is then recruited to the region and further decompacts the chromatin. SAGA is an acronym for <u>Spt</u>/<u>Ada</u>/<u>GCN5</u>/<u>A</u>cetyltransferase. *Spt, Ada,* and *GCN5* are the names of three genes that had been shown to be transcriptionally regulated by this histone acetyltransferase. A second transcription factor termed *SBP* (an acronym for a mating type <u>s</u>witching cell cycle <u>b</u>ox <u>p</u>rotein) binds to an enhancer in the promoter region and initiates transcription by recruiting general transcription factors and RNA polymerase.

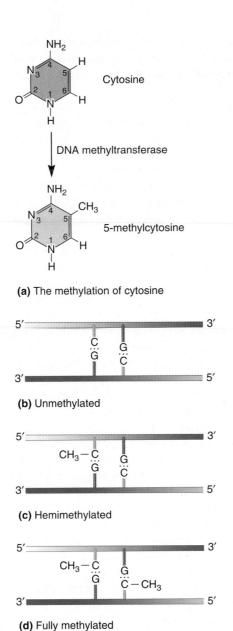

FIGURE 15.15   **DNA methylation on cytosine residues.**
(a) Methylation occurs via an enzyme known as DNA methylase or
DNA methyltransferase, which attaches a methyl group to the number
5 carbon on cytosine. A CG sequence can be (b) unmethylated,
(c) hemimethylated, or (d) fully methylated.

cytosine residues at the number 5 position of the cytosine base.
The sequence that is methylated is

$$
\begin{array}{c}
CH_3 \\
| \\
5'-CG-3' \\
3'-GC-5' \\
| \\
CH_3
\end{array}
$$

Note that this sequence contains cytosines in both strands. When
only one strand is methylated, this is called *hemimethylation,*

whereas the methylation of the cytosines in both strands is
termed *full methylation.*

DNA methylation usually inhibits the transcription of
eukaryotic genes, particularly when it occurs in the vicinity of the
promoter. In vertebrates and plants, many genes contain **CpG
islands** near their promoters. (Note: CpG refers to a dinucleotide
of C and G in DNA that is connected by a phosphodiester linkage.) These CpG islands are commonly 1,000 to 2,000 base pairs
in length and contain many CpG sites. In the case of **housekeeping genes,** which encode proteins that are required in most cells
of a multicellular organism, the cytosine bases in the CpG islands
are unmethylated. Therefore, housekeeping genes are not inhibited and tend to be expressed in most cell types. By comparison,
other genes are highly regulated and may be expressed only in a
particular cell type. These are **tissue-specific genes.** In some
cases, it has been found that the expression of such genes may be
silenced by the methylation of CpG islands. Overall, evidence is
accumulating that unmethylated CpG islands are correlated with
active genes, while suppressed genes contain methylated CpG
islands. In this way, DNA methylation may play an important
role in the silencing of tissue-specific genes.

There are two general ways that methylation can inhibit
transcription. First, methylation of CpG islands may prevent the
binding of transcription factors to the promoter region (fig.
15.16*a*). For example, methylated CG sequences could prevent the
binding of an activator protein to an enhancer element, presumably by the methyl group protruding into the major groove of the
DNA. The inability of an activator protein to bind to the DNA
would inhibit the initiation of transcription. However, CG methylation does not slow down the movement of RNA polymerase
along a gene. In vertebrates and plants, coding regions downstream from the core promoter usually contain methylated CG
sequences, but these do not hinder the elongation phase of transcription. This suggests that methylation must occur in the vicinity of the promoter to have an inhibitory effect on transcription.

A second way that methylation inhibits transcription is by
converting chromatin from an open to a closed conformation
(fig. 15.16*b*). Proteins known as **methyl-CpG-binding proteins**
bind methylated sequences. These proteins contain a domain
called the methyl-binding domain that specifically recognizes a
methylated CG sequence. Once bound to the DNA, the methyl-
CpG-binding protein recruits other proteins to the region that
cause the chromatin to become very compact. For example,
methyl-CpG-binding proteins may recruit histone deacetylase to
a methylated CpG island near a gene promoter. Histone deacety-
lation removes acetyl groups from the histone proteins and
thereby favors a chromatin conformation that is very compact.

DNA Methylation Is Heritable

Methylated DNA sequences are inherited during cell division.
Experimentally, if fully methylated DNA is introduced into a
plant or vertebrate cell, the DNA will remain fully methylated
even in subsequently produced daughter cells. However, if the
same sequence of nonmethylated DNA is introduced into a cell, it

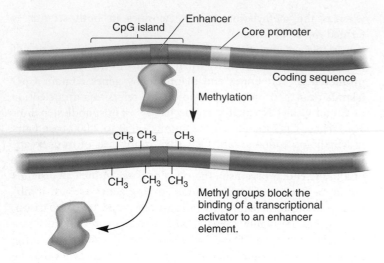

Methyl groups block the binding of a transcriptional activator to an enhancer element.

(a) Methylation inhibits the binding of transcriptional activators.

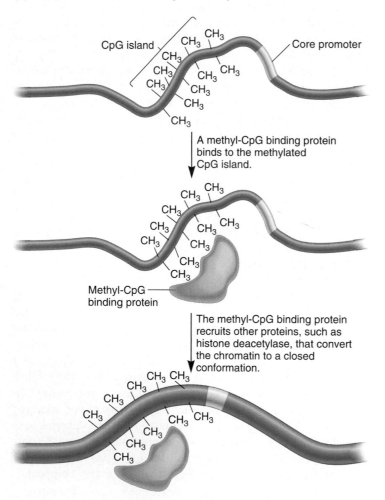

A methyl-CpG binding protein binds to the methylated CpG island.

Methyl-CpG binding protein

The methyl-CpG binding protein recruits other proteins, such as histone deacetylase, that convert the chromatin to a closed conformation.

(b) Methyl-CpG binding protein recruits other proteins that cause the region to become more compact.

FIGURE 15.16   Transcriptional silencing via methylation. (a) The methylation of a CpG island may inhibit the binding of transcriptional activators to the promoter region. **(b)** The binding of a methyl-CpG-binding protein to a CpG island may lead to the recruitment of proteins that suppress transcription. For example, a methyl-CpG-binding protein may recruit a histone deacetylase to the promoter region and thereby favor a closed conformation of chromatin.

will remain nonmethylated in the daughter cells. These observations indicate that the pattern of methylation is retained following DNA replication and, therefore, is inherited in future daughter cells.

Figure 15.17 illustrates a molecular model that explains how methylation can be passed from parent to daughter cell. Arthur Riggs, Robin Holliday, and J. E. Pugh originally suggested this model. The DNA in a particular cell becomes methylated in both strands by ***de novo* methylation** (i.e., new methylation). When this cell divides, it will replicate its DNA and synthesize complementary daughter strands. Initially, the newly made daughter strands contain nonmethylated cytosines. This hemimethylated DNA is efficiently recognized by **DNA methylase** (also called **DNA methyltransferase**), which makes it fully methylated. This process is called **maintenance methylation,** since it preserves the methylated condition in future cells. Overall, maintenance methylation appears to be an efficient process that routinely occurs within vertebrate and plant cells. By comparison, *de novo* methylation and demethylation are infrequent and highly regulated events. According to this view, the initial methylation or demethylation of a given gene can be regulated so that it occurs in a specific cell type or stage of development. Once methylation has occurred, it can then be transmitted from mother to daughter cells via maintenance methylation.

The methylation mechanism shown in figure 15.17 can explain the phenomenon of genomic imprinting, which is described in chapter 7. In this case, specific genes are methylated during oogenesis or spermatogenesis, but not both. Following fertilization, the pattern of methylation is maintained in the offspring. For example, if a gene is methylated only during spermatogenesis, the allele that is inherited from the father will be methylated in the somatic cells of the offspring, while the maternal allele will remain unmethylated. Along these lines, geneticists are also eager to determine how variations in DNA methylation patterns may be important for cell differentiation. It may be a key way to silence genes in different cell types. However, additional research will be necessary to understand how specific genes may be targeted for *de novo* methylation or demethylation during different developmental stages or in specific cell types.

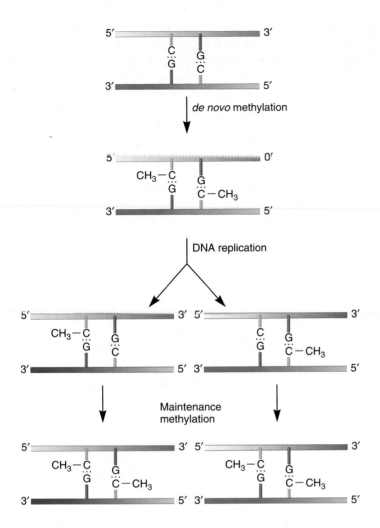

FIGURE 15.17 **A molecular model for the inheritance of DNA methylation.** The DNA initially becomes methylated *de novo*. The *de novo* methylation is a rare, highly regulated event. Once this occurs, DNA replication produces hemimethylated DNA molecules, which are then fully methylated by a maintenance DNA methylase. Maintenance methylation is a routine event that is expected to occur for all hemimethylated DNA.

TABLE 15.2

Gene Regulation via RNA Processing and Translation

Effect	Description
Alternative splicing	Certain pre-mRNAs can be spliced in more than one way, leading to polypeptides that have different amino acid sequences. Alternative splicing is often cell specific so that a protein can be fine-tuned to function in a particular cell type. It is an important form of gene regulation in multicellular eukaryotic species.
RNA editing	The sequence of RNAs can be altered after the RNA is made. This usually involves the addition or deletion of one or a few bases, or a change of C to U, or A to I. This does not appear to be a widespread phenomenon, though the study of RNA editing is still relatively recent.
RNA stability	The amount of RNA is greatly influenced by the half-life of RNA transcripts. A long polyA tail promotes stability due to the binding of polyA binding protein. Some RNAs with a relatively short half-life contain sequences that target them for rapid destruction. Some RNAs are stabilized by specific RNA-binding proteins that usually bind near the 3' end.
RNA interference	Double-stranded RNA can mediate the degradation of homologous mRNAs in the cell. This mechanism probably provides eukaryotic cells with protection from invasion by certain types of viruses. It may also prevent the movement of transposable elements.
General regulation of translation	The function of translational initiation factors may be regulated to permit or inhibit translation. This regulation affects all cellular mRNAs. Inhibition of translation is desirable if a cell has been exposed to a virus or to toxic materials.
Translational regulation of specific mRNAs	Some mRNAs are regulated via binding proteins that inhibit the ability of the ribosomes to initiate translation. These proteins usually bind at the 5' end of the mRNA and thereby prevent the ribosome from binding.

15.3 REGULATION OF RNA PROCESSING AND TRANSLATION

Thus far, we have considered a variety of mechanisms that regulate the level of gene transcription. In eukaryotic species, it is also common for gene expression to be regulated at the RNA level. The function of RNA can be controlled in several general ways (table 15.2). One way is pre-mRNA processing. Following transcription, a pre-mRNA transcript is processed before it becomes a functional mRNA. These processing events, which include splicing, capping, polyA tailing, and RNA editing, were described in chapter 12. We will see shortly how alternative splicing and RNA editing are regulated at the RNA level.

Another strategy for regulating gene expression is to influence the concentration of mRNA. As we have seen in chapter 15,

this can be accomplished by regulating the rate of transcription. When the transcription of a gene is increased, a higher concentration of the corresponding RNA results. In addition, RNA concentration is greatly affected by the stability or half-life of a particular RNA. Factors that increase RNA stability are expected to raise the concentration of that RNA molecule. Later in chapter 15, we will examine how sequences within mRNA molecules greatly affect their stability. In addition, a newly discovered mechanism of RNA degradation, known as RNA interference, involves double-stranded RNAs that direct the breakdown of specific RNAs within the cell.

Finally, another way to regulate RNA function is to control the ability of mRNAs to be translated. As we have seen in chapter 13, mRNA translation relies on the translational machinery (e.g., ribosomes, tRNA, etc.). In eukaryotes, one important mechanism for regulating translation is to alter the rate of translation via the ribosomal machinery. This occurs in two ways. One is to directly affect the function of translational initiation factors. This has the general effect of increasing or decreasing the translation of many mRNAs within the cell. Second, RNA-binding proteins can prevent ribosomes from initiating the translation process for specific mRNAs. In this section, we will examine both of these mechanisms for regulating mRNA translation.

Alternative Splicing Regulates Which Exons Occur in an RNA Transcript, Allowing Different Proteins to Be Made from the Same Structural Gene

When it was first discovered, the phenomenon of splicing seemed like a rather wasteful process. During transcription, energy is used to synthesize intron sequences. Likewise, energy is also used to remove introns via a large spliceosome complex. This observation troubled many geneticists, because evolution tends to select against wasteful processes. Therefore, instead of simply viewing splicing as a wasteful process, many geneticists expected to find that pre-mRNA splicing has one or more important biological roles. In recent years, one very important biological advantage has become apparent. This is **alternative splicing,** which refers to the phenomenon that a pre-mRNA can be spliced in more than one way.

To understand the biological effects of alternative splicing, remember that the sequence of amino acids within a polypeptide determines the structure and function of a protein. Alternative splicing produces two (or more) polypeptides with differences in their amino acid sequences, leading to possible changes in their functions. In most cases, the alternative versions of the protein will have similar functions, because much of their amino acid sequences are identical to each other. Nevertheless, alternative splicing produces differences in amino acid sequences that will provide each polypeptide with its own unique characteristics. The biological advantage of alternative splicing is that two (or more) different polypeptide sequences can be derived from a single gene. This allows an organism to carry fewer genes in its genome.

The degree of splicing and alternative splicing varies greatly among different species. Baker's yeast, for example, contains about 6,000 genes and approximately 300 (i.e., 5%) encode mRNAs that are spliced. Of these 5%, only a few have been shown to be alternatively spliced. Therefore, in this unicellular eukaryote, alternative splicing is not a major mechanism to generate protein diversity. By comparison, complex multicellular organisms seem to rely on alternative splicing to a great extent. Humans contain approximately 35,000 different genes and most of these contain one or more introns. Recent estimates suggest that a minimum of one-third of all human mRNAs are alternatively spliced. Furthermore, certain mRNAs are alternatively spliced to an extraordinary extent. In fact, certain mRNAs can be alternatively spliced to produce dozens or even hundreds of different mRNAs. This provides a much greater potential for human cells to create protein diversity.

Figure 15.18 considers an example of alternative splicing for a gene that encodes a protein known as α-tropomyosin. This

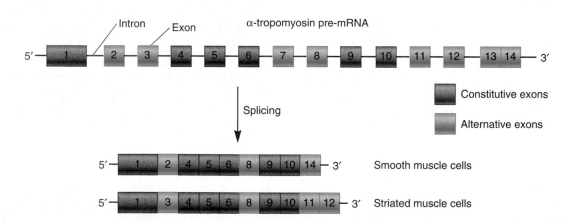

FIGURE 15.18   **Alternative ways that the rat α-tropomyosin pre-mRNA can be spliced.** The *top* part of this figure depicts the gene structure of the rat α-tropomyosin gene. Exons are shown as *colored boxes,* and introns are illustrated as connecting *black lines.* The *lower* part of the figure describes the final mRNA products in smooth and striated muscle cells. Note: Exon 8 is found in the final mRNA of smooth and striated muscle cells, but not in the mRNA of certain other cell types. The junction between exon 13 and 14 contains a 3' splice so that exon 13 can be separated from exon 14.

GENES→TRAITS   Alpha-tropomyosin functions in the regulation of cell contraction in muscle and nonmuscle cells. Alternative splicing of the α-tropomyosin gene provides a way to vary contractility in different types of cells by modifying the function of α-tropomyosin. As shown here, the alternatively spliced versions of α-tropomyosin mRNA produce α-tropomyosin proteins that differ slightly from each other in their structure (i.e., amino acid sequence). Presumably, these alternatively spliced versions of α-tropomyosin vary in function to meet the needs of the cell type in which they are found. For example, the sequence of exons 1–2–4–5–6–8–9–10–14 produces an α-tropomyosin protein that functions suitably in smooth muscle cells. Overall, alternative splicing affects the traits of an organism by allowing a single gene to encode several versions of a protein, each optimally suited to the cell type in which it is made.

protein functions in the regulation of cell contraction. It is located along the thin filaments found in smooth muscle cells (e.g., in the uterus and small intestine) and striated muscle cells (e.g., in cardiac and skeletal muscle). Alpha-tropomyosin is also synthesized in many types of nonmuscle cells but in lower amounts. Within a multicellular organism, different types of cells must regulate their contractibility in subtly different ways. One way that this may be accomplished is to produce different forms of α-tropomyosin by alternative splicing.

The intron-exon structure of the rat α-tropomyosin gene and two alternative ways that the pre-mRNA can be spliced are described in figure 15.18. The gene contains 14 exon regions. The exons shown in *red* are termed **constitutive exons.** These exons (i.e., 1, 4, 5, 6, 9, and 10) are always found in the mature mRNA from all cell types. Presumably, constitutive exons encode polypeptide segments of the α-tropomyosin protein that are necessary for its general structure and function. By comparison, **alternative exons,** shown in *green,* are not always found in the mRNA after splicing has occurred. The polypeptide sequences encoded by alternative exons may subtly change the function of α-tropomyosin to meet the needs of the cell type in which they are found. For example, figure 15.18 shows the predominant splicing products found in smooth muscle cells and striated muscle cells. When comparing smooth and striated muscle cells, exon 2 encodes a segment of the α-tropomyosin protein that slightly alters its function to make it suitable for smooth muscle cells. By comparison, exon 2 is not included in the α-tropomyosin mRNA found in striated muscle cells. Instead, this mRNA contains exon 3, which is presumably more suitable for that cell type.

It is important to emphasize that alternative splicing is not a random event. Rather, the specific pattern of splicing is regulated in any given cell. The molecular mechanism for the regulation of alternative splicing involves proteins known as **splicing factors.** Such splicing factors play a key role in the choice of particular splice sites. One such category of splicing factors is the **SR proteins.** These splicing factors contain a domain at their carboxyl terminal end that is rich in serine (S) and arginine (R) and is involved in protein-protein recognition. They also contain an RNA-binding domain at their amino terminal end.

As discussed in chapter 12, components of the spliceosome recognize the 5′ and 3′ splice sites and then remove the intervening intron. The key effect of splicing factors is to modulate the ability of the spliceosome to choose 5′ and 3′ splice sites. This can occur in two ways. Some splicing factors act as repressors that inhibit the ability of the spliceosome to recognize a splice site. For example, a splicing repressor could recognize a 3′ splice site and prevent the spliceosome from binding there. Instead, the spliceosome would choose the next available 3′ splice site. As shown in figure 15.19a, this would result in **exon skipping.** In this example, the splicing repressor would cause exon 2 to be skipped and not included in the mature mRNA. Alternatively, the role of other splicing factors is to enhance the ability of the

spliceosome to recognize particular splice sites. Figure 15.19b shows the effects of splicing factors that facilitate the recognition of a 3′ splice site and a 5′ splice site that flank exon 3. These splicing factors would promote the inclusion of exon 3 in the mature mRNA.

Alternative splicing in different tissues is thought to occur because each cell type has its own characteristic concentration of many kinds of splicing factors. Furthermore, much like transcription factors, splicing factors may be regulated by protein-protein interactions, covalent modifications, and the binding of small effector molecules. Overall, the differences in the composition of splicing factors, and the regulation of their activities, form the basis for alternative splicing decisions.

The Nucleotide Sequence of RNA Can Be Modified by RNA Editing

The term **RNA editing** refers to a change in the nucleotide sequence of an RNA molecule that involves additions or deletions of particular bases, or a conversion of one type of base to another (e.g., a cytosine to a uracil). In the case of mRNAs, editing can have various effects, such as generating start codons, generating stop codons, and changing the coding sequence for a polypeptide.

The phenomenon of RNA editing was first identified in trypanosomes, the protozoa that cause sleeping sickness. As with the discovery of RNA splicing, the initial finding of RNA editing was met with great skepticism. Since that time, however, RNA editing has been shown to occur in various organisms and in a variety of ways, although its functional significance is slowly emerging. Over the next decade or so, it will be interesting to see if RNA editing is a widespread occurrence or if it occurs only in a few selected cases. Table 15.3 describes several examples where RNA editing has been found.

The molecular mechanisms of RNA editing have been the subject of many investigations. In the specific case of trypanosomes, the editing process involves the participation of a **guide RNA.** This guide RNA can direct the insertion of one or more uracil nucleotides into an RNA or the deletion of one or more uracils. As shown in figure 15.20, the guide RNA has two important characteristics. First, its 5′ anchor is complementary to the RNA that is to be edited. And second, it has uracil nucleotides at its 3′ end. During the editing process, the 5′ anchor binds to the target RNA. The RNA is cleaved at a defined location by an endonuclease, and the 3′ end of the guide RNA becomes displaced from the target RNA. For insertions, an enzyme called terminal U-transferase then attaches uracil-containing nucleotides. For deletions, an enzyme called 3′-U-exonuclease removes one or more uracils. Finally, RNA ligase rejoins the two pieces of RNA.

A more widespread mechanism for RNA editing involves changes of one type of base to another. In this form of editing, a base in the RNA is deaminated (i.e., an amino group is removed

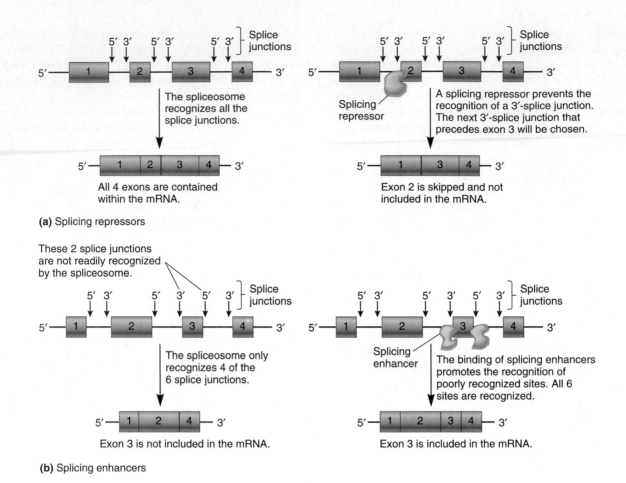

FIGURE 15.19   The roles of splicing factors during alternative splicing. (a) Splicing factors can act as repressors to prevent the recognition of splice sites. In this example, the presence of the splicing repressor causes exon 2 to be skipped (i.e., not included in the mRNA). **(b)** Other splicing factors can enhance the recognition of splice sites. In this example, the splicing enhancers promote the recognition of sites that flank a particular exon, thereby causing its inclusion in the mature mRNA.

TABLE 15.3

Examples of RNA Editing

Organism	Type of Editing	Found in:
Trypanosomes (protozoa)	Primarily additions but occasionally deletions of uracil nucleotides	Many mitochondrial mRNAs
Land plants	C-to-U conversion	Many mitochondrial and chloroplast mRNAs, tRNAs, and rRNAs
Slime mold	C additions	Many mitochondrial mRNAs
Mammals	C-to-U conversion	Apolipoprotein B mRNA, and NFI mRNA, which encodes a tumor-suppressor protein
	A-to-I conversion	Glutamate receptor mRNA, many tRNAs
Drosophila	A-to-I conversion	mRNA for calcium and sodium channels

from the base). When cytosine is deaminated, uracil is formed, and when adenine is deaminated, inosine is formed (fig. 15.21). Inosine is recognized as guanine during translation.

An example of RNA editing occurs in mammals involving an mRNA that encodes a protein called apolipoprotein B. The mRNA may be edited so that a single C is changed to a U. This converts a glutamine codon (CAA) to a stop codon (UAA) and thereby results in a shorter apolipoprotein. In this case, RNA editing produces an apolipoprotein B with an altered structure. Therefore, RNA editing can produce two proteins from the same gene, much like the phenomenon of alternative splicing described earlier in this chapter.

The Stability of mRNA Influences mRNA Concentration

In eukaryotes, the stability of mRNAs can vary considerably. Certain mRNAs have very short half-lives (namely, several minutes),

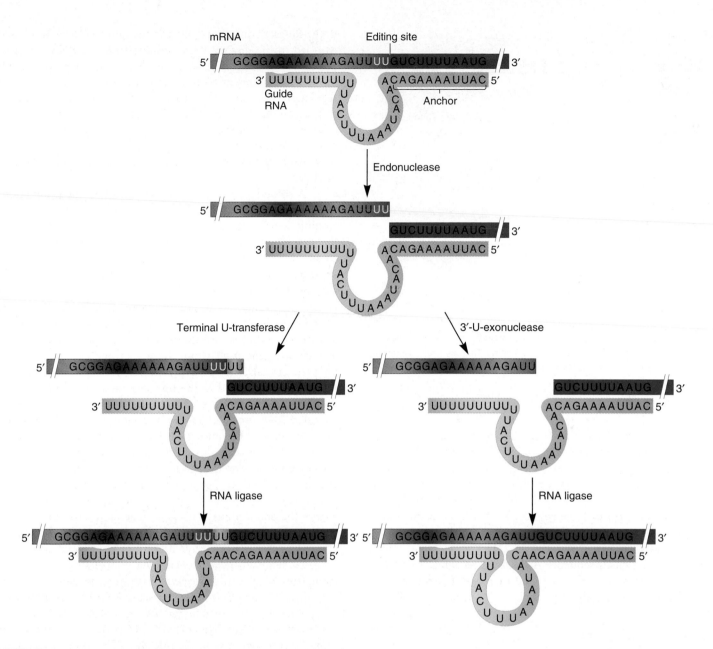

FIGURE 15.20   RNA editing in trypanosomes. The 5′ end of the guide RNA (the anchor) binds to the mRNA. The mRNA is then cleaved at the editing site by an endonuclease. On the *left* side, uracil residues are added by terminal U-transferase. On the *right* side, uracils are removed by 3′-U-exonuclease. Finally, the ends of the RNA are rejoined by RNA ligase.

whereas others can persist for several days. In some cases, the stability of an mRNA can be regulated so that its half-life is shortened or lengthened. A change in the stability of mRNA can greatly influence the cellular concentration of that mRNA molecule. In this way, factors that influence RNA stability can dramatically affect gene expression.

Various factors can play a role in mRNA stability. One important structural feature is the length of the polyA tail. As you may recall from chapter 12, most newly made mRNAs contain a polyA tail that averages 200 nucleotides in length. The polyA tail is recognized by the **polyA-binding protein.** The binding of this protein enhances RNA stability. However, as an mRNA ages, its polyA tail tends to be shortened by the action of cellular exonucleases. Once it becomes less than 10 to 30 adenosines in length, the polyA-binding protein can no longer bind, and the mRNA is rapidly degraded by exo- and endonucleases.

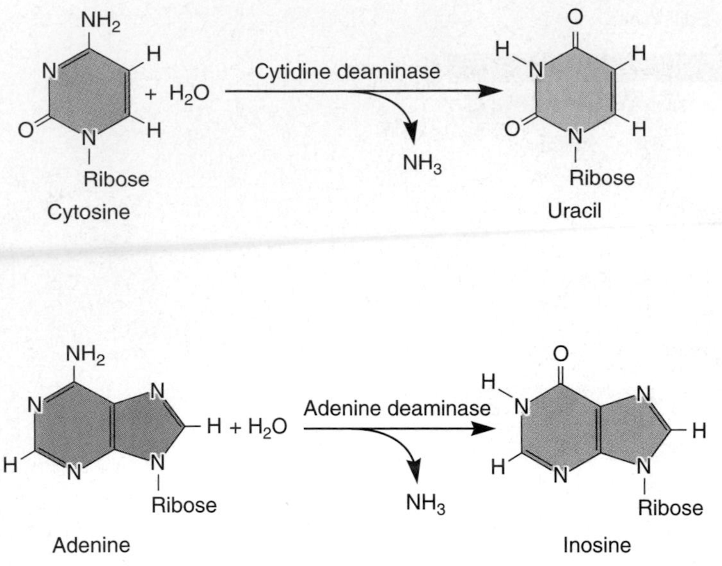

FIGURE 15.21 **RNA editing by deamination.** A cytidine deaminase can remove an amino group from cytosine, thereby creating uracil. An adenine deaminase can remove an amino group from adenine to make inosine.

Certain mRNAs, particularly those with short half-lives, contain sequences that act as destabilizing elements. While these destabilizing elements can be located anywhere within the mRNA, they are most commonly located at the 3′ end between the stop codon and the polyA tail. This region of the mRNA is known as the **3′-<u>un</u>translated <u>region</u> (3′-UTR).** An example of a destabilizing element is the **AU-rich element (ARE)** that is found in many short-lived mRNAs (fig. 15.22). This element, which contains the consensus sequence AUUUA, is recognized by cellular proteins that bind to the ARE and thereby influence whether or not the mRNA is rapidly degraded.

Double-Stranded RNA Can Mediate mRNA Degradation

Another way that specific RNAs can be targeted for degradation is a newly discovered mechanism involving double-stranded RNA. Research in plants and the nematode *Caenorhabditis elegans* led to the discovery that double-stranded RNA can silence the expression of particular genes. Studies of plant viruses were one avenue of research that identified this mechanism of gene regulation. Certain plant viruses produce double-stranded RNA as part of their life cycle. In addition, these plant viruses may carry genes that are very similar to genes that already exist in the genome of the plant cell. When such viruses infect plant cells, they silence the expression of the plant gene that is similar to the viral gene.

A second area of plant research involved the production of transgenic plants, a topic that is discussed in chapter 19. Using cloning techniques, it is possible to introduce cloned genes into the genome of plants. Surprisingly, it was often observed that when cloned genes were introduced in multiple copies, the expression of the gene was often silenced. We now know that this may be due to the formation of double-stranded RNA. When a cloned gene randomly inserts into a genome, it may, as a matter of random chance, happen to insert itself next to a promoter for a plant gene that is already present in the genome (fig. 15.23). The cloned gene itself would have a promoter to transcribe its template strand. If it randomly inserts next to a plant gene promoter that is oriented in the opposite direction, both strands of the cloned gene would be transcribed, thereby generating double-stranded RNA. As more copies of a cloned gene are introduced into a genome, it becomes more likely that the scenario described in figure 15.23 may occur. This event silences the expression of the cloned gene even though it is present in multiple copies. Furthermore, if the cloned gene is homologous to a plant gene that is already present in the plant cell, this phenomenon will also silence the endogenous plant gene. Therefore, the curious observation made by researchers was that increasing the number of copies of a cloned gene frequently led to gene silencing.

Evidence for gene silencing via double-stranded RNA also came from studies in *C. elegans*. To study gene expression in this organism, researchers injected antisense RNA (i.e., an RNA that is complementary to a specific mRNA) into oocytes and found that it silenced gene expression. That is not surprising since the antisense RNA would bind to the mRNA and prevent its translation. The surprising discovery came when Andrew Fire and Craig Mello and their colleagues injected double-stranded RNA into oocytes. Shockingly, the double-stranded RNA was 10 times more

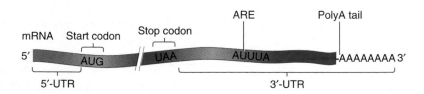

FIGURE 15.22 **The location of AU-rich elements (AREs) within mRNAs.** One or more AREs are commonly found within the 3′-UTRs (untranslated regions) of mRNAs with short half-lives. The 5′-UTR is the untranslated region of the mRNA that precedes the start codon.

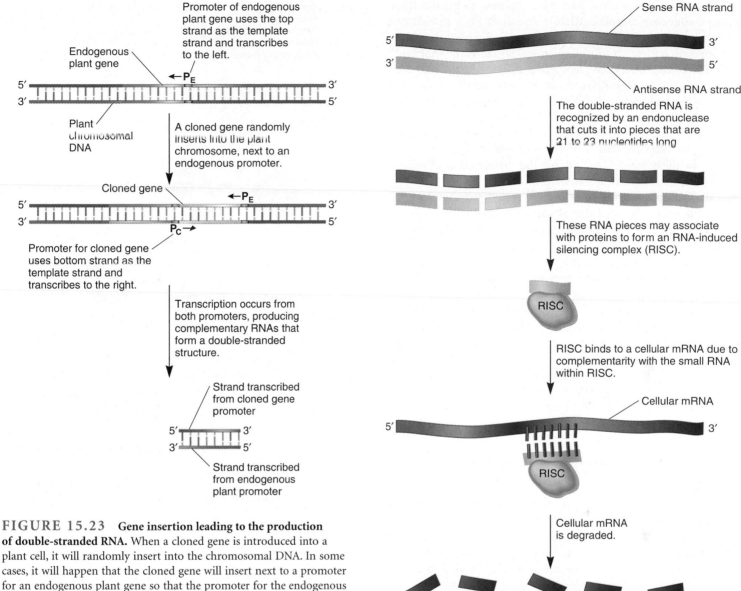

Promoter of endogenous plant gene uses the top strand as the template strand and transcribes to the left.

Endogenous plant gene

← P_E

5′
3′

Plant chromosomal DNA

A cloned gene randomly inserts into the plant chromosome, next to an endogenous promoter.

Cloned gene

← P_E

5′
3′

P_C →

Promoter for cloned gene uses bottom strand as the template strand and transcribes to the right.

Transcription occurs from both promoters, producing complementary RNAs that form a double-stranded structure.

Strand transcribed from cloned gene promoter

5′ ⎯⎯⎯ 3′
3′ ⎯⎯⎯ 5′

Strand transcribed from endogenous plant promoter

FIGURE 15.23   Gene insertion leading to the production of double-stranded RNA. When a cloned gene is introduced into a plant cell, it will randomly insert into the chromosomal DNA. In some cases, it will happen that the cloned gene will insert next to a promoter for an endogenous plant gene so that the promoter for the endogenous gene will transcribe the antisense strand of the cloned gene. Therefore, the two promoters will transcribe the sense strand and the antisense strand to produce complementary RNAs that will form double-stranded RNA.

Sense RNA strand

5′   3′
3′   5′

Antisense RNA strand

The double-stranded RNA is recognized by an endonuclease that cuts it into pieces that are 21 to 23 nucleotides long.

These RNA pieces may associate with proteins to form an RNA-induced silencing complex (RISC).

RISC

RISC binds to a cellular mRNA due to complementarity with the small RNA within RISC.

Cellular mRNA

5′   3′

RISC

Cellular mRNA is degraded.

RISC

FIGURE 15.24   Mechanism of RNA interference. Double-stranded RNA is recognized by an endonuclease that digests the RNA into small pieces, 21 to 23 nucleotides in length. These small RNAs associate with proteins to form an RNA-induced silencing complex (RISC). Short RNA strands that were part of the antisense strand will bind to complementary mRNAs. After RISC binding, endonucleases within the complex cut the RNA and thereby inactivate it. It should be emphasized that the short RNAs are RNA specific. For example, if the sense strand of the double-stranded RNA was the RNA that encodes β-globin, the short RNAs of the antisense strand would recognize only globin mRNAs and silence their expression.

potent at inhibiting the expression of the corresponding mRNA, compared to the injection of antisense RNA. They coined the term **RNA interference (RNAi)** to describe this phenomenon.

Since these studies, researchers have begun to unravel the molecular mechanism for RNA interference. A proposed mechanism for RNAi is shown in figure 15.24. The double-stranded RNA is first recognized by an endonuclease, called **dicer,** that cleaves the double-stranded RNA into short segments that are 21 to 23 nucleotides in length. These short RNAs then associate with cellular proteins to become part of a complex called the **RNA-induced silencing complex (RISC).** The short RNA strands that were part of the antisense strand will specifically bind to mRNAs that are complementary to its sequence. Therefore, the short antisense RNA directs the RISC to specific mRNAs in the cell. After the RISC binds to an mRNA, endonucleases within the complex cut the RNA and thereby inactivate it. In this way, the expression of the gene that encodes this mRNA is silenced.

RNA interference is widely found in eukaryotic species. It is believed to offer a host defense mechanism against certain viruses. In particular, it can inhibit the proliferation of viruses that have a double-stranded RNA genome and viruses that produce double-stranded RNAs as part of their life cycles. In addition, it has been suggested that RNAi may also play a role in the silencing of certain transposable elements. As discussed in chapter 17, transposable elements are DNA segments that have the capacity to move throughout the genome, an event termed *transposition.* Certain transposable elements produce double-stranded RNA intermediates as part of the transposition process. RNAi may inhibit their ability to move. In addition, transposable elements carry genes that are needed in the transposition process. The random insertion of many transposable elements in a cell may ultimately lead to gene silencing due to the scenario described in figure 15.23. The silencing of genes within transposable elements via RNAi would protect the organism against the potentially harmful effects of the transposition process.

Phosphorylation of Ribosomal Initiation Factors Can Alter the Rate of Translation

Let's now turn our attention to regulatory mechanisms that affect translation. Modulation of translational initiation factors is widely used to control fundamental cellular processes. Under certain conditions, it is advantageous for a cell to stop synthesizing proteins. For example, if a virus infects a cell, it is vital to inhibit protein synthesis so that the virus cannot manufacture viral proteins. Likewise, if critical nutrients are in short supply, it is beneficial for a cell to conserve its resources by inhibiting unnecessary protein synthesis.

As discussed in chapter 13, initiation factors are required to begin protein translation. The phosphorylation of many different initiation factors has been found to affect translation in eukaryotic cells. Two factors, eIF2 and eIF4F, appear to play a central role in controlling the initiation of translation. The functions of these two translational initiation factors are modulated by phosphorylation in opposite ways. When the α subunit of eIF2 (known as eIF2α) is phosphorylated, translation is inhibited, whereas the phosphorylation of eIF4F increases the rate of translation.

Figure 15.25 shows the events leading to translational inhibition by eIF2α. A variety of conditions can lead to a shutdown

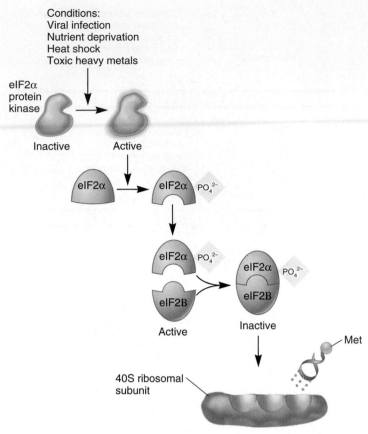

Translation is inhibited because the initiator tRNA^{Met} does not bind to the 40S subunit.

FIGURE 15.25 **The pathway that leads to the phosphorylation of eIF2α (eukaryotic initiation factor) and the inhibition of translation.**

of protein synthesis, including viral infection, nutrient deprivation, heat shock, and the presence of toxic heavy metals. These conditions promote the activation of protein kinases known as eIF2α protein kinases. Several eIF2α protein kinases have been identified. Once activated, an eIF2α protein kinase can phosphorylate eIF2α. The phosphorylation of eIF2α causes it to bind tightly to another initiation factor subunit called eIF2B. Functional eIF2B is necessary so that eIF2 can promote the binding of the initiator tRNA^{met} to the 40S subunit. However, when the phosphorylated eIF2α binds to eIF2B, it prevents eIF2B from functioning. Therefore, the initiator tRNA^{met} does not bind to the 40S subunit, and translation is inhibited.

A second important way to control translation is via the eIF4F translation factor that modulates the binding of mRNA to the ribosomal initiation complex. The function of eIF4F is stimulated by phosphorylation. A variety of conditions have been shown to cause eIF4F to become phosphorylated. These include the presence of growth factors, insulin, and other signaling molecules that promote cell proliferation. Conversely, conditions such as heat shock and viral infection decrease the level of eIF4F phosphorylation and thereby inhibit translation.

The Regulation of Iron Assimilation Is an Example of the Regulatory Effect of RNA-Binding Proteins on Translation

As we have just seen, the phosphorylation of translational initiation factors can modulate the translation of mRNA. Because these initiation factors are necessary to translate all of a cell's mRNA, this form of regulation affects the expression of many mRNAs. By comparison, particular mRNAs are sometimes regulated by RNA-binding proteins that directly affect translational initiation or RNA stability. The regulation of iron assimilation provides a well-studied example in which both of these phenomena occur. Before discussing the translational control, it is interesting to consider the biology of iron metabolism.

Iron is an essential element for the survival of living organisms because it is required for the function of many different enzymes. The pathway whereby mammalian cells take up iron is depicted in figure 15.26. Iron ingested by an animal is absorbed into the bloodstream and becomes bound to transferrin, a protein that carries iron through the bloodstream. The transferrin-Fe^{3+} complex is recognized by a transferrin receptor on the surface of cells; the complex binds to the receptor and then is transported into the cytosol by endocytosis. Once inside, the iron is then released from transferrin. At this stage, it may bind to cellular enzymes that require iron for their activity. Alternatively, if there is an overabundance of iron, the excess iron is stored within a hollow, spherical protein known as ferritin. The storage of excess iron within ferritin helps to prevent the toxic buildup of too much iron within the cell.

Since iron is a vital yet potentially toxic substance, mammalian cells have evolved an interesting way to regulate iron assimilation. The two mRNAs that encode ferritin and the transferrin receptor are both influenced by an RNA-binding protein known as the **iron regulatory protein (IRP)**. This protein binds to a regulatory element within the mRNA known as the **iron response element (IRE)**. The ferritin mRNA has an IRE in its 5′-untranslated region (5′-UTR). When IRP binds to this IRE, it inhibits the translation of the ferritin mRNA (fig. 15.27a). However, when iron is abundant in the cytosol, the iron binds directly to IRP and prevents it from binding to the IRE. Under these conditions, the ferritin mRNA is translated to make more ferritin

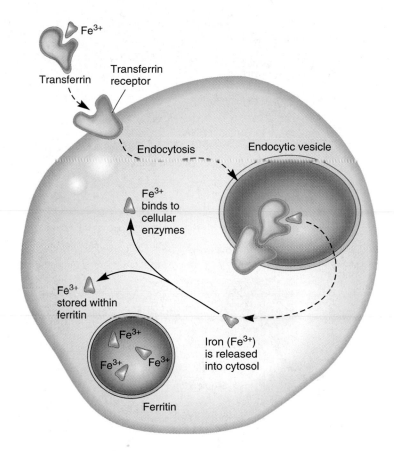

FIGURE 15.26 **The uptake of iron (Fe^{3+}) into mammalian cells.**

protein. The synthesis of ferritin prevents the toxic buildup of iron within the cytosol.

The transferrin receptor mRNA also contains iron response elements. However, the IREs in the transferrin receptor mRNA are located in the 3′-UTR. When IRP binds to these IREs, it does not inhibit translation. Instead, it increases the stability of the mRNA by blocking the action of RNA-degrading enzymes. This leads to increased amounts of transferrin receptor mRNA within the cell when the cytosolic levels of iron are very low (fig. 15.27b). Under these conditions, more transferrin receptor is made. This promotes the uptake of iron when it is in short supply. In contrast, when iron is abundant within the cytosol, IRP is removed from the transferrin receptor mRNA, and the mRNA becomes rapidly degraded. This leads to a decrease in the amount of transferrin receptor and thereby helps to prevent the uptake of too much iron into the cell.

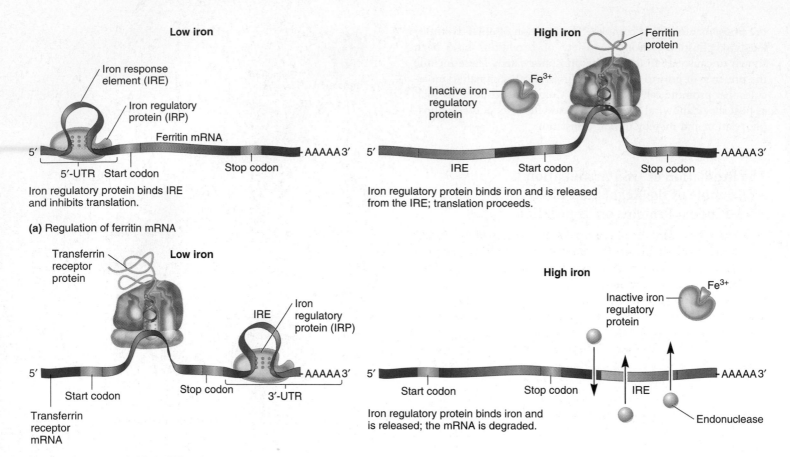

Low iron

Iron response
element (IRE)

Iron regulatory
protein (IRP)

Ferritin mRNA

5′ — 5′-UTR — Start codon — Stop codon — AAAAA 3′

Iron regulatory protein binds IRE
and inhibits translation.

(a) Regulation of ferritin mRNA

High iron — Ferritin protein

Inactive iron regulatory protein — Fe^{3+}

5′ — IRE — Start codon — Stop codon — AAAAA 3′

Iron regulatory protein binds iron and is released
from the IRE; translation proceeds.

Transferrin
receptor
protein   **Low iron**

IRE   Iron regulatory protein (IRP)

5′ — Start codon — Stop codon — 3′-UTR — AAAAA 3′

Transferrin
receptor
mRNA

Iron regulatory protein binds IRE and
enhances the stability of the mRNA.

(b) Regulation of transferrin receptor mRNA

High iron

Inactive iron regulatory protein — Fe^{3+}

5′ — Start codon — Stop codon — IRE — AAAAA 3′

Iron regulatory protein binds iron and
is released; the mRNA is degraded.   Endonuclease

FIGURE 15.27   The regulation of iron assimilation genes by IREs and IRP (iron regulatory elements and iron regulatory protein). (a) The binding of the iron regulatory protein (IRP) to the iron response element (IRE) in the 5′-UTR of ferritin mRNA inhibits translation. When Fe^{3+} binds to IRP, the protein is removed from the ferritin mRNA, so that translation can proceed. **(b)** The binding of IRP to IRE in the 3′-UTR of the transferrin receptor mRNA. In this case, the binding of IRP enhances the stability of the mRNA and leads to a higher concentration of this mRNA. Therefore, more transferrin receptor protein is made. When Fe^{3+} levels are high and this ion binds to IRP, the IRP dissociates from the IRE and the transferrin receptor mRNA is rapidly degraded.

GENES→TRAITS   This form of translational control allows cells to use iron appropriately. When the cellular concentration of iron is low, the translation of the transferrin receptor is increased, thereby enhancing the ability of cells to take up more iron. Also, the translation of ferritin mRNA is inhibited, which is not needed to store excess iron. By comparison, when the cellular concentration of iron is high, the translation of ferritin mRNA is enhanced. This leads to the synthesis of ferritin, an iron storage protein that prevents the toxic buildup of iron. Also, when iron is high, the transferrin receptor mRNA is degraded, which decreases further uptake of iron.

CONCEPTUAL SUMMARY

In chapter 15, we have surveyed a wide variety of mechanisms that allow eukaryotes to regulate gene expression. Many **regulatory transcription factors** have been identified that vary in their structures and modes of action. Such transcription factors recognize short DNA sequences called **regulatory elements** or **response elements** in the vicinity of genes and influence the events that occur at the core promoter. **Enhancers** are response elements that **up regulate** gene expression, whereas **silencers** have the opposite effect of **down regulating** transcription.

Response elements can be **bidirectional** and function at a fairly large distance away from the core promoter site. Regulatory transcription factors may interact with **TFIID** and **mediator** to influence the function of RNA polymerase. In chapter 15, we have considered the molecular mechanism of two regulatory transcription factors that respond to cell hormones. The **glucocorticoid receptor** is a regulatory transcription factor that binds glucocorticoid hormone directly. Once the hormone is bound, it activates several different genes by binding to glucocorticoid response elements (GREs) that are next to the genes. The **CREB protein** responds to intracellular levels of **cAMP.** The CREB protein binds to response elements known as **CREs.** When CREB becomes phosphorylated, it stimulates transcription by **transactivation.**

Gene regulation may also involve changes in DNA structure and chromatin compaction. The DNA itself can be altered by **gene amplification, rearrangement,** or **methylation.** The first two mechanisms are uncommon and were only briefly described in this chapter. For transcription to occur, the chromatin cannot be in a **closed conformation.** Instead, it must be in a loosely packed or **open conformation.** In the case of globin genes, a segment of DNA called the **locus control region (LCR)** plays a role in regulating the packing of DNA in this region. Transcriptional activators play a role in the recruitment of enzymes to the promoter region that convert the chromatin from a closed to an open conformation. These include **histone acetyltransferase** and **ATP-dependent remodeling enzymes.** In contrast, DNA methylation in vertebrates and higher plants appears to be an important mechanism to turn genes off in a **tissue-specific** or developmentally specific manner. **CpG islands** may be found in the vicinity of promoters, and their methylation is a way to silence transcription. Methylation may prevent the binding of transcription factors and/or promote compaction.

Gene regulation can also occur at the RNA level. For example, an RNA transcript can be regulated by **alternative splicing.** This RNA processing event influences the type of protein that is made from an mRNA transcript. **RNA editing** can change the sequence of an RNA transcript after it has been made. The stability of RNA can also be affected by particular sequences within the mRNA. Factors that promote RNA stability lead to a higher concentration of that RNA transcript. RNA can also be a target for degradation by double-stranded RNA, a mechanism termed **RNA interference (RNAi).** In addition, the rate of mRNA translation can be controlled in two ways. First, the translational machinery can be regulated by affecting the activity of translational initiation factors. Two particular factors, eIF2α and eIF4F, are modulated by phosphorylation in opposite ways. When eIF2α is phosphorylated, translation is inhibited, whereas the phosphorylation of eIF4F increases the rate of translation. Second, specific RNA-binding proteins, such as the **iron regulatory protein (IRP),** can affect the ability of certain mRNAs to be translated.

The regulation of gene expression in eukaryotes has been studied via many different techniques. In chapter 15, we have considered a few approaches to study this phenomenon. We saw, in one example, that Weintraub and Groudine were able to detect changes in the degree of DNA packing of the globin genes by using a DNase I sensitivity assay. This approach exploits the fact that tightly packed chromatin is less susceptible to DNase I digestion than is DNA in an open conformation. Furthermore, the identification of people who possess intact β-globin genes yet have thalassemia has suggested that a region near globin genes, known as the locus control region, is involved in controlling the degree of chromatin packing.

Other studies have focused on the identification of regulatory transcription factors that influence the expression of particular genes. By comparing the structures of many different transcription factors, researchers have found that they tend to have common domains that act to either bind DNA or interact with effector molecules. The sequences of many response elements have also been determined. In a few cases, the combined efforts of many research groups have elucidated the detailed molecular mechanisms for the regulation of particular genes. For example, in this chapter, we have examined how the glucocorticoid receptor and CREB protein regulate genes at the level of transcription. We have also considered how IRP regulates ferritin and transferrin receptor expression at the level of translation. The solved problems and experimental questions at the end of chapter 15 also serve to illustrate several techniques that have been used to understand the molecular mechanisms of eukaryotic gene regulation.

PROBLEM SETS & INSIGHTS

Solved Problems

S1. Describe how the tight packing of chromatin in a closed conformation may prevent gene transcription.

Answer: There are several possible ways that the tight packing of chromatin physically inhibits transcription. First, it may prevent transcription factors and/or RNA polymerase from binding to the major groove of the DNA. Second, it may prevent RNA polymerase from forming an open complex, which is necessary to begin transcription. Third, it could prevent looping in the DNA, which may be necessary to activate transcription.

S2. What are the two alternative ways that IRP can affect gene expression at the RNA level?

Answer: The ferritin mRNA has an IRE in its 5′-UTR. When IRP binds to this IRE, it inhibits the translation of the ferritin mRNA. This decreases the amount of ferritin protein, which is not needed when iron levels are low. However, when iron is abundant in the cytosol, the iron binds directly to IRP and prevents it from binding to the IRE. This allows the ferritin mRNA to be translated, producing more ferritin protein. IREs are also located in the transferrin receptor mRNA in the

3′-UTR. When IRP binds to these IREs, it increases the stability of the mRNA. This leads to an increase in the amount of transferrin receptor mRNA within the cell when the cytosolic levels of iron are very low. Under these conditions, more transferrin receptor is made to promote the uptake of iron, which is in short supply. When the iron is found in abundance within the cytosol, IRP is removed from the mRNA, and the mRNA becomes rapidly degraded. This leads to a decrease in the amount of the transferrin receptor.

S3. Eukaryotic response elements are often orientation independent and can function in a variety of locations. Explain the meaning of this statement.

Answer: Orientation independence means that the response element can function in the forward or reverse direction. In addition, response elements can function at a variety of locations that may be upstream or downstream from the core promoter. DNA looping is thought to bring the regulatory transcription factors bound at the response elements and the general transcription factors bound at the core promoter in close proximity with one another.

S4. To gain a molecular understanding of how the glucocorticoid receptor works, geneticists have attempted to "dissect" the protein to identify smaller domains that play specific functional roles. Using recombinant DNA techniques described in chapter 18, particular segments in the coding region of the glucocorticoid receptor gene can be removed. The altered gene can then be expressed in a living cell to see if functional aspects of the receptor have been changed or lost. For example, the removal of the portion of the gene encoding the carboxyl terminal half of the protein causes a loss of glucocorticoid binding. These results indicate that the carboxyl terminal portion contains a domain that functions as a glucocorticoid-binding site. The figure shown here illustrates the locations of several functional domains within the glucocorticoid receptor relative to the entire amino acid sequence.

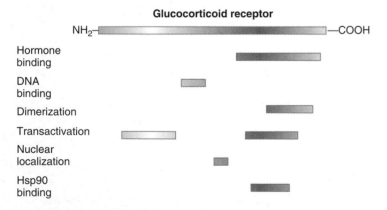

Based on your understanding of the mechanism of glucocorticoid receptor function described in figure 15.6, explain the functional roles of the different domains in the glucocorticoid receptor.

Answer: The hormone-binding domain is located in the carboxyl terminal half of the protein. This part of the protein also contains the region that is necessary for HSP90 binding and receptor dimerization. A nuclear localization sequence (NLS) is located near the center of the protein. After hormone binding, the NLS is exposed on the surface of the protein and allows it to be targeted to the nucleus. The DNA-binding domain, which contains zinc fingers, is also centrally located in the

primary amino acid sequence. Zinc fingers promote DNA binding to the major groove. Finally, two separate regions of the protein, one in the amino terminal half and one in the carboxyl terminal half, are necessary for the transactivation of RNA polymerase. If these domains are removed from the receptor, it can still bind to the DNA but it cannot activate transcription.

S5. A common approach to identify genetic sequences that play a role in the transcriptional regulation of a gene is the strategy that is sometimes called **promoter bashing.** This approach requires gene cloning methods, which are described in chapter 18. A clone is obtained that has the coding region for a structural gene as well as the region that is upstream from the core promoter. This upstream region is likely to contain genetic regulatory elements such as enhancers and silencers. The diagram shown here depicts a cloned DNA region that contains the upstream region, the core promoter, and the coding sequence for a protein that is expressed in human liver cells. The upstream region may be several thousand base pairs in length.

Upstream region Core promoter Coding sequence of liver-specific gene

To determine if promoter bashing has an effect on transcription, it is helpful to have an easy way to measure the level of gene expression. One way to accomplish this is to swap the coding sequence of the gene of interest with the coding sequence of another gene. For example, the coding sequence of the *lacZ* gene, which encodes β-galactosidase, is frequently swapped because it is easy to measure the activity of β-galactosidase because there is an assay for its enzymatic activity. The *lacZ* gene is called a "reporter gene" because it is easy to measure its activity. As shown here, the coding sequence of the *lacZ* gene has been swapped with the coding sequence of the liver-specific gene. In this new genetic construct, the expression and transcriptional regulation of the *lacZ* gene is under the control of the core promoter and upstream region of the liver-specific gene.

Upstream region Core promoter   *lacZ* gene coding sequence

Now comes the "bashing" part of the experiment. Different segments of the upstream region are deleted and then the DNA is transformed into living cells. In this case, the researcher would probably transform the DNA into liver cells, because those are the cells where the gene is normally expressed. The last step is to measure the β-galactosidase activity in the transformed liver cells.

In the diagram shown here, the upstream region and the core promoter have been divided into five regions, labeled A–E. One of these regions was deleted (i.e., bashed out), and then the rest of the DNA segment was transformed into liver cells.

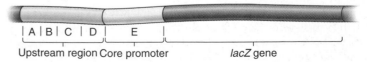

|A|B|C|D| E |

Upstream region Core promoter   *lacZ* gene

The data shown here are from this experiment.

Region Deleted	Percentage of* β-Galactosidase Activity
None	100
A	100
B	330
C	100
D	5
E	<1

*The amount of β-galactosidase activity in the cells carrying an undeleted upstream and promoter region was assigned a value of 100%. The amounts of activity in the cells carrying a deletion were expressed relative to this 100% value.

Explain what these results mean.

Answer: The amount of β-galactosidase activity found in liver cells that do not carry a deletion reflects the amount of expression under normal circumstances. If the core promoter (region E) is deleted, very little expression is observed. This is expected because a core promoter is needed for transcription. If enhancers are deleted, the activity should be less than 100%. It appears that one or more enhancers are found in region D. If a silencer is deleted, the activity should be above 100%. From the data shown it appears that one or more silencers are found in region B. Finally, if a deletion has no effect, there may not be any response elements there. This was observed for regions A and C.

Note: The deletion of an enhancer will have an effect on β-galactosidase activity only if the cell is expressing the regulatory transcription factor that binds to the enhancer and transactivates transcription. Likewise, the deletion of a silencer will have an effect only if the cell is expressing the repressor protein that binds to the silencer and inhibits transcription. In the problem described, the liver cells must be expressing the activators and repressors that recognize the response elements found in regions B and D, respectively.

Conceptual Questions

C1. Discuss the common points of control in eukaryotic gene regulation.

C2. Discuss the structure and function of response elements. Where are they located relative to the core promoter?

C3. What is meant by the term *transcription factor modulation?* List three general ways that this can occur.

C4. What do the terms *transactivation* and *transinhibition* mean? Explain how they occur at the molecular level.

C5. Are the following statements true or false?

A. An enhancer is a type of response element.

B. A core promoter is a type of response element.

C. Transcriptional regulatory proteins bind to response elements.

D. An enhancer may cause the down regulation of transcription.

C6. Transcription factors usually contain one or more motifs that play key roles in their function. What is the function of the following motifs?

A. Helix–turn–helix

B. Zinc finger

C. Leucine zipper

C7. The binding of an effector molecule, protein-protein interactions, and posttranslational modifications are three common ways to modulate the activities of transcription factors. Which of these three mechanisms are used by steroid receptors and the CREB protein?

C8. Describe the steps that occur for the glucocorticoid receptor to bind to a GRE.

C9. Let's suppose a mutation in the glucocorticoid receptor does not prevent the binding of the glucocorticoid hormone to the protein but does prevent the ability of the receptor to activate transcrip-

tion. Make a list of all the possible defects that may explain why transcription cannot be activated.

C10. Explain how phosphorylation affects the function of the CREB protein.

C11. A particular drug inhibits the protein kinase that is responsible for phosphorylating the CREB protein. How would this drug affect the following events?

A. The ability of the CREB protein to bind to CREs

B. The ability of extracellular hormone to enhance cAMP levels

C. The ability of the CREB protein to stimulate transcription

D. The ability of the CREB protein to dimerize

C12. The glucocorticoid receptor and the CREB protein are two examples of transcriptional activators. These proteins bind to response elements and transactivate transcription.

A. How would the function of the glucocorticoid receptor be shut off? (Note: The answer to this question is not directly described in chapter 15. You have to rely on your understanding of the functioning of other proteins that are modulated by the binding of effector molecules, such as the *lac* repressor.)

B. What type of enzyme would be needed to shut off the transactivation of transcription by the CREB protein?

C13. As discussed in chapter 15, transcription factors such as the glucocorticoid receptor and the CREB protein form homodimers and activate transcription. Other transcription factors form heterodimers. For example, a transcription factor known as myogenic bHLH forms a heterodimer with a protein called the E protein. This heterodimer activates the transcription of genes that promote muscle cell differentiation. However, when myogenic bHLH forms a heterodimer with a protein called the Id protein, transcriptional activation does not occur. (Note: Id stands for Inhibitor of differentiation.) Which of the following possibilities described would best explain this observation? Only one possibility is correct.

	Myogenic bHLH	E Protein	Id Protein
Possibility 1.			
DNA-binding domain:	Yes	No	No
Leucine zipper:	Yes	No	Yes
Possibility 2.			
DNA-binding domain:	Yes	Yes	No
Leucine zipper:	Yes	Yes	Yes
Possibility 3.			
DNA-binding domain:	Yes	No	Yes
Leucine zipper:	Yes	No	No

C14. An enhancer, located upstream from a gene, has the following sequence:

```
5′-GTAG-3′
3′-CATC-5′
```

This enhancer is orientation independent. Which of the following sequences would also work as an enhancer?

A. 5′-CTAC-3′
   3′-GATG-5′

B. 5′-GATG-3′
   3′-CTAC-5′

C. 5′-CATC-3′
   3′-GTAG-5′

C15. The DNA-binding domain of each CREB protein recognizes the sequence 5′–TGACGTCA–3′. As a matter of random chance, how often would you expect this sequence to occur in the human genome, which contains approximately 3 billion base pairs? Actually, there are only a few dozen genes that are activated by the CREB protein. Does the value of a few dozen agree with the number of random sites found in the human genome? If the number of random sites in the human genome is much higher than a few dozen, provide at least one explanation why the CREB protein is not activating more than a few dozen genes.

C16. Solved problem S4 shows the locations of domains in the glucocorticoid receptor relative to the amino and carboxyl terminal ends of the protein. Make a drawing that illustrates the binding of a glucocorticoid receptor dimer to the DNA. In your drawing label the amino and carboxyl ends, the hormone-binding domains, the DNA-binding domains, the dimerization domains, and the transactivation domains.

C17. The gene that encodes the enzyme called tyrosine hydroxylase is known to be up regulated by the CREB protein. Tyrosine hydroxylase is expressed in nerve cells and is involved in the synthesis of catecholamine, a neurotransmitter. The exposure of cells to adrenaline normally up regulates the transcription of the tyrosine hydroxylase gene. A mutant cell line was identified in which the tyrosine hydroxylase gene was not up regulated when exposed to adrenaline. List all the possible mutations that could explain this defect. How would you explain the defect if only the tyrosine hydroxylase gene was not up regulated by the CREB protein, but other genes having CREs were properly up regulated in response to adrenaline in this cell line?

C18. What is the predominant form of chromatin in the eukaryotic nucleus? What must happen to its structure for transcription to take place? Discuss what is believed to cause this change in chromatin structure.

C19. Explain how the acetylation of core histones may loosen chromatin packing.

C20. Figure 15.13 shows the nucleosome structure of the β-globin gene in an active and inactive state. Suggest two or more possible reasons why the arrangement shown in figure 15.13*a* would be transcriptionally inactive.

C21. Explain the function of a locus control region.

C22. Let's suppose you have found a patient with thalassemia in which the globin genes have been translocated to a different chromosome. The sequence of the β-globin gene is perfectly normal in this patient with a translocation, yet it is not expressed. Propose a reasonable explanation for this observation.

C23. What is DNA methylation? When we say that DNA methylation is heritable, what do we mean? How is it passed from a mother to a daughter cell?

C24. Let's suppose that a vertebrate organism carries a mutation that causes some cells that would normally differentiate into nerve cells to differentiate into muscle cells. A molecular analysis of this mutation revealed that it was in a gene that encodes a methylase. Explain how an alteration in a methylase could produce this phenotype.

C25. What is a CpG island? Where would you expect one to be located? How does the methylation of CpG islands affect gene expression?

C26. What is the function of a splicing factor? Explain how splicing factors can regulate the tissue-specific splicing of mRNAs.

C27. Figure 15.18 shows the products of alternative splicing for the α-tropomyosin pre-mRNA. Let's suppose that smooth muscle cells produce splicing factors that are not produced in other cell types. Explain where you think such splicing factors bind and how they influence the splicing of the α-tropomyosin pre-mRNA.

C28. Let's suppose a person was homozygous for a mutation in the IRP gene that changed the structure of the iron regulatory protein in such a way that it could not bind iron, but it could still bind to IREs. How would this mutation affect the regulation of ferritin and transferrin receptor mRNAs? Do you think such a person would need more iron in his/her diet compared to normal individuals? Do you think that excess iron in his/her diet would be more toxic compared to normal individuals? Explain your answer.

C29. In response to toxic substances (e.g., iron), eukaryotic cells often use translational or posttranslational regulatory mechanisms to prevent cell death, rather than using transcriptional regulatory mechanisms. Explain why.

C30. What are the advantages and disadvantages of mRNAs with a short half-life compared to mRNAs with a long half-life?

C31. What conditions lead to the phosphorylation of eIF2α? Discuss why a cell would want to shut down translation when these conditions occur.

C32. What is the relationship between mRNA stability and mRNA concentration? What factors affect mRNA stability?

C33. Describe how the binding of the iron regulatory protein affects the mRNAs for ferritin and the transferrin receptor. How does iron influence this process?

Experimental Questions

E1. What is a DNase I sensitive site? With regard to DNase I sensitivity, what do you expect will happen when a gene is converted from an inactive to a transcriptionally active state?

E2. In the experiment of figure 15.10, explain how the use of S1 nuclease makes it possible to determine whether or not the β-globin gene is DNase I sensitive. In this experiment, why is it necessary to precipitate the DNA in step 7?

E3. Explain how investigations of thalassemias ultimately led to studies that identified a locus control region.

E4. Researchers can isolate a sample of cells, such as skin fibroblasts, and grow them in the laboratory. This is called a cell culture. A cell culture can be exposed to a sample of DNA. If the cells are treated with agents that make their membranes permeable to DNA, the cells may take up the DNA and incorporate the DNA into their chromosomes. This process is called transformation or transfection. Experimenters have transformed human skin fibroblasts with methylated DNA and then allowed the fibroblasts to divide for several cellular generations. The DNA in the daughter cells was then isolated, and the segment that corresponded to the transformed DNA was examined. This DNA segment in the daughter cells was also found to be methylated. However, if the original skin fibroblasts were transformed with unmethylated DNA, the DNA found in the daughter cells was also unmethylated. Do fibroblasts express a *de novo* methylase, a maintenance methylase, or both? Explain your answer.

E5. Restriction enzymes, described in chapter 18, are enzymes that recognize a particular DNA sequence and cleave the DNA (along the DNA backbone) at that site. The restriction enzyme known as *Not*I recognizes the sequence

```
5'-GCGGCCGC-3'
3'-CGCCGGCG-5'
```

However, if the cytosines in this sequence have been methylated, *Not*I will not cleave the DNA at this site. For this reason, *Not*I is commonly used to investigate the methylation state of CpG islands.

A researcher has studied a gene, which we will call gene *T*, that is found in corn. This gene encodes a transporter that is involved in the uptake of phosphate from the soil. A CpG island is located near the core promoter of gene *T*. The CpG island has a single *Not*I site. The arrangement of gene *T* is shown here.

	CpG island	Core promoter	Coding sequence for gene *T*
Sal I	1,500 bp	*Not* I 3,800 bp	*Eco*RI

A *Sal*I restriction site is located upstream from the CpG island, and an *Eco*RI restriction site is located near the end of the coding sequence for gene *T*. The distance between the *Sal*I and *Not*I sites is 1,500 bp and the distance between the *Not*I and *Eco*RI sites is 3,800 bp. There are no other sites for *Sal*I, *Not*I, or *Eco*RI in this region.

Okay, now here is the question. Let's suppose a researcher has isolated DNA samples from four different tissues in a corn plant. These include the leaf, the tassel, a section of stem, and a section of root. The DNA was then digested with all three restriction enzymes, separated by gel electrophoresis, and then probed

with a DNA fragment that is complementary to the gene *T* coding sequence. The results are shown here.

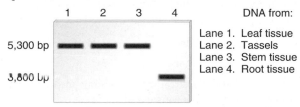

DNA from:
Lane 1. Leaf tissue
Lane 2. Tassels
Lane 3. Stem tissue
Lane 4. Root tissue

In which type of tissue is the CpG island methylated? Does this make sense based on the function of the protein encoded by gene *T*?

E6. You will need to understand solved problem S5 before answering this question. A muscle-specific gene was cloned and then subjected to promoter bashing as described in solved problem S5. As shown here, six regions, labeled A–F, were deleted, and then the DNA was transformed into muscle cells.

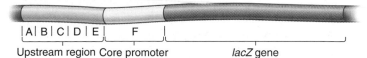

A	B	C	D	E	F	
Upstream region					Core promoter	*lacZ* gene

The data shown here are from this experiment.

Region Deleted	Percentage of β-Galactosidase Activity
None	100
A	20
B	330
C	100
D	5
E	15
F	<1

Explain these results.

E7. You will need to understand solved problem S5 before answering this question. A gene that is normally expressed in pancreatic cells was cloned and then subjected to promoter bashing as described in solved problem S5. As shown here, four regions, labeled A–D, were individually deleted, and then the DNA was transformed into pancreatic cells or into kidney cells.

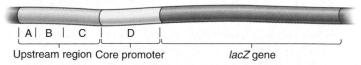

A	B	C	D	
Upstream region			Core promoter	*lacZ* gene

The data shown here are from this experiment.

Region Deleted	Cell Type Transformed	Percentage of β-Galactosidase Activity
None	Pancreatic	100
A	Pancreatic	5
B	Pancreatic	100
C	Pancreatic	100
D	Pancreatic	<1

(continued)

Region Deleted	Cell Type Transformed	Percentage of β-Galactosidase Activity
None	Kidney	<1
A	Kidney	<1
B	Kidney	100
C	Kidney	<1
D	Kidney	<1

If we assume that the upstream region has one silencer and one enhancer, answer the following questions:

A. Where are the silencer and enhancer located?

B. Why don't we detect the presence of the silencer in the pancreatic cells?

C. Why isn't this gene normally expressed in kidney cells?

E8. A gel retardation assay can be used to determine if a protein binds to a segment of DNA. When a segment of DNA is bound by a protein, its mobility will be retarded and the DNA band will appear higher in the gel. In the gel retardation assay shown here, a cloned gene fragment that is 750 bp in length contains a response element that is recognized by a transcription factor called protein X. Previous experiments have shown that the presence of hormone X results in transcriptional activation by protein X. The results of a gel retardation assay are shown here.

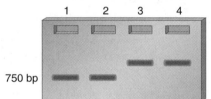

Lane 1. No additions
Lane 2. Plus hormone X
Lane 3. Plus protein X
Lane 4. Plus protein X and hormone X

Explain the action of hormone X.

E9. Chapter 18 describes a blotting method known as Northern blotting, which can be used to detect RNA that is transcribed from a particular gene. In this method, a specific RNA is detected using a short segment of cloned DNA as probe. The DNA probe, which is radioactive, is complementary to the RNA that the researcher wishes to detect. After the radioactive probe DNA binds to the RNA within a blot of a gel, the RNA is visualized as a dark (radioactive) band on an X-ray film. The method of Northern blotting can be used to determine the amount of a particular RNA transcribed in a given cell type. If one type of cell produces twice as much of a particular mRNA compared to another cell, the band appears twice as intense.

For this question, a researcher has a DNA probe that is complementary to the ferritin mRNA. This probe can be used to specifically detect the amount of ferritin mRNA on a gel. A researcher began with two flasks of human skin cells. One flask contained a very low concentration of iron, and the other flask had a high concentration of iron. The mRNA was isolated from these cells and then subjected to Northern blotting, using a probe that is complementary to the ferritin mRNA. The sample loaded in lane 1 was from the cells grown in a low concentration of iron, and the sample in lane 2 was from the cells grown in a high concentration of iron. Three Northern blots are shown here, but only one of them is correct. Based on your understanding of ferritin mRNA regulation, which blot (i.e., a, b, or c) would be your expected result? Explain. Which blot (i.e., a, b, or c) would be your expected result if the gel had been probed with a DNA segment that is complementary to the transferrin receptor mRNA?

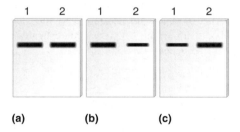

(a)   (b)   (c)

Questions for Student Discussion/Collaboration

1. Explain how DNA methylation could be used to regulate gene expression in a tissue-specific way. When and where would *de novo* methylation occur, and when would demethylation occur? What would occur in the germ line?

2. Enhancers can be almost anywhere and affect the transcription of a gene. Let's suppose you have a gene cloned on a piece of DNA, and the DNA fragment is 50,000 bp in length. Using cloning methods described in chapter 18, you can cut out short segments from this 50,000 bp fragment and then reintroduce the smaller fragments into a cell that can express the gene. You would like to know if there are any enhancers within the 50,000 bp region that may affect the expression of the gene. Discuss the *most efficient* strategy you can think of to trim your 50,000 bp fragment and thereby locate enhancers. You can assume that the coding sequence of the gene is in the center of the 50,000 bp fragment and that you can trim the 50,000 bp fragment into any size piece you want using molecular techniques described in chapter 18.

3. How are regulatory transcription factors and regulatory splicing factors similar in their mechanism of action? In your discussion, consider the domain structures of both types of proteins. How are they different?

Note: All answers appear at the website for this textbook; the answers to even-numbered questions are in the back of the textbook.

www.mhhe.com/brooker

Visit the Online Learning Center for practice tests, answer keys, and other learning aids for this chapter. Enhance your understanding of genetics with our interactive exercises, web links, news feeds, tutorial service, and much more.

GENE MUTATION AND DNA REPAIR

16

::

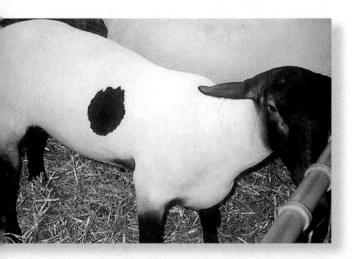

As we have seen throughout this textbook, the primary function of DNA is to store information for the synthesis of cellular proteins. A key aspect of the gene expression process is that the DNA itself does not normally change. This allows DNA to function as a permanent storage unit. However, on relatively rare occasions, a **mutation** can occur. The term *mutation* refers to a heritable change in the genetic material. This means that the structure of DNA has been changed permanently, and this alteration can be passed from mother to daughter cells during cell division. If a mutation occurs in reproductive cells, it may also be passed from parent to offspring.

The topic of mutation is centrally important in all fields of genetics, including molecular genetics, Mendelian inheritance, and population genetics. Mutations provide the allelic variation that we have discussed throughout this textbook. For example, phenotypic differences such as tall versus dwarf pea plants are due to mutations that alter the expression of particular genes. With regard to their phenotypic effects, mutations can be beneficial, neutral, or detrimental. On the positive side, mutations are essential to the continuity of life. They provide the variation that enables species to change and adapt to their environments. Mutations are the foundation for evolutionary change. On the negative side, however, new mutations are much more likely to be harmful rather than beneficial to the individual. The genes within each species have evolved to work properly. They have functional promoters, coding sequences, terminators, etc., that allow the genes to be expressed. Random mutations are more likely to disrupt these sequences rather than improve their function. For example, many inherited human diseases result from mutated genes.

In addition, diseases such as skin and lung cancer can be caused by environmental agents that are known to cause DNA mutations. For these and many other reasons, understanding the molecular nature of mutations is a deeply compelling area of research. In chapter 16, we will consider the nature of mutations and their consequences on gene expression at the molecular level.

Since mutations can be quite harmful, organisms have developed several ways to repair damaged DNA. DNA repair systems reverse DNA damage before it results in a mutation that could potentially have negative consequences. DNA repair systems have been studied extensively in many organisms, particularly *Escherichia coli,* yeast, and mammals. A variety of systems repair different types of DNA lesions. In chapter 16, we will examine the ways that several of these DNA repair systems operate.

16.1 CONSEQUENCES OF MUTATION

To understand why DNA mutations are beneficial or detrimental, we must appreciate how changes in DNA structure can ultimately affect DNA function. Much of our understanding of mutation has come from the study of experimental organisms, such as bacteria, yeast, and *Drosophila.* Researchers can expose these organisms to environmental agents that cause mutation and then study the consequences of the induced mutations. In addition, since these organisms have a short generation time, researchers can investigate the effects of mutation when they are passed from parent to offspring.

Changes in chromosome structure are referred to as **chromosome mutations,** and changes in chromosome number are called **genome mutations.** Both chromosome and genome mutations can usually be seen with the aid of a light microscope. As discussed in chapter 8, they are important occurrences within natural populations of eukaryotic organisms. By comparison, **single-gene mutations** are relatively small changes in DNA structure that occur within a particular gene. In chapter 16, we will be primarily concerned with the ways that mutations may affect the molecular and phenotypic expression of single genes. We will also consider how the timing of mutations during an organism's development has important consequences.

Gene Mutations Are Molecular Changes in the DNA Sequence of a Gene

A gene mutation occurs when the sequence of the DNA within a gene is altered in a permanent way. A gene mutation can change the base sequence within a gene, or it can involve a removal or addition of one or more nucleotides.

A **point mutation** is a change in a single base pair within the DNA. For example, the DNA sequence shown here has been altered by a **base substitution.**

$$5'-\text{AACGCTAGATC}-3' \rightarrow 5'-\text{AACGC}\textbf{G}\text{AGATC}-3'$$
$$3'-\text{TTGCGATCTAG}-3' 3'-\text{TTGCG}\textbf{C}\text{TCTAG}-5'$$

A change of a pyrimidine to another pyrimidine (C to T) or a purine to another purine (A to G) is called a **transition.** This type of mutation is more common than a **transversion,** in which

purines and pyrimidine are interchanged. The example just shown is a transversion (T to G change), not a transition.

Besides base substitutions, a short sequence of DNA may be deleted from or added to the chromosomal DNA:

$$5'-\text{AACGCTAGATC}-3' \rightarrow 5'-\text{AACGCTC}-3'$$
$$3'-\text{TTGCGATCTAG}-3' 3'-\text{TTGCGAG}-5'$$
(deletion of 4 bp)

$$5'-\text{AACGCTAGATC}-3' \rightarrow 5'-\text{AAC}\textbf{AGTC}\text{GCTAGATC}-3'$$
$$3'-\text{TTGCGATCTAG}-3' 3'-\text{TTG}\textbf{TCAG}\text{CGATCTAG}-5'$$
(addition of 4 bp)

As we will see next, small deletions or additions to the sequence of a gene can significantly affect its function.

Gene Mutations Can Alter the Coding Sequence Within a Gene

The occurrence of a mutation within the coding sequence of a structural gene can have various effects on the amino acid sequence of the polypeptide encoded by that gene. Table 16.1 describes the possible effects of point mutations. **Silent mutations** are those that do not alter the amino acid sequence of the polypeptide even though the nucleotide sequence has changed. Since the genetic code is degenerate, silent mutations can occur in certain bases within a codon, such as the third base, so that the specific amino acid is not changed. In contrast, **missense mutations** are base substitutions in which an amino acid change does occur. An example of a missense mutation occurs in the human disease known as sickle-cell anemia, which was discussed in chapter 4. This disease involves a mutation in the β-globin gene. In the most common form of this disease, a missense mutation alters the polypeptide sequence so that the sixth amino acid is changed from a glutamic acid to valine. This single amino acid substitution alters the structure and function of the hemoglobin protein. One consequence of this alteration is that the red blood cells sickle under conditions of low oxygen. A photo of this is shown in figure 16.1. It may seem surprising that a single amino acid substitution could have such a profound effect on the phenotype of cells and cause a serious disease.

Nonsense mutations involve a change from a normal codon to a termination codon. This causes the translation of the polypeptide to be terminated earlier than expected, producing a truncated polypeptide (see table 16.1). When a nonsense mutation occurs in a bacterial operon, it may inhibit the expression of downstream genes. This phenomenon, termed **polarity,** is described in solved problem S4 at the end of chapter 16. Finally, **frameshift mutations** involve the addition or deletion of a number of nucleotides that is not divisible by three. Since the codons are read in multiples of three, this shifts the reading frame so that a completely different amino acid sequence occurs downstream from the mutation.

Except for silent mutations, new mutations are more likely to produce polypeptides that have reduced rather than better function. For example, nonsense mutations will produce polypeptides that are substantially shorter and, therefore, unlikely to function properly. Likewise, frameshift mutations dramatically alter the amino acid sequence of polypeptides and are thereby

TABLE 16.1
Consequences of Point Mutations Within the Coding Sequence

Type of Change	Mutation in the DNA	Example
None	None	5′–A–T–G–A–C–C–G–A–C–C–C–G–A–A–A–G–G–G–A–C–C–3′* Met – Thr – Asp – Pro – Lys – Gly – Thr –
Silent	Base substitution	5′–A–T–G–A–C–C–G–A–C–C–C–C–A–A–A–G–G–G–A–C–C–3′ Met – Thr – Asp – Pro – Lys – Gly – Thr –
Missense	Base substitution	5′–A–T–G–C–C–C–G–A–C–C–C–G–A–A–A–G–G–G–A–C–C–3′ Met – Pro – Asp – Pro – Lys – Gly – Thr –
Nonsense	Base substitution	5′–A–T–G–A–C–C–G–A–C–C–C–G–T–A–A–G–G–G–A–C–C–3′ Met – Thr – Asp – Pro – STOP!
Frameshift	Addition/deletion	5′–A–T–G–A–C–C–G–A–C–G–C–C–G–A–A–A–G–G–G–A–C–C–3′ Met – Thr – Asp – Ala – Glu – Arg – Asp –

*DNA sequence in the coding strand. Note that this sequence is the same as the mRNA sequence except that the RNA contains uracil (U) instead of thymine (T).

likely to disrupt function. Missense mutations are less likely to alter function, since they involve a change of a single amino acid within polypeptides that typically contain hundreds of amino acids. When a missense mutation has no detectable effect on protein function, it is referred to as a **neutral mutation.** A missense mutation that substitutes an amino acid with a similar chemistry as the original amino acid is likely to be neutral. For example, a missense mutation that substitutes a glutamic acid for an aspartic acid is likely to be neutral because both amino acids are negatively charged and have similar side chain structures. Silent mutations are also considered neutral mutations.

Mutations can occasionally produce a polypeptide that has an enhanced ability to function. While these favorable mutations are relatively rare, they may result in an organism with a greater likelihood to survive and reproduce. If this is the case, natural selection may cause such a favorable mutation to increase in frequency within a population. This topic will be discussed later in this chapter and also in chapter 25.

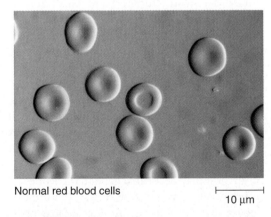

Normal red blood cells |— 10 µm —|

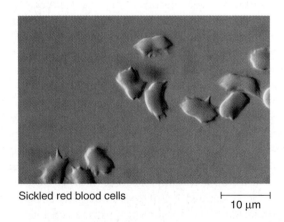

Sickled red blood cells |— 10 µm —|

(a) Micrographs of red blood cells

NORMAL: NH₂ – VALINE – HISTIDINE – LEUCINE – THREONINE – PROLINE – GLUTAMIC ACID – GLUTAMIC ACID

SICKLE CELL: NH₂ – VALINE – HISTIDINE – LEUCINE – THREONINE – PROLINE – VALINE – GLUTAMIC ACID

(b) A comparison of the amino acid sequence between normal β-globin and sickle-cell β-globin.

FIGURE 16.1 **Missense mutation in sickle-cell anemia. (a)** Normal red blood cells (*left*) and sickled red blood cells (*right*). **(b)** A comparison of the amino acid sequence of the normal β-globin polypeptide and the polypeptide encoded by the sickle-cell allele. This figure shows only a portion of the polypeptide sequence, which is 146 amino acids long. As seen here, a missense mutation changes the sixth amino from a glutamic acid to a valine. GENES→TRAITS   A missense mutation alters the structure of β-globin, which is a subunit of hemoglobin, the oxygen-carrying protein in the red blood cells. When an individual is homozygous for this allele, this missense mutation causes the red blood cells to sickle under conditions of low oxygen tension. The sickling phenomenon is a description of the trait at the cellular level. At the organismal level, the sickled cells clog the capillaries, thereby causing painful crises. The shortened life span of the red blood cells leads to symptoms of anemia.

Gene Mutations Are Also Given Names That Describe How They Affect the Wild-Type Genotype and Phenotype

Thus far, several genetic terms have been introduced that describe the molecular effects of mutations. Genetic terms are also used to describe the effects of mutations relative to a wild-type genotype. In a natural population, the **wild type** is a relatively common genotype. For example, the most common form of the β-globin gene is called the wild-type allele. For some genes, multiple alleles may be prevalent in a population, so there is not a single wild-type allele.

A **forward mutation** changes the wild-type genotype into some new variation. For example, in the sickle-cell allele of the β-globin gene, the sixth amino acid is changed from a glutamic acid to a valine. The mutation is "forward" in an evolutionary sense, since it has changed from the prevalent genotype in the population. If a forward mutation is beneficial, it may move evolution forward; otherwise, it will probably be eliminated from the population. A **reverse mutation** has the opposite effect. For example, if the sickle-cell allele mutated back to the wild-type allele (valine to glutamic acid), this would be a reverse mutation or a **reversion.**

Another way to describe a mutation is based on its influence on the wild-type phenotype. When a mutation alters the phenotypic characteristics of an organism, it is said to be a **variant.** In other words, a mutant gene is a variant when it has an effect on phenotype. Variants are often characterized by their differential ability to survive. As mentioned, a neutral mutation does not alter protein function, so it does not affect survival. A **deleterious mutation** will decrease the chances of survival. The extreme example of a deleterious mutation is a **lethal mutation,** which results in death to the cell or organism. On the other hand, a **beneficial mutation** will enhance the survival or reproductive success of an organism. In some cases, an allele may be either deleterious or beneficial depending on the genotype and/or the environmental conditions. An example is the sickle-cell allele. In the homozygous state, the sickle-cell allele lessens the chances of survival. However, when an individual is heterozygous for the sickle-cell allele and wild-type allele, this increases the chances of survival due to malarial resistance. Finally, some mutations are called **conditional mutants** because they affect the phenotype only under a defined set of conditions. Geneticists often study conditional mutants in microorganisms; a common example is a temperature-sensitive (*ts*) mutant. A bacterium that has a *ts* mutation grows normally in one temperature range (i.e., the permissive temperature range) but exhibits defective growth at a different temperature range (i.e., the nonpermissive temperature range). For example, an *E. coli* strain carrying a *ts* mutation may be able to grow at 37°C but not at 42°C, whereas the wild-type strain can grow at either temperature.

A second mutation will sometimes affect the phenotypic expression of a first mutation. For example, a forward mutation may cause an organism to grow very slowly. A second mutation at another site in the organism's DNA may restore the normal growth rate, converting the original mutant back to the wild-type condition. Geneticists call these second-site mutations **suppressors** or **suppressor mutations.** This name is meant to indicate that a suppressor mutation acts to suppress the phenotypic effects of another mutation. A suppressor mutation differs from a reversion, because it occurs at a DNA site that is distinct from that of the first mutation.

Suppressor mutations are classified according to their relative locations with regard to the mutation they suppress. When the second mutant site is within the same gene as the first mutation, it is termed an **intragenic suppressor.** Alternatively, a suppressor mutation can be in a different gene from the first mutation; this is called an **intergenic suppressor.**

There are two general types of intergenic suppressors: those that involve an ability to defy the genetic code and those that involve a mutant structural gene. Common examples of the first type are suppressor tRNA mutants, which have been identified in microorganisms. Suppressor tRNA mutants have a change in the anticodon region that causes the tRNA to behave contrary to the genetic code. For example, nonsense suppressors are mutant tRNAs that recognize a stop codon, and instead of stopping translation, put an amino acid into the growing polypeptide chain. This type of mutant tRNA can suppress a nonsense mutation in another gene. However, such bacterial strains grow poorly because they may suppress stop codons in normal genes.

A second type of intergenic suppressor mutation are those that occur within structural genes. These suppressor mutations usually involve a change in the expression of one gene that compensates for a loss-of-function mutation affecting another gene. For example, a first mutation may cause one protein to be partially or completely defective. An intergenic suppressor mutation in a different structural gene might overcome this defect by altering the structure of a second protein so that it could take over the functional role that the first protein cannot perform. Alternatively, intergenic suppressors may involve proteins that participate in a common cellular function. When a first mutation decreases the activity of a protein, a suppressor mutation could enhance the function of a second protein involved in this common function and thereby overcome the defect in the first protein. Interestingly, intergenic suppressors sometimes involve mutations in genetic regulatory proteins such as transcription factors. When a first mutation causes a protein to be defective, a suppressor mutation may occur in a gene that encodes a transcription factor. The mutant transcription factor transcriptionally activates other genes that can compensate for the loss-of-function mutation in the first gene.

Gene Mutations Can Occur Outside of the Coding Sequence and Still Influence Gene Expression

Thus far in chapter 16, we have focused our attention primarily on mutations in the coding regions of genes and their effects on gene expression. In chapters 12 through 15, we learned how various sequences outside the coding sequence play important roles during the process of gene expression. A mutation can occur within noncoding sequences and thereby affect gene expression (table 16.2). For example, a mutation may alter the sequence within the core promoter of a gene. If the mutant promoter

TABLE 16.2
Possible Consequences of Gene Mutations Outside of the Coding Sequence

Sequence	Effect of Mutation
Promoter	May increase or decrease the rate of transcription
Response element/operator site	May disrupt the ability of the gene to be properly regulated
5'-UTR/3'-UTR	May alter the ability of mRNA to be translated; may alter mRNA stability
Splice recognition sequence	May alter the ability of pre-mRNA to be properly spliced

sequence becomes more like the consensus sequence, the mutation may increase the rate of transcription. This is called an **up promoter mutation.** In contrast, a **down promoter mutation** occurs when a mutation causes the promoter to become less like the consensus sequence, decreasing its affinity for regulatory factors and decreasing the transcription rate.

In chapter 14, we considered how mutations could affect regulatory sequences. For example, mutations in the *lac* operator site, called *lacO^C* mutants, prevent the binding of the *lac* repressor protein. This causes the *lac* operon to be constitutively expressed even in the absence of lactose. Bacteria strains with *lacO^C* mutations are at a selective disadvantage compared to wild-type *E. coli* strains, because they waste their energy expressing the *lac* operon even when these proteins are not needed. As noted in table 16.2, mutations can also occur in other noncoding regions of a gene and alter gene expression in a way that may affect phenotype. For example, mutations in eukaryotic genes can alter splice junctions and affect the order and/or number of exons that are contained within mRNA. In addition, mutations that affect the untranslated regions of mRNA (i.e., 5'- and 3'-UTRs) may affect gene expression if they alter the stability of mRNA or its ability to be translated.

DNA Sequences Known as Trinucleotide Repeats Are Hotspots for Mutation

Researchers have discovered several human genetic diseases caused by an unusual form of mutation known as **trinucleotide repeat expansion (TNRE).** The term refers to the phenomenon in which a sequence of three nucleotides can readily increase in number from one generation to the next. In normal individuals, certain genes and chromosomal locations contain regions where trinucleotide sequences are repeated in tandem. These sequences are transmitted normally from parent to offspring without mutation. However, in persons with TNRE disorders, the length of a trinucleotide repeat has increased above a certain critical size and becomes prone to frequent expansion. This phenomenon is depicted here, where the trinucleotide repeat of CAG has expanded from 11 tandem copies to 18 copies.

CAGCAGCAGCAGCAGCAGCAGCAGCAGCAGCAG $n = 11$

to

CAGCAGCAGCAGCAGCAGCAGCAGCAGCAGCAGCAGCAGCA GCAGCAGCAG $n = 18$

Several human diseases have been discovered that involve these types of expansions. These include fragile X syndrome (FRAXA), FRAXE mental retardation, myotonic muscular dystrophy (DM), spinal and bulbar muscular atrophy (SBMA), Huntington disease (HD), and spinocerebellar ataxia (SCA1) (table 16.3). In some cases, the expansion is within the coding sequence of the gene. Typically, such an expansion is a CAG repeat. Since CAG encodes a glutamine codon, these repeats cause the encoded proteins to contain long tracks of glutamine. Although the reasons for the disease symptoms in TNRE disorders are not well understood, the presence of glutamine tracts causes the proteins to aggregate with each other. This aggregation of proteins or protein fragments carrying glutamine repeats is correlated with the progression of the disease. In other TNRE disorders, the expansions are located in the noncoding regions of genes. In the case of the two fragile X syndromes, the repeat produces CpG islands

TABLE 16.3
TNRE Disorders

Disease	SBMA	HD	SCA1	FRAXA	FRAXE	DM
Repeat Sequence	CAG	CAG	CAG	CGG	GCC	CTG
Location of Repeat	Coding sequence	Coding sequence	Coding sequence	5'-UTR	5'-UTR	3'-UTR
Number of Repeats in Normal Individuals	11–33	6–37	6–44	6–53	6–35	5–37
Number of Repeats in Affected Individuals	36–62	27–121	43–81	>200	>200	>200
Pattern of Inheritance	X linked	Autosomal dominant	Autosomal dominant	X linked	X linked	Autosomal dominant
Disease Symptoms	Neuro-degenerative	Neuro-degenerative	Neuro-degenerative	X breakage	Mental retardation	Muscle disease
Anticipation*	None	Male	Male	Female	None	Female

*Indicates the parent in which TNRE occurs most prevalently.

that become methylated. As discussed in chapter 15, methylation can lead to chromatin compaction and thereby silence gene transcription. For myotonic muscular dystrophy, it has been hypothesized that these expansions cause abnormal changes in RNA structure and thereby produce disease symptoms.

There are two particularly unusual features that TNRE disorders have in common. First, the severity of the disease tends to worsen in future generations. This phenomenon is called **anticipation.** A second perplexing feature of TNRE disorders is that the severity of the disease depends on whether the disease is inherited from the mother or father. In the case of Huntington disease, TNRE is likely to occur if the mutant gene is inherited from the father. In contrast, myotonic muscular dystrophy is more likely to get worse if it is inherited from the mother. Overall, TNRE is a newly discovered form of mutation that is receiving a lot of attention by the research community. It poses many challenging questions in molecular genetics. The phenomenon of TNRE also makes it particularly difficult for genetic counselors to advise couples as to the severity of these diseases if they are passed to their children.

At the DNA level, the cause of TNRE is not well understood. It has been speculated that the trinucleotide repeat produces alterations in DNA structure, such as stem-loop formation, and this may lead to errors in DNA replication. However, future research will be necessary to understand the underlying mechanism that causes TNRE. Nevertheless, it is well established that TNRE within certain genes alters the expression of the gene and thereby produces the disease symptoms.

Changes in Chromosome Structure Can Affect the Expression of a Gene

The mutations that we have previously considered in this chapter have been small changes in the DNA sequence of particular genes. A change in chromosome structure can also be associated with an alteration in the expression of a specific gene. Quite commonly, an inversion or translocation has no obvious phenotypic consequence. However, in 1925, Alfred Sturtevant was the first to recognize that chromosomal rearrangements in *Drosophila melanogaster* can influence phenotypic expression (namely, eye morphology). In some cases, a chromosomal rearrangement may affect a gene because the chromosomal breakpoint actually occurs within the gene itself. In other cases, a gene may be left intact, but its expression may be altered when it is moved to a new location. When this occurs, the change in gene location is said to have a **position effect.**

There are two common reasons for position effects. Figure 16.2 depicts a schematic example in which a piece of one chromosome has been inverted or translocated to a different chromosome. One possibility is that a gene may be moved next to regulatory sequences (i.e., silencers or enhancers) that influence the expression of the relocated gene (fig. 16.2a). Alternatively, a chromosomal rearrangement may reposition a gene from a euchromatic region to a chromosome that is very highly condensed (heterochromatic). When the gene is moved to a heterochromatic region, its expression may be turned off (fig. 16.2b). This second type of position effect may produce a variegated phenotype in which the expression of the gene is variable. For genes that affect

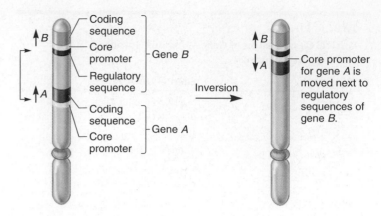

(a) Position effect due to regulatory sequences

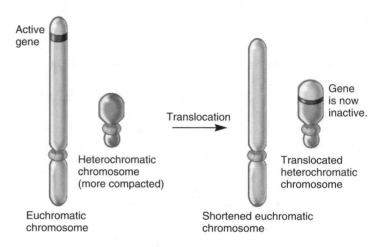

(b) Position effect due to translocation to a heterochromatic chromosome

FIGURE 16.2   Causes of position effects. (a) A chromosomal inversion has repositioned the core promoter of gene *A* next to the regulatory sequences for gene *B*. Since regulatory sequences are often bidirectional, the regulatory sequences for gene *B* may regulate the transcription of gene *A*. **(b)** A translocation has moved a gene from a euchromatic to a heterochromatic chromosome. This type of position effect prevents the expression of the relocated gene.

pigmentation, this produces a mottled appearance rather than an even color. Figure 16.3 shows a position effect that alters eye color in *Drosophila*. Figure 16.3a depicts a normal red-eyed fruit fly, and figure 16.3b shows a mutant fly in which a gene affecting eye color has been relocated to a heterochromatic chromosome. The variegated appearance of the eye occurs because the degree of heterochromatinization varies across different regions of the eye. In cells where heterochromatinization has turned off the eye color gene, a white phenotype occurs, while other cells allow this same region to remain euchromatic and produce a red phenotype.

Mutations Can Occur in Germ-Line or Somatic Cells

In this section, we have considered many different ways that mutations can affect gene expression. For multicellular organisms, the timing of mutations also plays an important role. A mutation can occur very early in life, such as in a gamete or a fertilized egg,

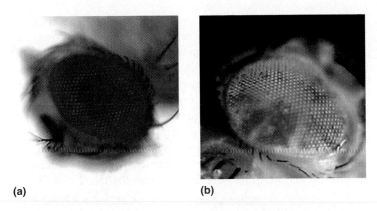

(a)   **(b)**

FIGURE 16.3 **A position effect that alters eye color in *Drosophila*.** (a) A normal red eye. (b) An eye in which an eye color gene has been relocated to a heterochromatic chromosome. This sometimes inactivates the gene and produces a variegated phenotype.

GENES→TRAITS   Variegated eye color occurs because the degree of heterochromatinization varies throughout different regions of the eye. In some cells, heterochromatinization occurs and turns off the eye color gene, thereby leading to the white phenotype. In other patches of cells, the region containing the eye color allele remains euchromatic, yielding a red phenotype.

or it may occur later in life, such as in the embryonic or adult stages. The exact time when mutations occur can be important with regard to the severity of the genetic effect and their ability to be passed from parent to offspring.

Geneticists classify the cells of animals into two types: the germ line and the somatic cells. The term **germ line** refers to cells that give rise to the gametes such as eggs and sperm. A germ-line mutation can occur directly in a sperm or egg cell, or it can occur in a precursor cell that produces the gametes. If a mutant gamete participates in fertilization, all the cells of the resulting offspring will contain the mutation (fig. 16.4a). Likewise, when an individual who has inherited a **germ-line mutation** produces gametes, the mutation may be passed along to future generations of offspring.

The **somatic cells** comprise all cells of the body excluding the germ-line cells. Examples include muscle cells, nerve cells, and so forth. Mutations can also occur within somatic cells at early or late stages of development. Figure 16.4b illustrates the consequences of a mutation that occurs during the embryonic stage. In this example, a **somatic mutation** has occurred within a single embryonic cell. As the embryo grows, this single cell will be the precursor for many cells of the adult organism. Therefore, in the adult, a patch of tissue will contain the mutation. The size of the patch will depend on the timing of the mutation. In general, the earlier the mutation occurs during development, the larger the patch. An individual who has somatic regions that are genotypically different from each other is called a **genetic mosaic**.

Figure 16.5 illustrates an individual who had a somatic mutation occur during an early stage of development. In this case, the person has a patch of gray hair while the rest of the hair is pigmented. Presumably, this individual initially had a single mutation occur in an embryonic cell that ultimately gave rise to a patch of scalp that produced the gray hair. Although a patch of gray hair is not a particularly harmful phenotypic effect, mutations during early stages of life can be quite harmful, especially if

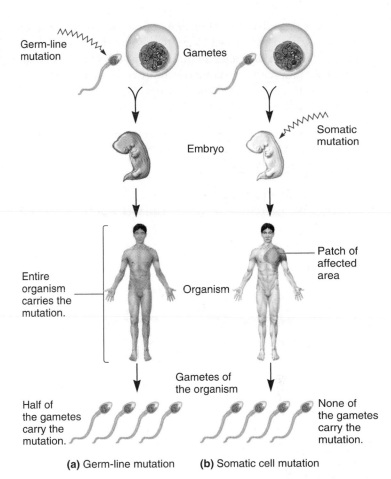

FIGURE 16.4 **The effects of somatic versus germ-line mutations.**

FIGURE 16.5 **Example of a somatic mutation.**

GENES→TRAITS   This individual, Bonnie Raitt, has a patch of gray hair, because a somatic mutation occurred in a single cell during embryonic development that prevented pigmentation of the hair. This cell continued to divide to produce a patch of gray hair.

they disrupt essential developmental processes. Therefore, even though it is prudent to avoid environmental agents that cause mutations during all stages of life, the possibility of somatic mutations is a rather compelling reason to avoid them during the very early stages of life such as fetal development, infancy, and early childhood.

16.2 OCCURRENCE AND CAUSES OF MUTATION

As we have seen, mutations can have a wide variety of effects on the phenotypic expression of genes. For this reason, geneticists have expended a great deal of effort identifying the causes of mutations. This has been a truly challenging task because a myriad of agents can alter the structure of DNA and thereby cause mutation. Geneticists categorize the cause of mutation in one of two ways. **Spontaneous mutations** are changes in DNA structure that result from abnormalities in biological processes, whereas **induced mutations** are caused by environmental agents (table 16.4).

Many causes of spontaneous mutations are examined in other chapters throughout this textbook. As discussed in chapter

TABLE 16.4
Causes of Mutations

Common Causes of Mutations	Description
Spontaneous	
Aberrant recombination	Abnormal crossing over may cause deletions, duplications, translocations, and inversions (see chapter 8).
Aberrant segregation	Abnormal chromosomal segregation may cause aneuploidy or polyploidy (see chapter 8).
Errors in DNA replication	A mistake by DNA polymerase may cause a point mutation (see chapter 11).
Toxic metabolic products	The products of normal metabolic processes may be chemically reactive agents that can alter the structure of DNA.
Transposable elements	Transposable elements can insert themselves into the sequence of a gene (see chapter 17).
Depurination	On rare occasions, the linkage between purines (i.e., adenine and guanine) and deoxyribose can spontaneously break. If not repaired, it can lead to mutation.
Deamination	Cytosine and 5-methyl cytosine can spontaneously deaminate to create uracil or thymine.
Tautomeric shifts	Spontaneous changes in base structure can cause mutations if they occur immediately prior to DNA replication.
Induced	
Chemical agents	Chemical substances may cause changes in the structure of DNA.
Physical agents	Physical phenomena such as UV light and X rays damage the DNA.

8, abnormalities in crossing over can produce chromosome mutations such as deletions, duplications, translocations, and inversions. Aberrant segregation of chromosomes during meiosis can cause genome mutations (i.e., changes in chromosome number). In chapter 11, we discovered that DNA polymerase can make a mistake during DNA replication by putting the wrong base in a newly synthesized daughter strand. Errors in DNA replication are usually infrequent except in certain viruses, such as HIV, that have relatively high rates of spontaneous mutations. In addition, normal metabolic processes may produce chemicals within the cell that can react directly with the DNA and alter its structure. Also, as will be described in chapter 17, transposable genetic elements can alter gene sequences by inserting themselves into genes. In chapter 16, we will examine how spontaneous changes in nucleotide structure can cause mutation. Overall, a distinguishing feature of spontaneous mutations is that their underlying cause originates within the cell.

By comparison, the cause of induced mutations originates outside the cell. Induced mutations are produced by environmental agents that enter the cell and then alter the structure of DNA. Agents that are known to alter the structure of DNA are called **mutagens.** Mutagens can be chemical substances or physical agents that ultimately lead to changes in DNA structure.

In this section, we will begin by examining the random nature of spontaneous mutations and general features of the mutation rate. We will also explore several mechanisms by which mutagens can alter the structure of DNA. Laboratory tests that can identify potential mutagens will then be described.

Spontaneous Mutations Are Random Events

For a few centuries, biologists have wondered whether genotypic changes occur purposefully as a result of environmental conditions or whether they are spontaneous events that may occur randomly in any gene of any individual. The question of whether such mutations are spontaneous occurrences or causally related to environmental conditions has an interesting history. In the nineteenth century, Jean Baptiste Lamarck proposed that physiological events (e.g., use and disuse) determine whether traits are passed along to offspring. For example, his theory suggested that an individual who practiced and became adept at a physical activity, such as the long jump, would pass that quality on to his/her offspring. Alternatively, Charles Darwin proposed that genetic variation occurs as a matter of chance and that natural selection results in the differential reproductive success of organisms that are better adapted to their environments. According to this view, those individuals who happen to contain beneficial mutations will be more likely to survive and pass these genes to their offspring. These opposing theories of the nineteenth century were tested in bacterial studies in the 1940s and 1950s. Two of these studies are described here.

Salvadore Luria and Max Delbrück were interested in the ability of bacteria to become resistant to infection by a bacteriophage called T1. When a population of *E. coli* cells is exposed to T1, a small percentage of bacteria are found to be resistant to T1 infection and pass this trait to their progeny. Luria and Delbrück were interested in whether such resistance, called *ton^r* (T one resistance), is due to the occurrence of spontaneous mutations or whether it is a physiological adaptation that occurs at a low rate within the bacterial population.

According to the physiological adaptation theory, the rate of adaptation should be a relatively constant value and depend on the exposure to the bacteriophage. Therefore, when comparing different populations of bacteria, the number of *ton^r* bacteria should be an essentially constant proportion of the total population. In contrast, the spontaneous mutation theory depends on the timing of mutation. If a *ton^r* mutation occurs early within the proliferation of a bacterial population, many *ton^r* bacteria will be found within that population. However, if it occurs much later in population growth, fewer *ton^r* bacteria will be observed. In general, a spontaneous mutation theory predicts a much greater fluctuation in the number of *ton^r* bacteria among different populations. This test, therefore, has become known as the **fluctuation test.**

To distinguish between the physiological adaptation and spontaneous mutation theories, Luria and Delbrück inoculated 20 individual tubes and one large flask with *E. coli* cells and grew them in the absence of T1 phage. The flask was grown to produce a very large population of cells, while each individual culture was grown to a smaller population of approximately 20 million cells. They then plated the individual cultures onto media containing T1 phage. Likewise, 10 subsamples, each consisting of 20 million bacteria, were removed from the large flask and plated onto media with T1 phage.

The results of the Luria and Delbrück experiment are shown in figure 16.6. Within the smaller individual cultures, a great fluctuation was observed in the number of *ton^r* mutants. These results are consistent with a spontaneous mutation theory in which the timing of a mutation during the growth of a culture greatly affects the number of mutant cells. For example, in tube 14 there were many *ton^r* bacteria. Luria and Delbrück reasoned that a mutation occurred randomly in one bacterium at an early stage of the population growth, before the bacteria were exposed to T1 on plates. This mutant bacterium then divided to produce many daughter cells that inherited the *ton^r* trait. In other tubes, such as 1 and 3, this spontaneous mutation did not occur, so that none of the bacteria had a *ton^r* phenotype. By comparison, the cells plated from the large flask tended toward a relatively constant and intermediate number of *ton^r* bacteria. Because the large growth flask had so many cells, several independent *ton^r* mutations were likely to have occurred during different stages of its growth. In a single flask, however, these independent events would be mixed together to give an average value of *ton^r* cells.

Randomly Occurring Mutations Can Give an Organism a Survival Advantage

Joshua and Ester Lederberg were also interested in the relationship between mutation and the environmental conditions that select for mutation. At the time of their studies, some scientists still

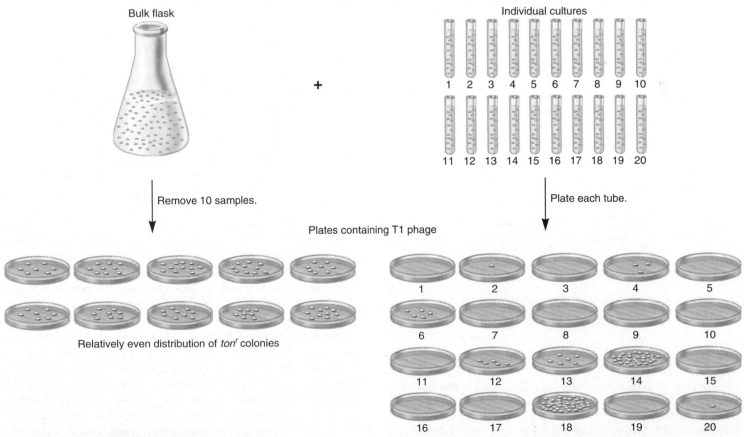

FIGURE 16.6 The Luria-Delbrück fluctuation test.

held the belief that selective conditions could promote the forma-tion of specific mutations allowing the organism to survive. These scientists thought that if bacteria were exposed to an antibiotic, for example, the presence of the antibiotic would actu-ally cause gene mutations that confer antibiotic resistance. This hypothesis, which is similar to Lamarck's adaptation hypothesis, was known as the **directed mutation hypothesis.** In contrast, the **random mutation theory,** which is consistent with a Darwinian viewpoint, indicates that mutations occur at random. According to this theory, environmental factors that affect survival simply select for the survival of those individuals that happen to possess a beneficial mutation.

An experimental approach to distinguish between these two possibilities was developed by the Lederbergs in the 1950s. They used a technique known as **replica plating.** As shown in figure 16.7, they plated a large number of bacteria onto a master plate that did not contain any selective agent (namely, T1). A sterile piece of velvet cloth was lightly touched to this plate in order to pick up a few bacterial cells from each colony. This replica was then

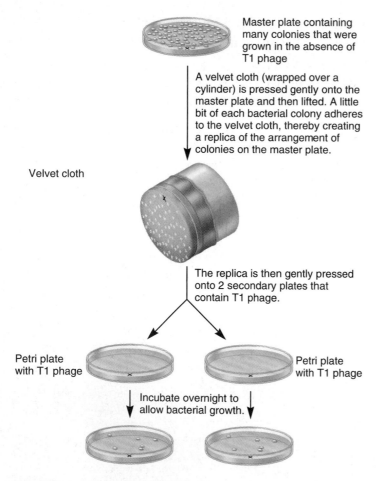

Master plate containing many colonies that were grown in the absence of T1 phage

A velvet cloth (wrapped over a cylinder) is pressed gently onto the master plate and then lifted. A little bit of each bacterial colony adheres to the velvet cloth, thereby creating a replica of the arrangement of colonies on the master plate.

Velvet cloth

The replica is then gently pressed onto 2 secondary plates that contain T1 phage.

Petri plate with T1 phage

Petri plate with T1 phage

Incubate overnight to allow bacterial growth.

FIGURE 16.7 **Replica plating.** Bacteria were first plated on a master plate under nonselective conditions. A sterile velvet cloth was used to make a replica of the master plate. This replica was gently pressed onto two secondary plates that contained a selective agent. In this case, the two secondary plates contained T1 bacteriophage. Only those mutant cells that are *ton*^r could grow to form visible colonies. Note: The *black x* indicates the alignment of the velvet and the plates.

transferred to secondary plates that contained an agent that selected for the growth of bacterial cells with a particular genotype.

In the example shown in figure 16.7, the secondary plates contained T1 bacteriophages. On these plates, only those mutant cells that are *ton*^r could grow. On the secondary plates, a few colonies were observed. Strikingly, they occupied the same loca-tion on each plate. These results indicated that the mutations conferring *ton*^r occurred randomly while the cells were growing on the (nonselective) master plate; the presence of the bacterio-phage in the secondary plates simply selected for the growth of previously occurring *ton*^r mutants. These results supported the random mutation theory. In contrast, the directed mutation hypothesis would have predicted that *ton*^r bacterial mutants would occur after the cells were transferred to the secondary plates. If that had been the case, they would not be expected to arise in identical locations on different secondary plates.

Mutation Rates and Frequencies Are Ways to Quantitatively Assess Mutation in a Population

Because mutations occur spontaneously among populations of living organisms, geneticists are greatly interested in learning how prevalent they are. The term **mutation rate** is the likelihood that a gene will be altered by a new mutation. It is commonly expressed as the number of new mutations in a given gene per generation. In general, the spontaneous mutation rate for a par-ticular gene is in the range of 1 in 100,000 to 1 in 1 billion, or 10^{-5} to 10^{-9} per cell generation. These numbers tell us that it is very unlikely that a particular gene will mutate due to natural causes. However, the mutation rate is not a constant number. The presence of mutagens within the environment can increase the rate of induced mutations to a much higher value than the spon-taneous mutation rate. In addition, mutation rates vary substan-tially from species to species and even within different strains of the same species. One explanation for this variation is the many different causes of mutations (refer back to table 16.4).

As we have learned more about mutation rate, it has been necessary to modify the view that mutations are a totally random process. Within the same individual, some genes mutate at a much higher rate than other genes. This is because some genes are larger than others (providing a greater chance for mutation), and their locations within the chromosome may be more suscep-tible to mutation. Even within a single gene, there are usually **hot spots,** which are select regions of a gene that are more likely to mutate than are other regions (refer back to fig. 6.20).

Before we end our discussion of mutation rate, it is helpful to distinguish the rate of new mutation from the concept of **muta-tion frequency.** The mutation frequency for a gene is the number of mutant genes divided by the total number of genes within a population. If 1 million bacteria were plated and 10 were found to be mutant, the mutation frequency would be 1 in 100,000, or 10^{-5}. As described earlier in the chapter, Luria and Delbrück showed that among the bacteria in the 20 tubes, the timing of mutations influenced the mutation frequency within any particular tube. Some tubes had a high frequency of mutation, while others did not. Therefore, the mutation frequency depends not only on the mutation rate but also on the timing of mutation, and on the likelihood that a mutation will be passed to future generations. The mutation frequency is an important genetic concept, particularly in the field of population genetics. As discussed in chapter 25, mutation frequencies may rise above the mutation rate due to evolutionary forces such as natural selection and genetic drift.

Spontaneous Mutations Can Arise by Depurination, Deamination, and Tautomeric Shifts

Thus far, we have considered the random nature of mutation and the quantitative difference between mutation rate and mutation frequency. We now turn our attention to the molecular changes in DNA structure that can cause mutation. Our first examples concern changes that can occur spontaneously, albeit at a low rate. The most common type of chemical change that occurs naturally is **depurination,** which involves the removal of a purine (guanine or adenine) from the DNA. The covalent bond between deoxyribose and a purine base is somewhat unstable and occasionally undergoes a spontaneous reaction with water that releases the base from the sugar (fig. 16.8a). This is termed an **apurinic site.** In a typical mammalian cell, approximately 10,000 purines are lost from the DNA in a 20-hour period at 37°C. The rate of loss is higher if the DNA is exposed to agents that cause certain types of base modifications such as the attachment of

alkyl groups (e.g., methyl or ethyl groups). Fortunately, as discussed later in this chapter, apurinic sites are recognized by DNA repair enzymes that repair the lesion. If the repair system fails, however, a mutation may result during subsequent rounds of DNA replication. At the apurinic site, there will not be a complementary base to specify the incoming base for the new strand. Rather, any of the four bases will be added to the new strand in the region that is opposite the apurinic site (fig. 16.8b). This may lead to a new mutation.

A second spontaneous lesion that may occur in DNA is the **deamination** of cytosines. The other bases are not readily deaminated. As shown in figure 16.9a, deamination involves the removal of an amino group from the cytosine base. This produces uracil. As discussed later in chapter 16, DNA repair enzymes can recognize uracil as an inappropriate base within DNA and subsequently remove it. However, if such repair does not take place, a mutation may result because uracil hydrogen bonds with adenine during DNA replication. Therefore, if a DNA template strand has uracil, a newly made strand will incorporate adenine into the daughter strand instead of guanine.

Figure 16.9b shows the deamination of 5-methyl cytosine. As discussed in chapter 15, the methylation of cytosine occurs in many eukaryotic species. It also occurs in prokaryotes. If 5-methyl cytosine is deaminated, the resulting base is thymine, which is a normal constituent of DNA. Therefore, this poses a problem for DNA repair enzymes since they cannot determine which is the incorrect base: the thymine that was produced by deamination or the guanine in the opposite strand (which originally base-paired with the methylated cytosine). For this reason, methylated cytosine bases tend to create hot spots for mutation. As an example, researchers analyzed 55 spontaneous mutations that occurred within the *lacI* gene of *E. coli* and determined that

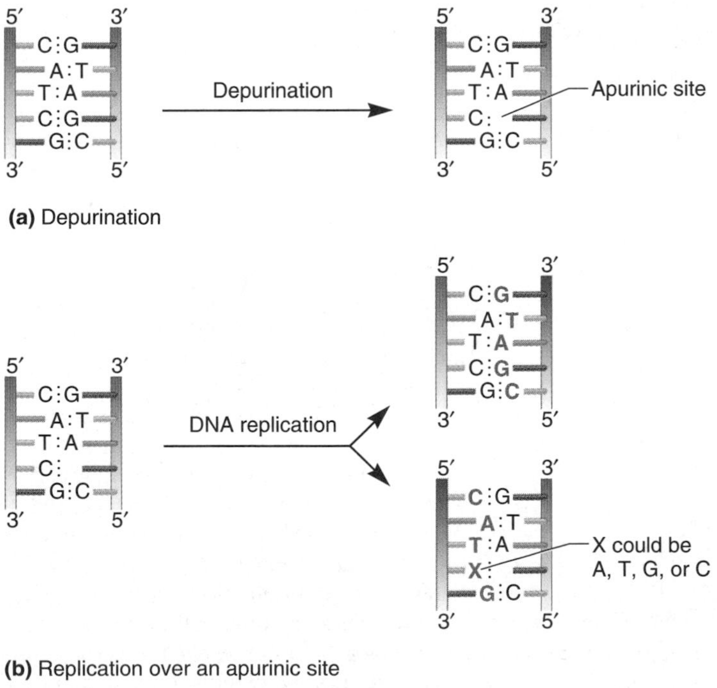

(a) Depurination

(b) Replication over an apurinic site

FIGURE 16.8   Spontaneous depurination. (a) The bond between guanine and deoxyribose is broken, thereby releasing the base. This leaves an apurinic site in the DNA. **(b)** If an apurinic site remains in the DNA as it is being replicated, any of the four nucleotides can be added to the newly made strand. Since three out of four (A, T, and G) are the incorrect nucleotide, this has a 75% chance of causing a mutation.

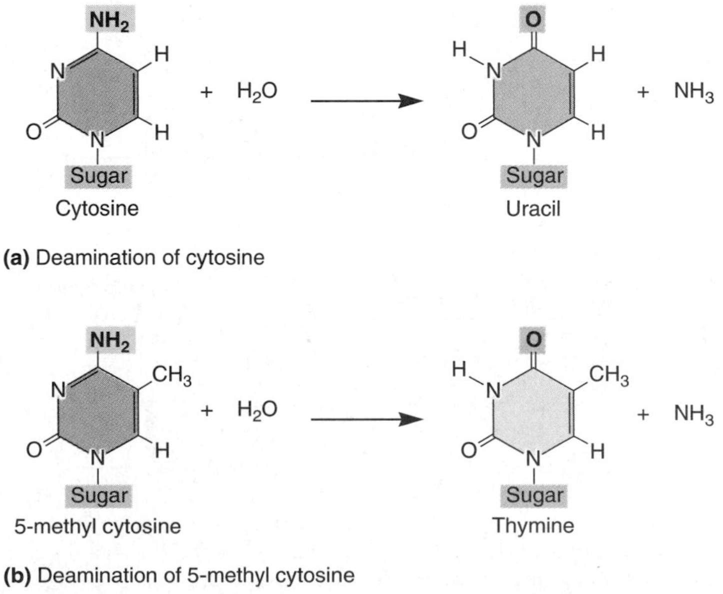

(a) Deamination of cytosine

(b) Deamination of 5-methyl cytosine

FIGURE 16.9   Spontaneous deamination of cytosine and 5-methyl cytosine. (a) The deamination of cytosine produces uracil. **(b)** The deamination of 5-methyl cytosine produces thymine.

44 of them involved changes at sites that were originally occupied by a methylated cytosine base.

A third way that mutations may arise spontaneously involves a temporary change in base structure called a **tautomeric shift.** This event is shown in figure 16.10a. The common, stable form of thymine and guanine is the keto form, while the common form of adenine and cytosine is the amino form. At a low rate, T and G can interconvert to an enol form, while A and C can change to an imino form. Though the relative amounts of the enol and imino forms of these bases are relatively small, they can

Common   **Rare**   **Common**   **Rare**

Guanine — Keto form → Enol form

Adenine — Amino form → Imino form

Thymine — Keto form → Enol form

Cytosine — Amino form → Imino form

(a) Tautomeric shifts that occur in the 4 bases found in DNA

Thymine (enol)   Guanine (keto)   Cytosine (imino)   Adenine (amino)

(b) Mis–base pairing due to tautomeric shifts

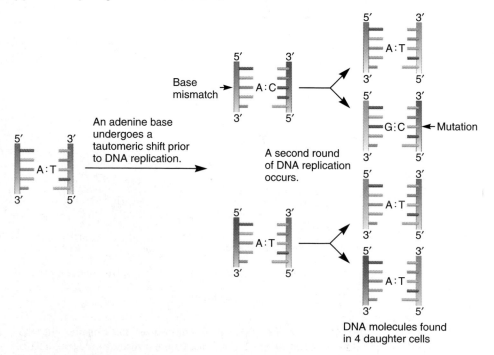

Base mismatch → A:C

An adenine base undergoes a tautomeric shift prior to DNA replication.

A second round of DNA replication occurs.

A:T

G:C ← Mutation

A:T

A:T

A:T

DNA molecules found in 4 daughter cells

(c) Tautomeric shifts and DNA replication can cause mutation

FIGURE 16.10 **Tautomeric shifts and their ability to cause mutation. (a)** The common forms of the bases are shown on the *left,* while the rare forms produced by a tautomeric shift are shown on the *right.* **(b)** In these two examples, the rare enol form of thymine pairs with the common keto form of guanine, and the rare imino form of cytosine pairs with the common amino form of adenine. **(c)** A tautomeric shift occurred in an adenine base just prior to replication, causing the formation of an AC base pair. If not repaired, a second round of replication will lead to the formation of a permanent GC mutation. Note: A tautomeric shift is a very temporary situation. During the second round of replication, the adenine base that shifted prior to the first round of DNA replication is likely to have shifted back to its normal form. Therefore, during the second round of replication, a thymine base will be found opposite this adenine.

cause mutation because these rare forms of the bases do not conform to the AT/GC rule of base pairing. Instead, if one of the bases is in the enol or imino form, hydrogen bonding will promote AC and GT base pairs, as shown in figure 16.10b.

For a tautomeric shift to cause a mutation, it must occur immediately prior to DNA replication. This phenomenon is shown in figure 16.10c. An adenine base in the template strand has undergone a tautomeric shift just prior to the replication of the complementary daughter strand. During replication, the daughter strand will incorporate a cytosine opposite this adenine, creating a base mismatch. This mismatch could be repaired via a mismatch repair system (discussed later in this chapter) or via the proofreading function of DNA polymerase. However, if these repair mechanisms fail, the next round of DNA replication will create a double helix with a GC base pair, while the correct base pair should be AT. As shown in the *right* side of figure 16.10c, one of four daughter cells will inherit this GC mutation.

EXPERIMENT 16A

X-Rays Were the First Environmental Agent Shown to Cause Induced Mutations

As mentioned previously in table 16.4, changes in DNA structure can also be caused by environmental agents. These agents are called mutagens, and the mutations they cause are referred to as induced mutations. In 1927, Hermann Müller devised an approach to show that X rays can cause induced mutations in *Drosophila melanogaster*. Müller reasoned that a mutagenic agent might cause some genes to become defective. His experimental approach focused on the ability of a mutagen to cause defects in X-linked genes that result in a recessive lethal phenotype.

To determine if X rays increase the rate of recessive, X-linked lethal mutations, Müller wanted to have an easy way to detect the occurrence of such mutations. He cleverly realized that he had available a laboratory strain of fruit flies that could make this possible. In particular, he conducted his crosses in such a way that a female fly that inherited a new mutation causing a recessive X-linked lethal allele would not be able to produce any male offspring. This made it very easy for him to detect lethal mutations; he had to count only the number of female flies that could not produce sons. In flies that had not been exposed to X rays, nearly all females were able to produce sons. In contrast, if X rays were acting as a mutagenic agent, Müller hypothesized that exposure to X rays would increase the numbers of females unable to produce sons.

To understand Müller's crosses, we need to take a closer look at a peculiar version of one of the X chromosomes in a strain of flies that he used in his crosses. This X chromosome, designated *ClB*, had three important genetic alterations.

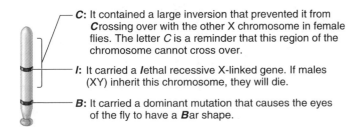

C: It contained a large inversion that prevented it from **C**rossing over with the other X chromosome in female flies. The letter **C** is a reminder that this region of the chromosome cannot cross over.

l: It carried a **l**ethal recessive X-linked gene. If males (XY) inherit this chromosome, they will die.

B: It carried a dominant mutation that causes the eyes of the fly to have a **B**ar shape.

Note: C and B are uppercase because they are inherited in a dominant manner, while *l* is lowercase because it is a recessive allele.

A female fly that has one copy of this X chromosome would have bar-shaped eyes, because bar is a dominant allele. Even though this X chromosome has a lethal allele, a female fly can survive if the corresponding gene on the other X chromosome is a normal allele. In Müller's experiments, the goal was to determine if exposure to X rays caused a mutation on the normal X chromosome (not the *ClB* chromosome) that created a lethal allele in any essential X-linked gene except for the gene that already had a lethal allele on the *ClB* chromosome (fig. 16.11). If a recessive lethal mutation occurred on the normal X chromosome, this female could survive because it would be heterozygous for recessive lethal mutations in two different genes. However, because each X chromosome would have a lethal mutation, this female would not be able to produce any living sons.

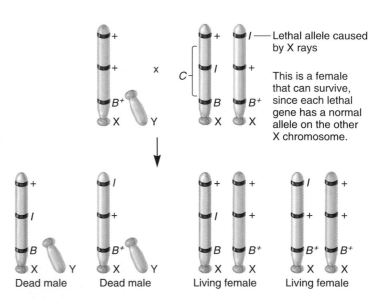

No living male offspring are produced from this cross.

FIGURE 16.11   A strategy to detect the presence of lethal X-linked mutations. X rays may cause a recessive lethal mutation to occur in the normal X chromosome. This female also contains another lethal allele in the *ClB* chromosome. Nevertheless, this female could survive because it would be heterozygous for recessive lethal mutations in two different genes. However, since each X chromosome would have a lethal mutation, this female would not be able to produce any living sons.

whether antioxidants can counteract the effects of mutagens and thereby lower the cancer rate.

Mutagens can alter the structure of DNA in various ways. Some mutagens act by covalently modifying the structure of nucleotides. For example, **nitrous acid** replaces amino groups with keto groups ($-NH_2$ to $=O$). This can change cytosine to uracil and adenine to hypoxanthine. When this mutated DNA replicates, the modified bases do not pair with the appropriate nucleotides in the newly made strand. Instead, uracil pairs with adenine, and hypoxanthine pairs with cytosine (fig. 16.13). Other chemical mutagens can also disrupt the appropriate pairing between nucleotides by alkylating bases within the DNA. During alkylation, methyl or ethyl groups are covalently attached to the bases. Examples of alkylating agents include **nitrogen mustards** and **ethyl methanesulfonate (EMS).**

Some mutagens exert their effects by directly interfering with the DNA replication process. For example, **acridine dyes,** such as **proflavin,** contain flat planar structures that interchelate into the double helix by sandwiching between adjacent base pairs, thereby distorting the helical structure. When DNA containing these mutagens is replicated, single-nucleotide additions and/or deletions can be incorporated into the newly made daughter strands.

Compounds such as **5-bromouracil (5BU)** and **2-aminopurine** are nucleotide base analogues that become incorporated into daughter strands during DNA replication. 5-bromouracil is a thymine analogue that can be incorporated into DNA instead of thymine. Like thymine, 5BU can base-pair with adenine. However, at a relatively high rate, 5BU will tautomerize and base-pair with guanine (fig. 16.14a). When this occurs during DNA replication, 5BU causes a mutation in which an AT base pair is changed to a G-5BU base pair (fig. 16.14b). This is a transition, since the adenine has been changed to a guanine, both of which are purines. During the next round of DNA replication,

FIGURE 16.13   **Mispairing of modified bases that have been deaminated using nitrous acid.** During DNA replication, uracil pairs with adenine, and hypoxanthine pairs with cytosine. This will create mutations in the newly replicated strand during DNA replication.

(a) Base pairing of 5BU with adenine or guanine

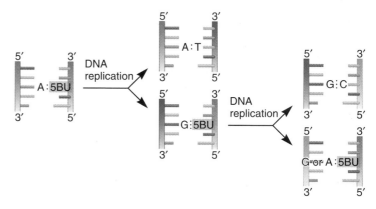

(b) How 5BU causes a mutation in a base pair during DNA replication

FIGURE 16.14   **Structure and effects of the mispairing of 5-bromouracil with guanine. (a)** In its keto form, 5BU bonds with adenine, while in its enol form, it bonds with guanine. **(b)** Guanine may be incorporated into a newly made strand by pairing with 5BU. After a second round of replication, the DNA will contain a CG base pair instead of the original AT base pair.

the template strand containing the guanosine base will create a GC base pair. In this way, 5-bromouracil can promote a change of an AT base pair into a GC base pair.

DNA molecules are also sensitive to physical agents such as radiation. In particular, radiation of short wavelength and high energy, known as ionizing radiation, can alter DNA structure. This type of radiation includes X rays and gamma rays. Ionizing radiation can penetrate deeply into biological materials, where it creates chemically reactive molecules known as free radicals. These molecules can alter the structure of DNA in a variety of ways. Exposure to high doses of ionizing radiation can cause base deletions, single nicks in DNA strands, cross-linking, and even chromosomal breaks.

Nonionizing radiation, such as UV light, contains less energy, and so it penetrates only the surface of material such as

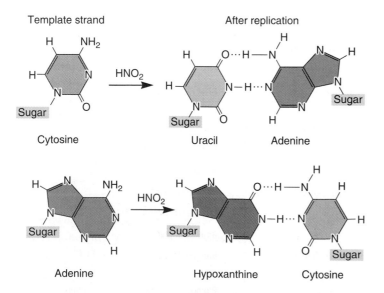

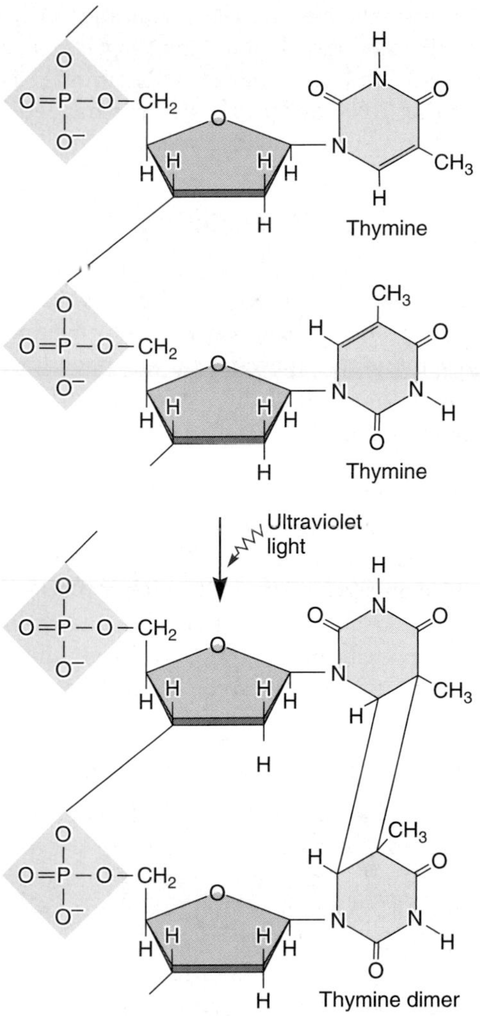

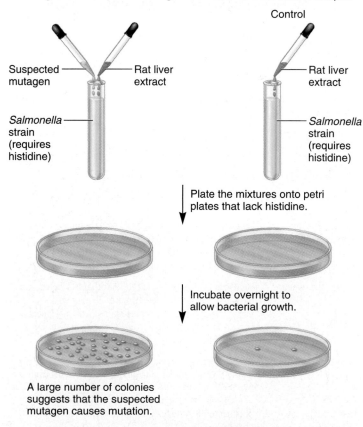

FIGURE 16.15   **Formation and structure of a thymine dimer.**

the skin. Nevertheless, UV light is known to cause DNA mutation. For example, as shown in figure 16.15, UV light causes the formation of cross-linked **thymine dimers.** A thymine dimer within a DNA strand may cause a mutation when that DNA strand is replicated. Plants, in particular, must have effective ways to prevent UV damage since they are exposed to sunlight throughout the day. As discussed later in chapter 16, several DNA repair systems are able to detect UV damage and repair it before a permanent mutation results.

Testing Methods Can Determine If an Agent Is a Mutagen

To determine if an agent is mutagenic, researchers use testing methods that can monitor whether or not an agent increases the rate of mutation. Many different kinds of tests have been used to evaluate mutagenicity. One commonly used test is the **Ames test,** which was developed by Bruce Ames. This test uses strains of a bacterium, *Salmonella typhimurium,* that cannot synthesize the amino acid histidine. These strains contain a point mutation within a gene that encodes an enzyme required for histidine

biosynthesis. The mutation renders the enzyme inactive. Therefore, the bacteria cannot grow on petri plates unless histidine has been added to the growth medium. However, a second mutation (e.g., a reversion) may occur that restores the ability to synthesize histidine. In other words, a second mutation can cause a reversion back to the wild-type condition. The Ames test monitors the rate at which this second mutation occurs and thereby indicates whether an agent increases the mutation rate above the spontaneous rate.

Figure 16.16 outlines the steps in the Ames test. The suspected mutagen is mixed with a rat liver extract and a *Salmonella* strain of bacteria that cannot synthesize histidine. A mutagen may require activation by cellular enzymes; the rat liver extract provides a mixture of enzymes that may activate a mutagen. This step improves the ability of the test to identify agents that may cause mutation in mammals. After the incubation period, a large number of bacteria are then plated on a minimal growth medium that does not contain histidine. The *Salmonella* strain is not expected to grow on these plates. However, if a mutation has occurred that allows the strain to synthesize histidine, it can grow on these plates to form a visible bacterial colony. To estimate the mutation rate, the colonies that grow on the minimal media are counted and compared to the total number of bacterial cells that were originally streaked on the plate. For example, if 10,000,000 bacteria were plated and 10 growing colonies were observed, the rate

FIGURE 16.16   **The Ames test for mutagenicity.**

of mutation is 10 out of 10,000,000; this equals 1 in 10^6, or simply 10^{-6}. As a control, bacteria that have not been exposed to the mutagen are also tested, since a low level of spontaneous mutations is expected to occur.

As discussed in solved problem S3 at the end of chapter 16, there are several *his⁻* strains of *Salmonella* that contain different kinds of mutations within the coding sequence. This makes it possible to determine if a mutagen causes transitions, transversions, or frameshift mutations.

16.3 DNA REPAIR

Because most mutations are deleterious, DNA repair systems are vital to the survival of all organisms. If DNA repair systems did not exist, environmentally induced and spontaneous mutations would be so prevalent that few species, if any, would survive. The necessity of DNA repair systems becomes evident when they are missing. Bacteria contain several different DNA repair systems. Yet when even a single system is absent, the bacteria have a much higher rate of mutation. In fact, the rate of mutation is so high that these bacterial strains are sometimes called mutator strains. Likewise, in humans, an individual who is defective in only a single DNA repair system may manifest various disease symptoms, including a higher risk of skin cancer. This increased risk is due to the inability to repair UV-induced mutations.

Living cells contain several DNA repair systems that can fix different types of DNA alterations (table 16.6). Each repair system is composed of one or more proteins that play specific roles in the repair mechanism. In most cases, DNA repair is a multistep process. First, one or more proteins in the DNA repair system detect an irregularity in DNA structure. Next, the abnormality is removed by the action of DNA repair enzymes. Finally,

TABLE 16.6
Common Types of DNA Repair Systems

System	Description
Direct repair	An enzyme recognizes an incorrect alteration in DNA structure and directly converts it back to a correct structure.
Base excision repair and nucleotide excision repair	An abnormal nucleotide or base is first recognized and removed from the DNA. A segment of DNA in this region is excised, and then the complementary DNA strand is used as a template to synthesize a normal DNA strand.
Mismatch repair	Similar to excision repair except that the DNA defect is a base pair mismatch in the DNA, not an abnormal nucleotide. The mismatch is recognized, and a segment of DNA in this region is removed. The parental strand is used to synthesize a normal daughter strand of DNA.
Recombinational repair	Occurs when DNA damage causes a gap in synthesis during DNA replication. The gap is exchanged between the abnormal DNA and the corresponding region in the normal, replicated double helix. After this occurs, it is possible to fill in the gap using the complementary DNA strand.

normal DNA is synthesized via DNA replication enzymes. In this section, we will examine several different repair systems that have been characterized in bacteria, yeast, and mammals. Their diverse ways of repairing DNA underscore the extreme necessity for the structure of DNA to be maintained properly.

Damaged Bases Can Be Directly Repaired

In a few cases, the covalent modification of nucleotides by mutagens can be reversed by specific cellular enzymes. As discussed earlier in this chapter, UV light causes the formation of thymine dimers. It was found that the yeast enzyme photolyase can repair thymine dimers by splitting the dimers, restoring DNA to its original condition (fig. 16.17a). This process directly restores the structure of DNA.

A protein known as an alkyltransferase can remove methyl or ethyl groups from guanine bases that have been mutagenized by agents such as nitrogen mustards and ethyl methanesulfonate. This protein is called an alkyltransferase because it transfers the methyl or ethyl group from the base to a cysteine side chain within the alkyltransferase protein (fig. 16.17b). Surprisingly, this permanently inactivates alkyltransferase—it can be used only once!

Base Excision Repair Removes a Damaged Base

A second type of repair system, called **base excision repair (BER),** involves the function of a category of enzymes known as **DNA-N-glycosylases.** This type of enzyme can recognize an abnormal base and cleave the bond between it and the sugar in the DNA backbone (fig. 16.18). Depending on the species, this repair system can eliminate abnormal bases such as uracil, 3-methyladenine, 7-methylguanine, and pyrimidine dimers.

Figure 16.18 illustrates the steps involved in DNA repair via N-glycosylase. In this example, the DNA contains a uracil residue in its sequence. This could have happened spontaneously or by the action of a chemical mutagen (e.g., nitrous acid) that deaminates cytosine to produce uracil. N-glycosylase recognizes a uracil within the DNA and cleaves the bond between the sugar and base. This releases the uracil base and leaves behind an apyrimidinic (AP) nucleotide. This abnormal nucleotide is recognized by a second enzyme, **AP-endonuclease,** which makes a cut on the 5′ side. DNA polymerase (which has a 5′ to 3′ exonuclease activity) removes the abnormal region and, at the same time, replaces it with normal nucleotides. This process is called **nick translation** (although nick replication would be a more accurate term). Finally, DNA ligase closes the nick.

Nucleotide Excision Repair Systems Remove Segments of Damaged DNA

An important general process for DNA repair is **nucleotide excision repair (NER).** This type of system can repair many different types of DNA damage, including UV-induced damage (namely, thymine dimers), chemically modified bases, missing bases, and certain types of cross-links. In NER, several nucleotides in the damaged strand are removed from the DNA, and the intact

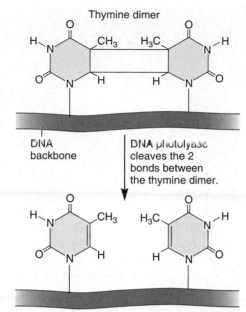

Thymine dimer

DNA
backbone

DNA photolyase
cleaves the 2
bonds between
the thymine dimer.

The normal structure of the 2 thymines is restored.

(a) Direct repair of a thymine dimer

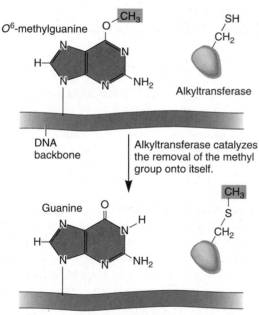

O^6-methylguanine

Alkyltransferase

DNA
backbone

Alkyltransferase catalyzes
the removal of the methyl
group onto itself.

Guanine

The normal structure of guanine is restored.

(b) Direct repair of a methylated base

FIGURE 16.17   **Direct repair of damaged bases in DNA.**
(a) The repair of thymine dimers by photolyase. **(b)** The repair of
methylguanine by the transfer of the methyl group to alkyltransferase.

strand is used as a template for resynthesis of a normal complementary strand. NER is found in all eukaryotes and prokaryotes,
although its molecular mechanism is better understood in
prokaryotic species.

In *E. coli*, the NER system requires four key proteins, designated UvrA, UvrB, UvrC, and UvrD, plus the help of DNA poly-

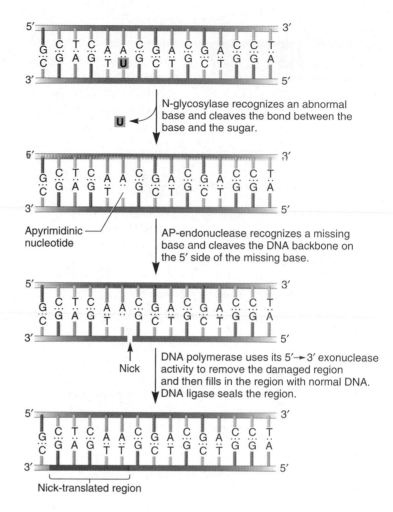

N-glycosylase recognizes an abnormal
base and cleaves the bond between the
base and the sugar.

Apyrimidinic
nucleotide

AP-endonuclease recognizes a missing
base and cleaves the DNA backbone on
the 5′ side of the missing base.

Nick

DNA polymerase uses its 5′→ 3′ exonuclease
activity to remove the damaged region
and then fills in the region with normal DNA.
DNA ligase seals the region.

Nick-translated region

FIGURE 16.18   **Base excision repair.** An abnormal base is
initially recognized by an enzyme known as N-glycosylase, which
cleaves the bond between the base and the sugar. Depending on
whether a purine or pyrimidine is removed, this creates an apurinic
or apyrimidinic site, respectively. In either case, the region is recognized
by AP-endonuclease, which makes a cut at the 5′ side of the abnormal
nucleotide. DNA polymerase then removes the abnormal region and
fills in with normal DNA. Finally, DNA ligase seals the repaired strand.

merase and DNA ligase. UvrA, B, C, and D recognize and remove
a short segment of a damaged DNA strand. They are named Uvr
because they are involved in Ultraviolet light repair of pyrimidine
dimers, although the UvrA–D proteins are also important in
repairing chemically damaged DNA.

Figure 16.19 outlines the steps involved in the *E. coli* NER
system. A protein trimer consisting of two UvrA molecules and
one UvrB molecule tracks along the DNA in search of damaged
DNA. Such DNA will have a distorted double helix, which is sensed
by the UvrA/UvrB complex. When a damaged segment is identified, the two UvrA proteins are released and UvrC binds to the site.
The UvrC protein makes incisions in the damaged strand on both
sides of the damaged site. Typically, the damaged strand is cut four
to five nucleotides away from the 3′ side of the damage and eight
nucleotides from the 5′ end. After this incision process, UvrD,
which is a helicase, recognizes the region and separates the two

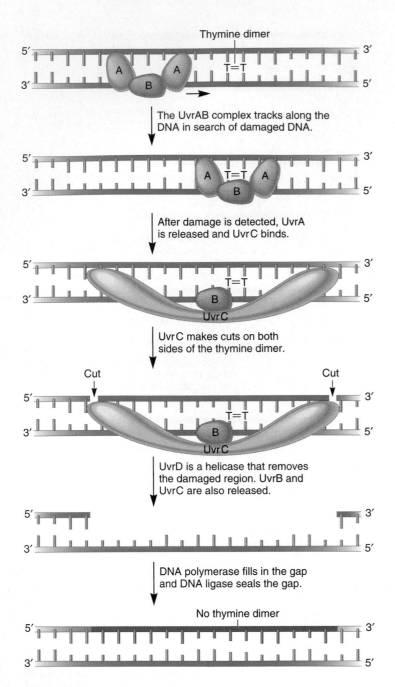

FIGURE 16.19 **Nucleotide excision repair in *E. coli*.**
UvrA/UvrB trimer scans along the DNA. When the trimer encounters a thymine dimer in the DNA, the UvrA dimer is released. UvrC binds to UvrB. UvrC make two cuts in the damaged DNA strand. One cut is made eight nucleotides 5′ to the site, the other approximately five nucleotides 3′ from the site. UvrC is released and UvrD binds to the site. UvrD is a helicase that causes the damaged strand to be removed. UvrB, C, and D are released. DNA polymerase resynthesizes a complementary strand. DNA ligase makes the final connection.

strands of DNA. This releases a short DNA segment that contains the damaged region. Following the excision of the damaged DNA, DNA polymerase fills in the gap using the undamaged strand as a template. Finally, DNA ligase makes the final covalent connection between the newly made DNA and the original DNA strand.

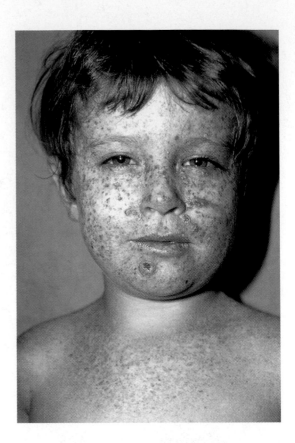

FIGURE 16.20   **An individual affected with xeroderma pigmentosum.**

GENES→TRAITS This disease involves a defect in genes that are involved with nucleotide excision repair. Xeroderma pigmentosum can be caused by defects in seven different NER genes. Affected individuals have an increased sensitivity to sunlight because of an inability to repair UV-induced DNA lesions. In addition, they may also have pigmentation abnormalities, many premalignant lesions, and a high predisposition to skin cancer.

Several human diseases have been shown to involve inherited defects in genes involved in nucleotide excision repair. These include xeroderma pigmentosum (XP), Cockayne syndrome (CS), and PIBIDS. (PIBIDS is an acronym for a syndrome with symptoms that include photosensitivity, ichthyosis [a skin abnormality], brittle hair, impaired intelligence, decreased fertility, and short stature.) A common characteristic in all three syndromes is an increased sensitivity to sunlight because of an inability to repair UV-induced lesions. Figure 16.20 shows a photograph of an individual with xeroderma pigmentosum. Such individuals have pigmentation abnormalities, many premalignant lesions, and a high predisposition to skin cancer. They may also develop early degeneration of the nervous system.

Genetic analyses of patients with XP, CS, and PIBIDS have revealed that these syndromes result from defects in a variety of different genes that encode NER proteins. For example, xeroderma pigmentosum can be caused by defects in seven different NER genes. In all cases, individuals have a defective nucleotide excision repair pathway. In recent years, several human NER genes have been successfully cloned and sequenced. Although more research is needed to completely understand the mechanisms of

DNA repair, the identification of NER genes has helped unravel the complexities of NER pathways in human cells.

Mismatch Repair Systems Recognize and Correct a Base Pair Mismatch

Thus far, we have considered several DNA repair systems that recognize abnormal nucleotide structures within DNA, including thymine dimers, alkylated bases, and the presence of uracil in the DNA. Another type of abnormality that should not occur in DNA is a **base mismatch.** As described in chapter 9, the structure of the DNA double helix obeys the AT/GC rule of base pairing. During the normal course of DNA replication, however, an incorrect nucleotide may be added to the growing strand by mistake. This creates a mismatch between a nucleotide in the parental and newly made strand. There are various DNA repair mechanisms that recognize and remove this mismatch. For example, as described in chapter 11, DNA polymerase has a 3′ to 5′ proofreading ability that can detect mismatches and remove them. However, if this proofreading ability fails, cells contain additional DNA repair systems that can detect base mismatches and fix them. An interesting DNA repair system is the **methyl-directed mismatch repair** system that exists in all species. Similar to defects in nucleotide excision repair systems, mutations in the human mismatch repair system are associated with particular types of cancer. For example, mutations in two human mismatch repair genes, *hMSH2* and *hMLH1*, play a role in the development of a type of colon cancer known as hereditary nonpolyposis colorectal cancer.

The molecular mechanism of mismatch repair has been studied extensively in *E. coli*. This system involves the participation of several proteins that detect the mismatch and specifically remove the segment from the newly made daughter strand. Keep in mind that the newly made daughter strand contains the incorrect base, while the parental strand is normal. Therefore, an important aspect of methyl-directed mismatch repair is that it specifically repairs the newly made strand rather than the original template strand. As shown in figure 16.21, three proteins in *E. coli*, designated MutL, MutH, and MutS, detect the mismatch and direct the removal of the mismatched base from the newly made strand. These proteins are named Mut because their absence leads to a much higher <u>mu</u>tation rate compared to normal strains of *E. coli*. A key characteristic of the MutH protein is that it can distinguish between the parental DNA strand and the daughter strand. Prior to DNA replication, the parental DNA has already been methylated. Immediately after DNA replication, it takes some time before a newly made daughter strand is methylated. Therefore, the MutH protein can distinguish between the parental strand and newly made strand because the newly made strand is not methylated.

The role of MutS is to locate mismatches. Once a mismatch is detected, MutS forms a complex with MutL. MutL acts as a linker that binds to MutH by a looping mechanism (fig. 16.21). This stimulates MutH, which is bound to a hemimethylated site, to make a cut in the nonmethylated DNA strand. After the strand

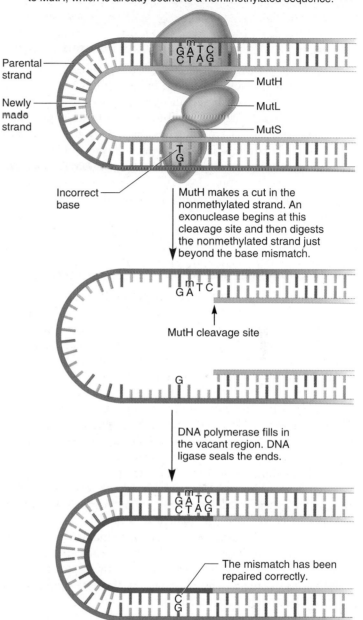

FIGURE 16.21   Methyl-directed mismatch repair in *E. coli*. MutS slides along the DNA and recognizes base mismatches in the double helix. MutL binds to MutS and acts as a linker between MutS and MutH. The DNA must loop for this interaction to occur. The role of MutH is to identify the methylated strand of DNA, which is the nonmutated parental strand. The methylated adenine is designated with an m. MutH then makes a cut in the nonmethylated strand, and an exonuclease digests the strand until it passes the MutS/MutL region. This leaves a gap, which is filled in by DNA polymerase and sealed by DNA ligase.

is cut, an exonuclease digests the nonmethylated DNA strand in the direction of the mismatch and proceeds just beyond the mismatch site. This leaves a gap in the daughter strand that is repaired by DNA polymerase and DNA ligase. As illustrated in figure 16.21, the net result is that the mismatch has been corrected by removing the incorrect region in the daughter strand and then resynthesizing the correct sequence using the parental DNA as a template.

Damaged DNA Can Be Repaired by Recombination

Certain types of DNA damage can halt the progression of DNA replication. For example, a thymine dimer inhibits DNA replication because it prevents base pairing in a newly made daughter strand. For this reason, a gap will be created in the region that cannot be replicated. DNA gaps are particularly harmful, since they may cause chromosomal breaks and further mutations. To avoid these consequences, the unreplicated gap may be repaired by genetic recombination. We will consider **recombinational repair** in chapter 16, and examine the molecular mechanisms of genetic recombination in chapter 17.

DNA replication produces a genetically identical pair of double-stranded DNA molecules. Recombinational repair occurs while the two DNA copies are being made. To understand this process, we must pay close attention to the relationship between newly made DNA copies during replication. In figure 16.22, the strands in the two double helices are labeled A, B, C, and D. Because DNA replication makes identical copies of genetic material, strands A and C have the same DNA sequence. Strands B and D also have the same sequence and are complementary to A and C. In figure 16.22, a thymine dimer had previously occurred in one of the parental strands (which is now labeled strand D). The thymines in a thymine dimer cannot hydrogen bond properly to incoming nucleotides during DNA replication. Therefore, a thymine dimer disrupts the DNA replication process by creating a gap in the newly made strand after the replication fork has passed. Recombinational repair fixes this gap. This occurs in a two-step process. First, the gap is replaced with the same region from the other parental DNA strand: a short segment of strand A replaces the corresponding segment of strand C. This removes the gap in strand C, but it creates a gap in strand A. The next step is to fill in the gap in strand A, using the normal strand B as a template. The net result is that neither of the replicated DNA double helices contain gaps, although the thymine dimer is still present in one of the double helices.

Actively Transcribed DNA Is Repaired More Efficiently Than Nontranscribed DNA

Before we end our discussion of DNA repair systems, it is interesting to mention that not all DNA is repaired at the same rate. In the 1980s, Philip Hanawalt and colleagues conducted experiments showing that actively transcribed genes in eukaryotes and prokaryotes are more efficiently repaired following radiation damage than is nontranscribed DNA. The targeting of DNA repair enzymes to actively transcribing genes may have several biological advantages. First, active genes are loosely packed and may be more vulnerable to DNA damage. Likewise, the process of transcription itself may make the DNA more susceptible to agents that cause damage. In addition, DNA regions that contain

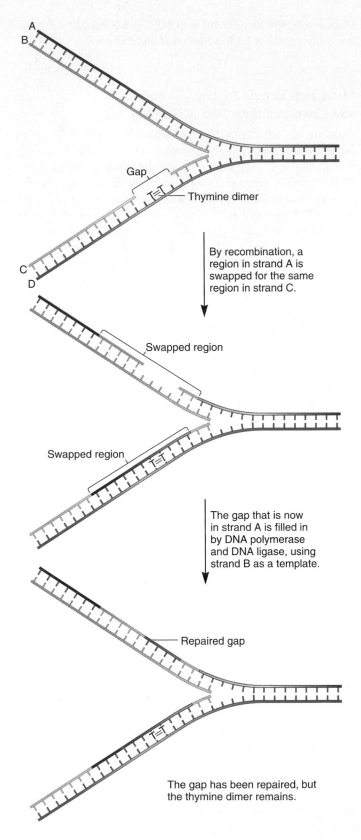

FIGURE 16.22   DNA repair by recombination. See text for a description.

actively transcribed genes are more likely to be important for survival than nontranscribed regions. This may be particularly true in differentiated cells, which no longer divide. In nondividing cells, gene transcription (rather than DNA replication) is of utmost importance.

In *E. coli*, a protein known as **transcription-repair coupling factor (TRCF)** is responsible for targeting the nucleotide excision repair system to actively transcribed genes having damaged DNA (fig. 16.23). In this example, RNA polymerase had been transcribing a gene until it encountered a DNA lesion and became stalled. TRCF is a helicase that displaces the stalled RNA polymerase from the damaged template strand. TRCF also has a binding site for UvrA. Therefore, after RNA polymerase has been removed, TRCF recruits the excision repair system to the damaged site by recognizing the UvrA/UvrB complex. After the UvrA/UvrB complex has bound to this site, TRCF is released from the damaged site. At this stage, the UvrA/UvrB complex can begin excision repair (as previously described in fig. 16.19).

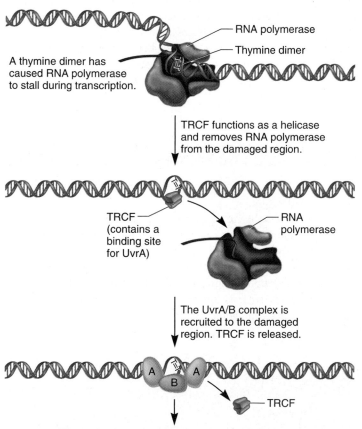

A thymine dimer has caused RNA polymerase to stall during transcription.

RNA polymerase
Thymine dimer

TRCF functions as a helicase and removes RNA polymerase from the damaged region.

TRCF (contains a binding site for UvrA)

RNA polymerase

The UvrA/B complex is recruited to the damaged region. TRCF is released.

A   A
B

TRCF

The region is repaired as described in figure 16.19.

FIGURE 16.23 **Targeting of the nucleotide excision repair system to an actively transcribing gene containing a DNA alteration.** When RNA polymerase encounters damaged DNA, it will stall over the damaged site. TRCF, which is a helicase, recognizes a stalled RNA polymerase and removes it from the damaged site. TRCF has a binding site for UvrA and thereby recruits the UvrA/UvrB complex to the region. This initiates the process of nucleotide excision repair.

In eukaryotes, the mechanism that couples DNA repair and transcription is not completely understood. Several different proteins have been shown to act as transcription-repair coupling factors. Some of these have been identified in people with high rates of mutation. As mentioned earlier, Cockayne syndrome (CS) involves defects in genes that play a role in nucleotide excision repair. Two genes, termed *CS-A* and *CS-B*, encode proteins that function as transcription-repair coupling factors. In addition, certain general transcription factors, such as TFIIH, play a role in both transcription and excision repair.

Damaged DNA May Be Replicated by Translesion DNA Polymerases

Despite the efficient action of numerous repair systems that remove lesions in DNA in an error-free manner, it is inevitable that some lesions may escape these repair mechanisms. Such lesions may be present when DNA is being replicated. If so, replicative DNA polymerases, such as polIII in *E. coli*, are highly sensitive to geometric distortions in DNA and are unable to replicate through DNA lesions. During the past decade, it has been discovered that cells are equipped with specialized DNA polymerases that will assist the replicative DNA polymerase during the process of **translesion synthesis (TLS).** These TLS polymerases, which are also described in chapter 11 (see table 11.4), contain an active site with a loose, flexible pocket that can accommodate aberrant structures in the template strand. When a replicative DNA polymerase encounters a damaged region, it is swapped with a TLS polymerase.

A negative consequence of translesion synthesis is low fidelity. Due to their flexible active site, TLS polymerases are much more likely to incorporate the wrong nucleotide in a newly made daughter strand. The mutation rate is typically in the range of 10^{-2} to 10^{-3}. By comparison, replicative DNA polymerases are highly intolerant of the geometric distortions imposed on DNA by the incorporation of incorrect nucleotides, and consequently, they incorporate wrong nucleotides with a very low frequency. In other words, they copy DNA with a high degree of fidelity.

In *E. coli*, translesion synthesis has been shown to occur under extreme conditions that promote DNA damage, an event termed the **SOS response.** Causes of the SOS response include high doses of UV light and other types of mutagens. It results in the up regulation of several genes that function to repair the DNA lesions, restore replication, and prevent premature cell division. When the SOS response occurs, the damaged DNA that has not been repaired is replicated by DNA polymerases II, IV, and V, which can replicate over damaged regions. As mentioned, this translesion DNA synthesis results in a high rate of mutation, usually in the range of 10^{-2} to 10^{-3}. Even though this is a harmful result, it nevertheless allows the bacteria to survive under conditions of extreme environmental stress. Furthermore, the high rate of mutation may provide genetic variability within the bacterial population, so that certain cells may become resistant to the harsh conditions.

CONCEPTUAL SUMMARY

Mutations are heritable changes in the genetic material. They can result from small changes in the genetic material (**single-gene mutations**), rearrangements in chromosome structure (**chromosome mutations**), or changes in chromosome number (**genome mutations**). Mutations can have various consequences on gene expression. If a mutation occurs within the coding sequence, it may alter the amino acid sequence of the encoded polypeptide. **Missense mutations** cause single amino acid substitutions, **nonsense mutations** result in truncated polypeptides, and **frameshift mutations** produce changes in the mRNA codon reading frame. In general, these types of mutations are more likely to be detrimental rather than beneficial to protein structure and function. Mutations can also occur outside the coding sequence and significantly affect gene expression. For example, mutations within promoters, response elements, splicing sequences, and untranslated sequences can have major consequences on the level of gene expression. If a mutation occurs within the **germ line,** it will occur in all the cells of an organism. Furthermore, a germ-line mutation can be passed from parent to offspring. By comparison, **somatic mutations** affect only a patch of the organism. The size of the patch depends on how early during development the mutation has occurred.

Mutations are random events, although **hot spots** for higher rates of mutation can occur at certain locations within genes.

Spontaneous mutations occur during normal cellular conditions. These can arise from **depurination, deamination,** and **tautomeric shifts.** In addition, a variety of environmental agents, known as **mutagens,** are known to induce mutations. These include many different types of **chemical** and **physical agents.** Mutagens can alter DNA structure by modifying bases, removing bases, causing errors in DNA replication, and producing breaks in DNA strands. Testing methods such as the **Ames test** can determine whether an agent is mutagenic.

Due to the prevalence of mutagens in the environment, organisms have evolved DNA repair systems that can detect and repair different types of DNA lesions. A few types of DNA repair systems can recognize altered bases and repair them directly. Alternatively, **base** and **nucleotide excision repair** systems recognize alterations in DNA structure and excise the abnormal region. Excision repair mechanisms are found in all organisms and represent a prominent, general system for DNA repair. Excision repair is particularly important in repairing actively transcribed genes. Certain proteins, such as the **transcription-repair coupling factor** (**TRCF**) of *E. coli*, can target the excision repair system to a damaged gene that is being actively transcribed. Finally, living organisms also possess **mismatch repair** and **recombinational repair systems** that operate after or during the DNA replication process.

EXPERIMENTAL SUMMARY

Experimentally, there are many ways to study the occurrence, causes, and consequences of mutation. Luria and Delbrück conducted a **fluctuation test** to show that mutations occur randomly in a population. Similarly, the Lederbergs used replica plating to demonstrate that mutations occur randomly, and that a selective agent such as T1 phage simply selects for the growth of organisms that happen to have incurred a mutation providing T1 resistance. However, there are **hot spots** where mutations may occur more frequently. In chapter 6, the work of Benzer showed that a

few sites within bacteriophage genes were more likely to mutate than were other sites.

Müller was the first scientist to establish that environmental agents such as X rays can cause mutation. His work showed that X rays dramatically increased the likelihood of recessive, X-linked lethal alleles. Since these studies, many mutagens have been discovered by biochemists, microbiologists, and geneticists. Testing methods, such as the **Ames test,** can ascertain whether or not an agent is a mutagen.

PROBLEM SETS & INSIGHTS

Solved Problems

S1. There are mutant tRNAs that act as nonsense and missense suppressors. At the molecular level, explain how you think these suppressors work.

Answer: A suppressor is a second-site mutation that suppresses the phenotypic effects of a first mutation. Intergenic suppressor mutations in tRNA genes can act as nonsense or missense suppressors. For example, let's suppose a first mutation puts a stop codon into a structural gene. A second mutation in a tRNA gene can alter the anticodon region of a tRNA so that the anticodon recognizes a stop codon but inserts an amino acid at this site. A missense suppressor is a mutation in a tRNA

gene that changes the anticodon so that it puts in the wrong amino acid at a normal codon that is not a stop codon. These mutant tRNAs are termed missense tRNAs. For example, a tRNA that normally recognizes glutamic acid may incur a mutation that changes its anticodon sequence so that it recognizes a glycine codon instead. Like nonsense suppressors, missense suppressors can be produced by mutations in the anticodon region of tRNAs so that the tRNA recognizes an incorrect codon. Alternatively, missense suppressors can also be produced by mutations in aminoacyl-tRNA synthetases that cause them to attach the incorrect amino acid to a tRNA.

S2. If the rate of mutation is 10^{-5}, how many new mutations would you expect in a population of one million bacteria?

Answer: If we multiply the mutation rate times the number of bacteria (10^{-5} times 10^6), we obtain a value of 10 new mutations in this population. This answer is correct, but it is an oversimplification of mutation rate. For any given gene, the mutation rate is based on a probability that an event will occur. Therefore, when we consider a particular population of bacteria, we should be aware that the actual rate of new mutation would vary. Even though the rate may be 10^{-5}, we would not be surprised if a population of 1 million bacteria had 9 or 11 new mutations instead of the expected number of 10. We would be surprised if it had 5,000 new mutations, since this value would deviate much too far from our expected number.

S3. In the Ames test, there are several *Salmonella* strains that contain different types of mutations within the gene that encodes an enzyme necessary for histidine biosynthesis. These mutations include transversions, transitions, and frameshifts. Why do you think it would be informative to test a mutagen with these different types of strains?

Answer: Different types of mutagens have different effects on DNA structure. For example, if a mutagen caused transversions, an experimenter would want to use a *Salmonella* strain in which a transversion would convert a *his⁻* strain into a *his⁺* strain. This type of strain would make it possible to detect the effects of the mutagen.

S4. In chapter 14, we discussed how bacterial genes can be arranged in an operon structure in which a polycistronic mRNA contains the coding sequences for two or more genes. For a normal operon, there is a relatively short distance between the stop codon in the first gene and the start codon in the next gene. In addition to the ribosomal-binding sequence at the beginning of the first gene, there is also a ribosomal-binding sequence at the beginning of the second gene, and at the beginning of all subsequent genes. After the ribosome has moved past the stop codon in the first gene, it quickly encounters a ribosomal-binding site in the next gene. This prevents the complete disassembly of the ribosome as would normally occur after a stop codon and leads to the efficient translation of the second gene.

Jacob and Monod, who studied mutations in the *lac* operon, discovered a mutation in the *lacZ* gene that had an unusual effect. This mutation resulted in a shorter version of the β-galactosidase protein, and it also prevented the expression of the *lacY* gene, even though the *lacY* gene was perfectly normal. Explain this mutation.

Answer: This mutation introduced a stop codon much earlier in the *lacZ* gene sequence. This type of mutation is referred to as a **polar mutation** and the phenomenon is called **polarity.** The explanation is that the early nonsense codon causes the ribosome to disassemble before it has a chance to reach the next gene sequence in the operon. The ribosomal-binding sequence that precedes the second gene is efficiently recognized only by a ribosome that has traveled to the normal stop codon in the first gene.

S5. A reverse mutation or a reversion is a mutation that returns a mutant codon back to a codon that gives a wild-type phenotype. At the DNA level, this can be an exact reversion or an equivalent reversion.

GAG (glutamic acid)	Forward → mutation	GTG (valine)	Exact → reversion	GAG (glutamic acid)
GAG (glutamic acid)	Forward → mutation	GTG (valine)	Equivalent → reversion	GAA (glutamic acid)
GAG (glutamic acid)	Forward → mutation	GTG (valine)	Equivalent → reversion	GAT (aspartic acid)

An equivalent reversion produces a protein that is equivalent to the wild type with regard to structure and function. This can occur in two ways. In some cases, the reversion produces the wild-type amino acid (in this case, glutamic acid), but it uses a different codon than the wild-type gene. Alternatively, an equivalent reversion may substitute an amino acid that is structurally similar to the wild-type amino acid. In our example, an equivalent reversion has changed valine to an aspartic acid. Since aspartic and glutamic acids are structurally similar, this type of reversion can restore the wild-type structure and function.

Now here is the question. The template strand within the coding sequence of a gene has the following sequence:

3'-TACCCCTTCGACCCCGGA-5'

This template produces an mRNA, 5'–AUGGGGAAGCUGGGG-CCA–3', that encodes a polypeptide with the sequence methionine–glycine–lysine–leucine–glycine–proline.

A forward mutation changes the template strand to 3'–TAC-CCCT<u>A</u>CGACCCCGGA–5'.

After the forward mutation, another mutation occurs to change this sequence again. Would the following second mutations be an exact reversion, an equivalent reversion, or neither?

A. 3'-TACCCCT<u>C</u>CGACCCCGGA-5'

B. 3'-TACCCCT<u>T</u>CGACCCCGGA-5'

C. 3'-TACCCCG<u>A</u>CGACCCCGGA-5'

Answer:

A. This is probably an equivalent reversion. The third codon, which encodes a lysine in the normal gene, is now an arginine codon. Arginine and lysine are both basic amino acids, so the polypeptide would probably function normally.

B. This is an exact reversion.

C. The third codon, which is a lysine in the normal gene, has been changed to a leucine codon. It is difficult to say if this would be an equivalent reversion or not. Lysine is a basic amino acid and leucine is a nonpolar amino acid. The protein may still function normally with a leucine at the third codon, or it may function abnormally. You would need to test the function of the protein to determine if this was an equivalent reversion or not.

Conceptual Questions

C1. For each of the following mutations, is it a transition, transversion, addition, or deletion? The original DNA strand is 5'–GGACTA-GATAC–3'. (Note: Only one strand is shown)

A. 5'-GAACTAGATAC-3'

B. 5'-GGACTAGAGAC-3'

C. 5'-GGACTAGTAC-3'

D. 5'-GGAGTAGATAC-3'

C2. Discuss the differences between gene, chromosome, and genome mutations.

C3. A gene mutation changes an AT base pair to a GC pair. This causes a gene to encode a truncated protein that is nonfunctional. An organism that carries this mutation cannot survive at high temperatures. Make a list of all the genetic terms that could be used to describe this type of mutation.

C4. What does a suppressor mutation suppress? What is the difference between an intragenic and intergenic suppressor?

C5. How would each of the following types of mutations affect the amount of functional protein that is expressed from a gene?

A. Nonsense

B. Missense

C. Up promoter mutation

D. Mutation that affects splicing

C6. X rays strike a chromosome in a living cell and ultimately cause the cell to die. Did the X rays produce a mutation? Explain why or why not.

C7. The lactose permease is encoded by the *lacY* gene of the *lac* operon. Let's assume a mutation occurred at codon 64 that changed the normal glycine codon into a valine codon. The mutant lactose permease is unable to function. However, a second mutation, which changes codon 50 from an alanine codon to a threonine codon, is able to restore function. Are the following terms appropriate or inappropriate to describe this second mutation?

A. Reversion

B. Intragenic suppressor

C. Intergenic suppressor

D. Missense mutation

C8. Nonsense suppressors tend to be very inefficient at their job of allowing readthrough of a stop codon. How would it affect the cell if they were efficient at their job?

C9. Are each of the following mutations silent, missense, nonsense, or frameshift? The original DNA strand is 5′–ATGGGACTAGATACC–3′. (Note: Only the coding strand is shown; the first codon is methionine)

A. 5′-ATGGGTCTAGATACC-3′

B. 5′-ATGCGACTAGATACC-3′

C. 5′-ATGGGACTAGTTACC-3′

D. 5′-ATGGGACTAAGATACC-3′

C10. In chapters 12 through 15, we discussed many sequences that are outside the coding sequence but are important for gene expression. Look up two of these sequences and write them out. Explain how a mutation could change these sequences and thereby alter gene expression.

C11. Explain two ways that a chromosomal rearrangement can cause a position effect.

C12. Is a random mutation more likely to be beneficial or harmful? Explain your answer.

C13. Which of the following mutations could be appropriately described as a position effect?

A. A point mutation at the −10 position in the promoter region prevents transcription.

B. A translocation places the coding sequence for a muscle-specific gene next to an enhancer that is turned on in nerve cells.

C. An inversion flips a gene from the long arm of chromosome 17 (which is euchromatic) to the short arm (which is heterochromatic).

C14. As discussed in chapter 22, most forms of cancer are caused by environmental agents that produce mutations in somatic cells. Is an individual with cancer considered a genetic mosaic? Explain why or why not.

C15. Discuss the consequences of a germ-line versus a somatic cell mutation.

C16. Draw and explain how alkylating agents alter the structure of DNA.

C17. Explain how a mutagen can interfere with DNA replication to cause a mutation. Give two examples.

C18. What type of mutation (transition, transversion, and/or frameshift) would you expect each of the following mutagens to cause?

A. Nitrous acid

B. 5-bromouracil

C. Proflavin

C19. Explain what happens to the sequence of DNA during trinucleotide repeat expansion. If someone was mildly affected with a TNRE disorder, what issues would be important when considering to have offspring?

C20. Distinguish between spontaneous and induced mutations. Which are more harmful? Which are avoidable?

C21. Are mutations random events? Explain your answer.

C22. Give an example of a mutagen that can change cytosine to uracil. Which DNA repair system(s) would be able to repair this defect?

C23. If a mutagen causes bases to be removed from nucleotides within DNA, what repair system would fix this damage?

C24. Trinucleotide repeat expansions (TNREs) are associated with several different human inherited diseases. Certain types of TNREs produce a long stretch of glutamines (a type of amino acid) within the encoded protein. This long stretch of glutamines somehow inhibits the function of the protein and thereby causes a disorder. In cases where a TNRE exerts its detrimental effect by producing a glutamine stretch, are the following statements true or false?

A. The TNRE is within the coding sequence of the gene.

B. The TNRE prevents RNA polymerase from transcribing the gene properly.

C. The trinucleotide sequence is CAG.

D. The trinucleotide sequence is CCG.

C25. With regard to TNRE, what is meant by the term *anticipation*?

C26. What is the difference between the mutation rate compared to the mutation frequency?

C27. Achondroplasia is a rare form of dwarfism. It is caused by an autosomal dominant mutation within a single gene. Among 1,422,000 live births, the number of babies born with achondroplasia was 31. Among those 31 babies, 18 of them had one parent with achondroplasia. The remaining babies had two unaffected parents. What is the mutation frequency for this disorder among these 1,422,000 babies? What is the mutation rate for achondroplasia?

C28. A segment of DNA has the following sequence:

```
TTGGATGCTGG
AACCTACGACC
```

A. What would be the sequence immediately after reaction with nitrous acid? Let the letter H represent hypoxanthine and U represent uracil.

B. Let's suppose this DNA was reacted with nitrous acid. The nitrous acid was then removed and the DNA was replicated for two generations. What would be the sequences of the DNA products after the DNA had replicated two times? Your answer should contain the sequences of four double helices.

C29. In the treatment of cancer, the basis for many types of chemotherapy and radiation therapy is that mutagens are more effective at killing dividing cells compared to nondividing cells. Explain why. What are possible harmful side effects of chemotherapy and radiation therapy?

C30. An individual contains a somatic mutation that changes a lysine codon into a glutamic acid codon. Prior to acquiring this mutation, the individual had been exposed to UV light, proflavin, and 5-bromouracil. Which of these three agents would be the most likely to have caused this somatic mutation? Explain your answer.

C31. Which of the following examples is likely to be caused by a somatic mutation?

A. A purple flower has a small patch of white tissue.

B. One child, in a family of seven, is an albino.

C. One apple tree, in a very large orchard, produces its apples two weeks earlier than any of the other trees.

D. A 60-year-old smoker develops lung cancer.

C32. How would nucleotide excision repair be affected in each case if one of the following proteins were missing? Describe the condition of the DNA that had been repaired in the absence of the protein.

A. UvrA

B. UvrC

C. UvrD

D. DNA polymerase

C33. During methyl-directed mismatch repair, why is it necessary to distinguish between the template strand and the newly made daughter strand? How is this accomplished?

C34. When does recombinational repair occur? What would happen if DNA damage occurred in exactly the same location in both parental DNA strands?

C35. When DNA-N-glycosylase recognizes thymine dimers, it only detects the thymine located on the 5′ side of the thymine dimer as being abnormal. Draw and explain the steps whereby a thymine dimer would be removed by the consecutive actions of DNA-N-glycosylase, AP-endonuclease, and DNA polymerase.

C36. Discuss the relationship between transcription and DNA repair. Why is it beneficial to repair actively transcribed DNA more efficiently than nontranscribed DNA?

C37. Three common ways to repair changes in DNA structure are nucleotide excision repair, mismatch repair, and recombinational repair. Which of these three mechanisms would be used to fix the following types of DNA changes?

A. A change in the structure of a base caused by a mutagen in a nondividing eukaryotic cell

B. A change in DNA sequence caused by a mistake made by DNA polymerase

C. A thymine dimer in the DNA of an actively dividing bacterial cell

C38. What is the underlying genetic defect that causes xeroderma pigmentosum? How can the symptoms of this disease be explained by the genetic defect?

C39. In *E. coli*, a methylase enzyme encoded by the *dam* gene recognizes the sequence 5′–GATC–3′ and attaches a methyl group to the N^6 position of adenine. *E. coli* strains, which have the *dam* gene deleted, are known to have a higher spontaneous mutation rate compared to normal strains. Explain why.

C40. Discuss the similarities and differences between nucleotide excision repair and methyl-directed mismatch repair.

Experimental Questions

E1. The Luria-Delbrück fluctuation test is consistent with the random mutation theory. How would the results have been different if the physiological adaptation hypothesis had been correct?

E2. Explain how the technique of replica plating supports a random mutation theory but conflicts with the adaptation hypothesis. Outline how you would use this technique to show that antibiotic resistance is due to random mutations.

E3. In the experiment of figure 16.12, the *ClB* chromosome carries a large paracentric inversion. This inversion prevents crossing over with the other X chromosome (which may carry an X-ray-induced lethal mutation). Does a paracentric inversion really prevent crossing over? Explain. You may wish to review chapter 8 to answer this question.

E4. As you may have noticed in the data table of figure 16.12, the daughters whose fathers were exposed to X rays were less likely to produce any offspring compared to the control daughters (whose fathers were not exposed to X rays). Among 1,015 daughters (whose fathers had been exposed to X rays), only 783 produced offspring. By comparison, among 1,011 control daughters, 947 were able to produce offspring. Suggest reasons why the daughters of irradiated fathers appear less likely to produce any offspring compared to the control daughters.

E5. In Müller's experiment with *ClB* chromosomes, is the experiment measuring the mutation rate within a single gene? Explain. If we divide 91 by 783, we obtain a mutation rate of 11.6%. In your own words, explain what this value means.

E6. Researchers were interested in whether two different chemicals were mutagens. Let's call them chemical A and chemical B. They followed the protocol described in the experiment of figure 16.12 and obtained the following results:

Treatment	Number of *ClB* Daughters Crossed to Normal Males	Number of Tubes Containing Offspring	Number of Tubes Lacking Male Offspring
Control	2,108	2,077	3
Chemical A	1,402	1,378	2
Chemical B	4,203	3,100	77

Would you conclude that chemical A and/or chemical B is a mutagen? Explain.

E7. From an experimental point of view, is it better to use haploid or diploid organisms for mutagen testing? Consider both the Ames test and the experiment of figure 16.12 when preparing your answer.

E8. How would you modify the Ames test to discover physical mutagens? Would it be necessary to add the rat liver extract? Explain why or why not.

E9. During an Ames test, bacteria were exposed to a potential mutagen. Also, as a control, another sample of bacteria was not exposed to the mutagen. In both cases, 10 million bacteria were plated and the following results were obtained:

No mutagen: 17 colonies

With mutagen: 2,017 colonies

Calculate the mutation rate in the presence and absence of the mutagen. How much does the mutagen increase the rate of mutation?

E10. Richard Boyce and Paul Howard-Flanders conducted an experiment that provided biochemical evidence that thymine dimers are removed from the DNA by a DNA repair system. In their studies, bacterial DNA was radiolabeled so that the amount of radioactivity reflected the amount of thymine dimers. The DNA was then subjected to UV light, causing the formation of thymine dimers. When radioactivity was found in the soluble fraction, thymine dimers had been excised from the DNA by a DNA repair system. But when the radioactivity was in the insoluble fraction, the thymine dimers had been retained within the DNA. The following table illustrates some of their results involving a normal strain of *E. coli* and a second strain that was very sensitive to killing by UV light:

Strain	Treatment	Radioactivity in the Insoluble Fraction (cpm)	Radioactivity in the Soluble Fraction (cpm)
Normal	No UV	<100	<40
Normal	UV-treated, incubated 2 hours at 37°C	357	940
Mutant	No UV	<100	<40
Mutant	UV-treated, incubated 2 hours at 37°C	890	<40

Adapted from: Boyce, R. P., and Howard-Flanders, P. *Proc. Natl. Acad. Sci. USA* 51, 293–300.

Explain the results found in this table. Why is the mutant strain sensitive to UV light?

Questions for Student Discussion/Collaboration

1. In *E. coli*, a variety of mutator strains have been identified in which the spontaneous rate of mutation is much higher than normal strains. Make a list of the types of abnormalities that could cause a strain of bacteria to become a mutator strain. Which abnormalities do you think would give the highest rate of spontaneous mutation?

2. Discuss the times in a person's life when it would be most important to avoid mutagens. Which parts of a person's body should be most protected from mutagens?

3. A large amount of research is aimed at studying mutation. However, there is not an infinite amount of research dollars. Where would you put your money for mutation research?

1. Testing of potential mutagens

2. Investigating molecular effects of mutagens

3. Investigating DNA repair mechanisms

4. Or some other place?

Note: All answers appear at the website for this textbook; the answers to even-numbered questions are in the back of the textbook.

Visit the Online Learning Center for practice tests, answer keys, and other learning aids for this chapter. Enhance your understanding of genetics with our interactive exercises, web links, news feeds, tutorial service, and much more.

RECOMBINATION AND TRANSPOSITION AT THE MOLECULAR LEVEL

17

::

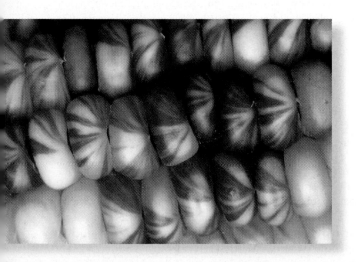

Genetic recombination is the process in which chromosomes are broken and then rejoined to form a new genetic combination, different from the original. A major category of genetic recombination is homologous recombination, which is an essential feature of all species. As its name suggests, **homologous recombination** occurs between DNA segments that are homologous to each other. Recombination enhances genetic diversity, helps to maintain genome integrity (i.e., DNA repair), and ensures the proper segregation of chromosomes.

In chapter 17, we will also examine ways that nonhomologous segments of DNA may recombine with each other. During **site-specific recombination,** nonhomologous DNA segments are recombined at specific sites. This type of recombination occurs within genes that encode antibody polypeptides and also occurs when certain viruses integrate their genomes into host cell DNA. Finally, we will end chapter 17 with a description of an unusual form of genetic recombination known as **transposition.** As we will learn, small segments of DNA called **transposons** can move themselves to multiple locations within the host's chromosomal DNA.

From a molecular viewpoint, homologous recombination, site-specific recombination, and transposition all are important mechanisms for DNA rearrangement. These processes involve a series of steps that direct the breakage and rejoining of DNA fragments. Various cellular proteins are necessary for these steps to occur properly. The past few decades have seen many exciting advances in our understanding of genetic recombination at the molecular level. In the first part of chapter 17, we will consider the general concepts of recombination and examine models that explain how recombination occurs.

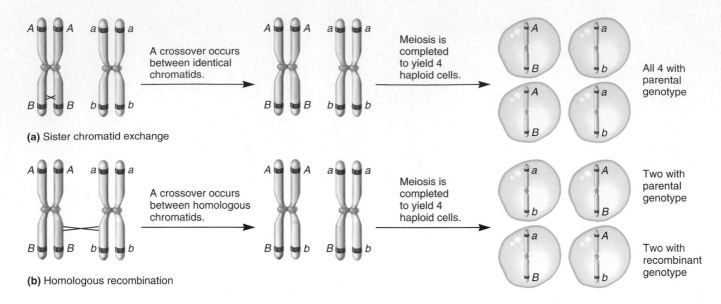

FIGURE 17.1   **Crossing over between eukaryotic chromosomes. (a)** Sister chromatid exchange occurs when genetically identical chromatids cross over. **(b)** Homologous recombination occurs when homologous chromosomes cross over. Homologous recombination may lead to a new combination of alleles, which is called a recombinant (or nonparental) genotype.

GENES→TRAITS   Homologous recombination is particularly important when we consider the relationships between multiple genes and multiple traits. For example, if the X chromosome in a female fruit fly carried alleles for red eyes and gray body and its homologue carried alleles for white eyes and yellow body, homologous recombination could produce recombinant chromosomes that carry alleles for red eyes and yellow body, or alleles for white eyes and gray body. Therefore, new combinations of two or more alleles can arise when homologous recombination takes place.

17.1 SISTER CHROMATID EXCHANGE AND HOMOLOGOUS RECOMBINATION

As described in chapters 3 and 5, chromosomes that have similar or identical sequences frequently participate in crossing over during meiosis I and occasionally during mitosis. As you may recall, crossing over involves the alignment of a pair of homologous chromosomes, followed by the breakage of two chromosomes at analogous locations, and the subsequent exchange of the corresponding segments (refer back to fig. 3.9). When crossing over occurs between sister chromatids, it is called **sister chromatid exchange (SCE).** Since sister chromatids are genetically identical

to each other, SCE does not produce a new combination of alleles (fig. 17.1a). Therefore, it is not considered a form of recombination. By comparison, it is also common for homologous chromosomes to cross over. This is homologous recombination, because two similar (but not identical) homologues have exchanged genetic material. As shown in figure 17.1b, homologous recombination may produce a new combination of alleles in the resulting chromosomes. Recombinant chromosomes contain a combination of alleles not found in the parental chromosomes.

In this section, we will begin with an experimental approach to detect sister chromatid exchange, overcoming the obstacle that the exchanged chromosomes are genetically identical to each other. We will then focus our attention on the molecular mechanisms that underlie homologous recombination.

The Staining of Harlequin Chromosomes Can Reveal Recombination Between Sister Chromatids

Our understanding of crossing over and genetic recombination has come from a variety of experimental approaches including genetic, biochemical, and cytological analyses. Chromosomal staining methods have made it possible to visualize the genetic exchange between eukaryotic chromosomes. In the 1970s, the Russian cytogeneticist A. F. Zakharov and colleagues expended much effort developing methods that improved our ability to identify chromo-

somes. They made the interesting observation that chromosomes labeled with the nucleotide analogue **5-bromodeoxyuridine (BrdU)** bind certain types of stain to a different degree compared to normal chromosomes. Paul Perry and Sheldon Wolff extended this approach to differentially stain sister chromatids and microscopically identify sister chromatid exchanges.

Before we consider the experiment of Perry and Wolff, let's examine how their staining procedure allowed them to accurately distinguish the two sister chromatids. In their approach, eukaryotic cells were grown in a laboratory and exposed to BrdU for

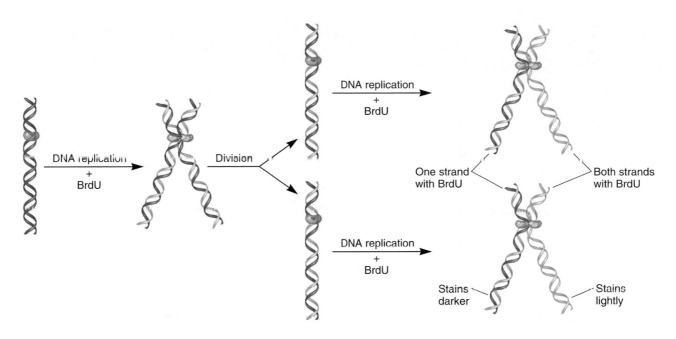

FIGURE 17.2   **Harlequin chromosomes.**

two rounds of DNA replication. After the second round of DNA replication, one of the sister chromatids contained one normal strand and one BrdU-labeled strand. The other sister chromatid had two BrdU-labeled strands (fig. 17.2). When treated with two dyes, Hoechst 33258 and Giemsa, the sister chromatid containing two strands with BrdU stains very weakly and appears light, whereas the sister chromatid with only one strand containing BrdU stains much more strongly and appears very dark. In this way, the two sister chromatids can be distinguished microscopically. Chromosomes stained in this way have been referred to as harlequin chromosomes, because they are reminiscent of a harlequin character's costume with its variegated pattern of light and dark patches.

The steps in Perry and Wolff's protocol are shown in figure 17.3. They began with a commonly used mammalian cell line and exposed the cells to BrdU for two rounds of DNA replication. Near the end of the second round, colcemid was added to prevent the completion of mitosis. The cells were treated with KCl to spread out the chromosomes, which were subsequently fixed, and then stained with Hoechst 33258 and Giemsa.

■ **THE HYPOTHESIS**

Crossing over may occur between sister chromatids.

■ **TESTING THE HYPOTHESIS** — **FIGURE 17.3**   **The staining of harlequin chromosomes reveals sister chromatid exchange.**

Starting material: A laboratory cell line of Chinese hamster ovary (CHO) cells.

	Experimental level	**Conceptual level**
1. Expose CHO cells to BrdU for two cell generations (approximately 24 hours). Note: CHO cells are Chinese hamster ovary cells, which is a commonly used mammalian cell line.	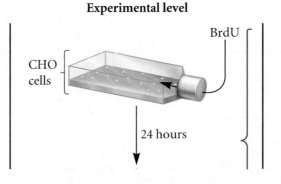	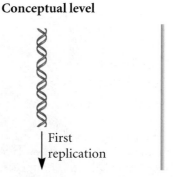

(*continued*)

2. Near the end of the growth, expose the cells to colcemid. This prevents the cells from completing mitosis following the second round of DNA replication.

3. Add 0.075 M KCl to spread the chromosomes and then methanol/acetic acid to fix the cells.

4. Stain with Hoechst 33258, rinse, and later stain with Giemsa. Note: This refinement in the staining procedure greatly improved the ability to discern the sister chromatids.

5. View under a microscope.

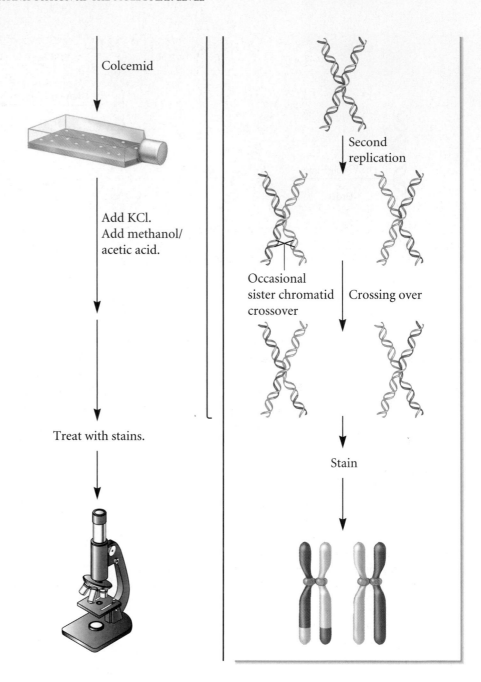

Colcemid

Add KCl.
Add methanol/
acetic acid.

Treat with stains.

Second replication

Occasional sister chromatid crossover

Crossing over

Stain

THE DATA

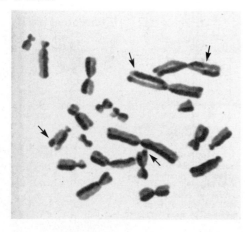

INTERPRETING THE DATA

A micrograph of the results is shown in the data of figure 17.3. As seen here, the chromosomes show the classic harlequin appearance, due to the differential staining of the sister chromatids. Furthermore, examples of sister chromatid exchange are clearly visible. The *arrows* depict regions where crossing over has taken place. In this study, Perry and Wolff found that SCEs occurred at a frequency of approximately 0.67 per chromosome. This method has provided an accurate (and dramatic) way to visualize genetic exchange between sister chromatids.

Many subsequent studies have used the harlequin staining method to study the effects of agents that may influence the frequency of genetic exchanges. Researchers have found that DNA damage caused by radiation and chemical mutagens tends to increase the level of genetic exchange. When cells are exposed to these types of mutagens, the technique of harlequin staining has revealed a substantial increase in the frequency of SCEs.

A self-help quiz involving this experiment can be found at the Online Learning Center.

The Holliday Model Describes a Molecular Mechanism for the Recombination Process

We now turn our attention to genetic exchange that occurs between homologous chromosomes. It is surprising that the first molecular model of homologous recombination did not come from a biochemical analysis of DNA or from electron microscopy studies. Instead, it was deduced from the outcome of genetic crosses in fungi.

As discussed in chapter 5, geneticists have learned a great deal from the analysis of fungal asci, because an ascus contains the products of a single meiosis. When two haploid fungi that differ at a single gene are crossed to each other, it is expected that the ascus will contain an equal proportion of each genotype. For example, if a pigmented strain of *Neurospora* producing orange spores is crossed to an albino strain producing white spores, the octad should contain four orange spores and four white spores (refer back to fig. 5.13). As early as 1934, H. Zickler noticed that unequal proportions of the spores sometimes occurred within asci. He occasionally observed octads with six orange spores and two white spores, or six white spores and two orange spores.

Zickler used the term **gene conversion** to describe this phenomenon. It occurred at too high a rate to be explained by new mutations. Subsequent studies by several researchers confirmed this phenomenon in yeast and *Neurospora*. When gene conversion occurs, one allele is converted to the allele on the homologous chromosome.

Based on studies involving gene conversion, Robin Holliday proposed a model in 1964 to explain the molecular steps that occur during homologous recombination. In chapter 17, we will first consider the steps in the Holliday model and then move on to more recent models. Later in chapter 17, we will examine how the Holliday model can explain the phenomenon of gene conversion.

The **Holliday model** is shown in figure 17.4a. At the beginning of the process depicted in this figure, two homologous chromatids are aligned with each other. According to the model, a break occurs at identical sites in one strand of the two homologous chromatids. The strands then invade the opposite helices and base pair with the complementary strands. This event is followed by a covalent linkage to create a **Holliday junction.** The cross in the Holliday junction can migrate in a lateral direction. As it does so, a DNA strand in one helix is swapped for a DNA strand in the other helix. This process is called **branch migration,** because the branch connecting the two double helices migrates laterally. Since the DNA sequences in the homologous chromosomes are similar but not identical, the swapping of the DNA strands during branch migration may produce regions in the double-stranded DNA that are called **heteroduplexes.** A heteroduplex is a DNA double helix that contains mismatches. In other words, since the DNA strands in this region are from homologous chromosomes, their sequences are not perfectly complementary, yielding mismatches.

A key issue involving recombination is the manner in which the Holliday structure eventually separates into two double-stranded molecules. As shown in figure 17.4a, the Holliday structure may or may not make a 180° turn. This is called isomerization, because the two structures before and after the 180° turn are structural isomers of each other. This means that they are chemically identical except for the relative locations of certain segments. The final steps in the recombination process are collectively called **resolution,** because they involve the breakage and rejoining of two DNA strands to create two separate chromosomes. In other words, the entangled DNA strands become resolved into two separate structures. Without isomerization, breakage can occur in the same two strands that were originally broken (see *top* of fig. 17.4a). If this happens, the strands are rejoined to produce a nonrecombinant pair of chromosomes. The only difference between this nonrecombinant pair and the original parental pair is the short heteroduplex region. Alternatively, if isomerization has occurred, the resolution phase can involve breakage of the two DNA strands that were not originally broken. In this case, the rejoining of the corresponding strands produces two recombinant chromosomes.

The Holliday model can account for the general properties of recombinant chromosomes that are formed during eukaryotic meiosis. As was mentioned, the original model was based on the results of crosses in fungi where the products of meiosis are contained within a single ascus. Nevertheless, molecular research in many other organisms has supported the central tenets of the Holliday model. A particularly convincing piece of evidence came from electron microscopy studies in which recombination structures could be visualized. Figure 17.4b shows an electron micrograph of two DNA fragments that are in the process of recombination. This structure has been called a chi (χ) form because its shape is similar to the Greek letter χ.

More Recent Models Have Refined the Molecular Steps of Homologous Recombination

As more detailed studies of genetic recombination have become available, certain steps in the Holliday model have been reconsidered. In particular, more recent models have modified the initiation phase of recombination. It is now known that the first step need not involve nicks at identical sites in one strand of each

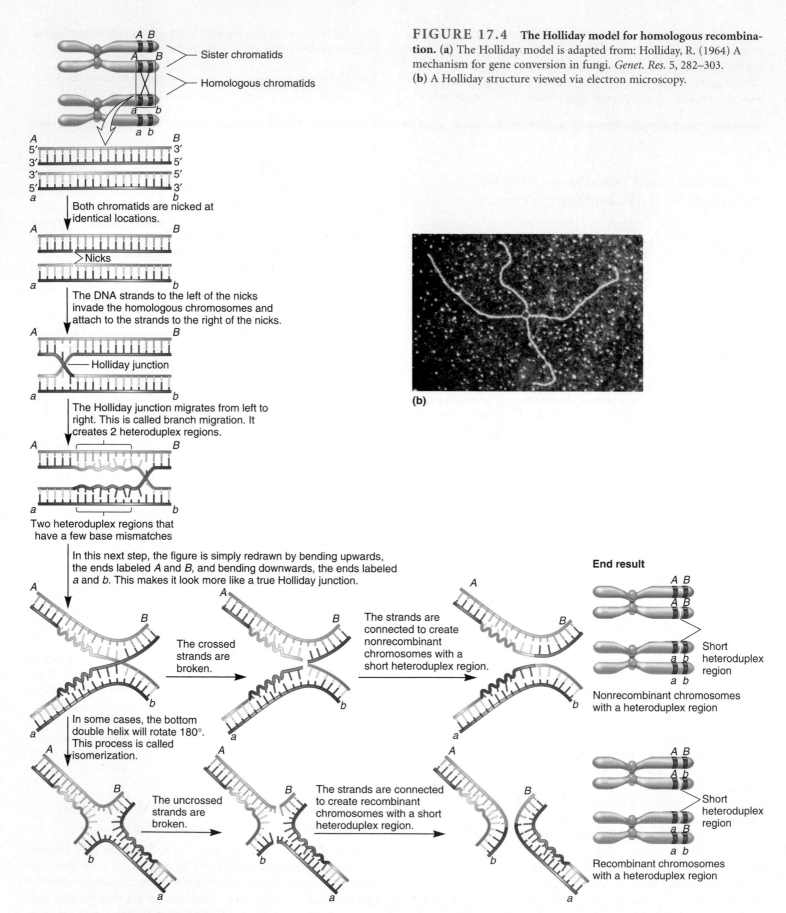

FIGURE 17.4 **The Holliday model for homologous recombination.** (a) The Holliday model is adapted from: Holliday, R. (1964) A mechanism for gene conversion in fungi. *Genet. Res.* 5, 282–303. (b) A Holliday structure viewed via electron microscopy.

Sister chromatids

Homologous chromatids

Both chromatids are nicked at identical locations.

Nicks

The DNA strands to the left of the nicks invade the homologous chromosomes and attach to the strands to the right of the nicks.

Holliday junction

The Holliday junction migrates from left to right. This is called branch migration. It creates 2 heteroduplex regions.

Two heteroduplex regions that have a few base mismatches

In this next step, the figure is simply redrawn by bending upwards, the ends labeled A and B, and bending downwards, the ends labeled a and b. This makes it look more like a true Holliday junction.

The crossed strands are broken.

The strands are connected to create nonrecombinant chromosomes with a short heteroduplex region.

In some cases, the bottom double helix will rotate 180°. This process is called isomerization.

The uncrossed strands are broken.

The strands are connected to create recombinant chromosomes with a short heteroduplex region.

End result

Short heteroduplex region

Nonrecombinant chromosomes with a heteroduplex region

Short heteroduplex region

Recombinant chromosomes with a heteroduplex region

(b)

(a) The Holliday model for homologous recombination

homologous chromatid. In fact, such an event is rather unlikely, Instead, it is more likely for a DNA helix to incur a break in both strands of one chromatid or a single nick. Both of these types of changes have been shown to initiate genetic recombination. Therefore, newer models have tried to incorporate these experimental observations. A model proposed by Matthew Meselson and Charles Radding hypothesized that a single nick in one DNA strand initiates recombination. A second model, proposed by Jack Szostak, Terry Orr-Weaver, Rodney Rothstein, and Franklin Stahl, suggests that a double-stranded break initiates the recombination process. This is called the **double-stranded break model.** Though recombination may occur via more than one mechanism, recent evidence suggests that double-stranded breaks promote homologous recombination during meiosis and during DNA repair.

Figure 17.5 shows the general steps in the double-stranded break model. As seen here, the top chromosome has suffered a double-stranded break. In *E. coli,* this break is recognized by a protein complex called RecBCD that further degrades the region

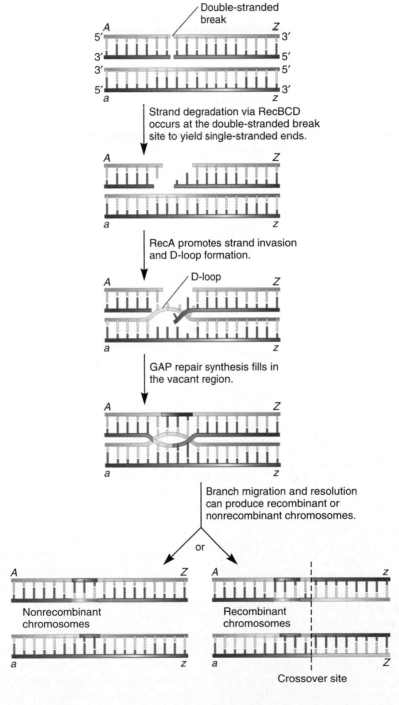

FIGURE 17.5   A simplified version of the double-stranded break model. For simplicity, this illustration does not include the formation of heteroduplex DNA.

and generates a single-stranded DNA segment that can invade the lower DNA double helix. This invasion is facilitated by a protein called RecA. Strand invasion generates a structure called a displacement loop, or D-loop. After the D-loop is formed, you will notice that there are two regions where there is a gap in the DNA. If you look carefully at figure 17.5, the gap exists in the top strand of both double helices. To fill in this region, DNA synthesis occurs in the relatively short gaps where a DNA strand is missing. This DNA synthesis is called **DNA gap repair synthesis.** Once this is completed, two Holliday junctions are produced. Depending on the way these are resolved, the end result is a recombinant or a nonrecombinant chromosome containing a short heteroduplex.

Various Proteins Are Necessary to Facilitate Homologous Recombination

The homologous recombination process requires the participation of many proteins that catalyze different steps in the recombination pathway. Homologous recombination is found in all species, and the types of proteins that participate in the steps outlined in figure 17.5 are very similar. The cells of any given species may have more than one molecular mechanism to carry out homologous recombination. The enzymology of this process is best understood in *Escherichia coli.* Table 17.1 summarizes some of the *E. coli* proteins that play critical roles in one recombination pathway that is found in this species. Though it is beyond the scope of this textbook, *E. coli* has other pathways to carry out homologous recombination.

TABLE 17.1

E. coli **Proteins That Play a Role in Homologous Recombination**

Protein	Description
RecBCD	A complex of three proteins that tracks along the DNA and recognizes double-stranded breaks. The complex partially degrades the double-stranded regions to generate single-stranded regions that can participate in strand invasion. RecBCD is also involved in loading RecA onto single-stranded DNA. In addition, RecBCD can create single-stranded breaks at chi sites.
Single-strand binding protein	Coats broken ends of chromosomes and prevents excessive strand degradation.
RecA	Binds to single-stranded DNA and promotes strand invasion, which enables homologous strands to find each other. It also promotes the displacement of the complementary strand to generate a D-loop.
RecG	Though its function is not entirely understood, RecG protein can promote branch migration of three-strand pre-Holliday junctions. It is thought to play a role in the initial formation of Holliday junctions.
RuvABC	This protein complex binds to Holliday junctions. RuvAB promotes branch migration. RuvC is an endonuclease that cuts the crossed or uncrossed strands to resolve Holliday junctions into separate chromosomes.

RecBCD is a protein complex composed of the RecB, RecC, and RecD proteins. (The term *Rec* indicates that these proteins are involved with recombination.) The RecBCD complex plays an important role in the initiation of recombination involving double-stranded breaks. If we look back at figure 17.5, RecBCD recognizes a double-stranded break within DNA and catalyzes DNA unwinding and strand degradation. The action of RecBCD produces single-stranded DNA ends that can participate in strand invasion and exchange. The single-stranded DNA ends are coated with single-stranded binding protein to prevent their further degradation. The RecBCD complex can also create breaks in the DNA at sites known as chi sequences. In *E. coli,* the chi sequence is 5′–GCTGGTGG–3′. As RecBCD tracks along the DNA, when it encounters a chi sequence, it cuts one of the DNA strands to the 3′ side of this sequence. This generates a single nick that can initiate homologous recombination.

The function of the RecA protein is to promote strand invasion. To accomplish this task, it can bind to the single-stranded ends of DNA molecules that are generated from the activity of RecBCD. A large number of RecA proteins bind to single-stranded DNA, forming a structure called a filament. During strand invasion, this filament makes contact with the unbroken chromosome. Initially, this contact is most likely to occur at nonhomologous regions. The contact point slides along the DNA until it reaches a homologous region. Some current models suggest that a homologous region is recognized by the formation of triplex DNA. Once a homologous site is located, RecA catalyzes the displacement of one DNA strand, and the invading single-stranded DNA quickly forms a normal double helix with the other strand. This causes the formation of a D-loop, as described previously in figure 17.5. RecA proteins mediate the movement of the invading strand and the displacement of the complementary strand. This occurs in such a way that the displaced strand invades the vacant region of the broken chromosome.

Proteins that bind specifically to Holliday junctions have also been identified. RecG and a complex of three proteins termed RuvABC specifically bind to Holliday junctions. Though the mechanisms of Holliday junction formation and resolution are not entirely understood, it appears that RecG, along with the help of RecA, plays a critical role in the initial formation of a Holliday junction. The RuvABC complex promotes branch migration and the cleavage of the Holliday junction, which is necessary for resolution.

Before ending this discussion regarding the molecular mechanism of homologous recombination, it is worthwhile for us to briefly consider recombinational events during meiosis in eukaryotic cells. As described in chapter 3, crossing over between homologous chromosomes is an important event during prophase of meiosis I. An intriguing question is how are crossover sites chosen between two homologous chromosomes? While this question is not entirely understood, molecular studies in two different yeast species, *Saccharomyces cerevisiae* and *Schizosaccharomyces pombe,* suggest that double-stranded breaks initiate the homologous recombination that occurs during meiosis. In other words, double-stranded breaks create sites where a crossover will occur. In *S. cerevisiae*, the formation of DNA double-stranded breaks that initiate meiotic recombination requires at least 10 different proteins. One particular protein, termed Spo11 protein, is thought to be instrumental in cleaving the DNA and thereby creating the double-stranded break. However, the roles of the other proteins, and the interactions among them, are not well understood. Once a double-stranded break is made, homologous recombination can then occur according to the model described previously in figure 17.3.

Gene Conversion May Result from DNA Gap Repair Synthesis or DNA Mismatch Repair

As mentioned earlier in chapter 17, genetic recombination can lead to an event where two different alleles become two identical alleles. Since one of the alleles has been converted to the other, this process is known as **gene conversion.** The Holliday model, as well as newer models, can account for the phenomenon of gene conversion.

There are two possible ways that gene conversion can occur. One way is via DNA gap repair synthesis. Figure 17.6 illustrates how gap repair synthesis can lead to gene conversion according to the double-stranded break model. The top chromosome, which carries the recessive *b* allele, has suffered a double-stranded break in this gene. A gap is created by the digestion of the DNA in the double helix. This digestion eliminates the *b* allele. The two template strands used in gap repair synthesis are from one homologous chromatid. This helix carries the dominant *B* allele. Therefore, after gap repair synthesis takes place, the top chromosome will contain the *B* allele, as will the bottom chromosome. Gene conversion has changed the recessive *b* allele to a dominant *B* allele.

A second mechanism that accounts for gene conversion is DNA mismatch repair, a topic that was described in chapter 16. To understand how this works, let's take a closer look at the heteroduplex structure that is formed during homologous recombination (refer back to fig. 17.4). A heteroduplex contains a DNA strand from each of the two original parental chromosomes. It is possible that the two parental chromosomes may contain an allelic difference within this region. In other words, this short region may contain DNA sequence differences. If this is the case, the heteroduplex region that is formed after branch migration will contain an area of base mismatch. Gene conversion occurs when recombinant chromosomes are repaired and result in two copies of the same allele.

The DNA repair of a heteroduplex that may cause gene conversion is shown in figure 17.7. The two parental chromosomes contained different alleles due to a single base-pair difference in their DNA sequences as shown at the *top* of the figure. During recombination, branch migration has occurred across this region, thereby creating two heteroduplexes with base mismatches. As was described in chapter 16, DNA mismatches will be recognized by DNA repair systems and repaired to a DNA double helix that obeys the AT/GC rule. These two mismatches can be repaired in four possible ways. As shown here, two possibilities produce no gene conversion, whereas the other two lead to gene conversion.

The top chromosome carries the recessive *b* allele, while the bottom chromosome carries the dominant *B* allele.

A double-stranded break occurs within the recessive gene.

A region adjacent to the double-stranded break is digested away, which eliminates the recessive *b* allele.

RecA promotes strand invasion and D-loop formation.

GAP repair synthesis uses the strands from the dominant *B* allele to fill in the region.

After the resolution phase is completed

Both chromosomes carry the *B* allele.

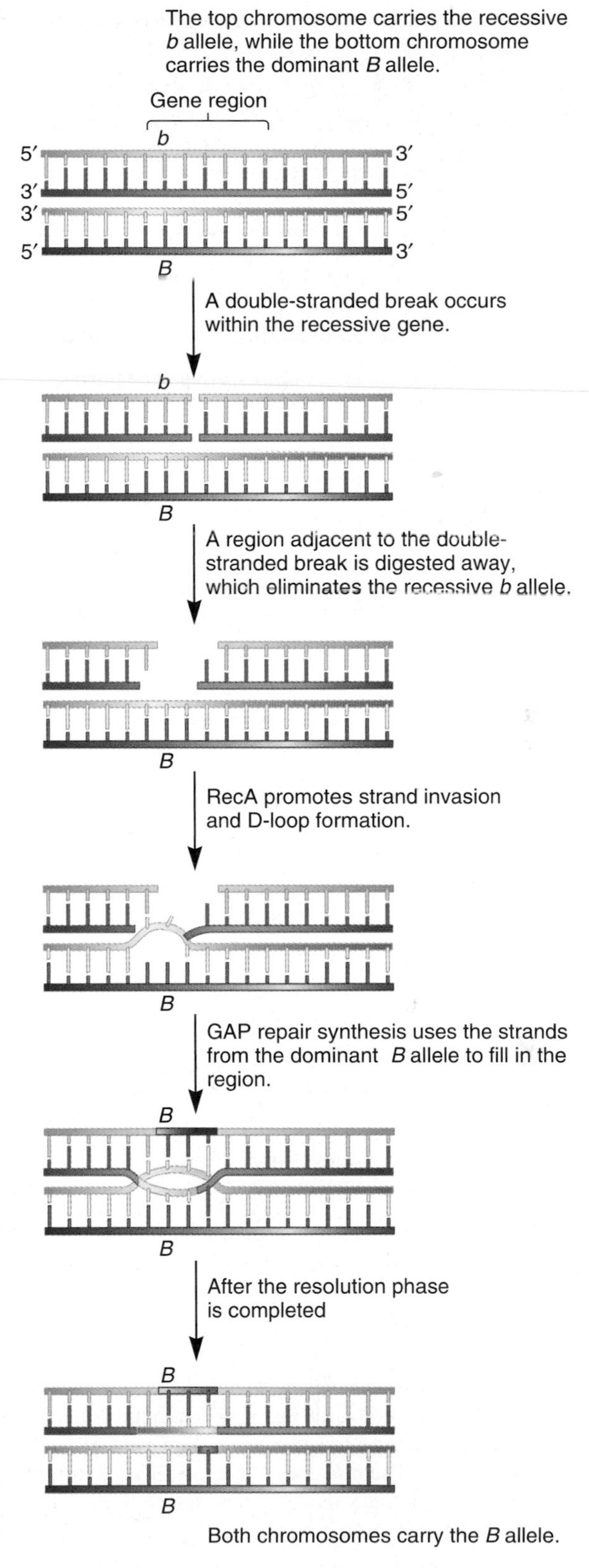

FIGURE 17.6 **Gene conversion by gap repair synthesis in the double-stranded break model.** A gene is found in two alleles, designated *B* and *b*. A double-stranded break occurs in the DNA encoding the *b* allele. Both of these DNA strands are digested away, thereby eliminating the *b* allele. A complementary DNA strand encoding the *B* allele migrates to this region and provides the template to synthesize a double-stranded region. Following resolution, both DNA double helices carry the *B* allele.

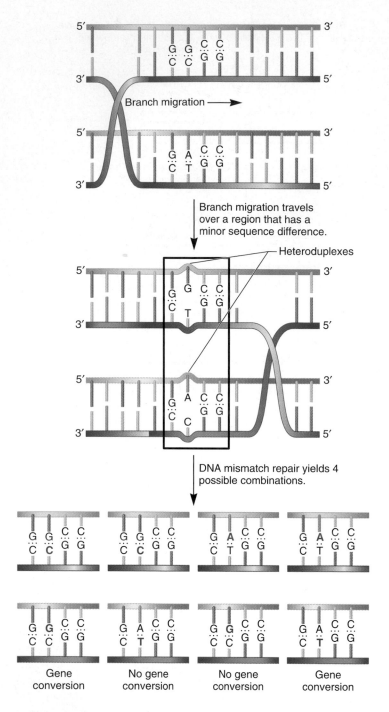

FIGURE 17.7   Gene conversion by mismatch DNA repair. A branch migrates past a homologous region that contains slightly different DNA sequences. This produces two heteroduplexes—DNA double helices with mismatches. The mismatches can be repaired in four possible ways by the mismatch DNA repair system described in chapter 16. Two of these ways result in gene conversion. The repaired base is shown in *red*.

17.2   SITE-SPECIFIC RECOMBINATION

Thus far in chapter 17, we have examined recombination between segments of DNA that are homologous, or identical, to each other. Site-specific recombination is another mechanism where DNA fragments can recombine to make new genetic combinations. During this process, two DNA segments with little or no homology align themselves at specific sites. The sites are relatively short DNA sequences (a dozen or so nucleotides in length) that provide a specific location where recombination will occur. Chromosome breakage and reunion occur at these defined sites to create a recombinant chromosome. These sites are recognized by specialized enzymes that catalyze the breakage and rejoining of DNA fragments within the sites.

Certain viruses use site-specific recombination to insert their viral DNA into their host cell's chromosome. This process has been examined extensively in bacteriophage λ. In addition, mammalian genes that encode antibody polypeptides are rearranged by site-specific recombination which enables the generation of a diverse array of antibodies. In this section, we will consider both mechanisms.

The Integration of Viral Genomes Can Occur by Site-Specific Recombination

The life cycle of some viruses involves the integration of viral DNA into host cell DNA. Certain bacteriophages, for example, can integrate their viral DNA into the bacterial chromosome, creating a **prophage.** This prophage can exist in a latent, or **lysogenic,** state for many generations. The integration of phage DNA is well understood for bacteriophage λ, which infects *E. coli.* The life cycles of phage λ were described in chapter 14. The integration occurs by a mechanism involving site-specific recombination. The steps leading to integration of the viral DNA are outlined in figure 17.8.

Integration of the λ DNA into the *E. coli* chromosome requires sequences known as **attachment sites.** As shown at the *top* of figure 17.8, a common core sequence within the attachment site sequences is identical in the λ DNA and the *E. coli* chromosome. An enzyme known as integrase is encoded by a gene in the λ DNA. Several molecules of integrase recognize the core sequences and bring them close together. Integrase then makes staggered cuts in both the λ and *E. coli* attachment sites. The strands are then exchanged, and the ends are ligated together. In this way, the phage DNA is integrated into the host cell chromosome. As a prophage, the λ DNA may remain latent for many generations. Certain conditions (e.g., when the bacterium is exposed to UV light) may act to stimulate the excision of the prophage from the host DNA, thereby reactivating the virus. Excision also requires integrase, which catalyzes the reverse reaction, as well as a second protein known as excisionase.

Antibody Diversity in the Immune System Is Produced by Site-Specific Recombination

As we have just seen, viruses can integrate their DNA into the host cell chromosome by a site-specific recombinational event.

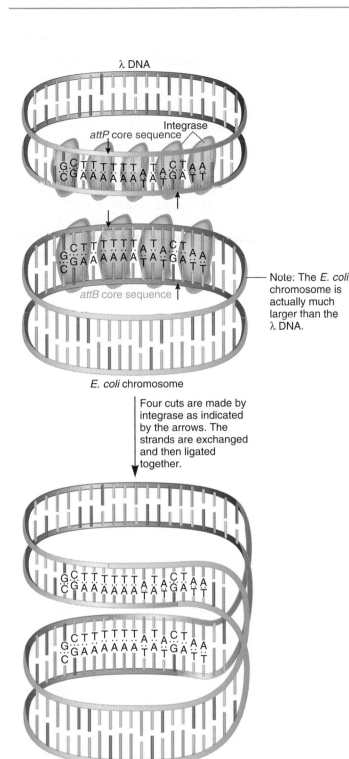

λ DNA

Integrase
attP core sequence

G C T T T T T T A   C T A A
C G A A A A A A T A T A G A T T

Note: The *E. coli*
chromosome is
actually much
larger than the
λ DNA.

G C T T T T T T A   C T
C G A A A A A A T A T G A
C   T T

attB core sequence

E. coli chromosome

Four cuts are made by
integrase as indicated
by the arrows. The
strands are exchanged
and then ligated
together.

G C T T T T T T A T A C T A A
C G A A A A A A T A T A G A T T

G C T T T T T T A T A C T A A
C G A A A A A A T A T G A T T
C   T T

λ DNA integrated into
E. coli chromosome

FIGURE 17.8   The integration of λ DNA into the *E. coli* chromosome. The site designated *attP* is the sequence in the λ DNA that attaches to the *attB* site in the *E. coli* chromosome. As noted here, the core sequences of *attP* and *attB* are identical to each other and thereby provide recognition sites for site-specific recombination.

This process requires a viral enzyme, integrase, that recognizes the sites and catalyzes the recombination reaction. A similar process occurs in certain cells of the immune system. The DNA sequences within antibody genes are rearranged by enzymes that recognize specific sites within those genes and catalyze the breakage and reunion of DNA segments. Before we discuss the details of this mechanism, it is interesting to consider the biology of antibodies.

Antibodies or **immunoglobulins (Igs)** are proteins produced by the B cells of the immune system. Their function is to recognize foreign substances (namely, viruses, bacteria, etc.) and target them for destruction. Antibodies recognize sites known as epitopes within the structures of foreign substances (also known as **antigens**). The recognition between an antibody and antigen is very specific, with each type of antibody recognizing a single epitope. Within the immune system, each B cell produces a single type of antibody. However, our bodies have millions of B cells. Site-specific recombination allows each B cell to produce an antibody with a different amino acid sequence. These differences in the amino acid sequences of antibody proteins enable them to recognize different epitopes. In this way, the immune system can identify an impressive variety of substances as being foreign antigens and thereby target them for destruction.

From a genetic viewpoint, the production of millions of different antibodies poses an interesting problem. If a distinct gene were needed to produce each different antibody polypeptide, the DNA would need to contain millions of different antibody genes. By comparison, consider that the entire human genome contains only about 35,000 different genes. To generate millions of antibody molecules with different polypeptide sequences, an unusual mechanism has evolved in which the DNA is cut and reconnected by site-specific recombination. We might call this "DNA splicing," although the term *splicing* is usually reserved for cutting and rejoining of RNA molecules. With this mechanism, only a few large antibody precursor genes are needed to produce millions of different antibodies. These precursor genes are spliced in many different ways to produce a vast array of polypeptides with differing amino acid sequences.

Figure 17.9 shows the site-specific recombination of an antibody precursor gene. Antibodies are tetrameric proteins composed of two heavy polypeptide chains and two light chains. One type of light chain is the κ (kappa) light chain, which is a component of a class of antibodies known as immunoglobulin G (IgG). The organization of this precursor gene of the κ light chain is shown at the top of figure 17.9. At the *left* side, there are approximately 300 regions known as variable (V) sequences or domains. In addition, there are four different joining (J) sequences and a single constant (C) sequence. Each variable domain or joining domain encodes a different amino acid sequence.

During the maturation of B cells, the κ light precursor gene is cut and rejoined so that one variable domain becomes adjacent to a joining domain. At the end of every V domain and the beginning of every J domain is located a **recombination signal sequence** that functions as a recognition site for site-specific recombination between the V and J regions. The recombination event is initiated by two proteins called **RAG1** and **RAG2.** RAG is

Organization of domains in light-chain gene

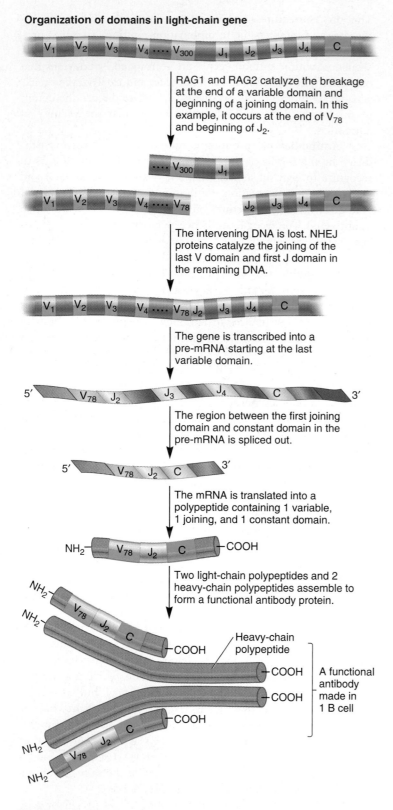

RAG1 and RAG2 catalyze the breakage at the end of a variable domain and beginning of a joining domain. In this example, it occurs at the end of V_{78} and beginning of J_2.

The intervening DNA is lost. NHEJ proteins catalyze the joining of the last V domain and first J domain in the remaining DNA.

The gene is transcribed into a pre-mRNA starting at the last variable domain.

The region between the first joining domain and constant domain in the pre-mRNA is spliced out.

The mRNA is translated into a polypeptide containing 1 variable, 1 joining, and 1 constant domain.

Two light-chain polypeptides and 2 heavy-chain polypeptides assemble to form a functional antibody protein.

Heavy-chain polypeptide

A functional antibody made in 1 B cell

FIGURE 17.9 Site-specific recombination within the precursor gene that encodes the κ light chain for immunoglobulin G (IgG) proteins.

an acronym for recombination-activating gene. These proteins recognize recombination signal sequences and generate two double-stranded breaks, one at the end of a V domain and one at the beginning of a J domain. For example, in the recombination event shown in figure 17.9, RAG1 and RAG2 have made cuts at the end of variable domain number 78 and the beginning of joining domain number 2. The intervening region is lost, and the two ends are then joined to each other. The connection phase of this process is catalyzed by a group of proteins termed **nonhomologous DNA end-joining (NHEJ) proteins.** It is interesting to note that the fusion process is not entirely precise so that a few nucleotides can be added or lost at the junction between the variable and joining domains. This imprecision further accentuates the diversity in antibody genes.

Following transcription, the fused VJ region is contained within a pre-mRNA transcript that is also spliced to connect the J and C domains. After this has occurred, a B cell will produce only the particular κ light chain encoded by the specific VJ fusion domain and the constant domain.

The heavy-chain polypeptides are produced by a similar recombination mechanism. In this case, there are about 500 variable domains and four joining domains. In addition, there are also 12 diversity (D) domains, which are found between the variable and joining domains. The recombination first involves the connection of a D and J domain, followed by the connection of a V and DJ domain. The same proteins that catalyze VJ fusion are involved in the recombination with the heavy-chain gene. Collectively, this process is called **V(D)J recombination.** The D is in parentheses because this type of domain is found only in the heavy-chain genes, not in the light-chain genes.

The recombination process within immunoglobulin genes produces an enormous diversity in polypeptides. Even though it occurs at specific junctions within the antibody gene, the recombination is fairly random with regard to the particular V and J domains that can be joined. If we assume that any of the 300 different variable sequences can be spliced next to any of the four joining sequences, this amounts to 1,200 possible combinations. Overall, the possible number of functional antibodies that can be produced by V(D)J recombination is rather staggering. The number of heavy-chain possibilities is $500 \times 12 \times 4 = 24,000$. Since any light chain–heavy chain combination is possible, this yields $1,200 \times 24,000 = 28,800,000$ possible antibody molecules from the random recombination within two precursor genes!

17.3 TRANSPOSITION

The last form of recombination that we will consider is transposition. In some ways, transposition resembles the site-specific recombination that we examined for phage λ. In that case, a segment of λ DNA was able to integrate itself into the *E. coli* chromosome. Transposition also involves the integration of small segments of DNA into the chromosome. Transposition, though, can occur at many different locations within the genome. The DNA segments that transpose themselves are known as **transposable** elements (TEs). TEs have sometimes been referred to as "jumping genes," because they are inherently mobile.

Transposable elements were first identified by Barbara McClintock in the early 1950s from her classic studies with corn plants. Since that time, geneticists have discovered many different types of TEs in organisms as diverse as bacteria, fungi, plants, and animals. The advent of molecular technology has allowed scientists to understand more about the characteristics of TEs that enable them to be mobile. In this section, we will examine the characteristics of TEs and explore mechanisms that explain how they move. We will also discuss the biological significance of TEs and their uses as experimental tools.

EXPERIMENT 17B

McClintock Found That Chromosomes of Corn Plants Contain Loci That Can Move

Barbara McClintock began her scientific career as a student at Cornell University. Her interests quickly became focused on the structure and function of the chromosomes of corn plants, an interest that continued for the rest of her life. She spent countless hours examining corn chromosomes under the microscope. She was technically gifted and, in addition, had a theoretical mind that could propose ideas that conflicted with conventional wisdom.

During her long career as a scientist, McClintock identified many unusual features of corn chromosomes. She noticed that one strain of corn had the strange characteristic that a particular chromosome, number 9, tended to break at a fairly high rate at the same site. McClintock termed this a **mutable site** or **locus.** This observation initiated a six-year study concerned with highly unstable chromosomal locations. In 1951, at the end of her study, McClintock proposed that these mutable sites are actually locations where transposable elements have been inserted into the chromosomes. At the time of McClintock's studies, such an idea was entirely unorthodox.

McClintock focused her efforts on the relationship between a mutable locus and its phenotypic effects on corn kernels. There were several genes on a single chromosome that could be used as chromosomal markers in her crosses. All of these genes affected the phenotype of corn kernels. Each gene existed in (at least) one dominant and one recessive allele that could be detected easily by examining the phenotypes of kernels. A few of these alleles are described here. The chromosome also contained a mutable locus that McClintock termed **Ds** (for dissociation) because the locus was known to frequently cause chromosomal breaks. In the chromosome shown here, the Ds locus is located next to several genes affecting kernel traits.

In this case, there are three genes that exist as two or more alleles:

1. C is an allele for normal kernel color (dark red), c is a recessive allele of the same gene that causes a colorless kernel, while C^I is a third allele of this gene that is dominant to both C and c and causes a colorless kernel.

2. Sh is an allele that produces normal endosperm, while sh is a second, recessive allele that causes shrunken endosperm. Note: The endosperm is the storage material in the kernel that is used by the plant embryo to provide energy for growth.

3. Wx is the allele that produces normal starch in the endosperm, while wx (waxy) is a recessive allele that produces a waxy-appearing phenotype.

McClintock produced strains of corn which demonstrated that the Ds locus could be transposed to different locations within the corn genome. For example, figure 17.10 illustrates a strain of corn in which it was possible to determine if Ds had moved, because the movement of Ds occasionally causes chromosome breakage. It is important to mention that the endosperm

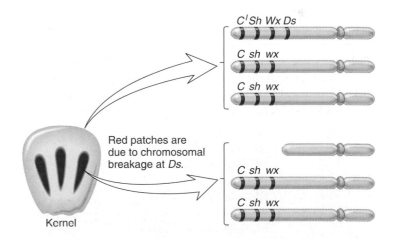

FIGURE 17.10   The sectoring trait in corn kernels.
GENES→TRAITS This kernel is expected to be colorless, because the C^I allele is dominant. On occasion, though, the movement of Ds may cause a chromosome break, thereby losing the C^I, Sh, and Wx alleles. As such a cell continues to divide, it will produce a patch of daughter cells that are red, waxy, and shrunken. Therefore, this sectoring trait arises from the loss of genes that occurs when the movement of Ds causes chromosome breakage.

of a kernel is triploid: it is derived from the fusion of two maternal haploid nuclei and one paternal haploid nucleus. The kernel shown in figure 17.10 was produced by a cross in which the pollen carried the top chromosome (i.e., C^I Sh Wx Ds) while the two maternal chromosomes are shown below (C sh wx). This kernel is expected to be colorless, because the C^I allele is dominant and causes a colorless phenotype. However, as the kernel develops, some chromosomes may break at the Ds locus and lose the distal part of this chromosome. This would produce a sector of cells having a different phenotype. In this case, the patch would be red, shrunken, and waxy.

By analyzing many kernels, McClintock was also able to identify cases in which Ds had moved to a new location. For example, if Ds had moved out of its original location and inserted between Sh and Wx, a break at Ds would produce the following combination:

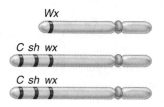

This genotype would produce patches on the kernel that are red and shrunken but not waxy. In this way, McClintock identified 20 independent cases in which the Ds element had moved to a new location within this chromosome. Overall, the results from many crosses were consistent with the idea that Ds can transpose itself throughout the corn genome. McClintock also found that a second locus, termed Ac (for activator), was necessary for the Ds site to move. It is now known that the Ac locus contains a gene that encodes an enzyme called transposase, which is necessary for Ds to move. We will discuss the function of transposase later in this chapter. Some strains of McClintock's corn contained the activator locus, while others did not.

During her studies, McClintock noticed a particularly exciting and unusual event. By making the appropriate cross, she intended to produce kernels with the following genotype:

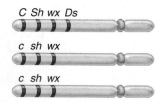

The kernels are expected to be red. Because the strain also contained the Ac locus, breakage would occur occasionally at the Ds locus to produce colorless patches. Among 4,000 kernels, she noticed one kernel with the opposite phenotype. It had a colorless background with red patches. This suggested that the background genotype was recessive, c, and it was mutable to become C. In this kernel, the Ds element had moved into the C gene.

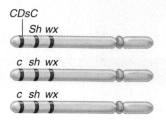

McClintock postulated that the colorless phenotype was due to a transposition of Ds into the C gene. She proposed that, when Ds is located within the C gene, it inactivated the C gene thereby resulting in the recessive colorless phenotype. However, when Ds occasionally transposed out of the gene, the C allele would be restored and a red patch would result. In this case, the formation of patches, or sectoring, is due to the movement of Ds out of its location within the C gene and the rejoining of the two ends, not simply due to chromosome breakage. According to this hypothesis, the red phenotype should be associated with two observations. First, Ds should have moved to a new location. Second, the restored C allele should no longer be mutable.

To further study this phenomenon, McClintock carried out the experiment shown in figure 17.11. As seen here, a cross produced kernels that were not entirely colorless. In most cases, there was red sectoring, because the Ds element left the C gene in a few cells after the kernel had started to develop. However, in the ear of corn shown in figure 17.11, one kernel was completely red. This suggests that the Ds element transposed out of the C gene during the formation of the haploid male gametophyte that produced the pollen nuclei that fertilized this particular red kernel. Therefore, all of the cells in this kernel would have a red phenotype. (Note: Every kernel in an ear of corn is equivalent to a distinct offspring, each being produced from the union of haploid cells from the male and female gametophytes.)

■ THE HYPOTHESIS

The transposition of the Ds element into the normal C gene prevents kernel pigmentation. When the Ds element transposes back out of the C gene, the normal C allele is restored and the kernel becomes red.

TESTING THE HYPOTHESIS — **FIGURE 17.11** Evidence for transposable elements in corn.

Starting material: The male pollen of the corn plant had a chromosome in which the *Ds* element had moved into the *C* gene. The male plant also contains the *Ac* locus.

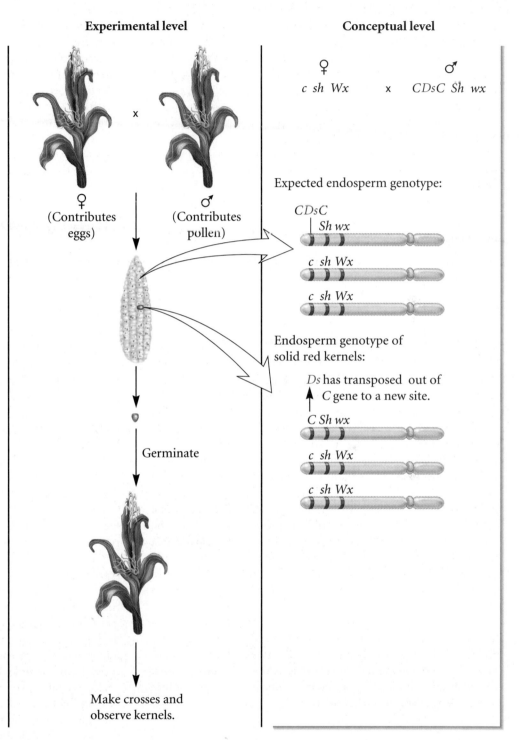

Experimental level

Conceptual level

1. Cross a plant that is homozygous for *c*, *sh*, and *Wx* to a plant that is homozygous for *CDsC*, *Sh*, *wx*.

♀   ♂
c sh Wx   x   *CDsC Sh wx*

♀
(Contributes eggs)

♂
(Contributes pollen)

Expected endosperm genotype:

CDsC
| *Sh wx*

c sh Wx

c sh Wx

2. In this cross, most kernels will have a white background with red sectoring. On rare occasions, *Ds* may have transposed during male gametophyte formation, producing a completely red kernel.

3. Identify those occasional kernels that are completely red.

Endosperm genotype of solid red kernels:

Ds has transposed out of
▲ *C* gene to a new site.

C Sh wx

c sh Wx

c sh Wx

4. Germinate the solid red kernels.

Germinate

5. Conduct crosses to determine the location of *Ds* in the plants derived from the solid red kernels. Note: This can be done using the chromosomal markers and the strategy that was described at the beginning of this experiment. When observing the results of these crosses, determine whether the *C* gene is still mutable. (Is red sectoring occurring?)

Make crosses and observe kernels.

■ THE DATA

Strain	Kernel Phenotype	Location of Ds	Mutability?
From parental cross (see step 2)	White background with red sectoring	Within the C gene	Yes, red sectoring occurred in strains containing Ac.
From red kernels (see step 3)	Red kernels	Ds had moved out of the C gene to another location.	No, the C gene was stable; no sectoring was observed.

■ INTERPRETING THE DATA

By conducting the appropriate crosses, McClintock found that in the progeny of the red kernels, the Ds locus had moved out of the C gene to another location (see data table of figure 17.11).

In addition, the "restored" C gene behaved normally. In other words, it was no longer a highly mutable gene exhibiting a sectoring phenotype. Taken together, the results are consistent with the idea that the Ds locus can move around the corn genome by transposition.

When McClintock published these results, they were met with great skepticism. Some geneticists of that time were unable to accept the idea that the genetic material was susceptible to frequent rearrangement. Instead, they wanted to believe that the genetic material was always very stable and permanent in its structure. Over the next several decades, the scientific community came to realize that transposable elements are a widespread phenomenon. Much like Gregor Mendel and Charles Darwin, Barbara McClintock was clearly ahead of her time. She was awarded the Nobel Prize in 1983, more than 30 years after her original discovery.

A self-help quiz involving this experiment can be found at the Online Learning Center.

Transposable Elements and Retroelements Move via One of Three Transposition Pathways

Since the pioneering studies of Barbara McClintock, many different transposable elements have been found in bacteria, fungi, plant, and animal cells. Three general types of transposition pathways have been identified (fig. 17.12). In **simple transposition,** the TE is removed from its original site and transferred to a new target site. This mechanism is also called a "cut-and-paste" mechanism because the element is cut out of its original site and pasted into a new site. Transposable elements that move via simple transposition are widely found in bacteria and eukaryotic species. By comparison, **replicative transposition** involves the replication of the TE and insertion of the newly made copy into a second site. In this case, one of the TEs remains in its original location and the other is inserted at another location. Replicative transposition is relatively uncommon and found only in bacterial species.

A third category of elements move via an RNA intermediate. This form of transposition is found only in eukaryotic species, where it is very common. These types of elements are known as **retroelements, retrotransposons,** or **retroposons.** Like replicative transposons, retroelements increase in number during retrotranspositional events.

Each Type of Transposable Element Has a Characteristic Pattern of DNA Sequences

As researchers have studied TEs from many species, they have found that DNA sequences within transposable elements are organized in several different ways. Figure 17.13 describes a few ways that TEs are organized, although many variations are possible. All TEs are flanked by direct repeats (DR). The TE itself is the region between two direct repeats. The simplest TEs, which are commonly found in bacteria, are known as **insertion sequences.** As shown in figure 17.13a, insertion sequences have two important characteristics. First, both ends of the insertion sequence contain **inverted repeats (IRs).** Inverted repeats are DNA sequences that are identical (or very similar) but run in opposite directions, such as the following:

```
5′-CTGACTCTT-3′     and     5′-AAGAGTCAG-3′
3′-GACTGAGAA-5′             3′-TTCTCAGTC-5′
```

Depending on the particular element, the lengths of inverted repeats range from 9 to 40 bp in length. In addition, insertion sequences may contain a central region that encodes the enzyme **transposase,** which catalyzes the transposition event.

Composite transposons contain additional genes that are not necessary for transposition *per se.* They commonly contain genes that confer a selective advantage to the organism. Composite transposons are prevalent in bacteria, where they often contain genes that provide resistance to antibiotics or toxic heavy metals. For example, the composite transposon shown in figure 17.13a contains two insertion sequences flanking a gene that confers antibiotic resistance. During transposition of a composite transposon, only the inverted repeats at the ends of the transposon are involved in the transpositional event. Whenever insertion sequences are found at both ends of a gene, they create a composite transposon.

Replicative transposons have a sequence organization that is similar to insertion sequences except that replicative transposons have a resolvase gene that is found between the inverted repeats (fig. 17.13b). As discussed later in this chapter, both transposase

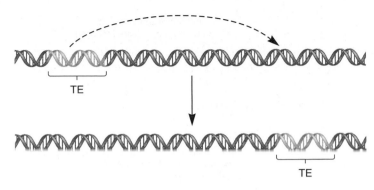

(a) Simple transposition

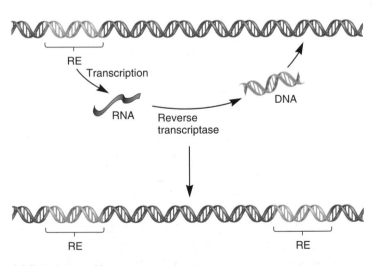

Replication

(b) Replicative transposition

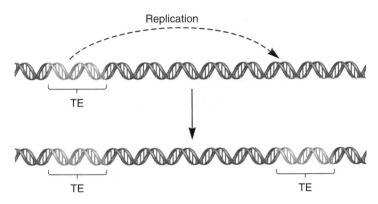

Transcription

RNA

DNA

Reverse transcriptase

(c) Retrotransposition

FIGURE 17.12   **Three mechanisms of transposition.**

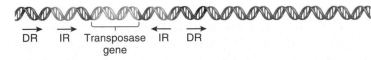

DR   IR   Transposase gene   IR   DR

Insertion sequence

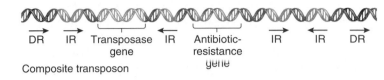

DR   IR   Transposase gene   IR   Antibiotic-resistance gene   IR   IR   DR

Composite transposon

(a) Elements that move by simple transposition

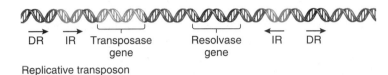

DR   IR   Transposase gene   Resolvase gene   IR   DR

Replicative transposon

(b) An element that moves by replicative transposition

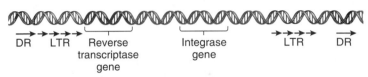

DR   LTR   Reverse transcriptase gene   Integrase gene   LTR   DR

Viral-like retroelement

DR   Reverse transcriptase gene   DR

Nonviral-like retroelement

(c) Elements that move by retrotransposition (via an RNA intermediate)

FIGURE 17.13   **Common organizations of transposable elements.** Direct repeats (DRs) are found within the host DNA. Inverted repeats (IRs) are at the ends of most transposable elements. Long terminal repeats (LTRs) are regions containing a large number of tandem repeats.

and resolvase are needed to catalyze the transposition of replicative transposons.

The organization of retroelements can be quite variable, and they are categorized based on their evolutionary relationship to retroviral sequences. Retroviruses are RNA viruses that make a DNA copy that integrates into the host's genome. The **viral-like**

retroelements are evolutionarily related to known retroviruses. These transposable elements have retained the ability to move around the genome, though, in most cases, mature viral particles are not produced. Viral-like retroelements contain **long terminal repeats (LTRs)** at both ends of the element (fig. 17.13c). The LTRs are typically a few hundred nucleotides in length. Like their viral counterparts, viral-like retroelements encode virally related proteins such as **reverse transcriptase** and **integrase** that are needed for the transposition process.

By comparison, nonviral-like retroelements appear less like retroviruses in their sequence, although some similarity, such as the occurrence of a reverse transcriptase gene, may be present (fig. 17.13c). Many nonviral-like retroelements, however, do not share any sequence similarity with known viruses. Instead, some nonviral-like retroelements are evolutionarily derived from normal eukaryotic genes. For example, the *Alu* family of repetitive sequences found in humans is derived from a single ancestral gene known as the 7SL RNA gene (a component of signal recognition particle, which was described in chapter 13). This gene sequence has been copied by retroposition to achieve the current number of greater than 500,000 copies. Incredibly, it constitutes approximately 5 to 6% of the human genome!

Transposable elements are considered to be **complete** (or **autonomous**) elements when they contain all the information necessary for transposition or retroposition to take place. However, TEs are often **incomplete** (or **nonautonomous**). An incomplete element typically lacks a gene such as transposase or reverse transcriptase that is necessary for transposition. The *Ds* element described in the experiment of figure 17.11 is an incomplete element that lacks a transposase gene. The complete version of this element is called an *Ac* locus or *Ac* element (i.e., <u>Ac</u>tivator element). As mentioned earlier, an *Ac* element provides a transposase gene that enables *Ds* to transpose. Therefore, incomplete TEs such as *Ds* can transpose only when the transposase gene is present at another region in the genome.

Transposase Catalyzes the Excision and Insertion of Transposable Elements

Now that we have an understanding of the sequence organization of transposable elements, we can examine the steps of the transposition process. The enzyme transposase catalyzes the removal of a TE from its original site in the chromosome and its subsequent insertion at another location. A general scheme for simple transposition is shown in figure 17.14. Transposase proteins recognize the inverted repeat sequences at the ends of the TE and bring them close together. The DNA is cleaved at the ends of the TE, excising it from its original site within the chromosome. This excision event is followed by an insertion process in which transposase cleaves the target DNA sequence at staggered recognition sites. The TE is then ligated to the target DNA.

As noted in figure 17.14, the ligation of the transposable element into its new site initially leaves short gaps in the target DNA. Notice that the DNA sequences in these gaps are complementary to each other (in this case, ATGCT and TACGA). Therefore, when they are filled in by DNA gap repair synthesis, the DNA base pair sequences at both ends of the TE are identical. These two sequences are called **direct repeats (DRs),** because they are in the same <u>direct</u>ion and they are <u>repeated</u> at both ends of the element. The occurrence of direct repeats is a common feature of all TEs (refer back to fig. 17.13).

Replicative Transposition Requires Both Transposase and Resolvase

Replicative transposition has been studied in several bacterial transposons and in bacteriophage μ (mu), which behaves like a transposon. The net result of replicative transposition is that a

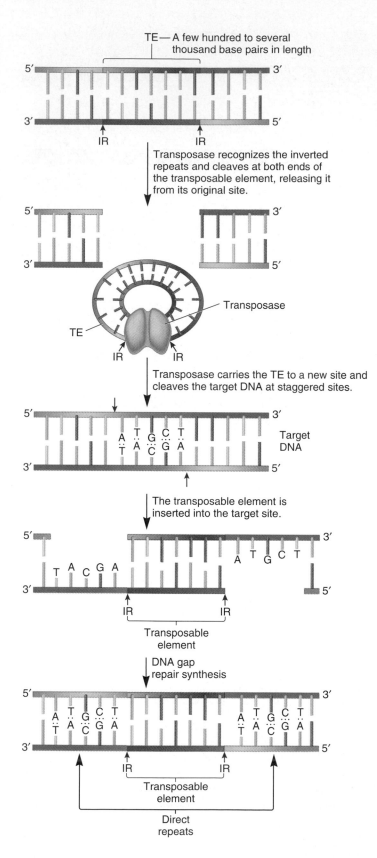

FIGURE 17.14   **Simple transposition.**

transposable element occurs at a new site and a TE also remains in its original location. Figure 17.15 describes a model for replicative transposition between two circular DNA molecules. One DNA molecule already has a TE, whereas the other does not. In this mechanism, transposase initially makes one cut at each end of the TE and two cuts in the target DNA. Note that this differs from simple transposition, in which the transposase makes four cuts and completely removes the TE from its original site (compare figs. 17.14 and 17.15). In replicative transposition, one strand of the TE is left at its original location.

Following ligation, both the host DNA and the transposable element have a long gap. Gap repair DNA synthesis copies the host DNA gap as well as the TE. This creates two copies of the TE within a large circular molecule known as a cointegrant. The enzyme resolvase catalyzes homologous recombination within the TEs so that the cointegrant can be resolved into two separate DNA molecules. One of these molecules contains the TE in its original location, and the other has a TE at a new location.

In the preceding discussion, we have seen that replicative transposons are duplicated when they transpose, whereas TEs that move via simple transposition are completely removed and placed in a new location. Thus, you might think that such TEs would have difficulty increasing in number. Interestingly, this is not the case. Cut-and-paste TEs can easily increase in number, because transposition often occurs around the time of DNA replication. For these TEs, an increase in number can happen in the following way (also see solved problem S3). After a replication fork has passed a region containing a TE, there will be two TEs behind the fork (i.e., one in each of the replicated regions). One of these TEs could then transpose from its original location into a region ahead of the replication fork. After the replication fork has passed this second region and DNA replication is completed, there will be two TEs in one of the chromosomes, and one TE in the other chromosome. In this way, cut-and-paste TEs can increase in number. We will discuss the biological significance of transposon proliferation later in this chapter.

Retroelements Use Reverse Transcriptase and Integrase for Retrotransposition

Thus far, we have considered how DNA elements (i.e., transposons) can move throughout the genome. By comparison, retroelements use an RNA intermediate in their transposition mechanism. As shown in figure 17.16, the movement of retroelements also requires two key enzymes, reverse transcriptase and integrase. In this example, the host cell already contains a retroelement known as the *Alu* sequence within its genome. This retroelement, which behaves like a gene, is transcribed into RNA. In a series of steps, reverse transcriptase uses this RNA as a template to synthesize a double-stranded DNA molecule. The ends of the double-stranded DNA are then recognized by integrase, which catalyzes the insertion of the DNA into the host chromosomal DNA. The integration of retroelements can occur at many locations within the genome. Furthermore, because a single retroelement can be copied into many RNA transcripts, retroelements may accumulate rapidly within a genome.

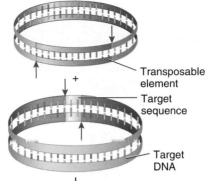

Transposable element

Target sequence

Target DNA

Transposase makes cuts at the arrows and the strands are exchanged and ligated together.

The gaps are filled in by DNA polymerase and sealed by DNA ligase.

Resolvase catalyzes recombination between the 2 elements. This resolves them into 2 separate structures, each with a copy of the transposable element.

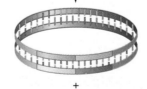

+

FIGURE 17.15 Replicative transposition. The transposase catalyzes the movement of a DNA strand carrying the transposable element to a new recipient site. DNA gap repair synthesis at the previous and new sites produces two double-stranded elements within a large circular molecule known as a cointegrant. Similar to the resolution step of homologous recombination described in figure 17.4, a resolvase separates the cointegrant into two separate structures that each contain a TE.

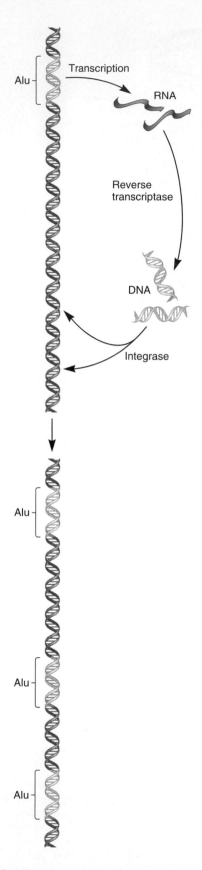

FIGURE 17.16 **The mechanism of retrotransposition.**

Transposable Elements May Have Important Influences on Mutation and Evolution

Over the past few decades, researchers have found that transposable elements probably occur in the genomes of all species. Table 17.2 describes a few TEs that have been studied in great detail. As discussed in chapter 10, the genomes of eukaryotic species typically contain moderately and highly repetitive sequences. In some cases, these repetitive sequences are due to the proliferation of TEs. In mammals, for example, **LINEs** are long interspersed elements that are usually 1,000 to 5,000 bp in length and found in 20,000 to 100,000 copies per genome. **SINEs** are short interspersed elements that are less than 500 bp in length. A specific example of a SINE is the *Alu* sequence, present in 500,000 to 1 million copies in the human genome.

The biological significance of transposons in the evolution of prokaryotic and eukaryotic species remains a matter of debate. According to the **selfish DNA hypothesis,** TEs exist because they contain the characteristics that allow them to multiply within the host cell DNA. In other words, they resemble parasites in the sense that they inhabit the host without offering any selective advantage. They can proliferate within the host as long as they do not harm the host to the extent that they significantly disrupt survival.

Alternatively, other geneticists have argued that most transpositional events are deleterious, and, therefore, TEs would be eliminated from the genome if they did not also offer a compensating advantage. Several potential advantages have been suggested. For example, TEs may cause greater genetic variability by promoting recombination. In addition, bacterial TEs often carry an antibiotic-resistance gene that provides the organism with a survival advantage. Researchers have also suggested that transposition may cause the insertion of exons into the coding sequences of structural genes. This phenomenon, called **exon shuffling,** may lead to the evolution of genes with more diverse functions.

While this controversy remains unresolved, it is clear that transposable elements can rapidly enter the genome of an organism and proliferate quickly. In *Drosophila melanogaster,* for example, a TE known as the P element was probably introduced into this species in the 1950s. Laboratory stocks of *D. melanogaster* collected prior to this time do not contain P elements. Remarkably, in the last 50 years, the P element has expanded throughout *D. melanogaster* populations worldwide. The only strains without the P element are laboratory strains collected prior to the 1950s. This observation underscores the surprising ability of TEs to infiltrate a population of organisms.

Transposable elements have a variety of effects on chromosome structure and gene expression (table 17.3). Since many of these outcomes are likely to be harmful, transposition is usually a highly regulated phenomenon that occurs only in a few individuals under certain conditions. Agents such as radiation, chemical mutagens, and hormones stimulate the movement of TEs. When it is not carefully regulated, transposition is likely to be potently detrimental. For example, in *D. melanogaster,* if M strain females

TABLE 17.2
Examples of Transposable Elements

Element	Type	Approximate Length (bp)	Description
Bacterial			
IS1	Insertion sequence	768	An insertion sequence that is commonly found in 5–8 copies in *E. coli.*
Mu	Replicative transposon	36,000	*Mu* is a true virus that can insert itself anywhere in the *E. coli* chromosome. Its name, *Mu,* is derived from its ability to insert into genes and mutate them.
Tn10	Composite transposon	9,300	One of many different bacterial transposons that carries antibiotic resistance.
Tn951	Composite transposon	16,600	A transposon that provides bacteria with genes that allow them to metabolize lactose.
Yeast			
Ty elements	Viral-like retroelement	6,200	A retroelement found in *S. cerevisiae* at about 35 copies per genome.
Drosophila			
P elements	Cut-and-paste transposon	500–3,000	A transposon that may be found in 30–50 copies in P strains of *Drosophila.* It is absent from M strains.
Copia-like elements	Viral-like retroelement	5,000–8,000	A family of *copia*-like elements is found in *Drosophila,* which vary slightly in their lengths and sequences. Typically, each family member is found at about 5–100 copies per genome.
Humans			
Alu sequence	Nonviral-like retroelement	300	As discussed in chapter 10, *Alu* sequences are SINEs that are abundantly interspersed throughout the human genome.
L1	Viral-like retroelement	6,500	A human LINE found in 50,000–100,000 copies in the human genome.
Plants			
Ac/Ds	Cut-and-paste transposon	4,500	*Ac* is an autonomous transposon found in corn and other plant species. It carries a transposase gene. *Ds* is a nonautonomous version that lacks a functional transposase gene.

TABLE 17.3
Possible Consequences of Transposition

	Cause
Chromosome Structure	
Chromosomal breakage	Excision of a TE.
Chromosomal rearrangements	Homologous recombination between TEs located at different positions in the genome.
Gene Expression	
Mutation	Incorrect excision of TEs.
Gene inactivation	Insertion of a TE into a gene.
Alteration in gene regulation	Transposition of a gene next to regulatory sequences or the transposition of regulatory sequences next to a gene.
Alteration in the exon content of a gene	Insertion of exons into the coding sequence of a gene via TEs. This phenomenon is called exon shuffling.
Gene duplications	Creation of a composite transposon that transposes to another site in the genome.

(which lack P elements) are crossed with P strain males (which contain numerous P elements), the hybrid offspring contain a variety of abnormalities, which include a high rate of mutation and chromosome breakage. This deleterious outcome, which is called **hybrid dysgenesis,** occurs because the P elements can transpose freely in the offspring.

Transposons Have Become Important Tools in Molecular Biology

The unique and unusual features of transposons have made them an important experimental tool in molecular biology. For researchers, the introduction of a transposon into a cell is a convenient way to abolish the expression of particular genes. If a transposon "hops" into a gene, it is likely to inactivate the gene's function. In addition, this phenomenon can be used to clone a particular gene in an approach known as **transposon tagging.** In this strategy, researchers use transposons in an attempt to clone novel genes.

An early example of transposon tagging involved an X-linked gene in *Drosophila* that affects eye color. As was described in chapters 3 and 4, this *Drosophila* X-linked gene can exist in the wild-type (red) allele and a loss-of-function allele that causes a white-eye phenotype. In 1981, Paul Bingham, in collaboration with Robert Levis and Gerald Rubin, used transposon tagging to clone this gene (fig. 17.17). Prior to their cloning work, a wild-type strain of *Drosophila* had been characterized that carried a transposable element called *copia.* From this red-eyed strain, a white-eyed strain was obtained in which the *copia* element had transposed to a region on the X chromosome that corresponded

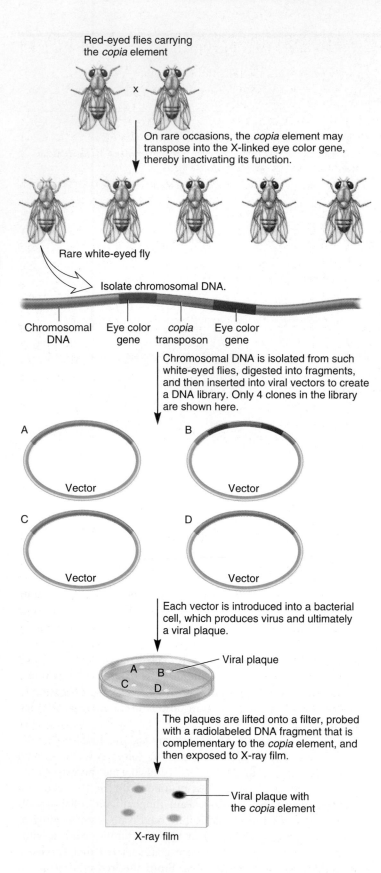

Red-eyed flies carrying the *copia* element

On rare occasions, the *copia* element may transpose into the X-linked eye color gene, thereby inactivating its function.

Rare white-eyed fly

Isolate chromosomal DNA.

Chromosomal DNA   Eye color gene   *copia* transposon   Eye color gene

Chromosomal DNA is isolated from such white-eyed flies, digested into fragments, and then inserted into viral vectors to create a DNA library. Only 4 clones in the library are shown here.

A   Vector

B   Vector

C   Vector

D   Vector

Each vector is introduced into a bacterial cell, which produces virus and ultimately a viral plaque.

Viral plaque

The plaques are lifted onto a filter, probed with a radiolabeled DNA fragment that is complementary to the *copia* element, and then exposed to X-ray film.

Viral plaque with the *copia* element

X-ray film

FIGURE 17.17   The procedure of transposon tagging. A transposon, known as *copia*, was introduced into a *Drosophila* strain with red eyes. On rare occasions, a white-eyed fly was produced from this red-eyed strain. In this case, the phenotype has occurred because the *copia* transposon inserted itself into the wild-type eye color gene and thereby inactivated it. Chromosomal DNA was isolated from the white-eyed fly, digested with a restriction enzyme, and cloned into a vector, creating a DNA library as described in chapter 18. The DNA library is probed with a radiolabeled fragment complementary to the *copia* element. A plaque carrying the TE shows up as a dark spot on an X-ray film. This plaque may also contain a version of the eye color gene that was disrupted by the TE.

GENES→TRAITS   A white-eyed fly may occur due to the insertion of a transposable element into a gene that confers eye color. As discussed in chapter 13, the wild-type eye color gene encodes a protein that is necessary for red pigment production. When a TE inserts into this gene, it disrupts the coding sequence and thereby causes the gene to produce a nonfunctional protein. Therefore, no red pigment can be made, and a white-eye phenotype results. In many cases, transposons affect the phenotypes of organisms by inactivating individual genes.

to where the eye color gene mapped. The researchers reasoned that the white-eye phenotype could be due to the insertion of the *copia* element into the wild-type gene, thereby inactivating it.

To clone the eye color gene, chromosomal DNA from this white-eyed strain was isolated, digested with restriction enzymes, and cloned into viral vectors. This procedure of creating a DNA library will be described in chapter 18. A DNA library is a collection of vectors that contain different pieces of chromosomal DNA. If a transposon has "jumped" into the eye color gene, vectors that contain this gene will also contain the transposon sequence. In other words, the presence of the transposon tags the eye color gene. Therefore, a radiolabeled fragment of DNA that is complementary to the transposon sequence can be used as a probe to identify plaques that also contain the eye color gene. The method of using a probe to screen a DNA library will also be described in chapter 18. In the example of figure 17.17, the method of transposon tagging was successful at cloning an eye color gene in *Drosophila*.

Sister chromatid exchange (SCE) involves the breakage and union of identical sister chromatids. It does not change the sequence of the genetic material. By comparison, **genetic recombination** occurs when segments of DNA are broken and reconnected to form new combinations. During **homologous recombination,** homologous DNA regions become aligned and exchange DNA segments. This enhances genetic variability by producing recombinant chromosomes with new combinations of alleles.

At the molecular level, several models have attempted to explain the steps in homologous recombination. The **Holliday model** was the first example of a molecular explanation for the recombination process. This model was able to account for the phenomenon of **gene conversion,** in which one allele of a gene is converted to another allele. More recently, other models, such as the Meselson/Radding and **double-stranded break model,** have more accurately described certain steps in the recombination pathway. In addition, much progress has been made toward identifying the many proteins that play important roles in recombination.

A second mechanism for recombination is known as **site-specific recombination,** because the breakage and rejoining of the DNA segments occurs at particular DNA sequences. Site-specific recombination is responsible for the integration of certain bacteriophages such as λ into the host genome. In addition, site-specific recombination within **immunoglobulin** genes is important in generating an astounding diversity in antibody polypeptides. During this process, regions in precursor immunoglobulin genes known as V, D, and J domains are connected to each other via recombination. This mechanism enables the production of a very diverse array of antibody proteins.

Transposition is a third way that DNA segments can rearrange. Short segments of DNA known as **transposable elements (TEs)** possess characteristics that enable them to be mobile. During **simple transposition,** a "cut-and-paste" mechanism occurs where the TE is cut out of its original site and ligated to a new site. **Transposase** catalyzes this reaction. By comparison, in **replicative transposition** a TE is duplicated, with one TE remaining in its original location and a new TE located at a new site. Finally, **retroposition** occurs via an RNA intermediate. In this case, the **retroelement** is transcribed into RNA and then copied into DNA by **reverse transcriptase.** This DNA is then integrated into the chromosome by **integrase.** From an evolutionary viewpoint, all mechanisms of transposition can cause many different types of mutations.

Crossing over, which may occur between sister chromatids or homologous chromosomes, can be detected by several techniques. **Sister chromatid exchange (SCE)** can be observed using staining methods that produce harlequin chromosomes, which distinguish the sister chromatids within a pair. As was discussed in chapter 5, **homologous recombination** can also be identified from the outcome of dihybrid crosses involving linked genes, where the frequency of recombinant offspring is a measure of the frequency of crossing over. At the molecular level, researchers have studied homologous recombination by identifying and characterizing the proteins and intermediates that facilitate the process. As shown in figure 17.4, **Holliday junctions** are a type of intermediate involved in homologous recombination.

Likewise, researchers have elucidated the mechanisms of site-specific recombination and transposition using genetic and molecular techniques. McClintock followed genetic crosses in corn to provide compelling evidence for the existence of **transposable elements (TEs).** Since that time, researchers have studied many different TEs. The sequencing of TEs has revealed certain patterns of DNA sequences that are required for **transposition** and **retrotransposition.** Researchers have also identified proteins that are necessary to promote the movement of TEs. In addition, molecular biologists have taken advantage of our knowledge of transposition and now routinely use transposons to inactivate and clone genes via **transposon tagging.**

PROBLEM SETS & INSIGHTS

Solved Problems

S1. Zickler was the first person to demonstrate gene conversion by observing unusual ratios in *Neurospora* octads. At first, it was difficult for geneticists to believe these results because they seemed to contradict the Mendelian concept that alleles do not physically interact with each other. However, work by Mary Mitchell provided convincing evidence that gene conversion actually takes place. She investigated three different genes in *Neurospora*. One *Neurospora* strain had three mutant alleles: *pdx-1* (pyridoxine-requiring), *pyr-1* (pyrimidine-requiring), and *col-4* (a mutation that affected growth morphology). The *pdx-l* gene had been previously shown to map in between the *pyr-1* and *col-4* genes. As shown here, this strain was crossed to a wild-type *Neurospora* strain.

$$pyr\text{-}1 \ pdx\text{-}1 \ col\text{-}4 \times pyr\text{-}1^+ \ pdx\text{-}1^+ \ col\text{-}4^+$$

She first analyzed many octads with regard to their requirement for pyridoxine. Out of 246 octads, two of them had an aberrant ratio in which two spores were *pdx-1* and six were *pdx-1*⁺. These same spores were then analyzed with regard to the other two genes. In both cases, the aberrant asci gave a normal 4:4 ratio of *pyr-1: pyr-1*⁺ and *col-4:col-4*⁺. Explain these results.

Answer: These results can be explained by gene conversion. The gene conversion took place in a limited region of the chromosome (within the *pdx-1* gene), but it did not affect the flanking genes (*pyr-1* and *col-4*) located on either side of the *pdx-1* gene. In the asci containing two *pdx-1* alleles and six *pdx-1*⁺ alleles, a crossover occurred during meiosis I in the region of the *pdx-1* gene. Gene conversion changed the *pdx-1* allele into the *pdx-1*⁺ allele. This gene conversion could have occurred by two mechanisms. If branch migration occurred across the *pdx-1* gene, a heteroduplex may have formed, and this could be repaired by mismatch DNA repair as described in figure 17.7. In the aberrant asci with two *pdx-1* and six *pdx-1*⁺ alleles, the *pdx-1* allele was converted to *pdx-1*⁺. Alternatively, gene conversion of *pdx-1* into *pdx-1*⁺ could have taken place via gap repair synthesis as described in figure 17.6. In this case, the *pdx-1* allele would have been digested away, and the DNA encoding the *pdx-1*⁺ allele would have migrated into the digested region and provided a template to make a copy of the *pdx-1*⁺ allele. Note: Since this pioneering work, additional studies have shed considerable light concerning the phenomenon of gene conversion. It occurs at a fairly high rate in fungi, approximately 0.1 to 1% of the time. It is not due to new mutations occurring during meiosis.

S2. Recombination involves the pairing of identical or homologous sequences, followed by crossing over and the resolution of the intertwined helices. On rare occasions, the direct repeats or the inverted repeats within a single transposable element can recognize each other and undergo genetic recombination. What are the consequences when the direct repeats recombine? What are the consequences when the inverted repeats recombine?

Answer:

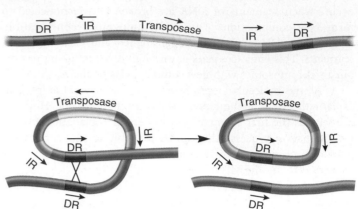

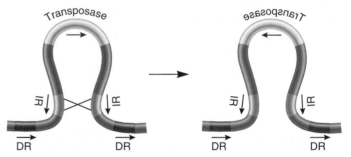

Most of the transposable element has been excised.

(a) Recombination between direct repeats

The sequence within the transposable element has been inverted. Note that the transposase gene has changed to the opposite direction.

(b) Recombination between inverted repeats

S3. Replicative transposons are duplicated when they transpose, while cut-and-paste TEs are simply removed and placed in a new location. Based on these mechanisms, you might think that cut-and-paste TEs would have difficulty increasing in number. Interestingly, this is not the case. The reason why is because transposition frequently occurs during DNA replication. Explain how the number of copies of a cut-and-paste TE can increase in number within a single chromosome if transposition occurs during chromosomal replication.

Answer: The diagram shown here illustrates how this can occur.

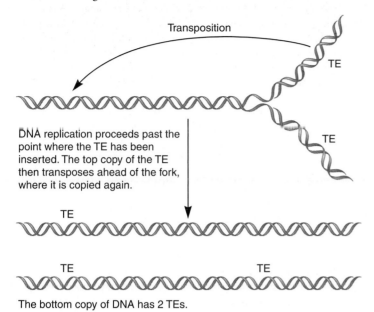

Transposition

DNA replication proceeds past the point where the TE has been inserted. The top copy of the TE then transposes ahead of the fork, where it is copied again.

TE

TE   TE

The bottom copy of DNA has 2 TEs.

As shown here, a chromosome containing a single transposon is in the process of DNA replication. After the replication fork has passed the transposon, one of the TEs transposes to a new location that has not yet replicated. Following the completion of DNA replication, one of the chromosomes has one TE while the other chromosome has two TEs. This second chromosome contains an additional TE compared to the parental chromosome. In this way, cut-and-paste TEs can increase in number.

Conceptual Questions

C1. Describe the similarities and differences between sister chromatid exchange and homologous recombination. Would you expect the same types of proteins to be involved in both processes? Explain.

C2. The molecular mechanism of sister chromatid exchange is similar to homologous recombination except that the two segments of DNA are sister chromatids instead of homologous chromatids. If branch migration occurs during sister chromatid exchange, will a heteroduplex be formed? Explain why or why not. Can gene conversion occur during sister chromatid exchange?

C3. A schematic drawing of an uncrossed Holliday junction is shown here. One chromatid is shown in *red* and the homologous chromatid is shown in *blue*. The red chromatid carries a dominant allele labeled *A,* and a recessive allele labeled *b,* whereas the blue chromatid carries a recessive allele labeled *a* and a dominant allele labeled *B.*

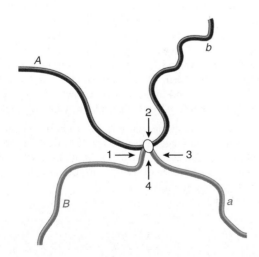

Where would the breakage of the crossed strands have to occur to get recombinant chromosomes? Would it have to occur at sites 1 and 3, or at sites 2 and 4? What would be the genotypes of the two recombinant chromosomes?

C4. Which steps in the double-stranded break model for recombination would be inhibited if the following proteins were missing? Explain the function of each protein required for the step that is inhibited.

A. RecBCD

B. RecA

C. RecG

D. RuvABC

C5. What are the two molecular mechanisms that can explain the phenomenon of gene conversion? Would both of these mechanisms occur in the double-stranded break model?

C6. Is homologous recombination an example of mutation? Explain.

C7. What are recombinant chromosomes? How do they differ from the original parental chromosomes from which they are derived?

C8. During homologous recombination (see fig. 17.4), the resolution steps can produce recombinant or nonrecombinant chromosomes. Explain how this can occur.

C9. What is gene conversion?

C10. Make a list of the differences between the Holliday model and the double-stranded break model.

C11. In recombinant chromosomes, where is gene conversion likely to have taken place: near the breakpoint or far away from the breakpoint? Explain.

C12. What are the events that RecA protein facilitates?

C13. According to the double-stranded break model, does gene conversion necessarily involve DNA mismatch repair? Explain.

C14. What type of DNA structure is recognized by RecG and RuvABC? Do you think these proteins recognize DNA sequences? Be specific about what type(s) of molecular recognition these proteins can perform.

C15. Briefly describe three ways that antibody diversity is produced.

C16. Describe the function of RAG1/RAG2 and NHEJ proteins.

C17. According to the scenario shown in figure 17.9, how many segments of DNA (one, two, or three) are removed during site-specific recombination within the gene that encodes the κ (kappa) light chain for IgG proteins? How many segments are spliced out of the pre-mRNA?

C18. Describe the role that integrase plays during the insertion of λ DNA into the host chromosome.

C19. If you were examining a sequence of chromosomal DNA, what characteristics would cause you to believe that the DNA contained a transposable element?

C20. According to the current model for replicative transposition, does the transposable element replicate before or after it transposes? Explain your answer.

C21. Why does transposition always produce direct repeats in the host DNA?

C22. Which types of transposable elements have the greatest potential for proliferation: cut-and-paste TEs, replicative TEs, or retroelements? Explain your choice.

C23. Do you consider transposable elements to be mutagens? Explain.

C24. Let's suppose that a species of mosquito has two different types of simple transposons that we will call X elements and Z elements. The X elements appear quite stable. When analyzing a population of 100 mosquitoes, every mosquito has six X elements, and they are always located in the same chromosomal locations among different individuals. In contrast, the Z elements seem to "move around" quite a bit. Within the same 100 mosquitoes, the number of Z elements ranges from 2 to 14, and the locations of the Z elements tend to vary considerably among different individuals. Explain how one simple transposon can be stable and another simple transposon can be mobile, within the same group of individuals.

C25. Your textbook describes five different types of transposable elements including insertion sequences, composite transposons, replicative transposons, viral-like retrotransposons, and nonviral-like retrotransposons. Which of these five types of TEs would have the following features?

A. Require transcription to transpose

B. Require transposase to transpose

C. Have direct repeats

D. Have inverted repeats

C26. What features distinguish a transposon from a retroelement? How are their sequences different and how are their mechanisms of transposition different?

C27. Solved problem S2 illustrates the consequences of crossing over between the direct and inverted repeats within a single transposable element. The drawing here shows the locations of two copies of the same transposable element within a single chromosome. The chromosome is depicted according to its G banding pattern. (Note: G bands are described in chapter 8.) The direct and inverted repeats are labeled 1, 2, 3, and 4, from left to right.

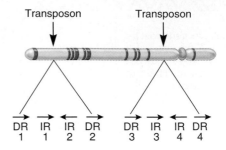

Draw the end result of recombination between the following sequences. Your drawing should include the banding pattern of the resulting chromosome.

A. DR-1 and DR-4

B. IR-1 and IR-4

C28. What is the difference between an autonomous versus a nonautonomous transposable element? Is it possible for nonautonomous TEs to move? If yes, explain how.

C29. An operon in the bacterium *Salmonella typhimurium* has the following arrangement:

DR IR   IR DR
   promoter→   *H2 rH1*

The promoter for this operon is contained within a transposable element. The *H2* gene encodes a protein that is part of the bacterial flagellum. The *rH1* gene encodes a repressor protein that represses the *H1* gene, which is found at another location in the bacterial chromosome. The *H1* gene also encodes a flagellar protein. When the promoter is found in the arrangement shown here, the *H2/rH1* operon is turned on. This results in flagella that contain the H2 protein. The H1 protein is not made because the rH1 repressor prevents the transcription of the *H1* gene. At a frequency of approximately 1 in 10,000 (which is much higher than the spontaneous mutation rate), this strain of bacterium can "switch" its expression so that *H2* is turned off, and *H1* is turned on. Bacteria that have *H1* turned on and *H2* turned off can also switch back to having *H2* turned on and *H1* turned off. This switch also occurs at a frequency of about 1 in 10,000. Based on your understanding of transposons and recombination, explain how switching occurs in *Salmonella typhimurium*. Hint: Take a look at solved problem S2.

C30. The occurrence of multiple transposons within the genome of organisms has been suggested as a possible cause of chromosomal rearrangements such as deletions, translocations, and inversions. How could the occurrence of transposons promote these kinds of structural rearrangements?

Experimental Questions

E1. With the harlequin staining technique, one sister chromatid appears to fluoresce more brightly than the other. Why?

E2. In the data shown here, harlequin staining was used to determine the rate of SCEs in the presence of a suspected mutagen.

	Frequency of SCEs/Chromosome
No mutagen	0.67
With suspected mutagen	14.7

Would you conclude that this substance is a mutagen?

E3. In the experiment described in experimental question E2, at what point would you need to add the mutagen: before the first round of DNA replication, after the first round but before the second round, or after the second round?

E4. Let's suppose that a researcher followed the protocol described in the experiment of figure 17.3 but exposed the cells to BrdU for three cell generations instead of two. Near the end of growth, the cells were exposed to colcemid to prevent them from completing mitosis following the third round of DNA replication. What would be the expected results if a parental cell contained a total of four chromosomes (two homologous pairs) and a single sister chromatid exchange occurred after the second replication? Your drawing should show four cells that contain four condensed chromosomes in each cell.

E5. Based on your understanding of the experiment of figure 17.3, does BrdU enhance the binding of Giemsa to the chromatids or inhibit the binding of Giemsa? Explain.

E6. Briefly explain how McClintock determined that *Ds* was occasionally moving from one chromosomal location to another. Discuss the type of data she examined to arrive at this conclusion.

E7. In the data table of figure 17.3, is the solid red phenotype due to chromosome breakage or the excision of a transposable element? Explain how you have arrived at your conclusion.

E8. In your own words, explain the term *transposon tagging*.

E9. Tumor-suppressor genes are normal human genes that prevent uncontrollable cell growth. Starting with a normal laboratory human cell line, describe how you could use transposon tagging to identify tumor-suppressor genes. Note: When a transposable element hops into a tumor-suppressor gene, it may cause uncontrolled cell growth. This is detected as a large clump of cells among a normal monolayer of cells.

E10. As discussed in the experiment of figure 17.11, the presence of a transposon can create a "mutable site" that is subject to frequent chromosomal breakage. Why do you think a transposon creates a mutable site? If chromosomal breakage occurs, do you think the transposon has moved somewhere else? How would you experimentally determine if it has?

E11. Rubin and Spradling devised a method of introducing a transposon into *Drosophila*. This approach has been important for the transposon tagging of many *Drosophila* genes. They began with a P element that had been cloned on a plasmid. (Note: Methods of cloning are described in chapter 18.) Using cloning methods, they inserted the wild-type allele for the *rosy* gene into the P element in this plasmid. The recessive allele, *rosy*, results in a rosy eye color, while the wild-type allele, *rosy+*, produces red eyes. This plasmid also has an intact transposase gene. The cloned DNA is shown here:

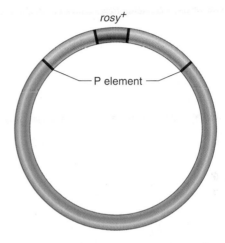

They used a micropipette to inject this DNA into regions of embryos that would later become reproductive cells. These embryos were originally homozygous for the recessive *rosy* allele. However, the P element carrying the *rosy+* allele could "hop" out of the plasmid and into a chromosome of the cells that were destined to become germ cells (i.e., sperm or egg cells). After they had matured to adults, these flies were then mated to flies that were homozygous for the recessive *rosy* allele. If offspring inherited a chromosome carrying the P element with the *rosy+* gene, such offspring would have red eyes. Therefore, the phenotype of red eyes provided a way to identify offspring that had a P element insertion.

Now here is the question. Let's suppose you were interested in identifying genes that play a role in wing development. Outline the experimental steps you would follow, using the clone with the P element containing the *rosy+* gene as a way to transposon tag genes that play a role in wing development. Note: You should assume that the inactivation of a gene involved in wing development would cause an abnormality in wing shape. Also keep in mind that most P element insertions inactivate genes and may be inherited in a recessive manner.

Questions for Student Discussion/Collaboration

1. Make a list of the similarities and differences among homologous recombination, site-specific recombination, and transposition.

2. If no genetic recombination of any kind could occur, what would be the harmful and beneficial consequences?

3. Based on your current knowledge of genetics, discuss whether or not you think the selfish DNA theory is correct.

Note: All answers appear at the website for this textbook; the answers to even-numbered questions are in the back of the textbook.

www.mhhe.com/brooker

Visit the Online Learning Center for practice tests, answer keys, and other learning aids for this chapter. Enhance your understanding of genetics with our interactive exercises, web links, news feeds, tutorial service, and much more.

RECOMBINANT DNA TECHNOLOGY

::

18

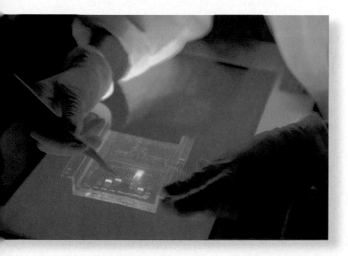

Recombinant DNA technology is the use of in vitro molecular techniques to isolate and manipulate fragments of DNA. In the early 1970s, the first successes in making recombinant DNA molecules were accomplished independently by two groups at Stanford University: David Jackson, Robert Symons, and Paul Berg; and Peter Lobban and A. Dale Kaiser. Both groups were able to isolate and purify pieces of DNA in a test tube and then covalently link two or more DNA fragments. In other words, they constructed chimeric molecules called **recombinant DNA molecules.** Shortly thereafter, it became possible to introduce such recombinant DNA molecules into living cells. Once inside a host cell, the recombinant molecules can be replicated to produce many identical copies of a gene. This achievement ushered in the era of **gene cloning.**

Recombinant DNA technology and gene cloning have enabled geneticists to probe relationships between gene sequences and phenotypic consequences and, thereby, have been fundamental to our understanding of gene structure and function. Most researchers in molecular genetics are familiar with recombinant DNA technology and apply it frequently in their work. Significant practical applications of recombinant DNA technology also have been developed. These include exciting advances such as **gene therapy,** screening for human diseases, recombinant vaccines, and the production of **transgenic** plants and animals in agriculture (in which the cloned gene from one species is transferred to some other species). In chapter 18, we will focus primarily on the use of recombinant DNA technology as a way to further our

Part V
Genetic Technologies

understanding of gene structure and function. We will examine how molecular techniques make it possible to clone genes, detect genes, and analyze DNA sequences. In chapter 19, we will consider modern applications that have arisen as a result of these technologies. Chapters 20 and 21 are devoted to **genomics,** which involves techniques to analyze many genes and even entire genomes.

18.1 GENE CLONING

Molecular biologists want to understand how the molecules within living cells contribute to cell structure and function. Because proteins are the workhorses of cells, and because proteins are the products of genes, most molecular biologists focus their attention on the structure and function of proteins, or the genes that encode them. In any particular laboratory, it is common to study just one or perhaps a few different genes or proteins. At the molecular level, this poses a daunting task. In eukaryotic species, for example, any given cell can express thousands of different proteins, making the study of any single gene or protein akin to a "needle in a haystack" exploration. To overcome this truly formidable obstacle, researchers frequently take the approach of cloning the genes that encode their proteins of interest. The term *gene cloning* refers to the phenomenon of isolating and making many copies of a gene. The laboratory methods that are necessary to clone a gene were devised during the early 1970s. Since then, many technical advances have enabled gene cloning to become a widely used procedure among scientists, including geneticists, cell biologists, biochemists, plant biologists, microbiologists, evolutionary biologists, clinicians, and biotechnologists.

Table 18.1 summarizes some of the common uses of gene cloning. In modern molecular biology, the myriad of uses for gene cloning is remarkable. For this reason, gene cloning has provided the foundation for critical technical advances in a variety of disciplines including molecular biology, genetics, cell biology, biochemistry, and medicine. In this section, we will examine the reagents and procedures that are used to clone genes. Later sections in chapter 18, as well as chapters 19 through 21, will consider many of the uses of gene cloning that are described in table 18.1.

Cloning Experiments Usually Involve Two Kinds of DNA Molecules: Chromosomal DNA and Vector DNA

Let's begin our discussion of gene cloning by considering a recombinant DNA technology in which a gene is removed from its native site within a chromosome and inserted into a smaller segment of DNA known as a **vector.** When introduced into a living cell, a vector can replicate and produce many identical copies of the inserted gene.

If a scientist wants to clone a particular gene, the source of the gene is the chromosomal DNA of the species that carries the gene. For example, if the goal is to clone the rat β-globin gene,

TABLE 18.1
Some Uses of Gene Cloning

Technique	Description
Gene sequencing	Cloned genes provide enough DNA to subject the gene to DNA sequencing (described later in chapter 18). The sequence of the gene can reveal the gene's promoter, regulatory sequences, and coding sequence. Gene sequencing is also important in the identification of alleles that cause cancer and inherited human diseases.
Mutagenesis	A cloned gene can be manipulated to change its DNA sequence. Mutations within genes can help to identify gene sequences such as promoters and regulatory elements. The study of a mutant gene can also help to elucidate its normal function and how its expression may affect the roles of other genes. Mutations in the coding sequence can reveal which amino acids are important for a protein's structure and function.
Gene probes	A cloned gene can be used as a probe to identify similar or identical genes or RNA. These methods, known as Southern and Northern blotting, are described later in chapter 18. Probes can also be used to localize genes within intact chromosomes (see chapter 20). DNA probes are used in techniques such as DNA fingerprinting to identify criminals (see chapter 19).
Expression of cloned genes	Cloned genes can be introduced into a different cell type or different species. The expression of cloned genes has many uses: *Research* 1. The expression of a cloned gene can help to elucidate its cellular function. 2. The coding sequence of a gene can be placed next to an active promoter and then introduced into a culture of cells that will express a large amount of the protein. This greatly aids in the purification of large amounts of protein that may be needed for biochemical or biophysical studies. *Biotechnology* 1. Cloned genes can be introduced into bacteria to make pharmaceutical products such as insulin (see chapter 19). 2. Cloned genes can be introduced into plants and animals to make transgenic species with desirable traits (see chapter 19). *Clinical* 1. Cloned genes are used in the procedure of gene therapy (see chapter 19).

this gene is found within the chromosomal DNA of rat cells. In this case, therefore, the rat's chromosomal DNA is one type of DNA that is needed in a cloning experiment. To prepare chromosomal DNA, an experimenter first obtains cellular tissue from the organism of interest. To clone the rat β-globin gene, a researcher would obtain tissue from a rat. The preparation of chromosomal

DNA then involves the breaking open of cells and the extraction and purification of the DNA using biochemical techniques such as chromatography and centrifugation (see the appendix).

As mentioned, a second type of DNA is used in a cloning experiment—a small, specialized DNA segment known as a vector. The purpose of vector DNA is to act as a carrier of the DNA segment that is to be cloned. (The term *vector* comes from a Latin term meaning carrier.) In cloning experiments, a vector may carry a small segment of chromosomal DNA, perhaps only a single gene. By comparison, a chromosome carries many more genes, perhaps a few thousand. Like a chromosome, a vector is replicated when it resides within a living cell; a cell that harbors a vector is called the **host cell.** When a vector is replicated within a host cell, the DNA that it carries is also replicated.

The vectors commonly used in gene cloning experiments were derived originally from two natural sources. Some vectors are **plasmids,** which are small circular pieces of DNA. As discussed in chapter 6, plasmids are found naturally in many strains of bacteria and occasionally in eukaryotic cells. Many naturally occurring plasmids carry genes that confer resistance to antibiotics or other toxic substances. These plasmids are called **R factors.** Many of the plasmids used in modern cloning experiments were derived from R factors.

Plasmids also contain a DNA sequence, known as an origin of replication, that is recognized by the replication enzymes of the host cell. It is the sequence of the origin of replication that determines the host cell specificity of a vector. Some plasmids have origins of replications with a broad host range. Such a plasmid can replicate in the cells of many different species. Alternatively, many vectors used in cloning experiments have a limited host cell range. In cloning experiments, researchers must choose a vector that replicates in the appropriate cell types for their experiments. If researchers want a cloned gene to be propagated in *Escherichia coli,* the vector they employ must have an origin of replication that is recognized by this species of bacterium.

Commercially available plasmids have been genetically engineered for effective use in cloning experiments. They contain unique sites where geneticists can easily insert pieces of DNA. Another useful feature of cloning vectors is that they often contain resistance genes that provide host bacteria with the ability to grow in the presence of a toxic substance. Such a gene is called a **selectable marker,** since the expression of the gene selects for the growth of the bacterial cells. Many selectable markers are genes that confer antibiotic resistance to the host cell. For example, the gene amp^R encodes an enzyme known as β-lactamase. This enzyme degrades ampicillin, an antibiotic that normally kills bacteria. Bacteria containing the amp^R gene can grow on media containing ampicillin, because they can degrade it. In a cloning experiment where the amp^R gene is found within the plasmid, the growth of cells in the presence of ampicillin identifies bacteria that contain the plasmid.

An alternative type of vector used in cloning experiments is a **viral vector.** As discussed in chapter 6, viruses can infect living cells and propagate themselves by taking control of the host cell's metabolic machinery. When a chromosomal gene is inserted into a viral genome, the gene will be replicated whenever the viral DNA is replicated. Therefore, viruses can be used as vectors to carry other pieces of DNA. When a virus is used as a vector, the researcher analyzes viral plaques rather than bacterial colonies. The characteristics of viral plaques are described in chapter 6 (see fig. 6.14).

Molecular biologists use hundreds of different vectors in cloning experiments. It is beyond the scope of this textbook to mention more than a few. Table 18.2 provides a general description of several different types of vectors that are commonly used to clone small segments of DNA. In addition, other types of vectors, such as **cosmids, BACs,** and **YACs,** are used to clone large pieces of DNA. These vectors will be described in chapter 20. Also, vectors designed to introduce genes into plants and animals will be discussed in chapter 19.

TABLE 18.2

Some Vectors Used in Cloning Experiments

Example	Type	Description
pBluescript	Plasmid	A type of vector like the one shown in figure 18.2. It is used to clone small segments of DNA and propagate them in *E. coli.*
YEp24	Plasmid	This is an example of a **shuttle vector,** which can replicate in two different species, *E. coli* and *Saccharomyces cerevisiae.* It carries an origin of replication from both species. Using a shuttle vector, a researcher can propagate this plasmid in two unrelated species.
λgt11	Viral	This vector is derived from the bacteriophage λ, which is described in chapter 14. λgt11 also contains a promoter from the *lac* operon. When fragments of DNA are cloned next to this promoter, the DNA is expressed in *E. coli.* This is an example of an **expression vector.** An expression vector is designed to clone the coding sequence of genes so they will be transcribed and translated correctly.
M13mp18	Viral	This virus produces single-stranded DNA as a natural part of its life cycle. For this reason, it is commonly used as a vector to determine the sequence of cloned DNA fragments, as described later in chapter 18.
SV40	Viral	This virus naturally infects mammalian cells. Genetically altered derivatives of the SV40 viral DNA are used as vectors for the cloning and expression of genes in mammalian cells that are grown in the laboratory.
Baculovirus	Viral	This virus naturally infects insect cells. In a laboratory, insect cells can be grown in liquid media. Unlike most other types of cells, insect cells often express large amounts of proteins that are encoded by cloned genes. When researchers want to make a large amount of a protein, they can clone the gene that encodes the protein into baculovirus and then purify the protein from insect cells.

Enzymes Are Used to Cut DNA into Pieces and Join the Pieces Together

A key step in a cloning experiment is the insertion of chromosomal DNA into a plasmid or viral vector. This requires the cutting and pasting of DNA fragments. To cut DNA, researchers use enzymes known as **restriction endonucleases** or **restriction enzymes.** The restriction enzymes used in cloning experiments bind to a specific base sequence and then cleave the DNA backbone at two defined locations, one in each strand. Figure 18.1 shows the action of a restriction enzyme. Discovered by

Werner Arber, Hamilton Smith, and Daniel Nathans in the 1960s and 1970s, they are made naturally by many different species of bacteria. Restriction enzymes protect bacterial cells from invasion by foreign DNA, particularly that of bacteriophages. Researchers have isolated and purified restriction enzymes from many bacterial species and now use them in their cloning experiments.

Currently, several hundred different restriction enzymes from various bacterial species have been identified and are available commercially to molecular biologists. Table 18.3 gives a few examples. Restriction enzymes usually recognize sequences that are **palindromic.** This means that the sequence is identical when

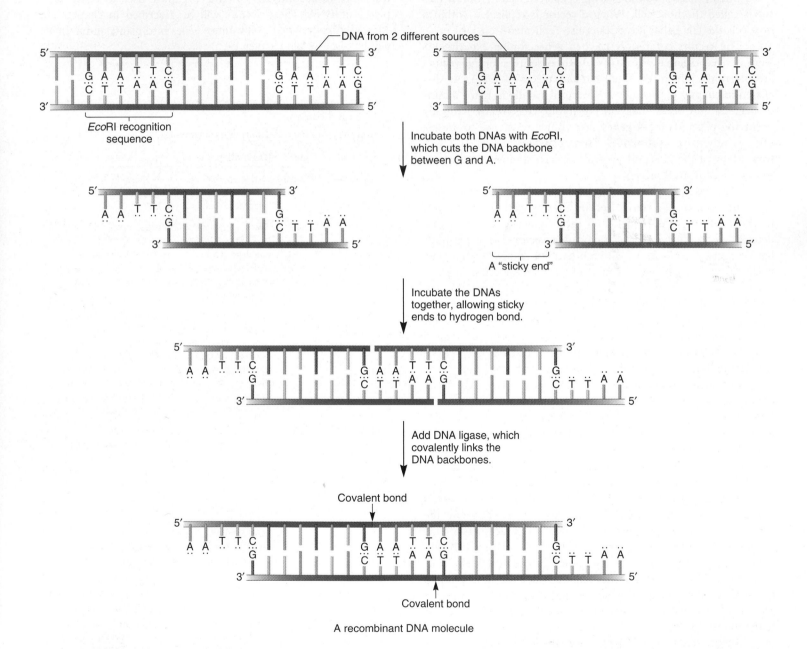

FIGURE 18.1   The action of a restriction enzyme and the production of recombinant DNA. The restriction enzyme *Eco*RI binds to a specific sequence, in this case 5′–GAATTC–3′. It then cleaves the DNA backbone, producing DNA fragments. The single-stranded ends of the DNA fragments can hydrogen bond with each other, because they have complementary sequences. The enzyme DNA ligase can then catalyze the formation of covalent bonds in the DNA backbones of the fragments.

TABLE 18.3
Some Restriction Enzymes Used in Gene Cloning

Restriction Enzyme*	Bacterial Source	Sequence Recognized†
BamHI	Bacillus amyloliquefaciens H	↓ 5′–GGATCC–3′ 3′–CCTAGG–5′ ↑
ClaI	Caryophanon latum	↓ 5′–ATCGAT–3′ 3′–TAGCTA–5′ ↑
EcoRI	E. coli RY13	↓ 5′–GAATTC–3′ 3′–CTTAAG–5′ ↑
NaeI	Nocardia aerocolonigenes	↓ 5′–GCCGGC–3′ 3′–CGGCCG–5′ ↑
PstI	Providencia stuartii	↓ 5′–CTGCAG–3′ 3′–GACGTC–5′ ↑
SacI	Streptomyces achromogenes	↓ 5′–GAGCTC–3′ 3′–CTCGAG–5′ ↑

*Restriction enzymes are named according to the species in which they are found. The first three letters are italicized because they indicate the genus and species names. Because a species may produce more than one restriction enzyme, the enzymes are designated I, II, III, etc., to indicate the order in which they were discovered in a given species.

†The arrows show the locations in the upper and lower DNA strands where the restriction enzymes cleave the DNA backbone. A complete list of restriction enzymes can be found at rebase.neb.com/rebase.

read in the opposite direction in the complementary strand. For example, the sequence recognized by EcoRI is 5′–GAATTC–3′ in the top strand. Read in the opposite direction in the bottom strand, this sequence is also 5′–GAATTC–3′.

Certain types of restriction enzymes are useful in cloning because they digest DNA into fragments with "sticky ends." This means, as shown in figure 18.1, that these DNA fragments will hydrogen bond to each other due to their complementary sequences. As shown here, the complementary sequences promote interactions between two different pieces of DNA. The ends of two different DNA pieces will hydrogen bond to each other because of their compatible sticky ends. However, not all restriction enzymes cut DNA to produce sticky ends. For example, as shown in table 18.3, the enzyme NaeI cuts DNA to produce blunt ends.

The hydrogen bonding between the sticky ends of DNA fragments promotes temporary interactions between two DNA fragments. However, this interaction is not stable, because it involves only a few hydrogen bonds between complementary bases. To establish a permanent connection between two DNA fragments, the sugar-phosphate backbones within the DNA strands must be covalently linked together. This linkage is catalyzed by **DNA ligase.** Figure 18.1 illustrates DNA ligase catalyzing covalent bond formation in the sugar-phosphate backbones of both DNA strands after the sticky ends have hydrogen bonded with each other.

Gene Cloning Involves the Insertion of DNA Fragments into Vectors, Which Are Then Propagated Within Host Cells

Now that we are familiar with the materials, let's outline the general strategy that is followed in a typical cloning experiment. The procedure shown in figure 18.2 seeks to clone a chromosomal gene into a plasmid vector that already carries the amp^R gene. To begin this experiment, the chromosomal DNA is isolated and digested with a restriction enzyme. This enzyme will cut the chromosomes into many small fragments. Many copies of the plasmid DNA are also cut at one site with the same restriction enzyme, so that the vector will have the same sticky ends as the chromosomal DNA fragments. The digested chromosomal DNA and plasmid DNA are mixed together and incubated under conditions that promote the binding of complementary sticky ends.

DNA ligase is then added to catalyze the covalent linkage between DNA fragments. In some cases, the two ends of the vector will simply ligate back together, restoring the vector to its original circular structure. This is called a **recircularized vector.** In other cases, a fragment of chromosomal DNA may become ligated to both ends of the vector; thus, a segment of chromosomal DNA has been inserted into the vector. The vector containing a piece of chromosomal DNA is referred to as a **hybrid vector.**

Following ligation, the DNA is introduced into living cells treated with agents that render them permeable to DNA molecules. Cells that are able to take up DNA from the extracellular medium are called **competent cells.** This step in the procedure is called **transformation,** when a plasmid vector is used, or **transfection,** when a viral vector is introduced into a host cell. In the experiment shown in figure 18.2, a plasmid is introduced into bacterial cells that were originally sensitive to ampicillin. The bacteria are then streaked on plates containing bacterial growth media and ampicillin. If a bacterium has taken up a plasmid carrying the amp^R gene, it will continue to divide and form a bacterial colony containing tens of millions of cells. Because each cell within a single colony is derived from the same original cell, all the cells within a colony contain the same type of plasmid DNA.

In the experiment shown in figure 18.2, the experimenter can distinguish between bacterial colonies that contain a recircularized vector versus those with a hybrid vector carrying a piece of chromosomal DNA. As shown here, the chromosomal DNA has been inserted into a region of the vector that contains the lacZ gene, which encodes the enzyme β-galactosidase. The insertion of chromosomal DNA into the vector disrupts the lacZ gene so that it is no longer able to produce a functional enzyme. By

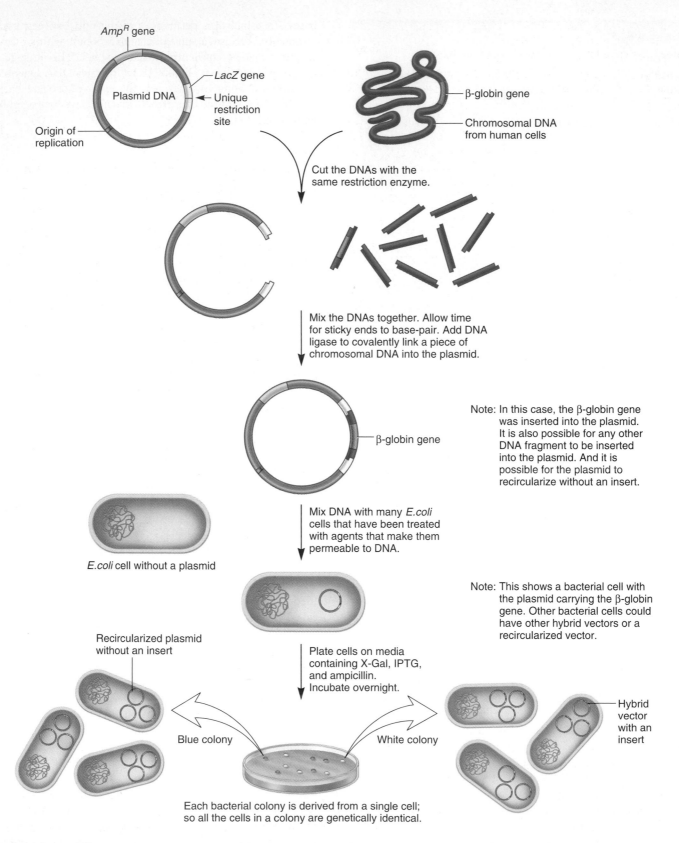

FIGURE 18.2   **The steps in gene cloning.** Note: X-Gal refers to the colorless compound 5-bromo-4-chloro-3-indolyl-β-D-galactoside. IPTG is an acronym for isopropyl-β-D-thiogalactopyranoside, which is a nonmetabolizable lactose analogue that can induce the *lac* promoter.

comparison, a recircularized vector has a functional *lacZ* gene. The functionality of *lacZ* can be determined by providing the growth media with a colorless compound, X-Gal, which is cleaved by β-galactosidase into a blue dye. Bacteria grown in the presence of X-Gal and IPTG (an inducer of the *lacZ* gene) will form blue colonies if they have a functional β-galactosidase and white colonies if they do not. In this experiment, therefore, bacterial colonies containing recircularized vectors will form blue colonies, while colonies containing hybrid vectors will be white.

In the example of figure 18.2, one of the white colonies contains a hybrid vector that carries a human gene; the segment containing the human gene is shown in *red*. The net result of gene cloning is to produce an enormous number of copies of a hybrid vector. During transformation, a single bacterial cell usually takes up a single copy of a hybrid vector. However, two subsequent events lead to the amplification of the cloned gene. First, because the vector has an origin of replication, the bacterial host cell replicates the hybrid vector to produce many copies per cell. Second, the bacterial cells divide approximately every 30 minutes. Following overnight growth, a population of many millions of bacteria will be obtained. Each of these bacterial cells will contain many copies of the cloned gene. For example, a bacterial colony may be comprised of 10 million cells, with each cell containing 50 copies of the hybrid vector. Therefore, this bacterial colony would contain 500 million copies of the cloned gene!

The preceding description has acquainted you with the steps required to clone a gene. A misleading aspect of figure 18.2 is that the digestion of the chromosomal DNA with restriction enzymes seems to yield only a few DNA fragments, one of which contains the gene of interest. In an actual cloning experiment, however, the digestion of the chromosomal DNA with a restriction enzyme produces tens of thousands of different pieces of chromosomal DNA, not just a single piece of chromosomal DNA that happens to be the gene of interest. Later in chapter 18, we will consider how to identify bacterial colonies containing the specific gene that a researcher wants to clone.

Recombinant DNA technology can also be used to clone fragments of DNA that do not code for genes. For example, sequences such as telomeres, centromeres, and highly repetitive sequences have been cloned by this procedure.

EXPERIMENT 18A

In the First Gene Cloning Experiment, Cohen, Chang, Boyer, and Helling Inserted a *kanamycin*[R] Gene into a Plasmid Vector

Now that we are familiar with the basic procedures followed in a cloning experiment, let's consider the first successful attempt at creating a recombinant DNA molecule and propagating that molecule in bacterial cells. This was accomplished in a collaboration between Stanley Cohen, Annie Chang, Herbert Boyer, and Robert Helling in 1973. Several prior important discoveries led to their ability to clone a gene. In 1970, H. Gobind Khorana found that DNA ligase could covalently link DNA fragments together. In 1972, Janet Mertz, Ronald Davis, and Vittorio Sgaramella discovered that the digestion of DNA with the restriction enzyme *Eco*RI produced sticky ends that enabled the DNA fragments to hydrogen bond. With these two observations in mind, Cohen, Chang, Boyer, and Helling realized that it might be possible to create recombinant DNA molecules by digesting DNA with *Eco*RI, allowing the fragments to hydrogen bond with each other, and then covalently linking these fragments together with DNA ligase.

One last condition was necessary for the researchers to succeed in cloning a gene. They needed to identify a vector that could independently replicate itself once inside a host cell. They chose a plasmid vector as their vehicle for gene cloning. In their collection of plasmids, Cohen, Chang, Boyer, and Helling found one small plasmid, designated pSC101, that had a single *Eco*RI site and carried a gene for tetracycline resistance.

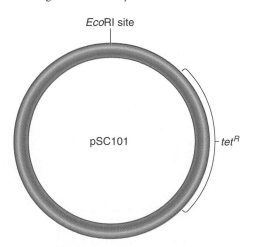

They reasoned that it should be possible to open up this plasmid with *Eco*RI and then insert another piece of DNA into this *Eco*RI site via the hydrogen bonding of sticky *Eco*RI ends.

As a source of a gene to insert into pSC101, they obtained a second plasmid, which they called pSC102.

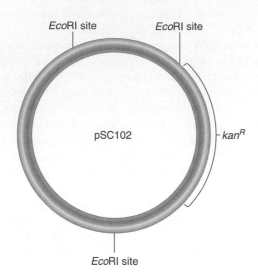

*Eco*RI site   *Eco*RI site

pSC102

⌐ *kan*^R

*Eco*RI site

This plasmid carried a gene, *kan*^R, that provides resistance to the antibiotic kanamycin. However, this second plasmid was cut at three locations by *Eco*RI, yielding three DNA fragments. One of these DNA fragments would be expected to carry the kanamycin-resistance gene (unless, unluckily, *Eco*RI happened to cut in the middle of the *kan*^R gene). Their goal in this experiment was to clone the *kan*^R gene by inserting the DNA fragment carrying this gene into the *Eco*RI site of pSC101 and then introducing the hybrid plasmid into bacterial cells.

As shown in figure 18.3, the researchers isolated plasmid DNA from bacterial strains carrying pSC101 or pSC102 (step 1). The purified plasmids were cut with *Eco*RI and then mixed together to allow the sticky ends to hydrogen bond with each other. Ligase was added to covalently seal the plasmids. It was hoped that a hybrid vector would be obtained in which pSC101 carried the kanamycin-resistance gene from pSC102. Such a plasmid would confer resistance to both tetracycline and kanamycin. To identify such a plasmid, the DNA sample was exposed to bacteria that were treated in such a way as to promote the uptake of DNA. As seen in figure 18.3 (step 7), some colonies were obtained that were resistant to both antibiotics. Four of these colonies were picked, and DNA was isolated from them. This DNA was then subjected to two different procedures. In step 9, some of the DNA was analyzed by density gradient centrifugation. In step 10, some of the DNA was digested with *Eco*RI and subjected to gel electrophoresis. As a control, samples of pSC101 and pSC102 were also subjected to these two procedures.

■ THE HYPOTHESIS

A piece of DNA carrying a gene can be inserted into a plasmid vector using recombinant DNA techniques. If this recombinant plasmid is introduced into a bacterial host cell, it will be replicated and transmitted to daughter cells, producing many copies of the recombinant plasmid.

■ TESTING THE HYPOTHESIS — FIGURE 18.3 The first success at gene cloning.

Starting material: Three different strains of *E. coli:* one strain that did not carry any plasmid and two strains that carried pSC101 or pSC102.

Experimental level   **Conceptual level**

1. Isolate and purify the two types of plasmid DNA:

Grow the bacterial cells containing the plasmids.

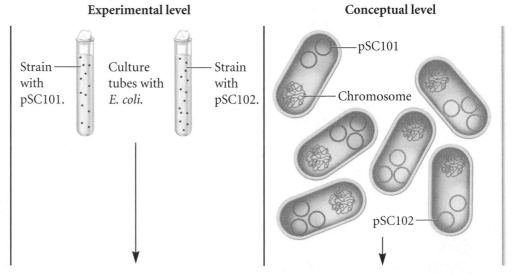

Strain with pSC101.   Culture tubes with *E. coli.*   Strain with pSC102.

pSC101

Chromosome

pSC102

(continued)

Break open the cells. One way to break open cells is to subject them to harsh sound waves, or sonication.

Isolate each type of plasmid DNA by density gradient centrifugation (see the appendix for a description of this procedure).

2. Digest the plasmid DNAs with *Eco*RI.

3. Mix together the two samples.

4. Add DNA ligase.

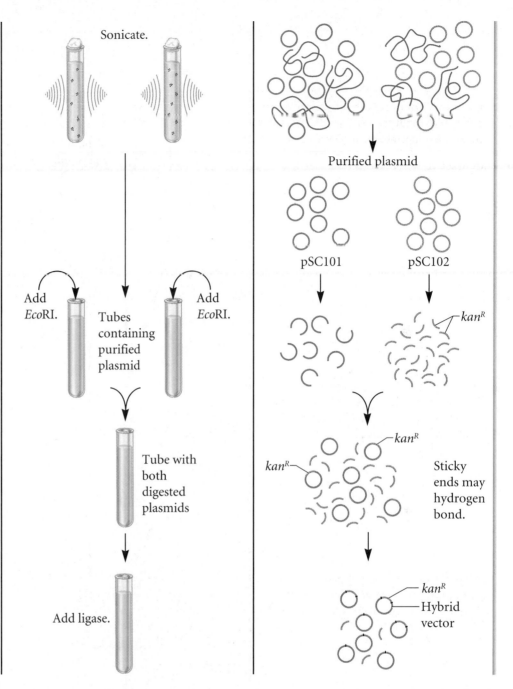

(*continued*)

5. Grow an *E. coli* strain that does not carry a plasmid. Treat the cells with $CaCl_2$ to make them permeable to DNA.

6. Add the ligated DNA samples to the bacterial cells. Most cells do not take up any plasmid but an occasional cell will take up one plasmid.

7. Plate the cells on growth media containing both tetracycline and kanamycin. Grow overnight to allow growth of visible bacterial colonies.

8. Pick four colonies from the plates. The plasmid found in these colonies is designated pSC105. Grow the colonies in liquid culture containing radiolabeled deoxyribonucleotides.

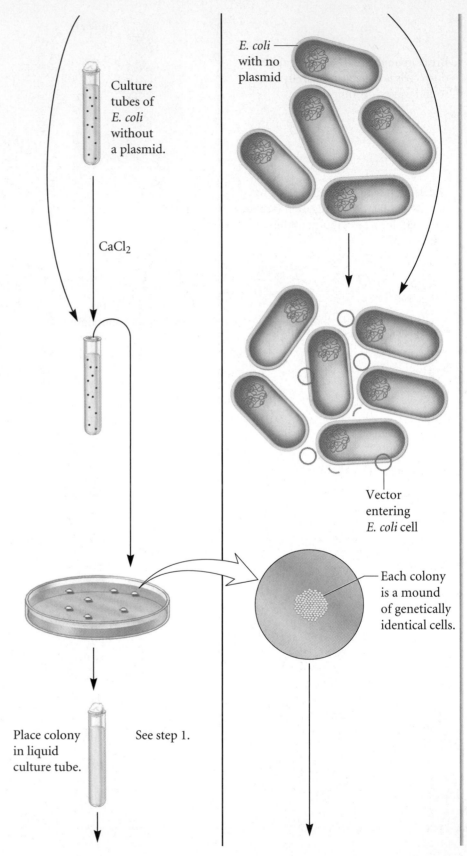

Culture tubes of *E. coli* without a plasmid.

$CaCl_2$

E. coli with no plasmid

Vector entering *E. coli* cell

Each colony is a mound of genetically identical cells.

Place colony in liquid culture tube.

See step 1.

(continued)

9. To isolate the radiolabeled plasmid DNA, break open the cells and subject the DNA to cesium chloride density gradient centrifugation. As described in the appendix, collect fractions and count them in a scintillation counter. (See data shown on the top *left.*)

As a control, subject pSC101 and pSC102 to the same procedure. (See data shown on the bottom *left.*)

10. Also, as described under Interpreting the Data, digest the plasmid DNAs with *Eco*RI, and subject them to gel electrophoresis. (See data shown on the *right.*)

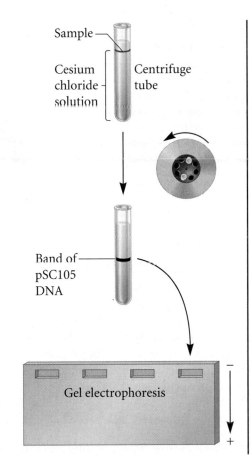

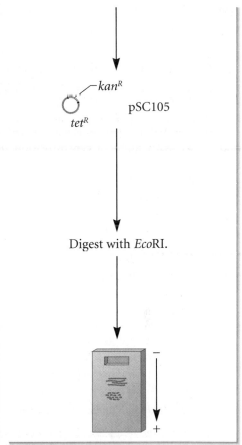

THE DATA

Results from step 9:

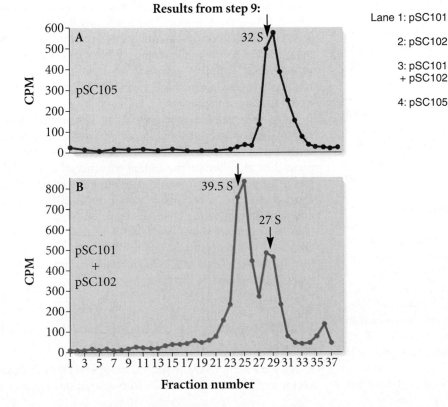

Fraction number

Results from step 10:

Lane 1: pSC101

2: pSC102

3: pSC101 + pSC102

4: pSC105

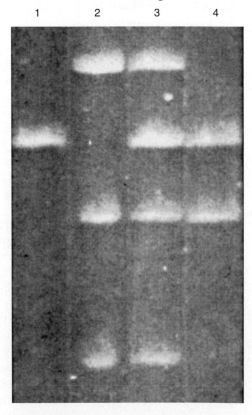

INTERPRETING THE DATA

With regard to their data, let's first consider the results of density gradient centrifugation. As shown in the control experiment, the pSC102 plasmid is larger than pSC101 and sediments at a density of 39.5 S, whereas the latter plasmid has a density of 27 S. (Note: The letter S refers to Svedberg units, a unit of centrifugation velocity. See the appendix.) In this control experiment, the pSC101 and pSC102 plasmids were mixed together and then subjected to density gradient centrifugation. This yielded two peaks at 27 S and 39.5 S. The *top* figure shows the results for pSC105 obtained from a bacterial colony that was resistant to both tetracycline and kanamycin. In this case, there is a plasmid with an intermediate density of 32 S rather than a mixture of two plasmids. These results indicate that this bacterial colony contained a recombinant plasmid, not a mixture of pSC101 and pSC102.

This conclusion is confirmed in the gel electrophoresis analysis. When digested with *Eco*RI, pSC101 yielded a single band, whereas pSC102 yielded three bands (see lanes 1 and 2). In lane 3, the experimenters had mixed together samples of pSC101 and pSC102 (isolated in step 1) and digested them with *Eco*RI. As expected, this produced four bands. Lane 4 shows the plasmid DNA, pSC105, from a bacterial colony that was tetracycline and kanamycin resistant. This plasmid showed the pSC101 band plus one other band that was found in pSC102. The results of lane 4 are consistent with the idea that the pSC105 plasmid is formed by

the insertion of one fragment from pSC102 into the single *Eco*RI site of pSC101. This recombinant plasmid is shown here.

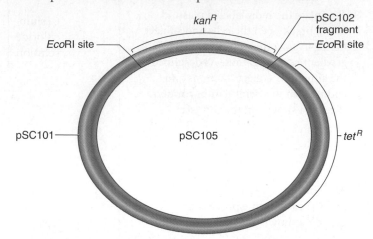

This experiment showed the scientific community that it is possible to create recombinant DNA molecules and then propagate them in bacterial cells. In other words, it is possible to clone genes. This hallmark achievement ushered in the era of gene cloning.

A self-help quiz involving this experiment can be found at the Online Learning Center.

cDNA Can Be Made from mRNA via Reverse Transcriptase

In our preceding discussion of gene cloning described in figure 18.2, chromosomal DNA and plasmid DNA were used as the material to clone genes. Alternatively, a sample of RNA can provide a starting point to clone DNA. As mentioned in chapter 17, the enzyme **reverse transcriptase** can use RNA as a template to make a complementary strand of DNA. This enzyme is encoded in the genome of retroviruses. It provides a way for retroviruses to convert their RNA genome into DNA molecules that can integrate into the host cell's chromosomes. Likewise, reverse transcriptase is encoded in viral-like retroelements and is needed in the retrotransposition of such elements.

Researchers can use purified reverse transcriptase in a strategy to clone genes, using mRNA as the starting material. This strategy is shown in figure 18.4. RNA is purified from a sample of cells. The RNA is mixed with primers composed of a string of thymine-containing nucleotides. Such an oligonucleotide is called a poly-dT primer. Since eukaryotic mRNAs contain a polyA tail, poly-dT primers will be complementary to the 3′ end of mRNAs. Deoxyribonucleotides and reverse transcriptase are then added to make a DNA strand that is complementary to the mRNA. Next, RNaseH, DNA polymerase, and DNA ligase are added. RNaseH partially digests the RNA, generating short RNAs that are used as primers by DNA polymerase to make a second DNA strand that

is complementary to the DNA strand made by reverse transcriptase. Finally, DNA ligase will seal any nicks in this second DNA strand. When DNA is made from RNA as the starting material, the DNA is called **complementary DNA (cDNA).** The term originally referred to the single strand of DNA that is complementary to the RNA template. However, it now refers to any DNA, whether it is single or double stranded, that is made using RNA as the starting material.

From a research perspective, an important advantage of cDNA is that it lacks introns. Since introns can be quite large, it is much simpler to insert cDNAs into vectors if researchers want to focus their attention on the coding sequence of a gene. For example, if the primary goal was to determine the coding sequence of a structural gene, a researcher would insert cDNA into a vector and then determine the DNA sequence of the insert as described later in chapter 18. Similarly, if a scientist wanted to express an encoded protein of interest in a cell that would not splice out the introns properly (e.g., in a bacterial cell), it would be necessary to make cDNA clones of the respective gene.

Restriction Mapping Is Used to Locate the Restriction Sites Within a Vector

As we have seen, DNA or gene cloning involves the digestion of vector and chromosomal DNA with restriction enzymes and the

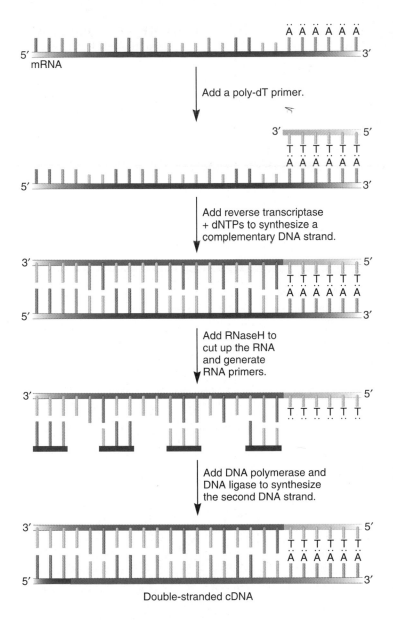

FIGURE 18.4 **Synthesis of cDNA.** A poly-dT primer anneals to the 3' end of mRNAs. Reverse transcriptase then catalyzes the synthesis of a complementary DNA strand (cDNA). RNaseH digests the mRNA into short pieces that are used as primers by DNA polymerase to synthesize the second DNA strand. The 5' to 3' exonuclease function removes all of the RNA primers except the one at the 5' end (because there is no primer upstream from this site). This RNA primer can be removed by the subsequent addition of an RNase. After the double-stranded cDNA is made, it can then be inserted into vectors as described in figure 18.2. Prior to vector insertion, short pieces of DNA, called linkers, are covalently attached to both ends of the cDNA. The DNA sequence of the linkers contains unique restriction sites, making it easy to cut the cDNA at both ends and insert it into a vector.

subsequent ligation of DNA fragments into vectors. In this type of procedure, the locations of restriction enzyme sites are crucial. In the vector, for example, it is desirable to have a unique restriction site for the insertion of chromosomal DNA. After successfully cloning of a large chromosomal DNA insert, a researcher may wish to use the hybrid vector to obtain clones having smaller pieces of the chromosomal DNA. The process of making smaller clones from a larger one is called **subcloning.** To make a subclone, the hybrid vector DNA must be cut with one or more restriction enzymes to produce smaller fragments of the insert DNA, which are then inserted into a new vector. Overall, cloning and subcloning methods require knowledge of the locations of restriction enzyme sites in vectors and hybrid vectors.

A common approach to examine the locations of restriction sites is known as **restriction mapping.** Figure 18.5 outlines the restriction mapping of a bacterial plasmid. To begin this experiment, the small circular plasmid DNA is isolated and purified from host cells. Samples of the purified DNA are then placed in separate test tubes that contain a particular restriction enzyme or combination of enzymes. The plasmid DNA is incubated with the restriction enzymes long enough for digestion to occur. The DNA fragments are then separated by gel electrophoresis. In lane 1 of this experiment, a different DNA sample, known as a set of molecular markers, is also subjected to gel electrophoresis. This sample contains a mixture of DNA fragments with molecular masses that are known from previous experiments. (These markers are purchased from commercial sources or can be prepared in the laboratory.) To determine the sizes of the fragments obtained by digesting the plasmid, the fragments in lanes 2–8 are compared to the known markers in lane 1.

The restriction map shown in figure 18.5 was deduced by comparing the sizes of fragments obtained from the single, double, and triple digestions. The starting plasmid is a circular molecule 4,363 bp in length. A single digestion with any of the three enzymes yields a single linear fragment of size 4,363 bp. This means that *Eco*RI, *Bam*HI, and *Pst*I cut the plasmid at a single site.

The double digestion with *Eco*RI and *Bam*HI yields two fragments, of about 380 bp and 3,980 bp. This result indicates that the *Eco*RI and *Bam*HI sites are approximately 380 bp apart in one direction along the circle and 3,980 bp apart along the circle in the opposite direction. Likewise, the pairwise combinations of *Eco*RI/*Pst*I and *Bam*HI/*Pst*I indicate how far apart these sites are along the circular plasmid.

Finally, the triple digestion confirms the locations of the single sites for *Eco*RI, *Bam*HI, and *Pst*I. Taken together, these results provide a map of the restriction sites within this plasmid. A similar approach can be used on a hybrid vector to determine the locations of restriction sites within a fragment of DNA that has been inserted into a plasmid.

Alternatively, another way to obtain a restriction map is via DNA sequencing, a technique described at the end of chapter 18. If

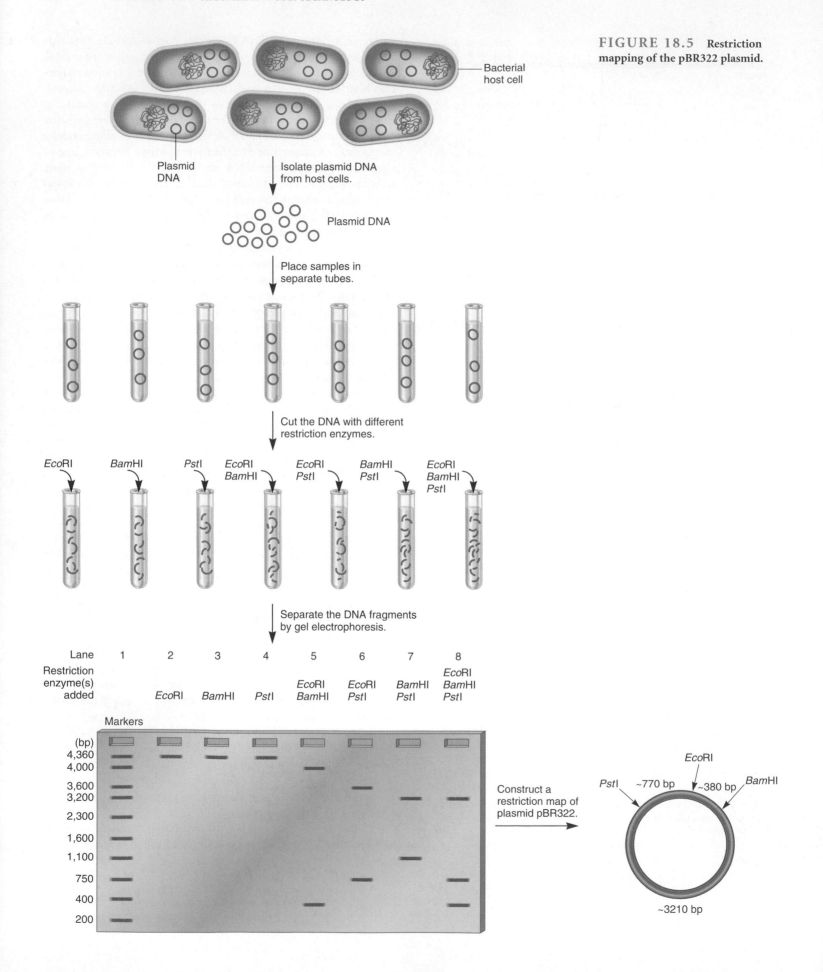

FIGURE 18.5   Restriction mapping of the pBR322 plasmid.

the DNA sequence of a plasmid and insert have been determined, computer programs can scan the sequence and identify sites that are recognized by particular restriction enzymes. For example, *Eco*RI recognizes the sequence 5′–GAATTC–3′. If the sequence of a hybrid vector contained this sequence, a computer program could identify this sequence as an *Eco*RI site. Such a program would scan the DNA sequence for the recognition sequences of many different restriction enzymes and generate a detailed restriction map of a hybrid vector.

A second use of restriction enzymes is gene mapping. Restriction digestion of chromosomal DNA yields a mixture of fragments that vary in length. Furthermore, due to variations in DNA sequence, different individuals exhibit variation in the sizes of particular DNA fragments that are generated by restriction digestion. This phenomenon, known as **restriction fragment length polymorphism (RFLP),** enables researchers to map particular genes within a species' genome. This approach is discussed in chapter 20.

Polymerase Chain Reaction (PCR) Can Also Be Used to Make Many Copies of DNA

In our previous discussions of gene cloning, the DNA of interest was inserted into a vector, which then was introduced into a host cell. The replication of the vector within the host cell, and the proliferation of the host cells, led to the production of many copies of the DNA. Another way to copy DNA, without the aid of vectors and host cells, is a technique called **polymerase chain reaction (PCR),** which was developed by Kary Mullis in 1985.

The PCR method is outlined in figure 18.6. In this example, the starting material contains a sample of DNA, known as the **template DNA.** The goal of PCR is to make many copies of the template DNA in a defined region. Several reagents are added to facilitate the synthesis of DNA. These include a high concentration of two oligonucleotide primers (which are complementary to sequences at the ends of the DNA fragment to be amplified), deoxyribonucleoside triphosphates (dNTPs), and a thermostable form of DNA polymerase called *Taq* **polymerase,** isolated from the bacterium *Thermus aquaticus*. A thermostable form of DNA polymerase is necessary, because PCR involves heating steps that would inactivate most other natural forms of DNA polymerase (which are thermolabile).

To make copies of the DNA, this double-stranded template is denatured by heat treatment and then the oligonucleotide primers bind to the template DNA as the temperature is lowered. The binding of the primers to the DNA is called **annealing.** Once the primers have annealed, *Taq* polymerase catalyzes the synthesis of complementary DNA strands. This doubles the amount of the template DNA. The sequential process of denaturation–annealing–synthesis is then repeated many times to double the amount of template DNA many times over. This method is called a chain reaction, because the products of each previous reaction (i.e.,

newly made DNA strands) are used as reactants (i.e., as template strands) in subsequent reactions.

PCR is carried out in a thermal reactor, known as a **thermocycler,** that automates the timing of each cycle. The experimenter mixes the DNA sample, an excess amount of primers, *Taq* polymerase, and dNTPs together in a single tube. The tube is placed in a thermocycler, and the experimenter sets the machine to operate within a defined temperature range and number of cycles. During each cycle, the thermocycler increases the temperature to denature the DNA strands and then lowers the temperature to allow annealing and DNA synthesis to take place. Typically, each cycle lasts 2 to 3 minutes and is then repeated. A typical PCR run is likely to involve 20 to 30 cycles of replication and takes a few hours to complete. The PCR technique can amplify the amount of DNA by a staggering amount. After 30 cycles of amplification, a DNA sample will increase 2^{30}-fold, which is approximately a billion-fold.

The PCR reaction shown in figure 18.6 seeks to amplify a particular DNA segment. As shown here, the sequences of the PCR primers are complementary to two specific sequences within the template DNA. Therefore, the two primers bind to these sites and the intervening region is replicated. To conduct this type of PCR experiment, the researcher must have prior knowledge about the sequence of the template DNA in order to design oligonucleotide primers complementary to two sites in the template sequence. When specific primers can be constructed, PCR can amplify a specific region of DNA from a very complex mixture of template DNA. For example, if a researcher uses two primers that anneal to the human β-globin gene, PCR can specifically amplify the β-globin gene from a DNA sample that contains all of the human chromosomes!

Alternatively, PCR can be used to amplify a sample of chromosomal DNA semispecifically or nonspecifically. As discussed in chapter 19, this approach is used in DNA fingerprinting analysis. In a semispecific PCR experiment, the primers recognize a repetitive DNA sequence found at several sites within the genome. When chromosomal DNA is used as a template, this will amplify many different DNA fragments. In a nonspecific approach, a mixture of short PCR primers with many different random sequences is used. These primers will anneal randomly throughout the genome and amplify most of the chromosomal DNA. Nonspecific DNA amplification is used to increase the total amount of DNA in very small samples, such as blood stains found at crime scenes.

PCR is also used to detect and quantitate the amount of specific RNAs in living cells. To accomplish this goal, RNA is isolated from a sample and mixed with reverse transcriptase and a primer that will bind to the 3′ end of the RNA of interest. This generates a single-stranded cDNA, which then can be used as template DNA in a conventional PCR reaction. This method, called **reverse transcriptase PCR (RT-PCR),** is extraordinarily sensitive. It can detect the expression of small amounts of RNA in a single cell.

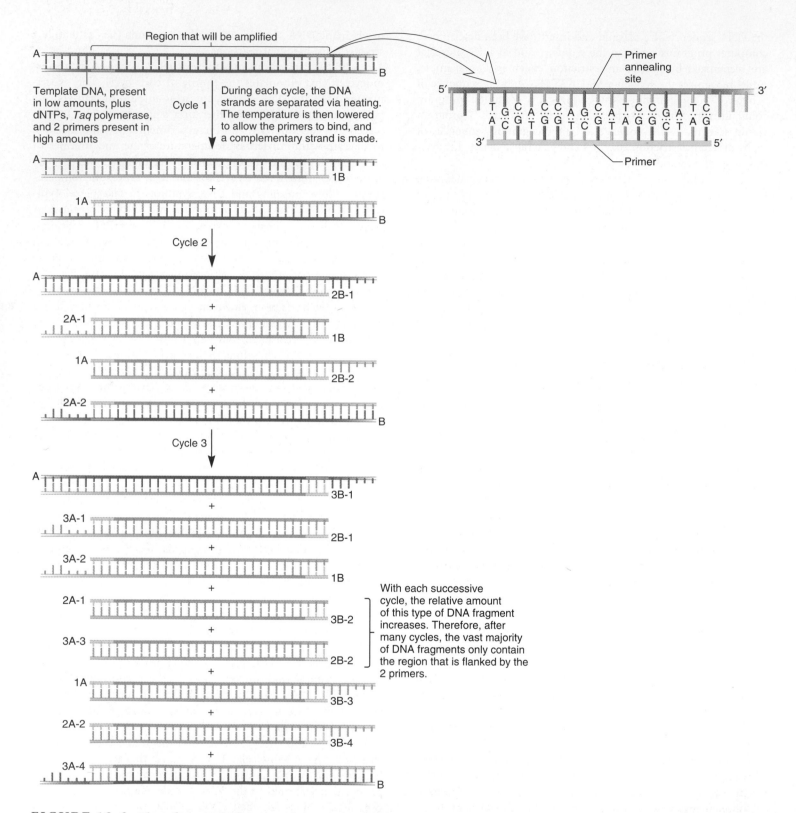

FIGURE 18.6   **The technique of polymerase chain reaction (PCR).** During each cycle, oligonucleotides that are complementary to the ends of the targeted DNA sequence bind to the DNA and act as primers for the synthesis of this DNA region. The primers used in actual PCR experiments are usually 15 to 20 nucleotides in length. The region between the two primers is typically hundreds of nucleotides in length, not just several nucleotides as shown here. The net result of PCR is the synthesis of many copies of DNA in the region that is flanked by the two primers.

18.2 DETECTION OF GENES AND GENE PRODUCTS

The advent of gene cloning has enabled scientists to investigate gene structure and function at the molecular level. However, unlike the first gene cloning experiment in which a single gene was removed from one plasmid (pSC102) and cloned into another plasmid (pSC101), molecular geneticists usually want to study genes that are originally within the chromosomes of living species. This presents a problem, because chromosomal DNA contains thousands of different genes. For this reason, researchers need methods to specifically identify a gene within a mixture of many other genes or DNA fragments. The term **gene detection** refers to methods that distinguish one particular gene among a mixture of thousands of other genes. We will see that the molecular identification of genes has greatly advanced our understanding of how genes function. Also, we will learn how gene detection methods are used to clone genes.

When studying gene expression at the molecular level, scientists also need techniques for the identification of gene products, such as the RNA that is transcribed from a particular gene or the protein that is encoded by an mRNA. In this section, we will consider the methodology and uses of a few common detection strategies.

A DNA Library Is a Collection of Many Cloned Fragments of DNA

In a typical cloning experiment (refer back to fig. 18.2), the treatment of the chromosomal DNA with restriction enzymes yields tens of thousands of different DNA fragments. Therefore, after the DNA fragments are ligated individually to vectors, the researcher has a collection of hybrid vectors, with each vector containing a particular fragment of chromosomal DNA. This collection of hybrid vectors is known as a **DNA library** (see fig. 18.7). When the starting material is chromosomal DNA, the library is called a **genomic library.** Similarly, it is also common for researchers to make a **cDNA library** that contains hybrid vectors with cDNA inserts. A cDNA library would be made if a researcher wanted to express the encoded protein of interest in a cell that would not splice out the introns properly. As mentioned earlier, because cDNA is produced from mature mRNA via reverse transcriptase, it lacks any introns.

In most cloning experiments, the ultimate goal is to clone a specific gene. For example, suppose that a geneticist wishes to clone the rat β-globin gene. To begin a cloning experiment, rat chromosomal DNA would be isolated from rat cells. This chromosomal DNA would be digested with a restriction enzyme, yielding thousands of DNA fragments. The chromosomal fragments would then be ligated to vector DNA and transformed into bacterial cells. Unfortunately, only a small percentage, perhaps one in a few thousand, of the hybrid vectors would actually contain the rat β-globin gene. For this reason, researchers must have some way to identify those rare bacterial colonies that happen to contain the cloned gene of interest, in this case, a colony that contains the rat β-globin gene.

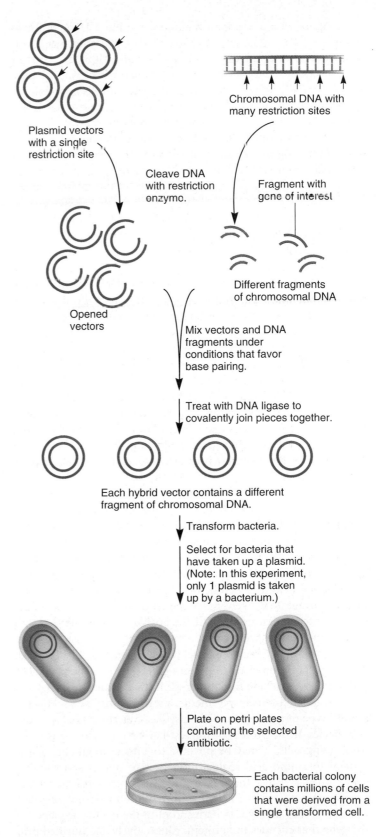

FIGURE 18.7   The construction of a DNA library. The digestion of chromosomal DNA produces many fragments. The fragment containing the gene of interest is highlighted in *red*. Following ligation, each vector contains a different piece of chromosomal DNA.

Figure 18.8 describes a method for using a DNA probe to identify a bacterial colony that has the rat β-globin gene. This procedure is referred to as **colony hybridization.** The master plate shown at the top of figure 18.8 has many bacterial colonies. Each bacterial colony is composed of bacterial cells containing a hybrid vector with a different piece of rat chromosomal DNA. The goal is to identify a colony that contains the rat β-globin gene. To do so, a nitrocellulose filter is laid gently onto the master plate containing many bacterial colonies. After the filter is lifted, some cells from each colony are attached to it. In this way, the nitrocellulose filter paper contains a replica of the colonies on the master plate.

The cells on the filter are then permeabilized and the DNA within the cells is denatured and fixed (i.e., adhered) to the nitrocellulose filter. The filter is then submerged in a solution containing a radiolabeled DNA probe. In this case, the probe is a DNA strand that is complementary to one of the DNA strands of the rat β-globin gene. The probe is given time to hybridize to the DNA on the filter, which has already been denatured into single strands. If a bacterial colony contains the rat β-globin gene, the probe will hybridize to the DNA in this colony. Most bacterial colonies are not expected to contain this gene. The unbound probe is then washed away, and the filter is placed next to X-ray film. If the DNA within a bacterial colony did hybridize to the probe, a dark spot will appear on the film in the corresponding location. Therefore, a dark spot identifies a bacterial colony that contains the rat β-globin gene. Following the identification of "hot" colonies, the experimenter can go back to the master plate (which contains living cells) to identify the colonies, pick them from the plate, and grow bacteria containing the cloned gene.

There are several different strategies a scientist might use to obtain a probe for a gene cloning experiment. In some cases, the gene of interest may have already been cloned. A piece of the cloned gene then can be used as a radiolabeled probe. However, in many circumstances, a scientist may attempt to clone a gene that has never been cloned before. In such a case, one strategy is to use a probe that likely has a sequence similar to that of the gene of interest. For example, if the goal is to clone the β-globin gene from a South American rodent, it is expected that this gene is similar to the rat β-globin gene, which has already been cloned. Therefore, the cloned rat β-globin gene can be used as a DNA probe to "fish out" the β-globin gene from a similar species.

In other cloning experiments, a scientist may be looking for a novel type of gene that no one else has ever cloned before from any species. If the protein of interest has been previously isolated from living cells, it may be possible to subject fragments of the protein to amino acid sequence analysis to obtain short amino acid sequences. Since the genetic code is known, a researcher can use this amino acid sequence information to design oligonucleotide probes that would correspond to the coding sequence of the gene that encodes the protein. Alternatively, the purified protein could be used to generate antibodies. In that case, the antibody would recognize clones that express the protein. Like DNA probes, labeled antibodies can be used as probes in a colony hybridization experiment. The use of antibodies as labeling tools will be discussed later in chapter 18.

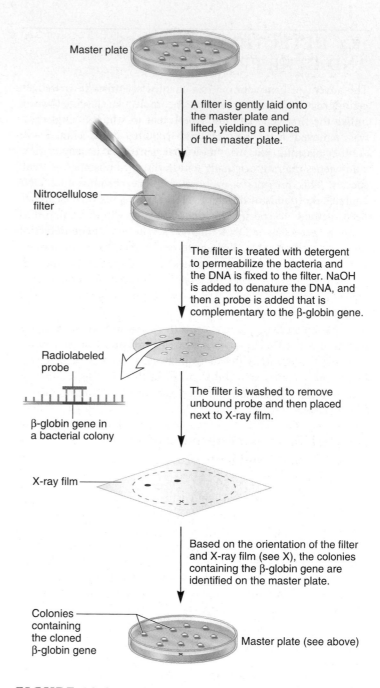

FIGURE 18.8   A colony hybridization experiment. A replica of the master plate is blotted onto a nitrocellulose filter. The filter is treated with detergent (SDS) to permeabilize the bacteria and the DNA is fixed to the filter. The filter is then treated with sodium hydroxide (NaOH) to separate the DNA into single strands. Radioactively labeled probes are added. Probes are single-stranded DNA with base sequences complementary to that of the gene of interest. The probe will hybridize with the desired gene from the bacterial cell DNA. The filter is washed to remove unbound probe and then exposed to X-ray film. Finally, the developed film is compared to the replica of the master plate to identify bacterial colonies carrying the gene of interest.

Southern Blotting Is Used to Detect DNA Sequences

Earlier in chapter 18, we learned that PCR can be used to amplify a specific gene from a sample of chromosomal DNA. This is a commonly used method to detect a specific gene. Likewise, the technique of **Southern blotting** can also detect the presence of a particular gene sequence within a mixture of many chromosomal DNA fragments. This method, developed by E. M. Southern in 1975, has several uses. It can determine the copy number of a gene within the genome of an organism. For example, Southern blotting has revealed that rRNA genes are found in multiple copies within a genome whereas many other structural genes are unique. In the study of human genetic diseases, Southern blotting can also detect small gene deletions that cannot be distinguished under the light microscope (see solved problem S4 at the end of chapter 18). A common use of Southern blotting is to identify gene families. As discussed in chapter 8, a gene family is a group of two or more genes derived from the same ancestral gene. The members of a gene family have similar but not identical DNA sequences; they are homologous genes. As we will see, Southern blotting can distinguish the homologous members of a gene family. Similarly, Southern blotting can identify homologous genes among different species. This is called a zoo blot.

Prior to a Southern blotting experiment, a gene of interest, or a fragment of a gene, has been cloned. This cloned DNA is labeled (e.g., radiolabeled) in vitro, and then the labeled DNA is used as a probe to detect the presence of the gene or a homologous gene within a mixture of many DNA fragments. The basis for a Southern blotting experiment is that two DNA fragments will bind to each other only if they have complementary sequences. In such an experiment, the labeled strands from the cloned gene will pair specifically with complementary DNA strands, even if the complementary strands are found within a mixture of many other DNA pieces.

Figure 18.9a shows the Southern blotting procedure. The goal of this experiment is to determine if the chromosomal DNA contains a base sequence complementary to a specific probe. To begin the experiment, the chromosomal DNA is isolated and digested with a restriction enzyme. Since the restriction enzyme cuts the chromosomal DNA at many different sites within the chromosomes, this step produces thousands of DNA pieces of different sizes. The chromosomal pieces are loaded onto a gel that separates them according to their size. The DNA pieces within the gel are denatured by soaking the gel in a NaOH solution and then the DNA is transferred (i.e., blotted) to a nitrocellulose (or nylon) filter. One way to accomplish this is via a transfer apparatus in which the gel is placed next to a nitrocellulose filter and then an electric field causes the DNA fragments to electrophoretically move from the gel into the filter (fig. 18.9b). An alternative type of transfer apparatus uses a vacuum rather than an electric field to pull the DNA fragments into the nitrocellulose filter. The DNA fragments are then tightly fixed to that filter. At this point, the filter contains many unlabeled DNA fragments that have been denatured and separated according to size.

The next step is to determine if any of these unlabeled fragments from the chromosomal DNA contain sequences comple-

mentary to the probe, which must be labeled. A common labeling method is to incorporate the radioisotope ^{32}P into the probe DNA, which labels the phosphate group in the DNA backbone. The filter, which has the unlabeled chromosomal DNA attached to it, is submerged in a solution containing the radiolabeled probe. If the radiolabeled probe and a fragment of chromosomal DNA are complementary, they will hydrogen bond to each other. Any unbound radiolabeled DNA is then washed away, and the filter is exposed to X-ray film. Locations where radiolabeled probe has bound will appear as dark bands on the X-ray film.

Important variables in the Southern blot procedure are the temperature and ionic strength of the hybridization and wash steps. If these steps are done at very high temperatures or at low salt concentrations, then the probe DNA and chromosomal fragment must be very complementary—nearly a perfect match—to hybridize. This condition is called **high stringency.** Conditions of high stringency are used to detect close or exact complementarity between the probe and a chromosomal DNA fragment. However, if the temperature is lower and/or the ionic strength is higher, DNA sequences that are not perfectly complementary may hybridize to the probe. This is called **low stringency.** Conditions of low stringency are used to detect homologous genes with DNA sequences that are similar but not identical to the cloned gene being used as a probe. In the results shown in figure 18.9a, conditions of high stringency reveal that the gene of interest is found only in a single copy in the genome. At low stringency, however, two other bands are detected. These results suggest that this gene is a member of a **gene family** composed of three distinct members.

Northern Blotting Is Used to Detect RNA

Let's now turn our attention to the technique known as **Northern blotting,** which is used to identify a specific RNA within a mixture of many RNA molecules. (Note: Even though Southern blotting was named after E. M. Southern, Northern blotting was not named after anyone called Northern! It was originally termed reverse-Southern blotting and later Northern blotting.) Northern blotting is used to investigate the transcription of genes at the molecular level. This method can determine if a specific gene is transcribed in a particular cell type (e.g., nerve versus muscle cells) or at a particular stage of development (e.g., fetal versus adult cells). Also, Northern blotting can reveal if a pre-mRNA transcript is alternatively spliced into two or more mRNAs of different sizes.

From a technical viewpoint, Northern blotting is rather similar to Southern blotting with a few important differences. As shown in figure 18.10, RNA (rather than chromosomal DNA) is extracted and purified from living cells. This RNA can be isolated from a particular cell type under a given set of conditions or during a particular stage of development. Any given cell will produce thousands of different types of RNA molecules, because cells express many genes at any given time. After the RNA is extracted from cells and purified, it is loaded onto a gel that separates the RNA transcripts according to their size. The RNAs within the gel are then blotted onto a nitrocellulose (or nylon) filter and probed with a radiolabeled fragment of DNA from a cloned gene. Using

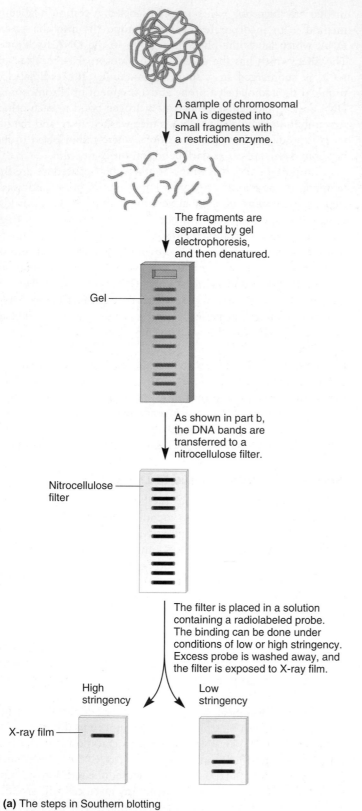

A sample of chromosomal DNA is digested into small fragments with a restriction enzyme.

The fragments are separated by gel electrophoresis, and then denatured.

Gel ——

As shown in part b, the DNA bands are transferred to a nitrocellulose filter.

Nitrocellulose filter ——

The filter is placed in a solution containing a radiolabeled probe. The binding can be done under conditions of low or high stringency. Excess probe is washed away, and the filter is exposed to X-ray film.

High stringency   Low stringency

X-ray film ——

(a) The steps in Southern blotting

FIGURE 18.9   **The technique of Southern blotting. (a)** Chromosomal DNA is isolated from a sample of cells and then digested with a restriction enzyme to yield many DNA fragments. The fragments are then separated by gel electrophoresis according to molecular mass. The bands within the gel are transferred to a nitrocellulose or nylon filter, a step known as blotting. The blot is then placed in a solution containing a radiolabeled probe, which is the product of gene cloning. When the probe DNA is hybridized to the mixture of chromosomal DNA fragments at high temperature and/or low ionic strength (i.e., high stringency), the probe recognizes only a single band. When hybridization is performed at lower temperature and/or higher ionic strength (i.e., low stringency), this band and two additional bands hybridize with the radiolabeled probe. **(b)** A transfer apparatus. The gel is placed next to a nitrocellulose filter, and the gel/filter are sandwiched between thick blotting paper, and between anode and cathode plates. When an electric field is applied, the DNA fragments move toward the anode plate to the nitrocellulose filter.

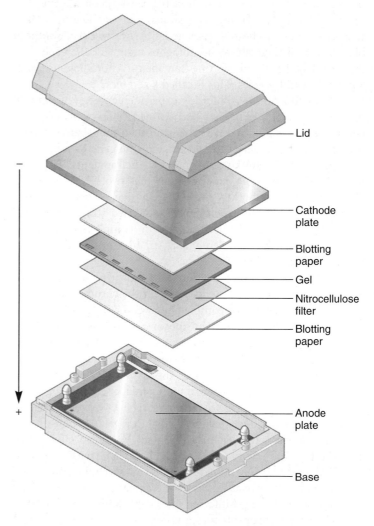

Lid

Cathode plate

Blotting paper

Gel

Nitrocellulose filter

Blotting paper

Anode plate

Base

(b) The transfer step in a Southern blotting experiment

Lane 1: Smooth muscle

Lane 2: Striated muscle

Lane 3: Brain cells

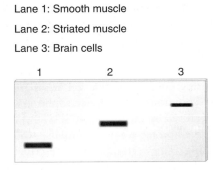

FIGURE 18.10 **The results of Northern blotting.** RNA is isolated from a sample of cells and then separated by gel electrophoresis. The separated RNA bands are blotted to a nitrocellulose or nylon filter and then placed in a solution containing a radioactive probe. Following autoradiography, RNA molecules that are complementary to the probe appear as dark bands on the X-ray film.

this method, RNAs that are complementary to the radiolabeled DNA fragment are detected as dark bands on an X-ray film. In the example of figure 18.10, the probe was complementary to an mRNA that encodes a protein called tropomyosin. In lane 1, the RNA was isolated from smooth muscle cells, in lane 2 from striated muscle cells, and in lane 3 from brain cells. As seen here, smooth and striated muscle cells produce a large amount of this mRNA. This is expected because tropomyosin plays a role in the regulation of cell contraction. By comparison, brain cells have much less of this mRNA. In addition, we see that the molecular masses of the three mRNA are slightly different among the three cell types. This indicates that the pre-mRNA is alternatively spliced to contain different combinations of exons.

As an alternative to Northern blotting, another way to detect a specific RNA is via RT-PCR. This method was discussed earlier in chapter 18.

Western Blotting Is Used to Detect Proteins

For structural genes, the net result of gene expression is the synthesis of proteins. A particular protein within a mixture of many different protein molecules can be identified by a third detection procedure, **Western blotting.** Similar to Northern blotting, Western blotting can determine if a specific protein is made in a particular cell type or at a particular stage of development. Technically, this procedure is also rather similar to Southern and Northern blotting.

In a Western blotting experiment, proteins are extracted from living cells (fig. 18.11). As with RNA, any given cell will produce many different proteins at any time, because it is expressing many structural genes. After the proteins have been extracted from the cells, they are loaded onto a gel that separates them by molecular mass. To perform the separation step, the proteins are first dissolved in sodium dodecyl sulfate (SDS), a detergent that denatures proteins and coats them with negative charges. The negatively charged proteins are then separated in a gel made of polyacrylamide. This method of separating proteins is called SDS-PAGE (polyacrylamide gel electrophoresis).

Lane 1: Red blood cells

Lane 2: Brain cells

Lane 3: Intestinal cells

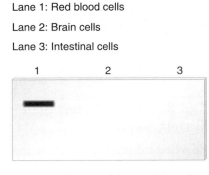

FIGURE 18.11 **The results of Western blotting.** Proteins are isolated from a sample of cells and separated according to their molecular mass via SDS-polyacrylamide gel electrophoresis. The proteins are then transferred to a nitrocellulose or nylon filter, and then the filter blot is placed in a solution containing a primary antibody. The primary antibody specifically recognizes the protein of interest, in this case, β-globin. A secondary antibody, which recognizes the constant domain of the primary antibody, is then added. The secondary antibody is also conjugated to alkaline phosphatase. When XP is added, alkaline phosphatase converts it to a black compound that stains the protein band. Therefore, a black band indicates where the primary antibody has recognized the protein of interest. In lane 1, proteins were isolated from mouse red blood cells. As seen here, the β-globin polypeptide is made in these cells. By comparison, lanes 2 and 3 were samples of brain cells and intestinal cells, which do not synthesize β-globin.

Following SDS-PAGE, the proteins within the gel are blotted to a nitrocellulose (or nylon) filter. The next step is to use a probe that will recognize a specific protein of interest. An important difference between Western blotting and either Southern or Northern blotting is the use of an **antibody** as a probe, rather than a cloned gene. Antibodies bind to sites on molecules known as **epitopes** or antigenic sites within **antigens.** In the case of proteins, an antigenic site is a short sequence of amino acids. Because the amino acid sequence is a unique feature of each protein, any given antibody will specifically recognize a particular protein. This is called the primary antibody for that protein.

After the primary antibody has been given sufficient time to recognize the protein of interest, any unbound primary antibody is washed away and a secondary antibody is added. A secondary antibody is an antibody that binds to the primary antibody. Secondary antibodies, which may be radiolabeled or conjugated to an enzyme, are used for convenience, as secondary antibodies are available commercially. In general, it is easier for researchers to purchase these antibodies rather than label their own primary antibodies. In a Western blotting experiment, the secondary antibody provides a way to detect the protein of interest in a gel blot. For example, it is common for the secondary antibody to be attached to the enzyme alkaline phosphatase. When the colorless dye, XP, is added to the blotting solution, alkaline phosphatase converts the dye to a black compound. Because the secondary antibody binds to the primary antibody, a protein band that is recognized by the primary antibody will become black. In the example shown in figure 18.11, the primary antibody recognizes β-globin. In lane 1, proteins were isolated from mouse red blood cells. The *black* band indicates that the β-globin polypeptide is made in these cells. By comparison, the proteins loaded into lanes 2 and 3 were from samples of brain cells and intestinal cells. The absence of any bands indicates that these cell types do not synthesize β-globin.

Techniques Can Be Used to Detect the Binding of Proteins to DNA Sequences

In addition to detecting the presence of genes and gene products using blotting techniques, researchers often want to study the binding of proteins to specific sites on a DNA molecule. For example, the molecular investigation of transcription factors requires methods that can identify interactions between transcription factor proteins and specific DNA sequences. A technically simple, widely used method for identifying this type of interaction is the **gel retardation assay** (also known as the **band shift assay**). This technique was used originally to study rRNA-protein interactions and quickly became popular after its success in studying protein-DNA interactions in the *lac* operon. Now it is commonly used as a technique to detect interactions between eukaryotic transcription factors and DNA response elements.

The technical basis for a gel retardation assay is that the binding of a protein to a DNA fragment will retard the fragment's ability to move within a polyacrylamide or agarose gel. During electrophoresis, DNA fragments are pulled through the gel matrix toward the bottom of the gel by the voltage gradient. Smaller fragments of DNA migrate more quickly through a gel matrix than do larger fragments. As you might expect, therefore, the binding of a protein to a DNA fragment will retard the DNA's rate of movement through the gel matrix, because a protein-DNA complex has a higher mass. When comparing a DNA fragment and a protein-DNA complex after electrophoresis, the complex will produce a band closer to the top of the gel. The protein-DNA complex is shifted to a higher band than the DNA alone, because the complex migrates more slowly to the bottom of the gel (fig. 18.12).

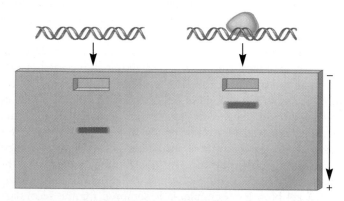

FIGURE 18.12   The results of a gel retardation assay. The binding of protein to a labeled fragment of DNA retards its rate of movement through a gel. For the results shown in the lane to the *right*, if the concentration of the DNA fragment was higher than the concentration of the protein, then there would be two bands: one band with protein bound (at a higher molecular mass) and one band without protein bound (corresponding to the band found in the *left* lane).

A gel retardation assay must be carried out under nondenaturing conditions. This means that the buffers and gel cannot cause the unfolding of proteins or the separation of the DNA double helix. This is necessary so that the proteins and DNA retain their proper structure and thereby can bind to each other. The nondenaturing conditions of a gel retardation assay differ from the more common SDS–gel electrophoresis, in which the proteins are denatured by the detergent SDS.

A second method for studying protein-DNA interactions is **DNA footprinting.** This technique was described originally by David Galas and Albert Schmitz in 1978. A DNA-footprinting experiment attempts to identify one or more regions of DNA that interact with a DNA-binding protein. In their original study, Galas and Schmitz identified a site in the *lac* operon, known as the *lac* operator site, that is bound by a DNA-binding protein called the *lac* repressor.

To understand the basis of a DNA-footprinting experiment, we need to consider the molecular interactions among three things: a fragment of DNA, DNA-binding proteins, and agents that can alter DNA structure. As an example, let's examine the binding of RNA polymerase to a bacterial promoter, a topic discussed in chapter 12. When the RNA polymerase holoenzyme binds to the promoter to form a closed complex, it binds tightly to the −35 and −10 promoter regions, but the protein covers up an even larger region of the DNA. Thus, holoenzyme bound at the promoter prevents other molecules from gaining access to this region of the DNA. The enzyme DNase I, which can cleave covalent bonds in the DNA backbone, is used as a reagent to determine if a DNA region has a protein bound to it. Galas and Schmitz reasoned that DNase I cannot cleave the DNA at locations where a protein is bound. In the foregoing example, it is expected that RNA polymerase holoenzyme will bind to a promoter and protect this DNA region from DNase I cleavage.

Figure 18.13 shows the results of a DNA-footprinting experiment. In this experiment, a sample of many identical DNA fragments, which are 150 bp in length, were radiolabeled at only one end. The sample of fragments was then divided into two tubes; tube A did not contain any holoenzyme, whereas tube B contained RNA polymerase holoenzyme. DNase I was then added to both tubes. The tubes were incubated long enough for DNase I to cleave the DNA at a single site in each DNA fragment. Each tube contained many 150 bp DNA fragments, and the cutting in any DNA strand by DNase I occurs randomly. Therefore, the DNase I treatment should produce a mixture of many smaller DNA fragments. A key point, however, is that DNase I cannot cleave the DNA in a region where RNA polymerase holoenzyme is bound. After DNase I treatment, the DNA fragments within the two tubes were separated by gel electrophoresis and DNA fragments containing the labeled end were detected by autoradiography.

In the absence of RNA polymerase holoenzyme (tube A), DNase I should cleave the 150 bp fragments randomly at any single location. Therefore, a continuous range of sizes occurs (fig. 18.13). In contrast, DNase I should not be able to cleave the DNA in the region where RNA polymerase holoenzyme is bound. Looking at the gel lane from tube B, there are no bands in the size range from 25 to 105 nucleotides. These bands are missing

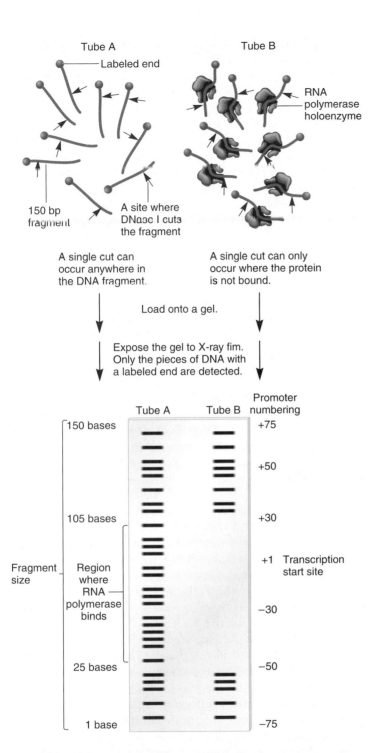

FIGURE 18.13 A DNA-footprinting experiment. Both tubes contained 150 bp fragments of DNA that were incubated with DNase I. Tube B also contained RNA polymerase holoenzyme. The binding of RNA polymerase holoenzyme protected a region of about 80 nucleotides (namely, the −50 region to the +30 region) from DNase I digestion. Note: The promoter numbering convention shown here is the same as that described previously in figure 12.3.

because DNase I cannot cleave the DNA within the region where the holoenzyme is bound. These results occur because the middle portion of the 150 bp fragment contains a promoter sequence that binds the RNA polymerase holoenzyme. The *right* side of the figure numbers the bases according to their position within the gene. (The site labeled +1 is where transcription begins.) As seen here, RNA polymerase covers up a fairly large region of about 80 nucleotides, from the −50 region to the +30 region. As discussed in chapter 12, the promoter sequence is found at the −35 and −10 regions. The observation that RNA polymerase shields an 80-nucleotide region has been obtained for many other bacterial promoters.

As illustrated in this experiment, DNA footprinting can identify the DNA region that interacts with a DNA-binding protein. In addition to RNA polymerase–promoter binding, DNA footprinting has been used to identify the sites of binding for many other types of DNA-binding proteins, such as eukaryotic transcription factors and histones. This technique has greatly facilitated our understanding of protein-DNA interactions.

18.3 ANALYSIS AND ALTERATION OF DNA SEQUENCES

As we have seen throughout this textbook, our knowledge of genetics can be largely attributed to an understanding of DNA structure and function. The feature that underlies all aspects of inherited traits is the DNA sequence. For this reason, analyzing and altering DNA sequences is a powerful approach in understanding genetics. In this last section, we will begin by examining a technique called **DNA sequencing.** This method enables researchers to determine the base sequence of DNA found in genes and other chromosomal regions. It is one of the most important tools for exploring genetics at the molecular level. DNA sequencing is practiced by scientists around the world. The amount of scientific information that is contained within experimentally determined DNA sequences has become enormous. In chapter 20, we will learn how researchers can determine the complete DNA sequence of entire genomes. In chapter 21, we will consider how computers play an essential role in the storage and analysis of genetic sequences.

Not only can researchers determine DNA sequences, they can also use another technique, known as **site-directed mutagenesis,** to change the sequence of cloned DNA segments. At the end of this section, we will examine how site-directed mutagenesis is conducted, and how it provides information regarding the function of genes.

The Dideoxy Method of DNA Sequencing Is Based on Our Knowledge of DNA Replication

Molecular geneticists often seek to determine nucleotide sequences as a first step toward understanding the function and expression of genes. For example, the investigation of genetic sequences has been vital to our understanding of promoters, regulatory elements, and the genetic code itself. Likewise, an examination of sequences has facilitated our understanding of origins of replication, centromeres, telomeres, and transposable elements. In this section, we will examine the technique of DNA sequencing, which is used to determine the base sequence within a DNA strand.

During the 1970s, two methods for DNA sequencing were devised. One method, developed by Alan Maxam and Walter Gilbert, involves the base-specific chemical cleavage of DNA. Another method, developed by Frederick Sanger and colleagues, is known as **dideoxy sequencing.** Because it has become the more popular method of DNA sequencing, we will consider the dideoxy method here.

The dideoxy procedure of DNA sequencing is based on our knowledge of DNA replication but uses a clever twist. As described in chapter 11, DNA polymerase connects adjacent deoxyribonucleotides by catalyzing a covalent linkage between the 5′ phosphate on one nucleotide and the 3′ –OH group on the previous nucleotide (refer back to fig. 11.10). Chemists, though, can synthesize deoxyribonucleotides that are missing the –OH group at the 3′ position (fig. 18.14). These synthetic nucleotides are called **dideoxyribonucleotides.** (Note: The prefix *dideoxy-* refers to the fact that there are two (di) removed (de) oxygens (oxy) compared to ribose; ribose has –OH groups at both the 2′ and 3′ positions). Sanger reasoned that if a dideoxyribonucleotide is added to a growing DNA strand, the strand can no longer grow, because the dideoxyribonucleotide is missing the 3′ –OH group. The incorporation of a dideoxyribonucleotide into a growing strand is therefore referred to as **chain termination.**

Before describing the steps of this DNA-sequencing protocol, we need to become acquainted with the DNA segments that are used in a sequencing experiment. Prior to DNA sequencing, the segment of DNA to be sequenced must be obtained in large amounts. This is accomplished using gene cloning or PCR techniques, which were described earlier in this chapter. In figure 18.15, the segment of DNA to be sequenced (which we will call the target DNA) was cloned into a vector at a defined location. In many DNA-sequencing experiments, the target DNA is cloned into the vector at a site adjacent to a primer-annealing site. The aim of the experiment is to determine the base sequence of the target DNA that has been inserted next to the primer-annealing site. In the experiment shown in figure 18.15, the vector DNA is from a virus called M13. When introduced into a host cell, the virus will produce single-stranded DNA as part of its life cycle.

With these ideas in mind, we can now describe the steps involved in DNA sequencing (fig. 18.15). First, a sample containing many copies of the single-stranded DNA is mixed with primers that will bind to the primer-annealing site. This annealing process is identical to hybridization, because the primer and primer-annealing site are complementary to each other. All four types of deoxyribonucleotides and DNA polymerase are then added to the annealed DNA fragments, and the mixture is divided into four separate tubes. In addition to the four deoxyribonucleotides, each of the four tubes has a low concentration of a different dideoxyribonucleotide. The tubes are then incubated to allow DNA polymerase to make strands complementary to the target DNA sequence. In the third tube, which in this example contains ddTTP, chain termination can occasionally occur at the sixth or thirteenth position of the newly synthesized DNA strand if a ddT becomes incorporated at either of these sites. Note that the complementary A base is found at the sixth and thirteenth position in the target DNA. Therefore, in this tube, we would expect to make some DNA strands that will terminate at the sixth or thirteenth positions. Likewise, in the second tube, a ddATP can cause chain termination only at the second, seventh, eighth, or eleventh positions because a T is found at the corresponding positions in the target strand.

Within the four tubes, mixtures of DNA strands of different lengths have been made. These DNA strands can be separated according to their lengths by running them on an acrylamide gel. The shorter strands move to the bottom of the gel more quickly than the longer strands. To detect the newly made DNA strands, small amounts of radiolabeled deoxyribonucleotides are added to each reaction. This enables the strands to be visualized as bands when the gel is exposed to X-ray film. In figure 18.15, the DNA strands in the four tubes were run in separate lanes on an acrylamide gel. Because we know which dideoxyribonucleotide was added to each tube, we also know which base is at the very end of each DNA strand separated on this gel. Therefore, it becomes possible to read the DNA sequence by reading which base is at the end of every DNA strand and matching this sequence with the length of the strand. Reading the base sequence, from bottom to top, is much like climbing a ladder of bands. Therefore, the sequence obtained by this method is referred to as a **sequencing ladder.**

An important innovation in the method of dideoxy sequencing is **automated sequencing.** Instead of having four separate tubes, with a single type of dideoxyribonucleotide in each tube, automated sequencing uses one tube containing all four types of dideoxyribonucleotides. However, each type of dideoxyribonucleotide (ddA, ddT, ddG, and ddC) has a different-colored fluorescent label attached. After incubating the target DNA with deoxyribonucleotides, the four types of fluorescent dideoxyribonucleotides, and DNA polymerase, the sample is then loaded into a single lane of a gel. A schematic example of a lane of automated sequencing is shown in figure 18.16a.

In automated sequencing, a sample is loaded at the top of a gel, and then the fragments are separated by electrophoresis. Theoretically, it would be possible to read this sequence directly from the gel. From a practical perspective, however, it is more efficient to automate the procedure using a laser and fluorescence

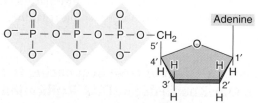

2′, 3′-Dideoxyadenosine triphosphate

FIGURE 18.14   The structure of a dideoxyribonucleotide. Note that the 3′ group is a hydrogen rather than an –OH group. For this reason, another nucleotide cannot be attached at the 3′ position.

FIGURE 18.15 **The protocol for DNA sequencing by the dideoxy method.** The method shown here begins with single-stranded DNA in which the target DNA has been inserted into a viral vector. Alternatively, double-stranded DNA can be used as the template DNA. However, for the primer to bind, the template must be denatured into single-stranded DNA at the beginning of the experiment.

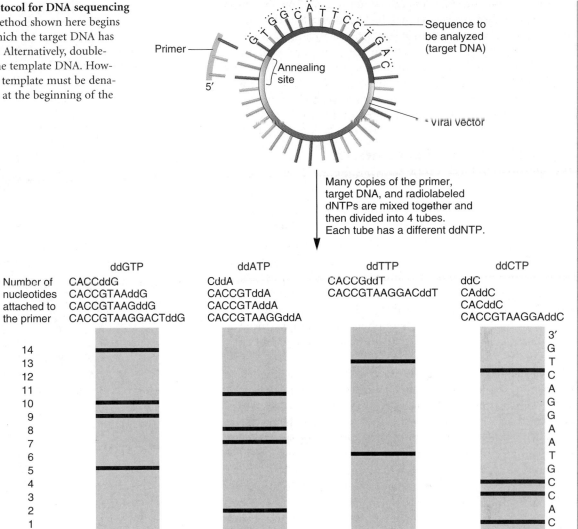

FIGURE 18.16 **Automated DNA sequencing.** (a) This diagram schematically depicts how the sample of DNA fragments each contain a fluorescently labeled dideoxyribonucleotide at their ends. (b) The mixture of DNA fragments is electrophoresed off the end of the gel. As each band is released from the bottom of the gel, the fluorescent dye is excited by a laser, and the fluorescence emission is recorded by a fluorescence detector. The detector reads the level of fluorescence at four wavelengths, corresponding to the four dyes. As shown in the printout, the peaks of fluorescence correspond to the DNA sequence that is complementary to the target DNA.

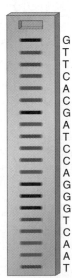

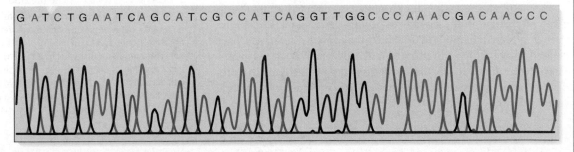

(a) Automated sequencing gel   **(b)** Output from automated sequencing detector. Electrophoresis is continued until each band emerges from the bottom of the gel. As each band comes off, a laser excites the fluorescent dye, and a fluorescence detector records the amount of fluorescence emission. The detector reads the level of fluorescence at four wavelengths, corresponding to the four dyes. An example of the printout from the fluorescence detector is shown in figure 18.16b. As seen here, the peaks of fluorescence correspond to the DNA sequence that is complementary to the target DNA.

In Vitro Site-Directed Mutagenesis Is a Technique to Alter DNA Sequences

As we have seen, the dideoxy technique provides a way to determine the base sequence of DNA. To understand how the genetic material functions, researchers often analyze mutations that alter the normal DNA sequence and thereby affect the expression of genes and the outcome of traits. For example, geneticists have discovered that many inherited human diseases, such as sickle-cell anemia and hemophilia, involve mutations within specific genes. These mutations provide insight into the function of the genes in normal individuals. Hemophilia, for example, involves deleterious mutations in genes that normally encode blood clotting factors.

Because the analysis of mutations can provide important information about normal genetic processes, researchers often wish to obtain mutant organisms. Mutations can arise spontaneously; Mendel's pea plants are a classic example of allelic strains with different phenotypes that arose from spontaneous mutations. In addition, experimental organisms can be treated with mutagens that increase the rate of mutations.

More recently, researchers have developed molecular techniques to make mutations within cloned genes or other DNA segments. One widely used method, known as **in vitro site-directed mutagenesis,** allows a researcher to produce a mutation at a specific site within a cloned DNA segment. With this technique, a DNA sequence can be altered in a specific way. For example, if a DNA sequence is 5′–AAATTTCTTTAAA–3′, a researcher can use site-directed mutagenesis to change it to 5′–AAATTTGTTTAAA–3′; in this case, the researcher deliberately changed the seventh base from a C to a G. Because the sequence of DNA has been altered at a specific site, this approach is called site-directed or site-specific mutagenesis. The site-directed mutant can then be introduced into a living organism to see how the mutation affects the expression of a gene, the function of a protein, and the phenotype of an organism.

The first successful attempts at site-directed mutagenesis involved changes in the sequences of viral genomes. These studies were conducted in the 1970s. Since that time, researchers have devised methods for the mutagenesis of cloned DNA segments. Mark Zoller and Michael Smith developed a protocol for the site-directed mutagenesis of DNA that has been cloned into a viral vector. It is shown in figure 18.17.

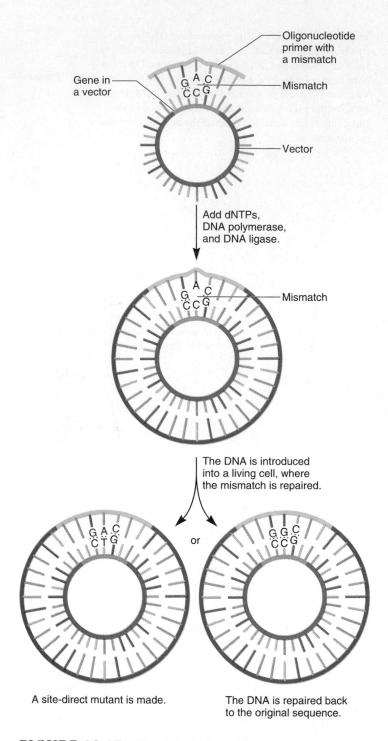

FIGURE 18.17   The method of site-directed mutagenesis.
GENES→TRAITS   To examine the relationship between genes and traits, researchers can alter gene sequences via site-directed mutagenesis. The altered gene can then be introduced into a living organism to examine how the mutation affects the organism's traits. For example, a researcher could introduce a nonsense mutation into the middle of the *lacY* gene in the *lac* operon. If this site-directed mutant were introduced into an *E. coli* bacterium that did not a have a normal copy of the *lacY* gene, the bacterium would be unable to use lactose. These results indicate that a functional *lacY* gene is necessary for bacteria to have the trait of lactose utilization.

Prior to this experiment, a DNA fragment was inserted into a viral vector such as M13 that synthesizes single-stranded DNA as a part of its natural life cycle. This single-stranded DNA can be isolated and used in an in vitro site-directed mutagenesis experiment. As in PCR, this single-stranded DNA is referred to as the template DNA, because it is used as a template to synthesize a complementary strand. Alternatively, double-stranded DNA can be used as the template DNA, but it must be denatured into single-stranded DNA at the start of the experiment.

As shown in figure 18.17, an oligonucleotide primer is allowed to hybridize or anneal to the template DNA. The primer, typically 20 or so nucleotides in length, is synthesized chemically. (A shorter version is shown in figure 18.17 for simplicity.) The scientist designs the base sequence of the primer. The primer has two important characteristics. First, most of the sequence of the primer is complementary to the site in the DNA where the mutation is to be made. However, a second feature is that the primer contains a region of mismatch where the primer and template DNA are not complementary. The mutation will occur in this mismatched region. For this reason, site-directed mutagenesis is sometimes referred to as oligonucleotide-directed mutagenesis.

After the primer and template have annealed, the complementary strand is synthesized by adding deoxyribonucleoside triphosphates (dNTPs), DNA polymerase, and DNA ligase. This yields a double-stranded molecule that contains a mismatch only at the desired location. This double-stranded DNA is then transfected into a bacterial cell. Within the cell, the DNA mismatch will likely be repaired (see chapter 17). Depending on which base is replaced, this may produce the mutant sequence or the original sequence. Clones containing the desired mutation can be identified by DNA sequencing and used for further studies.

After a site-directed mutation has been made within a cloned gene, its consequences are analyzed by introducing the mutant gene into a living cell or organism. As described earlier in chapter 18, hybrid vectors containing cloned genes can be introduced into bacterial cells. Following transformation or transfection into a bacterium, a researcher can study the differences in function between the mutant and wild-type genes and the proteins they encode. Similarly, mutant genes made via site-directed mutagenesis can be introduced into plants and animals, as will be described in chapter 19.

CONCEPTUAL AND EXPERIMENTAL SUMMARY

Recombinant DNA technology, the development of which began in the early 1970s, has revolutionized our understanding of molecular genetics. It is now possible to use **restriction enzymes** to cut DNA fragments out of their native sites within a chromosome. The DNA fragments can then be ligated to a **vector,** which propagates within a living **host cell.** This technology is known as **DNA cloning,** or if the fragment of DNA contains a gene, as **gene cloning.** In addition, **PCR** can produce many copies of a DNA fragment. DNA cloning has enabled researchers to study the structure and function of genes at the molecular level. Hundreds of thousands of different genes have been cloned from hundreds of different species. In addition, gene cloning has many practical applications, which will be described in chapter 19.

In some cases, a researcher needs to detect a specific gene or gene product. For example, **gene detection** procedures are usually needed to identify particular cloned genes in a **DNA library.** In addition, detection of gene products can provide information regarding the expression pattern of a gene in particular cell types or during specific stages of development. Three common methods of detection are **Southern blotting, Northern blotting,** and **Western blotting.** In Southern blotting, a DNA probe is used to detect the presence of a gene or other DNA sequences within a mixture of DNA fragments. Under conditions of **high stringency,** this technique can determine if a species has a particular gene; under conditions of **low stringency,** it can determine if a gene is a member of a **gene family.** The Northern blotting procedure is used to detect the transcription of RNA from a specific gene. In this technique, a DNA probe is used to detect RNA. Both Southern and Northern blotting rely on sequence homology between the DNA probe and the DNA or RNA in the sample. They are called **hybridization** techniques because the labeled DNA probe forms a hybrid with a specific molecule in the sample. Finally, Western blotting is used to detect the protein product from a particular gene. In this technique, an **antibody** is used as a probe because antibodies bind to proteins very specifically. In addition to detection methods, researchers can use a **gel retardation assay** and **DNA footprinting** as a way to study protein-DNA interactions.

Perhaps the greatest advance of recombinant DNA technology is that it has enabled researchers to determine the nucleotide base sequence of DNA. In the **dideoxy** method of **DNA sequencing,** dideoxyribonucleotides are used that terminate the action of DNA polymerase at specific locations in the growing DNA strands. When the terminated DNA strands are separated on the basis of their size by gel electrophoresis, this creates a ladder of DNA fragments that are terminated at defined locations. The DNA sequence can then be determined by reading the bases along the ladder. More recently, DNA sequencing has become automated by using dideoxyribonucleotides that are fluorescently labeled. In addition, researchers can answer questions concerning the functional importance of DNA sequences by intentionally altering the DNA sequence via **site-directed mutagenesis.** In this procedure, a mutagenic DNA primer, which is synthesized chemically, is used to alter the DNA sequence at a particular site in a cloned fragment of DNA.

PROBLEM SETS & INSIGHTS

Solved Problems

S1. RNA was isolated from four different cell types and probed with a radiolabeled cloned gene that is called gene *X*. The results are shown here.

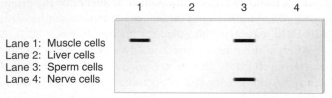

Lane 1: Muscle cells
Lane 2: Liver cells
Lane 3: Sperm cells
Lane 4: Nerve cells

Explain the results of this experiment.

Answer: In this Northern blot, a dark band appears in those lanes where RNA was isolated from muscle and sperm cells but not from liver and nerve cells. These results indicate that the muscle and sperm cells are transcribing gene *X*, but the liver and nerve cells are not. The muscle cells show a single band, while the sperm cells show this band plus a second band of lower molecular mass. An interpretation of these results is that the sperm cells can alternatively splice the RNA to produce a second RNA containing fewer exons.

S2. In the Western blotting experiment shown here, proteins were extracted from red blood cells obtained from tissue samples at different stages of human development. The primary antibody recognizes the *β*-globin polypeptide that is found in the hemoglobin protein.

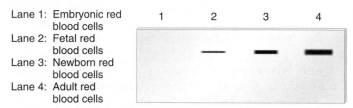

Lane 1: Embryonic red blood cells
Lane 2: Fetal red blood cells
Lane 3: Newborn red blood cells
Lane 4: Adult red blood cells

Explain these results.

Answer: As shown here, the amount of *β*-globin increases during development. There is little detectable *β*-globin produced during embryonic development. The amount increases significantly during fetal development and becomes maximal in the adult. These results indicate that the *β*-globin gene is "turned on" in later stages of development, leading to the synthesis of the *β*-globin polypeptide. This experiment illustrates how a Western blot can provide information concerning the relative amount of a specific protein within living cells.

S3. A DNA strand has the sequence 3′–ATACGACTAGTCGGGACC-ATATC–5′. If the primer in a dideoxy sequencing reaction anneals

just to the left of this sequence, draw what the sequencing ladder would look like.

Answer:

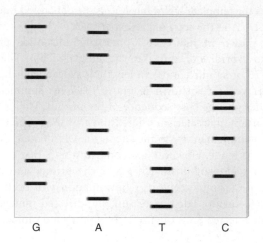

G   A   T   C

S4. The human genetic disease PKU involves a defect in a gene that encodes the enzyme phenylalanine hydroxylase. It is inherited as a recessive autosomal disorder. Using the normal phenylalanine hydroxylase gene as a probe, a Southern blot was carried out on a PKU patient, one of her parents, and a normal unrelated person. In this example, the DNA fragments were subjected to acrylamide gel electrophoresis, rather than agarose gel electrophoresis, since acrylamide gel electrophoresis is better able to detect small deletions within genes. The following results were obtained:

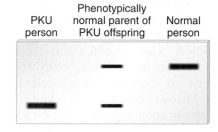

| PKU person | Phenotypically normal parent of PKU offspring | Normal person |

Suggest an explanation for these results.

Answer: In the affected person, the PKU defect is caused by a small deletion within the PKU gene. The parents are heterozygous for the normal gene and the deletion. The PKU-affected person carries only the deletion, which runs at a lower molecular mass than the normal gene.

Conceptual Questions

C1. Discuss three important advances that have resulted from gene cloning.

C2. What is a restriction enzyme? What structure does it recognize? What type of chemical bond does it cleave? Be as specific as possible.

C3. Write a double-stranded sequence that is 20 nucleotides long and is palindromic.

C4. What is cDNA? In eukaryotes, how would cDNA differ from genomic DNA?

C5. Explain and draw the structural feature of a dideoxyribonucleotide that causes chain termination.

Experimental Questions

E1. What is the functional significance of sticky ends in a cloning experiment? What types of chemical bonds make the ends sticky?

E2. Table 18.3 describes the cleavage sites of six different restriction enzymes. After these restriction enzymes have cleaved the DNA, five of them produce short single-stranded (sticky) ends that can hydrogen bond with complementary sticky ends, as shown in figure 18.1. The efficiency of sticky ends binding together depends on the number of hydrogen bonds; more hydrogen bonds makes the ends "stickier" and more likely to stay attached. Rank these five restriction enzymes in table 18.3 (from best to worst) with regard to the efficiency of their sticky ends binding to each other.

E3. Describe the important features of cloning vectors. Explain the purpose of selectable marker genes in cloning experiments.

E4. How does gene cloning produce many copies of a gene?

E5. In your own words, describe the series of steps necessary to clone a gene. Your answer should include the use of a probe to identify a bacterial colony that contains the cloned gene of interest.

E6. What is a hybrid vector? How is a hybrid vector constructed? Explain how X-Gal can be used in a method to identify hybrid vectors that contain segments of chromosomal DNA.

E7. In the experiment of figure 18.3, would the researchers have been successful if pSC102 had an *Eco*RI site in the middle of the kan^R gene? Explain why or why not.

E8. In the experiment of figure 18.3, would the results have been entirely convincing if they had done only gel electrophoresis (step 10) but not the density gradient centrifugation experiment (step 9)? Can you think of an alternative explanation for the results shown in step 10 that would not require the insertion of the kan^R gene into pSC101? In other words, can you think of an explanation for the results shown in step 10 that would not require the construction of a recombinant plasmid? Do the results shown in step 9 rule out your alternative explanation?

E9. A circular plasmid was digested with one or more restriction enzymes, run on a gel, and the following results were obtained:

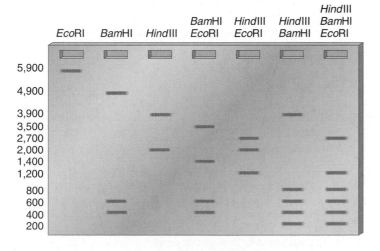

Construct a restriction map for this plasmid.

E10. If a researcher began with a sample that contained three copies of double-stranded DNA, how many copies would be present after 27 cycles of PCR?

E11. Why is a thermostable form of DNA polymerase (e.g., *Taq* polymerase) used in PCR? Is it necessary to use a thermostable form of DNA polymerase in the techniques of dideoxy DNA sequencing or site-directed mutagenesis?

E12. Reverse transcriptase is an enzyme that uses RNA as template to make a complementary strand of DNA. Experimentally, it is used to make cDNA. Reverse transcriptase can also be used in conjunction with PCR to amplify RNAs. This method is called reverse transcriptase PCR, or RT-PCR. In other words, it is possible to make many copies of double-stranded DNA using RNA as a template. Starting with a sample of RNA that contained the mRNA for the β-globin gene, explain how you could create many copies of the β-globin cDNA using RT-PCR.

E13. Let's suppose that you have recently cloned a gene, which we will call gene *X*, from corn. You use this cloned gene to probe genomic DNA from corn in a Southern blot experiment under conditions of low and high stringency. The following results were obtained:

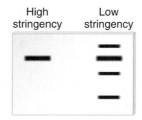

What do these results mean?

E14. What is a DNA library? Do you think that this is an appropriate name?

E15. Some vectors used in cloning experiments contain bacterial promoters that are adjacent to unique cloning sites. This makes it possible to insert a gene sequence next to the bacterial promoter and express the gene in bacterial cells. These are called expression vectors. If you wanted to express a eukaryotic protein in bacterial cells, would you clone genomic DNA or cDNA into the expression vector? Explain your choice.

E16. Southern and Northern blotting depend on the phenomenon of hybridization. In these two techniques, explain why hybridization occurs. Which member of the hybrid is labeled?

E17. In Southern, Northern, and Western blotting, what is the purpose of gel electrophoresis?

E18. What is the purpose of a Northern blot experiment? What types of information can it tell you about the transcription of a gene?

E19. Let's suppose an X-linked gene in mice exists as two alleles, which we will call *B* and *b*. Remember that X inactivation occurs in the somatic cells of female mammals (see chapter 7). Allele *B* encodes an mRNA that is 900 nucleotides long, while allele *b* contains a small deletion that shortens the mRNA to a length of 825 nucleotides. Draw the expected results of a Northern blot using mRNA isolated from somatic tissue of the following mice:

Lane 1. mRNA from an X^bY male mouse

Lane 2. mRNA from an X^bX^b female mouse

Lane 3. mRNA from an X^BX^b female mouse. Note: The sample taken from the female mouse is not from a clone of cells. It is from a tissue sample, like the one shown at the beginning of the experiment of figure 7.6.

E20. The method of Northern blotting can be used to determine the amount of a particular RNA transcribed in a given cell type and the size of the mRNA. Alternative splicing (discussed in chapter 15) can produce mRNAs from the same gene that have different lengths. A Northern blot is shown here using a DNA probe that is complementary to the mRNA encoded by a particular gene. The mRNA in lanes 1–4 was isolated from different cell types.

Lane 1: mRNA isolated from nerve cells
Lane 2: mRNA isolated from kidney cells
Lane 3: mRNA isolated from spleen cells
Lane 4: mRNA isolated from muscle cells

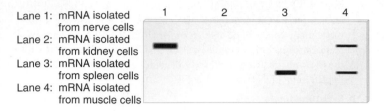

Explain these results.

E21. Southern blotting can be used to detect the presence of repetitive sequences, such as transposable elements, that are present in multiple copies within the chromosomal DNA of an organism. (Note: Transposable elements are described in chapter 17.) In the Southern blot shown here, chromosomal DNA was isolated from three different strains of Baker's yeast, digested with a restriction enzyme, run on a gel, blotted, and then probed with a radioactive DNA probe that is complementary to a transposable element called the *Ty* element.

Southern blot

Explain, in a general way, why the banding patterns are not the same in lanes 1, 2, and 3.

E22. In chapter 8, figure 8.10 describes the evolution of the globin gene family. All of the genes in this family are homologous to each other, though the degree of sequence similarity varies depending on the time of divergence. Genes that have diverged more recently have sequences that are more similar. For example, the α_1 and α_2 genes have DNA sequences that are more similar to each other compared to the α_1 and ξ genes. In a Southern blot experiment, the degree of sequence similarity can be discerned by varying the stringency of hybridization. At high temperature (i.e., high stringency), the probe will only recognize genes that are a perfect or very close match. At lower temperature, however, homologous

genes with lower degrees of similarities can be detected because slight mismatches are tolerated. If a Southern blot was conducted on a sample of human chromosomal DNA, and a probe was used that was a perfect match to the β-globin gene, rank the following genes (from those that would be detected at high stringency down to those that would only be detected at low stringency) as they would appear in a Southern blot experiment: Mb, α_1, β, γ_A, δ, and ε.

E23. In the Western blot shown here, polypeptides were isolated from red blood cells and muscle cells from two different individuals. One individual was normal, and the other individual suffered from a disease known as thalassemia, which involves a defect in hemoglobin. In the Western blot, the gel blot was exposed to an antibody that recognizes β-globin, which is one of the polypeptides that constitute hemoglobin.

Lane 1: Proteins isolated from normal red blood cells
Lane 2: Proteins isolated from the red blood cells of a thalassemia patient
Lane 3: Proteins isolated from normal muscle cells
Lane 4: Proteins isolated from the muscle cells of a thalassemia patient

Explain these results.

E24. Let's suppose a researcher was interested in the effects of mutations on the expression of a structural gene that encodes a polypeptide that is 472 amino acids in length. This polypeptide is expressed in leaf cells of *Arabidopsis thaliana*. Since the average molecular mass of an amino acid is 120 Daltons, this protein has a molecular mass of approximately 56,640 Daltons. Make a drawing that shows the expected results of a Western blot using polypeptides isolated from the leaf cells that were obtained from the following individuals:

Lane 1. A normal plant

Lane 2. A plant that is homozygous for a deletion that removes the promoter for this gene

Lane 3. A plant that is heterozygous in which one gene is normal and the other gene has a mutation that introduces an early stop codon at codon 112

Lane 4. A plant that is homozygous for a mutation that introduces an early stop codon at codon 112

Lane 5. A plant that is homozygous for a mutation that changes codon 108 from a phenylalanine codon into a leucine codon

E25. If you wanted to know if a protein was made during a particular stage of development, what technique would you choose?

E26. Explain the basis for using an antibody as a probe in a Western blot experiment.

E27. Starting with pig cells and a probe that is the human β-globin gene, describe how you would clone the β-globin gene from pigs. You may assume that you have available all the materials needed in a cloning experiment. How would you confirm that a putative clone really contained a β-globin gene?

E28. A cloned gene fragment contains a response element that is recognized by a regulatory transcription factor. Previous experiments have shown that the presence of a hormone results in transcriptional activation by this transcription factor. To study this effect, you conduct a gel retardation assay and obtain the following results:

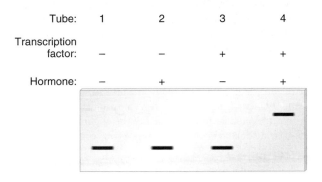

Tube: 1   2   3   4

Transcription
factor: –   –   +   +

Hormone: –   +   –   +

Explain the action of the hormone.

E29. Describe the rationale behind a gel retardation assay.

E30. Certain hormones, such as adrenaline, can increase the levels of cAMP within cells. Let's suppose you can pretreat cells with or without adrenaline and then prepare a cell extract that contains the CREB protein (see chapter 15 for a description of the CREB protein). You then use a gel retardation assay to analyze the ability of the CREB protein to bind to a DNA fragment containing a cAMP response element (CRE). Describe what the expected results would be.

E31. A gel retardation assay can be used to study the binding of proteins to a segment of DNA. In the experiment shown here, a gel retardation assay was used to examine the requirements for the binding of RNA polymerase II (from eukaryotic cells) to the promoter of a structural gene. The assembly of RNA polymerase II at the core promoter is described in chapter 12. In this experiment, the segment of DNA containing a promoter sequence was 1,100 bp in length. The fragment was mixed with various combinations of proteins and then subjected to a gel retardation assay.

Lane 1: No proteins added
Lane 2: TFIID
Lane 3: TFIIB
Lane 4: RNA polymerase II
Lane 5: TFIID + TFIIB
Lane 6: TFIID + RNA
        polymerase
Lane 7: TFIID +
        TFIIB + RNA
        polymerase II

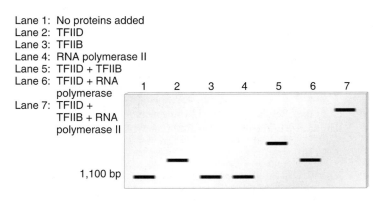

1   2   3   4   5   6   7

1,100 bp

Explain which proteins (TFIID, TFIIB, and/or RNA polymerase II) are able to bind to this DNA fragment by themselves. Which transcription factors (i.e., TFIID and/or TFIIB) are needed for the binding of RNA polymerase II?

E32. As described in chapter 15, certain regulatory transcription factors bind to the DNA and transactivate RNA polymerase II. When glucocorticoid binds to the glucocorticoid receptor (a regulatory transcription factor), this changes the conformation of the receptor and allows it to bind to the DNA. The glucocorticoid receptor binds to a DNA sequence called a glucocorticoid response element (GRE). In contrast, other regulatory transcription factors, such as the

CREB protein, do not require hormone binding in order to bind to DNA. The CREB protein can bind to the DNA in the absence of any hormone, but it will not transactivate RNA polymerase II unless the CREB protein is phosphorylated. (Phosphorylation is stimulated by certain hormones.) The CREB protein binds to a DNA sequence called a cAMP response element (CRE). With these ideas in mind, draw the expected results of gel retardation assay conducted on the following samples:

Lane 1. A 600 bp fragment containing a GRE, plus the glucocorticoid receptor

Lane 2. A 600 bp fragment containing a GRE, plus the glucocorticoid receptor, plus glucocorticoid hormone

Lane 3. A 600 bp fragment containing a GRE, plus the CREB protein

Lane 4. A 700 bp fragment containing a CRE, plus the CREB protein

Lane 5. A 700 bp fragment containing a CRE, plus the CREB protein, plus a hormone (such as adrenaline) that causes the phosphorylation of the CREB protein

Lane 6. A 700 bp fragment containing a CRE, plus the glucocorticoid receptor, plus glucocorticoid hormone

E33. In the technique of DNA footprinting, the binding of a protein to a region of DNA will protect that region from digestion by DNase I by blocking the ability of DNase I to gain access to the phosphodiester linkages in the DNA. In the DNA-footprinting experiment shown here, a researcher began with a sample of cloned DNA that was 400 bp in length. This DNA contained a eukaryotic promoter for RNA polymerase II. The assembly of RNA polymerase II at the core promoter is described in chapter 12. For the sample loaded in lane 1, no proteins were added. For the sample loaded in lane 2, the 400 bp fragment was mixed with RNA polymerase II plus TFIID and TFIIB.

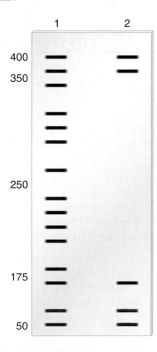

1   2

400
350

250

175

50

Which region of this 400 bp fragment of DNA is bound by RNA polymerase II and TFIID and TFIIB?

E34. Explain the rationale behind a DNA-footprinting experiment.

E35. DNA sequencing can help us to identify mutations within genes. The following is an experiment in which a normal gene and a mutant gene have been sequenced:

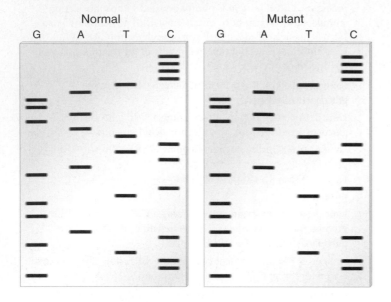

Locate and describe the mutation.

E36. A sample of DNA was subjected to automated DNA sequencing as shown here.

    A. What is the sequence of this DNA segment?

    B. Discuss the advantages of automated sequencing over conventional sequencing that uses radiolabeled nucleotides.

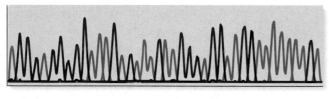

G = Black   T = Red
A = Green   C = Blue

E37. A gene has been cloned into M13 phage, which makes single-stranded DNA as part of its normal life cycle. A portion of the coding sequence of this gene is shown here:

```
5'-GCCCCCGATCTACATCATTACGGCGAT-3'
3'-CGGGGGCTAGATGTAGTAATGCCGCTA-5'
```

The bottom strand also is the strand that is packaged into the M13 phage. (The bottom strand corresponds to the template strand shown in figure 18.15.) This portion of the gene encodes a polypeptide with the amino acid sequence alanine-proline–aspartic acid–leucine–histidine–histidine–tyrosine–glycine–aspartic acid.

Using the method of site-directed mutagenesis, a researcher wants to change the leucine codon into an arginine codon, with an oligonucleotide that is 19 nucleotides long. What is the sequence of the oligonucleotide that should be made? You should designate the 5' and 3' ends of the oligonucleotide in your answer. Note: The mismatch should be in the middle of the oligonucleotide, and a one-base mismatch is preferable over a two- or three-base mismatch.

E38. Let's suppose you want to use site-directed mutagenesis to investigate a DNA sequence that functions as a response element for hormone binding. From previous work, you have narrowed down the response element to a sequence of DNA that is 20 bp in length with the following sequence:

```
5'-GGACTGACTTATCCATCGGT-3'
3'-CCTGACTGAATAGGTAGCCA-5'
```

As a strategy to pinpoint the actual response element sequence, you decide to make 10 different site-directed mutants and then analyze their effects by a gel retardation assay. What mutations would you make? What results would you expect to obtain?

E39. Site-directed mutagenesis can also be used to explore the structure and function of proteins. For example, changes can be made to the coding sequence of a gene to determine how alterations in the amino acid sequence affect the function of a protein. Let's suppose that you are interested in the functional importance of one glutamic acid residue within a protein you are studying. By site-directed mutagenesis, you make mutant proteins in which this glutamic acid codon has been changed to other codons. You then test the encoded mutant proteins for functionality. The results are as follows:

	Functionality
Normal protein	100%
Mutant proteins containing:	
Tyrosine	5%
Phenylalanine	3%
Aspartic acid	94%
Glycine	4%

From these results, what would you conclude about the functional significance of the glutamic acid residue within the protein?

Questions for Student Discussion/Collaboration

1. Discuss and make a list of some of the reasons that it would be informative for a geneticist to determine the amount of a gene product. Use specific examples of known genes (e.g., β-globin and other genes) when making your list.

2. Make a list of all the possible genetic questions that could be answered using site-directed mutagenesis.

Note: All answers appear at the website for this textbook; the answers to even-numbered questions are in the back of the textbook.

---

www.mhhe.com/brooker

Visit the Online Learning Center for practice tests, answer keys, and other learning aids for this chapter. Enhance your understanding of genetics with our interactive exercises, web links, news feeds, tutorial service, and much more.

BIOTECHNOLOGY

::

19

Biotechnology is broadly defined as technologies that involve the use of living organisms, or products from living organisms, as a way to benefit humans. Biotechnology is not a new topic. It began about 12,000 years ago when humans began to domesticate animals and plants for the production of food. Since that time, many species of microorganisms, plants, and animals have become routinely used for human benefit. More recently, the term *biotechnology* has become associated with molecular genetics. Since the 1970s, molecular genetic tools have provided new, improved ways to make use of living organisms to benefit humans. As discussed in chapter 18, recombinant DNA techniques can be used to genetically engineer microorganisms. In addition, recombinant methods enable the introduction of genetic material into plants and animals. As will be discussed in chapter 19, an organism that has integrated recombinant DNA into its genome is called **transgenic.**

In the 1980s, court rulings made it possible to patent recombinant microorganisms as well as transgenic plants and animals. This has provided great economic stimulus to the growth of many biotechnology industries. In chapter 19, we will examine how molecular techniques have expanded our knowledge of the genetic characteristics of commercially important species. We will also discuss examples in which recombinant microorganisms and transgenic plants and animals have been given characteristics that are useful to humans. These include recombinant bacteria that make human insulin, transgenic tomatoes with a longer shelf life, and transgenic livestock that produce human proteins in their milk. In addition, the controversial topics of **mammalian cloning** and **stem cell research** will be examined from a technical point of view. We will also touch upon some of the ethical issues that are associated with these technologies.

Finally, we will consider two technologies that are directly applied to humans. Advances in molecular genetics have spawned technologies based on the analysis of DNA among different individuals. **DNA fingerprinting** is now a common method to analyze the DNA from an individual. Like conventional fingerprinting, this technique can be used for identification purposes. It also can be applied to analyze possible genetic relationships between individuals. In the last section of chapter 19, we will learn how certain inherited diseases have the potential to be treated via **gene therapy.** In this relatively new approach, cloned genes are introduced into individuals with genetic diseases in an attempt to compensate for mutant genes.

19.1 USES OF MICROORGANISMS IN BIOTECHNOLOGY

Microorganisms are used to benefit humans in various ways (table 19.1). In this section, we will examine how molecular genetic tools have become increasingly important for influencing and improving our use of microorganisms. Such tools can produce recombinant microorganisms with genes that have been manipulated in vitro. This approach can improve strains of microorganisms currently in use, and it has even yielded strains that make products not normally produced by microorganisms. For example, several human genes have been introduced into

TABLE 19.1

Common Uses of Microorganisms

Application	Examples
Production of medicines	Antibiotics
	Synthesis of human insulin in recombinant *E. coli*
Food fermentation	Cheese, yogurt, vinegar, wine, and beer
Biological control	Control of plant diseases, insect pests, and weeds
	Symbiotic nitrogen fixation
	Prevention of frost formation
Bioremediation	Cleanup of environmental pollutants such as petroleum hydrocarbons and recalcitrant synthetics

bacteria to produce medically important products such as insulin and human growth hormone.

Overall, the use of recombinant microorganisms is an area of great research interest and potential. As discussed in this section, several recombinant strains are in current use, making profitable products. However, in some areas of biotechnology, the commercial use of recombinant strains has proceeded slowly. This is particularly true for applications in which recombinant microorganisms may be used to produce food products or where they are released into the environment. In such cases, safety concerns and negative public perceptions have slowed the commercial use of recombinant microorganisms. Nevertheless, molecular genetic research continues, and many biotechnologists expect an expanding use of recombinant microbes in the future.

EXPERIMENT 19A

Somatostatin Was the First Human Peptide Hormone Produced by Recombinant Bacteria

During the 1970s, geneticists became aware of the great potential of recombinant DNA technology to produce therapeutic agents to treat certain human diseases. Healthy individuals possess many different genes that encode short peptide and longer polypeptide hormones. Diseases can result when an individual is unable to produce these hormones. This can occur because the individual may have inherited a defective gene or because the cells that produce the hormone have been damaged.

In 1976, Robert Swanson and Herbert Boyer formed Genentech Inc. The aspiration of this company was to engineer bacteria to synthesize useful products, particularly peptide and polypeptide hormones. Their first contract was with Keiichi Itakura and Arthur Riggs. Their intent was to engineer a bacterial strain

that would produce somatostatin, a human hormone that functions to inhibit the secretion of a number of other hormones, including growth hormone, insulin, and glucagon. Somatostatin was chosen for purely technical reasons. It is very small (it contains only 14 amino acids) and can be detected easily via radioimmunoassay.

Before discussing the details of this experiment, let's consider the researchers' approach to constructing the somatostatin gene. To express somatostatin in bacteria, the coding sequence for somatostatin must be inserted next to a bacterial promoter that is contained within a plasmid. In this experiment, the researchers did not clone the human somatostatin gene and insert it into a bacterial plasmid. Instead, due to its small size, they chemically synthesized short oligonucleotides that would hydrogen bond with each other to form the coding sequence for somatostatin.

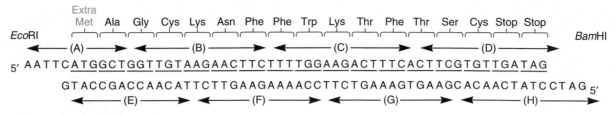

As shown here, eight separate oligonucleotides (labeled *A* through *H*) were synthesized chemically. Due to base complementarity within their sequences, the oligonucleotides hydrogen bonded to each other, forming a longer double-stranded DNA fragment with several important characteristics. First, its single-stranded ends allow it to be inserted into *Eco*RI and *Bam*HI restriction sites within plasmid DNA. The middle of this DNA fragment encodes the amino acid sequence of the somatostatin peptide hormone.

In addition, an extra methionine was added at the amino terminal end. This methionine provided a link between somatostatin and a bacterial protein (namely, β-galactosidase). This was necessary because the researchers learned, during the course of their experiments, that somatostatin made in bacteria is degraded rapidly by cellular proteases. To prevent this from happening, they linked the somatostatin sequence to the bacterial gene encoding β-galactosidase. When this linked gene is expressed in bacteria, a fusion protein is made between somatostatin and β-galactosidase. The fusion protein is not rapidly degraded. The researchers could then separate somatostatin from β-galactosidase by treatment with cyanogen bromide (CNBr), which cleaves polypeptides at the carboxyl terminal side of methionine.

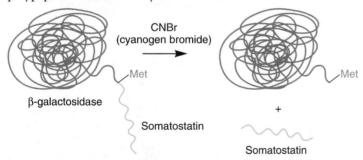

The steps in their protocol are shown in figure 19.1. As described here, the researchers made a synthetic somatostatin gene that was flanked by unique restriction sites and had a methionine codon at the beginning of the somatostatin-coding sequence. This gene was then inserted into a plasmid at the end of a *lacZ* gene, which encodes β-galactosidase. As a control, they also inserted the somatostatin gene in the wrong orientation (shown on the *right*). The plasmid with the wrong orientation should not make any somatostatin. The plasmids were then introduced into *E. coli* cells. The promoter on the plasmid, which was the *lac* promoter of the *lac* operon, was induced with IPTG (a nonmetabolizable lactose analogue). In the cells harboring the somatostatin gene in the correct orientation, this would induce the synthesis of a fusion protein containing β-galactosidase and somatostatin. The bacterial cells were then collected by centrifugation and exposed to formic acid and cyanogen bromide. As mentioned, the cyanogen bromide cleaves polypeptides next to methionine residues. Therefore, this treatment would break the link between β-galactosidase and somatostatin. The amount of somatostatin was then determined by a radioimmunoassay (see appendix).

■ THE HYPOTHESIS

It is possible to produce human somatostatin in a recombinant bacterium.

■ TESTING THE HYPOTHESIS — FIGURE 19.1 The production of human somatostatin in *E. coli*.

Starting material: A normal *E. coli* strain that was unable to synthesize somatostatin, and bacterial plasmids that carry the *amp^R* gene along with bacterial promoters.

	Experimental level	Conceptual level
1. Chemically synthesize eight oligonucleotides. When they are mixed together, the complementary oligonucleotides will hybridize to each other.	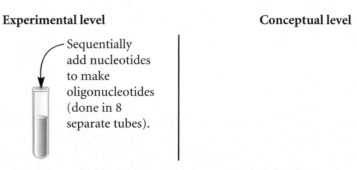 Sequentially add nucleotides to make oligonucleotides (done in 8 separate tubes).	

(*continued*)

2. Treat with DNA ligase to covalently link the oligonucleotides, which will form the molecule shown at the *right*.

3. Using recombinant techniques described in chapter 18 (fig. 18.2), insert this fragment into a plasmid by digesting the plasmid at unique sites for *Eco*RI and *Bam*HI.

4. Transform the plasmids into *E. coli* by treatment with CaCl₂. The transformed cells are spread on plates containing ampicillin. Grow overnight to obtain bacterial colonies that are ampicillin resistant because they contain the plasmid.

5. Grow these recombinant bacteria and activate the *lac* promoter with isopropyl thiodigalactoside (IPTG).

Note: The *lac* promoter in this plasmid is controlled by the *lac* repressor (see chapter 15). IPTG is an inducer that activates transcription of the β-galactosidase gene by removing the *lac* repressor.

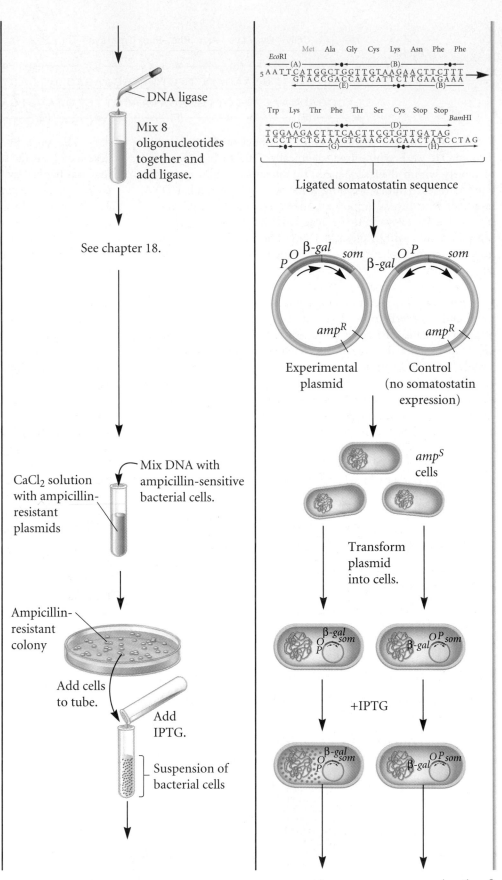

(*continued*)

6. Place in a tube and centrifuge to obtain a bacterial cell pellet.

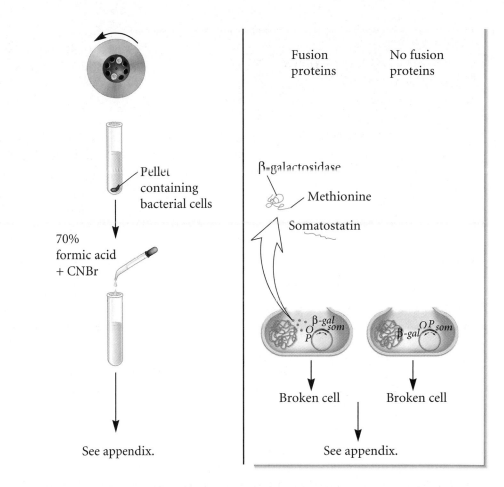

7. Resuspend the pellet in 70% formic acid and cyanogen bromide (5 mg/ml).

   Note: This breaks open the cells and cleaves polypeptides at methionine residues.

8. Determine the amount of somatostatin using a radioimmunoassay. (See the appendix for a description of this procedure.)

THE DATA

Plasmid Strain	Amount of Somatostatin (detected by RIA) (picograms of somatostatin/ milligram of bacterial proteins)
Correct orientation	8–320*
Incorrect orientation	<0.1–0.4

*The amount of somatostatin was determined in several independent experiments.

INTERPRETING THE DATA

As shown in the data table, recombinant bacteria carrying the somatostatin gene in the correct orientation produced this hormone. The amount of somatostatin varied from 8 to 320 picograms per milligram of bacterial proteins. This variability could be attributed to several factors, including protein degradation, incomplete cyanogen bromide cleavage, and unknown genetic changes in the plasmids during bacterial cell growth. In spite of this variability, the exciting result was the production of a human hormone in recombinant bacteria. By comparison, the plasmid with the incorrect orientation did not produce a significant amount of the hormone. This study was the first demonstration that recombinant bacteria could make products encoded by human genes. At the time, this was a major breakthrough that catalyzed the growth of the biotechnology industry!

A self-help quiz involving this experiment can be found at the Online Learning Center.

Many Important Medicines Are Produced by Recombinant Microorganisms

Since the pioneering studies described in the experiment of figure 19.1, recombinant DNA technology has developed bacterial strains that synthesize several other human peptides and proteins. A few examples are described in table 19.2.

In 1982, the U.S. Food and Drug Administration approved the sale of human insulin made by recombinant bacteria. In normal individuals, insulin is produced by the β cells of the pancreas. Insulin functions to regulate several physiological processes, particularly the uptake of glucose into fat and muscle cells. Persons with insulin-dependent diabetes cannot synthesize an adequate amount of insulin, due to a loss of their β cells. Today, these people can purchase genetically engineered human insulin to treat their disease. Prior to 1982, insulin was isolated from the pancreases removed from cattle and pigs. Unfortunately, in some

TABLE 19.2

Examples of Medical Agents Produced by Recombinant Microorganisms

Drug	Action	Treatment
Insulin	A hormone that promotes glucose uptake	For diabetic patients
Tissue plasminogen activator (TPA)	Dissolves blood clots	For heart attack victims and other arterial occlusions
Superoxide dismutase	Antioxidant	For heart attack victims to minimize tissue damage
Factor VIII	Blood clotting factor	For certain types of hemophilia patients
Renin inhibitor	Lowers blood pressure	For hypertension
Erythropoietin	Stimulates the synthesis of red blood cells	For anemia

cases, diabetic individuals became allergic to cow insulin. These allergic patients had to use expensive combinations of insulin from human cadavers and other animals. Now, of course, they can use human insulin made by recombinant bacteria.

As shown in figure 19.2, insulin is a hormone composed of two polypeptide chains, called the A and B chains. To make this hormone using bacteria, the coding sequences of the A and B chains are placed next to the coding sequence of a native *E. coli* protein, β-galactosidase. This creates a fusion protein comprising β-galactosidase and the A or B chain. As with somatostatin, this step is necessary because the A and B chains are rapidly degraded when expressed in bacterial cells by themselves. The fusion proteins, however, are not. After the fusion proteins are expressed in

bacteria, they can be purified and then treated with cyanogen bromide (CNBr) to separate β-galactosidase from the A or B chain. The A and B chains are then purified and mixed together under conditions in which they will fold and associate with each other to make a functional insulin hormone.

Bacterial Species Can Be Used as Biological Control Agents

The term **biological control** refers to the use of living organisms or their products to alleviate plant diseases or damage from environmental conditions. During the past 20 years, interest in the biological control of plant diseases and insect pests as an alternative to chemical pesticides has increased. Biological control agents can prevent disease in several ways. In some cases, nonpathogenic microorganisms are used to compete effectively against pathogenic strains for nutrients or space. In other cases, microorganisms may produce a toxin that inhibits other pathogenic microorganisms or insects without harming the plant.

Biological control can also involve the use of microorganisms living in the field. A successful example is the use of *Agrobacterium radiobacter* to prevent crown gall disease caused by *Agrobacterium tumefaciens*. *A. radiobacter* produces agrocin 84, an antibiotic that kills *A. tumefaciens*. *A. radiobacter* contains genes that confer resistance to the agrocin 84 it produces.

Molecular geneticists have determined that the genes responsible for agrocin 84 synthesis and resistance are located on a plasmid. Unfortunately, this plasmid occasionally is transferred from *A. radiobacter* to *A. tumefaciens* during interspecies conjugation. When this occurs, *A. tumefaciens* can gain resistance to agrocin 84. To prevent this from happening, researchers have identified *A. radiobacter* strains in which this plasmid has been altered genetically to prevent its transfer during conjugation. This

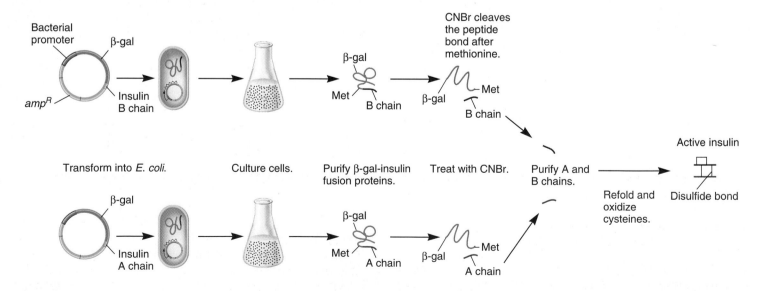

FIGURE 19.2 **The use of bacteria to make human insulin.**

GENES→TRAITS The synthesis of human insulin is not a trait that bacteria normally possess. However, genetic engineers can introduce the genetic sequences that encode the A and B chains of human insulin via recombinant DNA technology, yielding bacteria that make these polypeptides as fusion proteins with β-galactosidase. CNBr treatment releases the A and B polypeptides, which are then purified and oxidized to form functional human insulin.

conjugation-deficient strain is now used commercially worldwide to prevent crown gall disease. This strain still kills neighboring *A. tumefaciens* strains by making agrocin 84. Because it is conjugation deficient, it prevents the formation of *A. tumefaciens* strains that could acquire the plasmid and become resistant to killing.

The Release of Recombinant Microorganisms into the Environment Is Sometimes Controversial

As we have seen, genetically altered strains can have commercial applications in the field. Whether or not a microorganism is recombinant has become an important issue in the use of biological control agents that are released into the environment. If DNA is altered in vitro using molecular techniques and then reintroduced back into a microorganism, that microorganism is considered to be recombinant. In contrast, alterations in the genetic characteristics of a microorganism, such as the acquisition of naturally occurring plasmids or mutagenesis by chemical agents and radiation, do not produce bacteria that are classified as recombinant strains. This distinction is important from the perspective of governmental regulation and public perception even though the organisms produced by the two approaches may be genetically identical or nearly identical.

Knowledge from molecular genetic research is used to develop both nonrecombinant and recombinant strains with desirable characteristics. Each year, many new strains of nonrecombinant microorganisms are analyzed in field tests for the biological control of plant diseases and insect pests. By comparison, the release of recombinant microorganisms and their use in field tests have proceeded much more slowly. This slow progress is related to increased levels of governmental regulation and, in some cases, to negative public perception of recombinant microorganisms.

As an example of the controversial nature of this topic, let's consider the first field test of a recombinant bacterium. It involved the use of a genetically engineered strain of *Pseudomonas syringae* to control frost damage. Experiments by Steven Lindow at the University of California, Berkeley, and his colleagues showed that the formation of ice on the surface of plants is enhanced by the presence of certain bacterial species. These *Ice*$^+$ species synthesize cellular proteins that promote ice nucleation (i.e., the initiation of ice crystals). A way to prevent this from occurring is to use strains that cannot make ice-nucleation proteins (*Ice*$^-$ species). When applied to the surface of plants, an *Ice*$^-$ strain can compete with and thereby reduce the proliferation of *Ice*$^+$ bacteria. Using recombinant DNA technology, an *Ice*$^-$ strain of *P. syringae* was constructed in the early 1980s.

Lindow sought approval for field tests of an *Ice*$^-$ recombinant strain in Tulelake, California. For several years, these tests were delayed because of a lawsuit from the Foundation on Economic Trends in Washington, D.C. During that time, Lindow made great efforts to ensure the safety of this project by studying the local environment. He also consulted with local townspeople where the field test was to take place. Initially, the idea was well received by the local residents. Unfortunately, however, another company tested similar bacteria on the roof of an Oakland facil-

FIGURE 19.3   **The release of recombinant microorganisms in a field test.**

ity. The Environmental Protection Agency (EPA) had not approved this experiment. The media reported this incident, and it caused many townspeople to become apprehensive about the release of recombinant bacteria in Tulelake. Nevertheless, in 1987, approval was finally granted for the field testing of the recombinant *P. syringae* (fig. 19.3).

In the first test on several thousand strawberry plants, the plants were ripped out by vandals. In a second field test, the ability of *Ice*$^-$ bacteria to protect potato plants was tested. Again, some (but not all) of the plants were destroyed by vandals. The results of this field experiment showed that the *Ice*$^-$ bacteria did protect the potato plants from frost damage. In addition, soil sampling showed that the recombinant bacteria were contained at the field site and did not proliferate into surrounding areas. Even so, the release of recombinant microorganisms into the environment remains controversial. Since this first test, relatively few recombinant strains have been released.

Microorganisms Can Reduce Environmental Pollutants

The term **bioremediation** refers to the use of microorganisms to decrease pollutants in the environment. As its name suggests, this is a biological remedy for pollution. During bioremediation, enzymes produced by a microorganism modify a toxic pollutant by altering or transforming its structure. This event is called **biotransformation.** In many cases, biotransformation results in **biodegradation,** in which the toxic pollutant is degraded, yielding less complex, nontoxic metabolites. Alternatively, biotransformations without biodegradation can also occur. For example, toxic heavy metals can often be rendered less toxic by oxidation or reduction reactions carried out by microorganisms. Another way to alter the toxicity of organic pollutants is by promoting polymerization. In many cases, polymerized toxic compounds are less likely to leach from the soil and, therefore, are less environmentally toxic than their parent compounds.

Since the early 1900s, microorganisms have been used in the treatment and degradation of sewage. More recently, the field of bioremediation has expanded into the treatment of hazardous and refractory chemical wastes (i.e., chemicals that are difficult to degrade), usually associated with chemical and industrial activity. These pollutants include petroleum hydrocarbons, halogenated organic compounds, pesticides, herbicides, and organic solvents. Many new applications that use microorganisms to degrade these pollutants are being tested. The field of bioremediation has been fostered, to a large extent, by better knowledge of how pollutants are degraded by microorganisms, the identification of new and useful strains of microbes, and the ability to enhance bioremediation through genetic engineering.

Molecular genetic technology is key in identifying genes that encode enzymes involved in bioremediation. The characterization of the relevant genes greatly enhances our understanding of how microbes can modify toxic pollutants. In addition, recombinant strains created in the laboratory can be more efficient at degrading certain types of pollutants.

In 1980, in a landmark case (*Diamond v. Chakrabarty*), the U.S. Supreme Court ruled that a live, recombinant microorganism is patentable as a "manufacture or composition of matter." The first recombinant microorganism to be patented was an "oil-eating" bacterium that contained a laboratory-constructed plasmid. This strain can oxidize the hydrocarbons commonly found in petroleum. It grew faster on crude oil than did any of the natural isolates that were tested. However, it has not been a commercial success because this recombinant strain metabolizes only a limited number of toxic compounds; the number of compounds actually present in crude oil is over 3,000. Unfortunately, the recombinant strain did not degrade many higher-molecular-weight compounds, which tend to persist in the environment.

Currently, bioremediation should be considered a developing industry. This field will need well-trained molecular geneticists to conduct research aimed at elucidating the mechanisms whereby microorganisms degrade toxic pollutants. In the future, recombinant microorganisms may provide an effective way to decrease the levels of toxic chemicals within our environment. However, this approach will require careful studies to demonstrate that recombinant organisms are effective at reducing pollutants and are safe when released into the environment.

19.2 NEW METHODS FOR GENETICALLY MANIPULATING PLANTS AND ANIMALS

As mentioned at the beginning of chapter 19, transgenic organisms contain recombinant DNA that has been integrated into their genome. The production of transgenic plants and animals is a relatively new, exciting area of biotechnology. In recent years, a few transgenic species have reached the stage of commercialization. Many researchers believe that this technology holds great promise for innovations in agricultural quality and productivity. However, the degree to which this potential may be realized will

FIGURE 19.4   A comparison between a normal mouse and a transgenic mouse.

GENES→TRAITS   The transgenic mouse (on the *right*) carries a gene encoding human growth hormone. This introduction of the human gene into the mouse's genome causes it to grow larger.

depend, in part, on the public's concern about the release and consumption of transgenic species.

In some cases, transgenic organisms have been made in which a cloned gene from one species (the **transgene**) is transferred to some other species. A dramatic example of this is shown in figure 19.4. In the experiment shown here, the gene that encodes the human growth hormone was introduced into the genome of a mouse. The larger mouse shown on the *right* is a transgenic mouse that expresses the human growth hormone gene.

In this section, we will begin by examining the two mechanisms whereby cloned DNA becomes integrated into the chromosomal DNA of animal and plant cells. We will then explore the current techniques used to create transgenic animals and plants. Next, we will consider two very controversial topics: mammalian cloning and stem cell research. These topics have received enormous public attention due to the complex ethical issues that they raise.

The Integration of a Cloned Gene into a Chromosome Can Result in Gene Addition or Gene Replacement

In chapter 18, we considered methods used to clone genes. As described there, a common approach is to insert a chromosomal gene into a vector and then propagate the vector in living microorganisms such as bacteria or yeast cells. Cloned genes can also be introduced into plant and animal cells. However, to be

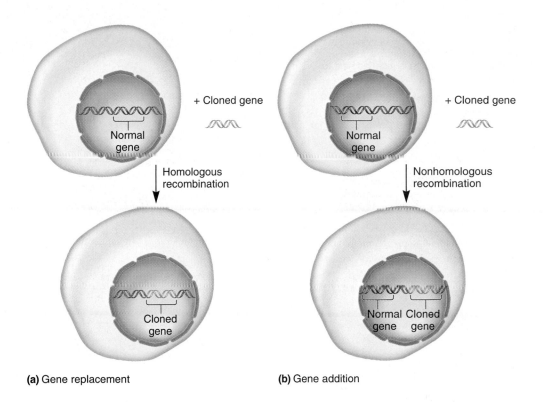

(a) Gene replacement   **(b)** Gene addition

FIGURE 19.5   **The introduction of a cloned gene into a cell can lead to gene replacement or gene addition.** (**a**) If the gene undergoes homologous recombination and replaces the normal chromosomal gene, this is gene replacement. (**b**) Alternatively, the cloned gene may recombine nonhomologously at some other chromosomal location, leading to gene addition.

inherited stably from generation to generation, the cloned gene must become integrated into one (or more) of the chromosomes that reside in the cell nucleus. The integration of cloned DNA into a chromosome occurs by recombination, which was described in chapter 17.

Figure 19.5 illustrates how a cloned gene can integrate into a chromosome by recombination. If the genome of the host cell carries the same gene, and if the cloned gene is swapped with the normal chromosomal gene by homologous recombination, then the cloned gene will replace the normal gene within the chromosome (fig. 19.5*a*). This is **gene replacement.** If the cloned gene has been rendered inactive by mutation and replaces the normal gene, it becomes possible to study how the loss of the normal gene function affects the organism. This is called a **gene knockout.** Alternatively, the cloned gene may recombine at another location within the genome by nonhomologous recombination (fig. 19.5*b*). When this occurs, both the normal and cloned genes will be present. This process is known as **gene addition.**

Molecular Biologists Can Produce Mice That Contain Gene Replacements

In bacteria and yeast, which have relatively small genomes, homologous recombination between cloned genes and the host cell chromosome occurs at a relatively high rate, so that gene replacement is commonly achieved. This is useful when a researcher or biotechnologist wants to make a gene mutation in vitro and then compare the effects of the normal and mutant genes on the phenotype of the organism. However, in more complex eukaryotes with very large genomes, the introduction of cloned genes into cells is much more likely to result in gene addition rather than gene replacement. For example, when a cloned gene is introduced into a mouse cell, it will undergo homologous recombination only 0.1% of the time. Most of the time (namely, 99.9%), gene addition occurs.

To produce mice with gene replacements, molecular biologists have devised laboratory procedures to preferentially select cells in which homologous recombination has occurred. This approach is shown in figure 19.6. The gene of interest (i.e., the normal gene) is found in a mouse chromosome and this same gene has been cloned so that it can be manipulated in vitro. The cloned gene, referred to as the target gene, is shown at the top of figure 19.6. The cloned gene is altered using two selectable marker genes. These selectable markers influence whether or not mouse embryonic cells can grow in the presence of certain drugs. First, the cloned gene is inactivated by inserting a neomycin-resistance gene (called *Neo^R*) into the center of its coding sequence. *Neo^R* provides cells with resistance to neomycin. Next, a thymidine kinase gene, designated *TK,* is inserted adjacent to the target gene but not within the target gene itself. This gene renders cells sensitive to killing by a drug called gancyclovir.

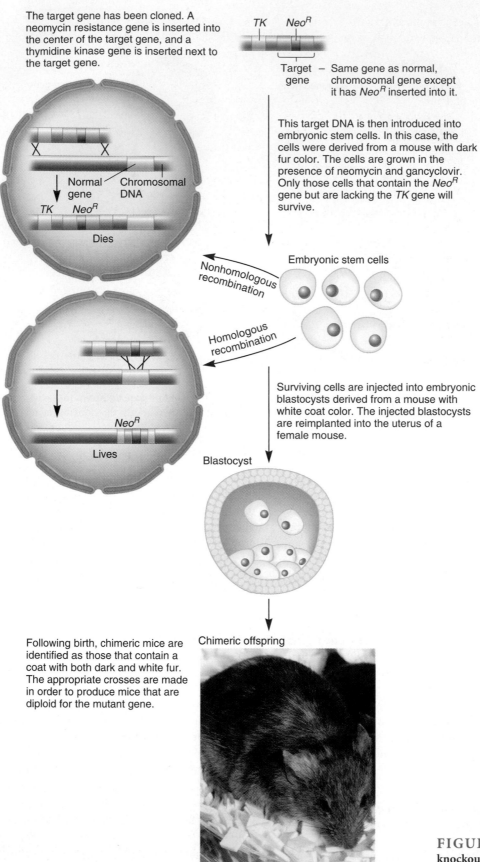

The target gene has been cloned. A neomycin resistance gene is inserted into the center of the target gene, and a thymidine kinase gene is inserted next to the target gene.

TK Neo^R

Target — Same gene as normal, gene   chromosomal gene except it has Neo^R inserted into it.

This target DNA is then introduced into embryonic stem cells. In this case, the cells were derived from a mouse with dark fur color. The cells are grown in the presence of neomycin and gancyclovir. Only those cells that contain the Neo^R gene but are lacking the TK gene will survive.

Normal   Chromosomal
gene   DNA

TK Neo^R

Dies

Nonhomologous recombination

Embryonic stem cells

Homologous recombination

Neo^R

Lives

Surviving cells are injected into embryonic blastocysts derived from a mouse with white coat color. The injected blastocysts are reimplanted into the uterus of a female mouse.

Blastocyst

Following birth, chimeric mice are identified as those that contain a coat with both dark and white fur. The appropriate crosses are made in order to produce mice that are diploid for the mutant gene.

Chimeric offspring

FIGURE 19.6 **Producing a gene knockout in mice.** The *bottom* shows a photograph of a chimeric mouse. Note the patches of black and white fur.

After the target gene has been modified in vitro, it is introduced into mouse embryonic stem cells, which can be grown in the laboratory. When the cells are grown in the presence of neomycin and gancyclovir, most nonhomologous recombinants will be killed because they will also carry the *TK* gene. In contrast, homologous recombinants in which the normal gene has been replaced with the target gene contain only the *Neo^R* gene, and so they will be resistant to both drugs. The surviving embryonic cells can then be injected into a blastocyst, an early embryo that is obtained from a pregnant mouse. In the example shown in figure 19.6, the embryonic cells are from a mouse with dark fur and the blastocysts are from a mouse with white fur. The embryonic cells can mix with the blastocyst cells to create a **chimera.** This is an organism that contains cells from two different individuals. To identify chimeras, the injected blastocysts are reimplanted into the uterus of a female mouse and allowed to develop. When this mouse gives birth, chimeras are identified easily because they contain patches of white and dark coat color (fig. 19.6).

Chimeric animals that contain a single-gene replacement can then be mated to other mice to produce offspring that carry the mutant gene. Since mice are diploid, it is usually necessary to make two or more subsequent crosses to create a strain of mice that contain both copies of the mutant target gene. Later in this chapter, we will consider some of the applications of gene replacements in research and medicine.

Agrobacterium tumefaciens Can Be Used to Make Transgenic Plants

As we have just seen, the introduction of cloned genes into embryonic cells can produce transgenic animals. The production of transgenic plants is somewhat easier, because certain plant cells are **totipotent,** which means that an entire organism can be regenerated from somatic cells. Therefore, a transgenic plant can be made by the introduction of cloned genes into somatic tissue, such as the tissue of a leaf. After the cells of a leaf have become transgenic, an entire plant can be regenerated by the treatment of the leaf with plant growth hormones, which cause it to form roots and shoots.

Molecular biologists can use the bacterium *Agrobacterium tumefaciens,* which naturally infects plant cells, to produce transgenic plants. A plasmid from the bacterium, known as the **Ti plasmid** (Tumor-inducing plasmid), induces tumor formation after a plant has been infected (fig. 19.7*a*). A segment of the plasmid DNA, known as **T DNA** (for transferred DNA), is transferred from the bacterium to the infected plant cells. The T DNA from the Ti plasmid becomes integrated into the chromosomal DNA of the plant cell by recombination. After this occurs, genes within the T DNA that encode plant growth hormones cause uncontrolled plant cell growth. This produces a cancerous plant growth known as a crown gall tumor (fig. 19.7*b*).

Since *A. tumefaciens* inserts its T DNA into the chromosomal DNA of plant cells, it can be used as a vector to introduce cloned genes into plants. Molecular geneticists have been able to

modify the Ti plasmid to make this an efficient process. The T DNA genes that cause tumorigenesis have been identified. Fortunately for genetic engineers, when these genes are deleted, the T DNA is still taken up into plant cells and integrated within the plant chromosomal DNA. However, a crown gall tumor does not form. In addition, geneticists have inserted selectable marker genes into the T DNA to allow selection of plant cells that have taken up the T DNA. A gene that provides resistance to the antibiotic kanamycin is a commonly used selectable marker. Finally, the Ti plasmids used in cloning experiments have been modified to contain unique restriction sites for the convenient insertion of any gene.

Figure 19.8 shows the general strategy for producing transgenic plants via T DNA-mediated gene transfer. A desired gene is cloned into a genetically engineered Ti plasmid and then transformed into *A. tumefaciens.* Plant cells are exposed to the transformed *A. tumefaciens.* After allowing time for infection, the plant cells are grown on a solid medium that contains kanamycin and carbenicillin. Carbenicillin kills *A. tumefaciens,* and kanamycin kills any plant cells that have not taken up the T DNA. Therefore, the only surviving cells are those plant cells that have integrated the T DNA into their genome. Since the T DNA also contains the cloned gene of interest, the selected plant cells are expected to have received this cloned gene as well. The cells are then transferred to a medium that contains the plant growth hormones necessary for the regeneration of entire plants. These plants can then be analyzed to verify that they are transgenic plants containing the cloned gene.

A. tumefaciens infects a wide range of plant species including most dicotyledonous plants, most gymnosperms, and some monocotyledonous plants. However, not all plant species are infected by this bacterium. Fortunately, other methods are available for introducing genes into plant cells. The second most common way to produce transgenic plants is an approach known as **biolistic gene transfer** (i.e., biological ballistics). In this method, plant cells are bombarded with high-velocity microprojectiles coated with DNA. When fired upon by this "DNA gun," the microprojectiles penetrate the cell wall and membrane and thereby enter the plant cell.

Other methods are also available to introduce DNA into plant cells. For example, DNA can enter plant cells by **microinjection** (i.e., by use of microscopic-sized needles) or may be targeted electrophoretically into cells by **electroporation** (i.e., by use of electrical current, which creates transient pores in the plasma membrane through which DNA can enter a cell). Because the rigid plant cell wall is a difficult barrier for DNA entry, other approaches involve the use of protoplasts, which are plant cells that have had their cell walls removed. DNA can be introduced into protoplasts using a variety of methods, including treatment with polyethylene glycol and calcium phosphate.

The production of transgenic plants has become routine practice for many agriculturally important plant species. These include alfalfa, corn, cotton, soybean, tobacco, and tomato. Some of the applications of transgenic plants will be described later in chapter 19.

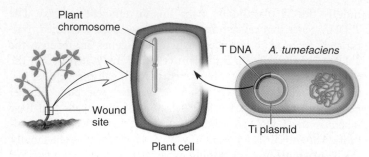

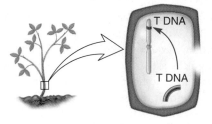

Agrobacterium tumefaciens is found within the soil. A wound on the plant enables the bacterium to infect the plant cells.

During infection, the T DNA within the Ti plasmid is transferred to the plant cell. The T DNA becomes integrated into the plant cell's DNA. Genes within the T DNA promote uncontrolled plant cell growth.

The growth of the recombinant plant cells produces a crown gall tumor.

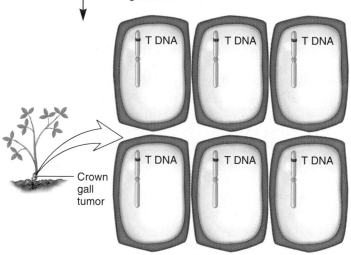

(a) The production of a crown gall tumor by *A. tumefaciens* infection

FIGURE 19.7 *Agrobacterium tumefaciens* infecting a pecan tree and causing a crown gall tumor.

(b) A crown gall on a pecan tree

Researchers Have Succeeded in Cloning Mammals from Somatic Cells

We now turn our attention to cloning as a way to genetically manipulate mammals. The term *cloning* has many different meanings. In chapter 18, we discussed gene cloning, which involves methods that produce many copies of a gene. The cloning of an entire organism is a different matter. **Organismal cloning** refers to methods that produce two or more genetically identical individuals. By accident, this happens in nature; identical twins are genetic clones that began from the same fertilized egg. Similarly, researchers can take mammalian embryos at an early stage of development (e.g., two-cell to eight-cell stage), separate the cells, implant them into the uterus, and obtain multiple births of genetically identical individuals.

In the case of plants, cloning is an easier undertaking. Plants can be cloned from somatic cells. In most cases, it is easy to take a cutting from a plant, expose it to growth hormones, and obtain a separate plant that is genetically identical to the original. However, this approach has not been possible with mammals. For several decades, scientists believed that chromosomes within the somatic cells of mammals had incurred irreversible genetic changes that render them unsuitable for cloning. However, this hypothesis has proven to be incorrect. In 1997, Ian Wilmut and his colleagues at the Roslin Institute created clones of sheep using the genetic material from somatic cells. As you may have heard, they named the first cloned lamb Dolly.

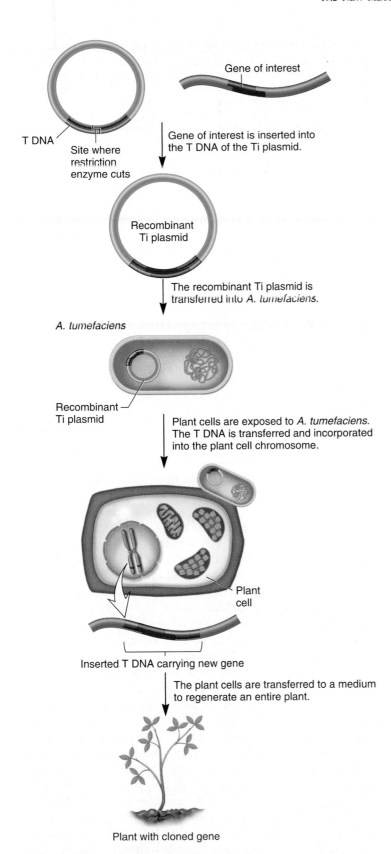

T DNA

Site where restriction enzyme cuts

Gene of interest

Gene of interest is inserted into the T DNA of the Ti plasmid.

Recombinant Ti plasmid

The recombinant Ti plasmid is transferred into *A. tumefaciens*.

A. tumefaciens

Recombinant Ti plasmid

Plant cells are exposed to *A. tumefaciens*. The T DNA is transferred and incorporated into the plant cell chromosome.

Plant cell

Inserted T DNA carrying new gene

The plant cells are transferred to a medium to regenerate an entire plant.

Plant with cloned gene

FIGURE 19.8 **The transfer of genes into plants using the Ti plasmid from *A. tumefaciens* as a vector.**

Figure 19.9 illustrates how Dolly was created. The researchers removed mammary cells from an adult female sheep and grew them in the laboratory. The researchers then extracted the nucleus from a sheep oocyte and fused a diploid mammary cell with the enucleated oocyte cell. Fusion was promoted by electrical pulses. After fusion, the zygote was implanted into the uterus of an adult sheep. One hundred and forty-eight days later, Dolly was born.

Evidence suggested that Dolly may have been "genetically older" than her actual age would have indicated. As mammals age, chromosomes in somatic cells tend to shorten from the telomeres (i.e., from the ends of the chromosome). Therefore, older individuals have shorter chromosomes in their somatic cells compared to younger ones. This shortening does not seem to occur in the cells of the germ line, however. When researchers analyzed the chromosomes in the somatic cells of Dolly when she was about 3 years old, the lengths of her chromosomes were consistent with a sheep that was significantly older, say 9 or 10 years old. The sheep that donated the somatic cell that produced Dolly was 6 years old, and her mammary cells had been grown in culture for several cell doublings before a mammary cell was fused with an oocyte. This led researchers to suspect that Dolly's shorter telomeres were a result of shortening in the somatic cells of the sheep that donated the nucleus. In 2003, the Roslin Institute announced the decision to euthanize 6-year-old Dolly after an examination showed progressive lung disease. Her death has raised concerns among experts that the techniques used to produce Dolly could have caused premature aging.

However, research in mice and cattle has shown the opposite results; the telomeres of these cloned animals appear to be the correct length. For example, cloning was conducted on mice via the method described in figure 19.9 for six consecutive generations. The cloned mice of the sixth generation had normal telomeres. Further research will be necessary to determine if cloning via somatic cells has an effect on the length of telomeres in subsequent generations.

Mammalian cloning is still at an early stage of development. Nevertheless, the breakthrough of creating Dolly has shown that it is technically possible. In recent years, cloning from somatic cells has been achieved in several mammalian species, including sheep, cows, mice, goats, and pigs. This provides the potential for many practical applications. With regard to livestock, cloning would enable farmers to use the somatic cells from their best individuals to create genetically homogeneous herds. This could be advantageous with regard to agricultural yield, although such a genetically homogeneous herd may be more susceptible to rare diseases.

Though some people are concerned about the practical uses of cloning agricultural species, a great majority have become very concerned with the possibility of human cloning. This prospect has raised serious ethical questions. Some people feel that it is morally wrong and threatens the basic fabric of parenthood and family. Others feel that it is a technology that offers a new avenue for reproduction, one that could be offered to infertile couples, for example. In the public sector, the sentiment toward human

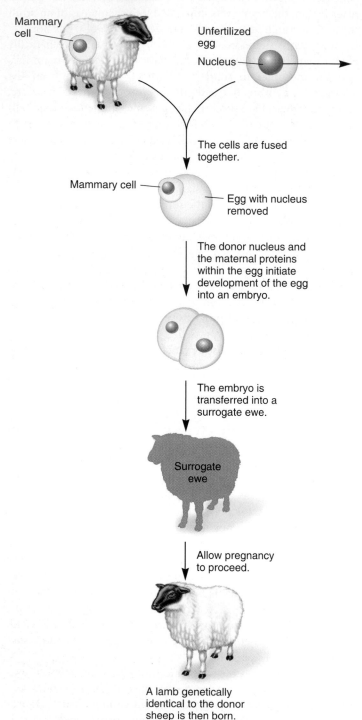

Donor sheep's mammary cell is extracted and grown in a tissue culture flask. Another sheep's unfertilized egg is extracted, and the nucleus is removed.

Mammary cell

Unfertilized egg

Nucleus

The cells are fused together.

Mammary cell

Egg with nucleus removed

The donor nucleus and the maternal proteins within the egg initiate development of the egg into an embryo.

The embryo is transferred into a surrogate ewe.

Surrogate ewe

Allow pregnancy to proceed.

A lamb genetically identical to the donor sheep is then born.

FIGURE 19.9 **Protocol for the successful cloning of sheep.**
GENES→TRAITS Dolly was (almost) genetically identical to the sheep that donated a mammary cell to create Dolly. Dolly and the donor sheep were (almost) genetically identical in the same way that identical twins are; they carried the same set of genes and looked remarkably similar. However, they may have had minor genetic differences due to possible differences in their mitochondrial DNA and may have exhibited some phenotypic differences due to maternal effect or imprinted genes.

cloning has been generally negative. Indeed, many countries have issued an all-out ban on human cloning, while others permit limited research in this area. In the future, our society will have to wrestle with the legal and ethical aspects of cloning as it applies not only to animals but also to people.

Stem Cells Have the Ability to Divide and Differentiate into Different Cell Types

Stem cells supply the cells that construct our bodies from a fertilized egg. In adults, stem cells also replenish worn-out or damaged cells. To accomplish this task, stem cells have two common characteristics. First, they have the capacity to divide, and second, they have the capacity to **differentiate** into one or more specialized cell types. As shown in figure 19.10, the two daughter cells that are produced from the division of a stem cell can have different fates. One of the cells may remain an undifferentiated stem cell, while the other daughter cell can differentiate into a specialized cell type. With this type of asymmetric division/differentiation pattern, the population of stem cells remains constant, yet the stem cells provide a population of specialized cells. In the adult, this type of mechanism is needed to replenish cells that have a finite life span, such as skin epithelial cells and red blood cells.

In mammals, stem cells are commonly categorized according to their developmental stage and their ability to differentiate (fig. 19.11). The ultimate stem cell is the fertilized egg, which, via multiple cellular divisions, can give rise to an entire organism. A fertilized egg is considered totipotent, because it can produce all the cell types in the adult organism. The early mammalian embryo contains **embryonic stem cells (ES cells)**, which are found in the inner cell mass of the blastocyst. The blastocyst is the stage of embryonic development prior to uterine implantation (i.e., the preimplantation embryo). Embryonic stem cells are **pluripotent,** which means that they can also differentiate into every or almost every cell type of the body. However, a single embryonic stem cell has lost the ability to produce an entire, intact individual. During the early fetal stage of development, the germ-line cells found in the gonads also are pluripotent. These cells are called **embryonic germ cells (EG cells).** Interestingly, certain types of human cancers called teratocarcinomas arise from cells that are pluripotent. These bizarre tumors contain a variety of tissues including cartilage, neuroectoderm, muscle, bone, skin, ganglionic structures, and primitive glands. Due to

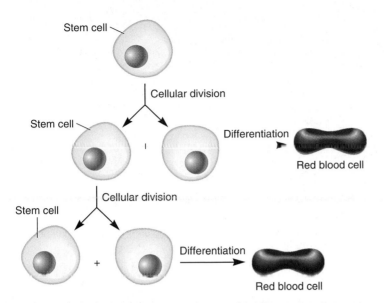

FIGURE 19.10   **Growth pattern of stem cells.** The two main traits that stem cells exhibit are an ability to divide and an ability to differentiate. When a stem cell divides, one of the two cells remains a stem cell while the other daughter cell differentiates into a specialized cell type.

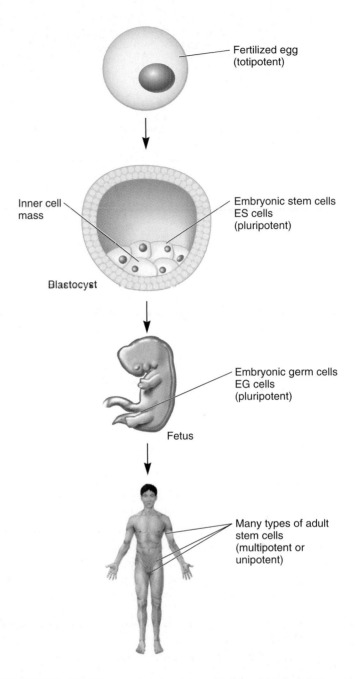

FIGURE 19.11   **Occurrence of stem cells at different stages of development.**

the seemingly embryonic origin of teratocarcinoma cells, these cells are termed **embryonic carcinoma cells (EC cells).**

As mentioned, adults also contain stem cells, but these are thought to be **multipotent** or **unipotent.** A multipotent stem cell can differentiate into several cell types but far fewer than an embryonic stem cell. For example, hematopoietic stem cells (HSCs) found in the bone marrow can supply cells that populate two different tissues, namely, the blood and lymphoid tissues (fig. 19.12). Furthermore, each of these tissues contains several cell types. Multipotent hematopoietic stems cells can follow a pathway in which cell division produces a myeloid progenitor cell, which can then differentiate into a red blood cell, megakaryocyte, basophil, monocyte, eosinophil, neutrophil, or dendritic cell. Alternatively, an HSC can follow a path in which it becomes a lymphoid progenitor cell, which then differentiates into a T cell, B cell, natural killer cell, or dendritic cell. Other stem cells found in the adult seem to be unipotent. For example, primordial germ cells in the testis only differentiate into a single cell type, the sperm.

Interest in stem cells centers around two main areas. Because stem cells have the capacity to differentiate into multiple cell types, the study of stem cells may help us to understand basic genetic mechanisms that underlie the process of development. The genetic events that govern early developmental stages are described in chapter 23. A second compelling reason why people have become interested in stem cells is the potential to treat human diseases or injuries that cause cell and tissue damage. This application has already become a reality in certain cases. For example, bone marrow transplants are used to treat patients with

certain forms of cancers. When bone marrow from a healthy person is injected into the body of a patient who has had his/her immune system wiped out via radiation, the stem cells within the transplanted marrow have the ability to proliferate and differentiate within the body of the patient.

Renewed interest in the use of stem cells in the potential treatment of many other diseases has been fostered by studies in 1998 that showed that embryonic stem cells can be extracted and successfully propagated from human blastocysts (ES cells) and germ cells of aborted fetuses (EG cells). As mentioned, ES and EG

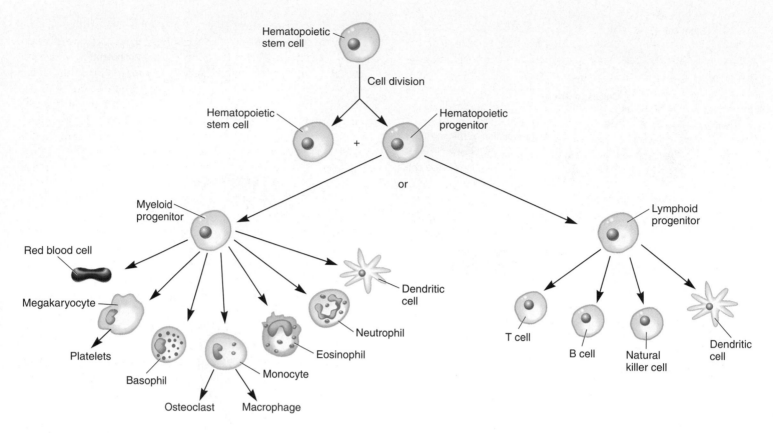

FIGURE 19.12   **Fates of hematopoietic stem cells.**

cells are pluripotent. Therefore, they have the capacity to produce many different kinds of tissue. As shown in table 19.3, embryonic cells could potentially be used to treat a wide variety of diseases associated with cell and tissue damage. By comparison, it would be difficult, based on our modern knowledge, to treat these diseases with adult stem cells because of the inability to locate most types of adult stem cells within the body and successfully grow

TABLE 19.3

Potential Uses of Stem Cells to Treat Diseases

Cell/Tissue Type	Disease Treatment
Nerve	Implantation of cells into the brain to treat Parkinson disease Treatment of injuries such as spinal cord injuries
Skin	Treatment of burn victims and other types of skin disorders
Cardiac	Repair of heart damage associated with heart attacks
Cartilage	Repair of joints damaged by injury or arthritis
Bone	Repair of damaged bone or replacement with new bone
Liver	Repair or replacement of liver tissue that has been damaged by injury or disease
Skeletal muscle	Repair or replacement of damaged muscle

them in the laboratory. Even hematopoietic stem cells are elusive. In the bone marrow, about 1 cell in 10,000 is a stem cell, yet that is enough to populate all of the blood and lymphoid cells of the body. The stem cells of most other adult tissues are equally elusive, if not more so. Furthermore, other types of stem cells in the adult body would be difficult to remove in sufficient numbers that would be required for transplantation. By comparison, ES and EG cells are easy to identify and they have the great advantage of rapid growth in the laboratory. For these reasons, ES and EG cells offer a greater potential for transplantation, based on our current knowledge of stem cell biology.

For ES or EG cells to be used in transplantation, it will be necessary to derive methods that cause them to differentiate into the appropriate type of tissue. For example, if the goal was to repair a spinal cord injury, ES or EG cells would need the appropriate cues that would cause them to differentiate into nerve tissue. At present, much research is needed to understand and potentially control the fate of ES or EG cells. Currently, researchers speculate that a complex variety of factors determines the developmental fates of stem cells. These probably include internal factors within the stem cells themselves, as well as external factors such as the properties of neighboring cells and the presence of hormones and growth factors in the environment.

From an ethical perspective, the primary issue that raises debate is the source of the stem cells for research and potential treatments. If adult stem cells could be identified and propagated in the laboratory, it is likely that an ethical dilemma would be avoided because most people do not have serious moral objections to current procedures such as bone marrow transplantation, discussed earlier. However, the ethical controversy revolves around sources of ES and EG cells. Most ES cells have been derived from human embryos that were produced from in vitro fertilization and were not used by the couples. Most EG cells are obtained from aborted fetuses, either those that were spontaneously aborted or those in which the decision to abort was not related to donating the fetal tissue to research. Some feel that it is morally wrong to use such tissue in research and/or the treatment of disease. Furthermore, some people fear that this technology could lead to intentional abortions for the sole purpose of obtaining fetal tissues for transplantation. Alternatively, others feel that the embryos and fetuses that have been the sources of ES and EG cells were not going to become living individuals, and, therefore, it is beneficial to study these cells and to use them in a positive way to treat human diseases and injury. It is not clear whether these two opposing viewpoints can reach a common ground. As a compromise, many governments are enacting laws that limit or prohibit the use of embryos or fetuses to obtain stem cells, yet permit the use of stem cell lines that are already available in research laboratories. In the United States, for example, federal law prohibits the use of government funding for research projects that involve the destruction of embryos to obtain stem cells. However, government-sponsored research can be done on stem cell lines that were created prior to this legislation.

19.3 APPLICATIONS OF TRANSGENIC PLANTS AND ANIMALS

In this section, we will examine applications arising from the ability of scientists to make transgenic plants and animals. In research, transgenic species can provide important information regarding the functional roles of genes. In addition, transgenic animals as model organisms are becoming invaluable tools for investigating the mechanisms and treatment of human diseases.

A relatively new area of research is the use of transgenic species in agriculture. For centuries, agriculture has relied on selective breeding programs to produce plants and animals with desirable characteristics. Traditional mating strategies yield offspring carrying genes that will provide them with worthwhile traits. For agriculturally important species, this often means the production of strains that are larger, have disease resistance, and yield high-quality food. It is now technologically possible to complement traditional breeding strategies with modern molecular genetic approaches. In this section, we will discuss the current and potential uses of transgenic organisms in agriculture.

Gene Replacements in Mice Can Be Used to Understand Gene Function and Human Disease

Earlier, in figure 19.6, we considered how a researcher can replace a normal mouse gene with one that has been inactivated by the insertion of an antibiotic-resistance gene. When a mouse is homo-

zygous for an inactivated gene, this is called a gene knockout. The inactive mutant gene has replaced both copies of the normal gene. In other words, the function of the normal gene has been knocked out. Gene replacements and gene knockouts have become powerful tools for understanding gene function.

In some cases, gene knockouts have shown that the function of a gene is critical within a particular tissue or during a specific stage of development. In many cases, however, a gene knockout produces no detectable phenotypic effect at all. This has led geneticists to conclude that mammalian genomes have a fair amount of **gene redundancy.** This means that when one type of gene is inactivated, another gene with a similar function may be able to compensate for the inactive gene.

A particularly exciting avenue of gene replacement research is its application in the study of human disease. For example, the disease cystic fibrosis (CF) is one of the most common and severe inherited human disorders. We will describe its symptoms later in chapter 19. In humans, the defective gene that causes CF has been identified. Likewise, the homologous gene in mice was later identified. Using the technique of gene replacement, researchers have produced mice that are homozygous for the same type of mutation that is found in humans with CF. This is called a **gene knock in,** because the function of the normal gene has been altered in a specific way. Such mice exhibit disease symptoms resembling those found in humans (namely, respiratory and digestive abnormalities). Therefore, these mice can be used as model organisms to study this human disease. Furthermore, these mouse models have been used to test the effects of various therapies in the treatment of the disease.

Biotechnology Holds Promise in Producing Transgenic Livestock

The technology for creating transgenic mice has been extended to other animals, and much research is underway to develop transgenic species of livestock, including fish, sheep, pigs, goats, and cattle. For some, the ability to modify the characteristics of livestock via the introduction of cloned genes is an exciting prospect. As discussed next, milk from transgenic animals has become an essential source of human proteins having therapeutic value. In addition, work is currently underway to produce genetically modified pigs that are expected to be resistant to rejection mechanisms that occur following organ transplantation to humans. These pigs may become a source of organs or cells for patients.

A novel avenue of research involves the production of medically important proteins in the mammary glands of livestock. This approach is sometimes called **molecular pharming.** As shown in table 19.4, several human proteins have been successfully produced in the milk of domestic livestock. This list is expanding at a rapid rate. Compared to the production of proteins in bacteria, one advantage is that certain proteins are more likely to function properly when expressed in mammals. This may be due to posttranslational modifications (e.g., attachment of carbohydrate groups) that occur in eukaryotes but not in bacteria. In addition, certain proteins may be degraded rapidly or folded improperly when expressed in bacteria. Furthermore, the

TABLE 19.4

Proteins That Can Be Produced in the Milk of Domestic Animals

Protein	Host	Use
Lactoferrin	Cattle	Used as an iron supplement in infant formula
Tissue plasminogen activator (TPA)	Goat	Dissolves blood clots
Antibodies	Cattle	Used to combat specific infectious diseases
α-1-antitrypsin	Sheep	Treatment of emphysema
Factor IX	Sheep	Treatment of certain inherited forms of hemophilia
Insulin-like growth factor	Cattle	Treatment of diabetes

yield of recombinant proteins in milk can be quite large. Dairy cows, for example, produce about 10,000 liters of milk per year per cow. In most cases, a transgenic cow can produce approximately 1 g/L of the transgenic protein in its milk.

To introduce a human gene into an animal so that the encoded protein will be secreted into its milk, the strategy is to clone the gene next to a milk-specific promoter. Eukaryotic genes often are expressed in a tissue-specific fashion. In mammals, certain genes are expressed specifically within the mammary gland so that their protein product will be secreted into the milk. Examples of milk-specific genes include genes that encode milk proteins such as β-lactoglobulin, casein, and whey acidic protein. To express a human gene that encodes a protein hormone into a domestic animal's milk, the promoter and regulatory sequences for a milk-specific gene are linked to the coding sequences for the human gene (fig. 19.13). In this way, the protein hormone encoded by the human gene will be expressed within the mammary gland and secreted into the milk. The milk can then be obtained from the animal, and the human hormone isolated.

Transgenic Plants Can Be Given Characteristics That Are Agriculturally Useful

Various traits can be modified in transgenic plants (table 19.5). Frequently, transgenic research has sought to produce plant strains resistant to insects, disease, and herbicides. For example, transgenic plants highly tolerant of particular herbicides have been made. The Monsanto Company has produced transgenic plant strains tolerant of glyphosate, the active agent in the herbicide Roundup™. Compared to nontransgenics, these plants grow quite well in the presence of glyphosate-containing herbicides (fig. 19.14). Another important approach is to make plant strains that are disease resistant. In many cases, virus-resistant plants have been developed by introducing a gene that encodes a viral coat protein. When the plant cells express the viral coat protein, they become resistant to infection by that pathogenic virus.

Human hormone gene

Using recombinant DNA technology (described in chapter 18), clone a human hormone gene next to a sheep β-lactoglobulin promoter. This promoter is functional only in mammary cells so that the protein product is secreted into the milk.

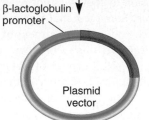

β-lactoglobulin promoter

Plasmid vector

Inject this DNA into a sheep oocyte. The plasmid DNA will integrate into the chromosomal DNA, resulting in the addition of the hormone gene into the sheep's genome.

Implant the fertilized oocyte into a female sheep, which then gives birth to a transgenic sheep offspring.

Transgenic sheep

Obtain milk from female transgenic sheep. The milk contains a human hormone.

Purify the hormone from the millk.

FIGURE 19.13 **Strategy for expressing human genes in a domestic animal's milk.** The β-lactoglobulin gene is normally expressed in mammary cells, whereas the human hormone gene is not. To express the human hormone gene in milk, the promoter from the milk-specific gene is linked to the coding sequence of the hormone gene. In addition to the promoter, a short signal sequence may also be necessary so that the protein will be secreted from the mammary cells and into the milk.

GENES→TRAITS   By using genetic engineering, researchers can give sheep the trait of producing a human hormone in their milk. This hormone can be purified from the milk and used to treat humans.

TABLE 19.5

Traits That Have Been Modified in Transgenic Plants

Trait	Examples
Plant Protection	
Resistance to viral, bacterial, and fungal pathogens	Transgenic plants that express the pokeweed antiviral protein are resistant to a variety of viral pathogens.
Resistance to insects	Transgenic plants that express the CryIA protein from *Bacillus thuringiensis* are resistant to a variety of insects.
Resistance to herbicides	Transgenic plants can express proteins that render them resistant to particular herbicides (see fig. 19.14).
Plant Quality	
Improvement in storage	Transgenic plants can express antisense RNA that silences a gene involved in ripening (see fig. 19.15).
Change in plant composition	Transgenic strains of canola have been altered with regard to oil composition; the seeds of the Brazil nut have been rendered methionine-rich via transgenic technology.
New Products	
Biodegradable plastics	Transgenic plants have been made that can synthesize polyhydroxyalkanoates, which are used as biodegradable plastics.
Vaccines	Transgenic plants have been modified to produce vaccines against many human and animal diseases including hepatitis B, cholera, and malaria.
Pharmaceuticals	Transgenic plants have been made that produce a variety of medicines including human interferon-α (to fight viral diseases and cancer), human epidermal growth factor (for wound repair), and human aprotinin (for transplantation surgery).
Antibodies	Human antibodies have been made in transgenic plants to battle various diseases such as non-Hodgkin's lymphoma.

FIGURE 19.14 **Transgenic plants that are resistant to glyphosate.**

GENES→TRAITS   This field of soybean plants has been treated with glyphosate. The plants on the *left* have been genetically engineered to contain a herbicide-resistance gene. They are resistant to killing by glyphosate. By comparison, the dead or stunted plants in the row with the *orange stick* do not contain this gene.

FIGURE 19.15 **The genetically engineered FlavrSavr tomato.**

GENES→TRAITS   This transgenic tomato plant has been genetically engineered to contain an artificial gene that encodes an antisense RNA to the gene encoding poly-galacturonase. The antisense RNA will bind to the mRNA that encodes polygalactur-onase because the two are complementary, thereby preventing this mRNA from being translated. The antisense RNA thus inhibits the expression of the polygalacturonase gene and prevents overripening. In this way, the FlavrSavr tomato has the trait of not overripening as quickly as traditional tomatoes.

Some transgenic plants have been approved for human consumption. The first example was the FlavrSavr tomato (fig. 19.15), which was developed by Calgene Inc. This transgenic tomato plant had been given a gene that encodes an antisense RNA. This antisense RNA is complementary to the mRNA that codes for the enzyme polygalacturonase. This enzyme, which is involved in ripening, digests sugar linkages within the pectin that is found in plant cell walls. The antisense RNA binds to the poly-galacturonase mRNA, preventing it from being translated. In addition, double-stranded RNA is targeted for degradation (RNA interference), as discussed in chapter 15. Therefore, the expression of the antisense RNA silences the expression of the polygalactur-onase gene. The practical advantage of this transgenic tomato is improved shelf life; these tomatoes do not spoil (i.e., become overripe) as quickly as traditional tomatoes. Therefore, to improve

their flavor, these tomatoes can be allowed to ripen on the vine for a longer period of time. This is an important consideration in the $5-billion annual U.S. tomato market. By comparison, other commercial tomatoes commonly are picked while they are green and ripen later.

19.4 DNA FINGERPRINTING

DNA fingerprinting, also known as **DNA profiling,** is a technology that identifies particular individuals using properties of their DNA. Like the human fingerprint, the DNA of each individual is a distinctive characteristic that provides a means of identification. The application of DNA fingerprinting to forensics has captured the most public attention. Samples of blood and semen can be subjected to DNA fingerprinting to determine whether a particular individual has been at a crime scene. In addition, DNA fingerprinting can determine if two individuals are related genetically. For example, DNA fingerprinting is routinely used in paternity testing. In this section, we will examine how DNA fingerprinting is conducted and consider several of its applications.

DNA Fingerprinting Provides a Unique Banding Pattern for Every Individual

When subjected to DNA fingerprinting, the chromosomal DNA gives rise to a series of bands on a gel (fig. 19.16). The order of bands is an individual's DNA fingerprint. It is the unique pattern of these bands that makes it possible to distinguish individuals.

In the case of humans, the development of DNA fingerprinting relied on the identification of DNA fragments that are quite variable among members of the human population. This naturally occurring variation causes each individual to have a unique DNA fingerprint. In the 1980s, Alec Jeffries and his colleagues found that certain locations within human chromosomes are particularly variable in their lengths. These loci contain tandemly repeated sequences called **minisatellites.** Within the human population, the number of tandem repeats at each minisatellite varies substantially. Therefore, minisatellites are also referred to as loci with a **variable number of tandem repeats (VNTRs).**

The occurrence of VNTRs is shown schematically in figure 19.17, where the chromosomal DNA from two individuals is compared. The diagram emphasizes two features of the chromosomal DNA. First, it shows the sites recognized by a particular restriction enzyme (designated R). As a matter of chance, these sites are interspersed throughout the genome of both individuals. Second, VNTRs (designated by a series of Vs) are also found. At some sites, the two individuals have VNTRs of similar or identical length. At other sites, the VNTRs of the two individuals differ substantially. Therefore, when the chromosomal DNAs are digested with the restriction enzyme, these differences yield a distinct pattern of DNA fragments when analyzed via gel electrophoresis.

Figure 19.17 depicts only a short segment of DNA. When the chromosomal DNA from a sample of actual cells is digested with a restriction enzyme, this would yield too many DNA fragments to analyze. In traditional DNA fingerprinting, DNA probes are used that hybridize specifically to the repeat sequence located

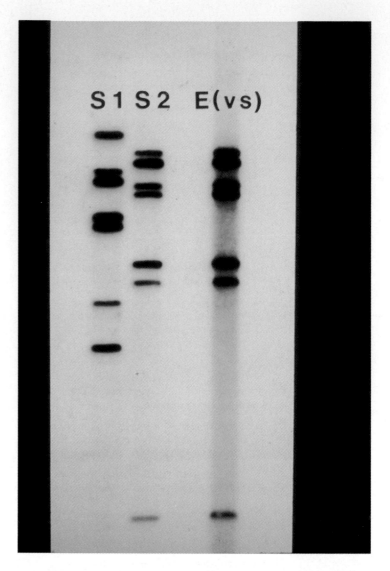

FIGURE 19.16   A comparison of two DNA fingerprints. The chromosomal DNA from two different individuals (Suspect 1, **S1,** and Suspect 2, **S2**) was subjected to DNA fingerprinting. The DNA at a crime scene, **E(vs),** was also subjected to DNA fingerprinting. Following the hybridization of a radiolabeled probe, their DNA appears as a series of bands on a gel. It is the dissimilarity in the pattern of these bands that distinguishes different individuals, much as the differences in physical fingerprint patterns can be used for identification. As seen here, **S2** matches the DNA found at the crime scene.

within selected VNTRs. DNA fingerprinting probes are made that recognize a selected VNTR sequence found at a relatively small number of sites (say, 5–30 sites) in the human genome. Using these probes, a DNA fingerprinting analysis examines the length of the DNA fragments at the sites of the chosen VNTR sequence. This, therefore, involves analysis of the 5 to 30 bands that correspond to the VNTR sites, as in figure 19.16.

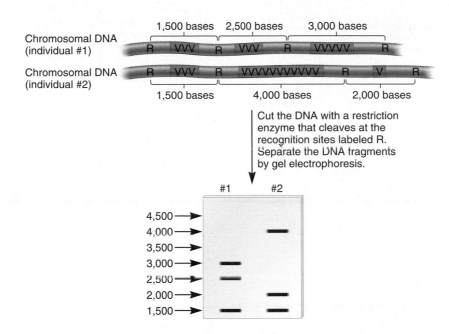

FIGURE 19.17 **A comparison of VNTRs between two individuals.** The letter *R* represents restriction sites found in both individuals. Groups of the letter *V* represent a sequence that is repeated a variable number of times between the two individuals. The variation in the number of repeats affects the sizes of the DNA fragments produced when the DNA is digested with the restriction enzyme that cleaves at the designated sites.

Figure 19.18*a* outlines the steps in a traditional DNA finger-printing experiment. This procedure is a Southern blot using a radiolabeled probe complementary to a selected VNTR sequence. The chromosomal DNA is isolated from a sample and digested with a restriction enzyme. The resulting DNA fragments are then separated by gel electrophoresis. The fragments in the gel are blotted onto a nitrocellulose filter, the DNA is denatured, and the filter is exposed to the radiolabeled probe. Because the probe is complementary to a selected VNTR sequence, it hybridizes to approximately 5 to 30 fragments of DNA that contain this sequence and thereby labels 5 to 30 bands. This is called a multi-locus probe (MLP). In Figure 19.18*a*, the results of a DNA finger-printing experiment on two samples are compared. Since the pattern of band sizes does not match between the two samples, it is concluded that the samples came from different individuals.

In the past decade, the technique of DNA fingerprinting has become automated, much like the automation that changed the procedure of DNA sequencing described in chapter 18. DNA fingerprinting is now done using PCR, which amplifies **short tandem repeats (STRs).** Like VNTRs, STRs are found in multiple sites in the genome of humans and other species and are variable among different individuals. The main difference between traditional VNTRs and STRs is size; STRs are much shorter, usually between 100 and 450 bp. STRs are also called **microsatellites,** while VNTRs are called minisatellites. As you may recall from chapter 10, tandem repeat sequences are called satellites because they sediment away from the rest of the chromosomal DNA during equilibrium gradient centrifugation.

Using primers, the STRs from a sample of DNA are amplified by PCR and then separated by gel electrophoresis according to their molecular masses. Like automated DNA sequencing, the amplified STR fragments are fluorescently labeled. A laser excites the fluorescent molecule within an STR, and a detector records the amount of fluorescence emission for each STR. As shown in figure 19.18*b,* this type of DNA fingerprint yields a series of peaks, each peak having a characteristic molecular mass. In this automated approach, the pattern of peaks is an individual's DNA fingerprint.

DNA Fingerprinting Is Used for Identification and Relationship Testing

Within the past decade or so, the uses of DNA fingerprinting have expanded in many ways. DNA fingerprinting is gaining acceptance as a precise method of identification. In medicine, it is used to identify different species of bacteria and fungi and even can distinguish among closely related strains of the same species. This is useful so that clinicians can treat patients with the appropriate antibiotics. A second common use is in the field of forensics. In many criminal cases, DNA fingerprinting has provided critical evidence that an individual was at a crime scene. Forensic DNA was first used in the U.S. court system in 1986.

When a sample taken from a crime scene matches the DNA fingerprint of an individual, the probability that this match could occur simply by chance can be calculated. Each STR length is given a probability score based on its observed frequency within a reference human population (e.g., Caucasian). A DNA fingerprint contains many peaks, and the probability scores for each peak can be multiplied together to arrive at the likelihood that a particular pattern of peaks would be observed. For example, if a DNA fingerprint contains 20 fluorescent peaks, and the probability

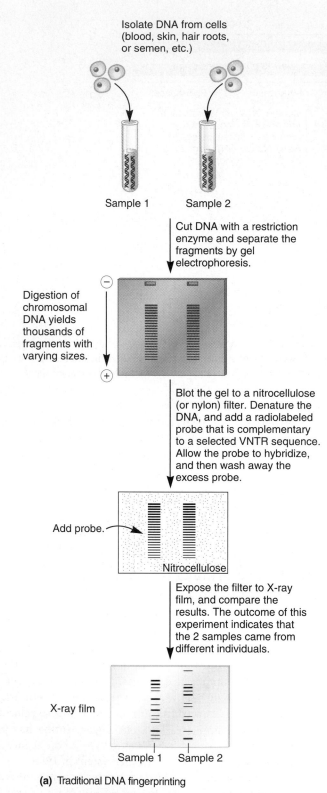

Isolate DNA from cells (blood, skin, hair roots, or semen, etc.)

Sample 1   Sample 2

Cut DNA with a restriction enzyme and separate the fragments by gel electrophoresis.

Digestion of chromosomal DNA yields thousands of fragments with varying sizes.

Blot the gel to a nitrocellulose (or nylon) filter. Denature the DNA, and add a radiolabeled probe that is complementary to a selected VNTR sequence. Allow the probe to hybridize, and then wash away the excess probe.

Add probe.

Nitrocellulose

Expose the filter to X-ray film, and compare the results. The outcome of this experiment indicates that the 2 samples came from different individuals.

X-ray film

Sample 1   Sample 2

(a) Traditional DNA fingerprinting

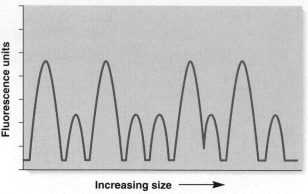

Fluorescence units

Increasing size ⟶

(b) Automated DNA fingerprinting

FIGURE 19.18 **Protocol for DNA fingerprinting. (a)** Traditional DNA fingerprinting. Chromosomal DNA is isolated from a sample of cells. This is often a blood sample, hair roots, or semen. In this experiment, we are comparing two different samples. The DNA is cut with a restriction enzyme, and then the DNA fragments are separated by gel electrophoresis. The DNA fragments within the gel are then transferred to a nitrocellulose (or nylon) filter. The DNA is denatured, and a radiolabeled probe is added that is complementary to a selected VNTR sequence. The probe is given sufficient time to hybridize to any complementary DNA fragments, and then any unbound probe is washed away. Finally, the filter is exposed to X-ray film. The outcome of this experiment indicates that the two samples came from different individuals. **(b)** Automated DNA fingerprinting. A sample of DNA is amplified using primers that recognize the ends of STRs. The STR fragments are fluorescently labeled and then separated by gel electrophoresis. The fluorescent molecules within each STR are excited with a laser, and the amount of fluorescence is measured via a fluorescence detector. A printout from the detector is shown here.

Another important application of DNA fingerprinting is relationship testing. Persons who are related genetically will have some bands or peaks in common. The number they share depends on the closeness of their genetic relationship. For example, an offspring is expected to receive half of his or her VNTRs from one parent and the rest from the other. Figure 19.19 schematically shows a traditional DNA fingerprint of an offspring, mother, and two potential fathers. In paternity testing, the offspring's DNA fingerprint is first compared to that of the mother. The bands that the offspring has in common with the mother are depicted in *purple*. The bands that are not similar between the offspring and the mother must have been inherited from the father. These bands are depicted in *red*. In figure 19.19, male 2 does not have many of the paternal bands. Therefore, he can be excluded as being the father of this child. However, male 1 has all of the paternal bands. Using a calculation similar to that just described, one can compute the likelihood that male 1 could have all of these bands by chance alone. This yields a rather small probability, and it is highly probable that he is the biological father.

of each peak is 1/4, the likelihood of having that pattern would be $(1/4)^{20}$, or roughly 1 in 1 trillion. Therefore, a match between two samples is rarely a matter of random chance. However, when evaluating DNA fingerprinting evidence, other factors such as the potential mishandling of samples and the misinterpretation of data must also be considered.

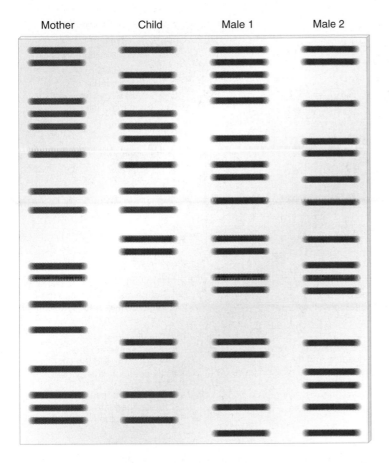

Mother   Child   Male 1   Male 2

FIGURE 19.19 **The use of DNA fingerprinting to establish paternity.** The bands are schematically colored *purple* for maternal and *red* for paternal. Because of matching bands with the offspring, male 1 is the likely father.

TABLE 19.6
Future Prospects in Gene Therapy

Type of Disease	Treatment of
Blood	Sickle-cell anemia, hemophilia, severe combined immunodeficiency syndrome (SCID)
Metabolic	Glycogen storage diseases, lysosomal storage diseases, phenylketonuria, and Lesch-Nyhan syndrome
Neurological	Duchenne muscular dystrophy, myotonic muscular dystrophy
Lung	Cystic fibrosis
Cancer	Brain tumors, breast cancer, colorectal cancer, malignant melanoma, ovarian cancer, and several other types of malignancies
Cardiovascular	Atherosclerosis, essential hypertension
Infectious	AIDS, possibly other viral diseases that involve latent infections

stage of development, success has been very limited in spite of a large amount of research. Relatively few patients have been treated with gene therapy. Nevertheless, a few results have been promising. Table 19.6 describes several types of diseases that are being investigated as potential targets for gene therapy. In this section, we will examine the approaches to gene therapy and how it may be used to treat human disease.

Gene Therapy Involves the Introduction of Cloned Genes into Human Cells

A key step in gene therapy is the introduction of a cloned gene into the cells of people. The techniques used to transfer a cloned gene into human cells can be categorized as nonviral and viral gene transfer methods. The most common nonviral technique involves the use of **liposomes,** which are lipid vesicles (fig. 19.20a). The DNA containing the gene of interest is complexed with liposomes that carry a positive charge (i.e., cationic liposomes). The DNA-liposome complexes are taken up by cells via endocytosis. An advantage of gene transfer via liposomes is that the liposomes do not elicit an immune response. A disadvantage is that the efficiency of gene transfer may be very low.

A second way to transfer genes into human cells is via viruses (fig. 19.20b). The genetic modification of certain viral genomes has led to the development of gene therapy vectors with a capacity to infect cells or tissues, much like the ability of wild-type viruses to infect cells. However, in contrast to wild-type viruses, gene therapy vectors have been genetically engineered so they have lost their capacity for replication in target cells. Nevertheless, when these viral vectors are packaged into viral coats using specialized packaging cell lines or helper viruses, the genetically engineered viruses are naturally taken up by cells and thereby promote the uptake of the cloned DNA of interest. Commonly used viruses for gene therapy include retroviruses, adenoviruses, and parvoviruses.

19.5 HUMAN GENE THERAPY

Throughout this textbook, we have considered examples in which mutant genes cause human diseases. As discussed in chapter 22, these include rare inherited disorders and common diseases such as cancer. Because mutant genes cause disease, geneticists are actively pursuing the goal of using normal, cloned genes to compensate for defects in mutant genes. Gene therapy is the introduction of cloned genes into living cells in an attempt to cure disease. It represents a new, potentially powerful means of treating a wide variety of illnesses.

Many current research efforts in gene therapy are aimed at alleviating inherited human diseases. There are more than 4,000 human genetic diseases that involve a single gene abnormality. Common examples include cystic fibrosis, sickle-cell anemia, and hemophilia. In addition, gene therapies have also been aimed at treating diseases such as cancer and cardiovascular disease that may occur later in life. Some scientists are even pursuing research that will use gene therapy to combat infectious diseases such as AIDS. Even though gene therapy is still at an early

FIGURE 19.20   **Methods of gene transfer used in gene therapy. (a)** Nonviral approach. In this example, the DNA containing the gene of interest is complexed with cationic liposomes. These complexes are taken into cells by endocytosis, in which a portion of the plasma membrane invaginates and creates an intracellular vesicle known as an endosome. After it is released from the endosome, the DNA may then integrate into the host chromosomal DNA via recombination. **(b)** Viral approach. In this example, the gene of interest is cloned into a retrovirus. During the viral life cycle, the genome of the virus is transcribed into RNA, which is packaged into a phage coat. When the retrovirus infects a cell, the RNA is reverse transcribed into double-stranded DNA, which then integrates into the host chromosome. Viruses used in gene therapy have been genetically altered so they cannot proliferate after entry into the host cell.

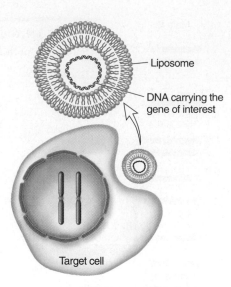

Liposome

DNA carrying the gene of interest

Target cell

DNA-liposome complex is taken into the target cell by endocytosis.

The liposome is degraded within the endosome and the DNA is released into the cytosol.

The DNA is imported into the cell nucleus.

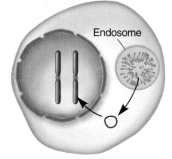

Endosome

By recombination, the DNA carrying the gene of interest is integrated into a chromosome of the target cell.

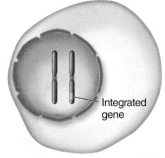

Integrated gene

(a) Nonviral approach

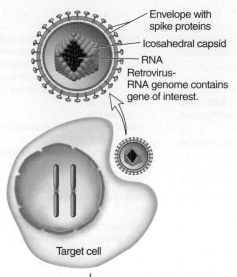

Envelope with spike proteins

Icosahedral capsid

RNA

Retrovirus-RNA genome contains gene of interest.

Target cell

Retrovirus is taken into the target cell via endocytosis.

The viral coat is disassembled in the endosome, and the RNA is released into the cytosol.

The RNA is reverse transcribed into DNA, which travels into the nucleus.

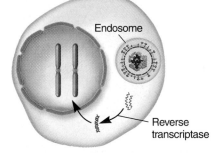

Endosome

Reverse transcriptase

By recombination, the viral DNA, carrying the gene of interest, is integrated into a chromosome of the target cell.

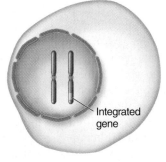

Integrated gene

(b) Viral approach

A key advantage of viral vectors is their ability to efficiently transfer cloned genes to a variety of human cell types. However, a major disadvantage of viral-mediated gene therapy is the potential to evoke an undesirable immune response when injected into a patient. The inflammatory responses induced by adenovirus particles, for example, can be very strong and even fatal in patients treated with these adenoviral constructs. Therefore, much effort has been aimed at preventing the inflammatory responses mediated by virus particles. These include the application of immunosuppressive drugs, and the generation of less immunogenic viral vectors by further genetic modification within the viral genome.

Adenosine Deaminase Deficiency Was the First Inherited Disease Treated with Gene Therapy

Adenosine deaminase (ADA) is an enzyme involved in purine metabolism. If both copies of the *ADA* gene are defective, deoxyadenosine will accumulate within the cells of the individual. At high concentration, deoxyadenosine is particularly toxic to lymphocytes in the immune system, namely, T cells and B cells. In affected individuals, the destruction of T and B cells leads to severe combined immunodeficiency disease (SCID). If left untreated, SCID is typically fatal at an early age (generally, 1 to 2 years old), because the compromised immune system of these individuals cannot fight infections.

Three approaches can be used to treat ADA deficiency. In some cases, it is possible to receive a bone marrow transplant from a compatible donor. A second method is to treat SCID patients with purified ADA that is coupled to polyethylene glycol (PEG). This PEG-ADA is taken up by lymphocytes and can correct the ADA deficiency. Unfortunately, these two approaches are not always available and/or successful. A third, more recent approach is to treat ADA patients with gene therapy.

On September 14, 1990, the first human gene therapy was approved for a young girl suffering from ADA deficiency. This work was carried out by a large team of researchers composed of R. Michael Blaese, Kenneth Culter, W. French Anderson, and several collaborators at the National Institutes of Health (NIH). Prior to this clinical trial, the normal gene for ADA had been cloned into a retroviral vector that can infect lymphocytes. The general aim of this therapy was to remove lymphocytes from the blood of the young girl with SCID, introduce the normal *ADA* gene into her cells, and then return them to her bloodstream.

Figure 19.21 outlines the protocol for the experimental treatment. Lymphocytes (i.e., T cells) were removed and cultured in a laboratory. The lymphocytes were then transfected with a nonpathogenic retrovirus that had been genetically engineered to contain the normal *ADA* gene. During the life cycle of a retrovirus, the retroviral genetic material is inserted into the host cell's DNA. Therefore, because this retrovirus contained the normal *ADA* gene, this gene also was inserted into the chromosomal DNA of the girl's lymphocytes. After this had occurred in the laboratory, the cells were reintroduced back into the patient. This is called an *ex vivo* approach because the genetic manipulations occur outside the body, yet the products are reintroduced into the body.

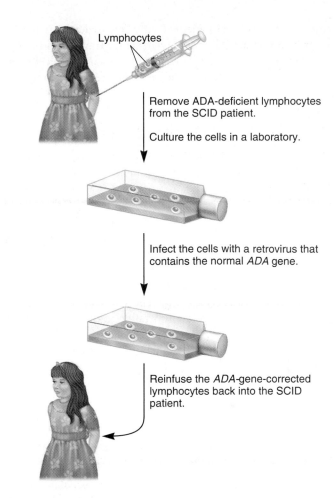

Lymphocytes

Remove ADA-deficient lymphocytes from the SCID patient.

Culture the cells in a laboratory.

Infect the cells with a retrovirus that contains the normal *ADA* gene.

Reinfuse the *ADA*-gene-corrected lymphocytes back into the SCID patient.

FIGURE 19.21 **The first human gene therapy, for adenosine deaminase (ADA) deficiency.**

In this clinical trial, two patients were enrolled, and a third patient was later treated in Japan. The results of this trial showed that the transfer of DNA into a large number of human cells is feasible. In at least one patient, T cells carrying the cloned gene were still detectable 8 to 10 years after they had been transferred. However, most of the circulating T cells were not found to contain the cloned gene. Because researchers were not able to withhold PEG-ADA treatment in these patients, it was not possible to determine whether or not gene transfer into T cells was of significant clinical benefit.

Another form of SCID, termed SCID-X1, is inherited as an X-linked trait. SCID-X1 is characterized by a block in T cell growth and differentiation. This block is caused by mutations in the gene encoding the γ_c cytokine receptor, which plays a key role in the recognition of signals that are needed to promote the growth, survival, and differentiation of T cells. Similar to the ADA trial shown in figure 19.21, a gene therapy trial for SCID-X1 was initiated in which a normal γ_c cytokine receptor gene was introduced into SCID-X1 patients' lymphocytes, and then the lymphocytes were reintroduced back into their bodies. At a 10-month follow-up, T cells expressing the normal γ_c cytokine receptor were detected in two patients. Most importantly, the T cell counts in these two patients had risen to levels that were comparable to normal individuals. This clinical trial was the first clear demonstration that gene therapy can provide clinical benefit. In this case, it seemed to provide a complete correction of the disease phenotype.

Aerosol Sprays May Be Used to Treat Cystic Fibrosis

Cystic fibrosis (CF) is one of the most common recessive inherited disorders with debilitating consequences. It is caused by a defect in a gene that encodes a protein that functions in the transport of ions across the plasma membrane of epithelial cells (e.g., cells lining the respiratory tract and the intestinal tract). A defect in membrane transport leads to an abnormality in salt and water balance, which causes a variety of symptoms, particularly an overaccumulation of mucus in the lungs. Even though great strides have been made in the treatment of the symptoms of CF, this disease remains associated with repeated lung infections and a shortened life. In most cases, mortality results from chronic lung infections.

CF has been the subject of much gene therapy research. As mentioned earlier in chapter 19, *CF* gene knockouts in mice have generated a CF animal model. These animal models have been used to test the efficacy of gene therapy approaches. More recently, clinical trials have tested the ability of gene therapy to improve the condition of patients suffering from CF. To implement CF gene therapy, it is necessary to deliver the normal *CF* gene to the lung cells. Unlike ADA gene therapy in which the lymphocytes can be treated ex vivo, it is not possible to remove lung epithelial cells and then put them back into the individual. Instead, it is necessary to design innovative approaches that can target the *CF* gene directly to the lung cells.

To achieve this goal, CF gene therapy methods have involved the use of an inhaled aerosol spray. In one protocol, the normal *CF* gene is cloned into an adenovirus. Adenoviruses normally infect lung epithelial cells and cause a lung infection. This adenovirus, however, has been engineered so that it will gain entry into the epithelial cells but not cause a lung infection. In addition, the adenovirus has been engineered to contain the normal *CF* gene. A second approach is liposome delivery of the *CF* gene. In this therapy, the normal *CF* gene is complexed with liposomes. When inhaled by the patient via an aerosol spray, the lung epithelial cells take up this liposome complex.

Like ADA gene therapy, CF gene therapy is at an early stage of development. It is hoped that gene therapy eventually will become an effective method to alleviate the symptoms associated with this disease.

CONCEPTUAL AND EXPERIMENTAL SUMMARY

The connection between **biotechnology** and molecular genetics has become an important interface. Molecular geneticists are analyzing the microorganisms that we use for food fermentation, **biological control, bioremediation,** and the production of medicines, enzymes, and fuels. This research yields a better understanding of the molecular mechanisms that underlie the beneficial features of these microorganisms. In many cases, this knowledge can be used to produce microbial strains better suited to serve humans. Some of these are recombinant strains that can make products not normally made by bacteria.

In addition to microorganisms, molecular tools are also being directed toward agriculturally important plants and animals. Several examples of **transgenic** species have been made and tested. A few are commercially available. In the future, it is expected that the production and commercial success of transgenic species will become more prevalent. This will provide exciting career opportunities for well-trained molecular geneticists.

Molecular genetic tools can also be directed toward humans. In some cases, such as **cloning** and **stem cell research,** ethical issues will have to be examined before these methods can be applied to humans. In other cases, the technologies are currently in use. For example, molecular genetics has produced a technology, **DNA fingerprinting,** that can be used as a tool for human identification. As a forensic tool, this technology can determine whether an individual may have been at a crime scene based on samples of blood, hair (roots), or semen. In addition, it can be used to evaluate possible genetic relationships among individuals. Very recently, recombinant DNA technology is being used to combat human diseases via **gene therapy.** In this approach, a cloned gene is introduced into somatic cells as a way to compensate for a defective gene. In the future, this may become an important way to alleviate many human disorders.

PROBLEM SETS & INSIGHTS

Solved Problems

S1. Which of the following would appropriately be described as a transgenic organism?

A. The sheep "Dolly," which was produced by cloning.

B. A sheep that produces human α-1-antitrypsin in its milk.

C. The FlavrSavr strain of tomato.

D. A hybrid strain of corn produced from crossing two inbred strains of corn. The inbred strains were not transgenic.

Answer:

    A. No. Dolly was not produced using recombinant techniques. There were not pieces of DNA that were cut and combined in a new way.

    B. Yes.

    C. Yes.

    D. No. The hybrids simply contain chromosomal genes from two different parental strains.

S2. Describe the strategy for producing human proteins in the milk of livestock.

Answer: Milk proteins are encoded by genes with promoters and regulatory sequences that direct the expression of these genes within the cells of the mammary gland. To get other proteins expressed in the mammary gland, the strategy is to link the promoter and regulatory sequences from a milk-specific gene to the coding sequence of the gene that encodes the human protein of interest. In some cases, it is also necessary to add a signal sequence to the amino terminal end of the target protein. A signal sequence is a short polypeptide that directs the secretion of a protein from a cell. If the target protein does not already have a signal sequence, it is possible to use a signal sequence from a milk-specific gene to promote the secretion of the target protein from the mammary cells and into the milk. During this process, the signal sequence is cleaved from the secreted protein.

S3. Explain how one type of probe can be used to identify multiple bands in the Southern blot of a DNA-fingerprinting experiment. Why do the lengths of these bands vary among different individuals?

Answer: The reason why the same probe can identify multiple bands is because the probe is complementary to a genetic sequence that is located at several different sites throughout the genome. Not only is the sequence located at several different sites, it is also tandemly repeated at each site. Within populations, there is variation in the number of these tandem repeats at each site. If the number of tandem repeats at a given site is high, this will yield a higher molecular weight band (i.e., a longer band) in a Southern blot. This type of site is termed a VNTR (for variable number of tandem repeats).

Conceptual Questions

C1. What is a recombinant microorganism? Discuss examples of recombinant microbes.

C2. A conjugation-deficient strain of *A. radiobacter* is used to combat crown gall disease. Explain how this bacterium prevents the disease and the advantage of a conjugation-deficient strain.

C3. What is bioremediation? What is the difference between biotransformation and biodegradation?

C4. What is a biological control agent? Briefly describe three examples.

C5. As described in table 19.2, several medical agents are now commercially produced by genetically engineered microorganisms. Discuss the advantages and disadvantages of making these agents this way.

C6. What is a mouse model for human disease?

C7. What is a transgenic organism? Describe three examples.

C8. What part of the *A. tumefaciens* DNA gets transferred to the genome of a plant cell during infection?

C9. Explain the difference between gene addition and gene replacement. Would the following descriptions be examples of gene addition or gene replacement?

    A. A mouse model to study cystic fibrosis

    B. Introduction of a pesticide-resistance gene into corn using the Ti plasmid of *A. tumefaciens*

C10. As described in chapter 7, not all inherited traits are determined by nuclear genes (i.e., genes located in the cell nucleus) that are expressed during the life of an individual. In particular, maternal effect genes and mitochondrial genes are notable exceptions. With these ideas in mind, let's consider the cloning of sheep (e.g., Dolly).

    A. With regard to maternal effect genes, would the phenotype of such a cloned animal be determined by the animal that donated the enucleated oocyte or by the animal that donated the somatic cell nucleus? Explain.

    B. Would the cloned animal inherit extranuclear traits from the animal that donated the oocyte or by the animal that donated the somatic cell? Explain.

    C. In what ways would you expect this cloned animal to be similar to or different from the animal that donated the somatic cell? Is it fair to call such an animal a "clone" of the animal that donated the nucleus?

C11. Discuss some of the worthwhile traits that can be modified in transgenic plants.

C12. Discuss the concerns that some people have with regard to the uses of genetically engineered organisms.

C13. If a person with adenosine deaminase deficiency (ADA) or cystic fibrosis was treated for their disease with gene therapy, and their therapy was successful, would the person pass the normal gene or the mutant gene to their offspring? Explain your answer.

Experimental Questions

E1. Recombinant bacteria can produce hormones that are normally produced in humans. Briefly describe how this is accomplished.

E2. In the experiment of figure 19.1, why did the plasmid with β-galactosidase in the wrong orientation fail to produce somatostatin?

E3. *Bacillus thuringiensis* can make toxins that kill insects. This toxin must be applied several times during the growth season to prevent insect damage. As an alternative to repeated applications, one strategy is to apply bacteria directly to leaves. However, *B. thuringiensis* does not survive very long in the field. Other bacteria, such as

Pseudomonas syringae, do. Propose a way to alter *P. syringae* so that it could be used in the battle against insects. Discuss advantages and disadvantages of this approach compared to the repeated applications of the insecticide from *B. thuringiensis*.

E4. In the experiment of figure 19.1, explain how the coding sequence for somatostatin was constructed. Why was it necessary to link this coding sequence to the sequence for β-galactosidase? How were these two polypeptides separated after the fusion protein was synthesized in *E. coli*?

E5. Explain how it is possible to select for homologous recombination in mice. What phenotypic marker is used to readily identify chimeric mice?

E6. To produce transgenic plants, plant tissue is exposed to *A. tumefaciens* and then grown in media containing kanamycin, carbenicillin, and plant growth hormones. Explain the purpose behind each of these three agents. What would happen if you left out the kanamycin?

E7. List and briefly describe five methods for the introduction of cloned genes into plants.

E8. What is a gene knockout? Is an animal or plant with a gene knockout a heterozygote or homozygote? What might you conclude if a gene knockout does not have a phenotypic effect?

E9. Nowadays, it is common for researchers to identify genes using cloning methods described in chapters 18 and 19. A gene can be identified according to its molecular features. For example, a segment of DNA can be identified as a gene because it contains the right combination of sequences: a promoter, exons, introns, and a terminator. Or a gene can be identified because it is transcribed into mRNA. In the study of plants and animals, it is relatively common for researchers to identify genes using molecular techniques without knowing the function of the gene. In the case of mice, the function of the gene can be investigated by making a gene knockout. If the knockout causes a phenotypic change in the mouse, this may provide an important clue regarding the function of a gene. For example, if a gene knockout produced an albino mouse, this would indicate that the gene that had been knocked out probably plays a role in pigment formation. The experimental strategy of first identifying a gene based on its molecular properties and then investigating its function by making a knockout is called **reverse genetics.** Explain how this approach is opposite (or "in reverse") to the conventional way that geneticists study the function of genes.

E10. According to the methods described in figure 19.6, can homologous recombination that results in gene replacement cause the integration of both the *TK* and *Neo^R* genes? Explain why or why not. Describe how the thymidine kinase gene and neomycin-resistance gene are used in a selection scheme that favors gene replacement.

E11. What is a chimera? How are chimeras made?

E12. Evidence (see *Nature* 399, 316–17) suggested that Dolly may have been "genetically older" than her actual age would have suggested. As mammals age, the chromosomes in somatic cells tend to shorten from the telomeres. Therefore, older individuals have shorter chromosomes in their somatic cells compared to younger ones. When researchers analyzed the chromosomes in the somatic cells of Dolly when she was about 3 years old, the lengths of her chromosomes were consistent with a sheep that was significantly older, say 9 or 10 years old. (Note: The sheep that donated the somatic cell that produced Dolly was 6 years old, and her mammary cells had been grown in culture for several cell doublings before a mammary cell was fused with an oocyte.)

A. Suggest an explanation why Dolly's chromosomes seemed so old.

B. Let's suppose that Dolly at age 11 gave birth to a lamb named Molly; Molly was produced naturally (by mating Dolly with a normal male). When Molly was 8 years old, a sample of somatic cells was analyzed. How old would you expect Molly's chromosomes to appear, based on the phenomenon of telomere shortening? Explain your answer.

C. Discuss how the observation of chromosome shortening, which was observed in Dolly, might affect the popularity of organismal cloning.

E13. When transgenic organisms are made, the transgene may integrate into multiple sites within the genome. Furthermore, the integration site may influence the expression of the gene. For example, if a transgene integrates into a heterochromatic (i.e., highly condensed) region of a chromosome, the transgene may not be expressed. For these reasons, it is important for geneticists to analyze transgenic organisms with regard to the number of transgene insertions and the expression levels of the transgenes. Chapter 18 describes three methods (Southern blot, Northern blot, and Western blot) that can be used to detect genes and gene products. Which of these techniques would you use to determine the number of transgenes in a transgenic plant or animal? Why is it important to know the number of transgenes? (Hint: You may want to use a transgenic animal or plant as breeding stock to produce many more transgenic animals or plants.) Which technique would you use to determine the expression levels of transgenes?

E14. What is molecular pharming? Compared to the production of proteins by bacteria, why might it be advantageous?

E15. What is organismal cloning? Are identical twins in humans considered to be clones? With regard to agricultural species, what are some potential advantages to organismal cloning?

E16. Researchers have identified a gene in humans that (when mutant) causes severe dwarfism and mental retardation. This disorder is inherited in an autosomal recessive manner, and the mutant allele is known to be a loss-of-function mutation. The same gene has been found in mice, although a mutant version of the gene has not been discovered in mice. To develop drugs and an effective therapy to treat this disorder in humans, it would be experimentally useful to have a mouse model. In other words, it would be desirable to develop a strain of mice that carry the mutant allele in the homozygous condition. Experimentally, how would you develop such a strain?

E17. Here are traditional DNA fingerprints of five people: a child, mother, and three potential fathers:

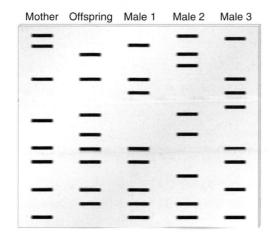

Mother   Offspring   Male 1   Male 2   Male 3

Which males can be ruled out as being the father? Explain your answer. If one of the males could be the father, explain the general strategy for calculating the likelihood that he could match the offspring's DNA fingerprint by chance alone.

E18. What is DNA fingerprinting? How can it be used in human identification?

E19. What roles do PCR and Southern blotting play in the analysis of DNA fingerprints?

E20. What is a VNTR? Discuss the relationship between VNTRs and DNA fingerprinting.

E21. When analyzing the DNA fingerprints of a father and his biological daughter, a technician examined 50 bands and found that 30 of them were a perfect match. In other words, 30 out of 50, or 60% of the bands, were a perfect match. Is this percentage too high, or would you expect a value of only 50%? Explain why or why not.

E22. What would you expect to be the minimum percentage of matching bands in a DNA fingerprint for the following pairs of individuals?

A. Mother and son

B. Sister and brother

C. Uncle and niece

D. Grandfather and grandson

E23. Treatment of adenine deaminase deficiency (ADA) is an example of ex vivo gene therapy. Why is this therapy called ex vivo? Can ex vivo gene therapy be used to treat all inherited diseases? Explain.

E24. Describe the targeting methods used in cystic fibrosis gene therapy. Provided the *CF* gene gets to the patient's lung cells, would you expect this to be a permanent cure for the patient, or would it be necessary to perform this gene therapy on a regular basis (say, monthly)?

E25. Many clinical trials are under way that involve the use of gene therapies to inhibit the growth of cancer cells. As discussed in chapter 22, oncogenes are mutant genes that are overexpressed and cause cancer. New gene therapies are aimed at silencing an oncogene by producing antisense RNA that recognizes the mRNA that is transcribed from an oncogene. Based on your understanding of antisense RNA (described in chapter 14), explain how this strategy would prevent the growth of cancer cells.

Questions for Student Discussion/Collaboration

1. Discuss the advantages and disadvantages of gene therapy. Because a limited amount of funding is available for gene therapy research, make a priority list of the three top diseases that you would fund. Discuss your choices.

2. A commercially available strain of *P. syringae* marketed as Frostban B is used to combat frost damage. This is a naturally occurring *Ice*⁻ strain. Discuss the advantages and disadvantages of using this strain compared with a recombinant version.

3. Make a list of the types of traits you would like to see altered in transgenic plants and animals. Suggest ways (i.e., what genes would you use?) to accomplish these alterations.

Note: All answers appear at the website for this textbook; the answers to even-numbered questions are in the back of the textbook.

www.mhhe.com/brooker

Visit the Online Learning Center for practice tests, answer keys, and other learning aids for this chapter. Enhance your understanding of genetics with our interactive exercises, web links, news feeds, tutorial service, and much more.

STRUCTURAL GENOMICS

::

20

CHAPTER OUTLINE

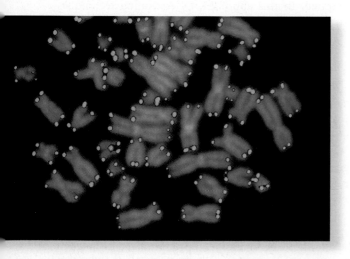

The term **genome** refers to the total genetic composition of an organism. For example, the nuclear genome of humans is composed of 22 different autosomes and two sex chromosomes. In addition, humans have a mitochondrial genome that is composed of a single circular chromosome.

As genetic technology has progressed over the past few decades, researchers have gained increasing ability to analyze the composition of genomes as a whole unit. The term **genomics** refers to the molecular analysis of the entire genome of a species. Genome analysis is a molecular dissection process applied to a complete set of chromosomes. Segments of chromosomes are cloned and analyzed in progressively smaller pieces, the locations of which are known on the intact chromosomes. This is the mapping phase of genome analysis. The mapping of the genome ultimately progresses to the determination of the complete DNA sequence, which provides the most detailed description of an organism's genome at the molecular level.

In 1995, a team of researchers headed by Craig Venter and Hamilton Smith obtained the first complete DNA sequence of the bacterial genome from *Hemophilus influenzae*. Its genome is composed of a single circular chromosome 1.83 million base pairs (bp) in length and contains approximately 1,743 genes (fig. 20.1). In 1996, the first entire DNA sequence of a eukaryote, *Saccharomyces cerevisiae* (baker's yeast), was completed. This work was carried out by a European-led consortium of more than 100 laboratories, including some in the United States, Japan, and Canada. The overall coordinator was Andre Goffeau in Belgium.

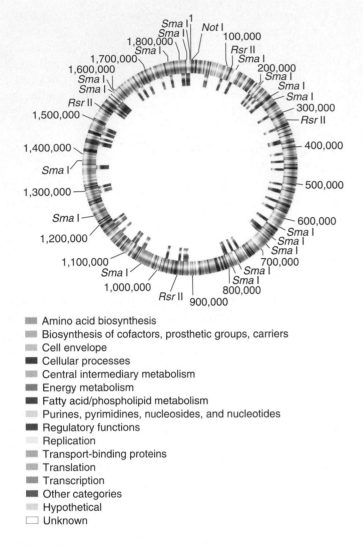

- Amino acid biosynthesis
- Biosynthesis of cofactors, prosthetic groups, carriers
- Cell envelope
- Cellular processes
- Central intermediary metabolism
- Energy metabolism
- Fatty acid/phospholipid metabolism
- Purines, pyrimidines, nucleosides, and nucleotides
- Regulatory functions
- Replication
- Transport-binding proteins
- Translation
- Transcription
- Other categories
- Hypothetical
- Unknown

FIGURE 20.1   A complete map of the genome of the bacterium *Hemophilus influenzae.* On the outside of the circular chromosome, the numbering of nucleotides begins at the *top* of the circle and proceeds clockwise in 100,000-nucleotide increments. The map also shows the locations of restriction enzyme sites for *Not*I, *Sma*I, and *Rsr*II. As shown in the key, the general categories of genes are color coded.

GENES→TRAITS   By mapping and DNA sequencing the entire genome of a species, researchers can identify all the genes it possesses. It is the expression of these many genes that ultimately determines an organism's inherited traits.

The yeast genome contains 16 linear chromosomes; these 16 chromosomes have a combined length of about 12 million bp and contain approximately 6,200 genes. Since that time, genome sequences from many prokaryotes and eukaryotes have been completed. These will be examined later.

In chapter 20, we will focus on methods that are aimed at elucidating the organization of the sequences within a species' genome. As mentioned, this process begins with the mapping of regions along an organism's chromosomes and ends with the determination of the complete DNA sequence. This process is called **structural genomics.** Once a genome sequence is known, it then becomes possible to examine, at the level of many genes,

how the components of a genome interact to produce the traits of an organism. This is called **functional genomics.** Ultimately, a long-term goal of researchers is to determine the roles of all cellular proteins, as well as the interactions that these proteins experience, to produce the characteristics of particular cell types, and the traits of complete organisms. This is the research area known as **proteomics.** In chapter 20, we will explore the steps of structural genomics. In chapter 21, we will consider the exciting prospects that functional genomics and proteomics have to offer regarding our future understanding of genetics.

20.1 CYTOGENETIC AND LINKAGE MAPPING

There are three common ways to determine the organization of DNA regions. These are cytogenetic, linkage, and physical mapping strategies. Before we discuss these approaches in detail, it is helpful to get an overview by comparing them. **Cytogenetic mapping** (also called **cytological mapping**) relies on the localization of gene sequences within chromosomes that are viewed microscopically. In most cases, each chromosome has a characteristic banding pattern, and genes are mapped cytologically relative to a band location. This type of mapping is more refined in organisms with polytene chromosomes, such as *Drosophila,* because these chromosomes have a very detailed banding pattern (see chapter 8). By comparison, in chapter 5, we considered how genetic crosses are conducted to map the relative locations of genes within a chromosome. Such genetic studies, which are called **linkage mapping,** use the frequency of genetic recombination between different genes to determine their relative spacing and order along a chromosome. In eukaryotes, linkage analysis involves crosses among organisms that are heterozygous for two or more genes. The number of recombinant offspring provides a relative measure of the distance between genes, which is computed in map units (or centiMorgans). Finally, genes can be **physically mapped.** In this approach, DNA cloning techniques are used to determine the location of and distance between genes and other DNA regions. In a physical map, the distances are computed as the number of nucleotide bp between genes.

Figure 20.2 compares the map distance between two genes, *sc* (scute, an abnormality in bristle formation) and *w* (white eye) in *Drosophila melanogaster* according to linkage, cytogenetic, and physical maps. In the linkage map, genetic crosses indicate that the two genes are approximately 1.5 map units (mu) apart. In the cytogenetic map, analysis of polytene chromosomes has shown that the *sc* gene is located at band 1B3 and the *w* gene is located at band 3C2. With regard to physical mapping, the two genes are approximately 1.5×10^6 bp apart from each other along the X chromosome. Correlations between linkage, cytogenetic, and physical maps often vary from species to species and from one region of the chromosome to another. For example, a distance of 1 mu may correspond to 10^6 bp in one region of the chromosome, but other regions may recombine at a much lower rate so that a distance of 1 mu may be a much longer physical segment of DNA.

Linkage map:

Cytogenetic map:

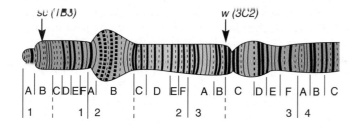

Physical map:

FIGURE 20.2 **A comparison of linkage, cytogenetic, and physical maps.** Each of these maps shows the distance between the *sc* and *w* genes in *Drosophila melanogaster*.

In this section, we will explore several techniques aimed at producing cytogenetic and linkage maps. Later in chapter 20, we will consider physical mapping and the current state of ongoing worldwide projects to analyze the genomes of humans and many other species.

Cytogenetic Mapping Provides a Way to Determine the Locations of Genes by Microscopically Examining the Chromosomes

As mentioned, the goal of cytogenetic mapping is to determine the location of a DNA region, such as a gene, as it occurs along an intact chromosome. It is commonly used in eukaryotes, which have much larger chromosomes. Cytologically, eukaryotic chromosomes can be distinguished from one another by their size, centromeric location, and banding patterns. By treating chromosomal preparations with particular dyes, a discrete banding pattern is obtained for each chromosome. Cytogeneticists use this banding pattern as a way to describe specific regions along a chromosome.

Cytogenetic mapping attempts to determine the location of a particular gene relative to a banding pattern of a chromosome. For example, the human gene that encodes phenylalanine hydroxylase, the enzyme that is defective in people with phenylketonuria (PKU), is located on chromosome 12, at the band designated q24. Cytogenetic mapping is often used as a first step in the localization of genes in plants and animals. However, since it relies on light microscopy, cytological analysis has a fairly crude

limit of resolution. In most species, cytogenetic mapping is accurate only within limits of approximately 5 million bp along a chromosome. In species that have polytene chromosomes, such as *Drosophila*, the resolution is much better. A common strategy used by geneticists is to roughly locate a gene by cytogenetic analysis and then determine its location more precisely by the physical mapping methods described later in chapter 20.

In situ Hybridization Can Localize Genes Along Particular Chromosomes

In situ **hybridization** can detect the location of a gene at a particular site within an intact chromosome. This method is widely used to cytologically map the locations of genes or other DNA sequences within large eukaryotic chromosomes. The term *in situ* (from the Latin for "in place") indicates that the procedure is conducted on chromosomes that are being held in place (i.e., adhered to a surface).

To map a gene via *in situ* hybridization, researchers use a probe to detect the location of the gene within a set of chromosomes. The gene of interest has been cloned previously as described in chapter 18. The DNA of the cloned gene is used as a probe to find where the same gene is located within a set of chromosomes. Because the cloned gene, which is a very small piece of DNA relative to a chromosome, will hybridize only to its complementary sequence on a particular chromosome, this technique provides a distinct localization for any cloned gene. For example, let's consider the gene that causes the white-eye phenotype in *Drosophila* when it carries a loss-of-function mutation. This gene has already been cloned. If a single-stranded piece of this cloned DNA is mixed with *Drosophila* chromosomes in which the DNA has been denatured, it will bind only to the X chromosome, at the location corresponding to the site of the eye color gene.

The most common method of *in situ* hybridization uses fluorescently labeled DNA probes. This specialized form of *in situ* hybridization is referred to as **fluorescence *in situ* hybridization (FISH)**. Figure 20.3 describes the steps of the FISH procedure. The cells are prepared using a technique that keeps the chromosomes intact. A DNA probe is then added. For example, the added DNA probe might be complementary to a specific gene. In this case, the goal of a FISH experiment is to determine the location of the gene within a particular set of chromosomes.

To detect where the probe has bound to a chromosome, the probe is subsequently tagged with a fluorescent molecule. A fluorescent molecule is one that absorbs light at a particular wavelength and then emits light at a longer wavelength. To detect the light emitted by a fluorescent probe, a fluorescence microscope is used. Such a microscope contains filters that allow the passage of light only within a defined wavelength range. The sample is illuminated at the absorption wavelength of the fluorescent molecule and then viewed through a filter that transmits light within the emission wavelength of the molecule. Since the fluorescent probe binds only to a specific sequence, it is seen as a colorfully glowing region against a nonglowing background. Because of this background, the slightest amount of color is readily detectable. This makes fluorescence microscopy a very sensitive technique. The

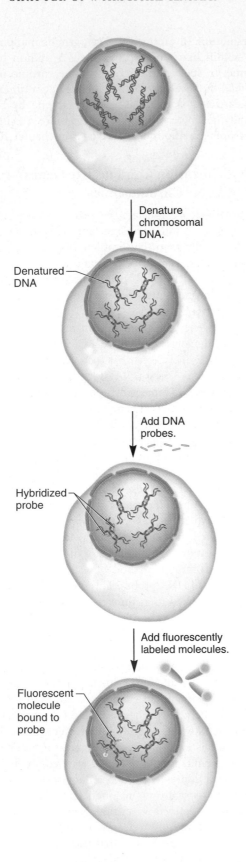

Denature chromosomal DNA.

Denatured DNA

Add DNA probes.

Hybridized probe

Add fluorescently labeled molecules.

Fluorescent molecule bound to probe

FIGURE 20.3   The technique of fluorescence *in situ* hybridization (FISH). The cells are treated with agents that make them swell and are then fixed on slides. The chromosomal DNA is denatured into single-stranded regions. The single-stranded probe is added. The DNA probe hybridizes to the denatured chromosomal DNA only at specific sites in the genome that are complementary to the probe. A fluorescently labeled molecule that binds to the probe DNA is added. (The probe DNA has been chemically modified to allow the fluorescent label to specifically attach to probe DNA.) The preparation is viewed under a fluorescence microscope. Note that the chromosomes are highly condensed metaphase chromosomes that have already replicated. Therefore, each X-shaped chromosome actually contains two copies of a particular chromosome. These are sister chromatids. Because the sister chromatids are identical, a probe that recognizes one sister chromatid will also bind to the other. In some cases, the sister chromatids are so close to each other that the fluorescence looks like a single colored spot. In other cases, you can see two distinguishable adjacent spots because the probe binds to both sister chromatids.

results of a FISH experiment are then compared to a sample of chromosomes that have been Giemsa stained so that the location of a probe can be mapped relative to the G banding pattern.

Figure 20.4 illustrates the results of an experiment involving six different DNA probes. The six probes were cloned pieces of DNA corresponding to six different fragments of DNA from human chromosome 5. In this experiment, each probe was labeled with a different combination of fluorescent molecules. This enabled researchers to distinguish the probes when they became bound to their corresponding locations on chromosome 5. In this experiment, computer-imaging methods were used to assign each fluorescently labeled probe a different color. In this way, fluorescence *in situ* hybridization discerns the sites along chromosome 5 corresponding to the six different probes. In a visual, colorful way, this technique was used here to determine the order and relative distances between several specific sites along a single chromosome. Overall, FISH is commonly used in genetics and cell biology. FISH has also gained widespread use in research and clinical applications.

Linkage Mapping Can Use Molecular Markers

Genetic linkage mapping relies on the frequency of recombinant offspring to determine the distance between sites that are located along the same chromosome. The linkage mapping methods described in chapters 5 and 6 used allelic differences between genes to map the relative locations of those genes along a chromosome. As an alternative to gene mapping, geneticists have realized that regions of DNA, which need not encode genes, can be used as markers along a chromosome. A **molecular marker** is a segment of DNA that is found at a specific site along a chromosome and has properties that enable it to be uniquely recognized

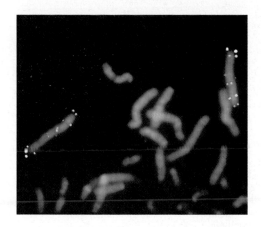

FIGURE 20.4   **The results of a fluorescence *in situ* hybridization experiment.** In this experiment, six different probes were used to locate six different sites along chromosome 5. The colors are due to computer imaging of fluorescence emission; they are not the actual colors of the fluorescent labels. There are two spots at each site because the chromosomes have replicated to form sister chromatids that lie side-by-side.

using molecular tools such as gel electrophoresis. As with alleles, the characteristics of molecular markers may vary from individual to individual. Therefore, the distances between linked molecular markers can be determined from the outcomes of crosses. In most species, it is much easier to identify many molecular markers rather than identify many allelic differences among individuals. For this reason, geneticists have increasingly turned to molecular markers as points of reference along genetic maps. As described in table 20.1, many different kinds of molecular markers are used by geneticists. These markers can be used in linkage mapping and also in cytogenetic and physical mapping studies. We will consider **restriction fragment length polymorphisms (RFLPs)** first.

As discussed in chapter 18, restriction enzymes recognize specific DNA sequences and cleave the DNA at those sequences. Along a very long chromosome, a particular restriction enzyme will recognize many sites. For example, a commonly used restriction enzyme, *Eco*RI, recognizes 5′-GAATTC-3′. Simply by chance, this six-nucleotide sequence is expected to occur (on average) every 4^{-6} times, or once in every 4,096 nucleotides. A chromosome composed of millions of nucleotides contains many *Eco*RI sites, which will be randomly distributed along the chromosome. Therefore, *Eco*RI will digest a chromosome into many smaller pieces of different lengths.

When comparing two different individuals, the digestion of chromosomal DNA by a given restriction enzyme may produce certain fragments that differ in their length, even though such fragments are found at the same chromosomal locations in the two individuals (fig. 20.5). We would say there is **polymorphism** in the population with regard to the length of a particular DNA fragment. This variation can arise for several reasons. Among different individuals, genetic changes such as small deletions and duplications may subtract or add segments of DNA to a particular region of the chromosome. This tends to occur more frequently if a region contains a repetitive sequence. Alternatively, a mutation may change the DNA sequence in a way that alters a recognition site for a restriction enzyme. For example, in figure 20.5, the elimination of an *Eco*RI site has occurred in the chromosome of individual 2. This has altered the sizes of certain DNA fragments when the chromosomal DNA is digested with the restriction enzyme.

TABLE 20.1

Common Types of Molecular Markers

Marker	Description
Restriction fragment length polymorphism (RFLP)	A site in a genome where the distance between two restriction sites varies among different individuals. These sites are identified by restriction enzyme digestion of chromosomal DNA, and the use of Southern blotting to identify the specific fragment.
Amplified restriction fragment length polymorphism (AFLP)	The same as an RFLP except that the fragment is amplified via PCR instead of isolating the chromosomal DNA and subjecting the chromosomal DNA to restriction digestion and Southern blotting.
Minisatellite, also called a site with a variable number of tandem repeats (VNTR)	A site in the genome that contains many repeat sequences. These sites are usually in the size range of several hundred to a few thousand base pairs. Because they are usually polymorphic, they were once used in DNA fingerprinting but have been largely superseded by microsatellites (see chapter 19).
Microsatellite, also called a short tandem repeat (STR)	A site in the genome that contains many short, tandem repeat sequences. These sites are usually in the size range of 100 to 500 bp, and they may be polymorphic within a population. They are isolated via PCR.
Single-nucleotide polymorphism (SNP)	A site in the genome where a single nucleotide is polymorphic among different individuals. These sites occur commonly in all genomes, and they are gaining greater use in the mapping of disease-causing alleles and in the mapping of genes that contribute to quantitative traits.
Sequence-tagged site (STS)	This is a general term to describe any molecular marker that is found at a unique site in the genome and is amplified by PCR. AFLPs, microsatellites, and SNPs can provide sequence-tagged sites within a genome.

Figure 20.6 considers the application of RFLP analysis along a short chromosomal region among three diploid individuals. Due to slight differences in their DNA sequences, these individuals vary with regard to the existence of an *Eco*RI site shown in *red*. In certain individuals, a mutation may change the sequence at this

FIGURE 20.5   Restriction fragment length polymorphisms (RFLPs).

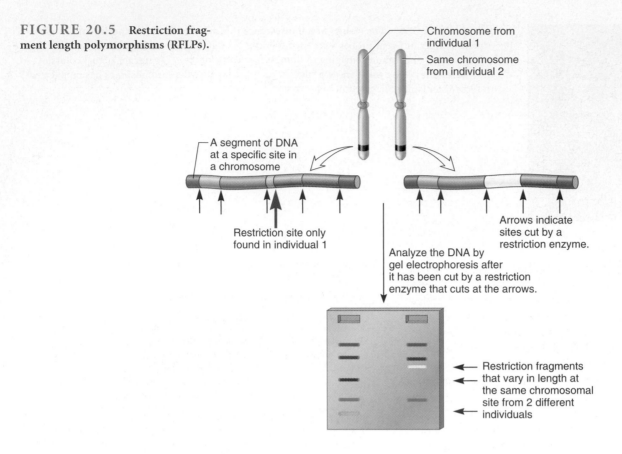

Chromosome from
individual 1

Same chromosome
from individual 2

A segment of DNA
at a specific site in
a chromosome

Restriction site only
found in individual 1

Arrows indicate
sites cut by a
restriction enzyme.

Analyze the DNA by
gel electrophoresis after
it has been cut by a restriction
enzyme that cuts at the arrows.

Restriction fragments
that vary in length at
the same chromosomal
site from 2 different
individuals

site so that it is no longer recognized by *Eco*RI. In a diploid species, each individual has two copies of this region. In individual 1, both of these *Eco*RI sites are present and in individual 2, both are absent. By comparison, individual 3 is a heterozygote. In one of the chromosomes the site is present, but in the other it is absent. In this experiment, the DNA was isolated and then cut by *Eco*RI. The DNA from individual 1 is cut at six sites by *Eco*RI, whereas individual 2 has only five locations where the DNA is cut. In individual 2, an *Eco*RI site is missing because the sequence of DNA in this region has been changed from GAATTC, so it is not recognized by *Eco*RI.

When subjected to gel electrophoresis, the change in the DNA sequence at a single *Eco*RI restriction site produces a variation in the sizes of homologous DNA fragments. The term *polymorphism* (meaning many forms) refers to the idea that the individuals within a population differ with regard to these particular DNA fragments. As noted in figure 20.6, not all restriction sites are polymorphic. The three individuals share many DNA fragments that are identical in size. When a DNA segment is identical among all members of a population, it is said to be **monomorphic** (meaning one form). By comparison, if there is genetic variation in a population for a certain DNA region, it is said to be polymorphic. As a practical rule of thumb, a DNA segment is considered monomorphic when over 99% of the individuals in the population have identical sequences at that segment.

The preceding discussion was meant as an overview of the phenomenon of RFLPs. However, the experiment of figure 20.6 is a technical oversimplification, because it considers only a very short region of chromosomal DNA. In an actual RFLP analysis, DNA samples containing all of the chromosomal DNA would be isolated from these three individuals. The digestion of the chromosomal DNA with *Eco*RI would then yield so many fragments that the results would be very difficult to analyze. To circumvent this problem, Southern blotting is used to identify specific RFLPs.

Figure 20.7 illustrates such an experiment that reconsiders the chromosomes from the three individuals described in figure 20.6. A cloned piece of DNA that corresponds to the region near the fourth *Eco*RI site is used as a radiolabeled probe in a Southern blot of the chromosomal DNA. When the blot is exposed to X-ray film, we will see only the DNA band that can hybridize to the radiolabeled probe. As shown in figure 20.7, individual 1 shows one band of 3,000 bp and individual 2 has a band that is 4,500 bp long. As discussed in regard to figure 20.6, this difference is due to the absence of an *Eco*RI site in individual 2.

Individual 3 has both bands because of heterozygosity for this *Eco*RI site. In other words, the heterozygote will have two bands of different lengths, whereas homozygotes will display only one band, because the RFLP procedure detects the molecular products of each chromosome.

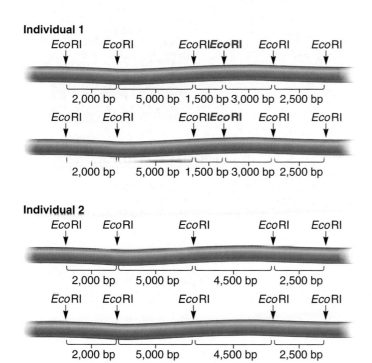

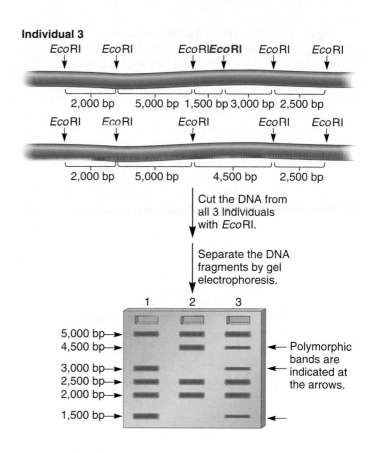

FIGURE 20.6   **An RFLP analysis of chromosomal DNA from three different individuals.** The *Eco*RI site shown in *red* is sometimes missing because a mutation has changed the DNA sequence in this region.

The Distance Between Two Linked RFLPs Can Be Determined

Now that we understand what an RFLP is, let's consider how RFLPs are mapped and used as chromosomal markers. In the genetic mapping experiments described in chapter 5, we learned how the results of genetic crosses are used to map the distance between two genes. In that type of analysis, the proportions of offspring with recombinant phenotypes were used to calculate the map distance between genes. Likewise, we can map the distance between two RFLPs by making crosses and analyzing the offspring. In an RFLP analysis, however, we do not look at the phenotypic characteristics of the offspring (e.g., white eyes or miniature wings). Instead, we look at their DNA bands on a gel. To do this, DNA is obtained from a tissue sample.

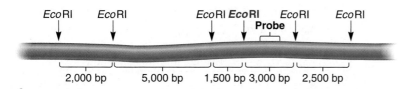

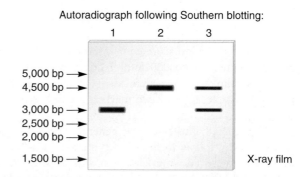

FIGURE 20.7   **Southern blot hybridization of a specific RFLP.** The *left* side shows the region of the chromosome that was used as a probe in a Southern blot hybridization. The radiolabeled probe is a DNA segment located just to the right of the fourth *Eco*RI site. In this experiment, a Southern blot was conducted on the chromosomal DNA from the same three individuals described in figure 20.6.

Figure 20.8 presents an example of how RFLP mapping is done. Let's suppose there are two RFLPs that are located at different sites in the genome. One RFLP is detected with probe 1 and yields either a 4,500 bp band or a 6,500 bp band. A second RFLP is detected with probe 2 and yields either a 2,000 bp band or a 1,500 bp band. In the experiment of figure 20.8, the goal is to determine whether or not the 4,500/6,500 and 2,000/1,500 sites are linked along the same chromosome and, if so, to find the map distance between them. The researcher begins with two strains that are homozygous: 4,500 and 2,000 versus 6,500 and 1,500. These strains are crossed to each other to produce a heterozygote. The heterozygote can then be crossed to a homozygote that carries only the 4,500 and 2,000 bands. This is analogous to a dihybrid test cross described in chapter 5 in which an F_1 heterozygote is crossed to a homozygote (see fig. 5.9). The results of this cross depend on whether the RFLPs are closely linked or not (see the *right* side of figure 20.8). If an offspring inherits pairs of RFLPs like the parents (4,500 bp and 2,000 bp or 6,000 bp and 1,500 bp), such offspring will show four bands of 4,500 bp, 2,000 bp, 6,500 bp, and 1,500 bp or two bands of 4,500 and 2,000 bp. The recombinant categories are 6,500, 4,500, and 2,000 bp or 4,500, 2,000, and 1,500 bp. If the RFLPs were not linked, we would expect a 1:1:1:1 ratio of the four types among the offspring. If the RFLPs were linked, we would expect a higher percentage of offspring with a parental combination of RFLPs. Figure 20.8 describes the results of 100 offspring. There are many more parental offspring, indicating that the RFLPs are closely linked.

In an actual analysis of RFLP data, researchers simultaneously analyze many RFLPs and decide on the likelihood of linkage between two RFLPs by using a statistical test called the **lod** (<u>l</u>ogarithm of <u>od</u>s) **score method.** This method was devised by Newton Morton in 1955. Although the theoretical basis of this approach is beyond the scope of this textbook, computer programs analyze pooled data from a large number of pedigrees or crosses involving many RFLPs. The programs determine the probability that any two markers exhibit a certain degree of linkage (i.e., are within a particular number of map units apart), and also the probability that the data would have been obtained if the two markers were unlinked. The lod score is then calculated as

$$\text{Lod score} = \log_{10} \frac{\text{Probability of a certain degree of linkage}}{\text{Probability of independent assortment}}$$

For example, if the lod score equals 3, the $\log_{10}$ of 1,000 equals 3. This means there is a 1,000-fold greater probability that the two markers are linked rather than assort independently. Traditionally, geneticists accept that two markers are linked if the lod score is 3 or higher.

Figure 20.8 illustrates the general type of approach that researchers can follow to map RFLP markers. If a lod score suggested that these two markers were linked, we can divide the recombinant offspring (7 + 9 = 16) by the total number (100) to obtain a map distance of 16 mu.

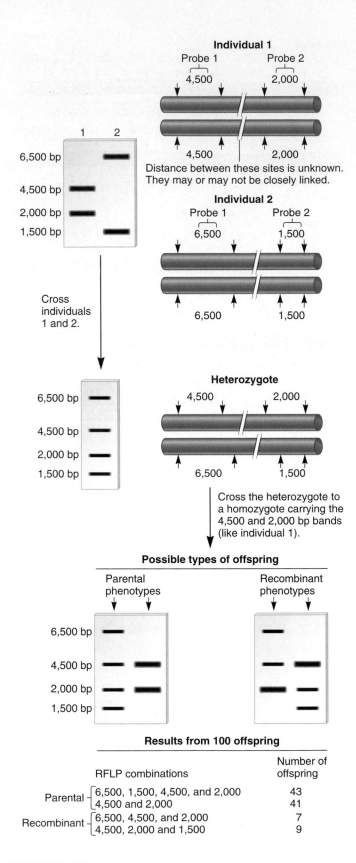

FIGURE 20.8   Linkage analysis of RFLP markers.

An RFLP Map Describes the Locations of Many Different RFLPs Throughout the Genome

As described in the preceding section, the map distance between two RFLPs can be determined by analyzing the transmission of RFLP markers from parents to offspring. Such an analysis can be conducted on many different RFLPs to determine their relative locations throughout a genome. RFLPs are quite common in virtually all species, so that geneticists can easily map the locations of many RFLPs within a genome. A genetic map composed of many RFLP markers is called an **RFLP map.** Figure 20.9 shows a simplified RFLP map of the plant *Arabidopsis thaliana* (a small plant in the mustard family), which is one of the favorite organisms of plant molecular geneticists. Many RFLPs have been mapped to different locations along the five *Arabidopsis* chromosomes.

An RFLP map, like the one shown in figure 20.9, can be used to locate functional genes within the genome. In the analysis of human disease-causing genes and in the genetic analysis of agriculturally important crops, RFLPs along with other markers described later are widely used as a way to locate genes along particular chromosomes. As an example, solved problem S2 at the end of chapter 20 illustrates how RFLP analysis can be used to map a gene that confers herbicide resistance.

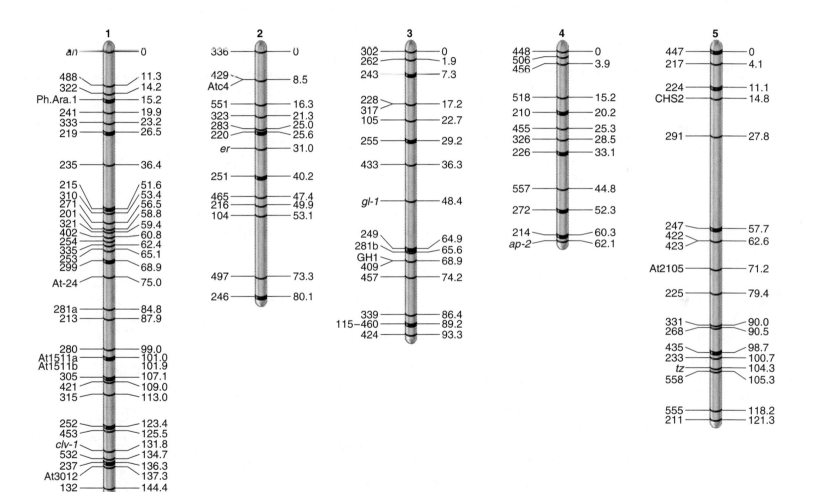

FIGURE 20.9 **An RFLP linkage map of *Arabidopsis thaliana*.** This plant has five different chromosomes. The *left* side of each chromosome describes the locations of RFLP markers. The numbers along the *right* side of each chromosome are the map distances in map units. For example, 219 and 235 are RFLPs located 9.9 mu apart on chromosome 1. The *top* marker at the end of each chromosome was arbitrarily assigned as the starting point (zero) for each chromosome. In addition, the map shows the locations of a few known genes (shown in *red*): *Ph.Ara.1* = phytochrome, *At-24* = nitrate reductase, *At1511a/b* = small RNA found in seeds, *clv-1* = clavata-1, *At3012* = alcohol dehydrogenase, *Atc4* = actin, *er* = erecta, *gl-1* = glabra-1, *GH1* = acetolactate synthase, *ap-2* = apelata-2, *CHS2* = chalcone synthase, *At2105* = 12S seed storage protein, and *tz* = thiazole-requiring allele.

GENES→TRAITS As a first step in mapping the locations of an organism's genes, researchers may initially determine the sites of RFLPs along the chromosomes. In this case, researchers determined the locations of many of these sites along the *Arabidopsis* chromosomes. By mapping these RFLP sites, it becomes easier to locate genes within the *Arabidopsis* genome. The identification of genes helps researchers to elucidate the relationship between genes and traits.

EXPERIMENT 20A

RFLP Analysis Can Be Used to Follow the Inheritance of Disease-Causing Alleles

As we have just seen, RFLP analysis can locate particular genes relative to an RFLP marker. This method may also be used to inform prospective parents of the likelihood that they are heterozygous for a recessive allele that may cause a genetic disease. This information would allow them to predict the likelihood of having children affected with the disorder. This genetic testing method is particularly useful for diseases in which the phenotype of the heterozygote may not differ from that of the normal homozygote.

The assumption behind this approach is that a disease-causing allele had its origin in a single individual known as a **founder,** who lived many generations ago. Since that time, the allele has spread throughout portions of the human population. In the case of recessive disorders, descendants of the founder who are affected with the disease have inherited two copies of the mutant allele. A second assumption is that the founder is likely to have had a polymorphic molecular marker that lies somewhere near the mutant allele. This is a reasonable assumption, because all people carry many polymorphic markers throughout their genomes. If a polymorphic marker lies very close to the disease-causing allele, it is unlikely that a crossover will occur in the intervening region. Therefore, such a polymorphic marker may be linked to the disease-causing allele for many generations. For this reason, an association between a particular polymorphic marker and a disease-causing allele may help to predict whether a person is a heterozygote for a recessive disease-causing allele.

As an example, let's consider the allele that causes sickle-cell anemia. A couple may want to know if they are heterozygous carriers of the recessive allele that causes sickle-cell anemia (see chapter 4 for a description of this disease). Other techniques are now used to distinguish sickle-cell homozygotes and heterozygotes. Nevertheless, the experiment described in figure 20.10 by Yuet Kan and Andree Dozy is a classic study conducted in 1978, which confirmed that RFLP markers can be used to predict the likelihood that an individual carries a disease-causing allele. The researchers conducted an RFLP analysis in the region of the β-globin gene. The normal β-globin gene encodes the β-globin polypeptide; this is a subunit of normal hemoglobin, designated hemoglobin A (Hb^A). A mutant form of the β-globin gene encodes an abnormal polypeptide, which results in the formation of hemoglobin S (Hb^S). Individuals who are homozygous for this mutant gene develop sickle-cell anemia. Kan and Dozy set out to determine if RFLP differences could be detected between chromosomes containing the normal and mutant forms of the β-globin gene. In other words, they wanted to discover if the restriction digestion of chromosomal DNA in the vicinity of the β-globin gene was different in the Hb^A and Hb^S forms of the gene.

■ THE HYPOTHESIS

A mutant allele may be closely linked to a particular RFLP. People who carry the mutant allele can be identified by the presence of this RFLP.

■ TESTING THE HYPOTHESIS — FIGURE 20.10   RFLP linkage to a disease-causing allele.

Starting materials: The β-globin gene had been cloned previously. This clone was used as a radiolabeled probe. The researchers obtained blood samples from several individuals, some of whom were known to be affected with sickle-cell anemia.

Experimental level	Conceptual level

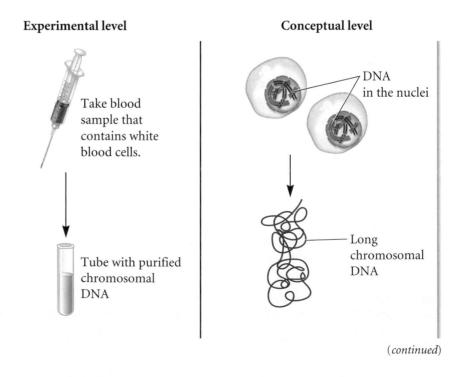

1. Take blood samples from 73 people. (Note: The *Hb* genotype was already known from a biochemical analysis of the subjects' hemoglobin.)

   Take blood sample that contains white blood cells.

   DNA in the nuclei

2. Isolate chromosomal DNA from the cells. This is similar to plasmid isolation. It involves breaking the cells and then using chromatography or centrifugation to purify the DNA. See experiment 19A.

   Tube with purified chromosomal DNA

   Long chromosomal DNA

(*continued*)

3. Digest the chromosomal DNA with the restriction enzyme *Hpa*I.

Add *Hpa*I.

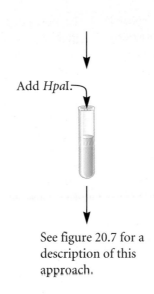

See figure 20.7 for a description of this approach.

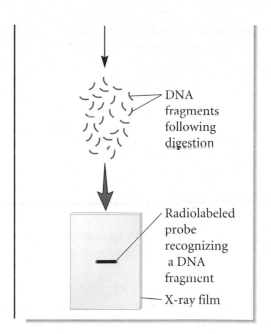

DNA fragments following digestion

Radiolabeled probe recognizing a DNA fragment

X-ray film

4. Use a radiolabeled DNA probe of the β-globin gene to conduct a Southern blot experiment. (Note: The procedure of Southern blotting is described in chapter 18.)

THE DATA

Results from 5 different individuals

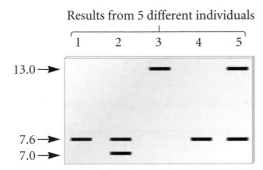

Summary of data from 73 people

Hemoglobin Composition	*Hpa*I β-globin RFLP						Frequency of the 13 kbp Fragment
	7.6/7.6	7.6/7.0	7.0/13	7.6/13	13/13	Total	
African*							
AA	8	6	0	1	0	15	3%
AS	5	1	1	9	0	16	31%
SS	0	0	0	4	11	15	87%
Caucasian							
AA	12	0	0	0	0	12	0%
Asian							
AA	15	0	0	0	0	15 / 73	0%

*Racial origin of the individuals used in this study.

Data from: Kan, Y.W., and Dozy, A.M. (1978) Polymorphism of DNA sequence adjacent to human β-globin structural gene: Relationship to sickle mutation. *Proc. Natl. Acad. Sci. USA* 75, 5631–5635.

INTERPRETING THE DATA

In this study, the genotypes of 73 individuals were already known from a biochemical analysis of their hemoglobin characteristics.

The goal of this research was to see if an RFLP analysis could also distinguish individuals carrying the mutant gene from those carrying the normal gene. As shown in figure 20.10, they began with samples of white blood cells from many individuals and isolated the chromosomal DNA. Next, the chromosomal DNA was digested with the restriction enzyme *Hpa*I. The DNA fragments were separated by gel electrophoresis and then subjected to Southern blotting, using a probe that was complementary to the β-globin gene. Three different RFLPs were found among a population of 73 individuals. These RFLPs produced fragments of molecular weights 7,000 bp, 7,600 bp, and 13,000 bp (or 7.6, 7.0, and 13 kbp [kilobase pairs], respectively).

With regard to the Hb^A and Hb^S alleles, the occurrence of these RFLPs was not random. Instead, the 13 kbp RFLP was usually found in persons who were known to have at least one copy of the Hb^S allele. For example, in approximately 87% of the homozygotes ($Hb^S Hb^S$), the Hb^S allele was associated with the 13 kbp fragment. This observation is consistent with the diagram shown here.

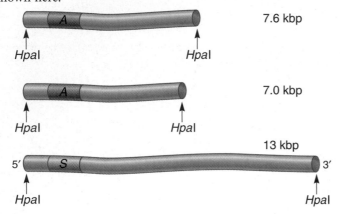

In most people with hemoglobin A, the site where the normal β-globin gene is located has a distance of 7.6 or 7.0 kbp between the *Hpa*I sites. For people carrying the Hb^S allele, the mutant β-globin gene is found in a chromosomal region where the *Hpa*I sites are usually 13 kbp apart.

In most but not all cases, the Hb^S allele was found to be closely linked to an RFLP of 13 kbp. This type of information can be used as a predictive tool in human genetic analysis. For example, if an individual was found to be heterozygous, 7.6/13, it would be fairly likely that the individual is a heterozygous, $Hb^A Hb^S$, carrier of the abnormal gene. By comparison, if an RFLP analysis revealed that an individual was 7.6/7.6, that person would more likely be homozygous for the normal Hb^A allele.

Since these initial studies of the β-globin gene, RFLP markers have been determined to be closely linked to many other alleles associated with human genetic diseases. The analysis of RFLP markers has been used frequently as a tool to predict if individuals are heterozygous carriers of recessive alleles that cause human genetic diseases. This approach has also been used in prenatal testing for a variety of genetic abnormalities.

A self-help quiz involving this experiment can be found at the Online Learning Center.

Genetic Mapping Also Uses Molecular Markers Called Microsatellites

To make a highly refined map of a genome, many different polymorphic sites must be identified and their transmission followed from parent to offspring over many generations. In the case of humans and many other species, however, relatively few polymorphic genes are available for this type of analysis. Therefore, linkage mapping in humans and other eukaryotic species often focuses on the transmission of molecular markers, such as RFLPs, which are not actually genes.

In most species, an even easier method of generating molecular markers has been discovered. These newer markers are short, simple sequences called **microsatellites** that are abundantly interspersed throughout a species' genome and are quite variable in length among different individuals. For example, the most common microsatellite encountered in humans is a sequence $(CA)_n$, where n may range from 5 to more than 50. In other words, this dinucleotide sequence can be tandemly repeated 5 to 50 or more times. These sequences are called microsatellites to distinguish them from the much larger tandemly repeated sequences called satellites (see chapter 10). The $(CA)_n$ microsatellite is found, on average, about every 10,000 bases in the human genome. Researchers have identified thousands of different DNA

segments that contain $(CA)_n$ microsatellites, located at many distinct sites within the human genome. Using primers complementary to the unique DNA sequences that flank a $(CA)_n$ region, a particular microsatellite can be specifically amplified by PCR. In other words, the PCR primers will amplify only a particular microsatellite but not the thousands of others that are interspersed throughout the genome (fig. 20.11).

When a pair of PCR primers amplifies a single DNA sequence within a genome, the amplified region is called a **sequence-tagged site (STS).** In a diploid species, there will be two copies of a given STS in any individual. When an STS contains a microsatellite, the two PCR products will be identical only if the region is the same length in both copies (i.e., if the individual is homozygous for the microsatellite). However, if an individual has two copies that differ in the number of repeats in the microsatellite sequence, the two PCR products obtained will be slightly different in length (as in fig. 20.11).

Like RFLPs, microsatellites that have length polymorphisms allow researchers to follow their transmission from parent to offspring. PCR amplification of particular microsatellites provides an important strategy in the genetic analysis of human pedigrees. This idea is shown in figure 20.12. Prior to this analysis, a unique segment of DNA containing a microsatellite had been identified. Using PCR primers complementary to the unique

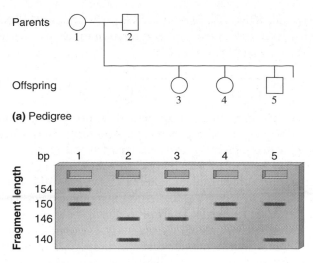

(a) Pedigree

(b) Electrophoretic gel of PCR products for the polymorphic STSs found in the family in (a).

FIGURE 20.12   Inheritance pattern of microsatellites in a human pedigree.

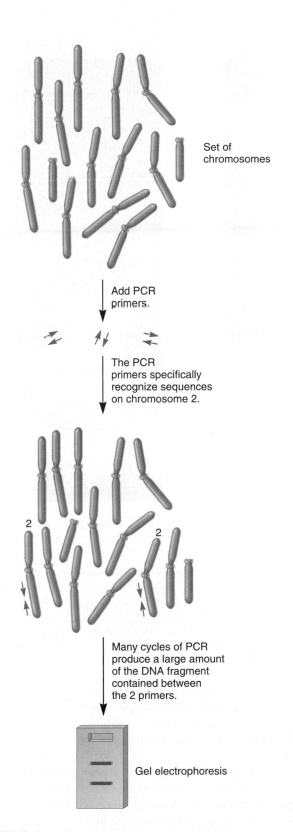

Set of chromosomes

Add PCR primers.

The PCR primers specifically recognize sequences on chromosome 2.

Many cycles of PCR produce a large amount of the DNA fragment contained between the 2 primers.

Gel electrophoresis

FIGURE 20.11   Identifying a microsatellite using PCR primers.

flanking segments, two parents and their three offspring can be tested for the inheritance of this microsatellite. A small sample of cells is obtained from each individual and subjected to PCR amplification, as described earlier in figure 20.11. The amplified PCR products are then analyzed by high-resolution gel electrophoresis, which can detect small differences in the lengths of DNA fragments. The mother's PCR products were 154 and 150 bp in length, while the father's were 146 and 140 bp. Their first offspring inherited the 154 bp product from the mother and the 146 bp from the father, the second inherited the 150 bp from the mother and the 146 bp from the father, and the third inherited the 150 bp from the mother and the 140 bp from the father. As shown in the figure, the transmission of polymorphic microsatellites is relatively easy to follow from generation to generation.

The simple pedigree analysis shown in figure 20.12 illustrates the general method used to follow the transmission of a single microsatellite that is polymorphic in length. In genetic linkage studies, the goal is to follow the transmission of many different microsatellites to determine those that are linked along the same chromosome versus those that are not. Those that are not linked will independently assort from generation to generation. Those that are linked tend to be transmitted to the same offspring. In a large pedigree, it is possible to identify cases where linked microsatellites have segregated due to crossing over. The frequency of crossing over provides a measure of the map distance, in this case between different microsatellites. This approach can help researchers to obtain a finely detailed genetic linkage map of the human chromosomes.

A key difference between RFLPs and microsatellites is that RFLPs are identified by restriction enzyme digestion and Southern blotting. This is technically more difficult than the PCR method that is used to produce microsatellites. A newer kind of molecular marker, termed **amplified restriction fragment length polymorphism** (**AFLP**) combines the approaches used to make

RFLPs and STSs. To identify AFLPs, chromosomal DNA is digested with one or two restriction enzymes, and then specific fragments are amplified via PCR.

20.2 PHYSICAL MAPPING

We now turn our attention to methods aimed at establishing a physical map of a species' genome. Physical mapping requires the cloning of many pieces of chromosomal DNA. The cloned DNA fragments are then characterized by size (i.e., length in nucleotide bp), the gene(s) they contain, and their relative locations along a chromosome. In recent years, physical mapping studies have led to the DNA sequencing of entire genomes.

As mentioned in chapter 10, eukaryotic genomes are very large; the human genome, for example, is approximately 3.2 billion bp in length and the *Drosophila* genome is roughly 137 million bp. When making a physical map of a genome, researchers must characterize many DNA clones that contain much smaller pieces of the genome. For any given species, the physical mapping of the entire genome is a massive undertaking carried out as a collaborative project by many genome research laboratories. In this section, we will examine the general strategies used in creating a physical map of a species' genome. We will also consider how physical mapping information can be used to clone genes.

Specific Chromosomes Can Be Isolated by Pulsed-Field Gel Electrophoresis or Chromosome Sorting

Eukaryotic genomes are composed of several different chromosomes. A first step in the dissection of a large genome involves techniques that separate individual chromosomes or large pieces of chromosomes. One technique that can accomplish this goal is **pulsed-field gel electrophoresis (PFGE),** developed by David Schwartz and Charles Cantor. Using this method, entire small chromosomes or large pieces of chromosomes are subjected to electrophoresis. As its name suggests, pulsed-field gel electrophoresis differs from conventional electrophoresis; the DNA is driven through the gel using alternating pulses of electrical current at different angles. This can be achieved using electrophoresis devices that have two sets of electrodes. The alternating pulses can achieve a good separation of DNA fragments that vary in length from 50,000 bp up to 10 million bp. Therefore, this method is used to separate large fragments of DNA or even small chromosomes.

For PFGE, individual chromosomes or large fragments of chromosomes must be isolated very carefully. To avoid mechanical breakage of the chromosomes, cells are embedded in agarose blocks (also called plugs). The agarose protects the chromosomal DNA from mechanical breakage. Once inside the plug, the cells are lysed, and if desired, the chromosomal DNA can be subjected to a restriction digestion while still in the plug. PFGE can be performed in conjunction with restriction digestion to obtain large DNA fragments. This can be achieved using a restriction enzyme that cuts very infrequently. To separate the large pieces of DNA, the plug is inserted into a well in an agarose gel and then subjected to PFGE. Figure 20.13*a* illustrates a separation of yeast

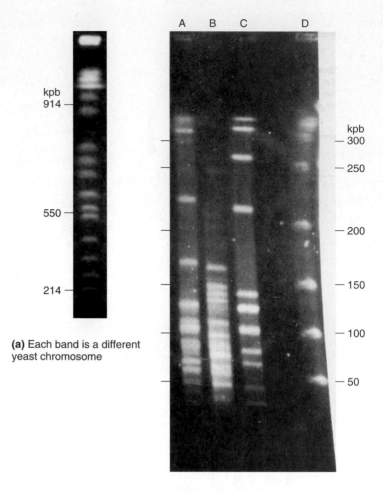

(a) Each band is a different yeast chromosome

(b) Each band is a fragment of a bacterial chromosome

FIGURE 20.13   **Pulsed-field gel electrophoresis (PFGE).** **(a)** Yeast chromosomes that have not been digested with a restriction enzyme. The numbers at the *left* indicate the size of the DNA fragments in kilobase pairs (kbp). **(b)** The *Hemophilus influenzae* chromosome digested with *Eag*I (lane A), *Nae*I (lane B), and *Sma*I (lane C). Lane D shows molecular weight markers.

chromosomes that have not been subjected to restriction digestion. Figure 20.13*b* shows fragments of DNA that have been obtained by digesting a bacterial chromosome with three different restriction enzymes that cut the bacterial chromosome infrequently.

PFGE has two important uses in physical mapping studies. First, it can be used to purify large pieces of DNA for the construction of libraries, which will be described later. For example, the DNA containing a particular yeast chromosome can be isolated from a gel slice and used to construct a chromosome-specific library. In addition, pulsed-field gels can be used to coarsely map the locations of genes. If an experimenter has cloned two different genes, these clones can be labeled and used as probes in a Southern blot of a pulsed-field gel. If both probes bind to the same band, they are located on the same chromosome in the genome or on the same large piece of a chromosome.

The chromosomes of bacteria, and some simpler eukaryotes such as yeast and protozoa, are small enough to be resolved by PFGE. However, the chromosomes of higher eukaryotes (e.g., mammals and plants) are usually too large to be resolved by this procedure. Instead, individual chromosomes can be isolated by **chromosome sorting,** a technique pioneered by scientists at the Los Alamos National Research Laboratory.

As shown in figure 20.14, the starting material for chromosome sorting is a mixture of chromosomes. The chromosomes are stained with two fluorescent dyes: Hoechst 33258, which binds preferentially to AT-rich DNA, and chromomycin A_3, which binds to GC-rich DNA. Each chromosome will have its own characteristic numbers of AT- and GC-rich regions, so that each chromosome will have a distinctive level of fluorescence. The stained chromosomes are forced through a small opening and pass through a laser beam that excites the fluorescent dyes. The stream containing the chromosomes is separated into individual droplets, which are given a negative charge if the fluorescence detector indicates they contain the desired chromosome. In this example, the goal is to separate one type of human chromosome from the rest of the chromosomes. If the fluorescence intensity in the droplet indicates that the drop contains the desired human chromosome, the droplet is given a negative charge and deflected away from the main stream and into a separate tube. In this way, a particular chromosome can be separated from a mixture of many different chromosomes. This device can separate chromosomes at the amazing rate of 1,000 to 2,000 per second!

A Physical Map of a Chromosome Is Constructed by Creating a Contiguous Series of Clones from a Chromosome-Specific Library

As we have just learned, both pulsed-field gel electrophoresis and chromosome sorting can be used to obtain a sample that contains a single type of chromosome or large fragment of a chromosome. The sample of DNA can then be digested into many smaller pieces with a restriction enzyme; these fragments are then cloned into vectors to create a **chromosome-specific library.** Such a library contains a collection of hybrid vectors with different pieces of DNA from the same chromosome.

The next step in physical mapping studies is to determine the relative locations of the cloned chromosomal pieces, as they would occur in an intact chromosome. In other words, the members of the library must be organized according to their actual locations along a chromosome. To obtain a complete physical map of a chromosome, researchers need a series of clones that contain overlapping pieces of chromosomal DNA. Such a collection of clones, known as a **contig,** contains a <u>contiguous</u> region of a chromosome that is found as overlapping regions within a group of vectors (fig. 20.15). As discussed later, cloning vectors known as YACs and cosmids commonly are used in the construction of a contig.

Different experimental strategies can be used to align the members of a contig. The general approach is to identify adjacent members that contain overlapping regions. Historically, Southern blotting (described in chapter 18) was first used to determine if

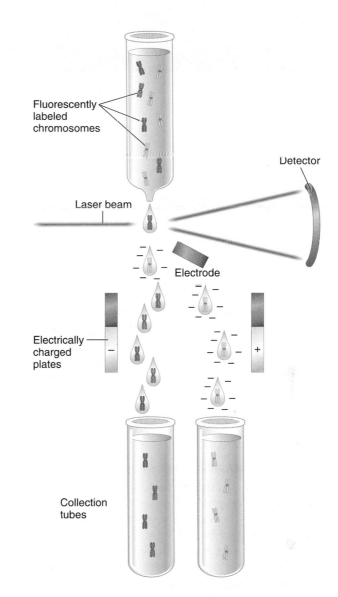

FIGURE 20.14 **Chromosome sorting.** A mixture of chromosomes is treated with two fluorescent stains, Hoechst 33258 and chromomycin A_3, to label the chromosomes. Each type of chromosome is already known to have its own unique level of fluorescence. The chromosome mixture leaves the nozzle in droplets with one chromosome per droplet. A laser beam strikes each droplet. A fluorescence detector identifies fluorescent chromosomes by the intensity and wavelength of light emitted by the chromosomes. An electrode negatively charges a droplet carrying the desired chromosome. As droplets pass between the electrically charged plates, the droplets with a negative charge move closer to the positive plate. In this way, the desired chromosomes are separated as they fall into different collection tubes.

two different clones contain an overlapping region. In the example shown in figure 20.15, the DNA from clone 1 could be radiolabeled in vitro and then used as a probe in a Southern blot of the other clones shown in this figure. Clone 1 would hybridize to clone 2, because they share identical DNA sequences in the overlapping region. Similarly, clone 2 could be used as a probe to show that it will hybridize to clone 1 and clone 3. By conducting Southern blots between many combinations of clones, researchers

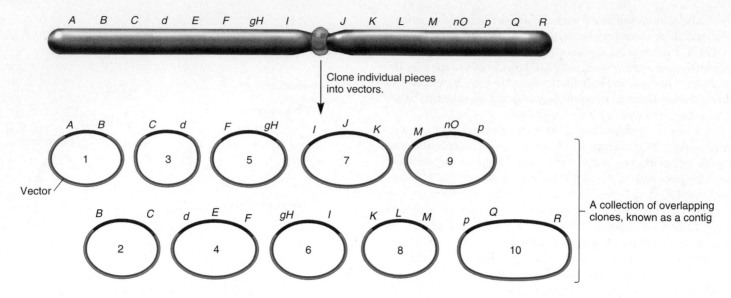

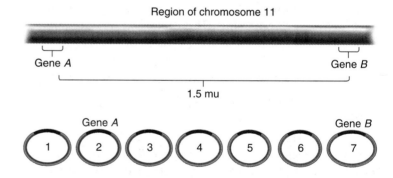

FIGURE 20.15   The construction of a contig. Large pieces of chromosomal DNA are cloned into vectors. The numbers denote the order of the members of the contig. For example, 3 is between 2 and 4. Note that the chromosome is labeled with letters that designate the locations of genes. The members of the contig have overlapping regions. For example, clone 3 ends with gene *d* and clone 4 begins with gene *d;* clone 4 ends with gene *F* and clone 5 begins with gene *F*.

can determine which clones have common overlapping regions and thereby order them, as they would occur along the chromosome.

Alternatively, other methods to order the members of a contig would involve the use of molecular markers. For example, if many STSs have already been identified along a chromosome via linkage and/or cytogenetic analysis, the STSs can be used as probes to order the members of a contig. Another approach is to subject the members of the chromosome-specific library to a restriction enzyme to produce a restriction digest pattern for each member. The patterns of fragments are then analyzed by computer programs, which can identify regions that are potentially overlapping.

An ultimate goal of physical mapping procedures is to obtain a complete contig for each type of chromosome within a full set. For example, in the case of humans, a complete physical map requires a contig for each of the 22 autosomes and for the X and Y chromosomes. A contig represents a physical map of a chromosome. Geneticists can also correlate cloned DNA fragments in a contig with locations along a chromosome obtained from linkage or cytogenetic mapping. This can be accomplished by identifying members of the contig with inserts that have already been mapped by linkage or cytological methods. For example, a member of a contig may contain a gene, RFLP, or STS that previously has been mapped by genetic linkage. Figure 20.16 considers a situation in which two members of a contig carry genes already mapped by linkage analysis to be approximately 1.5 mu apart on chromosome 11. In this example, clone 2 has an insert that carries gene *A*, while clone 7 has an insert that carries gene *B*. Since a contig is composed of overlapping members, a researcher can line up the contig along chromosome 11 starting with gene *A* and gene *B* as reference points. In this example,

FIGURE 20.16   The use of genetic markers to align a contig. In this example, gene *A* and gene *B* had been mapped previously to specific regions of chromosome 11. Gene *A* was found within the insert of clone 2, gene *B* within the insert of clone 7. This made it possible to align the contig using gene *A* and gene *B* as genetic markers (i.e., reference points) along chromosome 11.

genes *A* and *B* serve as markers that identify the location of specific members of the contig.

YAC Cloning Vectors Are Used to Make Contigs of Eukaryotic Chromosomes

For large eukaryotic genomes, it is much easier to create contigs when the cloning vector can accept chromosomal DNA inserts of very large size. In general, most plasmid and viral vectors can accommodate inserts only a few thousand to perhaps tens of thousands of nucleotides in length. If a plasmid or viral genome has a DNA insert that is too large, it will have difficulty with DNA replication and is likely to suffer deletions in the insert.

By comparison, other cloning vectors, known as a **yeast artificial chromosome (YAC)** and a **bacterial artificial chromosome (BAC),** can reliably contain inserted DNA fragments of much larger size. The first YACs were developed by David Burke, Georges Carle, and Maynard Olson in 1987. An insert within a YAC can be several hundred thousand to perhaps 2 million nucleotides in length. For a small human chromosome, a few hundred YACs would be sufficient to create a contig with fragments that span the entire length of the chromosome. By comparison, it would take thousands or even tens of thousands of hybrid plasmids to create such a contig. BACs, which are genetically engineered F factors, can typically contain inserts up to 500,000 bp. Bacterial F factors are described in chapter 6.

At the molecular level, YACs have structural similarities to normal eukaryotic chromosomes yet have characteristics that make them suitable for cloning. The general structure of a YAC vector is shown at the *top left* of figure 20.17. The YAC vector contains two telomeres (*TEL*), a centromere (*CEN*), a bacterial origin of replication (*ORI*), a yeast origin of replication (known as an *ARS* for autonomous replication sequence), selectable markers, and unique cloning sites that are each recognized by a single restriction enzyme. Without an insert, the circular form of this vector can replicate in *E. coli*. After a large fragment has been inserted, the linear form of the vector can replicate in yeast.

In the experiment shown in figure 20.17, the chromosomal DNA is digested with the restriction enzyme *Eco*RI at a low concentration so that only some of the restriction sites are cut. This partial digestion will result in only occasional cleavage of the chromosomal DNA to yield very large DNA fragments. In other words, most restriction sites that are normally recognized by the enzyme will not be cleaved, because there is not enough of the restriction enzyme. The circular YAC vector is also digested with *Eco*RI and a second restriction enzyme, *Bam*HI, to yield two arms of the YAC. The YAC arms are then ligated to the large fragments of chromosomal DNA and transformed into yeast cells. When ligation occurs in the desired way, a large piece of chromosomal DNA becomes ligated to both arms of the YAC. Since each arm contains a different selectable marker, it is possible to select for the growth of yeast cells that carry a YAC construct having both arms.

YAC and BAC cloning vectors are very useful in the construction of contigs that span long segments of chromosomes. They are commonly the first step in creating a rough physical

FIGURE 20.17   **The use of YAC vectors in DNA cloning.**

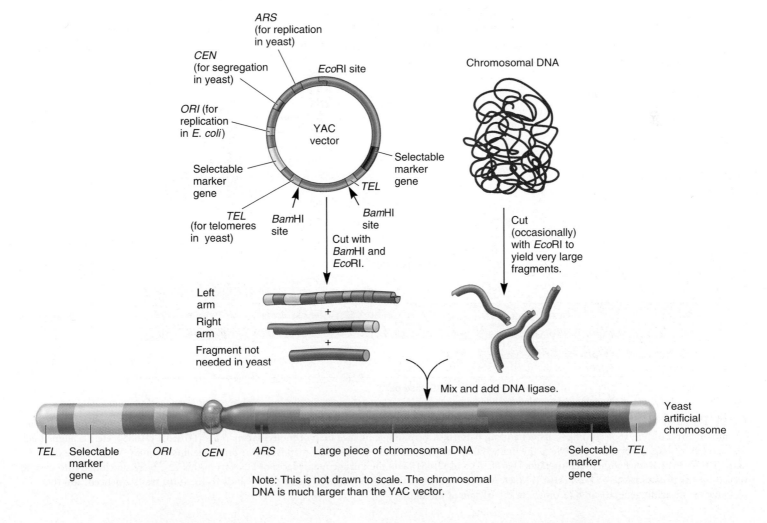

Note: This is not drawn to scale. The chromosomal DNA is much larger than the YAC vector.

map of a genome. While this is an important step in physical mapping, the large insert sizes make them difficult to use in gene cloning and sequencing experiments. Therefore, libraries containing hybrid vectors with smaller insert sizes are needed. Most commonly, a type of cloning vector called a **cosmid** is used. A cosmid is a hybrid between a plasmid vector and phage λ; its

DNA can replicate in a cell like a plasmid or be packaged into a protein coat like a phage. Cosmid vectors typically can accept DNA fragments that are tens of thousands of bp in length.

Figure 20.18 illustrates a comparison between cytogenetic, linkage, and physical maps of chromosome 16. Actually, this is a very simplified map of chromosome 16. A much more detailed

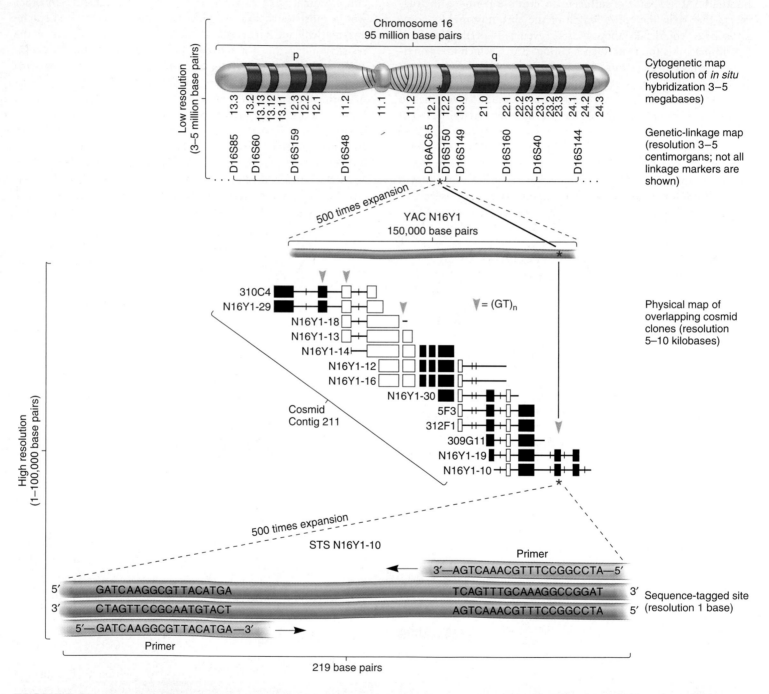

FIGURE 20.18   A correlation of cytogenetic, linkage, and physical maps of human chromosome 16. The *top* part shows the cytological map of human chromosome 16 according to its G banding pattern. A very simple linkage map of molecular markers (D16S85, D16S60, etc.) is aligned *below* the cytological map. A correlation between the linkage map and a segment of the physical map is shown *below* the linkage map. A YAC clone designated YAC N16Y1 is located between markers D16AC6.5 and D16S150 on the linkage map. Pieces of DNA from this YAC were subcloned into cosmid vectors and the cosmids (N16Y1-29, N16Y1-18, etc.) were aligned relative to each other. One of the cosmids (N16Y1-10) was sequenced, and this sequence was used to generate an STS shown at the *bottom* of the figure.

map is available, although it would take well over 10 pages of this textbook to print it! The *top* of this figure shows the banding pattern of this chromosome. Underneath the banded chromosome are molecular markers that have been mapped by linkage analysis. These same markers have been mapped cytologically. A complete contig of this chromosome has also been produced by generating a series of overlapping YACs. However, figure 20.18 shows the location of only one YAC within the contig. Cosmids within this region are shown below the YAC. In addition, an STS is found within the region and provides a molecular marker for cosmids N16Y1-19 and N16Y1-10, as well as YAC N16Y1.

Positional Cloning Can Be Achieved Using Chromosome Walking

The creation of a contig bears many similarities to a gene cloning strategy known as **positional cloning.** This term refers to a method in which a gene is cloned based on its mapped position along a chromosome. This approach has been successful in the cloning of many human genes, particularly those that cause genetic diseases when mutated. These include genes involved in cystic fibrosis, Huntington disease, and Duchenne muscular dystrophy.

A common method used in positional cloning is known as **chromosome walking.** To initiate this type of experiment, a gene's position relative to a marker must be known from mapping studies. For example, a gene may be known to be fairly close to a previously mapped gene or RFLP marker. This provides a starting point to molecularly "walk" toward the gene of interest.

Figure 20.19 considers a chromosome walk in which the goal is to locate a gene that we will call gene A. In this example, genetic mapping studies have revealed that gene A is relatively close to another gene, called gene B, that has been previously cloned. Gene A and gene B have been deduced from genetic crosses to be approximately 1 mu apart. To begin this chromosome walk, a cloned DNA fragment that contains gene B and flanking sequences can be used as a starting point to walk to gene A.

To walk from gene B to gene A, a series of library screening methods are followed. In this example, the starting materials are a cosmid library and a clone containing gene B. A small piece of DNA from the first cosmid vector containing gene B is inserted into another vector. This is called **subcloning.** The subclone is radiolabeled and used as a probe to screen a cosmid library. This will enable the researchers to identify a second clone that is closer to gene A. A subclone from this second clone is then used to screen the library a second time. This allows the researchers to identify a clone that is even closer to gene A. This repeated pattern of subcloning and library screening is used to walk toward gene A. The term *chromosome walking* is an appropriate description of this technique, because each clone takes you a step closer to the gene of interest. When starting at gene B in figure 20.19, a researcher would also want to have markers to the left of gene B to ensure that they were not walking in the wrong direction.

The number of steps required to reach the gene of interest depends on the distance between the starting and ending points,

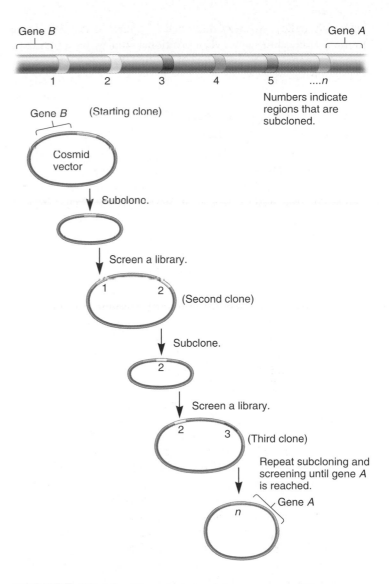

FIGURE 20.19 **The technique of chromosome walking.**

and on the sizes of the inserts in the library. If the two points are 1 mu apart, they are expected to be approximately 1 million bp apart, although the correlation between map units and physical distances can be quite variable. In a typical walking experiment, each clone might have an average insert size of 50,000 bp. Therefore, it will take more than 20 walking steps to reach the gene of interest. If chromosome walking is done in conjunction with DNA sequencing, this can be a laborious undertaking spanning several years. For this reason, researchers are eager to locate starting points in a chromosome walking experiment that are as close as possible to the gene of interest.

It may have occurred to you that the subcloning steps of chromosome walking are unnecessary if another researcher previously had created a contig in the region of the chromosome where the gene is located. In this case, the experimenter may analyze the members of the contig, rather than following the extremely tedious strategy described in figure 20.19. This desire to more easily clone genes via a positional strategy has provided a strong impetus for the creation of contigs that can be shared by many researchers.

The Human Genome Project Has Stimulated Genomic Research

In 1988, the National Institutes of Health in Bethesda, Maryland, established an Office of Human Genome Research, with James Watson as its first director. Discussions among scientists to undertake this project began in the mid-1980s. The Human Genome Project officially began on October 1, 1990. It has been the largest internationally coordinated undertaking in the history of biological research. From its outset, the goals of the Human Genome Project have been the following:

1. *To obtain a genetic linkage map of the human genome.* This has involved the identification of thousands of genetic markers and their localization along the autosomes and sex chromosomes.

2. *To obtain a physical map of the human genome.* This has required the cloning of many segments of chromosomal DNA into YACs and cosmids.

3. *To obtain the DNA sequence of the entire human genome.* The first (nearly complete) sequence was published in February 2001. This is considered a first draft and likely will be refined during subsequent years. The entire genome is approximately 3.2 billion nucleotide base pairs in length. If the entire human genome were typed in a textbook like this, it would be nearly 1 million pages long!

4. *To develop technology for the management of human genome information.* The amount of information that has been obtained from this project is staggering, to say the least. The Human Genome Project has developed user-friendly tools to provide scientists easy access to up-to-date information obtained from the project. The Human Genome Project also has developed analytical tools to interpret genome information.

5. *To analyze the genomes of other model organisms.* These include bacterial species (e.g., *Escherichia coli* and *Bacillus subtilis*), *Drosophila melanogaster* (fruit fly), *Caenorhabditis elegans* (a nematode), *Arabidopsis thaliana* (a simple plant), and *Mus musculus* (mouse).

6. *To develop programs focused on understanding and addressing the ethical, legal, and social implications of the results obtained from the Human Genome Project.* The Human Genome Project has sought to identify the major genetic issues that will affect members of our society and to develop policies to address these issues. For example, some people are worried that their medical insurance companies may discriminate against them if it is found that they carry a disease-causing or otherwise deleterious gene.

7. *To develop technological advances in genetic methodologies.* Some of the efforts of the Human Genome Project have involved improvements in molecular genetic technology such as gene cloning, contig construction, DNA sequencing, and so forth. The project has also been aimed at developing computer technology for data processing, storage, and the analysis of sequence information (see chapter 21).

A great benefit expected from the characterization of the human genome is the ability to identify our genes. Mutations in many different genes are known to be correlated with human diseases, which include cancer, heart disease, and many other abnormalities. The identification of mutant genes that cause inherited diseases has been a strong motivation for the Human Genome Project. A detailed genetic and physical map has made it profoundly easier for researchers to locate such genes. Furthermore, a complete DNA sequence of the human genome provides researchers with insight into the types of proteins encoded by these genes. The cloning and sequencing of disease-causing alleles is expected to play an increasingly important role in the future diagnosis and treatment of disease.

Many Genome Sequences Have Been Determined

In just a couple of decades, our ability to map and sequence genomes has improved dramatically. Motivation behind genome sequencing projects comes from a variety of sources. For example, basic research scientists can greatly benefit from a genome sequence. It allows a scientist to know which genes a given species has, and it aids in the cloning and characterization of such genes. This has been the impetus for genome projects involving experimental organisms such as *Escherichia coli, Saccharomyces cerevisiae, Drosophila melanogaster, Caenorhabditis elegans, Arabidopsis thaliana,* and the mouse. A second reason for genome sequencing has involved human disease. This is an important goal involving the sequencing of the human genome. It is hoped that it will aid in the identification of genes that (when mutant) play a role in disease. Likewise, the sequencing of many bacterial genomes has been related to infectious diseases. Dozens of bacterial genomes have been sequenced, and many of these are from species that have the potential to be pathogenic in humans. The sequencing of such bacterial genomes may help to elucidate the genes that play a role in the infectious process. Finally, the genomes of agriculturally important species have been the subject of genome sequencing projects. An understanding of a species' genome may aid in the development of new strains of livestock and plant species that have improved traits from an agricultural perspective.

Table 20.2 describes the results of several genome-sequencing projects that have been completed in the past decade or so. Newly completed genome sequences are emerging rapidly, particularly those of bacterial species. As we obtain more genome sequences, it becomes progressively more interesting to compare them to each other as a way to understand the process of evolution. As described in chapter 21, the field of functional genomics enables researchers to probe the roles of many genes as they interact to generate the phenotypic traits of the species that contain them.

TABLE 20.2

Examples of Genomes That Have Been Sequenced

Species	Genome Size (bp)	Number of Genes	Description
Prokaryotic			
Escherichia coli	4,639,000	4,289	A widespread inhabitant of the gut of animals; also a model research organism.
Mycoplasma genitalium	580,000	468	An inhabitant of the human genital tract with a very small genome.
Helicobacter pylori	1,668,000	1,590	A common inhabitant of the stomach that may cause gastritis, peptic ulcer, and gastric cancer.
Mycobacterium tuberculosis	4,412,000	~4,000	The bacterial species that causes tuberculosis.
Eukaryotic			
Saccharomyces cerevisiae (baker's yeast)	12,069,000	~6,200	This is one of the simplest eukaryotic species and has been extensively studied by researchers to understand eukaryotic cell biology and other molecular mechanisms.
Caenorhabditis elegans	~97,000,000	~19,000	A nematode worm that has been a model organism to study animal development.
Drosophila melanogaster	~137,000,000	~14,000	The fruit fly has been a model organism to study many genetic phenomena including development.
Arabidopsis thaliana	~142,000,000	~26,000	A model organism studied by plant biologists.
Oryza sativa, subspecies japonica and indica (rice)	~420,000,000–460,000,000	~40,000–50,000	A cereal grain with a relatively small genome. It is very important worldwide as a food crop.
Homo sapiens (humans)	~3,200,000,000	~35,000	The sequencing of the human genome will help us understand our inherited traits and may aid in the identification and treatment of diseases.

CONCEPTUAL AND EXPERIMENTAL SUMMARY

In chapter 20, we have considered a variety of methods aimed at the molecular analysis of entire **genomes.** This approach, termed **structural genomics,** begins with the mapping of the genome and progresses ultimately to a complete DNA sequence. As we have seen, the three main strategies to map the locations of genes and molecular markers are cytogenetic, linkage, and physical mapping.

Cytogenetic mapping relies on light microscopy to locate genes or other DNA markers along intact chromosomes. The most commonly used method of cytogenetic mapping is **fluorescence *in situ* hybridization (FISH),** which can locate genes with remarkable sensitivity. In this technique, a small DNA probe is hybridized to intact chromosomes to determine where that DNA sequence is located in the genome. **Linkage mapping** relies on the outcome of crosses to map the relative locations of genes or molecular markers along chromosomes. In chapter 5, we studied linkage mapping that involved crosses between individuals heterozygous for two or more genes. In chapter 20, we have considered how **molecular markers,** such as **restriction fragment length polymorphisms (RFLPs)** or **microsatellites,** can be mapped by a similar approach. Compared to conventional linkage analysis, molecular mapping uses techniques that examine the composition of an individual's chromosomes rather than relying on their external phenotypes. It is now common for geneticists to construct maps of entire genomes using molecular markers such as RFLPs and microsatellites. These maps can be used to determine the locations of functional genes by correlating the inheritance of the gene with the inheritance of a particular RFLP or microsatellite.

Physical mapping relies on the molecular analysis of an organism's genome via DNA cloning methods. As a start, it is common to isolate individual chromosomes or large fragments of chromosomes by **pulsed-field gel electrophoresis (PFGE)** or **chromosome sorting.** A **chromosome-specific library** can then be made using **yeast artificial chromosomes (YACs)** or **bacterial artificial chromosomes (BACs)** as cloning vectors, the large DNA inserts of which can be aligned to make a **contig.** A more refined contig is made by cloning smaller fragments of DNA into **cosmid** vectors. The methods of constructing a contig are similar in theory to **positional cloning.** To clone a gene by this method, the gene is first mapped to a region of the chromosome and a clone from that region is used as a starting point in a **chromosome walk** toward the gene of interest. This is accomplished by sequential hybridization methods. Positional cloning can be accomplished more easily if a contig has already been constructed in the region where the gene is located. The ultimate goal of physical mapping is the determination of the complete DNA sequence. This has been accomplished for a variety of species, and newly completed genome sequences are emerging at a rapid pace.

Solved Problems

S1. An RFLP marker is located 1 million bp away from a gene of interest. Your goal is to start at this RFLP marker and walk to this gene. The average insert size in the library is 55,000 bp and the average overlap at each end is 5,000 bp. Approximately how many steps will it take to get there?

Answer: Each step is only 50,000 bp (i.e., 55,000 minus 5,000) because you have to subtract the overlap between adjacent fragments, which is 5,000 bp. Therefore, it will take about 20 steps to go 1 million bp.

S2. When many RFLPs have been mapped within the genome of a plant, an RFLP analysis can be used to map a herbicide-resistance gene. For example, let's suppose that an agricultural geneticist has two strains, one that is herbicide resistant and one that is herbicide sensitive. The two strains differ with regard to many RFLPs. The sensitive and resistant strains are crossed, and the F_1 offspring are allowed to self-fertilize. The F_2 offspring are then analyzed with regard to their herbicide sensitivity and RFLP markers. The following results were obtained:

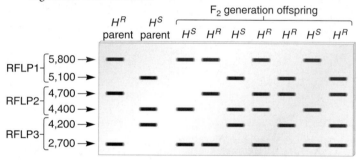

To which RFLP might the herbicide-resistance gene be linked?

Answer: The aim of this experiment is to correlate the presence of a particular RFLP with the herbicide-resistant phenotype. As shown in the data, the herbicide-resistant parent and all the herbicide-resistant offspring have an RFLP that is 4,700 bp in length. In an actual experiment, a more thorough lod analysis would be conducted to determine if linkage is considered likely. If so, the 4,700 bp RFLP may either contain the gene that confers herbicide resistance or, as is more likely, the two may be linked. If the 4,700 bp RFLP has already been mapped to a particular site in the plant's genome, the herbicide-resistance gene also maps to the same site or very close to it. This information may then be used by a plant breeder when making future crosses to produce herbicide-resistant plant strains.

S3. Does a molecular marker have to be polymorphic to be useful in mapping studies? Does a molecular marker have to be polymorphic to be useful in linkage mapping (i.e., involving family pedigree studies or genetic crosses)? Explain why or why not.

Answer: A molecular marker does not have to be polymorphic to be useful in mapping studies. Many sequence-tagged sites that are used in physical or cytogenetic studies are monomorphic. Monomorphic markers can provide landmarks in mapping studies.

In linkage studies, a marker must be polymorphic to be useful. Polymorphic molecular markers can be RFLPs or microsatellites. To compute map distances in linkage analysis, individuals must be heterozygous for two or more markers (or genes). For experimental organisms, heterozygotes are testcrossed to homozygotes, and then the number of recombinant offspring and nonrecombinant offspring are determined. For markers that do not assort independently (i.e., linked markers), the map distance is computed as the number of recombinant offspring divided by the total number of offspring, times 100.

S4. The distance between two molecular markers that are linked along the same chromosome can be determined by analyzing the outcome of crosses. This can be done in humans by analyzing the members of a pedigree. However, the accuracy of linkage mapping in human pedigrees is fairly limited because the number of people in most families is pretty small. As an alternative, researchers can analyze a population of sperm, produced from a single male, and compute linkage distance in this manner. As an example, let's suppose a male is heterozygous for two polymorphic sequence-tagged sites. STS-1 exists in two sizes: 234 bp and 198 bp. STS-2 also exists in two sizes: 423 bp and 322 bp. A sample of sperm was collected from this man, and individual sperm were placed into 40 separate tubes. In other words, there was one sperm in each tube. Believe it or not, one sperm has enough DNA to conduct PCR! Into each of the 40 tubes were added the primers that amplify STS-1 and STS-2, and then the samples were subjected to PCR. The following results were obtained.

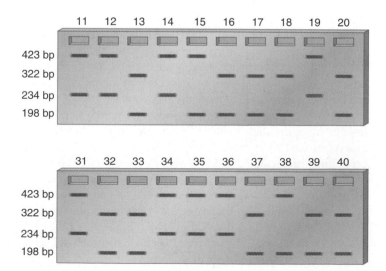

A. What is the arrangement of these two sequence-tagged sites in this individual?

B. What is the linkage distance between STS-1 and STS-2?

Answer: Keep in mind that mature sperm are haploid, so they will have only one copy of STS-1 and one copy of STS-2.

A. If we look at the 40 lanes, most of the lanes (i.e., 36 of them) have either the 234 bp and 423 bp STSs or the 198 bp and 322 bp. This is the arrangement of STSs in this male. One chromosome has STS-1 that is 234 bp and STS-2 that is 423 bp and the homologous chromosome has STS-1 that is 198 bp and STS-2 that is 322 bp.

B. There are four recombinant sperm, shown in lanes 15, 22, 25, and 38.

$$\text{Map distance} = \frac{4}{40} \times 100$$

$$= 10.0 \text{ mu}$$

Note: This is a relatively easy experiment compared to a pedigree analysis, which would involve contacting lots of relatives and collecting samples from each of them.

S5. The technique of chromosome walking involves the stepwise cloning of adjacent DNA fragments until the gene of interest is reached. This method was used to clone the gene that (when mutant) causes cystic fibrosis, an autosomal recessive disorder in humans. Linkage mapping studies first indicated that the *CF* gene was located between two markers designated MET and D7S8 (see next). Later mapping studies put the *CF* gene between IRP and D7S8. Walking was initiated from the D7S122 marker.

MET	D7S122	IRP	*CF* gene	D7S8

(from *Science* 245, 1059–69)

In this study, the researchers restriction mapped this region using two enzymes, *Not*I and *Xho*I. The locations of sites are shown here:

MET	D7S122	IRP	*CF* gene	D7S8

↑A↑   B ↑C↑ D ↑E↑F↑
N X   X X   X X N

N = *Not*I,   X = *Xho*I

How would the researchers know they were walking toward the *CF* gene and not away from it?

Answer: The general answer is that they had to determine if they were walking toward the IRP marker or toward the MET marker.

As depicted in the previous diagram, *Not*I would produce a large DNA fragment (including regions A–F) that contained both the D7S122 and IRP markers. Since the D7S122 and IRP markers were available as probes, they could confirm that this region contained both markers by Southern hybridization of this *Not*I fragment to both markers. *Xho*I cuts this region into smaller pieces. The piece labeled B carried the D7S122 marker, and the piece labeled D carried the IRP marker. Again, this could be confirmed by hybridization. This map enabled the researchers to determine which direction to walk. During their walking, they began at fragment B (which contained the D7S122 marker) and walked toward the adjacent fragment labeled C. They then walked to the fragment labeled D, which contains the IRP marker. In contrast, if they had walked toward the fragment labeled A, they would then have walked to the MET marker. This is the incorrect direction, because they knew from their linkage mapping that IRP is closer to the *CF* gene compared to MET.

Conceptual Questions

C1. A person with a rare genetic disease has a sample of her chromosomes subjected to *in situ* hybridization using a probe that is known to recognize band p11 on chromosome 7. Even though her chromosomes look cytologically normal, the probe does not bind to this person's chromosomes. How would you explain these results? How would you use this information to positionally clone the gene that is related to this disease?

C2. For each of the following decide if it could be appropriately described as a genome:

A. The *E. coli* chromosome

B. Human chromosome 11

C. A complete set of 10 chromosomes in corn

D. A copy of the single-stranded RNA packaged into HIV (human immunodeficiency virus)

C3. Which of the following statements are true about molecular markers?

A. All molecular markers are segments of DNA that carry specific genes.

B. A molecular marker is a segment of DNA that is located at a specific location in a genome.

C. We can follow the transmission of a molecular marker by analyzing the phenotype (e.g., the individual's bodily characteristics) of offspring.

D. We can follow the transmission of molecular markers using molecular techniques such as gel electrophoresis.

E. An STS is a molecular marker.

Experimental Questions

E1. Would the following methods be described as linkage, cytogenetic, or physical mapping?

   A. Fluorescence *in situ* hybridization (FISH)

   B. Conducting dihybrid crosses to compute map distances

   C. Chromosome walking

   D. Examination of polytene chromosomes in *Drosophila*

   E. Use of RFLPs in crosses

   F. Using YACs and cosmids to construct a contig

E2. In an *in situ* hybridization experiment, what is the relationship between the sequence of the probe DNA and the site on the chromosomal DNA where the probe binds?

E3. Describe the technique of *in situ* hybridization. Explain how it can be used to map genes.

E4. The cells from a malignant tumor were subjected to *in situ* hybridization using a probe that recognizes a unique sequence on chromosome 14. The probe was detected only once in each of these cells. Explain these results and speculate on their significance with regard to the malignant characteristics of these cells.

E5. The legend to figure 20.3 describes the technique of FISH. Why is it necessary to "fix" the cells (and the chromosomes inside of them) to the slides? What does it mean to fix them? Why is it necessary to denature the chromosomal DNA?

E6. Explain how the use of DNA probes with different fluorescence emission wavelengths can be used in a single experiment to map the locations of two or more genes. This method is called *chromosome painting*. Explain why this is an appropriate term.

E7. A researcher is interested in a gene that is found on human chromosome 21. Describe the expected results of a FISH experiment using a probe that is complementary to this gene. How many spots would you see if the probe was used on a sample from a normal individual versus an individual with Down syndrome?

E8. What is a contig? Explain how you would determine that two clones in a contig are overlapping.

E9. Contigs are often made using YAC and cosmid vectors. What are the advantages and disadvantages of these two types of vector? Which type of contig would you make first, a YAC or cosmid contig? Explain.

E10. Describe the molecular features of a YAC cloning vector. What is the primary advantage of a YAC compared to plasmid or viral vectors?

E11. In general terms, what is a polymorphism? Explain the molecular basis for a restriction fragment length polymorphism. How is an RFLP detected experimentally? Why are RFLPs useful in physical mapping studies? How can they be used to clone a particular gene?

E12. Five RFLPs designated 1A and 1B, 2A and 2B, 3A and 3B, 4A and 4B, and 5A and 5B are known to map along chromosome 4 of corn. A plant breeder has obtained a strain of corn that carries a pesticide-resistance gene that (from previous experiments) is known to map somewhere along chromosome 4. The plant breeder crosses this pesticide-resistance strain that is homozygous for RFLPs 1A, 2B, 3A, 4B, and 5A to a pesticide-sensitive strain that is

homozygous for 1B, 2A, 3B, 4A, and 5B. The F_1 generation plants were allowed to self-hybridize to produce the following F_2 plants:

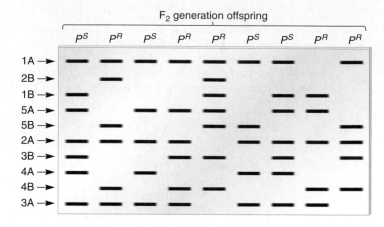

Based on these results, which RFLP does the pesticide-resistance gene map closest to?

E13. Let's suppose there are two different RFLPs in a species. RFLP 1 can be found in 4,500 bp and 5,200 bp lengths; RFLP 2 can be 2,100 bp and 3,200 bp. A homozygote, 4,500 bp and 2,100 bp, was crossed to a homozygote 5,200 bp and 3,200 bp. The F_1 offspring were then crossed to a homozygote containing the 4,500 bp and 2,100 bp fragments. The following results were obtained:

5,200, 4,500, and 2,100	40 offspring
5,200, 4,500, 3,200, 2,100	98 offspring
4,500 and 2,100	97 offspring
4,500, 2,100, and 3,200	37 offspring

   Conduct a chi square analysis to determine if these two RFLPs are linked. If so, calculate the map distance between them.

E14. Explain how a detailed RFLP map of an organism's genome can be helpful in mapping the location of functional genes. Describe an experimental strategy you would follow to map a gene near a particular RFLP.

E15. A woman has been married to two different men and produced five children. This group is analyzed with regard to three different STSs: STS-1 is 146 and 122 bp; STS-2 is 102 and 88 bp, and STS-3 is 188 and 204 bp. The mother is homozygous for all three STSs: STS-1 = 122, STS-2 = 88, and STS-3 = 188. Father 1 is homozygous for STS-1 = 122 and STS-2 = 102, and heterozygous for STS-3 = 188/204. Father 2 is heterozygous for STS-1 = 122/146, STS-2

= 88/102, and homozygous for STS-3 = 204. The five children have the following results:

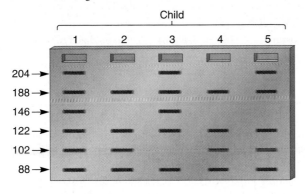

Child

Which children can you definitely assign to father 1 and father 2?

E16. An experimenter used primers to nine different STSs to test their presence along five different YAC clones. The results are shown here.

Alignment of STSs and YACs

	STSs								
	1	2	3	4	5	6	7	8	9
YACs									
1	−	−	−	−	−	+	+	−	+
2	+	−	−	−	+	−	−	+	−
3	−	−	−	+	−	+	−	−	−
4	−	+	+	−	−	−	−	+	−
5	−	−	+	−	−	−	−	−	+

Make a contig map that describes the alignment of the five YACs.

E17. In the Human Genome Project, researchers have collected linkage data from many crosses in which the male was heterozygous for markers and many crosses where the female was heterozygous for markers. The distance between the same two markers, computed in map units or centiMorgans, is different between males and females. In other words, the linkage maps for human males and females are not the same. Propose an explanation for this discrepancy. Do you think the sizes of chromosomes (excluding the Y chromosome) in human males and females are different? How could physical mapping resolve this discrepancy?

E18. Take a look at solved problem S4. Let's suppose a male is heterozygous for two polymorphic sequence-tagged sites. STS-1 exists in two sizes: 211 bp and 289 bp. STS-2 also exists in two sizes: 115 bp and 422 bp. A sample of sperm was collected from this man, and individual sperm were placed into 30 separate tubes. Into each of the 30 tubes were added the primers that amplify STS-1 and STS-2,

and then the samples were subjected to PCR. The following results were obtained:

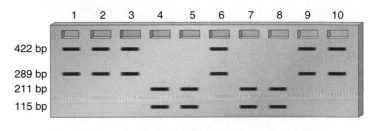

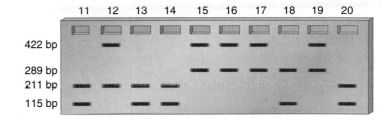

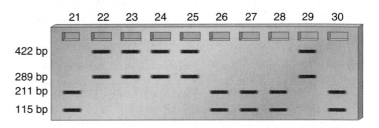

A. What is the arrangement of these two sequence-tagged sites in this individual?

B. What is the linkage distance between STS-1 and STS-2?

C. Could this approach of analyzing a population of sperm be applied to RFLPs?

E19. A gene affecting flower color in petunias is closely linked to an RFLP. A red allele for this gene is associated with a 4,000 bp RFLP, while a purple allele is linked to a 3,400 bp version of this same RFLP. A second gene in petunias affects flower size. An allele causing big flowers is linked to a 7,200 bp RFLP, while a small-flower allele is linked to the same RFLP that is 1,600 bp. A true-breeding strain with small, red flowers was crossed to a true-breeding strain with big, purple flowers. All the F_1 offspring had big, purple flowers. These F_1 offspring were then crossed to true-breeding petunias with small, red flowers. The following results were obtained:

Red, small	725
Red, big	111
Purple, small	109
Purple, big	729

Are these two genes linked to each other? If so, compute the map distance. What would be the expected outcome regarding the inheritance of the RFLPs among the offspring?

E20. An agricultural geneticist has studied a gene in alfalfa that affects pesticide resistance. It exists in three alleles that confer low, medium, and high levels of resistance. This gene has a significant impact on the yield of alfalfa, depending on seasonal variation in pest problems. This geneticist has followed the basic protocol in figure 20.6, using *Eco*RI as the enzyme to digest the chromosomal DNA. Unfortunately, after tireless efforts, and the analysis of thousands of offspring, it has not been possible to identify an RFLP that is associated with the three alleles of this pesticide-resistance gene. What should the geneticist do next? In other words, discuss ways to vary the RFLP method, or propose alternative approaches to identify molecular markers that may be linked to the pesticide-resistance gene.

E21. Figure 20.8 describes the transmission of two RFLPs that were linked and 16 mu apart. If these two RFLPs had not been linked and 100 offspring had been analyzed, what would have been the expected results?

E22. Explain why it is necessary to use the technique of Southern blotting for RFLP mapping.

E23. Compared to a conventional plasmid, what additional sequences are required in a YAC vector so that it can behave like an artificial chromosome? Describe the importance of each required sequence.

E24. In the data of figure 20.10, individuals who were homozygous for the sickle-cell allele had an 87% likelihood of carrying the 13.0 kbp RFLP. Based on this observation, is it likely that the Hb^S allele originated in an individual with a 13.0 kbp RFLP? How would you explain the observation that the Hb^S allele is found with the 7.6 kbp RFLP 13% of the time?

E25. In the experiment of figure 20.10, explain the conclusion that the 13.0 kbp fragment is more closely linked to Hb^S compared to the 7.0 and 7.6 fragments. Is it always linked? Why or why not? How is this type of information useful?

E26. Explain the technique of pulsed-field gel electrophoresis (PFGE). What special precautions are needed to prevent the mechanical breakage of the chromosomes? What are the uses of PFGE?

E27. When conducting physical mapping studies, place the following methods in their most logical order:

A. Clone large fragments of DNA to make a YAC library.

B. Determine the DNA sequence of subclones from a cosmid library.

C. Isolate whole chromosomes or large DNA fragments via pulsed-field gel electrophoresis or chromosome sorting.

D. Subclone YAC fragments to make a cosmid library.

E. Subclone cosmid fragments for DNA sequencing.

E28. Four cosmid clones, which we will call cosmid A, B, C, and D, were subjected to a Southern blot in pairwise combinations. The insert size of each cosmid was also analyzed. The following results were obtained:

Cosmid	Insert Size (bp)	Hybridized to?
A	6,000	C
B	2,200	C, D
C	11,500	A, B, D
D	7,000	B, C

Draw a map that shows the order of the inserts within these four cosmids.

E29. What is an STS (sequence-tagged site)? How are STSs generated experimentally? What are the uses of STSs? Explain how a microsatellite can produce a polymorphic STS.

E30. A human gene, which we will call gene *X*, is located on chromosome 11 and is found as a normal allele and a recessive disease-causing allele. The location of gene *X* has been approximated on the map shown here that contains four STSs, labeled STS-1, STS-2, STS-3, and STS-4.

STS-1	STS-2	STS-3	Gene *X*	STS-4

A. Explain the general strategy of positional cloning.

B. If you applied the approach of positional cloning to clone gene *X*, where would you begin? As you progressed in your cloning efforts, how would you know if you were walking toward gene *X* or away from gene *X*?

C. How would you know you had reached gene *X*? (Keep in mind that gene *X* exists as a normal allele and a disease-causing allele.)

E31. Describe how you would clone a gene by positional cloning. Explain how a (previously made) contig would make this task much easier.

E32. Describe two experimental strategies to obtain a purified preparation of a particular type of chromosome. Why are these techniques useful in genome analysis?

Questions for Student Discussion/Collaboration

1. How is it possible to obtain an RFLP linkage map? What kind of experiments would you conduct to correlate the RFLP linkage map with the positions of known genes that had already been cloned? Discuss the uses of RFLPs in genetic analyses.

2. What is a molecular marker? Give two examples. Discuss why it is easier to locate and map many molecular markers rather than functional genes.

3. Which goals of the Human Genome Project do you think are the most important? Why? Discuss the types of ethical problems that might arise as a result of identifying all of our genes.

Note: All answers appear at the website for this textbook; the answers to even-numbered questions are in the back of the textbook.

Visit the Online Learning Center for practice tests, answer keys, and other learning aids for this chapter. Enhance your understanding of genetics with our interactive exercises, web links, news feeds, tutorial service, and much more.

FUNCTIONAL GENOMICS, PROTEOMICS, AND BIOINFORMATICS

21

::

In chapter 20, we learned that the outcome of structural genomics is the mapping of an entire genome and, eventually, the determination of the complete DNA sequence. The amount of information that is found within a species' genome is enormous. The goal of **functional genomics** is to elucidate the roles of genetic sequences in a given species. In most cases, functional genomics is aimed at an understanding of gene function. At the genomic level, researchers can study genes as groups. For example, the information from a genome-sequencing project can help researchers study entire metabolic pathways. This provides a description of the ways in which gene products work together to carry out cellular processes. In addition, a study of genomic sequences can help to identify regions that play other functional roles. For example, a careful analysis of repetitive sequences in bacteria helped to identify DNA sequences that are recognized by primase during DNA replication.

Because most genes encode proteins, a goal of many molecular biologists is to understand the functional roles of the proteins that a species can produce. This field is called **proteomics,** and the entire collection of proteins that a given species can make is called its **proteome.** Similar to genomics, proteomics is aimed at understanding the interplay among many proteins as they function to create cells and, ultimately, the traits of a given species.

From a research perspective, functional genomics and proteomics can be broadly categorized in two ways: experimental and computational. The experimental approach involves the study of groups of genes or proteins using molecular techniques in the laboratory. The first two sections of chapter 21 will focus on these types of methods. By comparison, a computational strategy attempts to extract information within genetic sequences using a mathematical approach. This area, which is called **bioinformatics,** has become an important branch of science. The tools of bioinformatics are computers, computer programs, and genetic sequences (e.g., DNA sequences or amino acid sequences). In the last section of chapter 21, we will consider the field of bioinformatics and learn how this area has provided great insights into the subjects of functional genomics and proteomics.

21.1 FUNCTIONAL GENOMICS

Though the rapid sequencing of genomes, particularly the human genome, has generated great excitement in the field of genetics, many would argue that an understanding of genomic function is fundamentally more interesting. In the past, our ability to study genes involved many of the techniques described in chapter 18. Specific genes can be cloned using vectors or PCR, and their expression and regulation can be studied using methods such as Northern blotting, gel retardation assays, and site-directed mutagenesis. These approaches continue to provide a solid foundation for our understanding of gene function. More recently, genome-sequencing projects have enabled researchers to consider gene function at a more complex level. As mentioned, we now can examine groups of many genes simultaneously, as they work as integrated units to produce the characteristics of cells and the traits of complete organisms. This section of chapter 21 is aimed at an understanding of these types of techniques.

Expressed Genes Can Be Identified in a cDNA Library

As discussed in chapter 10, chromosomal DNA contains regions that encode genes, which are punctuated by regions that do not contain genes. Therefore, a basic goal of genomic research is to definitively identify the regions of DNA that are actually genes. To do so, one approach is to show that a given region is transcribed into RNA. This can be accomplished by the generation of a **cDNA library.** As described in chapter 18, cDNA (i.e., complementary DNA) is made using RNA as a starting material. A cDNA library is also called an **expressed sequence tag library (EST library)** because the sequences can also be used as markers in physical mapping studies. The members of an EST library can be subjected to DNA sequencing, and then EST sequences can be compared to a complete genome sequence. A particular EST sequence will match a genomic sequence indicating that the genomic region encodes a gene. If the genome of a given species has already been sequenced, only a short region of a cDNA clone must be sequenced to find a match. From the perspective of gene function, a cDNA or EST library provides a reliable identification of a region that is truly transcribed. The mere fact that a region is transcribed indicates that the corresponding genomic region is a gene.

The method of making a cDNA library can be extended to study gene regulation at the genomic level. The strategy is to isolate mRNA under different conditions and then identify particular mRNAs that are expressed only under one set of conditions. For example, in the experiment shown in figure 21.1, samples of

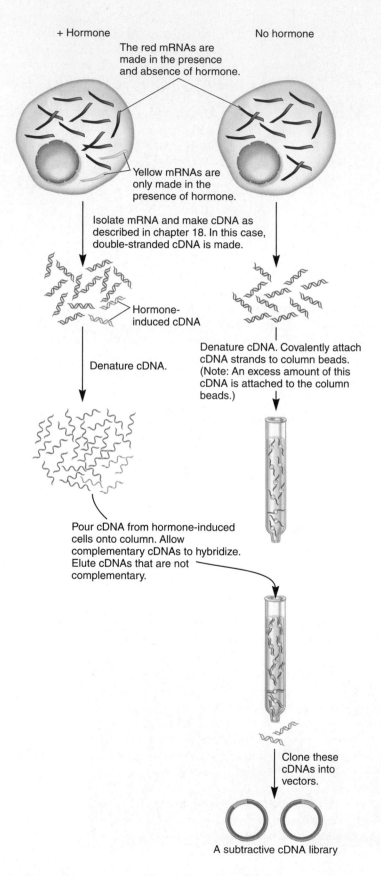

+ Hormone   No hormone

The red mRNAs are made in the presence and absence of hormone.

Yellow mRNAs are only made in the presence of hormone.

Isolate mRNA and make cDNA as described in chapter 18. In this case, double-stranded cDNA is made.

Hormone-induced cDNA

Denature cDNA.

Denature cDNA. Covalently attach cDNA strands to column beads. (Note: An excess amount of this cDNA is attached to the column beads.)

Pour cDNA from hormone-induced cells onto column. Allow complementary cDNAs to hybridize. Elute cDNAs that are not complementary.

Clone these cDNAs into vectors.

A subtractive cDNA library

FIGURE 21.1   Creation of subtractive cDNA library.

cells were incubated with and without a particular hormone. The goal is to identify those genes that are turned on by the presence of the hormone. The mRNAs from these cells are isolated and used to make two different groups of cDNAs, which we will call hormone-induced and hormone-uninduced. The hormone-uninduced cDNAs are tightly bound to a column, and the DNA is denatured. The cDNAs derived from the hormone-induced cells are then denatured and run over the column. Any cDNAs derived from the hormone-induced cells that match cDNAs from the hormone-uninduced cells will bind to the column, because they hybridize to their complementary cDNA. However, cDNAs that are only derived from the hormone-induced cells will not bind to the column, because there is no complementary cDNA partner. The cDNAs that do not bind to the column are eluted, and these are cloned into vectors to create a cDNA library termed a **subtractive cDNA library.** This approach is also called **subtractive hybridization.** The members of the subtractive cDNA library can be subjected to DNA sequencing to identify the genes that are turned on by the presence of the hormone.

A Microarray Can Identify Genes That Are Transcribed

Researchers have developed an exciting new technology, called **DNA microarrays** (also called **gene chips**), that makes it possible to monitor the expression of thousands of genes simultaneously. A DNA microarray is a small silica, glass, or plastic slide that is dotted with many different sequences of DNA, each corresponding to a short sequence within a known gene. For example, one spot in a microarray may correspond to a sequence within the β-globin gene while another spot could correspond to a different gene, such as a gene that encodes an iron transporter. A single slide contains tens of thousands of different spots in an area that is the size of a postage stamp. The relative location of each spot is known. These microarrays are typically produced using spotting technologies that are quite similar to the way that an inkjet printer works. In some microarrays, different DNA fragments, which were made synthetically (e.g., by PCR), are individually spotted onto the slide. The DNA fragments are typically 500 to 5,000 bp in length, and a few thousand to tens of thousands are spotted to make a single array. Alternatively, short oligonucleotides of different DNA sequences can be directly synthesized on the surface of the slide. In this process, the DNA sequence at a given spot is produced by selectively controlling the growth of the oligonucleotide using narrow beams of light. In this case, there can be hundreds of thousands of different spots on a single array. Overall, the technology of making DNA microarrays is quite amazing.

Once a DNA microarray has been made, it is used as a hybridization tool. This idea is shown in figure 21.2. In this experiment, RNA was isolated from a sample of cells and then used to make fluorescently labeled cDNA. The labeled cDNAs were then layered onto a DNA microarray. Those cDNAs that are complementary to the DNAs in the microarray will hybridize and thereby remain bound to the microarray. The array is then washed and placed in a scanning confocal fluorescence microscope that scans each pixel (the smallest element in a visual image). After correction for local background, the final fluorescence intensity for each spot is obtained by averaging across the pixels in each spot. This results in a group of fluorescent spots at defined locations in the microarray. If the fluorescence intensity in a spot is high, it means that there was a large amount of cDNA in the sample that hybridized to the DNA at this location. Because the DNA sequence of each spot is already known, a fluorescent spot identifies cDNAs that are complementary to those DNA sequences. Furthermore, because the cDNA was generated from mRNA, this technique identifies RNAs that have been made in a particular cell type under a given set of conditions.

The technology of DNA microarrays has found many important uses (table 21.1). Thus far, its most common use is to study gene expression patterns. This helps us to understand how genes are regulated in a cell-specific manner, and how environmental conditions can induce or repress the transcription of genes. In some cases, microarrays can even help to elucidate the

TABLE 21.1

Applications of DNA Microarrays

Application	Description
Cell-specific gene expression	A comparison of microarray data using cDNAs derived from RNA of different cell types can identify genes that are expressed in a cell-specific manner.
Gene regulation	Environmental conditions play an important role in gene regulation. A comparison of microarray data may reveal genes that are induced under one set of conditions and repressed under another set of conditions.
Elucidation of metabolic pathways	Genes that encode proteins that participate in a common metabolic pathway are oftentimes expressed in a parallel manner. This can be revealed from a microarray analysis (as described later in figure 21.3). This application overlaps with the study of gene regulation via microarrays.
Tumor profiling	Different types of cancer cells exhibit striking differences in their profiles of gene expression. This can be revealed by a DNA microarray analysis. This approach is gaining widespread use to subclassify tumors that are sometimes morphologically indistinguishable. This may provide information that can improve a patient's clinical treatment.
Genetic variation	A mutant allele may not hybridize to a spot on a microarray as well as a wild-type allele. Therefore, microarrays are gaining widespread use as a tool to detect genetic variation. This has been used to identify disease-causing alleles in humans and mutations that contribute to quantitative traits in plants and other species.
Microbial strain identification	Microarrays can distinguish between closely related bacterial species and subspecies.

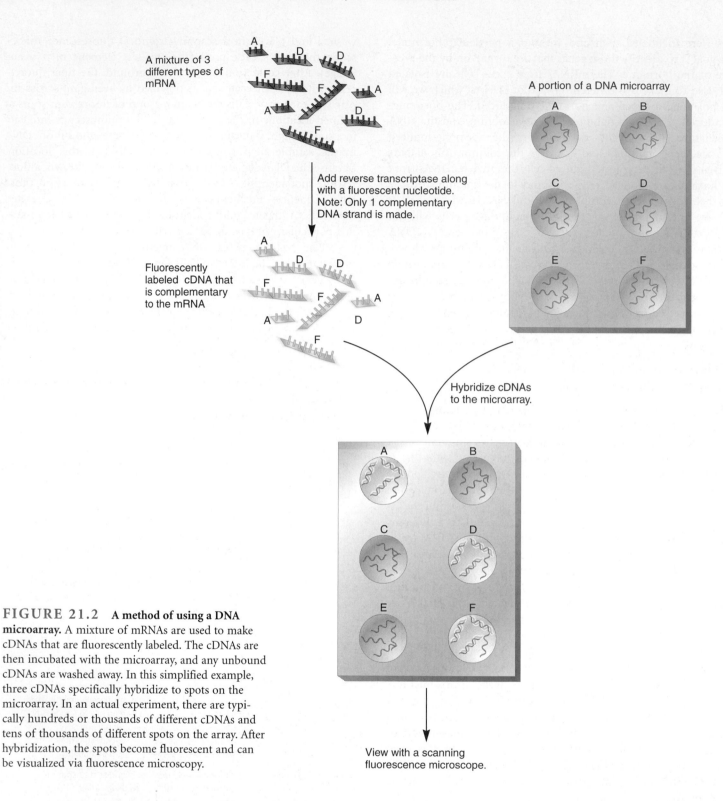

A mixture of 3 different types of mRNA

A portion of a DNA microarray

Add reverse transcriptase along with a fluorescent nucleotide. Note: Only 1 complementary DNA strand is made.

Fluorescently labeled cDNA that is complementary to the mRNA

Hybridize cDNAs to the microarray.

View with a scanning fluorescence microscope.

FIGURE 21.2 **A method of using a DNA microarray.** A mixture of mRNAs are used to make cDNAs that are fluorescently labeled. The cDNAs are then incubated with the microarray, and any unbound cDNAs are washed away. In this simplified example, three cDNAs specifically hybridize to spots on the microarray. In an actual experiment, there are typically hundreds or thousands of different cDNAs and tens of thousands of different spots on the array. After hybridization, the spots become fluorescent and can be visualized via fluorescence microscopy.

genes encoding proteins that participate in a complicated metabolic pathway. Microarrays can also be used as identification tools. For example, gene expression patterns can aid in the categorization of tumor types. Such identification can be important in the treatment of the disease. Instead of using labeled cDNA,

researchers can also hybridize labeled genomic DNA to a microarray. This can be used to identify mutant alleles in a population of individuals. In addition, this technology is proving useful in the correct identification of closely related bacterial species and subspecies.

EXPERIMENT 21A

The Coordinate Regulation of Many Genes Is Revealed by a DNA Microarray Analysis

One way that cells respond to environmental changes is via the coordinate regulation of genes. Under one set of environmental conditions, a particular set of genes may be induced, while under another set of conditions those same genes may be repressed. In the past, researchers have been able to study this type of gene regulation using tools that can analyze the expression of a few genes at a time. The advent of microarrays, however, makes it possible to study the expression of the whole genome under different sets of environmental conditions.

One of the first studies using this approach involved the analysis of the yeast genome. As mentioned in chapter 20, the genome of baker's yeast, *Saccharomyces cerevisiae,* was the first eukaryotic genome to be sequenced. It encodes approximately 6,200 different genes. An important process in the growth of yeast cells, as well as cells of other species, is the ability to metabolize carbon sources using different metabolic pathways. When yeast cells have glucose available, they metabolize the glucose to pyruvate. This process is called glycolysis. If oxygen is present, the pyruvate can be broken down via the tricarboxylic acid cycle (TCA cycle) that occurs in the mitochondrion. Therefore, when yeast are first given glucose and then allowed to metabolize it in the presence of oxygen, they first metabolize the carbohydrate via glycolysis and then, when the glucose is used up, they metabolize the products of glycolysis via the TCA cycle. The process of switching from glycolysis to the TCA cycle is called a diauxic shift, and it involves major changes in the expression of genes involved with carbohydrate metabolism. The goal of the experiment described in figure 21.3 was to identify genes that are induced and repressed as yeast cells shift from glycolysis to the TCA cycle. It was carried out by Joseph DeRisi, Vishwanath Iyer, and Patrick Brown in 1997.

As shown in figure 21.3, yeast cells were initially given glucose as their carbon source for growth and then allowed to grow for several hours. Over time, the glucose was used up, and the cells shifted from glycolysis to the TCA cycle. At various time points, samples of cells were removed, and the RNA was isolated. The RNA was then exposed to reverse transcriptase in the presence of deoxyribonucleotides, one of which was fluorescently labeled. This created fluorescently labeled cDNAs.

To determine the relative changes in RNA synthesis, two different fluorescent dyes were used. The RNA collected at the first time point (i.e., when glucose was at a high level) was used to make cDNA that contained a green fluorescent dye. The RNAs collected at later time points were used to make cDNAs that contained a red fluorescent dye. A sample of green cDNA (from the first time point) was then mixed with a sample of red cDNA (from later time points), and the mixture was hybridized to a microarray containing all 6,200 yeast genes. The fluorescence microscope can measure the amount of red fluorescence and green fluorescence at each spot in the microarray. The fluorescence ratio, red fluorescent units divided by green fluorescent units, provided a way to quantitatively determine how the expression of the genes was changing. For example, if the ratio was high, this means that a gene was being induced as glucose levels fall, because the amount of red cDNA would be higher than the amount of green cDNA. Alternatively, if the red : green ratio was low, this means that a gene was being repressed as glucose is used up.

■ THE HYPOTHESIS

A diauxic switch from glycolysis to the TCA cycle will involve the induction of certain genes and the repression of other genes.

■ TESTING THE HYPOTHESIS — FIGURE 21.3 **The use of a DNA microarray to study carbohydrate metabolism in yeast.**

Starting material: A commonly used strain of *Saccharomyces cerevisiae* (baker's yeast) was used in this study. The researchers had made a DNA microarray containing (nearly) all of the known yeast genes.

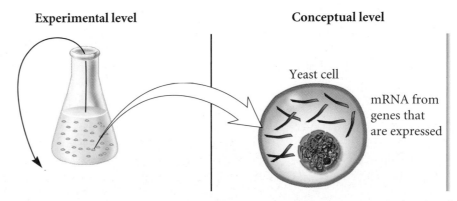

Experimental level   **Conceptual level**

1. Inoculate yeast into a media containing 2% glucose. Grow for up to 21 hours.

Yeast cell

mRNA from genes that are expressed

(continued)

2. Beginning at 9 hours after inoculation, take out samples of cells and isolate mRNA. This involves breaking open cells and running the cell contents over a poly-dT column under high-salt conditions. Since mRNA has a polyA tail, it will bind to this column while other cell components flow through. The purified mRNA can be then eluted by adding a buffer to the top of the column that contains a low concentration of salt. This breaks the interaction between the poly-dT and polyA tails. In this experiment, mRNA samples were isolated at approximately 2-hour intervals, beginning at 9 hours and ending at 21 hours.

3. Add reverse transcriptase and fluorescently labeled nucleotides to make complementary strands of fluorescently labeled cDNA. Note: The sample at 9 hours was used to make green cDNA while mRNA samples collected at later time points were used to make red cDNA.

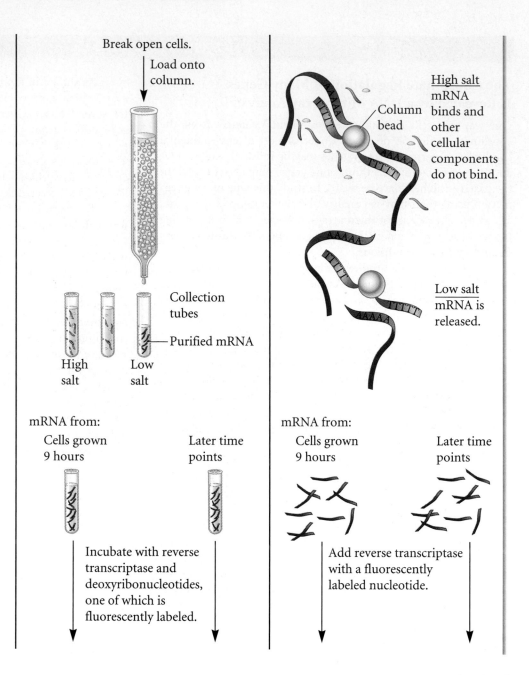

Break open cells.

Load onto column.

Collection tubes

Purified mRNA

High salt   Low salt

mRNA from:
Cells grown 9 hours   Later time points

Incubate with reverse transcriptase and deoxyribonucleotides, one of which is fluorescently labeled.

High salt
mRNA binds and other cellular components do not bind.

Column bead

Low salt
mRNA is released.

mRNA from:
Cells grown 9 hours   Later time points

Add reverse transcriptase with a fluorescently labeled nucleotide.

(continued)

4. At each time point, mix together the cDNA derived from that time (i.e., red cDNA) with cDNA derived from the 9-hour time point (i.e., green cDNA).

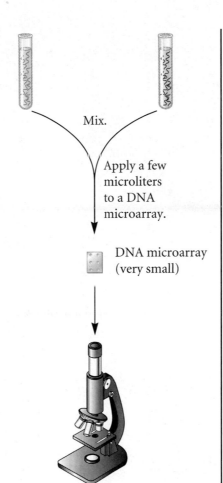

Mix.

Apply a few microliters to a DNA microarray.

DNA microarray (very small)

5. Hybridize the mixture to the yeast DNA microarray as described in figure 21.2.

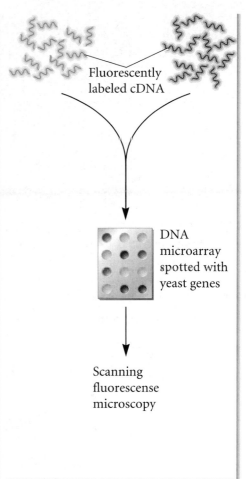

Fluorescently labeled cDNA

DNA microarray spotted with yeast genes

6. Examine the DNA microarray with a scanning fluorescence microscope. The data are then analyzed by a computer, which can correlate expression levels among different genes. (This is a cluster analysis.)

Scanning fluorescense microscopy

■ THE DATA

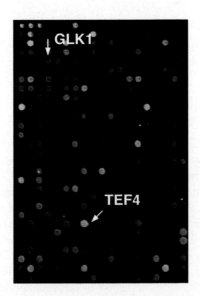

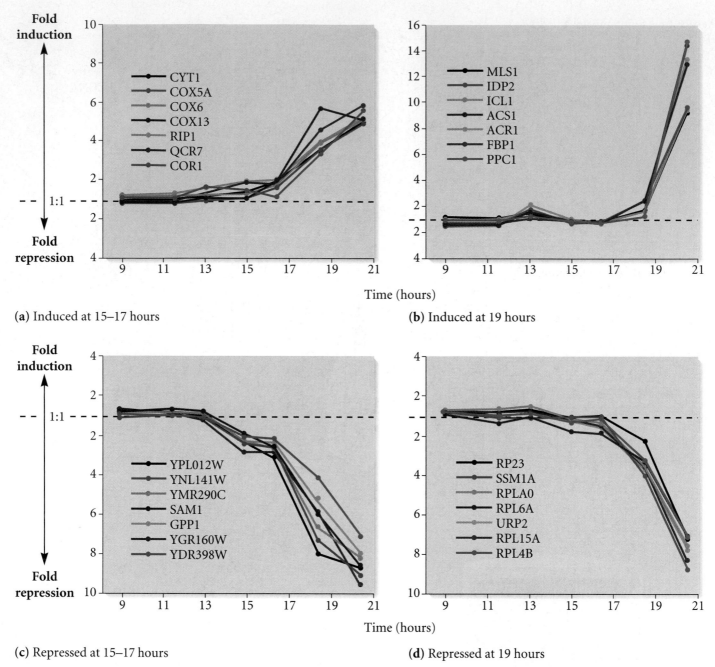

(a) Induced at 15–17 hours

(b) Induced at 19 hours

(c) Repressed at 15–17 hours

(d) Repressed at 19 hours

INTERPRETING THE DATA

A portion of one microarray is shown in the data of figure 21.3. As you can see here, the array shows many spots, some of which are relatively green and some of which are relatively red. The green spots would indicate genes that are expressed at higher levels during early stages of growth when glucose levels are high. An example is a gene designated *TEF4,* which is involved with protein synthesis. Red spots identify genes that are expressed when the glucose is depleted. The gene *GLK1* encodes an enzyme that phosphorylates glucose but is expressed only when glucose levels are low. In addition, many spots are not fluorescent, indicating that there was not much cDNA in the sample to hybridize to the DNA strands on the microarray. These spots would correspond to genes that are not greatly expressed under either condition. Judging from the numbers of red and green spots, the shift from glycolysis to the TCA cycle involved a great amount of gene induction and repression. By determining the red : green ratio at each spot, it was found that 710 genes were induced by at least a factor of two, while 1,030 genes were repressed by a factor of two or more. Thus, it was found that a diauxic shift involves a staggering amount of gene regulation. Of the 6,200 yeast genes, 1,740, or roughly 28%, appeared to be regulated as a result of a diauxic shift.

Because the gene sequence in each spot is known, the next step is to relate the levels of gene expression based on the microarray data to specific genes. A common goal is to identify genes whose pattern of expression seems to strongly correlate with each other. This is called a **cluster analysis.** The data of figure 21.3 illustrates how microarray data can be used to make this type of comparison. Though the sequences of these genes are known from the yeast genome-sequencing project, the functions of some of these genes are not known. Each time point involved the measurement of the red : green ratio at a particular spot. Parts a–d show the analysis of 28 genes (7 genes in each figure) over the course of 9 to 21 hours following the addition of glucose. The diauxic shift occurred at approximately 15 hours.

The genes shown in part a were induced at 15 to 17 hours of growth. Due to their coordinate regulation, these genes may be controlled by the same transcription factor(s) and may participate in a common metabolic response to the induction of the TCA cycle. In fact, several of the genes in part a have already been studied, and they are known to play a role in the ability of the mitochondrion to metabolize pyruvate. This makes sense, since the diauxic shift occurred around 15 hours of growth, which correlates with the time when the genes in part a were induced. By comparison, the genes in part b were induced later, after the TCA cycle had operated for a few hours and when the carbon sources were becoming depleted. Therefore, it would seem that the genes in part b were induced as a response to the operation of the TCA cycle or they were induced because the carbon sources in the media were low. Though further work needs to be done to elucidate the functions of some of the genes shown in parts a and b,

the data provide insights that suggest a common regulation and metabolic function for particular groups of genes. It would seem more likely that the transcription factor(s) that regulate the genes in part a would be different from those that regulate the genes in part b. Likewise, the genes in parts a and b may work in different cellular pathways.

Parts c and d illustrate a similar phenomenon except that the switch to the TCA cycle and the operation of the TCA cycle represses those genes. The genes shown in part c were repressed at the time of the diauxic shift, while those shown in part d were repressed after the TCA cycle had operated for a few hours. Most of the genes shown in part d were already known to function as ribosomal proteins. Therefore, as the carbon sources in the media were depleted, these results suggest that one of the cellular responses is to diminish the synthesis of ribosomes, which, in turn, would slow down the rate of protein synthesis. It would seem that the yeast cells were trying to conserve energy at this late stage of growth.

Overall, the data shown in the experiment of figure 21.3 illustrate how a microarray analysis can shed light on gene function at the genomic level. It provides great insight regarding gene regulation and may help to identify groups of proteins (i.e., clusters) that share a common cellular function.

A self-help quiz involving this experiment can be found at the Online Learning Center.

21.2 PROTEOMICS

Thus far in chapter 21, we have considered ways to characterize the genome of a given species and study its function. Because most genes encode proteins, a logical next step is to examine the functional roles of the proteins that a species can make. As mentioned, this field is called proteomics and the entire collection of a species' proteins is its proteome.

Though the study of a species' genome provides important information regarding its proteome, genomics represents only the first step in our comprehensive understanding of protein structure and function. Researchers often use genomic information to initiate proteomic studies, but such information must be followed up with research that involves the direct analysis of proteins. For example, as discussed in the preceding section, a DNA microarray may provide insights into the transcription of particular genes under a given set of conditions. However, the correlation between mRNA expression and protein abundance is often less than 0.5. This is because protein levels are greatly affected by the rate of mRNA translation and by the turnover rate of a given protein. Therefore, DNA microarray data must be confirmed using other methods such as a Western blotting (discussed in chapter 18), which directly determines the abundance of a protein in a given cell type.

A second way that genomic data can elucidate the workings of the proteome is via homology. As discussed later in chapter 21,

homology between the genes of different species can be used to predict protein function and/or structure, when the function or structure of the protein in one of the species is already known. However, homology may not provide direct information regarding the regulation of protein structure and function. Also, it may not reveal potential types of protein-protein interactions in which a given protein may participate. Therefore, even though homology is profoundly useful as a guiding tool, a full understanding of protein function must involve the direct analysis of proteins as they are found in living cells.

As we move into the twenty-first century, a key challenge facing molecular biologists is the study of proteomes. Much like genomic research, this will require the collective contributions of many research scientists, as well as improvements in technologies that are aimed at unraveling the complexities of the proteome. Nonetheless, it is exciting to realize that a new phase of molecular biology has just begun!

The Proteome Is Much Larger Than the Genome

From the sequencing and analysis of an entire genome, researchers can identify all the genes that a given species contains. The size of the proteome, however, is larger than the genome, and its actual size is somewhat more difficult to determine. The larger size of the proteome is rooted in a variety of cellular processes. As discussed in chapters 12 and 15, changes in pre-mRNA structures may ultimately affect the resulting amino acid sequence of a protein. The most important alteration that occurs commonly in eukaryotic species is **alternative splicing.** For many genes, a single pre-mRNA can be spliced into more than one version. The splicing is often cell specific, or it may be related to environmental conditions. As discussed in chapter 15, alternative splicing is widespread, particularly among higher eukaryotes. It can lead to the production of several or perhaps dozens of different polypeptide sequences from the same pre-mRNA. This greatly increases the number of potential proteins in the proteome. Similarly, the phenomenon of **RNA editing** can lead to changes in the coding sequence of an mRNA. However, RNA editing is much less common than alternative splicing.

A second process that greatly diversifies the composition of a proteome is the phenomenon of **posttranslational covalent modification.** Certain types of modifications are primarily involved with the assembly and construction of a functional protein. These alterations include proteolytic processing, disulfide bond formation, and the attachment of prosthetic groups, sugars, or lipids. These are typically irreversible changes that are necessary to produce a functional protein. In contrast, other types of modifications, such as phosphorylation, acetylation, and methylation, are often reversible modifications that transiently affect the function of a protein. Because a protein may be subject to several different types of covalent modification, this can greatly increase the forms of a particular protein that are found in a cell at any given time.

Two-Dimensional Gel Electrophoresis Is Used to Separate a Mixture of Cellular Proteins

As we have just learned, the proteome is usually much larger than a species' genome. Even so, any given cell within a complex multicellular organism will produce only a subset of the proteins that are found in the proteome of a species. For example, the human genome has approximately 35,000 different genes, yet a muscle cell makes only a few thousand types of proteins. The subset of proteins that a cell makes depends primarily on the cell type, the stage of development, and the environmental conditions. An objective of researchers in the field of proteomics is the identification and functional characterization of all the proteins that a cell type will make. Because cells produce thousands of different proteins, this is a daunting task. Nevertheless, along with genomic research, the past decade has seen important advances in our ability to isolate and identify cellular proteins.

The workhorse technique in the field of proteomics is **two-dimensional gel electrophoresis.** It is a separation technique that can distinguish hundreds or even thousands of different proteins within a cell extract. The steps in this procedure are shown in figure 21.4. As its name suggests, the technique involves two different gel electrophoresis procedures. A sample of cells is lysed, and the proteins are loaded onto the top of a tube gel that separates proteins according to their net charge at a given pH. A protein migrates to the point in the gel where its net charge is zero. This is termed *isoelectric focusing.* After the tube gel has run, it is placed on top of a slab gel that contains sodium dodecyl sulfate

(SDS). The SDS coats each protein uniformly, so that proteins in the slab gel are separated according to their molecular mass. Smaller proteins move toward the bottom of the gel more quickly than larger ones. After the slab gel has run, the proteins within the gel can be stained with a dye. As seen in figure 21.4, the end result is a collection of spots, each spot corresponding to a unique cellular protein. The resolving power of two-dimensional gel electrophoresis is extraordinary. Proteins that differ by a single charged amino acid can be resolved as two distinct spots using this method.

There are various ways to identify spots on a gel that may be interesting to researchers. One possibility is that a given cell type may show a few very large spots that are not found when proteins are analyzed from other cell types. The relative abundance of such spots may indicate that a particular protein is important to cell structure or function. Secondly, certain spots on a two-dimensional gel may be seen only under a given set of conditions. For example, a researcher may be interested in the effects of a hormone on the function of a particular cell type. Two-dimensional gel electrophoresis could be conducted on a sample in which the cells had not been exposed to the hormone versus a sample in which they had. This may reveal particular spots that are present only when the cells are exposed to a given hormone. Finally, abnormal cells often express proteins that are not found in normal cells. This is particularly true for cancer cells. A researcher may compare normal and cancer cells via two-dimensional gel electrophoresis to identify proteins that are expressed only in cancer cells.

Mass Spectrometry Is Used to Identify Proteins

Two-dimensional gel electrophoresis is often used as the first step in the separation of cellular proteins. The next goal is to correlate a given spot on a two-dimensional gel with a particular protein. To accomplish this goal, a spot on a two-dimensional gel can be cut out of the gel to obtain a tiny amount of the protein within the spot. In essence, the two-dimensional gel procedure purifies a small amount of the cellular protein of interest. The next step is to identify that protein. This can be accomplished via **mass spectrometry.**

Figure 21.5 describes how mass spectrometry can determine the amino acid sequence of a protein. As shown here, the technique actually determines the mass of peptide fragments that are produced by digesting a purified protein with an enzyme that cuts the protein into small peptide fragments. The peptides are mixed with an organic acid and dried onto a metal slide. The sample is then struck with a laser. This causes the peptides to become ejected from the slide in the form of an ionized gas, in which the peptide contains one or more positive charges. The charged peptides are then accelerated via an electric field and fly toward a detector. The time of flight is determined by their mass and net charge. A measurement of the flight time provides an extremely accurate way to determine the mass of a peptide.

An ultimate goal of mass spectrometry is to determine the amino acid sequence of a given peptide. This is accomplished using two mass spectrometers, a method called **tandem mass**

Load a mixture of proteins onto an isoelectric focusing tube gel.

pH 4.0

Proteins migrate until they reach the pH where their net charge is 0. At this point, a single band could contain 2 or more different proteins.

pH 10.0

Lay the tube gel onto an SDS-gel and separate proteins according to their molecular mass.

SDS-gel

pH 4.0   pH 10.0

200 kDa

10 kDa

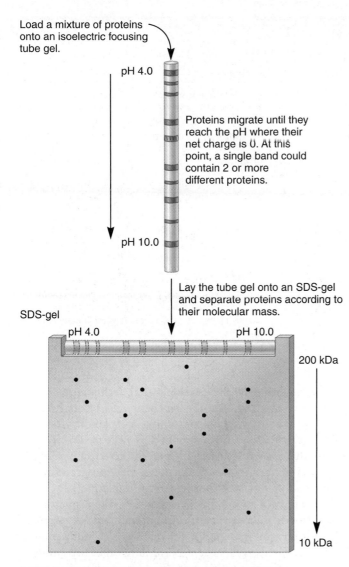

(a) The technique of 2-dimensional gel electrophoresis

pH 4.0   pH 10.0

200 kDa

10 kDa

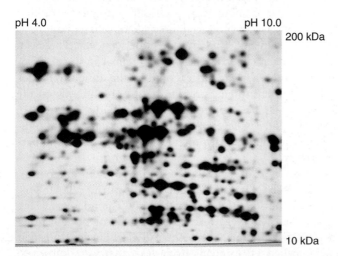

(b) A 2-dimensional gel that has been stained to visualize proteins

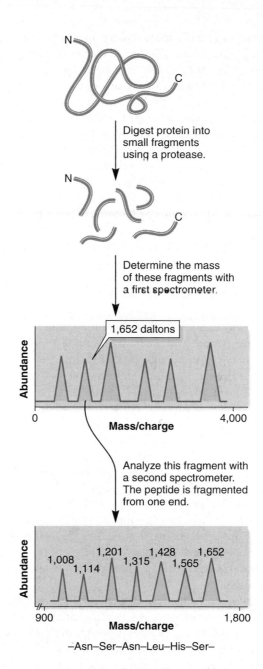

N   C

Digest protein into small fragments using a protease.

N

C

Determine the mass of these fragments with a first spectrometer.

1,652 daltons

Abundance

0   Mass/charge   4,000

Analyze this fragment with a second spectrometer. The peptide is fragmented from one end.

Abundance

1,008   1,114   1,201   1,315   1,428   1,565   1,652

900   Mass/charge   1,800

–Asn–Ser–Asn–Leu–His–Ser–

FIGURE 21.5 The use of tandem mass spectrometry to determine the amino acid sequence of a peptide.

FIGURE 21.4 Two-dimensional gel electrophoresis. (a) The technique involves two electrophoresis steps. First, a mixture of proteins is separated on an isoelectric focusing gel that has the shape of a tube. Proteins migrate to the point where their net charge is zero. This tube gel is placed into a long well on top of an SDS-polyacrylamide gel. This second gel separates the proteins according to their mass. In this diagram, only a few spots are seen, but an actual experiment would involve a mixture of hundreds or thousands of different proteins. (b) A photograph of an SDS-gel that has been stained for proteins. Each spot represents a unique cellular protein.

spectrometry. The first spectrometer measures the mass of a given peptide. This same peptide is then analyzed by a second spectrometer after the peptide has been digested into smaller fragments. The differences in the masses of the peaks in the spectrum from the second spectrometer reveal the amino acid sequence of the peptide, because the masses of all 20 amino acids are known. For example, as shown in figure 21.5, let's suppose that a peptide had a mass of 1,652 daltons. If one amino acid at the end were removed, and the smaller peptide had a mass that was 87 daltons less (i.e., 1,565 daltons), this would indicate that a serine is at one end of the peptide, because the mass of serine within a polypeptide chain is 87 daltons. If two amino acids were removed at one end and the mass was 224 daltons less, this would correspond to the removal of one serine (87 daltons) and one histidine (137 daltons). If three amino acids were removed, and the mass was decreased by 337 daltons, this would correspond to the removal of one serine (87 daltons), one histidine (137 daltons), and one leucine (113 daltons). Thus, from these measurements, we would conclude that the amino acid sequence from one end of the peptide was serine–histidine–leucine.

After a researcher has obtained a few short peptide sequences from a given protein, genomic information can readily predict the entire amino acid sequence of the protein. For example, if a peptide had the sequence serine–histidine–leucine–alanine–alanine–asparagine, one could determine the possible codon sequences that could encode such a peptide. More than one sequence is possible due to the degeneracy of the genetic code. Using computer software described later in chapter 21, the codon sequences would be used as query sequences to scan the entire genomic sequence. This program would locate a match between the predicted codon sequence and a specific gene within the genome. In this way, mass spectrometry makes it possible to identify the gene that encodes the entire protein. The gene sequence, in turn, can be used to predict the amino acid sequence of the whole protein.

It should also be mentioned that mass spectrometry can be used to identify protein covalent modifications. For example, if an amino acid within a peptide was phosphorylated, the mass of the peptide would be increased by the mass of a phosphate group. This increase in mass can be determined via mass spectrometry.

Protein Microarrays Can Be Used to Study Protein Expression and Function

Earlier in chapter 21, we learned about DNA microarrays, which have gained widespread use to study gene expression at the transcriptional level. The technology to make DNA microarrays is being applied to make **protein microarrays.** In this type of technology, proteins, rather than DNA molecules, are spotted onto a glass or silica slide. The development of protein microarrays is a bit more challenging because proteins are much more easily damaged by the manipulations that occur during microarray formation. For example, the three-dimensional structure of a protein may be severely damaged by drying, which usually occurs during the formation of a microarray. This has created additional chal-

TABLE 21.2

Applications of Protein Microarrays

Application	Description
Protein expression	An antibody microarray can measure protein expression, since each antibody in a given spot recognizes a specific amino acid sequence. This can be used to study the expression of proteins in a cell-specific manner. It can also be used to determine how environmental conditions affect the levels of particular proteins.
Protein function	The substrate specificity and enzymatic activities of groups of proteins can be analyzed by exposing a functional protein microarray to a variety of substrates.
Protein-protein interactions	The ability of two proteins to interact with each other can be determined by exposing a functional protein microarray to fluorescently labeled proteins.
Pharmacology	The ability of drugs to bind to cellular proteins can be determined by exposing a functional protein microarray to different kinds of labeled drugs. This can help to identify the targets within a cell to which a given drug may bind.

lenges to researchers who are developing the technology of protein microarrays. In addition, the synthesis and purification of proteins tend to be more time-consuming compared to the production of DNA, which can be amplified by PCR or directly synthesized on the microarray itself. In spite of these technical difficulties, however, the last few years have seen progress in the production and uses of protein microarrays (table 21.2).

There are two common approaches to protein microarray analysis: **antibody microarrays** and **functional protein microarrays.** The purpose of an antibody microarray is to study protein expression. As discussed in chapter 18, antibodies are proteins that recognize antigens. One type of antigen that an antibody can recognize is a short peptide sequence found within another protein. Therefore, an antibody can specifically recognize a cellular protein. Researchers can produce thousands of different antibodies, each one recognizing a different peptide sequence. These can be spotted onto a microarray. Cellular proteins can then be isolated, fluorescently labeled, and exposed to the antibody microarray. When a given protein is recognized by an antibody on the microarray, it will be captured by the antibody and remain bound to that spot. The level of fluorescence at a given spot indicates the amount of a cellular protein that is recognized by a particular antibody.

The other type of array is a functional protein microarray. To make this type of array, researchers must purify cellular proteins and then spot them onto a microarray. The microarray can then be analyzed with regard to specific kinds of protein function. In 2000, for example, Heng Zhu, Michael Snyder, and their colleagues purified 119 proteins from yeast that were known to function as protein kinases. These kinds of proteins attach phosphate groups onto other cellular proteins. A microarray was made consisting of different possible protein substrates of these 119 kinases, and then the array was exposed to each of the 119 kinases in the presence of radiolabeled ATP. By following the incorporation of phosphate into the array, they determined the substrate specificity of each kinase. On a much larger scale, the same group of researchers purified 5,800 different yeast proteins and spotted them onto a microarray. The array was then exposed to fluorescently labeled calmodulin, which is a regulatory protein that binds calcium ions. Several proteins in the microarray were found to bind calmodulin. Some of these were already known to be regulated by calmodulin. Interestingly, other proteins in the array that had not been previously known to bind calmodulin were identified.

21.3 BIOINFORMATICS

Geneticists use computers to collect, store, manipulate, and analyze data. Molecular genetic data, which comes in the form of a DNA, RNA, or amino acid sequence, is particularly amenable to computer analysis. The ability of computers to analyze data at a rate of millions or even billions of operations per second has made it possible to solve problems concerning genetic information that were thought intractable a few decades ago. In recent years, the marriage between genetics and biocomputing has yielded an important branch of science known as bioinformatics. Several scientific journals are largely devoted to this topic. Computer analysis of genetic sequences usually relies on three basic components: a computer, a computer program, and some type of data. In genetic research, the data are typically a particular genetic sequence or several sequences that a researcher or clinician wants to study. For example, this could be a DNA sequence derived from a cloned DNA fragment. A sequence of interest may be relatively short or thousands to millions of nucleotides in length. Experimentally, DNA sequences and related data are obtained using the techniques described in chapters 18 and 20.

In this section, we will first consider the fundamental concepts that underlie the analysis of genetic sequences. We will then consider how these methods are used to provide insights regarding functional genomics and proteomics. Chapter 26 will consider applications of bioinformatics in the area of evolutionary biology. In addition, the website for this textbook provides students with the opportunity to run computer programs themselves. This type of hands-on learning will help you to see how the computer has become a valuable tool to analyze genetic data. Furthermore, at our textbook website, you will be provided with information regarding other websites that allow students to learn about gene sequences and run a myriad of programs.

Sequence Files Are Analyzed by Computer Programs

A **computer program** is a defined series of operations that can analyze data in a desired way. For example, a computer program might be designed to take a DNA sequence and translate it into an amino acid sequence. A first step in the computer analysis of genetic data is the creation of a **computer data file** to store the data. This file is simply a collection of information in a form suitable for storage and manipulation on a computer. In genetic

studies, a computer data file typically might contain an experimentally obtained DNA, RNA, or amino acid sequence. For example, a file could contain the DNA sequence of one strand of the *lacY* gene from *Escherichia coli,* as shown here. The numbers to the left represent the base number in the sequence file.

```
    1  ATGTACTATT  TAAAAAACAC  AAACTTTTGG  ATGTTCGGTT  TATTCTTTTT
   51  CTTTTACTTT  TTTATCATGG  GAGCCTACTT  CCCGTTTTTC  CCGATTTGGC
  101  TACATGACAT  CAACCATATC  AGCAAAGTG   ATACGGGTAT  TATTTTGCC
  151  GCTATTTCTC  TGTTCTCGCT  ATTATTGGAA  GGGGTGTTTG  GTGTGGTTG
  201  TGACAAACTC  GGGCTGCGCA  AATACCTGCT  GTGGATTATT  ACGGCATGT
  251  TAGTCATGTT  TGCGCCGTTC  TTTATTTTTA  TCTTCGGGCC  ACTGTTACAA
  301  TACAACATTT  TAGTAGGATC  GATTGTTGGT  GGTATTTATC  TAGGCTTTTG
  351  TTTTAACGCC  GGTGCGCCAG  CAGTAGAGGC  ATTTATTGAG  AAAGTCAGCC
  401  GTCGCAGTAA  TTTCGAATTT  GGTCGCGCGC  GGATGTTTGG  CTGTGTTGGC
  451  TGGGCGCTGT  GTGCCTGAT   TGTCGGCATC  ATGTTCACCA  TCAATAATCA
  501  GTTTGTTTTC  TGGCTGGGCT  CTGGCTGTGC  ACTCATCCTC  GCCGTTTTAC
  551  TCTTTTTCGC  CAAAACGGAT  GCGCCCTCTT  CTGCCACGGT  TGCCAATGCG
  601  GTAGGTGCCA  ACCATTCGGC  ATTTAGCCTT  AAGTGGCAC   TGGAACTGTT
  651  CAGACAGCCA  AAACTGTGGT  TTTGTCTACT  GTATGTTATT  GGCGTTTCCT
  701  GCACCTACGA  TGTTTTTGAC  CAACAGTTTG  CTAATTTCTT  TACTTCGTTC
  751  TTTGCTACCG  GTGAACAGGG  TACGCGGGTA  TTTGGCTACG  TAACGACAAT
  801  GGGCGCAATTA CTTAACGCCT  CGATTATCTT  CTTGCGCCA   CTGATCATTA
  851  ATCGCATCGG  TGGGAAAAAC  GCCCTGCCTG  TGGCTGGCAC  TATTATCTCT
  901  CTACGTATTA  TTGGCTCATC  GTTCGCCACC  TCAGCGCTGG  AAGTGGTTAT
  951  TCTGAAAACG  CTGCATATGT  TTGAAGTACC  GTTCCTGCTG  GTGGGCTGCT
1,001  TTAAATATAT  TACCAGCCAG  TTTGAAGTGC  GTTTTCAGC   GACGATTTAT
1,051  CTGGTCTGTT  TCTGCTTCTT  TAAGCAACTG  GCATGATTT   TTATGTCTGT
1,101  ACTGGCGGGC  AATATGTATG  AAAGCATCGG  TTTCCAGGGC  GCTTATCTGG
1,151  TGCTGGGTCT  GGTGGCGCTG  GGCTTCACCT  TAATTTCCGT  GTTCACGCTT
1,201  AGCGGCCCCG  GCCCGCTTTC  CCTGCTGCGT  CGTCAGGTGA  ATGAAGTCGC
1,251  TTAA
```

To store data in a computer data file, a scientist creates the file (if it does not already exist) and enters the data. This may be done simply by using a keyboard to enter (i.e., type) the data into the file. For example, a scientist could obtain a DNA sequence by reading a sequencing ladder and then keyboarding the data into a computer file. Entry via a keyboard can be used to create a small computer data file, say for short genetic sequences less than a few thousand nucleotides in length. For very long genetic sequences, like those obtained in genome-sequencing projects, this would be a tedious and error-prone process. Nowadays, it is more common for genetic sequence data to be directly entered into a computer file by laboratory instruments (e.g., densitometers and fluorometers). These are designed with the capability to read data such as a sequencing ladder and enter the information directly into a computer file.

The purpose of making a computer file that contains a genetic sequence is to take advantage of the swift speed with which computers can analyze this information. Genetic sequence data in a computer file can be investigated in many different ways corresponding to the myriad of questions a researcher might ask about the sequence and its functional significance. These include:

1. Does a sequence contain a gene?
2. Where are functional sequences such as promoters, regulatory sites, and splice sites located within a particular gene?
3. Does a sequence encode a polypeptide? If so, what is the amino acid sequence of the polypeptide?
4. Does a sequence predict certain structural features for DNA, RNA, or proteins? For example, is a DNA sequence likely to be in a B-DNA conformation? What is the secondary structure of an RNA sequence or polypeptide sequence?

5. Is a sequence homologous to any other known sequences?

6. What is the evolutionary relationship between two or more genetic sequences?

To answer these and many other questions, computer programs have been written to analyze genetic sequences in particular ways. These programs have been devised by theoreticians who understand basic genetic principles and can design computational strategies for analyzing genetic sequences. When constructing a computer program, a theoretician has a goal that the program is meant to fulfill. For example, a theoretician may wish to write a program to translate a DNA sequence into an amino acid sequence. Based on knowledge of the genetic code, a theoretician can devise computational procedures, or algorithms, that relate a DNA sequence to an amino acid sequence. In a computer program, these procedures are executed as a stepwise "plan of operations" that manipulates the data in a sequence file. For a computer to perform the plan of operation, the program instructions must be written in a programming language the computer can decipher. After this has been accomplished, the program is tested to see if it works. Usually, errors or bugs are found in the program that prevent it from working. After these bugs are overcome, in a process called debugging, the program is ready for use. An ideal computer program is one that accurately manipulates data and at the same time is easy to use (user-friendly).

As mentioned, a few programs that manipulate genetic sequences are available at the website for this textbook. After you have read this section, it is useful to visit the website and try some of these programs for yourself. In addition, the website provides links to other websites that specialize in the analysis of genetic sequences.

As an example, let's consider a computer program aimed at translating a DNA sequence into an amino acid sequence and see how it might work in practice. The operation of the program as it would appear on a computer screen is shown in figure 21.6. The geneticist (i.e., the user) has a DNA sequence file that he or she may want to have translated into an amino acid sequence. The user is sitting at a computer; this computer is connected to a program that can translate a DNA sequence into an amino acid sequence. In this hypothetical example, the program is named TRANSLATION. To begin running the program, the user types the name of the program, TRANSLATION, or clicks on the name of the program. As it runs, the program asks a series of questions. These are depicted on the screen in black. The typed answers of the user are shown in red. The first question asks which sequence file the user wants translated. In this case, the user wants the *lacY* gene sequence translated into an amino acid sequence. The name of this file is lacY.SEQ. The program then asks which codon translation table to use. As you may recall from chapter 13, some species have slight variations in their usage of the genetic code. In this case, the user selects the standard genetic code. Next, the program requires the parameters of the sequence file that is to be translated. The user decides to begin the translation at the first nucleotide in the sequence file and end the translation at nucleotide number 1,254. The user wants the program to trans-

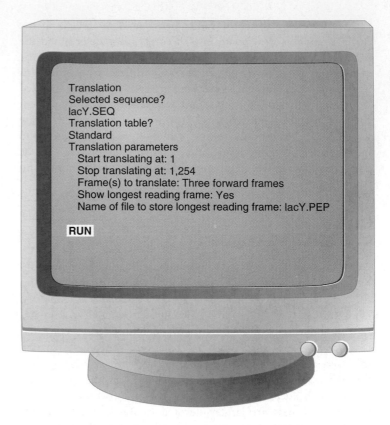

```
Translation
Selected sequence?
lacY.SEQ
Translation table?
Standard
Translation parameters
    Start translating at: 1
    Stop translating at: 1,254
    Frame(s) to translate: Three forward frames
    Show longest reading frame: Yes
    Name of file to store longest reading frame: lacY.PEP

RUN
```

FIGURE 21.6   **The operation of a computer program as it would appear on a computer screen.**

late the sequence in all three forward reading frames and to show the longest reading frame (i.e., the longest amino acid sequence that is found among the three possible reading frames). This translated sequence is saved in a file that the user wants to be named lacY.PEP. (Note: The symbol PEP provides a reminder that it is a poly**pep**tide sequence.) The user clicks on the RUN button and translation program proceeds to translate the *lacY* sequence and stores the amino acid sequence in a file called lacY.PEP. The contents of the lacY.PEP file are shown here.

```
  1 MYYLKNTNFW MFGLFFFFYF FIMGAYFPFF PIWLHDINHI SKSDTGIIFA
 51 AISLFSLLFQ PLFGLLSDKL GLRKYLLWII TGMLVMFAPF FIFIFGPLLQ
101 YNILVGSIVG GIYLGFCFNA GAPAVEAFIE KVSRRSNFEF GRARMFGCVG
151 WALCASIVGI MFTINNQFVF WLGSGCALIL AVLLFFAKTD APSSATVANA
201 VGANHSAFSL KLALELFRQP KLWFLSLYVI GVSCTYDVFD QQFANFFTSF
251 FATGEQGTRV FGYVTTMGEL LNASIMFFAP LIINRIGGKN ALLLAGTIMS
301 VRIIGSSFAT SALEVVIKLT LHMFEVPFLL VGCFKYITSQ FEVRFSATIY
351 LVCFCFFKQL AMIFMSVLAG NMYESIGFQG AYLVLGLVAL GFTLISVFTL
401 SGPGPLSLLR RQVNEVA
```

In this file, which was created by the computer program TRANSLATION, each of the 20 amino acids is given a single-letter abbreviation. The advantages of running this program are speed and accuracy. It can translate a relatively long genetic sequence within seconds. By comparison, it would probably take you a few hours to look each codon up in the genetic code table and write the sequence out in the correct order. If you visit the textbook website or links at the website and actually run a program like

TRANSLATION, you will discover that such a program can translate a genetic sequence into six reading frames (i.e., 3 forward and 3 reverse). This is useful if a researcher does not know where the start codon is located and/or does not know the direction of the coding sequence.

In genetic research, computer programs (i.e., software) are frequently sold as parts of large software packages. A genetics software package contains many computer programs that can analyze genetic sequences in different ways. For example, one program can translate a DNA sequence into an amino acid sequence, another program can locate introns within genes, and there are many other capabilities. These software packages commonly are sold to universities, hospitals, and industry. At these locations, a central computer with substantial memory and high-speed computational abilities is responsible for running the software. Individuals with personal computers can connect to this central computer. In this way, many individuals can use the programs within a software package. In addition, many programs are freely available on the Internet. As mentioned, links to some of these other websites can be found at the website for this textbook.

The Scientific Community Has Collected Sequence Files and Stored Them in Large Computer Databases

In chapter 18, we considered how researchers can clone and sequence genes. Likewise, in chapter 20, we learned how scientists are investigating the genetic sequences of entire genomes from several species, including humans. The amount of genetic information that is generated by researchers has become enormous. The Human Genome Project, for example, has produced more data than any other undertaking in the history of biology. With these advances, scientists realize that another critical use of computers is to store the staggering amount of data produced from genetic research.

When a large number of computer data files are collected and stored in a single location, this is called a **database.** In addition to genetic sequences, the files within databases are **annotated,** which means they contain a concise description of a sequence (e.g., a gene sequence), the name of the organism from which this sequence was obtained, and the function of the encoded protein, if it is known. The file may also describe other features of significance and provide a published reference that contains the sequence.

The scientific community has collected the genetic information from thousands of research labs and created several large databases. Table 21.3 describes some of the major genetic databases in use worldwide. These databases enable researchers to access and compare genetic sequences that are obtained by many laboratories. Later in chapter 21, we will learn how researchers can use databases to analyze genetic sequences.

The databases described in table 21.3 collect genetic information from many different species. Scientists have also created more specialized databases, called **genome databases,** that focus on the genetic characteristics of a single species. Genome databases have been created for species of bacteria (e.g., *E. coli*), yeast

TABLE 21.3

Examples of Major Computer Databases

Type	Description
Nucleotide sequence	DNA sequence data are collected into three internationally collaborating databases: GenBank (a USA database), EMBL (European Molecular Biology Laboratory Nucleotide Sequence Database), and DDBJ (DNA Databank of Japan). These databases receive sequence and sequence annotation data from genome projects, sequencing centers, individual scientists, and patent offices. These databases are accessed via the Internet and on CD-ROM.
Amino acid sequence	Amino acid sequence data are collected into a few international databases including Swissprot (Swiss protein database), PIR (Protein Information Resource), Genpept (translated peptide sequences from the GenBank database), and TrEMBL (Translated sequences from the EMBL database).
Three-dimensional structure	PDB (Protein Data Bank) collects the three-dimensional structures of biological macromolecules with an emphasis on protein structure. These are primarily structures that have been determined by X-ray crystallography and NMR, but some models are included in the database. These structures are stored in files that can be viewed on a computer with the appropriate software.
Protein motifs	Prosite is a database containing a collection of sequence motifs that are characteristic of a protein family, domain structure, or posttranslational modification. Pfam is a database of protein families with multiple sequence alignments.

From Persson, Bengt (2000) Bioinformatics in protein analysis. *EXS* 88, 215–231.

(e.g., *Saccharomyces cerevisiae*), worms (e.g., *Caenorhabditis elegans*), fruit flies (e.g., *Drosophila melanogaster*), plants (e.g., *Arabidopsis thaliana*), and mammals (e.g., mice and humans). The primary aim of genome databases is to organize the information from sequencing and mapping projects for a single species. Genome databases identify the known genes within an organism and describe their map locations in the genome. In addition, a genome database may provide information concerning gene alleles, bibliographic information, a directory of researchers who study the species, and other pertinent information.

Different Computational Strategies Can Identify Functional Genetic Sequences

At the molecular level, the function of the genetic material is based largely on specific genetic sequences that play distinct roles. For example, codons are three-base sequences that specify particular amino acids, and promoters are sequences that provide a binding site for RNA polymerase to initiate transcription. Computer programs can be designed to scan very long sequences and locate meaningful features within the sequence. To illustrate this concept in a hypothetical way, let's first consider the following sequence file, which contains an alphabetic sequence of 54 letters:

Sequence file:

```
GJTYLLAMAQLHEOGYLTOBWENTMNMTORXXXTGOODNTHEQA
LLYTLSTORE
```

We will now compare how three different computer programs can analyze this sequence to identify meaningful features. The goal of our first program is to locate all the English words within this sequence. If we ran this program, we would obtain the following result:

```
GJTYLLAMAQLHEOGYLTOBWENTMNMTORXXXTGOODNTHEQA
LLYTLSTORE
```

In this case, a computer program has identified locations where the sequence of letters forms a word. Several words (which are underlined) have been located within this sequence.

A second computer program could be aimed at locating a series of words that are organized in the correct order to form a grammatically logical English sentence. If we used our sequence file and ran this program, we would obtain the following result:

```
GJTYLLAMAQLHEOGYLTOBWENTMNMTORXXXTGOODNTHEQA
LLYTLSTORE
```

The second program has identified five words that form a logical sentence.

Finally, a computer program might be used to identify patterns of letters, rather than words or sentences. For example, a computer program could locate a pattern of five letters that occurs in both the forward and reverse directions. If we applied this program to our sequence file, we would obtain the following:

```
GJTYLLAMAQLHEOGYLTOBWENTMNMTORXXXTGOODNTHEQA
LLYTLSTORE
```

In this case, the program has identified a pattern where five letters are found in both the forward and reverse directions.

In the three previous examples, we can distinguish between **sequence recognition** (as in our first example) and **pattern recognition** (as in our third example). In sequence recognition, the program has the information that a specific sequence of symbols has a specialized meaning. This information must be supplied to the computer program. For example, the first program would have access to the information from a dictionary with all known English words. With this information, the first program can identify sequences of letters that make words. By comparison, the third program does not rely on specialized sequence information. Rather, it is looking for a pattern of symbols that can occur within any group of symbol arrangements.

Overall, the simple programs we have considered illustrate three general types of identification strategies:

1. *Locate specialized sequences within a very long sequence.* A specialized sequence is called a **sequence element.** The computer program has a list of predefined sequence elements and can identify such elements within a sequence of interest.

2. *Locate an organization of sequences.* As shown in the second program, this could be an organization of sequence elements. Alternatively, it could be an organization of a pattern of sequences.

3. *Locate a pattern of sequences.* The third program is an example.

The great power of computer analysis is that these types of operations can be performed with great speed and accuracy on sequences of symbols that may be enormously long.

Now that we understand the general ways that computer programs identify sequences, let's consider specific examples. As we have discussed throughout this textbook, there are many short nucleotide sequences that play specialized roles in the structure or function of genetic material. In general, a genetic sequence with a particular function is called a sequence element or **sequence motif.** Table 21.4 lists examples. A geneticist may want to locate a short sequence element within a longer nucleotide sequence in a data file. For example, a sequence of chromosomal DNA might be tens of thousands of nucleotides in length, and a geneticist may want to know whether a sequence element, such as a TATA box, is found at one or more sites within the chromosomal DNA. To do so, a researcher could visually examine the long chromosomal DNA sequence in search of a TATA sequence. Of course, this would be tedious and prone to error. By comparison, the appropriate computer program can locate a sequence element within seconds. Therefore, computers are very useful for this type of application.

TABLE 21.4

Short Sequence Elements That Can Be Identified by Computer Analysis

Type of Sequence	Examples*
Promoter	Many *E. coli* promoters contain TTGACA (−35 site) and TATAAT (−10 site). Eukaryotic core promoters may contain CAAT boxes, GC boxes, TATA boxes, etc.
Response elements	Glucocorticoid response element (AGR† ACA), cAMP response element (GTGACGTRA).
Start codon	ATG
Stop codons	TAA, TAG, TGA
Splice site	GTRAGT————————YNYTRAC(Y)ₙAG
Polyadenylation signal	AATAAA
Highly repetitive sequences	Relatively short sequences that are repeated many times throughout a genome.
Transposable elements	Usually characterized by a pattern in which direct repeats flank inverted repeats.

*The sequences shown in this table would be found in the DNA. For gene sequences, only the coding strand is shown. In the transcribed RNA, U would be substituted for T.

†R = purine (A or G), Y = pyrimidine (T or C), N = A, T, G, or C, U in RNA = T in DNA.

By comparing the amino acid sequences and known functions of proteins in thousands of cases, researchers have also found amino acid motifs that carry out specialized functions within proteins. For example, researchers have determined that the amino acid motif asparagine–X–serine (where X is any amino acid except proline) within eukaryotic proteins is a glycosylation site (i.e., it may have a carbohydrate attached to it). The Prosite database (refer back to table 21.3) contains a collection of all amino acid sequence motifs known to be functionally important. Researchers can use computer programs to determine whether an amino acid sequence contains any of the motifs found in the Prosite database. This may help them to understand the role of a newly found protein of unknown function.

Several Computer-Based Approaches Can Identify Structural Genes Within a Nucleotide Sequence

As described in chapter 12, a structural gene is composed of nucleotide sequences organized in a particular way. A typical gene contains a promoter, followed by a start codon, a coding sequence, a stop codon, and a transcriptional termination site. In addition, most genes contain regulatory sequences (e.g., eukaryotic response elements or prokaryotic operator sites), and eukaryotic genes are likely to contain introns. After researchers have sequenced a long segment of chromosomal DNA, they frequently want to know if the sequence contains any genes. In an attempt to answer this question, geneticists can use computer programs that are aimed at identifying genes in long genomic DNA sequences.

Computer programs can employ different strategies to locate genes. A **search by signal** approach relies on known sequences such as promoters, start and stop codons, and splice sites to help predict whether or not a DNA sequence contains a structural gene. The program tries to locate an organization of known sequence elements that normally are found within a gene. It would try to locate a region that contains a promoter sequence, followed by a start codon, a coding sequence, a stop codon, and a transcriptional terminator.

A second strategy is a **search by content** approach. The goal here is to identify sequences with a nucleotide content that differs significantly from a random distribution. Within structural genes, this occurs primarily due to codon usage. Although there are 64 codons, most organisms display a **codon bias** within structural genes. This means that certain codons are used much more frequently than others. For example, UUA, UUG, CUU, CUC, CUA, and CUG all specify leucine. In yeast, however, the UUG codon is used for that purpose 80% of the time. Codon bias allows organisms to more efficiently rely on a smaller population of tRNA molecules. A search by content strategy, therefore, attempts to locate coding regions by identifying regions where the nucleotide content displays certain types of bias.

Another way to locate coding regions within a DNA sequence is to examine translational reading frames. As discussed in chapter 13, each codon contains three nucleotides. In a new DNA sequence, researchers must consider that the reading of codons (in groups of three nucleotides) could begin with the first nucleo-

tide (reading frame 1), the second nucleotide (reading frame 2), or the third nucleotide (reading frame 3). An **open reading frame (ORF)** is a region of a nucleotide sequence that does not contain any stop codons. Because most proteins are several hundred amino acids in length, a relatively long reading frame is required to encode them. In prokaryotic species, long ORFs are contained within the chromosomal gene sequences. In eukaryotic genes, however, the chromosomal coding sequence may be interrupted by introns. As described earlier in chapter 21, one way to determine eukaryotic ORFs is to clone and sequence cDNA, which is complementary to mRNA. Alternatively, a computer program can translate a genomic DNA sequence in all three reading frames, seeking to identify a long ORF. In figure 21.7, a DNA sequence has been translated in all three reading frames. Only one of the three reading frames (3) contains a very long open reading frame without any stop codons, suggesting that this DNA sequence encodes a protein. In figure 21.7, it was assumed that the reading frame was from left to right. In an uncharacterized genetic sequence, it is possible that a reading frame could proceed from right to left. Therefore, six reading frames are possible in a newly discovered genetic sequence.

Before ending this section it should be pointed out that computer programs are not always accurate in their prediction of gene sequences. In particular, it is oftentimes difficult for programs to predict the correct start codon and the precise intron/exon boundaries. In some cases, computer programs may even suggest that a region encodes a gene when it does not. Therefore, while a bioinformatic approach is a relatively easy tool to identify potential genes, it should not be viewed as a definitive method. The confirmation that a DNA region encodes an actual gene requires laboratory experimentation to show that it is truly transcribed into RNA.

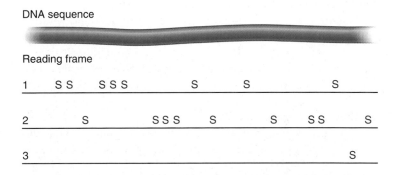

FIGURE 21.7 **Translation of a DNA sequence in all three reading frames.** The three lines represent the translation of a gene sequence in each of three forward reading frames; the reading frames proceed from left to right. The letter *S* indicates the location of a stop codon. Reading frame 3 has a very long open reading frame, suggesting that it may be the reading frame for a structural gene. Reading frames 1 and 2 are not likely to be the reading frames for a structural gene, because they contain many stop codons. During the cloning of DNA, the orientation of a gene may become flipped so that the coding sequence is inverted. Therefore, when analyzing many cloned DNA fragments, six reading frames (i.e., three forward and three reverse) are evaluated. Only the three forward frames are shown here.

Computer Programs Can Identify Homologous Sequences

We now turn our attention to the uses of computer technology to identify genes that are evolutionarily related. The ability to sequence DNA allows geneticists to examine evolutionary relationships at the molecular level. This has become an extremely powerful tool in the field of genomics. When comparing genetic sequences, researchers frequently find two or more sequences that are similar. For example, the sequence of the *lacY* gene that encodes the lactose permease in *E. coli* is similar to that of the gene that encodes the lactose permease in a closely related bacterium, *Klebsiella pneumoniae.* As shown here, when segments of the two *lacY* genes are lined up, approximately 78% of their bases are a perfect match.

```
           151                                                    200
E. coli        TCTTTTTCTT TTACTTTTTT ATCATGGGAG CCTACTTCCC GTTTTTCCCG
K. pneumoniae  TCTTTTTCTT TTACTATTTC ATTATGTCAG CCTACTTTCC TTTTTTTCCG

           201                                                    250
               ATTTGGCTAC ATGACATCAA CCATATCAGC AAAAGTGATA CGGGTATTAT
               GTGTGGCTGG CGGAAGTTAA CCATTTAACC AAAACCGAGA CGGGTATTAT
```

In this case, the two sequences are similar because the genes are **homologous** to each other. This means they have been derived from the same ancestral gene. This idea is shown schematically in figure 21.8. An ancestral *lacY* gene was located in a bacterium that preceded the evolutionary divergence between *E. coli* and *K. pneumoniae.* After these two bacteria had diverged from each other, their *lacY* genes accumulated distinct mutations that produced somewhat different base sequences for this gene. Therefore, in these two species of bacteria, the *lacY* genes are similar but not identical. When two homologous genes are found in different species, these genes are termed **orthologs.**

Two or more homologous genes can also be found within a single organism. These are termed **paralogous** genes or **paralogs.** As discussed in chapter 8, this can occur because abnormal gene duplication events can produce multiple copies of a gene and ultimately lead to the formation of a gene family. A **gene family** consists of two or more copies of homologous genes within the genome of a single organism.

It is important not to confuse the terms *homology* and *similarity.* **Homology** implies a common ancestry. **Similarity** means

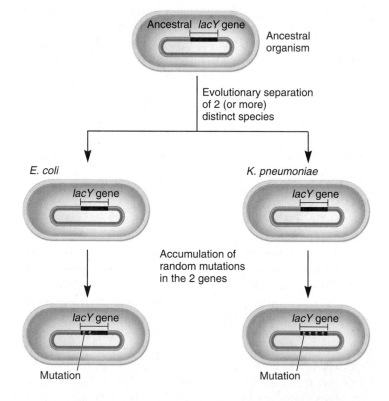

FIGURE 21.8 **The origin of homologous *lacY* genes in *Escherichia coli* and *Klebsiella pneumoniae.*** This figure emphasizes a single gene within an ancestral organism. During evolution, the ancestral organism diverged into two different species, *E. coli* and *K. pneumoniae.* After this divergence, the *lacY* gene in the two separate species accumulated mutations, yielding *lacY* genes with different sequences.

GENES→TRAITS   After two organisms diverge evolutionarily, their genes will accumulate different random mutations. This example concerns the *lacY* gene, which encodes lactose permease. In both species, the function of lactose permease is to transport lactose into the cell. The *lacY* gene in these two species has accumulated different mutations that slightly alter the amino acid sequence of the protein. Researchers have determined that these two species transport lactose at significantly different rates. Therefore, the changes in gene sequences have affected the ability of these two species to use lactose, an important trait for survival.

that two sequences have similar sequences. In many cases, such as the *lacY* example, similarity is due to homology. However, this is not always the case. Short genetic sequences may be similar to each other even though two genes are not related evolutionarily. For example, many nonhomologous bacterial genes contain similar promoter sequences at the −35 and −10 regions.

A Simple Dot Matrix Can Compare the Degree of Similarity Between Two Sequences

To evaluate the similarity between two sequences, a matrix can be constructed. In a general way, figure 21.9 illustrates the use of a simple dot matrix. In figure 21.9*a*, the sequence GENETICSIS-COOL is compared to itself. Each point in the grid corresponds to one position of each sequence. The matrix allows all such pairs to be compared simultaneously. Dots are placed where the same letter occurs in both sequences. The key observation is that regions of similarity are distinguished by the occurrence of many dots

along a diagonal line within the matrix. In contrast, figure 21.9*b* compares two unrelated sequences: GENETICSISCOOL and THECOURSEISFUN. In this comparison, no diagonal lines are seen. In some cases, two sequences may be related to each other but differ in length. Figure 21.9*c* compares the sequences GENETICSISCOOL and GENETICSISVERYCOOL. In this example, the second sequence is four letters longer than the first. In the dot matrix, two diagonal lines occur. To align these two lines, a gap must be created in the first sequence. Figure 21.9*d* shows the insertion of a gap that aligns the two sequences. Now the two diagonal lines fall along the same line.

Overall, figure 21.9 illustrates two important features of matrix methods. First, regions of homology are recognized by a series of dots that lie along a diagonal line. Second, gaps can be inserted to align sequences that are of unequal length but still have some similarity. These same concepts hold true when genetic sequences are compared to each other. Unfortunately, for long genetic sequences, a simple dot matrix approach is not adequate.

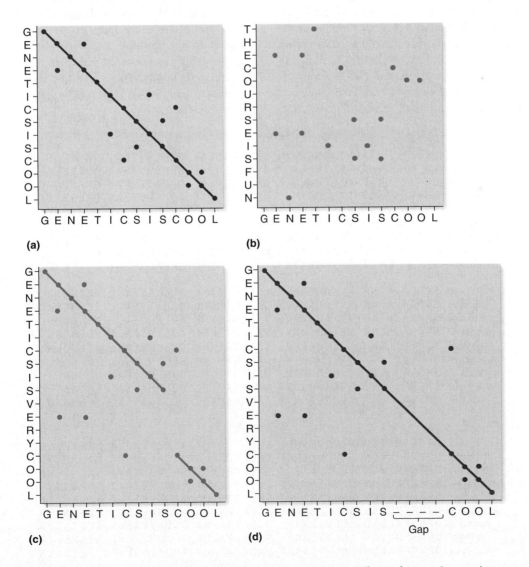

FIGURE 21.9   The use of dot matrices to evaluate similarity between two sequences. When a dot matrix contains many points that fall on a diagonal line, this indicates regions of similarity.

Instead, dynamic programming methods are used to identify similarities between genetic sequences. This approach was originally proposed by Saul Needleman and Christian Wunsch. Dynamic programming methods are theoretically similar to a dot matrix, but they involve mathematical operations that are beyond the scope of this textbook.

In their original work of 1970, Needleman and Wunsch demonstrated that whale myoglobin and human β-hemoglobin have similar sequences. Since then, this approach has been extended to compare more than two genetic sequences. Newer computer programs can align several genetic sequences and sensibly put in gaps. This produces a **multiple sequence alignment.** You may run such a program at the website for this textbook.

To illustrate the usefulness of multiple sequence alignment, let's use the general methods of Needleman and Wunsch and apply them to the globin gene family. As you may recall from chapter 8, hemoglobin is a protein found in red blood cells; it is responsible for carrying oxygen through the bloodstream. In humans, nine paralogous globin genes are functionally expressed. (There are also several pseudogenes that are not expressed and one myoglobin gene.) The nine globin genes fall into two categories: the α-chains and the β-chains. The α-chain genes are α_1, α_2, θ, and ξ; the β-chain genes are β, δ, γ_A, γ_G, and ε. Each hemoglobin protein is composed of two α-chains and two β-chains.

Because the globin genes are expressed at different stages of human development, the composition of hemoglobin changes during the course of growth. For example, the ξ and ε genes are expressed during early embryonic development, whereas the α and β genes are expressed in the adult. Insights into the structure and function of the hemoglobin polypeptide chains can be gained by comparing their sequences. In the computer-based experiment described in figure 21.10, the sequences of the human globin polypeptides are compared in a multiple sequence alignment using dynamic programming methods.

An inspection of a multiple sequence alignment may reveal important features concerning the similarities and differences within a gene family. In this alignment, dots are shown where it is necessary to create gaps to keep the sequences aligned. As we can see, the sequence similarity is very high between α_1, α_2, θ, and ξ. In fact, the α_1 and α_2 amino acid sequences are identical. This suggests that the four types of α chains likely carry out very similar functions. Likewise, the β chains encoded by the β, δ, γ_A, γ_G, and ε genes are very similar to each other. In the globin gene family, the α chains are much more similar to each other than they are to the β chains, and vice versa.

In general, amino acids that are highly conserved within a gene family are more likely to be important functionally. The *arrows* in the multiple sequence alignment point to histidine amino acids that are conserved in all nine members of the hemoglobin gene family. These histidine residues are involved in the necessary function of binding the heme molecule to the globin polypeptides.

```
                    1                                                          50
    β beta          VHLTPEEEKSA VTALWGKV..  NVDEVGGEAL GRLLVVYPWT QRFFESFGDL
    δ delta         VHLTPEEEKSA VNALWGKV..  NVDAVGGEAL GRLLVVYPWT QRFFESFGDL
   γA gamma-A       GHFTEEDKAT  ITSLWGKV..  NVEDAGGEAL GRLLVVYPWT QRFFESFGDL
   γG gamma-G       GHFTEEDKAT  ITSLWGKV..  NVEDAGGEAL GRLLVVYPWT QRFFESFGDL
    ε epsilon       VHFTAEEKAA  VTSLWSKM..  NVEEAGGEAL GRLLVVYPWT QRFFESFGDL
   α1 alpha-1       VLSPADKTN   VKAAWGKVGA HAGEGAEAL  ERMFLSFPTT KTYFPHF.DL
   α2 alpha-2       VLSPADKTN   VKAAWGKVGA HAGEGAEAL  ERMFLSFPTT KTYFPHF.DL
    θ theta         ALSAEDRAL   VRALWKKLGS NVGVYTTEAL ERTFLAFPAT KTYFSHL.DL
    ξ zeta          SLTKTERTI   IVSMWAKIST QADTIGTETL ERLFLSHPQT KTYFPHF.DL

                    51                                                         100
    β beta          STPDAVMGNP KVKAHGKKVL GAFSDGLAHL DNLKGTFATL SELHCDKLHV
    δ delta         SSPDAVMGNP KVKAHGKKVL GAFSDGLAHL DNLKGTFSQL SELHCDKLHV
   γA gamma-A       SSASAIMGNP KVKAHGKKVL TSLGDAIKHL DDLKGTFAQL SELHCDKLHV
   γG gamma-G       SSASAIMGNP KVKAHGKKVL TSLGDAIKHL DDLKGTFAQL SELHCDKLHV
    ε epsilon       SSPSAILGNP KVKAHGKKVL TSFGDAIKNM DNLKPAFAKL SELHCDKLHV
   α1 alpha-1       SHGSA..... QVKGHGKKVA DALTNAVAHV DDMPNALSAL SDLHAHKLRV
   α2 alpha-2       SHGSA..... QVKGHGKKVA DALTNAVAHV DDMPNALSAL SDLHAHKLRV
    θ theta         SPGSS..... QVKAHGQKVA DALSLAVERL DDLPHALSAL SHLHACQLRV
    ξ zeta          HPGSA..... QLRAHGSKVV AAVGDAVKSI DDIGGALSKL SELHAYQLRV
                                       ↑                              ↑
                    101                                                        148
    β beta          DPENFRLLGN VLVCVLAHHF GKEFTPPVQA AYQKVVAGVA NALAHKYH
    δ delta         DPENFRLLGN VLVCVLARNF GKEFTPQMQA AYQKVVAGVA NALAHKYH
   γA gamma-A       DPENFRLLGN VLVCVLAIHF GKEFTPEVQA SWQKMVTAVA SALSSRYH
   γG gamma-G       DPENFRLLGN VLVCVLAIHF GKEFTPEVQA SWQKMVTAVA SALSSRYH
    ε epsilon       DPENFRLLGN VMVIILATHF GKEFTPEVQA AWQKLVSAVA IALAHKYH
   α1 alpha-1       DPVNFKLLSH CLLVTLAAHL PAEFTPAVHA SLDKFLASVS TVLTSKYR
   α2 alpha-2       DPVNFKLLSH CLLVTLAAHL PAEFTPAVHA SLDKFLASVS TVLTSKYR
    θ theta         DPASFQLLGH CLLVTLARHL PGDFSPALQA SLDKFLSHSVI SALVSEYR
    ξ zeta          DPVNFKLLSH CLLVTLAARF PADFTAEAHA AWDKFLSVVS SVLTEKYR
```

FIGURE 21.10   A multiple sequence alignment among selected members of the globin gene family in humans.

Overall, the alignment shown in figure 21.10 illustrates the type of information that can be derived from a multiple sequence alignment. In this case, multiple sequence alignment has shown that a group of nine genes falls into two closely related subgroups. The alignment has also identified particular amino acids within the proteins' sequences that are highly conserved. This conservation is consistent with an important role in protein function.

A Database Can Be Searched to Identify Homologous Sequences

Homologous genes usually carry out similar or identical functions. As we have just considered, the members of the globin gene family are all involved with carrying and transporting oxygen. Likewise, the *lacY* genes in *E. coli* and *K. pneumoniae* both encode lactose permeases that transport lactose across the bacterial cell membrane.

In general, there is a strong correlation between homology and function. In many cases, the first indication of the function of a newly determined sequence is through homology to known sequences in a database. An example is the gene that is altered in cystic fibrosis patients. After this gene was identified in humans, a database search revealed that it is homologous to several genes found in other species. Moreover, a few of the homologous genes were already known to encode proteins that function in the transport of ions and small molecules across the plasma membrane. This observation provided an important clue that cystic fibrosis involves a defect in ion transport.

The ability of computer programs to identify homology between genetic sequences provides a powerful tool for predicting the function of genetic sequences. In 1990, Stephen Althschul, David Lipman, and their colleagues developed an approach called a **b**asic **l**ocal **a**lignment **s**earch **t**ool (**BLAST**). The BLAST program has been described by many geneticists as the single most important bioinformatic tool. This type of computer program can start with a particular genetic sequence and then locate homologous sequences within a large database. You may run this program at the website for this textbook. Because there are only four bases but 20 amino acids, homology among protein sequences is easier to identify than is DNA sequence homology. Among proteins, sequences that diverged more than 2.5 billion years ago can still be correlated. By comparison, it becomes difficult to identify homologous DNA sequences that diverged more than 100 million years ago.

Table 21.5 describes the results of a database search that started with the amino acid sequence for human δ-(delta) globin. This sequence was used as a "query sequence" by the BLAST program to search the Swissprot database, which contains hundreds of thousands of different protein sequences. With a high-speed computer, an amino acid sequence of a given protein (i.e., the query sequence) can be compared to hundreds of thousands of polypeptide sequences in the Swissprot database. The BLAST program can determine which sequences are the closest matches. As an example, table 21.5 shows the results using human δ-globin as the query sequence. Since the human δ-globin sequence is

TABLE 21.5

An Example of a Database Search

Sequence in the Database	Species	% Identity with Human δ-globin*
δ-globin	Human	100.0
δ-globin	Chimpanzee	99.3
β-globin	Hanuman langur	95.2
δ-globin	Colobus polykomos	95.9
β-globin	Common gibbon	94.5
β-globin	Green monkey	94.5
β-globin	Colobus polykomos	93.8
β-globin	Mandrill	93.2
δ-globin	Black-handed spider monkey	93.8
β-globin	Japanese macaque	93.2

*The *right* column describes the percentages of amino acids that are identical to human δ-globin. The entries in this table are listed in the order of their degree of similarity. In some cases (e.g., compare hanuman langur and colobus polykomos), a sequence with a lower percent identity may be judged as more similar to human δ-globin because the alignment may require fewer gaps in the sequence.

already in the Swissprot database, its closest match is to itself. This search identified many other sequences that are very similar to human δ-globin. These include β-globin and δ-globin sequences from several primate species.

The results shown in table 21.5 illustrate the remarkable computational abilities of current computer technology. In minutes, the human δ-globin sequence can be compared to hundreds of thousands of different sequences.

Genetic Sequences Can Be Used to Predict the Structure of RNA and Proteins

Another topic in which bioinformatics has impacted functional genomics and proteomics is the area of structure prediction. The function of macromolecules such as DNA, RNA, and proteins relies on their structure. The three-dimensional structure of these macromolecules, in turn, depends on the linear sequences of their building blocks. In the case of DNA and RNA, this means a linear sequence of nucleotides; proteins are composed of a linear sequence of amino acids. Currently, the three-dimensional structure of macromolecules is determined primarily through the use of biophysical techniques such as X-ray crystallography and nuclear magnetic resonance (NMR). These methods are technically difficult and very time-consuming. DNA sequencing, by comparison, requires much less effort. Therefore, since the three-dimensional structure of macromolecules depends ultimately on the linear sequence of their building blocks, it would be far easier if we could predict the structure (and function) of DNAs, RNAs, and proteins from their sequence of building blocks.

As discussed in chapter 9, RNA molecules typically are folded into a secondary structure, which commonly contains double-stranded regions. These secondary structure regions are further folded and twisted to adopt a tertiary conformation. As mentioned in many places throughout the textbook, such structural features of RNA molecules are functionally important. For example, the folding of RNA into secondary structures such as stem-loops affects transcriptional termination and other regulatory events. Therefore, geneticists are interested in the secondary and tertiary structures that RNA molecules can adopt.

Many approaches are available for investigating RNA structure. In addition to biophysical and biochemical techniques, computer modeling of RNA structure has become an important tool. Modeling programs can consider different types of information. For example, the known characteristics of RNA secondary structure, such as the ability to form double-stranded regions, can provide parameters for use in a modeling program.

A comparative approach can also be used in RNA structure prediction. This method assumes that RNAs of similar function and sequence have a similar structure. For example, the genes that encode certain types of RNAs, such as the 16S rRNAs, have been sequenced from many different species. Among different species, the 16S rRNAs have similar but not identical sequences. Computer programs can compare many different 16S rRNA sequences to aid in the prediction of secondary structure. Figure 21.11 illustrates a secondary structural model for 16S rRNA based on a comparative sequence analysis. This large RNA contains 45 stem-loop regions. As you can imagine, it would be rather difficult to deduce such a model without the aid of a computer!

Structure prediction is also used in the area of proteomics. As described in chapter 12, proteins contain repeating secondary structural patterns known as α helices and β sheets. Several computer-based approaches attempt to predict secondary structure from the primary amino acid sequence. These programs base their predictions on different types of parameters. Some programs rely on the physical and energetic properties of the amino acids and the polypeptide backbone. More commonly, however, secondary structure predictions are based on the statistical frequency of amino acids within secondary structures that have been crystallized.

For example, Peter Chou and Gerald Fasman have compiled X-ray crystallographic data to calculate the likelihood that an amino acid will be found in an α helix or a β sheet. Certain amino acids, such as glutamate and alanine, are likely to be found in an α helix; others, such as valine and isoleucine, are more likely to be found in a β-sheet structure. Such information can be used to predict whether a sequence of amino acids within a protein is likely to be folded in an α-helix or β-sheet conformation. Secondary structure prediction is correct for approximately 60 to 70% of all sequences. While this degree of accuracy is promising, it is generally not sufficient to predict protein secondary structure reliably. Therefore, one must be cautious in interpreting the results of a secondary structure prediction program.

In recent years, an exciting computer methodology known as a neural net has been applied to protein secondary structure prediction. A computer neural network is a large number of calculation units organized into interconnected layers; this structure

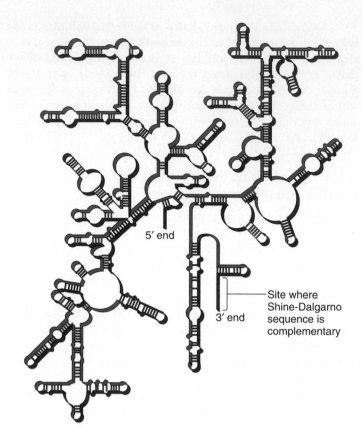

5′ end

Site where Shine-Dalgarno sequence is complementary

3′ end

FIGURE 21.11   A secondary structural model for *E. coli* 16S rRNA.

is reminiscent of the organization of neurons in the brain. The input layer receives data and may (or may not) transmit that information to the next layer. A neural network can adjust the parameters that define the interconnections among its units in response to data; they can thus be trained to identify complex patterns coming from the input data. For example, the amino acid sequences of proteins with known crystal structures can be used to train a network (i.e., adjust its parameters) to predict secondary structures for new amino acid sequences. Thus far, neural networks have yielded small improvements in secondary structure predictions. In the future, a combination of innovative predictive approaches and increased information concerning the biophysical properties of amino acids may make secondary structure prediction a reliable strategy.

The three-dimensional structure of a protein is extremely difficult to predict solely from its amino acid sequence. However, researchers have had some success in predicting tertiary structure using a comparative approach. This strategy requires that the protein of interest be homologous to another protein, the tertiary structure of which already has been solved by X-ray crystallography. In this situation, the crystal structure of the known protein can be used as a starting point to model the three-dimensional structure of the protein of interest. This approach is known as homology-based modeling, knowledge-based modeling, or comparative homology. As an example, two research groups predicted the structure of a protein encoded by HIV. The protein, known as HIV protease, is homologous to other proteases with structures that have been solved by X-ray crystallography. Two similar models of the HIV protease were predicted before its actual crystal structure was determined. Both models turned out to be fairly accurate representations of the actual structure, which was later solved by X-ray crystallography.

CONCEPTUAL AND EXPERIMENTAL SUMMARY

The field of **functional genomics** attempts to elucidate the roles of genetic sequences in a given species. This involves the analysis of groups of genes and even the entire genome. A **cDNA** or **EST library** can aid in the identification of genes, because the cDNAs are derived from cellular mRNA (a product of gene transcription). A **subtractive cDNA library** can identify genes that are expressed only under a particular environmental condition. **DNA microarrays** are small slides that contain many spots of DNA fragments. Some arrays contain every known gene within a species' genome. The arrays are used as hybridization tools to determine which genes are expressed in a given cell type or under a particular set of conditions. Usually, fluorescently labeled cDNA is hybridized to the array, although genomic DNA or RNA could also be used. By comparing the results of DNA microarray analyses at different time points (as in figure 21.3) or under different sets of environmental conditions, researchers can identify genes that are coordinately regulated. This is termed a **cluster analysis.** These genes often encode proteins that participate in common cellular functions.

Proteome is the term used to describe all the proteins that a species can produce. The field of **proteomics** is aimed at a comprehensive understanding of protein function. Due to RNA processing and the posttranslational modifications of proteins, the proteome may be much larger than the genome. A common way to isolate proteins is via **two-dimensional gel electrophoresis.** Spots from a two-dimensional gel can be removed, and the amino acid sequences of peptide fragments can be determined by **tandem mass spectrometry.** If the genome sequence is known, the amino acid sequence of peptides can be used to scan the genome and identify the chromosomal gene. The gene sequence provides the sequence of the entire protein. Recently, microarray analysis has been applied to proteins. An **antibody microarray** consists of a collection of antibodies that recognize short peptide sequences. This type of array can assess the level of protein expression from a given cell type under particular environmental conditions. A **functional protein microarray** consists of many different cellular proteins. This array can probe the function of proteins, such as their substrate specificity, interactions with drugs, or interactions with other cellular proteins.

The computer has become an important tool in genetic studies and has played a large role in creating the field of **bioinformatics.** Genetic sequences contained within **computer data files** can be analyzed by many different programs. For example, short sequence elements can be located within very long genetic sequences. Likewise, various strategies can be used to find structural genes within DNA sequences. In addition to analyzing genetic sequences, computers are used as storage facilities for genetic information. The scientific community has created large **databases** that contain genetic data from thousands of laboratories.

Homology has become an important concept in sequence analysis. Two genes are **homologous** if they have been derived from a common ancestral gene. Homologous genes have similar sequences and are very likely to carry out similar or identical functions. Computer programs can produce a **multiple sequence alignment** in which several homologous gene sequences are compared. In addition, geneticists may take newly identified sequences and search a database to identify homologous sequences using the **BLAST** program. In many cases, the identification of homologous genes within a database can provide important clues concerning the function of a recently determined sequence.

The three-dimensional structure of DNA, RNA, and proteins ultimately depends on the sequence of their building blocks. Therefore, much research is being directed toward predicting the structure and function of these macromolecules from their genetic sequences. Computer programs have been developed to predict the secondary and/or tertiary structure of DNA, RNA, and proteins. These programs rely on known biochemical and biophysical properties of the macromolecules and may also incorporate data concerning homologous sequences of known structure. Some predictive programs are more reliable than others. Nevertheless, future research is expected to improve the predictive capabilities of such computer programs.

PROBLEM SETS & INSIGHTS

Solved Problems

S1. When a cell experiences a change in its environment, it may activate one or more genes as a way to respond to the changes. For example, when confronted with a toxic substance, a cell may turn on genes that encode proteins that can degrade or export the toxin. Let's suppose a researcher is interested in the ability of mouse liver cells to protect themselves against toxic heavy metals, such as mercury. To understand the cellular response, the researcher grows mouse liver cells in the laboratory and divides them into two samples. One of the samples is exposed to mercury while the other is not. Samples of cDNA were made from these cells, and then a subtractive cDNA library was made as described in figure 21.1. In this case, the cDNAs derived from the cells that were not exposed to mercury were the cDNAs bound to the column. The results showed that the subtractive library contained seven different types of cDNAs.

A. What do these results mean?

B. What would you do next, assuming that the entire mouse genome is contained within a database?

Answer:

A. The results mean that seven different genes are turned on when liver cells are exposed to mercury.

B. Take the seven different cDNAs and subject them to DNA sequencing. Next, the cDNA sequences are used (via computer programs) to find their genomic matches. This will make it possible to determine the entire gene sequences. Then take the entire coding sequence, for each of the seven genes, and look for homology between each mouse gene sequence and sequences with a database. In other words, you would do a database search, using each of the seven sequences as the query sequence. If you discovered that any of the seven genes were homologous to other genes whose function is already known, this would give you direct information regarding the probable function of particular genes. For example, if one of the genes turned out to be homologous to a gene in bacteria that encodes a protein that is already known to function in the export of metal ions, this would suggest that one way that mouse liver cells try to avoid the toxic effects of mercury is to try to pump it out.

S2. To answer this question, you will need to look back at the evolution of the globin gene family, which is shown in chapter 8, figure 8.10. Throughout the evolution of this gene family, mutations have occurred, and these mutations have caused the modern-day globin polypeptides to have similar but significantly different amino acid sequences. If we look at the sequence alignment in figure 21.10, we can make logical guesses regarding the timing of mutations, based on a comparison of the amino acid sequences of family members. What is/are the most probable time(s) that mutations occurred to produce the following amino acid differences? Note: You will have to examine the alignment in figure 21.10 and the evolutionary timescale in figure 8.10 to answer this question.

A. Val-111 and Cys-111

B. Met-112 and Leu-112

C. Ser-141, Asn-141, Ile-141, and Thr-141

Answer:

A. We do not know if the original globin gene encoded a cysteine or valine at codon 111. The mutation could have changed cysteine to valine or valine to cysteine. The mutation probably occurred after the duplication that produced the α-globin family and β-globin family (about 300 million years ago) but before the gene duplications that occurred in the last 200 millions years to produce the multiple copies of the globin genes on chromosome 11 and chromosome 16. Therefore, all of the globin genes on chromosome 11 have a valine at codon 111 while all of the globin genes on chromosome 16 have a cysteine.

B. Met-112 occurs only in the ε-globin polypeptide; all of the other globin polypeptides contain a leucine at position 112. Therefore, the primordial globin gene probably contained a leucine codon at position 112. After the gene duplication that produced the ε-globin gene, a mutation occurred that changed this leucine codon into a methionine codon. This would have occurred since the evolution of primates (i.e., within the last 10 or 20 million years).

C. When we look at the possible codons at position 141 (i.e., Ser-141, Asn-141, Ile-141, and Thr-141), we notice that a serine codon is found in θ-globin, ζ-globin, and γ-globin. Since the θ- and ζ-globin genes are found on chromosome 16 and the γ-globin genes are found on chromosome 11, it is probable that serine is the primordial codon, and that the other codons (asparagine, isoleucine, and threonine) arose later by mutation of the serine codon. If this is correct, the Thr-141 codon arose after the gene duplications that produced the θ- and ζ-globin genes. And the Asn-141 and Ile-141 mutations arose after the gene duplications that produced the γ-globin genes. Therefore, the Thr-141, Asn-141, and Ile-141 arose since the evolution of primates (i.e., within the last 10 or 20 million years).

S3. Using a comparative sequence analysis, the secondary structures of rRNAs have been predicted. Among many homologous rRNAs, one stem-loop usually has the following structure:

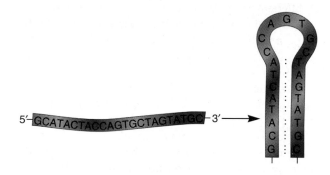

You have sequenced a homologous rRNA from a new species and have obtained most of its sequence, but you cannot read the last five bases on your sequencing gel.

```
5'-GCATTCTACCAGTGCTAG?????-3'
```

Of course, you will eventually repeat this experiment to determine the last five bases. However, before you get around to doing this, what do you expect will be the sequence of the last five bases?

Answer: AATGC–3'

This will also form a similar stem-loop structure.

S4. How can codon bias be used to search for structural genes within uncharacterized genetic sequences?

Answer: Most species exhibit a bias in the codons they use within the coding sequence of structural genes. This causes the base content within coding sequences to differ significantly from that of noncoding DNA regions. By knowing the codon bias for a particular species, researchers can use a computer to locate regions that display this bias and thereby identify what are likely to be the coding regions of structural genes.

Conceptual Questions

C1. Discuss the meaning of the following terms: structural genomics, functional genomics, and proteomics.

C2. Discuss the reasons why the proteome is larger than the genome of a given species.

C3. What is a database? What types of information are stored within a database? Where does the information come from? Discuss the objectives of a genome database.

C4. Besides the examples listed in table 21.4, list five short sequences that a geneticist might want to locate within a DNA sequence.

C5. Discuss the distinction between sequence recognition and pattern recognition.

C6. A multiple sequence alignment of five homologous proteins is shown here:

```
  1                                                     50
1 MLAFLNQVRK PTLDLPLEVR RKMWFKPFM. QSYLVVFIGY LTMYLIRKNF
2 MLAFLNQVRK PTLDLALDVR RKMWFKPFM. QSYLVVFIGY LTMYLIRKNF
3 MLPFLKAPAD APL.MTDKYE IDARYRYWRR HILLTIWLGY ALFYFTRKSF
4 MLSFLKAPAN APL.ITDKHE VDARYRYWRR HILITIWLGY ALFYFTRKSF
5 MLSIFKPAPH KAR.LPAA.E IDPTYRRLRW QIFLGIFFGY AAYYLVRKNF

  51                                                    100
1 NIAQNDMIST YGLSMTQLGM IGLGFSITYG VGKTLVSYYA DGKNTKQFLP
2 NIAQNDMIST YGLSMTELGM IGLGFSITYG VGKTLVSYYA DGKNTKQFLP
3 NAAVPEILAN GVLSRSDIGL LATLFYITYG VSKFVSGIVS DRSNARYFMG
4 NAAAPEILAS GILTRSDIGL LATLFYITYG VSKFVSGIVS DRSNARYFMG
5 ALAMPYLVEQ .GFSRGDLGF ALSGISIAYG FSKFIMGSVS DRSNPRVFLP
```

Discuss some of the interesting features that this alignment reveals.

C7. What is the difference between similarity and homology?

C8. Which of the following statements uses the term *homologous* correctly?

A. The two X chromosomes in female mammalian cells are homologous to each other.

B. The α-tubulin gene in *Saccharomyces cerevisiae* is homologous to the α-tubulin gene in *Arabidopsis thaliana*.

C. The promoter of the *lac* operon is homologous to the promoter of the *trp* operon.

D. The *lacY* gene of *E. coli* and *Klebsiella pneumoniae* are approximately 60% homologous to each other.

C9. When comparing (i.e., aligning) two or more genetic sequences, it is sometimes necessary to put in gaps. Explain why. Discuss two changes (i.e., two types of mutations) that could happen during the evolution of homologous genes that would explain the occurrence of gaps in multiple sequence alignments.

Experimental Questions

E1. Explain how a subtractive cDNA library can provide information regarding gene regulation.

E2. Take a look at solved problem S1 regarding a subtractive DNA library. Would you want to load a relatively small amount of cDNA from the cells that had been exposed to mercury, compared to the amount of cDNA that is attached to the column? Explain. What would happen if you ran too much cDNA over the column from the cells that had been exposed to mercury?

E3. With regard to DNA microarrays, answer the following questions:

A. What is attached to the slide? Be specific about the number of spots, the lengths of DNA fragments, and the origin of the DNA fragments.

B. What is hybridized to the microarray?

C. How is hybridization detected?

E4. In the experiment of figure 21.3, explain how the ratio of red : green fluorescence provides information regarding gene regulation.

E5. What is meant by the term *cluster analysis*? How is this approach useful?

E6. For two-dimensional gel electrophoresis, what physical properties of proteins promote their separation in the first dimension and the second dimension?

E7. Can two-dimensional gel electrophoresis be used as a purification technique? Explain.

E8. Explain how tandem mass spectroscopy can be used to determine the sequence of a peptide. Once a peptide sequence is known, how is this information used to determine the sequence of the entire protein?

E9. Describe the two general types of protein microarrays. What are their possible applications?

E10. Discuss the strategies that can be used to identify a structural gene using bioinformatics.

E11. What is a motif? Why is it useful for computer programs to identify functional motifs within amino acid sequences?

E12. Discuss why it is useful to search a database to identify sequences that are homologous to a newly determined sequence.

E13. The secondary structure of 16S rRNA has been predicted using a computer-based sequence analysis. In general terms, discuss what type of information is used in a comparative sequence analysis, and explain what assumptions are made concerning the structure of homologous RNAs.

E14. Discuss the basis for secondary structure prediction in proteins. How reliable is it?

E15. To reliably predict the tertiary structure of a protein based on its amino acid sequence, what type of information must be available?

E16. In figure 21.6, we considered a program that can translate a DNA sequence into a polypeptide sequence. A researcher has a sequence file that contains the amino acid sequence of a polypeptide and runs a program that is opposite to the translation program. This other program is called "backtranslate." It can take an amino acid sequence file and determine the sequence of DNA that would encode such a polypeptide. How does this program work? In other words, what are the genetic principles that underlie this program? What type of sequence file would this program generate: a nucleotide sequence or an amino acid sequence? Would the backtranslate program produce only a single sequence file? Explain why or why not.

E17. As mentioned in experimental question E16, we considered a computer program that can translate a DNA sequence into a polypeptide sequence. Instead of running this program, a researcher could simply look the codons up in a genetic code table and determine the sequence by hand. What are the advantages of running this program compared to the old-fashioned way of doing it by hand?

E18. To identify the following types of genetic occurrences, would a program use sequence recognition, pattern recognition, or both?

A. Whether a segment of *Drosophila* DNA contains a P element (which is a specific type of transposable element)

B. Whether a segment of DNA contains a stop codon

C. In a comparison of two DNA segments, whether there is an inversion in one segment compared to the other segment

D. Whether a long segment of bacterial DNA contains one or more genes

E19. The goal of many computer programs is to identify "sequence elements" within a long segment of DNA. What is a sequence element? Give two examples. How is the specific sequence of a sequence element determined? In other words, is it determined by the computer program or by genetic studies? Explain.

E20. Take a look at the multiple sequence alignment in figure 21.10 of the globin polypeptides from amino acids 101–148.

A. Which of these amino acids are likely to be most important for globin structure and function? Explain why.

B. Which are likely to be least important?

E21. See solved problem S2 before answering this question. Based on the sequence alignment in figure 21.10, what is/are the most probable time(s) that mutations occurred in the human globin gene family to produce the following amino acid differences:

A. His-119 and Arg-119

B. Gly-121 and Pro-121

C. Glu-103, Val-103, and Ala-103

E22. Take a look at solved problem S2 and the codon table found in chapter 13 (table 13.2). Assuming that a mutation involving a single-base change is more likely than a double-base change, propose how the Asn-141, Ile-141, and Thr-141 codons arose. In your answer, describe which of the six possible serine codons is/are likely to be the primordial serine codon of the globin gene family, and how that codon changed to produce the Asn-141, Ile-141, and Thr-141 codons.

E23. Membrane proteins often have transmembrane regions that span the membrane in an α-helical conformation. These transmembrane segments are about 20 amino acids long and usually contain amino acids with nonpolar (i.e., hydrophobic) amino acid side chains. Researchers can predict whether a polypeptide sequence has transmembrane segments based on the occurrence of segments that contain 20 nonpolar amino acids. To do so, each amino acid is assigned a hydropathy value, based on the chemistry of its amino acid side chain. Amino acids with very nonpolar side chains are given a high value, whereas amino acids that are charged and/or polar are given low values. The hydropathy values usually range from about +4 to −4.

Computer programs have been devised that scan the amino acid sequence of a polypeptide and calculate values based on the hydropathy values of the amino acid side chains. You can run such a program at the website for this textbook. The program usually scans a window of seven amino acids and assigns an average hydropathy value. For example, the program would scan amino acids 1–7 and give an average value, then it would scan 2–8 and give a value, then it would scan 3–9 and give a value, and so on, until it reached the end of the polypeptide sequence (i.e., until it reached the carboxyl terminus).

The program then produces a figure, known as a hydropathy plot, which describes the average hydropathy values throughout the entire polypeptide sequence. An example of a hydropathy plot is shown here.

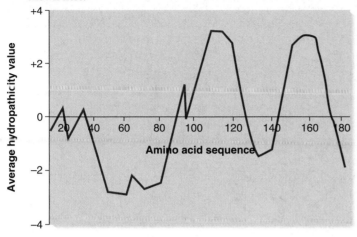

A. How many transmembrane segments are likely in this polypeptide?

B. Draw the structure of this polypeptide if it were embedded in the plasma membrane. Assume that the amino terminus is found in the cytoplasm of the cell.

E24. Explain how a computer program can predict RNA secondary structure. What is the underlying genetic concept that is used by the program to predict secondary structure?

E25. Are the following statements about protein structure prediction true or false?

A. The prediction of secondary structure relies on information regarding the known occurrence of amino acid residues in α helices or β sheets from X-ray crystallographic data.

B. The prediction of secondary structure is highly accurate, nearly 100% correct.

C. To predict the tertiary structure of a protein based on its amino acid sequence, it is necessary that the protein of interest is homologous to another protein whose tertiary structure is already known.

Questions for Student Discussion/Collaboration

1. Let's suppose that you are in charge of organizing and publicizing a genomic database for the mouse genome. Make a list of innovative strategies that you would initiate to make the mouse genome database useful and effective.

2. Let's suppose that a 5-year-old told you that she was interested in pursuing a career studying the three-dimensional structure of proteins. (Okay, so she's a bit precocious.) Would you advise her to become a geneticist, a mathematical theoretician, or a biophysicist?

3. If you have access to the necessary computer software, make a sequence file and analyze it in the following ways: What is the translated sequence in all three reading frames? What is the longest open reading frame? Is the sequence homologous to any known sequences? If so, does this provide any clues about the function of the sequence?

Note: All answers appear at the website for this textbook; the answers to even-numbered questions are in the back of the textbook.

www.mhhe.com/brooker

Visit the Online Learning Center for practice tests, answer keys, and other learning aids for this chapter. Enhance your understanding of genetics with our interactive exercises, web links, news feeds, tutorial service, and much more.

MEDICAL GENETICS AND CANCER

::

22

Genetic information is highly personal and unique. Our genes underlie every aspect of human health, both in function and dysfunction. Obtaining a detailed understanding of how genes work together and interact with environmental factors ultimately will allow us to appreciate the differences between the events in normal cellular processes versus those that occur in disease pathogenesis. Such knowledge will have a profound impact on the way many diseases are defined, diagnosed, treated, and prevented. It is expected that this will bring about revolutionary changes in medical practices. In fact, such changes are already beginning. Currently, several hundred genetic tests are in clinical use, with many more under development. Most of these tests detect mutations associated with rare genetic disorders that follow Mendelian inheritance patterns. These include Duchenne muscular dystrophy, cystic fibrosis, sickle-cell anemia, and Huntington disease. In addition, genetic tests are available to detect the predisposition to develop certain forms of cancer.

Approximately 4,000 genetic diseases afflict people, but this is almost certainly an underestimate. Given enough time and effort, scientists can learn how to prevent or treat a great many of them. Most of the genetic disorders discussed in the first part of chapter 22 are the direct result of a mutation in one gene. Even so, one of the most difficult problems facing scientists is to learn how genes contribute to diseases that have a complex pattern of inheritance involving several genes. These include common medical disorders such as diabetes, asthma, and mental illness. In these cases, a single mutant gene cannot determine whether a person has a disease or not. Instead, a number of genes may each

PART VI

Genetic Analysis of Individuals and Populations

make a subtle contribution to a person's susceptibility to a disease. Unraveling these complexities will be a challenge for some time to come. The availability of the human genome sequence, discussed in chapter 20, will help in this undertaking.

In chapter 22, we will focus our attention on ways that mutant genes contribute to human disease. In the first part of the chapter, we will explore the molecular basis of several genetic disorders and their patterns of inheritance. We will also examine how genetic testing can determine if an individual carries a defective allele. The second part of chapter 22 will concern cancer, a disease that involves the uncontrolled growth of somatic cells. We will examine the underlying genetic basis for this disease and discuss the roles that many different genes may play in the development of cancer.

22.1 GENETIC ANALYSIS OF HUMAN DISEASES

The genetic analysis of humans is a topic that is hard to resist. Almost everyone who looks at a newborn is tempted to speculate whether the baby resembles the mother, the father, or perhaps a distant relative. In this section, we will focus primarily on human genetic diseases rather than common traits found in the general population. Even so, the study of human genetic diseases provides insights regarding our traits. The disease hemophilia illustrates this point. As you may know, hemophilia is a condition in which an individual's blood will not clot properly. By analyzing people with this disorder, researchers have identified genes that participate in the process of blood clotting. An analysis of people with hemophilia has elucidated a clotting pathway involving several different proteins. Therefore, when we study the inheritance of genetic diseases, we often learn a great deal about the genetic basis for normal physiological processes as well.

Thousands of human diseases have an underlying genetic basis. Therefore, human genetic analysis is of great medical importance. In this section, we will examine the causes and inheritance patterns of human genetic diseases that result from defects in single genes. As we will learn, the mutant genes that cause these diseases often follow simple Mendelian inheritance patterns.

A Genetic Basis for a Human Disease May Be Suggested from a Variety of Observations

When we view the characteristics of human beings, we usually think that some traits are inherited while others are caused by environmental factors. For example, when the facial features of two related individuals look strikingly similar, we think that this similarity has a genetic basis. The profound resemblance between identical twins is a perfect example. By comparison, other traits are governed by the environment. If we see a person with purple hair, we likely suspect that he or she has used hair dye as opposed to carrying a genetic abnormality.

For common human traits, as well as for diseases, geneticists would like to know the relative contributions from genetics and the environment. Is a disease caused by a pathogenic micro-

organism, a toxic agent in the environment, or a faulty gene? Unlike the case with experimental organisms, we cannot conduct human crosses to elucidate the genetic basis for diseases. Instead, we must rely on analyses of families that already exist. As described in the following list, several observations are consistent with the idea that a disease is caused, at least in part, by the inheritance of mutant genes. When the occurrence of a disease correlates with several of these observations, a geneticist will become increasingly confident that it has a genetic basis.

1. *When an individual exhibits a disease, this disorder is more likely to occur in genetic relatives than in the general population.* For example, if someone has cystic fibrosis, it is more likely that his or her relatives will also have this disease than will randomly chosen members of the general population.
2. *Identical twins share the disease more often than nonidentical twins.* Identical twins, also called **monozygotic twins (MZ),** are genetically identical to each other, because they were formed from the same sperm and egg. By comparison, nonidentical twins, also called fraternal or **dizygotic twins,** are formed from separate pairs of sperm and egg cells. Dizygotic twins share, on average, 50% of their genetic material. When a disorder has a genetic component, both identical twins are more likely to exhibit the disorder than are both nonidentical twins.

Geneticists evaluate the **concordance** for a disorder, which is the degree to which it is inherited, by calculating the percentage of twin pairs in which both twins exhibit the disorder relative to pairs where only one twin shows the disorder. Theoretically, for diseases that are caused by a single gene, concordance among identical twins should be 100%. For dizygotic twins, concordance for dominant disorders is expected to be 50%, assuming that only one parent is a heterozygous individual. For recessive diseases, concordance among dizygotic twins would be 25% if we assume that both parents are heterozygous carriers. However, the actual concordance values that are observed for most single-gene disorders are usually less than such theoretical values because some disorders are not completely penetrant, and because one twin may have a disorder due to a new mutation that occurred after fertilization.

3. *The disease does not spread to individuals sharing similar environmental situations.* Inherited disorders cannot spread from person to person. The only way genetic diseases can be transmitted is from parent to offspring during sexual reproduction.
4. *Different populations tend to have different frequencies of the disease.* Due to evolutionary forces, which we will discuss in chapter 25, the frequencies of traits usually vary among different populations of humans. For example, the frequency of the disease sickle-cell anemia is highest among African populations and relatively low in other parts of the world (see fig 25.11).
5. *The disease tends to develop at a characteristic age.* Many genetic disorders exhibit a characteristic **age of onset** at

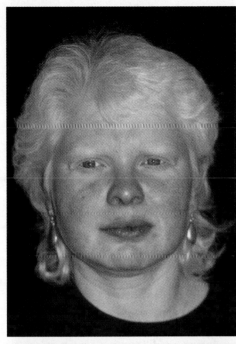

FIGURE 22.1 **The albino phenotype in humans and other mammals.**

GENES→TRAITS   Certain enzymes (encoded by genes) are necessary for the production of pigment. A homozygote with two defective alleles in these pigmentation genes exhibits an albino phenotype. This phenotype can occur in humans and other animals.

which the disease appears. Many mutant genes exert their effects during embryonic and fetal development, so that their effects are apparent at birth. Other genetic disorders tend to develop much later in life.

6. *The human disorder may resemble a disorder that is already known to have a genetic basis in an animal.* In animals,

where we can conduct experiments, various traits are known to be governed by genes. For example, the albino phenotype is found in humans as well as in many animals (fig. 22.1).

7. *A correlation is observed between a disease and a mutant human gene or a chromosomal alteration.* A particularly convincing piece of evidence that a disease has a genetic basis is the identification of altered genes or chromosomes that occur only in people exhibiting the disorder. When comparing two individuals, one with a disease and one without, we expect them to have differences in their genetic material, if the disorder has a genetic component. We can discover alterations in gene sequences by gene cloning and DNA sequencing techniques (as described in chapter 18); we can detect alterations in chromosome structure and number by the microscopic examination of chromosomes as described in chapter 8.

Inheritance Patterns of Human Diseases May Be Determined via Pedigree Analysis

When a human disorder is caused by a mutation in a single gene, the pattern of inheritance can be deduced by analyzing human pedigrees. To use this method, a geneticist must obtain data from many large pedigrees containing several individuals who exhibit the disorder. To appreciate the basic features of pedigree analysis, we will examine a few large pedigrees that involve diseases inherited in different ways. You may wish to review figure 2.10 for the organization and symbols of pedigrees.

The pedigree shown in figure 22.2 concerns a genetic disorder called Tay-Sachs disease (TSD). TSD was first described by Warran Tay, a British ophthalmologist, and Bernard Sachs, an American neurologist, in the 1880s. Affected individuals appear healthy at birth but then develop neurodegenerative symptoms at 4 to 6 months of age. The primary characteristics are cerebral degeneration, blindness, and loss of motor function. Individuals with TSD typically die in the third or fourth year of life. This disease is particularly prevalent in Ashkenazi (eastern European) Jewish populations, in which it has a frequency of about 1 in 3,600 births. It is over 100 times less frequent in most other human populations.

At the molecular level, the mutation that causes TSD is in a gene that encodes the enzyme hexosaminidase A (hexA). HexA is responsible for the breakdown of a category of lipids called GM_2-gangliosides. This lipid is prevalent in the cells of the central nervous system. A defect in the ability to break down this lipid, as in TSD, leads to an excessive accumulation of this lipid in nerve cells and eventually causes neurodegeneration and the symptoms just described.

As illustrated in figure 22.2, Tay-Sachs disease is inherited in an autosomal recessive manner. The four common features of autosomal recessive inheritance are as follows:

1. *Frequently, an affected offspring will have two unaffected parents.* For rare recessive traits, the parents are usually unaffected. For deleterious alleles that cause early death or

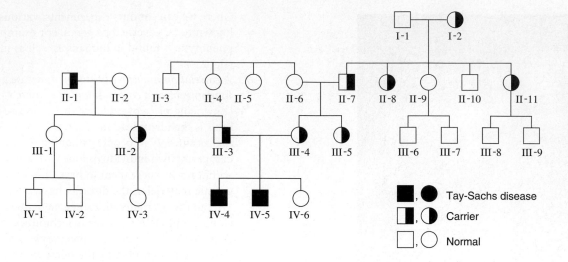

FIGURE 22.2   **A family pedigree of Tay-Sachs disease.**

infertility, the two parents must be unaffected. This is always the case in TSD.

2. *When two unaffected heterozygotes have children, the percentage of affected children is (on average) 25%.*

3. *Two affected individuals will have 100% affected children.* This observation can be made only when a recessive trait produces fertile, viable individuals. In the case of TSD, the affected individual dies in early childhood, and so it is not possible to observe crosses between two affected people.

4. *The trait occurs with the same frequency in both sexes.*

Autosomal recessive inheritance is a common mode of transmission for genetic disorders, particularly those that involve defective enzymes. The heterozygous carrier has 50% of the normal enzyme, which is sufficient for a normal phenotype. Hundreds of human genetic diseases are inherited in this way. In many cases, mutant genes that cause disease have been cloned and/or mapped. Several of these are described in table 22.1.

Now let's examine a human pedigree involving an autosomal dominant trait (fig. 22.3). In this example, the affected indi-

viduals have a disorder called Huntington disease (also called Huntington chorea). The major symptom of this disease, which usually occurs during middle age, is a degeneration of certain types of neurons in the brain, leading to personality changes, dementia, and early death. In 1993, the gene for Huntington disease was

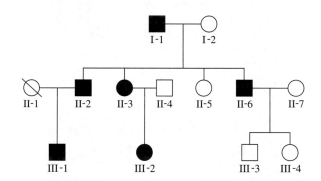

FIGURE 22.3   **A family pedigree of Huntington disease.**

TABLE 22.1
Examples of Human Disorders Inherited in an Autosomal Recessive Manner

Disorder	Chromosome Location	Gene Product	Description
Adenosine deaminase deficiency	20q	Adenosine deaminase	Defective immune system and skeletal and neurological abnormalities
Albinism (type I)	11q	Tyrosinase	Inability to synthesize melanin; white skin, hair, etc.
Cystic fibrosis	7q	CF transmembrane regulator	Water imbalance in tissues of the pancreas, intestine, sweat glands, and lungs due to impaired ion transport; leads to lung damage
Phenylketonuria	12q	Phenylalanine hydroxylase	Foul-smelling urine and neurological abnormalities; may be remedied by diet modification at birth
Sickle-cell anemia	11p	β-globin	Anemia, blockages in blood circulation
Tay-Sachs disease	15q	Hexosaminidase A	Similar to Sandhoff disease, progressive neurodegeneration

TABLE 22.2

Examples of Human Diseases Inherited in an Autosomal Dominant Manner

Trait/Disease	Chromosome Location	Gene Product	Description
Achondroplasia	4p	Fibroblast growth factor receptor	A common form of dwarfism associated with a defect in the growth of long bones
Osteoporosis	7q	Collagen (type 1 A2)	Brittle, weakened bones
Familial hypercholesterolemia	19p	LDL receptor	Characterized by very high serum levels of low-density lipoprotein; a predisposing factor in heart disease
Huntington disease	4p	Huntingtin	Neurodegeneration that occurs relatively late in life, usually in middle age
Neurofibromatosis I	17q	Tumor-suppressor gene	Individuals may exhibit spots of abnormal pigmentation (cafe-au-lait spots) and growth of noncancerous tumors in the nervous system

identified and sequenced. It encodes a protein called huntingtin that is expressed in neurons but is also found in some cells not affected in Huntington disease. In persons with this disorder, a mutation has added a polyglutamine tract (i.e., many glutamines in a row) to the amino acid sequence of the huntingtin protein. This causes an aggregation of the protein in neurons. However, additional research will be needed to elucidate the molecular relationship between the abnormality in the huntingtin protein and the disease symptoms. Five common features of autosomal dominant inheritance are as follows:

1. *An affected offspring usually has one or both affected parents.* However, this is not always the case. Some dominant traits show incomplete penetrance (see chapter 4), so that a heterozygote may not exhibit the trait even though it may be passed to offspring who do exhibit the trait. Also, a dominant mutation may occur during gametogenesis, so that two unaffected parents may produce an affected offspring.
2. *An affected individual, with only one affected parent, is expected to produce 50% affected offspring (on average).*
3. *Two affected, heterozygous individuals will have (on average) 25% unaffected offspring.*
4. *The trait occurs with the same frequency in both sexes.*
5. *For most dominant, disease-causing alleles, the homozygote is more severely affected with the disorder. In some cases, a dominant allele may be lethal in the homozygous condition.*

Numerous dominant diseases have been identified in humans (table 22.2). In these disorders, 50% of the normal protein is not sufficient to produce a normal phenotype. Though many dominant diseases may involve defects in enzymes, it is also common for them to involve alterations in structural proteins or membrane receptors. For example, one inherited form of osteoporosis is a missense mutation in a gene encoding a subunit of collagen, a constituent of bone and cartilage. In this genetic disorder, a mutation in the type 1 A2 collagen gene substitutes a serine residue for a glycine, producing a kink in collagen, which disrupts its structure. This alteration in collagen conformation leads to weakened bones.

A third pattern of inheritance common in humans is X-linked recessive inheritance (table 22.3). With regard to recessive genetic diseases, X-linked inheritance poses a special problem for males. Most X-linked genes lack a counterpart on the Y chromosome; males thus have a single copy of (i.e., are hemizygous for) these genes. Therefore, a female heterozygous for an X-linked recessive gene will pass this trait to half of her sons. This idea is shown in a Punnett square for hemophilia. In this example, X^{h-A} is the chromosome that carries a mutant allele causing hemophilia.

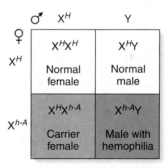

TABLE 22.3

Examples of Human Disorders Inherited in an X-linked Recessive Manner

Trait/Disease	Gene Product	Description
Duchenne muscular dystrophy	Dystrophin	A progressive degeneration of muscles that begins in early childhood
Hemophilia A	Clotting factor VIII	Defect in blood clotting
Hemophilia B	Clotting factor IX	Defect in blood clotting
Testicular feminization	Androgen receptor	Missing the male steroid hormone receptor; XY individuals have external features that are feminine but internally have undescended testes and no uterus

As mentioned previously, hemophilia is a disorder in which the blood cannot clot properly when a wound occurs. For individuals with this trait, a minor cut may bleed for a very long time, and small bumps can lead to large bruises because internal broken capillaries may leak blood profusely before they are repaired. For hemophiliacs, common injuries pose a threat of severe internal or external bleeding. Hemophilia A (also called classical hemophilia) is caused by a defect in an X-linked gene that encodes the clotting protein factor VIII. This disease has also been called the "Royal disease," because it has affected many members of European royal families. The pedigree shown in figure 22.4 illustrates the prevalence of hemophilia A among the descendants of Queen Victoria of England. The pattern of X-linked recessive inheritance is revealed by the following observations:

1. *Males are much more likely to exhibit the trait.*
2. *The mothers of affected males often have brothers or fathers who are affected with the same trait.*
3. *The daughters of affected males will produce, on average, 50% affected sons.*

Many Genetic Disorders Are Heterogeneous

Hemophilia can be used to illustrate another concept in genetics called **heterogeneity.** This term refers to the phenomenon in

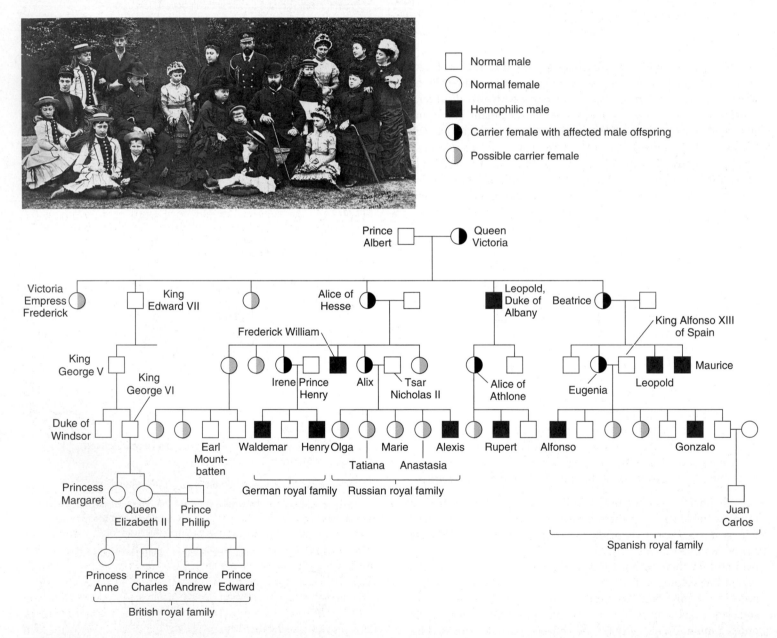

FIGURE 22.4 **The inheritance pattern of hemophilia A in the royal families of Europe.** Pictured are Queen Victoria and Prince Albert of Great Britain, with some of their descendants.

which a particular type of disease may be caused by mutations in different genes. For example, blood clotting involves the participation of several different proteins. These proteins take part in a cellular cascade that leads to the formation of a clot. Therefore, a defect in any of the proteins needed for clot formation can cause hemophilia. Hemophilia B is caused by a defect in the blood clotting protein factor IX; it is also X linked. In addition, several other forms of hemophilia are caused by defects in autosomal genes that also encode clotting factor proteins.

In hemophilia, genetic heterogeneity arises from the participation of several proteins in a common cellular process. Another mechanism that may lead to genetic heterogeneity occurs when proteins are composed of two or more different subunits, with each subunit being encoded by a different gene. The disease thalassemia presents an example of genetic heterogeneity caused by a mutation in a protein composed of multiple subunits. This potentially life-threatening disease involves defects in the ability of the red blood cells to transport oxygen. The underlying cause is an alteration in hemoglobin. In adults, hemoglobin is a tetrameric protein composed of two α chains and two β chains; α-globin and β-globin are encoded by separate genes (namely, the α- and β-globin genes). Two main types of thalassemia have been discovered in human populations: α-thalassemia, in which the α-globin subunit is defective; and β-thalassemia, in which the β-globin subunit is defective.

Unfortunately, genetic heterogeneity may greatly confound pedigree analysis. For example, a human pedigree might contain individuals with X-linked hemophilia and other individuals with autosomally linked hemophilia; a geneticist who assumed that all affected individuals had defects in the same gene would be unable to explain the resulting pattern of inheritance. For disorders such as hemophilia and thalassemia, pedigree analysis is not a major problem, since the biochemical basis for these diseases is well understood. However, for rare diseases that are poorly understood at the molecular level, genetic heterogeneity may profoundly obscure the pattern of inheritance.

Genetic Testing Can Identify Many Inherited Human Diseases

Because genetic abnormalities occur in the human population at a significant level, people have sought ways to determine whether individuals possess disease-causing alleles. The term **genetic testing** refers to the use of testing methods to discover if an individual carries a genetic abnormality. Table 22.4 describes several different testing strategies. The term **genetic screening** refers to population-wide genetic testing.

In many cases, single-gene mutations that affect the function of cellular proteins can be examined at the protein level. If a gene encodes an enzyme, biochemical assays to measure that enzyme's activity may be available. As mentioned earlier, Tay-Sachs disease involves a defect in the enzyme hexA. Enzymatic assays for this enzyme involve the use of an artificial substrate in

TABLE 22.4

Testing Methods for Genetic Abnormalities

Method	Description
Protein Level	
Biochemical	As described in chapter 22 for Tay-Sachs disease, the enzymatic activity of a protein can be assayed in vitro.
Immunological	The presence of a protein can be detected using antibodies that specifically recognize that protein. An example of this type of technique, known as Western blotting, is described in chapter 18.
DNA or Chromosomal Level	
RFLP linkage	As described in the experiment of figure 20.10, certain inherited alleles are linked to RFLPs.
Altered restriction site	A gene mutation may alter the presence of a particular restriction enzyme site. A sample of DNA from an individual can be subjected to restriction enzyme digestion and gel electrophoresis (see chapter 18) to see whether or not the restriction enzyme site is present.
DNA sequencing	If the normal gene has already been identified and sequenced, it is possible to design PCR primers that can amplify the gene from a sample of cells. The amplified DNA segment can then be subjected to DNA sequencing as described in chapter 18.
In situ hybridization	A DNA probe that hybridizes to a particular gene can be used to determine if the gene is present in an individual. This approach can be used to determine if a gene or segment of a gene has been deleted or altered. The technique of fluorescence in situ hybridization is described in chapter 20.
Karyotyping	The chromosomes from a sample of cells can be stained and then analyzed microscopically for abnormalities in chromosome structure and number.

which 4-methylumbelliferone (MU) is covalently linked to N-acetylglucosamine (GlcNAc). HexA cleaves this covalent bond and releases MU, which is fluorescent.

$$\text{MU–GlcNAc} \xrightarrow{\text{HexA}} \text{MU} + \text{GlcNAc}$$
$$\text{(nonfluorescent)} \qquad \text{(fluorescent)}$$

To perform this assay, a small sample of cells is collected and incubated with MU–GlcNAc, and the fluorescence is measured with a device called a fluorometer. Individuals affected with Tay-Sachs will produce little or no fluorescence, whereas individuals who are homozygous for the normal *hexA* allele produce a high level of fluorescence. Heterozygotes, who have 50% hexA activity, are expected to produce intermediate levels of fluorescence.

An alternative approach is to detect single-gene mutations at the DNA level. To apply this testing strategy, researchers must have previously identified the mutant gene using molecular techniques. While this is often a difficult task, progress in the cloning and sequencing of human genes is advancing rapidly. The cloning of several human genes, such as those involved in Duchenne muscular dystrophy, cystic fibrosis, and Huntington disease, has made it possible to test for affected individuals or those who may be carriers of these diseases. Table 22.4 describes several ways to test for gene mutations by analyzing DNA sequences. The laboratory techniques involved were described in chapters 18 and 20.

As discussed in chapter 8, many human genetic abnormalities involve changes in chromosome structure and/or number. In fact, changes in chromosome number are the most common class of human genetic abnormality. Most of these result in spontaneous abortions. However, approximately 1 in 200 live births are aneuploid or have unbalanced chromosomal alterations. About 5% of infant and childhood deaths are related to such genetic abnormalities. With regard to testing, changes in chromosome number and many changes in chromosome structure can be detected by karyotyping the chromosomes with a light microscope.

In the United States, genetic screening for certain disorders has become common medical practice. For example, pregnant women over 35 years old often have tests conducted to see if their fetuses are carrying chromosomal abnormalities. As discussed in chapter 8, these tests are indicated because the rate of such defects increases with the age of the mother. Another example is the widespread screening for phenylketonuria (PKU). An inexpensive test can determine if newborns have this disease. Those who test positive can then be given a low-phenylalanine diet to avoid PKU's devastating effects.

Genetic screening also has been conducted on specific populations in which a genetic disease is prevalent. For example, in 1971, community-based screening for heterozygous carriers of Tay-Sachs disease was begun among specific Ashkenazi Jewish populations. Over the course of one generation, the incidence of TSD births was reduced by 90%. For most rare genetic abnormalities, however, genetic screening is not routine practice. Rather, genetic testing is performed only when a family history reveals a strong likelihood that a couple may produce an affected child. This typically involves a couple that already has an affected child or has other relatives afflicted with a genetic disease.

Genetic testing can be performed prior to birth. There are two common ways of obtaining cellular material from a fetus for the purpose of genetic testing: **amniocentesis** and **chorionic villus sampling.** In amniocentesis, a doctor removes amniotic fluid containing fetal cells using a needle that is passed through the abdominal wall (fig. 22.5). The cells are cultured for several weeks and then karyotyped to determine the number of chromosomes per cell and whether changes in chromosome structure have occurred. In chorionic villus sampling, a small piece of the chorion (the fetal part of the placenta) is removed, and a karyotype is prepared directly from the collected cells. Chorionic villus sampling can be performed earlier during pregnancy than amniocentesis (around the eighth to tenth week, compared with the fourteenth to sixteenth week for amniocentesis). Weighed against this advantage, however, is that chorionic villus sampling may pose a slightly greater risk of causing a miscarriage.

Genetic testing and screening is a medical practice with many social and ethical dimensions. For example, people who learn they are carriers of genetic diseases, such as Huntington disease or familial breast cancer (described later in chapter 22), are affected profoundly by the news. Some argue that people have a right to know about their genetic makeup; others assert that it does more harm than good. Another issue is privacy. Will an understanding of our genetic makeup subject us to discrimination by employers or medical insurance companies? During the twenty-first century, we will gain an ever-increasing awareness of our genetic makeup and the underlying causes of genetic diseases. As a society, it will be necessary, yet very difficult, to establish guidelines for the uses of genetic testing.

Prions Are Infectious Particles That Alter Protein Function Posttranslationally

Before we end this section on inherited diseases, let's consider an unusual mechanism in which agents known as **prions** cause disease. As shown in table 22.5, prions cause several types of neurodegenerative diseases affecting humans and livestock. Recent evidence has shown that prions also exist in yeast. In 1982, Stanley Prusiner was the first to propose that prions act as infectious agents composed entirely of protein. The term emphasizes the

TABLE 22.5

Neurodegenerative Diseases Caused by Prions

Disease	Description
Infectious Diseases	
Kuru	A human disease that was once common in New Guinea. It begins with a loss of coordination, usually followed by dementia. Infection was spread by cannibalism, a practice that ended in 1958.
Scrapie	A disease of sheep and pigs characterized by intense itching in which the animals tend to scrape themselves against trees, etc., followed by neurodegeneration.
Mad cow disease	Begins with changes in posture and temperament, followed by loss of coordination and neurodegeneration.
Human Inherited Diseases	
Creutzfeldt-Jakob disease	Characterized by loss of coordination and dementia. On rare occasions, this disease may be spread by direct contact of bodily fluid between an infected and uninfected individual.
Gerstmann-Straussler-Scheinker disease	Characterized by loss of coordination and dementia.
Familial fatal insomnia	Begins with sleeping and autonomic nervous system disturbances followed by insomnia and dementia.

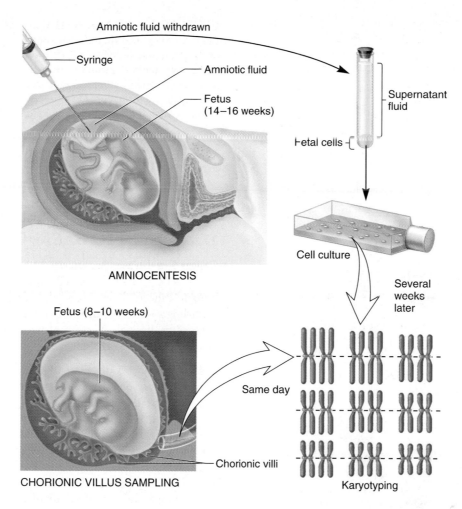

FIGURE 22.5   Techniques to determine genetic abnormalities before birth. In amniocentesis, amniotic fluid is withdrawn and fetal cells are collected by centrifugation. The cells are then allowed to grow in laboratory culture media for several weeks prior to karyotyping. In chorionic villus sampling, a small piece of the chorion is removed. These cells can be prepared directly for karyotyping.

prion's unusual character as a proteinaceous infectious agent. Before the discovery of prions, all known infectious agents such as viruses and bacteria contained their own genetic material (either DNA or RNA).

Prion-related diseases arise from the ability of the prion protein to exist in two conformational states: a normal form PrP^C, which does not cause disease, and an abnormal form, PrP^{Sc}, which does. The gene encoding the prion protein is found in humans, and the protein is expressed at low levels in certain types of cells such as nerve cells. The abnormal conformation of the prion protein can come from two sources. An individual can be "infected" with the abnormal protein by direct contact with an affected individual or by eating meat from an organism with the disease. Alternatively, some people carry alleles of the *PrP* gene that cause the normal prion protein to convert spontaneously to the abnormal conformation at a very low rate. These individuals have an inherited predisposition to develop a prion-related dis-

ease. An example of an inherited prion disease is familial fatal insomnia (table 22.5).

Let's now consider the molecular mechanism by which prions cause disease. As noted, the prion protein can exist in two conformations, PrP^{Sc} and PrP^C. As shown in figure 22.6, the abnormal conformation, PrP^{Sc}, acts as a catalyst to convert normal prion proteins within the cell to the abnormal conformation. Therefore, once an abnormal prion protein is found within a cell, the normal prion proteins already present will be converted to the abnormal conformation. As a prion disease progresses, the PrP^{Sc} protein is deposited as dense aggregates in the cells of the brain and peripheral nervous tissues. This deposition is correlated with the disease symptoms affecting the nervous system. Some of the abnormal prion protein is also excreted from infected cells, where it can travel through the bloodstream and be taken up by uninfected cells. In this way, a prion disease can spread through the body just like many viral diseases.

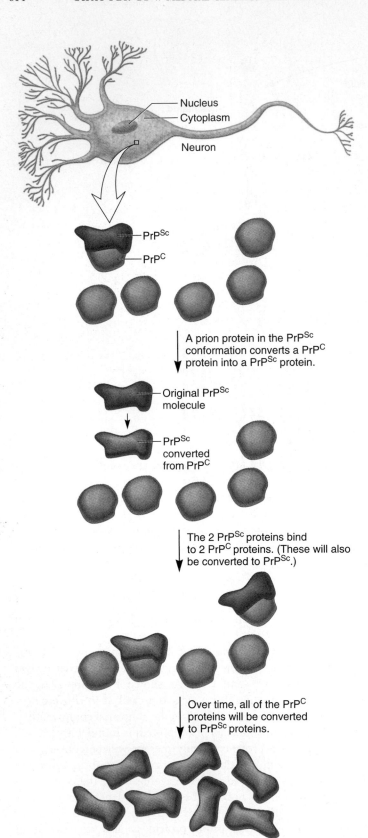

A prion protein in the PrP^{Sc} conformation converts a PrP^{C} protein into a PrP^{Sc} protein.

The 2 PrP^{Sc} proteins bind to 2 PrP^{C} proteins. (These will also be converted to PrP^{Sc}.)

Over time, all of the PrP^{C} proteins will be converted to PrP^{Sc} proteins.

FIGURE 22.6 **A proposed molecular mechanism of prion diseases.** A healthy neuron normally contains only the PrP^{C} conformation of the prion protein. When a PrP^{Sc} protein is found within the cell, it will catalyze the conversion of PrP^{C} to PrP^{Sc}. Then, both of the proteins in the PrP^{Sc} conformation will bind to PrP^{C} proteins and convert them to the PrP^{Sc} conformation. Over time, the PrP^{Sc} conformation will accumulate to high levels, leading to symptoms of the prion diseases.

22.2 GENETIC BASIS OF CANCER

Cancer is a disease characterized by uncontrolled cell division. It is a genetic disease at the cellular level. More than 100 kinds of human cancers have been identified, and they are classified according to the type of cell that has become cancerous. Though cancer is a diverse collection of many diseases, some characteristics are common to all cancers.

1. Most cancers originate in a single cell. This single cell, and its line of daughter cells, undergoes a series of genetic changes that accumulate during cell division. In this regard, a cancerous growth can be considered to be **clonal** in origin. A hallmark of a cancer cell is that it divides to produce two daughter cancer cells.
2. At the cellular and genetic levels, cancer usually is a multistep process that begins with a precancerous genetic change (i.e., a **benign** growth) and following additional genetic changes, progresses to cancerous cell growth (fig. 22.7).
3. Once a cellular growth has become **malignant** (i.e., cancerous), the cancer cells are **invasive** (i.e., they can invade healthy tissues) and **metastatic** (i.e., they can migrate to other parts of the body).

Approximately 1 million Americans are diagnosed with cancer each year; about half that number will die from the disease. In 5 to 10% of cancers, a higher predisposition to develop the cancer is an inherited trait. We will examine some inherited forms of cancer later in chapter 22. Most cancers, though, perhaps 90 to 95%, are not passed from parent to offspring. Rather, cancer is usually an acquired condition that typically occurs later in life. While some cancers are caused by spontaneous mutations and viruses, at least 80% of all human cancers are related to exposure to agents that promote genetic changes in somatic cells. These environmental agents, such as UV light and certain chemicals, are mutagens that alter the DNA. If the DNA is permanently modified in somatic cells, such changes may be transmitted during cell division. These DNA alterations can lead to effects on gene expression that ultimately affect cell division and thereby lead to cancer. An environmental agent that causes cancer in this manner is called a **carcinogen.**

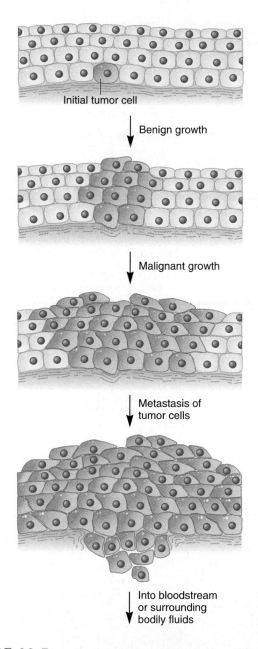

Initial tumor cell

Benign growth

Malignant growth

Metastasis of
tumor cells

Into bloodstream
or surrounding
bodily fluids

FIGURE 22.7   Progression of cellular growth leading to cancer.
GENES→TRAITS   In a healthy individual, an initial gene mutation converts a
normal cell into a tumor cell. This tumor cell divides to produce a benign tumor. Additional genetic changes in the tumor cells may occur, leading to malignant growth. At
a later stage in malignancy, the tumor cells will invade surrounding tissues, and some
malignant cells may metastasize by traveling through the bloodstream to other parts
of the body. As a trait, cancer can be viewed as a series of genetic changes that
eventually lead to uncontrolled cell growth.

In this section, we will begin by considering some early
experimental observations that suggested genes play a role in
cancer. We will then explore how genetic abnormalities, which
affect the functions of particular cellular proteins, can lead to
cancer.

Certain Viruses Can Cause Cancer by Carrying Viral Oncogenes into the Cell

As mentioned, most types of cancers are caused by mutagens
that alter the structure and expression of genes. A few viruses,
however, are known to cause cancer in plants, animals, and
humans. We begin our discussion here, because early studies of
cancer-causing viruses identified the first genes that play a role
in cancer. Many of these viruses can also infect normal laboratory-
grown cells and convert them into malignant cells. The process
of converting a normal cell into a malignant cell is called **transformation.**

Most cancer-causing viruses are not very potent at inducing cancer. An organism must be infected for a long period of
time before a tumor actually develops. Furthermore, most viruses
are inefficient at transforming or are unable to transform normal
cells grown in the laboratory. By comparison, a few types of
viruses rapidly induce tumors in animals and efficiently transform cells in culture. These are called **acutely transforming
viruses (ACVs).** About 40 ACVs have been isolated from chickens, turkeys, mice, rats, cats, and monkeys. The first virus of this
type, the Rous sarcoma virus (RSV), was isolated from chicken
sarcomas by Peyton Rous in 1911.

During the 1970s, RSV research led to the first identification of a gene that promotes cancer—an **oncogene.** Researchers
investigated RSV by infecting chicken fibroblast cells. This causes
the chicken fibroblasts to grow like cancer cells. Researchers identified mutant RSV strains that infected and proliferated within
chicken cells without transforming them into malignant cells.
These RSV strains were determined to contain a defective viral
gene designated *src* (for *sar̲coma*, the type of cancer it causes).
The *src* gene (also known as a v-*src*, because it is found within a
viral genome) was the first example of a viral oncogene, a gene in
a viral genome that promotes cancer.

Since the viral oncogene of RSV is not necessary for viral
replication, researchers were curious as to why the virus has this
v-*src* oncogene. Harold Varmus and Michael Bishop, in collaboration with Peter Vogt, soon discovered that normal host cells contain a copy of the *src* gene in their chromosomes. This normal
copy of the *src* gene in the host is termed c-*src*, for c̲ellular *src*; it
is encoded in the chromosomal DNA of the host cell. The c-*src*
gene does not cause cancer. However, once incorporated into a
viral genome, this gene can become a viral oncogene that promotes cancer. There are two possible explanations of this phenomenon. First, the many copies of the virus made during viral
replication may lead to overexpression of the *src* gene. Alternatively, the v-*src* gene may accumulate additional mutations that
convert it to an oncogene.

RSV has acquired the *src* gene by capturing it from a host
cell's chromosome. This can occur during the RSV life cycle. RSV
is a retrovirus, with an RNA genome. During its life cycle, a retrovirus uses reverse transcriptase to make a DNA copy of its
genome, which becomes integrated as a **provirus** into the host cell

genome. This integration may occur next to a proto-oncogene. During transcription of the proviral DNA, the neighboring proto-oncogene may be included in the RNA transcript. This RNA transcript can then recombine with an RNA retroviral genome within the cell to yield a retrovirus that contains an oncogene.

Since these early studies of RSV, many other retroviruses carrying oncogenes have been investigated. The characterization of such oncogenes has led to the identification of several genes with oncogenic potential. Besides retroviruses, several viruses with DNA genomes cause tumors (table 22.6). Some of these are known to cause cancer in humans.

TABLE 22.6

Examples of Viruses That Cause Cancer

Virus	Description
Retroviruses	
Rous sarcoma virus	Causes sarcomas in chickens.
Simian sarcoma virus	Causes sarcomas in monkeys.
Abelson leukemia virus	Causes leukemia in mice.
Hardy-Zuckerman-4 feline sarcoma virus	Causes sarcomas in cats.
DNA tumor viruses	
Hepatitis B, SV40, polyomavirus	Causes liver cancer in several species including humans. Does not cause cancer in their natural hosts but can transform cells in culture.
Papillomavirus	Causes benign tumors and malignant carcinomas in several species including humans. Causes cervical cancer in humans.
Adenovirus	Does not cause cancer in its natural host but can transform cells in culture.
Herpesvirus	Causes carcinoma in frogs and T-cell lymphoma in chickens. A human herpesvirus, Epstein-Barr virus, is a causative agent in Burkitt's lymphoma, which occurs primarily in immunosuppressed individuals such as AIDS patients.

EXPERIMENT 22A

DNA Isolated from Malignant Mouse Cells Can Transform Normal Mouse Cells into Malignant Cells

The study of retroviruses and other tumor-producing viruses led to the identification of a few dozen viral oncogenes. These were the first genes implicated in causing cancer. However, most cancers are caused not by viruses but by environmental mutagens that alter the expression of normal cellular genes. Therefore, researchers also wanted to identify cellular genes that have been altered in a way that leads to malignancy. Methods developed in the study of viral oncogenes were to prove valuable in this search. In particular, Miroslav Hill and Jana Hillova showed in 1971 that the purified DNA of RSV could be taken up by chicken fibroblasts and would transform them into malignant cells.

In 1979, Robert Weinberg and his colleagues wanted to determine if chromosomal DNA purified from cells that have become malignant due to exposure to mutagens can transform normal cells into malignant cells. If so, this would be the first step in the identification of chromosomally located genes that had been converted to oncogenes. Before we discuss this experiment, let's consider how a researcher can identify malignant cells in a laboratory. A widely used assay relies on the ability to recognize a clump of transformed cells as a distinct **focus** that grows over a monolayer of cells on a culture dish (fig. 22.8). Unlike normal

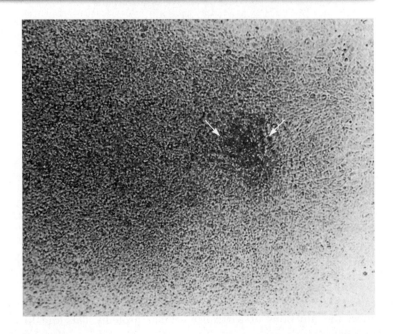

FIGURE 22.8   **The malignant growth of cells.** The growth of normal cells leads to the formation of a monolayer, while certain transformed cells grow as a focus (see *arrows*) or a raised pile of cells.

cells, which grow as a simple monolayer on culture dishes, a malignant focus piles up to form a mass of cells. The focus of cells is derived from a single cell that had become malignant and then proliferated. At the microscopic level, the malignant cells also have altered morphologies.

Now that we understand how malignant cells are identified in a laboratory, let's examine the steps of Weinberg's experiment. They began with several malignant cell lines that had been previously characterized. They also had normal cell lines. As shown in figure 22.9, they isolated the DNA from the malignant and normal cells and then mixed this DNA with normal cells. Calcium

phosphate was added to promote the uptake of the DNA by the normal cells. The cells were incubated for 2 to 3 weeks and then the researchers examined the plates for the growth of malignant foci of cells. In other words, they looked to see if any of the normal cells had been transformed into malignant cells.

■ THE HYPOTHESIS

Cellular DNA isolated from malignant cells will be taken up by normal cells and will transform them into malignant cells.

■ TESTING THE HYPOTHESIS — **FIGURE 22.9** Identification of chromosomal oncogenes.

Starting material: Several mouse cell lines. Some of the cell lines were normal, whereas others were malignant due to exposure to chemical or physical mutagens. It was known that none of the cell lines in this experiment were infected with oncogenic viruses.

Experimental level

1. Extract the chromosomal DNA from normal or malignant cell lines.

Homogenize the cells.

Isolate the nuclei by differential centrifugation. See the appendix.

Dissolve the nuclear membrane with detergent, and isolate the chromosomal DNA.

2. Mix the DNA with normal mouse fibroblast cells that are growing on a tissue culture plate.

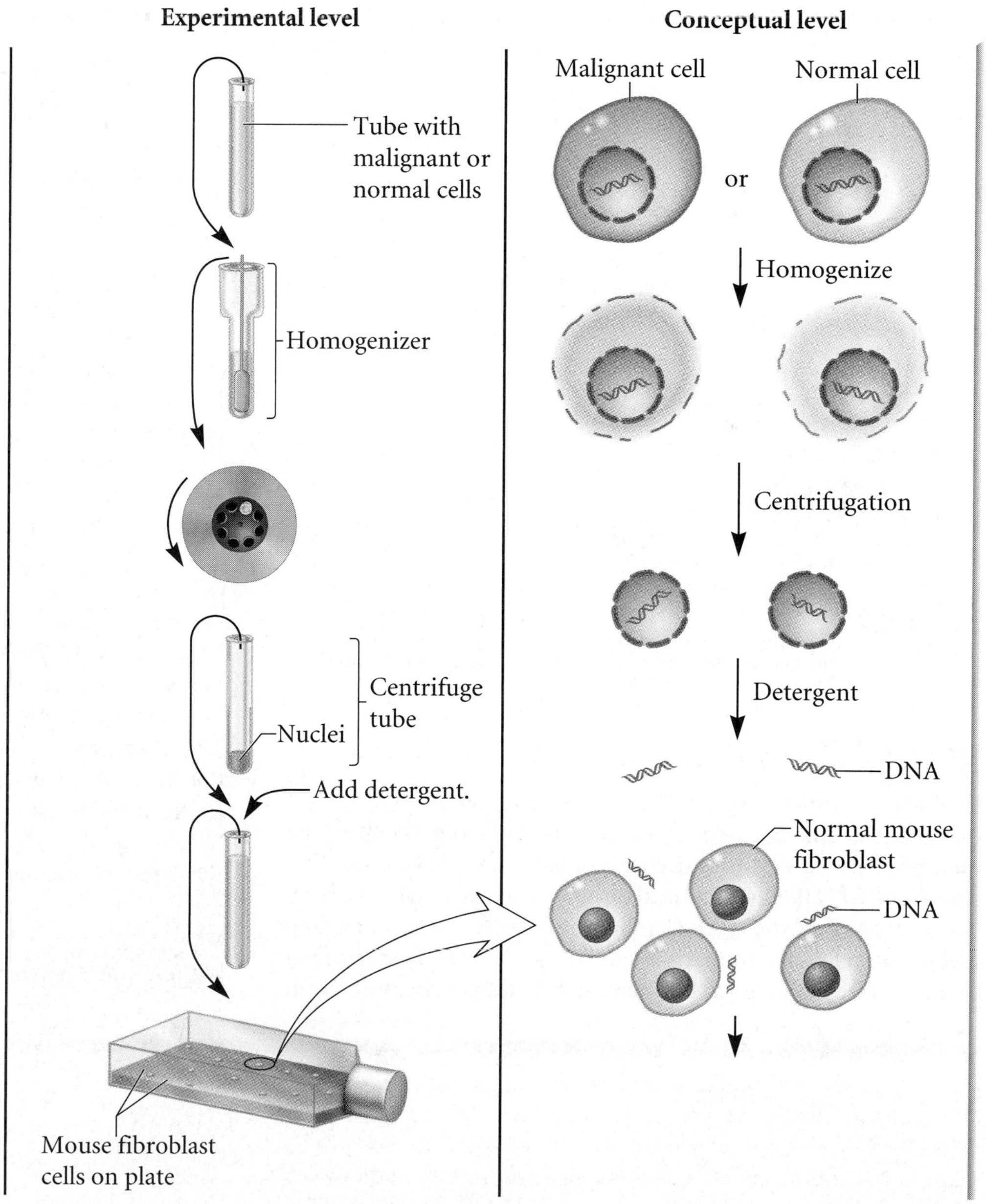

Tube with malignant or normal cells

Homogenizer

Nuclei

Centrifuge tube

Add detergent.

Mouse fibroblast cells on plate

Conceptual level

Malignant cell   Normal cell

or

Homogenize

Centrifugation

Detergent

DNA

Normal mouse fibroblast

DNA

(continued)

3. Add a buffer containing calcium ions and phosphate ions. This buffer makes the cells permeable to DNA.

4. Incubate for 14–20 days.

5. Examine the plates under a light microscope for cells growing as transformed foci.

Note: Normal cells in culture grow as a monolayer and very rarely form foci.

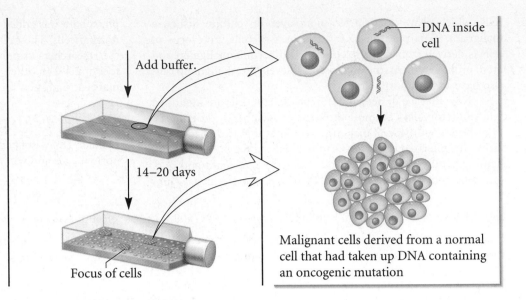

Add buffer.

DNA inside cell

14–20 days

Focus of cells

Malignant cells derived from a normal cell that had taken up DNA containing an oncogenic mutation

■ THE DATA

Source of DNA	Recipient Cells	Number of Malignant Foci Found on 12 Plates
Malignant Cell Lines		
MC5-5-0	NIH3T3 (normal fibroblasts)	48*
MCA16	"	5
MB66 MCA ad 36	"	8
MB66 MCA ACL6	"	0
MB66 MCA ACL13	"	0
Normal Cell Lines		
NIH3T3	"	<1
C3H10T1/2	"	0

*In this experiment, 2 of the plates were contaminated, so this is 48 foci on 10 plates.

■ INTERPRETING THE DATA

As shown in the data, the DNA isolated from some (but not all) malignant cell lines could transform normal mouse cells, which then proliferated and produced malignant foci. These results are consistent with the hypothesis that oncogenes had been taken up and expressed in the normal mouse cells, converting them into malignant cells. By comparison, the DNA isolated from normal cells did not cause a significant amount of transformation. From these experiments, it is not clear why the DNA from two of the malignant cell lines (namely, MB66 MCA ACL6 and MB66 MCA ACL13) could not transform the normal mouse cells. One possibility is that some malignancies are caused by dominant oncogenes, while others involve genes that act recessively. Recessive genes would be unable to transform normal mouse cells that already contain the normal (nonmalignant) dominant allele. Later in chapter 22, we will learn how another category of genes involved in cancer, called tumor-suppressor genes, act recessively.

At about the same time as Weinberg's work, Geoffrey Cooper and his colleagues showed that the DNA from normal cells could be isolated in vitro and activated to become oncogenic. Taken together, these early studies provided the first demonstration that organisms contain cellular oncogenes that cause malignancy. Just 2 years later, in 1981, the laboratories of Weinberg and Cooper identified the first cellular oncogene in humans. They isolated chromosomal DNA from human bladder carcinoma cells and used it to transform mouse cells in vitro. This chromosomal DNA contained a human oncogene that was the result of a mutation. The identity of the gene was eventually determined by transposon tagging as described in solved problem S2 at the end of chapter 22. This observation paved the way for the isolation of many human cellular oncogenes.

A self-help quiz involving this experiment can be found at the Online Learning Center.

Many Oncogenes Have Abnormalities That Affect Proteins Involved in Cell Division Pathways

As researchers began to identify oncogenes, they wanted to understand how these abnormal genes cause cancer. In parallel with cancer research, cell biologists have studied the roles that normal cellular proteins play in cell division. In eukaryotic species, the cell cycle is regulated in part by polypeptide hormones known as growth factors. As shown in figure 22.10, a growth factor binds to cell surface receptors and initiates a cascade of cellular events that lead eventually to cell division. In this example, a growth hormone known as epidermal growth factor (EGF) binds to its receptor, leading to the activation of an intracellular signaling pathway. This pathway, also known as a signal cascade, transmits a change in gene transcription. In other words, the transcription of specific genes is activated in response to the growth hormone. Once these genes are transcribed and translated, the gene products function to promote the progression through the cell cycle. Figure 22.10 is just one example of a pathway between a growth factor and gene activation. Eukaryotic species produce many different growth factors, and the signaling pathways are often more complex than the one shown here.

Most, but not all, oncogenes encode proteins that function in cell growth signaling pathways (table 22.7). These include growth factors, growth factor receptors, proteins involved in the intracellular signaling pathways, and transcription factors.

An oncogene may promote cancer by keeping a cell growth signaling pathway in a permanent "on" position. This can occur in two ways. In some cancers, the oncogene is overexpressed, yielding too much of the encoded protein. Research groups headed by Robert Gallo and Mark Groudine showed that a *myc* gene, designated c-*myc*, was amplified about 10-fold in a human promyelocytic leukemia cell line. Since that time, it has been found that *myc* genes are overexpressed in many forms of cancer, including breast cancer, lung cancer, and colon carcinoma. Presumably, the overexpression of this transcription factor leads to the transcriptional activation of genes that promote cell division.

Rather than producing too much of a given protein, certain oncogenes produce a functionally aberrant protein. For example, mutations that alter the amino acid sequence of the Ras protein have been shown to cause functional abnormalities. The Ras protein is a GTPase, which hydrolyzes GTP to GDP + P_i (fig. 22.11). Therefore, after it has been activated, the Ras protein returns to its inactive state by hydrolyzing GTP. Mutations that convert normal *ras* into an oncogenic *ras* either decrease the GTPase activity of the Ras protein or they increase the rate of exchange of bound GDP for GTP. Both of these functional changes result in a greater amount of the active GTP-bound form of the Ras protein. In this way, these mutations keep the signaling pathway turned on.

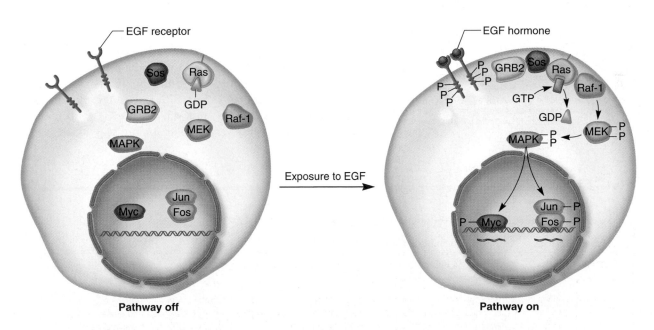

Pathway off   **Pathway on**

Exposure to EGF

FIGURE 22.10 **Activation of a cell signaling pathway by a growth hormone.** In this example, epidermal growth factor (EGF) binds to two EGF receptors, causing them to dimerize and phosphorylate each other. An intracellular protein called GRB2 is attracted to the phosphorylated EGF receptor, and it is subsequently bound by another protein called Sos. The binding of Sos to GRB2 enables Sos to activate a protein called Ras. This activation involves the binding of GTP and the release of GDP. The activated Ras/GTP complex then activates Raf-1, which is a protein kinase. Raf-1 phosphorylates MEK, and then MEK phosphorylates MAPK. More than one MAPK may be involved. Finally, the phosphorylated form of MAPK activates transcription factors. This leads to the transcription of genes, which encode proteins that promote cell division.

TABLE 22.7
Examples of Oncogenes

Gene	Cellular Function
Growth Factors*	
sis	Platelet-derived growth factor
int-2	Fibroblast growth factor
Receptors	
erbB	Growth factor receptor for EGF (epidermal growth factor)
trk	Growth factor receptor for CSF-1 (cytostatic factor)
fms	Growth factor receptor for NGF (nerve growth factor)
K-sam	Growth factor receptor for FGF (fibroblast growth factor)
Intracellular Signaling Proteins	
ras	GTP/GDP-binding protein
raf	Serine kinase
src	Tyrosine kinase
abl	Tyrosine kinase
gsp	G-protein α subunit
Transcription Factors	
jun	Transcription factor
fos	Transcription factor
myc	Transcription factor
gli	Transcription factor
erbA	Steroid receptor (which functions as a transcription factor)

*The genes described in this table are found in humans as well as other eukaryotic species. Many of these genes were initially identified in retroviruses. Most of the genes have been given three-letter names that are abbreviations for the type of cancer the oncogene causes or the type of virus in which the gene was first identified.

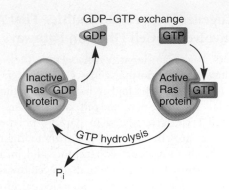

FIGURE 22.11 **Functional cycle of the Ras protein.** When GTP is bound to Ras, this activates the function of Ras and promotes cell division. The hydrolysis of GTP to GDP and P_i converts the active form to an inactive form.

Genetic Changes in Cellular Proto-Oncogenes Convert Them to Cellular Oncogenes

A **proto-oncogene** is a normal cellular gene that has the potential to incur a mutation to become an oncogene. A fundamental issue in cancer biology is to identify the specific genetic alterations that convert normal proto-oncogenes into abnormal oncogenes. By isolating and studying oncogenes at the molecular level, researchers have discovered four main ways this occurs (table 22.8). These changes can be categorized as missense mutations, gene amplifications (i.e., increase in copy number), chromosomal translocations, and viral integrations.

As mentioned previously, changes in the structure of the Ras protein can cause it to become permanently activated. This is caused by a missense mutation in the *ras* gene. The human genome contains four different but evolutionarily related *ras* genes: *ras*H, *ras*N, *ras*K-4a, and *ras*K-4b. All four genes encode proteins with very similar amino acid sequences containing a

TABLE 22.8
Genetic Changes That Convert Proto-Oncogenes into Oncogenes

Type of Change	Description and Examples
Missense mutation	A change in the amino acid sequence of a proto-oncogene protein may cause it to function in an abnormal way. Missense mutations can convert *ras* genes into oncogenes.
Gene amplification	The copy number of a proto-oncogene may be increased by gene duplication. *Myc* genes have been amplified in human leukemias; breast, stomach, lung, and colon carcinomas; and neuroblastomas and glioblastomas. *ErbB* genes have been amplified in glioblastomas, squamous cell carcinomas, and breast, salivary gland, and ovarian carcinomas.
Chromosomal translocations	A piece of chromosome may be translocated to another chromosome and affect the expression of genes at the breakpoint site. In Burkitt's lymphoma, a region of chromosome 8 is translocated to either chromosome 2, 14, or 22. The breakpoint in chromosome 8 causes the overexpression of the c-*myc* gene.
Viral integration	When a virus integrates into the chromosome, it may enhance the expression of nearby proto-oncogenes. In avian lymphomas, the integration of the avian leukosis virus can enhance the transcription of the c-*myc* gene.

Adapted from Cooper, G. M. (1995) *Oncogenes,* 2d ed. Jones and Bartlett Publishers, Boston.

total of 188 or 189 amino acids. Missense mutants in these normal *ras* genes are associated with particular forms of cancer. For example, a missense mutation in *ras*H that changes a glycine to a valine is responsible for the conversion of *ras*H into an oncogene:

	1	2	3	4	5	6	7	8	9	10	11	12	13	188	189
Normal	Met	Thr	Glu	Tyr	Lys	Leu	Val	Val	Val	Gly	Ala	Gly	Gly	Leu	Ser
Human *ras*H	ATG	ACG	GAA	TAT	AAG	CTG	GTG	GTG	GTG	GGC	GCC	GGC	GGT........	CTC	TCC
												↓ GTC			
Oncogenic *ras*H	Met	Thr	Glu	Tyr	Lys	Leu	Val	Val	Val	Gly	Ala	Val	Gly.........	Leu	Ser

Experimentally, chemical carcinogens have been shown to cause these missense mutations and thereby lead to cancer.

Another genetic event that occurs in cancer cells is an abnormal increase in the copy number of a proto-oncogene. Gene amplification does not normally happen in mammalian cells, but it is a common occurrence in cancer cells. As mentioned previously, Gallo and Groudine discovered that c-*myc* was amplified in a human leukemia cell line. Many human cancers are associated with the amplification of particular oncogenes. In some cases, the extent of oncogene amplification is correlated with the progression of tumors to increasing malignancy. These include the amplification of N-*myc* in neuroblastomas and *erbB-2* in breast carcinomas. In other types of malignancies, gene amplification is more random and may be a secondary event that increases the expression of oncogenes previously activated by other genetic changes.

A third type of genetic alteration that can lead to cancer is a chromosomal translocation. While structural abnormalities are common in cancer cells, very specific types of chromosomal translocations have been identified in certain types of tumors. In 1960, Peter Nowell discovered that chronic myelogenous leukemia was correlated with the presence of a shortened version of chromosome 22, which he called the Philadelphia chromosome after the city where it was discovered. Rather than a deletion, this shortened chromosome is the result of a reciprocal translocation between chromosomes 9 and 22. Later studies revealed that this translocation activates a proto-oncogene, *abl*, in an unusual way (fig. 22.12). The reciprocal translocation involves breakpoints within the *bcr* and *abl* genes. Following the reciprocal translocation, the first part of the *bcr* gene fuses with the *abl* gene. This yields an oncogene that encodes an abnormal fusion protein, which contains the polypeptide sequences encoded from both genes. However, the gene is controlled by the *bcr* promoter.

In other forms of cancer, chromosomal translocations cause an overexpression of an oncogene. In Burkitt's lymphoma, for example, a region of chromosome 8 is translocated to either chromosome 2, 14, or 22. The breakpoint in chromosome 8 is near the c-*myc* gene, and the sites on chromosomes 2, 14, and 22 correspond to locations of different immunoglobulin genes that are normally expressed in lymphocytes. The translocation of the c-*myc* gene near the immunoglobulin genes leads to the overexpression of the c-*myc* gene and thereby promotes malignancy in lymphocytes.

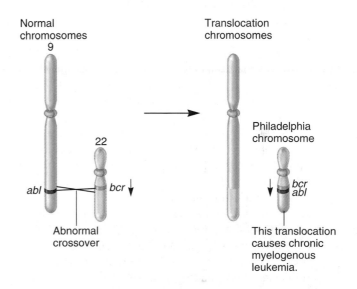

FIGURE 22.12 **The reciprocal translocation commonly found in people with chronic myelogenous leukemia.**

GENES→TRAITS   In healthy individuals, the *bcr* gene is located on chromosome 22 and the *abl* gene is on chromosome 9. In certain forms of myelogenous leukemia, a reciprocal translocation causes the first part of *bcr* to fuse with *abl*. This combined gene encodes an abnormal fusion protein, which leads to leukemia.

Tumor-Suppressor Genes Play a Role in Preventing the Proliferation of Cancer Cells

Thus far, we have considered how oncogenes promote cancer. An oncogene is an abnormally activated gene that leads to uncontrolled cell growth. We will now turn our attention to a second category of genes called **tumor-suppressor genes.** As the name suggests, the role of a tumor-suppressor gene is to prevent cancerous growth. Therefore, when a tumor-suppressor gene becomes inactivated by mutation, it becomes more likely that cancer will occur.

The first identification of a human tumor-suppressor gene involved studies of retinoblastoma, a tumor that occurs in the retina of the eye. Some people have an inherited predisposition to develop this disease within the first few years of life. By comparison, the noninherited form of retinoblastoma, which is caused by environmental agents, is more likely to occur later in life.

Based on these differences, Alfred Knudson proposed a "two-hit" model for retinoblastoma. According to this idea, retinoblastoma requires two mutations to occur. People with the inherited form already have received one mutation from one of their parents. They need only one additional mutation to develop the disease. Since there are more than 1 million cells in the retina, it is relatively likely that a mutation may occur in one of these cells at an early age, leading to the disease. However, people with the noninherited form of the disease must have two mutations in the same retinal cell to cause the disease. Since two rare events are much less likely to occur than a single such event, the noninherited form of this disease is expected to occur much later in life, and only rarely. Therefore, this hypothesis explains the different populations typically affected by the inherited and noninherited forms of retinoblastoma.

Since Knudson's original hypothesis in 1971, molecular studies have confirmed the two-hit hypothesis. The gene that causes this disease is designated *rb* (for retinoblastoma). *Rb* is found on the long arm of chromosome 13. Most people have two normal copies of this gene. Persons with hereditary retinoblastoma have inherited one functionally defective copy. Therefore, in nontumorous cells throughout the body, they have one normal copy and one defective copy of *rb*. However, in retinal tumor cells, the normal *rb* gene has also suffered the second hit (i.e., a mutation), which renders it defective. These observations verify the two-hit theory of Knudson.

More recent studies have revealed how the Rb protein prevents the proliferation of cancer cells (fig. 22.13). The Rb protein regulates a transcription factor called E2F, which activates genes required for cell cycle progression. The binding of the Rb protein to E2F inhibits its activity and prevents cell division. When a normal cell is supposed to divide, cellular proteins called cyclins bind to cyclin-dependent protein kinases. This activates the kinases, which then leads to the phosphorylation of the Rb protein. The phosphorylated form of the Rb protein is released from E2F, thereby allowing E2F to activate genes needed to progress through the cell cycle. By comparison, it is easy to see how the cell cycle becomes unregulated without functional Rb protein. When both copies of Rb are defective, the E2F protein is always active. This explains why uncontrolled cell division can occur.

The Vertebrate *p53* Gene Is a Master Tumor-Suppressor Gene That Senses DNA Damage

After the *rb* gene, the second tumor-suppressor gene discovered was the *p53* gene. *p53* is the most commonly altered gene in human cancers. About 50% of all human cancers are associated with defects in this gene. This includes malignant tumors of the lung, breast, esophagus, liver, bladder, and brain as well as sarcomas, lymphomas, and leukemias. For this reason, an enormous amount of research has been aimed at elucidating the function of the p53 protein.

A primary role of the p53 protein is to determine if a cell has incurred DNA damage. If so, p53 will promote three types of cellular pathways that are aimed at preventing the proliferation of cells with damaged DNA. First, when confronted with a DNA-damaging agent, the cell can try to repair its DNA. This may prevent the accumulation of mutations that activate oncogenes or inactivate tumor-suppressor genes. Second, if a cell is in the process of dividing, it can arrest itself in the cell cycle. By stopping its cell cycle, the cell has more time to repair its DNA and avoid producing two mutant daughter cells. The third, and most drastic event, is that a cell can initiate a series of events called **apoptosis,** or programmed cell death. In response to DNA-damaging agents, a cell may program its own death.

Apoptosis is an active process that involves cell shrinkage, chromatin condensation, and DNA degradation. This process is facilitated by proteases known as **caspases,** which are activated during apoptosis. These types of proteases are sometimes called the "executioners" of the cell. Caspases digest selected cellular proteins such as microfilaments, which are components of the intracellular cytoskeleton. This causes the cell to break down into small vesicles that are eventually phagocytosized by cells of the immune system. Apoptosis occurs in a small number of cells as a

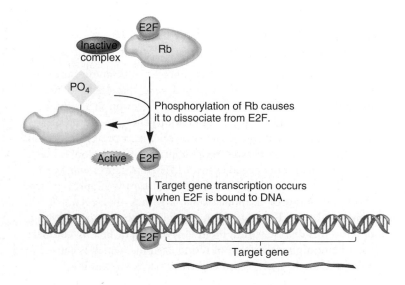

FIGURE 22.13   **Interactions between the Rb and E2F proteins.** The binding of the Rb protein to E2F inhibits the ability of E2F to function. This prevents cell division. For cell division to occur, cyclins bind to cyclin-dependent protein kinases, which then phosphorylate the Rb protein. The phosphorylated Rb protein is released from E2F. The free form of E2F can activate target genes needed to progress through the cell cycle.

Environmental mutagen
causes DNA damage
(double-strand breaks).

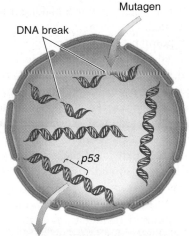

Mutagen

DNA break

p53

Induction of the *p53* gene leads to the
synthesis of the p53 protein, which functions
as a transcription factor. This transcription
factor can:

• Activate genes that promote DNA repair.

• Activate genes that arrest cell division and
  may generally repress other genes that are
  required for cell division.

• Activate genes that promote apoptosis.

FIGURE 22.14   **Central role of *p53* in preventing the proliferation of cancer cells.** Expression of the *p53* gene, which encodes a transcription factor, is induced by agents that cause double-stranded DNA
breaks. The p53 transcription factor may activate genes that promote
DNA repair, activate genes that arrest cell division, repress genes
required for cell division, and activate genes that promote apoptosis.

normal process of embryonic development. In addition, it is beneficial for an adult organism to kill an occasional cell with cancer-
causing potential.

Figure 22.14 summarizes how *p53* plays a central role in all
three processes. As shown here, the expression of the *p53* gene is
induced by agents that damage the DNA. The inducing signal
appears to be double-stranded DNA breaks. The p53 protein
functions as a transcription factor. It contains a DNA-binding
domain and a transcriptional activation domain. Experimental
studies have shown that p53 can activate the transcription of several specific target genes. As shown in figure 22.14, it can activate
genes that promote DNA repair, arrest the cell cycle, and promote
apoptosis. In addition, p53 appears to act as a negative regulator
by interacting with general transcription factors. This decreases
the general expression of many other structural genes. This inhibition may also help prevent the cell from dividing.

Overall, p53 activates the expression of a few specific cellular genes and, at the same time, inhibits the expression of many
other genes in the cell. Through the regulation of gene transcrip-

tion, the p53 protein can prevent the proliferation of cells that
have incurred DNA damage.

Many Types of Tumor-Suppressor Genes Can Promote Cancer When Their Function Is Lost

During the past three decades, researchers have identified many
tumor-suppressor genes that contribute to the development and
progression of cancer (table 22.9). Some tumor-suppressor genes
encode proteins that have direct effects on the regulation of cell
division. An example would be the *rb* gene. As mentioned earlier,
the Rb protein negatively regulates E2F. If both copies of the *rb*
gene are inactivated, the growth of cells will be accelerated.
Therefore, loss of function of these kinds of negative regulators
has a direct impact on the abnormal cell division rates seen in

TABLE 22.9

Functions of Selected Tumor-Suppressor Genes

Gene	Function
rb	As described in figure 22.13, the Rb protein is a negative regulator of E2F. This represses the transcription of genes required for DNA replication and cell division.
p53	p53 is a transcription factor that positively regulates a few specific target genes and negatively regulates others in a general manner. It acts as a sensor of DNA damage. It can prevent the progression through the cell cycle and also can promote apoptosis.
p16	A protein that negatively regulates cyclin-dependent kinases. This kinase promotes the transition from the G_1 phase of the cell cycle to the S phase.
Pten	A protein phosphatase that functions to maintain a low level of a phospholipid (PIP-3) within the cell. PIP-3 is a signaling molecule that promotes cell division.
NF1	The NF1 protein stimulates Ras to hydrolyze its GTP to GDP. Loss of NF1 function causes the Ras protein to be overactive, which promotes cell division.
APC	A negative regulator of a cell-signaling pathway called the Wnt pathway. The Wnt pathway leads to the activation of genes that promote cell division.
VHL	Functions in the targeting of certain proteins for (ubiquitin-mediated) protein degradation. When these proteins are not degraded, the cell cycle may proceed in an unregulated manner.
BRCA-1, BRCA-2	Brca1 and Brca2 proteins are both involved in the cellular defense against DNA damage, although the precise function of the proteins is still not known. It is not clear whether they play a role in sensing DNA damage, or may act to facilitate DNA repair.
WT1	A transcription factor that regulates genes in kidney cells and gonadal cells. The genes that WT1 regulates are poorly understood.
MTS1	The MTS1 protein acts as an inhibitor of cyclin-dependent protein kinases. The activities of these protein kinases promote the progression through the cell cycle.

cancer cells. In other words, when the Rb protein is lost, a cell becomes more likely to divide.

Alternatively, other tumor-suppressor genes play a role in the proper maintenance of the genome. The proteins they encode function in the sensing of genome integrity. Such proteins are vital for the detection of abnormalities such as DNA breaks and improperly segregated chromosomes. When such abnormalities are detected, the proteins participate in regulatory pathways that prevent cell division. Many of the proteins in this category are called **checkpoint proteins** because their role is to <u>check</u> the integrity of the genome and prevent cells from progressing past a certain <u>point</u> in the cell cycle if genetic abnormalities are detected. Figure 22.15 shows a simplified diagram of the eukaryotic cell cycle. As mentioned earlier, proteins called cyclins and cyclin-dependent protein kinases (Cdks) are responsible for advancing a cell through the four phases of the cell cycle. For example, an activated G_1-cyclin/Cdk is necessary to advance from the G_1 phase to the S phase. The formation of activated cyclin/Cdk complexes is regulated via checkpoint proteins.

There are several checkpoints in the cell cycle of human cells. Figure 22.15 shows three of the major checkpoints. Both the G_1 and G_2 checkpoints involve the functions of proteins that can sense if the DNA has incurred damage. If so, these checkpoint proteins, such as p53, can prevent the formation of active cyclin/Cdk complexes. This would stop the progression of the cell cycle. A checkpoint also exists in metaphase. This checkpoint is monitored by proteins that can sense if a chromosome is not correctly attached to the spindle apparatus, making it likely for it to be improperly segregated.

Overall, checkpoint proteins prevent the division of cells that may have incurred DNA damage or harbor abnormalities in chromosome attachment. This provides a mechanism to stop the accumulation of genetic abnormalities that could produce cancer cells within the body. When checkpoint genes are lost, cell division may not be directly accelerated. However, the loss of checkpoint protein function makes it more likely that undesirable genetic changes will occur that could cause cancerous growth.

Along these lines, it should also be mentioned that the genes encoding DNA repair enzymes are inactivated in certain forms of cancer. The loss of a DNA repair enzyme makes it more likely for a cell to accumulate mutations that could create an oncogene and/or eliminate the function of tumor-suppressor genes. However, many geneticists do not classify the genes that encode DNA repair enzymes to be tumor-suppressor genes because the encoded proteins do not play a role in the regulation of cell division.

Thus far, we have considered the functions of proteins that are encoded by tumor-suppressor genes. Cancer biologists would also like to understand how tumor-suppressor genes are inactivated, because this knowledge may aid in the prevention of cancer. It has been proposed that there are three common ways that the function of tumor-suppressor genes can be lost. First, a mutation can occur specifically within a tumor-suppressor gene to inactivate its function. For example, a mutation could wreck the promoter of a tumor-suppressor gene or introduce an early stop codon in the coding sequence. Either of these would prevent the expression of a functional protein. A second way to inhibit tumor-suppressor genes is via DNA methylation. As discussed in chapter 15, transcription is inhibited when the methylation of CpG islands occurs near a promoter region. The methylation of CpG islands near the promoters of tumor-suppressor genes has been found in many types of tumors, suggesting that this form of gene inactivation plays an important role in the formation and/or progression of malignancy. However, it is not clear why tumor-suppressor genes are aberrantly methylated in cancer cells. Additional research will be necessary to understand why this happens. Finally, aneuploidy is frequently associated with many types of cancer. As discussed in chapter 8, aneuploidy involves the loss or addition of one or more chromosomes so that the total number of chromosomes is not an even multiple of a chromosome set. It has been proposed that chromosome loss may contribute to the progression of cancer if the lost chromosome carries one or more tumor-suppressor genes.

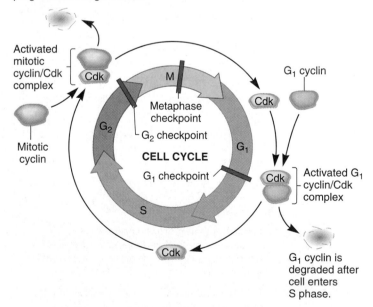

Mitotic cyclin is degraded as cell progresses through mitosis.

Activated mitotic cyclin/Cdk complex

Cdk

Mitotic cyclin

G_1 cyclin

Cdk

M

Metaphase checkpoint

G_2 checkpoint

G_2

CELL CYCLE

G_1

G_1 checkpoint

S

Cdk

Cdk

Activated G_1 cyclin/Cdk complex

G_1 cyclin is degraded after cell enters S phase.

FIGURE 22.15 **Checkpoints in the cell cycle.** This is a simplified diagram of the cell cycle of humans. Progression through the cell cycle requires the formation of activated cyclin/Cdk complexes. There are several different types of cyclin proteins, which are typically degraded after the cell has progressed to the next phase. The formation of activated cyclin/Cdk complexes is regulated by checkpoint proteins. If these proteins detect DNA damage, they will prevent the formation of activated cyclin/Cdk complexes. In addition, other checkpoint proteins can detect chromosomes that are incorrectly attached to the spindle apparatus and prevent further progression through metaphase.

Most Forms of Cancer Involve Multiple Genetic Changes Leading to Malignancy

The discovery of oncogenes and tumor-suppressor genes, along with molecular techniques that can detect genetic alterations, has enabled researchers to study the progression of certain forms of cancer at the molecular level. Many cancers begin with a benign genetic alteration that, over time and with additional mutations, leads to malignancy. Furthermore, a malignancy can continue to accumulate genetic changes that make it even more difficult to treat.

In 1990, Eric Fearon and Bert Vogelstein proposed a series of genetic changes that lead to colorectal cancer, the second most common cancer in the United States. As shown in figure 22.16, colorectal cancer is derived from cells in the mucosa of the colon. The loss of function of *APC*, a tumor-suppressor gene on chromosome 5, leads to an increased proliferation of mucosal cells and the development of a benign polyp, a noncancerous growth. Additional genetic changes involving the loss of other tumor-suppressor genes and the activation of an oncogene (namely, *ras*) leads eventually to the development of a carcinoma. In figure 22.16, the genetic changes that lead to colon cancer seem to occur in an orderly sequence. While the growth of a tumor often begins with mutations in *APC*, the order of mutations shown in figure 22.16 is not absolute. It is the total number of genetic changes, not their exact order, that is important.

Inherited Forms of Cancers May Be Caused by Defects in Tumor-Suppressor Genes and DNA Repair Genes

As we have just seen, cancer cells usually are generated by multiple genetic alterations that cause the activation of oncogenes and the inactivation of tumor-suppressor genes. As mentioned earlier in chapter 22, about 5 to 10% of all cancers involve inherited (germ-line) mutations. People who have inherited such mutations have a predisposition to develop cancer. This does not mean they will definitely get cancer, but they are more likely to develop the disease than are individuals in the general population. When individuals have family members who have developed certain forms of cancer, they may be tested to determine if they also carry a mutant gene. For example, von Hippel-Lindau disease and familial adenomatous polyposis are examples of syndromes for which genetic testing to identify at-risk family members is considered the standard of care.

Most inherited forms of cancer involve a defect in a tumor-suppressor gene (table 22.10). In these cases, the individual is heterozygous, with one normal and one inactive allele. As described earlier for retinoblastoma, a mutation in the remaining normal allele may occur in a somatic cell, thereby producing a cell with two inactive copies of the tumor-suppressor gene. At the phenotypic level, a predisposition for developing cancer is inherited in a dominant fashion, since a heterozygote exhibits this tendency. However, the actual development of cancer is recessive, because a second somatic mutation that results in two abnormal alleles (in the same cell) is necessary to promote malignancy.

As noted in table 22.10, not all hereditary forms of cancer are due to defective tumor-suppressor genes. For example, multiple endocrine neoplasia type 2 is due to the activation of an oncogene. In addition, several types of hereditary cancers are associated with defects in DNA repair enzymes. Two different genes (*MSH*2 and *MLH*1) identified in nonpolyposis (i.e., non-polyp-forming)

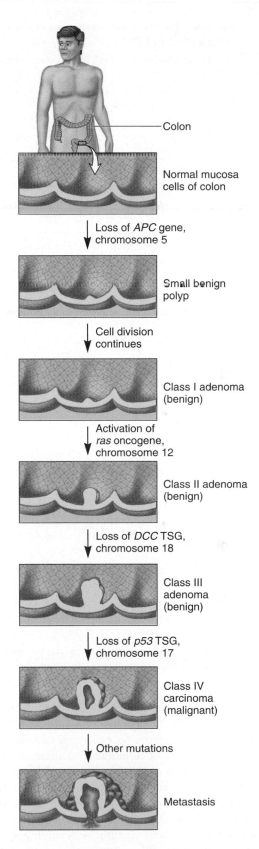

FIGURE 22.16 **Multiple genetic changes leading to colorectal cancer.**

TABLE 22.10

Inherited Mutant Genes That Confer a High Predisposition to Develop Cancer

Gene	Type of Cancer*
Tumor-Suppressor Genes	
VHL	von Hippel-Lindau disease, which is typically characterized by a clear cell renal carcinoma
APC	Familial adenomatous polyposis and familial colon cancer
rb	Retinoblastoma
p53	Li-Fraumeni syndrome, which is characterized by a wide spectrum of tumors including soft-tissue and bone sarcomas, brain tumors, adenocortical tumors, and premenopausal breast cancers
Pten	Cowden syndrome, which is characterized by the presence of multiple hamartomas in the skin, breast, thyroid, gastrointestinal tract, and central nervous system and an increased risk of breast and thyroid carcinomas
BRCA-1	Familial breast cancer
BRCA-2	Familial breast cancer
WT1	Wilm's tumor, which is a nephroblastoma
NF1	Neurofibromatosis
MTS1	Hereditary malignant melanoma
Oncogenes	
RET	Multiple endocrine neoplasia type 2
DNA Repair Genes	
Mismatch repair:	
MSH2	Nonpolyposis colorectal cancer
MLH1	Nonpolyposis colorectal cancer
Nucleotide excision repair: (can involve mutations in at least seven different genes designated XP-A to XP-G)	UV-sensitive forms of cancer such as basal cell carcinoma

*Many of the genes described in this table are mutated in more than one type of cancer. The cancers listed are those in which it has been firmly established that a predisposition to develop the disease is commonly due to germ-line mutations in the designated gene.

colorectal cancer are defects in enzymes required for mismatch DNA repair. A defect in DNA repair may lead to a higher rate of mutation and thereby increase the likelihood of cancer-causing mutations. Likewise, as discussed in chapter 16, some human diseases, such as xeroderma pigmentosum, involve inherited defects in genes that are involved with nucleotide excision repair. These individuals have many premalignant lesions and a predisposition to develop skin cancer.

CONCEPTUAL SUMMARY

Throughout chapter 22, we have examined many human diseases that have a genetic basis, such as Tay-Sachs disease, Huntington disease, hemophilia, and thalassemia. In some cases, mutant genes are passed from parents to offspring, causing inherited abnormalities. The pattern of inheritance of these mutant genes can often be deduced from a pedigree analysis, although this may be difficult if the disease has a heterogeneous genetic origin. In the case of single-gene defects, some genetic diseases are recessive, some are dominant, and some are X linked. In addition, irregularities in chromosome structure and/or number can cause phenotypic abnormalities. These types of genetic defects are more common in humans than defects in single genes.

Cancer is also a human disease with a genetic basis. Cancer can be caused by oncogenic viruses, such as RSV and other **acutely transforming viruses (ACVs),** or by the inheritance of mutant genes. Spontaneous mutations can also cause cancer. However, most forms of cancer are due to environmentally induced mutations in somatic cells that lead to uncontrolled cell growth. These mutations may involve two types of genes: **oncogenes** and **tumor-suppressor genes.** An oncogene, which is derived from a normal **proto-oncogene,** is an abnormally activated gene that stimulates cell growth. At the molecular level, a proto-oncogene may be converted to an oncogene by a missense mutation, a chromosome translocation, gene amplification, or viral integration. By comparison, a tumor-suppressor gene normally inhibits cell growth, but if rendered inactive, unconstrained cell growth may ensue. A master tumor-suppressor gene, called *p53*, plays a critical role in monitoring DNA damage and preventing the division of cells that have been damaged.

EXPERIMENTAL SUMMARY

Researchers cannot experiment with humans, as they can with model organisms such as *Drosophila* and mice. Nevertheless, researchers can gain knowledge of human genetic disorders by analyzing the occurrence of diseases within pedigrees and conducting in vitro tests on human cell samples.

Often, when a human disease is newly identified, researchers would like to know if it is caused, wholly or in part, by a mutation in a gene. Various observations point to a genetic cause. These include a higher frequency of affected individuals among family members and in identical versus fraternal twins, the inability of the disease to be spread by contact, different rates of the disease among different populations, a characteristic age of onset, a similar disease of known genetic origin in animals, and a correlation between a disease and a mutant gene.

Once genetics has become established as a cause of disease, researchers attempt to determine the pattern of transmission from parents to offspring. Many, but not all, genetic diseases follow a simple Mendelian pattern of inheritance. A pedigree analysis may indicate whether a disease is transmitted recessively or dominantly, and if it is X linked or autosomal.

To understand the relationship between genetics and disease, researchers investigate how mutant genes lead to the disease symptoms at the cellular and organismal levels. There are many experimental approaches to investigate this question; most of these have not been described in chapter 22. At the molecular level, the ultimate goal is to identify the function of the protein encoded by the normal gene and the defect resulting from the disease-causing mutation. In chapter 22, we have considered many examples where a disease-causing allele has been identified, and the function of the normally encoded protein is known. With this information, molecular and cell biologists can begin to unravel how a gene mutation affects protein function, cellular

processes, and ultimately the phenotype of the individual. Much research will continue to be aimed at understanding how mutant genes lead to abnormalities in people. In addition, our identification of an increasing number of mutant genes has provided an impetus to determine via **genetic tests** if individuals are carriers of a genetic lesion and have the potential of transmitting a genetic lesion to their children.

Cancer is a collection of human diseases that receives a large amount of attention. Experimentally, our understanding of cancer has been aided by the characterization of viruses that can cause cancer in animals or transform laboratory-grown cells. The study of these tumor-causing viruses has led to the identification of many oncogenes that play a central role in promoting cancerous cell growth. In addition, experiments in which DNA is used to transform cells in vitro have also enabled researchers to identify oncogenes.

Researchers continue to investigate the specific ways that mutations alter cancer-causing genes and the combinations of mutations that lead to malignant growth. By studying the DNA at the cellular and molecular level, scientists have discovered that an oncogene can be created by a missense mutation, a chromosome translocation, gene amplification, or viral integration, which lead to an abnormally high quantity and/or activity of a growth-related protein. By comparison, mutations that inactivate tumor-suppressor genes may render an individual more vulnerable to the proliferation of cancer.

In the future, scientists will continue to explore how oncogenes and tumor-suppressor genes influence the cell division cycle. They would ultimately like to understand this relationship in hopes of discovering how we can prevent cancer after these mutations have occurred.

PROBLEM SETS & INSIGHTS

Solved Problems

S1. The pedigree shown here concerns a human disease known as familial hypercholesterolemia.

This disorder is characterized by an elevation of serum cholesterol in the blood. Though relatively rare, this genetic abnormality

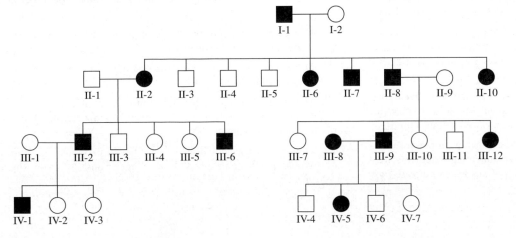

can be a contributing factor to heart attacks. At the molecular level, this disease is caused by a defective gene that encodes a protein called the low-density lipoprotein receptor (LDLR). In the bloodstream, serum cholesterol is bound to a carrier protein known as low-density lipoprotein (LDL). LDL binds to LDLR so that cells can absorb cholesterol. When LDLR is defective, it becomes more difficult for the cells to absorb cholesterol. This explains why the levels of blood cholesterol remain high. Based on the pedigree, what is the most likely pattern of inheritance of this disorder?

Answer: The pedigree is consistent with a dominant pattern of inheritance. An affected individual always has an affected parent. Also, individuals III-8 and III-9, who are both affected, produced unaffected offspring. If this trait was recessive, two affected parents should always produce affected offspring. However, because the trait is dominant, two heterozygous parents can produce homozygous unaffected offspring. The ability of two affected parents to have unaffected offspring is a striking characteristic of dominant inheritance. On average, we would expect that two heterozygous parents should produce 25% unaffected offspring. In the family containing IV-4, IV-5, IV-6, and IV-7, there are actually three out of four unaffected offspring. This higher-than-expected proportion of unaffected offspring is not too surprising because the family is a very small group and may deviate substantially from the expected value due to random sampling error.

S2. One way to identify a human cellular oncogene is to use human DNA from a malignant cell to transform a mouse cell. A mouse cell that has been transformed by human DNA will have a human oncogene incorporated into its genome; it also is likely to have *Alu* sequences that are closely linked to this human oncogene. (Note: As discussed in chapter 10, the human genome contains *Alu* sequences interspersed every 5,000 to 6,000 bp. *Alu* sequences are not found in the mouse genome.) Discuss how the *Alu* sequence can provide a way to clone human oncogenes.

Answer: One approach is transposon tagging, described in chapter 17. When a mouse cell is transformed with a human DNA fragment containing an oncogene, that fragment is likely to contain an *Alu* sequence

as well. To clone the human oncogene, the chromosomal DNA can be isolated from the transformed mouse cells, digested with a restriction enzyme, and cloned into vectors to create a library of DNA fragments. The members of the library that carry the human oncogene can be identified using a probe complementary to the *Alu* sequence because this sequence is not found in the mouse genome. Using this strategy, researchers have identified several human cellular oncogenes. In human bladder carcinoma, for example, the human cellular oncogene called *ras* has been identified this way.

S3. Oncogenes sometimes result from the genetic rearrangements (e.g., translocations) that produce gene fusions. An example is the Philadelphia chromosome that causes the *bcr* gene to fuse with the *abl* gene. Suggest two different reasons why a gene fusion could create an oncogene.

Answer: An oncogene is derived from a genetic change that abnormally activates the expression of a gene that plays a role in cell division. When a genetic change creates a gene fusion, this can abnormally activate the expression of the gene in two ways.

The first way is at the level of transcription. The promoter and part of the coding sequence of one gene may become fused with the coding sequence of another gene. For example, the promoter and part of the coding sequence of the *bcr* gene may fuse with the coding sequence of the *abl* gene. After this has occurred, the *abl* gene may be abnormally activated because it is now under the control of the *bcr* promoter, rather than its own normal promoter. If the *bcr* promoter is much stronger than the *abl* promoter, or if it is turned on in different cells compared to the *abl* promoter, this may lead to the overexpression of the abl protein compared to its normal level of expression.

A second way that a gene fusion can cause abnormal activation is at the level of protein structure. The fusion protein has parts of two different polypeptides. For example, the abl/bcr fusion protein has parts of the abl polypeptide and the bcr polypeptide. Perhaps the abl portion of the fusion protein affects the structure of the bcr portion of the polypeptide in such a way that the bcr portion becomes abnormally active, or vice versa.

Conceptual Questions

C1. With regard to pedigree analysis, make a list of the patterns that distinguish recessive, dominant, and X-linked genetic diseases from each other.

C2. Explain, at the molecular level, why human genetic diseases often follow a simple Mendelian pattern of inheritance, whereas most normal traits (e.g., the shape of your nose or the size of your liver) are governed by multiple gene interactions.

C3. Many genetic disorders are heterogeneous. What does this mean? Give two examples of genetic heterogeneity. How does genetic heterogeneity confound a pedigree analysis?

C4. In general, why do changes in chromosome structure and/or number affect an individual's phenotype? Explain why some changes in chromosome structure, such as reciprocal translocations, do not.

C5. We often speak of diseases such as PKU and achondroplasia as having a "genetic basis." Explain whether the following statements are accurate with regard to the genetic basis of any human disease (not just PKU and achondroplasia).

A. An individual must inherit two copies of a mutant allele to have disease symptoms.

B. A genetic predisposition or genetic basis means that an individual has inherited one or more alleles that make it more likely for them to develop disease symptoms compared to other individuals.

C. A genetic predisposition to develop a disease may be passed from parents to offspring.

D. The genetic basis for a disease is always more important than the environment.

C6. Figure 22.1 illustrates albinism in humans, whales, and the wildebeest. Describe two other genetic disorders that are found in both humans and animals.

C7. Discuss why a genetic disease might have a particular age of onset. Would an infectious disease have an age of onset? Explain why or why not.

C8. Gaucher disease (type I) is due to a defect in a gene that encodes a protein called acid β-D-glucosidase. This enzyme plays a role in carbohydrate metabolism within the lysosome. The gene is located on the long arm of chromosome 1. Persons who inherit two defective copies of this gene exhibit Gaucher disease. The symptoms include an enlarged spleen, bone lesions, and changes in skin pigmentation. Let's suppose a phenotypically normal woman, whose father had Gaucher disease, has a child with a phenotypically normal man, whose mother had Gaucher disease.

A. What is the probability that this child will have the disease?

B. What is the probability that this child will have two normal copies of this gene?

C. If this couple has five children, what is the probability that one of them will have Gaucher disease and four will be phenotypically normal?

C9. Ehler-Danlos syndrome is a relatively rare disorder that is due to a mutation in a gene that encodes a protein called collagen (type 3 A1). Collagen is a protein found in the extracellular matrix that plays an important role in the formation of skin, joints, and other connective tissues. Persons afflicted with this syndrome have extraordinarily flexible skin and very loose joints. The pedigree shown here contains several members affected with Ehler-Danlos syndrome, shown with black symbols. Based on this pedigree, does this syndrome appear to be an autosomal recessive, autosomal dominant, X-linked recessive, or X-linked dominant trait? Explain your reasoning.

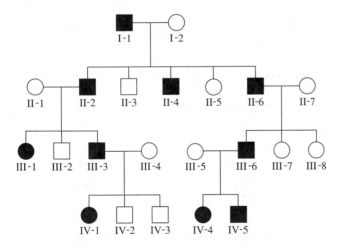

C10. Hurler syndrome is due to a mutation in a gene that encodes a protein called α-L-iduronidase. This protein functions within the lysosome as an enzyme that breaks down mucopolysaccharides (a type of polysaccharide that has many acidic groups attached). When this enzyme is defective, excessive amounts of dermatan sulfate and heparin sulfate accumulate within the lysosomes, especially in liver cells and connective tissue cells. This leads to symptoms such as an enlarged liver and spleen, bone abnormalities,

cloudy corneas, heart problems, and severe neurological problems. The pedigree shown here contains three members affected with Hurler syndrome, indicated with black symbols. Based on this pedigree, does this syndrome appear to be an autosomal recessive, autosomal dominant, X-linked recessive, or X-linked dominant trait? Explain your reasoning.

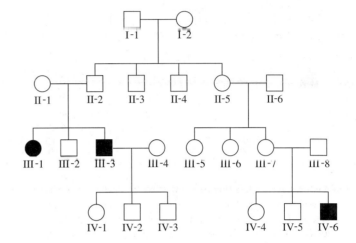

C11. Like Hurler syndrome, Fabry disease involves an abnormal accumulation of substances within lysosomes. However, the lysosomes of people with Fabry disease show an abnormal accumulation of lipids. The defective enzyme is α-galactosidase A, which is a lysosomal enzyme that functions in lipid metabolism. This defect causes cell damage, especially to the kidneys, heart, and eyes. The gene that encodes α-galactosidase A is found on the X chromosome. Let's suppose a phenotypically normal couple produces two sons with Fabry disease and one normal daughter. What is the probability that the daughter (when she grows up) will have an affected son?

C12. Achondroplasia is a rare form of dwarfism. It is caused by an autosomal dominant mutation that affects the gene that encodes a fibroblast growth factor receptor. Among 1,422,000 live births, the number of babies born with achondroplasia was 31. Among those 31 babies, 18 of them had one parent with achondroplasia. The remaining babies had two unaffected parents. How do you explain these 13 babies, assuming that the mutant allele has 100% penetrance? What are the odds that these 13 individuals will pass this mutant gene to their offspring?

C13. Lesch-Nyhan syndrome is due to a mutation in a gene that encodes a protein called hypoxanthine-guanine phosphoribosyltransferase (HPRT). HPRT is an enzyme that functions in purine metabolism. Persons afflicted with this syndrome have severe neurodegeneration and loss of motor control. The pedigree shown here contains several members with Lesch-Nyhan syndrome. Affected members are shown with black symbols. Based on this pedigree, does this syndrome appear to be an autosomal recessive, autosomal dominant,

X-linked recessive, or X-linked dominant trait? Explain your reasoning.

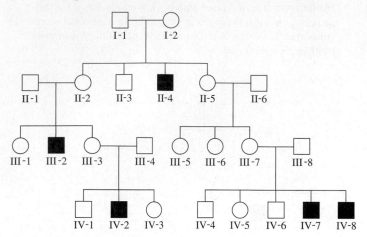

C14. Marfan syndrome is due to a mutation in a gene that encodes a protein called fibrillin-1. It is inherited as a dominant trait. The fibrillin-1 protein is the main constituent of extracellular microfibrils. These microfibrils can exist as individual fibers or associate with a protein called elastin to form elastic fibers. The gene that encodes fibrillin-1 is located on the long arm of chromosome 15. Individuals tend to be tall and thin and may have abnormalities in the skeletal, eye, and cardiovascular systems. It has been suggested that Abraham Lincoln may have been afflicted with this disorder. Let's suppose a phenotypically normal woman has a child with a man who has Marfan syndrome.

A. What is the probability that this child will have the disease?

B. If this couple has three children, what is the probability that none of them will have Marfan syndrome?

C15. Sandhoff disease is due to a mutation in a gene that encodes a protein called hexosaminidase B. This disease has symptoms that are similar to Tay-Sachs disease. Weakness begins in the first 6 months of life. Individuals exhibit early blindness and progressive mental and motor deterioration. The pedigree shown here contains several members with Sandhoff disease. Affected members are shown with black symbols.

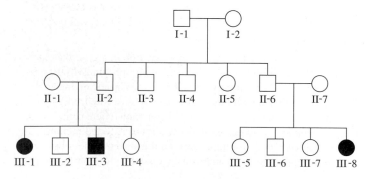

A. Based on this pedigree, does this syndrome appear to be an autosomal recessive, autosomal dominant, X-linked recessive, or X-linked dominant trait? Explain your reasoning.

B. What is the likelihood that II-1, II-2, II-3, II-4, II-5, II-6, and II-7 carry a mutant allele for the hexosaminidase B gene?

C16. What is a prion? Explain how a prion relies on normal cellular proteins to cause a disease such as mad cow disease.

C17. Some people have a genetic predisposition for developing prion diseases. Examples are described in table 22.5. In the case of Gerstmann-Straussler-Scheinker disease, the age of onset is typically at 30 to 50 years, and the duration of the disease (which leads to death) is about 5 years. Explain how someone can live for a relatively long time (i.e., 30 to 50 years) without symptoms and then succumb to the disease in a relatively short time (i.e., 5 years).

C18. Familial fatal insomnia (described in table 22.5) is inherited as an autosomal dominant trait. Researchers have identified the *PrP* gene, located on the short arm of chromosome 20, and tried to understand the relationship between the *PrP* gene sequence and the molecular basis of familial fatal insomnia.

A. It has been found that a rare mutation at codon 178, changing an aspartic acid to an asparagine, can cause this disease. In addition, codon 129 seems to play a role. The normal human population is polymorphic at codon 129, which may encode valine or methionine. If codon 178 is the normal aspartic acid codon, it does not seem to matter if valine or methionine is found at position 129. In other words, if codon 178 is aspartic acid, Met-129 and Val-129 codons are not associated with disease symptoms. However, if codon 178 is a mutant codon that specifies asparagine, then it does matter. Familial fatal insomnia seems to require a methionine at codon 129 (and an asparagine codon at position 178 in the same gene). Suggest a possible reason why this is the case.

B. Also, researchers have compared the sequences of the *PrP* gene in many people with familial fatal insomnia. Keep in mind that this is a dominant autosomal trait, so that people with this disorder have one mutant copy of the *PrP* gene and one normal copy. People with familial fatal insomnia must have one abnormal copy of the gene that contains Met-129 and Asn-178. The second copy of the gene has Asp-178, and it may have Met-129 or Val-129. Although the studies are fairly recent, some results suggest that people having a Met-129 codon in this second copy of the *PrP* gene causes the disease to develop more rapidly than people who have Val-129 in the second copy. Propose an explanation why disease symptoms may occur more rapidly when Met-129 is found in the second copy of the *PrP* gene.

C19. What is the difference between an oncogene and a tumor-suppressor gene? Give two examples.

C20. What is a proto-oncogene? What are the typical functions of proteins encoded by proto-oncogenes? At the level of protein function, what are the general ways that proto-oncogenes can be converted into oncogenes?

C21. What is a retroviral oncogene? Is it necessary for viral infection and proliferation? How have retroviruses acquired oncogenes?

C22. A genetic predisposition to develop cancer is usually inherited as a dominant trait. At the level of cellular function, are the alleles involved actually dominant? Explain why some individuals who have inherited these dominant alleles do not develop cancer during their lifetimes.

C23. Describe the types of genetic changes that commonly convert a proto-oncogene into an oncogene. Give three examples. Explain how the genetic changes are expected to alter the activity of the gene product.

C24. Relatively few inherited forms of cancer involve the inheritance of mutant oncogenes. Instead, most inherited forms of cancer are defects in tumor-suppressor genes or DNA repair genes. Give two or more reasons why we seldom see inherited forms of cancer involving activated oncogenes.

C25. The *rb* gene encodes a protein that inhibits E2F, a transcription factor that activates several genes involved in cell division. Mutations in *rb* are associated with certain forms of cancer, such as retinoblastoma. Under each of the following conditions, would you expect cancer to occur?

A. One copy of *rb* is defective; both copies of *E2F* are normal.

B. Both copies of *rb* are defective; both copies of *E2F* are normal.

C. Both copies of *rb* are defective; one copy of *E2F* is defective.

D. Both copies of *rb* and *E2F* are defective.

C26. A *p53* knockout mouse (i.e., both copies of *p53* are defective) has been produced by researchers. This type of mouse appears normal at birth. However, it is highly sensitive to UV light. Based on your knowledge of *p53*, explain the normal appearance at birth and the high sensitivity to UV light.

C27. With regard to cancer cells, which of the following statements are true?

A. Cancer cells are clonal, which means that they are derived from a single mutant cell.

B. To become cancerous, cells usually accumulate multiple genetic changes that eventually result in uncontrolled growth.

C. Most cancers are caused by oncogenic viruses.

D. Cancer cells have lost the ability to properly regulate cell division.

C28. When the DNA of a human cell becomes damaged, this leads to the activation of the *p53* gene. What is the general function of the p53 protein (i.e., is it an enzyme, transcription factor, cell cycle protein, etc.)? Describe three ways that the synthesis of the p53 protein affects cellular function. Why does an organism want these three things to happen when a cell's DNA has been damaged?

Experimental Questions

E1. Which of the following experimental observations would suggest that a disease has a genetic basis?

A. The frequency of the disease is less likely in relatives that live apart compared to relatives that live together.

B. The frequency of the disease is unusually high in a small group of genetically related individuals who live in southern Spain.

C. The disease symptoms usually begin around the age of 40.

D. It is more likely that both monozygotic twins will be affected by the disease compared to dizygotic twins.

E2. At the beginning of chapter 22, we discussed the types of experimental observations that suggest a disease is inherited. Which of these observations do you find the least convincing? The most convincing? Explain your answer.

E3. What is meant by the term *genetic testing*? What is different between testing at the protein level versus testing at the DNA level? Describe five different techniques used in genetic testing.

E4. A particular disease is found in a group of South American Indians. During the 1920s, many of these people migrated to Central America. In the Central American group, the disease is never found. Discuss whether or not you think the disease has a genetic component. What types of further observations would you make?

E5. Chapter 18 describes a blotting method known as Western blotting that can be used to detect a polypeptide that is translated from a particular mRNA. In this method, a particular polypeptide or protein is detected by an antibody that specifically recognizes a segment of its amino acid sequence. After the antibody binds to the polypeptide within a gel, a secondary antibody (which is labeled) is used to visualize the polypeptide as a dark band. For example, an antibody that recognizes α-galactosidase A could be used to specifically detect the amount of α-galactosidase A protein on a gel. α-galactosidase A is the enzyme that is defective in people with Fabry disease. It is inherited as an X-linked recessive disorder. Amy, Nan,

and Pete are brother and sisters, and Pete has Fabry disease. Aileen, Jason, and Jerry are brothers and sister, and Jerry has Fabry disease. Amy, Nan, and Pete are not related to Aileen, Jason, and Jerry. Amy, Nan, and Aileen are concerned that they could be carriers of a defective α-galactosidase A gene. A sample of cells was obtained from each of these six individuals and subjected to Western blotting, using an antibody against α-galactosidase A. Samples were also obtained from two unrelated normal females (lanes 7 and 8). The results are shown here.

Samples from:

Lane 1. Amy
Lane 2. Nan
Lane 3. Pete
Lane 4. Aileen
Lane 5. Jason
Lane 6. Jerry
Lane 7. Normal female
Lane 8. Normal female

Note: Due to X inactivation in females, the amount of expression of genes on the single X chromosome in males is equal to the amount of expression from genes on both X chromosomes in females.

A. Explain the type of mutation (i.e., missense, nonsense, promoter, etc.) that causes Fabry disease in Pete and Jerry.

B. What would you tell Amy, Nan, and Aileen regarding the likelihood that they are carriers of the mutant allele and the probability of having affected offspring?

E6. An experimental assay for the blood clotting protein called factor IX is available. A blood sample was obtained from each member of the pedigree shown here. The amount of factor IX protein is shown within the symbol of each member. The number is the percentage observed in normal individuals who do not carry a mutant copy of the gene.

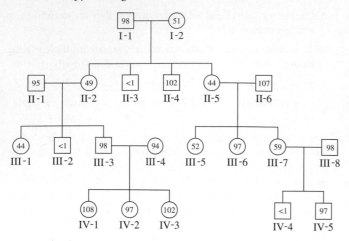

What are the likely genotypes of each member of this pedigree?

E7. Let's suppose a cell line has become malignant because it has accumulated mutations that inactivate two different tumor-suppressor genes. A researcher follows the protocol described in the experiment of figure 22.9 and isolates DNA from this mutant cell line. The DNA is used to transform normal NIH3T3 cells. What results would you expect? Would you expect a high number of malignant foci, or not? Explain your answer.

E8. What is a transformed cell? Describe three different methods to transform cells in a laboratory.

E9. In the experiment of figure 22.9, what would be the results if the DNA sample had been treated with either DNase, RNase, or protease prior to the treatment with calcium and phosphate ions?

E10. Explain how the experimental study of oncogenic viruses has increased our understanding of cancer.

E11. Discuss ways to distinguish whether a particular form of cancer involves an inherited predisposition or is due strictly to (postzygotic) somatic mutations. In your answer, consider that only one mutation may be inherited, but the cancer might develop only after several somatic mutations.

E12. The codon change (Gly-12 to Val-12) in normal human *rasH* that converts it to oncogenic *rasH* has been associated with many types of cancers. For this reason, researchers would like to develop drugs to inhibit oncogenic *rasH*. Based on your molecular understanding of the Ras protein, what types of drugs might you develop? In other words, what would be the structure of the drugs, and how would they inhibit Ras protein? How would you test the efficacy of the drugs? What might be some side effects?

E13. Describe how normal mammalian cells grow on solid media in a laboratory and how cancer cells grow. Is a malignant focus derived from a single cancer cell or from many independent cancer cells that happen to be in the same region of a tissue culture plate?

Questions for Student Discussion/Collaboration

1. Make a list of the benefits that may arise from genetic testing as well as possible negative consequences. Discuss the items on your list.

2. Our government has finite funds to devote to cancer research. Discuss which aspects of cancer biology you would spend the most money pursuing.

   A. Identifying and characterizing oncogenes and tumor-suppressor genes

   B. Identifying agents in our environment that cause cancer

   C. Identifying viruses that cause cancer

   D. Devising methods aimed at killing cancer cells in the body

   E. Informing the public of the risks involved in exposure to carcinogens

In the long run, which of these areas would you expect to be the most effective in decreasing mortality due to human cancer?

Note: All answers appear at the website for this textbook; the answers to even-numbered questions are in the back of the textbook.

www.mhhe.com/brooker

Visit the Online Learning Center for practice tests, answer keys, and other learning aids for this chapter. Enhance your understanding of genetics with our interactive exercises, web links, news feeds, tutorial service, and much more.

DEVELOPMENTAL GENETICS

::

23

Multicellular organisms, such as plants and animals, begin their lives with a fairly simple organization (namely, a fertilized egg) and then proceed step by step to a much more complex arrangement. As this occurs, cells divide, migrate, and change their characteristics, as they become highly specialized units within a multicellular individual. In an adult, each cell plays its own particular role for the good of the entire individual. In animals, for example, muscle cells allow an organism to move, while intestinal cells facilitate the absorption of nutrients. This division of labor among the various cells and organs of the individual works collectively to promote its survival.

Developmental genetics, currently one of the hottest fields in molecular biology, is concerned with the roles that genes play in orchestrating the changes that occur during development. In chapter 23, we will examine how the sequential actions of genes provide a program for the development of an organism from a fertilized egg to an adult. The last couple of decades have seen staggering advances in our understanding of developmental genetics at the molecular level. Scientists have chosen a few experimental organisms, such as the fruit fly, nematode, frog, mouse, and *Arabidopsis,* and worked toward the identification and characterization of the genes required for running their developmental programs. In certain organisms, notably the fruit fly, most of the genes that play a critical role in the early stages of development have been identified. Researchers are now exploring how the proteins encoded by these genes control the course of development. In chapter 23, we will consider several examples in which geneticists understand how the actions of genes govern the developmental process.

23.1 INVERTEBRATE DEVELOPMENT

We will begin our discussion of multicellular development by considering two model organisms, *Drosophila melanogaster* and the nematode *Caenorhabditis elegans*, that have been pivotal in our understanding of developmental genetics. As we first saw in chapter 3, *Drosophila* has been a favorite subject of geneticists since 1910, when Thomas Hunt Morgan isolated his first white-eyed mutant. The fruit fly has been used to determine many of the fundamental principles of genetics, including the chromosome theory of inheritance, the random mutation theory, and linkage mapping, to mention a few.

In developmental biology, *Drosophila* is also useful, for a variety of reasons. First, researchers have identified many mutant strains with altered developmental pathways. The techniques for generating and analyzing mutants in this organism are more advanced than in any other animal. Second, at the embryonic and larval stages, *Drosophila* is large enough to conduct transplantation experiments, yet small enough to determine where particular genes are expressed at critical stages of development.

By comparison, *C. elegans* is used by developmental geneticists because of its simplicity. The adult organism is a small transparent worm composed of only about 1,000 somatic cells. Starting with the fertilized egg, the pattern of cell division and the fate of each cell within the embryo are completely known.

In this section, we will begin by examining the general features of *Drosophila* development. We will then focus our attention on embryonic development (embryogenesis), because it is during this stage that the overall body plan is determined. We will see how the expression of particular genes and the localization of gene products within the embryo influence the developmental process. We will then briefly consider development in *C. elegans*. In this organism, we will examine how the timing of gene expression plays a key role in determining the developmental fate of particular cells.

The Early Stages of Embryonic Development Determine the Pattern of Structures in the Adult Organism

Multicellular development in plants and animals follows a body plan or pattern. The term **pattern** refers to the spatial arrangement of different regions of the body. At the cellular level, the body pattern is due to the arrangement of cells and their specialization.

The progressive growth of a fertilized egg into an adult organism involves four types of cellular events: cell division, cell movement, cell differentiation, and cell death. The coordination of these four events leads to the formation of a body with a particular pattern. As we will see, the temporal expression of genes and the localization of gene products at precise regions in the fertilized egg and early embryo are the critical phenomena that underlie this coordination.

Figure 23.1 illustrates the general sequence of events in *Drosophila* development. The oocyte is the most critical cell in determining the pattern of development in the adult organism. It is an elongated cell with preestablished axes and a well-defined cytoplasmic organization (fig. 23.1a). After fertilization takes place, the zygote goes through a series of nuclear divisions that are not accompanied by cytoplasmic division. Initially, the resulting nuclei are scattered throughout the yolk, but eventually they migrate to the periphery. This is the syncytial blastoderm stage (fig. 23.1b).

After the nuclei have lined up along the cell membrane, individual cells are formed as portions of the cell membrane surround each nucleus; this creates a cellular blastoderm (fig. 23.1c). This structure is composed of a sheet of cells on the outside with yolk in the center. In this arrangement, the cells are distributed asymmetrically. At the posterior end are a group of cells called the pole cells. These are the primordial germ cells that eventually give rise to gametes in the adult organism.

After blastoderm formation is complete, dramatic changes occur during gastrulation (fig. 23.1d). This stage involves a great deal of **cell migration,** which produces three cell layers known as the ectoderm, mesoderm, and endoderm. In general, the ectoderm remains on the outside of the gastrula, the endoderm is on the inside, and the mesoderm is wedged in the middle.

As this process occurs, the embryo begins to be subdivided into detectable units. Initially, shallow grooves divide the embryo into 14 **parasegments.** However, this is a transient condition. A short time later, these grooves disappear, and new boundaries are formed that divide the embryo into morphologically discrete **segments.** Figure 23.1e shows the segmented pattern of a *Drosophila* embryo about 10 hours after fertilization. Later in this section, we will explore how the coordination of gene expression underlies the formation of these parasegments and segments.

At the end of **embryogenesis,** a larva hatches from the egg (fig. 23.1f) and begins feeding on its own. In *Drosophila*, there are three larval stages, separated by molts. During molting, the larva sheds its cuticle, a hardened extracellular shell that is secreted by the epidermis.

After the third larval stage, *Drosophila* proceeds through a process known as metamorphosis. This occurs during the pupal stage. Groups of cells called imaginal disks were produced earlier in development. During metamorphosis, these imaginal disks grow and differentiate into the structures found in the adult fly (e.g., head, wings, legs, abdomen). The adult fly then emerges from the pupal case.

In metazoa (i.e., animals that are more complex than unicellular protozoa), the final result of development is an adult body organized along three axes: the **dorso-ventral axis,** the **antero-posterior axis,** and the **right-left axis** (fig. 23.1g). An additional axis, used mostly for designating limb parts, is the **proximo-distal axis.**

Although many interesting developmental events occur during the three larval stages and the pupal stage, we will focus most of our attention on the events that occur during embryonic development. Even before hatching, the embryo develops the basic body plan that will be found in the adult organism. In other words, during the early stages of development, the embryo is divided into segments that correspond to the segments of the larva

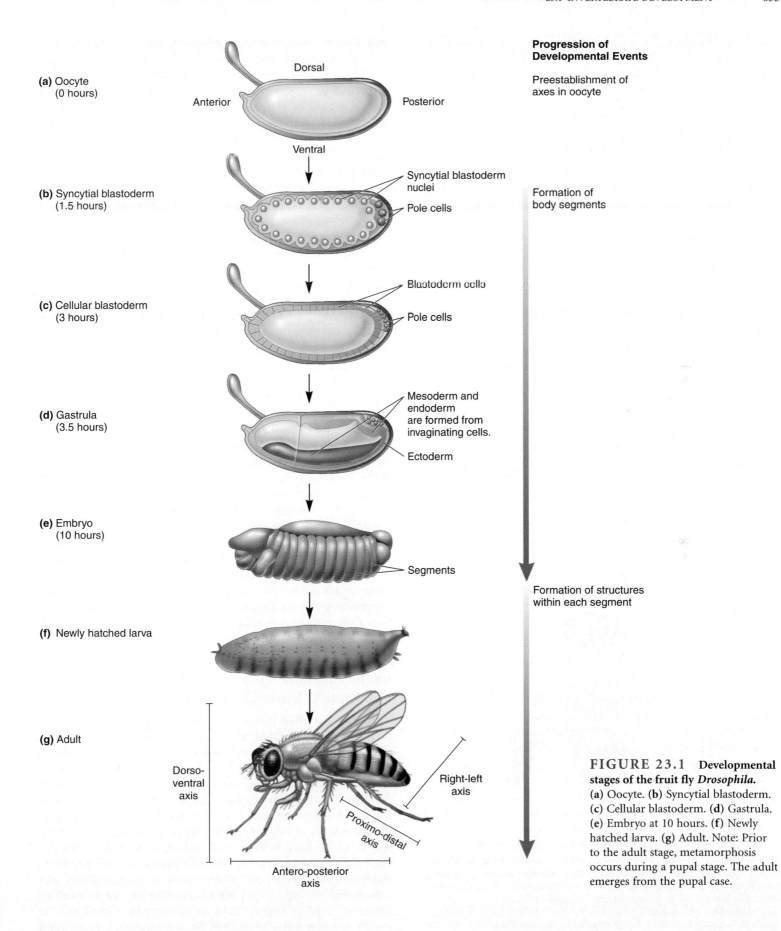

Progression of Developmental Events

Preestablishment of axes in oocyte

Formation of body segments

Formation of structures within each segment

(a) Oocyte
(0 hours)

Dorsal

Anterior

Posterior

Ventral

(b) Syncytial blastoderm
(1.5 hours)

Syncytial blastoderm nuclei

Pole cells

(c) Cellular blastoderm
(3 hours)

Blastoderm cells

Pole cells

(d) Gastrula
(3.5 hours)

Mesoderm and endoderm are formed from invaginating cells.

Ectoderm

(e) Embryo
(10 hours)

Segments

(f) Newly hatched larva

(g) Adult

Dorso-ventral axis

Right-left axis

Proximo-distal axis

Antero-posterior axis

FIGURE 23.1   Developmental stages of the fruit fly *Drosophila*. (a) Oocyte. (b) Syncytial blastoderm. (c) Cellular blastoderm. (d) Gastrula. (e) Embryo at 10 hours. (f) Newly hatched larva. (g) Adult. Note: Prior to the adult stage, metamorphosis occurs during a pupal stage. The adult emerges from the pupal case.

and adult. Therefore, an understanding of how these segments form in the embryo is critical to our understanding of pattern formation.

The Study of *Drosophila* Mutants with Disrupted Developmental Patterns Has Identified Genes That Control Development

Mutations that alter the course of *Drosophila* development have greatly contributed to our understanding of the normal process of development. For example, figure 23.2 shows an example of mutations in a gene called *Ultrabithorax*. This mutant fly has four wings instead of two; the halteres (balancing organs that resemble miniature wings), which are found on the third thoracic segment, are changed into wings, normally found on the second thoracic segment. The term *bithorax* refers to the observation that the characteristics of the second thoracic segment are duplicated.

Edward Lewis, a pioneer in the genetic study of development, became interested in the bithorax phenotype and began investigating it in 1946. He discovered that the mutant chromosomal region actually contained a complex of three genes involved in specifying developmental pathways in the fly. A gene that plays a central role in specifying the final identity of a body region is called a **homeotic gene.** We will discuss particular examples of homeotic genes later in chapter 23.

During the 1960s and 1970s, interest in the relationship between genetics and embryology blossomed as biologists began to appreciate the role of genetics at the molecular and cellular levels. It soon became clear that the genome of multicellular organisms contains groups of genes that initiate a program of development involving networks of gene regulation. By identifying mutant alleles that disrupt development, geneticists have begun to unravel the pattern of gene expression that underlies the normal pattern of multicellular development.

Early in development, a category of genes known as **segmentation genes** plays a role in the formation of body segments. The expression of segmentation genes in specific regions of the embryo causes it to become segmented, as shown earlier in figure 23.1e. In the 1970s, Christiane Nusslein-Volhard and Eric Wieschaus undertook a systematic search for *Drosophila* mutants with disrupted segmentation patterns. It was their pioneering effort that identified most of the genes required for the embryo to develop a segmented pattern. They identified three classes of segmentation genes: **gap genes, pair-rule genes,** and **segment-polarity genes.**

Figure 23.3 illustrates a few of the interesting phenotypic effects that Nusslein-Volhard and Wieschaus observed when a particular segmentation gene is defective. The *gray boxes* indicate the regions that are missing in the resulting larvae. When a mutation inactivates a gap gene, a contiguous section of the larva is missing (fig. 23.3a). In other words, there is a gap of several segments. For example, when the *Krüppel* gene is defective, about eight segments are missing from the larva. By comparison, a defect in a pair-rule gene causes alternating parasegments to be deleted (fig. 23.3b). For example, when the *even-skipped* gene is defective, the odd-numbered parasegments are missing. Finally, segment-polarity mutations cause portions of segments to be missing either an anterior or a posterior region. Figure 23.3c shows a mutation in a segmentation gene known as *gooseberry*. When this gene is defective, the anterior portion of each segment is missing from the larva. In the case of segment-polarity mutants, the segments adjacent to the deleted regions exhibit a mirror-image duplication.

Overall, the phenotypic effects of mutant segmentation genes provide geneticists with important clues regarding the roles of these genes in the developmental process of segmentation. Later in this section, we will examine when and where the segmentation genes are expressed and how their pattern of expression leads to the segmentation of the embryo.

The Generation of a Body Pattern Depends on the Positional Information That Each Cell Receives During Development

The identification of mutant alleles that disrupt the developmental process has permitted great insight into the genes controlling pattern formation. Before we consider how these genes function, let's consider a central concept in developmental biology known as **positional information.** For an organism to develop into a segmented pattern with unique morphological and cellular features, each cell of the body must somehow become the appropriate cell type based on its position relative to the other cells. A cell may respond to positional information in various ways. For example, positional information may stimulate a cell to divide into two daughter cells, or it may cause a cell to differentiate into a particular cell type. Positional information may cause a cell or group of cells to migrate in a particular direction from one region of the embryo to another. Finally, positional information

FIGURE 23.2   The bithorax mutation in *Drosophila*.

GENES→TRAITS   A normal fly contains two wings on the second thoracic segment and two halteres on the third thoracic segment. However, this mutant fly contains multiple mutations in a homeotic gene called *Ultrabithorax*. (The function of the *Drosophila* homeotic genes is discussed later in chapter 23.) In this fly, the third thoracic segment has the same characteristics as the second thoracic segment, thereby producing a fly with four wings instead of the normal number of two.

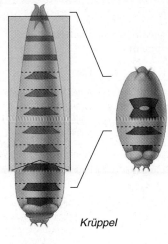

Krüppel

(a) Gap

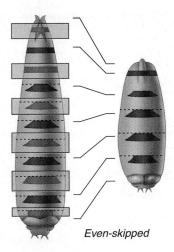

Even-skipped

(b) Pair-rule

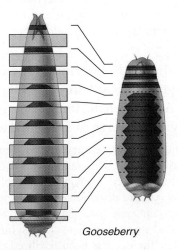

Gooseberry

(c) Segment-polarity

FIGURE 23.3 **Phenotypic effects in *Drosophila* larvae that have mutations in segmentation genes.** Effects shown are caused by defects in **(a)** gap genes, **(b)** pair-rule genes, and **(c)** segment-polarity genes.

may cause a cell to die. This process, known as **apoptosis,** is a necessary event during normal development. For example, in the later stages of development of the retina in *Drosophila*, excess cells are eliminated by apoptosis. Also, during mammalian development, individual fingers and toes are created when embryonic cells that lie between each future finger and toe undergo apoptosis.

Molecules that convey positional information and promote developmental changes are known as **morphogens.** A morphogen influences the developmental fate of a cell. It provides the positional information that stimulates a cell to divide, differentiate, migrate, or die. For example, as we will learn later, when regions of a *Drosophila* embryo are exposed to a high concentration of the morphogen known as Bicoid, they differentiate into structures characteristic of the anterior region of the body.

Within an oocyte and during embryonic development, morphogens typically are distributed along a concentration gradient (fig. 23.4). In other words, the concentration of a morphogen varies from low to high in different regions of the organism. A key feature of morphogens is that they act in a concentration-dependent manner. At a high concentration, a morphogen will restrict a cell into a particular developmental pathway; at a lower concentration, it will not. There is often a critical **threshold concentration** above which the morphogen will exert its effects but below which it is ineffective.

During the earliest stages of development, several morphogenic gradients are preestablished within the oocyte (fig. 23.4*a*). At the cellular blastoderm stage, the zygote subdivides into many smaller cells. Due to the preestablished gradient of morphogens within the oocyte, these smaller cells have higher or lower concentrations of morphogens, depending on their location. In this way, the morphogen gradients in the oocyte can provide positional information that is important in establishing the general polarity of an embryo along two main axes: the antero-posterior axis and the dorso-ventral axis. This topic will be described in greater detail later in chapter 23.

A morphogen gradient can also be established by cell secretion and transport. A certain cell or group of cells may synthesize and secrete a morphogen at a specific stage of development. After secretion, the morphogen may be transported to neighboring cells. The concentration of the morphogen is usually highest near the cells that secrete it (fig. 23.4*b*). The morphogen may then influence the developmental fate of the cells that are exposed to it. The process by which a cell or group of cells governs the developmental fate of neighboring cells is known as **induction.**

In addition to morphogens, positional information is conveyed by **cell adhesion** (fig. 23.4*c*). Each cell makes its own collection of surface receptors; these receptors cause it to adhere to other cells and/or to the extracellular matrix (ECM), which consists primarily of carbohydrates and fibrous proteins. Such receptors are known as **cell adhesion molecules (CAMs).** A cell may gain positional information via the combination of contacts it makes with other cells or with the ECM.

The phenomenon of cell adhesion, and its role in multicellular development, was first recognized by H. V. Wilson in 1907. He took multicellular sponges and disaggregated them into individual

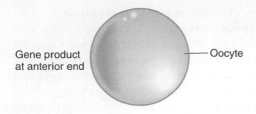

Gene product at anterior end — Oocyte

(a) Asymmetric distribution of morphogens in the oocyte

Cellularized embryo

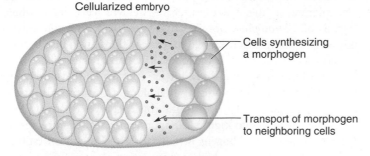

Cells synthesizing a morphogen

Transport of morphogen to neighboring cells

(b) Asymmetric synthesis and extracellular distribution of a morphogen

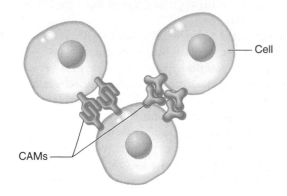

Cell

CAMs

(c) Cell-to-cell contact conveys positional information

FIGURE 23.4 Three molecular mechanisms of positional information.

cells. Remarkably, the cells actively migrated until they adhered to one another to form a new sponge, complete with the chambers and canals that characterize a sponge's internal structure! When sponge cells from different species were mixed, they sorted themselves properly, adhering only to cells of the same species. Overall, these results indicate that cell adhesion plays an important role in governing the position that a cell will adopt during development.

Figure 23.4 provides an overview of the general phenomena that underlie positional information during embryonic development. With these ideas in mind, we can begin to examine how specific genes in *Drosophila* encode morphogens that convey positional information. In this organism, the establishment of the body axes and division of the fly into segments involves the participation of a few dozen genes. Table 23.1 lists many of the important genes governing pattern formation during embryonic development. These genes are often given interesting names based on the observed phenotype when they are mutant. It is

TABLE 23.1
Examples of *Drosophila* Genes That Play a Role in Pattern Development

Description	Examples
Some genes play a role in determining the axes of development. Also, certain genes govern the formation of the extreme terminal (anterior and posterior) regions.	Anterior: *bicoid, exuperantia, hunchback, swallow, staufen* Posterior: *nanos, cappuccino, oskar, pumilio, spire, staufen, tudor, vasa* Terminal: *torso, torsolike, Trunk, NTF-1* Dorso-ventral: *Toll, cactus, dorsal easter, gurken, nudel, pelle, pipe, snake, spatzle*
Some genes play a role in promoting the subdivision of the embryo into segments. These are called segmentation genes. As described, there are three types of segmentation genes known as gap genes, pair-rule genes, and segment-polarity genes.	Gap genes: *empty spiracles, giant, huckebein, hunchback, knirps, Krüppel, tailless, orthodenticle* Pair-rule genes: *even-skipped, hairy, runt, fushi tarazu, paired* Segment-polarity genes: *frizzled, frizzled-2, engrailed, patched, smoothened, hedgehog, wingless*
Some genes play a role in determining the fate of particular segments. These are known as homeotic genes, and *Drosophila* has two clusters of homeotic genes known as the *Antennapedia* complex and the *bithorax* complex.	*Antennapedia* complex: *labial, proboscipedia, Deformed, Sex combs reduced, Antennapedia P* *Bithorax* complex: *Ultrabithorax, abdominal A, Abdominal B*

Adapted from Kalthoff, K. (2001) *Analysis of Biological Development*, 2d ed. McGraw-Hill, New York.

beyond the scope of this textbook to examine how all of these genes exert their effects during embryonic development. Instead, we will consider a few examples that illustrate how the expression of a particular gene and the localization of its gene product have a defined effect on the pattern of development.

The Gene Products of Maternal Effect Genes Are Deposited Asymmetrically into the Oocyte and Establish the Antero-posterior and Dorso-ventral Axes at a Very Early Stage of Development

The first stage in *Drosophila* embryonic pattern development is the establishment of the body axes. This occurs before the embryo becomes segmented. In fact, the morphogens necessary to establish these axes are distributed prior to fertilization. During oogenesis, certain gene products, which are important in early developmental stages, are deposited asymmetrically within the egg. Later, after the egg has been fertilized and development begins, these gene products will establish independent developmental programs that govern the formation of the body axes of the embryo. The products of several different genes ensure that proper development will occur.

As shown in figure 23.5, a few gene products act as key morphogens, or receptors for morphogens, that initiate changes in embryonic development. As shown here, these gene products are deposited asymmetrically in the egg. For example, the prod-

FIGURE 23.5 **The establishment of the axes of polarity in the *Drosophila* embryo.** This figure shows some of the gene products that are critical in the establishment of the antero-posterior, terminal, and dorso-ventral axes. (**a**) *Bicoid* mRNA is distributed in the anterior end of the oocyte and promotes the formation of anterior structures. (**b**) *Nanos* mRNA is localized to the posterior end and promotes the formation of posterior structures. (**c**) The Torso receptor protein is found in the membrane at either end of the oocyte and causes the formation of structures that are found only at the ends of the organism. (**d**) The Toll receptor protein is activated at the ventral side of the embryo and establishes the dorso-ventral axis.

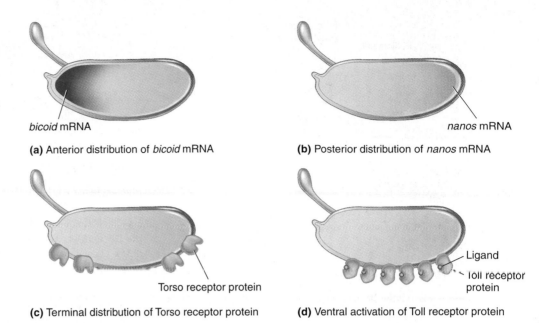

(**a**) Anterior distribution of *bicoid* mRNA

bicoid mRNA

(**b**) Posterior distribution of *nanos* mRNA

nanos mRNA

(**c**) Terminal distribution of Torso receptor protein

Torso receptor protein

(**d**) Ventral activation of Toll receptor protein

Ligand

Toll receptor protein

uct of the *bicoid* gene is necessary to initiate development of the anterior structures of the organism. During oogenesis, the mRNA for *bicoid* accumulates in the anterior region of the oocyte. In contrast, the mRNA from the *nanos* gene accumulates in the posterior end. Later in development, the *nanos* mRNA will be translated into protein, which functions to influence posterior development. *Nanos* is required for the formation of the abdomen.

In addition to *bicoid* and *nanos,* the development of the structures at the extreme anterior and posterior ends of the embryo are regulated in part by a receptor protein called Torso. This receptor, which is activated only at the anterior and posterior ends of the egg, is necessary for the formation of the terminal ends of the embryo. The dorso-ventral axis is governed, in part, by a receptor protein known as Toll. This receptor is synthesized throughout the embryo, but is activated by ligand binding only along the ventral midline of the embryo.

Let's now take a closer look at the molecular mechanism of one of these modulators, namely *bicoid*. The *bicoid* gene got its name because a larva defective in this gene develops with two posterior ends (fig. 23.6). This allele exhibits a maternal effect pattern of inheritance (see chapter 7). A female fly that is phenotypically normal (because its mother was heterozygous for the normal *bicoid* allele), but genotypically homozygous for an inactive *bicoid* allele (because it inherited the inactive allele from its mother and father), will produce 100% affected offspring even when mated to a male that is homozygous for the normal *bicoid* allele. In other words, the genotype of the mother determines the phenotype of the offspring. This occurs because the *bicoid* gene product is provided to the oocyte via the nurse cells.

In the ovaries of female flies, the nurse cells are localized asymmetrically toward the anterior end of the oocyte. During oogenesis, gene products are transferred into the oocyte via cell-to-cell connections called cytoplasmic bridges. Maternally encoded gene products enter one side of the oocyte (fig. 23.7a). This side will eventually become the anterior end of the embryo. The *bicoid* gene is actively transcribed in the nurse cells, and *bicoid* mRNA is transported into the anterior end of the oocyte. The 3′

end of *bicoid* mRNA contains a signal that is recognized by binding proteins thought necessary for the transport of this mRNA into the oocyte. After it enters the oocyte, the *bicoid* mRNA is trapped near the anterior side.

Figure 23.7b shows an *in situ* hybridization experiment in which a *Drosophila* egg was examined via a probe complementary to the *bicoid* mRNA. The technique of *in situ* hybridization was described in chapter 20. As seen here, the *bicoid* mRNA is highly concentrated near the anterior pole of the egg cell. When the *bicoid* mRNA subsequently is translated, a gradient of Bicoid protein is established as shown in figure 23.7c.

After fertilization occurs, the Bicoid protein functions as a transcription factor. A remarkable feature of this protein is that its ability to influence gene expression is tuned exquisitely to its concentration. Depending on the distribution of the Bicoid protein, this transcription factor will activate only genes in certain regions of the embryo. For example, Bicoid stimulates a gene

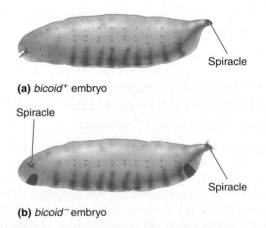

(**a**) *bicoid*⁺ embryo

Spiracle

Spiracle

(**b**) *bicoid*⁻ embryo

Spiracle

FIGURE 23.6 **The *bicoid* mutation in *Drosophila.*** (**a**) A normal *bicoid*⁺ embryo. (**b**) A *bicoid*⁻ embryo in which both ends of the larva develop posterior structures. For example, both ends develop a spiracle, which normally is found only at the posterior end.

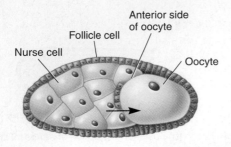

(a) Transport of maternal effect gene products into the oocyte

(b) *In situ* hybridization of *bicoid* mRNA

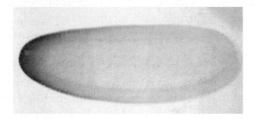

(c) Immunostaining of Bicoid protein

FIGURE 23.7 **Asymmetrical localization of gene products during oogenesis in *Drosophila*.** (a) The nurse cells transport gene products into the anterior end of the developing oocyte. (b) An *in situ* hybridization experiment showing that the *bicoid* mRNA is trapped near the anterior end. (c) The *bicoid* mRNA is translated into protein soon after fertilization. The location of the Bicoid protein is revealed by immunostaining using an antibody that specifically recognizes this protein.

called *hunchback* in the anterior half of the embryo, but its concentration is too low in the posterior half to activate the *hunchback* gene.

Gap, Pair-Rule, and Segment-Polarity Genes Act Sequentially to Divide the *Drosophila* Embryo into Segments

After the antero-posterior, dorso-ventral, and terminal regions of the embryo have been established by maternal effect genes, the next developmental process organizes the embryo transiently into parasegments and then permanently into segments. The segmentation pattern of the embryo is shown in figure 23.8. This pattern of positional information will be maintained, or "remembered," throughout the rest of development. In other words, each segment of the embryo will give rise to unique morphological features in the adult. For example, T2 will become a thoracic seg-

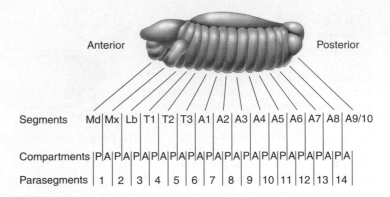

Segments	Md	Mx	Lb	T1	T2	T3	A1	A2	A3	A4	A5	A6	A7	A8	A9/10
Compartments	P A	P A	P A	P A	P A	P A	P A	P A	P A	P A	P A	P A	P A	P A	P A
Parasegments	1	2	3	4	5	6	7	8	9	10	11	12	13	14	

FIGURE 23.8 **A comparison of segments and parasegments in the *Drosophila* embryo.** Note that the parasegments and segments are out of register. The posterior (P) and anterior (A) regions are shown for each segment.

ment with a pair of legs and a pair of wings, and A8 will become a segment of the abdomen.

Figure 23.8 shows the overlapping relationship between parasegments and segments. As seen here, the boundaries of the segments are out of register with the boundaries of the parasegments. An appreciation of this feature is critical to our understanding of segmentation. From the viewpoint of genes, we will see that the parasegments are the locations where gene expression is controlled spatially. The anterior compartment of each segment coincides with the posterior region of a parasegment; the posterior compartment of a segment coincides with an anterior region of the next parasegment. The pattern of gene expression that occurs in the anterior region of one parasegment and the posterior region of an adjacent parasegment results in the formation of a segment.

Now that we have a general understanding of the way the *Drosophila* embryo is subdivided, we can examine how particular genes cause it to become segmented into this pattern. As mentioned, the genes that play a role in the formation of body segments are called segmentation genes; there are three classes of segmentation genes: gap genes, pair-rule genes, and segment-polarity genes. The expression and activation patterns of these genes in specific regions of the embryo cause it to become segmented.

A partial, simplified scheme of the genetic hierarchy that leads to a segmented pattern in the *Drosophila* embryo is shown in figure 23.9. This figure presents the general sequence of events that occurs during the early stages of embryonic development. However, many more genes are actually involved in this process (refer back to table 23.1). As described in this figure, the following steps occur:

1. Maternal effect gene products, such as *bicoid* mRNA, are deposited asymmetrically into the oocyte. These gene products form a gradient that will later influence the formation of axes, such as the antero-posterior axis.

2. After fertilization, maternal effect gene products activate **zygotic genes.** In contrast to maternal effect genes, which are expressed during oogenesis, zygotic genes are expressed

after fertilization. The first zygotic genes to be activated are the gap genes. As shown in step 2 of figure 23.9, gap genes are activated as broad bands within particular regions of the embryo. These bands do not correspond to parasegments or segments within the embryo.

3. The gap genes and maternal effect genes then activate the pair-rule genes. The photograph shown in step 3 of figure 23.9 illustrates the alternating pattern of *even-skipped* gene expression. Note that pair-rule genes are expressed in particular parasegments. *Even-skipped* is expressed in the odd-numbered parasegments. Solved problem S4 at the end of chapter 23 examines how the *even-skipped* gene can be expressed in this pattern of alternating stripes.

4. Once the pair-rule genes are activated in an alternating banding arrangement, their gene products then regulate the segment-polarity genes. As shown in step 4 of figure

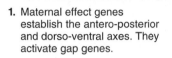

1. Maternal effect genes establish the antero-posterior and dorso-ventral axes. They activate gap genes.

Asymmetric localization of maternal effect gene products

Examples

Asymmetric localization of Bicoid protein. Other maternal effect gene products (not shown) are also asymmetrically localized.

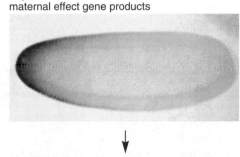

2. Gap gene products act as genetic regulators of pair-rule genes. They bind to stripe-specific enhancers that are located adjacent to pair-rule genes.

Gap gene expression

Gap gene expression occurs as broad bands in the embryo. In this photo, Krüppel protein is shown in *red* and Hunchback in *green*. Their region of overlap is *yellow*. Other gap genes (not shown) are also expressed.

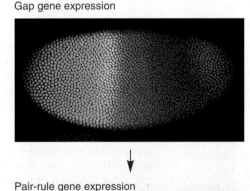

3. The expression of pair-rule genes in a stripe defines the boundary of a parasegment. Pair-rule gene products regulate the expression of segment-polarity genes.

Pair-rule gene expression

Pair-rule genes are expressed in alternating stripes. Each stripe corresponds to a parasegment. In this photo, the Even-skipped gene product is expressed in the light bands that correspond to odd-numbered parasegments.

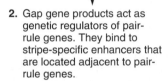
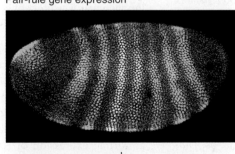

4. Segment-polarity genes define the anterior or posterior compartment of each parasegment.

Segment-polarity gene expression

Segment-polarity genes are expressed in either an anterior or posterior compartment. This photo shows the product of the *engrailed* gene in the anterior compartment of each parasegment.

FIGURE 23.9 Overview of pattern formation in *Drosophila*. Note: When comparing steps 3 and 4, the embryo has undergone a 180° turn, folding back on itself.

23.9, the segment-polarity gene *engrailed* is expressed in the anterior region of each parasegment. Another segment-polarity gene, *wingless*, is expressed in the posterior region. Later in development, the anterior end of one parasegment and the posterior end of another parasegment will develop into a segment with particular morphological characteristics.

The Expression of Homeotic Genes Controls the Phenotypic Characteristics of Segments

Thus far, we have considered how the *Drosophila* embryo becomes organized along axes and then into a segmented body pattern. Now we will examine how each segment develops its unique morphological features. Geneticists often use the term **cell fate** to describe the ultimate morphological features that a cell or group of cells will adopt. For example, the fate of the cells in segment T2 in the *Drosophila* embryo is to develop into a thoracic segment containing two legs and two wings. In *Drosophila*, the cells in each segment of the body have their fate determined at a very early stage of embryonic development, long before the morphological features become apparent.

Our understanding of developmental fate has been greatly aided by the identification of mutant genes that alter cell fates. In animals, the first mutant of this type was described by the German entomologist G. Kraatz in 1876. He observed a sawfly (*Climbex axillaris*) in which part of an antenna was replaced with a leg. During the late nineteenth century, the English zoologist William Bateson collected many of these types of observations and published them in 1894 in a book entitled *Materials for the Study of Variation Treated with Especial Regard to Discontinuity in the Origin of Species*. In this book, Bateson coined the term **homeotic** to describe mutant alleles in which one body part is replaced by another.

As mentioned earlier in chapter 23, Edward Lewis began to study strains of *Drosophila* having homeotic mutations. This work, which began in 1946, was the first systematic study of homeotic genes. Each homeotic gene controls the fate of a particular region of the body. *Drosophila* contains two clusters of homeotic genes called the *bithorax* complex and the *Antennapedia* complex. Figure 23.10 shows the organization of genes within these complexes. The Antennapedia complex contains five genes, designated *lab, pb, Dfd, Scr,* and *Antp*. The bithorax complex has three genes, *Ubx, abd-A,* and *Abd-B*. Both of these complexes are located on chromosome 3, but a large segment of DNA separates them.

As noted in figure 23.10, the order of these genes along chromosome 3 correlates with the antero-posterior axis of the body. For example, *lab* is expressed in the anterior segment and governs the formation of mouth structures. The normal *Antp* gene is expressed strongly in the thoracic region during embryonic development and controls the formation of thoracic structures. Transcription of the *Abd-B* gene occurs in the posterior region of the embryo; this gene controls the formation of the posterior-most abdominal segments.

The role of homeotic genes in determining the identity of particular segments has been revealed by mutations that alter their function. As shown in figure 23.11, the antennapedia muta-

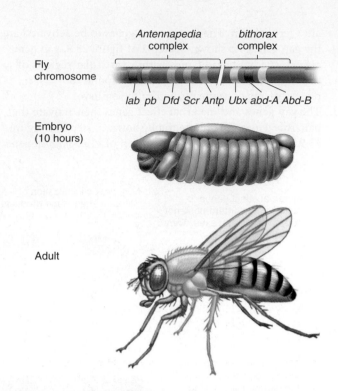

FIGURE 23.10 **Expression pattern of homeotic genes in *Drosophila*.** The order of homeotic genes, *labial* (*lab*), *proboscipedia* (*pb*), *Deformed* (*Dfd*), *Sex combs reduced* (*Scr*), *Antennapedia* (*Antp*), *Ultrabithorax* (*Ubx*), *abdominal A* (*abd-A*), and *Abdominal B* (*Abd-B*), correlates with the order of expression in the embryo. The expression pattern of four of these genes is shown. *Lab* (*purple*) is expressed in the region that will eventually give rise to mouth structures. *Dfd* (*green*) is expressed in the region that will form much of the head. *Antp* (*blue*) is expressed in embryonic segments that give rise to thoracic segments, and *Abd-B* (*yellow*) is expressed in posterior segments that will form the abdomen. The order of gene expression, from anterior to posterior, parallels the order of genes along the chromosome.

tion is a **gain-of-function mutation** in the *Antp* gene that causes it to be expressed in an additional place in the embryo. In this case, the *Antp* gene is also expressed abnormally in the anterior segment that normally gives rise to the antennae. In other words, there has been a gain of *Antp* function in this segment. The abnormal expression of *Antp* in this region causes the antennae to be converted into legs!

Investigators have also studied many loss-of-function alleles in homeotic genes. When a particular homeotic gene is defective, the region that it normally governs will usually be controlled by the homeotic gene that acts in the adjacent anterior region. For example, the *Ubx* gene normally functions within parasegments 5 and 6. If this gene is missing, this section of the fly becomes converted to the structures that are found in parasegment 4.

The homeotic genes are part of the genetic hierarchy that produces the morphological characteristics of the fly. They are regulated in a very complex way. Their expression is controlled by gap genes and pair-rule genes, and they are also regulated by interactions among themselves. Other groups of genes stabilize the patterns of homeotic gene expression after the initial pattern is established by segmentation genes. A group of genes known as

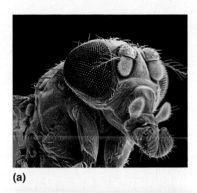

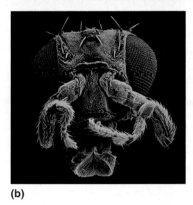

FIGURE 23.11   **The Antennapedia mutation in *Drosophila.***

GENES→TRAITS   **(a)** A normal fly with antennae. **(b)** This mutant fly has a gain-of-function mutation in which the *Antp* gene is expressed in the embryonic segment that normally gives rise to antennae. The expression of *Antp* causes this region to have legs rather than antennae.

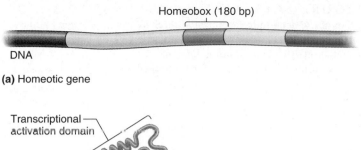

(a) Homeotic gene

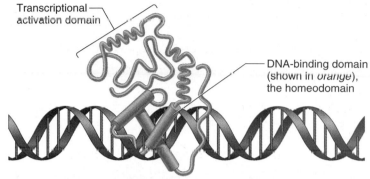

(b) Homeotic protein bound to DNA

FIGURE 23.12   **Molecular features of homeotic proteins.**
(a) A homeotic gene (shown in *tan*) contains a 180 bp sequence called the homeobox (shown in *orange*). **(b)** When a homeotic gene is expressed, it produces a protein that functions as a transcription factor. The homeobox encodes a region of the protein called a homeodomain, which binds to the major groove of DNA. These DNA-binding sites are found within genetic regulatory elements (i.e., enhancers). These enhancers are found in the vicinity of promoters that are turned on by homeotic proteins. For this to occur, the homeotic protein also contains a transcriptional activation domain, which will activate the transcription of a gene after the homeodomain has bound to the DNA.

the *Polycomb* genes represses the expression of homeotic genes in regions of the embryo where they should not act. This is accomplished by the remodeling of chromatin structure to convert it to a closed conformation that is more compact and unable to be transcribed. Alternatively, in regions where the homeotic genes are active, the products of *Trithorax* genes maintain an open conformation that is capable of transcription. Overall, the concerted actions of many gene products cause the homeotic genes to be expressed only in the appropriate region of the embryo, as shown in figure 23.10.

Because they are part of a genetic hierarchy, it is not too surprising that homeotic genes encode transcription factors. The coding sequence of homeotic genes contains a 180 bp consensus sequence, known as a **homeobox** (fig. 23.12). This sequence was first discovered in the *Antp* and *Ubx* genes, and it has since been found in all *Drosophila* homeotic genes and in some other genes affecting pattern development, such as *bicoid*. The protein domain encoded by the homeobox is called a **homeodomain.** The arrangement of α helices within the homeodomain promotes the binding of the protein to the major groove of DNA. In this way, homeotic proteins can bind to DNA in a sequence-specific manner. In addition to DNA-binding ability, homeotic proteins also contain a transcriptional activation domain that functions to activate the genes to which the homeodomain can bind.

The transcription factors encoded by homeotic genes activate the next category of genes, sometimes called **realizator genes.** These genes produce the morphological characteristics of each segment. Much current research attempts to identify realizator genes and determine how their expression in particular regions of the embryo leads to morphological changes in the embryo, larva, and adult.

In some cases, realizator genes also encode transcription factors. Presumably, such realizator genes control the expression of other sets of genes that will alter the morphological characteristics of cells. In other cases, realizator genes encode proteins involved in cell-to-cell signaling pathways. The activation of these signaling pathways is thought to play a key role in the ability of cells and groups of cells to adopt their correct morphologies. It is expected that research during the next few decades will shed considerable light on the pathways by which realizator genes control morphological changes in the fruit fly.

The Developmental Fate of Each Cell in the Nematode *Caenorhabditis elegans* Is Known

We now turn our attention to another invertebrate, *C. elegans,* that has been the subject of numerous studies in developmental genetics. The embryo develops within the eggshell and hatches when it reaches a size of 550 cells. After hatching, it continues to grow and mature as it passes through four successive molts. It takes about 3 days for a fertilized egg to develop into an adult worm.

FIGURE 23.13   **A cell lineage diagram of the nematode *Caenorhabditis elegans.*** This partial lineage diagram illustrates how the cells divide to produce different regions of the adult. The fate of the intestinal cell lineage is shown in greater detail than that of other cell lineages. A complete lineage diagram is known for this organism, although its level of detail is beyond the scope of this textbook.

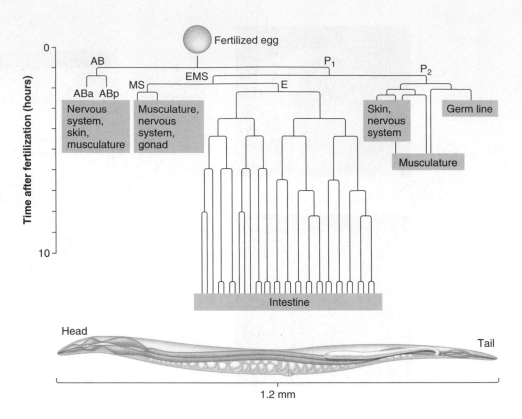

With regard to sex, *C. elegans* can be a male (and only produce sperm) or a hermaphrodite (capable of producing sperm and egg cells). An adult male is composed of 1,031 somatic cells and produces about 1,000 sperm. A hermaphrodite consists of 959 somatic cells and produces about 2,000 gametes (both sperm and eggs).

A remarkable feature of this organism is that the pattern of cellular development is extremely invariant from worm to worm. In the early 1960s, Sydney Brenner pioneered the effort to study the pattern of cell division in *C. elegans.* To do so, a researcher can identify a particular cell at an embryonic stage, follow that cell as it divides, and observe where its descendant cells will be located in the adult.

Because *C. elegans* is transparent and composed of relatively few cells, researchers can follow cell division step by step, beginning with a fertilized egg and ending with an adult worm. An illustration that depicts how cell division proceeds is called a **lineage diagram.** It describes the cell division patterns and fates

of any cell's descendants. Figure 23.13 shows a partial lineage diagram for a *C. elegans* hermaphrodite. At the first cell division, the egg divides to produce two cells, called AB and P_1. AB then divides into two cells, ABa and ABp; and P_1 divides into two cells, EMS and P_2. As noted in figure 23.13, the EMS cell then divides into two cells, called MS and E. The cellular descendants of the E cell give rise to the worm's intestine. In other words, the fate of the E cell's descendants is to develop into intestinal cells. This diagram also illustrates the concept of a **cell lineage.** This term refers to a series of cells that are derived from each other by cell division. For example, the EMS cell, E cell, and the intestinal cells are all part of the same cell lineage.

Having a lineage diagram for an organism is an important experimental advantage. It allows researchers to investigate how gene expression in any cell, at any stage of development, may affect the outcome of a cell's fate. In the experiment described next, we will see how the timing of gene expression is an important parameter in the fate of a cell's descendants.

Heterochronic Mutations Disrupt the Timing of Developmental Changes in *C. elegans*

Our discussion of *Drosophila* development has focused on how the spatial expression and localization of gene products can lead to a particular pattern of embryonic development. Another important issue in development is timing. The cells of a multicellular

organism must know when to divide and when to differentiate into a particular cell type. If the timing of these processes is not coordinated, certain tissues will develop too early or too late, disrupting the developmental process.

In *C. elegans,* the timing of developmental events can be examined carefully at the cellular level. As mentioned, the fate of each cell has been determined. Using a microscope, a researcher can focus on a particular cell within this transparent worm and watch it divide into two cells, then four cells, and so forth. Therefore, a scientist can judge whether a cell is behaving as it should during the developmental process.

To identify genes that play a role in the timing of cell fates, researchers have searched for mutant alleles that disrupt the normal timing process. In a collaboration in the late 1970s, H. Robert Horvitz and John Sulston set out to identify mutant alleles in *C. elegans* that disrupt cell fates or the timing of cell fates.

Prior to this work, they did not know what phenotypic effects to expect from a mutation that altered the fate of cells within a cell lineage. Using a microscope, they screened thousands of worms for altered morphologies that might indicate an abnormality in development. During this screening process, one of the phenotypic abnormalities they found was an egg-laying defective phenotype. They reasoned that since the egg-laying system depended on a large number of cell types (vulval cells, muscle cells, and nerve cells), an abnormality in any of the cell lineages leading to these cell types might cause an inability to lay eggs.

In *C. elegans,* an egg-laying defective phenotype is easy to identify, because the hermaphrodite will be able to fertilize its own eggs but will be unable to lay them. When this occurs, the eggs actually hatch within the hermaphrodite's body. This leads to the death of the hermaphrodite as it becomes filled with hatching worms. This egg-laying defective phenotype, in which the hermaphrodite becomes filled with its own offspring, is called a "bag of worms." Eventually, the newly hatched larva eat their way out and can be saved for further study.

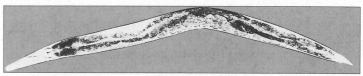

Normal *C. elegans*

C. elegans with egg-laying defective phenotype, the "bag of worms"

In their initial study, published in 1980, the egg-laying defective phenotype yielded several mutant strains that were defective in particular cell lineages. A few years later, in the experiment described in figure 23.14, Victor Ambros and H. Robert Horvitz took this same approach and were able to identify genes that play a key role in the timing of cell fate. They began with wild-type *C. elegans* and three mutant lines designated *n536, n355,* and *n540*. All three of the mutant lines showed an egg-laying defect. Right after hatching, they observed the fates of particular cells via microscopy. In other words, a researcher would spend hours looking at larvae under the microscope and recording the patterns of cell division. The patterns in the mutants and wild type were then used to create lineage diagrams from particular cells.

■ THE HYPOTHESIS

Mutations that cause an egg-laying defective phenotype may affect the timing of cell lineages.

■ TESTING THE HYPOTHESIS — FIGURE 23.14 **Identification of heterochronic mutations in *C. elegans*.**

Starting material: Prior to this work, many laboratories had screened thousand of *C. elegans* worms and identified many different mutant strains that were egg-laying defective. (Note: There are many different genes that when mutated may cause an egg-laying defective phenotype. Only some of them are expected to be genes that alter the timing of cell fate within a particular cell lineage.)

	Experimental level	**Conceptual level**
1. Obtain a large number of *C. elegans* strains that have an egg-laying defective phenotype. The wild-type strain was also studied as a control.	Normal adult worm: — Intestine Single row of eggs — Egg-laying mutant: — Intestine Many eggs — crowded inside	

(*continued*)

2. Right after hatching, observe the fate of particular cells via microscopy. This involves long hours of viewing specific cells within a worm and watching to see if they divide at the appropriate time. Typically, the viewer looks at the cell nuclei (which are relatively easy to see in this transparent worm) and keeps track of when they divide. A researcher can watch and time the division of the cell nuclei as a way to monitor cell division. In this example, a researcher began watching a cell called the T cell and monitored its division pattern, and the pattern of subsequent daughter cells, during the first and second larval stages. These patterns were examined in both wild-type and egg-laying defective worms.

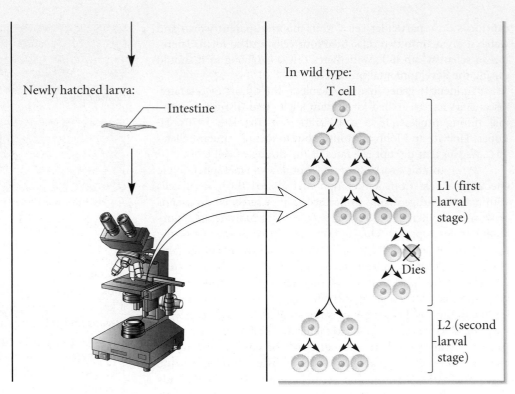

THE DATA

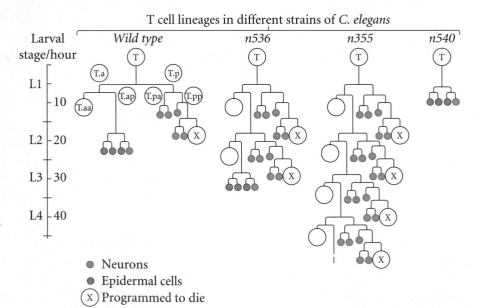

T cell lineages in different strains of *C. elegans*

● Neurons
● Epidermal cells
ⓧ Programmed to die

INTERPRETING THE DATA

Lineage diagrams involving cells derived from a cell called a T cell are shown in the data of figure 23.14. As seen here, the wild-type strain follows a particular pattern of cell division for the T cell lineage. Each division event occurs at specific times during the L1 and L2 larval stages. In the normal strain, the T cell divides during the L1 larval stage to produce a T.a and T.p cell. The T.a cell also divides during L1 to produce a T.aa and T.ap cell. The T.p cell divides during L1 to produce T.pa and T.pp. These cells also divide during L1, eventually producing five neurons (labeled in *blue*) and one cell that is programmed to die (designated with an X). During the L2 larval stage, the T.ap cell resumes division to produce four cells: three epidermal cells (labeled in *red*) and one neuron.

The other T cell lineages are from worms that carry mutations that cause an egg-laying defect. It is now known that these three mutants are located in a gene called *lin-14*. The allele designated *n536* has caused the reiteration of the normal events of L1 during the L2 larval stage. In L2, the only cell of this lineage that is supposed to divide is T.ap. In worms carrying the *n536* allele, however, this cell behaves as if it were a T cell, rather than a T.ap cell. It produces a group of cells that are identical to what a T cell produces during the L1 stage. In the L3 stage, the cells in the *n536* strain behave as if they were in L2. Besides the egg-laying defect, the phenotypic outcome of this irregularity in the timing of cell fates is a worm that has a few more cells and goes through five or six larval stages instead of the normal four.

A more severe allele that causes multiple reiterations is the *n355* allele. This strain continues to reiterate the normal events of L1 during the L2, L3, and L4 stages. In contrast, the *n540* allele has an opposite effect on the T cell lineage. During the L1 larval stage, the T cell behaves as if it were a T.ap cell in the L2 stage. In this case, it skips the divisions and cell fates of the L1 and proceeds directly to cell fates that occur during the L2 stage.

The types of mutations described here are called **heterochronic mutations.** This term refers to the fact that the timing of fates for particular cell lineages is not synchronized with the development of the rest of the organism. More recent molecular data have shown that this is due to an irregular pattern of gene expression. In wild-type worms, the Lin-14 protein accumulates during the L1 stage and promotes the T cell division pattern shown for the wild type. During L2, the Lin-14 protein diminishes to negligible levels. The *n536* and *n355* alleles are examples of gain-of-function mutations. In these alleles, the Lin-14 protein persists during later larval stages. For the *n536* allele, it is made one cell division too late, while the *n355* allele continues to be expressed for several cell divisions. By comparison, the *n540* allele is a loss-of-function mutation. This allele causes Lin-14 to be inactive during L1, so that it cannot promote the normal L1 pattern of cell division and cell fate.

Overall, the results described in this experiment are consistent with the idea that the precise timing of *lin-14* expression during development is necessary to correctly control the fates of particular cells in *C. elegans*. Mutations that alter the expression of *lin-14* lead to phenotypic abnormality, including the inability to lay eggs. This detrimental phenotypic consequence illustrates the importance of the correct timing for cell division and differentiation during development.

A self-help quiz involving this experiment can be found at the Online Learning Center.

23.2 VERTEBRATE DEVELOPMENT

Embryologists have studied the morphological features of development in many vertebrate species. Historically, amphibians and birds have been studied extensively, because their eggs are rather large and easy to manipulate. For example, certain developmental stages of the chicken and frog (*Xenopus laevis*) have been described in great detail. In more recent times, the successes obtained in *Drosophila* have shown the great power of genetic analyses in elucidating the underlying molecular mechanisms that govern biological development. With this knowledge, many researchers are attempting to understand the genetic pathways that govern the development of the more complex body structure found in vertebrate organisms.

Several vertebrate species have been the subject of developmental studies. These include the mouse, the chicken, *Xenopus,* and the small aquarium zebrafish (*Brachydanio rerio*). In this section, we will primarily discuss the genes that are important in mammalian development, particularly those that have been characterized in the mouse. Among mammals, the most extensive genetic analyses have been performed on the mouse. As we will see, several genes affecting its developmental pathways have been cloned and characterized. In this section, we will examine how these genes affect the course of mouse development.

Researchers Have Identified Homeotic Genes in Vertebrates

In most vertebrates, which have long generation times and produce relatively few offspring, it is not practical to screen large numbers of embryos or offspring in search of mutant phenotypes with developmental defects. As an alternative, the most successful way of identifying genes that affect vertebrate development has been the use of molecular techniques to identify vertebrate genes similar to those that control development in simpler organisms such as *Drosophila.*

As discussed in chapters 21 and 26, species that are evolutionarily related to each other often contain genes with similar DNA sequences. When two or more genes have similar sequences because they are derived from the same ancestral gene, they are called **homologous genes.** Homologous genes found in different species are termed **orthologs.** Because they have similar sequences, a DNA strand from one gene will hybridize to a complementary strand of a homologue.

The general approach of using cloned *Drosophila* genes as probes to identify homologous vertebrate genes has been quite successful. Using this method, researchers have found complexes of homeotic genes in many vertebrate species that bear striking similarities to those in the fruit fly. In the mouse, these groups of adjacent homeotic genes are called **Hox complexes.** As shown in figure 23.15, the mouse has four *Hox* complexes, designated *HoxA* (on chromosome 6), *HoxB* (on chromosome 11), *HoxC* (on chromosome 15), and *HoxD* (on chromosome 2). There are a total of 38 genes in the four complexes. There are 13 different gene types within the four *Hox* complexes, although none of the four complexes contains representatives of all 13 types of genes.

Remarkably, several of the homeotic genes in fruit flies and mammals are strikingly similar. Among the first six types of genes, five of them are homologous to genes found in the *Antennapedia* complex of *Drosophila*. Among the last seven, three are homologous

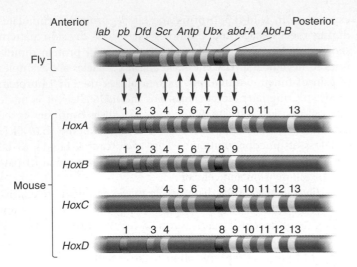

FIGURE 23.15   A comparison of homeotic genes in *Drosophila* and the mouse. The mouse contains four gene clusters, *HoxA–D,* that correspond to certain homeotic genes found in *Drosophila*. Thirteen different types of homeotic genes are found in the mouse, although each *Hox* gene cluster does not contain all 13 genes. In this drawing, orthologous genes are aligned in columns. For example, *lab* is the orthologue to *HoxA-1, HoxB-1,* and *HoxD-1; Ubx* is orthologous to *HoxA-7* and *HoxB-7.*

to the genes of the *bithorax* complex. These results indicate that there are fundamental similarities in the ways that fruit flies and mammals undergo embryonic development. This suggests that there is a "universal body plan" for animal development.

Like the *bithorax* and *Antennapedia* complexes in *Drosophila,* the arrangement of *Hox* genes along the mouse chromosome reflects their pattern of expression from the anterior to the posterior end (fig. 23.16*a*). This phenomenon is seen in more detail in figure 23.16*b,* which shows the expression pattern for a group of *HoxB* genes in a mouse embryo. Overall, these results are consistent with the idea that the *Hox* genes play a role in determining the fates of segments along the antero-posterior axis.

Currently, researchers are trying to understand the functional roles of the genes within the *Hox* complexes in vertebrate development. In the fly, great advances in developmental genetics have been made by studying mutant alleles in genes that control development. In mice, however, few natural mutations have been identified that affect development. This has made it difficult to understand the role that genetics plays in the development of the mouse and other vertebrate organisms.

To circumvent this problem, geneticists are taking an approach known as **reverse genetics.** In this strategy, researchers first identify the wild-type gene using cloning methods. In this case, the *Hox* genes in vertebrates have been cloned using *Drosophila* genes as probes. The next step is to create a mutant version of a *Hox* gene in vitro. This mutant allele is then reintroduced into a mouse using techniques described in chapter 19. When the function of the wild-type gene is thereby eliminated, this is called a **gene knockout.** In this way, researchers can determine how the mutant allele affects the phenotype of the mouse.

The term *reverse genetics* reflects the idea that the experimental steps occur in an order opposite to that in the conventional approach used in *Drosophila* and other organisms. In the fly, mutant alleles were identified by their phenotype first, and then they were cloned. In the mouse, the genes were cloned first, the mutations were made in vitro, and then these were introduced into the mouse to observe their phenotypic effects.

In recent years, several laboratories have used a reverse genetics approach to understand how the *Hox* genes affect vertebrate development. In *Drosophila,* loss-of-function alleles for homeotic genes usually show an anterior transformation. This means that the segment where the defective homeotic gene is expressed now exhibits characteristics that resemble the adjacent anterior segment. Similarly, certain gene knockouts (e.g., *HoxA-2, B-4,* and *C-8*) also show anterior transformations within particular regions of the mouse. However, knockouts of other *Hox* genes (e.g., *A-11*) have posterior transformations, and knockouts of *A-3* and *A-1* exhibit abnormalities in morphology but no clear homeotic transformations. Interestingly, a *HoxA-5* knockout in mice shows evidence of both anterior and posterior transformations, which is also seen in *Drosophila* when its ortholog, *Scr,* is knocked out. Overall, the current picture indicates that the *Hox* genes in vertebrates play a key role in patterning the anteroposterior axis. Nevertheless, additional research will be necessary to understand the individual roles that each of the 38 *Hox* genes plays during embryonic development.

Genes That Encode Transcription Factors Also Play a Key Role in Cell Differentiation

Throughout most of chapter 23, we have focused our attention on patterns of gene expression that occur during the very early stages of development. These genes control the basic body plan of the organism. As this process occurs, cells become **determined.** This term refers to the phenomenon that a cell will be destined to become a particular cell type. In other words, its fate has been predetermined to eventually become a particular type of cell such as a nerve cell. This occurs long before a cell becomes **differentiated.** This later term means that a cell's morphology and function have changed, usually permanently, into a highly specialized cell type. For example, an undifferentiated mesodermal cell may differentiate into a specialized muscle cell, or an ectodermal cell may differentiate into a nerve cell.

At the molecular level, the profound morphological differences between muscle cells and nerve cells arise from gene regulation. Though nerve and muscle cells contain the same genetic material (i.e., the same set of genes), they regulate the expression of their genes in very different ways. Certain genes that are transcriptionally active in muscle cells are completely inactive in nerve cells, and vice versa. Therefore, nerve and muscle cells express different proteins, which affect the morphological and physiological characteristics of the respective cells in distinct ways. In this manner, differential gene regulation underlies cell differentiation.

In *Drosophila,* we learned earlier that a hierarchy of gene regulation is responsible for establishing the body pattern. Mater-

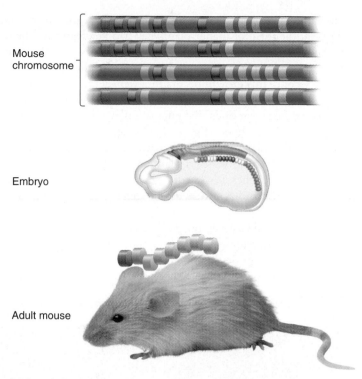

(a) Correlation between *Hox* gene arrangement and expression

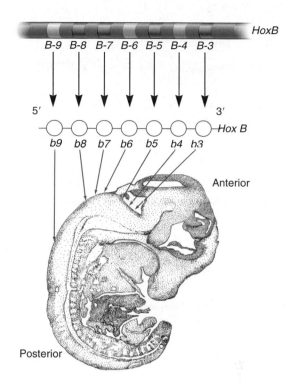

(b) Anterior expression boundaries for a series of *HoxB* genes

FIGURE 23.16   **Expression pattern of *Hox* genes in the mouse.** **(a)** A schematic illustration of the *Hox* gene expression in the embryo and the corresponding regions in the adult. **(b)** A more detailed description of *HoxB* expression in a mouse embryo. The arrows indicate the anterior-most boundaries for the expression of *HoxB-3* to *HoxB-9*. The order of *Hox* gene expression, from anterior to posterior, parallels the order of genes along the chromosome. (From K. Kalthoff, *Analysis of Biological Development*, 2d ed. ©2001 McGraw-Hill, Inc.)

nal effect genes control the expression of gap genes, which control the expression of pair-rule genes, and so forth. A similar type of hierarchy is thought to underlie cell differentiation. Researchers have identified specific genes that cause cells to differentiate into particular cell types. These genes trigger undifferentiated cells to differentiate and follow their proper cell fates.

In 1987, Harold Weintraub and his colleagues identified a gene, which they called *MyoD*. This gene plays a key role in skeletal muscle cell differentiation. Experimentally, when the cloned *MyoD* gene was expressed in fibroblast cells in a laboratory, the fibroblasts differentiated into skeletal muscle cells. This result was particularly remarkable because fibroblasts normally differentiate into osteoblasts (bone cells), chrondrocytes (cartilage cells), adipocytes (fat cells), and smooth muscle cells, but in vivo they never differentiate into skeletal or cardiac muscle cells.

Since this initial discovery, researchers have found that *MyoD* belongs to a small group of genes that initiate muscle development. Besides *MyoD,* these include *Myogenin, Myf5,* and *Mrf4.* All four of these genes encode transcription factors that contain a **basic domain** and a **helix-loop-helix domain (bHLH).** The basic domain is responsible for DNA binding and the activation of skeletal muscle-cell-specific genes. The HLH domain is necessary for dimer formation between transcription factor proteins. Because of their common structural features and their role in muscle differentiation, MyoD, Myogenin, Myf5, and Mrf4 are

called **myogenic bHLH** proteins. They are found in all vertebrates and have been identified in several invertebrates, such as *Drosophila* and *C. elegans.* In all cases, the myogenic bHLH genes are activated during skeletal muscle cell development.

At the molecular level, two key features enable myogenic bHLH proteins to promote muscle cell differentiation. First, the basic domain binds specifically to a muscle-cell-specific enhancer sequence; this sequence is adjacent to genes that are expressed only in muscle cells (fig. 23.17). Therefore, when bHLH proteins are activated, they can bind to these enhancers and activate the expression of many different muscle-specific genes. They may exert their effects via chromatin remodeling or via the transactivation of RNA polymerase. In this way, myogenic bHLH proteins function as master switches that activate the expression of many muscle-specific genes. When the encoded proteins are synthesized, they change the characteristics of an undifferentiated cell into those of a highly specialized skeletal muscle cell.

Another important aspect of myogenic bHLH proteins is that their activity is regulated by dimerization. As shown in figure 23.17, heterodimers may be activating or inhibitory. When a heterodimer forms between a myogenic bHLH protein and an E protein, which also contains a bHLH domain, the heterodimer binds to the DNA and activates gene expression (fig. 23.17*a*). However, when a heterodimer forms between a myogenic bHLH protein and a protein called Id (for <u>i</u>nhibitor of <u>d</u>ifferentiation),

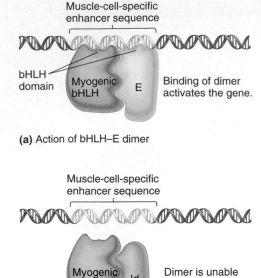

(a) Action of bHLH–E dimer

(b) Action of bHLH–Id dimer

FIGURE 23.17   Regulation of muscle-cell-specific genes by myogenic bHLH proteins. (a) A heterodimer formed from a myogenic bHLH protein and an E protein can bind to a muscle-cell-specific enhancer sequence and activate gene expression. **(b)** When a myogenic bHLH protein forms a heterodimer with an Id protein, it cannot bind to the DNA and therefore does not activate gene transcription.

the heterodimer cannot bind to DNA, because the Id protein lacks a basic domain (fig. 23.17b). The Id protein is produced during early stages of development and prevents myogenic bHLH proteins from promoting muscle differentiation too soon. At later stages of development, the amount of Id protein falls, and myogenic bHLH proteins can then combine with E proteins to induce muscle differentiation.

23.3 PLANT DEVELOPMENT

In developmental plant biology, the model organism for genetic analysis is *Arabidopsis thaliana*. Unlike most higher plants, which have long generation times and large genomes, *Arabidopsis* has a generation time of about 2 months and a genome size of 14×10^7 bp, which is similar to *Drosophila* and *C. elegans*. A flowering *Arabidopsis* plant is small enough to be grown in the laboratory, and it produces a large number of seeds. Like *Drosophila*, *Arabidopsis* can be subjected to mutagens such as X rays to generate mutations that alter developmental processes. The small genome size of this organism makes it relatively easy to map these mutant alleles and eventually clone the relevant genes (as described in chapters 18 and 20).

The morphological patterns of growth are markedly different between plants and animals. As described previously, animal embryos become organized along antero-posterior, dorso-ventral,

and lateral axes, and then they subdivide into segments. By comparison, the form of higher plants has two key features. The first is the root-shoot axis. Most plant growth occurs via cell division near the tips of the shoots and the bottoms of the roots.

Second, this growth occurs in a well-defined radial pattern. For example, early in *Arabidopsis* growth, a rosette of leaves is produced from leaf buds that emanate in a spiral pattern directly from the main shoot (fig. 23.18). Later, the shoot generates branches that will also produce leaf buds as they grow. Overall, the radial pattern in which a plant shoot gives off the buds that give rise to branches, leaves, and flowers is an important mechanism that determines much of the general morphology of the plant.

At the cellular level too, plant development differs markedly from animal development. For example, cell migration does not occur during plant development. In addition, the development of a plant does not rely on morphogens that are deposited asymmetrically in the oocyte. In plants, an entirely new individual can be

FIGURE 23.18   *Arabidopsis.* The plant is relatively small, making it easy to grow many of them in the laboratory.

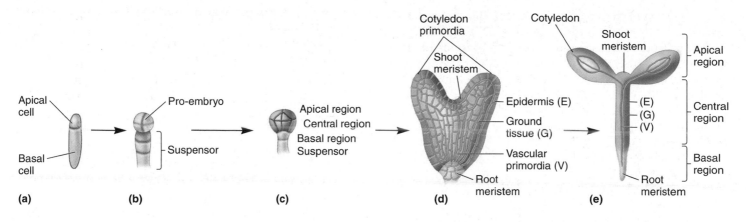

FIGURE 23.19 **Developmental steps in the formation of a plant embryo.** (a) The two-cell stage consists of the apical cell and basal cell. (b) The eight-cell stage consists of a pro-embryo and a suspensor. The suspensor gives rise to extraembryonic tissue, which is needed for seed formation. (c) At this stage of embryonic development, the three main regions of the embryo (i.e., apical, central, and basal) have been determined. (d) At the heart stage, all of the plant tissues have begun to form. Note that the shoot meristem is located between the future cotyledons, while the root meristem is on the opposite side. (e) A seedling.

regenerated from most types of somatic cells. In other words, most plant cells are **totipotent,** meaning that they have the ability to produce an entire individual. By comparison, animal development invariably relies on the organization within an oocyte as a starting point for development.

In spite of these apparent differences, the underlying molecular mechanisms of pattern development in plants still share similarities with those in animals. In this section, we will consider a few examples in which genes encoding transcription factors play a key role in plant development.

Plant Growth Occurs from Meristems That Are Formed During Embryonic Development

Figure 23.19 illustrates a common sequence of events that takes place in the development of seed plants such as *Arabidopsis*. After fertilization, the first cellular division is asymmetrical and produces a smaller cell, called the apical cell, and a larger basal cell (fig. 23.19a). The apical cell will give rise to most of the embryo, and it will later develop into the shoot of the plant. The basal cell will give rise to the root, along with extraembryonic tissue, which is required for seed formation. At the heart stage, which is composed of only about 100 cells, the basic organization of the plant has been established. As shown in figure 23.19d, the **shoot meristem** will arise from a group of cells located between the cotyledons. As its name suggests, these cells are the precursors that will produce the shoot of the plant. The **root meristem** is located at the opposite side; it will create the root.

A meristem contains an organized group of actively dividing stem cells. As discussed in chapter 19, stem cells retain the ability to divide and differentiate into multiple cell types. As they grow, meristems produce offshoots of proliferating cells. On the shoot meristem, for example, these offshoots or buds give rise to structures such as leaves and flowers. The organization of the shoot meristem is shown in figure 23.20. It is organized into three areas called the **organizing center,** the **central zone,** and the

peripheral zone. The role of the organizing center is to ensure the proper organization of the meristem and preserve the correct number of actively dividing stem cells. The central zone is an area where undifferentiated stem cells are always maintained. The peripheral zone contains dividing cells that will eventually differentiate into plant structures. For example, the peripheral zone may form a bud, which will produce a leaf or flower.

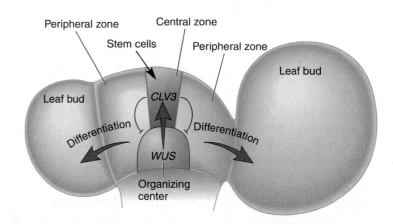

FIGURE 23.20 **Organization of the shoot meristem.** The role of the organizing center is to ensure the proper organization of the meristem and preserve the correct number of actively dividing stem cells. The central zone is where undifferentiated stem cells are always maintained and the peripheral zone contains dividing cells that will eventually differentiate into plant structures. The organization of the shoot meristem is controlled by the *WUS* and *CLV3* genes, which are acronyms for *Wuschel* and *clavata*, respectively. The *WUS* gene is expressed in the organizing center and induces the cells in the central zone to become undifferentiated stem cells. These stem cells turn on the *CLV3* gene, which encodes a secreted protein that binds to receptors in the cells of the peripheral zone and prevents them from expressing the *WUS* gene. This limits the area of *WUS* gene expression to the underlying organizing center and thereby maintains a small population of stem cells at the growing tip.

In *Arabidopsis,* the organization of the shoot meristem is controlled by two critical genes termed *WUS* and *CLV3*. The *WUS* gene enocodes a transcription factor that is expressed in the organizing center (fig. 23.20). The expression of the *WUS* gene induces the adjacent cells in the central zone to become undifferentiated stem cells. These stem cells then turn on the *CLV3* gene, which encodes a secreted protein. The CLV3 protein binds to receptors in the cells of the peripheral zone, and this prevents them from expressing the *WUS* gene. This limits the area of *WUS* gene expression to the underlying organizing center and thereby maintains a small population of stem cells at the growing tip. The shoot meristem in *Arabidopsis* contains only about 100 cells. The inhibition of *WUS* expression in the peripheral cells also allows them to embark on a path of cell differentiation so they can produce structures such as leaves and flowers.

The shoot meristem produces all aerial parts of the plant, which include the stem as well as lateral structures such as leaves and flowers. The root meristem only gives rise to the root. In the seedling shown earlier in figure 23.19e, three main regions are observed. The **apical region** produces the leaves and flowers of the plant. The **central region** (not to be confused with the central zone) creates the stem. It is the radial pattern of cells in the central region that causes the radial growth observed in plants. Finally, the **basal region** produces the roots. Each of these three regions develops differently as indicated by their unique cell division patterns and distinct morphologies. In addition, by analyzing mutants that disrupt the developmental process, researchers have discovered that these three regions express different sets of genes. G. Jürgens and his colleagues began a search to identify a category of genes, known as the **apical-basal-patterning genes,** that are important in early stages of development. As described in table 23.2, defects in apical-basal-patterning genes cause dramatic effects in one of the three main regions. For example, the *gurke* gene is necessary for apical development. When it is defective, the embryo lacks apical structures. Currently, a great amount of effort is directed toward the identification of genes that govern pattern formation in these three regions of *Arabidopsis* and other plants.

TABLE 23.2

Examples of *Arabidopsis* Genes That Affect the Development of the Apical, Central, or Basal Region

Gene	Description
Apical	
Gurke	Minor loss-of-function alleles in the *Gurke* gene produce seedlings with highly reduced or no cotyledons. Complete loss-of-function eliminates the entire shoot.
Pin1	Encodes a putative auxin exporter that is expressed in the peripheral zone. It plays a role in the location of bud sites, which give rise to leaves and flowers.
Aintegumenta	Encodes a transcription factor that is also expressed in the peripheral zone. Its expression maintains the proliferative cell state during the growth of lateral buds.
Central	
Fackel	Encodes an enzyme involved in phytosterol synthesis. Loss-of-function alleles show a severe defect in stem growth.
Scarecrow	Encodes a transcription factor protein that plays a role in the asymmetric division that produces the radial pattern of growth in the stem. Note: The scarecrow protein also affects cell division patterns in roots and plays a role in sensing gravity.
Basal	
Monopterous	Encodes a transcription factor. When this gene is defective, the plant embryo cannot initiate the formation of root structures, but root structures can be formed postembryonically under the correct growth conditions. This gene seems to be required for organizing root formation in the embryo but is not required for root formation per se.
Hobbit	Encodes a subunit of a protein that functions as a cell cycle checkpoint during anaphase. Hobbit function may be required to couple cell division to cell differentiation in the meristem or to restrict the response to growth hormones such as auxin. Loss-of-function alleles in this gene are incapable of forming a root.
Overall Organization	
Gnom	Plays a role in the stable fixation of the apical-basal axis of the *Arabidopsis* embryo.

Plant Homeotic Genes Control Flower Development

Although the term *homeotic* was coined by William Bateson to describe homeotic mutations in animals, the first known homeotic genes were described in plants. In ancient Greece and Rome, for example, double flowers in which stamens were replaced by petals were noted. In current research, geneticists have been studying these types of mutations to better understand developmental pathways in plants. Many homeotic mutations affecting flower development have been identified in *Arabidopsis* and also in the snapdragon (*Antirrhinum majus*).

Examples in *Arabidopsis* are shown in figure 23.21. Part *a* shows a normal *Arabidopsis* flower. It is composed of four concentric whorls of structures. The outer whorl contains four sepals, which protect the flower bud before it opens. The second whorl is composed of four petals, and the third whorl contains six stamens. The stamens are the structures that make the male

gametophyte, pollen. Finally, the innermost whorl contains two carpels, which are fused together. The carpel produces the female gametophyte. The homeotic mutants shown in figure 23.21 have undergone transformations of particular whorls. For example, in figure 23.21*b*, the sepals have been transformed into carpels, the petals into stamens.

By analyzing the effects of many different homeotic mutations in *Arabidopsis,* Elliot Meyerowitz and his colleagues proposed the **ABC model** for flower development. In this model, three classes of genes, called *A*, *B*, and *C*, govern the formation of sepals, petals, stamens, and carpels. More recently, a fourth category of genes called the *Sepallata* genes (*SEP* genes) have been found to be required for this process. In *Arabidopsis,* figure 23.22 illustrates how these genes affect normal flower development. In the outermost whorl (whorl 1), the gene *A* products are made. This promotes sepal formation. In whorl 2, gene *A*, gene *B*, and

(a) Normal flower

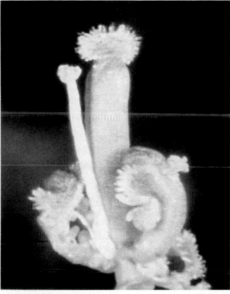

(b) Single homeotic mutant

(c) Triple mutant

FIGURE 23.21   Examples of homeotic mutations in *Arabidopsis*.

GENES→TRAITS   (a) A normal flower. It is composed of four concentric whorls of structures: sepals, petals, stamens, and carpel. (b) A homeotic mutant in which the sepals have been transformed into carpels, and the petals have been transformed into stamens. (c) A triple mutant in which all of the whorls have been changed into leaves.

SEP gene products are made, which promotes petal formation. In whorl 3, the expression of gene *B*, gene *C*, and *SEP* genes causes stamens to be made. Finally, in whorl 4, gene *C* and *SEP* genes promote carpel formation.

Now let's consider what happens in certain homeotic mutants. In the original ABC model, it was proposed that genes *A* and *C* repress each other's expression, and gene *B* functions independently. In a mutant defective in gene *A* expression, gene *C* will also be expressed in whorls 1 and 2. This produces a carpel-stamen-stamen-carpel arrangement. When gene *B* is defective, a flower cannot make petals or stamens. Therefore, a gene *B* defect yields a flower with a sepal-sepal-carpel-carpel arrangement. When gene *C* is defective, gene *A* is expressed in all four whorls. This results in a sepal-petal-petal-sepal pattern. If the expression of *SEP* genes is defective, the flower consists entirely of sepals, which is the origin of the gene's name.

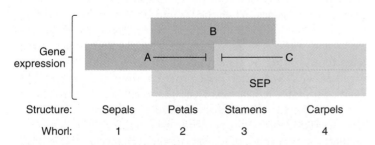

FIGURE 23.22   The ABC model of homeotic gene action in *Arabidopsis*. Note: This is a revised model based on the recent identification of *SEP* genes.

Overall, it appears that the types of genes described in figure 23.22 promote cell differentiation that leads to sepal, petal, stamen, or carpel structures. But what happens if genes *A*, *B*, and *C* are all defective? As shown in figure 23.21c, this produces a "flower" that is composed of leaves! These results indicate that the leaf structure is the default pathway and that the *A*, *B*, *C*, and *SEP* genes cause development to deviate from a leaf structure in order to make something else. In this regard, the sepals, petals, stamens, and carpels can be viewed as modified leaves. Johann Goethe originally proposed this idea over 200 years ago.

In *Arabidopsis*, there are two types of gene *A* (*apetala1* and *apetala2*), two types of gene *B* (*apetala3* and *pistillata*), one type of gene *C* (*agamous*), and three *SEP* genes (*SEP1*, *SEP2*, and *SEP3*). All of these plant homeotic genes encode transcription factor proteins. The proteins contain a DNA-binding domain and a dimerization domain. However, the *Arabidopsis* homeotic genes do not contain a sequence similar to the homeobox found in animal homeotic genes. Instead, most of them (except for *apetala2*) contain a **MADS box** (an acronym of the first four plant genes of this type that were identified: *MCM1*, *AG*, *DEF*, and *SRF*). In the transcription factor proteins, the MADS box promotes the binding of the transcription factor to specific DNA sequences.

Like the *Drosophila* homeotic genes, plant homeotic genes are part of a hierarchy of gene regulation. Genes that are expressed within the flower bud primordium produce proteins that activate the expression of these homeotic genes. Once they are transcriptionally activated, the homeotic genes then regulate the expression of other genes, the products of which promote the formation of sepals, petals, stamens, or carpels.

23.4 SEX DETERMINATION IN ANIMALS AND PLANTS

To end our discussion of development, it is interesting to consider how genetics plays a role in the development of male and female individuals. In the animal kingdom, the existence of two sexes is nearly universal. In plants, most species produce male and female haploid reproductive cells on the same individual. However, some plant species are sexually dimorphic. Such species produce two distinct types of individuals that have different abilities to produce male and female gametophytes. We will consider such types of plants in this section.

The underlying factors that distinguish female versus male development vary widely. In animals, sex determination is often caused by differences in chromosomal composition (table 23.3). In worms and flies, it is the ratio of X chromosomes to autosomes that determines sex. By comparison, it is the presence of the Y chromosome in mammals that causes maleness. Similarly, plants that are sexually dimorphic typically have sex chromosomes in which the plant that produces the male gametophyte has two types of sex chromosomes, similar to the XY system in mammals. In birds, it is the female that is heteromorphic for the sex chromosomes. ZW birds are female, while ZZ birds are male. Finally, sex determination is not always caused by two types of sex chromosomes. In bees, for example, males are monoploid while females are diploid. In many animal species, variation in chromosome composition is not the underlying factor that distinguishes female versus male development. Sex determination in many species of reptiles and fish is controlled by environmental factors such as temperature. For example, in the American alliga-

tor (*Alligator mississippiensis*), temperature controls sex determination. Eggs incubated at 33°C result in 100% male individuals, while eggs incubated at lower or higher temperatures such as 30°C or 34.5°C result in 100% or 95% females, respectively.

The adoption of one of two sexual fates is an event that has been studied in great detail in several species. It has been discovered that sex determination is a process that is controlled genetically by a hierarchy of genes that exert their effects in early embryonic development. In this section of chapter 23, we will consider features of these hierarchies in *Drosophila, C. elegans,* mammals, and plants.

In *Drosophila,* Sex Determination Involves a Regulatory Cascade That Includes Alternative Splicing

In diploid fruit flies, X0 flies develop into males while XX flies become females. It is the ratio of the number of X chromosomes to the diploid autosomes (1.0 in females versus 0.5 in males) that determines sex. Although male fruit flies usually carry a Y chromosome, it is not necessary for male development. The mechanism of sex determination begins in early embryonic development and involves a regulatory cascade composed of several genes. Females and males follow one of two alternative pathways, which are depicted in figure 23.23. Genes or gene products that are functionally active are shown in *tan,* while those that are not expressed or expressed in a functionally inactive form are depicted in *blue.*

Let's begin with the pathway that produces female flies. In females, the higher ratio of X chromosomes results in the embryonic expression of a gene designated *SXL* (fig. 23.23*a*). The *SXL* gene product functions as a splicing factor. In female embryos, the *SXL* gene product enhances its own expression by splicing its own RNA, an event termed an autoregulatory loop. In addition, it splices the RNA from two other genes called *MSL-2* and *TRA.* The Sxl protein promotes the splicing of the *MSL-2* mRNA in a way that introduces an early stop codon in the coding sequence and thereby produces a shortened version of the Msl-2 protein that is functionally inactive. This prevents dosage compensation in the female. By comparison, the Sxl protein promotes the splicing of *TRA* mRNA to produce an mRNA that is translated into a functional protein. Therefore, *SXL* activates *TRA.* The *TRA* gene product and a constitutively expressed product from a gene called *TRA-2* are also splicing factors. In the female, they cause the alternative splicing of the mRNAs that are expressed from the *DSX* and *FRU* genes. When the tra and tra-2 proteins cause these mRNAs to be alternatively spliced, the mRNAs are designated DSX^F and FRU^F. The female-specific *FRU* mRNA is not translated to a sex-specific gene product. However, the DSX^F mRNA, together with two other gene products from the *IX* and *HER* genes, promotes female sexual development and controls some aspects of female-specific behavior via the central nervous system. The DSX^F protein is known to be a transcription factor that regulates certain genes that promote these changes.

In X0 or XY flies, the *SXL* gene is not activated, due to the lower ratio between the single X chromosome and the diploid

TABLE 23.3
Mechanisms of Sex Determination in Selected Species

Species	Mechanism
Drosophila, C. elegans	Ratio of X chromosomes to autosomes determines sex. An X:A ratio of 1.0 results in females (or hermaphrodites in *C. elegans*) while an X:A ratio of 0.5 produces males.
Mammals	The presence of the Y chromosome causes maleness.
Birds	Males are ZZ and females are ZW.
Bees	Males are monoploid, while females (workers and queen bees) are diploid.
Certain species of reptiles, fish, and turtles	Environmental conditions, usually temperature, influence the ratio of male and female offspring.
Plants	The male plant, which produces pollen, is heteromorphic. In some sexually dimorphic species, the males are XY. In other species, there are not distinct sex chromosomes, but the male plant is heterozygous for a dominant gene that suppresses the female pathway.

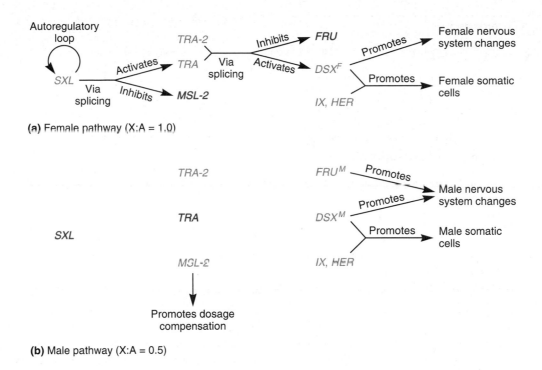

(a) Female pathway (X:A = 1.0)

(b) Male pathway (X:A = 0.5)

FIGURE 23.23 **Sex determination pathway for *Drosophila melanogaster*.** Genes or gene products that are functionally expressed are shown in *tan*, while those that are not expressed are shown in *blue*. The gene names are acronyms for the phenotypes that result from mutations that cause loss-of-function or aberrant expression. These are as follows: *SXL* (sex lethal), *MSL* (male sex lethal), *TRA* (transformer), *DSX* (double sex), *FRU* (fruitless), *IX* (intersex), and *HER* (hermaphrodite).

autosomes (fig. 23.23*b*). This permits the expression of *MSL-2*, which promotes dosage compensation. In fruit flies, dosage compensation is accomplished by turning up the expression of X-linked genes in the male to a level that is twofold higher. Therefore, even though the male has only one X chromosome, the expression of X-linked genes in the male and female are approximately equal. The absence of *SXL* expression in male embryos also promotes the development of maleness. Without *SXL*, the *TRA* mRNA is not properly spliced, so that *TRA* is not expressed. Without the tra protein, the *DSX* and *FRU* mRNAs are spliced in a different way (i.e., the default pathway) to produce mRNAs designated *DSX^M* and *FRU^M*. Along with *IX* and *HER* gene products, the dsx^M protein promotes male development as well as male-specific behavior. Like the dsx^F protein, the dsx^M protein is a transcription factor that regulates certain genes. Since *DSX^M* is spliced differently from *DSX^F*, the dsx^M protein's structure is different from the dsx^F protein, and this difference alters the regulation pattern of dsx^M. In addition, the *FRU^M* gene product is necessary for the regulation of genes that are involved in male-specific behaviors.

In *C. elegans* the Ratio of X Chromosomes to Autosomes Initiates a Regulatory Cascade That Determines Sex

C. elegans has two sexes: hermaphrodites and males. The hermaphrodites produce sperm during larval development that are stored in a structure called the spermathecae. In adulthood, hermaphrodites are anatomically female. They produce oocytes that are fertilized when they are forced through the spermathecae by muscular contractions. For fertilization to occur, one possibility is that a sperm that fertilizes an oocyte may be produced from the same worm. In other words, the hermaphrodite is capable of self-fertilization. Alternatively, a hermaphrodite can mate with a male worm, which produces only sperm. Such sperm are also stored in the spermathecae and can cross-fertilize eggs.

The sexual identity of *C. elegans*, hermaphrodite versus male, is a trait that is determined very early in embryonic development. Like *Drosophila*, it is controlled by the activities of many genes that interact in a regulatory cascade that determines the sex of the worm and controls dosage compensation. In *C. elegans*, the expression of genes on the X chromosome in the hermaphrodite, which carries two X chromosomes, is decreased to 50% compared to males, which carry one X chromosome.

Like *Drosophila*, the ratio of the X chromosomes to the autosomes is the factor that causes the regulatory cascade to follow one of two alternative pathways. In figure 23.24, genes or gene products that are expressed are shown in *tan*, while those that are silenced are depicted in *blue*. Let's first consider the events that occur during early embryonic development in the hermaphrodite. Since hermaphrodites have two X chromosomes, this results in a twofold higher ratio between the X chromosomes and autosomes, compared to males. This higher ratio results in an enhanced expression of *fox-1* and *sex-1*, which are located on

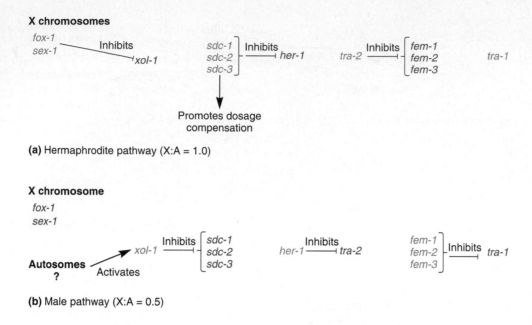

(a) Hermaphrodite pathway (X:A = 1.0)

(b) Male pathway (X:A = 0.5)

FIGURE 23.24   **Sex determination pathway for *C. elegans*.** Genes or gene products that are expressed at relatively high levels are shown in *tan*, while those that are not expressed or are expressed at low levels are shown in *blue*. The gene names are acronyms that usually refer to the effects of mutations. These are as follows: *fox* (feminizing on X), *sex* (sex determination), *xol* (XO lethal), *sdc* (sex determination and dosage compensation), *her* (hermaphrodite), *tra* (transformer), and *fem* (feminization).

the X chromosome. Presumably, because it is the ratio of X chromosomes to autosomes that is critical, there are genes on the autosomes that are involved, but these have yet to be definitively identified. In the XX hermaphrodite, the expression of *fox-1* and *sex-1* inhibits the expression of *xol-1*. The protein product of the *xol-1* gene inhibits the expression of three genes called *sdc-1*, *sdc-2*, and *sdc-3*. Since *xol-1* activity is inhibited in hermaphrodites, this permits the expression of the *sdc* genes. The expression of *sdc-1*, *sdc-2*, and *sdc-3* has two effects. First, these three genes are necessary for dosage compensation and second, they are necessary to permit the downstream expression of *tra-1*, which is needed to promote hermaphrodite development. This second effect occurs via a chain of events that are also shown in figure 23.24*a*. The *sdc* gene products inhibit the expression of *her-1*. When *her-1* is inhibited this permits the expression of *tra-2*. When *tra-2* is active, this inhibits the activities of *fem-1*, *fem-2*, and *fem-3*. When the *fem* gene products are inhibited, this permits the expression of *tra-1*. It is the *tra-1* gene product that promotes hermaphrodite development and inhibits male development.

Figure 23.24*b* describes the pathway for male development. Because males contain a single X chromosome, the expression of *fox-1* and *sex-1* is insufficient to inhibit the expression of *xol-1*. When *xol-1* is expressed, its protein product inhibits the expression of the *sdc* genes. This prevents dosage compensation and permits the expression of *her-1*. The *her-1* gene product then inhibits *tra-2*. This permits the expression of the *fem* genes, which inhibit the expression of *tra-1*. Without *tra-1* expression, the worm develops into a male instead of a hermaphrodite.

Experimentally, it is has been shown that XX worms that are lacking a functional *tra-1* gene develop into males.

In Mammals, the *Sry* Gene on the Y Chromosome Determines Maleness

In mammals, such as humans, mice, and marsupials, it is the presence of the Y chromosome that determines maleness. In cases of abnormal sex chromosome composition such as XXY, an individual develops into a male. The *Sry* gene, which is located on the Y chromosome, causes the sex determination pathway to follow a male developmental scheme. The *Sry* gene encodes a protein that contains a DNA-binding domain called an HMG box, which is found in a broad category of DNA-binding proteins known as the high mobility group. In several of these proteins, the HMG box is known to cause DNA bends. Thus far, the ability of the Sry protein to promote male sex determination is not well understood, but it may act as a transcription factor and/or promote changes in DNA architecture, much like the DNA remodeling proteins discussed in chapters 10 and 15.

Similar to fruit flies and worms, sex determination in mammals involves a cascade that is initiated in early embryonic development. However, the details of the pathway are not completely elucidated. Figure 23.25 illustrates a putative pathway in mammals. Two genes, designated *SFI* and *WTI*, are expressed in the early embryo, prior to sexual differentiation. In the male, the expression of the *Sry* gene activates the expression of a gene called *SOX9*. Like the *Sry* gene, *SOX9* encodes a protein with an

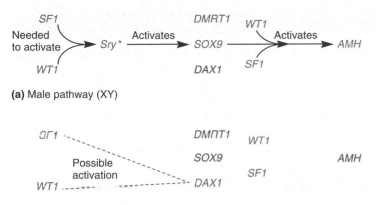

(a) Male pathway (XY)

(b) Female pathway (XX)

*The *Sry* gene is located on the Y chromosome, so it is only found in males.

FIGURE 23.25 **Sex determination pathway for mammals.**
Genes or gene products that are expressed at relatively high levels are shown in *tan*, while those that are not expressed or are inactive are shown in *blue*. The regulation of the *DAX1* gene is not well understood, but it may involve activation via *SF1* and/or *WT1* gene products. The gene names are acronyms that usually refer to the effects of mutations and/or their relatedness to other genes or chromosomal regions. These are as follows: *SF1* (steroidogenic factor-1), *WT1* (Wilm's tumor gene), *Sry* (sex determining region of the Y chromosome), *DMRT1* (doublesex- and mab-3-related transcription factor), *SOX9* (Sry-like HMG box), *DAX1* (dosage-sensitive sex-reversal, adrenal hypoplasia congenital, X chromosome), and *AMH* (anti-Müllerian hormone).

HMG domain. In addition, the SOX9 protein is larger than the Sry protein because it also contains two transcriptional activation domains, indicating that it is a transcriptional activator. *SOX9* expression is necessary for male development, since a loss-of-function allele of *SOX9* results in female development in XY animals. Furthermore, when researchers produced female mice with three copies of the *SOX9* gene, such individuals developed into males.

The downstream targets of the SOX9 protein that promote the male developmental pathway and/or prevent the female pathway are not well understood. Though further research is needed, it is thought that the expression of *SOX9*, together with *WTI* and *SFI*, result in the activation of the *AMH* gene. AMH is an acronym for anti-Müllerian hormone. In females, the Müllerian duct differentiates into the oviduct, uterus, and upper parts of the vagina, but in males the Müllerian duct regresses due to the action of anti-Müllerian hormone. Males that lack the *AMH* gene develop testes but are infertile due to the persistence of Müllerian-derived structures that interfere with sperm transfer. Therefore, although the *AMH* gene is important for the proper development of male characteristics, it is not required for testis determination.

Two other genes, designated *DAX1* and *DMRT1*, also play a role in sex determination. The *DAX1* gene is X linked, and its gene product is thought to prevent male development. In XX animals, its expression remains high, in contrast to XY males. XY animals with two copies of the *DAX1* gene develop into females when *Sry* gene expression is low due to a weak allele in the *Sry* gene. This suggests that *DAX1* can inhibit the effects of the *Sry*

gene. However, the *DAX1* gene is not needed for female development because XX mice that are lacking the *DAX1* gene develop as normal females. *DAX1* encodes a hormone-receptor protein. Further research is needed to understand the pathways that are activated via this receptor. Finally, the *DMRT1* gene in mammals is evolutionarily related to the *mab-3* gene in *C. elegans* and the *DSX* gene in *Drosophila*. These genes encode transcription factors that are involved in the differentiation of the gonads. In mammals, *DMRT1* expression is gonad-specific, and it is expressed at higher levels in the testes. XY mice that are lacking the *DMRT1* gene develop as males, but they are infertile due to severe defects in testis structure and the inability to produce sperm cells.

In Sexually Dimorphic Plants, the Male Plant Is Usually Heteromorphic

As mentioned earlier, most species of plants, about 95%, are sexually monomorphic. Only a single type of individual is produced that can make both male and female gametophytes. By comparison, sexually dimorphic species have two separate types of individuals. The most common types are **dioecious,** in which one kind of individual produces male gametophytes while the other produces female gametophytes. Other rare exceptions are possible. A few plants species are gynodioecious, which produce hermaphrodite plants or female plants. Alternatively, a few species are androdioecious and produce hermaphrodite plants or male plants.

The genetics of sex determination in dioecious plant species is beginning to emerge. Since *Arabidopsis* is sexually monomorphic, researchers have had to turn to other species to study sex determination in plants. For example, the white campion, *Silene latifolia*, has been the subject of numerous investigations. In this species, sex chromosomes, designated X and Y, are responsible for sex determination. The male plant is XY, while the female plant is XX. In other dioecious species, cytological examination of the chromosomes does not always reveal distinct types of sex chromosomes. Nevertheless, the male plants appear to be the heterozygote. This has been determined because male plants are often "inconstant," which means that they produce an occasional fruit. When a male plant undergoes such selfing, the seeds from the fruit produce a 3:1 ratio of male to female plants. This is the expected result if the male parent was heterozygous for a dominant gene that promotes maleness.

Researchers are beginning to identify loci that are important for sex determination in plants, though further studies are needed to unravel the genetic hierarchy that promotes sexual development. A locus designated Su^F plays a role in this process. The Su^F locus is believed to contain a gene that acts as a dominant suppressor of the female pathway. It is responsible for producing the 3:1 ratio when male plants undergo selfing. In addition, other loci that appear to control early and late anther development have been discovered. Along with Su^F, these loci are linked on the Y chromosome in *S. latifolia*. Eventually, the discovery of more genes that control sexual development will allow researchers to propose alternative pathways that promote male and female development in dioecious plant species.

CONCEPTUAL SUMMARY

In chapter 23, we have examined the role that genetics plays in the **development** of multicellular organisms. Each organism has its own developmental program, driven by a hierarchy of genes that encode transcription factors and other types of regulatory proteins. In *Drosophila*, the developmental program begins in the oocyte. It is here that a few key gene products act as **morphogens** that are asymmetrically located or activated. This provides **positional information** that initiates development of the body plan or pattern in this organism. The maternal effect gene products serve to organize the **antero-posterior** and **dorso-ventral** axes soon after fertilization. Once these axes have been established, the next step is to divide the embryo into **segments.**

Three categories of **segmentation genes,** known as **gap genes, pair-rule genes,** and **segment-polarity genes,** act sequentially to promote segmentation. The pair-rule genes control the identity of each parasegment, and the segment-polarity genes divide each parasegment into anterior and posterior regions. A segment, which is a morphological feature, is composed of the anterior region of one parasegment and the posterior region of an adjacent parasegment. The role of **homeotic** genes is to dictate the morphological characteristics of each segment. To do so, they activate **realizator genes,** which express proteins that determine the morphological characteristics of cells.

In *C. elegans,* the availability of a complete **lineage diagram** has allowed geneticists to identify **heterochronic mutations** that alter the timing of **cell fate.** These mutations lead to abnormalities in morphology. This phenomenon illustrates the importance of coordinated cell division in determining cell fate during development.

In vertebrates, much less is known about the programs that promote development. Nevertheless, they appear to bear many similarities to the programs in simpler invertebrates. For example, the *Hox* complex in mice contains groups of genes that are homologous to homeotic genes in *Drosophila.* Furthermore, experiments with gene knockouts of particular *Hox* genes indicate that they act as homeotic genes in vertebrates.

A well-studied feature of vertebrate development is **cell differentiation.** Researchers have identified genes involved in the differentiation of cells into highly specialized cell types. For example, the myogenic bHLH proteins play a critical role in the differentiation of skeletal muscle cells.

Morphologically, plant development differs markedly from animal development. Plants are organized along a root-shoot axis and growth occurs from **meristems.** Plant geneticists have identified some of the genes that play a role in the general organization of the plant embryo. Geneticists have also discovered many homeotic genes that dictate the morphology of particular structures later in development. Four classes of homeotic genes, *A, B, C,* and *SEP,* are involved in the formation of the four whorls that make up a flower. These results show that plant development is dictated by a genetic program that bears some similarities to those found in animals.

Sex determination in plants and animals relies on a genetic hierarchy that begins in early embryonic development. Different factors, such as sex chromosomes, temperature, etc., can provide the basis for choosing between the male- and female-specific pathway. The hierarchy in any given species may involve a variety of control mechanisms including transcriptional regulation, chromatin remodeling, and alternative splicing. The outcome of such regulation is to promote the development of male or female structures.

EXPERIMENTAL SUMMARY

In chapter 23, we have seen that a genetic approach (namely, the identification of mutant alleles) has been essential to our understanding of development. Researchers, particularly those studying invertebrates and *Arabidopsis,* have identified many different mutant alleles that alter key steps in the developmental process. In *Drosophila,* for example, these include mutations in maternal effect genes, segmentation genes, and homeotic genes. The molecular characterization of these genes has shown that many of them encode transcription factors that influence the expression of other genes. The study of developmental mutations has enabled scientists to piece together a genetic hierarchy that underlies many developmental programs.

The characterization of developmental mutants has also been correlated with the spatial localization of gene products using techniques such as *in situ* hybridization and immunostaining. This enables researchers to determine when and where a gene product is made during oogenesis or embryogenesis. This approach has shown that some mutations that alter development are gain-of-function alleles, which cause a gene to be expressed in the wrong place or at the wrong time. Other mutations are loss-of-function alleles, which cause a defect in gene expression. In *C. elegans,* the availability of a **lineage diagram** has also allowed researchers to examine how **heterochronic mutations** can affect the timing of developmental steps.

The key experimental advantage of studying invertebrates is the ability to identify mutations that alter developmental steps. This approach is not as easy in vertebrates, but a **reverse genetics** approach is proving to be successful. Using invertebrate genes as probes, researchers have identified vertebrate genes, such as the *Hox* genes in mice, that are homologous to the homeotic genes of invertebrates. To understand their role, the wild-type *Hox* genes have been mutated in vitro and reintroduced into mice replacing the normal gene. The results of this method are consistent with the idea that the *Hox* genes function as the mouse homologues of the *Drosophila* homeotic genes. However, further research will be needed to understand the individual roles of these 38 genes.

In plants, a genetic approach is also proving effective in elucidating the development process. Early in development, a category of genes, known as the **apical-basal-patterning genes,** appear necessary for the formation of the apical and basal regions of the embryo. These genes were identified by investigating the effects of mutations on the developing plant embryo.

Similarly, mutations in plant homeotic genes have been discovered by their ability to abnormally transform one plant structure into another (e.g., sepals into petals). A comparison of single, double, and triple homeotic mutations has provided the framework for the **ABC model** of flower development.

PROBLEM SETS & INSIGHTS

Solved Problems

S1. Discuss and distinguish the functional roles of the maternal effect genes, gap genes, pair-rule genes, and segment-polarity genes in *Drosophila.*

Answer: These genes are involved in pattern formation of the *Drosophila* embryo. The asymmetric distribution of maternal effect gene products in the oocyte establishes the antero-posterior and dorso-ventral axes. These gene products also control the expression of the gap genes, which are expressed as broad bands in certain regions of the embryo. The overlapping expression of maternal effect genes and gap genes controls the pair-rule genes, which are expressed in alternating stripes. A stripe corresponds to a parasegment. Within each parasegment, the expression of segment-polarity genes defines an anterior and posterior compartment. With regard to morphology, an anterior compartment of one parasegment and the posterior compartment of an adjacent parasegment will form a segment of the fly.

S2. With regard to genes affecting development, what are the phenotypic effects of gain-of-function mutations versus loss-of-function mutations?

Answer: Gain-of-function mutations cause a gene to be expressed in the wrong place, at the wrong time, or in an abnormal way. When they are expressed in the wrong place, that region may develop into an inappropriate structure. For example, when *Antp* is abnormally expressed in an anterior segment, this segment develops legs in place of antennae. When gain-of-function mutations cause a gene to be expressed at the wrong time, this can also disrupt the development process. Gain-of-function heterochronic alleles cause cell lineages to be reiterated and thereby alter the course of development. By comparison, loss-of-function mutations result in a defect in the expression of a gene. This usually will disrupt the developmental process, because the cells in the region where the gene is supposed to be expressed will not be directed to develop along the correct pathway.

S3. Mutations in genes that control the early stages of development are often lethal (e.g., see fig. 23.6*b*). To circumvent this problem, developmental geneticists may try to isolate *temperature-sensitive developmental mutants* or *ts alleles*. If an embryo carries a *ts* allele, it will develop correctly at the permissive temperature (e.g., 25°C) but will fail to develop if incubated at the nonpermissive temperature (e.g., 30°C). In most cases, *ts* alleles are missense mutations that slightly alter the amino acid sequence of a protein, causing a change in its structure that prevents it from working properly at the nonpermissive temperature. *Ts* alleles are particularly useful because they can provide insight regarding the stage of development when the protein is necessary. Researchers can take groups of embryos that carry a *ts* allele and expose them to the permissive and nonpermissive temperature at different stages of development.

In the experiment described next, embryos were divided into five groups and exposed to the permissive or nonpermissive temperature at different times after fertilization.

Time After Fertilization (hours)	Group:	1	2	3	4	5
0–1		25°C	25°C	25°C	25°C	25°C
1–2		25°C	30°C	25°C	25°C	25°C
2–3		25°C	25°C	30°C	25°C	25°C
3–4		25°C	25°C	25°C	30°C	25°C
4–5		25°C	25°C	25°C	25°C	30°C
5–6		25°C	25°C	25°C	25°C	25°C
SURVIVAL:		Yes	Yes	Yes	No	Yes

Explain these results.

Answer: By varying the temperature during different stages of development, researchers can pinpoint the stage when the function of the protein encoded by this *ts* allele is critical. As shown, embryos fail to survive if they are subjected to the nonpermissive temperature between 3 and 4 hours after fertilization, but they do survive if subjected to the nonpermissive temperature at other times of development. These results indicate that this protein plays a crucial role at the 3–4-hour stage of development.

S4. An intriguing question in developmental genetics is, How can a particular gene, such as *even-skipped,* be expressed in a multiple banding pattern as seen in figure 23.9? Another way of asking this question is, How is the positional information within the broad bands of the gap genes able to be deciphered in a way that causes the pair-rule genes to be expressed in this alternating banding pattern? The answer lies in a complex mechanism of genetic regulation. Certain pair-rule genes have several **stripe-specific enhancers** that are controlled by multiple transcription factors. A stripe-specific enhancer is typically a short segment of DNA, 300 to 500 bp in length, that contains binding sequences that are recognized by several different transcription factors. This term is a bit misleading since a stripe-specific enhancer is a regulatory region that contains both enhancer and silencer elements.

As an example, Michael Levine and his colleagues have investigated stripe-specific enhancers that are located near the promoter of the *even-skipped* gene. A segment of DNA, termed the stripe 2 enhancer, controls the expression of the *even-skipped* gene; this enhancer is responsible for the expression of the *even-skipped* gene in stripe 2, which corresponds to parasegment 3 of the embryo. The stripe 2 enhancer is a segment of DNA that contains binding sites for four transcription factors that are the products of the *Krüppel, bicoid, hunchback,* and *giant* genes. The Hunchback and Bicoid transcription factors bind to this enhancer and activate the transcription of the *even-skipped* gene. In contrast, the transcription factors encoded by the *Krüppel* and *giant* genes bind to the stripe 2 enhancer and repress transcription. The figure shown next describes the concentrations of these four transcription factor proteins in the region of parasegment 3 (i.e., stripe 2) in the *Drosophila* embryo.

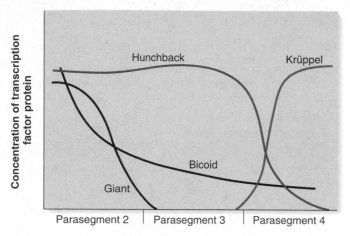

Region of *Drosophila* embryo

To study stripe-specific enhancers, researchers have constructed artificial genes in which the enhancer is linked to a reporter gene; the expression of the reporter gene is easy to detect.

The next figure shows the results of an experiment in which an artificial gene was made by putting the stripe 2 enhancer next to the β-galactosidase gene. This artificial gene was introduced into *Drosophila* and then embryos containing this gene were analyzed for β-galactosidase activity. If a region of the embryo is expressing β-galactosidase, the region will stain darkly because β-galactosidase converts a colorless compound into a dark blue compound.

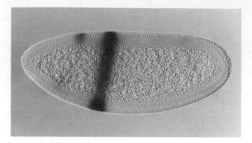

Explain these results.

Answer: As shown in the first figure to this problem, the concentrations of the Hunchback and Bicoid transcription factors are relatively high in the region of the embryo corresponding to stripe 2 (which is parasegment 3). The levels of Krüppel and Giant are very low in this region. Therefore, the high levels of activators and low levels of repressors cause the *even-skipped* gene to be transcribed. In this experiment, β-galactosidase was made only in stripe 2 (i.e., parasegment 3). These results show that the stripe-2-specific enhancer controls gene expression only in parasegment 3. Because we know that the *even-skipped* gene is expressed as several alternating stripes (as seen in fig. 23.9), the *even-skipped* gene must contain other stripe-specific enhancers that allow it to be expressed in these other alternating parasegments.

Conceptual Questions

C1. In the case of multicellular animals, what are the four types of cellular processes that occur so that a fertilized egg can develop into an adult organism? Briefly discuss the role of each process.

C2. The arrangement of body axes of the fruit fly are shown in figure 23.1g. Are the following statements true or false with regard to body axes in the mouse?

A. Along the antero-posterior axis, the head is posterior to the tail.

B. Along the dorso-ventral axis, the vertebrae of the back are dorsal to the stomach.

C. Along the dorso-ventral axis, the feet are dorsal to the hips.

D Along the proximo-distal axis, the toes on the hind legs are distal to the hips of the hind legs.

C3. If you observed fruit flies with the following developmental abnormalities, would you guess that a mutation has occurred in a segmentation gene or a homeotic gene? Explain your guess.

A. Three abdominal segments were missing.

B. One abdominal segment had legs.

C. A fly with the correct number of segments had two additional thoracic segments and two fewer abdominal segments.

C4. Which of the following statements are true with regard to positional information in *Drosophila*?

A. Morphogens are a type of molecule that conveys positional information.

B. Morphogenetic gradients are established only in the oocyte, prior to fertilization.

C. Cell adhesion molecules also provide a way for a cell to obtain positional information.

C5. Discuss the morphological differences between the parasegments and segments of *Drosophila*. Discuss the evidence, providing specific examples, that suggests the parasegments of the embryo are the subdivisions for the organization of gene expression.

C6. Here are schematic diagrams of mutant larvae.

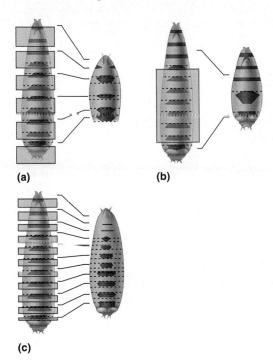

(a)   (b)

(c)

The *left* side of each pair shows a wild-type larva, with *gray boxes* showing the sections that are missing in the mutant larva. Which type of gene is defective in each larva: a gap gene, a pair-rule gene, or a segment-polarity gene?

C7. Describe what a morphogen is and how it exerts its effects. What do you expect will happen when a morphogen is expressed in the wrong place in an embryo? List five examples of morphogens that function in *Drosophila*.

C8. What is the meaning of the term *positional information?* Discuss three different ways that cells can obtain positional information. Which of these three ways do you think is the most important for the formation of a segmented body pattern in *Drosophila?*

C9. Gradients of morphogens can be preestablished in the oocyte. Also, later in development, morphogens can be secreted from cells. How are these two processes similar and different?

C10. Discuss how the anterior portion of the antero-posterior axis is established. What aspects of oogenesis are critical in establishing this axis? What do you think would happen if the *bicoid* mRNA was not trapped at the anterior end but instead diffused freely throughout the oocyte?

C11. Describe the function of the Bicoid protein. Explain how its ability to exert its effects in a concentration-dependent manner is a critical feature of its function.

C12. With regard to development, what are the roles of the maternal effect genes versus the zygotic genes? Which types of genes are needed earlier in the development process?

C13. Discuss the role of homeotic genes in development. Explain what happens to the phenotype of a fly when a gain-of-function homeotic gene mutation occurs in an abnormal region of the embryo. What are the consequences of a loss-of-function mutation in such a gene?

C14. Describe the molecular features of the homeobox and homeodomain. Explain how these features are important in the function of homeotic genes.

C15. What would you predict to be the phenotype of a *Drosophila* larva whose mother was homozygous for a loss-of-function allele in the *nanos* gene?

C16. Based on the photographs in figure 23.11, in which segments would the *Antp* gene normally be expressed?

C17. If a mutation in a homeotic gene produced the following phenotypes, would you expect the mutation to be a loss-of-function or a gain-of-function allele? Explain your answer.

A. An abnominal segment has antennae attached to it.

B. The most anterior abdominal segment resembles the most posterior thoracic segment.

C. The most anterior thoracic segment resembles the most posterior abdominal segment.

C18. Explain how loss-of-function alleles in the following categories of genes would affect the morphologies of *Drosophila* larvae:

A. Gap genes

B. Pair-rule genes

C. Segment polarity genes

C19. What is the difference between a maternal effect gene versus a zygotic gene? Of the following genes that play a role in *Drosophila* development, which would be maternal effect genes and which would be zygotic? Explain your answer.

A. *nanos*

B. *Antp*

C. *bicoid*

D. *lab*

C20. Cloning of mammals (such as Dolly) is described in chapter 19. Based on your understanding of animal development, explain why an enucleated egg is needed to clone mammals. In other words, what features of the oocyte are essential for animal development?

C21. A hypothetical cell lineage is shown here.

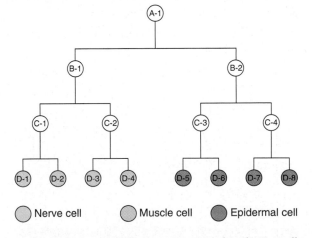

Nerve cell   Muscle cell   Epidermal cell

A gene, which we will call gene *X,* is activated in the B-1 cell so that the B-1 cell will progress through the proper developmental stages to produce three nerve cells (D-1, D-2, and D-3) and one muscle cell (D-4). Gene *X* is normally inactivated in A-1, C-1, and

C-2 cells as well as the four D cells. Draw the expected cell lineages if a heterochronic mutation had the following effects:

A. Gene *X* is turned on one cell division too early.

B. Gene *X* is turned on one cell division too late.

C22. What is a heterochronic mutation? How does it affect the phenotypic outcome of an organism? What phenotypic effects would you expect if a heterochronic mutation affected the cell lineage that determines the fates of intestinal cells?

C23. Discuss the similarities and differences between the *bithorax* and *Antennapedia* complexes in *Drosophila* and the *Hox* gene complexes in mice.

C24. What is cell differentiation? Discuss the role of bHLH proteins in the differentiation of muscle cells. Explain how they work at the molecular level. In your answer, indicate how protein dimerization is a key feature in gene regulation.

C25. The *myoD* gene in mammals plays a role in muscle cell differentiation, while the *Hox* genes are homeotic genes that play a role in the differentiation of particular regions of the body. Explain how the functions of these genes are similar and different.

C26. What is a totipotent cell? In the following examples, which cells would be totipotent? In the case of multicellular organisms (humans and corn), explain which cell types would be totipotent.

A. Humans

B. Corn

C. Yeast

D. Bacteria

C27. What is a meristem? Explain the role of meristems in plant development.

C28. Discuss the morphological differences between plants and animals. How are they different at the cellular level? How are they similar at the genetic level?

C29. Predict the phenotypic consequences of each of the following mutations:

A. *apetala1* defective

B. *pistillata* defective

C. *apetala1* and *pistillata* defective

Experimental Questions

E1. Researchers have used the cloning methods described in chapter 18 to clone the *bicoid* gene and express large amounts of the Bicoid protein. The Bicoid protein was then injected into the posterior end of a zygote immediately after fertilization. What phenotypic results would you expect to occur? What do you think would happen if the Bicoid protein was injected into a segment of a larva?

E2. Compare and contrast the experimental advantages of *Drosophila* and *C. elegans* in the study of developmental genetics.

E3. What is meant by the term *cell fate*? What is a lineage diagram? Discuss the experimental advantage of having a lineage diagram. What is a cell lineage?

E4. Explain why a lineage diagram is necessary to determine if a mutation is heterochronic.

E5. Explain the rationale behind the use of the "bag of worms" phenotype as a way to identify heterochronic mutations.

E6. Here are the results of cell lineage analyses of hypodermal cells in wild-type and mutant strains of *C. elegans*.

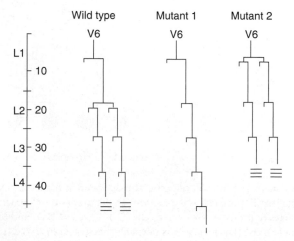

Explain the nature of the mutations in the altered strains.

E7. Take a look at solved problem S3 before answering this question. *Drosophila* embryos carrying a *ts* mutation were exposed to the permissive (25°C) or nonpermissive (30°C) temperature at different stages of development. Explain these results.

Time After Fertilization (hours)	Group: 1	2	3	4	5
0–1	25°C	25°C	25°C	25°C	25°C
1–2	25°C	30°C	25°C	25°C	25°C
2–3	25°C	25°C	30°C	25°C	25°C
3–4	25°C	25°C	25°C	30°C	25°C
4–5	25°C	25°C	25°C	25°C	30°C
5–6	25°C	25°C	25°C	25°C	25°C
SURVIVAL:	Yes	No	No	Yes	Yes

E8. All of the homeotic genes in *Drosophila* have been cloned. As discussed in chapter 18, cloned genes can be manipulated in vitro. They can be subjected to cutting and pasting, site-directed mutagenesis, etc. After *Drosophila* genes have been altered in vitro, they can be inserted into a *Drosophila* transposon vector (i.e., a P element vector) and then the genetic construct containing the altered gene within a P-element can be injected into *Drosophila* embryos. The P element will then transpose into the chromosomes and thereby introduce one or more copies of the altered gene into the *Drosophila* genome. This method is termed *P element transformation*.

With these ideas in mind, how would you make a mutant gene with a "gain-of-function" in which the *Antp* gene would be expressed where the *abd-A* gene is normally expressed? What phenotype would you expect for flies that carried this altered gene?

E9. You will need to understand solved problem S4 before answering this question. If the artificial gene containing the stripe 2 enhancer and the β-galactosidase gene was found within an embryo that also contained the following loss-of-function mutations, what results would you expect? In other words, would there be a stripe or not? Explain why.

A. *Krüppel*

B. *bicoid*

C. *hunchback*

D. *giant*

E10. Two techniques that are commonly used to study the expression patterns of genes that play a role in development are Northern blotting and *in situ* hybridization. As described in chapter 18, Northern blotting can be used to detect RNA that is transcribed from a particular gene. In this method, a specific RNA is detected by using a short segment of cloned DNA as a probe. The DNA probe, which is radioactive, is complementary to the RNA that the researcher wishes to detect. After the radioactive probe DNA binds to the RNA within a blot of a gel, the RNA is visualized as a dark (radioactive) band on an X-ray film. For example, a DNA probe that is complementary to the *bicoid* mRNA could be used to specifically detect the amount and size of the mRNA in a blot.

A second technique, termed *in situ* hybridization, can be used to identify the locations of genes on chromosomes. This technique can also be used to locate gene products within oocytes, embryos, and larvae. For this reason, it has been commonly used by developmental geneticists to understand the expression patterns of genes during development. The photograph in figure 23.7*b* is derived from the application of the *in situ* hybridization technique. In this case, the probe was complementary to *bicoid* mRNA.

Now here is the question. Suppose a researcher has three different *Drosophila* strains that have loss-of-function mutations in the *bicoid* gene. We will call them *bicoid-A, bicoid-B,* and *bicoid-C;* the wild type is designated *bicoid⁺*. To study these mutations, phenotypically normal female flies that are homozygous for the *bicoid* mutation were obtained, and their oocytes were analyzed using these two techniques. A wild-type strain was also analyzed as a control. In other words, RNA was isolated from some of the oocytes and analyzed by Northern blotting, and some oocytes were subjected to *in situ* hybridization. In both cases, the probe was complementary to the *bicoid* mRNA. The results are shown here.

Northern blot

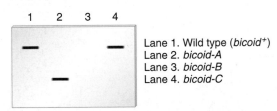

Lane 1. Wild type (*bicoid⁺*)
Lane 2. *bicoid-A*
Lane 3. *bicoid-B*
Lane 4. *bicoid-C*

In situ hybridization

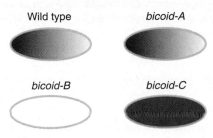

A. How can phenotypically normal female flies be homozygous for a loss-of-function allele in the *bicoid* gene?

B. Explain the type of mutation (e.g., deletion, point mutation, etc.) in each of the three strains. Explain how the mutation may cause a loss of normal function for the *bicoid* gene product.

C. Discuss how the use of both techniques provides more definitive information compared to the application of just one of the two techniques.

E11. Explain one experimental strategy you could follow to determine the functional role of the mouse *HoxD-3* gene.

E12. In the experiment of figure 23.14, suggest reasons why the *n536, n355,* and *n540* strains have an egg-laying defect.

E13. Another way to study the role of proteins (e.g., transcription factors) that function in development is to microinject the mRNA that encodes a protein, or the purified protein itself, into an oocyte or embryo, and then determine how this affects the subsequent development of the embryo, larva, and adult. For example, if Bicoid protein is injected into the posterior region of an oocyte, the resulting embryo will develop into a larva that has anterior structures at both ends. Based on your understanding of the function of these developmental genes, what would be the predicted phenotype if the following proteins or mRNAs were injected into normal oocytes?

A. *Nanos* mRNA injected into the anterior end of an oocyte

B. Antp protein injected into the posterior end of an embryo

C. *Toll* mRNA injected the dorsal side of an early embryo

E14. Why have geneticists been forced to use reverse genetics to study the genes involved in vertebrate development? Explain how this strategy differs from traditional genetic analyses like those done by Mendel.

Questions for Student Discussion/Collaboration

1. Compare and contrast the experimental advantages and disadvantages of *Drosophila, C. elegans,* mammals, and *Arabidopsis*.

2. It seems that developmental genetics boils down to a complex network of gene regulation. Try to draw how this network is structured for *Drosophila*. How many genes do you think are necessary to describe a complete developmental network for the fruit fly? How many genes do you think are needed for a network to specify one segment? Do you think it is more difficult to identify genes that are involved in the beginning, middle, or end of this network? Knowing what you know about *Drosophila* development, suppose you were trying to identify all the genes needed for development in a chicken. Would you first try to identify genes that are necessary for early development, or would you begin by identifying genes involved in cell differentiation?

3. At the molecular level, how do you think a gain-of-function mutation in a developmental gene might cause it to be expressed in the wrong place or at the wrong time? Explain what type of DNA sequence would be altered.

Note: All answers appear at the website for this textbook; the answers to even-numbered questions are in the back of the textbook.

www.mhhe.com/brooker

Visit the Online Learning Center for practice tests, answer keys, and other learning aids for this chapter. Enhance your understanding of genetics with our interactive exercises, web links, news feeds, tutorial service, and much more.

QUANTITATIVE GENETICS

::

24

Quantitative genetics concerns the study of traits that can be described numerically. Such traits are usually controlled by more than one gene. In humans, quantitative traits include height, the shape of our noses, and the rate at which we metabolize food, to name a few examples. The features of these traits can be characterized and analyzed with numbers. Quantitative genetics is an important branch of genetics for several reasons. In agriculture, most of the key characteristics of interest to plant and animal breeders are quantitative traits. These include traits such as weight, fruit size, resistance to disease, and the ability to withstand harsh environmental conditions. As we will see later in chapter 24, genetic techniques have improved our ability to develop strains of agriculturally important species with desirable quantitative traits.

Another important reason to study quantitative genetics is its relationship to evolution. Many of the traits that allow a species to adapt to its environment are quantitative. Examples include the long neck of a giraffe, the swift speed of the cheetah, and the sturdy branches of trees in windy climates. The importance of quantitative traits in the evolution of species will be discussed in chapter 26. In chapter 24, we will examine how genes and the environment contribute to the phenotypic expression of quantitative traits.

24.1 QUANTITATIVE TRAITS

When we compare characteristics among members of the same species, the differences are often quantitative rather than qualitative. A **quantitative trait** is any trait that varies measurably in a given species. Humans, for example, all have the same basic anatomical features (two eyes, two ears, etc.), but they differ in quantitative ways. People vary with regard to height, weight, the shape of facial features, pigmentation, and many other characteristics. As shown in table 24.1, quantitative traits can be categorized as anatomical, physiological, and behavioral. In addition, many human diseases exhibit characteristics and inheritance patterns analogous to those of quantitative traits.

In many cases, quantitative traits are easily measured and described numerically. Height and weight can be measured in centimeters and kilograms. Speed can be measured in kilometers per hour, and metabolic rate can be assessed as the grams of glucose burned per minute. Behavioral traits can also be quantified. A mating call can be evaluated with regard to its duration, sound level, and pattern. The ability to learn a maze can be described as the time and/or repetitions it takes to learn the skill. Finally, complex diseases such as diabetes can also be studied and described via numerical parameters. For example, the severity of the disease can be assessed by the age of onset or by the amount of insulin needed to prevent adverse symptoms.

From a scientific viewpoint, the measurement of quantitative traits is essential when comparing individuals or evaluating groups of individuals. It is not very informative to say that two people are tall. Instead, we are better informed if we know that one person is 6 feet 4 inches and the other is 6 feet 7 inches. In this branch of genetics, the measurement of a quantitative trait is how we describe the phenotype.

In the early 1900s, Francis Galton in England and his student Karl Pearson showed that many traits in humans and domesticated animals are quantitative in nature. To understand the underlying genetic basis of these traits, they founded what became known as the **biometric field** of genetics. During this period, Galton and Pearson developed various statistical tools for studying the variation of quantitative traits within groups of individuals; many of these tools are still in use today. In this sec-

tion, we will examine how quantitative traits are measured and how statistical tools are used to analyze their variation within groups.

Quantitative Traits Exhibit a Continuum of Phenotypic Variation That May Follow a Normal Distribution

In part II of this textbook, we considered many traits that fell into discrete categories. For example, fruit flies might have white eyes or red eyes, and pea plants might have wrinkled or smooth seeds. The alleles that govern these traits affect the phenotype in a qualitative way. In analyzing crosses involving these types of traits, each offspring can be put into a particular phenotypic category. Such attributes are called **discontinuous traits.**

In contrast, **quantitative traits** show a continuum of phenotypic variation within a group of individuals. For such traits, it is often impossible to place organisms into a discrete phenotypic class. For example, figure 24.1a is a classic photograph showing the range of heights of 175 students at the Connecticut Agricultural College. Though there is a minimum and maximum height, the range of heights between these minimum and maximum values is fairly continuous.

Because quantitative traits do not naturally fall into a small number of discrete categories, an alternative way to describe them is a **frequency distribution.** To construct a frequency distribution, the trait is divided arbitrarily into a number of convenient, discrete phenotypic categories. For example, in figure 24.1, the range of heights is partitioned into 1-inch intervals. Then a graph is made that shows the numbers of individuals found in each of several phenotypic categories.

Figure 24.1b shows a frequency distribution for the heights of students pictured in part a. The measurement of height is plotted along the x-axis, and the number of individuals who exhibit that phenotype is plotted on the y-axis. The values along the x-axis are divided into the discrete 1-inch intervals that define the phenotypic categories, even though height is essentially continuous within a group of individuals. For example, in figure 24.1a, 22 students were between 64.5 and 65.5 inches in height, which is plotted as the point (65 inches, 22 students) on the graph in figure 24.1b. This type of analysis can be conducted on any group of individuals who vary with regard to a quantitative trait.

The line in the frequency distribution depicts a **normal distribution curve,** a distribution for an infinite sample in which the trait of interest varies in a symmetric way around an average value. The distribution of measurements of many biological characteristics is approximated by a symmetrical bell curve like the line in figure 24.1b. We will consider the significance of this type of distribution next.

Statistical Methods Are Used to Evaluate a Frequency Distribution Quantitatively

Statistical tools are used to analyze a normal distribution in a number of ways. One measure that you are probably familiar with is a parameter called the **mean.** The mean is the sum of all

---

TABLE 24.1

Types of Quantitative Traits

Trait	Examples
Anatomical traits	Height, weight, number of bristles in *Drosophila*, ear length in corn, and the degree of pigmentation in flowers and skin
Physiological traits	Metabolic traits, speed of running and flight, ability to withstand harsh temperatures, and milk production in mammals
Behavioral traits	Mating calls, courtship rituals, ability to learn a maze, and the ability to grow or move toward light
Complex diseases	Diabetes, hypertension, arthritis, obesity

Number of individuals	1	0	0	1	5	7	7	22	25	26	27	17	11	17	4	4	1
Height in inches	58	59	60	61	62	63	64	65	66	67	68	69	70	71	72	73	74

(a)

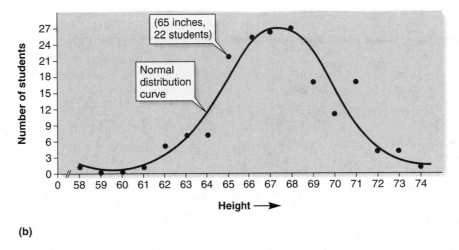

(b)

FIGURE 24.1 **Normal distribution of a quantitative trait.** (a) A classic photograph showing the distribution of heights in 175 students at the Connecticut Agricultural College. (b) A frequency distribution for the heights of students shown in (a).

the values in the group divided by the number of individuals in the group. It is computed using the following formula:

$$\bar{X} = \frac{\Sigma X}{N}$$

where

$\bar{X}$ is the mean

ΣX is the sum of all the values in the group

N is the number of individuals in the group

A more generalized form of this equation can be used.

$$\bar{X} = \frac{\Sigma f_i X}{N}$$

where

$\bar{X}$ is the mean

$\Sigma f_i X$ is the sum of all the values in the group; each value in the group is multiplied by its frequency (f_i) in the group.

N is the number of individuals in the group

For example, suppose a bushel of corn had ears with the following lengths (rounded to the nearest centimeter): 15, 14, 13, 14, 15, 16, 16, 17, 15, and 15. Then

$$\bar{X} = \frac{4(15) + 2(14) + 13 + 2(16) + 17}{10}$$

$$\bar{X} = 15 \text{ cm}$$

In genetics, we are often interested in the amount of phenotypic variation that exists in a group. As we will see later in chapter 24, and in chapters 25 and 26, variation lies at the heart of breeding experiments and evolution. Without variation, selective breeding is not possible, and evolution cannot favor one phenotype over another. A common way to evaluate variation within a population is with a statistic called the **variance.** The variance is the sum of the squared deviations from the mean divided by the degrees of freedom (df equals N-1; see chapter 2 for a review of df).

$$V_X = \frac{\Sigma f_i (X - \bar{X})^2}{N - 1}$$

where

V_X is the variance

$X - \bar{X}$ is the difference between each value and the mean

N equals the number of observations

For example, if we use the values given previously for the lengths of ears of corn, the variance in this group is calculated as follows:

$$\Sigma f_i (X - \bar{X})^2 = 4(15 - 15)^2 + 2(14 - 15)^2$$
$$+ (13 - 15)^2 + 2(16 - 15)^2 + (17 - 15)^2$$

$$\Sigma f_i (X - \bar{X})^2 = 0 + 2 + 4 + 2 + 4$$

$$\Sigma f_i (X - \bar{X})^2 = 12 \text{ cm}^2$$

$$V_X = \frac{\Sigma f_i (X - \bar{X})^2}{N - 1}$$

$$V_X = \frac{12 \text{ cm}^2}{9}$$

$$V_X = 1.33 \text{ cm}^2$$

Because the variance is computed from squared deviations, it is a statistic that may be difficult to understand intuitively. For example, weight can be measured in grams; the corresponding variance would be measured in square grams. Even so, variances are centrally important in the analysis of quantitative traits because they are **additive** under certain conditions. This means that the variances for different factors that contribute to a quantitative trait, such as genetic and environmental factors, can be added together to predict the total variance for that trait. Later in chapter 24, we will examine how this property is useful in predicting the outcome of genetic crosses.

To gain an intuitive grasp for variation, we can take the square root of the variance. This statistic is called the **standard deviation (SD).** Again, using the same values for length, the standard deviation is:

$$SD = \sqrt{V_X} = \sqrt{1.33}$$

$$SD = 1.15 \text{ cm}$$

If the values in a population follow a normal distribution, then it is easier to appreciate the amount of variation by considering the standard deviation. Figure 24.2 illustrates the relationship between the standard deviation and the percentages of individuals who deviate from the mean. Approximately 68% of all individuals have values within one standard deviation from the mean, either in the positive or negative direction. About 95% are within two standard deviations, and 99.7% are within three standard deviations. When a quantitative characteristic follows a normal distribution, less than 0.3% of the individuals will have values that are more or less than three standard deviations from the mean of the population. In our corn example, three standard deviations equals 3.45 cm. Therefore, we would expect that fewer than 0.3% of the ears of corn would be less than 11.55 cm or greater than 18.45 cm, assuming that length follows a normal distribution.

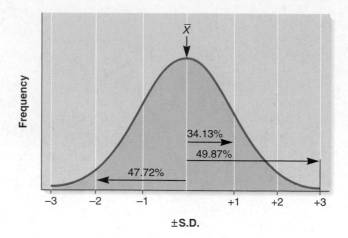

FIGURE 24.2   **The relationship between the standard deviation and the proportions of individuals in a normal distribution.** For example, 34.13% of the individuals in a population are between the mean and one standard deviation above the mean; 47.72% are between the mean and two standard deviations below the mean.

Some Statistical Methods Compare Two Variables to Each Other

In many biological problems, it is useful to compare two different variables. For example, we may wish to compare the occurrence of two different phenotypic traits. Do obese animals have larger hearts? Are brown eyes more likely to occur in people with dark skin pigmentation? A second type of comparison is between traits and environmental factors. Does insecticide resistance occur more frequently in areas that have been exposed to insecticides? Is heavy body weight more prevalent in colder climates? Finally, a third type of comparison is between traits and genetic relationships. Do tall parents tend to produce tall offspring? Do women with diabetes tend to have brothers with diabetes?

To gain insight into such questions, a statistic known as the **correlation coefficient** is often applied. To calculate this statistic, we first need to determine the **covariance,** which describes the degree of variation between two variables within a group. The covariance is similar to the variance, except that we multiply together the deviations of two different variables rather than squaring the deviations from a single factor.

$$CoV_{(X,Y)} = \frac{\Sigma[(X - \bar{X})(Y - \bar{Y})]}{N - 1}$$

where

X represents the values for one variable and $\bar{X}$ is the mean value in the group

Y represents the values for another variable and $\bar{Y}$ is the mean value in that group

N is the total number of pairs of observations

As an example, let's consider the weight of cattle at 5 years of age between parents and offspring. A farmer might be interested in this relationship to determine if genetic variation plays a role in the weight of cattle. The data here describe the 5-year weights for 10 different cows and their female offspring.

Mother's Weight (kg)	Offspring's Weight (kg)	$X - \bar{X}$	$Y - \bar{Y}$	$(X - \bar{X})(Y - \bar{Y})$
570	568	−26	−30	780
572	560	−24	−38	912
599	642	3	44	132
602	580	6	−18	−108
631	586	35	−12	−420
603	642	7	44	308
599	632	3	34	102
625	580	29	−18	−522
584	605	−12	7	−84
575	585	21	−13	273
$\bar{X} = 596$	$\bar{Y} = 598$			$\Sigma = 1,373$
$SD_X = 21.1$	$SD_Y = 30.5$			

$$CoV_{(X,Y)} = \frac{\Sigma[(X - \bar{X})(Y - \bar{Y})]}{N - 1}$$

$$CoV_{(X,Y)} = \frac{1,373}{10 - 1}$$

$$CoV_{(X,Y)} = 152.6$$

After we have calculated the covariance, we can evaluate the strength of the association between the two variables by calculating a **correlation coefficient (r)**.

$$r_{(X,Y)} = \frac{CoV_{(X,Y)}}{SD_X SD_Y}$$

This value, which ranges between +1 and −1, indicates how two factors vary in relation to each other. A value of +1 is a perfect correlation. It means that two factors vary in a completely predictable way relative to each other; as one factor increases, the other will increase with it. A value of zero indicates there is no detectable way in which the two factors vary relative to each other; the values of the two factors are not related. Finally, an inverse correlation, in which the correlation coefficient is negative, indicates that the two factors tend to vary in opposite ways to each other; as one factor increases, the other will decrease.

Let's use the data of 5-year weights for mother and offspring to calculate a correlation coefficient.

$$r_{(X,Y)} = \frac{152.6}{(21.1)(30.5)}$$

$$r_{(X,Y)} = 0.237$$

The result is a positive correlation between the 5-year weights of mother and offspring. In other words, the positive correlation value suggests that heavy mothers tend to have heavy offspring and lighter mothers have lighter offspring.

However, after a correlation coefficient has been calculated, one must evaluate whether the r value represents a true association between the two variables or whether it could be simply due to chance. To accomplish this, we can test the hypothesis that there is no real correlation (i.e., the null hypothesis): The r value differs from zero only as a matter of random sampling error. This is the same approach followed in the chi square analysis described in chapter 2. Like the chi square value, the significance of the correlation coefficient is directly related to sample size and the degrees of freedom (df). In testing the significance of correlation coefficients, df equals $N - 2$, which is one less than the degrees of freedom of variance (i.e., df for variance equals $N - 1$). Table 24.2 shows the relationship between the r values and degrees of freedom at the 5% and 1% significance levels.

This approach, though, is valid only if several assumptions are met. First, the values of X and Y in the study must have been obtained by an unbiased sampling of the entire population. In addition, this approach assumes that the scores of X and Y follow a normal distribution (like fig. 24.1) and that the relationship between X and Y is linear.

To illustrate the use of table 24.2, let's consider the correlation we have just calculated for 5-year weights of mother and offspring. In this case, we obtained a value of 0.237 for r, and the value of N was 10. Under these conditions, df equals 8. According to table 24.2, it is fairly likely that this value could have occurred as a matter of random sampling error. Therefore, we cannot conclude that the positive correlation is due to a true association

TABLE 24.2

Values of r at the 5% and 1% Significance Levels

Degrees of Freedom (df)	5%	1%	Degrees of Freedom (df)	5%	1%
1	.997	1.000	24	.388	.496
2	.950	.990	25	.381	.487
3	.878	.959	26	.374	.478
4	.811	.917	27	.367	.470
5	.754	.874	28	.361	.463
6	.707	.834	29	.355	.456
7	.666	.798	30	.349	.449
8	.632	.765	35	.325	.418
9	.602	.735	40	.304	.393
10	.576	.708	45	.288	.372
11	.553	.684	50	.273	.354
12	.532	.661	60	.250	.325
13	.514	.641	70	.232	.302
14	.497	.623	80	.217	.283
15	.482	.606	90	.205	.267
16	.468	.590	100	.195	.254
17	.456	.575	125	.174	.228
18	.444	.561	150	.159	.208
19	.433	.549	200	.138	.181
20	.423	.537	300	.113	.148
21	.413	.526	400	.098	.128
22	.404	.515	500	.088	.115
23	.396	.505	1000	.062	.081

Note: df equals $N - 2$.

From Spence, J. T. et al. (1976) *Elementary Statistics*, Prentice-Hall, Englewood Cliffs, New Jersey.

between the weights of mother and offspring (i.e., we cannot reject the null hypothesis).

In an actual experiment, however, a researcher would examine many more pairs of mothers and offspring, perhaps 500 to 1,000. If a correlation of 0.237 was observed for $N = 1,000$, the value would be significant at the 1% level. We would therefore reject the null hypothesis that weights are not associated with each other. Instead, we would conclude that there is a real association between the weights of mothers and their offspring. In fact, these kinds of experiments have been done for cattle weights, and the correlations between mothers and offspring have often been found to be significant.

Whenever a statistically significant correlation is obtained, we must be very cautious in interpreting its meaning. An r value that is statistically significant need not imply a cause-and-effect relationship. When parents and offspring display a significant correlation for a trait, we should not jump to the conclusion that genetics is the underlying cause of the positive association. In many cases, parents and offspring share similar environments, so that the positive association might be rooted in environmental factors. In general, correlations are quite useful in identifying positive or negative associations between two variables. We should use caution, however, because this statistic, by itself, cannot prove that the association is due to a cause and effect. When it has been established that the variables are related due to cause and effect (i.e., one variable affects the outcome of another variable), it is valid to use a **regression analysis** to predict how much one variable will change in response to the other variable. This approach is described in solved problem S4 at the end of chapter 24.

24.2 POLYGENIC INHERITANCE

In the preceding section, we saw that quantitative traits tend to show a continuum of variation and can be analyzed with various statistical tools. At the beginning of the 1900s, there was great debate concerning the inheritance of quantitative traits. The biometric school, founded by Francis Galton and Karl Pearson, argued that these types of traits are not controlled by discrete genes that affect phenotypes in a predictable way. To some extent, the biometric school favored a blending theory of inheritance, which had been proposed many years earlier. Alternatively, the followers of Mendel, led by William Bateson in England and William Castle in the United States, held firmly to the idea that traits are governed by genes, which are inherited as discrete units. As we know now, Bateson and Castle were correct. However, as we will see in this section, the difficulty of studying quantitative traits lies in the fact that these traits are influenced by multiple genes and substantial environmental factors.

Most quantitative traits are polygenic and exhibit a continuum of phenotypic variation. The term **polygenic inheritance** refers to the transmission of traits that are governed by two or more different genes. The locations on chromosomes that harbor genes that affect the outcome of quantitative traits are called **quantitative trait loci (QTLs).** As discussed later, QTLs are chro-

mosomal regions that are identified by genetic mapping. Depending on the precision of such mapping, it is likely that a QTL will contain many genes. In some cases, only a single gene in such a region may play an important role in a quantitative trait. Alternatively, a QTL may be a region that contains two or more closely linked genes that affect quantitative traits.

Just a few years ago, it was extremely difficult for geneticists to determine the inheritance patterns for genes underlying polygenic traits, particularly those determined by three or more genes having multiple alleles for each gene. Recently, however, molecular genetic tools (described in chapters 19 and 20) have greatly enhanced our ability to find regions in the genome where QTLs are likely to reside. This has been a particularly exciting advance in the field of quantitative genetics. In some cases, the identification of QTLs may allow the improvement of quantitative traits in agriculturally important species.

Polygenic Inheritance and Environmental Factors Create Overlaps Between Genotypes and Phenotypes

The first experiment demonstrating that continuous variation is related to polygenic inheritance was conducted by the Swedish geneticist Herman Nilsson-Ehle in 1909. He studied the inheritance of red pigment in the hull of wheat, *Triticum aestivum* (fig. 24.3*a*). When true-breeding plants with white hulls were crossed to a variety with red hulls, the F_1 generation had an intermediate color. When the F_1 generation was allowed to self-fertilize, there was great variation in redness in the F_2 generation, ranging from white, light red, intermediate red, medium red, and dark red. An unsuspecting observer might conclude that this F_2 generation displayed a continuous variation in hull color. However, as shown in figure 24.3*b*, Nilsson-Ehle carefully categorized the colors of the hulls and discovered that they followed a 1:4:6:4:1 ratio. He concluded that this species is diploid for two different genes that control hull color, each gene existing in a red or white allelic form. He hypothesized that these two loci must contribute additively to the color of the hull.

Later, it was discovered that a third gene also affects hull color. The strains that Nilsson-Ehle had used in his original experiments must have been homozygous for the white allele of this third gene. It makes sense that wheat would be diploid for three genes that affect hull color because we now know that *T. aestivum* is a hexaploid species. As discussed in chapter 8, it was derived from three diploid species. Therefore, *T. aestivum* has six copies of many genes.

As we have just seen, Nilsson-Ehle categorized wheat hull colors into several discrete genotypic categories. However, for many polygenic traits, this is difficult or impossible. In general, as the number of genes controlling a trait increases and the influence of the environment increases, the categorization of phenotypes into discrete genotypic classes becomes increasingly difficult if not impossible. Therefore, a Punnett square cannot be used to analyze most quantitative traits. Instead, statistical methods, which are described in chapter 24, must be employed.

♀ \ ♂	R1R2	R1r2	r1R2	r1r2
R1R2	R1R1R2R2 — Dark red	R1R1R2r2 — Medium red	R1r1R2R2 — Medium red	R1r1R2r2 — Intermediate red
R1r2	R1R1R2r2 — Medium red	R1R1r2r2 — Intermediate red	R1r1R2r2 — Intermediate red	R1r1r2r2 — Light red
r1R2	R1r1R2R2 — Medium red	R1r1R2r2 — Intermediate red	r1r1R2R2 — Intermediate red	r1r1R2r2 — Light red
r1r2	R1r1R2r2 — Intermediate red	R1r1r2r2 — Light red	r1r1R2r2 — Light red	r1r1r2r2 — White

(a) Red and white hulls of wheat

(b) *R1r1R2r2* x *R1r1R2r2*

FIGURE 24.3   **The Nilsson-Ehle experiment studying how continuous variation is related to polygenic inheritance in wheat. (a)** Red and white pigments in the hulls of wheat, *Triticum aestivum*. **(b)** Nilsson-Ehle carefully categorized the colors of the hulls in the F$_2$ generation and discovered that they followed a 1:4:6:4:1 ratio.

GENES→TRAITS   In this example, there are two genes, with two alleles each (red and white), that govern hull color. Offspring can display a range of colors, depending on how many copies of the red allele they inherit. If an offspring is homozygous for the red allele of both genes, it will have very dark red hulls. By comparison, if it carries three red alleles and one white allele, it will be medium red (which is not quite as deep in color). In this way, this polygenic trait can exhibit a range of phenotypes from very dark red to completely white.

Figure 24.4 illustrates how genotypes and phenotypes may overlap for polygenic traits. In this example, the environment (sunlight, soil conditions, and so forth) may affect the phenotypic outcome of a trait in plants (namely, seed weight). Figure 24.4a considers a situation in which seed weight is controlled by one gene with light (*w*) and heavy (*W*) alleles. A heterozygous plant (*Ww*) is allowed to self-fertilize. When the weight is only slightly influenced by variation in the environment, as seen on the *left*, the heavy, light, and intermediate seeds fall into separate, well-defined categories. When the environmental variation has a greater impact on seed weight, as shown on the *right*, there is more phenotypic variation in seed weight within each genotypic class. The variance in the frequency distribution on the *right* is much higher. Even so, it is still possible to classify most individuals into the three main categories.

By comparison, figure 24.4b illustrates a situation in which seed weight is governed by three genes instead of one, each existing in light and heavy alleles. When the environmental variation is low and/or plays a minor role in the outcome of this trait, the expected 1:6:15:20:15:6:1 ratio is observed. As shown in the *upper* illustration in *b*, nearly all the individuals fall within a phenotypic category that corresponds to their genotypes. When the environment has a more significant effect on phenotype, as shown in the *lower* illustration, the situation becomes more ambiguous. For example, individuals with five *W* alleles and one *w* allele have phenotypes that overlap with individuals having six *W* alleles or four *W* alleles and two *w* alleles. Therefore, it becomes difficult to categorize each phenotype into a unique genotypic class. Instead, the trait displays a continuum ranging from light to heavy seed weight.

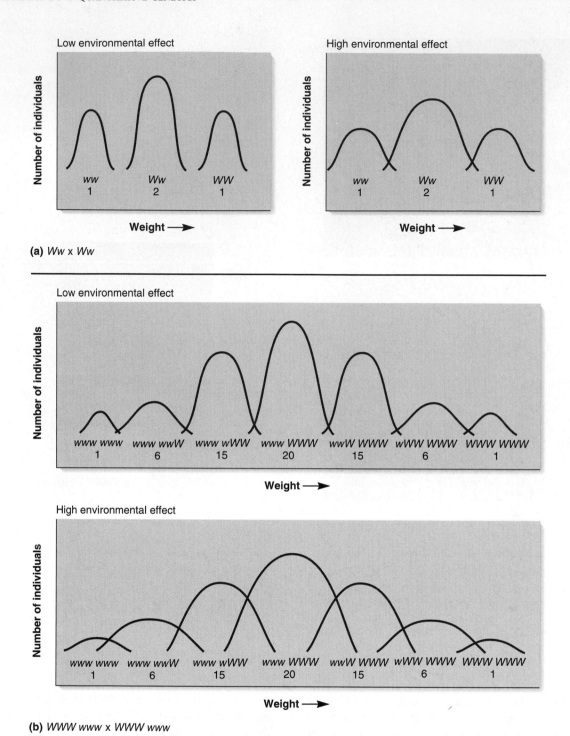

FIGURE 24.4 How genotypes and phenotypes may overlap for polygenic traits. (a) Situations in which seed weight is controlled by one gene, existing in light (*w*) and heavy (*W*) alleles. (b) Situations in which seed weight is governed by three genes instead of one, each existing in light and heavy alleles.

GENES→TRAITS   The ability of geneticists to correlate genotype and phenotype depends on how many genes are involved and how much the environment causes the phenotype to vary. In (a), a single gene influences weight. In the graph on the *left* side, the environment does not cause much variation in weight. This makes it easy to distinguish the three genotypes. There is no overlap in the weights of *ww*, *Ww*, and *WW* individuals. In the graph on the *right* side, the environment causes more variation in weight. In this case, a few individuals with *ww* genotypes will have the same weight as a few individuals with *Ww* genotypes; and a few *Ww* genotypes will have the same weight as *WW* genotypes. As shown in (b), it becomes even more difficult to distinguish genotype based on phenotype when three genes are involved. The overlaps are minor when the environment does not cause much weight variation. However, when the environment causes substantial phenotypic variation, the overlaps between genotypes and phenotypes are very pronounced and greatly confound genetic analysis.

Polygenic Inheritance Explains
DDT Resistance in *Drosophila*

As we have just learned, the phenotypic overlap for a quantitative trait may be so great that it may not be possible to establish discrete genotypic classes. This is particularly true if many genes contribute to the trait. As an alternative, another way to identify the genes affecting polygenic inheritance is to look for linkage between genes affecting quantitative traits and genes affecting discontinuous traits. This approach was first studied in *Drosophila melanogaster,* because many alleles had been identified and mapped to particular chromosomes.

In 1957, James Crow conducted one of the earliest studies to link quantitative genes to discontinuous genes. Crow, who was interested in evolution, spent time studying insecticide resistance in *Drosophila.* He noted, "Insecticide resistance is an example of evolutionary change, the insecticide acting as a powerful selective sieve for concentrating resistant mutants that were present in low frequencies in the population." His aim was to determine the genetic basis for insecticide resistance in *Drosophila melanogaster.* Many mutations were already known in this species, and these could serve as **genetic markers** for each of the four different chromosomes. Dominant alleles are particularly useful because they allow the experimenter to determine which chromosomes are inherited from either parent. The general strategy in identifying QTLs is to cross two strains that have different genetic markers and also differ with regard to the quantitative trait of interest. This produces an F_1 generation that is heterozygous for the markers and usually exhibits an intermediate phenotype for the quantitative trait. The next step is to **backcross** the F_1 offspring to the parental strains. This backcross produces a population of F_2 offspring that differ with regard to their combinations of parental chromosomes. A few offspring may contain all of their chromosomes from one parental strain or the other, but most offspring will contain a few chromosomes from one parental strain and the rest from the other strain. The genetic markers on the chromosomes provide a way to determine whether particular chromosomes were inherited from one parental strain or the other.

To illustrate how genetic markers work, figure 24.5 considers a situation in which two strains differ with regard to a quantitative trait, resistance to DDT, and also differ with regard to dominant alleles on chromosome 3. One strain is resistant to DDT and carries a dominant allele that causes minute bristles (M), while a different strain is sensitive to DDT and carries a dominant allele that causes a rough eye (R). The wild-type alleles, which are recessive, produce normal bristles (m) and smooth eyes (r). At the start of this experiment, it is not known if alleles affecting DDT resistance are located on this chromosome. If offspring from a backcross inherit both copies of chromosome 3 from the DDT-resistant strain, they will have minute bristles and smooth eyes. If they inherit both copies from the DDT-sensitive strain, they will have normal bristles and rough eyes. By comparison, a fly with minute bristles and rough eyes inherited one copy of chromosome 3 from the DDT-resistant

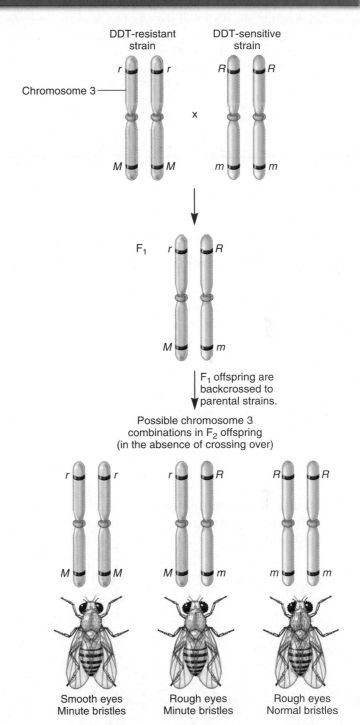

FIGURE 24.5   The use of gene markers to map a QTL affecting DDT resistance. One strain is DDT resistant. On chromosome 3, it also carries a dominant allele that causes minute bristles (M). The other strain is DDT sensitive and carries a dominant allele that causes a roughness to the eye (R). The wild-type alleles, which are recessive, produce normal bristles (m) and smooth eyes (r). F_2 offspring can have either both copies of chromosome 3 from the DDT-resistant strain, both from the sensitive strain, or one of each. This can be discerned by the phenotypes of the F_2 offspring.

strain and one copy from the DDT-sensitive strain. The transmission of the other *Drosophila* chromosomes can also be followed in a similar way. Therefore, the phenotypes of the offspring from the backcross provide a way to discern whether particular chromosomes were inherited from the DDT-resistant or DDT-sensitive strain.

Figure 24.6 shows the protocol followed by James Crow. He began with a DDT-resistant strain that had been produced by exposing flies to DDT for many generations. This DDT-resistant strain was crossed to a sensitive strain. As described previously in figure 24.5, the two strains had allelic markers that made it possible to determine the origins of the different *Drosophila* chromosomes. In this study, only chromosomes X, 2, and 3 were marked

with alleles. Chromosome 4 was neglected due to its very small size. The F_1 flies were backcrossed to both parental strains, and then the F_2 progeny were examined in two ways. First, their phenotypes were examined to determine whether particular chromosomes were inherited from the DDT-resistant or DDT-sensitive strain. Next, the flies were exposed to filter paper impregnated with DDT. It was then determined if the flies survived this exposure for 18 to 24 hours.

■ **THE HYPOTHESIS**

DDT resistance is a polygenic trait.

■ **TESTING THE HYPOTHESIS — FIGURE 24.6** Polygenic inheritance of DDT-resistance alleles in *Drosophila melanogaster.*

Starting material: A DDT-resistant and DDT-sensitive strain of fruit flies.

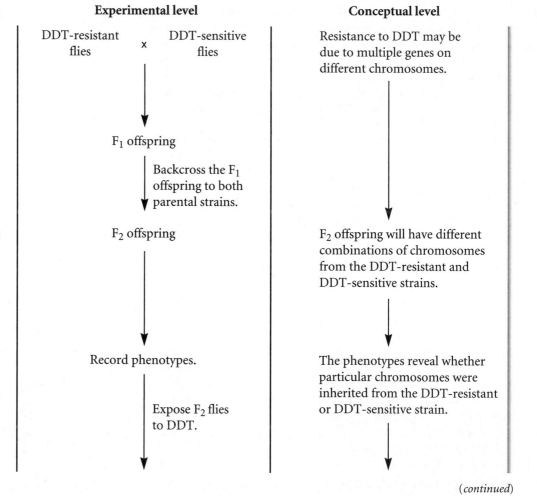

	Experimental level	Conceptual level
1. Cross the DDT-resistant strain to the sensitive strain. In each strain, chromosomes X, 2, and 3 were marked with alleles that provided easily discernible phenotypes.	DDT-resistant flies x DDT-sensitive flies	Resistance to DDT may be due to multiple genes on different chromosomes.
2. Take the F_1 flies and backcross to both parental strains.	F_1 offspring Backcross the F_1 offspring to both parental strains.	
3. Identify the origin of the chromosomes in each fly according to their phenotypes.	F_2 offspring	F_2 offspring will have different combinations of chromosomes from the DDT-resistant and DDT-sensitive strains.
4. Expose each fly to DDT on a filter paper for 18–24 hours.	Record phenotypes. Expose F_2 flies to DDT.	The phenotypes reveal whether particular chromosomes were inherited from the DDT-resistant or DDT-sensitive strain.

(continued)

5. Record the number of survivors.

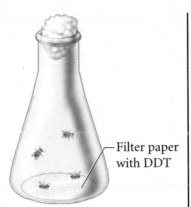

Filter paper
with DDT

The goal is to determine if survival is correlated with the inheritance of specific chromosomes from the DDT-resistant strain.

■ THE DATA

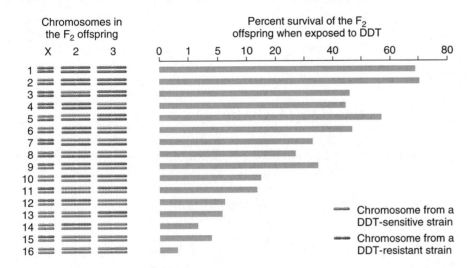

Chromosomes in the F$_2$ offspring

Percent survival of the F$_2$ offspring when exposed to DDT

— Chromosome from a
   DDT-sensitive strain

— Chromosome from a
   DDT-resistant strain

■ INTERPRETING THE DATA

The results of this analysis are shown in the data of figure 24.6. As a matter of chance, some offspring contained all of their chromosomes from one parental strain or the other, but most offspring contained a few chromosomes from one parental strain and the rest from the other. The data in figure 24.6 show that each copy of the X chromosome, and chromosomes 2 and 3, confer a significant amount of insecticide resistance. When a fly contained all of its X chromosomes, and chromosomes 2 and 3 from the insecticide-resistant parental strain, it had maximal insecti-

cide resistance. Even when only a single chromosome was derived from the insecticide-sensitive strain, the resistance to DDT was less than maximal. These results are consistent with the hypothesis that insecticide resistance is a polygenic trait involving multiple genes that reside on the X chromosome and on chromosomes 2 and 3.

A self-help quiz involving this experiment can be found at the Online Learning Center.

Quantitative Trait Loci (QTLs) Are Now Mapped by Linkage to Molecular Markers

In the previous experiment, we have seen how the minimum number of genes affecting a quantitative trait (i.e., pesticide resistance) was determined by the linkage of such unknown genes to known genes on the *Drosophila* chromosomes. In the past few years, newer research techniques have identified molecular markers, such as RFLPs and microsatellites, that serve as reference points along chromosomes. This topic is discussed in chapter 20. These markers have been used to construct detailed genetic maps of several species' genomes. Once a genome map is obtained, it becomes much easier to determine the number of genes that affect a quantitative trait. In addition to model organisms such as *Drosophila, Arabidopsis, Caenorhabditis elegans,* and mice, detailed molecular maps have been obtained for many species of agricultural importance. These include crops such as corn, rice, and tomatoes, as well as livestock such as cattle, pigs, sheep, and chickens.

To map the genes in a eukaryotic species, researchers now determine their locations by identifying molecular markers that are close to such genes. In 1989, Eric Lander and David Botstein extended this technique to identify QTLs that govern a quantitative trait. The basis of **QTL mapping** is the association between genetically determined phenotypes (e.g., quantitative traits) and molecular markers such as RFLPs and microsatellites. The general strategy for QTL mapping is shown in figure 24.7. This figure depicts two different strains of a diploid species with four chromosomes per set. The strains differ in two important ways. First, they must differ with regard to a quantitative trait of interest. In this example, the strain on the *left* produces large fruit, while the strain on the *right* produces small fruit. The unknown genes affecting this trait are designated with the letter *X*. A *black X* indicates a site promoting large fruit, and a *blue X* is the same site that promotes small fruit. Secondly, the two strains differ with regard to many molecular markers. In figure 24.7, these markers are designated 1A and 1B, 2A and 2B, and so forth. The markers 1A and 1B would mark the same chromosomal location in this species, namely the upper tip of chromosome 1. However, the two markers would be distinguishable in the two strains at the molecular level. For example, 1A might be a microsatellite that is 148 bp, while 1B might be 212 bp.

With these ideas in mind, the protocol shown in figure 24.7 begins by mating the two inbred strains to each other and then backcrossing the F_1 offspring to the parental strains. This would produce an F_2 generation with a great degree of variation. The F_2 offspring would then be characterized in two ways. First, they would be examined for their fruit size, and second, a sample of cells would be analyzed to determine which molecular markers were found in their chromosomes. The goal is to find an association between particular molecular markers and fruit size. For example, 2A would be strongly associated with large size, while 2B would be strongly associated with small size. By comparison, 9A and 9B would not be associated with large or small size, because a QTL affecting this trait is not in the vicinity of this marker. Overall, QTL mapping involves the analysis of a large number of markers and offspring. The data are analyzed by computer programs that can statistically associate the phenotype (e.g., fruit size) with particular markers. Because markers are found throughout the genome of a species, this provides a way to identify the locations of several different genes that possess allelic differences that may affect the outcome of a quantitative trait.

As an actual example of QTL mapping, in 1988, Andrew Paterson and his colleagues examined quantitative trait inheritance in the tomato. They studied a domestic strain of tomato and a South American green-fruited variety. These two strains differed in their RFLPs, and they also exhibited dramatic differences in three agriculturally important characteristics: fruit mass, soluble solids content (in the fruit), and fruit pH. The researchers crossed the two strains and then backcrossed the offspring to the domestic tomato. A total of 237 plants were then examined with regard to 70 known RFLP markers. In addition, between 5 and 20 tomatoes from each plant were analyzed with regard to fruit mass, soluble solids content, and fruit pH. Using this approach, they were able to map the genes contributing variation in these traits to particular intervals along the tomato chromosomes. They identified six loci causing variation in fruit mass, four affecting soluble solids content, and five with effects on fruit pH.

24.3 HERITABILITY

As we have just seen, recent approaches in molecular mapping have enabled researchers to identify the genes that contribute to a quantitative trait. The other key factor that affects the phenotypic outcomes of quantitative traits is the environment. All traits of biological organisms are influenced by genetics and the environment, and this is particularly pertinent in the study of quantitative traits.

A geneticist, however, can never actually determine the relative amount of a quantitative trait that is controlled by genetics or the amount that is governed by the environment. When you think about it, a researcher cannot raise an organism without an environment and discover the amount of a trait that is governed by genetics. Likewise, it is impossible to rear an organism without its genes and conclude how much the environment contributes to the outcome of a trait. Instead, the focus of our analysis of quantitative traits is to explain how variation, both genetic and environmental, will affect the phenotypic results.

The term **heritability** refers to the amount of phenotypic variation within a group of individuals that is due to genetic variation. If all the phenotypic variation in a group was due to genetic variation, the heritability would have a value of 1. If all the variation was due to environmental effects, the heritability would equal 0. For most groups of organisms, the heritability for a given trait lies between these two extremes. For example, both genes and diet affect the size that an individual will attain. Some individuals will inherit genes that tend to make them large, and a proper diet will also promote larger size. Other individuals will inherit genes that make them small, and an inadequate diet may contribute to small size. Taken together, both genetics and the environment influence the phenotypic results.

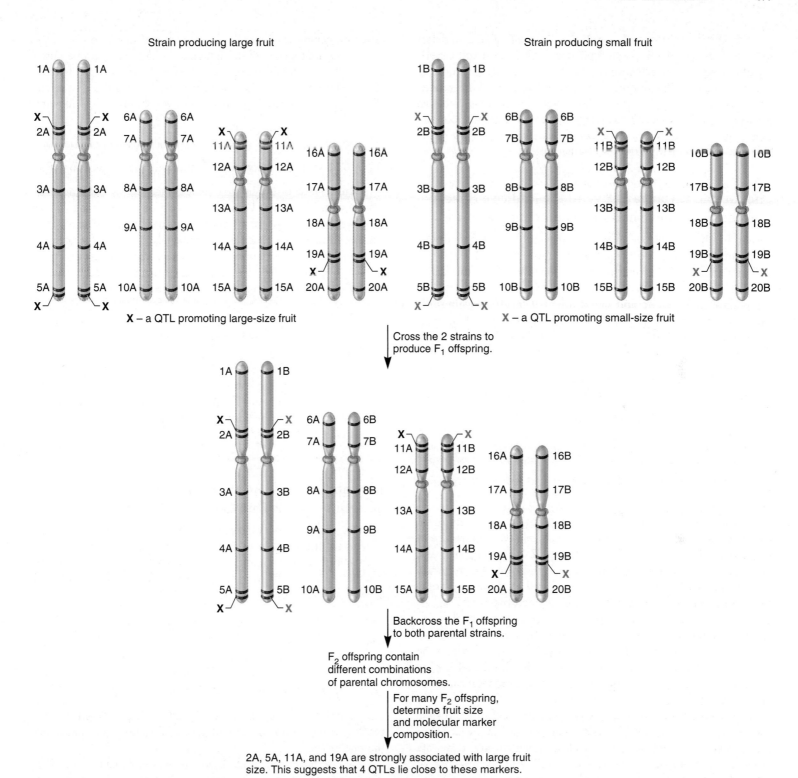

Strain producing large fruit

Strain producing small fruit

X – a QTL promoting large-size fruit

X – a QTL promoting small-size fruit

Cross the 2 strains to produce F₁ offspring.

Backcross the F₁ offspring to both parental strains.

F₂ offspring contain different combinations of parental chromosomes.

For many F₂ offspring, determine fruit size and molecular marker composition.

2A, 5A, 11A, and 19A are strongly associated with large fruit size. This suggests that 4 QTLs lie close to these markers.

FIGURE 24.7 **The general strategy for QTL mapping via molecular markers.** Two different inbred strains have four chromosomes per set. The strain on the *left* produces large fruit, while the strain on the *right* produces small fruit. The goal of this mapping strategy is to locate the unknown genes affecting this trait, which are designated with the letter X. A *black* X indicates a site promoting large fruit and a *blue* X is a site promoting small fruit. The two strains differ with regard to many molecular markers designated 1A and 1B, 2A and 2B, and so forth. The two strains are mated and then the F₁ offspring are backcrossed to the parental strains. Many F₂ offspring are then examined for their fruit size and to determine which molecular markers are found in their chromosomes. The data are analyzed by computer programs that can statistically associate the phenotype (e.g., fruit size) with particular markers. Since markers are found throughout the genome of this species, this provides a way to locate many different genes that may affect the outcome of a single quantitative trait. In this case, the analysis would predict four QTLs promoting heavier fruit weight that would be linked to regions of the chromosomes containing the following markers: 2A, 5A, 11A, and 19A.

In the study of quantitative traits, a primary goal is to determine how much of the phenotypic variation arises from genetic variation and how much comes from environmental variation. In this section, we will examine how geneticists analyze the genetic and environmental components that affect quantitative traits. As we will see, this approach has been applied with great success in breeding strategies to produce domesticated species with desirable and commercially valuable characteristics.

Phenotypic Variance Is Due to the Additive Effects of Genetic Variance and Environmental Variance

Earlier in chapter 24, we examined the amount of phenotypic variation within a group by calculating the variance. In studying quantitative trait variation, the first step is to partition this variation into components that are attributable to different causes. If we assume that genetic and environmental factors are the only two components that determine a trait, and if genetic and environmental factors are independent of each other, then the total variance for a trait in a group of individuals is

$$V_T = V_G + V_E$$

where

V_T is the total variance. It reflects the amount of variation that is measured at the phenotypic level.

V_G is the relative amount of variance due to genetic variation.

V_E is the relative amount of variance due to environmental variation.

The partitioning of variance into genetic and environmental components allows us to estimate their relative importance in influencing the variation within a group. If V_G is very high and V_E is very low, genetics plays a greater role in promoting variation within a group. Alternatively, if V_G is low and V_E is high, environmental causes underlie much of the phenotypic variation. As described later in chapter 24, a livestock breeder might want to apply selective breeding if V_G for an important (quantitative) trait is high. In this way, the characteristics of the herd may be improved. Alternatively, if V_G is negligible, it would make more sense to investigate (and manipulate) the environmental causes of phenotypic variation.

With experimental and domesticated animals, one possible way to determine V_G and V_E is by comparing the variation in traits between genetically identical and genetically disparate groups. For example, researchers have followed the practice of **inbreeding** to develop genetically homogeneous strains of mice. Inbreeding in mice involves many generations of brother-sister matings, which eventually produces strains that are **monomorphic** for all of their genes. The term *monomorphic* means that all the members of a population are homozygous for the same allele of a given gene. Within such an inbred strain of mice, V_G equals zero. Therefore, all phenotypic variation is due to V_E. When studying quantitative traits such as weight, an experimenter might want to know the genetic and environmental variance for a

different, genetically heterogeneous group of mice. To do so, the genetically homogeneous and heterogeneous mice could be raised under the same environmental conditions and their weights measured. The phenotypic variance for weight could then be calculated as described earlier. Let's suppose we obtained the following results:

$V_T = 0.30$ sq oz for the group of genetically homogeneous mice

$V_T = 0.52$ sq oz for the group of genetically heterogeneous mice

In the case of the homogeneous mice, $V_T = V_E$, because V_G equals zero. Therefore, V_E equals 0.30 sq oz. To estimate V_G for the heterogeneous group of mice, we could assume that V_E (i.e., the environmentally produced variance) is the same for them as it is for the homogeneous mice, because the two groups were raised in identical environments. This assumption allows us to calculate the genetic variance for the heterogeneous mice.

$$V_T = V_G + V_E$$
$$0.52 = V_G + 0.30$$
$$V_G = 0.22 \text{ sq oz}$$

This result tells us that some of the phenotypic variance in the genetically heterogeneous group is due to the environment (namely, 0.30 sq oz) and some (0.22 sq oz) is due to genetic variation in alleles that affect weight.

Heritability Is the Relative Amount of Phenotypic Variation That Is Due to Genetic Variation

Another way to view variance is to focus our attention on the genetic contribution to phenotypic variation. Heritability is the proportion of the phenotypic variance that is attributable to genetic variation. If we assume again that environment and genetics are the only two components, then

$$H_B^2 = V_G/V_T$$

where

H_B^2 is the heritability in the broad sense

V_G is the variance due to genetics

V_T is the total phenotypic variance

The heritability defined here, H_B^2, is called the **broad sense heritability.** It takes into account all genetic variation that may affect the phenotype.

As we have seen throughout this textbook, genes can affect phenotypes in various ways. As described earlier in chapter 24, the Nilsson-Ehle experiment showed that the alleles determining hull color in wheat affect the phenotype in an additive way. Alternatively, alleles affecting other traits may show a dominant/recessive relationship. In this case, the alleles are not strictly additive, because the heterozygote has a phenotype closer to, or perhaps the same as, the homozygote containing two copies of the dominant allele. In addition, another complicating factor is epistasis (which was described in chapter 4). The alleles for one gene may influence the phenotypic expression of the alleles of another gene. To account for these differences, geneticists usually subdivide V_G into these three different genetic categories:

$$V_G = V_A + V_D + V_I$$

where

V_A is the variance due to additive alleles

V_D is the variance due to alleles that follow a dominant/recessive pattern of inheritance

V_I is the variance due to genes that interact in an epistatic manner

In analyzing quantitative traits, geneticists often focus on V_A and neglect the contributions of V_D and V_I. This is done for scientific as well as practical reasons. For many quantitative traits, the additive effects of alleles often dominate the phenotypic outcome in genetic crosses. In addition, when the alleles behave additively, we can predict the outcomes of crosses based on the quantitative characteristics of the parents. The heritability of a trait due to the additive effects of alleles is called the **narrow sense heritability:**

$$h_N^2 = V_A/V_T$$

The narrow sense heritability (h_N^2) may be an inaccurate measure of the actual heritability if V_D and V_I are not small values. However, for many quantitative traits, geneticists have found that the value of V_A is very large compared to V_D and V_I. In such cases, the broad sense and the narrow sense heritabilities are similar to each other.

There are several ways to estimate narrow sense heritability. A common strategy is to measure a quantitative trait among groups of genetically related individuals. For example, agriculturally important traits, such as egg weight in poultry, can be analyzed in this way. To calculate the heritability, one would determine the observed egg weights between individuals whose genetic relationships are known, such as a mother and her female offspring. These data could then be used to compute a correlation between the parent and offspring using the methods described

earlier in chapter 24. The narrow sense heritability is then calculated as

$$h_N^2 = r_{obs}/r_{exp}$$

where

r_{obs} is the observed phenotypic correlation between related individuals

r_{exp} is the expected correlation based on the known genetic relationship

In our example, r_{obs} is the observed phenotypic correlation between parent and offspring. In particular research studies, the observed phenotypic correlation for egg weights between mothers and daughters has been found to be about 0.25 (although this varies among strains). The expected correlation, r_{exp}, is based on the known genetic relationship. A parent and child share 50% of their genetic material, so that r_{exp} equals 0.50. So,

$$h_N^2 = r_{obs}/r_{exp}$$
$$= 0.25/0.50 = 0.50$$

Note: For siblings, $r_{exp} = 0.50$; for identical twins, $r_{exp} = 1.0$; and for an uncle-niece relationship, $r_{exp} = 0.25$.

According to this calculation, about 50% of the phenotypic variation in egg weight is due to genetic variation; the other half is due to the environment.

When calculating heritabilities from correlation coefficients, keep in mind that this computation assumes that genetics and the environment are independent variables. This is not always the case. The environments of parents and offspring are often more similar to each other than they are to those of unrelated individuals. There are several ways to avoid this confounding factor. First, in human studies, one may analyze the heritabilities from correlations between adopted children and their biological parents. Alternatively, one can examine a variety of relationships (uncle-niece, identical twins versus fraternal twins, etc.) and see if the heritability values are roughly the same in all cases. This approach was applied in the study to be described next.

EXPERIMENT 24B

Heritability of Dermal Ridge Count in Human Fingerprints Is Very High

Fingerprints are inherited as a quantitative trait. It has long been known that identical twins have fingerprints that are very similar, whereas fraternal twins show considerably less agreement. Galton was the first researcher to study fingerprint patterns, but this trait became more amenable to genetic studies when Kristine Bonnevie developed a method for counting the number of ridges within a human fingerprint.

As shown in figure 24.8, human fingerprints can be categorized as having an arch, a loop, or a whorl (or a combination of these patterns). The primary difference among these patterns is the number of triple junctions, each known as a triradius (fig. 24.8b and c). At a triradius, a ridge emanates in three directions. An arch has zero triradii, a loop has one, and a whorl has two. In Bonnevie's method of counting, a line is drawn from a triradius to the center of the fingerprint. The ridges that touch this line are then counted. (Note: The triradius ridge itself is not counted, and the last ridge is not counted if it forms the center

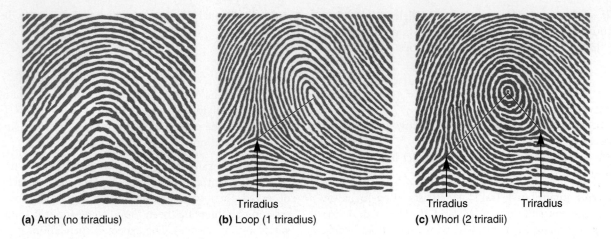

(a) Arch (no triradius)   **(b)** Loop (1 triradius)   **(c)** Whorl (2 triradii)

FIGURE 24.8 **Human fingerprints and the ridge count method of Bonnevie.** (a) This print has an arch rather than a triradius. The ridge count is zero. (b) This print has one triradius. A *straight line* is drawn from the triradius to the center of the print. The number of ridges dissecting this straight line is 13. (c) This print has two triradii. *Straight lines* are drawn from both triradii to the center. There are 16 ridges touching the *left line* and 7 touching the *right line,* giving a total ridge count of 23.

of the fingerprint.) With this method, one can obtain a ridge count for all 10 fingers. Bonnevie conducted a study on a small population and found that ridge count correlations were relatively high in genetically related individuals.

Sarah Holt, who was also interested in the inheritance of this quantitative trait, carried out a more exhaustive study of ridge counts in a British population. In groups of 825 males or 825 females, the ridge count on all 10 fingers varied from 0 to 300, with mean values of approximately 145 for males (standard deviation equaled 51.1) and 127 for females (standard deviation 52.5).

Based on these encouraging results, Holt decided to conduct a more detailed analysis of ridge counts by examining the fingerprint patterns of a large group of people and their close relatives. In the experiment of figure 24.9, the ridge counts for pairs of related individuals were determined by the method described in figure 24.8. The correlation coefficients for ridge counts were then calculated among the pairs of related individuals. To estimate the narrow sense heritability, the observed correlations were then divided by the expected correlations based on the known genetic relationships.

■ THE HYPOTHESIS

Dermal ridge count has a genetic component. The goal of this experiment is to determine the contribution of genetics in the variation of dermal ridge counts.

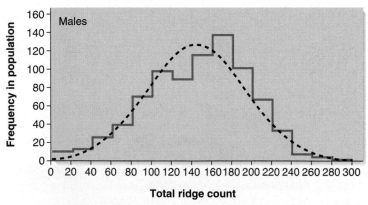

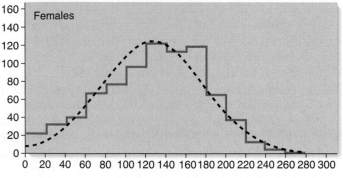

TESTING THE HYPOTHESIS — **FIGURE 24.9** Heritability of human fingerprint patterns.

Starting material: A group of human subjects from Great Britain.

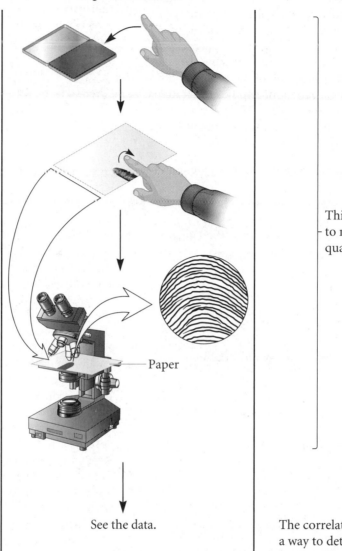

Experimental level

1. Take a person's finger and blot it onto an ink pad.

2. Roll the person's finger onto a recording surface to obtain a print.

3. With a low-power binocular microscope, count the number of ridges using the Bonnevie method described previously.

Paper

4. Calculate the correlation coefficients between different pairs of individuals as described earlier in chapter 24.

See the data.

Conceptual level

This is a method to measure a quantitative trait.

The correlation coefficient provides a way to determine the heritability for the quantitative trait.

THE DATA

Type of Relationship	Number of Pairs Examined	Correlation Coefficient	Heritability r_{obs}/r_{exp}
Parent-child	810	0.48 ± 0.04*	0.96
Parent-parent	200	0.05 ± 0.07	—†
Sibling-sibling	642	0.50 ± 0.04	1.00
Identical twins	80	0.95 ± 0.01	0.95
Fraternal twins	92	0.49 ± 0.08	0.98
			0.97 average heritability

*± Standard error of the mean.

†Note: We cannot calculate a heritability value, because the value for r_{exp} is not known. Nevertheless, the value for r_{obs} is very low, suggesting that there is a negligible correlation between unrelated individuals.

Adapted from Holt, S. B. (1961) Quantitative Genetics of Finger-Print Patterns. *British Med. Bull.* 17, 247–250.

INTERPRETING THE DATA

As seen in the data table, it becomes apparent that genetics plays the major role in explaining the variation in this trait. Genetically unrelated individuals (namely, parent-parent relationships) have a negligible correlation for this trait. By comparison, individuals who are genetically related have a substantially higher correlation. When the observed correlation coefficient is divided by the expected correlation coefficient based on the known genetic relationships, the average heritability value is 0.97, which is very close to 1.0. These values indicate that nearly all of the variation in fingerprint pattern is due to genetic variation. Significantly, fraternal and identical twins have substantially different correlation coefficients, even though we expect that they have been raised in very similar environments.

These results support the idea that genetics is playing the major role in promoting variation and that the results are not biased heavily by environmental similarities that may be associated with genetically related individuals. From an experimental viewpoint, the results show us how the determination of correlation coefficients between related and unrelated individuals can provide insight regarding the relative contributions of genetics and environment to the variation of a quantitative trait.

A self-help quiz involving this experiment can be found at the Online Learning Center.

Heritability Values Are Relevant Only to Particular Groups Raised in a Particular Environment

Table 24.3 describes heritability values that have been calculated for particular populations. Unfortunately, heritability is a widely misunderstood concept. Heritability describes the amount of phenotypic variation due to genetic variation *for a particular population raised in a particular environment.* The words *variation, particular population,* and *particular environment* cannot be overemphasized. For example, in one population of cattle the heritability for milk production may be 0.35, while in another group (with less genetic variation) the heritability may be 0.1. Second, if a group displays a heritability of 1.0 for a particular trait, this does not mean that the environment is unimportant in affecting the outcome of the trait. A heritability value of 1.0 only means that the amount of variation within this group is due to genetics. Perhaps the group has been raised in a relatively homogeneous environment, so that the environment has not caused a significant amount of variation. Nevertheless, the environment may be quite important. It just is not causing much variation within this particular group.

As a hypothetical example, let's suppose that we take a species of rodent and raise a group on a poor diet; we find their weights range from 1.5 to 2.5 pounds, with a mean weight of 2 pounds. We allow them to mate and then raise their offspring on a healthy diet of rodent chow. The weights of the offspring range from 2.5 to 3.5 pounds, with a mean weight of 3 pounds. In this hypothetical experiment, we might find a positive correlation in which the small parents tended to produce small offspring, and the large parents produce large offspring. The correlation of weights between parent and offspring might be, say, 0.5. In this case, the heritability for weight would be calculated as r_{obs}/r_{exp}, which equals 0.5/0.5, or 1.0. The value of 1.0 means that all the variation within the groups is due to genetics. The offspring vary from 2.5 to 3.5 pounds because of genetic variation, and also the parents range from 1.5 to 2.5 pounds because of genetics. However, as we see here, environment has played an important role. Presumably, the mean weight of the offspring is higher because of their better diet. This example is meant to emphasize the point that heritability tells us only the relative contributions of genetics and environment in influencing phenotypic *variation* in a *specific group* under a *specific set of conditions.* Heritability cannot indicate the relative importance of these two factors in determining the outcomes of traits. When a heritability value is high, it does not mean that a change in the environment cannot have a major impact on the outcome of the trait.

TABLE 24.3

Examples of Heritabilities for Quantitative Traits

Trait	Heritability Value*
Humans	
Stature	0.65
IQ testing ability	0.60
Cattle	
Body weight	0.65
Butterfat, %	0.40
Milk yield	0.35
Mice	
Tail length	0.40
Body weight	0.35
Litter size	0.20
Poultry	
Body weight	0.55
Egg weight	0.50
Egg production	0.10

*As emphasized in the textbook, these values apply to particular groups raised in particular environments. The value for IQ testing ability is an average value from many independent studies. The other values were taken from Falconer, D. S. (1989) *Introduction to Quantitative Genetics,* 3d ed. Longman, Essex, England.

With regard to the roles of genetics and environment (i.e., nature vs. nurture), the topic of human intelligence has been hotly debated. As a trait, intelligence is difficult to define or to measure. Nevertheless, performance on an IQ test has been taken by some people as a reflection of intelligence ever since Alfred Binet's test was first introduced in the United States in 1916. Even though such tests may have inherent bias and consider only a limited subset of human cognitive abilities, IQ tests still remain a method of assessing intelligence. By comparing IQ scores among related and unrelated individuals, various studies have attempted to estimate heritability values in selected human populations. These values have ranged from 0.3 to 0.8. A heritability value of around 0.6 is fairly common among many studies. Such a value indicates that over half of the heritability for IQ testing ability is due to genetic factors.

It is important to consider what a value of 0.6 means, and what it does not mean. It means that 60% of the variation in IQ testing ability is due to genetic variation in a selected population raised in a particular environment. It does not mean that 60% of an individual's IQ testing ability is due to genetics and 40% is due to the environment. It cannot be overemphasized that heritability is meaningless at the level of a single individual. Furthermore, even at the populational level, a heritability value of 0.6 does not mean that 60% of the IQ testing ability is due to genetics and 40% is due to the environment. Rather, it means that in the selected population that was examined, 60% of the variation in IQ testing ability is due to genetics while 40% is due to the environment. Heritability is strictly a populational value that pertains to variation. As mentioned already, it is not possible to determine the relative contributions of genetics and environment in producing a given trait, either at the individual or populational level. Indeed, such a perception is absurd because genetics and environment are both essential for biological existence.

Selective Breeding of Species Can Alter Quantitative Traits Dramatically

The term **selective breeding** refers to programs and procedures designed to modify phenotypes in species of economically important plants and animals. This phenomenon, also called **artificial selection,** is related to natural selection, which will be discussed in chapter 25. In fact, in forming his theory of natural selection, Charles Darwin was influenced by his observations of selective breeding by pigeon fanciers and other breeders. The primary difference between artificial and natural selection is how the parents are chosen. Natural selection is due to natural variation in reproductive success; in artificial selection, the breeder chooses individuals that possess traits that are desirable from a human perspective.

For centuries, humans have been practicing selective breeding to obtain domestic species with interesting or agriculturally useful characteristics. The common breeds of dogs and cats have been obtained by selective breeding strategies (fig. 24.10). As shown here, it is very striking how selective breeding can modify the quantitative traits in a species. When comparing a greyhound

Greyhound

German shepherd

Bulldog

Cocker spaniel

FIGURE 24.10   Some common breeds of dogs that have been obtained by selective breeding.

GENES→TRAITS   By selecting parents carrying the alleles that influence certain quantitative traits in a desired way, dog breeders have produced breeds with distinctive sets of traits. For example, the bulldog has alleles that give it short legs and a flat face. By comparison, the corresponding genes in a German shepherd are found in alleles that produce longer legs and a more pointy snout. All the dogs shown in this figure carry the same kinds of genes (e.g., many genes that affect their sizes, shapes, and fur color). However, the alleles for many of these genes are different among these dogs, thereby producing breeds with strikingly different phenotypes.

with a bulldog, the magnitude of the differences is fairly amazing. They hardly look like members of the same species.

Likewise, most of the food we eat is obtained from species that have been modified profoundly by selective breeding strategies. This includes products such as grains, fruit, vegetables, meat, milk, and juices. Figure 24.11 illustrates how certain characteristics in the wild mustard plant (*Brassica oleracea*) have been modified by selective breeding to create several species of domesticated crops. This plant is native to Europe and Asia, and plant breeders began to modify its traits approximately 4,000 years ago. As seen here, certain quantitative traits in the domestic strains differ considerably from those of the original wild species.

The phenomenon that underlies selective breeding is variation. Within a group of individuals, there may be allelic variation that affects the outcomes of quantitative traits. The fundamental strategy of the selective breeder is to choose parents that will pass on advantageous alleles to their offspring. In general, this means that the breeder will choose parents with desirable phenotypic characteristics. For example, if a breeder wants large cattle, the largest members of the herd will be chosen as parents for the next generation. These large cattle will transmit an array of alleles to their offspring that confer large size. The breeder will often choose genetically related individuals (e.g., brothers and sisters)

FIGURE 24.11 **Crop plants developed by selective breeding of the wild mustard plant.**

GENES→TRAITS   The wild mustard plant carries a large amount of genetic (i.e., allelic) variation, which was used by plant breeders to produce modern strains that are agriculturally desirable. For example, by selecting for alleles that promote the formation of large lateral buds, the strain of Brussels sprouts was created. By selecting for alleles that alter the leaf morphology, kale was developed. Although these six agricultural plants look quite different from each other, they carry many of the same alleles as the wild mustard. However, they differ in alleles that affect the formation of flowers, buds, stems, and leaves.

Wild mustard plant

Strain	Modified trait
Kohlrabi	Stem
Kale	Leaves
Broccoli	Flower buds and stem
Brussels sprouts	Lateral leaf buds
Cabbage	Terminal leaf bud
Cauliflower	Flower buds

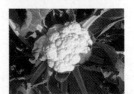

as the parental stock. As mentioned, the practice of mating between genetically related individuals is known as **inbreeding.** Some of the consequences of inbreeding will also be described in chapter 25.

Figure 24.12 illustrates a common outcome when selective breeding is conducted for a quantitative trait. Figure 24.12a shows the results of a program begun at the Illinois Experiment Station in 1896, even before the rediscovery of Mendel's laws. This experiment began with 163 ears of corn with an oil content ranging from 4 to 6%. In each of 80 succeeding generations, corn plants were divided into two separate groups. In one group, members with the highest oil content were chosen as parents of the next generation. In the other group, members with the lowest oil content were chosen. After many generations, the oil content in the first group rose to over 18%; in the other group, it dropped to less than 1%. These results show that selective breeding can modify quantitative traits in a very directed manner.

Similar results have been obtained for many other quantitative traits. Figure 24.12b shows an experiment by K. Mather, in which flies were selected on the basis of their bristle number. The starting group had an average of 40 bristles for females and 35 bristles for males. After eight generations, the group selected for high bristle number had an average of 46 bristles for females and 40 for males, while the group selected for low bristle number had an average of 36 bristles for females and 30 for males.

When comparing the curves in figure 24.12, some general observations can be made. Quantitative traits are often at an intermediate value in unselected populations. Therefore, artificial selection can increase or decrease the magnitude of the trait. Oil content can go up or down, and bristle number can increase or decrease. Nevertheless, there is a limit to how far this can continue. Presumably, the starting population possesses a large amount of genetic variation, which contributes to the diversity in phenotypes. In other words, at the beginning of these experiments, the heritability for the trait is fairly high. By carefully choosing the parents, each succeeding generation has a higher proportion of the desirable alleles. However, after many generations, the population will eventually become monomorphic for all or most of the desirable alleles that affect the trait of interest. At this point, additional selective breeding will have no effect. When this occurs, the heritability for the trait is near zero, because nearly all genetic variation has been eliminated from the population.

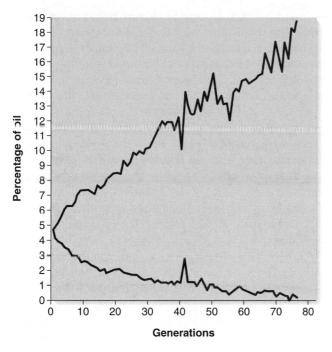

(a) Results of selective breeding for and against oil content in corn

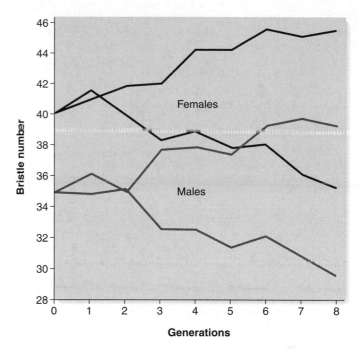

(b) Results of selective breeding for low and high bristle number in flies

FIGURE 24.12 **Common results of selective breeding for a quantitative trait.** (a) Selection for high and low oil content in corn. (b) Selection for high and low bristle number in flies.

In artificial selection experiments, the response to selection is the most common way to estimate the narrow sense heritability in a starting population. The heritability measured in this way is also called the **realized heritability.** It is calculated as

$$h_N^2 = \frac{R}{S}$$

where

R is the response to selection in the offspring
S is the selection differential in the parents

Here,

$$R = \bar{X}_O - \bar{X}$$

$$S = \bar{X}_P - \bar{X}$$

where
$\bar{X}$ is the mean of the starting population
$\bar{X}_O$ is the mean of the offspring
$\bar{X}_P$ is the mean of the parents

So,

$$h_N^2 = \frac{\bar{X}_O - \bar{X}}{\bar{X}_P - \bar{X}}$$

As an example, let's suppose we began with a population of fruit flies in which the average bristle number for both sexes was 37.5. The parents chosen from this population had an average bristle number of 40. The offspring of the next generation had an average bristle number of 38.7. With these values, the realized heritability is

$$h_N^2 = \frac{38.7 - 37.5}{40 - 37.5}$$

$$h_N^2 = \frac{1.2}{2.5} = 0.48$$

This result tells us that about 48% of the phenotypic variation is due to the additive effects of alleles.

An important aspect of narrow sense heritabilities is their ability to predict the outcome of selective breeding. Solved problem S1 at the end of chapter 24 illustrates this idea.

Heterosis May Be Explained by Dominance or Overdominance

As we have just seen, selective breeding can be used to alter the phenotypes of domesticated species in a highly directed way. An unfortunate consequence of inbreeding, however, is that it tends to decrease the overall genetic variation within a group and may unknowingly promote homozygosity for deleterious alleles. In agriculture, it is widely observed that when two different inbred strains are crossed to each other, the resulting offspring are more vigorous (e.g., larger, longer-lived) than either of the inbred parental strains. This phenomenon is called **heterosis** or **hybrid vigor.** In modern agricultural breeding practices, many strains of plants and animals are hybrids produced by crossing two different

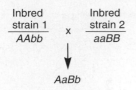

The recessive alleles (*a* and *b*) are slightly harmful in the homozygous condition.

The hybrid offspring is more vigorous, because the harmful effects of the recessive alleles are masked by the dominant alleles.

The Dominance Hypothesis

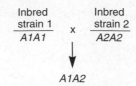

Neither the *A1* nor *A2* allele is recessive.

The hybrid offspring is more vigorous, because the heterozygous combination of alleles exhibits overdominance. This means that the *A1A2* heterozygote is more vigorous than either the *A1A1* or *A2A2* homozygote.

The Overdominance Hypothesis

FIGURE 24.13   **Mechanisms to explain heterosis. (a)** The dominance hypothesis. **(b)** The overdominance hypothesis.

inbred lines. In fact, much of the success in agricultural breeding programs is founded in heterosis. In rice, for example, hybrid strains have a 15 to 20% yield advantage over the best conventional inbred varieties under similar cultivation conditions.

The genetic basis for heterosis has been debated for more than 90 years, and the controversy is still not resolved. As shown in figure 24.13, there are two major hypotheses to explain heterosis. In 1908, Charles Davenport proposed the dominance hypothesis in which the effects of dominant alleles explain the favorable outcome in a heterozygote. He suggested that highly inbred strains have become homozygous for one or more recessive genes that are somewhat deleterious (but not lethal). This undesirable homozygosity can happen randomly as result of inbreeding or genetic drift (as described in chapter 25). Because the homozygosity occurs by chance, two different inbred strains are likely to be homozygously recessive for different genes. Therefore, when they are crossed to each other, the resulting hybrids are heterozygous and do not suffer the consequences of homozygosity for deleterious recessive alleles. In other words, the dominance of the beneficial alleles explains the observed heterosis.

Also in 1908, George Shull and Edward East proposed a second hypothesis, known as the overdominance hypothesis. As described in chapter 4, overdominance occurs when the heterozygote is more vigorous than either corresponding homozygote. According to this idea, heterosis occurs because the resulting hybrids are heterozygous for one or more genes that display overdominance. A classic case of overdominance involves the β-globin allele for sickle-cell anemia. In areas where malaria is prevalent, the heterozygote has a higher survival rate than both the wild-type homozygote and the homozygote affected with sickle-cell anemia (see chapter 4).

As mentioned earlier in chapter 24, the recent advent of QTL mapping has made it possible to identify the QTLs that underlie heterosis. In corn, Charles Stuber and his colleagues have found that most QTLs for grain yield support the overdominance hypothesis. By comparison, Steven Tanksley, working with colleagues in China, found that heterosis in rice seems to be due to dominance rather than overdominance. Though much research has been devoted to this question, a general consensus regarding the relative importance of dominance versus overdominance has yet to be reached. Over the next few years, it will be interesting to see if heterosis among agriculturally important species is usually explained by dominance, overdominance, or a combination of the two.

Finally, it should be pointed out that overdominance is very difficult to distinguish from **pseudo-overdominance,** a phenomenon initially suggested by James Crow. Pseudo-overdominance is really the same as dominance, except that two genes or more are very closely linked. For example, a QTL may be identified in a mapping experiment to be close to a particular molecular marker. However, it is possible that a QTL could contain two genes, both affecting the same quantitative trait. For example, at a single QTL, two genes, *a* and *B*, may be closely linked in one strain, while *A* and *b* would be closely linked in another strain. The hybrid is really heterozygous (*AaBb*) for two different genes, but this may be difficult to discern in mapping experiments because the genes are so close together. If a researcher assumed that there was only one gene at a QTL, the overdominance hypothesis would be favored, whereas if two genes were really present, the dominance hypothesis would be correct. Therefore, without very fine mapping, which is rarely done for QTLs, it is hard to distinguish between overdominance and pseudo-overdominance.

CONCEPTUAL SUMMARY

Quantitative genetics is the study of traits that vary in a continuous, quantitative way in populations. Such **quantitative traits** can be anatomical, physiological, or behavioral, and some diseases exhibit characteristics and inheritance patterns analogous to those of quantitative traits. Typically, such traits are determined by two or more genes and are said to be **polygenic.**

Within populations, quantitative traits commonly exhibit a continuum of phenotypic variation that may follow a **normal distribution.** Statistical methods can be used to analyze such a distribution. The **mean** describes the average value among the population; the **standard deviation (SD)** provides a measure of the amount of variation in the group. The **variance** is a measure of the squared deviations from the mean. Variances are useful statistics because they are additive. Also, some statistical methods compare two variables with each other. The **correlation coefficient** evaluates the strength of association between two variables.

In genetics, it is often used to see if there are phenotypic correlations between genetically related individuals.

The chromosomal locations of genes that influence quantitative traits are called **quantitative trait loci (QTLs).** The reasons most quantitative traits show a continuum are that each trait is governed by several genes existing in two or more alleles, and the environment contributes a substantial amount of variation to the phenotypic outcome.

All traits of biological organisms are influenced by genetics and the environment, particularly quantitative traits. Phenotypic variance is due to the additive effects of genetic variance and environmental variance. **Heritability** is the fraction of the phenotypic variance that is attributable to genetic variation; it describes the amount of phenotypic variation that is due to genetic variation for a particular population raised in a particular environment. **Broad sense heritability** takes into account all of the genetic categories that could affect the phenotype. By comparison, the heritability of a trait due to the additive effects of alleles is called the **narrow sense heritability.** For many quantitative traits, the broad sense and narrow sense heritabilities are similar.

EXPERIMENTAL SUMMARY

Geneticists measure the quantitative traits of an individual and describe them numerically. In a population, experimentally obtained numerical values from each individual are used to calculate statistics such as the mean, standard deviation, variance, and correlation.

More than one method can be used to determine the number of genes or QTLs that influence a quantitative trait. For example, QTLs can be identified via their linkage to genes on chromosomes. More recently, molecular markers are used to determine the number of QTLs for a given quantitative trait. In some cases, this approach enables plant and animal breeders to understand the genetic nature of quantitative traits and to develop useful breeding strategies.

A central issue for quantitative geneticists is the relative contributions of genetics and the environment that underlie the phenotypic variation seen in quantitative traits. In chapter 24, we have considered two experimental methods to measure the heritability of quantitative traits. One method is to compare the correlation coefficients among related and unrelated individuals. A second method is **selective breeding,** which is frequently used to improve quantitative traits in agriculturally important species. Quantitative traits are often at an intermediate value in unselected populations. Therefore, artificial selection can increase or decrease the magnitude of the trait. The response to selection can be used to estimate the heritability in the starting population. The heritability measured in this way is called the **realized heritability.**

Selective breeding typically involves the repeated breeding of related individuals to produce strains that are highly inbred. An unfortunate consequence of **inbreeding** is that it decreases the overall genetic variation within a group and may inadvertently promote homozygosity for deleterious alleles. When two different inbred strains are crossed to each other, the resulting offspring are often more vigorous than either of the inbred parental strains. This phenomenon, called **heterosis** or hybrid vigor, may be due to the dominant effects of masking harmful recessive alleles or to overdominance of alleles in the heterozygous state.

PROBLEM SETS & INSIGHTS

Solved Problems

S1. The narrow sense heritability for potato weight in a starting population of potatoes is 0.42, and the mean weight is 1.4 lb. If a breeder crosses two individuals with average potato weights of 1.9 and 2.1 lb, respectively, what is the predicted average weight of potatoes in the offspring?

Answer: The mean weight of the parents is 2.0 lb. To solve for the mean weight of the offspring:

$$h_N^2 = \frac{\bar{X}_O - \bar{X}}{\bar{X}_P - \bar{X}}$$

$$0.42 = \frac{\bar{X}_O - 1.4}{2.0 - 1.4}$$

$$\bar{X}_O = 1.65 \text{ lb}$$

S2. A farmer wants to increase the average body weight in a herd of cattle. She begins with a herd having a mean weight of 595 kg and chooses individuals to breed that have a mean weight of 625 kg. Twenty offspring were obtained, having the following weights in kilograms: 612, 587, 604, 589, 615, 641, 575, 611, 610, 598, 589, 620, 617, 577, 609, 633, 588, 599, 601, and 611. Calculate the realized heritability in this herd with regard to body weight.

Answer:

$$h_N^2 = \frac{R}{S}$$

$$= \frac{\bar{X}_O - \bar{X}}{\bar{X}_P - \bar{X}}$$

We already know the mean weight of the starting herd (595 kg) and the mean weights of the parents (625 kg). First, we need to calculate the mean weights of the offspring.

$$\bar{X}_O = \frac{\text{Sum of the offspring's weights}}{\text{Number of offspring}}$$

$$\bar{X}_O = 604 \text{ kg}$$

$$h_N^2 = \frac{604 - 595}{625 - 595}$$

$$= 0.3$$

S3. The following are data that describe the 6-week weights of mice and their offspring of the same sex:

Parent (g)	Offspring (g)
24	26
21	24
24	22
27	25
23	21
25	26
22	24
25	24
22	24
27	24

Calculate the correlation coefficient.

Answer: To calculate the correlation coefficient, we first need to calculate the means and standard deviations for each group:

$$\bar{X}_{parents} = \frac{24 + 21 + 24 + 27 + 23 + 25 + 22 + 25 + 22 + 27}{10} = 24$$

$$\bar{X}_{offspring} = \frac{26 + 24 + 22 + 25 + 21 + 26 + 24 + 24 + 24 + 24}{10} = 24$$

$$SD_{parents} = \sqrt{\frac{0 + 9 + 0 + 9 + 1 + 1 + 4 + 1 + 4 + 9}{9}}$$

$$= 2.1$$

$$SD_{offspring} = \sqrt{\frac{4 + 0 + 4 + 1 + 9 + 4 + 0 + 0 + 0 + 0}{9}}$$

$$= 1.6$$

Next, we need to calculate the covariance.

$$CoV_{(parents,\ offspring)} = \frac{\Sigma[(X_p - \bar{X}_p)(X_o - \bar{X}_o)]}{N - 1}$$

$$= \frac{0 + 0 + 0 + 3 + 3 + 2 + 0 + 0 + 0 + 0}{9}$$

$$= 0.9$$

Finally, we calculate the correlation coefficient.

$$r_{(parent,\ offspring)} = \frac{CoV_{(p,o)}}{SD_p SD_o}$$

$$r_{(parent,\ offspring)} = \frac{0.9}{(2.1)(1.6)}$$

$$r_{(parent,\ offspring)} = 0.27$$

S4. As described in chapter 24, the correlation coefficient provides a way to determine the strength of association between two variables. When the variables are related due to cause and effect (i.e., one variable affects the outcome of another variable), it is valid to use a regression analysis to predict how much one variable will change in response to the other variable. This is easier to understand if we plot the data for two variables. The graph shown here compares mothers' and offspring's body weights in cattle. The line running through the data points is called a **regression line.** It is the straight line that is closest to all the data points. It is the minimal distance away from the squared vertical distance of all the points.

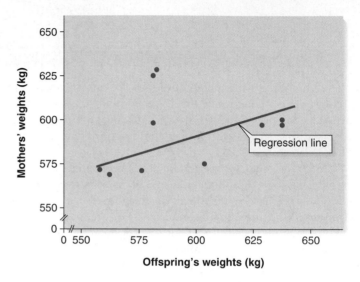

For many types of data, particularly those involving quantitative traits, the regression line can be represented by the equation

$$Y = bX + a$$

where

b is the regression coefficient
a is a constant
In this example, X is the value of an offspring's weight and Y is the value of its mother's weight.

In the equation shown, the value of b, known as the **regression coefficient,** represents the slope of the regression line. The value of a is the y-intercept (i.e., the value of Y when X equals zero). The equation shown is very useful because it allows us to predict the value of X at any given value of Y, and vice versa. To do so, we first need to determine the values of b and a. This can be accomplished in the following manner:

$$b = \frac{CoV_{(X,Y)}}{V_x}$$

$$a = \bar{Y} - b\bar{X}$$

Once the values of a and b have been computed, we can use them to predict the values of X or Y by using the equation

$$Y = bX + a$$

For example, if b = 0.5, a = 2, and Y = 31, this equation can be used to compute a value of X that equals 58. It is important to keep in mind that this equation predicts the average value of X. As we see in the preceding figure, the data points tend to be scattered around the regression line; there are significant deviations between the data points and the line. The equation predicts the values that are the most likely to occur. In an actual experiment, however, there will be some deviation between the predicted values and the experimental values due to random sampling error.

Now here is the question: using the data found in chapter 24 regarding weight in cattle, what is the predicted weight of an off-spring if its mother weighed 660 lb?

Answer: We first need to calculate a and b.

$$b = \frac{CoV_{(X,Y)}}{V_x}$$

We need to use the data on page 669 to calculate V_x, which is the variance for the mothers' weights. The variance equals 445.1. The covariance is already calculated on page 669; it equals 152.6.

$$b = \frac{152.6}{445.1}$$

$$b = 0.34$$

$$a = \bar{Y} - b\bar{X}$$

$$a = 598 - (0.34)(596) = 395.4$$

Now we are ready to calculate the predicted weight of the offspring using the equation

$$Y = bX + a$$

In this problem, $X = 660$ pounds

$$Y = 0.34(660) + 395.4$$

$$Y = 619.8 \text{ lb}$$

The average weight of the offspring is predicted to be 619.8 lb.

S5. Genetic variance can be used to estimate the number of genes affecting a quantitative trait by using the following equation:

$$n = \frac{D^2}{8V_G}$$

where

n is the number of genes affecting the trait

D is the difference between the mean values of the trait in two strains that have allelic differences at every gene that influences the trait

V_G is the genetic variance for the trait; it is calculated using data from both strains

For this method to be valid, several assumptions must be met. In particular, the alleles of each gene must be additive, each gene con-

tributes equally to the trait, all the genes assort independently, and the two strains are homozygous for alternative alleles of each gene. For example, if three genes affecting a quantitative trait exist in two alleles each, one strain could be *AA bb CC* and the other would be *aa BB cc*. In addition, the strains must be raised under the same environmental conditions. Unfortunately, these assumptions are not typically met with regard to most quantitative traits. Even so, when one or more assumptions are invalid, the calculated value of n is smaller than the actual number. Therefore, this calculation can be used to estimate the minimum number of genes that affect a quantitative trait.

Now here is the question. The average bristle numbers in two strains of flies were 35 and 42. The genetic variance for bristle number calculated for both strains was 0.8. What is the minimum number of genes that affect bristle number?

Answer: We apply the equation described previously.

$$n = \frac{D^2}{8V_G}$$

$$n = \frac{(35 - 42)^2}{8(0.8)}$$

$$n = 7.7 \text{ genes}$$

Because genes must come in whole numbers, and because this calculation is a minimum estimate, one would conclude that there must be at least eight genes that affect bristle number.

S6. Are the following statements regarding heritability true or false?

A. Heritability applies to a specific population raised in a particular environment.

B. Heritability in the narrow sense takes into account all types of genetic variance.

C. Heritability is a measure of the amount that genetics contributes to the outcome of a trait.

Answer:

A. True.

B. False. Narrow sense heritability considers only the effects of additive alleles.

C. False. Heritability is a measure of the amount of phenotypic *variation* that is due to genetic variation; it applies to the variation of a specific population raised in a particular environment.

Conceptual Questions

C1. Give several examples of quantitative traits. How are these quantitative traits described within groups of individuals?

C2. At the molecular level, explain why quantitative traits often exhibit a continuum of phenotypes within a population. How does the environment help produce this continuum?

C3. What is a normal distribution? Discuss this curve with regard to quantitative traits within a population. What is the relationship between the standard deviation and the normal distribution?

C4. Explain the difference between a continuous trait and a discontinuous trait. Give two examples of each. Are quantitative traits likely to be continuous or discontinuous? Explain why.

C5. What is a frequency distribution? Explain how the graph is made for a quantitative trait that is continuous.

C6. The variance for weight in a particular herd of cattle is 424 lb². The mean weight happens to be 424 lb. How heavy would an animal have to be if it was in the top 2.5% of the herd? The bottom 0.13%?

C7. Two different varieties of potatoes both have the same mean weight of 1.5 lb. One group has a very low variance, and the other has a much higher variance.

A. Discuss the possible reasons for the differences in variance.

B. If you were a potato farmer, would you rather raise a variety with a low or high variance? Explain your answer from a practical point of view.

C. If you were a potato breeder, and you wanted to develop potatoes with a heavier weight, would you choose the variety with a low or high variance? Explain your answer.

C8. If an r value equals 0.5 and $N = 4$, would you conclude that there is a positive correlation between the two variables? Explain your answer. What if $N = 500$?

C9. What does it mean when a correlation coefficient is negative? Can you think of examples?

C10. When a correlation coefficient is statistically significant, what do you conclude about the two variables? What do the results mean with regard to cause and effect?

C11. What is polygenic inheritance? Discuss the issues that make polygenic inheritance difficult to study.

C12. What is a quantitative trait locus (QTL)? Does a QTL contain one gene or many genes? What technique is commonly used to identify QTLs?

C13. Let's suppose that weight in a species of mammals is polygenic, and each gene exists as a heavy and light allele. If the allele frequencies in the population were equal for both types of allele (i.e., 50% heavy alleles and 50% light alleles), what percentage of individuals would be homozygous for the light alleles at all of the genes affecting this trait, if the trait was determined by the following number of genes?

A. Two

B. Three

C. Four

C14. The broad sense heritability for a trait equals 1.0. In your own words, explain what this value means. Would you conclude that the environment is unimportant in the outcome of this trait? Explain your answer.

C15. Compare and contrast the dominance and overdominance hypotheses. Based on your knowledge of mutations and genetics, which do you think is more likely?

C16. What is hybrid vigor (also known as heterosis)? Give examples that you might find in a vegetable garden.

C17. From an agricultural point of view, discuss the advantages and disadvantages of selective breeding. It is common for plant breeders to take two different, highly inbred strains, which are the product of many generations of selective breeding, and cross them to make hybrids. How does this approach overcome some of the disadvantages of selective breeding?

C18. Many beautiful varieties of roses have been produced, particularly in the last few decades. These newer varieties often have very striking and showy flowers, making them desirable as horticultural specimens. However, breeders and novices alike have noticed that some of these newer varieties do not have very fragrant flowers compared to the older, more traditional varieties. From a genetic point of view, suggest an explanation why some of these newer varieties with superb flowers are not as fragrant.

C19. In your own words, explain the meaning of the term *heritability*. Why is a heritability value valid only for a particular population of individuals raised in a particular environment?

C20. What is the difference between broad sense heritability and narrow sense heritability? Why is narrow sense heritability such a useful concept in the field of agricultural genetics?

C21. The heritability for egg weight in a group of chickens on a farm in Maine is 0.95. Are the following statements regarding heritability true or false? If a statement is false, explain why.

A. The environment in Maine has very little impact on the outcome of this trait.

B. Nearly all of the phenotypic variation in this group of chickens is due to genetic variation.

C. The trait is polygenic and likely to involve a large number of genes.

D. Based on the observation of the heritability in the Maine chickens, it is reasonable to conclude that it is likely that the heritability for egg weight in a group of chickens on a farm in Montana is also very high.

C22. In a fairly large population of people living in a commune in the southern United States, everyone cares about good nutrition. All the members of this population eat very nutritional foods, and their diets are very similar to each other. With regard to height, how do you think this commune population would compare to the general population in the following categories?

A. Mean height

B. Heritability for height

C. Genetic variation for alleles that affect height

C23. When artificial selection is practiced over many generations, it is common for the trait to reach a plateau in which further selection has little effect on the outcome of the trait. This phenomenon is illustrated in figure 24.12. Explain why.

C24. Discuss whether a natural population of wolves or a domesticated population of German shepherds is more likely to have a higher heritability for the trait of size.

C25. With regard to heterosis, would the following statements be consistent with the dominance hypothesis, the overdominance hypothesis, or both?

A. Strains that have been highly inbred have become monomorphic for one or more recessive alleles that are somewhat detrimental to the organism.

B. Hybrid vigor occurs because highly inbred strains are monomorphic for many genes, while hybrids are more likely to be heterozygous for those same genes.

C. If a gene exists in two alleles, hybrids are more vigorous because heterozygosity for the gene is more beneficial than homozygosity of either allele.

Experimental Questions

E1. Here are data for height and weight among 10 male college students.

Height (cm)	Weight (kg)
159	48
162	50
161	52
175	60
174	64
198	81
172	58
180	74
161	50
173	54

A. Calculate the correlation coefficients for this group.

B. Is the correlation coefficient statistically significant? Explain.

E2. The abdomen length (in millimeters) was measured in 15 male *Drosophila* and the following data were obtained: 1.9, 2.4, 2.1, 2.0, 2.2, 2.4, 1.7, 1.8, 2.0, 2.0, 2.3, 2.1, 1.6, 2.3, and 2.2. Calculate the mean, standard deviation, and variance for this population of male fruit flies.

E3. You will need to understand solved problem S5 before answering this question. The average weights for two varieties of cattle were 514 kg and 621 kg. The genetic variance for weight calculated for both strains was 382 kg^2. What is the minimum number of genes that affect weight variation in these two varieties of cattle?

E4. Using the same strategy as the experiment of figure 24.6, the following data are the survival of F_2 offspring obtained from backcrosses to insecticide-resistant and control strains:

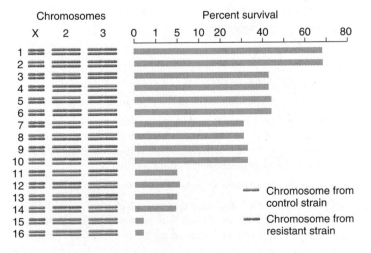

Interpret these results with regard to the locations of QTLs.

E5. In one strain of cabbage, you conduct an RFLP analysis with regard to head weight; you determine that seven QTLs affect this trait. In another strain of cabbage, you find that only four QTLs affect this trait. Note that both strains of cabbage are from the same species, although they may have been subjected to different degrees of inbreeding. Explain how one strain can have seven QTLs and another strain four QTLs for exactly the same trait. Is the second strain missing three genes?

E6. From an experimental viewpoint, what does it mean to say that an RFLP is associated with a trait? Let's suppose that two strains of pea plants differ in two RFLPs that are linked to two genes governing pea size. RFLP-1 is found in 2,000 bp and 2,700 bp bands, and RFLP-2 is found in 3,000 bp and 4,000 bp bands. The plants producing large peas have RFLP-1 (2,000 bp) and RFLP-2 (3,000 bp); those producing small peas have RFLP-1 (2,700 bp) and RFLP 2 (4,000 bp). A cross is made between these two strains, and the F_1 offspring are allowed to self-fertilize. Five phenotypic classes are observed: small peas, small-medium peas, medium peas, medium-large peas, and large peas. We assume that each of the two genes makes an equal contribution to pea size, and that the genetic variance is additive. Draw a gel, and explain what RFLP banding patterns you would expect to observe for these five phenotypic categories. Note: Certain phenotypic categories may have more than one possible banding pattern.

E7. Let's suppose that two strains of pigs differ in 500 RFLPs. One strain is much larger than the other. The pigs are crossed to each other, and the members of the F_1 generation are also crossed among themselves to produce an F_2 generation. Three distinct RFLPs are associated with F_2 pigs that are larger. How would you interpret these results?

E8. Outline the steps you would follow to determine the number of genes that influence the yield of rice. Describe the results you might get if rice yield is governed by six different genes.

E9. A researcher has two highly inbred strains of mice. One strain is susceptible to infection by a mouse leukemia virus, while the other strain is resistant. Susceptibility/resistance is a polygenic trait. The two strains were crossed together and all the F_1 mice were resistant. The F_1 mice were then allowed to interbreed, and 120 F_2 mice were obtained. Among these 120 mice, 118 were resistant to the viral pathogen, and two were sensitive. Discuss how many different genes may be involved in this trait. How would your answer differ if none of the F_2 mice had been susceptible to the leukemia virus? Hint: You should assume that the inheritance of one viral-resistance allele is sufficient to confer resistance.

E10. In a wild strain of tomato plants, the phenotypic variance for tomato weight is 3.2 g^2. In another strain of highly inbred tomatoes raised under the same environmental conditions, the phenotypic variance is 2.2 g^2. With regard to the wild strain:

A. Estimate V_G.

B. What is H_B^2?

C. Assuming all the genetic variance is additive, what is h_N^2?

E11. The average thorax length in a *Drosophila* population is 1.01 mm. You want to practice selective breeding to make larger *Drosophila*. To do so, you choose 10 parents (five males and five females) of the following sizes: 0.97, 0.99, 1.05, 1.06, 1.03, 1.21, 1.22, 1.17, 1.19, and 1.20. You mate them and then analyze the thorax sizes of 30 offspring (half male and half female).

0.99, 1.15, 1.20, 1.33, 1.07, 1.11, 1.21, 0.94, 1.07, 1.11, 1.20, 1.01, 1.02, 1.05, 1.21, 1.22, 1.03, 0.99, 1.20, 1.10, 0.91, 0.94, 1.13, 1.14, 1.20, 0.89, 1.10, 1.04, 1.01, 1.26

Calculate the realized heritability in this group of flies.

E12. In a strain of mice, the average 6-week body weight is 25 g and the narrow sense heritability for this trait is 0.21.

A. What would be the average weight of the offspring if parents with a mean weight of 27 g were chosen?

B. What weight of parents would you have to choose to obtain off-spring with an average weight of 26.5 g?

E13. Two tomato strains, A and B, both produce fruit that weighs, on average, 1 lb each. All of the variance is due to V_G. When these two strains are crossed to each other, the F_1 offspring display heterosis with regard to fruit weight, with an average weight of 2 lb. You take these F_1 offspring and backcross them to strain A. You then grow several plants from this cross and measure the weights of their fruit. What would be the expected results for each of the following scenarios?

A. Heterosis is due to a single overdominant gene.

B. Heterosis is due to two dominant genes, one in each strain.

C. Heterosis is due to two overdominant genes.

D. Heterosis is due to dominance of several genes each from strains A and B.

E14. You will need to understand solved problem S4 before answering this question. The variance for fathers (in square inches) was 112, the variance for sons was 122, and the covariance was 144. The mean height for fathers was 68 in., and the mean height for sons was 69 in. If a father had a height of 70 in., what is the most probable height of his son?

E15. In a particular strain of pigs, the correlation among littermates for weight is 0.15. What is the narrow sense heritability for this trait?

E16. A danger in computing heritability values from studies involving genetically related individuals is the possibility that these individuals share more similar environments than do unrelated individuals. In the experiment of figure 24.9, which data are the most compelling evidence that ridge count is not caused by genetically related individuals sharing common environments? Explain.

E17. A large, genetically heterogeneous group of tomato plants was used as the original breeding stock by two different breeders, named Mary and Hector. Each breeder was given 50 seeds and began an artificial selection strategy, much like the one described in figure 24.12. The seeds were planted, and the breeders selected the 10 plants with the highest mean tomato weights as the breeding stock for the next generation. This process was repeated over the course of 12 growing seasons, and the following data were obtained:

Mean Weight of Tomatoes (lb)

Year	Mary's Tomatoes	Hector's Tomatoes
1	0.7	0.8
2	0.9	0.9
3	1.1	1.2
4	1.2	1.3
5	1.3	1.3

(continued)

Mean Weight of Tomatoes (lb)

Year	Mary's Tomatoes	Hector's Tomatoes
6	1.4	1.4
7	1.4	1.5
8	1.5	1.5
9	1.5	1.5
10	1.5	1.5
11	1.5	1.5
12	1.5	1.5

A. Explain these results.

B. Another tomato breeder, named Martin, got some seeds from Mary's and Hector's tomato strains (after 12 generations), grew the plants, and then crossed them to each other. The mean weight of the tomatoes in these hybrids was about 1.7 lb. For a period of 5 years, Martin subjected these hybrids to the same experimental strategy that Mary and Hector had followed, and he obtained the following results:

Mean Weight of tomatoes (lb)

Year	Martin's Tomatoes
1	1.7
2	1.8
3	1.9
4	2.0
5	2.0

Explain Martin's data. Is heterosis occurring? Why was Martin able to obtain tomatoes that are heavier than 1.5 lb, while Mary's and Hector's strains appeared to plateau at this weight?

E18. The correlations for height were determined for 15 pairs of individuals with the following genetic relationships:

Mother/daughter: 0.36

Mother/granddaughter: 0.17

Sister/sister: 0.39

Sister/sister (fraternal twins): 0.40

Sister/sister (identical twins): 0.77

What is the average heritability for height in this group of females?

E19. An animal breeder had a herd of sheep with a mean weight of 254 lb at 3 years of age. He chose animals with mean weights of 281 lb as parents for the next generation. When these offspring reached 3 years of age, their mean weights were 269 lb.

A. Calculate the narrow sense heritability for weight in this herd.

B. Using the heritability value that you calculated in part A, what weight of animals would you have to choose to get offspring that weigh 275 lb (at 3 years of age)?

E20. The trait of blood pressure in humans has a frequency distribution that is similar to a normal distribution. The following graph shows the ranges of blood pressures for a selected population of people. The *red line* depicts the frequency distribution of the systolic pressures for the entire population. Several individuals with high blood pressure were identified, and the blood pressures of their relatives were determined. This frequency distribution is depicted with a *blue line*. (Note: The *blue line* does not include the people who were identified with high blood pressure; it includes only their relatives.)

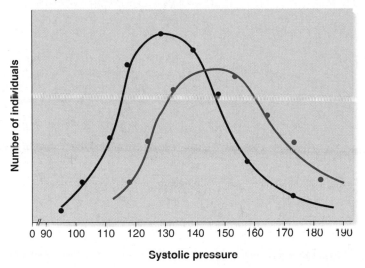

What do these data suggest with regard to a genetic basis for high blood pressure? What statistical approach could you use to determine the heritability for this trait?

Questions for Student Discussion/Collaboration

1. Discuss why heritability is an important phenomenon in agriculture. Discuss how it is misunderstood.

2. From a biological viewpoint, speculate as to why many traits seem to fit a normal distribution. Students with a strong background in math and statistics may want to explain how a normal distribution is generated, and what it means. Can you think of biological examples that do not fit a normal distribution?

3. What is heterosis? Discuss whether it is caused by a single gene or several genes. Discuss the two major hypotheses proposed to explain heterosis. Which do you think is more likely to be correct?

Note: All answers appear at the website for this textbook; the answers to even-numbered questions are in the back of the textbook.

www.mhhe.com/brooker

Visit the Online Learning Center for practice tests, answer keys, and other learning aids for this chapter. Enhance your understanding of genetics with our interactive exercises, web links, news feeds, tutorial service, and much more.

POPULATION GENETICS

::

25

The central issue in **population genetics** is genetic variation. Population geneticists want to know the extent of genetic variation within populations, why it exists, and how it changes over the course of many generations. Population genetics emerged as a branch of genetics in the 1920s and 1930s. Its mathematical foundations were developed by theoreticians who extended the principles of Mendel and Darwin by deriving formulae to explain the occurrence of genotypes within populations. These foundations can be largely attributed to three scientists: Sir Ronald Fisher, Sewall Wright, and J. B. S. Haldane. As we will see, support for their mathematical theories was provided by several researchers who analyzed the genetic composition of natural and experimental populations. More recently, population geneticists have used techniques to probe genetic variation at the molecular level. In addition, the staggering improvement in computer technology has aided population geneticists in the analysis of their genetic theories and data. In chapter 25, we will explore the genetic variation that occurs in populations and consider the reasons that cause variation to be prevalent.

25.1 GENES IN POPULATIONS

Population genetics may seem like a significant departure from other topics in this textbook, but it is a direct extension of our understanding of Mendel's laws of inheritance, molecular genetics, and the ideas of Darwin. The focus is shifted away from the individual and toward the population of which the individual is a member. Conceptually, all of the alleles of every gene in a population make up the **gene pool.** In this regard, each member of the population is viewed as receiving its genes from its parents, which, in turn, are members of the gene pool. Furthermore, individuals that reproduce contribute to the gene pool of the next generation. Population geneticists study the genetic variation within the gene pool and how such variation changes from one generation to the next. The emphasis is often on allelic variation. In this introductory section, we will examine some of the general features of populations and gene pools.

A Population Is a Group of Interbreeding Individuals That Shares a Gene Pool

In genetics, the term *population* has a very specific meaning. A **population** is a group of individuals of the same species that can interbreed with one another. Many species occupy a wide geographic range and are divided into discrete populations. For example, distinct populations of a given species may be located on different continents. Or populations on the same continent could be divided by a large mountain range.

A large population usually is composed of smaller groups called **subpopulations, local populations,** or **demes.** The members of a subpopulation are far likelier to breed among themselves than with other members of the general population. Subpopulations are often separated from each other by moderate geographic barriers. As shown in figure 25.1, two populations of

FIGURE 25.1   Two local populations of Douglas fir. A wide river bottom separates the two local populations. There is a much higher frequency of cross-fertilization among members of the same local population than between the two different local populations.

Douglas fir (*Pseudotsuga menziesii*) are separated by a wide river bottom, where the fir trees cannot grow. Trees on opposite sides of this river bottom constitute local populations. Interbreeding is much more apt to occur among members of each local population than between members of neighboring populations. On relatively rare occasions, however, pollen can be blown across the river bottom, which allows interbreeding between these two local populations.

Populations typically are dynamic units that change from one generation to the next. A population may change its size, geographic location, and genetic composition. With regard to size, natural populations commonly go through cycles of "feast or famine," during which the population swells or shrinks. In addition, natural predators or disease may periodically decrease the size of a population to significantly lower levels; the population later may rebound to its original size. Populations or individuals within populations may migrate to a new site and establish a distinct population in this location. The environment at this new geographic location may differ from the original site. As population sizes and locations change, their genetic composition generally changes as well. As described later in chapter 25, population geneticists have developed mathematical theories that predict how the gene pool will change in response to fluctuations in size, migration, and new environments.

Some Genes Are Monomorphic, and Others Are Polymorphic

In population genetics, the term **polymorphism** (meaning many forms) refers to the observation that many traits display variation within a population. Historically, polymorphism first referred to the variation in traits that are observable with the naked eye. Polymorphisms in color and pattern have long attracted the attention of population geneticists. These include studies of melanism in the peppered moth and of variation in snail color, which are discussed later in chapter 25. Figure 25.2 illustrates a striking example of polymorphism in the Hawaiian happy-face spider (*Theridion grallator*). The three individuals shown in this figure are from the same species, but they differ in alleles that affect color and pattern.

At the DNA level, polymorphism is due to two or more alleles that influence the phenotype of the individual that inherits them. In other words, it is due to genetic variation. Geneticists also use the term **polymorphic** to describe a gene that commonly exists as two or more alleles in a population. By comparison, a **monomorphic** gene exists predominantly as a single allele in a population. By convention, when a single allele is found in at least 99% of all cases, the gene is considered monomorphic. Said another way, a polymorphic gene must have one or more additional alleles that make up at least 1% of the alleles in the population.

During the 1960s, molecular techniques became available that could assess variation in alleles that influence enzyme structure. John Hubby and Richard Lewontin studied allelic variation in populations of the fruit fly *Drosophila pseudoobscura*. This species of fruit fly was chosen because, unlike *D. melanogaster*, *D. pseudoobscura* does not share its habitat with humans, and so it is

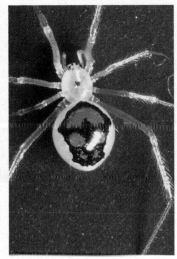

FIGURE 25.2   Polymorphism in the Hawaiian happy-face spider.

GENES→TRAITS   These three spiders are members of the same species and carry the same genes. However, several genes that affect pigmentation patterns are polymorphic, meaning that there is more than one allele for each gene within the population. This polymorphism within the Hawaiian happy-face spider population produces members that look quite different from each other.

considered truly wild. In addition, it has a wide geographic distribution in North and Central America. Hubby and Lewontin studied many structural genes that were known to encode enzymes. At the molecular level, genetic variation in a structural gene may result in the production of an enzyme with slight differences in its amino acid sequence compared to the wild-type enzyme. Based on previous work, Hubby and Lewontin knew that these differences in amino acid sequence could be detected as changes in the mobility of the enzymes during gel electrophoresis. Two enzymes with alterations in their gel mobilities due to small differences in their amino acid sequences are called **allozymes.** In other words, they are <u>all</u>eles of the same gene that encodes an en<u>zyme.</u>

In their study, Hubby and Lewontin looked for genetic variation in 18 different enzymes in five different fruit fly populations. On average, 30% of the enzymes were found as two or more allozymes. This means that the genes encoding these enzymes had slightly different DNA sequences, yielding enzymes with slightly different amino acid sequences. This study tends to underestimate genetic variability, since some amino acid substitutions do not alter protein mobility during gel electrophoresis. Nevertheless, the important conclusion from this work was that the genetic variability within these populations is quite high. Around the time of this work, Harry Harris also found that approximately 30% of human genes encoding enzymes are polymorphic. More recently, many population geneticists have investigated genetic variation using DNA sequencing methods described in chapter 18.

In most natural populations, a substantial percentage of genes are polymorphic. Examples are shown in figure 25.3. However, in small populations that are near extinction, genetic variation is expected to be low because the gene pool is derived from a small number of individuals. An extreme example is the African cheetah population, which has a genetic variation near zero. Later

in chapter 25, we will examine how small population size contributed to this problem. By comparison, approximately 30% of human genes are polymorphic. Because humans have approximately 35,000 different genes, this means that roughly 10,000 genes can be found as two or more different alleles.

Keep in mind that polymorphism refers to the diversity of genes in a population. Within a single individual, genetic variation will be less, because an individual may be homozygous for polymorphic genes. For example, the ABO blood type is determined by three alleles, designated I^A, I^B, and i. A person with type O blood is homozygous, ii, and therefore has less genetic variation than the population as a whole. While approximately 30% of

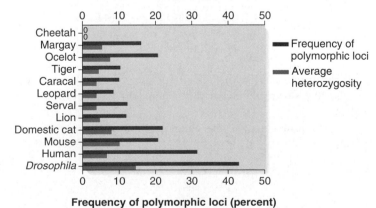

FIGURE 25.3   A comparison of polymorphisms in different species. The *upper red bars* indicate the percentage of genes that are polymorphic within an entire species; the *lower blue bars* indicate the average percentage of genes that are heterozygous within any given individual (average heterozygosity). This information was derived from the analysis of allozymes using gel electrophoresis.

all genes are polymorphic in human populations, less than 10% of all genes are heterozygous within any given individual.

Population Genetics Is Concerned with Allele and Genotype Frequencies

As we have seen, population geneticists want to understand the prevalence of polymorphic genes within populations. Much of their work evaluates the frequency of alleles in a quantitative way. Two fundamental calculations are central to population genetics: **allele frequencies** and **genotype frequencies.** The allele and genotype frequencies are defined as

$$\text{Allele frequency} = \frac{\text{Number of copies of an allele in a population}}{\text{Total number of all alleles for that gene in a population}}$$

$$\text{Genotype frequency} = \frac{\text{Number of individuals with a particular genotype in a population}}{\text{Total number of all individuals in a population}}$$

Though these two frequencies are related, a clear distinction between them must be kept in mind. As an example, let's consider a population of 100 pea plants with the following genotypes:

64 tall plants with the genotype TT

32 tall plants with the genotype Tt

4 dwarf plants with the genotype tt

When calculating an allele frequency, homozygous individuals have two copies of an allele, whereas heterozygotes only have one. For example, in tallying the t allele, each of the 32 heterozygotes has one copy of the t allele, and each dwarf plant has two copies. The allele frequency for t equals

$$t = \frac{32 + (2)(4)}{(2)(64) + (2)(32) + (2)(4)}$$

$$= \frac{40}{200} = 0.2, \text{ or } 20\%$$

This result tells us that the allele frequency of t is 20%. In other words, 20% of the alleles for this gene in the population are the t allele.

Let's now calculate the genotype frequency of tt (dwarf) plants.

$$tt = \frac{4}{64 + 32 + 4}$$

$$= \frac{4}{100} = 0.04, \text{ or } 4\%$$

We see that 4% of the individuals in this population are dwarf plants.

Allele and genotype frequencies are always less than or equal to 1 (i.e., less than or equal to 100%). If a gene is monomorphic, the allele frequency for the single allele will equal

or be close to a value of 1.0. For polymorphic genes, if we add up the frequencies for all the alleles in the population, we should obtain a value of 1.0. In our pea plant example, the allele frequency of t equals 0.2. The frequency of the other allele, T, equals 0.8. If we add the two together, we obtain a value of $0.2 + 0.8 = 1.0$. In the next section, we will learn how allele and genotype frequencies within populations are related.

25.2 THE HARDY-WEINBERG EQUILIBRIUM

Now that we have a general understanding of genes in populations, we can begin to relate these concepts to mathematical expressions as a way to examine whether allele and genotype frequencies will change over the course of many generations. In 1908, Godfrey Harold Hardy and Wilhelm Weinberg independently derived a simple mathematical expression that predicted stability of allele and genotype frequencies from one generation to the next. This expression, known as the **Hardy-Weinberg equation,** relates allele and genotype frequencies within a population. It is also called an **equilibrium,** because (under a given set of conditions, described later) the allele and genotype frequencies do not change over the course of many generations. This relationship established a framework on which to understand changes in gene frequencies within a population when such an equilibrium is violated.

In subsequent decades, as the field of population genetics developed, it became apparent that genetic stability is not always present in natural populations. Furthermore, genetic change underlies the theory of evolution. Therefore, population geneticists turned their efforts largely toward understanding how genetic stability is altered. In this section, we will examine the Hardy-Weinberg equilibrium and the conditions that must be met for it to be valid. At the end of chapter 25, we will learn how natural populations frequently violate the Hardy-Weinberg equilibrium and thereby cause allele frequencies to change.

The Hardy-Weinberg Equation Can Be Used to Calculate Genotype Frequencies Based on Allele Frequencies

The Hardy-Weinberg equation is a simple mathematical expression that relates genotype and allele frequencies. Let's first examine the mathematical components of the Hardy-Weinberg equation and then look at the conditions necessary for an equilibrium to be achieved.

Let's begin by considering a situation in which a gene is polymorphic and exists as two different alleles, A and a. If the allele frequency of A is denoted by the variable p, and the allele frequency of a by q, then

$$p + q = 1$$

For example, if $p = 0.8$, then q must be 0.2. In other words, if the allele frequency of A equals 80%, the remaining 20% of alleles must be a, because together they equal 100%.

For a gene that exists in two alleles, the Hardy-Weinberg equation states that

$$(p + q)^2 = 1$$

$$p^2 + 2pq + q^2 = 1 \text{ (Hardy-Weinberg equation)}$$

If this equation is applied to a gene that exists in alleles designated A and a, then

p^2 equals the genotype frequency of AA

$2pq$ equals the genotype frequency of Aa

q^2 equals the genotype frequency of aa

If $p = 0.8$ and $q = 0.2$, and if the population is in Hardy-Weinberg equilibrium, then

$$AA = p^2 = (0.8)^2 = 0.64$$

$$Aa = 2pq = 2(0.8)(0.2) = 0.32$$

$$aa = q^2 = (0.2)^2 = 0.04$$

In other words, if the allele frequency of A is 80% and the allele frequency of a is 20%, the genotype frequency of AA is 64%, Aa 32%, and aa 4%.

To see the relationship between allele frequencies and genotypes, figure 25.4 compares the Hardy-Weinberg equation with the Punnett square approach. As seen here, the Hardy-Weinberg equation results from the way gametes combine randomly with each other to produce offspring. In a population, the frequency of a gamete carrying a particular allele is equal to the allele frequency in that population. For example, the frequency of a gamete carrying the A allele equals 0.8.

We can use the product rule to determine the frequency of genotypes. For example, the frequency of producing an AA homozygote is $0.8 \times 0.8 = 0.64$, or 64%. Likewise, the probability of inheriting both a alleles is $0.2 \times 0.2 = 0.04$, or 4%. In our Punnett square, there are two different ways to produce heterozygotes (fig. 25.4). An offspring could inherit the A allele from its father and a from its mother, or A from its mother and a from its father.

Therefore, the frequency of heterozygotes is $pq + pq$, which equals $2pq$; in our example, this is $2(0.8)(0.2) = 0.32$, or 32%.

The Hardy-Weinberg equation predicts an equilibrium (i.e., unchanging allele and genotype frequencies) if certain conditions are met in a population. With regard to the gene of interest, these are as follows:

1. The population is so large that allele frequencies do not change due to random sampling effects.
2. Regarding the gene of interest, the members of the population mate with each other without regard to their phenotypes and genotypes.
3. There is no migration between different populations.
4. There is no survival or reproductive advantage for any of the genotypes. In other words, no natural selection occurs.
5. No new mutations occur in the gene of interest.

According to the equilibrium, the Hardy-Weinberg equation provides a quantitative relationship between allele and genotype frequencies in a population. Figure 25.5 describes this relationship for different allele frequencies of A and a. As expected, when the allele frequency of A is very low, the aa genotype predominates; when the A allele frequency is high, the AA homozygote is most frequent in the population. When the allele frequencies of A and a are intermediate in value, the heterozygote predominates.

In reality, no population satisfies the Hardy-Weinberg equilibrium completely. Nevertheless, in large natural populations with little migration and negligible natural selection, the Hardy-Weinberg equilibrium may be nearly approximated for certain genes. In addition, the Hardy-Weinberg equation can be extended to situations in which a single gene exists in three or more alleles. Solved problem S5 at the end of chapter 25 considers a problem in which a gene exists in three alleles.

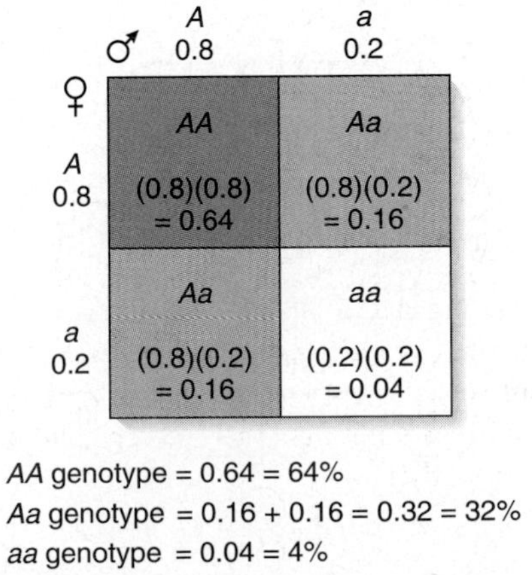

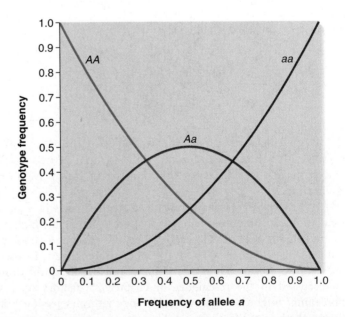

FIGURE 25.5 **The relationship between allele frequencies and genotype frequencies according to the Hardy-Weinberg equilibrium.** This graph assumes that A and a are the only two alleles for this gene.

AA genotype = 0.64 = 64%

Aa genotype = 0.16 + 0.16 = 0.32 = 32%

aa genotype = 0.04 = 4%

FIGURE 25.4 **A comparison between the Hardy-Weinberg equation and the Punnett square approach.**

As discussed in chapter 2, the χ^2 test can determine if observed and expected data are in agreement. Therefore, we can use a χ^2 test to see if a population really exhibits a Hardy-Weinberg equilibrium for a particular gene. To do so, it is necessary to distinguish the homozygotes and heterozygotes, either phenotypically or at the molecular level. This is necessary so that we can determine both the allele and genotype frequencies. As an example, we can consider a human blood type called the MN type. In this case, the blood type is determined by two codominant alleles, M and N. In an Inuit population in East Greenland, it was found that among 200 people, 168 were MM, 30 were MN, and 2 were NN. We can use these observed data to calculate the expected data, if we hypothesize that the population is in equilibrium.

$$\text{Allele frequency of } M = \frac{2(168) + 30}{400} = 0.915$$

$$\text{Allele frequency of } N = \frac{2(2) + 30}{400} = 0.085$$

Expected frequency of $MM = p^2 = (0.915)^2 = 0.837$
Expected number of MM individuals $= 0.837 \times 200 = 167.4$
$\qquad$ (or **167** rounded to the nearest individual)

Expected frequency of $NN = q^2 = (0.085)^2 = 0.007$
Expected number of mm individuals $= 0.007 \times 200 = 1.4$ (or **1** rounded to the nearest individual)

Expected frequency of $MN = 2pq = 2(0.915)(0.085) = 0.0156$
Expected number of MN individuals $= 0.156 \times 200 = 31.2$ (or **31** rounded to the nearest individual)

$$\chi^2 = \frac{(O_1 - E_1)^2}{E_1} + \frac{(O_2 - E_2)^2}{E_2} + \frac{(O_3 - E_3)^2}{E_3}$$

$$\chi^2 = \frac{(168 - 167)^2}{167} + \frac{(30 - 31)^2}{31} + \frac{(2 - 1)^2}{1}$$

$$= 1.04$$

With 2 degrees of freedom, the calculated chi square value is well within the acceptable range. Therefore, we accept the hypothesis that the alleles for this gene are in Hardy-Weinberg equilibrium.

When researchers have investigated other genes in various populations, it sometimes happens that an unacceptably high chi square value is obtained, indicating that the allele and genotype frequencies are not in a Hardy-Weinberg equilibrium. In these cases, we would say that such a population is in **disequilibrium.** As discussed in the remainder of chapter 25, factors such as inbreeding, migration, genetic drift, and natural selection may disrupt the Hardy-Weinberg equilibrium. Therefore, when population geneticists discover that a population is not in equilibrium, they try to determine if one or several of these four factors are causing the disequilibrium.

Nonrandom Mating May Occur in Natural and Human Populations

As mentioned earlier, one of the conditions required to establish the Hardy-Weinberg equilibrium is random mating. This means that individuals choose their mates irrespective of their genotypes and phenotypes. In many cases, particularly in human populations, this condition is violated frequently.

When two individuals are more likely to mate due to similar phenotypic characteristics, this is known as **assortative mating.** The opposite situation, where dissimilar phenotypes mate preferentially, is called **disassortative mating.** In addition, individuals may choose a mate who is part of the same genetic lineage. The mating of two genetically related individuals, such as cousins, is called **inbreeding.** This sometimes occurs in human societies and is more likely to take place in nature when population size becomes very limited. In chapter 24, we also learned that inbreeding is a useful strategy for developing agricultural breeds or strains with desirable characteristics. Conversely, **outbreeding,** which involves mating between unrelated individuals, can create hybrids that are heterozygous for many genes.

In the absence of other evolutionary processes, inbreeding and outbreeding do not affect allele frequencies in a population. However, these patterns of mating do disrupt the balance of genotypes that is predicted by the Hardy-Weinberg equilibrium. Let's first consider inbreeding in a family pedigree. Figure 25.6 illustrates a human pedigree involving a mating between cousins. Individuals III-2 and III-3 are cousins and have produced the son labeled IV-1. He is said to be **inbred,** because his parents are genetically related to each other.

During inbreeding, the gene pool is smaller, because the parents are related genetically. Wright and Fisher developed methods to quantify the degree of inbreeding. A **coefficient of inbreeding (F)** can be computed by analyzing the degree of relatedness within a pedigree. As an example, let's determine the coefficient of inbreeding for individual IV-1. To begin this problem, we must first identify all the common ancestors that this individual has. A common ancestor is anyone who is an ancestor to both of an individual's parents. In figure 25.6, IV-1 has one common

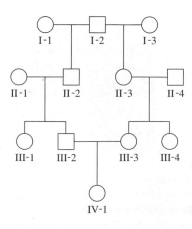

FIGURE 25.6 **A human pedigree containing inbreeding.** Individual IV-1 is the result of a consanguineous mating.

ancestor, I-2, his great-grandfather. I-2 is the grandfather of III-2 and III-3.

Our next step is to determine the inbreeding paths. An inbreeding path for an individual is the shortest path through the pedigree that includes both parents and the common ancestor. In a pedigree, there is an inbreeding path for each common ancestor. The length of each inbreeding path is calculated by adding together all of the individuals in the path except the individual of interest. In this case, there is only one path because there is only one common ancestor. To add the members of the path, we begin with individual IV-1, but we do not count him. We then move to his father (III-2); to his grandfather (II-2); to I-2, his great-grandfather (the common ancestor); back down to his other grandmother (II-3); and finally to his mother (III-3). This path has five members. Finally, to calculate the inbreeding coefficient, we use the following formula:

$$F = \Sigma (1/2)^n (1 + F_A)$$

where

F is the inbreeding coefficient of the individual of interest

n is the number of individuals in the inbreeding path, excluding the inbred offspring

F_A is the inbreeding coefficient of the common ancestor

Σ indicates that we add together $(1/2)^n(1 + F_A)$ for each inbreeding path

In this case, there is only one common ancestor. Also, we do not know anything about the heritage of the common ancestor, and so we assume that F_A is zero. Thus, in our example of figure 25.6,

$$F = \Sigma (1/2)^n (1 + 0)$$

$$= (1/2)^5 = 1/32 = 3.125\%$$

Our inbreeding coefficient, 3.125%, tells us the probability that a gene in the inbred individual (IV-1) is homozygous due to its inheritance from a common ancestor (I-2). In this case, therefore, each gene in individual IV-1 has a 3.125% chance of being homozygous because he has inherited the same allele twice from his great-grandfather (I-2), once through each parent. For example, let's suppose that the common ancestor (I-2) is heterozygous for the allele that causes cystic fibrosis. There is a 3.125% probability that the inbred individual (IV-1) is homozygous for this gene because he has inherited both copies from his great-grandfather. He has a 1.56% probability of inheriting both normal alleles and a 1.56% probability of inheriting both mutant alleles.

The coefficient of inbreeding is also called the **fixation coefficient.** That is why the inbreeding coefficient is denoted by the letter F (for *F*ixation). It is the probability that an allele will be fixed in the homozygous condition. The term *fixation* signifies that the homozygous individual can pass only one type of allele to their offspring.

In other pedigrees, an individual may have two or more common ancestors. In this case, the inbreeding coefficient F is calculated as the sum of the inbreeding paths. Such an example is described in solved problem S2 at the end of chapter 25.

Besides pedigrees, the effects of inbreeding can also be considered within a population of individuals. For example, let's con-

sider the situation in which the frequency of $A = p$ and the frequency of $a = q$. In a given population, the genotype frequencies are determined in the following way:

$p^2 + Fpq$ equals the frequency of AA

$2pq(1 - F)$ equals the frequency of Aa

$q^2 + Fpq$ equals the frequency of aa

Let's suppose that $p = 0.8$, $q = 0.2$, and $F = 0.25$. We can calculate the frequencies of the AA, Aa, and aa genotypes under these conditions as follows:

$$AA = p^2 + Fpq = (0.8)^2 + (0.25)(0.8)(0.2) = 0.68$$

$$Aa = 2pq(1 - F) = 2(0.8)(0.2)(1 - 0.25) = 0.24$$

$$aa = q^2 + Fpq = (0.2)^2 + (0.25)(0.8)(0.2) = 0.08$$

There will be 68% AA homozygotes, 24% heterozygotes, and 8% aa homozygotes. If inbreeding had not been practiced (i.e., $F = 0$), the genotype frequencies of AA would be p^2, which equals 64%, and aa would be q^2, which equals 4%. The frequency of heterozygotes would be $2pq$, which equals 32%. When comparing these numbers, we see that inbreeding raises the proportions of homozygotes and decreases the proportion of heterozygotes. In natural populations, the inbreeding coefficient tends to become larger as a population becomes smaller, because each individual has a more limited choice in mate selection.

Inbreeding can have both positive and negative consequences in a population. From an agricultural viewpoint, it results in a higher proportion of homozygotes, which may exhibit a desirable trait. For example, an animal breeder may use inbreeding to produce animals that are larger because they have become homozygous for alleles promoting larger size. On the negative side, many genetic diseases are inherited in a recessive manner (see chapter 22). For these rare recessive disorders, inbreeding increases the likelihood that an individual will be homozygous and therefore afflicted with the disease.

25.3 FACTORS THAT CHANGE ALLELE FREQUENCIES IN POPULATIONS

The Hardy-Weinberg equilibrium and the high prevalence of polymorphisms within natural populations seem, at first glance, to contradict each other. On the one hand, the Hardy-Weinberg equilibrium predicts that allele frequencies will not change from one generation to the next. On the other hand, research has shown that many genes are polymorphic, indicating that two or more alleles can attain a significant percentage within a population. This leads to a central question in population genetics: How do genetic polymorphisms originate, and why are they maintained in natural populations?

For very rare, detrimental alleles, the explanation is straightforward. Random mutations are far likelier to produce deleterious alleles than beneficial alleles, because the intricate functioning of proteins tends to be disrupted rather than improved by a haphazard change. As we will learn in this section, natural selection promotes the elimination of harmful alleles in a population. Thus, the low rate of mutation produces detrimental alleles, and the process of natural selection works to eliminate them. The opposing actions of these two processes maintain detrimental alleles at very low frequencies.

However, many genetic polymorphisms exist in which two or more alleles are found at substantial frequencies in a population. In these cases, the allele percentages commonly observed in genetic polymorphisms cannot be explained by new mutations; the allele frequencies are too high, the rate of new mutations too low. To illustrate this point, let's use our previous example of a polymorphic gene existing in alleles A and a at respective frequencies of 0.8 and 0.2. One possibility is that such a population is derived from an ancestral population that was originally monomorphic, containing only the A allele in its gene pool. Over time, new mutations could have converted the A allele into the a allele in some members of the population. However, to reach a frequency of 0.2 for the a allele, it is unlikely to be the result of the gradual accumulation of new mutations of A to a, because the rate of such mutations is exceedingly low. Therefore, though new mutations provide a source of allelic variation, they do not occur fast enough to generate the allele frequencies observed in many polymorphisms. After that new mutation occurred, other evolutionary forces must have increased the frequency of the a allele from near zero to a frequency of 0.2. But what are these other evolutionary forces, and how do they work?

We can divide evolutionary processes into two categories: neutral and adaptive (table 25.1). **Neutral processes** (or neutral

forces) alter allele frequencies in a random manner. In other words, neutral processes change allele frequencies without any regard to the survival of the individual. They occur as a matter of chance. The two key neutral processes that can change allele frequencies are migration and genetic drift. By comparison, **adaptive forces** are processes in which alleles that confer greater survival or reproductive success increase in frequency. Natural selection produces individuals that are well adapted to their environment. Those individuals who happen to possess these beneficial alleles are more likely to survive, reproduce, and contribute to the gene pool of the next generation. In this section, we will examine the contributions of neutral and adaptive processes in creating and maintaining genetic variation.

Mutations Provide the Source of Genetic Variation

As discussed in chapters 8 and 16, mutations involve changes in gene sequences, chromosome structure, and/or chromosome number. They are random events that occur spontaneously at a low rate or are caused by mutagens at a higher rate. In chapter 25, we will be concerned primarily with the effects of gene mutations on the frequency of new alleles within a population. Even so, as we will learn in chapter 26, changes in chromosome structure and number are also important events in the evolutionary process.

The Russian geneticist Sergei Tshetverikov was the first to suggest that mutational variability provides the raw material for evolution but does not constitute evolution itself. In other words, mutation can provide new alleles to a population but does not act as a major force in dictating the final balance of allele frequencies. Tshetverikov concluded that populations in nature absorb mutations "like a sponge" and retain them in a heterozygous condition, thereby providing a source of variability for future change.

In population genetics, it is most useful to consider how new mutations affect the survival and reproductive potential of the individual that inherits them. A new mutation may be beneficial, neutral, or deleterious. For structural genes that encode proteins, the effects of new mutations will depend on their impact on protein function. Neutral and deleterious mutations are far likelier to occur than beneficial mutations. For example, alleles can be altered in many different ways that render an encoded protein defective. As discussed in chapter 16, deletions, frameshift mutations, missense mutations, and nonsense mutations all may cause a gene to express a protein that is nonfunctional or less functional than the wild-type protein. Neutral mutations can also occur in several different ways. For example, a neutral mutation can change the wobble base without affecting the amino acid sequence of the encoded protein, or it can be a missense mutation that has no effect on protein function. Such mutations are point mutations at specific sites within the coding sequence. Neutral mutations can also occur within noncoding sequences of genes, such as introns. By comparison, beneficial mutations are relatively uncommon. To be advantageous, a new mutation could alter the amino acid sequence of a protein to yield a better-functioning product. While such mutations do occur, they are expected to be very rare for a population in a stable environment.

TABLE 25.1

Forces That Alter Allele Frequencies

Description	Examples
Mutation	Mutation introduces new alleles into populations but at a very low rate. New mutations may be beneficial, neutral, or deleterious. For alleles to rise to a significant percentage in a population, it is necessary for other genetic forces (i.e., genetic drift, migration, and/or natural selection) to operate.
Neutral Forces **Random genetic drift**	A random change in allele frequencies due to random sampling error. In other words, allele frequencies may change as a matter of chance from one generation to the next. This is much more likely to occur in a small population.
Migration	Migration can occur between two different populations that have different allele frequencies. The introduction of migrants into the recipient population may change the allele frequencies of the recipient population.
Adaptive Forces **Natural selection**	This occurs when the environment selects for individuals that possess certain alleles. Natural selection can favor the survival of members with beneficial alleles or disfavor the survival of individuals with deleterious alleles.

The **mutation rate** is the probability that a gene will be altered by a new mutation. It is expressed as the number of new mutations in a given gene per generation, commonly in the range of 1 in 100,000 to 1 in 1,000,000, or 10^{-5} to 10^{-6} per generation. However, studies of mutation rates typically follow the change of a normal (functional) gene to a deleterious (nonfunctional) allele. The mutation rate producing beneficial alleles is expected to be substantially less.

It is clear that new mutations provide genetic variability, but population geneticists also want to know how the mutation rate affects the allele frequencies in a population over time. To appreciate this idea, let's take the simple case where a gene exists in a functional allele, A; the allele frequency of A is denoted by the variable p. A deleterious mutation can convert the A allele into a nonfunctional allele, a. The allele frequency of a is designated by q. The conversion of the A allele into the a allele by mutation will occur at a rate that we call u. If we assume that the rate of the reverse mutation (a to A) is negligible, the increase in the frequency of the a allele after one generation will be

$$\Delta q = up$$

For example, let's consider the following conditions:

$p = 0.8$ (i.e., frequency of A is 80%)

$q = 0.2$ (i.e., frequency of a is 20%)

$u = 10^{-5}$ (i.e., the mutation rate of converting A to a)

$\Delta q = (10^{-5})(0.8) = (0.00001)(0.8) = 0.000008$

Therefore, in the next generation,

$$q_{n+1} = 0.2 + 0.000008 = 0.200008$$

$$p_{n+1} = 0.8 - 0.000008 = 0.799992$$

As we can see from this calculation, new mutations do not significantly alter the allele frequencies in a single generation.

We can use the following equation to calculate the change in allele frequency after any number of generations:

$$(1 - u)^t = \frac{p_t}{p_0}$$

where

u is the mutation rate of the conversion of A to a

t is the number of generations

p_0 is the allele frequency of A in the starting generation

p_t is the allele frequency of A after t generations

As an example, let's suppose that the allele frequency of A is 0.8, $u = 10^{-5}$, and we want to know what the allele frequency will be after 1,000 generations ($t = 1,000$). Plugging these values into the preceding equation and solving for p_t,

$$(1 - 0.00001)^{1,000} = \frac{p_t}{0.8}$$

$$p_t = 0.792$$

Therefore, after 1,000 generations the frequency of A has dropped only from 0.8 to 0.792. Again, these results point to how slowly the mutation rate changes allele frequencies. In natural populations, the rate of new mutation is rarely a significant catalyst in shaping allele frequencies. Instead, other processes such as migration, genetic drift, and natural selection have far greater effects on allele frequencies. We will examine how these other factors work in the remainder of chapter 25.

In Small Populations, Allele Frequencies Can Be Altered by Random Genetic Drift

In the 1930s, Sewall Wright played a key role in developing the concept of **random genetic drift,** which refers to random changes in allele frequencies due to sampling error. In other words, allele frequencies may drift from generation to generation as a matter of chance. Over the long run, genetic drift favors either the loss or the fixation of an allele. The rate at which an allele is lost or becomes fixed at 100% depends on the population size. Figure 25.7 illustrates the potential consequences of genetic drift in one large ($N = 1,000$) and five small ($N = 20$) populations. At the beginning of this hypothetical simulation, all of these populations have identical allele frequencies: $A = 0.5$ and $a = 0.5$. In the five small populations, this allele frequency fluctuates substantially from generation to generation. Eventually, one of the alleles is eliminated and the other is fixed at 100%. At this point, an allele has become monomorphic and cannot fluctuate any further. By comparison, the allele frequencies in the large population fluctuate much less, because random sampling error is expected to have a smaller impact. Nevertheless, genetic drift will lead to homozygosity even in large populations, but this will take many more generations to occur.

Now let's ask two questions:

1. How many new mutations do we expect in a natural population?
2. How likely is it that any new mutation will be either fixed in, or eliminated from, a population due to random genetic drift?

With regard to the first question, the average number of new mutations depends on the mutation rate (u) and the number of individuals in a population (N). If each individual has two copies of the gene of interest, the expected number of new mutations in this gene is

$$\text{Expected number of new mutations} = 2Nu$$

From this, we see that a new mutation is more likely to occur in a large population than in a small one. This makes sense, because the larger population has more copies of the gene to be mutated. If a new mutation does occur, it may be eliminated or fixed in a population. Due to genetic drift, the probability of fixation of a newly arising allele is

$$\text{Probability of fixation} = 1/2N \text{ (assuming equal numbers of males and females contribute to the next generation)}$$

In other words, the probability of fixation is the same as the allele frequency in the population. For example, if $N = 20$, the

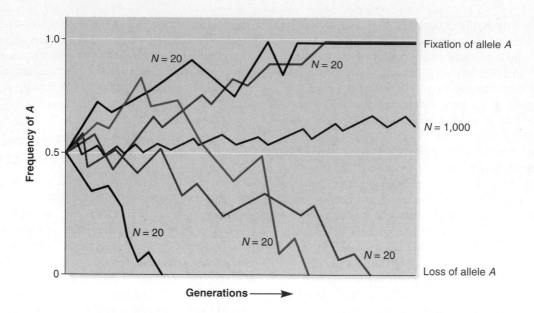

FIGURE 25.7   A hypothetical simulation of random genetic drift. In all cases, the starting allele frequencies are $A = 0.5$ and $a = 0.5$. The *colored lines* illustrate five populations in which $N = 20$; the *black line* shows a population in which $N = 1,000$.

probability of fixation equals $1/(2 \times 20)$, or 2.5%. Conversely, a new allele may be lost from the population.

$$\text{Probability of elimination} = 1 - \text{probability of fixation}$$

$$= 1 - 1/2N$$

As you may have noticed, the value of N has opposing effects with regard to mutations and their eventual fixation in a population. When N is very large, new mutations are much more likely to occur. Each new mutation, however, has a greater chance of being eliminated from the population due to random genetic drift. On the other hand, when N is small, the probability of new mutations is also small, but if they occur, the likelihood of fixation is relatively large.

Now that we have an appreciation for the phenomenon of genetic drift, we can ask a third question:

3. If fixation does occur, how many generations is it likely to take?

Again, this will depend on the number of individuals in the population.

$$\bar{t} = 4N$$

where

$\bar{t}$ equals the average number of generations to achieve fixation

N equals the number of individuals in the population, assuming that males and females contribute equally to each succeeding generation

As you may have expected, allele fixation will take much longer in large populations. If there are 1 million breeding members in a population, it will take 4 million generations, perhaps an insurmountable period of time, to reach fixation. In a small

group of 100 individuals, however, fixation will take only 400 generations. As we will see in chapter 26, the drifting of neutral alleles among different populations and species provides a way to measure the rate of evolution and can be used to determine evolutionary relationships.

Our preceding discussion of random genetic drift has emphasized two important points. First, genetic drift ultimately operates in a directional manner with regard to allele frequency. Over the long run, it leads to either allele fixation or elimination. Second, its impact is more significant in smaller populations. In nature, there are several interesting ways that small population size and genetic drift affect the genetic composition of a species. For example, some species occupy wide ranges in which small populations become geographically isolated from the rest of the species. The allele frequencies within these small populations are more susceptible to genetic drift. Because this is a random process, small isolated populations tend to be more genetically disparate in relation to other populations.

A second example of genetic drift is called the **bottleneck effect.** In nature, a population can be reduced dramatically in size by events such as earthquakes, floods, drought, and human destruction of habitat. Such events may randomly eliminate most of the members of the population without regard to genetic composition. The period of the bottleneck, when the population size is very small, may be influenced by genetic drift. First, the surviving members may have allele frequencies that differ from those of the original population. In addition, allele frequencies are expected to drift substantially during the generations when the population size is small. In extreme cases, alleles may even be eliminated. Eventually, the bottlenecked population may regain its original size (fig. 25.8a). However, the new population will have less genetic variation than the original large population. As mentioned earlier in chapter 25, the African cheetah population

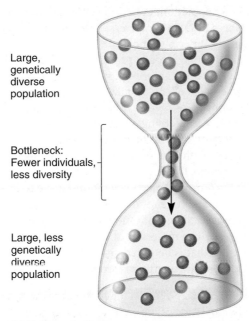

(a) Bottleneck effect

(b) An African cheetah

FIGURE 25.8 **The bottleneck effect, an example of genetic drift.** (a) A representation of the bottleneck effect. (b) The African cheetah. The modern species is monomorphic for nearly all of its genes. This low genetic variation is due to a genetic bottleneck that is thought to have occurred about 10,000 to 12,000 years ago.

has lost nearly all its genetic variation (fig. 25.8*b*). This is due to a bottleneck effect. An analysis by population geneticists has suggested that a severe bottleneck occurred approximately 10,000 to 12,000 years ago, reducing the population size to near extinction. The population eventually rebounded, but the bottleneck reduced the genetic variation to very low levels.

A third interesting case of genetic drift is the **founder effect.** This refers to the phenomenon in which a small group of individuals separates from a larger population and establishes a colony in a new location. For example, a few individuals may migrate from a large continental population and become the founders of an island population. The founder effect has two important consequences. First, the founding population is expected to have less genetic variation than the original population from which it was

derived. Second, as a matter of chance, the allele frequencies in the founding population may differ markedly from those of the original population.

Population geneticists have studied many examples where isolated populations have been started from a few members of another population. In the 1960s, Victor McKusick studied allele frequencies in the Old Order Amish of Lancaster County, Pennsylvania. At that time, this was a group of about 8,000 people, descended from just three couples that immigrated to the United States in 1770. Among this population of 8,000, a genetic disease known as the Ellis-vanCreveld syndrome (a recessive form of dwarfism) was found at a frequency of 0.07, or 7%. By comparison, this disorder is extremely rare in other human populations, even the population from which the founding members had originated. The high frequency of dwarfism in the Lancaster County population is a chance occurrence due to the founder effect.

Migrations Between Two Populations Can Alter Allele Frequencies

We have just seen how migration to a new location by a relatively small group can result in a founding population with an altered genetic composition due to genetic drift. In addition, migration between two different established populations can alter allele frequencies. For example, a species of birds may occupy two geographic regions that are separated by a large body of water. On rare occasions, the prevailing winds may allow birds from the western population to fly over this body of water and become members of the eastern population. If the two populations have different allele frequencies, and if migration occurs in sufficient numbers, this may alter the allele frequencies in the eastern population.

After migration has occurred, the new (eastern) population is called a **conglomerate.** To calculate the allele frequencies in the conglomerate, we need two kinds of information. First, we must know the original allele frequencies in the donor and recipient populations. Second, we must know the proportion of the conglomerate population that is due to migrants. With these data, we can calculate the change in allele frequency in the conglomerate population using the following equation:

$$\Delta p_C = m(p_D - p_R)$$

where

Δp_C is the change in allele frequency in the conglomerate population

p_D is the allele frequency in the donor population

p_R is the allele frequency in the original recipient population

m is the proportion of migrants that make up the conglomerate population.

$$m = \frac{\text{Number of donor individuals in the conglomerate population}}{\text{Total number of individuals in the conglomerate population}}$$

As an example, let's suppose the allele frequency of *A* is 0.7 in the donor population and 0.3 in the recipient population. A group of 20 individuals migrates and joins the recipient population, which originally had 80 members. Thus,

$$m = \frac{20}{20 + 80}$$

$$= 0.2$$

$$\Delta p_C = m(P_D - P_R)$$

$$= 0.2(0.7 - 0.3)$$

$$= 0.08$$

We can now calculate the allele frequency in the conglomerate:

$$p_C = p_R + \Delta p_C$$

$$= 0.3 + 0.08 = 0.38$$

Therefore, in the conglomerate population, the allele frequency of A has changed from 0.3 (its value before migration) to 0.38. This increase in allele frequency arises from the higher allele frequency of A in the donor population. Because population geneticists are concerned with allele frequencies rather than the migration of individuals, this phenomenon is called **gene flow.** It occurs whenever individuals migrate between populations having different allele frequencies.

In our previous example, we considered the consequences of a unidirectional migration from a donor to a recipient population. In nature, it is common for individuals to migrate in both directions. This bidirectional migration has two important consequences. Depending on its rate, migration tends to reduce differences in allele frequencies between neighboring populations. In fact, population geneticists can analyze allele frequencies in two different populations to evaluate the rate of migration between them. Populations that frequently mix their gene pools via migration tend to have similar allele frequencies, whereas isolated populations are expected to be more disparate. In addition, migration can enhance genetic diversity within a population. As discussed earlier in chapter 25, new mutations are relatively rare events. Therefore, a particular mutation may arise only in one population. Migration may then introduce this new allele into neighboring populations.

Natural Selection Favors the Survival of the Fittest

In the 1850s, Charles Darwin and Alfred Russel Wallace independently proposed the theory of **natural selection.** We will discuss the phenotypic consequences of natural selection in greater detail in chapter 26. According to this idea, there is a "struggle for existence." The conditions found in nature result in the selective survival and reproduction of individuals whose characteristics make them well adapted to their environment. These surviving individuals are more likely to reproduce and contribute offspring to the next generation. More recently, population geneticists have realized that natural selection can be related not only to differential survival but also to mating efficiency and fertility.

A modern description of natural selection can relate our knowledge of molecular genetics to the phenotypes of individuals.

1. Within a population, there is genetic variation arising from differences in DNA sequences. Distinct alleles may encode proteins of differing function.
2. Some alleles may encode proteins that enhance an individual's survival or reproductive capability as compared to other members of the population. For example, an allele may produce a protein that is more efficient at a higher temperature, conferring on the individual a greater probability of survival in a hot climate.
3. Individuals with beneficial alleles are more likely to survive and contribute to the gene pool of the next generation.
4. Over the course of many generations, allele frequencies of many different genes may change through natural selection, thereby significantly altering the characteristics of a species. The net result of natural selection is a population that is better adapted to its environment and/or more successful at reproduction.

Natural selection acts on phenotypes (which are derived from an individual's genotype). With regard to quantitative traits, there are three ways that natural selection may operate (fig. 25.9):

1. **Directional selection** favors the survival of one extreme of a phenotypic distribution that is better adapted to an environmental condition. For example, in a dimly lit forest, directional selection would favor mice with darker fur, because they would be better camouflaged and less susceptible to predation.
2. **Disruptive** (or **diversifying**) **selection** favors the survival of two (or more) different phenotypic classes of individuals. For example, disruptive selection favors the survival of mice with dark fur and light fur that occupy a heterogeneous environment. Disruptive selection is likely to occur in species that occupy diverse environments so that some members of the species will survive in each type of environment.
3. **Stabilizing selection** favors the survival of individuals with intermediate phenotypes. For example, mice with moderately colored fur may survive better in an environment that has an intermediate level of light, such as a tall grassland.

Haldane, Fisher, and Wright developed mathematical relationships to explain the theory of natural selection. As our knowledge of natural selection increased, it became apparent that it operates in many different ways. In chapter 25, we will consider a few examples of natural selection involving a single gene that exists in two alleles. In reality, however, natural selection acts on populations of individuals, in which many genes are polymorphic and each individual contains thousands or tens of thousands of different genes.

To begin our quantitative discussion of natural selection, we must examine the concept of **Darwinian fitness.** Fitness is the relative likelihood that a phenotype will survive and contribute to the gene pool of the next generation as compared to other phenotypes. Although this property often correlates with physical fitness, the two ideas should not be confused. Darwinian fitness is a

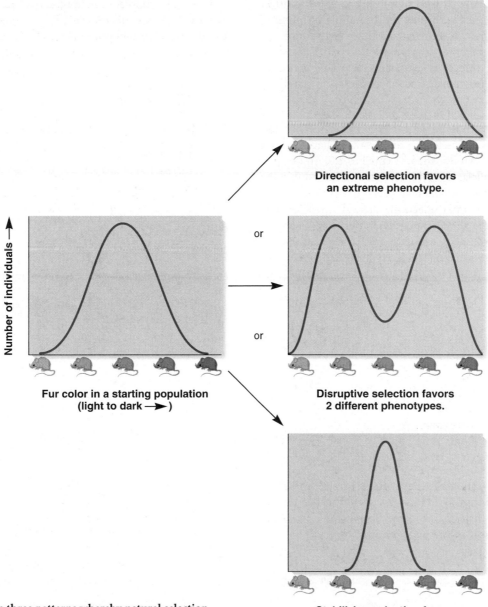

FIGURE 25.9 The three patterns whereby natural selection acts on phenotypes.

measure of reproductive superiority. An extremely fertile phenotype may have a higher Darwinian fitness than a less fertile phenotype that appears more physically fit.

To consider Darwinian fitness, let's use our example of two genes existing in the *A* and *a* alleles. We can assign fitness values to each of the three genotypic classes according to their relative reproductive potential. For example, let's suppose that the average reproductive success of the three genotypes are as follows:

AA produces five offspring.

Aa produces four offspring.

aa produces one offspring.

By convention, the genotype with the highest reproductive ability is given a fitness value of 1.0. Fitness values are denoted by the variable *W*. The fitness values of the other genotypes are assigned values relative to this 1.0 value.

$$\text{Fitness of } AA\text{: } W_{AA} = 1.0$$

$$\text{Fitness of } Aa\text{: } W_{Aa} = 4/5 = 0.8$$

$$\text{Fitness of } aa\text{: } W_{aa} = 1/5 = 0.2$$

Keep in mind that differences in reproductive achievement among these three genotypes could stem from various reasons. A common reason is that the fittest phenotype is more likely to survive. Of course, an individual must survive into adulthood to reproduce. A second possibility is that the most fit phenotype is more likely to mate. For example, a bird with brightly colored feathers may have an easier time attracting a mate than a bird with duller plumage. Finally, a third possibility is that the fittest phenotype may be more fertile. It may produce a higher number of gametes or gametes that are more successful at fertilization.

An opposite parameter that population geneticists use in their calculations is the **selection coefficient (s).** This measures the degree to which a genotype is selected against.

$$s = 1 - W$$

By convention, the genotype with the highest fitness has an *s* value of zero. Genotypes at a selective disadvantage have *s* values that are greater than zero but less than or equal to 1.0. An extreme case is a recessive lethal allele. It would have an *s* value of 1.0 in the homozygote, while the *s* value in the heterozygote could be zero.

In the case of directional selection, allele frequencies may change in a step-by-step, generation-per-generation way. To appreciate how this occurs, let's take a look at how fitness affects the Hardy-Weinberg equilibrium and allele frequencies. Again, let's suppose that a gene exists in two alleles, *A* and *a*. The three fitness values are

$$W_{AA} = 1.0$$

$$W_{Aa} = 0.8$$

$$W_{aa} = 0.2$$

In the next generation, we expect that the Hardy-Weinberg equilibrium will be modified in the following way:

Frequency of *AA:* $p^2 W_{AA}$
Frequency of *Aa:* $2pq W_{Aa}$
Frequency of *aa:* $q^2 W_{aa}$

In a population that is changing due to natural selection, these three terms may not add up to 1.0, as they would in the Hardy-Weinberg equilibrium. Instead, the three terms sum to a value known as the **mean fitness of the population.**

$$p^2 W_{AA} + 2pq W_{Aa} + q^2 W_{aa} = \overline{W}$$

Dividing both sides of the equation by the mean fitness of the population,

$$\frac{p^2 W_{AA}}{\overline{W}} + \frac{2pq W_{Aa}}{\overline{W}} + \frac{q^2 W_{aa}}{\overline{W}} = 1$$

Using this equation, we can calculate the expected genotype and allele frequencies after one generation of natural selection.

Frequency of *AA* genotype: $\dfrac{p^2 W_{AA}}{\overline{W}}$

Frequency of *Aa* genotype: $\dfrac{2pq W_{Aa}}{\overline{W}}$

Frequency of *aa* genotype: $\dfrac{q^2 W_{aa}}{\overline{W}}$

Allele frequency of *A:* $p_A = \dfrac{p^2 W_{AA}}{\overline{W}} + \dfrac{pq W_{Aa}}{\overline{W}}$

Allele frequency of *a:* $q_a = \dfrac{q^2 W_{aa}}{\overline{W}} + \dfrac{pq W_{Aa}}{\overline{W}}$

As an example, let's suppose that the starting allele frequencies are *A* = 0.5 and *a* = 0.5, and use fitness values of 1.0, 0.8, and 0.2 for the three genotypes, *AA*, *Aa*, and *aa*, respectively. We begin by calculating the mean fitness of the population.

$$p^2 W_{AA} + 2pq W_{Aa} + q^2 W_{aa} = \overline{W}$$

$$\overline{W} = (0.5)^2(1) + 2(0.5)(0.5)(0.8) + (0.5)^2(0.2)$$

$$\overline{W} = 0.25 + 0.4 + 0.05 = 0.7$$

After one generation of selection:

Frequency of *AA* genotype: $\dfrac{p^2 W_{AA}}{\overline{W}} = \dfrac{(0.5)^2(1)}{0.7} = 0.36$

Frequency of *Aa* genotype: $\dfrac{2pq W_{Aa}}{\overline{W}} = \dfrac{2(0.5)(0.5)(0.8)}{0.7} = 0.57$

Frequency of *aa* genotype: $\dfrac{q^2 W_{aa}}{\overline{W}} = \dfrac{(0.5)^2(0.2)}{0.7} = 0.07$

Allele frequency of *A:* $p_A = \dfrac{p^2 W_{AA}}{\overline{W}} + \dfrac{pq W_{Aa}}{\overline{W}}$

$$= \frac{(0.5)^2(1)}{0.7} + \frac{(0.5)(0.5)(0.8)}{0.7} = 0.64$$

Allele frequency of *a:* $q_a = \dfrac{q^2 W_{aa}}{\overline{W}} + \dfrac{pq W_{Aa}}{\overline{W}}$

$$= \frac{(0.5)^2(0.2)}{0.7} + \frac{(0.5)(0.5)(0.8)}{0.7} = 0.36$$

After one generation, the allele frequency of *A* has increased from 0.5 to 0.64, while the frequency of *a* has decreased from 0.5 to 0.36. This is because the *AA* genotype has the highest fitness, while the *Aa* and *aa* genotypes have lower fitness values. Another interesting feature of natural selection is that it raises the mean fitness of the population. If we assume that the individual fitness values are constant, the mean fitness of this next generation is

$$\overline{W} = p^2 W_{AA} + 2pq W_{Aa} + q^2 W_{aa}$$

$$= (0.64)^2(1) + 2(0.64)(0.36)(0.8) + (0.36)^2(0.2)$$

$$= 0.80$$

The mean fitness of the population has increased from 0.7 to 0.8. This population is better adapted to its environment than the previous one. Another way of viewing this calculation is that the subsequent population has a greater reproductive potential than the previous one. We could perform the same types of calculations to find the allele frequencies and mean fitness value in the next generation. If we assume that the individual fitness values remain constant, the frequencies of *A* and *a* in the next generation are 0.85 and 0.15, respectively, and the mean fitness increases to 0.931. As we can see, the general trend is to increase *A*, decrease *a*, and increase the mean fitness of the population.

In the previous example, we considered the effects of natural selection by beginning with allele frequencies at intermediate levels (namely, *A* = 0.5 and *a* = 0.5). Figure 25.10 illustrates

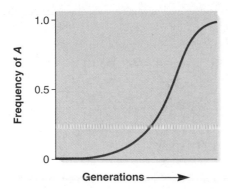

FIGURE 25.10 **The fate of a beneficial dominant allele that is introduced as a new mutation into a population.** This allele is beneficial in the homozygous condition: $W_{AA} = 1.0$. The corresponding heterozygote, Aa ($W_{Aa} = 0.8$), and recessive homozygote, aa ($W_{aa} = 0.2$), have lower fitness values.

what would happen if a new mutation introduced the A allele into a population that was originally monomorphic for the a allele. As before, the AA homozygote has a fitness of 1.0, the Aa heterozygote 0.8, and the recessive aa homozygote 0.2. Initially, the A allele is at a very low frequency in the population. If it is not lost initially due to genetic drift, its frequency slowly begins to rise and then, at intermediate values, rises much more rapidly.

Eventually, this type of natural selection may lead to fixation of the beneficial allele. However, this new beneficial allele is in a precarious situation when its frequency is very low. As mentioned earlier in chapter 25, random genetic drift is likely to eliminate new neutral mutations due to sampling error. Genetic drift can also eradicate beneficial alleles, although this is a little less likely because natural selection favors them.

Balanced Polymorphisms May Exist Due to Heterozygote Superiority or Heterogeneous Environments

As we have just seen, selection for an allele that is beneficial in a homozygote may lead to its eventual fixation in a population. Likewise, genetic drift also eliminates or fixes neutral alleles. So how do we explain the high levels of polymorphisms observed for genes in natural populations?

One possibility is that many of these polymorphisms involve neutral alleles not subject to natural selection. The genes become polymorphic due to genetic drift, and there has not been enough time for fixation to occur. As we have seen, fixation can require vast amounts of time. Similarly, directional selection for one of two alleles may be occurring, but again, there may not have been enough time for fixation of the allele that confers higher fitness. In other cases, however, a polymorphism may have reached an equilibrium where opposing selective forces have reached a balance; the population is not evolving toward fixation or allele elimination. This situation, known as a **balanced polymorphism,** can occur for several different reasons.

In some cases, a balanced polymorphism occurs when the heterozygote is at a selective advantage. The higher fitness of the heterozygote is balanced by the lower fitness values of both corresponding homozygotes. Let's consider the following case of relative fitness:

$$W_{AA} = 0.7$$
$$W_{Aa} = 1.0$$
$$W_{aa} = 0.4$$

This is an example of **heterozygote advantage,** also called **overdominance.** The selection coefficients are:

$$s_{AA} = 1 - 0.7 = 0.3$$
$$s_{Aa} = 1 - 1.0 = 0$$
$$s_{aa} = 1 - 0.4 = 0.6$$

Under these conditions, the population will reach an equilibrium in which

$$\text{Allele frequency of } A = \frac{s_{aa}}{s_{AA} + s_{aa}}$$
$$= \frac{0.6}{0.3 + 0.6} = 0.67$$
$$\text{Allele frequency of } a = \frac{s_{AA}}{s_{AA} + s_{aa}}$$
$$= \frac{0.3}{0.3 + 0.6} = 0.33$$

Balanced polymorphisms can sometimes explain the high frequency of alleles that are deleterious in a homozygous condition. A classic example is the Hb^S allele of the human β-globin gene, which was described in chapter 4. A homozygous $Hb^S Hb^S$ individual displays sickle-cell anemia, a disease that leads to the sickling of the red blood cells. The $Hb^S Hb^S$ homozygote has a lower fitness than a homozygote with two normal copies of the β-globin gene, $Hb^A Hb^A$. However, the heterozygote, $Hb^A Hb^S$, has the highest level of fitness in areas where malaria is endemic (see fig. 25.11). Compared to normal, $Hb^A Hb^A$, homozygotes, the heterozygotes have a 10 to 15% better chance of survival if infected by the malarial parasite, *Plasmodium falciparum*. Therefore, the Hb^S allele is maintained in populations where malaria is prevalent, even though the allele is detrimental in the homozygous state.

Besides sickle-cell anemia, several other gene mutations that cause human disease in the homozygous state are thought to be prevalent because of heterozygote advantage. These include cystic fibrosis, in which the heterozygote is resistant to diarrheal disease (such as cholera); PKU, in which the heterozygous fetus is resistant to abortion caused by a fungal toxin; and Tay-Sachs disease, in which the heterozygote is resistant to tuberculosis.

Balanced polymorphism also may occur when a species occupies a region that contains heterogeneous environments. One environment may favor a particular allele, while a neighboring environment may favor a different allele. Figure 25.12*a* shows a photograph of land snails, *Cepaea nemoralis,* that live in woods

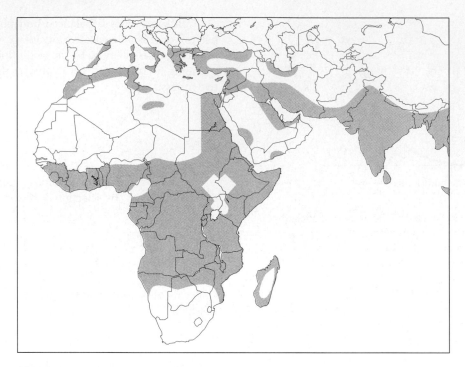

(a) Malaria prevalence

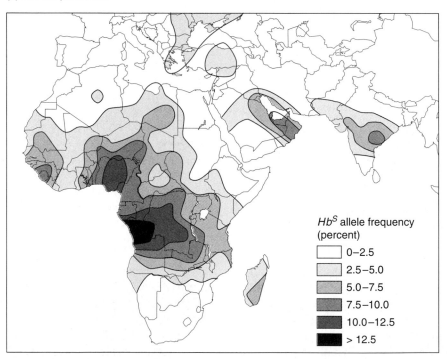

(b) Hb^S allele frequency

FIGURE 25.11   **The geographic relationship between malaria and the frequency of the sickle-cell allele in human populations. (a)** The geographic prevalence of malaria in Africa and surrounding areas. **(b)** The frequency of the Hb^S allele in the same areas.

GENES→TRAITS   The sickle-cell allele of the β-globin gene is maintained in human populations as a balanced polymorphism. In areas where malaria is prevalent, the heterozygote carrying one copy of the Hb^S allele has a greater fitness than either of the corresponding homozygotes ($Hb^A Hb^A$ and $Hb^S Hb^S$). Therefore, even though the $Hb^S Hb^S$ homozygotes suffer the detrimental consequences of sickle-cell anemia, this negative aspect is balanced by the beneficial effects of malarial resistance in the heterozygotes.

(a) Land snails

Habitat	Brown	Pink	Yellow
Beechwoods	0.23	0.61	0.16
Deciduous woods	0.05	0.68	0.27
Hedgerows	0.05	0.31	0.64
Rough herbage	0.004	0.22	0.78

(b) Frequency of snail color

FIGURE 25.12   **Polymorphism in the land snail *Cepaea nemoralis*. (a)** This species of snail can exist in several different colors and banding patterns. **(b)** Coloration of the snails is correlated with the specific environments where they are located. (Taken from Cain, A. J., and Sheppard, P. M. (1954) Natural selection in *Cepaea. Genetics* 39, 89–116.)

GENES→TRAITS   Snail coloration is an example of genetic polymorphism due to heterogeneous environments; the genes governing shell coloration are polymorphic. The predation of snails is correlated with their ability to be camouflaged in their natural environment. Snails in the beechwoods, where the soil is dark, have the highest frequency of brown shell color. Pink snails are usually found in the leaf litter of forest floors in deciduous woods. And yellow snails are prevalent in the sunny, unwooded areas of hedgerows and rough herbage.

and open fields. This snail is polymorphic in color and banding patterns. In 1954, A. J. Cain and P. M. Sheppard found that snail color was correlated with the environment. As shown in figure 25.12*b*, the highest frequency of brown shell color was found in snails in the beechwoods, where there are wide expanses of dark soil. Pink snails are most common in the leaf litter of forest floors in deciduous woods. The yellow snails are most abundant in the sunny, unwooded areas of hedgerows and rough herbage. It has been proposed that disruptive selection has favored the survival of these different phenotypes. Migration between the snail populations keeps the polymorphism in balance among these different environments.

EXPERIMENT 25A

Industrial Melanism in the Moth *Biston betularia* Is a Modern Example of Natural Selection

The concept of natural selection provides a framework for explaining how allele frequencies are maintained in populations; it also explains how species become adapted to the environments in which they live. As scientists, we tend to examine the ultimate result of natural selection. In other words, we observe species in their native environments, and in some cases, it seems obvious how certain characteristics provide an organism with a survival advantage. The smell of a skunk and the ability of a cactus to retain water are classic examples. To support the theory of natural selection, population geneticists would like to witness the process in a species over time. However, since it is such a slow process, this is not an easy task.

A first example of modern natural selection occurred in the decades that followed the Industrial Revolution (which began in the latter half of the eighteenth century). As a result of industrial-ization, large areas of the earth's surface are blanketed by the fallout of smoke particles. In and around industrial areas, this fallout is dramatically more evident than in undeveloped areas. The smoke particles tend to kill vegetative lichens on the trunks and boughs of trees. Rain washes the pollutants down the tree trunks until they are bare and black. By comparison, tree trunks in unpolluted areas are much lighter and are speckled with growing lichens.

During the early 1900s, scientists in Europe and North America began to notice the proliferation of "melanic" moth varieties that are darker in color compared to the previously abundant lighter varieties. An example is shown in figure 25.13*a*. This moth, *Biston betularia*, is polymorphic. One variety called *carbonaria* is very dark, while the *typical* variety is much lighter in color. As seen here, *carbonaria* is easily seen on a lichen-covered tree trunk while the *typical* variety is readily visible on dark trunks. These observations led geneticists to propose that the Industrial Revolution was selecting for the survival of *carbonaria*. It was hypothesized that the darkened color of the *carbonaria* variety made it less

(a) *Typical* variety (left) and *carbonaria* variety (right) of *Biston betularia*

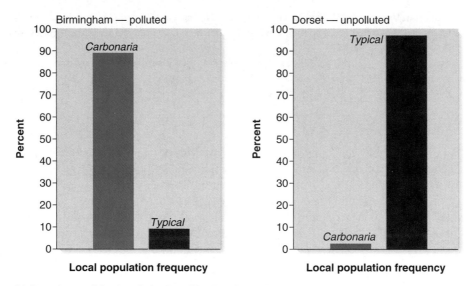

(b) Prevalence of the 2 varieties in polluted and unpolluted woods

FIGURE 25.13   Phenotypes and prevalence of *Biston betularia*. (a) The *typical* variety is seen on the *left,* while the *carbonaria* is on the *right.* (b) A comparison of the prevalence of the *typical* and *carbonaria* varieties in polluted and unpolluted areas.

susceptible to predation by birds. To support this idea, a field study in England showed that *carbonaria* is more prevalent in polluted areas of Birmingham, while the *typical* variety is more prevalent in unpolluted areas (fig. 25.13*b*).

In the experiment described in figure 25.14, H. Bernard Kettlewell, a British biologist, set out to test the hypothesis that the coloration polymorphism in *B. betularia* is due to natural selection involving predation by birds. To conduct this experiment, he collected many *carbonaria* and *typical* moths and marked their undersides. The moths were released into polluted or unpolluted woods. Researchers sat in a blind and watched to see if the moths were subject to predation by birds. Over the course of several days, the surviving moths were recaptured by

attracting them to a mercury vapor light. Therefore, this experimental protocol yielded two ways to evaluate natural selection. First, the predation rate was determined, and second, the recapture rate provided an independent measure of the survival of the two varieties of moths in the polluted and unpolluted woods.

THE HYPOTHESIS

The coloration of *B. betularia* affects the probability that birds will see and eat them. This natural selection leads to the survival of those moths with coloration more closely matching that on which they rest, whether that be in polluted or unpolluted woods.

■ TESTING THE HYPOTHESIS — FIGURE 25.14 Natural selection in *Biston betularia.*

Starting material: Samples of *B. betularia,* both *carbonaria* and *typical,* were collected from the wild. In some cases, they were bred in the laboratory, to be released later into the wild.

Experimental level	Conceptual level

1. Mark the underside of *carbonaria* and *typical* moths with cellulose paint.

Paintbrush

Marking provides identification of released moths versus those already present.

2. Release equal numbers of marked *carbonaria* and *typical* moths in either polluted woods near an industrial area of Birmingham or in an unpolluted woods in Dorset, England.

Release moths.

3. On several consecutive days, recapture moths that have been marked. This can be done by attracting them to a mercury vapor light. Record the number of recaptured moths.

Recapture moths:

Mercury vapor lights

Attracted

Recapture: Only surviving moths will be recaptured.

Record number of *carbonaria* and *typical* moths.

Compare the numbers of *carbonaria* and *typical* moths:

Natural selection may be occuring due to differences in predation by birds.

Bird predation may be higher for moths that are not well camouflaged.

4. Also, sit in a blind and observe the moths on a tree trunk. Record the numbers and types of sitting moths that are eaten by birds.

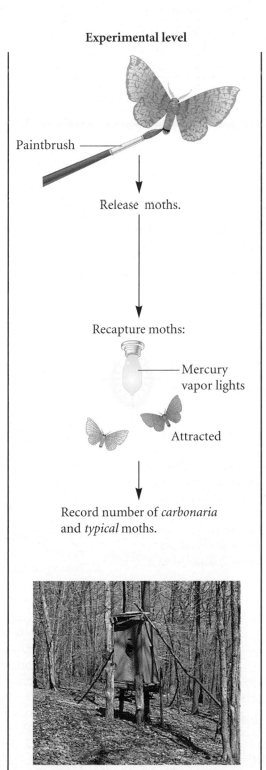

THE DATA

	Number Eaten by Birds		Percentage of Marked Moths That Were Recaptured	
Location	Carbonaria	Typical	Carbonaria	Typical
Dorset:				
unpolluted woods	43	15	7.0	12.5
Birmingham:				
polluted woods	26	164	27.5	13.1

Data taken from Kettlewell, H. B. D. (1956) Further Selection Experiments on Industrial Melanism in the *Lepidoptera*. *Heredity* 10, 287–301.

INTERPRETING THE DATA

As shown in the data, the *carbonaria* moths were more likely to be eaten in unpolluted woods by birds. For example, there was approximately a 3:1 ratio (i.e., 43 to 15) with regard to predation of *carbonaria* versus *typical* moths in unpolluted woods, while there was approximately a 1:6 ratio (i.e., 26 to 164) of predation of *carbonaria* versus *typical* moths in polluted woods. A lower percentage of *carbonaria* moths were recaptured in Dorset. Conversely, the *typical* moths were much more likely to be eaten in a polluted woods. Again, this observation was supported by the lower percentage of recapture of *typical* moths in Birmingham. Taken together, these results supported the idea that *carbonaria* are more likely to survive in polluted woods while the *typicals* are more likely to survive in unpolluted woods. In other words, the consequences of the Industrial Revolution have resulted in the natural selection for the *carbonaria* variety. As stated by Kettlewell, "These experiments showed that birds act as selective agents and that the melanic forms of *betularia* are at a cryptic advantage in an industrial area such as Birmingham."

A self-help quiz involving this experiment can be found at the Online Learning Center.

Genetic Load Is the Negative Consequence of Genetic Variation

As we have just seen, genetic variation in a population can lead to adaptation via natural selection. In general, we often think of genetic variation in a positive light. It allows species to adapt to their environment and provides the raw material for evolution. Furthermore, genetic diversity adds interest and beauty to our perception of biology. On the negative side, however, certain kinds of genetic variation are detrimental to the survival of a population. The term **genetic load** (**L**) refers to genetic variation that decreases the average fitness of a population compared to a (theoretical) maximum or optimal value.

$$L = \frac{(W_{max} - \overline{W})}{W_{max}}$$

where

L equals the genetic load
$\overline{W}$ equals the mean fitness of a population
W_{max} equals the maximal fitness of a population

Many factors contribute to the genetic load in populations. Several of these are listed here.

1. *Mutation.* Most new mutations are deleterious. Therefore, the accumulation of new mutations in a population contributes to the genetic load.
2. *Segregation.* As mentioned earlier in chapter 25, certain mutations are beneficial in the heterozygous condition but the homozygotes are disadvantaged. Therefore, the segregation of alleles to create homozygotes causes a genetic load in the population.
3. *Recombination.* Sometimes, linked genes occur in optimal combinations. For example, two genes may be found in *A* and *a* alleles and *B* and *b* alleles. Let's suppose that the *AB* and *ab* combinations are beneficial, while the *Ab* and *aB* combinations are less fit. Although natural selection may select for chromosomes containing the *AB* and *ab* pairs of genes, recombination (i.e., crossing over) may create *Ab* or *aB* combinations and thereby decrease the fitness of the population.
4. *Heterogeneous environments.* As mentioned earlier in chapter 25 (namely, for snail coloration), some species occupy two or more types of environments. This creates polymorphisms that are suited to different environments. Some members of the population will have genotypes that are not well suited to their environment. For example, a darkly colored snail may be located in a sunny environment.
5. *Meiotic drive/gamete selection.* Certain types of alleles and chromosomal rearrangements are transmitted preferentially to gametes even though they may be detrimental to the survival of the organism. For example, some alleles favor gamete viability even though they may not be beneficial in the adult.
6. *Maternal-fetal incompatibility.* In a few cases, the genotype of the mother may interact in an unfavorable way with that of a fetus. A classic example is the Rh factor, a type of red blood cell antigen. If a mother is Rh⁻ and a fetus is

Rh$^+$, the mother's body will occasionally mount an immunological reaction against the fetus, thereby diminishing its chances for survival.

7. *Finite population.* In a small population, genetic drift causes allele frequencies to deviate from their optimal fitness values.

8. *Migration.* When individuals from one environment migrate to a new environment, they usually do not possess the optimal combination of alleles for survival. Therefore, migration tends to decrease the fitness of a population.

CONCEPTUAL SUMMARY

In chapter 25, we have surveyed some of the critical issues in **population genetics.** The primary focus of this field is an understanding of genetic variation within **populations.** A substantial proportion (e.g., at least 30% in humans) of genes in natural populations are **polymorphic,** existing in two or more alleles. Population geneticists want to understand why this variation in the gene pool exists and how it may change.

According to the **Hardy-Weinberg equilibrium,** allele frequencies remain stable as long as several conditions are met. These include no new mutations, random mating, large population size, no migration, and no natural selection. While these conditions are never truly met, they can be approximated in large, well-adapted populations. Therefore, the Hardy-Weinberg equilibrium predicts that **genotype frequencies** will remain relatively constant when certain conditions are met.

However, we know that many processes in nature promote genetic change. As we will see in chapter 26, these forces lead to the evolution of species. The initial source of genetic variation is **mutation.** Other evolutionary forces may act to increase the frequency of some new alleles. In small populations, **neutral processes** such as **random genetic drift** can alter **allele frequencies** dramatically. Simply due to random sampling error, a new neu-

tral mutation may be eliminated from a population or it may become fixed in it. This genetic drift can occur via a **founder effect,** when a small population moves to a new site, or a **bottleneck effect,** when a population is reduced in size due to negative environmental events. Allele frequencies can also be altered by **migration.** When two populations differ in their allele frequencies, migration can alter the existing allele frequencies. This is known as **gene flow.**

Finally, **adaptive forces,** such as **natural selection,** make a species better adapted to its environment. Natural selection favors individuals with the greatest reproductive potential (i.e., the greatest **Darwinian fitness**). These may be individuals who are more likely to survive, more likely to mate, or more fertile. When natural selection favors one homozygote, it selects for the fixation of the favorable allele. However, when the heterozygote has the highest fitness, or when two or more phenotypes in an environmentally diverse region have the highest relative fitness values, a **balanced polymorphism** will result. In chapter 26, we will learn how the combined evolutionary processes of genetic drift, migration, and natural selection lead to the formation of new species. We will also examine how these processes change the sequences of genes among evolutionarily related species.

EXPERIMENTAL SUMMARY

A central approach in population genetics is to measure variation experimentally and then examine the mathematical relationships between genetic variation and the occurrence of genotypes and phenotypes within populations. In chapter 25, we have focused on how population geneticists explain phenotypic variation via mathematical theories. Experimentally, phenotypic variation can be measured by tallying the organisms within a population that display particular phenotypes. For example, we can count the number of dark snails and light snails that live in a particular location. This provides a measure of the phenotypic variation within this population. We also learned that variation can be studied at the molecular level by identifying **allozymes.** In chapter 26, we will see that many other molecular approaches, particularly DNA sequencing, allow population geneticists to measure the genetic variation within populations.

After variation is measured, population geneticists attempt to develop and apply mathematical relationships that provide insight into the causes and dynamics of the variation. Although geneticists understand that mutation is the source of new variation, their mathematical formulae indicate that the rate of new mutations is too low to explain the experimentally observed variation in natural populations. Instead, other processes, such as genetic drift, migration, and natural selection, must alter the frequency of new mutations after they have occurred. The mathematical theories of population geneticists tell us that these forces may be **directional,** leading to the eventual fixation or elimination of an allele, or they may promote a **balanced polymorphism** within a population. These theories also show us that variation may have negative consequences, known as the **genetic load.**

PROBLEM SETS & INSIGHTS

Solved Problems

S1. The phenotypic frequency of individuals who cannot taste phenylthiocarbamide (PTC) is approximately 0.3. The inability to taste this bitter substance is due to a recessive allele. If we assume that there are only two alleles in the population (namely, tasters, T, and nontasters, t) and that the population is in Hardy-Weinberg equilibrium, calculate the frequencies of these two alleles.

Answer: Let p = allele frequency of the taster allele and q = the allele frequency of the nontaster allele. The frequency of nontasters is 0.3. This is the frequency of the genotype tt, which in this case is equal to q^2.

$$q^2 = 0.3$$

To determine the frequency q of the nontaster allele, we take the square root of both sides of this equation:

$$q = 0.55$$

With this value, we can calculate the frequency p of the taster allele.

$$p = 1 - q$$

$$= 1 - 0.55 = 0.45$$

S2. In the pedigree shown here, answer the following questions with regard to individual VII-1:

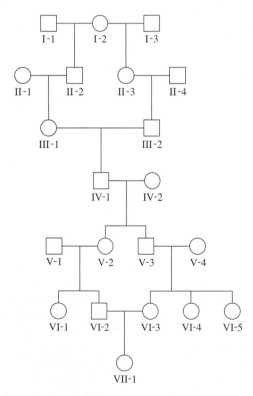

A. Who are the common ancestors of her parents?

B. What is the inbreeding coefficient?

Answer:

A. The common ancestors are IV-1 and IV-2. They are the grandparents of VI-2 and VI-3, who are the parents of VII-1.

B. The inbreeding coefficient is calculated using the formula

$$F = \Sigma(1/2)^n(1 + F_A)$$

In this case there are two common ancestors, IV-1 and IV-2. Also, IV-1 is inbred, because I-2 is a common ancestor to both of IV-1's parents. So, the first step is to calculate F_A, the inbreeding coefficient for this common ancestor. The inbreeding loop for IV-1 contains five people, III-1, II-2, I-2, II-3, and III-2. Therefore,

$$n = 5$$

$$F_A = (1/2)^5 = 0.03$$

Now we can calculate the inbreeding coefficient for VII-1. Each inbreeding loop contains five people: VI-2, V-2, IV-1, V-3, and VI-3; and VI-2, V-2, IV-2, V-3, and VI-3. Thus,

$$F = (1/2)^5 (1 + 0.03) + (1/2)^5 (1 + 0)$$

$$= 0.032 + 0.031 = 0.063$$

S3. As discussed in chapter 25, John Hubby and Richard Lewontin studied genetic variation in *Drosophila pseudoobscura*. They focused their attention on 18 different structural genes that were known to encode enzymes. The activity of these enzymes had already been well characterized by biochemists. Hubby and Lewontin wanted to know how much polypeptide sequence variation there was in these enzymes within different populations of *D. pseudoobscura*. At the molecular level, genetic variation in a structural gene may produce enzymes that have slight differences in their amino acid sequences. Based on previous work, Hubby and Lewontin knew that very small differences between the amino acid sequences of two enzymes could be detected by gel electrophoresis. For example, two alleles for a particular enzyme may differ only by a single amino acid. However, this amino acid difference may influence the mobility of the enzyme when it is subjected to gel electrophoresis. This is particularly evident when the amino acid alteration affects the overall charge on the protein. The 20 amino acids differ with regard to their side chain chemistry. The amino acid side chains are positively charged, negatively charged, or neutral. An amino acid substitution that changes a neutral amino acid to a charged one, or vice versa, will affect the net charge on the protein. The data from their study are shown here.

Population	Number of Monomorphic Genes	Number of Polymorphic Genes	Percentage of Polymorphic Genes
Strawberry Canyon	12	6	33
Wildrose	13	5	28
Cimarron	13	5	28
Mather	12	6	33
Flagstaff	13	5	28
			30 (average)

Discuss these data.

Answer: In this study, Hubby and Lewontin looked for genetic variation in 18 different enzymes in five different populations of fruit flies. On average, 30% of the enzymes were found as two or more allozymes. This means that the genes encoding these enzymes have DNA sequence differences resulting in alleles that cause the encoded proteins to have slightly different amino acid sequences. It should be pointed out that this study tends to underestimate genetic variability because some amino acid substitutions will not alter protein mobility during gel electrophoresis. Nevertheless, the conclusion from this work is that the genetic variability within these populations is quite high.

S4. The Hardy-Weinberg equilibrium provides a way to predict genotype frequency based on allele frequency. In the case of mammals, males are hemizygous for X-linked genes while females have two copies. Among males, the frequency of any X-linked trait will equal the frequency of males with the trait. For example, if an allele frequency for a X-linked disease-causing allele was 5%, then 5% of all males would be affected with the disorder. Female genotype frequencies are computed using the Hardy-Weinberg equation.

As a specific example, let's consider the human X-linked trait known as hemophilia A (see chapter 22 for a description of this disorder). In human populations, the allele frequency of the hemophilia A allele is approximately 1 in 10,000, or 0.0001. The other allele for this gene is the normal allele. Males can be affected or unaffected, whereas females can be affected, unaffected carriers, or unaffected noncarriers.

A. What are the allele frequencies for the mutant and normal allele in the human population?

B. Among males, what is the frequency of affected individuals?

C. Among females, what is the frequency of affected individuals and heterozygous carriers?

D. Within a population of 100,000 people, what is the expected number of affected males? In this same population, what is the expected number of carrier females?

Answer: Let p represent the normal allele and q represent the allele that causes hemophilia.

A. X^H normal allele, frequency = 0.9999 = p

   X^h hemophilia allele, frequency = 0.0001 = q

B. X^hY genotype frequency of affected males = q = 0.0001

C. X^hX^h genotype frequency of affected females = q^2 = $(0.0001)^2$
   = 0.00000001

   X^HX^h genotype frequency of carrier females = $2pq$
   = 2(0.9999)(0.0001) = 0.0002

D. We will assume that this population is composed of 50% males and 50% females.

Number of affected males = 50,000 × 0.0001 = 5

Number of carrier females = 50,000 × 0.0002 = 10

S5. The Hardy-Weinberg equation can be modified to include situations of three or more alleles. In its standard (two-allele) form, the Hardy-Weinberg equation reflects the Mendelian notion that each individual inherits two copies of each allele, one from both parents. For a two-allele situation, it is written as

$(p + q)^2 = 1$ (Note: The number 2 in this equation reflects the idea that the genotype is due to the inheritance of two alleles, one from each parent.)

This equation can be expanded to include three or more alleles. For example, let's consider a situation in which a gene exists as three alleles: *A1, A2,* and *A3.* The allele frequency of *A1* is designated by the letter p, *A2* by the letter q, and *A3* by the letter r. Under these circumstances, the Hardy-Weinberg equation becomes

$$(p + q + r)^2 = 1$$
$$p^2 + q^2 + r^2 + 2pq + 2pr + 2qr = 1$$

where

p^2 is the genotype frequency of *A1A1*

q^2 is the genotype frequency of *A2A2*

r^2 is the genotype frequency of *A3A3*

$2pq$ is the genotype frequency of *A1A2*

$2pr$ is the genotype frequency of *A1A3*

$2qr$ is the genotype frequency of *A2A3*

Now here is the question. As discussed in chapter 4, the gene that affects human blood type can exist in three alleles. In a Japanese population, the allele frequencies are

$$I^A = 0.28$$
$$I^B = 0.17$$
$$i = 0.55$$

Based on these allele frequencies, calculate the different possible genotype frequencies and blood type frequencies.

Answer: If we let p represent I^A, q represent I^B, and r represent i, then,

p^2 is the genotype frequency of I^AI^A, which is type A blood = $(0.28)^2$ = 0.08

q^2 is the genotype frequency of I^BI^B, which is type B blood = $(0.17)^2$ = 0.03

r^2 is the genotype frequency of ii, which is type O blood = $(0.55)^2$ = 0.30

$2pq$ is the genotype frequency of I^AI^B, which is type AB blood = 2(0.28)(0.17) = 0.09

$2pr$ is the genotype frequency of I^Ai, which is type A blood = 2(0.28)(0.55) = 0.31

$2qr$ is the genotype frequency of I^Bi, which is type B blood = 2(0.17)(0.55) = 0.19

Type A = 0.08 + 0.31 = 0.39, or 39%

Type B = 0.03 + 0.19 = 0.22, or 22%

Type O = 0.30, or 30%

Type AB = 0.09, or 9%

S6. Let's suppose that pigmentation in a species of insect is controlled by a single gene existing in two alleles, D for dark and d for light. The heterozygote Dd is intermediate in color. In a heterogeneous environment, the allele frequencies are $D = 0.7$ and $d = 0.3$. This polymorphism is maintained because the environment contains some dimly lit forested areas and some sunny fields. During a hurricane, a group of 1,000 insects is blown to a completely sunny area. In this environment, the fitness values are $DD = 0.3$, $Dd = 0.7$, and $dd = 1.0$. Calculate the allele frequencies in the next generation.

Answer: The first step is to calculate the mean fitness of the population.

$$p^2 W_{DD} + 2pq W_{Dd} + q^2 W_{dd} = \overline{W}$$

$$\overline{W} = (0.7)^2(0.3) + 2(0.7)(0.3)(0.7) + (0.3)^2(1.0)$$

$$= 0.15 + 0.29 + 0.09 = 0.53$$

After one generation of selection, we get

Allele frequency of D: $p_D = \dfrac{p^2 W_{DD}}{\overline{W}} + \dfrac{pq W_{Dd}}{\overline{W}}$

$$= \dfrac{(0.7)^2(0.3)}{0.53} + \dfrac{(0.7)(0.3)(0.7)}{0.53}$$

$$= 0.55$$

Allele frequency of d: $q_d = \dfrac{q^2 W_{dd}}{\overline{W}} + \dfrac{pq W_{Dd}}{\overline{W}}$

$$= \dfrac{(0.3)^2(1.0)}{0.53} + \dfrac{(0.7)(0.3)(0.7)}{0.53}$$

$$= 0.45$$

After one generation, the allele frequency of D has decreased from 0.7 to 0.55, while the frequency of d has increased from 0.3 to 0.45.

Conceptual Questions

C1. What is the gene pool? How is a gene pool described in a quantitative way?

C2. In genetics, what does the term *population* mean? Pick any species you like and describe how its population might change over the course of many generations.

C3. What is a genetic polymorphism? What is the source of genetic variation?

C4. For each of the following, state whether it is an example of an allele, genotype, and/or phenotype frequency:

A. Approximately 1 in 2,500 Caucasians is born with cystic fibrosis.

B. The percentage of carriers of the sickle-cell allele in West Africa is approximately 13%.

C. The number of new mutations for achondroplasia is approximately 5×10^{-5}.

C5. The term *polymorphism* can refer to both genes and traits. Explain the meaning of a polymorphic gene and a polymorphic trait. If a gene is polymorphic, does the trait that the gene affects also have to be polymorphic? Explain why or why not.

C6. Cystic fibrosis is a recessive autosomal trait. In certain Caucasian populations, the number of people born with this disorder is about 1 in 2,500. Assuming a Hardy-Weinberg equilibrium for this trait:

A. What are the frequencies for the normal and *CF* alleles?

B. What are the genotypic frequencies of homozygous normal, heterozygous, and homozygous affected individuals?

C. Assuming random mating, what is the probability that two phenotypically normal heterozygous carriers will choose each other as mates?

C7. Does inbreeding affect allele frequencies? Why or why not? How does it affect genotype frequencies? With regard to rare recessive diseases, what are the consequences of inbreeding in human populations?

C8. For a gene existing in two alleles, what are the allele frequencies when the heterozygote frequency is at its maximum value? What if there are three alleles?

C9. In a population, the frequencies of two alleles are $B = 0.67$ and $b = 0.33$. The genotype frequencies are $BB = 0.50$, $Bb = 0.37$, and $bb = 0.13$. Do these numbers suggest inbreeding? Explain why or why not.

C10. The ability to roll your tongue is inherited as a recessive trait. The frequency of the rolling allele is approximately 0.6 and the dominant (nonrolling) allele is 0.4. What is the frequency of individuals who can roll their tongues?

C11. In the pedigree shown here, answer the following questions for individual VI-1:

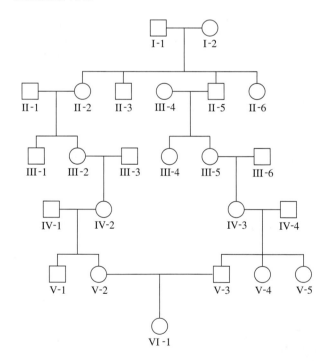

A. Is this individual inbred?

B. If so, who are her common ancestor(s)?

C. Calculate the inbreeding coefficient for VI-1.

D. Are the parents of VI-1 inbred?

C12. A family pedigree is shown here.

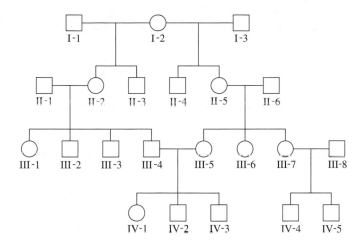

A. What is the inbreeding coefficient for individual IV-3?

B. Based on the data shown in this pedigree, is individual IV-4 inbred?

C13. A family pedigree is shown here.

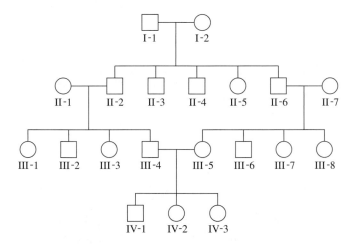

A. What is the inbreeding coefficient for individual IV-2? Who is/are her common ancestors?

B. Based on the data shown in this pedigree, is individual III-4 inbred?

C14. What evolutionary forces cause genetic polymorphisms to attain their levels of allele frequencies in natural populations? Discuss the relative importance of each type of process.

C15. In the term *genetic drift*, what is drifting? Why is this an appropriate term to describe this phenomenon?

C16. What is the difference between a neutral and an adaptive evolutionary process? Describe two or more examples of each. At the molecular level, explain how mutations can be neutral or adaptive.

C17. Let's suppose the mutation rate for converting a *B* allele into a *b* allele is 10^{-4}. The current allele frequencies are $B = 0.6$ and $b = 0.4$. How long will it take for the allele frequencies to equal each other, assuming that no genetic drift is taking place?

C18. Why is genetic drift more significant in small populations? Why does it take longer for genetic drift to cause allele fixation in large populations than in small ones?

C19. A group of four birds flies to a new location and initiates the formation of a new colony. Three of the birds are homozygous *AA*, and one bird is heterozygous *Aa*.

A. What is the probability that the *a* allele will become fixed in the population?

B. If fixation occurs, how long will it take?

C. How will the growth of the population, from generation to generation, affect the answers to parts A and B? Explain.

C20. Describe what happens to allele frequencies during the bottleneck effect. Discuss the relevance of this effect with regard to species that are approaching extinction.

C21. When two populations frequently intermix due to migration, what are the long-term consequences with regard to allele frequencies and genetic variation?

C22. Discuss the similarities and differences among directional, disruptive, and stabilizing selection.

C23. What is Darwinian fitness? What types of characteristics can promote high fitness values? Give several examples.

C24. What is the intuitive meaning of the mean fitness of a population? How does its value change in response to natural selection?

C25. What is the genetic load of a population? How does genetic variation contribute to the genetic load?

C26. Antibiotics are commonly used to combat bacterial and fungal infections. During the past several decades, however, antibiotic-resistant strains of microorganisms have become alarmingly prevalent. This has undermined the ability of physicians to treat many types of infectious disease. Discuss how the following processes that alter allele frequencies may have contributed to the emergence of antibiotic-resistant strains:

A. Random mutation

B. Genetic drift

C. Natural selection

C27. With regard to genetic drift, are the following statements true or false? If a statement is false, explain why.

A. Over the long run, genetic drift will lead to allele fixation or loss.

B. When a new mutation occurs within a population, genetic drift is more likely to cause the loss of the new allele rather than the fixation of the new allele.

C. Genetic drift promotes genetic diversity in large populations.

D. Genetic drift is more significant in small populations.

C28. Two populations of antelope are separated by a mountain range. It is known that the antelope may occasionally migrate from one population to the other. Migration can occur in either direction. Explain how migration will affect the following phenomena:

A. Genetic diversity in the two populations

B. Allele frequencies in the two populations

C. Genetic drift in the two populations

C29. Do the following examples describe directional selection, disruptive selection, or stabilizing selection?

A. Polymorphisms in snail color and banding pattern as described in figure 25.12

B. Thick fur among mammals exposed to cold climates

C. Birth weight in humans

D. Sturdy stems and leaves among plants exposed to windy climates

Experimental Questions

E1. Experimentally, how are allozymes typically detected in population genetics studies? Does this technique underestimate or overestimate the frequency of allozymes in populations? Explain. What technique should be able to detect all allozymes within a population?

E2. Let's suppose that a species of sparrow is found in two different geographic locations. By gel electrophoresis, you analyze 24 different enzymes to see if they are polymorphic. In one location, 2 out of 24 exist as two or more allozymes. In the other location, 11 out of 24 exist as two or more allozymes. With regard to the historical events that produced these two populations, propose at least two different hypotheses that would be consistent with these data.

E3. You will need to be familiar with the techniques described in chapter 18 to answer this question. Gene polymorphisms can be detected using a variety of cellular and molecular techniques. Which techniques would you use to detect gene polymorphisms at the following levels?

A. DNA level

B. RNA level

C. Polypeptide level

E4. The β-globin gene can exist in two different alleles termed Hb^A (the wild-type allele) and Hb^S (the allele that causes sickle-cell anemia in the homozygote). The Hb^A and Hb^S polypeptides can be considered allozymes of each other. The amino acid sequences of the Hb^A and Hb^S polypeptides are identical to each other except that the sixth amino acid is glutamic acid in the Hb^A polypeptide and valine in Hb^S. There are total of 146 amino acids in the β-globin polypeptide. In the gel shown here, a blood sample was collected from three individuals. The negative end of the gel is at the top.

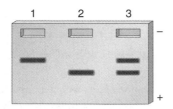

For each lane of this gel, explain whether the sample came from an Hb^AHb^A, Hb^AHb^S, or Hb^SHb^S individual.

E5. You will need to understand solved problem S5 to answer this question. As described in chapter 4, the gene for coat color in rabbits can exist in four alleles termed C (full coat color), c^{ch} (chinchilla), c^h (Himalayan), and c (albino). In a population of rabbits in Hardy-Weinberg equilibrium, the allele frequencies are

$$C = 0.34$$

$$c^{ch} = 0.17$$

$$c^h = 0.44$$

$$c = 0.05$$

A. What is the frequency of albino rabbits?

B. Among 1,000 rabbits, how many would you expect to have a Himalayan coat color?

C. Among 1,000 rabbits, how many would be heterozygotes with a chinchilla coat color?

E6. In a large herd of 5,468 sheep, 76 animals have yellow fat, compared to the rest of the members of the herd, which have white fat. Yellow fat is inherited as a recessive trait. It is assumed that this herd is in Hardy-Weinberg equilibrium.

A. What are frequencies of the white and yellow fat alleles in this population?

B. Approximately how many sheep with white fat are heterozygous carriers of the yellow allele?

E7. The human MN blood group is determined by two codominant alleles, M and N. The following data were obtained from various human populations:

Population	Place	Percentages		
		MM	MN	NN
Inuit	East Greenland	83.5	15.6	0.9
Navajo Indians	New Mexico	84.5	14.4	1.1
Finns	Karajala	45.7	43.1	11.2
Russians	Moscow	39.9	44.0	16.1
Aborigines	Queensland	2.4	30.4	67.2

Data from Speiss, E. B. (1990) *Genes in Populations*, 2d ed., Wiley-Liss, New York.

A. Calculate the allele frequencies in these five populations.

B. Which populations appear to be in Hardy-Weinberg equilibrium?

C. Which populations do you think have had significant intermixing due to migration?

E8. You will need to understand solved problem S5 before answering this question. In an island population, the following data were obtained regarding the numbers of people with each of the four blood types:

Type O   721

Type A   932

Type B   235

Type AB   112

Is this population in Hardy-Weinberg equilibrium? Explain your answer.

E9. In a donor population, the allele frequencies for the normal (Hb^A) and sickle-cell alleles (Hb^S) are 0.9 and 0.1, respectively. A group of 550 individuals migrates to a new population containing 10,000 individuals; in the recipient population, the allele frequencies are $Hb^A = 0.99$ and $Hb^S = 0.01$.

A. Calculate the allele frequencies in the conglomerate population.

B. Assuming that the donor and recipient populations are each in Hardy-Weinberg equilibrium, calculate the genotype frequencies in the conglomerate population prior to further mating between the donor and recipient populations.

C. What will be the genotype frequencies of the conglomerate population in the next generation, assuming that it achieves Hardy-Weinberg equilibrium in one generation?

E10. A recessive lethal allele has achieved a frequency of 0.22 due to genetic drift in a very small population. Based on natural selection, how would you expect the allele frequencies to change in the next three generations? Note: Your calculation can assume that genetic drift is not altering allele frequencies in either direction.

E11. Among a large population of 2 million gray mosquitoes, one mosquito is heterozygous for a body color gene; this mosquito has one gray allele and one blue allele. There is no selective advantage or disadvantage between gray and blue body color. All the other mosquitoes carry the gray allele.

A. What is the probability of fixation of the blue allele?

B. If fixation happens to occur, how many generations is it likely to take?

C. Qualitatively, how would the answers to parts A and B be affected if the blue allele conferred a slight survival advantage?

E12. As discussed in the experiment of figure 25.14, the dark form (*carbonaria*) of *Biston betularia* has a survival advantage in industrialized regions. The *carbonaria* allele is dominant to the *typical* (light) allele. In a particular forest within an industrialized region, the frequency of the *carbonaria* allele is 0.7 and the *typical* allele is 0.3. The fitness for the *carbonaria* moth is 1.0 compared with 0.47 for the *typical* moth. What would be the allele frequencies after one generation of selection?

E13. Resistance to the poison warfarin is a genetically determined trait in rats. Homozygotes carrying the resistance allele (*WW*) have a lower fitness because they suffer from vitamin K deficiency, but heterozygotes (*Ww*) do not. However, the heterozygotes are still resistant to warfarin. In an area where warfarin is applied, the heterozygote has a survival advantage. Due to warfarin resistance, the heterozygote is more fit than the normal homozygote (*ww*). The heterozygote is also more fit than the warfarin-resistant homozygote (*WW*), because the homozygote suffers from vitamin K deficiency. If the relative fitness values for *Ww*, *WW*, and *ww* individuals are 1.0, 0.37, and 0.19 in areas where warfarin is applied, calculate the allele frequencies at equilibrium. How would this equilibrium be affected if the rats were no longer exposed to warfarin?

E14. Why was it necessary in the experiment of figure 25.14 for Kettlewell to mark the undersides of the moths he released?

E15. Carry out a chi square analysis to determine if Kettlewell's results from figure 25.14 are significant. Carefully state the hypothesis you are testing, and then determine whether the chi square value allows you to accept or reject it.

E16. Based on the data described in figure 25.14, estimate the fitness of *carbonaria* moths in the Dorset woods. You should assume a fitness value of 1.0 for the *typical* moths in the Dorset woods. Note: You can calculate a fitness value based on the number of moths eaten by birds and also calculate a fitness value based on the percentage of moths that were recaptured. Which fitness value do you think is a more accurate estimate of the true fitness?

E17. The data in figure 25.12 describe the frequency of brown, pink, and yellow snails in different habitats. What would be the relative fitness values of the three types of snail in the following habitats? Note: The type with the highest fitness should be given a value of 1.0.

A. Beechwoods

B. Deciduous woods

C. Hedgerows

E18. Describe, in as much experimental detail as possible, how you would test the hypothesis that snail color distribution is due to predation.

Questions for Student Discussion/Collaboration

1. Discuss examples of assortative and disassortative mating in natural populations, human populations, and agriculturally important species.

2. Discuss the role of mutation in the origin of genetic polymorphisms. Suppose that a genetic polymorphism has two alleles at frequencies of 0.45 and 0.55. Describe three different scenarios to explain these observed allele frequencies. You can propose that the alleles are neutral, beneficial, or deleterious.

3. Most new mutations are detrimental, yet rare beneficial mutations can be adaptive. Discuss whether you think it is more important for natural selection to select against detrimental alleles or to select in favor of beneficial alleles. Which do you think is more significant in human populations?

Note: All answers appear at the website for this textbook; the answers to even-numbered questions are in the back of the textbook.

---

www.mhhe.com/brooker

Visit the Online Learning Center for practice tests, answer keys, and other learning aids for this chapter. Enhance your understanding of genetics with our interactive exercises, web links, news feeds, tutorial service, and much more.

EVOLUTIONARY GENETICS

::

Biological evolution is a heritable change in one or more characteristics of a population or species across many generations. Evolution can be viewed on a small scale as it relates to a single gene or it can be viewed on a larger scale as it relates to the formation of new species. At the end of chapter 25, we examined several factors that cause allele frequencies to change in populations. This process, also known as **microevolution,** concerns the changing composition of gene pools with regard to particular alleles over measurable periods of time. As we have seen, several evolutionary forces, such as mutation, inbreeding, genetic drift, migration, and natural selection, affect the allele and genotypic frequencies within natural populations. Thus, on a microevolutionary scale, evolution can be viewed as a change in allele frequency over time.

The goal of chapter 26 is to relate phenotypic changes that occur during evolution to the underlying genetic changes that cause them to happen. In the first part, we will be concerned with evolution on a larger scale, which leads to the origin of new species. This phenomenon was central to the development of evolutionary theory. In more advanced courses, you may also learn about evolution on an even grander scale. The term **macroevolution** refers to evolutionary changes that create new species and groups of species. It concerns the diversity of organisms established over long periods of time through the accumulated evolution and extinction of many species. Macroevolution involves relatively large changes in form and function that are sufficient to produce new species and higher taxa.

Evolutionary biologists often focus on traits that fall into three broad phenotypic categories:

1. *Adaptation.* Traits that make a species better able to survive in its natural environment.
2. *Reproductive success.* Traits that promote fertility, the ability to find a mate, etc.
3. *Reproductive isolation.* Traits that prevent two different species from successfully interbreeding.

In the second part of chapter 26, we will link molecular genetics to the evolution of species. The development of techniques for analyzing chromosomes and DNA sequences has greatly enhanced our understanding of evolutionary processes at the molecular level. The term **molecular evolution** refers to the molecular changes in genetic material that underlie the phenotypic changes associated with evolution. The topic of molecular evolution is a fitting way to end our discussion of genetics, since it integrates the ongoing theme of this textbook—the relationship between molecular genetics and traits—in the grandest and most profound ways.

Theodosius Dobzhansky, an influential evolutionary scientist, once said, "Nothing in biology makes sense except in the light of evolution." The extraordinarily diverse and seemingly bizarre array of species on our planet can be explained naturally within the context of evolution. An examination of molecular evolution allows us to make sense of the existence of these species at both the populational and the molecular levels.

26.1 ORIGIN OF SPECIES

Charles Darwin, a British naturalist who was born in 1809, was the first person to theorize that existing species have evolved from preexisting species. Like many great scientists, Darwin had a broad background in science, which enabled him to see connections among different disciplines. His thinking was influenced by theories of geology indicating that the earth is very old and that slow geological processes can lead eventually to substantial changes in the earth's characteristics.

A second important influence on Darwin was his own experimental observations. His famous voyage on the *HMS Beagle,* which lasted from 1832 to 1836, involved a careful examination of many different species. He observed the similarities among many discrete species, yet noted the differences that enabled them to be adapted to their environmental conditions. He was particularly struck by the distinctive adaptations of island species. For example, the finches found on the Galápagos Islands had unique phenotypic characteristics as compared to similar finches found on the mainland.

Finally, a third important impact on Darwin was a paper published in 1798, *Essay on the Principle of Population,* by Thomas Malthus, an English economist. Malthus asserted that the population size of humans can, at best, increase arithmetically due to increased land usage and improvements in agriculture, while the reproductive potential of humans can increase geometrically. He

argued that famine, war, and disease will limit population growth, especially among the poor.

With these three ideas in mind, Darwin had largely formulated his theory of evolution by the mid-1840s. He then spent several years studying barnacles without having published his ideas. The geologist Charles Lyell, who had greatly influenced Darwin's thinking, strongly encouraged Darwin to publish his theory of evolution. In 1856, Darwin began to write a long book to explain his ideas. In 1858, however, Alfred Wallace, a naturalist working in the East Indies, sent Darwin an unpublished manuscript to read prior to its publication. In it, Wallace proposed the same ideas concerning evolution. Darwin therefore quickly excerpted some of his writings on this subject, and two papers, one by Darwin and one by Wallace, were published in the Proceedings of the Linnaean Society of London. These papers were not widely recognized. A short time later, however, Darwin finished his book, *On the Origin of Species by Means of Natural Selection,* which expounded his ideas in greater detail and with experimental support. This book, which received high praise from many scientists and scorn from others, started a great debate concerning evolution. Although some of his ideas were incomplete due to the fact that the genetic basis of traits was not understood at that time, Darwin's work represents one of the most important contributions to our understanding of biology.

The theory of evolution was called "the theory of descent with modification through variation and natural selection" by Charles Darwin. As its name suggests, it is based on two fundamental principles: genetic variation and natural selection. A modern interpretation of evolution can view these principles with regard to speciation and microevolution.

1. *Genetic variation at the species level:* As we have seen in chapter 25, genetic variation is a consistent feature of most natural populations. Darwin observed that many species exhibit a great amount of phenotypic variation. Although the theory of evolution preceded Mendel's pioneering work in genetics, Darwin (as well as many other people before him) observed that offspring resemble their parents more than they do unrelated individuals. Therefore, he assumed that some phenotypic variation is passed from parent to offspring. However, the genetic basis for the inheritance of traits was not understood at that time.

   At the microevolutionary level: Genetic variation can involve allelic differences in genes, changes in chromosome structure, and alterations in chromosome number. These differences are caused by random mutations. Alternative alleles may affect the functions of the proteins they encode and thereby affect the phenotype of the organism. Likewise, changes in chromosome structure and number may affect gene expression and thereby influence the phenotype of the individual.

2. *Natural selection at the species level:* Darwin agreed with Malthus that most species produce many more offspring than survive and reproduce. This creates a "struggle for existence" that results in a "survival of the fittest." Over the

course of many generations, those individuals who happen to possess the most favorable traits will dominate the composition of the population. The ultimate result of natural selection is to make a species better adapted to its environment and/or more efficient at reproduction.

At the microevolutionary level: Some alleles encode proteins that provide the individual with a selective advantage. Over time, natural selection may change the allele frequencies of genes and thereby lead to the fixation of beneficial alleles and the elimination of detrimental alleles.

In this section, we will examine the features of evolution as it occurs in natural populations over time.

A Biological Species Is a Group of Reproductively Isolated Individuals

Perhaps it is now obvious that the study of evolution requires a dialogue among naturalists, taxonomists, and geneticists. However, after Darwin's death, and during the first three decades of the twentieth century, naturalists and taxonomists hardly interacted with geneticists. This situation changed dramatically due to the efforts of Theodosius Dobzhansky.

Dobzhansky was a naturalist and a taxonomist (with an interest in insects) who started working with Thomas Hunt Morgan (a pioneer in studying genetic mutations in *Drosophila* as described in chapters 3 and 5). Dobzhansky's profound perception was that an understanding of evolution must be rooted in the discreteness and discontinuity among different species. A key contribution of Dobzhansky was the idea that discrete species exist due to **isolating mechanisms.** In the late 1920s, he proposed that species are reproductively isolated from other species, which means they cannot successfully interbreed with other species. Thus, reproductive isolation is an important feature of a species' phenotypic traits. In 1937, Dobzhansky also published a book, *Genetics and the Origin of Species,* which laid the foundation for our modern understanding of evolution, sometimes called **Neo-Darwinean theory.** This book focused on adaptive changes in phenotype, particularly those that promote survival. Unlike Darwin (who thought that environmental variation was somehow responsible for genetic variation), Dobzhansky argued that random mutations provide the basis of variation, but he agreed with Darwin that natural selection is responsible for creating adaptive changes in species. In addition, Dobzhansky acknowledged that forces such as genetic drift and migration could also play important roles.

Ernst Mayr expanded on the ideas of Dobzhansky to provide a biological definition of a species. According to Mayr's **biological species concept,** a **species** is defined as a group of individuals whose members have the potential to interbreed with one another in nature to produce viable, fertile offspring but who cannot successfully interbreed with members of other species. There are several different ways to achieve reproductive isolation (table 26.1). These can be classified as **prezygotic mechanisms,** which prevent the formation of a zygote, and **postzygotic mecha-**

TABLE 26.1
Types of Reproductive Isolation Between Different Species

Prezygotic Mechanisms

Habitat isolation Species may occupy different habitats so that they never come in contact with each other.

Temporal isolation Species have different mating or flowering seasons, mate at different times of day, or become sexually active at different times of the year.

Sexual isolation Sexual attraction between males and females of different animal species is limited due to differences in behavior, physiology, or morphology.

Mechanical isolation The anatomical structures of genitalia prevent mating between different species.

Gametic isolation Gametic transfer takes place, but the gametes fail to unite with each other. This can occur because the male and female gametes fail to attract, because they are unable to fuse, or because the male gametes are inviable in the female reproductive tract of another species.

Postzygotic Mechanisms

Hybrid inviability The egg of one species is fertilized by the sperm from another species, but the fertilized egg fails to develop past early embryonic stages.

Hybrid sterility The interspecies hybrid survives, but it is sterile. For example, the mule is a cross between a female horse (*Equus caballus*) and a male donkey (*Equus asinus*).

Hybrid breakdown The F_1 interspecies hybrid is viable and fertile, but succeeding generations (i.e., F_2, etc.) become increasingly inviable. This is usually due to the formation of less fit genotypes by genetic recombination.

nisms, which prevent the development of a viable and fertile individual after fertilization has taken place. Reproductive isolation in nature may be circumvented when species are kept in captivity. For example, different species of the genus *Drosophila* rarely mate with each other in nature. In the laboratory, however, it is fairly easy to produce interspecies hybrids.

Hugh Paterson proposed an alternative way to view a species known as the **species recognition concept.** This idea agrees with the biological species concept in that it recognizes that species are reproductive communities that are isolated from other species. The difference in the species recognition concept concerns the mechanisms by which new species arise. According to Paterson, the sexual recognition of members of the same species acts as a positive force in evolution. There is variation in the ability of mates to locate each other and participate in a productive fertilization. Therefore, natural selection favors the survival of members that can recognize each other and mate successfully. In this regard, recognition can have a very broad meaning. It could involve factors such as visual recognition, behavioral recognition, or even cellular recognition (e.g., sperm and egg cells).

The species recognition concept is consistent with many examples of secondary sexual traits that lead to reproductive isolation. The term **sexual selection,** originally described by Darwin, refers to "the advantage that certain individuals have over

Blue-throated *Lampornis clemenciae*   Broad-billed *Cyanthus latirostris*   Rufous hummingbird *Selasphorus rufus*

FIGURE 26.1   **Species of male hummingbirds exhibit great differences in their feather plumage.**
GENES→TRAITS   One way that natural selection works is by promoting the sexual recognition of members of the same species. Natural selection favors the survival of members with phenotypes that make it easier for them to find each other and mate successfully. The male hummingbirds shown carry alleles that affect their plumage size and coloration. These distinctive features make it easier for males and females of the same species to recognize and mate with each other.

others of the same sex and species solely in respect of reproduction." In many species of animals, sexual selection affects male characteristics more intensely than it does female. Unlike females, who tend to be fairly uniform in their reproductive success, male success tends to be more variable, with some males mating with many females and others not mating at all. There are many examples of male traits that function solely for the purpose of recognition by the female of the species. Figure 26.1 illustrates the morphological differences among several species of male hummingbirds. These differences allow the females to recognize and choose an appropriate mate. By comparison, the females of these species look rather similar to each other.

Speciation Usually Occurs via a Branching Process Called Cladogenesis

By examining the fossil record, evolutionary biologists have found two different patterns of speciation (fig. 26.2). During **anagenesis** (fig. 26.2*a*; from the Greek, *ana*, meaning "up," and *genesis*, meaning "origin"), a single species is transformed into a different species over the course of many generations. During this process, the characteristics of the species change due to both neutral evolutionary forces and adaptive forces promoted by natural selection. As a result of natural selection, the new species may be better adapted to survive in its original environment, or the

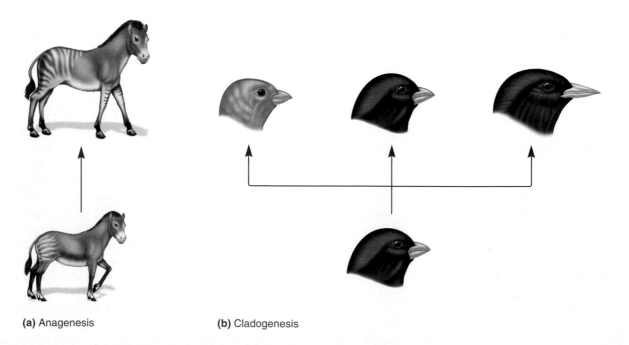

(a) Anagenesis   **(b)** Cladogenesis

FIGURE 26.2   **A comparison between anagenesis and cladogenesis, two patterns of speciation. (a)** Anagenesis is the change of one species into another; **(b)** cladogenesis involves a process whereby one original species is separated into two or more different species.

environment may have changed so that the new species is better adapted to the new surroundings.

By comparison, **cladogenesis** (fig. 26.2b; from the Greek *clados,* meaning "branch") involves the division of a species into two or more species. This is the most common form of speciation. Although cladogenesis is usually thought of as a branching process, it commonly occurs as a budding process in which a single species divides into the original species plus a new species with different characteristics. If we view evolution as a tree, the new species buds from the original species and develops characteristics that prevent it from interbreeding with the original one.

Divergent Evolution Can Be Allopatric, Parapatric, or Sympatric

The divergence of one species into two or more discrete species is the most common form of speciation. Depending on the geographic locations of the evolving population(s), speciation is categorized as allopatric, parapatric, or sympatric. Table 26.2 presents an overview of these types of speciation patterns. The following discussion will describe examples.

Allopatric speciation (from the Greek *allos,* "other," and Latin *patria,* "homeland") is thought to be the most prevalent

A. harrisi   *A. leucurus*

FIGURE 26.3 **An example of allopatric speciation: two closely related species of antelope squirrels that occupy opposite rims of the Grand Canyon.**

GENES→TRAITS   Harris's antelope squirrel (*Ammospermophilus harrisi*) is found on the south rim of the Grand Canyon, while the white-tailed antelope squirrel (*Ammospermophilus leucurus*) is found on the north rim. These two species evolved from a common species that existed before the canyon was formed. After it was formed, the two separated populations accumulated genetic changes due to mutation, natural selection, and genetic drift that eventually led to the formation of two distinct species.

way for a species to diverge. It occurs when members of a species become geographically separated from the other members. This form of speciation can occur by the geographic subdivision of large populations via geological processes. For example, a mountain range may emerge and split a species that occupies the lowland regions, or a creeping glacier may divide a population. Figure 26.3 shows an interesting example in which geological separation promoted speciation. Two species of antelope squirrels occupy opposite rims of the Grand Canyon. On the south rim is Harris's antelope squirrel (*Ammospermophilus harrisi*), while a closely related white-tailed antelope squirrel (*Ammospermophilus leucurus*) is found on the north rim. Presumably, these two species evolved from a common species that existed before the canyon was formed. Over time, the accumulation of genetic changes in the two populations led to the formation of two morphologically distinct species. Interestingly, birds that can easily fly across the canyon have not diverged into different species on the opposite rims.

Allopatric speciation can also occur via a second mechanism, known as the **founder effect,** which is thought to be more rapid and frequent than allopatric speciation caused by geological events. The founder effect, which was discussed in chapter 25, occurs when a small group migrates to a new location that is geographically separated from the main population. For example, a storm may force a small group of birds from the mainland to a distant island. In this case, migration between the island and the mainland populations is a very infrequent event. In a relatively short period of time, the founding population on the island may evolve into a new species. Several evolutionary forces may contribute to this rapid evolution. First, as discussed in chapter 25, genetic drift may quickly lead to the random fixation of certain alleles and the elimination of other alleles from the population. Another factor is natural selection. The environment on an island may differ significantly from the mainland environment.

TABLE 26.2

Common Genetic Mechanisms That Underlie Allopatric, Parapatric, and Sympatric Speciation

Type of Speciation	Common Genetic Mechanisms Responsible for Speciation
Allopatric—two large populations are separated by geographic barriers.	Many small genetic differences may accumulate over a long period of time, leading to reproductive isolation. Some of these genetic differences may be adaptive while others will be neutral.
Allopatric—a small founding population separates from the main population.	Genetic drift and shifts in adaptive peaks may lead to the rapid formation of a new species. If the group has moved to an environment that is different from its previous environment, natural selection is expected to favor beneficial alleles and eliminate harmful alleles.
Parapatric—two populations occupy overlapping ranges so that a limited amount of interbreeding occurs.	A new combination of alleles or chromosomal rearrangement may rapidly limit the amount of gene flow between neighboring populations because hybrid offspring have a very low fitness.
Sympatric—within a population occupying a single habitat in a continuous range, a small group evolves into a reproductively isolated species.	An abrupt genetic change leads to reproductive isolation. For example, a mutation may affect gamete recognition. In plants, the formation of a tetraploid often leads to the formation of new species because the interspecies hybrid (e.g., diploid × tetraploid) is triploid and sterile.

FIGURE 26.4   The concept of adaptive peaks. This illustration uses a geographic metaphor to symbolically represent genotypes. It suggests that certain genotypic combinations are favorable and form peaks of high fitness values, while other genotypic combinations are unfavorable and form valleys of low fitness values. If the environmental conditions change, the adaptive peaks may change as well.

To explain how natural selection may act on the phenotypic effects of many alleles, Sewall Wright developed the concept of **adaptive peaks** to describe combinations of many alleles that provide an optimal fitness for individuals in a stable environment. As shown in figure 26.4, several different combinations of alleles may lead to highly fit individuals. If a small change occurs in the frequency of alleles at one or a few loci, it will drive the population off an adaptive peak and into a less fit valley; natural selection, though, will tend to push the population back to the original peak.

When, however, a small group migrates to a new environment, the adaptive landscape is likely to be changed. Allele combinations that were highly adaptive in the original environment may be much less so in the new environment. If this occurs, natural selection may lead to a shift toward a new adaptive peak in which the frequencies of many alleles have been changed. In other words, natural selection will act to change the traits of the organism to make it better adapted to its new environment. Similarly, in a species occupying a large geographic range, the adaptive peaks of local populations may differ due to variation in the environmental conditions within the range.

Parapatric speciation (from the Greek *para,* "beside") occurs when members of a species are separated partially or when a species is very sedentary. In these cases, the geographic separation is not complete. For example, a mountain range may divide a species into two populations but with breaks in the range where the two groups are connected physically. In these zones of contact, the members of two populations can interbreed, although this tends to occur infrequently. Likewise, parapatric speciation may occur among very sedentary species even though no geographic isolation exists. Certain organisms are so sedentary that 100 to 1,000 meters may be sufficient to limit the interbreeding

between neighboring groups. Plants, terrestrial snails, rodents, grasshoppers, lizards, and many flightless insects may speciate in a parapatric manner.

Prior to parapatric speciation, the zones where two populations can interbreed are known as **hybrid zones.** For speciation to occur, the amount of gene flow within the hybrid zones must become very limited. In other words, there must be selection against the offspring produced in the hybrid zone. One way that this can happen is that each of the two parapatric populations may accumulate different neutral chromosomal rearrangements, such as inversions and balanced translocations. As discussed in chapter 8, if a hybrid individual has one chromosome with a large inversion and one that does not carry the inversion, crossing over during meiosis can lead to the production of grossly abnormal chromosomes. Therefore, such an individual is substantially less fertile. By comparison, an individual that is homozygous for two normal chromosomes or for two chromosomes carrying the same inversion will be fertile, because crossing over can proceed normally.

Finally, **sympatric speciation** (from the Greek *sym,* "together") occurs when members of a species initially occupy the same habitat within the same range. In plants, a common way for sympatry to occur is the formation of polyploids. As discussed in chapter 8, complete nondisjunction of chromosomes during gamete formation can increase the number of chromosome sets within the same species (autopolyploidy) or between different species (allopolyploidy). Polyploidy is so frequent in plants that it is a major form of speciation. In ferns and flowering plants, about 30 to 35% of the species are polyploid. By comparison, polyploidy is much less common in animals, but it can occur. For example, roughly 30 species of reptiles and amphibians have been identified that are polyploids derived from diploid relatives.

The formation of a polyploid can abruptly lead to reproductive isolation. As an example, let's consider the probable events that led to the formation of a natural species of common hemp nettle known as *Galeopsis tetrahit.* This species is thought to be an allotetraploid derived from two diploid species, *Galeopsis pubescens* and *Galeopsis speciosa.* These two diploid species contain 16 chromosomes each ($2n = 16$), while *G. tetrahit* contains 32. Figure 26.5 illustrates what would happen in crosses between the allotetraploid and the diploid species. The allotetraploid crossed to another allotetraploid produces an allotetraploid. The allotetraploid is fertile, because all of its chromosomes occur in homologous pairs that can segregate evenly during meiosis. However, a cross between an allotetraploid and a diploid produces an offspring that is monoploid for one chromosome set and diploid for the other set. These offspring are expected to be sterile, because they will produce highly aneuploid gametes that have incomplete sets of chromosomes. This hybrid sterility causes the allotetraploid to be reproductively isolated from the diploid species.

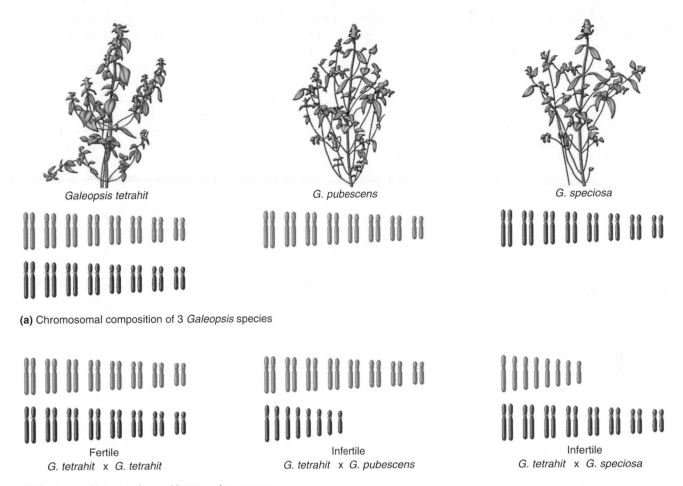

(a) Chromosomal composition of 3 *Galeopsis* species

Fertile
G. tetrahit x *G. tetrahit*

Infertile
G. tetrahit x *G. pubescens*

Infertile
G. tetrahit x *G. speciosa*

(b) Outcome of intraspecies and interspecies crosses

FIGURE 26.5 **A comparison of crosses between three natural species of hemp nettle with different ploidy levels.** *Galeopsis tetrahit* is an allotetraploid that is thought to be derived from *Galeopsis pubescens* and *Galeopsis speciosa*. If *G. tetrahit* is mated with the other two species, the F_1 hybrid offspring will be monoploid for one chromosome set and diploid for the other set. The F_1 offspring are likely to be sterile, because they will produce highly aneuploid gametes.

EXPERIMENT 26A

Müntzing Was Able to Re-Create an Allotetraploid Species

In studying biological evolution, we cannot look into the past and know with complete certainty how events occurred. An alternative way to explore past events is called **retrospective testing.** In this strategy, a researcher formulates a hypothesis and then collects observations as a way to confirm or refute the hypothesis. Therefore, this approach relies on the observation and testing of already existing materials, which themselves are the result of past events. The careful study or manipulation of these materials may then be used to support or refute ideas concerning how evolution may have occurred in the past.

An interesting example of this approach is provided in a study by Arne Müntzing, a Swedish botanist. In the 1930s, Müntzing first proposed that *Galeopsis tetrahit* is an allotetraploid that arose from an interspecies cross between *G. pubescens* and *G. speciosa* (fig. 26.5). To confirm this hypothesis, he attempted to "reenact" evolution by re-creating an "artificial" *G. tetrahit* by crossing *G. pubescens* and *G. speciosa* in a way that produced an allotetraploid.

As shown in figure 26.6, Müntzing began with *G. pubescens* and *G. speciosa* as his starting material. He knew that each of these two species occasionally produces diploid gametes (by nondisjunction), which could fuse together and produce an allotetraploid in a single step. However, he expected that this would be an extremely rare event. Therefore, rather than hoping for such a rare event to occur, he decided to make the allotetraploid in a two-step procedure.

Müntzing's first goal was to produce a strain that was diploid for *G. speciosa* chromosomes and monoploid for the *G. pubescens* chromosomes. To develop such a strain, he first crossed *G. pubescens* to *G. speciosa,* producing F_1 alloploids. Müntzing assumed that the F_1 alloploids would have a higher frequency of nondisjunction because they would contain one set of chromosomes from two different species. He hoped that the $F_1 \times F_1$ cross would tend to produce gametes that exhibited nondisjunction. An occasional offspring might be perfectly diploid for the chromosomes of one species (e.g., *G. speciosa*) and perfectly monoploid for the other species (e.g., *G. pubescens*). Out of 200 F_2 offspring he examined, he found one F_2 strain that contained 16 chromosomes from *G. speciosa* and 8 chromosomes from *G. pubescens* (which was named the intermediate F_2 strain).

Due to its unusual chromosomal composition, Müntzing knew that the intermediate F_2 strain would usually make highly aneuploid gametes that could not produce viable seeds. He rea-

soned that the intermediate F_2 strain might occasionally produce a gamete that would be the result of complete nondisjunction. Such a gamete would be diploid for the *G. speciosa* chromosomes and monoploid for the *G. pubescens* chromosomes. If this gamete fused with a normal haploid gamete from *G. pubescens,* this would result in a tetraploid offspring containing two sets of chromosomes from each species—the hypothesized composition of *G. tetrahit.* Therefore, to obtain such an offspring, he crossed the intermediate F_2 strain to a normal *G. pubescens* strain. Because the intermediate F_2 strain usually produced aneuploid gametes, it was almost unable to make any viable seeds. However, Müntzing did obtain one seed that had the chromosomal composition he wanted: it had a diploid number of chromosomes from both species. This seed was the artificial *G. tetrahit* Müntzing had hoped for.

■ THE HYPOTHESIS

G. tetrahit is an allopolyploid that contains a diploid chromosomal composition from *G. pubescens* and *G. speciosa.*

■ TESTING THE HYPOTHESIS — FIGURE 26.6 Creation of an allotetraploid species from two diploid species.

Starting material: Specimens of *G. pubescens, G. speciosa,* and *G. tetrahit.*

1. Cross *G. pubescens* to *G. speciosa.*

   Note: In most cases, this will produce alloploids that have 16 chromosomes.

2. Make an F_1 x F_1 cross. The fertility of the alloploids is very low, because the *G. pubescens* chromosomes usually do not pair properly with the *G. speciosa* chromosomes during meiosis. For this reason, the F_2 offspring obtained are expected to have abnormal numbers of chromosomes.

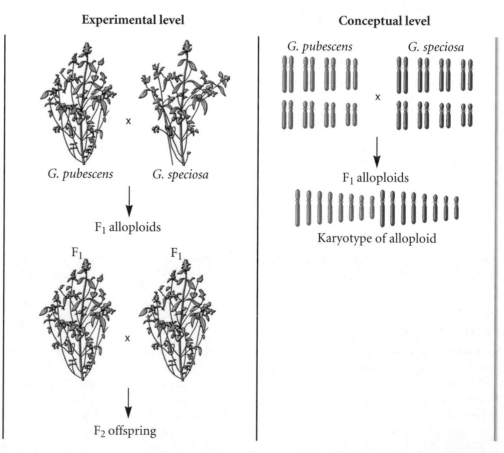

Experimental level

G. pubescens   *G. speciosa*

x

F_1 alloploids

F_1 F_1

x

F_2 offspring

Conceptual level

G. pubescens   *G. speciosa*

x

F_1 alloploids

Karyotype of alloploid

(continued)

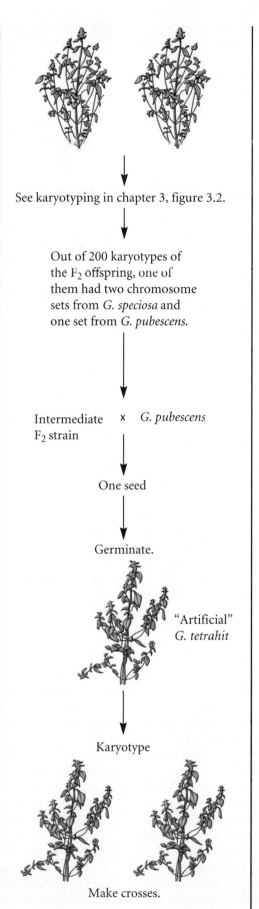

3. Examine the chromosome compositions of 200 F$_2$ offspring.

See karyotyping in chapter 3, figure 3.2.

4. Among 200 offspring examined, one of them had 16 chromosomes from *G. speciosa* and 8 chromosomes from *G. pubescens*. Presumably, this occurred because of nondisjunction of the *G. speciosa* chromosomes. Let's call this strain the "intermediate F$_2$ strain."

Out of 200 karyotypes of the F$_2$ offspring, one of them had two chromosome sets from *G. speciosa* and one set from *G. pubescens*.

5. Cross the intermediate F$_2$ strain to *G. pubescens*.

Intermediate   x   *G. pubescens*
F$_2$ strain

6. Only one seed was obtained! It was diploid for the chromosomes of both species. Presumably, it arose from a haploid *G. pubescens* gamete fusing with a gamete from the intermediate F$_2$ strain that had resulted from complete nondisjunction. This allotetraploid strain was called "artificial" *G. tetrahit*.

One seed

Germinate.

"Artificial" *G. tetrahit*

7. Examine the morphological features of the "artificial" *G. tetrahit* and compare them to the natural species.

Karyotype

8. Test the ability of the "artificial" *G. tetrahit* to be crossed with itself and with the natural *G. tetrahit* species.

Make crosses.

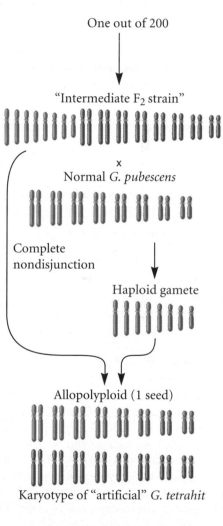

One out of 200

"Intermediate F$_2$ strain"

x

Normal *G. pubescens*

Complete nondisjunction

Haploid gamete

Allopolyploid (1 seed)

Karyotype of "artificial" *G. tetrahit*

To see if the natural and "artificial" *G. tetrahit* look the same

To see if the natural and "artificial" *G. tetrahit* produce fertile offspring

■ THE DATA

	Artificial G. tetrahit	Natural G. tetrahit	G. pubescens	G. speciosa
Chromosomal composition	32	32	16	16
Mean flower size (mm)	19	19	29	27
Calyx length (mm)	15	15	12	18
Degree of stem hairiness	Moderate	Moderate	Low	High

Fertility: The artificial *G. tetrahit* could produce fertile offspring when it was crossed to another artificial *G. tetrahit* and it could produce fertile offspring when it was crossed to a natural strain of *G. tetrahit*. Microscopically, normal chromosome pairing was observed during meiosis in the artificial and natural *G. tetrahit* hybrid.

■ INTERPRETING THE DATA

To determine if his data were compatible with the hypothesis, Müntzing examined the morphological characteristics of the plants and their ability to interbreed. Based on morphological characteristics, the artificial *G. tetrahit* species appeared to confirm his hypothesis. It had traits that were similar to the natural *G. tetrahit* species but different from *G. pubescens* and *G. speciosa*. Furthermore, the artificial and natural *G. tetrahit* strains could be mated to each other to produce fertile offspring. Overall, Müntzing came to the conclusion that "not only the artificial tetraploid but probably also natural *G. tetrahit* represents a synthesis of *pubescens*- and *speciosa*-genomes. Consequently, this case might be regarded as the first instance where a species already existing in nature [as a result of evolution] has been synthesized from the genomes of two other species."

A self-help quiz involving this experiment can be found at the Online Learning Center.

Evolution Can Proceed Gradually or Be Punctuated by Periods of Rapid Change

As we have seen, many different genetic mechanisms give rise to new species. For this reason, the rates of evolutionary change are not constant, although the degree of inconstancy has been debated since the time of Darwin. Even Darwin himself suggested that evolution can occur at fast and slow paces. Figure 26.7 illustrates contrasting views concerning the rates of evolutionary change. These views are not mutually exclusive but represent two different ways to consider the tempo of evolution. The notion of **gradualism** suggests that each new species evolves continuously over long spans of time (fig. 26.7*a*). The principal idea is that large phenotypic differences that produce the emergence of new species are due to the accumulation of many small genetic changes. By comparison, the concept of **punctuated equilibrium,** proposed by Niles Eldrege and Stephen Gould, suggests that the tempo of evolution is more sporadic (fig. 26.7*b*). According to this proposal, species exist relatively unchanged for many generations. During this period, the species is in equilibrium with its environment. These long periods of equilibrium are punctuated by relatively short periods (i.e., on a geological timescale) during which evolution occurs at a far more rapid rate.

In reality, neither of the views presented in figure 26.7 fully account for evolutionary change. The phenomenon of punctuated equilibrium seems to be supported by the fossil record. Paleontologists rarely find a gradual transition of fossil forms. Instead, it is much more common to observe species appearing as new forms rather suddenly in a layer of rocks, persisting relatively unchanged for a very long period of time, and then suddenly becoming extinct. It is presumed that the transition period during which a previous species evolved into a new, morphologically distinct species was so short that few, if any, of the transitional members were preserved as fossils. Even so, these rapid periods of change were likely followed by long periods of equilibrium that involved the additional accumulation of many small genetic changes, consistent with gradualism.

As discussed earlier, rapid evolutionary change can be explained by genetic phenomena. As we have seen throughout this textbook, single-gene mutations can have dramatic effects on phenotypic characteristics. Therefore, only a small number of new mutations may be required to alter phenotypic characteristics, eventually producing a group of individuals that make up a new species. Likewise, genetic events such as changes in chromosome structure (e.g., inversions and translocations) or chromosome number may abruptly create individuals with new phenotypic traits. On an evolutionary timescale, these types of events can be rather rapid, since one or only a few genetic changes can have a major impact on the phenotype of the organism.

In conjunction with genetic changes, species may also be subjected to sudden environmental shifts that quickly drive the gene pool in a particular direction via natural selection. For example, a small group may migrate to a new environment in which different alleles provide better adaptation to the surroundings. Alternatively, a species may be subjected to a relatively sudden environmental event that has a major impact on survival. There may be a change in climate, or a new predator may infiltrate the geographic range of the species. To ensure survival, these kinds of events may necessitate the rapid evolution of the gene pool by natural selection so that members of the population can survive the climatic change or have phenotypic characteristics that allow them to avoid the predator.

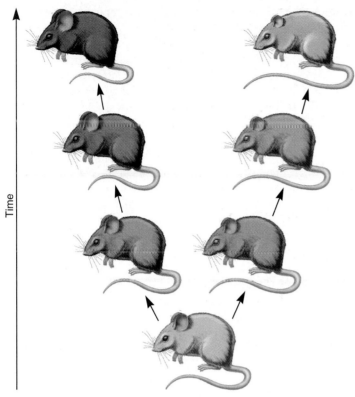

(a) Gradualism

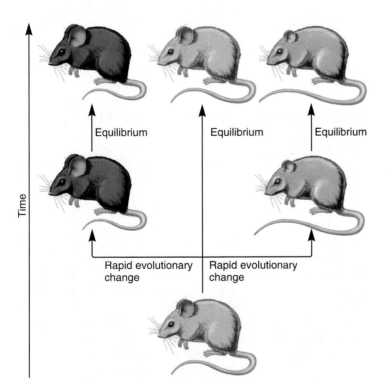

(b) Punctuated equilibrium

Overall, the fossil record and known genetic phenomena tend to support the idea that the tempo of evolution can be quite variable. In some cases, rapid evolutionary change has taken place and led to the formation of new species. During other periods, smaller phenotypic changes may occur over a longer timescale. In conjunction with phenotypic changes, the gradual accumulation of neutral mutations has been revealed by molecular analyses of DNA. We will explore this topic in the next section.

26.2 EVOLUTION AT THE MOLECULAR LEVEL

Historically, taxonomists have categorized different species primarily on the basis of phenotypes. Initially, morphological differences were used to construct evolutionary trees in which species that are more similar in appearance tended to be placed closer together on the tree. In addition, species have been categorized based on differences in physiology, biochemistry, and even behavior. However, because evolution involves genetic changes, it makes sense to categorize species based on the properties of their genetic material (i.e., their genotypes). Those species that are closely related evolutionarily are expected to have greater similarities in their genetic material than are distantly related species.

The advent of molecular approaches for analyzing DNA and gene products has revolutionized the field of evolution. In the past few decades, molecular genetics has greatly facilitated our understanding of speciation and evolution. Differences in nucleotide sequences are quantitative and can be analyzed using mathematical principles in conjunction with computer programs. Evolutionary changes at the DNA level can be objectively compared among different species to establish evolutionary relationships. Furthermore, this approach can be used to compare any two existing organisms, no matter how greatly they differ in their morphological traits. For example, we can compare DNA sequences between humans and bacteria, or between plants and fruit flies. Such comparisons would be very difficult at a morphological level.

There are many ways to compare the genetic material and gene products among different species. Table 26.3 lists the most common molecular approaches that geneticists use to evaluate evolutionary relationships. In this section, we will focus most of our attention on the evolution of gene sequences.

FIGURE 26.7 **A comparison of gradualism and punctuated equilibrium. (a)** During gradualism, the morphology of a species gradually changes due to the accumulation of small genetic changes. **(b)** During punctuated equilibrium, species exist essentially unchanged for long periods of time, during which they are in equilibrium with their environment. These equilibrium periods are punctuated by short periods of evolutionary change, during which morphological features may change rapidly.

TABLE 26.3
Experimental Approaches to Compare Species at the Molecular Level

Technique	Explanation
DNA Level	
DNA sequencing	The degrees of similarities and differences between DNA sequences are used to establish evolutionary relationships. This approach, along with the analysis of the deduced amino acid sequences encoded by structural genes, is the most common and reliable method that is used.
Hybridization	Genetic material from two different species is allowed to hybridize. The rate of hybridization will be faster for closely related species.
Restriction mapping	Segments of DNA are isolated from different species and subjected to restriction mapping. Closely related species will have more similar restriction maps.
Cytogenetic analysis	The chromosomes of different species are analyzed microscopically. Closely related species will have identical or similar chromosome numbers and banding patterns.
Protein Level	
Deduced amino acid sequence	Technically, DNA sequencing is much easier than amino acid sequencing. Therefore, to determine the amino acid sequence of a protein, the coding region of a structural gene is determined, and the amino acid sequence is deduced using the genetic code. The amino acid sequences of a given protein will be more similar between closely related species.
Immunological methods	Antibodies that recognize specific macromolecules, usually on the cell surface, are tested on different species. Antibodies that recognize a protein of one species will often recognize the same protein of closely related species but not distantly related species.

Homologous Genes Are Derived from a Common Ancestral Gene

Two genes are said to be **homologous** if they are derived from the same ancestral gene. As described in chapter 21, genes can exhibit **interspecies homology** and **intraspecies homology.** During evolution, a single species may become divided into two or more different species. When two homologous genes are found in different species, these genes are termed **orthologs.** In chapter 23, we considered orthologs of *Hox* genes. In that case, several homologous genes were identified in the fruit fly and the mouse. In addition, two or more homologous genes can also be found within a single species. These are termed **paralogous** genes or **paralogs.** As discussed in chapter 8, this can occur because abnormal gene duplication events can produce multiple copies of a gene and ultimately lead to the formation of a **gene family.** A gene family consists of two or more paralogs within the genome of a single organism.

Figure 26.8 shows examples of orthologs and paralogs in the globin gene family. Hemoglobin is an oxygen-carrying pro-

tein found in all vertebrate species. It is composed of two different subunits, encoded by the α-globin and β-globin genes. Figure 26.8 shows the deduced amino acid sequences of these genes. The sequences are homologous between humans and horses because of their evolutionary relationship. We would say that the human and horse α-globin genes are orthologs of each other, as are the human and horse β-globin genes. The α-globin and β-globin genes in humans are paralogs of each other.

If you scan the sequences shown in figure 26.8, you may notice that the sequences of the orthologs are more similar to each other than the paralogs. For example, the sequences of human and horse β-globins are much more similar than the sequences of the human α-globin and human β-globin. This suggests that the gene duplications that created the α-globin and β-globin genes occurred long before the evolutionary divergence that produced different species of mammals. For this reason, there was a greater amount of time for the α- and β-globin genes to accumulate changes compared to the amount of time that has elapsed since the evolutionary divergence of mammalian species. This idea is schematically shown in figure 26.9.

Based on the analysis of genetic sequences, evolutionary biologists have estimated that the gene duplication that created the α-globin and β-globin gene lineages occurred approximately 400 million years ago, while the speciation events that created different species of mammals occurred less than 200 million years ago. Therefore, the α-globin and β-globin genes have had much more time to accumulate changes relative to each other. There has been much less time for the accumulation of random mutations with regard to the α-globin genes among different mammalian species or the β-globin genes among different mammalian species.

Variation in Gene Sequences Can Be Used to Construct Phylogenetic Trees

A **phylogenetic tree** is a diagram that describes the evolutionary relationships among different species. In 1963, Linus Pauling and Emile Zukerkandl were the first to suggest that molecular data should be used to establish evolutionary relationships. With advances in DNA-sequencing technology, the amount of phylogenetic information in genetic sequences is immense. Molecular evolutionists want to analyze the information within genetic sequences to obtain a better understanding of evolutionary links.

Nucleotide and amino acid sequence data are particularly well suited to the construction of evolutionary trees because genetic sequences change over the course of many generations due to the accumulation of mutations. Therefore, when comparing homologous genes in different organisms, the DNA sequences from closely related organisms are more similar to each other than are the sequences from distantly related species. In a sense, the relatively constant rate of neutral mutation acts as a **molecular clock** on which to measure evolutionary time. According to this idea, neutral mutations will occur at a rate that is proportional to the rate of mutation per generation. On this basis, the genetic divergence between species that is due to neutral mutations reflects the time elapsed since their last common ancestor.

Hemoglobin

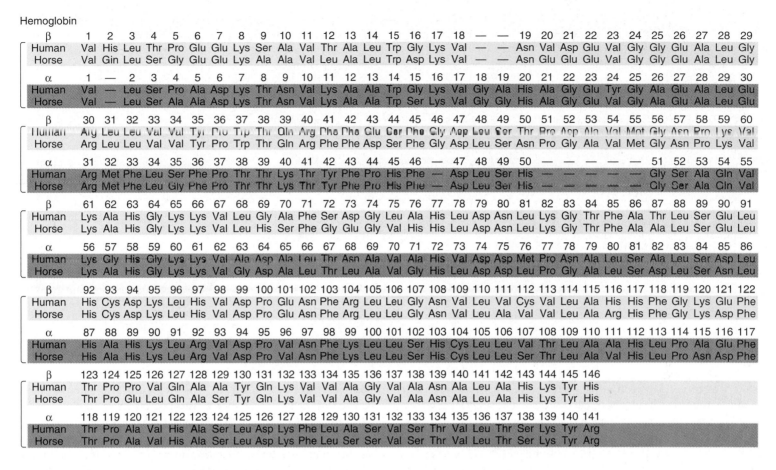

β	1	2	3	4	5	6	7	8	9	10	11	12	13	14	15	16	17	18	—	—	19	20	21	22	23	24	25	26	27	28	29
Human	Val	His	Leu	Thr	Pro	Glu	Glu	Lys	Ser	Ala	Val	Thr	Ala	Leu	Trp	Gly	Lys	Val	—	—	Asn	Val	Asp	Glu	Val	Gly	Gly	Glu	Ala	Leu	Gly
Horse	Val	Gln	Leu	Ser	Gly	Glu	Glu	Lys	Ala	Ala	Val	Leu	Ala	Leu	Trp	Asp	Lys	Val	—	—	Asn	Glu	Glu	Glu	Val	Gly	Gly	Glu	Ala	Leu	Gly

α	1	—	2	3	4	5	6	7	8	9	10	11	12	13	14	15	16	17	18	19	20	21	22	23	24	25	26	27	28	29	30
Human	Val	—	Leu	Ser	Pro	Ala	Asp	Lys	Thr	Asn	Val	Lys	Ala	Ala	Trp	Gly	Lys	Val	Gly	Ala	His	Ala	Gly	Glu	Tyr	Gly	Ala	Glu	Ala	Leu	Glu
Horse	Val	—	Leu	Ser	Ala	Ala	Asp	Lys	Thr	Asn	Val	Lys	Ala	Ala	Trp	Ser	Lys	Val	Gly	Gly	His	Ala	Gly	Glu	Val	Gly	Ala	Glu	Ala	Leu	Glu

β	30	31	32	33	34	35	36	37	38	39	40	41	42	43	44	45	46	47	48	49	50	51	52	53	54	55	56	57	58	59	60
Human	Arg	Leu	Leu	Val	Val	Tyr	Pro	Trp	Thr	Gln	Arg	Phe	Phe	Glu	Ser	Phe	Gly	Asp	Leu	Ser	Thr	Pro	Asp	Ala	Val	Met	Gly	Asn	Pro	Lys	Val
Horse	Arg	Leu	Leu	Val	Val	Tyr	Pro	Trp	Thr	Gln	Arg	Phe	Phe	Asp	Ser	Phe	Gly	Asp	Leu	Ser	Asn	Pro	Gly	Ala	Val	Met	Gly	Asn	Pro	Lys	Val

α	31	32	33	34	35	36	37	38	39	40	41	42	43	44	45	46	—	47	48	49	50	—	—	—	—	—	51	52	53	54	55
Human	Arg	Met	Phe	Leu	Ser	Phe	Pro	Thr	Thr	Lys	Thr	Tyr	Phe	Pro	His	Phe	—	Asp	Leu	Ser	His	—	—	—	—	—	Gly	Ser	Ala	Gln	Val
Horse	Arg	Met	Phe	Leu	Gly	Phe	Pro	Thr	Thr	Lys	Thr	Tyr	Phe	Pro	His	Phe	—	Asp	Leu	Ser	His	—	—	—	—	—	Gly	Ser	Ala	Gln	Val

β	61	62	63	64	65	66	67	68	69	70	71	72	73	74	75	76	77	78	79	80	81	82	83	84	85	86	87	88	89	90	91
Human	Lys	Ala	His	Gly	Lys	Lys	Val	Leu	Gly	Ala	Phe	Ser	Asp	Gly	Leu	Ala	His	Leu	Asp	Asn	Leu	Lys	Gly	Thr	Phe	Ala	Thr	Leu	Ser	Glu	Leu
Horse	Lys	Ala	His	Gly	Lys	Lys	Val	Leu	His	Ser	Phe	Gly	Glu	Gly	Val	His	His	Leu	Asp	Asn	Leu	Lys	Gly	Thr	Phe	Ala	Ala	Leu	Ser	Glu	Leu

α	56	57	58	59	60	61	62	63	64	65	66	67	68	69	70	71	72	73	74	75	76	77	78	79	80	81	82	83	84	85	86
Human	Lys	Gly	His	Gly	Lys	Lys	Val	Ala	Asp	Ala	Leu	Thr	Asn	Ala	Val	Ala	His	Val	Asp	Asp	Met	Pro	Asn	Ala	Leu	Ser	Ala	Leu	Ser	Asp	Leu
Horse	Lys	Ala	His	Gly	Lys	Lys	Val	Gly	Asp	Ala	Leu	Thr	Leu	Ala	Val	Gly	His	Leu	Asp	Asp	Leu	Pro	Gly	Ala	Leu	Ser	Asp	Leu	Ser	Asn	Leu

β	92	93	94	95	96	97	98	99	100	101	102	103	104	105	106	107	108	109	110	111	112	113	114	115	116	117	118	119	120	121	122
Human	His	Cys	Asp	Lys	Leu	His	Val	Asp	Pro	Glu	Asn	Phe	Arg	Leu	Leu	Gly	Asn	Val	Leu	Val	Cys	Val	Leu	Ala	His	His	Phe	Gly	Lys	Glu	Phe
Horse	His	Cys	Asp	Lys	Leu	His	Val	Asp	Pro	Glu	Asn	Phe	Arg	Leu	Leu	Gly	Asn	Val	Leu	Ala	Val	Val	Leu	Ala	Arg	His	Phe	Gly	Lys	Asp	Phe

α	87	88	89	90	91	92	93	94	95	96	97	98	99	100	101	102	103	104	105	106	107	108	109	110	111	112	113	114	115	116	117
Human	His	Ala	His	Lys	Leu	Arg	Val	Asp	Pro	Val	Asn	Phe	Lys	Leu	Leu	Ser	His	Cys	Leu	Leu	Val	Thr	Leu	Ala	Ala	His	Leu	Pro	Ala	Glu	Phe
Horse	His	Ala	His	Lys	Leu	Arg	Val	Asp	Pro	Val	Asn	Phe	Lys	Leu	Leu	Ser	His	Cys	Leu	Leu	Ser	Thr	Leu	Ala	Val	His	Leu	Pro	Asn	Asp	Phe

β	123	124	125	126	127	128	129	130	131	132	133	134	135	136	137	138	139	140	141	142	143	144	145	146
Human	Thr	Pro	Pro	Val	Gln	Ala	Ala	Tyr	Gln	Lys	Val	Val	Ala	Gly	Val	Ala	Asn	Ala	Leu	Ala	His	Lys	Tyr	His
Horse	Thr	Pro	Glu	Leu	Gln	Ala	Ser	Tyr	Gln	Lys	Val	Val	Ala	Gly	Val	Ala	Asn	Ala	Leu	Ala	His	Lys	Tyr	His

α	118	119	120	121	122	123	124	125	126	127	128	129	130	131	132	133	134	135	136	137	138	139	140	141
Human	Thr	Pro	Ala	Val	His	Ala	Ser	Leu	Asp	Lys	Phe	Leu	Ala	Ser	Val	Ser	Thr	Val	Leu	Thr	Ser	Lys	Tyr	Arg
Horse	Thr	Pro	Ala	Val	His	Ala	Ser	Leu	Asp	Lys	Phe	Leu	Ser	Ser	Val	Ser	Thr	Val	Leu	Thr	Ser	Lys	Tyr	Arg

FIGURE 26.8 **A comparison of the α- and β-globin polypeptides from humans and horses.** This figure shows the deduced amino acid sequences that were obtained by sequencing the exon portions of the corresponding genes. The gaps indicate where additional amino acids are found in the sequence of myoglobin, another member of this gene family.

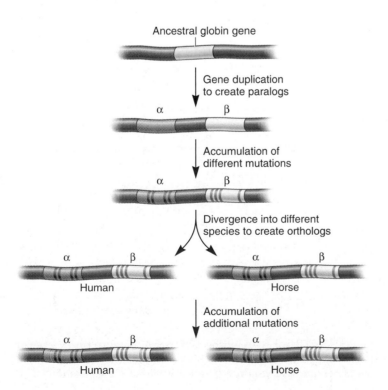

FIGURE 26.9 **Evolution of paralogous and orthologous genes.** In this schematic example, the primordial globin gene duplicated to create the α- and β-globin genes, which are paralogous genes. (Note: Actually, gene duplications occurred several times to create a large globin gene family, but this is a simplified example that shows only one gene duplication event. Also, chromosomal rearrangements have placed the α- and β-globin genes on different chromosomes.) Over time, these two paralogs accumulated different mutations that are designated with *red* and *blue lines.* At a later point in evolution, a species divergence occurred to create two different mammalian species such as horses and humans. Over time, the orthologs may also accumulate different mutations that are designated with *green lines.* As noted here, the orthologs have fewer differences compared to the paralogs because the gene duplication occurred prior to the species divergence.

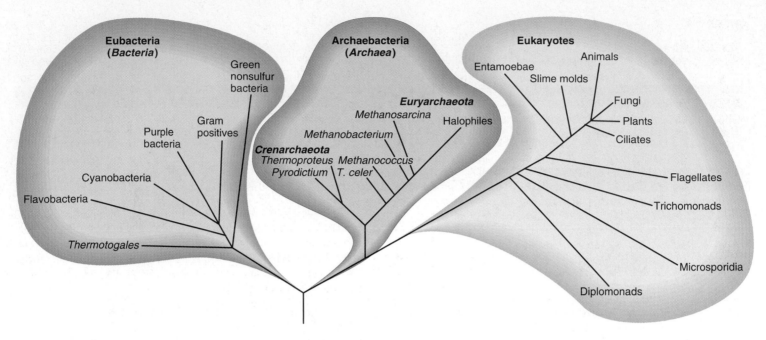

FIGURE 26.10 A phylogenetic tree of all life on earth based on 16S rRNA data.

For molecular evolutionary studies, the DNA sequences of many genes have been obtained from a wide range of sources. Several different types of gene sequences have been used to construct evolutionary trees. One very commonly analyzed gene is that encoding 16S rRNA, a small subunit rRNA. This gene has been sequenced from thousands of different species. It is as reliable a molecular measure of phylogenetic relationships among organisms as is now available. Because rRNA is universal in all living organisms, its function was established at an early stage in the evolution of life on this planet, and its sequence has changed fairly slowly. Presumably, most mutations in this gene are deleterious, so that few neutral or beneficial alleles can occur. This limitation causes this gene sequence to change very slowly during evolution. Furthermore, 16S rRNA is a rather large molecule, and so it contains a large amount of sequence information.

Figure 26.10 illustrates a phylogenetic tree of all life on earth based on a sequence analysis of the gene encoding the 16S rRNA. Carl Woese has championed this three-domain phylogenetic tree. The current model links all life on our planet. It proposes three main evolutionary branches: the **eubacteria,** the **archaebacteria,** and the **eukaryotes.** From these types of genetic analyses, it has become apparent that all living organisms are connected through a complex evolutionary tree.

By comparing sequence similarities among many genes in a myriad of species, researchers have found that the rates of sequence change differ dramatically for various genes. Slowly changing genes such as the gene that encodes 16S rRNA are useful for evaluating distant evolutionary relationships and also in assessing evolutionary relationships among bacterial species that evolve fairly rapidly due to their short generation times.

By comparison, other genes have changed more rapidly because of a greater tolerance of neutral mutations. In addition, the mitochondrial genome and DNA sequences within introns tend to have less negative selective pressure, and so their sequences change frequently during evolution. More rapidly changing DNA sequences have been used to elucidate more recent evolutionary relationships, particularly among eukaryotic species that have long generation times and tend to evolve more slowly. In these cases, slowly evolving genes may not be very useful for establishing evolutionary relationships because two closely related species likely have identical or nearly identical DNA sequences for such genes. Instead, it is easier to find sequence differences among closely related species when the DNA sequences are more rapidly changing. For example, figure 26.11 illustrates

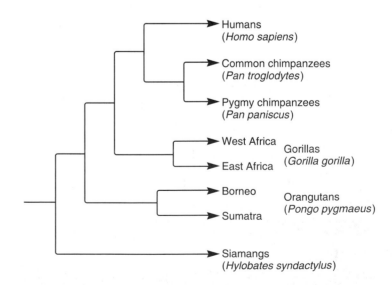

FIGURE 26.11 A phylogenetic tree of closely related species of hominoids, including humans. This tree is based on a comparison of mitochondrial gene sequences of cytochrome oxidase subunit II.

possible evolutionary relationships among hominoid species, including humans. This tree was derived by comparing DNA sequences in a mitochondrial gene that encodes a protein called cytochrome oxidase subunit II.

The construction of phylogenetic trees, like the ones shown in figures 26.10 and 26.11, is based on the amount of sequence similarity among different species. Many methods are used to construct phylogenetic trees. Some of these require computer programs that are beyond the scope of this textbook. However, a relatively simple method to construct a phylogenetic tree is the **UPGMA method** (<u>u</u>nweighted <u>p</u>air <u>g</u>roup <u>m</u>ethod with <u>a</u>rithmetic mean). This method assumes that after two species have diverged from each other, the rate of accumulation of mutations in a homologous region of DNA is approximately the same in the two species. If this assumption is met, closely related sequences (i.e., those that have diverged more recently) have greater sequence similarity compared to distantly related species.

As an example of the UPGMA method, we can consider the nucleotide similarities among five species of primates. In a homologous region containing 10,000 bp, the following numbers of sequence differences were found:

	Human	Chimpanzee	Gorilla	Orangutan	Rhesus Monkey
Human	0	145	151	298	751
Chimpanzee	145	0	157	294	755
Gorilla	151	157	0	304	739
Orangutan	298	294	304	0	710
Rhesus monkey	751	755	739	710	0

To construct a tree, we begin with the pair that is the most closely related. In this case, it would be humans and chimpanzees, because that pair has the fewest number of sequence differences. The number of nucleotide changes per 10,000 nucleotides equals 145. We divide 145 by 2 to yield the average number of changes that occurred in each species since the time they diverged. This number equals 72.5. In other words, about 72.5 changes occurred in humans and about 72.5 (different) changes occurred in chimpanzees since they diverged from a common ancestor, and that is why there are 145 nucleotide differences today. It is common to express this value as the percentage of nucleotide differences. So,

$$\text{Percentage of nucleotide differences} \atop \text{(humans vs. chimpanzees)} = \frac{72.5}{10,000} \times 100$$

$$= 0.725$$

Next, we consider the species that is the most closely related to humans and chimpanzees. That would be gorilla. There are 151 differences between gorillas and humans, and 157 between gorillas and chimpanzees. The average number between gorillas and humans/chimpanzees equals 151 + 157 divided by 2, which equals 154. We divide 154 by 2 to yield the average number that occurred in each species since the time they diverged from a common ancestor. This number equals 77. In other words, about 77 changes occurred in humans or chimpanzees and about 77

(different) changes occurred in gorillas since they diverged from a common ancestor, and that is why there are 154 nucleotide differences today. To express this value as the percentage of nucleotide differences:

$$\text{Percentage of nucleotide substitutions} \atop \text{(gorillas vs. humans/chimpanzees)} = \frac{77}{10,000} \times 100$$

$$= 0.77$$

Next, we consider the species that is the most closely related to humans, chimpanzees, and gorillas. That would be orangutan. There are 298 differences between orangutans and humans, 294 between orangutans and chimpanzees, and 304 between orangutans and gorillas. The average number between orangutans and humans/chimpanzees/gorillas equals 298 + 294 + 304 divided by 3, which equals 299. We divide 299 by 2 to yield the average number of changes that occurred in each species since the time they diverged from a common ancestor. This number approximately equals 149. To express this value as the percentage of nucleotide differences:

$$\text{Percentage of nucleotide substitutions} \atop {\text{(orangutans vs. humans/} \atop \text{chimpanzees/gorillas)}} = \frac{149}{10,000} \times 100$$

$$= 1.49$$

Finally, we consider the most distantly related species, which is the Rhesus monkey. There are 751 differences between Rhesus monkeys and humans, 755 between Rhesus monkeys and chimpanzees, 739 between Rhesus monkeys and gorillas, and 710 between Rhesus monkeys and orangutans. The average number between Rhesus monkeys compared to the humans, chimpanzees, gorillas, or orangutans equals 751 + 755 + 739 + 710 divided by 4, which equals 739. We divide 739 by 2 to yield the average number of changes that occurred in each species since the time they diverged from a common ancestor. This number approximately equals 369. To express this value as the percentage of nucleotide differences:

$$\text{Percentage of nucleotide substitutions} \atop {\text{(Rhesus monkeys vs. humans/} \atop \text{chimpanzees/gorillas/orangutans)}} = \frac{369}{10,000} \times 100$$

$$= 3.69$$

With these data, we can construct the phylogenetic tree shown in figure 26.12. This illustration shows the relative evolutionary relationships among these primate species.

Horizontal Gene Transfer Also Contributes to the Evolution of Species

The types of evolutionary trees depicted previously in figures 26.10 to 26.12 are examples of **vertical evolution.** This term means that species evolve from preexisting species by the accumulation of gene mutations, and by changes in chromosome structure and number. Vertical evolution involves genetic changes

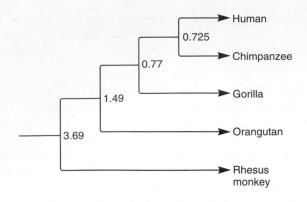

FIGURE 26.12 **Construction of a phylogenetic tree using the UPGMA method.**

in a series of ancestors that form a lineage. In addition to vertical evolution, however, species accumulate genetic changes by another process called **horizontal gene transfer.** This involves the exchange of genetic material among different species.

Figure 26.13 illustrates an example of horizontal gene transfer. In this case, a eukaryotic cell has engulfed a bacterium by endocytosis. During the degradation of the bacterium, a gene happens to escape to the nucleus of the cell, where it recombines

into a chromosome of the host cell. In other words, a gene has been transferred from a bacterial species to a eukaryotic species. By analyzing gene sequences among many different species, researchers have discovered that horizontal gene transfer is a widespread phenomenon. It can occur from prokaryotes to eukaryotes (as in fig. 26.13), from eukaryotes to prokaryotes, between different species of prokaryotes, and between different species of eukaryotes.

Gene transfer among bacterial species is relatively widespread. As discussed in chapter 6, there are three natural mechanisms of gene transfer known as conjugation, transformation, and transduction. When analyzing the genomes of bacterial species, it has become clear that a sizable fraction of their genes are derived from horizontal gene transfer. For example, in *E. coli* and *Salmonella typhimurium,* roughly 17% of their genes have been acquired via horizontal gene transfer during the past 100 million years. The functions of these acquired genes are quite varied, though they commonly involve functions that are readily acted upon by natural selection. Table 26.4 summarizes some categories of bacterial genes that frequently are transferred horizontally. These include genes that confer antibiotic resistance, the ability to degrade toxic compounds, and pathogenicity. Geneticists have suggested that much of the speciation that occurs in prokaryotic species is the

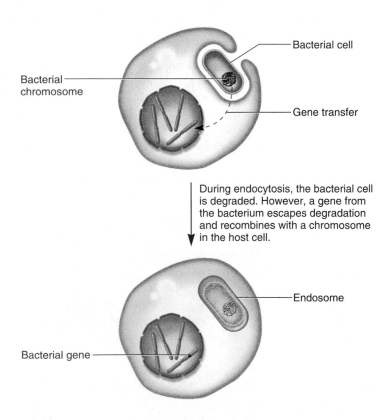

During endocytosis, the bacterial cell is degraded. However, a gene from the bacterium escapes degradation and recombines with a chromosome in the host cell.

FIGURE 26.13 **Horizontal gene transfer from a bacterium to a eukaryote.** In this example, a bacterium is engulfed by a eukaryotic cell, and a bacterial gene is transferred to the nucleus.

TABLE 26.4

Functions of Genes That Are Commonly Transferred Among Bacterial Species

General Function	Examples
Antibiotic resistance	During the past 50 years, the widespread use of antibiotics in medicine has dramatically selected for horizontal gene transfer events. Genes conferring resistance to many antibiotics, such as tetracycline, ampicillin, and streptomycin, are commonly found in many bacterial species.
Biodegradation pathways	To survive in particular environments containing toxic compounds, bacteria must contain enzymes that degrade the toxins. The genes encoding enzymes in biodegradative pathways are frequently transferred horizontally among different species.
Pathogenicity	Certain traits enable bacteria to become pathogenic. For example, the ability to adhere to the surface of intestinal cells enables some bacteria to colonize the intestine and cause an infection. The genes encoding proteins that facilitate colonization are subject to horizontal gene transfer.
Survival in extreme environments	Many species of bacteria exist in environments that require special proteins in order to survive. For example, bacteria that exist in hot springs require proteins that are stable at high temperatures. The genes encoding such proteins are often transferred horizontally among different species that occupy the same extreme environment.

result of horizontal gene transfer. In many cases, it may be the acquisition of new genes that allows a new survival strategy and thus the formation of a new species.

Among eukaryotes, horizontal gene transfer is also significant. In humans, for example, many bacterial genes have been acquired via two routes. First, as discussed in chapter 7, mitochondria were derived from an endosymbiotic relationship between a primordial eukaryotic cell and a bacterium. During the subsequent evolution of eukaryotes, most of the genes in the mitochondrial genome have been transferred to the chromosomes found in the cell nucleus. In addition, the first draft of the human genome sequence indicates that about 100 genes in the human genome were acquired directly from bacterial species (not through the mitochondrial genome). These 100 genes account for approximately 0.3% of the human genome and appear to have been acquired relatively recently on an evolutionary timescale. Therefore, when we view evolution, it is not simply a matter of one species evolving into one or more new species via the accumulation of mutations and chromosomal changes. In addition, the horizontal transfer of genes among different species enables species to acquire new traits that foster the evolutionary process.

EXPERIMENT 26B

Scientists Can Examine the Relationships Between Living and Extinct Flightless Birds by Analyzing Ancient DNA and Then Comparing DNA Sequences

The vast majority of our knowledge concerning molecular evolution has come from the analysis of DNA samples collected from living species. Using this approach, we can infer the prehistoric changes that gave rise to present-day DNA sequences. As an alternative, scientists have discovered that it is occasionally possible to obtain DNA sequence information from species that have lived in the past. In 1984, the first successful attempt at determining DNA sequences from extinct species was accomplished by groups at the University of California at Berkeley and the San Diego Zoo, including Russell Higuchi, Barbara Bowman, Mary Freiberger, Oliver Rynder, and Allan Wilson. They obtained a sample of dried muscle from a museum specimen of the quagga (*Equus quagga*), a zebralike species that became extinct in 1883. This piece of muscle tissue was obtained from an animal that had died 140 years ago. A sample of its skin and muscle had been preserved in salt in the Museum of Natural History at Mainz, Germany. The researchers were able to extract DNA from the sample, clone pieces of it into vectors, and then sequence hybrid vectors containing the quagga DNA. This pioneering study opened the field of **ancient DNA analysis,** also known as **molecular paleontology.**

Since the mid-1980s, many researchers have become excited about the information that might be derived from sequencing DNA obtained from older specimens. Currently there is debate concerning how long DNA can remain significantly intact after an organism has died. Over time, the structure of DNA is degraded by hydrolysis and the loss of purines. Nevertheless, under certain conditions (e.g., cold temperature, low oxygen, and so forth), DNA samples may be stable for as long as 50,000 to 100,000 years.

In most studies involving prehistoric specimens (in particular, those that are much older than the salt-preserved quagga sample), the ancient DNA is extracted from bone, dried muscle, or preserved skin. These samples are often obtained from museum specimens that have been gathered by archaeologists. However, it is unlikely that enough DNA will be extracted to enable a researcher to directly clone the DNA into a vector. Since 1985, however, the advent of PCR technology, which is described in chapter 18, has made it possible to amplify the very small amounts of DNA using PCR primers that flank a region within the 12S rRNA gene, a slowly changing gene. In recent years, this approach has been used to elucidate the phylogenetic relationships between modern and extinct species.

In the experiment described in figure 26.14, Alan Cooper, Cecile Mourer-Chauvire, Geoffrey Chambers, Arndt von Haeseler, Allan Wilson, and Svante Paabo investigated the evolutionary relationships between some extinct and modern species of flightless birds. Two groups of flightless birds, the kiwis and the moas, existed in New Zealand during the Pleistocene era. The moas are now extinct, although 11 species were formerly present. In this study, the researchers investigated the phylogenetic relationships among four extinct species of moas that were available as museum samples, kiwis of New Zealand, and several other (nonextinct) species of flightless birds. These included the emu and the cassowary (found in Australia and New Guinea), the ostrich (found in Africa and formerly Asia), and two rheas (found in South America).

The samples from the various species were subjected to PCR to amplify the 12S rRNA gene. This provided enough DNA to subject the gene to DNA sequencing. The sequences of the genes were aligned using computer programs described in chapter 21.

■ THE HYPOTHESIS

Because DNA is a relatively stable molecule, it can be PCR-amplified from a preserved sample of a deceased organism and subjected to DNA sequencing. A comparison of DNA sequences with modern species may help elucidate the phylogenetic relationships between extinct and modern species.

TESTING THE HYPOTHESIS — FIGURE 26.14 DNA analysis of phylogenetic relationships among flightless birds, including extinct moas.

Starting material: Tissue samples from four extinct species of moas were obtained from museum specimens. Tissue samples were also obtained from three species of kiwis, one ostrich, one cassowary, one emu, and two species of rhea.

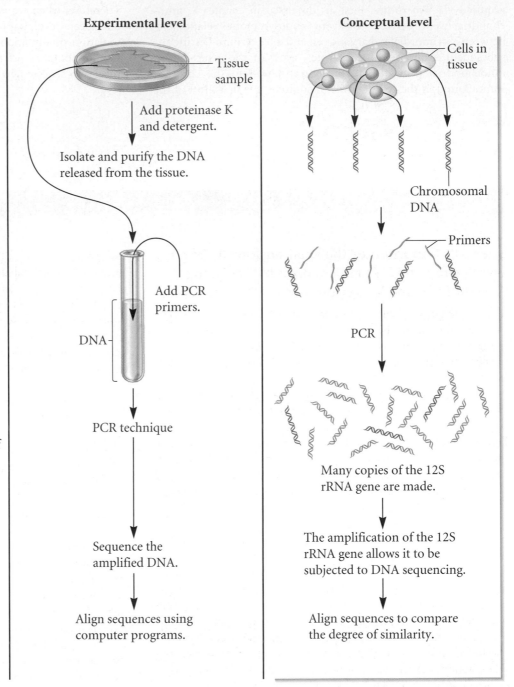

Experimental level   **Conceptual level**

1. For soft tissue samples, treat with proteinase K (which digests protein) and a detergent that dissolves cell membranes. This releases the DNA from the cells.

Tissue sample

Add proteinase K and detergent.

Isolate and purify the DNA released from the tissue.

Cells in tissue

Chromosomal DNA

2. Individually, mix the DNA samples with a pair of PCR primers that are complementary to the 12S rRNA gene.

Add PCR primers.

DNA

Primers

PCR

3. Subject the samples to PCR. See chapter 18 (fig. 18.6) for a description of PCR.

PCR technique

Many copies of the 12S rRNA gene are made.

4. Subject the amplified DNA fragments to DNA sequencing. See chapter 18 for a description of DNA sequencing.

Sequence the amplified DNA.

The amplification of the 12S rRNA gene allows it to be subjected to DNA sequencing.

5. Align the DNA sequences to each other. Methods of DNA sequence alignment are described in chapter 21.

Align sequences using computer programs.

Align sequences to compare the degree of similarity.

THE DATA

```
MOA 1      GCTTAGCCCTAAATCCAGATACTTACCCTACACAAGTATCCGCCCGAGAACTACGAGCACAAACGCTTAAAACTCTAAGGACTTGGCGGTGCCCCAAACCCACCTAGAGGAGCCTGTTCTATAATCGATAATCCACGATA
MOA 2      · · · · · · · · · · · · · · · · · · · · · · · · · · · · · · · · · · · · · · · · · · · · · · · · · · · · · · · · · · · · · · · · · · · · · · · · · · · · · · · · · · · · · · · · · · · · · · · · · · · · · · · · · · · · · · · · · · · ·
MOA 3      · · · G · · · · · · · · · T · · · · · · · · · · · · · · · · · · · · · · · · · · · T · · · · · · · · · · · · · · · · · · · · · · · · · · · · · · · · · · · · · · · · · · · · · · · · · · · · · · · · · · · · · · · · · · · · · · · · · · · ·
MOA 4      · · · · · · · · · · · · · · · · · · C · · · · · · · · · · · · · · · · · · · · · · · · · · · · · · · · · · · · · · · · · · · · · · · · · · · · · · · · · · · · · · · · · · · · · · · · · · · · · · · · · · · · · · · · · C · · · · · T · ·
KIWI 1     · · · · · · T · G · · · · GT · · CT · · · C · · · · · · · · · · · · · · · · · · · · · · · · · · · · · · T · · · · · · · · · · · · · · · · · · · · · · · · · · · · · · · · · · · C · · · · · ·
KIWI 2     · · · · · · T · G · G · · · · AT · · CT · · · C · · · · · · · · · · · · · · · · · · · · · · · · T · · · · · · · · · · · · · · · · · · · · · · · · · · · · · · · · · C · · · · · ·
KIWI 3     · · · · · · T · G · G · · G · AT · · · C · · · C · · · · · · · · · · · · · · · · · · · · · · · · T · · · · · · · · · · · · · · · · · · · · · · · · · · · · C · · · · · ·
EMU        · · · · · TT · · · · · C · · T · · CAG · C · · · · · · · T · · · · · · · · · · · · · · · · · · · · · · · · T · · · · · · · · · · · · · · · · · · · · · · · C · · · · · ·
CASSOWARY  · · · · · TT · · · · CG · TA · · CTG · · · · · · · · · · · · · · · · · · · · · · · · · · · · · · · · · · · · · · · · · · · · · · · · · · · · · · · · · · · C · · · · · ·
OSTRICH    · · · · · T · · AT · · · · · C · CT · · · · · · · · · · · · · · · · · · · · · · · · · · · · · · · · T · · · · · · · · · · · · · · · · · · · · · · · · · · · · · · · · T
RHEA 1     · · · · · · · · · · · · T · · · · · · · CT · · · · · · · · · · · · · · · · · · · · · · · · T · · · · · · · · · · · · · · · · · · · · · · · · · · · · · · · · · · · · · · ·
RHEA 2     · · · · · · · · · C · · · · · · · C · · C · · · · · · · · · · · · · · · · · · · · · · · · · · · · · · · · · · · · · · · · · · · · · · · · · · · · · · C · · · · · ·

MOA 1      CACCCGACCATCCCTCGCCCGT-GCAGCCTACATACCGCCGTCCCCAGCCCGCCT--AATGAAAG-AACAATAGCGAGCACAACAGCCCTCCCCCGCTAACAAGACAGGTCAAGGTATAGCATATGAGATGGAAGAAATG
MOA 2      · · · · · · · · · · · · · · · · · · A · · · · · · · · · · · · · · · · · · · · · · — · · · · · · · · · · · · · · · · · TCA-- · · · · · · · · · · · · · · · · · · · · · · · · · · · · · · · · · · · · · · · · · · · · · · · · · ·
MOA 3      · · · · · · T · T · · A · · · · · · · · · · · · · · · · · · — · · · · · · · · · · · · · · · · · TA--- · T · · · · · · · · · · · · · · · · · · · · · · · · · · · · · · · · · · · · · · · ·
MOA 4      · · · · · · T · T · · A · - · · · · · · · · · · · · · · · · —— · · · · · · · · · · · T · · · AC-- · · · · · · · · · · · · · · · · · · · · · · · · · · · · · · · · · · · · · · · ·
KIWI 1     · · · · A · · · T · T · · AAC-A · · · T · · · · · · · · G · · T · · · AA · · · G · · · — · · · C · · · A · · · · · TA · - · A · · · · · · · · · · · · · · · · C · · · · · · · · · ·
KIWI 2     · · · · A · · · T · T · · AAC-A · · · T · · · · · · · · G · · T · · · AA · · · G · · · — · · · C · · · A · · · · · TA · - · A · · · · · · · · · · · · · · · · C · · · · · · · · · ·
KIWI 3     · · · · A · · · T · T · · AAC-A · · · · · · · · · · · G · · · · · · AA · · · · · GC · · · · · · TACA- · A · · · · · · · · · · · · · · · · CC · C · · · · · G · · ·
EMU        · · · · AG · · T · T · · AA- · A · · · · · · · · · · · G · · · · · · · — · · · · · · · · · · · T · · · AC--TT · · · · · · · · · · · · · · · · · · G · · · · · · · · ·
CASSOWARY  · · · · A · · · T · T · · AA · TA · · · · · · · · · G · · · · · · · · · · —— · G · · · G · · · · · · · · · T · · · AC--T · · · · · · · · · · · · · · · · · G · · · · · · · · ·
OSTRICH    · · · · A · · C · · T · · A--T · · · · · · · · · · G · · · · · · C--- · G · · · G · · · · · · · · · · · · T · · · A--- · · · · · · · · · · · · · · · · · GAG · · · · · ·
RHEA 1     · · · · A · · · T · · T · · A · - · · · · · · · · · · · · · · · · · · TA · G · · · · · · C · · AG · T · T · · TA--- · · · · · · · · · · · · · · · · · · G · · · · · · · · ·
RHEA 2     · · · · · T · T · · A · - · · · · · · · · · · · · · · · TA · · · G · · · · · A · · · T · T · · TA--- · · · G · · · · · · · · · · · · · · · · · · · · · · · · · · · · · · · ·

MOA 1      GGCTACATTTTCTAACATAGAACACCC------------ACGAAAGAGAAGGTGAAACCCTCCTCAAAAGGCGGATTTAGCAGTAAAATAGAACAAGAATGCCTATTTTAAGCCCGGCCCTGGGGC
MOA 2      · · · · · · · · · · · · · · · · · · · · · · · · · · · · · · · · · · · · A · · · I · · · G · · · · · · · · · · · · · · · · · · · · · · · · · · · · · · · · · · · · T · · · · · · · · ·
MOA 3      · · · · · · · · · · · · · · T------------ · · · · · · · · · · · G · · · · · · · · · · · · · · · G · · · · · C · · · C · · · T · · · · · · · · · · ·
MOA 4      · · · · · · · · · · · · · · · · · · · · · ------------ · · · · · · · · · · · · · G · · · · · · · · · · · · C · · · C · · · · · T · · · · · · · · · · ·
KIWI 1     · · · · · · · A · · · · · T · T------------ · · · · A · GGT · · · T · C · · T · G · · · · · · C · · · T · · GA · T · · · · --T · · · A · · · ·
KIWI 2     · · · · · · · A · · · · · T · T------------ · · · · A · GGT · · · T · C · · T · G · · · · · · C · · · T · · GA · T · · · · --T · · · A · · · ·
KIWI 3     · · · · · · · A · · · · · T · T------------ · · · · A · GGTA · · · T · C · · T · G · · · A · · · C · · · T · · · A · T · · · · --- · · · A · · · ·
EMU        · · · · · · · · · T · T · ------------ · · · · · AG · T · · · · T · AC · T · · G · · · · · · C · · · T · · GA · T · · · · A--T · · T · A · · ·
CASSOWARY  · · · · · · · · · T · · ------------ · · · · A · G · T · · · T · A · · T · G · · · · · · · C · · · T · · GA · T · · · · A--- · · · · A · · · ·
OSTRICH    · · · · · · · · · T · A------------ · · · · · G · TA · · T · A · · · G · · · · · · · T · · · GA · T · · · · --T · · T · A · · ·
RHEA 1     · · · · · · · · TC · · · · A · ------------ · G · · · GGCA · · · -AC · · · CG · · · · · G · · G · TC · · A · · · C · · · --- · · · · A · · · ·
RHEA 2     · · · · · · · · GTC · · · · G · ------------ · · · · GGCA · · · -AC · · · CG · · · · · G · G · G · TC · · A · · C · C · · · --- · · · · A · · · ·
```

![] INTERPRETING THE DATA

The data of figure 26.14 illustrate a multiple sequence alignment of the amplified DNA sequences. The first line shows the DNA sequence of one extinct moa species. Underneath it are the sequences of the other species. When the other sequences are identical to the first sequence, a dot is placed in the corresponding position. When the sequences are different, the nucleotide base (A, T, G, or C) is placed there. In a few regions, the genes are different lengths. In these cases, a dash is placed at the corresponding position.

As you can see from the large number of dots, the sequences among these flightless birds are very similar. To establish evolutionary relationships, we need to focus on the few differences that occur. Some surprising results were obtained. The sequences from the kiwis (a New Zealand species) are actually more similar to the sequence from the ostrich (an African species) than they are to those of the moas, which were once found in New Zealand. Likewise, the kiwis are more similar to the emu and cassowary (found in Australia and New Guinea) than they are to the moas. Contrary to their original expectations, the authors concluded that the kiwis are more closely related to Australian and African flightless birds than they are to the moas. They proposed that New Zealand was colonized twice by ances-

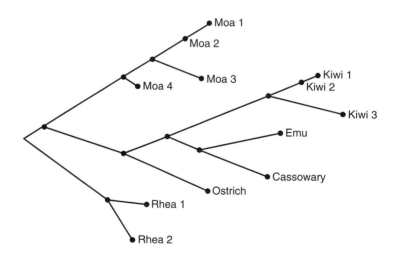

FIGURE 26.15 A revised phylogenetic tree of moas, kiwis, emus, cassowaries, ostriches, and rheas.

tors of flightless birds. As shown in figure 26.15, the researchers constructed a new evolutionary tree that illustrates the relationships among these modern and extinct species.

FIGURE 26.16   **Extinct organisms from which DNA sequences have been obtained.** Right panel, from *bottom right* to *top right:* quagga, marsupial wolf, sabre-toothed cat, moa, mammoth, cave bear, blue antelope, giant ground sloth, and Aurochs. Left panel, from *bottom left* to *top right:* mastodon, New Zealand coot, South Island piopio, Steller's sea cow, Neanderthal man, Aptornis defossor, Shasta ground sloth, pig-footed bandicoot, moa-nalo, and Myotragus balearicus. (From Hofreiter et al. (2001) *Nature Reviews Genetics,* volume 2, p. 357.)

Since these early studies, sequences of ancient DNA have been derived from a variety of species. Figure 26.16 shows some extinct organisms from which DNA sequences have been determined. Many of these samples were tens of thousands of years old. For example, the sample of a Neanderthal man was approximately 30,000 years old. The oldest samples are likely to be in the range of 50,000 to 100,000 years old.

A self-help quiz involving this experiment can be found at the Online Learning Center.

Genetic Variation at the Molecular Level Is Associated with Neutral Changes in Gene Sequences

In chapter 25, we learned that genetic variation is prevalent among natural populations. During the last few decades, a great debate has occurred among population and evolutionary geneticists. The debate centers on the reason for genetic variation in natural populations. Is it due primarily to mutations that are favored by natural selection or to random genetic events?

A **nonneutral mutation** is one that affects the phenotype of the organism and can be acted on by natural selection. A nonneutral mutation may only subtly alter the phenotype of an organism, or it may have a major impact. According to Darwin, natural selection is the agent that leads to evolutionary change in populations. It selects for the survival of the fittest and thereby promotes the establishment of beneficial alleles and the elimination of deleterious ones. Therefore, many geneticists have assumed that natural selection is the dominant force in changing the genetic composition of natural populations, thereby leading to variation.

Motoo Kimura proposed the **neutral theory of evolution.** According to this idea, most genetic variation observed in natural populations is due to the accumulation of neutral mutations. **Neutral mutations** do not affect the phenotype of the organism; neutral alleles are not acted on by natural selection. For example, a mutation within a structural gene that changes a glycine codon from GGG to GGC would not affect the amino acid sequence of the encoded protein. Because neutral mutations do not affect phenotype, they spread throughout a population according to their frequency of appearance and to genetic drift. This theory has been called the "survival of the luckiest" and also **non-Darwinian evolution** to contrast it with Darwin's "survival of the fittest." Kimura agreed with Darwin that natural selection is responsible for adaptive changes in a species during evolution. His main argument is that most modern variation in gene sequences is explained by neutral variation rather than adaptive variation. Kimura, along with his colleague Tomoko Ohta, suggested five principles that govern the evolution of genes at the molecular level:

1. For each protein, the rate of evolution, in terms of amino acid substitutions, is approximately constant with regard to neutral substitutions that do not affect protein structure or function.

   Evidence: As an example, the amount of genetic variation between the coding sequence of the human α-globin and β-globin genes is approximately the same as the difference between the α-globin and β-globin genes in the horse (shown previously in fig. 26.8). This type of comparison holds true among many different genes compared among many different species.

2. Proteins that are functionally less important for the survival of an organism, or parts of a protein that are less important for its function, tend to evolve faster than more important proteins or regions of a protein. In other words, during evolution, less important proteins will accumulate amino acid substitutions more rapidly than important proteins.

   Evidence: Certain proteins are critical for survival, and their structure is exquisitely tuned to their function.

Examples are the histone proteins necessary for nucleosome formation in eukaryotes. Histone genes tolerate very few mutations and have evolved extremely slowly. By comparison, fibrinopeptides, which bind to fibrinogen to form a blood clot, evolve very rapidly. Presumably, the sequence of amino acids in this polypeptide is not very important for allowing it to aggregate and form a clot. Another example concerns the amino acid sequences of enzymes. It is known that amino acid substitutions are very rare within the active site (which is critical for function) but are more frequent in other parts of the protein.

3. Amino acid substitutions that do not disrupt the existing structure and function of a protein (conservative substitutions) occur more frequently in evolution than disruptive amino acid changes.

   Evidence: When examining the rate of change of the coding sequence within structural genes, nucleotide substitutions are more likely to occur in the wobble base than in the first or second base within a codon. Mutations in the wobble base are often silent (i.e., do not change the amino acid sequence of the protein). In addition, conservative substitutions (i.e., a substitution with a similar amino acid, such as a nonpolar amino acid for another nonpolar amino acid) are fairly common. By comparison, nonconservative substitutions are less frequent, although they do occur. Nonsense and frameshift mutations are very rare within the coding sequences of genes. Also, intron sequences evolve more rapidly than exon sequences.

4. Gene duplication must always precede the emergence of a gene having a new function.

   Evidence: When a single copy of a gene exists in a species, it usually plays a functional role similar to that of the homologous gene found in another species. Gene duplications have created gene families in which each family member can evolve somewhat different functional roles. Examples include the globin family described in chapter 8 and the *Hox* genes described in chapter 23.

5. Selective elimination of definitely deleterious mutations and random fixation of selectively neutral or very slightly deleterious alleles occur far more frequently in evolution than Darwinian selection of advantageous mutants.

   Evidence: As mentioned in principle 3, silent and conservative mutations are much more common than nonconservative substitutions. Presumably these nonconservative mutations usually have a negative effect on the phenotype of the organism, so that they are effectively eliminated from the population by natural selection. On rare occasions, however, an amino acid substitution due to a mutation may have a beneficial effect on the phenotype. For example, a nonconservative mutation in the β-globin gene produces Hb^S, which gives an individual resistance to malaria in the heterozygous condition.

In general, the DNA sequencing of hundreds of thousands of different genes from thousands of species has provided compelling support for these five principles of gene evolution at the molecular level. However, the argument is by no means resolved. Some geneticists, called **selectionists,** oppose the neutralist theory. These geneticists often can offer persuasive theoretical arguments in favor of natural selection as the primary factor promoting genetic variation. In any case, the argument is largely a quantitative rather than a qualitative one. Each school of thought accepts that genetic drift and natural selection both play key roles in evolution. The neutralists argue that most genetic variation arises from neutral genetic mutations and genetic drift, whereas the selectionists argue that beneficial mutations and natural selection are primarily responsible.

Speciation Is Associated with Changes in Chromosome Structure and Number

In this section, we have emphasized mutations that alter the DNA sequences within genes. In addition to gene mutations, however, other types of changes, such as gene duplications, inversions, translocations, and changes in chromosome number, are important features of evolution.

As discussed earlier in chapter 26, changes in chromosome structure and/or number may not always be adaptive, but they can lead to reproductive isolation and the origin of new species. As an example of variation in chromosome structure among closely related species, figure 26.17 compares the banding pattern of the three largest chromosomes in humans and the corresponding chromosomes in chimpanzees, gorillas, and orangutans. The banding patterns are strikingly similar, because these species are closely related evolutionarily. However, there are some interesting differences. Humans have one large chromosome 2, but this chromosome is divided into two separate chromosomes in the other three species. This explains why humans have 23 types of chromosomes while the apes have 24. This may have occurred by a fusion of the two smaller chromosomes during the development of the human lineage. Another interesting change in chromosome structure is seen in chromosome 3. The banding patterns among humans, chimpanzees, and gorillas are very similar, but the orangutan has a large inversion that flips the arrangement of bands in the centromeric region.

With the advent of molecular techniques, researchers can analyze the chromosomes of two or more different species and identify regions that contain the same groups of linked genes, which are called **synteny groups.** Within a particular synteny group, the same types of genes are found in the same order. In 1995, Graham Moore and colleagues analyzed the locations of molecular markers along the chromosomes of several cereal grasses including rice (*Orzya sativa*), wheat (*Tricticum aestivum*), maize (*Zea mays*), foxtail millet (*Setaria itallica*), sugarcane (*Saccharum officinarum*), and sorghum (*Sorghum vulgare*). From this analysis, they were able to identify several large synteny groups that are common to most of these species (fig. 26.18). As an example, let's compare rice (12 chromosomes per set) to wheat (7

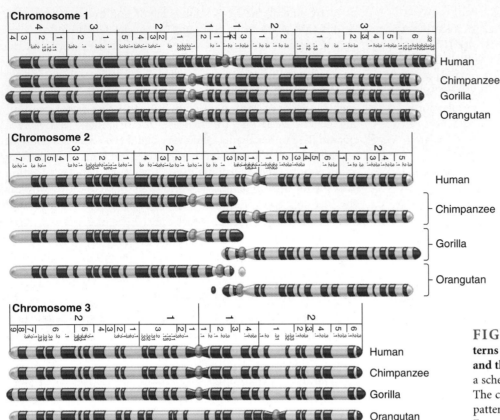

FIGURE 26.17   A comparison of banding patterns among the three largest human chromosomes and the corresponding chromosomes in apes. This is a schematic drawing of late prophase chromosomes. The conventional numbering system of the banding patterns is shown next to the human chromosomes. From *top* to *bottom*, the chromosomes are from humans, chimpanzees, gorillas, and orangutans.

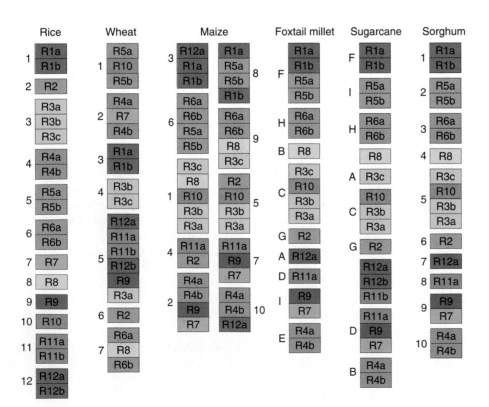

FIGURE 26.18   Synteny groups in cereal grasses. The synteny groups are named according to their relative arrangement in rice, which is shown on the *left*. In some cases, a synteny group or portion of a group may have incurred an inversion, but these are not shown in this figure. Also, the genome of maize contains two copies of most synteny groups, suggesting that it is derived from an ancestral polyploid species.

chromosomes per set). In rice, chromosome 6 contains two synteny groups designated R6a and R6b, while chromosome 8 contains a single synteny group, R8. In wheat, chromosome 7 consists of R6a and R6b at either end, while R8 is sandwiched in the middle. One possible explanation for this difference could be a chromosomal rearrangement during the evolution of wheat in which R8 became inserted into the middle of a chromosome containing R6a and R6b. Overall, the evolution of grass species has maintained most of the same types of genes among evolutionarily related species, but many chromosomal rearrangements have occurred. These types of rearrangements would promote reproductive isolation.

CONCEPTUAL SUMMARY

Biological evolution involves heritable changes in one or more characteristics in a population or species over the course of many generations. In the first part of chapter 26, we were concerned primarily with how evolution results in the formation of new **species.** To become a separate species, a population must be reproductively isolated from all other species. This enables their gene pool to evolve as a single unit. There are several ways that reproductive isolation can occur. The **biological species concept** emphasizes reproductive isolation as an important event that leads to speciation, whereas the **species recognition concept** emphasizes **sexual selection** as a positive force during speciation.

Speciation is usually a branching process (**cladogenesis**), although **anagenesis** (transformation of a single species) occasionally occurs. Depending on geographic barriers, divergent speciation may be **allopatric, parapatric,** or **sympatric.** Allopatric speciation, which involves geographic barriers, is thought to be the most widespread form of speciation. It can occur slowly, due to the gradual formation of a geographic barrier, or rapidly, due to the founder effect. Parapatric speciation is similar to allopatric, except that the geographic barriers are less complete, with ever-decreasing levels of interbreeding taking place in **hybrid zones.** Sympatric speciation does not require geographic barriers. Instead, an abrupt genetic event may lead to reproductive isolation. In plants, the formation of polyploids is a common form of sympatric speciation. Polyploids are reproductively isolated because they produce sterile hybrids when crossed to nonpolyploid species. In general, the fossil record suggests that speciation is often a rapid process that punctuates long periods during which the species is in equilibrium with its environment. This punctuated equilibrium hypothesis is contrasted with **gradualism,** which proposes a slower but steady rate of evolution due to the accumulation of many small genetic changes.

Molecular evolution is the study of the molecular changes in the genetic material during evolution. There are many ways to investigate the genetic material, but the analysis of DNA sequences and (the deduced) amino acid sequences are the most commonly used methods and the most informative. **Homologous genes** are derived from a common ancestral gene. **Interspecies homology** can be used to reconstruct **phylogenetic trees.** When two species are closely related evolutionarily, they tend to have gene sequences that are more similar to each other.

Kimura and Ohta proposed five principles of molecular evolution that are consistent with our present knowledge of gene sequences. The **neutral theory of evolution** argues that most variation in gene sequences is neutral. In the case of structural genes, **neutral mutations** are not expected to significantly alter the structure and function of the encoded protein; many **nonneutral mutations** are highly deleterious and are thus eliminated quickly from the gene pool; other nonneutral mutations are adaptive and are acted on by natural selection to change the characteristics of a species. According to the neutral theory of evolution, though, these adaptive changes represent a small proportion of the total number of genetic changes that occur during evolution. In addition to changes at the gene level, it is also common for changes in chromosome structure and number to occur. These events are often instrumental in leading to reproductive isolation.

EXPERIMENTAL SUMMARY

In a broad sense, evolutionary biologists would like to know how genetic changes have led to the phenotypic characteristics of present-day species. At the heart of this question is speciation: How do we explain the existence of so many different species? To answer this question, biologists have tried to understand how two different, but closely related, species have become reproductively isolated. By observing species in nature, they have identified several prezygotic and postzygotic mechanisms that prevent the production of viable, interspecies offspring. Researchers have also correlated reproductive isolation with experimentally observable genetic changes. For example, parapatric speciation may be associated with the accumulation of chromosomal inversions that prevent the production of fertile offspring within hybrid zones of contact. Sympatric speciation can occur via the formation of allotetraploids. As discussed in the experiment of figure 26.6, this speciation mechanism can be confirmed by producing "artificial" polyploids.

Evolutionary biologists are also interested in the rates of evolutionary change. Experimentally, the fossil record suggests that it is more common for evolution to follow a pattern of punctuated equilibrium, in which a species is well adapted to its environment for a long period of time (the equilibrium), which is punctuated by short periods of rapid evolutionary change. These short periods involve abrupt genetic changes (e.g., allopolyploidism or new alleles that have a dramatic effect on phenotype), or there may be a short period of strong selective pressure (e.g., the founder effect, a new predator in the region, or a significant environmental change).

At the molecular level, evolution is due to alterations in DNA sequences and changes in chromosome structure and number. Geneticists have found a great amount of variation within most species. As discussed in table 26.3, there are several experimental methods for detecting genetic variation at the DNA and protein levels. The earliest methods involved the study of allozymes; DNA sequencing is now the most common way to study molecular evolution. By comparing homologous gene sequences within a species and among different species, evolutionary biologists have been able to probe the relationship between organismal evolution and changes in DNA sequences. They have also con-

structed evolutionary trees that describe the phylogenetic relationships among many species. More recently, it has even been possible to analyze DNA sequences from some extinct species.

With all of this sequence information available, researchers have debated the origin of genetic variation in modern species. The data suggest that most genetic variation occurs through neutral changes that accumulate due to genetic drift. Assuming that the rate of new mutation is essentially constant, the accumulation of neutral changes provides a molecular clock to measure the timescale of evolution. However, not all variation is neutral. Adaptive changes, which are acted on by natural selection, must also occur and alter the phenotypic characteristics of species over time, thereby leading to the evolution of new, better-adapted species.

PROBLEM SETS & INSIGHTS

Solved Problems

S1. A codon for leucine is UUA. A mutation causing a single-base substitution in a gene can change this codon in the transcribed mRNA into GUA (valine), AUA (isoleucine), CUA (leucine), UGA (stop), UAA (stop), UCA (serine), UUG (leucine), UUC (phenylalanine), or UUU (phenylalanine). According to the neutral theory, which of these mutations would you expect to see within the genetic variation of a natural population? Explain.

Answer: The neutral theory proposes that neutral mutations will accumulate to the greatest extent in a population. Leucine is a nonpolar amino acid. For a UUA codon, single-base changes of CUA and UUG are silent, and so they would be the most likely to occur in a natural population. Likewise, conservative substitutions to other nonpolar amino acids such as isoleucine (AUA), valine (GUA), and phenylalanine (UUC and UUU) may not affect protein structure and function, and so they may also occur and not be eliminated rapidly by natural selection. The polar amino acid serine (UCA) is a nonconservative substitution; one would predict that it is more likely to disrupt protein function. Therefore, it may be less likely to be found. Finally, the stop codons, UGA and UAA, would be expected to diminish or eliminate protein function, particularly if they occur early in the coding sequence. These types of mutations are selected against and, therefore, are not usually found in natural populations.

S2. Explain why homologous genes have sequences that are similar but not identical.

Answer: Homologous genes are derived from the same ancestral gene. Therefore, as a starting point, they had identical sequences. Over time, however, each gene accumulates random mutations that the other homologous genes did not acquire. These random mutations change the gene from its original sequence. Therefore, much of the sequence between homologous genes remains identical, but some of the sequence will be altered due to the accumulation of independent random mutations.

S3. Explain why plants are more likely to evolve by sympatric speciation compared to animals.

Answer: The most common way for sympatric speciation to occur is by the formation of polyploids. For example, if one species is diploid ($2n$),

nondisjunction could produce an individual that is tetraploid ($4n$). If the tetraploid individual were a plant, and if the plant were monoecious (i.e., produces both pollen and egg cells), the plant could multiply to produce many tetraploid offspring. These offspring would be reproductively isolated from the diploid plants that are found in the same geographic area. This isolation occurs because the offspring of a cross between a diploid and tetraploid plant would be infertile. For example, if the pollen from a diploid plant fertilized the egg from a tetraploid plant, this would produce triploid offspring. The triploid offspring might be viable, but it is very likely that triploids would be sterile because they would produce highly aneuploid gametes. The gametes would be aneuploid because there are an odd number of homologous chromosomes. In this case there would be three copies of each homologous chromosome, and these could not be equally distributed into gametes. Therefore, when a tetraploid is produced, it can immediately create its own unique species that is reproductively isolated from other species.

In contrast, polyploidy rarely occurs in animals. Perhaps the main reason is because animals usually cannot tolerate polyploidy. Tetraploid animals typically die during early stages of development. In addition, most species of animals have male and female sexes. To develop a tetraploid species from a diploid species, nondisjunction would have to produce both a male and female offspring that could reproduce with each other. The chance of producing one tetraploid offspring is relatively rare. The chances of producing two tetraploid offspring that happen to be male and female, and happen to mate with each other to produce many offspring, would be extremely rare.

S4. As described in figure 26.17, evolution is associated with changes in chromosome structure and number. As seen here, chromosome 2 in humans is divided into two distinct chromosomes in chimpanzees, gorillas, and orangutans. In addition, chromosome 3 in the orangutan has a large inversion that is not found in the other three primates. Discuss the potential role of these types of changes in the evolution of these primate species.

Answer: As discussed in chapter 8, changes in chromosome structure, such as inversions and balanced translocations, may not have any phenotypic effects. Likewise, the division of a single chromosome into two distinct chromosomes may not have any phenotypic effect as long as the total amount of genetic material remains the same. Overall, the types of changes in chromosome structure and number shown in figure 26.17 may not have caused any changes in the phenotypes of primates. However, the changes would be expected to promote reproductive isolation. For example, if a gorilla mated with an orangutan, the offspring would be an inversion heterozygote for chromosome 3. As shown in figure 8.12, an inversion heterozygote may produce chromosomes that have too much or too little genetic material. This is particularly likely if the inversion is fairly large (such as the one shown in figure 26.17). The inheritance of too much or too little genetic material is likely to be detrimental or even lethal. For this reason, the hybrid offspring of a gorilla and orangutan would probably not be fertile. (Note: In reality, there are several other reasons why interspecies matings between gorillas and orangutans do not produce viable offspring.)

Overall, the primary effect of changes in chromosome structure and number, like the ones shown in figure 26.17, is to promote reproductive isolation. Once two populations become reproductively isolated, they will accumulate different mutations, and over the course of many generations, this will lead to two different species that have distinct characteristics.

It should be noted that changes in chromosome number in plants are more likely to have effects that abruptly lead to the formation of new species. This idea is discussed in solved problem S3.

Conceptual Questions

C1. Discuss the two principles on which evolution is based.

C2. Evolution, which is the genetic change in a population of organisms over time, is often described as the unifying theme in biology. Discuss how evolution is unifying at the molecular and cellular levels.

C3. What is meant by the term *reproductive isolation?* Give several examples. Compare and contrast the biological species concept and the species recognition concept with regard to reproductive isolation.

C4. What is sexual selection? For most animals, which sex is acted upon by natural selection, the male or female? Give examples. Describe and give examples of how sexual selection can alter the traits of animals.

C5. Would the following examples of reproductive isolation be considered a prezygotic or postzygotic mechanism?

A. Horses and donkeys can interbreed to produce mules, but the mules are infertile.

B. Three species of the orchid genus *Dendrobium* produce flowers 8 days, 9 days, and 11 days after a rainstorm. The flowers remain open for one day.

C. Two species of fish release sperm and egg into seawater at the same time, but the sperm of one species do not fertilize the eggs of the other species.

D. *Hyla chrysocelis* (diploid) and *H. versicolor* (tetraploid) can produce viable offspring, but the offspring are sterile.

C6. Distinguish between anagenesis and cladogenesis. Which type of speciation is more prevalent? Why?

C7. Describe three or more genetic mechanisms that may lead to the rapid evolution of a new species. Which of these genetic mechanisms are influenced by natural selection, and which are not?

C8. Explain the type of speciation (allopatric, sympatric, or parapatric) that is most likely to occur under each of the following conditions:

A. A pregnant female rat is transported by an ocean liner to a new continent.

B. A meadow containing several species of grasses is exposed to a pesticide that promotes nondisjunction.

C. In a very large lake containing several species of fish, the water level gradually falls over the course of several years. Eventually, the large lake becomes subdivided into smaller lakes, some of which are connected by narrow streams.

C9. Alloploids are created by crosses involving two different species. Explain why alloploids are reproductively isolated from the two original species from which they were derived. Explain why alloploids are usually sterile, whereas allotetraploids (containing a diploid set from each species) are commonly fertile.

C10. Discuss the evidence in favor of the punctuated equilibrium theory of evolution. What mechanisms could account for this pattern of evolution? In contrast, what type of genetic changes are consistent with gradualism?

C11. Discuss whether the phenomenon of reproductive isolation applies to bacteria, which reproduce asexually. How would a geneticist divide bacteria into separate species?

C12. Which of the following morphological traits would seem to be the result of sexual selection?

A. Male red-winged blackbirds have red bands on their wings.

B. Male insects usually have clasping organs that are used during reproduction.

C. Female mallards are usually brown and are better camouflaged compared to males.

D. Male mallards have green heads.

C13. Discuss the major differences between allopatric, sympatric, and parapatric speciation.

C14. Discuss the major goals in the field of molecular evolution.

C15. The following are two DNA sequences from homologous genes:

```
TTGCATAGGCATACCGTATGATATCGAAAACTAGAAAAATAGGGCG
ATAGCTA

GTATGTTATCGAAAAGTAGCAAAATAGGGCGATAGCTACCCAGACT
ACCGGAT
```

The two sequences, however, do not begin and end at the same location. Try to line them up according to their homologous regions.

C16. Why do some genes evolve at a fast rate and others at a slow rate? Explain how differences in the rate of molecular evolution can be useful with regard to establishing distant evolutionary relationships versus close evolutionary relationships.

C17. Which would you expect to exhibit a faster rate of evolutionary change, the nucleotide sequence of a gene or the amino acid sequence of the encoded polypeptide of the same gene? Explain your answer.

C18. When comparing the coding region of structural genes among closely related species, it is commonly found that certain regions of the gene have evolved more rapidly (i.e., have tolerated more changes in sequence) compared to other regions of the gene. Explain why different regions of a structural gene evolve at different rates.

C19. Plant seeds contain storage proteins that are encoded by plant genes. When the seed germinates, these proteins are rapidly hydrolyzed (i.e., the covalent bonds between amino acids within the polypeptides are broken), which releases amino acids for the developing seedling. Would you expect the genes that encode plant storage proteins to evolve slowly or rapidly, compared to genes that encode enzymes? Explain your answer.

C20. Figure 26.10 shows a phylogenetic tree of all life on earth based on 16S rRNA data. Based on your understanding of molecular genetics (in chapter 26 and other chapters), describe three or more observations that suggest that all life-forms on earth evolved from a common ancestor.

C21. Take a look at the α-globin and β-globin sequences in figure 26.8. Which sequences are more similar, the α-globin in humans and the α-globin in horse, or the α-globin in humans and the β-globin in humans? Based on your answer, would you conclude that the gene duplication that gave rise to the α-globin and β-globin genes occurred before or after the divergence of humans and horses? Explain your reasoning.

C22. Compare and contrast the neutral theory of evolution versus the Darwinian theory (i.e., selectionist) of evolution. Explain why the neutral theory of evolution is sometimes called non-Darwinian evolution.

C23. For each of the following examples, discuss whether it would be the result of neutral mutation or mutation that has been acted upon by natural selection, or both:

A. When comparing sequences of homologous genes, differences in the coding sequence are most common at the wobble base (i.e., the third base in each codon).

B. As described in the experiment of figure 25.14, the *carbonaria* moths are more prevalent in polluted woods while the *typical* moths are more prevalent in the unpolluted woods.

C. For a structural gene, the regions that encode portions of the polypeptide that are vital for structure and function are less likely to incur mutations compared to other regions of the gene.

D. When comparing the sequences of homologous genes, introns usually have more sequence differences compared to exons.

C24. As discussed in chapter 25, genetic variation is prevalent in natural populations. This variation is revealed in the electrophoresis of allozymes and in the DNA sequencing of genes. According to the neutral theory, discuss the relative importance of natural selection against detrimental mutations, natural selection in favor of beneficial mutations, and neutral mutations, in accounting for the genetic variation we see in natural populations.

C25. If you were comparing the karyotypes of species that are closely related evolutionarily, what types of similarities and differences would you expect to find?

C26. Would the rate of deleterious or beneficial mutations be a good molecular clock? Why or why not?

Experimental Questions

E1. In chapter 8, we discussed the use of colchicine to promote nondisjunction. If this drug had been available to Arne Müntzing, how might he have conducted the experiment described in figure 26.6 differently?

E2. In the experiment of figure 26.6, what do the results mean considering that the artificial *G. tetrahit* could produce fertile offspring with the natural *G. tetrahit*? What would the results have meant if these two strains had not produced fertile offspring?

E3. As described in the experiment of figure 26.6, Arne Müntzing concluded that *G. tetrahit* evolved from an allotetraploid that contains a diploid chromosomal composition of *G. pubescens* and *G. speciosa*. This conclusion was based on the phenotypic similarity between the natural and artificial *G. tetrahit* strains. Would this conclusion have been invalid if the natural and artificial *G. tetrahit* strains had shown significant phenotypic differences? Explain your answer.

E4. Two populations of snakes are separated by a river. The snakes cross the river only on rare occasions. The snakes in the two populations look very similar to each other, except that the members of the population on the eastern bank of the river have a yellow spot on the top of their head, while the members of the western population have an orange spot on the top of their head. Discuss two experimental methods that you might follow to determine whether the two populations are members of the same species or members of different species.

E5. Sympatric speciation by allotetraploidy has been proposed as a common mechanism for speciation. Let's suppose you were interested in the origin of certain grass species in Southern California. Experimentally, how would you go about determining if some of the grass species are the result of allotetraploidy?

E6. Two diploid species of closely related frogs, which we will call species A and species B, were analyzed with regard to genes that encode an enzyme called hexokinase. Species A has two distinct copies of this gene: *A1* and *A2*. In other words, this diploid species is *A1A1 A2A2*. The other species has three copies of the hexokinase gene, which we will call *B1, B2,* and *B3*. A diploid individual of species B would be *B1B1 B2B2 B3B3*. These hexokinase genes from the two species were subjected to DNA sequencing, and the percentage of sequence identity was compared among these genes. The results are shown here:

Percentage of DNA Sequence Identity

	A1	A2	B1	B2	B3
A1	100	62	54	94	53
A2	62	100	91	49	92
B1	54	91	100	67	90
B2	94	49	67	100	64
B3	53	92	90	64	100

If we assume that hexokinase genes were never lost in the evolution of these frog species, how many distinct hexokinase genes do you think there were in the most recent ancestor that preceded the divergence of these two species? Explain your answer. Also explain why species B has three distinct copies of this gene while species A only has two.

E7. A homologous DNA region, which was 20,000 bp in length, was sequenced among four different species. The following number of nucleotide differences were obtained:

	Species A	Species B	Species C	Species D
Species A	0	443	719	465
Species B	443	0	744	423
Species C	719	744	0	723
Species D	465	423	723	0

Construct a phylogenetic tree that describes the evolutionary relationships among these four species. Your tree should include values that show the percentage of nucleotide differences.

E8. A researcher sequenced a portion of a bacterial gene and obtained the following sequence, beginning with the start codon, which is underlined:

ATG CCG GAT TAC CCG GTC CCA AAC AAA ATG
ATC GGC CGC CGA ATC TAT CCC

The bacterial strain that contained this gene has been maintained in the laboratory and grown serially for many generations. Recently, another person working in the laboratory isolated DNA from the bacterial strain and sequenced the same region. The following results were obtained.

ATG CCG GAT TAT CCG GTC CCA AAT AAA ATG
ATC GGC CGC CGA ATC TAC CCC

Explain why these sequencing differences may have occurred.

E9. F_1 hybrids between two species of cotton, *Gossypium barbadense* and *G. hirsutum*, are very vigorous plants. However, F_1 crosses produce many seeds that do not germinate and a low percentage of very weak F_2 offspring. Suggest two reasons for these observations.

E10. A species of antelope contains 20 chromosomes per set. The species is divided by a mountain range into two separate populations, which we will call the eastern and western population. When comparing the karyotypes between these two populations, it was discovered that the members of the eastern population are homozygous for a large inversion within chromosome 14. How would this inversion affect the interbreeding between the two populations? Could such an inversion play an important role in speciation?

E11. Explain why molecular techniques were needed as a way to provide evidence for the neutral theory of evolution.

E12. Ancient samples often contain minute amounts of DNA. What technique can be used to increase the amount of DNA in an ancient sample? Explain how this technique is performed and how it increases the amount of a specific region of DNA.

E13. In the experiment of figure 26.14, explain how we know that the kiwis are more closely related to the emu and cassowary than to the moas. Cite particular regions in the sequences that support your answer.

E14. In chapter 20, we learned about a technique called *in situ* hybridization, during which a cloned piece of DNA is hybridized to a set of chromosomes. Let's suppose that we cloned a piece of DNA from *G. pubescens* and used it as a probe for *in situ* hybridization. What would you expect to happen if we hybridized it to the *G. speciosa*, the natural *G. tetrahit*, or the artificial *G. tetrahit* strains? Draw your expected results.

E15. A team of researchers has obtained a dinosaur bone (*Tyrannosaurus rex*) and has attempted to extract ancient DNA from it. Using primers to the 12S rRNA gene, they have used PCR and obtained a DNA segment that yields a sequence homologous to crocodile DNA. Other scientists are skeptical that this sequence is really from the dinosaur. Instead, they believe that it may be due to contamination from more recent DNA, such as the remains of a reptile that lived much more recently. What criteria might you use to establish the credibility of the dinosaur sequence?

Questions for Student Discussion/Collaboration

1. The raw material for evolution is random mutation. Discuss whether or not you view evolution as a random process.

2. Compare the forms of speciation that are slow with those that occur more rapidly. Make a list of the slow and fast forms. With regard to mechanisms of genetic change, what features do slow and rapid speciation have in common? What features are different?

3. Do you think that Darwin would object to the neutral theory of evolution?

Note: All answers appear at the website for this textbook; the answers to even-numbered questions are in the back of the textbook.

Visit the Online Learning Center for practice tests, answer keys, and other learning aids for this chapter. Enhance your understanding of genetics with our interactive exercises, web links, news feeds, tutorial service, and much more.

APPENDIX
EXPERIMENTAL TECHNIQUES

OUTLINE

- Methods of Cell Growth
- Microscopy
- Separation Methods
- Methods to Measure Concentrations of Molecules and to Detect Radioisotopes and Antigens

METHODS OF CELL GROWTH

Researchers often grow cells in a laboratory as a way to study them. This is known as a **cell culture.** Cell culturing offers several technical advantages. The primary advantage is that the growth medium is defined and can be controlled. Minimal growth medium contains the bare essentials for cell growth: salts, a carbon source, an energy source, essential vitamins, amino acids, and trace elements. In their experiments, geneticists often compare strains that can grow in minimal media and mutant strains that cannot grow unless the medium is supplemented with additional components. A rich growth medium contains many more components than are required for growth.

Researchers also add substances to the culture medium for other experimental reasons. For example, radioactive isotopes can be added to the culture medium to radiolabel cellular macromolecules. Or an experimenter could add a hormone to the growth medium and then monitor the cells' response to the hormone. In all of these cases, cell culturing is advantageous because the experimenter can control and vary the composition of the growth medium.

The first step in creating a cell culture is the isolation of a cell population that the researchers wish to study. For bacteria, such as *Escherichia coli,* and eukaryotic microorganisms, such as yeast and *Neurospora,* the researchers simply obtain a sample of cells from a colleague or a stock center. For animal or plant tissues, the procedure is a bit more complicated. When cells are contained within a complex tissue, they must first be dispersed by treating the tissue with agents that separate it into individual cells to create a cell suspension.

Once a desired population of cells has been obtained, researchers can grow them in a laboratory (i.e., in vitro) either suspended in a liquid growth medium or attached to a solid surface such as agar. Both methods have been used commonly in the experiments considered throughout this textbook. Liquid culture is often used when researchers want to obtain a large quantity of cells and isolate individual cellular components, such as nuclei or

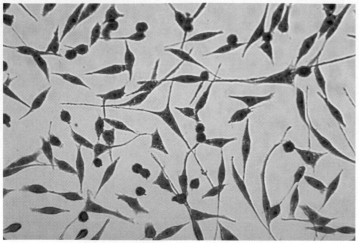

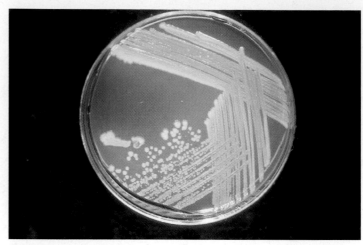

(a) Fibroblast (animal cell) culture

(b) Bacterial colonies

FIGURE A.1   **Growth of cells on solid growth media.** (a) Normal fibroblasts grow as a monolayer on the growth medium. (b) Bacterial cells form colonies that are a clonal population of cells derived from a single cell.

DNA. By comparison, figure A.1 shows micrographs of animal cells and bacteria cells that are grown on solid growth media. As discussed in experiment 22A, solid media are used to study cancer cells, because such cells can be distinguished by the formation of foci in which malignant cells pile up on top of each other. In gene cloning experiments with bacteria and yeast, solid media are also used. Each colony of cells is a clone of cells that is derived from a single cell that divided many times (fig. A.1*b*). As discussed in chapter 18, a solid medium is used in the isolation of individual clones that contain a desired gene.

MICROSCOPY

As you probably already know, **microscopy** is a technique to observe things that are not visible (or are hardly visible) with the naked eye. A key concept in microscopy is resolution, which is defined as the minimum distance between two objects that can be seen as separate from each other. The ability to resolve two points as being separate depends on several factors, including the wavelength of the illumination source (light or electron beam), the medium in which the sample is immersed, and the structural features of the microscope (which are beyond the scope of this textbook).

As shown in figure A.2, there are two widely used kinds of microscopes, the optical (light) microscope and the transmission electron microscope (TEM). The light microscope is used to resolve cellular structures to a limit of approximately 0.3 μm. (For comparison, a typical bacterium is about 1 μm long.) At this resolution, the individual cell organelles in eukaryotic cells can be discerned easily, and chromosomes are also visible. Karyotyping is accomplished via light microscopy after the chromosomes have been treated with stains. A variation of light microscopy known as fluorescence microscopy is often used to highlight a particular feature of a chromosome or cellular structure. The technique of fluorescence *in situ* hybridization (FISH; see chapter 20) makes use of

this type of microscope. Also, optical modifications in certain light microscopes (e.g., phase contrast and differential interference) can be used to exaggerate the differences in densities between neighboring cells or cell structures. These kinds of light microscopes are useful in monitoring cell division in living (unstained) cells or in transparent worms (as in experiment 23A).

The structural details of large macromolecules such as DNA and ribosomes are not observable by light microscopy. The coarse topology of these macromolecules can be determined by electron microscopy. Electron microscopes have a limit of resolution of about 2 nm, which is about 100 times finer than the best light microscopes. The primary advantage of electron microscopy over light microscopy is its better resolution. Disadvantages include a much higher expense and more extensive sample preparation. In transmission electron microscopy, the sample is bombarded with an electron beam. This requires that the sample be dried, fixed, and usually coated with a heavy metal that absorbs electrons.

SEPARATION METHODS

Biologists often wish to take complex systems and separate them into less complex components. For example, the cells within a complex tissue can be separated into individual cells, or the macromolecules within cells can be separated from the other cellular components. In this section, we will focus primarily on methods aimed at separating and ultimately purifying macromolecules.

Disruption of Cellular Components

In many experiments described in this textbook, researchers have obtained a sample of cells and then wish to isolate particular components within the cell. For example, a researcher may want to purify a protein that functions as a transcription factor. To do so, the researcher would begin with a sample of cells that synthe-

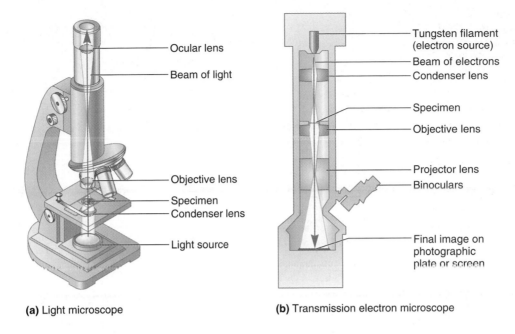

(a) Light microscope

- Ocular lens
- Beam of light
- Objective lens
- Specimen
- Condenser lens
- Light source

(b) Transmission electron microscope

- Tungsten filament (electron source)
- Beam of electrons
- Condenser lens
- Specimen
- Objective lens
- Projector lens
- Binoculars
- Final image on photographic plate or screen

FIGURE A.2   Design of (a) optical (light) and (b) transmission electron microscopes.

TABLE A.1
Common Methods of Cell Disruption

Method	Description
Sonication	The exposure of cells to intense sound waves, which breaks the cell membranes.
French press	The passage of cells through a small aperture under high pressure, which breaks the cell membranes and cell wall.
Homogenization	Cells are placed in a tube that contains a pestle. When the pestle is spun, the cells are squeezed through the small space between the pestle and the glass wall of the tube, thereby breaking them.
Osmotic shock	The transfer of cells into a hypoosmotic medium. The cells take up water and eventually burst.

size this protein and then break open the cells using one of the methods described in table A.1. In eukaryotes, the breakage of cells releases the soluble proteins from the cell; it also dissociates the cell organelles that are bounded by membranes. This mixture of proteins and cell organelles can then be isolated and purified by centrifugation and chromatographic methods, which are described next.

Centrifugation

Centrifugation is a method that is commonly used to separate cell organelles and macromolecules. A **centrifuge** contains a motor, which causes a rotor holding centrifuge tubes to spin very rapidly. As the rotor spins, particles move toward the bottom of the centrifuge tube; the rate at which they move depends on several factors, including their densities, sizes, and shapes and the viscosity of the medium. The rate at which a macromolecule or cell organelle sediments to the bottom of a centrifuge tube is called its **sedimentation coefficient,** which is normally expressed in Svedberg units (S). A sedimentation coefficient has the units of seconds: $1\text{ S} = 1 \times 10^{-13}$ seconds.

When a sample contains a mixture of macromolecules or cell organelles, it is likely that different components will sediment at different rates. This phenomenon, known as **differential centrifugation,** is shown in figure A.3. As seen here, particles with large sedimentation coefficients reach the bottom of the tube more quickly than those with smaller coefficients. There are two ways that researchers can use differential centrifugation as a separation technique. One way is to separate the **supernatant** from the **pellet** following centrifugation. The pellet is a collection of particles found at the bottom of the tube, and the supernatant is the liquid found above the pellet. In figure A.3, when the experimenter had subjected the sample to a low-speed spin, most of the particles with large sedimentation coefficients would be found in the pellet while most of the particles with small and intermediate coefficients would be found in the supernatant. A high-speed spin of the supernatant would then separate the small and intermediate particles. Therefore, differential centrifugation provides a way to segregate these three types of particles.

An alternative way to separate particles using centrifugation is to collect fractions. A **fraction** is a portion of the liquid contained within a centrifuge tube. The collection of fractions is done when the solution within the centrifuge tube contains a gradient. For example, as shown in figure A.4, the solution at the top of the tube has a lower concentration of cesium chloride (CsCl) than that at the bottom. In this experiment, a sample containing a mixture of cell organelles is layered on the top of the gradient and then centrifuged. In this example, the DNA and RNA separate from each other, because they have different sedimentation

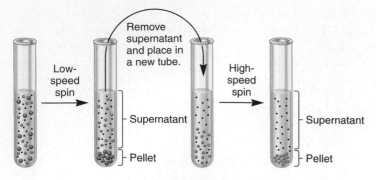

- Particles with large sedimentation coefficients
- Particles with intermediate sedimentation coefficients
- Particles with small sedimentation coefficients

FIGURE A.3   The method of differential centrifugation. A sample containing a mixture of particles with different sedimentation coefficients is placed in a centrifuge tube. The tube is subjected to a low-speed spin that pellets the particles with large sedimentation coefficients. After a high-speed spin of the supernatant, the particles with an intermediate sedimentation coefficient are found in the pellet, while those with a small sedimentation coefficient are in the liquid supernatant.

coefficients. The experimenter then punctures the bottom of the tube and collects fractions. The DNA fragments, which are heavier, come out of the tube in the earlier fractions; the RNA molecules will be collected in later fractions.

A type of gradient centrifugation that may also be used to separate macromolecules and organelles is **equilibrium density centrifugation.** In this method, the particles will sediment through the gradient, reaching a position where the density of the particle matches the density of the solution. At this point, the particle is at equilibrium and so does not move any farther toward the bottom of the tube.

Chromatography and Gel Electrophoresis

Chromatography is a method to separate different macromolecules and small molecules based on their chemical and physical properties. In this method, a sample is dissolved in a liquid solvent and exposed to some type of matrix, such as a column containing beads or a thin strip of paper. The degree to which the molecules interact with the matrix depends on their chemical and physical characteristics. For example, a positively charged molecule will bind tightly to a negatively charged matrix, while a neutral molecule will not.

Figure A.5 illustrates how column chromatography can be used to separate molecules that differ with respect to charge. Prior to this experiment, a column is packed with beads that are positively charged. There is plenty of space between the beads for molecules to flow from the top of the column to the bottom. However, if the molecules are negatively charged, they will spend some of their time binding to the positive charges on the surface of the beads. In the example shown in figure A.5, the *red* proteins are positively charged and, therefore, flow rapidly from the top of

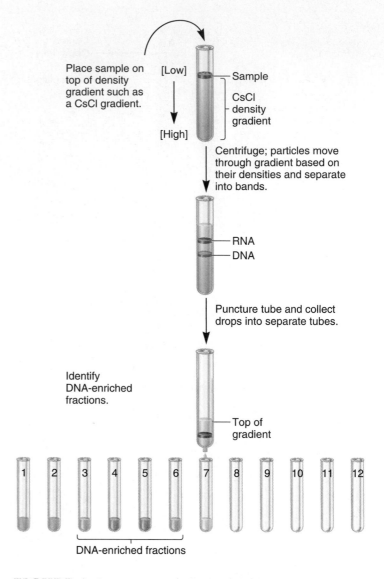

FIGURE A.4   Gradient centrifugation and the collection of fractions.

the column to the bottom. They emerge in the fractions that are collected early in this experiment. The *blue* proteins, however, are negatively charged and tend to bind to the beads. The binding of the blue proteins to the beads can be disrupted by changing the ionic strength or pH of the solution that is added to the column. Eventually, the blue proteins will be eluted (i.e., leave the column) in later fractions.

There are many variations of chromatography used by researchers to separate molecules and macromolecules. The type shown in figure A.5 is called ion-exchange chromatography, because its basis for separation depends on the charge of the molecules. In another type of column chromatography, known as gel filtration chromatography, the beads are porous. Small molecules are temporarily trapped within the beads, while large molecules flow between the beads. In this way, gel filtration separates molecules on the basis of size. To separate different types of macromolecules, such as proteins, researchers may use another type of bead; this bead has a preattached molecule that binds

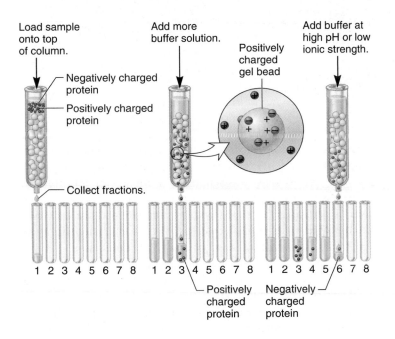

Load sample onto top of column.

Negatively charged protein

Positively charged protein

Collect fractions.

Add more buffer solution.

Positively charged gel bead

Positively charged protein

Negatively charged protein

Add buffer at high pH or low ionic strength.

1 2 3 4 5 6 7 8   1 2 3 4 5 6 7 8   1 2 3 4 5 6 7 8

FIGURE A.5 Ion-exchange chromatography.

specifically to the protein they want to purify. For example, if a transcription factor binds a particular DNA sequence as part of its function, the beads within a column may have this DNA sequence preattached to them. Therefore, the transcription factor will bind tightly to the DNA attached to these beads, while all other proteins will be eluted rapidly from the column. This form of chromatography is called affinity chromatography, because the beads have a special affinity for the macromolecule of interest.

Besides column chromatography, in which beads are packed into a column, there are other ways to make a matrix. In paper chromatography, molecules pass through a matrix composed of paper. The rate of movement of molecules through the paper depends on their degree of interaction with the solvent and paper. In thin-layer chromatography, a matrix is spread out as a very thin layer on a rigid support such as a glass plate. In general, paper and thin-layer chromatography are effective at separating small molecules, whereas column chromatography is used to separate macromolecules such as DNA fragments or proteins.

Gel electrophoresis combines chromatography and electrophoresis to separate molecules and macromolecules. As its name suggests, the matrix used in gel electrophoresis is composed of a gel. As shown in figure A.6, samples are loaded in wells

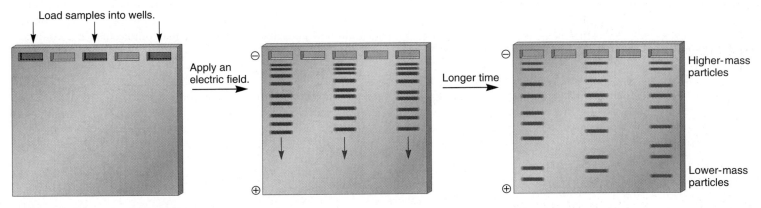

Load samples into wells.

Apply an electric field.

Longer time

Higher-mass particles

Lower-mass particles

(a) Separation of a mixture of particles by gel electrophoresis

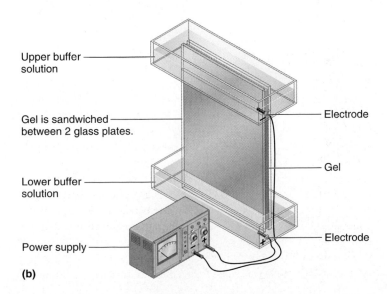

Upper buffer solution

Gel is sandwiched between 2 glass plates.

Lower buffer solution

Power supply

Electrode

Gel

Electrode

(b)

FIGURE A.6 Acrylamide gel electrophoresis of DNA fragments.

at one end of the gel, and an electric field is applied across the gel. This electric field causes charged molecules to migrate from one side of the gel to the other. The migration of molecules in response to an electric field is called **electrophoresis.** In the examples of gel electrophoresis found in this textbook, the macromolecules within the sample migrate toward the positive end of the gel. In most forms of gel electrophoresis, a mixture of macromolecules is separated according to their molecular masses. Small proteins or DNA fragments move to the bottom of the gel more quickly than larger ones. Since the samples are loaded in rectangular wells at the top of the gel, the molecules within the sample are separated into bands within the gel. These bands of separated macromolecules can be visualized with stains. For example, ethidium bromide is a stain that binds to DNA and RNA and can be seen under ultraviolet light.

The two most commonly used gels are polymers made from agarose or acrylamide. Proteins typically are separated on polyacrylamide gels, whereas DNA fragments are separated on agarose gels. Occasionally, researchers use polyacrylamide gels to separate DNA fragments that are relatively small (namely, less than 1,000 bp in length).

METHODS TO MEASURE CONCENTRATIONS OF MOLECULES AND TO DETECT RADIOISOTOPES AND ANTIGENS

To understand the structure and function of cells, researchers often need to detect the presence of molecules and macromolecules within living cells and to measure their concentrations. In this section, we will consider a variety of methods to detect and measure the concentrations of biological molecules and macromolecules.

Spectroscopy

Macromolecules found in living cells, such as proteins, DNA, and RNA, are fairly complex molecules that can absorb radiation (i.e., light). Likewise, small molecules such as amino acids and nucleotides can also absorb light. A device known as a **spectrophotometer** is used by researchers to determine how much radiation at various wavelengths a sample can absorb. The amount of absorption can be used to determine the concentration of particular molecules within a sample, because each type of molecule or macromolecule has its own characteristic wavelength(s) of absorption, called its absorption spectrum.

A spectrophotometer typically has two light sources, which can emit ultraviolet or visible light. As shown in figure A.7, the light source is passed through a monochromator, which emits the light at a desired wavelength. This incident light then strikes a sample contained within a cuvette. Some of the incident light is absorbed, and some is not. The unabsorbed light passes through the sample and is detected by the spectrophotometer. The amount of light that strikes the detector is subtracted from the amount of incident light, yielding the measure of absorption. In this way, the spectrophotometer provides an absorption reading for the sample. This reading can be used to calculate the concentration of particular molecules or macromolecules in a sample.

Detection of Radioisotopes

A radioisotope is an unstable form of an atom that decays to a more stable form by emitting α, β, or γ rays, which are types of ionizing radiation. In research, radioisotopes that are β and/or γ

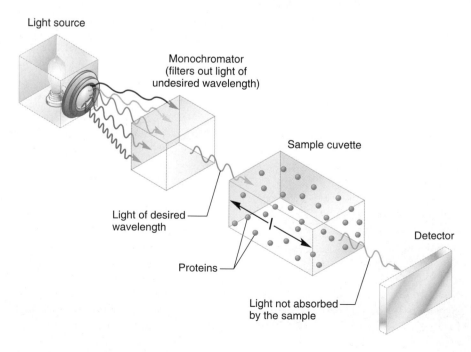

FIGURE A.7   **Design of a spectrophotometer.**

TABLE A.2
Some Useful Isotopes in Genetics

Isotope	Stable or Radioactive	Emission	Half-life
^2H	Stable		
^3H	Radioactive	β	12.3 years
^{13}C	Stable		
^{14}C	Radioactive	β	5,730 years
^{15}N	Stable		
^{18}O	Stable		
^{24}Na	Radioactive	β (and γ)	15 hours
^{32}P	Radioactive	β	14.3 days
^{35}S	Radioactive	β	87.4 days
^{45}Ca	Radioactive	β	164 days
^{59}Fe	Radioactive	β (and γ)	45 days
^{131}I	Radioactive	β (and γ)	8.1 days

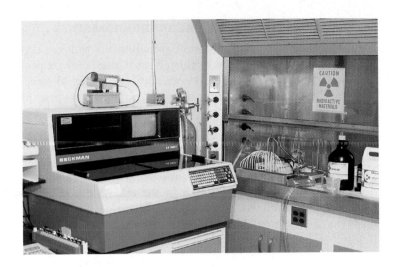

FIGURE A.8   A scintillation counter.

emitters are commonly used. A β ray is an emitted electron, and a γ ray is an emitted photon. Some radioisotopes commonly used in biological experiments are shown in table A.2.

Experimentally, radioisotopes are used frequently, because they are easy to detect. Therefore, if a particular compound is radiolabeled, its presence can be detected specifically throughout the course of the experiment. For example, if a nucleotide is radiolabeled with ^{32}P, a researcher can determine whether the isotope becomes incorporated into newly made DNA or whether it remains as the free nucleotide. Researchers commonly use two different methods to detect radioisotopes: scintillation counting and autoradiography.

The technique of **scintillation counting** permits a researcher to count the number of radioactive emissions from a sample containing a population of radioisotopes. In this approach, the sample is dissolved in a solution (called the scintillant) that contains organic solvents and one or more compounds known as fluors. When radioisotopes emit ionizing radiation, the energy is absorbed by the fluors in the solvent. This excites the fluor molecules, causing their electrons to be boosted to higher energy levels. The excited electrons return to lower, more stable energy levels by releasing photons of light. When a fluor is struck by ionizing radiation, it also absorbs the energy and then releases a photon of light within a particular wavelength range. The role of a device known as a scintillation counter is to count the photons of light that are emitted by the fluor. Figure A.8 shows a scintillation counter. To use this device, a researcher dissolves his or her sample in a scintillant and then places the sample in a scintillation vial. The vial is then placed in the scintillation counter, which detects the amount of radioactivity. The scintillation

counter has a digital meter that displays the amount of radioactivity in the sample, and it also provides a printout of the amount of radioactivity in counts per minute. A scintillation counter contains several rows for the loading and analysis of many scintillation vials. After they have been loaded, the scintillation counter will count the amount of radioactivity in each vial and provide the researcher with a printout of the amount of radioactivity in each vial.

A second way to detect radioisotopes is via **autoradiography.** This technique is not as quantitative as scintillation counting, because it does not provide the experimenter with a precise measure of the amount of radioactivity in counts per minute. However, autoradiography has the great advantage that it can detect the location of radioisotopes as they are found in macromolecules or cells. For example, autoradiography is used to detect a particular band on a gel or to map the location of a gene within an intact chromosome.

To conduct autoradiography, a sample containing a radioisotope is fixed and usually dried. If it is a cellular sample, it also may be thin sectioned. The sample is then pressed next to X-ray film (in the dark) and placed in a lightproof cassette. When a radioisotope decays, it will emit a β or γ ray, which may strike a thin layer of photoemulsion next to the film. The photoemulsion contains silver salts such as AgBr. When a radioactive particle is emitted and strikes the photoemulsion, a silver grain is deposited on the film. This produces a dark spot on the film, which correlates with the original location of the radioisotope in the sample. In this way, the dark image on the film reveals the location(s) of the radioisotopes in the sample. Figure 11.19 shows how autoradiography can be used to visualize the process of bacterial chromosome replication. In this case, radiolabeled nucleotides were incorporated into the DNA, making it possible to picture the topology of the chromosome as two replication forks pass around the circular chromosome.

Detection of Antigens by Radioimmunoassay

Antibodies, also known as **immunoglobulins,** are proteins that are used to ward off infection by foreign substances; they are produced by cells of the immune system. Antibodies bind to structures on the surface of foreign substances known as **epitopes;** the foreign substance is called an **antigen.** A particular antibody binds to a particular antigen with a very high degree of specificity. For this reason, antibodies have been used extensively by researchers to detect particular antigens. For example, a human protein such as hemoglobin can be injected into a rabbit. Human hemoglobin is a foreign substance in the rabbit's bloodstream. Therefore, the rabbit will make antibodies that specifically recognize human hemoglobin and are designed to destroy it. Researchers can isolate and purify these antibodies from a sample of the rabbit's blood and then use them to detect human hemoglobin in their experiments.

A **radioimmunoassay** is a method to measure the amount of an antigen in a biological sample. The steps in this method are shown in figure A.9a. The researcher begins with two tubes that have a known amount of radiolabeled antigen (shown in *blue*). An unknown amount of the same antigen, which is not radiolabeled (shown in *orange*), is added to the tube on the *right*. The nonradiolabeled antigen comes from a biological sample; the goal of this experiment is to determine how much of this antigen is contained within the sample. Next, a known amount of antibody is added to each of the two tubes. The amount of the antibody is less than the amount of the antigen, and so the nonlabeled and radiolabeled antigens compete with each other for binding to the antibody. After binding, a precipitating agent such as an anti-immunoglobulin antibody is added, and the precipitate is centrifuged to the bottom of the tube. The radioactivity in the precipitate is then determined by scintillation counting.

To calculate the amount of antigen in the sample being assayed, the researcher must determine the percentage of antibody that has bound to nonlabeled antigen. To do so, a second component of the experiment is to develop a standard curve in which a fixed amount of radiolabeled antigen is mixed with varying amounts of unlabeled antigen (fig. A.9b). Using this standard curve, a researcher can determine how much antigen is found in the unknown sample. For example, as shown in the *dashed line,* if the unknown sample had about 45% of the antibody bound, then the concentration of antigen in the sample would be between 50 and 75 nanomolar.

Radioimmunoassays are used to determine the concentrations of many different kinds of antigens. This includes small molecules such as hormones (as in experiment 19A) or macromolecules such as proteins.

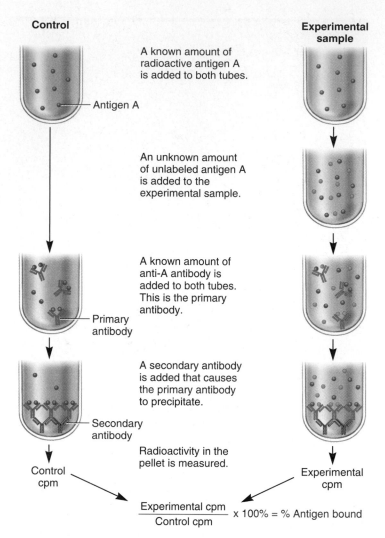

(a) A radioimmunoassay

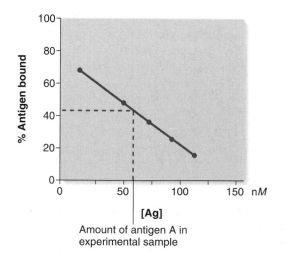

(b) Standard curve

FIGURE A.9 **The method of radioimmunoassay (a) and the construction of a standard curve (b).** In **(b),** the *dashed line* corresponds to the amount of antigen (Ag) bound by an unknown sample. This amounts to a concentration between 50 and 75 nanomolar of antigen.

SOLUTIONS TO EVEN-NUMBERED PROBLEMS

::

CHAPTER 1

Conceptual Questions

C2. A chromosome is a very long polymer of DNA. A gene is a specific sequence of DNA within that polymer; the sequence of bases creates a gene and distinguishes it from other genes. Genes are located in chromosomes, which are found within living cells.

C4. At the molecular level, a gene (which is a sequence of DNA) is first transcribed into RNA. The genetic code within the RNA is used to synthesize a protein with a particular amino acid sequence. This second process is called translation.

C6. Genetic variation involves the occurrence of genetic differences within members of the same species or different species. Within any population, there may be variation in the genetic material. There may be variation in particular genes so that some individuals carry one allele and other individuals carry a different allele. An example would be differences in coat color among mammals. There also may be variation in chromosome structure and number. In plants, differences in chromosome number can affect disease resistance.

C8. You could pick almost any trait. For example, flower color in petunias would be an interesting choice. Some petunias are red and others are purple. There must be different alleles in a flower color gene that affect this trait in petunias. In addition, the amount of sunlight, fertilizer, and water also affects the intensity of flower color.

C10. A DNA sequence is a sequence of nucleotides. Each nucleotide may have one of four different bases (i.e., A, T, G, or C). When we speak of a DNA sequence, we focus on the sequence of bases.

C12. A. A gene is a segment of DNA. For most genes, the expression of the gene results in the production of a functional protein. The functioning of proteins within living cells affects the traits of an organism.

B. A gene is a segment of DNA that usually encodes the information for the production of a specific protein. Genes are found within chromosomes. There are many genes within a single chromosome.

C. An allele is an alternative version of a particular gene. For example, suppose a plant has a flower color gene. One allele could produce a white flower, while a different allele could produce an orange flower. The white allele is an allele of the flower color gene.

D. A DNA sequence is a sequence of nucleotides. The information within a DNA sequence (which is transcribed into an RNA sequence) specifies the amino acid sequence within a protein.

C14. A. How genes and traits are transmitted from parents to offspring.

B. How the genetic material functions at the molecular and cellular levels.

C. Why genetic variation exists in populations, and how it changes over the course of many generations.

Experimental Questions

E2. This would be used primarily by molecular geneticists. The sequence of DNA is a molecular characteristic of DNA. In addition, as we will learn throughout this textbook, the sequence of DNA is interesting to transmission and population geneticists as well.

E4. A. Transmission geneticists. Dog breeders are interested in how genetic crosses affect the traits of dogs.

B. Molecular geneticists. This is a good model organism to study genetics at the molecular level.

C. Both transmission geneticists and molecular geneticists. Fruit flies are easy to cross and study the transmission of genes and traits from parents to offspring. Molecular geneticists have also studied many genes in fruit flies to see how they function at the molecular level.

D. Population geneticists. Most wild animals and plants would be the subject of population geneticists. In the wild, you cannot make controlled crosses. But you can study genetic variation within populations and try to understand its relationship to the environment.

E. Transmission geneticists. Agricultural breeders are interested in how genetic crosses affect the outcome of traits.

CHAPTER 2

Conceptual Questions

C2. In the case of plants, cross-fertilization occurs when the pollen and eggs come from different plants while in self-fertilization they come from the same plant.

C4. A homozygote that has two copies of the same allele.

C6. Diploid organisms contain two copies of each type of gene. When they make gametes, only one copy of each gene is found in a gamete. Two alleles cannot stay together within the same gamete.

C8. Genotypes: 1:1 *Tt* and *tt*

Phenotypes: 1:1 Tall and dwarf

C10. *c* is the recessive allele for constricted pods, *Y* is the dominant allele for yellow color. The cross is *ccYy* × *CcYy*. Follow the directions for setting up a Punnett square, as described in chapter 2. The genotypic ratio is 2 *CcYY* : 4 *CcYy* : 2 *Ccyy* : 2 *ccYY* : 4 *ccYy* : 2 *ccyy*. This 2:4:2:2:4:2 ratio could be reduced to a 1:2:1:1:2:1 ratio.

The phenotypic ratio is 6 inflated pods, yellow seeds : 2 inflated pods, green seeds: 6 constricted pods, yellow seeds : 2 constricted pods, green seeds. This 2:6:6:2 ratio could be reduced to a 1:3:3:1 ratio.

C12. Offspring with a nonparental phenotype are consistent with the idea of independent assortment. If two different traits were always transmitted together as unit, it would not be possible to get non-parental phenotypic combinations. For example, if a true-breeding parent had two dominant traits and was crossed to a true-breeding parent having the two recessive traits, the F_2 generation could not have offspring with one recessive and one dominant phenotype. However, because independent assortment can occur, it is possible for F_2 offspring to have one dominant and one recessive trait.

C14. A. Barring a new mutation during gamete formation, the chance is 100% because they must be heterozygotes in order to produce a child with a recessive disorder.

B. Construct a Punnett square. There is a 50% chance of heterozygous children.

C. Use the product rule. The chance of being phenotypically normal is 0.75 (i.e., 75%), so the answer is 0.75 × 0.75 × 0.75 = 0.422, which is 42.2%.

D. Use the binomial expansion equation where $n = 3$, $x = 2$, $p = 0.75$, $q = 0.25$. The answer is 0.422, or 42.2%.

C16. First construct a Punnett square. The chances are 75% of producing a solid pup and 25% of producing a spotted pup.

A. Use the binomial expansion equation where $n = 5$, $x = 4$, $p = 0.75$, $q = 0.25$. The answer is 0.396 = 39.6% of the time.

B. You can use the binomial expansion equation for each litter. For the first litter, $n = 6$, $x = 4$, $p = 0.75$, $q = 0.25$; for the second litter, $n = 5$, $x = 5$, $p = 0.75$, $q = 0.25$. Because the litters are in a specified order, we use the product rule and multiply the probability of the first litter times the probability of the second litter. The answer is 0.070, or 7.0%.

C. To calculate the probability of the first litter, we use the product rule and multiply the probability of the first pup (0.75) times the probability of the remaining four. We use the binomial expansion equation to calculate the probability of the remaining four, where $n = 4$, $x = 3$, $p = 0.75$, $q = 0.25$. The probability of

the first litter is 0.316. To calculate the probability of the second litter, we use the product rule and multiply the probability of the first pup (0.25) times the probability of the second pup (0.25) times the probability of the remaining five. To calculate the probability of the remaining five, we use the binomial expansion equation, where $n = 5$, $x = 4$, $p = 0.75$, $q = 0.25$. The probability of the second litter is 0.025. To get the probability of these two litters occurring in this order, we use the product rule and multiply the probability of the first litter (0.316) times the probability of the second litter (0.025). The answer is 0.008, or 0.8%.

D. Because this is a specified order, we use the product rule and multiply the probability of the firstborn (0.75) times the probability of the second born (0.25) times the probability of the remaining four. We use the binomial expansion equation to calculate the probability of the remaining four pups, where $n = 4$, $x = 2$, $p = 0.75$, $q = 0.25$. The answer is 0.040, or 4.0%.

C18. A. Use the product rule:

$$(1/4)(1/4) = 1/16$$

B. Use the binomial expansion equation:

$$n = 4, p = 1/4, q = 3/4, x = 2$$

$$P = 0.21 = 21\%$$

C. Use the product rule:

$$(1/4)(3/4)(3/4) = 0.14, \text{ or } 14\%$$

C20. A. 1/4

B. 1, or 100%

C. $(3/4)(3/4)(3/4) = 27/64 = 0.42$, or 42%

D. Use the binomial expansion equation where

$$n = 7, p = 3/4, q = 1/4, x = 3$$

$$P = 0.058, \text{ or } 5.8\%$$

E. The probability that the first plant is tall is 3/4. To calculate the probability that among the next four, any two will be tall, we use the binomial expansion equation, where $n = 4$, $p = 3/4$, $q = 1/4$, and $x = 2$.

The probability *P* equals 0.21.

To calculate the overall probability of these two events:

$$(3/4)(0.21) = 0.16, \text{ or } 16\%$$

C22. It violates the law of segregation because there are two copies of one gene in the gamete. The two alleles for the *A* gene did not segregate from each other.

C24. Based on this pedigree, it is likely to be dominant inheritance because an affected child always has an affected parent. In fact, it is a dominant disorder.

C26. It is impossible for the F_1 individuals to be true-breeding because they are all heterozygotes.

C28. 2 *TY*, *tY*, 2 *Ty*, *ty*, *TTY*, *TTy*, 2 *TtY*, 2 *Tty*

It may be tricky to think about, but you get 2 *TY* and 2 *Ty* because either of the two *T* alleles could combine with *Y* or *y*. Also, you get 2 *TtY* and 2 *Tty* because either of the two *T* alleles could combine with *t* and then combine with *Y* or *y*.

C30. The genotype of the F_1 plants is *Tt Yy Rr*. According to the laws of segregation and independent assortment, the alleles of each gene will segregate from each other, and the alleles of different genes will randomly assort into gametes. A *Tt Yy Rr* individual could make eight types of gametes: *TYR, TyR, Tyr, TYr, tYR, tyR, tYr,* and *tyr,* in equal proportions (i.e., 1/8 of each type of gamete). To determine genotypes and phenotypes, you could make a large Punnett square that would contain 64 boxes. You need to line up the eight possible gametes across the top and along the side, and then fill in the 64 boxes. Alternatively, you could use one of the two approaches described in solved problem S3. The genotypes and phenotypes would be:

1 *TT YY RR*
2 *TT Yy RR*
2 *TT YY Rr*
2 *Tt YY RR*
4 *TT Yy Rr*
4 *Tt Yy RR*
4 *Tt YY Rr*
8 *Tt Yy Rr*   = 27 tall, yellow, round

1 *TT yy RR*
2 *Tt yy RR*
2 *TT yy Rr*
4 *Tt yy Rr*   = 9 tall, green, round

1 *TT YY rr*
2 *TT Yy rr*
2 *Tt YY rr*
4 *Tt Yy rr*   = 9 tall, yellow, wrinkled

1 *tt YY RR*
2 *tt Yy RR*
2 *tt YY Rr*
4 *tt Yy Rr*   = 9 dwarf, yellow, round

1 *TT yy rr*
2 *Tt yy rr*   = 3 tall, green, wrinkled

1 *tt yy RR*
2 *tt yy Rr*   = 3 dwarf, green, round

1 *tt YY rr*
2 *tt Yy rr*   = 3 dwarf, yellow, wrinkled

1 *tt yy rr*   = 1 dwarf, green, wrinkled

C32. The wooly-haired male is a heterozygote, because he has the trait and his mother did not. (He must have inherited the normal allele from his mother.) Therefore, he has a 50% chance of passing the wooly allele to his offspring; his offspring have a 50% of passing the allele to their offspring; and these grandchildren have a 50% chance of passing the allele to their offspring (the wooly-haired man's great-grandchildren). Because this is an ordered sequence of independent events, we use the product rule: $0.5 \times 0.5 \times 0.5 = 0.125$, or 12.5%. Because there are not other Scandinavians on the island, there is an 87.5% chance of the offspring being normal (because they could not inherit the wooly-hair allele from anyone else). We use the binomial expansion equation to determine the likelihood that one out of eight great-grandchildren will have wooly hair, where $n = 8$, $x = 1$, $p = 0.125$, $q = 0.875$. The answer is 0.393, or 39.3%, of the time.

C34. Use the product rule. If the woman is heterozygous, there is a 50% chance of having an affected offspring: $(0.5)^7 = 0.0078$, or 0.78%, of the time. This is a pretty small probability. If the woman has an eighth child who is unaffected, however, she has to be a heterozygote, since it is a dominant trait. She would have to pass a normal allele to an unaffected offspring. The answer is 100%.

Experimental Questions

E2. The experimental difference depends on where the pollen comes from. In self-fertilization, the pollen and eggs come from the same plant. In cross-fertilization, they come from different plants.

E4. According to Mendel's law of segregation, the genotypic ratio should be 1 homozygote dominant : 2 heterozygotes : 1 homozygote recessive. This data table considers only the plants with a dominant phenotype. The genotypic ratio should be 1 homozygote dominant : 2 heterozygotes. The homozygote dominants would be true-breeding while the heterozygotes would not be true-breeding. This 1:2 ratio is very close to what Mendel observed.

E6. All three offspring had black fur. The ovaries from the albino female could only produce eggs with the dominant black allele (because they were obtained from a true-breeding black female). The actual phenotype of the albino mother does not matter. Therefore, all offspring would be heterozygotes (*Bb*) and have black fur.

E8. If we construct a Punnett square according to Mendel's laws, we expect a 9:3:3:1 ratio. Since a total of 556 offspring were observed, the expected number of offspring are

$556 \times 9/16 = 313$ round, yellow
$556 \times 3/16 = 104$ wrinkled, yellow
$556 \times 3/16 = 104$ round, green
$556 \times 1/16 = 35$ wrinkled, green

If we plug the observed and expected values into the chi square equation, we get a value of 0.51. There are four categories, so our degrees of freedom equals $n - 1$, or 3. If we look up our value in the chi square table (see table 2.1), it is well within the range of expected error if the hypothesis is correct. Therefore, we accept the hypothesis. In other words, the results are consistent with the law of independent assortment.

E10. A. If we let c^+ represent normal wings and c represent curved wings, and e^+ represents gray body and e represents ebony body:

Parental Cross: $cce^+e^+ \times c^+c^+ee$.

F_1 generation is heterozygous c^+ce^+e.

An F_1 offspring crossed to flies with curved wings and ebony bodies is

$$c^+ce^+e \times ccee$$

The F_2 offspring would be a 1:1:1:1 ratio of flies.

$$c^+ce^+e : c^+cee : cce^+e : ccee$$

B. The phenotypic ratio of the F_2 flies would be a 1:1:1:1 ratio of flies.

normal wings, gray body : normal wings, ebony bodies : curved wings, gray bodies : curved wings, ebony bodies

C. From part B, we expect 1/4 of each category. There are a total of 444 offspring. The expected number of each category is $1/4 \times 444$, which equals 111.

$$\chi^2 = \frac{(114 - 111)^2}{111} + \frac{(105 - 111)^2}{111} + \frac{(111 - 111)^2}{111} + \frac{(114 - 111)^2}{111}$$

$$\chi^2 = 0.49$$

With 3 degrees of freedom, a value of 0.49 or greater is likely to occur between 95 and 80% of the time. So we accept our hypothesis.

E12. Follow through the same basic chi square strategy as before. We expect 3/4 of the dominant phenotype and 1/4 of the recessive phenotype.

The observed and expected values are as follows (rounded to the nearest whole number):

Observed*	Expected	$\frac{(O - E)^2}{E}$
5,474	5,493	0.066
1,850	1,831	0.197
6,022	6,017	0.004
2,001	2,006	0.012
705	697	0.092
224	232	0.276
882	886	0.018
299	295	0.054
428	435	0.113
152	145	0.338
651	644	0.076
207	215	0.298
787	798	0.152
277	266	0.455
		$\chi^2 = 2.15$

*Due to rounding, the observed and expected values may not add up to precisely the same number.

Because $n = 14$, there are 13 degrees of freedom. If we look up this value in the chi square table, we have to look between 10 and 15 degrees of freedom. In either case, we would expect such a low value or higher to occur greater than 99% of the time. Therefore, we accept the hypothesis.

E14. The dwarf parent with terminal flowers must be homozygous for both genes since it is expressing these two recessive traits: *ttaa*, where *t* is the recessive dwarf allele, and *a* is the recessive allele for terminal flowers. The phenotype of the other parent is dominant for both traits. However, since this parent was able to produce dwarf offspring with axial flowers, it must have been heterozygous for both genes: *TtAa*.

E16. Our hypothesis is that blue flowers and purple seeds are dominant traits and they are governed by two genes that assort independently. According to this hypothesis, the F_2 generation should yield a ratio of 9 blue flowers, purple seeds : 3 blue flowers, green seeds : 3 white flowers, purple seeds : 1 white flower, green seeds. Because there are a total of 300 offspring produced, the expected numbers would be

$9/16 \times 300 = 169$ blue flowers, purple seeds

$3/16 \times 300 = 56$ blue flowers, green seeds

$3/16 \times 300 = 56$ white flowers, purple seeds

$1/16 \times 300 = 19$ white flowers, green seeds

$$\chi^2 = \frac{(103 - 169)^2}{169} + \frac{(49 - 56)^2}{56} + \frac{(44 - 56)^2}{56} + \frac{(104 - 19)^2}{19}$$

$$\chi^2 = 409.5$$

If we look up this value in the chi square table under 3 degrees of freedom, the value is much higher than would be expected 1% of the time by chance alone. Therefore, we reject the hypothesis. The idea that the two genes are assorting independently seems to be incorrect. The F_1 generation supports the idea that blue flowers and purple seeds are dominant traits.

Questions for Student Discussion/Collaboration

2. If we construct a Punnett square, the following probabilities will be obtained:

tall with axial flowers 3/8
dwarf with terminal flowers 1/8

To calculate the probability of being tall with axial flowers or dwarf with terminal flowers, we use the sum rule.

$$3/8 + 1/8 = 4/8 = 1/2$$

We use the product rule to calculate the ordered events of the first three offspring being tall, axial or dwarf, terminal, and the fourth offspring being tall, axial.

$$(1/2)(1/2)(1/2)(3/8) = 3/64 = 0.047 = 4.7\%$$

CHAPTER 3

Conceptual Questions

C2. The term *homologue* refers to the members of a chromosome pair. Homologues are usually the same size and carry the same types and order of genes. They may differ in that the genes they carry may be different alleles.

C4. Metaphase is the organization phase and anaphase is the separation phase.

C6. In metaphase I of meiosis, each pair of chromatids is attached to only one pole via the kinetochore microtubules. In metaphase of mitosis, there are two attachments (i.e., to both poles). If the attachment was lost, a chromosome would probably be lost and degraded because it would not migrate to a pole. Therefore, it would not become enclosed in a nuclear membrane after telophase. If left out in the cytoplasm, it would eventually be degraded.

C8. The reduction occurs because there is a single DNA replication event but two cell divisions. Because of the nature of separation during anaphase I, each cell receives one copy of each type of chromosome.

C10. It means that the arrangement of the maternally derived and paternally derived chromosomes is random during metaphase I. Refer to figure 3.17.

C12. There are three pairs of chromosomes. So the possible number of arrangements equals 2^3, which is 8.

C14. It would be much lower because pieces of maternal chromosomes would be mixed with the paternal chromosomes. Therefore, it is unlikely to inherit a chromosome that was completely paternally derived.

C16. During interphase, the chromosomes are greatly extended. In this conformation, they might get tangled up with each other and not sort properly during meiosis and mitosis. The condensation process probably occurs so that the chromosomes can easily align along the equatorial plate during metaphase without getting tangled up.

C18. During prophase II, your drawing should show four replicated chromosomes (i.e., four structures that look like Xs). Each chromosome is one homologue. During prophase of mitosis, there should be eight replicated chromosomes (i.e., eight Xs). During prophase of mitosis, there are pairs of homologues. The main difference is that prophase II has a single copy of each of the four chromosomes whereas prophase of mitosis has four pairs of homologues. At the end of meiosis I, each daughter cell received only one copy of a homologous pair, not both. This is due to the alignment of homologues during metaphase I and their separation during anaphase I.

C20. DNA replication does not take place during interphase II. The chromosomes at the end of telophase I have already replicated (i.e., they are found in pairs of sister chromatids). During meiosis II, the sister chromatids separate from each other yielding individual chromosomes.

C22. A. 20

   B. 10

   C. 30

   D. 20

C24. A. Dark males and light females; reciprocal: all dark offspring

   B. All dark offspring; reciprocal: dark females and light males

   C. All dark offspring; reciprocal: dark females and light males

   D. All dark offspring; reciprocal: dark females and light males

C26. To produce sperm, a spermatogonial cell first goes through mitosis to produce two cells. One of these remains a spermatogonial cell and the other progresses through meiosis. In this way, the testis continues to maintain a population of spermatogonial cells.

C28. A. $A B C, A B c, A b C, A b c, a B C, a b C, a B c, a b c$

   B. $A B C, A b C$

   C. $A B C, A B c, a B C, a B c$

   D. $A b c, a b c$

C30. The outcome is that all the daughters will be affected and all the sons will be unaffected. The ratio is 1:1.

C32. First set up the following Punnett Square:

Male gametes

	♂ X^H	Y
♀ X^H	$X^H X^H$	$X^H Y$
X^h	$X^H X^h$	$X^h Y$

Female gametes

There is a $1/4$ probability of each type of offspring.

A. 1/4

B. $(3/4)(3/4)(3/4)(3/4) = 81/256$

C. 3/4

D. The probability of an affected offspring is 1/4 and the probability of an unaffected offspring is 3/4. For this problem, you use the binomial expansion where $x = 2$, $n = 5$, $p = 1/4$, and $q = 3/4$. The answer is 0.26, or 26%, of the time.

C34. A. No. A male inherits the Y chromosome from his father, so he could not inherit an X chromosome from his paternal grandfather.

   B. Yes. A male could inherit a maternal X chromosome from his maternal grandfather. It would first be transmitted from his grandfather to his mother, and then from his mother to him.

   C. Yes. A female could inherit a maternal X chromosome from her maternal grandmother. It would first be transmitted from her grandmother to her mother, and then from her mother to her.

   D. Yes. A female could inherit a maternal X chromosome from her maternal great-great-grandmother. It would first be transmitted from her great-great-grandmother to a maternal great-grandparent (male or female). It could then be transmitted to a grandparent, and then to the mother, and then from her mother to her.

C36. A. The woman's mother must have been a heterozygote. So there is a 50% chance that the woman is a carrier. If she has children, 1/4 (i.e., 25%) will be affected sons if she is a carrier. However, there is only a 50% chance that she is a carrier. So we multiply 50% times 25%, which equals $0.5 \times 0.25 = 0.125$, or a 12.5% chance.

   B. If she already had a color-blind son, then we know she must be a carrier, so the chance is 25%.

   C. The woman is heterozygous and her husband is hemizygous for the color-blind allele. This couple will produce 1/4 offspring that are color-blind daughters. The rest are 1/4 carrier daughters, 1/4 normal sons, and 1/4 color-blind sons. Answer is 25%.

C38. A. The fly is a male because the ratio of X chromosomes to sets of autosomes is 1/2, or 0.5.

   B. The fly is female because the ratio is 1.0.

   C. The fly is male because the ratio is 0.5.

   D. The fly is female because the ratio is 1.0.

Experimental Questions

E2. Perhaps the most convincing observation was that all of the white-eyed flies of the F_2 generation were males; not a single white-eyed female was observed. This suggests a link between sex determination and the inheritance of this trait. Because sex determination in fruit flies is determined by the number of X chromosomes, this suggests a relationship between the inheritance of the X chromosome and the inheritance of this trait.

E4. The basic strategy is to set up a pair of reciprocal crosses. The phenotype of sons is usually the easiest way to discern the two patterns. If it is Y linked, the trait will only be passed from father to son. If it is X linked, the trait will be passed from mother to son.

E6. The 3:1 sex ratio occurs because the female produces 50% gametes that are XX (and must produce female offspring) and 50% that are X (and produce half male and half female offspring). The original female had one X chromosome carrying the red allele and two other X chromosomes carrying the eosin allele. Set up a Punnett square assuming that this female produces the following six types of gametes: $X^{w^+}X^{w-e}$, $X^{w^+}X^{w-e}$, $X^{w-e}X^{w-e}$, X^{w^+}, X^{w-e}, X^{w-e}. The male of this cross is X^wY.

Male gametes

	♂ X^w	Y
♀ $X^{w^+}X^{w-e}$	$X^{w^+}X^{w-e}X^w$ Red, female	$X^{w^+}X^{w-e}Y$ Red, female
$X^{w^+}X^{w-e}$	$X^{w^+}X^{w-e}X^w$ Red, female	$X^{w^+}X^{w-e}Y$ Red, female
$X^{w-e}X^{w-e}$	$X^{w-e}X^{w-e}X^w$ Eosin female	$X^{w-e}X^{w-e}Y$ Eosin female
X^{w^+}	$X^{w^+}X^w$ Red, female	$X^{w^+}Y$ Red, male
X^{w-e}	$X^{w-e}X^w$ Light-eosin female	$X^{w-e}Y$ Light-eosin male
X^{w-e}	$X^{w-e}X^w$ Light-eosin female	$X^{w-e}Y$ Light-eosin male

Female gametes (left axis label)

E8. If we use the data from the F_1 mating (i.e., F_2 results), there were 3,470 red-eyed flies. We would expect a 3:1 ratio between red- and white-eyed flies. Therefore, assuming that all red-eyed offspring survived, there should have been about 1,157 (i.e., 3,470/3) white-eyed flies. However, there were only 782. If we divide 782 by 1,157, we get a value of 0.676, or 67.6% have survived.

E10. You could karyotype other members of the family and see if affected members always carry the abnormal chromosome.

E12. These rare female flies would be XXY. Both X chromosomes would carry the white allele and the miniature allele. These female flies would have miniature wings because they would have inherited both X chromosomes from their mother.

E14. In X-linked recessive inheritance, it is much more common for males to be affected. In autosomal recessive inheritance, there is an equal chance of males and females being affected (unless there is a sex influence as described in chapter 4). For X-linked dominant inheritance, affected males would produce 100% affected daughters and not transmit the trait to their sons. This would not be true for autosomal dominant traits, where there is an equal chance of males and females being affected.

Questions for Student Discussion/Collaboration

2. It is not possible to give a direct answer, but the point is for students to be able to draw chromosomes in different configurations and understand the various phases. The chromosomes may or may not be

A. in homologous pairs;

B. connected as sister chromatids;

C. associated in bivalents;

D. lined up in metaphase;

E. moving toward the poles;

And so on.

CHAPTER 4

Conceptual Questions

C2. Sex-influenced traits are influenced by the sex of the individual even though the gene that governs the trait may be autosomally inherited. Pattern baldness in people is an example. Sex-limited traits are an extreme example of sex influence. The expression of a sex-limited trait is limited to one sex. For example, colorful plumage in certain species of birds is limited to the male sex. Sex-linked traits involve traits whose genes are found on the sex chromosomes. Examples in humans include hemophilia and color blindness.

C4. If the normal allele is dominant, it tells you that one copy of the gene produces a saturating amount of the protein encoded by the gene. Having twice as much of this protein, as in the normal homozygote, does not alter the phenotype. If the allele is incompletely dominant, this means that one copy of the normal allele is not saturating.

C6. There would be a ratio of 1 normal : 2 star-eyed individuals.

C8. The amount of protein produced from a single gene is not a saturating amount. Therefore, additional copies produce more of the protein, whose functional consequences can be increased.

C10. Types O and AB provide an unambiguous genotype. Type O can only be ii, and type AB can only be I^AI^B. It is possible for a couple to produce children with all four blood types. The couple would have to be I^Ai and I^Bi. If you construct a Punnett square, you will see that they can produce children with AB, A, B, and O blood types.

C12. A. 1/4

B. 0

C. $(1/4)(1/4)(1/4) = 1/64$

D. Use the binomial expansion:

$$P = \frac{n!}{x! \, (n - x)!} \, p^x q^{(n-x)}$$

$$n = 3, p = 1/4, q = 1/4, x = 2$$

$$P = 3/64 = 0.047 \text{ or } 4.7\%$$

C14. Let's begin with the assumption that the recessive alleles encode enzymes that are completely defective. If so, we could explain solved problem S1 in the following manner. Gene A might encode an enzyme that converts a colorless pigment precursor into a (black) pigmented molecule. If an animal is aa, no pigment is made and it becomes white. Gene C could encode an enzyme that converts some of this black pigment into a brown pigment to produce the agouti phenotype. If an animal is *cc*, no brown pigment is made, so the animal stays black.

C16. We know that the parents must be heterozygotes for both genes.

The genotypic ratio of their children is 1 *BB* : 2 *Bb* : 1 *Bb*

The phenotypic ratio depends on sex.

1 *BB* bald male : 1 *BB* bald female : 2 *Bb* bald males : 2 *Bb* nonbald females : 1 *bb* nonbald male : 1 *bb* nonbald female

A. 50%

B. 1/8

C. (3/8)(3/8)(3/8)= 27/512 = 0.05 or 5%

C18. A. The male offspring would be hemizygous for apricot and have 3% pigment. The female offspring would be heterozygous for apricot and white and also have 3% pigment.

B. Half of the male offspring would be hemizygous for coral and have 4% pigment, and the other half of the male offspring would be hemizygous for apricot and have 3% pigment. All of the female offspring would be heterozygous and have one copy of the dominant red allele. Therefore, all of the female offspring would have 100% pigment.

C. Half of the males would have red eyes with 100% pigment and the other half of the males would be hemizygous for the apricot allele and have 3% pigment. Half of the females would be heterozygous for the red allele and white allele and would have red eyes with 100% pigment (red is dominant). The other half of the females would be heterozygous for the white allele and apricot allele and have about 3% pigment.

D. All of the males would be hemizygous for the coral allele and have about 4% pigment. All of the females would be heterozygous for the coral and apricot alleles and have about 7% pigment.

C20. Set up a Punnett square, but keep in mind that the eosin gene is X linked.

	♂ $CX^{w\text{-}e}$	CY
♀		
CX^w	$CCX^{w\text{-}e}X^w$	CCX^wY
c^aX^w	$Cc^aX^{w\text{-}e}X^w$	Cc^aX^wY

Because *C* is dominant, the phenotypic ratios are two light-eosin females and two white males.

C22. A. Could be.

B. No, because an unaffected father has an affected daughter.

C. No, because two unaffected parents have affected children.

D. No, because an unaffected father has an affected daughter.

E. No, because both sexes exhibit the trait.

F. Could be.

C24. You would look at the pattern within families over the course of many generations. For a recessive trait, 25% of the offspring within a family are expected to be affected if both parents are unaffected carriers, and 50% of the offspring would be affected if one parent was affected. You could look at many families and see if these 25% and 50% values are approximately true. Incomplete penetrance would not necessarily predict such numbers. Also, for very rare alleles, incomplete penetrance would probably have a much higher frequency of affected parents producing affected offspring. For rare recessive disorders, it is most likely that both parents are heterozygous carriers. Finally, the most informative pedigrees would be situations in which two affected parents produce children. If they can produce an unaffected offspring, this would indicate incomplete penetrance. If all of their offspring were affected, this would be consistent with recessive inheritance.

C26. Because this is a rare trait, we assume the other parent is homozygous for the normal allele. The probability of a heterozygote passing the allele to his/her offspring is 50%. The probability of an affected offspring expressing the trait is 80%. We use the product rule to determine the likelihood of these two independent events.

$$(0.5)(0.8) = 0.4, \text{ or } 40\% \text{ of the time}$$

C28. This is an example of incomplete dominance. The heterozygous horses are palominos. For example, if *C* represents chestnut and *c* represents cremello, the chestnut horses are *CC*, the cremello horses are *cc*, and the palominos are *Cc*.

Experimental Questions

E2. Chinchilla 1 is heterozygous $c^{ch}c$.
Chinchilla 2 is heterozygous $c^{ch}c^h$.
Chinchilla 3 is heterozygous $c^{ch}c$.
Chinchilla 4 is probably $c^{ch}c^{ch}$ because it always produces chinchilla offspring when mated to chinchilla 1 or 2, which are heterozygous. However, it is possible that chinchilla 4 is also heterozygous $c^{ch}c^h$ or $c^{ch}c$.
Chinchilla 5 is $c^{ch}c^h$ or $c^{ch}c$.

As noted, we are not sure about the genotypes of chinchillas 4 and 5.

E4. The first offspring must be homozygous for the horned allele. The father's genotype is still ambiguous; he could be heterozygous or homozygous for the horned allele. The mother's genotype must be heterozygous because her phenotype is polled (she cannot be homozygous for the horned allele) but she produced a horned daughter (who must have inherited a horned allele from its mother).

E6. It is a sex-limited trait where *W* (white) is dominant but expressed only in females. In this cross of two yellow butterflies, the male is *Ww* but is still yellow because the white phenotype is limited to females. The female is *ww* and yellow. The offspring would be 50% *Ww* and 50% *ww*. However, all the males would be yellow. Half of the females would be white (*Ww*) and half would be yellow (*ww*). Overall, this would yield 50% yellow males, 25% yellow females, and 25% white females.

E8. One parent must be *RRPp*. The other parent could be *RRPp* or *RrPp*. All the offspring would inherit (at least) one dominant *R* allele. With regard to the other gene, 3/4 would inherit at least one copy of the dominant *P* allele. These offspring would have a walnut comb. The other 1/4 would be homozygous *pp* and have a rose comb (because they would also have a dominant *R* allele).

E10. Let's use the letters A and B for these two genes. Gene A exists in two alleles, which we will call A and a. Gene B exists in two alleles, B and b. The uppercase alleles are dominant to the lower-case alleles. The true-breeding long-shaped squash is $aabb$ and the true-breeding disk-shaped is $AABB$. The F_1 offspring are $AaBb$. You can construct a Punnett square to determine the outcome of self-fertilization of the F_1 plants. The Punnett square will have 16 boxes.

To get the disk-shaped phenotype, an offspring must inherit at least one dominant allele from both genes.

$1\ AABB + 2\ AaBB + 2\ AABb + 4\ AaBb = 9$ disk-shaped offspring

To get the round phenotype, an offspring must inherit at least one dominant allele for one of the two genes but must be homozygous recessive for only one of the two genes.

$1\ aaBB + 1\ AAbb + 2\ aaBb + 2\ Aabb = 6$ round-shaped offspring

To get the long phenotype, an offspring must inherit all recessive alleles: $1\ aabb$

E12. You would expect the alleles with intermediate pigmentation to exhibit a gene dosage effect because these are the ones that are not making a maximum amount of pigment. Alleles such as ivory, pearl, and apricot would be the best candidates. The alleles producing a darker reddish phenotype may not produce a gene dosage effect because a large amount of pigment is already made. To test this idea, you could set up a series of crosses, like the ones described in conceptual question C18, and then analyze the amount of pigment in the eyes of the offspring.

E14. In this cross, we expect that there will be a 9:7 ratio between red and white. In other words, 9/16 will be red and 7/16 will be white. Because there are a total of 345 plants, the expected values are

$9/16 \times 345 = 194$ red

$7/16 \times 345 = 151$ white

$$\chi^2 = \Sigma\ \frac{(O - E)^2}{E}$$

$$\chi^2 = \frac{(201 - 194)^2}{194} + \frac{(144 - 151)^2}{151}$$

$$\chi^2 = 0.58$$

With 1 degree of freedom, our chi square value is too small to reject our hypothesis. Therefore, we accept that it may be correct.

E16. Because the results of the first cross produce offspring with red eyes, it suggests that the vermilion allele and purple allele are not alleles of the same gene. The results of the second cross indicate that the vermilion allele is X linked, since all the male offspring had vermilion eyes, just like their mothers. Let pr^+ represent the red allele of the first gene and pr the purple allele. Let X^{v+} represent the red allele of the second gene and X^v the vermilion allele.

For the first cross: the male parents are $pr^+pr^+X^vY$ and the female parents are $prpr\ X^{v+}X^{v+}$. The F_1 offspring are shown in this Punnett square.

For the second cross: the male parents are $prprX^{v+}Y$ and the female parents are $pr^+pr^+X^vX^v$. The F_1 offspring are shown in this Punnett square.

Overall, the results of these two crosses indicate that the purple allele is autosomal, and the vermilion allele is X linked. To confirm this idea, the vermilion males could be crossed to homozygous wild-type females. The F_1 females should have red eyes. If these F_1 females are crossed to wild-type males, half of their sons should have vermilion eyes, if the gene is X linked. To verify that the purple allele is in an autosomal gene, a purple male could be crossed to a homozygous wild-type female. The F_1 offspring should all have red eyes. If these F_1 offspring are allowed to mate with each other, they should produce 1/4 purple offspring in the F_2 generation. In this case, the F_2 generation offspring with purple eyes would be both male and female.

Questions for Student Discussion/Collaboration

2. The easiest way to solve this problem is to take one trait at a time. With regard to combs, all the F_1 generation would be $RrPp$, or walnut comb. With regard to shanks, they would all be feathered, because they would inherit one dominant copy of a feathered allele. With regard to hen- or cock-feathering: 1 male cock-feathered : 1 male hen-feathered : 2 females hen-feathered. Overall then, we would have a 1:1:2 ratio of

walnut comb/feathered shanks/cock-feathered males
walnut comb/feathered shanks/hen-feathered males
walnut comb/feathered shanks/hen-feathered females

CHAPTER 5

Conceptual Questions

C2. An independent assortment hypothesis is used because it enables us to calculate the expected values based on Mendel's ratios. Using the observed and expected values, we can calculate whether or not the deviations between the observed and expected values are too large to occur as a matter of chance. If the deviations are very large, we reject the hypothesis of independent assortment.

C4.

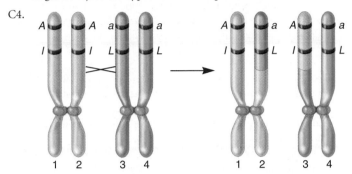

If the chromosomes labeled 2 and 4 move into one daughter cell, that will lead to a patch that is albino and has long hair. The other cell will receive chromosomes 1 and 3, which will produce a patch that has dark, short hair.

C6. A single crossover produces *A B C*, *A b c*, *a B C*, and *a b c*.

A. Between 2 and 3, between genes *B* and *C*

B. Between 1 and 4, between genes *A* and *B*

C. Between 1 and 4, between genes *B* and *C*

D. Between 2 and 3, between genes *A* and *B*

C8. The likelihood of scoring a basket would be greater if the basket was larger. Similarly, the chances of a crossover initiating in a region between two genes is proportional to the size of the region between the two genes. There are a finite number (usually a few) that occur between homologous chromosomes during meiosis, and the likelihood that a crossover will occur in a region between two genes depends on how big that region is.

C10. The pedigree suggests a linkage between the dominant allele causing nail-patella syndrome and the I^B allele of the ABO blood type gene. In every case, the individual who inherits the I^B allele also inherits this disorder.

C12. <u>*Ass-1*   43   *Sdh-1*   5   *Hdc*   9   *Hao-1*   6   *Odc-2*   8   *Ada-1*</u>

C14. The inability to detect double crossovers causes the map distance to be underestimated. In other words, there are more crossovers occurring in the region than we realize. When we have a double crossover, we do not get a recombinant offspring (in a dihybrid cross). Therefore, the second crossover cancels out the effects of the first crossover.

C16. The key feature is that all the products of a single meiosis are contained within a single sac. The spores in this sac can be dissected out, and then their genetic traits can be analyzed individually. In fungi, the traits most commonly analyzed are nutritional growth requirements (as described in chapter 5), as well as pigmentation and antibiotic resistance.

C18. In an unordered ascus, the products of meiosis are free to move around. In an ordered octad (or tetrad), they are lined up accord-

ing to their relationship to each other during meiosis and mitosis. An ordered octad can be used to map the distance between a single gene and its centromere.

C20. They would be higher with respect to gene *A*. First-division segregation patterns occur when there is not a crossover between the centromere and the gene of interest. Because gene *A* is closer to the centromere compared to gene *B*, it would be less likely to have a crossover between gene *A* and the centromere. This would make it more likely to observe first-division segregation.

Experimental Questions

E2. They could have used a strain with two abnormal chromosomes. In this case, the recombinant chromosomes would either look normal or have abnormalities at both ends.

E4. A gene on the Y chromosome in mammals would only be transmitted from father to son. It would be difficult to genetically map Y-linked genes because a normal male has only one copy of the Y chromosome, so you do not get any crossing over between two Y chromosomes. Occasionally, abnormal males (XYY) are born with two Y chromosomes. If such males were heterozygous for alleles of Y-linked genes, one could examine the normal male offspring of XYY fathers and determine if crossing over has occurred.

E6. The answer is explained in solved problem S5. We cannot get more than 50% recombinant offspring because the pattern of multiple crossovers can yield an average maximum value of only 50%. When a testcross does yield a value of 50% recombinant offspring, it can mean two different things. Either the two genes are on different chromosomes or the two genes are on the same chromosome but at least 50 mu apart.

E8. If two genes are at least 50 mu apart, you would need to map genes in between them to show that the two genes were actually in the same linkage group. For example, if gene *A* was 55 mu from gene *B*, there might be a third gene (e.g., gene *C*) that was 20 mu from *A* and 35 mu from *B*. These results would indicate that *A* and *B* are 55 mu apart, assuming dihybrid testcrosses between genes *A* and *B* yielded 50% recombinant offspring.

E10. Sturtevant used the data involving the following pairs: *y* and *w*, *w* and *v*, *v* and *r*, and *v* and *m*.

E12. A. Because they are 12 mu apart, we expect 12% (or 120) recombinant offspring. This would be approximately 60 *Aabb* and 60 *aaBb* plus 440 *AaBb* and 440 *aabb*.

B. We would expect 60 *AaBb*, 60 *aabb*, 440 *Aabb*, and 440 *aaBb*.

E14. Due to the large distance between the two genes, they will assort independently even though they are actually on the same chromosome. According to independent assortment, we expect 50% parental and 50% recombinant offspring. Therefore, this cross will produce 150 offspring in each of the four phenotypic categories.

E16. A. If we hypothesize two genes independently assorting, then the predicted ratio is 1:1:1:1. There are a total of 390 offspring. The expected number of offspring in each category is about 98. Plugging the figures into our chi square formula,

$$\chi^2 = \frac{(117-98)^2}{98} + \frac{(115-98)^2}{98} + \frac{(78-98)^2}{98} + \frac{(80-98)^2}{98}$$

$$\chi^2 = 3.68 + 2.95 + 4.08 + 3.31$$

$$\chi^2 = 14.02$$

Looking up this value in the chi square table under 3 degrees of freedom, we reject our hypothesis, since the chi square value is above 7.815.

B. Map distance:

$$\text{Map distance} = \frac{78 + 80}{117 + 115 + 78 + 80}$$

$$= 40.5 \text{ mu}$$

Because the value is relatively close to 50 mu, it is probably a significant underestimate of the true distance between these two genes.

E18. The percentage of recombinants for the green, yellow and wide, narrow is 7%, or 0.07; there will be 3.5% of the green, narrow and 3.5% of the yellow, wide. The remaining 93% parentals will be 46.5% green, wide and 46.5% yellow, narrow. The third gene assorts independently. There will be 50% long and 50% short with respect to each of the other two genes. To calculate the number of offspring out of a total of 800, we multiply 800 by the percentages in each category.

(0.465 green, wide)(0.5 long)(800) = 186 green, wide, long
(0.465 yellow, narrow)(0.5 long)(800) = 186 yellow, narrow, long
(0.465 green, wide)(0.5 short)(800) = 186 green, wide, short
(0.465 yellow, narrow)(0.5 short)(800) = 186 yellow, narrow, short
(0.035 green, narrow)(0.5 long)(800) = 14 green, narrow, long
(0.035 yellow, wide)(0.5 long)(800) = 14 yellow, wide, long
(0.035 green, narrow)(0.5 short)(800) = 14 green, narrow, short
(0.035 yellow, wide)(0.5 short)(800) = 14 yellow, wide, short

E20. Let's use the following symbols: G for green pods, g for yellow pods, S for green seedlings, s for bluish green seedlings, C for normal plants, c for creepers.

The parental cross is $GG\ SS\ CC$ crossed to $gg\ ss\ cc$.

The F_1 plants would all be $Gg\ Ss\ Cc$. If the genes are linked, the alleles G, S, and C would be linked on one chromosome and the alleles g, s, and c would be linked on the homologous chromosome.

The testcross is F_1 plants, which are $Gg\ Ss\ Cc$, crossed to $gg\ ss\ cc$.

To measure the distances between the genes, we can separate the data into gene pairs.

Pod color, seedling color

2,210 green pods, green seedlings—nonrecombinant
296 green pods, bluish green seedlings—recombinant
2,198 yellow pods, bluish green seedlings—nonrecombinant
293 yellow pods, green seedlings—recombinant

$$\text{Map distance} = \frac{296 + 293}{2,210 + 296 + 2,198 + 293} \times 100 = 11.8 \text{ mu}$$

Pod color, plant stature

2,340 green pods, normal—nonrecombinant
166 green pods, creeper—recombinant
2,323 yellow pods, creeper—nonrecombinant
168 yellow pods, normal—recombinant

$$\text{Map distance} = \frac{166 + 168}{2,340 + 166 + 2,323 + 168} \times 100 = 6.7 \text{ mu}$$

Seedling color, plant stature

2,070 green seedlings, normal—nonrecombinant
433 green seedlings, creeper—recombinant
2,056 bluish green seedlings, creeper—nonrecombinant
438 bluish green seedlings, normal—recombinant

$$\text{Map distance} = \frac{433 + 438}{2,070 + 433 + 2,056 + 438} \times 100 = 17.4 \text{ mu}$$

The order of the genes is seedling color, pod color, plant stature or you could say the opposite order. Pod color is in the middle. If we use the two shortest distances to construct our map:

S	11.8	G	6.7	C

E22. According to a hypothesis of independent assortment, we would expect an equal proportion of all four phenotypes. (You should construct a Punnett square if this is not apparent.) There are a total number of 1,074 offspring. Therefore, the expected number of each of the four categories is 1/4 × 1,074, which equals 268 (rounded to the nearest whole number). We use the value of 268 as our expected value in the chi square calculation.

$$\chi^2 = \frac{(368 - 268)^2}{268} + \frac{(160 - 268)^2}{268} + \frac{(194 - 268)^2}{268} + \frac{(352 - 268)^2}{268}$$

$$\chi^2 = 37.3 + 43.5 + 20.4 + 26.3$$

$$\chi^2 = 127.5$$

If we look up the value of 127.5 in our chi square table, with 3 degrees of freedom, the value lies far beyond the 0.01 probability level. Therefore, it is very unlikely to get such a large deviation if our hypothesis of independent assortment is correct. Therefore, we reject our hypothesis and conclude that the genes are linked. In the parental (true-breeding) generation, black is linked to chinchilla and brown is linked to Himalayan. Therefore, the recombinant offspring are black Himalayan and brown chinchilla.

To compute map distance:

$$\frac{160 + 194}{368 + 160 + 194 + 352} \times 100 = 33.0 \text{ mu}$$

E24. A.

	Parent						Parent			
b	7	A	4	C	×	B	7	a	4	c
b	7	A	4	C		B	7	a	4	c

					Offspring
b	7	A	4	C	
B	7	a	4	c	

B. A heterozygous F_2 offspring would have to inherit a chromosome carrying all of the dominant alleles. In the F_1 parent (of the F_2 offspring), a crossover in the 7 mu region between genes b and A (and between B and a) would yield a chromosome that was $B\ A\ C$ and $b\ a\ c$. If an F_2 offspring inherited the $B\ A\ C$ chromosome from its F_1 parent and the $b\ a\ c$ chromosome from the homozygous parent, it would be heterozygous for all three genes.

C. If you look at the answer to part B, a crossover between genes b and A (and between B and a) would yield $B\ A\ C$ and $b\ a\ c$ chromosomes. If an offspring inherited the $b\ a\ c$ chromosome from its F_1 parent and the $b\ a\ c$ chromosome from its homozygous parent, it would be homozygous for all three genes. The chances of a crossover in this region are 7%. However, half of this 7% crossover event yields chromosomes that are *B A C* and the other half yields chromosomes that are *b a c*. Therefore, the chances are 3.5% of getting homozygous F$_2$ offspring.

E26. A. The first thing to do is to determine which asci are parental ditypes (PD), nonparental ditypes (NPD), and tetratypes (T). A parental ditype will contain a 2:2 combination of spores with the same genotypes as the original haploid parents. The combination of the 502 asci are the parental ditypes. The nonparental ditypes are those containing a 2:2 combination of genotypes that are unlike the parentals. The combination of four asci fits this description. Finally, the tetratypes contain a 1:1:1:1 arrangement of genotypes, half of which have a parental genotype and half of which do not. There are 312 tetratypes in this case. Computing the map distance:

$$\text{Map distance} = \frac{\text{NPD} + (1/2)(\text{T})}{\text{Total number of asci}} \times 100$$

$$= \frac{4 + (1/2)(312)}{818}$$

$$= 19.6 \text{ mu}$$

If we use the more accurate equation:

$$\text{Map distance} = \frac{\text{T} + 6\text{NPD}}{\text{Total number of asci}} \times 0.5 \times 100$$

$$= \frac{312 + (6)(4)}{818}$$

$$= 20.5 \text{ mu}$$

B. The frequency of single crossovers is 0.205 if we use the more accurate equation.

C. Nonparental ditypes are produced from a double crossover. To compute the expected number, we multiply 0.205 × 0.205 = 0.042, or 4.2%. Since we had a total of 818 asci, we would expect 34.3 asci to be the product of a double crossover. However, as described in figure 5.17, only 1/4 of them would be a non-parental ditype. Therefore, we multiply 34.3 by 1/4, obtaining a value of 8.6 nonparental ditypes due to a double crossover. Since we observed only four, this calculation tells us that positive interference is occurring.

E28. A. Types   B. Number

pro-1 pro-1 pro-1 pro-1 pro$^+$ pro$^+$ pro$^+$ pro$^+$	402
pro$^+$ pro$^+$ pro$^+$ pro$^+$pro-1 pro-1 pro-1 pro-1	402
pro$^+$ pro$^+$pro-1 pro-1 pro$^+$ pro$^+$ pro-1 pro-1	49
pro-1 pro-1 pro$^+$ pro$^+$ pro-1 pro-1 pro$^+$ pro$^+$	49
pro$^+$ pro$^+$pro-1 pro-1 pro-1 pro-1 pro$^+$ pro$^+$	49
pro-1 pro-1 pro$^+$ pro$^+$ pro$^+$ pro$^+$ pro-1 pro-1	49

Questions for Student Discussion/Collaboration

2. The X and Y chromosomes are not completely distinct linkage groups. One might describe them as overlapping linkage groups having some genes in common, but most genes of which are not common to both.

CHAPTER 6

Conceptual Questions

C2. It is not a form of sexual reproduction whereby two distinct parents produce gametes that unite to form a new individual. However, conjugation is similar to sexual reproduction in the sense that the genetic material from two cells are somewhat mixed. In conjugation, there is not the mixing of two genomes, one from each gamete. Instead, there is a transfer of genetic material from one cell to another. This transfer can alter the combination of genetic traits in the recipient cell.

C4. An *F$^+$* strain contains a separate, circular piece of DNA that has its own origin of transfer. An *Hfr* strain has its origin of transfer integrated into the bacterial chromosome. An *F$^+$* strain can transfer only the DNA contained on the F factor. If given enough time, an *Hfr* strain can actually transfer the entire bacterial chromosome to the recipient cell.

C6. Sex pili promote the binding of donor and recipient cells and provide a passageway for the transfer of genetic material from the donor to the recipient cell.

C8. Though exceptions are common, interspecies genetic transfer via conjugation is not as likely because the cell surfaces do not interact correctly. Interspecies genetic transfer via transduction is also not very likely because each species of bacteria is sensitive to particular bacteriophages. The correct answer is transformation. A consequence of interspecies genetic transfer is that new genes can be introduced into a bacterial species from another species. For example, interspecies genetic transfer could provide the recipient bacterium with a new trait such as resistance to an antibiotic. Evolutionary biologists call this horizontal gene transfer, while the passage of genes from parents to offspring is termed vertical gene transfer.

C10. Cotransduction is the transduction of two or more genes. The distance between the genes determines the frequency of cotransduction. When two genes are close together, the cotransduction frequency would be higher compared to two genes that are relatively farther apart.

C12. If a site that frequently incurred a breakpoint was between two genes, the cotransduction frequency of these two genes would be much less than expected. This is because the site where the breakage occurred would separate the two genes from each other.

C14. The transfer of conjugative plasmids such as F factor DNA.

C16. A. If it occurred in a single step, transformation is the most likely mechanism because conjugation does not usually occur between different species, particularly distantly related species, and different species are not usually infected by the same bacteriophages.

B. It could occur in a single step, but it may be more likely to have involved multiple steps.

C. The use of antibiotics selects for the survival of bacteria that have resistance genes. If a population of bacteria is exposed to an antibiotic, those carrying resistant genes will survive and their relative numbers will increase in subsequent generations.

C18. The term *allele* means alternative forms of the same gene. Therefore, mutations in the same gene among different phages are alleles of each other; the mutations may be at different positions within the same gene. When we map the distance between mutations in the same gene, we are mapping the distance between the mutations that create different alleles of the same gene. An intragenic map describes the locations of mutations within the same gene.

Experimental Questions

E2. Mix the two strains together and then put some of them on plates containing streptomycin and some of them on plates without streptomycin. If mated colonies are present on both types of plates, then the phe^+ and thr^+ genes were transferred to the bio^+ met^+ phe^- thr^- strain. If colonies are found only on the plates that lack streptomycin, then the bio^+ and met^+ genes are being transferred to the bio^- met^- phe^+ thr^+ strain. This answer assumes a one-way transfer of genes from a donor to a recipient strain.

E4. An interrupted mating experiment is a procedure in which two bacterial strains are allowed to mate, and then the mating is interrupted at various time points. The interruption occurs by agitation of the solution in which the bacteria are found. This type of study is used to map the locations of genes. It is necessary to interrupt mating so that you can vary the time and obtain information about the order of transfer; which gene transferred first, second, etc.

E6. Mate unknown strains A and B to the F^- strain in your lab that is resistant to streptomycin and cannot use lactose. This is done in two separate tubes (i.e., strain A plus your F^- strain in one tube, and strain B plus your F^- strain in the other tube). Plate the mated cells on growth media containing lactose plates plus streptomycin. If you get growth of colonies, the unknown strain had to be the F^+ strain that had lactose utilization genes on its F factor.

E8. A. If we extrapolate these lines back to the x-axis, the $hisE$ intersects at about 3 minutes and the $pheA$ intersects at about 24 minutes. These are the values for the times of entry. Therefore, the distance between these two genes is 21 minutes (i.e., 24 minus 3).

   B. _____

   ↑   4   ↑   17   ↑

   $hisE$ $pabB$ $pheA$

E10. One possibility is that you could treat the P1 lysate with Dnase I, an enzyme that digests DNA. (Note: If DNA were digested with Dnase I, the function of any genes within the DNA would be destroyed.) If the DNA were within a P1 phage, it would be protected from Dnase I digestion. This would allow you to distinguish between transformation (which would be inhibited by Dnase I) versus transduction (which would not be inhibited by Dnase I). Another possibility is that you could try to fractionate the P1 lysate. Naked DNA would be smaller than a P1 phage carrying DNA. You could try to filter the lysate to remove naked DNA, or you could subject the lysate to centrifugation and remove the lighter fractions that contain naked DNA.

E12. Cotransduction frequency = $(1 - d/L)^3$

   For the normal strain:

   Cotransduction frequency = $(1 - 0.7/2)^3 = 0.275$, or 27.5%

   For the new strain:

   Cotransduction frequency = $(1 - 0.7/5)^3 = 0.64$, or 64%

   The experimental advantage is that you could map genes that are farther than 2 minutes apart. You could map genes that are up to 5 minutes apart.

E14. Cotransduction frequency = $(1 - d/L)^3$

$$0.53 = (1 - d/2 \text{ minutes})^3$$

$$(1 - d/2 \text{ minutes}) = \sqrt[3]{0.53}$$

$$(1 - d/2 \text{ minutes}) = 0.81$$

$$d = 0.38 \text{ minutes}$$

E16. A. We first need to calculate the cotransformation frequency, which equals 2/70, or 0.029.

   Cotransformation frequency = $(1 - d/L)^3$

$$0.029 = (1 - d/2 \text{ minutes})^3$$

$$d = 1.4 \text{ minutes}$$

   B. Cotransformation frequency = $(1 - d/L)^3$

$$= (1 - 1.4/4)^3$$

$$= 0.27$$

   As you may have expected, the cotransformation frequency is much higher when the transformation involves larger pieces of DNA.

E18. Benzer could use this observation as a way to evaluate if intragenic recombination had occurred. If two rII mutations recombined to make a wild-type gene, the phage would produce plaques in this $E.\ coli\ K12(\lambda)$ strain.

E20. $rIIA$: L47, L92

   $rIIB$: L33, L40, L51, L62, L65, and L91

E22. Benzer first determined the individual nature of each gene by showing that mutations within the same gene did not complement each other. He then could map the distance between two mutations within the same gene. The map distances defined each gene as a linear, divisible unit. In this regard, the gene is divisible due to crossing over.

Questions for Student Discussion/Collaboration

2. Epistasis is another example. When two different genes affect flower color, the white allele of one gene may be epistatic to the purple allele of another gene. A heterozygote (e.g., $CcPp$) for both genes would have purple flowers. In other words, the two purple alleles (C and P) are complementing the two white alleles (c and p). Other similar examples are discussed in chapter 4 in section 4.2 on Gene Interactions.

CHAPTER 7

Conceptual Questions

C2. A maternal effect gene is one in which the genotype of the mother determines the phenotype of the offspring. At the cellular level, this happens because maternal effect genes are expressed in the diploid nurse cells and then the gene products are transported into the oocyte. These gene products play key roles in the early steps of embryonic development.

C4. The genotype of the mother must be bic^- bic^-. That is why it produces abnormal offspring. Since the mother is alive and able to produce offspring, its mother (the maternal grandmother) must have been bic^+ bic^- and passed the bic^- allele to its daughter (the mother in this problem). The maternal grandfather also must have passed the *bic⁻* allele to its daughter. The maternal grand-father could be either *bic⁺ bic⁻* or *bic⁻ bic⁻*.

C6. The mother must be heterozygous. She is phenotypically abnormal because her mother must have been homozygous for the abnormal recessive allele. However, because she produces all normal off-spring, she must have inherited the normal dominant allele from her father. She produces all normal offspring because this is a maternal effect gene, and the gene product of the normal domi-nant allele is transferred to the oocyte.

C8. Maternal effect genes exert their effects because the gene products are transferred from nurse cells to oocytes. The gene products, mRNA and proteins, do not last a very long time before they are eventually degraded. Therefore, they can exert their effects only during early stages of embryonic development.

C10. Dosage compensation refers to the phenomenon that the genes on the sex chromosomes are expressed at similar levels even though the two sexes have different numbers of sex chromosomes. It may not always be necessary, but in many species it seems necessary so that the balance of gene expression between the autosomes and sex chromosomes is similar between the two sexes.

C12. Mammals: X inactivation in females

Fruit flies: double transcription on the X chromosome in males

Worms: decrease transcription of the X chromosome in hermaph-rodites to 50%

C14. X inactivation begins with the counting of *Xic*s. If there are two X chromosomes, one is targeted for inactivation. During embryo-genesis, this inactivation begins at the *Xic* locus and spreads to both ends of the X chromosome until it becomes a highly con-densed Barr body. The *TsiX* gene may play a role in the choice of the X chromosome that remains active. This may occur by its abil-ity to bind to *Xist* mRNA. The *Xist* gene, which is located in the *Xic* region, remains transcriptionally active on the inactivated X chromosome. It is thought to play an important role in X inactiva-tion by coating the inactive X chromosome. After the inactivation is established, it is maintained in the same X chromosome in somatic cells during subsequent cell divisions. In germ cells, how-ever, the X chromosomes are not inactivated so that an egg can transmit either copy of an active (noncondensed) X chromosome.

C16. A. One

B. Zero

C. Two

D. Zero

C18. The offspring inherited X^B from its mother and X^O and Y from its father. It is an XXY animal, which is male (but somewhat feminized).

C20. Erasure and reestablishment of the imprint occur during gameto-genesis. It is necessary to erase the imprint because each sex will either imprint or not imprint both alleles of a gene. In somatic cells, the two alleles for a gene are imprinted dissimilarly, depend-ing on the sex from which they were inherited.

C22. A person born with paternal uniparental disomy 15 would have Angelman syndrome, because this individual would not have an active copy of the *AS* gene; the paternally inherited copies of the *AS* gene are silenced. This individual (if he/she reproduced) would have normal offspring, because he/she does not have a deletion in either copy of chromosome 15.

C24. In some species, such as marsupials, X inactivation depends on the sex. This is similar to imprinting. Also, once X inactivation occurs during embryonic development, it is remembered throughout the rest of the life of the organism. Again, this is similar to imprinting. X inactivation in placental mammals is different from genomic imprinting in that it is not sex dependent. The X chromosome that is inactivated could be inherited from the mother or the father. There was no marking process on the X chromosome that occurred during gametogenesis. In contrast, genomic imprinting always involves a marking process during gametogenesis.

C26. The term *reciprocal cross* refers to two parallel crosses that involve the same genotypes of the two parents, but their sexes are opposite in the two crosses. For example: female *BB* × male *bb* and a recip-rocal cross in which a female *bb* × male *BB*. Autosomal inheritance gives the same result because the autosomes are transmitted from parent to offspring in the same way for both sexes. For extranuclear inheritance, the mitochondria and plastids are not transmitted via the gametes in the same way for both sexes. For maternal inheri-tance, the reciprocal crosses would show that the gene is always inherited from the mother.

C28. The phenotype of a petite mutant is that it forms small colonies on growth media that contain an energy source that does not require mitochondrial function. These mutants are unable to grow on an energy source that requires mitochondrial function. Because nuclear and mitochondrial genes are necessary for mitochondrial function, it is possible for a petite mutation to involve a gene in the nucleus or in the mitochondrial genome. Neutral petites lack most of their mitochondrial DNA, while suppressive petites usually lack small segments of the mitochondrial genetic material.

C30. The mitochondrial and chloroplast genomes are composed of a circular chromosome found in one or more copies. These copies are located in a region of the organelle known as the nucleoid. The number of genes per chromosome varies from species to species. Mitochondria tend to have fewer genes compared to chloroplasts. See table 7.3 for examples of the variation among mitochondrial and chloroplast genomes.

C32. A. Yes.

B. Yes.

C. No, it is determined by a gene in the chloroplast genome.

D. No, it is determined by a mitochondrial gene.

C34. Biparental extranuclear inheritance would resemble Mendelian inheritance in that offspring could inherit alleles of a given gene from both parents. It differs, however, when you think about it from the perspective of heterozygotes. For a Mendelian trait, the law of segregation tells us that a heterozygote passes one allele for a given gene to an offspring, but not both. In contrast, if a parent has a mixed population of mitochondria (e.g., some carrying a mutant gene and some carrying a normal gene), that parent could pass both types of genes (mutant and normal) to a single offspring, because more than one mitochondrion could be contained within a sperm or egg cell.

Experimental Questions

E2. The first type of observation was cytological. The presence of the Barr body in female cells was consistent with the idea that one of the X chromosomes was highly condensed. The second type was genetic. A variegated phenotype that is found only in females is consistent with the idea that certain patches express one allele and other patches express the other allele. This variegated phenotype would occur only if the inactivation happened at an early stage of embryonic development and was inherited permanently thereafter.

E4. The pattern of inheritance is consistent with imprinting. In every cross, the allele that is inherited from the father is expressed in the offspring, while the allele inherited from the mother is not.

E6. We assume that the snails in the large colony on the second island are true-breeding, *DD*. Let the male snail from the deserted island mate with a female snail from the large colony. Then let the F_1 snails mate with each other to produce an F_2 generation. Then let the F_2 generation mate with each other to produce an F_3 generation. Here are the expected results:

Female *DD* × Male *DD*

All F_1 snails coil to the right.
All F_2 snails coil to the right.
All F_3 snails coil to the right.

Female *DD* × Male *Dd*

All F_1 snails coil to the right.
All F_2 snails coil to the right because all of the F_1 females are *DD* or *Dd*.
15/16 of F_3 snails coil to the right, 1/16 of F_3 snails coil to the left (because 1/16 of the F_2 females are *dd*).

Female *DD* × Male *dd*.

All F_1 snails coil to the right.
All F_2 snails coil to the right because all of the F_1 females are *Dd*.
3/4 of F_3 snails coil to the right, 1/4 of F_3 snails coil to the left (because 1/4 of the F_2 females are *dd*).

E8. Let's first consider the genotypes of male A and male B. Male A must have two normal copies of the *Igf-2* gene. We know this because male A's mother was *Igf-2 Igf-2*; the father of male A must have been a heterozygote *Igf-2 Igf-2m* because half of the litter that contained male A also contained dwarf offspring. But since male A was not dwarf, it must have inherited the normal allele from its father. Therefore, male A must be *Igf-2 Igf-2*. We cannot be completely sure of the genotype of male B. It must have inherited the normal *Igf-2* allele from its father because male B is phenotypically normal. We do not know the genotype of male B's mother, but she could be either *Igf-2m Igf-2m* or *Igf-2 Igf-2m*. In either case, the mother of male B could pass the *Igf-2m* allele to an offspring, but we do not know for sure if she did. So, male B could be either *Igf-2 Igf-2m* or *Igf-2 Igf-2*.

For the *Igf-2* gene, we know that the maternal allele is inactivated. Therefore, the genotypes and phenotypes of females A and B are irrelevant. The phenotype of the offspring is determined only by the allele that is inherited from the father. Because we know that male A has to be *Igf-2 Igf-2*, we know that it can produce only normal offspring. Because both females A and B both produced dwarf offspring, male A cannot be the father. In contrast, male B could be either *Igf-2 Igf-2* or *Igf-2 Igf-2m*. Because both females gave birth to dwarf babies (and since male A and male B were the only two male mice in the cage), we conclude that male B must be *Igf-2 Igf-2m* and is the father of both litters.

E10. In mice, one of the two X chromosomes is inactivated; that is why females and males produce the same total amount of mRNA for most X-linked genes. In fruit flies, the expression of a male's X-linked genes is turned up twofold. In *C. elegans,* the expression

of hermaphrodite X-linked genes is turned down twofold. Overall, the total amount of expression of X-linked genes is the same in males and females (or hermaphrodites) of these three species. In fruit flies and *C. elegans,* heterozygous females and hermaphrodites express 50% of each allele compared to a homozygous male, so that heterozygous females and hermaphrodites produce the same total amount of mRNA from X-linked genes compared to males. Note: In heterozygous females of mice, fruit flies, and worms, there is 50% of each gene product (compared to hemizygous males and homozygous females).

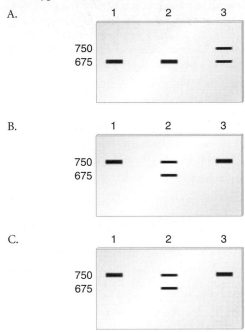

E12. In the absence of UV light, we would expect all sm^r offspring. With UV light, we would expect some sm^s offspring.

Questions for Student Discussion/Collaboration

2. An infective particle is something in the cytoplasm that contains its own genetic material and is not an organelle. Some symbiotic infective particles, such as those found in killer paramecia, are similar to mitochondria and chloroplasts since they contain their own genomes and are known to be bacterial in origin. The observation that these endosymbiotic relationships can initiate in modern species tells us that endosymbiosis can spontaneously happen. Therefore, it is reasonable that it happened a long time ago and led to the evolution of mitochondria and plastids.

CHAPTER 8

Conceptual Questions

C2. Small deletions and duplications are less likely to affect phenotype simply because they usually involve fewer genes. If a small deletion did have a phenotypic effect, you would conclude that a gene or genes in this region are required in two functional copies in order to have a normal phenotype.

C4. A gene family is a group of genes that are derived from the process of gene duplications. They have similar sequences, but the sequences have some differences due to the accumulation of mutations over many generations. The members of a gene family usually encode proteins with similar but specialized functions. The specialization may occur in different cells or at different stages of development.

C6. It has a pericentric inversion.

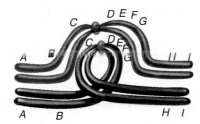

C8. There are four products from meiosis. One would be a normal chromosome and one would contain the inversion shown in the drawing to conceptual question C8. The other two chromosomes would be dicentric or acentric with the following order of genes:

centromere		centromere	
A ↓ B C D E F G H I J D C B	↓	A	Dicentric
M L K J I H G F E K L M			Acentric

C10. In the absence of crossing over, alternate segregation would yield half the cells with two normal chromosomes and half with a balanced translocation. For adjacent-1 segregation, two cells would be

  A B C D E + *A I J K L M*

And the other two cells would be:

  H B C D E + *H I J K L M*

C12. One of the parents may carry a balanced translocation between chromosomes 5 and 7. The phenotypically abnormal offspring has inherited an imbalanced translocation due to the segregation of translocated chromosomes during meiosis.

C14. A deficiency and an unbalanced translocation are more likely to have phenotypic effects because they create genetic imbalances. For a deficiency, there are too few copies of several genes, and for an unbalanced translocation, there are too many.

C16. You should draw out the inversion loop (as is done in fig. 8.12a). The crossover occurred between *P* and *U*.

C18. This person has a total of 46 chromosomes. However, this person would be considered aneuploid rather than euploid. This is because one of the sets is missing a sex chromosome and one set has an extra copy of chromosome 21.

C20. It may be related to genetic balance. In aneuploidy, there is an imbalance in gene expression between the chromosomes found in their normal copy number versus those that are either too many or too few. In polyploidy, the balance in gene expression is still maintained.

C22. The male offspring is the result of nondisjunction during oogenesis. The female produced an egg without any sex chromosomes. The male parent transmitted a single X chromosome carrying the red allele. This produces an X0 male offspring with red eyes.

C24. Trisomies 13, 18, and 21 survive because the chromosomes are small and probably contain fewer genes compared to the larger chromosomes. Individuals with abnormal numbers of X chromosomes can survive because the extra copies are converted to transcriptionally inactive Barr bodies. The other aneuploidies are lethal

because they cause a great amount of imbalance between the level of gene expression on the normal diploid chromosomes relative to the chromosomes that are trisomic or monosomic.

C26. Endopolyploidy means that a particular somatic tissue is polyploid even though the rest of the organism is not. The biological significance is not entirely understood although it has been speculated that an increase in ploidy may enable the cell to make more gene products that the cell needs.

C28. In certain types of cells, such as salivary cells, the homologous chromosomes pair with each other and then replicate about nine times to produce a polytene chromosome. The centromeres from each type of chromosome associate with each other at the chromocenter. This structure has six arms that arise from one arm of two telomeric chromosomes (the X and 4) and two arms each from chromosomes 2 and 3.

C30. The turtles are two distinct species that appear phenotypically identical. The turtles with 48 chromosomes are polyploid relatives (i.e., tetraploids) of the species with 24 chromosomes. In animals, it is somewhat hard to imagine how this could occur because animals cannot self-fertilize, so there had to be two animals (i.e., one male and one female) that became tetraploids. It is easy to imagine how one animal could become a tetraploid; complete nondisjunction could occur during the first cell division of a fertilized egg, thereby creating a tetraploid cell that continued to develop into a tetraploid animal. This would have to happen independently (i.e., in two individuals of opposite sex) to create a tetraploid species. If you mated a tetraploid turtle with a diploid turtle, the offspring would be triploid and probably phenotypically normal. However, the triploid offspring would be sterile because they would make highly aneuploid gametes.

C32. Polyploid, triploid, and euploid should not be used.

C34. The boy carries a translocation involving chromosome 21: probably a translocation in which nearly all of chromosome 21 is translocated to chromosome 14. He would have one normal copy of chromosome 14, one normal copy of chromosome 21, and the translocated chromosome that contains both chromosome 14 and chromosome 21. This boy is phenotypically normal because the total amount of genetic material is normal, although the total number of chromosomes is 45 (because chromosome 14 and chromosome 21 are fused into a single chromosome). His sister has Down syndrome because she has inherited the translocated chromosome, but she also must have one copy of chromosome 14 and two copies of chromosome 21. She has the equivalent of three copies of chromosome 21 (i.e., two normal copies and one copy fused with chromosome 14). This is why she has Down syndrome. One of the parents of these two children is probably normal with regard to karyotype (i.e., the parent has 46 normal chromosomes). The other parent would have a karyotype that would be like the phenotypically normal boy.

C36. Nondisjunction is a mechanism whereby the chromosomes do not segregate equally into the two daughter cells. This can occur during meiosis to produce cells with altered numbers of chromosomes, or it can occur during mitosis to produce a genetic mosaic individual. A third way to alter chromosome number is by interspecies crosses to produce an alloploid.

C38. A mutation occurred during early embryonic development to create the blue patch of tissue. One possibility is a mitotic nondisjunction in which the two chromosomes carrying the *b* allele went to one cell and the two chromosomes carrying the *B* allele went to the other daughter cell. A second possibility is that the chromosome carrying the *B* allele could be lost. A third possibility is that the *B* allele could have been deleted. This would cause the recessive *b* allele to exhibit pseudodominance.

C40. Homeologous chromosomes are chromosomes from two species that are evolutionarily related to each other. For example, chromosome 1 in chimpanzees and gorillas is homeologous; it carries the same types of genes.

C42. In general, Turner syndrome could be due to nondisjunction during oogenesis or spermatogenesis. However, the Turner individual with color blindness is due to nondisjunction during spermatogenesis. The sperm lacked a sex chromosome, due to nondisjunction, and the egg carried an X chromosome with the recessive color blindness allele. This X chromosome had to be inherited from the mother, because the father was not color-blind. The mother must be heterozygous for the recessive color blind allele, and the father is hemizygous for the normal allele. Therefore, the mother must have transmitted a single X chromosome carrying the color-blind allele to her offspring indicating that nondisjunction did not occur during oogenesis.

Experimental Questions

E2. Colchicine interferes with the spindle apparatus and thereby causes nondisjunction. At high concentrations, it can cause complete nondisjunction and produce polyploid cells.

E4. The primary purpose is to generate strains that are homozygous for all of their genes. Because the pollen is haploid, it has only one copy of each gene. The strain can later be made diploid (and homozygous) by treatment with colchicine.

E6. First, you would cross the two strains together. It is difficult to predict the phenotype of the offspring. Nevertheless, you would keep crossing offspring to each other and backcrossing them to the parental strains until you obtained a great-tasting tomato strain that was resistant to heat and the viral pathogen. You could then make this strain tetraploid by treatment with colchicine. If you crossed the tetraploid strain with your great-tasting diploid strain that was resistant to heat and the viral pathogen, you may get a triploid that had these characteristics. This triploid would probably be seedless.

E8. A polytene chromosome is formed when a chromosome replicates many times, and the chromatids lie side by side as shown in figure 8.21. The homologous chromosomes also lie side by side. Therefore, if there is a deletion, there will be a loop. The loop is the segment that is not deleted from one of the two homologues.

E10. The order, from most intense to least intense:

1 X chromosome-*bar*, 1 X chromosome-*ultra-bar*
1 X chromosome-*normal* and 1 X chromosome-*ultra-bar*
2 X chromosomes-*bar*
1 X chromosome-*normal* and 1 X chromosome-*bar*

It is interesting to note that the second and third genotypes of flies both have four copies of the gene. However, the fly having three copies on 1 X chromosome (1 X chromosome-*ultra-bar* and one copy on the other X chromosome) would produce more mRNA because of the position effect. That is why it has fewer facets (45) compared to a *bar* homozygote, which has 70 facets.

E12. Because the plant giving the pollen is heterozygous for many genes, some of the pollen grains may be haploid for recessive alleles, which are nonbeneficial or even lethal. However, some pollen grains may inherit only the dominant (beneficial) alleles and grow quite well.

Student Discussion/Collaboration Questions

2. There are lots of possibilities. The students could look in agriculture and botany textbooks to find many examples. In the insect world, there are interesting examples of euploidy affecting sex determination. Among amphibians and reptiles, there are also several examples of closely related species that have euploid variation.

4. 1. Polyploids are often more robust and disease resistant.

    2. Allopolyploids may have useful combinations of traits.

    3. Hybrids are often more vigorous; they can be generated from monoploids.

    4. Strains with an odd number of chromosome sets (e.g., triploids) are usually seedless.

CHAPTER 9

Conceptual Questions

C2. The transformation process is described in chapter 6.
   1. A fragment of DNA binds to the cell surface.
   2. It penetrates the cell wall/cell membrane.
   3. It enters the cytoplasm.
   4. It recombines with the chromosome.
   5. The genes within the DNA are expressed (i.e., transcription and translation).
   6. The gene products create a capsule. That is, they are enzymes that synthesize a capsule using cellular molecules as building blocks.

C4. The building blocks of a nucleotide are a sugar (ribose or deoxyribose), a nitrogenous base, and phosphate. In a nucleotide, the phosphate is already linked to the 5′ position on the sugar. When two nucleotides are hooked together, a phosphate on one nucleotide forms a covalent bond with a hydroxyl group at the 3′ position on another nucleotide.

C6. It is a phosphate group connecting two sugars at the 3′ and 5′ positions as shown in figure 9.11.

C8. `3′-CCGTAATGTGATCCGGA-5′`

C10. A drawing with 10 bp per turn is like figure 9.17 in the textbook. To make 15 bp per turn, you would have to add 5 more base pairs, but the helix should still make only one complete turn.

C12. The nucleotide bases occupy the major and minor grooves. Phosphate and sugar are found in the backbone. If a DNA-binding protein did not recognize a nucleotide sequence, it is probably not binding in the grooves, although it could in a nonspecific way. As an alternative, it is probably binding to the DNA backbone (i.e., sugar-phosphate sequence).

C14. The structure is shown in figure 9.8. You begin at the carbon that is to the right of the ring oxygen and number them in a clockwise direction. Antiparallel means that the backbones are running in the opposite direction. In one strand, the sugar carbons are oriented in a 3′ to 5′ direction while in the other strand they are oriented in a 5′ to 3′ direction.

C16. Double-stranded RNA is more like A DNA than B DNA. See the text for a discussion of A-DNA structure.

C18. Its nucleotide base sequence.

C20. G = 32%, C = 32%, A = 18%, T = 18%

C22. One possibility is a sequential mechanism. First, the double helix could unwind and copy itself via a semiconservative mechanism described in chapter 11. This would produce two double helices. Next, the third strand (bound in the major groove) could copy itself via a semiconservative mechanism. This new strand could be copied to make a copy that is identical to the strand that lies in the major groove. At this point, you would have two double helices and two strands that could lie in the major groove. These could assemble to make two triple helices.

C24. Lysines and arginines, and also polar amino acids.

C26. This DNA molecule contains 280 bp. There are 10 base pairs per turn, so there are 28 complete turns.

C28. A hydroxyl group is at the 3′ end and a phosphate group is at the 5′ end.

C30. Not necessarily. The AT/GC rule is required only of double-stranded DNA molecules.

C32. The first thing we need to do is to determine how many base pairs are in this DNA molecule. The linear length of 1 base pair is 0.34 nm, which equals 0.34×10^{-9} m. One centimeter equals 10^{-2} meters.

$$\frac{10^{-2}}{0.34 \times 10^{-9}} = 2.9 \times 10^7 \text{ bp}$$

There are approximately 2.9×10^7 bp in this DNA molecule, which equals 5.8×10^7 nucleotides. If 15% are adenine, then 15% must also be thymine. This leaves 70% for cytosine and guanine. Since cytosine and guanine bind to each other, there must be 35% cytosine and 35% guanine. If we multiply 5.8×10^7 times 0.35, we get

$$(5.8 \times 10^7)(0.35) = 2.0 \times 10^7 \text{ cytosines, or about 20 million cytosines}$$

C34. The methyl group is not attached to one of the atoms that hydrogen bonds with guanine, so methylation would not directly affect hydrogen bonding. It could indirectly affect hydrogen bonding if it perturbed the structure of DNA. Methylation may affect gene expression because it could alter the ability of proteins to recognize DNA sequences. For example, a protein might bind into the major groove by interacting with a sequence of bases that includes one or more cytosines. If the cytosines are methylated, this may prevent a protein from binding into the major groove properly. Alternatively, methylation could enhance protein binding. In chapter 7, we considered DNA-binding proteins that were influenced by the methyl-

ation of DMRs (differentially methylated regions) that occur during genomic imprinting.

Experimental Questions

E2. A. There are different possible reasons why most of the cells were not transformed.

  1. Most of the cells did not take up any of the type IIIS DNA.
  2. The type IIIS DNA was usually degraded after it entered the type IIR bacteria.
  3. The type IIIS DNA was usually not expressed in the type IIR bacteria.

  B. The antibody/centrifugation steps were used to remove the bacteria that had not been transformed. It enabled the researchers to determine the phenotype of the bacteria that had been transformed. If this step was omitted, there would have been so many colonies on the plate it would have been difficult to identify any transformed bacterial colonies, since they would have represented a very small proportion of the total number of bacterial colonies.

  C. They were trying to demonstrate that it was really the DNA in their DNA extract that was the genetic material. It was possible that the extract was not entirely pure and could contain contaminating RNA or protein. However, treatment with RNase and protease did not prevent transformation, indicating that RNA and protein were not the genetic material. In contrast, treatment with DNase blocked transformation, confirming that DNA is the genetic material.

E4. A. There are several possible explanations why 30% of the DNA is in the supernatant. One possibility is that not all of the DNA was injected into the bacterial cells. Alternatively, some of the cells may have been broken during the shearing procedure, thereby releasing the DNA.

  B. If the radioactivity in the pellet had been counted instead of the supernatant, the following figure would be produced:

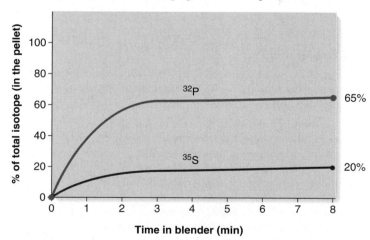

  C. ^{32}P and ^{35}S were chosen as radioisotopes to label the phages because phosphorous is found in nucleic acids, while sulfur is found only in proteins.

  D. There are multiple reasons why less than 100% of the phage protein is removed from the bacterial cells during the shearing process. Perhaps the shearing just is not strong enough to remove all of the phages. Perhaps the tail fibers remain embedded in the bacterium and only the head region is sheared off.

E6. This is really a matter of opinion. The Avery, MacLeod, and McCarty experiment seems to indicate directly that DNA is the genetic material since DNase prevented transformation while RNase and protease did not. However, one could argue that the DNA is required for the rough bacteria to take up some other contaminant in the DNA preparation. It would seem that the other contaminant would not be RNA or protein. The Hershey and Chase experiments indicate that DNA is being injected into bacteria, although quantitatively the results are not entirely convincing. Some ^{35}S-labeled protein was not sheared off, so the results do not definitely rule out the possibility that protein could be the genetic material. But the results do indicate that DNA is the more likely candidate.

E8. A. The purpose of chromatography was to separate the different types of bases.

B. It was necessary to separate the bases and determine the total amount of each type of base. In a DNA strand, all the bases are found within a single molecule, so it is difficult to measure the total amount of each type of base. When the bases are removed from the strand, each type can be purified, and then the total amount of each type of base can be measured with a spectrophotometer.

C. His results would probably not be very convincing if done on a single species. The strength of his data was that all species appeared to conform to the AT/GC rule, suggesting that this is a consistent feature of DNA structure. In a single species, the observation that A = T and G = C could occur as a matter of chance.

Questions for Student Discussion/Collaboration

2. Again, there are lots of possibilities. You could use a DNA-specific chemical and show that it causes heritable mutations. Perhaps you could inject an oocyte with a piece of DNA and produce a mouse with a new trait.

CHAPTER 10

Conceptual Questions

C2. Viruses also need sequences that enable them to be replicated. These sequences are equivalent to the origins of replication found in bacterial and eukaryotic chromosomes.

C4. A bacterium with two nucleoids is similar to a diploid eukaryotic cell because it would have two copies of each gene. The bacterium is different, however, with regard to alleles. A eukaryotic cell can have two different alleles for the same gene. For example, a cell from a pea plant could be heterozygous, *Tt*, for the gene that affects height. By comparison, a bacterium with two nucleoids has two identical chromosomes. Therefore, a bacterium with two nucleoids is homozygous for its chromosomal genes.

Note: As we will learn in chapter 14, on rare occasions a bacterium can contain another piece of DNA, called an F′ factor, that can carry a few genes. The alleles on an F′ factor can be different from the alleles on the bacterial chromosome.

C6. A. One loop is 40,000 bp. One base pair is 0.34 nm, which equals 0.34×10^{-3} μm. If we multiply the two together:

$$(40,000)(0.34 \times 10^{-3}) = 13.6 \ \mu m$$

B. Circumference = πD

$$13.6 \ \mu m = \pi D$$

$$D = 4.3 \ \mu m$$

C. No, it is too big to fit inside of *E. coli*. Supercoiling is needed to make the loops more compact.

C8. These drugs would diminish the amount of negative supercoiling. Negative supercoiling is needed to compact the chromosomal DNA, and it also aids in strand separation. Bacteria might not be able to survive and/or transmit their chromosomes to daughter cells if their DNA was not compacted properly. Also, because negative supercoiling aids in strand separation, these drugs would make it more difficult for the DNA strands to separate. Therefore, the bacteria would have a difficult time transcribing their genes and replicating their DNA, because both processes require strand separation. As discussed in chapter 11, DNA replication is needed to make new copies of the genetic material to transmit from mother to daughter cells. If DNA replication was inhibited, the bacteria could not grow and divide into new daughter cells. As discussed in chapters 12–15, gene transcription is necessary for cells to make proteins. If gene transcription was inhibited, the bacteria could not make many proteins that are necessary for survival.

C10.

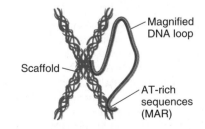

C12. The centromere is the attachment site for the kinetochore, which attaches to the spindle. If a chromosome is not attached to the spindle, it is free to "float around" within the cell, and it may not be near a pole when the nuclear membrane re-forms during telophase. If a chromosome is left outside of the nucleus, it is degraded during interphase. That is why the chromosome without a centromere is not likely to be found in daughter cells.

C14. Highly repetitive DNA, as its name suggests, is a DNA sequence that is repeated many times. It can be tandemly repeated or interspersed. Tandemly repeated DNA often has a base content that is significantly different from the rest of the chromosomal DNA so it sediments as a satellite band. In DNA renaturation studies, highly repetitive DNA renatures at a much faster rate because it is found at a higher concentration.

C16. During interphase (i.e., G_1, S, and G_2), the euchromatin is found primarily as a 30 nm fiber in a radial loop configuration. Most interphase chromosomes also have some heterochromatic regions where the radial loops are more highly compacted. During M phase, each chromosome becomes entirely heterochromatic. This is needed for the proper sorting of the chromosomes during nuclear division.

C18.

C20. During interphase, the chromosomes are found within the cell nucleus. They are less tightly packed and are transcriptionally active. Segments of chromosomes are anchored to the nuclear matrix. During M phase, the chromosomes become highly condensed and the nuclear membrane is fragmented into vesicles. The chromosomal DNA remains anchored to a scaffold, formed from the nuclear matrix. The chromosomes eventually become attached to the spindle apparatus via microtubules that attach to the kinetochore, which is attached to the centromere.

C22. There are 146 bp around the core histones. If the linker region is 54 bp, we expect 200 bp of DNA (i.e., 146 + 54) for each nucleosome and linker region. If we divide 46,000 bp by 200 bp we get 230. Because there are two molecules of H2A for each nucleosome, there would be 460 molecules of H2A in a 46,000 bp sample of DNA.

C24. The role of the core histones is to form the nucleosomes. In a nucleosome, the DNA is wrapped 1.65 times around the core histones. Histone H1 binds to the linker region. It may play a role in compacting the DNA into a 30 nm fiber.

C26. The answer is B and E. A Barr body is composed of a type of chromatin called heterochromatin. Heterochromatin is highly compacted. Euchromatin is not so compacted. A Barr body is not composed of euchromatin. The term *genome* refers to all the types of chromosomes that make up the genetic composition of an individual. A Barr body is just one chromosome, the X chromosome.

Experimental Questions

E2. This type of experiment gives the relative proportions of highly repetitive, moderately repetitive, and unique DNA sequences within the genome. The highly repetitive sequences renature at a fast rate, the moderately repetitive sequences renature at an intermediate rate, and the unique sequences renature at a slow rate.

E4. The amphibian probably has fewer structural genes than the mammal. The extra DNA is due to highly repetitive DNA sequences, which do not encode genes.

E6. 1. The repeating nucleosome structure was revealed from DNase I digestion studies.

2. Purification studies showed that the biochemical composition is an octamer of histones.

3. More recently, crystallography has shown the precise structure of the nucleosome.

4. Microscopy has revealed information about the 30 nm fiber and the attachment of chromatin to the nuclear matrix.

In general, it is easier to understand the molecular structure of something when it forms a regular repeating pattern. The eukaryotic chromosome has a repeating pattern of nucleosomes. The bacterial chromosome seems to be more irregular in its biochemical composition.

E8.

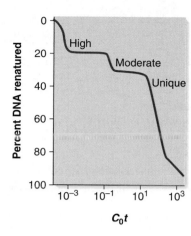

E10. You would expect all of the DNA in the sample to renature at a fast rate because it is a purified sample of highly repetitive DNA.

E12. You would get DNA fragments of about 446 to 496 bp (i.e., 146 bp plus 300 to 350).

E14. Histones are positively charged and DNA is negatively charged. They bind to each other by these ionic interactions. Salt is composed of positively charged ions and negatively charged ions. For example, when dissolved in water, NaCl becomes individual ions of Na$^+$ and Cl$^-$. When chromatin is exposed to a salt such as NaCl, the positively charged Na$^+$ ions could bind to the DNA and the negatively charged Cl$^-$ ions could bind to the histones. This would prevent the histones and DNA from binding to each other.

Questions for Student Discussion/Collaboration

2. This is a matter of opinion. It seems strange to have so much DNA that seems to have no obvious function. It is a waste of energy. Perhaps it has a function that we do not know about yet. On the other hand, evolution does occasionally cause bad things to accumulate within genomes, such as genes that cause diseases in the homozygous condition, etc. Perhaps this is just another example of the negative consequences of evolution.

CHAPTER 11

Conceptual Questions

C2. Bidirectionality refers to the idea that two replication forks emanate from one origin of replication. There are four DNA strands being made in two directions radiating outward from the origin.

C4. A.
```
           TTGGHTGUTGG
           HHUUTHUGHUU
```
B.
```
           TTGGHTGUTGG
           HHUUTHUGHUU
                ↓
    TTGGHTGUTGG  CCAAACACCAA
    AACCCACAACC  HHUUTHUGHUU
                ↓
TTGGHTGUTGG  TTGGGTGTTGG  CCAAACACCAA  CCAAACACCAA
AACCCACAACC  AACCCACAACC  GGTTTGTGGTT  HHUUTHUGHUU
```

C6. If we assume there are 4,600,000 bp of DNA, and that DNA replication is bidirectional at a rate of 750 nucleotides per second:

If there were just a single replication fork:

4,600,000/750 = 6,133 seconds, or 102.2 minutes

Because replication is bidirectional: 102.2/2 = 51.1 minutes

Actually, this is an average value based on a variety of growth conditions. Under optimal growth conditions, replication can occur substantially faster.

With regard to errors, if we assume an error rate of one mistake per 100,000,000 nucleotides:

4,600,000 × 1,000 bacteria

= 4,600,000,000 nucleotides of replicated DNA

4,600,000,000/100,000,000 = 46 mistakes

When you think about it, this is pretty amazing. In this population, DNA polymerase would cause only 46 single mistakes in a total of 1,000 bacteria, each containing 4.6 million bp of DNA.

C8. DNA polymerase would slide from right to left. The new strand would be

3′-CTAGGGCTAGGCGTATGTAAATGGTCTAGTGGTGG-5′

C10. A. When looking at figure 11.5, the first and third DnaA boxes are running in the same direction and the second and fourth DnaA boxes are running in the opposite direction. Once you realize that, you can see that the sequences are very similar to each other.

B. According to the direction of the first DnaA box, the consensus sequence is

TGTGGATAA
ACACCTATT

C. This sequence is nine nucleotides long. Because there are four kinds of nucleotides (i.e., A, T, G, and C), the chance of this sequence occurring by random chance is 4^{-9}, which equals once every 262,144 nucleotides. Because the *E. coli* chromosome is more than 10 times longer than this, it is fairly likely that this consensus sequence occurs elsewhere. The reason why there are not multiple origins, however, is because the origin has four copies of the consensus sequence very close together. The chances of having four copies of this consensus sequence occurring close together (as a matter of random chance) are very small.

C12. First, according to the AT/GC rule, a pyrimidine always hydrogen bonds with a purine. A transition still involves a pyrimidine hydrogen bonding to a purine, but a transversion causes a purine to hydrogen bond with a purine or a pyrimidine to hydrogen bond with a pyrimidine. The structure of the double helix makes it much more difficult for this later type of hydrogen bonding to occur. Second, the induced-fit phenomenon of the catalytic site of DNA polymerase makes it unlikely for DNA polymerase to catalyze covalent bond formation if the wrong nucleotide is bound to the template strand. A transition mutation creates a somewhat bad interaction between the bases in opposite strands, but it is not as bad as the fit caused by a transversion mutation. In a transversion, a purine is opposite another purine or a pyrimidine is opposite a pyrimidine. This is a very bad fit. And finally, the proofreading function of DNA polymerase is able to detect and remove an incorrect nucleotide that has been incorporated into the growing strand. A transversion is going to cause a larger distortion in the structure of the double helix and make it more likely to be detected by the proofreading function of DNA polymerase.

C14. Primase and DNA polymerase are able to knock the single-strand binding proteins off the template DNA.

C16. A. The right Okazaki fragment was made first. It is farthest away from the replication fork. The fork (not seen in this diagram) would be to the left of the three Okazaki fragments, and moving from right to left.

B. The RNA primer in the right Okazaki fragment would be removed first. DNA polymerase would begin by elongating the DNA strand of the middle Okazaki fragment and remove the right RNA primer with its 5′ to 3′ exonuclease activity. DNA polymerase I would use the 3′ end of the DNA of the middle Okazaki fragment as a primer to synthesize DNA in the region where the right RNA primer is removed. If the middle fragment was not present, DNA polymerase could not fill in this DNA (because it needs a primer).

C. You only need DNA ligase at the right arrow. DNA polymerase I begins at the end of the left Okazaki fragment and synthesizes DNA to fill in the region as it removes the middle RNA primer. At the left arrow, DNA polymerase I is simply extending the length of the left Okazaki fragment. No ligase is needed here. When DNA polymerase I has extended the left Okazaki fragment through the entire region where the RNA primer has been removed, it hits the DNA of the middle Okazaki fragment. This occurs at the right arrow. At this point, the DNA of the middle Okazaki fragment has a 5′ end that is a monophosphate. DNA ligase is needed to connect this monophosphate with the 3′ end of the region where the middle RNA primer has been removed.

D. As mentioned in the answer to part C, the 5′ end of the DNA in the middle Okazaki fragment is a monophosphate. It is a monophosphate because it was previously connected to the RNA primer by a phosphoester bond. At the location of the right arrow, there was only one phosphate connecting this deoxyribonucleotide to the last ribonucleotide in the RNA primer. For DNA polymerase to function, the energy to connect two nucleotides comes from the hydrolysis of the incoming triphosphate. In this location shown at the right arrow, however, the nucleotide is already present at the 5′ end of the DNA, and it is a monophosphate. DNA ligase needs energy to connect this nucleotide with the left Okazaki fragment. It obtains energy from the hydrolysis of ATP.

C18. 1. It recognizes the origin of replication.

2. It initiates the formation of a replication bubble.

3. It recruits helicase to the region.

C20. The picture would depict a ring of helicase proteins traveling along a DNA strand and breaking the hydrogen bonding between the two helices, as shown in figure 11.6.

C22. The leading strand is primed once, at the origin, and then DNA polymerase III makes it continuously in the direction of the replication fork. In the lagging strand, many short pieces of DNA are made. This requires many RNA primers and DNA polIII. The primers are removed by polI, which then fills in the gaps with DNA. DNA ligase then covalently connects the Okazaki fragments together. Having the enzymes within a complex provides coordination among the different steps in the replication process and thereby allows it to proceed more efficiently and faster.

C24. A processive enzyme is one that remains clamped to one of its substrates. In the case of DNA polymerase, it remains clamped to the template strand as it makes a new daughter strand. This is important to ensure a fast rate of DNA synthesis.

C26. It is necessary to have the correct number of chromosomes per cell. If DNA replication is too slow, daughter cells may not receive any chromosomes. If it is too fast, they may receive too many chromosomes.

C28. The opposite strand is made in the conventional way by DNA polymerase using the "telomerase added strand" as a template.

C30. The ends labeled *B* and *C* could not be replicated by DNA polymerase. DNA polymerase makes a strand in the 5′ to 3′ direction using a template strand that is running in the 3′ to 5′ direction. Also, DNA polymerase requires a primer. At the ends labeled *B* and *C*, there is no place (upstream) for a primer to be made.

C32. As shown in figure 11.23, the first step involves a binding of telomerase to the telomere. The 3′ overhang binds to the complementary RNA in telomerase. For this reason, a 3′ overhang is necessary for telomerase to replicate the telomere.

Experimental Questions

E2. A. You would probably still see a band of DNA, but you would only see a heavy band.

   B. You would probably not see a band because the DNA would not be released from the bacteria. The bacteria would sediment to the bottom of the tube.

   C. You would not see a band. Ethidium bromide stains the DNA and you are actually seeing the ethidium bromide when the gradient is exposed to UV light.

E4. You would need to add a primer (or primase), dNTPs, and DNA polymerase. If the DNA were double stranded, you would also need helicase. Adding single-strand binding protein and topoisomerase may also help.

E6. This is a critical step because you need to separate the radioactivity in the free nucleotides from the radioactivity in the newly made DNA strands. If you used an acid that precipitated free nucleotides and DNA strands, all of the radioactivity would be in the pellet. You would get the same amount of radioactivity in the pellet no matter how much DNA was synthesized into newly made strands. For this reason, the perchloric step is very critical. It separates radioactivity in the free nucleotides from radioactivity in the newly made strands.

E8. A. The left end is the 5′ end. If you flip the sequence of the first primer around, you will notice that it is complementary to the right end of the template DNA. The 5′ end of the first primer binds to the 3′ end of the template DNA.

   B. The sequence would be 3′–CGGGGCCATG–5′. It could not be used because the 3′ end of the primer is at the end of the template DNA. There wouldn't be any place for nucleotides to be added to the 3′ end of the primer and bind to the template DNA strand.

E10. A. Heat is used to separate the DNA strands, so you do not need helicase.

   B. Each primer must be a sequence that is complementary to one of the DNA strands. There are two types of primers, and each type binds to one of the two complementary strands. The arrowheads are at the 3′ end of the primers.

   C. A thermophilic DNA polymerase is used because DNA polymerases isolated from nonthermophilic species would be permanently inactivated during the heating phase of the PCR cycle. Remember that DNA polymerase is a protein, and most proteins are denatured by heating. However, proteins from thermophilic organisms have evolved to withstand heat. That is how thermophilic organisms survive at high temperatures.

   D. With each cycle, the amount of DNA is doubled. Because there are initially 10 copies of the DNA, there will be 10×2^{27} copies after 27 cycles. $10 \times 2^{27} = 1.34 \times 10^9 = 1.34$ billion copies of DNA. As you can see, PCR can amplify the amount of DNA by a staggering amount!

Questions for Student Discussion/Collaboration

2. Basically, the idea is to add certain combinations of enzymes and see what happens. You could add helicase and primase to double-stranded DNA along with radiolabeled ribonucleotides. Under these conditions, you would make short (radiolabeled) primers, and if you added DNA polymerase and deoxyribonucleotides, you would also make DNA strands. Alternatively, if you added DNA polymerase plus an RNA primer, you would not need to add primase. Or if you used single-stranded DNA as a template, rather than double-stranded DNA, you would not need to add helicase.

CHAPTER 12

Conceptual Questions

C2. The formation of the open complex, and the release of sigma factor.

C4. GGCATTGTCA

C6. The most highly conserved positions are the first, second, and sixth. In general, when promoter sequences are conserved, they are more likely to be important for binding. That explains why changes are not found at these positions; if a mutation altered a conserved position, the promoter would probably not work very well. By comparison, changes are occasionally tolerated at the fourth position and frequently at the third and fifth positions. The positions that tolerate changes are less important for binding by sigma factor.

C8. This will not affect transcription. However, it will affect translation by preventing the initiation of polypeptide synthesis.

C10. Sigma factor can slide along the major groove of the DNA. In this way, it is able to recognize base sequences that are exposed in the groove. When it encounters a promoter sequence, hydrogen bonding between the bases and the sigma factor protein can promote a tight and specific interaction.

C12. DNA-G/RNA-C
    DNA-C/RNA-G
    DNA-A/RNA-U
    DNA-T/RNA-A

The template strand is 3′–CCGTACGTAATGCCGTAGTGTGATC-CCTAG–5′ and the coding strand is 5′–GGCATGCATTACGGCAT-CACACTAGGGATC–3′. The promoter would be to the left (in the 3′ direction) of the template strand.

C14. Transcriptional termination occurs when the hydrogen bonding is broken between the DNA and the part of the newly made RNA transcript that is located in the open complex.

C16. Helicase and ρ protein bind to a nucleic acid strand and travel in the 5′ to 3′ direction. When they encounter a double-stranded region, they break the hydrogen bonds between complementary strands. ρ protein is different from DNA helicase in that it moves along an RNA strand, while DNA helicase moves along a DNA strand. The purpose of DNA helicase function is to promote DNA replication while the purpose of ρ protein function is to promote transcriptional termination.

C18. A. Mutations that alter the AU-rich region by introducing Gs and Cs, and mutations that prevent the formation of the stem-loop (hairpin) structure.

B. Mutations that alter the termination sequence and mutations that alter the ρ recognition site.

C. Eventually, somewhere downstream from the gene, there would be found another transcriptional termination sequence and transcription would terminate there. This second termination sequence might be found randomly or it might be at the end of an adjacent gene.

C20. Eukaryotic promoters are somewhat variable with regard to the pattern of sequence elements that may be found. In the case of structural genes that are transcribed by RNA polymerase II, it is common to have a TATA box, which is about 25 bp upstream from a transcriptional start site. The TATA box is important in the identification of the transcriptional start site and the assembly of RNA polymerase and various transcription factors. The transcriptional start site defines where transcription actually begins.

C22. It is primarily an accessibility problem. When the DNA is tightly wound around histones, it becomes difficult for large proteins, like RNA polymerase and transcription factors, to recognize the correct base sequence in the DNA and to catalyze the movement of the open complex. It is thought that a partial or complete disruption of the nucleosome structure is necessary for transcription to occur.

C24. Hydrogen bonding is usually the predominant type of interaction when proteins and DNA follow an assembly and disassembly process. In addition, ionic bonding and hydrophobic interactions could occur. Covalent interactions would not occur. High temperature and high salt concentrations tend to break hydrogen bonds. Therefore, high temperature and high salt would inhibit assembly and stimulate disassembly.

C26. In bacteria, the 5′ end of the tRNA is cleaved by RNaseP. The 3′ end is cleaved by a different endonuclease, and then a few nucleotides are digested away by an exonuclease that removes nucleotides one at a time until it reaches a CCA sequence.

C28. A ribozyme is an enzyme whose catalytic part is composed of RNA. Examples are RNaseP and self-splicing group I and II introns. It is thought that the spliceosome may have catalytic RNAs as well.

C30. Self-splicing means that an RNA molecule can splice itself without the aid of a protein. Group I and II introns can be self-splicing, although proteins can also facilitate the process.

C32. A transesterification reaction involves the breakage of one ester bond and formation of another ester bond.

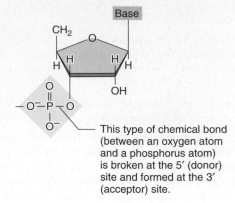

This type of chemical bond (between an oxygen atom and a phosphorus atom) is broken at the 5′ (donor) site and formed at the 3′ (acceptor) site.

C34. This is called trimming or processing rather than splicing because none of the mature rRNA fragments are ever ligated (i.e., connected) to each other after the cleavage of the primary rRNA transcript.

C36. A. U1 and U4/U6

B. U5

C. U2

C38. The 60-nucleotide sequence would be found in the closed loop. The closed loop is the region between the 5′ splice site and the branch site.

Experimental Questions

E2. An R loop is a loop of DNA that occurs when RNA is hybridized to double-stranded DNA. While the RNA is hydrogen bonding to one of the DNA strands, the other strand does not have a partner to hydrogen bond with so it bubbles out as a loop. RNA is complementary to the template strand, so that is the strand it binds to.

E4. The 1,100-nucleotide band would be observed from a normal individual (lane 1). A deletion that removed the −50 to −100 region would greatly diminish transcription, so the homozygote would produce hardly any of the transcript (just a faint amount as shown in lane 2) and the heterozygote would produce roughly half as much of the 1,100-nucleotide transcript (lane 3) compared to a normal individual. A nonsense codon would not have an effect on transcription; it affects only translation. So the individual with this mutation would produce a normal amount of the 1,100-nucleotide transcript (lane 4). A mutation that removed the splice acceptor site would prevent splicing. Therefore, this individual would produce a 1,550-nucleotide transcript (actually, 1,547 to be precise, 1,550 minus 3). The Northern blot is shown here:

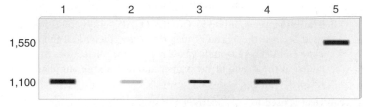

E6. A. It would not be retarded because ρ protein would not bind to the mRNA that is encoded by a gene that is terminated in a ρ-independent manner. The mRNA from such genes does not contain the sequence near the 3′ end that acts as a recognition site for the binding of ρ protein.

B. It would be retarded because ρ-protein would bind to the mRNA.

C. It would be retarded because U1 would bind to the pre-mRNA.

D. It would not be retarded because U1 would not bind to mRNA that has already had its introns removed. U1 binds only to pre-mRNA.

F8. A. mRNA molecules would bind to this column because they have a polyA tail. A polyA tail is complementary to poly-dT, so the two would hydrogen bond to each other. To purify mRNAs, one begins with a sample of cells; the cells need to be broken open by some technique such as homogenization or sonication. This would release the RNAs and other cellular macromolecules. The large cellular structures (e.g., organelles, membranes, etc.) could be removed from the cell extract by a centrifugation step. The large cellular structures would be found in the pellet, while soluble molecules such as RNA and proteins would stay in the supernatant. At this point, you would want the supernatant to contain a high salt concentration and neutral pH. The supernatant would then be poured over the poly-dT column. The mRNAs would bind to the poly-dT column and other molecules (i.e., other types of RNAs and proteins) would flow through the column. The mRNAs would bind to the poly-dT column via hydrogen bonds. To break the hydrogen bonds between the mRNAs and poly-dT, you could add a solution that contains a low salt concentration and/or a high pH. This would release the mRNAs, which would then be collected in a low salt/high pH solution as it dripped from the column.

B. The basic strategy is to attach an oligonucleotide (i.e., a short sequence of DNA) to the column matrix that is complementary to the type of RNA that you want to purify. For example, if an rRNA contained a sequence 5'–AUUCCUCCA–3', a researcher could chemically synthesize oligonucleotides with the sequence 3'–TAAGGAGGT–5' and attach these oligonucleotides to the column matrix. To purify rRNA, one would use this 3'–TAAGGAGGT–5' column and follow the general strategy described in part A.

Questions for Student Discussion/Collaboration

2. RNA transcripts come in two basic types: those that function as RNA (e.g., tRNA, rRNA, etc.) versus those that are translated (i.e., mRNA). As described in chapter 12, they play a myriad of functional roles. RNAs that form complexes with proteins carry out some interesting roles. In some cases, the role is to bind other types of RNA molecules. For example, rRNA in bacteria plays a role in binding mRNA. In other cases, the RNA plays a catalytic role. An example is RNaseP.

CHAPTER 13

Conceptual Questions

C2. When we say the genetic code is degenerate, it means that more than one codon can specify the same amino acid. For example, GGG, GGC, GGA, and GGU all specify glycine.

In general, the genetic code is nearly universal, because it is used in the same way by viruses, prokaryotes, fungi, plants, and animals. As discussed in table 13.3, there are a few exceptions, which occur primarily in protozoa and organellar genetic codes.

C4. A. This mutant tRNA would recognize glycine codons in the mRNA but would put in tryptophan amino acids where gly amino acids are supposed to be in the polypeptide chain.

B. This mutation tells us that the aminoacyl-tRNA synthetase is primarily recognizing other regions of the tRNA molecule besides the anticodon region. In other words, tryptophanyl-tRNA synthetase (i.e., the aminoacyl-tRNA synthetase that attaches tryptophan) primarily recognizes other regions of the tRNA^{trp} sequence (i.e., other than the anticodon region), such as the T- and D-loops. If aminoacyl-tRNA synthetases recognized only the anticodon region, we would expect glycyl-tRNA synthetase to recognize this mutant tRNA and attach glycine. That is not what happens.

C6. A. The answer is three. There are six leucine codons: UUA, UUG, CUU, CUC, CUA, and CUG. The anticodon AAU would recognize UUA and UUG. You would need two other tRNAs to *efficiently* recognize the other four leucine codons. These could be GAG and GAU or GAA and GAU.

B. The answer is one. There is only one codon, AUG, so you need only one tRNA with the anticodon UAC.

C. The answer is three. There are six serine codons: AGU, AGC, UCU, UCC, UCA, and UCG. You would need only one tRNA to recognize AGU and AGC. This tRNA could have the anticodon UCG or UCA. You would need two tRNAs to efficiently recognize the other four tRNAs. These could be AGG and AGU or AGA and AGU.

C8. 3'–CUU–5' or 3'–CUC–5'

C10. It can recognize 5'–GGU–3', 5'–GGC–3', and 5'–GGA–3'. All of these specify glycine.

C12. All tRNA molecules have some basic features in common. They all have a cloverleaf structure with three stem-loop structures. The second stem-loop contains the anticodon sequence that recognizes the codon sequence in mRNA. At the 3' end, there is an acceptor site, with the sequence CCA, that serves as an attachment site for an amino acid. Most tRNAs also have base modifications that occur within their nucleotide sequences.

C14. The role of aminoacyl-tRNA synthetase enzymes is to specifically recognize tRNA molecules and attach the correct amino acid to them. They are sometimes described as the second genetic code because the specificity of their attachment is a critical step in deciphering the genetic code. For example, if a tRNA has a 3'–GGG–5' anticodon, it will recognize a 5'–CCC–3' codon, which should specify proline. It is essential that the prolyl-tRNA-synthetase recognizes this tRNA and attaches proline to the 3' end. The other aminoacyl-tRNA synthetases should not recognize this tRNA.

C16. Bases that have been chemically modified can occur at various locations throughout the tRNA molecule. The significance of all of these modifications is not entirely known. However, within the anticodon region, base modification alters base pairing to allow the anticodon to recognize two or more different bases within the codon.

C18. No, it is not. Due to the wobble rules, the 5' base in the anticodon of a tRNA can sometimes recognize two or more bases in the third (3') position of the mRNA. Therefore, any given cell type synthesizes far fewer than 61 types of tRNAs.

C20. The assembly process is very complex at the molecular level. In eukaryotes, 33 proteins and one rRNA assemble to form a 40S

...eins and three rRNAs assemble to form a 60S ...mbly occurs within the nucleolus.

...face of the 30S subunit and at the interface

...n the 50S subunit

...om the 50S subunit

D. To the 30S subunit

C24. Most bacterial mRNAs contain a Shine-Dalgarno sequence, which is necessary for the binding of the mRNA to the small ribosomal subunit. This sequence, UAGGAGGU, is complementary to a sequence in the 16S rRNA. Due to this complementarity, these sequences will hydrogen bond to each other during the initiation stage of translation.

C26. The ribosome binds at the 5′ end of the mRNA and then scans in the 3′ direction in search of an AUG start codon. If it finds one that reasonably obeys the Kozak's rules, it will begin translation at that site. Aside from an AUG start codon, the two other features are a purine at the −3 position and a guanosine at the +4 position.

C28. The A site is the acceptor site. It is the location where a tRNA initially "floats in" and recognizes a codon in the mRNA. The only exception is the initiator tRNA that binds to the P site. The P site is the next location where the tRNA moves. When it first moves to the P site, it carries with it the polypeptide chain. In each round of elongation, the polypeptide chain is transferred from the tRNA in the P site to the amino acid attached to the tRNA in the A site. The third site is the E site. During translocation, the uncharged tRNA in the P site is transferred to the E site. It exits or is released from this site.

C30. Sorting signals provide an address that sends the protein to the correct location (i.e., compartment) within the cell. Proteins destined for the ER, Golgi, lysosomes, plasma membrane, or secretion have an SRP sorting signal at their amino terminal end. Nuclear proteins have an NLS sequence, etc. These sorting signals are recognized by cellular proteins/complexes that then act in a way to traffic the proteins to their correct destination.

C32. The initiation phase involves the binding of the Shine-Dalgarno sequence to the rRNA in the small ribosomal subunit. The elongation phase involves the binding of anticodons in tRNA to codons in mRNA.

C34. A. E site and P site (Note: A tRNA without an amino acid attached is only briefly found in the P site, just before translocation occurs.)

B. P site and A site (Note: A tRNA with a polypeptide chain attached is only briefly found in the A site, just before translocation occurs.)

C. Usually the A site, except the initiator tRNA can be found in the P site

C36. The tRNAs bind to the mRNA because their anticodon and codon sequences are complementary. When the ribosome translocates in the 5′ to 3′ direction, the tRNAs remain bound to their complementary codons, and the two tRNAs shift from the A site and P site to the P site and E site. If the ribosome tried to move in the 3′ direction, it would have to dislodge the tRNAs and drag them to a new position where they would not (necessarily) be complementary to the mRNA.

C38. 52

Experimental Questions

E2. A. There could have been other choices, but this template would be predicted to contain a cysteine codon, UGU, but would not contain any alanine codons.

B. You do not want to use ^{35}S because the radiolabel would be removed during the Raney nickel treatment.

C. There would not be a significant amount of radioactivity incorporated into newly made polypeptides with or without Raney nickel treatment. The only radiolabeled amino acid in this experiment was cysteine, which became attached to tRNA^{cys}. When exposed to Raney nickel, these cysteines were converted to alanine but only after they were already attached to tRNA^{cys}. If there were not any cysteine codons in the mRNA template, the tRNA^{cys} would not recognize this mRNA. Therefore, we would not expect to see much radioactivity in the newly made polypeptides.

E4. The initiation phase of translation is very different between bacteria and eukaryotes, so they would not be translated very efficiently. A bacterial mRNA would not be translated very efficiently in a eukaryotic translation system because it does not have a cap structure. A eukaryotic mRNA would probably not have a Shine-Dalgarno sequence near its 5′ end, so it would not be translated very efficiently in a bacterial translation system.

E6. Looking at the figure, the 5′ end of the template DNA strand is toward the right side. The 5′ ends of the mRNAs are farthest from the DNA, and the 3′ ends of the mRNAs are closest to the DNA. The start codons are slightly downstream from the 5′ ends of the RNAs.

E8.

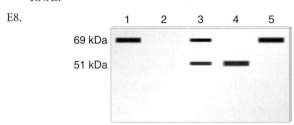

E10. A. If codon usage were significantly different between kangaroo and yeast cells, this would inhibit the translation process. For example, if the preferred leucine codon in kangaroos was CUU, translation would probably be slow in a yeast translation system. We would expect the yeast translation system to primarily contain leucine tRNAs with an anticodon sequence that is AAC because this tRNA^{leu} would match the preferred yeast leucine codon, which is UUG. In a yeast translation system, there probably would not be a large amount of tRNA with an anticodon of GAA, which would match the preferred leucine codon, CUU, of kangaroos. For this reason, kangaroo mRNA would not be translated very well in a yeast translation system, but it probably would be translated to some degree.

B. The advantage of codon bias is that a cell can rely on a smaller population of tRNA molecules to efficiently translate its proteins. A disadvantage is that mutations, which do not change the amino acid sequence but do change a codon (e.g., UUG to UUA), may inhibit the production of a polypeptide if a preferred codon is changed to a nonpreferred codon.

Questions for Student Discussion/Collaboration

2. This could be a very long list. There are similarities along several lines.

   1. There is a lot of molecular recognition going on, either between two nucleic acid molecules or between proteins and nucleic acid molecules. Students may see these as similarities or differences, depending on their point of view.

   2. There is biosynthesis going on in both processes. Small building blocks are being connected. This requires an input of energy.

   3. There are genetic signals that determine the beginning and ending of these processes.

   There are also many differences.

   1. Transcription produces an RNA molecule with a similar structure to the DNA, while translation produces a polypeptide with a structure that is very different from RNA.

   2. Depending on your point of view, it seems that translation is more biochemically complex, requiring more proteins and RNA molecules to accomplish the task.

CHAPTER 14

Conceptual Questions

C2. In bacteria, gene regulation greatly enhances the efficiency of cell growth. It takes a lot of energy to transcribe and translate genes. Therefore, a cell is much more efficient and better at competing in its environment if it expresses genes only when the gene product is needed. For example, a bacterium will express only the genes that are necessary for lactose metabolism when a bacterium is exposed to lactose. When the environment is missing lactose, these genes are turned off. Similarly, when tryptophan levels are high within the cytoplasm, the genes that are required for tryptophan biosynthesis are repressed.

C4. A. Regulatory protein

   B. Effector molecule

   C. DNA segment

   D. Effector molecule

   E. Regulatory protein

   F. DNA segment

   G. Effector molecule

C6. A *cis*-mutation is within a genetic regulatory sequence, such as an operator site, that affects the binding of a genetic regulatory protein. A *cis*-mutation affects only the adjacent genes that the genetic regulatory sequence controls. A *trans*-mutation is usually in a gene that encodes a genetic regulatory protein. A *trans*-mutation can be complemented in a merozygote experiment by the introduction of a normal gene that encodes the regulatory protein.

C8. A. No transcription would take place. The *lac* operon could not be expressed.

   B. No regulation would take place. The operon would be continuously turned on.

   C. The rest of the operon would function normally but none of the transacetylase would be made.

C10. Diauxic growth refers to the phenomenon whereby a cell firs[t] up one type of sugar (e.g., glucose) before it begins to metabo[lize a] second sugar (e.g., lactose). In this case, it is caused by gene regu[la]tion. When a bacterial cell is exposed to both sugars, the uptake o[f] glucose causes the cAMP levels in the cell to fall. When this occurs, the catabolite activator protein is removed from the *lac* operon so that it is not able to be (maximally) activated.

C12. A mutation that prevented the *lac* repressor from binding to the operator would make the *lac* operon constitutive only in the absence of glucose. However, this mutation would not be entirely constitutive because transcription would be inhibited in the presence of glucose. The disadvantage of constitutive expression of the *lac* operon is that the bacterial cell would waste a lot of energy transcribing the genes and translating the mRNA when lactose was not present.

C14. A. Without $araO_2$ the repression of the *ara* operon could not occur. The operon would be constitutively expressed at high levels because AraC protein could still activate transcription of the *ara* operon by binding to *araI*. The presence of arabinose would have no effect. Note: The binding of arabinose to AraC is not needed to form an AraC dimer at *araI*. The dimer is able to form because the loop has been broken. This point may be figured out if you notice that an AraC dimer is bound to $araO_1$ in the presence and absence of arabinose (see fig. 14.12).

   B. Without $araO_1$, the AraC protein would be overexpressed. It would probably require more arabinose to alleviate repression. In addition, activation might be higher because there would be more AraC protein available.

   C. Without *araI*, transcription of the *ara* operon cannot be activated. You might get a very low level of constitutive transcription.

   D. Without $araO_2$ the repression of the *ara* operon could not occur. However, without *araI*, transcription of the *ara* operon cannot be activated. You might get a very low level of constitutive transcription.

C16. A. Attenuation will not occur because loop 2–3 will form.

   B. Attenuation will occur because 2–3 cannot form, so 3–4 will form.

   C. Attenuation will not occur because 3–4 cannot form.

   D. Attenuation will not occur because 3–4 cannot form.

C18. The addition of Gs and Cs into the U-rich sequence would prevent attenuation. The U-rich sequence promotes the dissociation of the mRNA from the DNA, when the terminator stem-loop forms. This causes RNA polymerase to dissociate from the DNA and thereby causes transcriptional termination. The UGGUUGUC sequence would probably not dissociate because of the Gs and Cs. Remember that GC base pairs have three hydrogen bonds and are more stable than AU base pairs, which only have two hydrogen bonds.

C20. It takes a lot of cellular energy to translate mRNA into a protein. A cell wastes less energy if it prevents the initiation of translation rather than a later stage such as elongation or termination.

C22. One mechanism is that histidine could act as corepressor that shuts down the transcription of the histidine synthetase gene. A second mechanism would be that histidine could act as an inhibitor via feedback inhibition. A third possibility is that histidine inhibits the ability of the mRNA encoding histidine synthetase to be translated.

gene that encodes an antisense RNA. If the
ne synthetase protein was identical in the pres-
nce of extracellular histidine, a feedback inhibition
is favored, because this affects only the activity of the
synthetase enzyme, not the amount of the enzyme. The
two mechanisms would diminish the amount of this protein.

The two proteins are similar in that both bind to a segment of
DNA and repress transcription. They are different in three ways.
(1) They recognize different effector molecules (i.e., the *lac* repressor recognizes allolactose and the *trp* repressor recognizes tryptophan. (2) Allolactose causes the *lac* repressor to release from the
operator, while tryptophan causes the *trp* repressor to bind to its
operator. (3) The sequences of the operator sites that these two
proteins recognize are different from each other. Otherwise, the *lac*
repressor could bind to the *trp* operator and the *trp* repressor could
bind to the *lac* operator.

C26. In the lytic cycle, the virus directs the bacterial cell to make more
virus particles until eventually the cell lyses and releases them. In
the lysogenic cycle, the viral genome is incorporated into the host
cell's genome as a prophage. It remains there in a dormant state
until some stimulus causes it to erupt into the lytic cycle.

C28. The O_R region contains three operator sites and two promoters.
P_{RM} and P_R transcribe in opposite directions. The λ repressor will
first bind to O_{R1} and then O_{R2}. The binding of the λ repressor to
O_{R1} and O_{R2} inhibits transcription from P_R and thereby switches
off the lytic cycle. Early in the lysogenic cycle, the λ repressor protein concentration may become so high that it will occupy O_{R3}.
Later, when the λ repressor concentration begins to drop, it will
first be removed from O_{R3}. This allows transcription from P_{RM} and
maintains the lysogenic cycle. By comparison, the cro protein has
its highest affinity for O_{R3}, and so it binds there first. This blocks
transcription from P_{RM} and thereby switches off the lysogenic
cycle. The cro protein has a similar affinity for O_{R2} and O_{R1}, and so
it may occupy either of these sites next. It will bind to both O_{R2}
and O_{R1}. This turns down the expression from P_R, which is not
needed in the later stages of the lytic cycle.

C30. It would first increase the amount of cro protein so that the lytic
cycle would be favored.

C32. Neither cycle could be followed. As shown in figure 14.19, N protein is needed to make a longer transcript from P_L for the lysogenic
cycle and also to make a longer transcript from P_R for the lytic
cycle.

C34. If the F^- strain is lysogenic for phage λ, the λ repressor is already
being made in that cell. If the F^- strain receives genetic material
from an *Hfr* strain, you wouldn't expect it to have an effect on the
lysogenic cycle, which is already established in the F^- cell. However,
if the *Hfr* strain is lysogenic for λ and the F^- strain isn't, the *Hfr*
strain could transfer the integrated λ DNA (i.e., the prophage) to
the F^- strain. The cytoplasm of the F^- strain would not contain
any λ repressor. Therefore, this λ DNA could choose between the
lytic and lysogenic cycle. If it follows the lytic cycle, the F^- recipient bacterium will lyse.

Experimental Questions

E2. In samples loaded in lanes 1 and 4, we expect the repressor to bind
to the operator because there is no lactose present. In the sample

loaded into lane 4, the CAP protein could still bind cAMP because
there is no glucose. However, there really is no difference between
lanes 1 and 4, so it does not look like the CAP can activate transcription when the *lac* repressor is bound. If we compare samples
loaded into lanes 2 and 3, the *lac* repressor would not be bound in
either case, and the CAP would not be bound in the sample loaded
into lane 3. There is less transcription in lane 3 compared to lane 2,
but because there is some transcription seen in lane 3, we can conclude that the removal of the CAP (because cAMP levels are low) is
not entirely effective at preventing transcription. Overall, the
results indicate that the binding of the *lac* repressor is much more
effective at preventing transcription of the *lac* operon compared to
the removal of the catabolite activator protein.

E4. A. Yes, if you do not sonicate, then β-galactosidase will not be
released from the cell, and not much yellow color will be
observed. (Note: You may observe a little yellow color because
some of β-ONPG may be taken into the cell.)

B. No, you should still get yellow color in the first two tubes even if
you forgot to add lactose because the unmated strain does not
have a functional *lac* repressor.

C. Yes, if you forgot to add β-ONPG you could not get yellow color
because the cleavage of β-ONPG by β-galactosidase is what produces the yellow color.

E6. You could mate a strain that has an F′ factor carrying a normal *lac*
operon and a normal *lacI* gene to this mutant strain. Because the
mutation is in the operator site, you would still continue to get
expression of β-galactosidase, even in the absence of lactose.

E8. In this case, things are more complex because AraC acts as a
repressor and an activator protein. If AraC were missing due to
mutation, there would not be repression or activation of the *ara*
operon in the presence or absence of arabinose. It would be
expressed constitutively at low levels. The introduction of a
normal *araC* gene into the bacterium on an F′ factor would
restore normal regulation (i.e., a *trans*-effect).

E10. Antisense RNA prevents the translation of the mRNA to which it is
complementary. If the polypeptide that is encoded by the mRNA
has an important function during early embryonic stages, the
embryo may not develop properly. In this case, one may observe
gross developmental abnormalities. For example, if the polypeptide
is important in the formation of anterior structures, one may see
that the head does not develop properly when the antisense RNA is
injected into the fertilized oocyte. In contrast, if the polypeptide
does not play an important role during the early stages of development, the embryo should develop properly.

Questions for Student Discussion/Collaboration

2. A DNA loop may inhibit transcription by preventing RNA polymerase from recognizing the promoter. Or it may inhibit transcription by preventing the formation of the open complex. The bend
may expose the base sequence that the sigma factor of RNA polymerase is recognizing. The bend may "open up" the major groove
so that this base sequence is more accessible to the binding by
sigma factor and RNA polymerase.

CHAPTER 15

Conceptual Questions

C2. Response elements are relatively short genetic sequences that are recognized by regulatory transcription factors. After the regulatory transcription factor(s) has bound to the response element, it will affect the rate of transcription, either activating it or repressing it, depending on the action of the regulatory protein. Response elements are typically located in the upstream region near the promoter, but they can be located almost anywhere (i.e., upstream and downstream) and even quite far from the promoter.

C4. Transactivation occurs when a regulatory transcription factor binds to a response element and activates transcription. Transactivators may interact with TFIID and/or mediator to promote the assembly of RNA polymerase and general transcription factors at the promoter region. They also could alter the structure of chromatin so that RNA polymerase and transcription factors are able to gain access to the promoter. Transinhibition occurs when a regulatory transcription factor inhibits transcription. Transinhibitors or repressors also may interact with TFIID and/or mediator to inhibit RNA polymerase.

C6. A. DNA binding

B. DNA binding

C. Protein dimerization

C8. For the glucocorticoid receptor to bind to a GRE, the cell must be exposed to the hormone and it must enter the cell. It binds to the receptor, which releases HSP90. After HSP90 is released, the receptor dimerizes and then enters the nucleus. Once inside the nucleus, the dimer will recognize a pair of GREs and bind to them, thereby leading to the activation of specific genes that have two GREs next to them.

C10. Phosphorylation of the CREB protein causes it to act as a transactivator. The unphosphorylated CREB protein can still bind to CREs, but it does not stimulate transcription.

C12. A. Eventually, the glucocorticoid hormone will be degraded by the cell. The glucocorticoid receptor binds the hormone with a certain affinity. The binding is a reversible process. Once the concentration of the hormone falls below the affinity of the hormone for the receptor, the receptor will no longer have the glucocorticoid hormone bound to it. When the hormone is released, the glucocorticoid receptor will change its conformation, and it will no longer bind to the DNA.

B. An enzyme known as a phosphatase will eventually cleave the phosphate groups from the CREB protein. When the phosphates are removed, the CREB protein will stop transactivating transcription.

C14. The enhancer found in A would work, but the ones found in B and C would not. The sequence that is recognized by the transcriptional activator is 5′–GTAG–3′ in one strand and 3′–CATC–5′ in the opposite strand. This is the same arrangement found in A. In B and C, however, the arrangement is 5′–GATG–3′ and 3′–CATC–5′. In the arrangement found in B and C, the two middle bases (i.e., A and T) are not in the correct order.

C16.

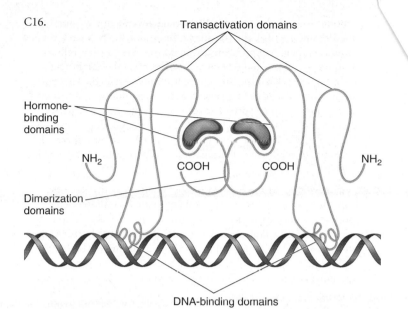

This is one hypothetical drawing of the glucocorticoid receptor. As discussed in your textbook, it forms a homodimer. The dimer shown here has rotational symmetry. If you flip the left side around, it has the same shape as the right side. The hormone-binding, DNA-binding, dimerization, and transactivation domains are labeled. The glucocorticoid hormone is shown in *orange.*

C18. The 30 nm fiber is the predominant form of chromatin during interphase. The chromatin must be converted from a closed conformation to an open conformation for transcription to take place. This "opening" process involves less packing of the chromatin and may involve changes in the locations of histone proteins. Transcriptional activators recruit histone acetyltransferase and ATP-dependent remodeling enzymes to the region, which leads to a conversion to the open conformation.

C20. The binding of the DNA to the core histones could

1. prevent transcriptional activators from recognizing enhancer elements that are required for transcriptional activation.

2. prevent RNA polymerase from binding to the core promoter.

3. prevent RNA polymerase from forming an open complex in which the DNA strands separate.

C22. The translocation breakpoint occurred between the β-globin gene and the locus control region. Therefore, the β-globin gene is not expressed because it needs the locus control region to allow the chromatin to be in an accessible conformation.

C24. Perhaps the methylase is responsible for methylating and inhibiting a gene that causes a cell to become a muscle cell. The methylase is inactivated by the mutation.

C26. The function of splicing factors is to influence the selection of splice sites in RNA. In certain cell types, the concentration of particular splicing factors is higher than in other tissues. The high concentration of particular splicing factors, and the regulation of their activities, may promote the selection of particular splice sites and thereby lead to tissue-specific splicing.

C28. This person would be unable to make ferritin, because the IRP would always be bound to the IRE. The amount of transferrin receptor mRNA would be high, even in the presence of high amounts of iron, because the IRP would always remain bound to the IRE and stabilize the transferrin receptor mRNA. Such a person would not have any problem taking up iron into his/her cells. In fact, this person would take up a lot of iron via the transferrin receptor, even when the iron concentrations were high. Therefore, he/she would not need more iron in the diet. However, excess iron in the diet would be very toxic for two reasons. First, the person cannot make ferritin, which prevents the toxic buildup of iron in the cytosol. Second, when iron levels are high, the person would continue to synthesize the transferrin receptor, which functions in the uptake of iron.

C30. A disadvantage of mRNAs with a short half-life is that the cells probably waste a lot of energy making them. If a cell needs the protein encoded by a short-lived mRNA, the cell has to keep transcribing the gene that encodes the mRNA because the mRNAs are quickly degraded. An advantage of short-lived mRNAs is that the cell can rapidly turn off protein synthesis. If a cell no longer needs the polypeptide encoded by a short-lived mRNA, it can stop transcribing the gene, and the mRNA will be quickly degraded. This will shut off the synthesis of more proteins rather quickly. With most long-lived mRNAs, it will take much longer to shut off protein synthesis after transcription has been terminated.

C32. If mRNA stability is low, this means that it is degraded more rapidly. Therefore, low stability results in a low mRNA concentration. The length of the polyA tail is one factor that affects stability. A longer tail makes mRNA more stable. Certain mRNAs have sequences that affect their half-lives. For example, AREs are found in many short-lived mRNAs. The AREs are recognized by cellular proteins that cause the mRNAs to be rapidly degraded.

Experimental Questions

E2. S1 nuclease cuts single-stranded DNA but not double-stranded DNA. If a gene is in an open conformation and cut by DNase I, it would not hybridize to the probe that is complementary to it. In this case, the single-stranded probe would be chewed up by the S1 nuclease. In contrast, if the gene is in a closed conformation and resistant to DNase I, the probe will hybridize to the gene and will not be digested with S1 nuclease. Therefore, the ability of S1 nuclease to digest the probe tells you whether or not the gene is being cleaved by DNase I.

The precipitation step is needed to separate the DNA fragments from free nucleotides. If the radioactive probe is bound to a complementary DNA strand, it will be protected from degradation by S1 nuclease; it will be found in the pellet. This occurs when the globin gene is in a closed conformation. In contrast, if the radioactive probe does not bind to a complementary DNA strand, it will be digested into free nucleotides; the free nucleotides will remain in the supernatant. This occurs when the globin gene is in an open conformation.

E4. These results indicate that the fibroblasts have a maintenance methylase because they can replicate and methylate DNA if it has already been methylated. However, the cells do not express a *de novo* methylase that recognizes this DNA segment, because the DNA in the daughter cells remains unmethylated if the donor DNA was unmethylated.

E6. Based on these results, there are enhancers that are located in regions A, D, and E. When these enhancers are deleted, the level of transcription is decreased. There also appears to be a silencer in region B, because a deletion of this region increases the rate of transcription. There do not seem to be any response elements in region C, or at least not any that function in muscle cells.

E8. The results indicate that protein X binds to the DNA fragment and retards its mobility (lanes 3 and 4). However, the hormone is not required for DNA binding. Because we already know that the hormone is needed for transactivation, it must play some other role. Perhaps, the hormone activates a signaling pathway that leads to the phosphorylation of the transcription factor, and phosphorylation is necessary for transactivation. This situation would be similar to the CREB protein, which is activated by phosphorylation. The CREB protein can bind to the DNA whether or not it is phosphorylated.

Questions for Student Discussion/Collaboration

2. Probably the most efficient method would be to systematically make deletions of progressively smaller sizes. For example, you could begin by deleting 20,000 bp on either side of the gene and see if that affects transcription. If you found that only the deletion on the 5′ end of the gene had an effect, you could then start making deletions from the 5′ end, perhaps in 10,000 bp or 5,000 bp increments until you localized response elements. You would then make smaller deletions in the putative region until it was down to a hundred or a few dozen nucleotides. At this point, you might conduct site-directed mutagenesis, as described in chapter 18, as a way to specifically identify the response element sequence.

CHAPTER 16

Conceptual Questions

C2. A gene mutation is a relatively small mutation that is localized to a particular gene. A chromosome mutation is a large enough change in the genetic material so that it can be seen with the light microscope. This would affect several genes. Genome mutations are changes in chromosome number.

C4. A suppressor mutation suppresses the phenotypic effects of some other mutation. Intragenic suppressors are within the same gene as the first mutation. Intergenic suppressors are in some other gene.

C6. The X rays did not produce a mutation because a mutation is a heritable change in the genetic material. In this case, the X rays have killed the cell, so changes in DNA structure cannot be passed from cell to cell or from parent to offspring.

C8. An efficient nonsense suppressor would probably inhibit cell growth because all of the genes that have their stop codons in the correct location would make proteins that would be too long. This would waste cellular energy, and in some cases, the elongated protein may not function properly.

C10. Here are two possible examples:

The consensus sequences for many bacterial promoters are −35: 5′–TTGACA–3′ and −10: 5′–TATAAT–3′. Most mutations that alter the consensus sequence would be expected to decrease the rate of transcription. For example, a mutation that changed the −35 region to 5′–GAGACA–3′ would decrease transcription.

The sequence 5′–TGACGTCA–3′ is recognized by the CREB transcription factor. If this sequence was changed to 5′–TGAGGTCA–3′ the CREB protein would not recognize it very well, and the adjacent gene would not be regulated properly in the presence of cAMP.

C12. Random mutations are more likely to be harmful. The genes within each species have evolved to work properly. They have functional promoters, coding sequences, terminators, etc., that allow the genes to be expressed. Mutations are more likely to disrupt these sequences. For example, mutations within the coding sequence may produce early stop codons, frameshift mutations, and missense mutations that result in a nonfunctional polypeptide. On rare occasions, however, mutations are beneficial; they may produce a gene that is expressed better than the original gene or produce a polypeptide that functions better. As discussed in chapter 25, beneficial mutations may be acted upon by natural selection over the course of many generations and eventually become the prevalent allele in the population.

C14. Yes, a person with cancer is a genetic mosaic. The cancerous tissue contains gene mutations that are not found in noncancerous cells of the body.

C16. The drawing should show the attachment of a methyl or ethyl group to a base within the DNA. The presence of the alkyl group disrupts the proper base pairing between the alkylated base and the normal base in the opposite DNA strand.

C18. A. Nitrous acid causes A→G and C→T mutations, which are transition mutations.

B. 5-bromouracil causes G→A mutations, which are transitions.

C. Proflavin causes small additions or deletions, which may result in frameshift mutations.

C20. A spontaneous mutation originates within a living cell. It may be due to spontaneous changes in nucleotide structure, errors in DNA replication, or products of normal metabolism that may alter the structure of DNA. The causes of induced mutations originate from outside the cell. They may be physical agents, such as UV light or X rays, or chemicals that act as mutagens. Both spontaneous and induced mutations may cause a harmful phenotype such as a cancer. In many cases, induced mutations are avoidable if the individual can prevent exposure to the environmental agent that acts as a mutagen.

C22. Nitrous acid can change a cytosine to uracil. Excision repair systems could remove the defect and replace it with the correct base.

C24. A. True.

B. False, the TNRE is not within the promoter, it is within the coding sequence.

C. True, CAG is a codon for glutamine.

D. False, CCG is a codon for proline.

C26. The mutation rate is the number of new mutations per gene per generation. The mutation frequency is the number of copies of a mutant gene within a population, divided by the total number of copies (mutant and nonmutant) of that gene. The mutation frequency may be much higher than the mutation rate if new mutations accumulate within a population over the course of many generations.

C28.

A.
```
TTGGHTGUTGG

HHUUTHUGHUU
```
↓ First round of replication
```
TTGGHTGUTGG   CCAAACACCAA

AACCCACAACC   HHUUTHUGHUU
```

↓ Second round of replication

B.
```
TTGGHTGUTGG   TTGGGTGTTGG   CCAAACACCAA   CCAAACACCAA

AACCCACAACC   AACCCACAACC   GGTTTGTGGTT   HHUUTHUGHUU
```

C30. The answer is 5-bromouracil. If this chemical is incorporated into DNA, it can change an AT base pair into a CG base pair. A lysine codon, AAG, could be changed into a glutamic acid codon, CAG, with this chemical. By comparison, UV light is expected to produce thymine dimers, which would not lead to a mutation creating a glutamic acid codon, and proflavin is expected to cause frameshift mutations.

C32. A. If UvrA was missing, the repair system would not be able to identify a damaged DNA segment.

B. If UvrC was missing, the repair system could identify the damaged segment, but it would be unable to make cuts in the damaged DNA strand.

C. If UvrD was missing, the repair system could identify the damaged segment and make cuts in the damaged strand, but it could not unwind the damaged and undamaged strands and thereby remove the damaged segment.

D. If DNA polymerase was missing, the repair system could identify the damaged segment, make cuts in the damaged strand, and unwind the damaged and undamaged strands to remove the damaged segment, but it could not synthesize a complementary DNA strand using the undamaged strands as a template.

C34. Recombinational repair occurs when the DNA is being replicated. If DNA damage occurred in exactly the same location in both replicated helices, recombinational repair could not work because it relies on the use of a strand in one replicated helix to act as a template to repair the complementary strand of the other replicated helix.

C36. In *E. coli*, the TRCF recognizes when RNA polymerase is stalled on the DNA. This stalling may be due to DNA damage such as a thymine dimer. The TRCF removes RNA polymerase and recruits the excision DNA repair system to the region, thereby promoting the repair of the template strand of DNA. It is beneficial to preferentially repair transcribed DNA because it is functionally important. It is a DNA region that encodes a gene.

C38. The underlying genetic defect that causes xeroderma pigmentosum is a defect in one of the genes that encode a polypeptide involved with nucleotide excision repair. These individuals are defective in repairing DNA abnormalities such as thymine dimers, abnormal bases, etc. Therefore, they are very sensitive to environmental agents such as UV light. Since they are defective at repair, UV light is more likely to cause mutations in these people, compared to unaffected individuals. For this reason, people with XP develop pigmentation abnormalities and premalignant lesions and have a high predisposition to skin cancer.

C40. Both types of repair systems recognize an abnormality in the DNA and excise the abnormal strand. The normal strand is then used as a template to synthesize a complementary strand of DNA. The systems differ in the types of abnormalities they detect. The mismatch repair system detects base pair mismatches, while the excision repair system recognizes thymine dimers, chemically modified bases, missing bases, and certain types of cross-links. The mismatch repair system operates immediately after DNA replication, allowing it to distinguish between the daughter strand (which contains the wrong base) and the parental strand. The excision repair system can operate at any time in the cell cycle.

Experimental Questions

E2. When cells from a master plate were replica plated onto two plates containing selective media with the T1 phage, T1-resistant colonies were observed at the same locations on both plates. These results indicate that the mutations occurred randomly while on the master plate (in the absence of T1) rather than occurring as a result of exposure to T1. In other words, mutations are random events, and selective conditions may promote the survival of mutant strains that occur randomly.

To show that antibiotic resistance is due to random mutation, one could follow the same basic strategy except the secondary plates would contain the antibiotic instead of T1 phage. If the antibiotic resistance arose as a result of random mutation on the master plate, one would expect the antibiotic-resistant colonies to appear at the same locations on two different secondary plates.

E4. Perhaps the X rays also produce mutations that make the *ClB* daughters infertile. Many different types of mutations could occur in the irradiated males and be passed to the *ClB* daughters. Some of these mutations could prevent the *ClB* daughter from being fertile. These mutations could interfere with oogenesis, etc. Such *ClB* daughters would be unable to have any offspring.

E6. You would conclude that chemical A is not a mutagen. The percentage of *ClB* daughters (whose fathers had been exposed to chemical A) that did not produce sons was similar to the control (compare 3 out of 2,108 with 2 out of 1,402). In contrast, chemical B appears to be a mutagen. The percentage of *ClB* daughters (whose fathers had been exposed to chemical B) that did not produce sons was much higher than the control (compare 3 out of 2,108 with 77 out of 4,203).

E8. You would expose the bacteria to the physical agent. You could also expose the bacteria to the rat liver extract, but it is probably not necessary for two reasons. First, a physical mutagen is not something that a person would eat. Therefore, the actions of digestion via the liver are probably irrelevant, if you are concerned that the agent might be a mutagen. Second, the rat liver extract would not be expected to alter the properties of a physical mutagen.

E10. The results suggest that the strain is defective in excision repair. If we compare the normal and mutant strains that have been incubated for 2 hours at 37°C, much of the radioactivity in the normal strain has been transferred to the soluble fraction because it has been excised. In the mutant strain, however, less of the radioactivity has been transferred to the soluble fraction, suggesting that it is not as efficient at removing thymine dimers.

Questions for Student Discussion/Collaboration

2. The worst time to be exposed to mutagens would be at very early stages of embryonic development. An early embryo is most sensitive to mutation because it will affect a large region of the body. Adults must also worry about mutagens for several reasons. Mutations in somatic cells can cause cancer, a topic to be discussed in chapter 22. Also, adults should be careful to avoid mutagens that may affect the ovaries or testes since these mutations could be passed along to offspring.

CHAPTER 17

Conceptual Questions

C2. Branch migration will not create a heteroduplex during sister chromatid exchange because the sister chromatids are genetically identical. There should not be any mismatches between the complementary strands. Gene conversion cannot take place because the sister chromatids carry alleles that are already identical to each other.

C4. The steps are described in figure 17.6.

   A. The ends of the broken strands would not be recognized and degraded.

   B. RecA protein would not recognize the single-stranded ends, and strand invasion of the homologous double helix would not occur.

   C. Holliday junctions would not form.

   D. Branch migration would not occur without these proteins. And resolution of the intertwined helices would not occur.

C6. Usually, the overall net effect is not to create any new mutations in particular genes. However, homologous recombination does rearrange the combinations of alleles along particular chromosomes. This can be viewed as a mutation, since the sequence of a chromosome has been altered in a heritable fashion.

C8. It depends on which way the breaks occur in the DNA strands during the resolution phase. If the two breaks occur in the crossed DNA strands, nonrecombinant chromosomes result. If the two breaks occur in the uncrossed strands, the result is a pair of recombinant chromosomes.

C10. Holliday model—proposes two breaks, one in each chromatid, and then both strands exchange a single strand of DNA. Double-stranded break model—proposes two breaks, both in the same chromatid. As in the Holliday model, single-strand migration occurs between both homologues. The double-stranded break model also proposes strand degradation and gap repair synthesis. Finally, the double-stranded break model generates two Holliday junctions, not just one.

C12. The RecA protein binds to single-stranded DNA and forms a filament. The formation of the filament promotes the sliding of the filament along another DNA region until it recognizes homologous sequences. This recognition process may involve the formation of triplex DNA. After recognition has occurred, RecA protein mediates the movement of the invading DNA strand and the displacement of the original strand.

C14. RecG and RuvABC bind to Holliday junctions. They do not necessarily recognize a DNA sequence but, instead, recognize a region of crossing over (i.e., a four-stranded structure).

C16. The function of the RAG1 and RAG2 proteins is to recognize the recombination signal sequences and make double-stranded cuts. In the case of V/J recombination, a cut is made at the end of one V region and the beginning of one J region. The NHEJ proteins recognize these ends and join them together. This is a form of DNA splicing. This creates different combinations of the V, J, (D), and constant regions, thereby creating a large amount of diversity in the encoded antibodies.

C18. Integrase is an enzyme that recognizes the attachment site sequences within the λ DNA and the *E. coli* chromosome. It brings the λ DNA close to the chromosome and then makes staggered cuts in the attachment sites. The strands are exchanged, and then integrase catalyzes the covalent attachment of the strands to each other. In this way, the λ DNA is inserted at a precise location within the *E. coli* chromosome.

C20. In figure 17.15, the TE has transposed prior to DNA replication. The transposon is single stranded at both locations. DNA replication (i.e., gap repair synthesis) copies the single-stranded elements into double-stranded elements. Therefore, the end result is two double-stranded TEs at two distinct locations.

C22. Retroelements have the greatest potential for proliferation because the element is transcribed into RNA as an intermediate. Many copies of this RNA could be transcribed and then copied into DNA by reverse transcriptase. Theoretically, many copies of the element could be inserted into the genome in a single generation.

C24. Keep in mind that each type of transposase recognizes only the inverted repeats of a particular type of transposable element. The mosquitoes must express a transposase that recognizes the Z elements. This explains why the Z elements are mobile. This Z element transposase must not recognize the inverted repeats of the X elements; the inverted repeats of the X elements and Z elements must have different sequences. This same group of mosquitoes must not express a transposase that recognizes the X elements. This explains why the X elements are very stable.

C26. A transposon is excised as a segment of DNA, and then it transposes to a new location via transposase. A retroelement is transcribed as RNA, and then reverse transcriptase makes a double-stranded copy of DNA. Integrase then inserts this DNA copy into the chromosome. All transposable elements have direct repeats. Transposons that move as DNA have inverted repeats, and they may encode transposase (as well as other genes). Retroelements do not have inverted repeats, but they may or may not have LTRs. Autonomous retroelements encode reverse transcriptase and integrase.

C28. An autonomous element has the genes that are necessary for transposition. For example, a cut-and-paste transposon that was autonomous would also have the transposase gene. A nonautonomous element does not have all the genes that are necessary for transposition. However, if a cell contains an autonomous element and a nonautonomous element of the same type, the nonautonomous element can move. For example, if a *Drosophila* cell contained two P elements, one autonomous and one nonautonomous, the transposase expressed from the autonomous P element could recognize the nonautonomous P element and catalyze its transposition.

C30. The formation of deletions and inversions is illustrated in the answer to conceptual question C27. A deletion occurs when there is recombination between the direct repeats in two different elements. The region between the elements is deleted. An inversion can happen when recombination occurs between inverted repeats between two different elements. This occurs when the inverted repeats are in the opposite orientation. A translocation can result when recombination occurs between transposable elements that are located on different (nonhomologous) chromosomes. In other words, the TEs

on different chromosomes align themselves and a crossover occurs. This will produce a translocation, as shown here.

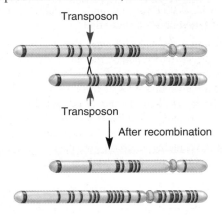

Experimental Questions

E2. You would conclude that the substance is a mutagen. Substances that damage DNA tend to increase the level of genetic exchange such as sister chromatid exchange.

E4. The drawing here shows the progression through three rounds of BrdU exposure. After one round, all of the chromosomes would be

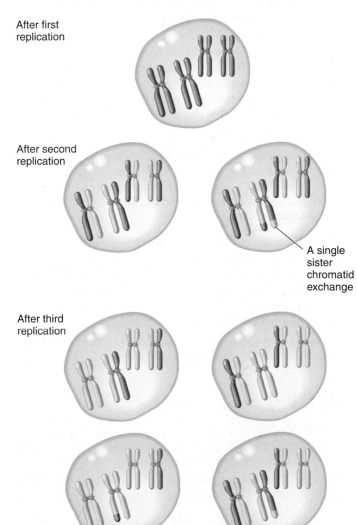

dark. After two rounds, all of the chromosomes would be harlequin. After three rounds, the number of light sister chromatids would be twice as much as the number of light sister chromatids found after two rounds of replication.

E6. When McClintock started with a colorless strain containing *Ds*, she identified 20 cases where *Ds* had moved to a new location to produce red kernels. This identification was possible because the 20 strains had a higher frequency of chromosomal breaks at a specific site, and because of the mutability of particular genes. She also had found a strain where *Ds* had inserted into the red-color-producing gene, so that its transposition out of the gene would produce a red phenotype. Overall, her analysis of the data showed that the sectoring (i.e., mutability) phenotype was consistent with the transposition of *Ds*.

E8. Transposon tagging is an experimental method that is aimed at cloning genes. In this approach, a transposon is introduced into a strain, and the experimenter tries to identify individuals in which the gene of interest has lost its function. In many cases, the loss of function has occurred because the transposon has been inserted into the gene. If so, one can make a library and then use a labeled transposon probe to identify clones in which the transposon has been inserted. This provides a way to clone the gene of interest. Later, one would use the transposon-inserted clone to screen a library from an individual with an active version of the gene. In this way, one could then identify a normal copy (nontransposon-inserted) of the gene.

E10. A transposon creates a mutable site because the excision of a transposon causes chromosomal breakage if the two ends are not reconnected properly. After it has moved out of its original site (causing chromosomal breakage), it may be inserted into a new site somewhere else. You could experimentally determine this by examining a strain that has incurred a chromosomal breakage at the first mutable site. You could microscopically examine many cells that had such a broken chromosome and see if a new mutable site had been formed. This new mutable site would be the site into which the transposon had moved. On occasion, there would be chromosomal breakage at this new mutable site, which could be observed microscopically.

Questions for Student Discussion/Collaboration

2. Harmful consequences: The level of genetic diversity would be decreased, because linked combinations of alleles would not be able to recombine. You would not be able to produce antibody diversity in the same way. Gene duplication could not occur, so the evolution of new genes would be greatly inhibited.

   Beneficial consequences: you would not get (as many) translocations, inversions, and the accumulation of selfish DNA.

CHAPTER 18

Conceptual Questions

C2. A restriction enzyme recognizes a DNA sequence and then cleaves a (covalent) phosphoester bond in each of two DNA strands.

C4. The term *cDNA* refers to DNA that is derived from mRNA. Compared to genomic DNA, it would lack introns.

Experimental Questions

E2. Remember that AT base pairs form two hydrogen bonds while GC base pairs form three hydrogen bonds. The order (from stickiest to least sticky) would be:

$$Bam\text{HI} = Pst\text{I} = Sac\text{I} > Eco\text{RI} > Cla\text{I}$$

E4. In conventional gene cloning, many copies are made because the vector replicates to a high copy number within the cell, and the cells divide to produce many more cells. In PCR, the replication of the DNA to produce many copies is facilitated by primers, nucleotides, and *Taq* polymerase.

E6. A hybrid vector is a vector that has a piece of "foreign" DNA inserted into it. The foreign DNA came from somewhere else, like the chromosomal DNA of some organism. To construct a hybrid vector, the vector and source of foreign DNA are digested with the same restriction enzyme. The complementary ends of the fragments are allowed to hydrogen bond to each other (i.e., sticky ends are allowed to bind), and then DNA ligase is added to create covalent phosphoester bonds. If all goes well, a piece of the foreign DNA will become ligated to the vector, thereby creating a hybrid vector.

   As described in figure 18.2, the insertion of foreign DNA can be detected using X-Gal. As seen here, the insertion of the foreign DNA causes the inactivation of the *lacZ* gene. The *lacZ* gene encodes the enzyme β-galactosidase, which is necessary to convert X-gal to a blue compound. If the *lacZ* gene is inactivated by the insertion of foreign DNA, the bacterial colonies will be white. If the vector has simply recircularized, and the *lacZ* gene remains intact, the colonies will be blue.

E8. If the *Eco*RI fragment containing the *kan^R* gene also had an origin of replication, it is possible that this fragment could circularize and become a very small plasmid. The electrophoresis results would be consistent with the idea that the same bacterial cells could contain two different plasmids: pSC101 and a small plasmid corresponding to the segment of DNA (now circularized) that carried the *kan^R* gene. However, the results shown in step 9 rule out this possibility. The density gradient centrifugation showed a single peak, corresponding to a plasmid that had an intermediate size between pSC101 and pSC102. In contrast, if our alternative explanation had been correct (i.e., that bacterial cells contain two plasmids), there would be two peaks from the density gradient centrifugation. One peak would correspond to pSC101 and the other peak would indicate a very small plasmid (i.e., smaller than pSC101 and pSC102).

E10. 3×2^{27}, which equals 4.0×10^8, or about 400 million copies.

E12. Initially, the mRNA would be mixed with reverse transcriptase and nucleotides to create a complementary strand of DNA. Reverse transcriptase also needs a primer. This could be a primer that is known to be complementary to the β-globin mRNA. Alternatively, mature mRNAs have a polyA tail, so one could add a primer that consists of many Ts. This is called a poly-dT primer. After the complementary DNA strand has been made, the sample would then be mixed with primers, *Taq* polymerase, and nucleotides, etc., and subjected to the standard PCR protocol. Note: the PCR reaction would have two kinds of primers. One primer would be complementary to the 5′ end of the mRNA and would be unique to the β-globin sequence. The other primer would be complementary to the 3′ end. This second primer could be a poly-dT primer or it could be a unique primer that would bind slightly upstream from the polyA-tail region.

E14. A DNA library is a collection of hybrid vectors that contain different pieces of DNA from a source of chromosomal DNA. Because it is a diverse collection of many different DNA pieces, the name *library* seems appropriate.

E16. Hybridization occurs due to the hydrogen bonding of complementary sequences. Due to the chemical properties of DNA and RNA strands, they form double-stranded regions when the base sequences are complementary. In a Southern and Northern experiment, the probe is labeled.

E18. The purpose of a Northern blot experiment is to determine if a gene is transcribed into RNA using a piece of cloned DNA as a probe. It can tell you if a gene is transcribed in a particular cell or at a particular stage of development. It can also tell you if a pre-mRNA is alternatively spliced.

E20. It appears that this mRNA is alternatively spliced to create a high molecular mass and a lower molecular mass product. Nerve cells produce a very large amount of the larger mRNA, whereas spleen cells produce a moderate amount of the smaller mRNA. Both types are produced in small amounts by the muscle cells. It appears that kidney cells do not transcribe this gene.

E22. 1. β (detected at the highest stringency)

2. δ

3. γ_A and ε

4. α_1

5. *Mb* (detected only at the lowest stringency)

Note: At the lowest stringency, all of the globin genes would be detected. At the highest stringency, only the β-globin gene would be detected.

E24. The Western blot is shown here. The sample in lane 2 came from a plant that was homozygous for a mutation that prevented the expression of this polypeptide. Therefore, no protein was observed in this lane. The sample in lane 4 came from a plant that is homozygous for a mutation that introduces an early stop codon into the coding sequence. As seen in lane 4, the polypeptide is shorter than normal (13.3 kDa). The sample in lane 3 was from a heterozygote that expresses about 50% of each type of polypeptide. Finally, the sample in lane 5 came from a plant that is homozygous for a mutation that changed one amino acid to another amino acid. This type of mutation, termed a missense mutation, may not be detectable on gel. However, a single amino acid substitution could affect polypeptide function.

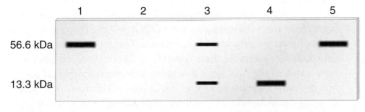

E26. The products of structural genes are proteins with a particular amino acid sequence. Antibodies can specifically recognize proteins due to their amino acid sequence. Therefore, an antibody can detect whether or not a cell is making a particular type of protein.

E28. In this case, the transcription factor binds to the response element when the hormone is present. Therefore, the hormone promotes the binding of the transcription factor to the DNA and thereby promotes transactivation.

E30. The levels of cAMP affect the phosphorylation of CREB, and this affects whether or not it can transactivate transcription. However, CREB can bind to CREs whether or not it is phosphorylated. Therefore, in a gel retardation assay, we would expect CREB to bind to CREs and retard their mobility using a cell extract from cells that were or were not retreated with adrenalin.

E32. The glucocorticoid receptor will bind to GREs if glucocorticoid hormone is also present. The glucocorticoid receptor does not bind to CREs. The CREB protein will bind to CREs (with or without hormone), but it will not bind to GREs. The expected results are shown here. In this drawing, the binding of CREB protein to the 700 bp fragment results in a complex with a higher mass compared to the glucocorticoid receptor binding to the 600 bp fragment.

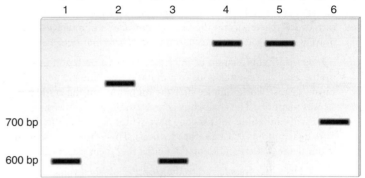

E34. The rationale behind a footprinting experiment has to do with accessibility. If a protein is bound to the DNA, it will cover up the part of the DNA where it is bound. This region of the DNA will be inaccessible to the actions of chemicals or enzymes that cleave the DNA, such as DNase-I.

E36. A. AGGTCGGTTGCCATCGCAATAATTTCTGCCTGAACCCAATA

B. Automated sequencing has several advantages. First, the reactions are done in a single tube as opposed to four tubes. Second, the detector can "read" the sequence and provide the researcher with a printout of the sequence. This is much easier than looking at an X-ray film and writing the sequence out by hand. It also avoids human error. Finally, automated sequencing does not require the use of radioisotopes, which are more expensive and require more laboratory precautions, compared to fluorescently labeled compounds.

E38. There are lots of different strategies one could follow. For example, you could mutate every other base and see what happens. It would be best to make very nonconservative mutations such as a purine for a pyrimidine or a pyrimidine for a purine. If the mutation prevents protein binding in a gel retardation assay, then the mutation is probably within the response element. If the mutation has no effect on protein binding, it probably is outside the response element.

Questions for Student Discussion/Collaboration

2. 1. Does a particular amino acid within a protein sequence play a critical role in the protein's structure or function?

2. Does a DNA sequence function as a promoter?

3. Does a DNA sequence function as a regulatory site?

4. Does a DNA sequence function as a splicing junction?

5. Is a sequence important for correct translation?

6. Is a sequence important for RNA stability?

And many others.

CHAPTER 19

Conceptual Questions

C2. *A. radiobacter* synthesizes an antibiotic that kills *A. tumefaciens*. The genes, which are necessary for antibiotic biosynthesis and resistance, are plasmid encoded and can be transferred during interspecies matings. If *A. tumefaciens* received this plasmid during conjugation, it would be resistant to killing. Therefore, the conjugation-deficient strain prevents the occurrence of *A. tumefaciens*–resistant strains.

C4. A biological control agent is an organism that prevents the harmful effects of some other agent in the environment. Examples include *Bacillus thuringiensis,* a bacterium that synthesizes compounds that act as toxins to kill insects, *Ice⁻* bacteria that inhibit the proliferation of *Ice⁺* bacteria, and the use of *Agrobacterium radiobacter* to prevent crown gall disease caused by *Agrobacterium tumefaciens*.

C6. A mouse model is a strain of mice that carries a mutation in a mouse gene that is analogous to a mutation in a human gene that causes disease. For example, after the mutation causing cystic fibrosis was identified, the analogous gene was mutated in the mouse. Mice with mutations in this gene have symptoms similar to the human symptoms (though not identical). These mice can be used to study the disease and to test potential therapeutic agents.

C8. The T DNA gets transferred to the plant cell; it then is incorporated into the plant cell's genome.

C10. A. With regard to maternal effect genes, the phenotype would depend on the animal that donated the oocyte. It is the cytoplasm of the oocyte that accumulates the gene products of maternal effect genes.

B. The extranuclear traits depend on the mitochondrial genome. Mitochondria are found in the oocyte and in the somatic cell. So, theoretically, both cells could contribute extranuclear traits. In reality, however, researchers have found that the mitochondria in Dolly were from the animal that donated the egg. It is not clear why she had no mitochondria from the mammary cell.

C. The cloned animal would be genetically identical to the animal that donated the nucleus with regard to traits that are determined by nuclear genes, which are expressed during the lifetime of the organism. The cloned animal would/could differ from the animal that donated the nucleus with regard to traits that are determined by maternal effect genes and extracytoplasmic genes. Such an animal is not a true clone, but it is likely that it would greatly resemble the animal that donated the nucleus, because the vast majority of genes are found in the cell nucleus.

C12. Some people are concerned with the release of genetically engineered microorganisms into the environment. The fear is that such organisms may continue to proliferate and it may not be possible to "stop them." A second fear is the use of genetically engineered organisms in the food we eat. Some people are worried that genetically engineered organisms may pose an unknown health risk. A third issue is ethics. Some people feel that it is morally wrong to tamper with the genetics of organisms. This opinion may also apply to genetic techniques such as cloning, stem cell research, and gene therapy.

Experimental Questions

E2. The plasmid with the wrong orientation would not work because the coding sequence would be in the wrong direction relative to the promoter sequence. Therefore, the region containing the somatostatin sequence would not be transcribed into RNA.

E4. To construct the coding sequence for somatostatin, the researchers synthesized eight oligonucleotides, labeled A–H. When these oligonucleotides were mixed together, they would hydrogen bond to each other due to their complementary sequences. The addition of ligase would covalently link the DNA backbones. The *left* side of the coding sequence had an *Eco*RI site and the *right* side had a *Bam*HI site; these sites made it possible to insert this sequence at the end of the β-galactosidase gene. You may also notice that an ATG codon (AUG in the mRNA) precedes the first alanine codon in somatostatin. This AUG codon specifies methionine, which allows the somatostatin to be cleaved from β-galactosidase using cyanogen bromide. This was necessary because somatostatin, by itself, is rapidly degraded in *E. coli*, whereas the fusion protein is not.

E6. A kanamycin-resistance gene is contained within the T DNA. Exposure to kanamycin selects for the growth of plant cells that have incorporated the T DNA into their genome. The carbenicillin kills the *A. tumefaciens*. The phytohormones promote the regeneration of an entire plant from somatic cells. If kanamycin were left out, it would not be possible to select for the growth of cells that had taken up the T DNA.

E8. The term *gene knockout* refers to an organism in which the function of a particular gene has been eliminated. For autosomal genes in eukaryotes, a gene knockout is a homozygote for a defect in both copies of the gene. If a gene knockout has no phenotypic effect, perhaps the gene is redundant. In other words, there may be multiple genes within the genome that can carry out the same function. Another reason why a gene knockout may not have a phenotypic effect is because of the environment. As an example, let's say a mouse gene is required for the synthesis of a vitamin. If the researchers were providing food that contained the vitamin, the knockout mouse that was lacking this gene would have a normal phenotype; it would survive just fine. Sometimes, researchers have trouble knowing the effects of a gene knockout until they modify the environmental conditions in which the animals are raised.

E10. Gene replacement occurs by homologous recombination. For homologous recombination to take place, two crossovers must occur, one at each end of the target gene. After homologous recombination, only the *Neo^R* gene, which is inserted into the target gene, can be incorporated into the chromosomal DNA of the embryonic stem cell. In contrast, nonhomologous recombination can involve two crossovers anywhere in the cloned DNA. Because the *TK* gene and target gene are adjacent to each other, nonhomologous recombination usually transfers both the *TK* gene and the *Neo^R* gene. If both genes are transferred to a stem cell, it will die because the cells are grown in the presence of gancyclovir. The product of the *TK* gene kills the cell under these conditions. In contrast, cells that have acquired the *Neo^R* gene due to homologous recombination but not the *TK* gene will survive. Stem cells, which have not taken up any of the cloned DNA, will die because they will be killed by the neomycin. In this way, the presence of gancyclovir and neomycin selects for the growth of stem cells that have acquired the target gene by homologous recombination.

E12. A. Dolly's chromosomes may seem so old because they were already old when they were in the nucleus that was incorporated into the enucleated egg. They had already become significantly shortened in the mammary cells. This shortening was not repaired by the oocyte.

B. Dolly's age does not matter. Remember that shortening does not occur in germ cells. However, Dolly's eggs are older than they seem by about 6 or 7 years, because Dolly's germ-line cells received their chromosomes from a sheep that was 6 years old, and the cells were grown in culture for a few doublings before a mammary cell was fused with an enucleated egg. Therefore, the calculation would be: 6 or 7 years (the age of the mammary cells that produced Dolly's germ-line cells) plus 8 years (the age of Molly), which equals 14 or 15 years. However, only half of Molly's chromosomes would appear to be 14 or 15 years old. The other half of her chromosomes, which she inherited from her father, would appear to be 8 years old.

C. Chromosome shortening is a bit disturbing, because it suggests that aging has occurred in the somatic cell, and this aging is passed to the cloned organism. If cloning was done over the course of many generations, this may eventually have a major impact on the life span of the cloned organism. It may die much earlier than a noncloned organism. However, chromosome shortening may not always occur. It does not seem to occur in mice, which were cloned for six consecutive generations.

E14. The term *molecular pharming* refers to the practice of making transgenic animals that will synthesize (human) products in their milk. It can be advantageous when bacterial cells are unable to make a functional protein product from a human gene. For example, some proteins are posttranslationally modified by the attachment of carbohydrate molecules. This type of modification does not occur in bacteria, but it may occur correctly in transgenic animals. Also, dairy cows produce large amounts of milk, which may improve the yield of the human product.

E16. You would first need to clone the normal mouse gene. Cloning methods are described in chapter 18. After the normal gene was cloned, you would then follow the protocol shown in figure 19.6. The normal gene would be inactivated by the insertion of the Neo^R gene, and the *TK* gene would be cloned next to it. This DNA segment would be introduced into mouse embryonic stem cells and grown in the presence of neomycin and gancyclovir. This selects for homologous recombinants. The surviving embryonic stem cells would be injected into early embryos, which would then develop into chimeras. The chimeric mice would be identified by their patches of light and dark fur. At this point, if all has gone well, a portion of the mouse is heterozygous for the normal gene and the gene that has the Neo^R insert. This mouse would be bred, and then brothers and sisters from the litter would be bred to each other. Southern blots would be conducted to determine if the offspring carried the gene with the Neo^R insert. At first, one would identify heterozygotes that had one copy of the inserted gene. These heterozygotes would be crossed to each other to obtain homozygotes. The homozygotes are gene knockouts because the function of the gene has been "knocked out" by the insertion of the Neo^R gene. Perhaps, these mice would be dwarf and exhibit signs of mental retardation. At this point, the researcher would have a mouse model to study the disease.

E18. DNA fingerprinting is a method of identification based on the properties of DNA. VNTR and STR sequences are variable with regard to size in natural populations. This variation can be seen when DNA fragments are subjected to gel electrophoresis. Within a population, any two individuals (except for identical twins) will display a different pattern of DNA fragments, which is called their DNA fingerprint.

E20. A VNTR is a sequence that is repeated several times within a genome and is variable in its length. In natural populations, it is common to find length variation. Therefore, any two individuals (who are not genetically identical) will differ with regard to the sizes of many of their VNTRs. When a gel is run and the VNTRs are seen in a Southern blot, the pattern of VNTR sizes provides a fingerprint of the individual's DNA. This fingerprint is a unique feature of each individual.

E22. The minimum percentage of matching bands is based on the genetic relationships.

A. 50%

B. 50% (on average, but occasionally it could be less)

C. 25% (on average, but occasionally it could be less)

D. 25% (on average, but occasionally it could be less)

E24. In cystic fibrosis gene therapy, an aerosol spray, containing the normal *CF* gene in a retrovirus or liposome, is used to get the normal *CF* gene into the lung cells. The epithelial cells on the surface of the lung will take up the gene. However, these surface epithelial cells have a finite life span, so it is necessary to have repeated applications of the aerosol spray. Ideally, scientists hope to devise methods whereby the normal *CF* gene will be able to penetrate more deeply into the lung tissue.

Questions for Student Discussion and Collaboration

2. From a genetic viewpoint, the recombinant and nonrecombinant strains are very similar. The main difference is their history. The recombinant strain has been subjected to molecular techniques to eliminate a particular gene. The nonrecombinant strain has the advantage of good public relations. People are less worried about releasing nonrecombinant strains into the environment.

CHAPTER 20

Conceptual Questions

C2. A. Yes

B. No, this is only one chromosome in the genome

C. Yes

D. Yes

Experimental Questions

E2. They are complementary to each other.

E4. Because normal cells contain two copies of chromosome 14, one would expect that a probe would bind to complementary DNA sequences on both of these chromosomes. If a probe recognized only one of two chromosomes, this means that one of the copies of chromosome 14 has been lost, or it has suffered a deletion in the region where the probe binds. With regard to cancer, the loss of this genetic material may be related to the uncontrollable cell growth.

E6. After the cells and chromosomes have been fixed to the slide, it is possible to add two or more different probes that recognize different sequences (i.e., different sites) within the genome. Each probe has a different fluorescence emission wavelength, so it can be identified by its color. Usually, a researcher will use computer imagery that recognizes the wavelength of each probe and then assigns that region a bright color. The color seen by the researcher is not the actual color emitted by the probe; it is a secondary color assigned by the computer. This can be done with two or more different probes, as a way to color the regions of the chromosomes that are recognized by the probes. In a sense, the probes, with the aid of a computer, are "painting" the regions of the chromosomes that are recognized by a probe. An example of chromosome painting is shown in figure 20.4. In this example, human chromosome 5 is painted with six different colors.

E8. A contig is a collection of clones that contain overlapping segments of DNA that span a particular region of a chromosome. To determine if two clones are overlapping, one could conduct a Southern blotting experiment. In this approach, one of the clones is used as a probe. If it is overlapping with the second clone, it will bind to it in a Southern blot. Therefore, the second clone is run on a gel and the first clone is used as a probe. If the band corresponding to the second clone is labeled, this means that the two clones are overlapping.

E10. YAC cloning vectors have the replication properties of a chromosome and the cloning properties of a plasmid. To replicate like a chromosome, the YAC vector contains an origin of replication and centromeric and telomeric sequences. Therefore, in a yeast cell, a YAC can behave as a chromosome. Like a plasmid, YACs also contain selectable markers and convenient cloning sites for the insertion of large segments of DNA. The primary advantage is the ability to clone very large pieces of DNA.

E12. The resistance gene appears to be linked to RFLP 4B.

E14. For most organisms, it is usually easy to locate many RFLPs throughout the genome. The RFLPs can be used as molecular markers to make a map of the genome. This is done using the strategy described in figure 20.8. To map a functional gene, one would also follow the same general strategy described in figure 20.8, except that the two strains would also have an allelic difference in the gene of interest. The experimenter would make crosses, such as dihybrid crosses, and determine the number of parental and recombinant offspring based on the alleles and RFLPs they had inherited. If an allele and RFLP are linked, there will be a much lower percentage (i.e., less than 50%) of the recombinant offspring.

E16.

Deduced Outcome

STSs

4 6   7 9   3 2   8 5 1

YAC

3   ⟵——————⟶

1   ⟵——————⟶

5   ⟵——————⟶

4   ⟵——————⟶

2   ⟵——————⟶

E18. A. One homologue contains the STS-1 that is 289 bp and STS-2 that is 422 bp while the other homologue contains STS-1 that is 211 bp and STS-2 that is 115 bp. This is based on the observation that 28 of the sperm have either the 289 bp and 422 bp bands or the 211 bp and 115 bp bands.

   B. There are two recombinant sperm; see lanes 12 and 18. Since there are two recombinant sperm out of a total of thirty:

$$\text{Map distance} = \frac{2}{30} \times 100$$

$$= 6.7 \text{ mu}$$

   C. In theory, this method could be used. However, there is not enough DNA in one sperm to carry out an RFLP analysis unless the DNA is amplified by PCR.

E20. One possibility is that the geneticist could try a different restriction enzyme. Perhaps there is sequence variation in the vicinity of the pesticide-resistance gene that affects the digestion pattern of a restriction enzyme other than *Eco*RI. There are hundreds of different restriction enzymes that recognize a myriad of different sequences.

   Alternatively, the geneticist could give up on the RFLP approach and try to identify one or more sequence-tagged sites that are in the vicinity of the pesticide-resistance gene. In this case, the geneticist would want to identify STSs that are also microsatellites. As described in figure 20.12, the transmission of microsatellites can be followed in genetic crosses. Therefore, if the geneticist could identify microsatellites in the vicinity of the pesticide-resistance gene, this would make it possible to predict the outcome of crosses. For example, let's suppose a microsatellite linked to the pesticide-resistance gene existed in three forms: 234 bp, 255 bp, and 311 bp. And let's also suppose that the 234 bp form was linked to the high-resistance allele, the 255 bp form was linked to the moderate-resistance allele, and the 311 bp form was linked to the low-resistance allele. According to this hypothetical example, the geneticist could predict the level of resistance in an alfalfa plant by analyzing the inheritance of these microsatellites.

E22. When chromosomal DNA is isolated and digested with a restriction enzyme, this produces thousands of DNA fragments of different sizes. This makes it impossible to see any particular band on a gel, if you simply stained the gel for DNA. Southern blotting allows you to detect one or more RFLPs that are complementary to the radioactive probe that is used.

E24. Based on these results, it is likely that the sickle-cell allele originated in an individual with the 13.0 kbp RFLP. This would explain why the *Hb^S* allele is usually transmitted with the 13.0 kbp RFLP. On occasion, however, a crossover could occur in the region between the β-globin gene and the distal *Hpa*I site.

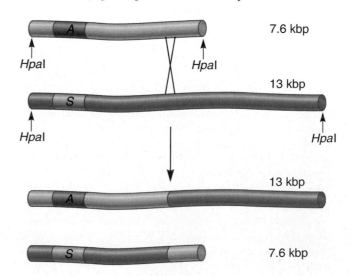

After crossing over, the *Hb^S* allele is now linked to a 7.6 kbp RFLP.

E26. PFGE is a method of electrophoresis that is used to separate small chromosomes and large DNA fragments. The electrophoresis devices used in PFGE have two sets of electrodes. The two sets of electrodes produce alternating pulses of current, and this facilitates the separation of large DNA fragments.

It is important to handle the sample gently to prevent the breakage of the DNA due to mechanical forces. Cells are first embedded in agarose blocks, and then the blocks are loaded into the wells of the gel. The agarose keeps the sample very stable and prevents shear forces that might mechanically break the DNA. After the blocks are in the gel, the cells within the blocks are lysed, and if desired, restriction enzymes can be added to digest the DNA. For PFGE, a restriction enzyme that cuts very infrequently might be used.

PFGE can be used as a preparative technique to isolate and purify individual chromosomes or large DNA fragments. PFGE can also be used, in conjunction with Southern blotting, as a mapping technique.

E28. Note: The insert of cosmid B is contained completely within the insert of cosmid C.

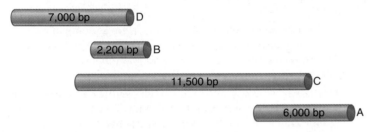

E30. A. The general strategy is shown in figure 20.19. The researcher begins at a certain location and then walks toward the gene of interest. You begin with a clone that has a marker that is known to map relatively close to the gene of interest. A piece of DNA at the end of the insert is subcloned and then used in a Southern blot to identify an adjacent clone in a cDNA library. This is the first "step." The end of this clone is subcloned to make the next step. And so on. Eventually, after many steps, you will arrive at your gene of interest.

B. In this example, you would begin at STS-3. If you walked a few steps and happened upon STS-2, you would know that you were walking in the wrong direction.

C. This is a difficult aspect of chromosome walking. Basically, you would walk toward gene X using DNA from a normal individual and DNA from an individual with a mutant gene X. When you have found a site where the sequences are different between the normal and mutant individual, you may have found gene X. You would eventually have to confirm this by analyzing the DNA sequence of this region and determining that it encodes a functional gene.

E32. Two techniques to purify chromosomes are pulsed-field gel electrophoresis and chromosome sorting. They are useful because the isolation of a single chromosome can efficiently lead to the construction of a contig for that particular chromosome. Also, the isolation of a particular chromosome can lead to the mapping of markers or genes on that chromosome.

Questions for Student Discussion/Collaboration

2. A molecular marker is a segment of DNA, not usually encoding a gene, which has a known location within a particular chromosome.

It marks the location of a site along a chromosome. RFLPs and STSs are examples. It is easier to use these types of markers because they can be readily identified by molecular techniques such as restriction digestion analysis and PCR. Locating functional genes is usually more difficult because this relies on conventional linkage mapping approaches whereby allelic differences in the gene are mapped by making crosses or following a pedigree. For monomorphic genes, this approach does not work.

CHAPTER 21

Conceptual Questions

C2. There are two main reasons why the proteome is larger than the genome. The first reason involves the processing of pre-mRNA, a phenomenon that occurs primarily in eukaryotic species. RNA splicing and editing can alter the codon sequence of mRNA and thereby produce alternative forms of proteins that have different amino acid sequences. The second reason for protein diversity is posttranslational modifications. There are many ways that a given protein's structure can be covalently modified by cellular enzymes. These include proteolytic processing, disulfide bond formation, glycosylation, attachment of lipids, phosphorylation, methylation, and acetylation, to name a few.

C4. Centromeric sequences, origins of replication, telomeric sequences, repetitive sequences, and enhancers. Other examples are possible.

C6. There are a few interesting trends. Sequences 1 and 2 are similar to each other, as are sequences 3 and 4. There are a few places where amino acid residues are conserved among all five sequences. These amino acids may be particularly important with regard to function.

C8. A. Correct.

B. Correct.

C. This is not correct. These are short genetic sequences that happen to be similar to each other. The *lac* operon and *trp* operon are not derived from the same primordial operon.

D. This is not correct. The two genes are homologous to each other. It is correct to say that their sequences are 60% identical.

Experimental Questions

E2. As described in solved problem S1, one reason for making a subtractive DNA library is to determine which RNAs are produced when environmental conditions change. You want to load a small amount of cDNA on the column that was derived from cells that had been exposed to mercury. You want all of the cDNA made in both the presence and absence of mercury to bind to the column. Remember that the cDNA, which is derived from mRNA that is made in the absence of mercury, is already bound to the column. You want the cDNA that is made in the presence of mercury to bind to this cDNA, if it is complementary. If too much of this cDNA is loaded, all of the cDNAs will have complementary cDNA bound to them, and some of them will not bind to the column, even though they may be complementary to the cDNAs. In other words, if you load too much cDNA (derived from the mercury-exposed cells), you will have saturated the binding sites for cDNAs that are made in both the presence and absence of mercury. You do not want this to happen, because you want only the cDNAs that are derived from mRNAs that are specifically expressed in the presence of mercury to flow through the column. These cDNAs are not complementary to any cDNAs that are attached to the column.

E4. The cDNA that is labeled with a green dye is derived from mRNA that was obtained from cells at an early time point, when glucose levels were high. The other samples of cDNA were derived from cells collected at later time points, when glucose levels were falling, and when the diauxic shift was occurring. These were labeled with a red dye. The green fluorescence provides a baseline for gene expression when glucose is high. At later time points, if the red : green ratio is high (i.e., greater than one), this means that a gene is induced as glucose levels fall, because there is more red cDNA compared to green cDNA. If the ratio is low (i.e., less than one), this means that a gene is being repressed.

E6. In the first dimension (i.e., in the tube gel), proteins are separated according to the isoelectric point. This is the pH at which a given protein's net charge is zero. In the second dimension (i.e., the slab gel), proteins are coated with SDS and separated according to their molecular mass.

E8. In tandem mass spectroscopy, the first spectrometer determines the mass of a peptide fragment from a protein of interest. The second spectrometer determines the masses of progressively smaller pieces that are derived from that peptide. Because the masses of each amino acid are known, the molecular masses of these smaller fragments reveal the amino acid sequence of the peptide. With peptide sequence information, it is possible to use the genetic code and produce putative DNA sequences that could encode such a peptide. These sequences, which are degenerate due to the degeneracy of the genetic code, are used as query sequences to search a genomic database. This will (hopefully) locate a match. The genomic sequence can then be analyzed to determine the entire coding sequence for the protein of interest.

E10. The first strategy is a search by signal approach, which relies on known sequences such as promoters, start and stop codons, and splice sites to help predict whether or not a DNA sequence contains a structural gene. It would try to identify a region that contains a promoter sequence, then a start codon, a coding sequence, and a stop codon. A second strategy is a search by content approach. The goal is to identify sequences whose nucleotide content differs significantly from a random distribution, which is usually due to codon bias. A search by content approach attempts to locate coding regions by identifying regions where the nucleotide content displays a bias. A third method to locate structural genes is to search for long open reading frames within a DNA sequence. An open reading frame is a sequence that does not contain any stop codons.

E12. By searching a database, one can identify genetic sequences that are homologous to a newly determined sequence. In most cases, homologous sequences carry out identical or very similar functions. Therefore, if one identifies a homologous member of a database whose function is already understood, this provides an important clue regarding the function of the newly determined sequence.

E14. The basis for secondary structure prediction is that certain amino acids tend to be found more frequently in α helices or β sheets. This information is derived from the locations of amino acids within proteins that have already been crystallized. Predictive methods are perhaps 60 to 70% accurate, which is not very good.

E16. The backtranslate program works by knowing the genetic code. Each amino acid has one or more codons (i.e., three-base

sequences) that are specified by the genetic code. This program would produce a sequence file that was a nucleotide base sequence. The backtranslate program would produce a degenerate base sequence because the genetic code is degenerate. For example, lysine can be specified by AAA or AAG. The program would probably store a single file that had degeneracy at particular positions. For example, if the amino acid sequence was lysine–methionine–glycine–glutamine, the program would produce the following sequence:

```
5'-AA(A/G)ATGGG(T/C/A/G)CA(A/G)
```

The bases found in parentheses are the possible bases due to the degeneracy of the genetic code.

E18. A. To identify a specific transposable element, a program would use sequence recognition. The sequence of P elements is already known. The program would be supplied with this information and scan a sequence file looking for a match.

B. To identify a stop codon, a program would use sequence recognition. There are three stop codons that are specific three-base sequences. The program would be supplied with these three sequences and scan a sequence file to identify a perfect match.

C. To identify an inversion (of any kind), a program would use pattern recognition. In this case, the program would be looking for a pattern in which the same sequence was running in opposite directions in a comparison of the two sequence files.

D. A search by signal approach uses both sequence recognition and pattern recognition as a means to identify genes. It looks for an organization of sequence elements that would form a functional gene. A search by content approach identifies genes based on patterns, not on specific sequence elements. This approach looks for a pattern in which the nucleotide content is different from a random distribution. The third approach to identify a gene is to scan a genetic sequence for long open read frames. This approach is a combination of sequence recognition and pattern recognition. The program is looking for specific sequence elements (i.e., stop codons) but it is also looking for a pattern in which the stop codons are far apart.

E20. A. The amino acids that are most conserved (i.e, the same in all of the family members) are most likely to be important for structure and/or function. This is because a mutation that changed the amino acid might disrupt structure and function, and these kinds of mutations would be selected against. Completely conserved amino acids are found at the following positions: 101, 102, 105, 107, 108, 116, 117, 123 (Note: Asp or Glu are found here, and these two amino acids are very similar), 124, 130, 134, 139, 143, and 147.

B. The amino acids that are least conserved are probably not very important because changes in the amino acid does not seem to inhibit function. (If it did inhibit function, natural selection would eliminate such a mutation.) At one location, position 118, there are five different amino acids.

E22. As described in part C of solved problem S2, a serine codon was likely to be the ancestral codon. If we look at the codon table, an AGU or AGC codon for serine could change into an Asn, Thr, or Ile codon by a single-base change. In contrast, UCU, UCC, UCA, and UCG codons, which also code for serine, could not change into Asn or Ile codons by a single-base change. Therefore, the two likely scenarios are shown next. The mutated base is underlined.

The mutations would actually occur in the DNA, although the sequences of the RNA codons are shown here.

<u>Ancestral codon</u>

A<u>C</u>U (Thr)   ←   AGU (Ser)   →A<u>A</u>U (Asn)

↓

A<u>U</u>U (Ile)

<u>Ancestral codon</u>

A<u>C</u>C (Thr)   ←   AGC (Ser)   →A<u>A</u>C (Asn)

↓

A<u>U</u>C (Ile)

E24. RNA secondary structure is based on the ability of complementary sequences (i.e., sequences that obey the AU/GC rule) to form a double helix. The program employs a pattern recognition approach. It looks for complementary sequences based on the AU/GC rule.

Questions for Student Discussion/Collaboration

2. This is a very difficult question. In 20 years, we may have enough predictive information so that the structure of macromolecules can be predicted from their genetic sequences. If so, it would be better to be a mathematical theoretician with some genetics background. If not, it is probably better to be a biophysicist with some genetics background.

CHAPTER 22

Conceptual Questions

C2. When a disease-causing allele affects a trait, it is causing a deviation from normality, but the gene involved is not usually the only gene that governs the trait. For example, an allele causing hemophilia prevents the normal blood clotting pathway from operating correctly. It follows a simple Medelian pattern because a single gene affects the phenotype. Even so, it is known that normal blood clotting is due to the actions of many genes.

C4. Changes in chromosome number, and unbalanced changes in chromosome structure, tend to affect phenotype because they create an imbalance of gene expression. For example, in Down syndrome, there are three copies of chromosome 21 and therefore three copies of all the genes on chromosome 21. This leads to a relative overexpression of genes that are located on chromosome 21 compared to the other chromosomes. Balanced translocations and inversions often are without phenotypic consequences because the total amount of genetic material is not altered, and the level of gene expression is not significantly changed.

C6. There are lots of possible answers; here are a few. Dwarfism occurs in people and dogs. Breeds like the dachshund and basset hound are types of dwarfism in dogs. There are diabetic people and mice. There are forms of inherited obesity in people and mice. Hip dysplasia is found in people and dogs.

C8. A. Because a person must inherit two defective copies of this gene and it is known to be on chromosome 1, the mode of transmission is autosomal recessive. Both members of this couple must be heterozygous because they have one affected parent (who had to transmit the mutant allele to them) and their phenotypes are normal (so they must have received the normal allele from their other parent). Because both parents are heterozygotes, there is a

1/4 chance of producing a homozygote with Gaucher disease. If we let G represent the normal allele, and g the mutant allele:

B. From this Punnett square, we can also see that there is a 1/4 chance of producing a homozygote with both normal copies of the gene.

C. We need to apply the binomial expansion to solve this problem. See chapter 2 for a description of this calculation. In this problem, $n = 5$, $x = 1$, $p = 0.25$, $q = 0.75$. The answer is 0.396, or 39.6%.

C10. The mode of transmission is autosomal recessive. All of the affected individuals do not have affected parents. Also, the disorder is found in both males and females. If it were X-linked recessive, individual III-1 would have to have an affected father, which she does not.

C12. The 13 babies have acquired a new mutation. In other words, during spermatogenesis or oogenesis, or after the egg was fertilized, a new mutation occurred in the fibroblast growth factor gene. These 13 individuals have the same chances of passing the mutant allele to their offspring compared to the 18 individuals who inherited the mutant allele from a parent. The chance is 50%.

C14. Because this is a dominant trait, the mother must have two normal copies of the gene, and the father (who is affected) is most likely to be a heterozygote. (Note: the father could be a homozygote, but this is extremely unlikely because the dominant allele is very rare.) If we let M represent the mutant Marfan allele, and m the normal allele, the following Punnett square can be constructed:

A. There is a 50% chance that this couple will have an affected child.

B. We use the product rule. The odds of having an unaffected child are 50%. So if we multiply $0.5 \times 0.5 \times 0.5$, this equals 0.125, or a 12.5% chance of having three unaffected offspring.

C16. A prion is a protein that behaves like an infectious agent. The infectious form of the prion protein has an abnormal conformation. This abnormal conformation is termed PrP^{Sc}, while the normal conformation of the protein is termed PrP^{C}. A prion protein in the PrP^{Sc} conformation can bind to a prion protein in the PrP^C conformation and convert it to the PrP^{Sc} form. An accumulation of the PrP^{Sc} form is what causes the disease symptoms. In the case of mad cow disease, an animal is initially exposed to a small amount of the prion protein in the PrP^{Sc} conformation. This PrP^{Sc} protein then converts the normal PrP^C proteins, which are normally found within the cells of the animal, into the PrP^{Sc} conformation. The disease symptoms rely on the conversion of the endogenous PrP^C conformation into the PrP^{Sc} form.

C18. A. Keep in mind that a conformational change is a stepwise process. It begins in one region of a protein and involves a series of small changes in protein structure. The entire conformational change from PrP^C to PrP^{Sc} probably involves many small changes in protein structure that occur in a stepwise manner. Perhaps the conformational change (from PrP^C to PrP^{Sc}) begins in the vicinity of position 178 and then proceeds throughout the rest of the protein. If there is a methionine at position 129, the complete conformational change (from PrP^C to PrP^{Sc}) can take place. However, perhaps the valine at position 129 somehow blocks one of the steps that are needed to complete the conformational change. (Note: This answer is purely speculative. The actual biochemical mechanism is not known.)

B. Based on the answer to part A, once the PrP^{Sc} conformational change is completed, a PrP^{Sc} protein can bind to another prion protein in the PrP^C conformation. Perhaps it begins to convert it (to the PrP^{Sc} conformation) by initiating a small change in protein structure in the vicinity of position 178. The conversion would then proceed in a stepwise manner until the PrP^{Sc} conformation has been achieved. If a valine is at position 129, this could somehow inhibit one of the steps that are needed to complete the conformational change. If an individual had Val-129 in the polypeptide encoded by the second *PrP* gene, half of their prion proteins would be less sensitive to conversion by PrP^{Sc}, compared to individuals who had Met-129. This would explain why individuals with Val-129 in half of the prion proteins would have disease symptoms that would progress more slowly.

C20. A proto-oncogene is a normal cellular gene that typically plays a role in cell division. It can be altered by mutation to become an oncogene and thereby cause cancer. At the level of protein function, a proto-oncogene can become an oncogene by synthesizing too much of a protein or synthesizing the same amount of a protein that is abnormally active.

C22. The predisposition to develop cancer is inherited in a dominant fashion because the heterozygote has the higher predisposition. The mutant allele is actually recessive at the cellular level. But, because we have so many cells in our bodies, it becomes relatively likely that a defective mutation will occur in the single normal gene and lead to a cancerous cell. Some heterozygous individuals may not develop the disease as a matter of chance. They may be lucky and not get a defective mutation in the normal gene. Or, perhaps their immune system is better at destroying cancerous cells once they arise.

C24. If an oncogene was inherited, it may cause uncontrollable cell growth at an early stage of development and thereby cause an embryo to develop improperly. This could lead to an early spontaneous abortion and thereby explain why we do not observe individuals with inherited oncogenes. Another possibility is that inherited oncogenes may adversely affect gamete survival, which would make it difficult for them to be passed from parent to offspring. A third possibility would be that oncogenes could affect the fertilized zygote in a way that would prevent the correct implantation in the uterus.

C26. The role of p53 is to sense DNA damage and prevent damaged cells from proliferating. Perhaps, prior to birth, the fetus is in a protected environment so that DNA damage may be minimal. In other words, the fetus may not really need p53. After birth, agents such as UV light may cause DNA damage. At this point, p53 is important. A *p53* knockout is more sensitive to UV light because it cannot repair its DNA properly in response to this DNA-damaging agent, and it cannot kill cells that have become irreversibly damaged.

C28. The p53 protein is a regulatory transcription factor; it binds to DNA and influences the transcription rate of nearby genes. This transcription factor (1) activates genes that promote DNA repair; (2) activates genes that arrest cell division and represses genes that are required for cell division; and (3) activates genes that promote apoptosis. If a cell is exposed to DNA damage, it has a greater potential to become malignant. Therefore, an organism wants to avoid the proliferation of such a cell. When exposed to an agent that causes DNA damage, a cell will try to repair the damage. However, if the damage is too extensive, the p53 protein will stop the cell from dividing and program it to die. This helps to prevent the proliferation of cancer cells in the body.

Experimental Questions

E2. Perhaps the least convincing is the higher incidence of the disease in particular populations. Because populations living in specific geographic locations are exposed to their own unique environment, it is difficult to distinguish genetic versus environmental causes for a particular disease. The most convincing evidence might be the higher incidence of a disease in related individuals and/or the ability to correlate a disease with the presence of a mutant gene. Overall, however, the reliability that a disease has a genetic component should be based on as many observations as possible.

E4. You would probably conclude that it is less likely to have a genetic component. If it were rooted primarily in genetics, it would be likely to be found in the Central American population. Of course, there is a chance that very few or none of the people who migrated to Central America were carriers of the mutant gene. This is somewhat unlikely for a large migrating population. By comparison, one might suspect that an environmental agent that is present in South America but not present in Central America may underlie the disease. Researchers could try to search for this environmental agent (e.g., pathogenic organism, etc.).

E6. Males I-1, II-1, II-4, II-6, III-3, III-8, and IV-5 have a normal copy of the gene. Males II-3, III-2, and IV-4 are hemizygous for an inactive mutant allele. Females III-4, III-6, IV-1, IV-2, and IV-3 have two normal copies of the gene, whereas females I-2, II-2, II-5, III-1, III-5, and III-7 are heterozygous carriers of a mutant allele.

E8. A transformed cell is one that has become malignant. In a laboratory, this can be done in three ways. First, the cells could be treated with a mutagen that would convert a proto-oncogene into an oncogene. Second, cells could be exposed to the DNA from a malignant cell line. Under the appropriate conditions, this DNA can be taken up by the cells and integrated into their genome so that they become malignant. A third way to transform cells is by exposure to an oncogenic virus.

E10. By comparing oncogenic viruses with strains that have lost their oncogenicity, researchers have been able to identify particular genes that cause cancer. This has led to the identification of many onco-genes. From this work, researchers have also learned that normal cells contain proto-oncogenes that usually play a role in cell divi-sion. This suggests that oncogenes exert their effects by upsetting the cell division process. In particular, it appears that oncogenes are abnormally active and keep the cell division cycle in a permanent "on" position.

E12. One possible category of drugs would be GDP analogues (i.e., compounds that resemble the structure of GDP). Perhaps one could find a GDP analogue that binds to the Ras protein and locks it in the inactive conformation.

> One way to test the efficacy of such a drug would be to incubate the drug with a type of cancer cell that is known to have an overactive Ras protein, and then plate the cells on solid media. If the drug locked the Ras protein in the inactive conforma-tion, it should inhibit the formation of malignant growth or malig-nant foci.

> There are possible side effects of such drugs. First, they might block the growth of normal cells, since Ras protein plays a role in normal cell proliferation. Second (if you have taken a cell biology course), there are many GTP/GDP-binding proteins in cells, and the drugs could somehow inhibit cell growth and func-tion by interacting with these proteins.

Questions for Student Discussion/Collaboration

2. There is no clearly correct answer to this question, but it should stimulate a large amount of discussion.

CHAPTER 23

Conceptual Questions

C2. A. False, the head is anterior to the tail.

B. True.

C. False, the feet are posterior to the hips. Along the dorso-ventral axis, they are about the same.

D. True.

C4. A. True.

B. False, because gradients are also established after fertilization during embryonic development.

C. True.

C6. A. This is a mutation in *runt,* which is a pair-rule gene.

B. This is a mutation in *knirps,* which is a gap gene.

C. This is a mutation in *patched,* which is a segment-polarity gene.

C8. Positional information refers to the phenomenon whereby the spa-tial locations of morphogens and CAMs provide a cell with infor-mation regarding its position relative to other cells. In *Drosophila,* the formation of a segmented body pattern relies initially on the spatial location of maternal gene products. These gene products lead to the sequential activation of the segmentation genes.

C10. The anterior portion of the antero-posterior axis is established by the action of Bicoid. During oogenesis, the mRNA for Bicoid enters the anterior end of the oocyte and is sequestered there to establish an anterior (high) to posterior (low) gradient. Later, when the mRNA is translated, the Bicoid protein in the anterior region establishes a genetic hierarchy that leads to the formation of ante-

rior structures. If Bicoid was not trapped in the anterior end, it is likely that anterior structures would not form.

C12. Maternal gene products influence the formation of the main body axes including the antero-posterior, dorso-ventral, and terminal regions. They are needed very early in development. Zygotic genes, particularly the three classes of the segmentation genes, are neces-sary after the axes have been established. The segmentation genes are expressed (after fertilization) during embryogenesis.

C14. The coding sequence of homeotic genes contains a 180 bp consen-sus sequence known as a homeobox. The protein domain encoded by the homeobox is called a homeodomain. The homeodomain contains three conserved sequences that are folded into α-helical conformations. The arrangement of these α helices promotes the binding of the protein to the major groove of the DNA. Helix III is called the recognition helix because it recognizes a particular nucleotide sequence within the major groove. In this way, homeotic proteins are able to bind to DNA in a sequence-specific manner and thereby activate particular genes.

C16. It would normally be expressed in the three thoracic segments that have legs (T1, T2, and T3).

C18. A. When a mutation inactivates a gap gene, a contiguous section of the larva is missing.

B. When a mutation inactivates a pair-rule gene, some regions that are derived from alternating parasegments are missing.

C. When a mutation inactivates a segment-polarity gene, portions are missing at either the anterior or posterior end of the seg-ments.

C20. Proper development in mammals is likely to require the products of maternal effect genes that play a key role in initiating embryonic development. The adult body plan is merely an expansion of the embryonic body plan. Because the starting point for the develop-ment of an embryo is the oocyte, this explains why an enucleated oocyte is needed to clone mammals. The oocyte is needed to estab-lish the embryonic body plan.

C22. A heterochronic mutation is one that alters the timing when a gene (involved in development) is normally expressed. The gene may be expressed too early or too late, which causes certain cell lineages to be out of sync with the rest of the animal. If a heterochronic muta-tion affected the intestine, the animal may end up with too many intestinal cells if it is a gain-of-function mutation or too few if it is a loss-of-function mutation. In either case, the effects might be detrimental because the growth of the intestine must be coordi-nated with the growth of the rest of the animal.

C24. Cell differentiation is the specialization of a cell into a particular cell type. In the case of skeletal muscle cells, the bHLH proteins play a key role in the initiation of cell differentiation. When bHLH proteins are activated, they are able to bind to enhancers and acti-vate the expression of many different muscle-specific genes. In this way, myogenic bHLH proteins turn on the expression of many muscle-specific proteins. When these proteins are synthesized, they change the characteristics of the cell into those of a muscle cell. Myogenic bHLH proteins are regulated by dimerization. When a heterodimer forms between a myogenic bHLH protein and an E protein, it activates gene expression. However, when a heterodimer forms between myogenic bHLH proteins and a protein called Id, the heterodimer is unable to bind to DNA. The Id protein is pro-duced during early stages of development and prevents myogenic bHLH proteins from promoting muscle differentiation too soon. At later stages of development, the amount of Id protein falls, and myogenic bHLH proteins can combine with E proteins to induce muscle differentiation.

C26. A totipotent cell is a cell that has the potential to create a complete organism.

A. In humans, a fertilized egg is totipotent, and the cells during the first few embryonic divisions are totipotent. However, after several divisions, embryonic cells lose their totipotency and, instead, are determined to become particular tissues within the body.

B. In plants, most living cells are totipotent.

C. Because yeast are unicellular, one cell is a complete individual. Therefore, yeast cells are totipotent; they can produce new individuals by cell division.

D. Because bacteria are also unicellular, one cell is a complete individual. Therefore, bacteria are totipotent; they can produce new individuals by cell division.

C28. Animals begin their development from an egg and then form antero-posterior and dorso-ventral axes. The formation of an adult organism is an expansion of the embryonic body plan. Plants grow primarily from two meristems, a shoot and root meristem. At the cellular level, plant development is different in that it does not involve cell migration, and most plant cells are totipotent. Animals require the organization within an oocyte to begin development. At the genetic level, however, animal and plant development are similar in that they involve a genetic hierarchy of transcription factors that govern pattern formation and cell specialization.

Experimental Questions

E2. *Drosophila* is more advanced from the perspective that many more mutant alleles have been identified that alter development in specific ways. The hierarchy of gene regulation is particularly well understood in the fruit fly. *C. elegans* has the advantage of simplicity and a complete knowledge of cell fate. This enables researchers to explore how the timing of gene expression is critical to the developmental process.

E4. To determine that a mutation is affecting the timing of developmental decisions, a researcher needs to know the normal time or stage of development when cells are supposed to divide, and what types of cells will be produced. With this information (i.e., a lineage diagram), one can then determine if particular mutations alter the timing when cell division occurs.

E6. Mutant 1 is a gain-of-function allele; it keeps reiterating the L1 pattern of division. Mutant 2 is a loss-of-function allele; it skips the L1 pattern and immediately follows an L2 pattern.

E8. As discussed in chapter 15, most eukaryotic genes have a core promoter that is adjacent to the coding sequence; regulatory elements that control the transcription rate at the promoter are typically upstream from the core promoter. Therefore, to get the *Antp* gene product expressed where the *abd-A* gene product is normally expressed, you would link the upstream genetic regulatory region of the *abd-A* gene to the coding sequence of the *Antp* gene. This construct would be inserted into the middle of a P element (see next). The construct shown here would then be introduced into an embryo by P element transformation.

P element	*abd-A* regulatory region	/	*Antp* coding sequence	P element

The *Antp* gene product is normally expressed in the thoracic region and produces segments with legs, as illustrated in figure 23.11. Therefore, because the *abd-A* gene product is normally expressed in the anterior abdominal segments, one might predict that the genetic construct shown above would produce a fly with legs attached to the segments that are supposed to be the anterior abdominal segments. In other words, the anterior abdominal segments would probably resemble thoracic segments with legs.

E10. A. The female flies must have had mothers that were heterozygous for a (dominant) normal allele and the mutant allele. Their fathers were either homozygous for the mutant allele or heterozygous. The female flies inherited a mutant allele from both their father and mother. Nevertheless, because their mother was heterozygous for the normal (dominant) allele and mutant allele, and because this is a maternal effect gene, their phenotype is based on the genotype of their mother. The normal allele is dominant, so they have a normal phenotype.

B. *Bicoid-A* appears to have a deletion that removes part of the sequence of the gene and thereby results in a shorter mRNA. *Bicoid-B* could also be a deletion that removes all of the sequence of the *bicoid* gene or it could be a promoter mutation that prevents the expression of the *bicoid* gene. *Bicoid-C* seems to be a point mutation that does not affect the amount of the *bicoid* mRNA.

With regard to function, all three mutations are known to be loss-of-function mutations. *Bicoid-A* probably eliminates function by truncating the Bicoid protein. The Bicoid protein is a transcription factor. The *bicoid-A* mutation probably shortens this protein and thereby inhibits its function. The *bicoid-B* mutation prevents expression of the *bicoid* mRNA. Therefore, none of the Bicoid protein would be made, and this would explain the loss of function. The *bicoid-C* mutation seems to prevent the proper localization of the *bicoid* mRNA in the oocyte. There must be proteins within the oocyte that recognize specific sequences in the *bicoid* mRNA and trap it in the anterior end of the oocyte. This mutation must change these sequences and prevent these proteins from recognizing the *bicoid* mRNA.

C. If we only used the technique of Northern blotting, we would not have understood how *bicoid-C* was abnormal. Likewise, if we had only used the technique of *in situ* hybridization, we would not have understood how *bicoid-A* was abnormal.

E12. An egg-laying defect is somehow related to an abnormal anatomy. The *n540* strain has fewer neurons compared to a normal worm. Perhaps the *n540* strain is unable to lay eggs because it is missing neurons that are needed for egg laying. The *n536* and *n355* strains have an abnormal abundance of neurons. Perhaps this overabundance also interferes with the proper neural signals that are needed for egg laying.

E14. Geneticists who are interested in mammalian development have used reverse genetics because it has been difficult for them to identify mutations in developmental genes based on phenotypic effects in the embryo. This is because it is difficult to screen a large number of mammalian embryos in search of abnormal ones that carry mutant genes. It is easy to have thousands of flies in a laboratory, but it is not easy to have thousands of mice. Instead, it is easier to clone the normal gene based on its homology to invertebrate genes and then make mutations in vitro. These mutations can be introduced into a mouse to create a gene knockout. This strategy is opposite to that of Mendel, who characterized genes by first identifying phenotypic variants (e.g., tall vs. dwarf, green seeds vs. yellow seeds, etc.).

Questions for Student Discussion/Collaboration

2. In this problem, the students should try to make a flow diagram that begins with maternal effect genes, then gap genes, pair rule genes, and segment-polarity genes. These genes then lead to homeotic genes and finally realizator genes. It is almost impossible to make an accurate flow diagram because there are so many gene interactions, but it is instructive to think about developmental genetics in this way. It is probably easier to identify mutant phenotypes that affect later stages of development because they are less likely to be lethal. However, modern methods can screen for conditional mutants as described in solved problem S3. To identify all of the genes necessary for chicken development, you may begin with early genes, but this assumes that you have some way to identify them. If they had been identified, you would then try to identify the genes that they stimulate or repress. This could be done using molecular methods described in chapters 14, 15, and 18.

CHAPTER 24

Conceptual Questions

C2. At the molecular level, quantitative traits often exhibit a continuum of phenotypic variation because they are usually influenced by multiple genes that exist as multiple alleles. A large amount of environmental variation will also increase the phenotypic overlaps among different genotypic categories.

C4. A discontinuous trait is one that falls into discrete categories. Examples include brown eyes versus blue eyes in humans or purple versus white flowers in pea plants. A continuous trait is one that does not fall into discrete categories. Examples include height in humans and fruit weight in tomatoes. Most quantitative traits are continuous; the trait falls within a range of values. The reason why quantitative traits are continuous is because they are usually polygenic and greatly influenced by the environment. As shown in figure 24.4b, this tends to create ambiguities between genotypes and a continuum of phenotypes.

C6. To be in the top 2.5% is about two standard deviation units. If we take the square root of the variance, the standard deviation would be 20.6 lb. To be in the top 2.5% an animal would have to weigh 41.2 lb heavier than the mean, which equals 465.2 lb. To be in the bottom 0.13%, an animal would have to be three standard deviations lighter, which would be 61.8 lb lighter than the mean or 362.2 lb.

C8. There is a positive correlation, but it could have occurred as a matter of chance alone. You would need to conduct more experimentation to determine if there is a significant correlation such as examining a greater number of pairs of individuals. If $N = 500$, the correlation would be statistically significant and you would conclude that the correlation did not occur as a matter of random chance. However, you could not conclude cause and effect.

C10. When a correlation coefficient is statistically significant, it means that the association is likely to have occurred for reasons other than random sampling error. It may indicate cause and effect but not necessarily. For example, large parents may have large offspring due to genetics (cause and effect). However, the correlation may be related to the sharing of similar environments rather than cause and effect.

C12. Quantitative trait loci are sites within chromosomes that contain genes that affect a quantitative trait. It is possible for a QTL to contain one gene, or it may contain two or more closely linked genes. QTL mapping, which involves linkage to known molecular markers, is commonly used to determine the locations of QTLs.

C14. If the broad sense heritability equals 1.0, it means that all of the variation in the population is due to genetic variation rather than environmental variation. It does not mean that the environment is unimportant in the outcome of the trait. Under another set of environmental conditions the trait may have turned out quite differently.

C16. Hybrid vigor is the phenomenon in which an offspring produced from two inbred strains is more vigorous than the corresponding parents. Tomatoes and corn are often the products of hybrids.

C18. When a species is subjected to selective breeding, the breeder is focusing his/her attention on improving one particular trait. In this case, the rose breeder is focused on the size and quality of the flowers. Because the breeder usually selects a small number of individuals (e.g., the ones with best flowers) as the breeding stock for the next generation, this may lead to a decrease in the allelic diversity at other genes. For example, several genes affect flower fragrance. In an unselected population, these genes may exist as "fragrant alleles" and "nonfragrant alleles." After many generations of breeding for large flowers, the fragrant alleles may be lost from the population, just as a matter of random chance. This is a common problem of selective breeding. As you select for an improvement in one trait, you may inadvertently diminish the quality of an unselected trait.

Other people have suggested that the lack of fragrance may be related to flower structure and function. Perhaps the amount of energy that a flower uses to make beautiful petals somehow diminishes its capacity to make fragrance.

C20. Broad sense heritability takes into account all genetic factors that affect the phenotypic variation in a trait. Narrow sense heritability considers only alleles that behave in an additive fashion. In many cases, the alleles affecting quantitative traits appear to behave additively. More importantly, if a breeder assumes that the heritability of a trait is due to the additive effects of alleles, it is possible to predict the outcome of selective breeding. This is also termed the *realized heritability*.

C22. A. Because of their good nutrition, you may speculate that they would grow to be taller.

B. If the environment is rather homogeneous, then heritability values tend to be higher because the environment contributes less to the amount of variation in the trait. Therefore, in the commune, the heritability might be higher, because they uniformly practice good nutrition. On the other hand, because the commune is a smaller population, the amount of genetic variation might be less, so this would make the heritability lower. However, because the problem states that the commune population is large, we would probably assume that the amount of genetic variation is similar to the general population. Overall, the best guess would be that the heritability in the commune population is higher because of the uniform nutrition standards.

C. As stated in part B, the amount of variation would probably be similar, because the commune population is large. As a general answer, larger populations tend to have more genetic variation. Therefore, the general population probably has a bit more variation, but it may not be much more than the commune population.

C24. A natural population of animals is more likely to have a higher genetic diversity compared to a domesticated population. This is because domesticated populations have been subjected to many generations of selective breeding, which decreases the genetic diversity. Therefore, V_G is likely to be higher for the natural population. The other issue is environment. It is difficult to say which group would have a more homogeneous environment. In general, natural populations tend to have a more heterogeneous environment, but not always. If the environment is more heterogeneous, this tends to cause more phenotypic variation, which makes V_E higher.

$$\text{Heritability} = V_G/V_T$$

$$= V_G/(V_G + V_E)$$

When V_G is high, this increases heritability. When V_E is high, this decreases heritability. In the natural wolf population, we would expect that V_G would be high. In addition, we would guess that V_E might be high as well (but that is less certain). Nevertheless, if this were the case, the heritability of the wolf population might be similar to the domestic population. This is because the high V_G in the wolf population is balanced by its high V_E. On the other hand, if V_E is not that high in the wolf population, or if it is fairly high in the domestic population, then the wolf population would have a higher heritability for this trait.

Experimental Questions

E2. To calculate the mean, we add the values together and divide by the total number.

Mean =

$$\frac{1.9 + 2(2.4) + 2(2.1) + 3(2.0) + 2(2.2) + 1.7 + 1.8 + 2(2.3) + 1.6}{15}$$

Mean = 2.1

The variance is the sum of the squared deviations from the mean divided by $N - 1$. The mean value of 2.1 must be subtracted from each value, and then the square is taken. These 15 values are added together and then divided by 14 (which is $N - 1$).

$$\text{Variance} = \frac{0.85}{14}$$

$$= 0.061$$

The standard deviation is the square root of the variance.

Standard deviation = 0.25

E4. The results are consistent with the idea that there are QTLs for this trait on chromosomes 2 and 3 but not on the X chromosome.

E6. When we say that an RFLP is associated with a trait, we mean that a gene that influences a trait is closely linked to an RFLP. At the chromosomal level, the gene of interest is so closely linked to the RFLP that a crossover almost never occurs between them.

Note: Each plant inherits four RFLPs, but it may be homozygous for one or two of them.

Small: 2,700 and 4,000 (homozygous for both)

Small-medium: 2,700 (homozygous), 3,000, and 4,000; or 2,000, 2,700, and 4,000 (homozygous)

Medium: 2,000 and 4,000 (homozygous for both); or 2,700 and 3,000 (homozygous for both); or 2,000, 2,700, 3,000, and 4,000

Medium-large: 2,000 (homozygous), 3,000, and 4,000; or 2,000, 2,700, and 3,000 (homozygous)

Large: 2,000 and 3,000 (homozygous for both)

E8. Let's assume that there is an extensive molecular marker map for the rice genome. We would begin with two strains of rice, one with a high yield and one with a low yield, that greatly differ with regard to the molecular markers that they carry. We would make a cross between these two strains to get F_1 hybrids. We would then backcross the F_1 hybrids to either of the parental strains and then examine hundreds of offspring with regard to their rice yields and molecular markers. In this case, our expected results would be that six different markers in the high-producing strain would be correlated with offspring that produce higher yields. We might get fewer than six bands if some of these genes are closely linked and associate with the same marker. We also might get fewer than six if the two parental strains have the same marker that is associated with one or more of the genes that affect yield.

E10. A. If we assume that the highly inbred strain has no genetic variance:

$$V_G \text{ (for the wild strain)} = 3.2 \text{ g}^2 - 2.2 \text{ g}^2 = 1.0 \text{ g}^2$$

B. $H_B^2 = 1.0 \text{ g}^2/3.2 \text{ g}^2 = 0.31$

C. It is the same as H_B^2 so it also equals 0.31.

E12. A.

$$h_N^2 = \frac{\bar{X}_O - \bar{X}}{\bar{X}_P - \bar{X}}$$

$$0.21 = (\bar{X}_O - 25 \text{ g})/(27 \text{ g} - 25 \text{ g})$$

$$\bar{X}_O - 25 \text{ g} = 2 \text{ g}(0.21)$$

$$\bar{X}_O = 25.42 \text{ g}$$

B.

$$0.21 = (26.5 \text{ g} - 25 \text{ g})/(\bar{X}_P - 25 \text{ g})$$

$$(\bar{X}_P - 25 \text{ g})(0.21) = 1.5 \text{ g}$$

$$\bar{X}_P = 32.14 \text{ g parents}$$

However, because this value is so far from the mean, there may not be 32.14 g parents in the population of mice that you have available.

E14. We first need to calculate a and b. In this calculation, X represents the height of fathers and Y represents the height of sons.

$$b = \frac{144}{112} = 1.29$$

$$a = 69 - (1.29)(68) = -18.7$$

For a father who is 70 in. tall,

$$Y = (1.29)(70) + (-18.7) = 71.6$$

The most likely height of the son would be 71.6 in.

E16. The identical and fraternal twins, who probably share very similar environments, but who differ in the amount of genetic material they share, are a strong argument against an environmental bias. The differences in the observed correlations (0.49 vs 0.99) are consistent with the differences in the expected correlations (0.5 vs 1.0).

E18. $h_N^2 = r_{obs}/r_{exp}$

The value for r_{exp} comes from the known genetic relationships.

Mother/daughter	$r_{obs} = 0.36$	$r_{exp} = 0.5$	$h_N^2 = 0.72$
Mother/granddaughter	$r_{obs} = 0.17$	$r_{exp} = 0.25$	$h_N^2 = 0.68$
Sister/sister	$r_{obs} = 0.39$	$r_{exp} = 0.5$	$h_N^2 = 0.78$
Twin sisters (fraternal)	$r_{obs} = 0.40$	$r_{exp} = 0.5$	$h_N^2 = 0.80$
Twin sisters (identical)	$r_{obs} = 0.77$	$r_{exp} = 1.0$	$h_N^2 = \underline{0.77}$
			$\overline{0.75}$

The average heritability is 0.75.

E20. These data suggest that there might be a genetic component to blood pressure, because the relatives of people with high blood pressure also seem to have high blood pressure themselves. Of course, more extensive studies would need to be conducted to determine the role of environment. To calculate heritability, the first thing to do is to calculate the correlation coefficient between relatives to see if it is statistically significant. If it is, then you could follow the approach described in the experiment of figure 24.9. You would determine the correlation coefficients between genetically related individuals as a way to determine the heritability for the trait. In this approach, heritability equals r_{obs}/r_{exp}. It would be important to include genetically related pairs that were raised apart (e.g., uncles and nieces) to see if they had a similar heritability value compared to genetically related pairs raised in the same environment (e.g., brothers and sisters). If their values were similar, this would give you some confidence that the heritability value is due to genetics and not due to fact that relatives often share similar environments.

Questions for Student Discussion/Collaboration

2. Most traits depend on the influence of many genes. Also, genetic variation is a common phenomenon in most populations. Therefore, most individuals have a variety of alleles that contribute to a given trait. For quantitative traits, some alleles may make the trait bigger and other alleles may make the trait turn out smaller. If a population contains many different genes and alleles that govern a quantitative trait, most individuals will have an intermediate phenotype because they will have inherited some large and some small alleles. Fewer individuals will inherit a predominance of large alleles or a predominance of small alleles. An example of a quantitative trait that does not fit a normal distribution is snail pigmentation. The dark snails and light snails are favored rather than the intermediate colors because they are less susceptible to predation.

CHAPTER 25

Conceptual Questions

C2. A population is a group of interbreeding individuals. Let's consider a squirrel population in a forested area. Over the course of many generations, several things could happen to this population that may change its gene pool. A forest fire, for example, could dramatically decrease the number of individuals and thereby cause a bottleneck. This would decrease the genetic diversity of the population. A new predator may enter the region and natural selection may select for the survival of squirrels that are best able to evade the predator. Another possibility is that a group of squirrels within the population may migrate to a new region and found a new squirrel population.

C4. A. Phenotype frequency and genotype frequency

B. Genotype frequency

C. Allele frequency

C6. A. The genotype frequency for the *CF* homozygote is 1/2,500, or 0.004. This would equal q^2. The allele frequency is the square root of this value, which equals 0.02. The frequency of the corresponding normal allele equals $1 - 0.02 = 0.98$.

B. CF homozygote is 0.004. Normal homozygote is $(0.98)^2 = 0.96$. Heterozygote is $2(0.98)(0.02)$, which equals 0.039.

C. If a person is known to be a heterozygous carrier, the chances that this particular person will happen to choose another as a mate is equal to the frequency of heterozygous carriers in the population, which equals 0.039, or 3.9%. The chances that two randomly chosen individuals will choose each other as mates equals $0.039 \times 0.039 = 0.0015$, or 0.15%.

C8. For two alleles, the heterozygote is at a maximum when they are 0.5 each. For three alleles, the two heterozygotes are at a maximum when each allele is 0.33.

C10. Because this is a recessive trait, only the homozygotes for the rolling allele will be able to roll their tongues. If p equals the rolling allele and q equals the nonrolling allele, the Hardy-Weinberg equation predicts that the frequency of homozygotes who can roll their tongues would be p^2. In this case, $p^2 = (0.6)^2 = 0.36$, or 36%.

C12. A. The inbreeding coefficient is calculated using the formula

$$F = \Sigma(1/2)^n(1 + F_A)$$

In this case there is one common ancestor, I-2. Because we have no prior history on I-2, we assume she is not inbred, which makes $F_A = 0$. The inbreeding loop for IV-3 contains five people, III-4, II-2, I-2, II-5, and III-5. Therefore, $n = 5$.

$$F = (1/2)^5(1 + 0) = 1/32 = 0.031$$

B. Based on the data shown in this pedigree, individual IV-4 is not inbred.

C14. Migration, genetic drift, and natural selection are the driving forces that alter allele frequencies within a population. Natural selection acts to eliminate harmful alleles and promote beneficial alleles. Genetic drift involves random changes in allele frequencies that may eventually lead to elimination or fixation of alleles. It is thought to be important in the establishment of neutral alleles in a population. Migration is important because it introduces new alleles into neighboring populations. According to the neutral theory, genetic drift is largely responsible for the variation seen in natural populations.

C16. A neutral force is one that alters allele frequencies without any regard to whether the changes are beneficial or not. Genetic drift and migration are the two main ways this can occur. An example is the founder effect, when a group of individuals migrate to a new location such as an island. Adaptive forces increase the reproductive success of a species. Natural selection is the adaptive force that tends to eliminate harmful alleles from a population and increase the frequency of beneficial alleles. An example would be the long neck of the giraffe, which enables it to feed in tall trees. At the molecular level, beneficial mutations may alter the coding sequence of a gene and change the structure and function of the protein in a way that is beneficial. For example, the sickle-cell anemia allele alters the structure of hemoglobin, and in the heterozygous condition, this affects the sensitivity of red blood cells to the malaria pathogen.

C18. Genetic drift is due to sampling error, and the degree of sampling error depends on the population size. In small populations, the relative proportion of sampling error is much larger. If genetic drift is moving an allele toward fixation, it will take longer in a large population because the degree of sampling error is much smaller.

C20. During the bottleneck effect, allele frequencies are dramatically altered due to genetic drift. In extreme cases, some alleles are lost while others may become fixed at 100%. The overall effect is to decrease the genetic diversity within the population. This may make it more difficult for the species to respond in a positive way to changes in the environment. Species that are approaching extinction are also facing a bottleneck as their numbers decrease. The loss of genetic diversity may make it even more difficult for the species to rebound.

C22. Directional selection favors the phenotype at one phenotypic extreme. Over time, natural selection is expected to favor the fixation of alleles that cause these phenotypic characteristics. Disruptive selection favors two or more phenotypic categories. It will lead to a population with a balanced polymorphism for the trait. Stabilizing selection favors individuals with intermediate phenotypes. It will also tend to promote polymorphism, since the favored individuals may be heterozygous for particular alleles.

C24. The intuitive meaning of the mean fitness of a population is the relative likelihood that members of a population will reproduce. If the mean fitness is high, it is likely that an average member will survive and produce offspring. Natural selection increases the mean fitness of a population.

C26. A. Random mutation is the source of genetic variation that may lead to antibiotic resistance. A random mutation may create an antibiotic-resistance allele. This could occur in different ways. Two possibilities are:

1. Many antibiotics exert their effects by binding to an essential cellular protein within the microorganism and inhibiting its function. A random mutation could occur in the gene that encodes such an essential cellular protein; this could alter the structure of the protein in a way that would prevent the antibiotic from binding to the protein or inhibiting its function.

2. As another possibility, microorganisms, which are killed by antibiotics, possess many enzymes, which degrade related compounds. A random mutation could occur in a gene that encodes a degradative enzyme so that the enzyme now recognizes the antibiotic and degrades it.

B. When random mutations occur, they may be lost due to genetic drift. This is particularly likely when the frequency of the mutation is very low in a large population. Alternatively (and much less likely), a random mutation that confers antibiotic resistance could become fixed in a population.

C. If a random mutation occurs that confers antibiotic resistance, and if the mutation is not lost by genetic drift, natural selection will favor the growth of microorganisms that carry the antibiotic-resistance allele if the organisms are exposed to the antibiotic. Therefore, if antibiotics are widely used, this will kill microorganisms that are sensitive and favor the proliferation of ones that happen to carry antibiotic-resistance alleles.

C28. A. Migration will increase the genetic diversity in both populations. A random mutation could occur in one population to create a new allele. This new allele could be introduced into the other population via migration.

B. The allele frequencies between the two populations will tend to be similar to each other, due to the intermixing of their alleles.

C. Genetic drift depends on population size. When two populations intermix, this has the effect of increasing the overall population size. In a sense, the two smaller populations behave somewhat like one big population. Therefore, the effects of genetic drift are lessened when the individuals in two populations can migrate. The net effect is that allele loss and allele fixation are less likely to occur due to genetic drift.

Experimental Questions

E2. 1. One hypothesis is that the population having only two allozymes was founded from a small group that left the other population. When the small founding group left, it had less genetic diversity than the original population.

2. The population with more genetic diversity may be in a more diverse environment so it may select for a greater variety of phenotypes.

3. It may just be a matter of chance that one population had accumulated more neutral alleles than the other.

E4. Glutamic acid is a negatively charged amino acid and valine is neutral. The Hb^A polypeptide has a glutamic acid at the sixth position while Hb^S has a valine. Therefore, the Hb^A polypeptide will move a little more quickly toward the positive end of the gel.

Lane 1—$Hb^S Hb^S$

Lane 2—$Hb^A Hb^A$

Lane 3—$Hb^A Hb^S$

E6. A. Let W represent the white fat allele and w represent the yellow fat allele. Assuming a Hardy-Weinberg equilibrium, we can let p^2 represent the genotype frequency of WW animals, and then Ww would be $2pq$ and ww would be q^2. The only genotype frequency we know is that of the ww animals.

$$ww = q^2 = \frac{76}{5,468}$$

$$q^2 = 0.014$$

$q = 0.12$, which is the allele frequency of w

$$p = 1 - q$$

$p = 0.88$, which is the allele frequency of W

B. The heterozygous carriers are represented by $2pq$. If we use the values of p and q, which were calculated in part A:

$$2pq = 2(0.88)(0.12) = 0.21$$

Approximately 21% of the animals would be heterozygotes with white fat.

If we multiply 0.21 times the total number of animals in the herd:

$$0.21 \times 5,468 = 1,148 \text{ animals}$$

E8. The first thing we need to do is to determine the allele frequencies. Let p represent i, q represent I^A, and r represent I^B.

p^2 is the genotype frequency of ii

q^2 is the genotype frequency of $I^A I^A$

r^2 is the genotype frequency of $I^B I^B$

$2pq$ is the genotype frequency of $I^A i$

$2pr$ is the genotype frequency of $I^B i$

$2qr$ is the genotype frequency of $I^A I^B$

$$p^2 = \frac{721}{721 + 932 + 235 + 112}$$

$$p^2 = 0.36$$

$$p = 0.6$$

Next, we can calculate the allele frequency of I^A. Keep in mind that there are two genotypes ($I^A I^A$ and $I^A i$) that result in type A blood.

$$q^2 + 2pq = \frac{932}{721 + 932 + 235 + 112}$$

$$q^2 + 2(0.6)q = 0.47$$

$$q = 0.31$$

Now it is easy to solve for r,

$$p + q + r = 1$$

$$0.6 + 0.31 + r = 1$$

$$r = 0.09$$

Based on these allele frequencies, we can compare the observed and expected values. To determine the expected values, we multiply the genotype frequencies times 2,000, which was the total number of individuals in this population.

p^2 is the genotype frequency of $ii = (0.6)^2(2,000) = 720$

q^2 is the genotype frequency of $I^A I^A = (0.31)^2(2,000) = 192$

r^2 is the genotype frequency of $I^B I^B = (0.09)^2(2,000) = 16$

$2pq$ is the genotype frequency of $I^A i = 2(0.31)(0.6)(2,000) = 744$

$2pr$ is the genotype frequency of $I^B i = 2(0.09)(0.6)(2,000) = 216$

$2qr$ is the genotype frequency of $I^A I^B = 2(0.31)(0.09)(2,000) = 112$

	Expected Numbers	Observed Numbers
Type O	720	721
Type A	192 + 744 = 936	932
Type B	16 + 216 = 232	235
Type AB	112	112

The observed and expected values agree quite well. Therefore, it does appear that this population is in Hardy-Weinberg equilibrium.

E10. Let's assume that the relative fitness values are 1.0 for the dominant homozygote and the heterozygote and 0 for the recessive homozygote. The first thing we need to do is to calculate the mean fitness for the population.

$$p^2 W_{AA} + 2pq W_{Aa} + q^2 W_{aa} = \overline{W}$$

$$(0.78)^2 + 2(0.78)(0.22) = \overline{W}$$

$$\overline{W} = 0.95$$

The genotype frequency in the next generation for AA equals

$$\frac{p^2 W_{AA}}{\overline{W}}$$

$$\frac{(0.78)^2}{0.95} = 0.64$$

$$p = \sqrt{0.64} = 0.80 \quad \text{and} \quad q = 0.20$$

For the second generation, we first need to calculate the mean fitness of the population, which now equals 0.96. Using the preceding equation, the genotype frequency of AA in the second generation equals 0.67 and the allele frequency equals 0.816. The frequency of the recessive allele in the second generation would equal 0.184 and the mean fitness would now equal 0.967. The genotype frequency of AA in the third generation would be 0.688 and the allele frequency would be 0.83. The frequency of the recessive allele would be 0.17.

E12. If we let C represent the *carbonaria* allele and c represent the *typical* allele:

$$W_{CC} = 1.0$$

$$W_{Cc} = 1.0$$

$$W_{cc} = 0.47$$

In the next generation, we expect that the Hardy-Weinberg equilibrium will be modified by the following amount:

$$p^2 W_{CC} + 2pq W_{Cc} + q^2 W_{cc}$$

In a population that is changing due to natural selection, these three genotypes will not add up to 1.0 as in the Hardy-Weinberg equilibrium. Instead, the three genotypes will add up to the mean fitness of the population.

$$p^2 W_{CC} + 2pq W_{Aa} + q^2 W_{aa} = \overline{W}$$

$$(0.7)^2(1.0) + 2(0.7)(0.3)(1.0) + (0.3)^2(0.47) = \overline{W}$$

$$\overline{W} = 0.95$$

After one generation of selection:

Allele frequency of C: $\quad p = \dfrac{p^2 W_{CC}}{\overline{W}} + \dfrac{pq W_{Cc}}{\overline{W}}$

$$p = \frac{(0.7)^2(1.0)}{0.95} + \frac{(0.7)(0.3)(1.0)}{0.95}$$

$$p = 0.74$$

Allele frequency of c: $\quad q = \dfrac{q^2 W_{cc}}{\overline{W}} + \dfrac{pq W_{Cc}}{\overline{W}}$

$$q = \frac{(0.3)^2(0.47)}{0.95} + \frac{(0.7)(0.3)(1.0)}{0.95}$$

$$q = 0.27$$

After one generation, the allele frequency of C has increased from 0.7 to about 0.74 while the frequency of c has decreased from 0.3 to about 0.27. This is because the homozygous, cc, genotype has a lower fitness compared to the heterozygous, Cc, and homozygous, CC, genotypes.

E14. Each area that he tested had its own endogenous population of moths. For example, the polluted areas had many more darkly colored moths, so we would expect to capture many more of these simply because there are more of them in the first place. Kettlewell wanted to release an equal number of moths of both types and then recapture them as a way to examine how well each type of moth could survive in polluted and unpolluted environments.

E16. Fitness based on the number eaten by birds:

The number of moths eaten by birds is really a measure of the selection coefficient (s), not a measure of fitness. $s = 1 - W$

We first need to compare the *carbonaria* and the *typicals*.

$$carbonaria = \frac{43}{43 + 15}$$

$$carbonaria = 0.74$$

$$typical = \frac{15}{43 + 15}$$

$$typical = 0.26$$

If we wish to give the *typical* moths a fitness value of 1.0, this means the selection coefficient for *typical* moths must be zero. Therefore, to calculate the selection coefficient for the *carbonaria* moths:

$$s_{carbonaria} = 0.74 - 0.26 = 0.48$$

$$W_{carbonaria} = 1 - 0.48 = 0.52$$

Fitness based on the number of moths recaptured:

$$W_{carbonaria} = \frac{7.0}{12.5}$$

$$W_{carbonaria} = 0.56$$

The two values (0.52 and 0.56) agree reasonably well. The fitness value based on recapture data is probably more reliable because it seems to be an unbiased measure of the survival rate. The fitness based on the number eaten by birds is somewhat biased because it assumes that this is the only factor that affects the survival of the two types of moths. However, there could be other factors. For example, animals other than birds may eat moths.

E18. One could follow an analogous protocol as conducted by Kettlewell. You could mark snails with a dye and release equal numbers of dark and light snails into dimly lit forested regions and sunny fields. At a later time, recapture the snails and count them. It would be important to have a method of unbiased recapture because the experimenter would have an easier time locating the light snails in a forest and the dark snails in a field. Perhaps one could bait the region with something that the snails like to eat and only collect snails that are at the bait. In addition to this type of experiment, one could also sit in a blind and observe predation as it occurs.

Questions for Student Discussion/Collaboration

2. Mutation is responsible for creating new alleles, but the rate of new mutations is so low that it cannot explain allele frequencies in this range. Let's call the two alleles B and b and assume that B was the original allele and b is more recent allele that arose as a result of mutation. Three scenarios explain the allele frequencies:

   1. The b allele is neutral and reached its present frequency by genetic drift. It has not reached elimination or fixation yet.

   2. The b allele is beneficial and its frequency is increasing due to natural selection. However, there has not been enough time to reach fixation.

   3. The Bb heterozygote is at a selective advantage, leading to a balanced polymorphism.

CHAPTER 26

Conceptual Questions

C2. Evolution is unifying because all living organisms on this planet evolved from the same primordial organism. At the molecular level, all organisms have a great deal in common. With the exception of a few viruses, they all use DNA as their genetic material. This DNA is found within chromosomes, and the sequence of the DNA is organized into units called genes. Most genes are structural genes that encode the amino acid sequence of polypeptides. Polypeptides fold to form functional units called proteins. At the cellular level, all living organisms also share many similarities. For example, living cells share many of the same basic features including a plasma membrane, ribosomes, enzymatic pathways, etc. In addition, as discussed in chapter 7, the mitochondria and chloroplasts of eukaryotic cells are evolutionarily derived from bacterial cells.

C4. Sexual selection is a form of natural selection that favors traits that make it more likely for an organism to reproduce. In many species of animals, natural selection selects for male traits that make it more likely to find a mate and successfully produce offspring. For example, the bright plumage of male birds makes it easier for the female to identify the males of their own species. Many species of deer have antlers so that males may spar with each other as a way to achieve reproductive superiority.

C6. Anagenesis is the evolution of one species into another, while cladogenesis is the divergence of one species into two or more species. Cladogenesis is more prevalent. There may be many reasons why. It is common for an abrupt genetic change such as alloploidy to produce a new species from a preexisting one. Also, migrations of a few members of species into a new region may lead to the formation of a new species in the new region (i.e., allopatric speciation).

C8. A. Allopatric

   B. Sympatric

   C. At first, it may involve parapatric speciation with a low level of intermixing. Eventually, when smaller lakes are formed, allopatric speciation will occur.

C10. The main evidence in favor of punctuated equilibrium is the fossil record. Paleontologists rarely find a gradual transition of fossil forms. The transition period in which environmental pressure and genetic changes cause a previous species to evolve into a new species is thought to be so short that few, if any, of the transitional members would be preserved as fossils. Therefore, the fossil record primarily contains representatives from the long equilibrium periods. Also, rapid evolutionary change is consistent with known genetic phenomena, including single-gene mutations that have dramatic effects on phenotypic characteristics, the founder effect, and genetic events such as changes in chromosome structure (e.g., inversions and translocations) or chromosome number, which may abruptly create individuals with new phenotypic traits. In some cases, however, gradual changes are observed in certain species over long periods of time. In addition, the gradual accumulation of mutations is known to occur from the molecular analyses of DNA.

C12. A. Yes, it may help the females identify the males of this species.

B. Yes, it helps the males to reproduce.

C. No, it may help the females survive, but it probably does not help the males to identify the females.

D. Yes, it may help the females identify the males of this species.

C14. One of the major goals in the field of molecular evolution is to understand, at the molecular level, how changes in the genetic material have led to the formation of present-day species. Along these same lines, molecular biologists would like to understand why genetic variation is prevalent within a single species, and also to examine the degree of differences in genetic variation among different species. Comparisons of the genetic material at the molecular level can help to elucidate evolutionary relationships. The field of molecular evolution also is aimed at understanding how genes change at the molecular level. Geneticists would like to know the rate of genetic change and whether changes are neutral or adaptive.

C16. The rate at which a gene evolves depends on whether mutations in the gene will affect the function of the encoded RNA or protein. If a gene can tolerate many mutations without inhibiting the function of the encoded RNA or protein, it will evolve rapidly. Some genes, however, are unable to tolerate many mutations because most mutations inhibit function. When examining evolutionary relationships between species, it is helpful to have a moderate number of differences between the species, but not too many and not too few. Rapidly evolving genes are useful to examine relationships between closely related species because they will probably have a significant number of differences. More distantly related species can be compared by analyzing slowly evolving genes.

C18. Some regions of a polypeptide are particularly important for the structure or function of a protein. For example, a region of a polypeptide may form the active site of an enzyme. The amino acids that are found within the active site are likely to be exquisitely located for the binding of the enzyme's substrate and/or for catalysis. Changes in the amino acid sequence of the active site usually have a detrimental effect on the enzyme's functions. Therefore, these types of polypeptide sequences (like those found in active sites) are not likely to change. If they did change, natural selection would prevent the change from being transmitted to future generations. In contrast, other regions of a polypeptide are less important. These other regions would be more tolerant of changes in amino acid sequence and therefore would evolve more rapidly. When comparing related protein sequences, regions that are important for function can often be identified based on less sequence variation.

C20. There are lots of possibilities. A few are listed here.

1. DNA is used as the genetic material.

2. The semiconservative mechanism of DNA replication is the same.

3. The genetic code is fairly universal.

4. Certain genes are found in all forms of life (such as 16S rRNA genes).

5. Gene structure and organization is pretty similar among all forms of life.

6. RNA is transcribed from genes.

7. mRNA is used as a messenger to synthesize polypeptides.

C22. Both theories propose that random mutations occur in populations. The primary difference between the neutral theory and the selectionist theory is the relative contributions of neutral mutations and nonneutral mutations for explaining present-day variation. Are most forms of genetic variation neutral or brought about by natural selection? Both sides agree that natural selection has been an important force in shaping the phenotypes of all species. It is natural selection that favors the traits that allow organisms to survive in their environments. Within each species, however, there is still a great deal of variation (i.e., not all members of the same species are genetically identical). The neutral theory would argue that most of this variation is neutral. Random mutations occur, which have no effect on the phenotype of the individual, and genetic drift causes the allele frequency to rise to significant levels. The neutral theory suggests that most (but certainly not all) variation can be explained in this manner. In contrast, the selectionist theory suggests that most of the variation seen in natural populations is due to natural selection. The neutral theory of evolution is sometimes called non-Darwinian evolution because it isn't based on natural selection, which was a central tenet in Darwin's theory.

C24. Natural selection plays a dominant role in the elimination of deleterious mutations and the fixation of beneficial mutations. Genetic drift also would affect the frequencies of deleterious and beneficial alleles, particularly in small populations. Neutral alleles are probably much more frequent than beneficial alleles. Neutral alleles are not acted upon by natural selection. Nevertheless, they can become prevalent within a population due to genetic drift. The neutral theory suggests that most of the genetic variation in natural populations is due to the accumulation of neutral mutations. A large amount of data supports this theory, although some geneticists still oppose it.

C26. The rate of deleterious and beneficial mutations would probably not be a good molecular clock. Their rate of formation might be relatively constant, but their rate of elimination or fixation would probably be quite variable. These alleles are acted upon by natural selection. As environmental conditions change, the degree to which natural selection would favor beneficial alleles and eliminate deleterious alleles would also change. For example, natural selection favors the sickle-cell allele in regions where malaria is prevalent but not in other regions. Therefore, the prevalence of this allele does not depend solely on its rate of formation and random genetic drift.

Experimental Questions

E2. Because the artificial and natural *G. tetrahit* strains can produce fertile offspring, it means that the chromosomes in these strains form homologous sets. When the offspring make gametes, each chromosome has a homologous partner to pair with. Therefore, the offspring are able to make gametes that have a complete set of chromosomes rather than making highly aneuploid gametes that would be inviable. If the two strains had not produced fertile offspring, it would mean that reproductive isolating mechanisms exist between them. Such mechanisms could have been due to mutations that occurred after the formation of the natural species. Alternatively, it could suggest that the natural species was not really an allotetraploid of *G. speciosa* and *G. pubescens*.

E4. Some possible experimental approaches are listed here.

1. You could karyotype some members of each population. Different species can often be distinguished by changes in chromosome structure and number.

2. You could see if eastern and western snakes could mate with each other in captivity. If they cannot, this would suggest reproductive isolation. If they can, they still might be different species, if mating never occurs in nature.

3. You could sit in a blind (a box where someone cannot be seen by animals) and watch both populations to see if eastern and western snakes ever mate with each other in nature.

E6. If we compare the percentages of sequence identity among the five genes, we see that *A1* is very similar to *B2*, and *A2* is very similar to *B1* and *B3*. This suggests that the most recent ancestor had two copies of the gene prior to the formation of these two species. After the formation of species A and B, one of the *B* genes must have duplicated again in species B to form *B1* and *B3*.

E8. The mutations that have occurred in this sequence are neutral mutations. In all cases, the wobble base has changed, and this change would not affect the amino acid sequence of the encoded polypeptide. Therefore, a reasonable explanation is that the gene has accumulated random neutral mutations over the course of many generations. This observation would be consistent with the neutral theory of evolution. A second explanation would be that one of these two researchers made a few experimental mistakes when determining the sequence of this region.

E10. Inversions do not affect the total amount of genetic material. Usually, inversions do not affect the phenotype of the organism. Therefore, if members of the two populations were to interbreed, the offspring would probably be viable because they would have inherited a normal amount of genetic material from each parent. However, such offspring would be inversion heterozygotes. As described in chapter 8 (see fig. 8.12), crossing over during meiosis may create chromosomes that have too much or too little genetic material. If these unbalanced chromosomes are passed to the next generation of offspring, the offspring may not survive. For this reason, inver-

sion heterozygotes (that are phenotypically normal) may not be very fertile because many of their offspring will die. Because inversion heterozygotes are less fertile, this would tend to keep the eastern and western populations reproductively isolated. Over time, this would aid in the independent evolution of the two populations and would ultimately promote the evolution of the two populations into separate species.

E12. The technique of PCR is used to amplify the amount of DNA in a sample. To accomplish this, one must use oligonucleotide primers that are complementary to the region that is to be amplified. For example, as described in the experiment of figure 26.14, PCR primers that were complementary to and flank the 12S rRNA gene can be used to amplify the 12S rRNA gene. The technique of PCR is described in chapter 18.

E14. We would expect the probe to hybridize to the natural *G. tetrahit* and also the artificial *G. tetrahit* because both of these strains contain two sets of chromosomes from *G. pubescens*. We would expect two bright spots in the *in situ* experiment. Depending on how closely related *G. pubescens* and *G. speciosa* are, the probe may also hybridize (to two sites) in the *G. speciosa* genome, but this is difficult to predict *a priori*. If so, the *G. tetrahit* species would show four spots.

Questions for Student Discussion/Collaboration

2. The founder effect and allotetraploidy are examples of rapid forms of evolution. In addition, some single-gene mutations may have a great impact on phenotype and lead to the rapid evolution of new species by cladogenesis. Geological processes may promote the slower accumulation of alleles and alter a species' characteristics more gradually. In this case, it is the accumulation of many phenotypically minor genetic changes that ultimately leads to reproductive isolation. Slow and fast mechanisms of evolution have the common theme that they result in reproductive isolation. This is a prerequisite for the evolution of new species. Fast mechanisms tend to involve small populations and a small number of genetic changes. Slower mechanisms may involve larger populations and involve the accumulation of a large number of genetic changes that each contributes in a small way to reproductive isolation.

G L O S S A R Y

::

A

A an abbreviation for adenine.

acentric describes a chromosome without a centromere.

acridine dye a type of chemical mutagen that causes frameshift mutations.

acrocentric describes a chromosome with the centromere significantly off center, but not at the very end.

activator a transcriptional regulatory protein that increases the rate of transcription.

acutely transforming virus (ACT) a virus that readily transforms normal cells into malignant cells, when grown in a laboratory.

adaptive peak a combination of many alleles that provides an optimal fitness for individuals in a stable environment.

adaptor hypothesis a hypothesis that proposes a tRNA has two functions: recognizing a three-base codon sequence in mRNA and carrying an amino acid that is specific for that codon.

adenine a purine base found in DNA and RNA. It base-pairs with thymine in DNA.

age of onset for alleles that cause genetic diseases, the time of life at which disease symptoms appear.

alkaptonuria a human genetic disorder involving the accumulation of homogentisic acid due to a defect in homogentisic acid oxidase.

allele an alternative form of a specific gene.

allele frequency the number of copies of a particular allele in a population divided by the total number of all alleles for that gene in the population.

allodiploid an organism that contains one set of chromosomes from two different species.

allopatric speciation (Greek, *allos,* "other"; Latin, *patria,* "homeland") an evolutionary phenomenon in which speciation occurs when members of a species become geographically separated from the other members.

alloploid an organism that contains chromosomes from two (or more) different species.

allopolyploid an organism that contains two (or more) sets of chromosomes from two (or more) species.

allosteric enzyme an enzyme that contains two binding sites—a catalytic site and a regulatory site.

allotetraploid an organism that contains two sets of chromosomes from two different species.

allozymes two or more enzymes (encoded by the same type of gene) with alterations in their amino acid sequences, which may affect their gel mobilities.

α helix a type of secondary structure found in proteins.

alternative exon an exon that is not always found in mRNA. It is only found in certain types of alternatively spliced mRNAs.

alternative splicing refers to the phenomenon in which a pre-mRNA can be spliced in more than one way.

amber a stop codon with the sequence UAG.

Ames test a test using strains of a bacterium, *Salmonella typhimurium,* to determine if a substance is a mutagen.

amino acid a building block of polypeptides and proteins. It contains an amino group, a carboxyl group, and a side chain.

aminoacyl site (A site) a site on the ribosome where a charged tRNA initially binds.

aminoacyl-tRNA synthetase an enzyme that catalyzes the attachment of a specific amino acid to the correct tRNA.

2-aminopurine a base analogue that acts as a chemical mutagen.

amino terminus the location of the first amino acid in a polypeptide chain. The amino acid at the amino terminus still retains a free amino group that is not covalently attached to the second amino acid.

amniocentesis a method of obtaining cellular material from a fetus for the purpose of genetic testing.

amplified restriction fragment length polymorphism (AFLP) a RFLP that is amplified via PCR.

anabolic enzyme an enzyme involved in connecting organic molecules to create larger molecules.

anagenesis (Greek, *ana,* "up," and *genesis,* "origin") the evolutionary phenomenon in which a single species is transformed into a different species over the course of many generations.

anaphase the fourth stage of M phase. As anaphase proceeds, half of the chromosomes move to one pole and the other half move to the other pole.

ancient DNA analysis analysis of DNA that is extracted from the remains of extinct species.

aneuploid not euploid. Refers to a variation in chromosome number such that the total number of chromosomes is not an exact multiple of a set or *n* number.

annealing the process in which two complementary segments of DNA bind to each other.

annotated in files involving genetic sequences, annotation is a description of the known function and features of the sequence, as well as other pertinent information.

antero-posterior axis in animals, the axis that runs from the head (anterior) to the tail or base of the spine (posterior).

anther culture the generation of monoploid plants by heat-shock treatment of anthers.

antibodies proteins produced by the B cells of the immune system that recognize foreign substances (namely, viruses, bacteria, and so forth) and target them for destruction.

antibody microarray a small silica, glass, or plastic slide that is dotted with many different antibodies, which recognize particular amino acid sequences within proteins.

anticipation the phenomenon in which the severity of an inherited disease tends to get worse in future generations.

anticodon a three-nucleotide sequence in tRNA that is complementary to a codon in mRNA.

antigens foreign substances that are recognized by antibodies.

anti-oncogene see *tumor-suppressor gene.*

antiparallel an arrangement in a double helix in which one strand is running in the 5′ to 3′ direction while the other strand is 3′ to 5′.

antisense RNA an RNA strand that is complementary to a strand of mRNA.

AP-endonuclease a DNA repair enzyme that recognizes a DNA region that is missing a base, and makes a cut in the DNA backbone near that site.

apical–basal-patterning gene one of several plant genes that play a role in embryonic development.

apical region in plants, the region that produces the leaves and flowers.

apoptosis programmed cell death.

apurinic site a site in DNA that is missing a purine base.

archaebacteria one of the three domains of life. Archaebacteria, also known as *archae,* are prokaryotic species. They tend to live in extreme environments, and are much less common than bacteria (i.e., eubacteria).

ARS elements DNA sequences found in yeast that function as origins of replication.

artificial selection see *selective breeding.*

ascus (pl. **asci**) a sac that contains haploid spores of fungi (i.e., yeast or molds).

asexual reproduction   the way that some unicellular organisms produce new individuals. In this process, a preexisting cell divides to produce two new cells.

A site   see *aminoacyl site.*

assortative mating   breeding in which individuals with similar phenotypes preferentially mate with each other.

AT/GC rule   in DNA, the phenomenon in which an adenine base in one strand always hydrogen bonds with a thymine base in the opposite strand, and a guanine base always hydrogen bonds with a cytosine.

ATP-dependent chromatin remodeling   see *chromatin remodeling.*

attachment site   a site in a host cell chromosome where a virus will integrate during site-specific recombination.

attenuation   a mechanism of genetic regulation, seen in the *trp* operon, in which a short RNA is made but its synthesis is terminated before RNA polymerase can transcribe the rest of the operon.

AU-rich element (ARE)   a sequence found in many short-lived mRNAs that contains the consensus sequence AUUUA.

automated sequencing   the use of fluorescently labeled dideoxyribonucleotides and a fluorescence detector to sequence DNA.

autonomous transposable element   a transposable element that contains all of the information necessary for transposition or retroposition to take place.

autopolyploid   a polyploid produced within a single species due to nondisjunction.

autosomes   chromosomes that are not sex chromosomes.

B

BAC   see *bacterial artificial chromosome.*

backbone   the portion of a DNA or RNA strand that is composed of the repeated covalent linkage of the phosphates and sugar molecules.

backcross   in genetics, this usually refers to a cross of F_1 hybrids to individuals that have genotypes of the parental generation.

bacterial artificial chromosome (BAC)   a cloning vector that propagates in bacteria and is used to clone large fragments of DNA.

bacteriophages (or **phages**)   viruses that infect bacteria.

balanced polymorphism   when natural selection favors the maintenance of two or more alleles in a population.

band shift assay   see *gel retardation assay.*

Barr body   a structure in the interphase nuclei of somatic cells of female mammals that is a highly condensed X chromosome.

basal region   in plants, the region that produces the roots.

basal transcription apparatus   the minimum number of proteins that is needed to transcribe a gene.

base   a nitrogen-containing molecule that is a portion of a nucleotide in DNA or RNA. Examples of bases are adenine, thymine, guanine, cytosine, and uracil.

base excision repair   a type of DNA repair in which a modified base is removed from a DNA strand. Following base removal, a short region of the DNA strand is removed, which is then resynthesized using the complementary strand as a template.

base mismatch   when two bases opposite each other in a double helix do not conform to the AT/GC rule. For example, if A is opposite C, that would be a base mismatch.

base pair   the structure in which two nucleotides in opposite strands of DNA hydrogen bond with each other. For example, an AT base pair is a structure in which an adenine-containing nucleotide in one DNA strand hydrogen bonds with a thymine-containing nucleotide in the complementary strand.

base substitution   a point mutation in which one base is substituted for another base.

benign   refers to a noncancerous tumor that is not invasive and cannot metastasize.

β sheet   a type of secondary structure found in proteins.

bidirectional replication   the phenomenon in which two DNA replication forks emanate in both directions from an origin of replication.

bilateral gynandromorph   an animal in which one side is phenotypically male and the other side is female.

binary fission   the physical process whereby a bacterial cell divides into two daughter cells. During this event, the two daughter cells become divided by the formation of a septum.

biodegradation   the breakdown of a larger molecule into a smaller molecule via cellular enzymes.

bioinformatics   literally, this term means the study of biological information. Recently, this term has been associated with the analysis of genetic sequences, using computers and computer programs.

biolistic gene transfer   the use of microprojectiles to introduce DNA into plant cells.

biological control   the use of microorganisms or products from microorganisms to alleviate plant diseases or damage from undesirable environmental conditions (e.g., frost damage).

biological evolution   the accumulation of genetic changes in a species or population over the course of many generations.

biological species concept   definition of a species as a group of individuals whose members have the potential to interbreed with one another in nature to produce viable, fertile offspring, but who cannot interbreed successfully with members of other species.

bioremediation   the use of microorganisms to decrease pollutants in the environment.

biotechnology   technologies that involve the use of living organisms, or products from living organisms, as a way to benefit humans.

biotransformation   the conversion of one molecule into another via cellular enzymes. This term is often used to describe the conversion of a toxic molecule into a nontoxic molecule.

bivalent   a structure in which two pairs of homologous sister chromatids have synapsed (i.e., aligned) with each other.

blending theory of inheritance   an early, incorrect theory of heredity. According to this view, the seeds that dictate hereditary traits are able to blend together from generation to generation. The blended traits would then be passed to the next generation.

bottleneck effect   a type of genetic drift that occurs when most members of a population are eliminated without any regard to their genetic composition.

box   in genetics, a term used to describe a sequence with a specialized function.

branch migration   the lateral movement of a Holliday junction.

broad sense heritability   heritability that takes into account all genetic factors.

5-bromodeoxyuridine   a base analogue that can be incorporated into chromosomes during DNA replication. The presence of this analogue can affect the ability of the chromosomes to absorb certain dyes. This is the basis for the staining of harlequin chromosomes.

5-bromouracil   a base analogue that acts as a chemical mutagen.

C

C   an abbreviation for cytosine.

cancer cell   a cell that has lost its normal growth control. Cancer cells are invasive (i.e., they can invade normal tissues) and metastatic (i.e., they can migrate to other parts of the body).

CAP   an abbreviation for the catabolite activator protein, a genetic regulatory protein found in bacteria.

capping   the covalent attachment of a 7-methyl-guanosine nucleotide to the 5′ end of mRNA in eukaryotes.

CAP site   the sequence of DNA that is recognized by CAP.

carbohydrate   organic molecules with the general formula $C(H_2O)$. An example of a simple carbohydrate would be the sugar, glucose. Large carbohydrates are composed of multiple sugar units.

carboxyl terminus   the location of the last amino acid in a polypeptide chain. The amino acid at the carboxyl terminus still retains a free carboxyl group that is not covalently attached to another amino acid.

carcinogen   an agent that can cause cancer.

caspases   proteolytic enzymes that play a role in apoptosis.

catabolic enzyme   an enzyme that is involved in the breakdown of organic molecules into small units.

catabolite activator protein   see *CAP.*

catabolite repression   the phenomenon in which a catabolite (such as glucose) represses the expression of certain genes (such as the *lac* operon).

catenane   interlocked circular molecules.

cDNA   see *complementary DNA.*

cDNA library   a DNA library made from a collection of cDNAs.

cell adhesion   when the surfaces of cells bind to each other or to the extracellular matrix.

cell adhesion molecule (CAM)   a molecule (e.g., surface protein or carbohydrate) that plays a role in cell adhesion.

cell culture   refers to the growth of cells in a laboratory.

cell cycle   in eukaryotic cells, a series of stages through which a cell progresses in order to divide. The phases are G for growth, S for synthesis (of the genetic material), and M for mitosis. There are two G phases, G_1 and G_2.

cell fate   the final morphological features that a cell or group of cells will adopt.

cell fusion   describes the process in which individual cells are mixed together and made to fuse with each other.

cell lineage   a series of cells that are descended from a cell or group of cells by cell division.

cell plate   the structure that forms between two daughter plant cells that leads to the separation of the cells by the formation of an intervening cell wall.

centiMorgans (cM)   (same as a map unit) a unit of map distance obtained from genetic crosses. Named in honor of Thomas Hunt Morgan.

central region   in plants, it is the region that creates the stem. It is the radial pattern of cells in the central region that causes the radial growth observed in plants.

central zone   in plants, an area in the meristem where undifferentiated stem cells are always maintained.

centromere   a segment of eukaryotic chromosomal DNA that provides an attachment site for the kinetochore.

centrosome   a cellular structure from which microtubules emanate.

chain termination   refers to the stoppage of growth of a DNA strand, RNA strand, or polypeptide sequence.

character   in genetics, this word has the same meaning as trait.

Chargaff's rule   see *AT/GC rule.*

charged tRNA   a tRNA that has an amino acid attached to its 3′ end by an ester bond.

checkpoint protein   a protein that monitors the conditions of DNA and chromosomes and may prevent a cell from progressing through the cell cycle if an abnormality is detected.

chiasma (pl. **chiasmata**)   the site where crossing over occurs between two chromosomes. It resembles the Greek letter chi, χ.

chimera   an organism composed of cells that are embryonically derived from two different individuals.

chi square (χ^2) test   a commonly used statistical method to determine the goodness of fit. This method can be used to analyze population data in which the members of the population fall into different categories. It is particularly useful for evaluating the outcome of genetic crosses, because these usually produce a population of offspring that differ with regard to phenotypes.

chorionic villus sampling   a method to obtain cellular material from a fetus for the purpose of genetic testing.

chromatid   following chromosomal replication in eukaryotes, the two copies remain attached to each other in the form of sister chromatids.

chromatin   the association between DNA and proteins that is found within chromosomes.

chromatin remodeling   a change in chromatin structure that alters the degree of compaction and/or the spacing of nucleosomes.

chromocenter   the central point where polytene chromosomes aggregate.

chromomere   a dark band within a polytene chromosome.

chromosome   the structures within living cells that contain the genetic material. Genes are physically located within the structure of chromosomes. Biochemically, chromosomes contain a very long segment of DNA, which is the genetic material, and proteins, which are bound to the DNA and provide it with an organized structure.

chromosome mutation   a substantial change in chromosome structure that may affect more than a single gene.

chromosome sorting   a technique in which chromosomes are dispersed into individual droplets, and the droplets carrying the desired chromosome are sorted into a separate tube.

chromosome specific library   a collection of clones that contain different DNA fragments from a single chromosome.

chromosome territory   in the cell nucleus, each chromosome occupies a nonoverlapping region called a chromosome territory.

chromosome theory of inheritance   a theory of Sutton and Boveri, which indicated that the inheritance patterns of traits can be explained by the transmission patterns of chromosomes during gametogenesis and fertilization.

chromosome walking   a common method used in positional cloning in which a mapped gene or RFLP marker provides a starting point to molecularly walk toward a gene of interest via overlapping clones.

***cis*-acting element**   a sequence of DNA, such as a regulatory element, that exerts a *cis* effect.

***cis*-effect**   an effect on gene expression due to genetic sequences immediately adjacent to the gene.

cistron   refers to the smallest genetic unit that produces a positive result in a complementation experiment. A cistron is equivalent to a gene.

cladogenesis (Greek, *clados*, "branch")   during evolution, a form of speciation that involves the division of a species into two or more species.

cleavage furrow   a constriction that causes the division of two animal cells during cytokinesis.

clone   the general meaning of this term is to make many copies of something. In genetics, this term has several meanings: (1) a single cell that has divided to produce a colony of genetically identical cells; (2) an individual who has been produced from a somatic cell of another individual, such as the sheep Dolly; (3) many copies of a DNA fragment that are propagated within a vector or produced by PCR.

closed complex   the complex between transcription factors, RNA polymerase, and a bacterial promoter before the DNA has denatured to form an open complex.

closed conformation   a tightly packed conformation of chromatin that cannot be transcribed.

cluster analysis   the analysis of microarray data to determine if certain groups (i.e., clusters) of genes are expressed under the same conditions.

cM   an abbreviation for centiMorgans; also see *map unit.*

coding strand   the strand in DNA that is not used as a template for mRNA synthesis.

codominance   a pattern of inheritance in which two alleles are both expressed in the heterozygous condition. For example, a person with the genotype $I^A I^B$ will have the blood type AB, and will express both surface antigens A and B.

codon   a sequence of three nucleotides in mRNA that functions in translation. A start codon, which usually specifies methionine, initiates translation, and a stop codon terminates translation. The other codons specify the amino acids within a polypeptide sequence according to the genetic code.

codon bias   the phenomenon that, in a given species, certain codons are used more frequently than other codons.

coefficient of inbreeding (F)   see *fixation coefficient.*

cohesin   a protein complex that facilitates the alignment of sister chromatids.

colinearity   the correspondence between the sequence of codons in the DNA coding strand and the amino acid sequence of a polypeptide.

colony hybridization   a technique in which a probe is used to identify bacterial colonies that contain a hybrid vector with a gene of interest.

common ancestor   someone who is an ancestor to both of an individual's parents.

competence factors   proteins that are needed for bacterial cells to become naturally transformed by extracellular DNA.

competent cells   cells that can be transformed by extracellular DNA.

complementary   describes sequences in two DNA strands that match each other according to the AT/GC rule. For example, if one strand has the sequence of ATGGCGGATTT, then the complementary strand must be TACCGCCTAAA.

complementary DNA (cDNA)   DNA that is made from an RNA template by the action of reverse transcriptase.

complementation   a phenomenon in which the presence of two different mutant alleles in the same organism produces a wild-type phenotype. It usually happens because the two mutations are in different genes, so that the organism carries one copy of each mutant allele and one copy of each wild-type allele.

complete transposable element   see *autonomous transposable element.*

composite transposon   a transposon that contains additional genes, such as antibiotic resistance genes, that are not necessary for transposition *per se.*

computer data file   a file (a collection of information) stored by a computer.

computer program   a series of operations that can analyze data in a defined way.

concordance   in genetics, the degree to which pairs of individuals (e.g., identical twins or fraternal twins) exhibit the same trait.

condensation   a change in chromatin structure to become more compact.

condensin   a protein complex that plays a role in the condensation of interphase chromosomes to become metaphase chromosomes.

conditional alleles   alleles in which the phenotypic expression depends on the environmental conditions. An example are temperature-sensitive alleles, which only affect the phenotype at a particular temperature.

conglomerate   a population composed of members of an original population plus new members that have migrated from another population.

conjugation   a form of genetic transfer between bacteria that involves direct physical interaction between two bacterial cells. One bacterium acts as donor and transfers genetic material to a recipient cell.

conjugative plasmid   a plasmid that can be transferred to a recipient cell during conjugation.

consensus sequence   the most commonly occurring bases within a sequence element.

constitutive exon   an exon that is always found in mRNA following splicing.

constitutive gene   a gene that is not regulated and has essentially constant levels of expression over time.

constitutive heterochromatin   regions of chromosomes that are always heterochromatic and are permanently transcriptionally inactive.

contig   a series of clones that contain overlapping pieces of chromosomal DNA.

control element   see *regulatory sequence or element.*

corepressor   a small effector molecule that binds to a repressor protein, thereby causing the repressor protein to bind to DNA and inhibit transcription.

core promoter   a DNA sequence that is absolutely necessary for transcription to take place. It provides the binding site for general transcription factors and RNA polymerase.

correlation coefficient (r)   a statistic with a value that ranges between -1 and 1. It describes how two factors vary with regard to each other.

cosmid a vector that is a hybrid between a plasmid vector and phage λ. Cosmid DNA can replicate in a cell like a plasmid or be packaged into a protein coat like a phage. Cosmid vectors can accept fragments of DNA that are typically tens of thousands of base pairs in length.

cotransduction the phenomenon in which bacterial transduction transfers a piece of DNA carrying two closely linked genes.

cotransformation the phenomenon in which bacterial transformation transfers a piece of DNA carrying two closely linked genes.

cotranslational events that occur during translation.

cotranslational sorting refers to the sorting of proteins into the ER. The protein is actually translated into the ER lumen or ER membrane.

covariance a statistic that describes the degree of variation between two variables within a group.

cpDNA an abbreviation for chloroplast DNA.

CpG island a group of CG sequences that may be clustered near a promoter region of a gene. The methylation of the cytosine residues inhibits transcription.

CREB protein (cAMP response element-binding protein) a regulatory transcription factor that becomes activated in response to specific cell-signaling molecules.

cross a mating between two distinct individuals. An analysis of their offspring may be conducted to understand how traits are passed from parent to offspring.

cross-fertilization same meaning as *cross*. It requires that the male and female gametes come from separate individuals.

crossing over a physical exchange of chromosome pieces that most commonly occurs during prophase of meiosis I.

C-terminus see *carboxyl terminus.*

cyclin a type of protein that plays a role in the regulation of the eukaryotic cell cycle.

cyclin-dependent protein kinases (CDKs) enzymes that are regulated by cyclins and can phosphorylate other cellular proteins by covalently attaching a phosphate group.

cytogenetic mapping the mapping of genes or genetic sequences using microscopy.

cytogenetics the field of genetics that involves the microscopic examination of chromosomes.

cytokinesis the division of a single cell into two cells. The two nuclei produced in M phase are segregated into separate daughter cells during cytokinesis.

cytological mapping see *cytogenetic mapping.*

cytoplasmic inheritance (also known as *extranuclear inheritance*) refers to the inheritance of genetic material that is not found within the cell nucleus.

cytosine a pyrimidine base found in DNA and RNA. It base-pairs with guanine in DNA.

D

Dam methylase an enzyme in bacteria that attaches methyl groups to the adenine base in DNA that is found within the sequence GATC.

Darwinian fitness the relative likelihood that a phenotype will survive and contribute to the gene pool of the next generation as compared with other phenotypes.

database a computer storage facility that stores many data files such as those containing genetic sequences.

deamination the removal of an amino group from a molecule. For example, the removal of an amino group from cytosine produces uracil.

deficiency condition in which a segment of chromosomal material is missing.

degeneracy in genetics, this term means that more that one codon specifies the same amino acid. For example, the codons GGU, GGC, GGA, and GGG all specify the amino acid glycine.

degrees of freedom in a statistical analysis, the number of categories that are independent of each other.

deletion condition in which a segment of DNA is missing.

deletion mapping the use of strains carrying deletions within a defined region to map a mutation of unknown location.

deme see *subpopulation.*

de novo methylation the methylation of DNA that has not been previously methylated. This is usually a highly regulated event.

deoxyribonucleic acid (DNA) the genetic material. It is a double-stranded structure, with each strand composed of repeating units of deoxyribonucleotides.

deoxyribose the sugar found in DNA.

depurination the removal of a purine base from DNA.

determined cell a cell that is destined to differentiate into a specific cell type.

developmental genetics the area of genetics concerned with the roles of genes in orchestrating the changes that occur during development.

diauxic growth the sequential use of two sugars by a bacterium.

dicentric describes a chromosome with two centromeres.

dicentric bridge the region between the two centromeres in a dicentric chromosome.

dideoxyribonucleotides a nucleotide used in DNA sequencing that is missing the $3'$–OH group. If a dideoxyribonucleotide is incorporated into a DNA strand, it stops any further growth of the strand.

dideoxy sequencing a method of DNA sequencing that uses dideoxyribonucleotides to terminate the growth of DNA strands.

differentially methylated region (DMR) in the case of imprinting, a site that is methylated during spermatogenesis or oogenesis, but not both.

differentiated cell a cell that has become a specialized type of cell within a multicellular organism.

dihybrid cross a cross in which an experimenter follows the outcome of two different traits.

dihybrid testcross a cross in which an experimenter crosses an individual that is heterozygous for two genes to an individual that is homozygous recessive for the same two genes.

dimeric DNA polymerase a complex of two DNA polymerase proteins that move as a unit during DNA replication.

dioecious in plants, a species in which male and female gametophytes are produced on a single (sporophyte) individual.

diploid a organism or cell that contains two copies of each type of chromosome.

directional selection natural selection that favors an extreme phenotype. This usually leads to the fixation of the favored allele.

direct repeat (DR) short DNA sequences that flank transposable elements in which the DNA sequence is repeated in the same direction.

disassortative mating breeding in which individuals with unlike phenotypes preferentially mate with each other.

discontinuous trait a trait in which each offspring can be put into a particular phenotypic category.

disruptive selection natural selection that favors both extremes of a phenotypic category. This results in a balanced polymorphism.

dizygotic twins also known as fraternal twins; twins formed from separate pairs of sperm and egg cells.

DNA the abbreviation for *deoxyribonucleic acid.*

dnaA box sequence serves as a recognition site for the binding of the DnaA protein, which is involved in the initiation of bacterial DNA replication.

DnaA protein a protein that binds to the dnaA box sequence at the origin of replication in bacteria, and initiates DNA replication.

DNA fingerprinting a technology to identify a particular individual based on the properties of their DNA.

DNA footprinting a method to study protein–DNA interactions in which the binding of a protein to DNA protects the DNA from digestion by DNase I.

DNA gap repair synthesis the synthesis of DNA in a region where a DNA strand has been previously removed, usually by a DNA repair enzyme or by an enzyme involved in homologous recombination.

DNA gyrase also known as topoisomerase II; an enzyme that introduces negative supercoils into DNA using energy from ATP. Gyrase can also relax positive supercoils when they occur.

DNA helicase an enzyme that separates the two strands of DNA.

DNA library a collection of many hybrid vectors, each vector carrying a particular fragment of DNA from a larger source. For example, each hybrid vector in a DNA library might carry a small segment of chromosomal DNA from a particular species.

DNA ligase an enzyme that catalyzes a covalent bond between two DNA fragments.

DNA methylation the phenomenon in which an enzyme covalently attaches a methyl group ($-CH_3$) to a base (usually adenine or cytosine) in DNA.

DNA methyltransferase the enzyme that attaches methyl groups to adenine or cytosine bases.

DNA microarray a small silica, glass, or plastic slide that is dotted with many different sequences of DNA, corresponding to short sequences within known genes.

DNA-N-glycosylase an enzyme that can recognize an abnormal base and cleave the bond between it and the sugar in the DNA backbone.

DNA polymerase an enzyme that catalyzes the covalent attachment of nucleotides together to form a strand of DNA.

DNA primase an enzyme that synthesizes a short RNA primer for DNA replication.

DNA profiling see *DNA fingerprinting.*

DNA replication the process in which original DNA strands are used as templates for the synthesis of new DNA strands.

DNase an enzyme that cuts the sugar-phosphate backbone in DNA.

DNase I an endonuclease that cleaves DNA.

DNA sequencing a method to determine the base sequence in a segment of DNA.

DNA supercoiling the formation of additional coils in DNA due to twisting forces.

domain a segment of a protein that has a specific function.

dominant   describes an allele that determines the phenotype in the heterozygous condition. For example, if a plant is *Tt* and has a tall phenotype, the *T* (tall) allele is dominant over the *t* (dwarf) allele.

dorso-ventral axis   in animals, the axis from the spine (dorsal) to the stomach (ventral).

dosage compensation   refers to the phenomenon that in species with sex chromosomes, one of the sex chromosomes is altered so that males and females will have similar levels of gene expression, even though they do not contain the same complement of sex chromosomes.

double helix   the arrangement in which two strands of DNA (and sometimes RNA) interact with each other to form a double-stranded helical structure.

double-stranded break model   a model for homologous recombination in which the event that initiates recombination is a double-stranded break in one of the double-helices.

down promoter mutation   a mutation in a promoter that inhibits the rate of transcription.

down regulation   genetic regulation that leads to a decrease in gene expression.

duplication   the copying of a segment of DNA.

E

editosome   a complex that catalyzes RNA editing.

egg cell   also known as an ovum; it is a female gamete that is usually very large and nonmotile.

electroporation   the use of electric current that creates transient pores in the plasma membrane of a cell to allow entry of DNA.

embryogenesis   an early stage of animal and plant development, leading to the production of an embryo with organized tissue layers, and a body plan organization.

embryonic carcinoma cell (EC cell)   a type of pluripotent stem cell found in a specific type of human tumor.

embryonic germ cell (EG cell)   a type of pluripotent stem cell found in the gonads of the fetus.

embryonic stem cell (ES cell)   a type of pluripotent stem cell found in the early blastocyst.

empirical approach   a strategy in which experiments are designed to determine quantitative relationships as a way to derive laws that govern biological, chemical, or physical phenomena.

empirical laws   laws that are discovered using an empirical (observational) approach.

endonuclease   an enzyme that can cut in the middle of a DNA strand.

endopolyploidy   in a diploid individual, the phenomenon in which certain cells of the body may be polyploid.

endosymbiosis   a symbiotic relationship in which the symbiont actually lives inside ("*endo*") the larger of the two species.

endosymbiosis theory   the theory that the ancient origin of plastids and mitochondria was the result of certain species of bacteria taking up residence within a primordial eukaryotic cell.

enhancer   a DNA sequence that functions as a regulatory element. The binding of a regulatory transcription factor to the enhancer increases the level of transcription.

enzyme   a protein that functions to accelerate chemical reactions within the cell.

enzyme adaptation   the phenomenon in which a particular enzyme only appears within a living cell after the cell has been exposed to the substrate for that enzyme.

epigenetic inheritance   an inheritance pattern in which a modification to a nuclear gene or chromosome alters gene expression in an organism, but the expression is not changed permanently over the course of many generations.

episome   a segment of bacterial DNA that can exist as an F factor and also integrate into the chromosome.

epistasis   an inheritance pattern where one gene can mask the phenotypic effects of a different gene.

epitope   the structure on the surface of an antigen that is recognized by an antibody.

E site   see *exit site.*

essential gene   a gene that is essential for survival.

EST library   see *expressed sequence tagged library.*

ethyl methanesulfonate (EMS)   a type of chemical mutagen that alkylates bases (i.e., attaches methyl or ethyl groups).

eubacteria   one of the three domains of life. Eubacteria, also known as *bacteria,* are prokaryotic species.

euchromatin   DNA that is not highly compacted and may be transcriptionally active.

eukaryotes (Greek, "true nucleus")   one of the three domains of life. A defining feature is organisms whose cells contain nuclei bounded by cell membranes. Some simple eukaryotic species are single-celled protists and yeast; more complex multicellular species include fungi, plants, and animals.

euploid   describes an organism in which the chromosome number is an exact multiple of a chromosome set.

evolution   see *biological evolution.*

excisionase   an enzyme that excises a prophage from a host cell's chromosome.

exit site (E site)   a site on the ribosome from which an uncharged tRNA exits.

exon   a segment of RNA that is contained within the RNA after splicing has occurred. In mRNA, the coding sequence of a polypeptide is contained within the exons.

exon shuffling   the phenomenon that exons have been transferred between different genes during evolution. Some researchers believe that transposable elements have played a role in this phenomenon.

exon skipping   when an exon is spliced out of a pre-mRNA.

exonuclease   an enzyme that digests an RNA or DNA strand from the end.

expressed sequence tagged (EST) library   a DNA library containing many clones that have different cDNA inserts.

expression vector   a cloning vector that contains a promoter so that the gene of interest will be transcribed into RNA when the vector is introduced into a host cell.

expressivity   the degree to which a trait is expressed. For example, flowers with deep red color would have a high expressivity of the red allele.

extranuclear inheritance (also known as *cytoplasmic inheritance*)   refers to the inheritance of genetic material that is not found within the nucleus.

F

F   see *fixation coefficient.*

facultative heterochromatin   heterochromatin that is derived from the conversion of euchromatin to heterochromatin.

fate map   a diagram that depicts how cell division proceeds in an organism.

feedback inhibition   the phenomenon in which the final product of a metabolic pathway inhibits an enzyme that acts early in the pathway.

fertilization   the union of sperm and egg to begin the life of a new organism.

F factor   a fertility factor found in certain strains of bacteria in addition to their circular chromosome. Strains of bacteria that contain an F factor are designated F^+; strains without F factors are F^-. F factors carrying pieces of chromosomal DNA are known as F' factors.

F_1 generation   the offspring produced from a cross of the parental generation.

F_2 generation   the offspring produced from a cross of the F_1 generation.

fidelity   a term used to describe the accuracy of a process. If there are few mistakes, a process has a high fidelity.

fine structure mapping   also known as *intragenic mapping;* the aim of fine structure mapping is to ascertain the distances between two (or more) different mutations within the same gene.

fitness   see *Darwinian fitness.*

fixation coefficient   the likelihood that an individual will be homozygous for a given gene because both copies were inherited from a common ancestor.

fluorescence *in situ* hybridization (FISH)   *in situ* hybridization in which the probe is fluorescent.

focus   with regard to cancer cells, it means a clump of raised cells that grow without regard to contact inhibition.

footprinting   see *DNA footprinting.*

fork   see *replication fork.*

forward mutation   a mutation that changes the wild-type genotype into some new variation.

founder effect   changes in allele frequencies that occur when a small group of individuals separates from a larger population and establishes a colony in a new location.

frameshift mutation   a mutation that involves the addition or deletion of nucleotides not in a multiple of three and thereby shifts the reading frame of the codon sequence downstream from the mutation.

frequency distribution   a graph that describes the numbers of individuals that are found in each of several phenotypic categories.

functional genomics   the study of gene function at the genome level. It involves the study of many genes simultaneously.

G

G   an abbreviation for guanine.

gain-of-function mutation   a mutation that causes a gene to be expressed in an additional place where it is not normally expressed, or during a stage of development when it is not normally expressed.

gamete   a reproductive cell (usually haploid) that can unite with another reproductive cell to create a zygote. Sperm and egg cells are types of gametes.

gametogenesis   the production of gametes (i.e., sperm or egg cells).

gametophyte   the haploid generation of plants.

gap gene   one category of segmentation genes.

G bands   the chromosomal banding pattern that is observed when the chromosomes have been treated with the chemical dye Giemsa.

gel retardation assay   a technique to study protein–DNA interactions in which the binding of protein to a DNA fragment retards it mobility during gel electrophoresis.

gene   a unit of heredity that may influence the outcome of an organism's traits.

gene addition   the addition of a cloned gene into a site in a chromosome of a living cell.

gene amplification   an increase in the copy number of a gene.

gene chip   see *DNA microarray.*

gene cloning   the production of many copies of a gene using molecular methods such as PCR or the introduction of a gene into a vector that replicates in a host cell.

gene conversion   the phenomenon in which one allele is converted to another allele due to genetic recombination and DNA repair.

gene detection   an experimental method, such as Southern blotting, that is used to detect the presence of a particular gene.

gene dosage effect   when the number of copies of a gene affects the phenotypic expression of a trait.

gene duplication   an increase in the copy number of a gene. Can lead to the evolution of gene families.

gene expression   the process in which the information within a gene is accessed, first to synthesize RNA (and proteins), and eventually to affect the phenotype of the organism.

gene family   two or more different genes within a single species that are homologous to each other because they were derived from the same ancestral gene.

gene flow   changes in allele frequencies due to migration.

gene interaction   when two or more different genes influence the outcome of a single trait.

gene knockout   when both copies of a normal gene have been replaced by an inactive mutant gene.

gene mutation   a relatively small mutation that only affects a single gene.

gene pool   the totality of all genes within a particular population.

general transcription factor   one of several proteins that are necessary for basal transcription at the core promoter.

gene rearrangement   a rearrangement in segments of a gene, as occurs in antibody precursor genes.

gene redundancy   the phenomenon in which an inactive gene is compensated for by another gene with a similar function.

gene regulation   the phenomenon in which the level of gene expression can vary under different conditions.

gene replacement   the swapping of a cloned gene made experimentally with a normal chromosomal gene found in a living cell.

gene therapy   the introduction of cloned genes into living cells in an attempt to cure or alleviate disease.

genetic code   the correspondence between a codon (i.e., a sequence of three bases in an mRNA molecule) and the functional role that the codon plays during translation. Each codon specifies a particular amino acid or the end of translation.

genetic cross   a mating between two individuals and the analysis of their offspring in an attempt to understand how traits are passed from parent to offspring.

genetic drift   random changes in allele frequencies due to sampling error.

genetic load (L)   genetic variation that decreases the average fitness of a population as compared with a (theoretical) maximum or optimal value.

genetic mapping   any method used to determine the linear order of genes as they are linked to each other along the same chromosome. This term is also used to describe the use of genetic crosses to determine the linear order of genes.

genetic marker   any genetic sequence that is used to mark a specific location on a chromosome.

genetic mosaic   see *mosaicism.*

genetic recombination   (1) the process in which chromosomes are broken and then rejoined to form a novel genetic combination. (2) the process in which alleles are assorted and passed to offspring in combinations that are different from the parents.

genetics   the study of heredity.

genetic screening   the use of testing methods to determine if an individual is a heterozygous carrier for or has a genetic disease.

genetic testing   the analysis of individuals with regard to their genes or gene products. In many cases, the goal is to determine if an individual carries a mutant gene.

genetic transfer   describes the physical transfer of genetic material from one bacterial cell to another.

genetic variation   genetic differences among members of the same species or among different species.

genome   all of the chromosomes and DNA sequences that an organism can possess.

genome database   a database that focuses on the genetic sequences and characteristics of a single species.

genome mutation   a change in chromosome number.

genomic clone   a clone made from the digestion and cloning of chromosomal DNA.

genomic imprinting   a pattern of inheritance that involves a change in a single gene or chromosome during gamete formation. Depending on whether the modification occurs during spermatogenesis or oogenesis, imprinting governs whether an offspring will express a gene that has been inherited from its mother or father.

genomic library   a DNA library made from chromosomal DNA fragments of a single species.

genomics   the molecular analysis of the entire genome of a species.

genotype   the genetic composition of an individual, especially in terms of the alleles for particular genes.

genotype frequency   the number of individuals with a particular genotype in a population divided by the total number of individuals in the population.

germ cells   the gametes (i.e., sperm and egg cells).

germ line   a lineage of cells that gives rise to gametes.

germ-line mutation   a mutation in a cell of the germ line.

glucocorticoid receptor   a type of steroid receptor that functions as a regulatory transcription factor.

goodness of fit   the degree to which the observed data and expected data are similar to each other. If the observed and predicted data are very similar, the goodness of fit is high.

gradualism   an evolutionary hypothesis suggesting that each new species evolves continuously over long spans of time. The principal idea is that large phenotypic differences that cause the divergence of species are due to the accumulation of many small genetic changes.

grande   normal (large-sized) yeast colonies.

grooves   in DNA, the indentations where the atoms of the bases are in contact with the surrounding water. In B-DNA there is a smaller minor groove and a larger major groove.

group I intron   a type of intron found in self-splicing RNA that uses free guanosine in its splicing mechanism.

group II intron   a type of intron found in self-splicing RNA that uses an adenine nucleotide within the intron itself in its splicing mechanism.

guanine   a purine base found in DNA and RNA. It base-pairs with cytosine in DNA.

guide RNA   in trypanosome RNA editing, an RNA molecule that directs the addition of uracil residues into the mRNA.

gyrase   see *DNA gyrase.*

H

haploid   describes the phenomenon that gametes contain half the genetic material found in somatic cells. For a species that is diploid, a haploid gamete contains a single set of chromosomes.

Hardy–Weinberg equilibrium   the phenomenon that under certain conditions allele frequencies will be maintained in a stable condition and genotypes can be predicted according to

$$p^2 + 2pq + q^2 = 1 \text{ (Hardy–Weinberg equation)}$$

helicase   see *DNA helicase.*

helix–loop–helix domain   a domain found in transcription factors that enables them to bind to the DNA.

hemizygous   describes the single copy of an X-linked gene in the male. A male mammal is said to be hemizygous for X-linked genes.

heritability   the amount of phenotypic variation within a particular group of individuals that is due to genetic factors.

heterochromatin   highly compacted DNA. It is usually transcriptionally inactive.

heterochronic mutation   a mutation that alters the timing of expression of a gene, and thereby alters the outcome of cell fates.

heterodimer   when two polypeptides encoded by different genes bind to each other to form a dimer.

heteroduplex   a double-stranded region of DNA that contains one or more base mismatches.

heterogametic sex   in species with two types of sex chromosomes, the heterogametic sex is the gender that produces two types of gametes. For example, in mammals, the male is the heterogametic sex, because a sperm can contain either an X or a Y chromosome.

heterogamous   describes a species that produces two morphologically different types of gametes (i.e., sperm and eggs).

heterogeneity   refers to the phenomenon that a particular type of disease may be caused by mutations in two or more different genes.

heterogeneous nuclear RNA (hnRNA)   same as pre-mRNA.

heterokaryon   a cell produced from cell fusion that contains two separate nuclei.

heteroplasmy   when a cell contains variation in a particular type of organelle. For example, a plant cell could contain some chloroplasts that make chlorophyll and other choroplasts that do not.

heterosis   the phenomenon in which hybrids display traits superior to either corresponding parental strain. Heterosis is usually different from overdominance, because the hybrid may be heterozygous for many genes, not just a single gene, and because the superior phenotype may be due to the masking of deleterious recessive alleles.

heterozygote   an individual who is heterozygous.

heterozygote advantage   a pattern of inheritance in which a heterozygote is more vigorous than either of the corresponding homozygotes.

heterozygous   describes a diploid individual who has different copies (i.e., two different alleles) of the same gene.

Hfr strain (for **High frequency of recombination**)   a bacterial strain in which an F factor has become integrated into the bacterial chromosome. During conjugation, an Hfr strain can transfer segments of the bacterial chromosome.

high stringency   refers to highly selective hybridization conditions that promote the binding of DNA or RNA fragments that are perfect or almost perfect matches.

histone acetyltransferase   an enzyme that attaches acetyl groups to the N-terminus of histone proteins.

histone deacetylase   an enzyme that removes acetyl groups from the N-terminus of histone proteins.

histones   a group of proteins involved in forming the nucleosome structure of eukaryotic chromatin.

hnRNA   an abbreviation for heterogeneous nuclear RNA.

holandric gene   a gene on the Y chromosome.

Holliday junction   a site where an unresolved crossover has occurred between two homologous chromosomes.

Holliday model   a model to explain the molecular mechanism of homologous recombination.

homeobox   a 180 base pair consensus sequence found in homeotic genes.

homeodomain   the protein domain encoded by the homeobox. The homeodomain promotes the binding of the protein to the DNA.

homeologous   describes the analogous chromosomes from evolutionarily related species.

homeotic gene   a gene that functions in governing the developmental fate of a particular region of the body.

homoallelic   describes two or more alleles in different organisms that are due to mutations at exactly the same base within a gene.

homodimer   when two polypeptides encoded by the same gene bind to each other to form a dimer.

homogametic sex   in species with two types of sex chromosomes, the homogametic sex is the gender that produces only one type of gamete. For example, in mammals, the female is the homogametic sex, because an egg can only contain an X chromosome.

homologous   in the case of genes, this term describes two genes that are derived from the same ancestral gene. Homologous genes have similar DNA sequences. In the case of chromosomes, the two homologues of a chromosome pair are said to be homologous to each other.

homologous recombination   the exchange of DNA segments between homologous chromosomes.

homologue   one of the chromosomes in a pair of homologous chromosomes.

homozygous   describes a diploid individual who has two identical copies of the same allele.

horizontal gene transfer   the transfer of genes between different species.

host cell   a cell that is infected with a virus or bacterium.

hot spots   sites within a gene that are more likely to be mutated than other locations.

housekeeping gene   a gene that encodes a protein required in most cells of a multicellular organism.

Hox **genes**   mammalian genes that play a role in development. They are homologous to homeotic genes found in *Drosophila*.

Human Genome Project   a worldwide collaborative project that aims to provide a detailed map of the human genome, and obtain a complete DNA sequence of the human genome.

hybrid   (1) an offspring obtained from a hybridization experiment; (2) a cell produced from a cell fusion experiment in which the two separate nuclei have fused to make a single nucleus.

hybrid cell   a cell produced from the fusion of two different cells, usually from two different species.

hybrid dysgenesis   a syndrome involving defective *Drosophila* offspring, due to the phenomenon that P elements can transpose freely.

hybridization   (1) the mating of two organisms of the same species with different characteristics; (2) the phenomenon in which two single-stranded molecules renature together to form a hybrid molecule.

hybrid vector   a cloning vector that contains an insert of foreign DNA.

hybrid vigor   see *heterosis*.

hypothesis testing   using statistical tests to determine if the data from genetic crosses are consistent with a hypothesis regarding a particular pattern of inheritance.

I

illegitimate recombination   see *nonhomologous recombination*.

immunoglobulin (IgG)   see *antibodies*.

imprinting   see *genomic imprinting*.

inborn error of metabolism   a genetic disease that involves a defect in a metabolic enzyme.

inbreeding   the practice of mating between genetically related individuals.

incomplete dominance   a pattern of inheritance in which a heterozygote that carries two different alleles exhibits a phenotype that is intermediate to the corresponding homozygous individuals. For example, an *Rr* heterozygote may be pink, while the *RR* and *rr* homozygotes are red and white, respectively.

incomplete penetrance   a pattern of inheritance in which a dominant allele does not always control the phenotype of the individual.

incomplete transposable element   see *nonautonomous transposable element*.

induced mutation   a mutation caused by environmental agents.

inducer   a small effector molecule that binds to a genetic regulatory protein and thereby increases the rate of transcription.

infective particle   genetic material found within the cytoplasm of eukaryotic cells that differs from the genetic material normally found in cell organelles.

inhibitor   a small effector molecule that binds to an activator protein, causing the protein to be released from the DNA and thereby inhibiting transcription.

insertion sequences   the simplest transposable elements. They are commonly found in bacteria.

in situ **hybridization**   a technique used to cytologically map the locations of genes or other DNA sequences within large eukaryotic chromosomes. In this method, a complementary probe is used to detect the location of a gene within a set of chromosomes.

integrase   an enzyme that functions in the integration of viral DNA or retroelements into the host chromosome.

interference   see *positive interference*.

intergenic region   in a chromosome, a region of DNA that lies between two different genes.

intergenic suppressor   a suppressor mutation that is in a different gene from the gene that contains the first mutation.

interphase   the series of phases G_1, S, and G_2, during which a cell spends most of its life.

interrupted mating   a method used in conjugation experiments in which the length of time that the bacteria spend conjugating is stopped by a blender treatment or other type of harsh agitation.

interstitial deficiency   when an internal segment is lost from a linear chromosome.

intervening sequence   also known as an *intron*. A segment of RNA that is removed during RNA splicing.

intragenic mapping   see *fine structure mapping*.

intragenic suppressor   a suppressor mutation that is within the same gene as the first mutation that it suppresses.

intrinsic termination   transcriptional termination that does not require the function of the rho protein.

intron   intervening sequences that are found between exons. Introns are spliced out of the RNA prior to translation.

invasive   refers to a tumor that can invade surrounding tissue.

inversion   a change in the orientation of genetic material along a chromosome such that a segment is flipped in the reverse direction.

inversion heterozygote   a diploid individual that carries one normal chromosome and a homologous chromosome with an inversion.

inversion loop   the loop structure that is formed when the homologous chromosomes of an inversion heterozygote attempt to align themselves (i.e., synapse) during meiosis.

inverted repeats   DNA sequences found in transposable elements that are identical (or very similar) but run in the opposite directions.

iron regulatory protein   a translational regulatory protein that recognizes iron response elements that are found in specific mRNAs. It may inhibit translation or stabilize the mRNA.

iron response element   an RNA sequence that is recognized by the iron regulatory protein.

isoacceptor tRNAs   two different tRNAs that can recognize the same codon.

isogamous   describes a species that makes morphologically similar gametes.

isolating mechanism   a mechanism that favors reproductive isolation. These can be prezygotic or postzygotic.

K

karyotype   a photographic representation of all the chromosomes within a cell. It reveals how many chromosomes are found within an actively dividing somatic cell.

kinetochore   a group of cellular proteins that attach to the centromere during meiosis and mitosis.

knockout   see *gene knockout*.

Kozac's rules   a set of rules that describes the most favorable types of bases that flank a eukaryotic start codon.

L

lagging strand   a strand during DNA replication that is synthesized as short Okazaki fragments in the direction away from the replication fork.

lariat   an excised intron structure composed of a circle and a tail.

law of independent assortment   see *Mendel's law of independent assortment*

law of segregation   see *Mendel's law of segregation*

leading strand   a strand during DNA replication that is synthesized continuously toward the replication fork.

lesion-replicating polymerase   a type of DNA polymerase that can replicate over a DNA region that contains an abnormal structure (i.e., a lesion).

lethal allele   an allele that may cause the death of an organism.

library   see *DNA library*.

ligase   see *DNA ligase*.

lineage diagram   in developmental biology, a depiction of a cell and all of its descendants that are produced by cell division.

LINEs   in mammals, long interspersed elements that are usually 1 to 5 kbp in length and found in 20,000 to 100,000 copies per genome.

linkage   refers to the occurrence of two or more genes along the same chromosome.

linkage group   a group of genes that are linked together because they are found on the same chromosome.

linkage mapping   the mapping of genes or other genetic sequences along a chromosome by analyzing the outcome of crosses.

lipid   a general name given to an organic molecule that is insoluble in water. Cell membranes contain a large amount of lipid.

liposome   a vesicle that is surrounded by a phospholipid bilayer.

local population   see *subpopulation*.

locus (pl. **loci**)   the physical location of a gene within a chromosome.

locus control region (LCR)   a segment of DNA that is involved in the regulation of chromatin opening and closing.

lod score method   a method that analyzes pooled data from a large number of pedigrees or crosses to determine the probability that two genetic markers exhibit a certain degree of linkage. A lod score value of >3 or higher is usually accepted as strong evidence that two markers are linked.

long terminal repeat (LTR) sequences   sequences containing many short segments that are tandemly repeated. They are found in retroviruses and virallike retroelements.

loop domain   a segment of chromosomal DNA that is anchored by proteins, so that it forms a loop.

loss-of-function allele   an allele of a gene that encodes an RNA or protein that is nonfunctional or compromised in function.

loss-of-function mutation   a change in a genetic sequence that creates a loss-of-function allele.

low stringency   refers to hybridization conditions in which DNA or RNA fragments that have some mismatches are still able to recognize and bind to each other.

LTRs   see *long terminal repeat sequences*.

Lyon hypothesis   a hypothesis to explain the pattern of X inactivation seen in mammals. Initially, both X chromosomes are active. However, at an early stage of embryonic development, one of the two X chromosomes is randomly inactivated in each somatic cell.

lysogenic cycle   a type of growth cycle for a phage in which the phage integrates its genetic material into the chromosome of the bacterium. This integrated phage DNA can exist in a dormant state for a long time, during which no new bacteriophages are made.

lytic cycle   a type of growth cycle for a phage in which the phage directs the synthesis of many copies of the phage genetic material and coat proteins. These components then assemble to make new phages. When synthesis and assembly is completed, the bacterial host cell is lysed and the newly made phages are released into the environment.

M

macroevolution   evolutionary changes above the species level involving relatively large changes in form and function that are sufficient to produce new species and higher taxa.

macromolecule   a large organic molecule composed of smaller building blocks. Examples include DNA, RNA, proteins, and large carbohydrates.

MADS box   a domain found in several transcription factors that play a role in plant development.

maintenance methylation   the methylation of hemimethylated DNA following DNA replication.

major groove   a wide indentation in the DNA double helix in which the bases have access to water.

malignant   describes a tumor composed of cancerous cells.

mammalian cloning   experimentally, this refers to the use of somatic cell nuclei and enucleated eggs to create a clone of a mammal.

map distance   the relative distance between sites (e.g., genes) along a single chromosome.

map unit (mu)   a unit of map distance obtained from genetic crosses. One map unit is equivalent to 1% recombinant offspring in a test cross.

mass spectrometry   a technique to accurately measure the mass of molecule, such as a peptide fragment. The technique is described in chapter 21.

maternal effect   an inheritance pattern for certain nuclear genes in which the genotype of the mother directly determines the phenotypic traits of her offspring.

maternal inheritance   inheritance of DNA that occurs through the cytoplasm of the egg.

matrix-attachment region (MAR)   a site in the chromosomal DNA that is anchored to the nuclear matrix or scaffold.

maturase   an enzyme that enhances the rate of splicing of Group I and II introns.

mean   the sum of all the values in a group divided by the number of individuals in the group.

mediator   a protein complex that interacts with RNA polymerase II and various regulatory transcription factors. Depending on its interactions with regulatory transcription factors, mediator may stimulate or inhibit RNA polymerase II.

meiosis   a form of nuclear division in which the sorting process results in the production of haploid cells from a diploid cell.

meiotic nondisjunction   the event in which chromosomes do not segregate equally during meiosis.

Mendelian inheritance   the common pattern of inheritance observed by Mendel, which involves the transmission of eukaryotic genes that are located on the chromosomes found within the cell nucleus.

Mendel's law of independent assortment   two different genes will randomly assort their alleles during gamete formation (if they are not linked).

Mendel's law of segregation   the two copies of a gene segregate from each other during transmission from parent to offspring.

meristem   in plants, an organized group of actively dividing cells.

merozygote   a partial diploid strain of bacteria containing F′ factor genes.

messenger RNA (mRNA)   a type of RNA that contains the information for the synthesis of a polypeptide.

metacentric   describes a chromosome with the centromere in the middle.

metaphase   the third phase of M phase. The chromosomes align along the center of the spindle apparatus and the formation of the spindle apparatus is complete.

metastatic   describes cancer cells that migrate to other parts of the body.

methylation   see *DNA methylation*.

methyl-CpG-binding protein   a protein that binds to a CpG island when it is methylated.

methyl-directed mismatch repair   a DNA repair system in *E. coli* that detects a mismatch and specifically removes the segment from the newly made daughter strand.

methyltransferase   see *DNA methyltransferase*.

microarray   see *DNA microarray, protein microarray,* or *antibody microarray*.

microevolution   changes in the gene pool with regard to particular alleles that occur over the course of many generations.

microsatellite   short simple sequences (typically a couple hundred base pairs in length) that are interspersed throughout a genome and are quite variable in length among different individuals. They can be amplified by PCR.

microscopy   the use of a microscope to view cells or subcellular structures.

minichromosomes   a structure formed from many copies of circular DNA molecules.

minisatellite   a repetitive sequence that was formerly used in DNA fingerprinting. Its use has been largely superceded by smaller repetitive sequences called microsatellites.

minor groove   a narrow indentation in the DNA double helix in which the bases have access to water.

minute   a unit of measure in bacterial conjugation experiments. This unit refers to the relative time it takes for genes to first enter a recipient strain during conjugation.

missense mutation   a base substitution that leads to a change in the amino acid sequence of the encoded polypeptide.

mitochondrial DNA   the DNA found within mitochondria.

mitosis   a type of nuclear division into two nuclei, such that each daughter cell will receive the same complement of chromosomes.

mitotic nondisjunction   an event in which chromosomes do not segregate equally during mitosis.

mitotic recombination   recombination that occurs during mitosis.

mitotic spindle apparatus (also known as the mitotic spindle)   the structure that organizes and separates the chromosomes during M phase of the eukaryotic cell cycle.

molecular clock   refers to the phenomenon that the rate of neutral mutations can be used as a tool to measure evolutionary time.

molecular evolution   the molecular changes in the genetic material that underlie the phenotypic changes associated with evolution.

molecular genetics   an examination of DNA structure and function at the molecular level.

molecular marker a segment of DNA that is found at a specific site in the genome and has properties that enable it to be uniquely recognized using molecular tools such as gel electrophoresis.

molecular paleontology the analysis of DNA sequences from extinct species.

molecular pharming a recombinant technology that involves the production of medically important proteins in the mammary glands of livestock.

monoallelic expression in the case of imprinting, refers to the phenomenon that only one of the two alleles of a given gene are transcriptionally expressed.

monohybrid an individual produced from a monohybrid cross.

monohybrid cross a cross in which an experimenter is following the outcome of only a single trait.

monomorphic a term used to describe a gene that is found as only one allele in a population.

monoploid an organism with a single set of chromosomes within its somatic cells.

monosomic a diploid cell that is missing a chromosome (i.e., $2n - 1$).

monozygotic twins twins that are genetically identical to each other because they were formed from the same sperm and egg.

morph a form or phenotype in a population. For example, red eyes and white eyes are different eye color morphs.

morphogen a molecule that conveys positional information and promotes developmental changes.

mosaicism when the cells of part of an organism differ genetically from the rest of the organism.

motif the name given to a domain or amino acid sequence that functions in a similar manner in many different proteins.

M phase a general name given to nuclear division that can apply to mitosis or meiosis. It is divided into prophase, prometaphase, metaphase, anaphase, and telophase.

mRNA see *messenger RNA*.

mtDNA an abbreviation for *mitochondrial DNA*.

multiple alleles when the same gene exists in two or more alleles within a population.

multiple sequence alignment an alignment of two or more genetic sequences based on their homology to each other.

multipotent a type of stem cell that can differentiate into several different types of cells.

mutagen an agent that causes alterations in the structure of DNA.

mutant alleles alleles that have been created by altering a wild-type allele by mutation.

mutation a permanent change in the genetic material that can be passed from cell to cell or from parent to offspring.

mutation frequency the number of mutant genes divided by the total number of genes within the population.

mutation rate the likelihood that a gene will be altered by a new mutation.

myogenic bHLH protein a type of transcription factor involved in muscle cell differentiation.

N

n an abbreviation that designates the number of chromosomes in a set. In humans, $n = 23$ and an individual has $2n = 46$ chromosomes.

narrow sense heritability heritability that takes into account only those genetic factors that are additive.

natural selection refers to the process whereby differential fitness acts on the gene pool. When a mutation creates a new allele that is beneficial, the allele may become prevalent within future generations because the individuals possessing this allele are more likely to survive and/or reproduce and pass the beneficial allele to their offspring.

negative control transcriptional regulation by repressor proteins.

neutral mutation a mutation that has no detectable effect on protein function and/or no detectable effect on the survival of the organism.

neutral theory of evolution the theory that most genetic variation observed in natural populations is due to the accumulation of neutral mutations.

nick translation the phenomenon in which DNA polymerase uses its 5′ to 3′ exonuclease activity to remove a region of DNA and, at the same time, replaces it with new DNA.

nitrous acid a type of chemical mutagen that deaminates bases, thereby changing amino groups to keto groups.

nonautonomous transposable element a transposable element that lacks a gene such as transposase or reverse transcriptase that is necessary for transposition.

noncoding strand the strand of DNA within a structural gene that is complementary to the mRNA. The noncoding strand is used as a template to make mRNA.

noncomplementation the phenomenon in which two mutant alleles in the same organism do not produce a wild-type phenotype.

non-Darwinian evolution see *neutral theory of evolution*.

nondisjunction event in which chromosomes do not segregate properly during mitosis or meiosis.

nonessential genes genes that are not absolutely required for survival, although they are likely to be beneficial to the organism.

nonhomologous DNA end-joining protein (NHEJ) a protein that joins the ends of DNA fragments that are not homologous. This occurs during site-specific recombination of immunoglobulin genes.

nonhomologous recombination the exchange of DNA between nonhomologous segments of chromosomes or plasmids.

nonneutral mutation a mutation that affects the phenotype of the organism and can be acted on by natural selection.

nonparental refers to combinations of alleles or traits that are not found in the parental generation.

nonsense codon a stop codon.

nonsense mutation a mutation that involves a change from a normal codon to a stop codon.

normal distribution a distribution for an infinite sample in which the trait of interest varies in a symmetrical way around an average value.

Northern blotting a technique used to detect a specific RNA within a mixture of many RNA molecules.

N-terminus see *amino terminus*.

nuclear genes genes that are located on chromosomes found in the cell nucleus of eukaryotic cells.

nuclear matrix (or nuclear scaffold) a group of proteins that anchor the loops found in eukaryotic chromosomes.

nucleic acid RNA or DNA. A macromolecule that is composed of repeating nucleotide units.

nucleoid a darkly staining region that contains the genetic material of mitochondria, chloroplasts, or bacteria.

nucleolus a region within the nucleus of eukaryotic cells where the assembly of ribosomal subunits occurs.

nucleoprotein a complex of DNA (or RNA) and protein.

nucleoside structure in which a base is attached to a sugar, but no phosphate is attached to the sugar.

nucleosome the repeating structural unit within eukaryotic chromatin. It is composed of double-stranded DNA wrapped around an octamer of histone proteins.

nucleotide the repeating structural unit of nucleic acids, composed of a sugar, phosphate, and base.

nucleotide excision repair (NER) a DNA repair system in which several nucleotides in the damaged strand are removed from the DNA and the undamaged strand is used as a template to resynthesize a normal strand.

nucleus a membrane-bounded organelle in eukaryotic cells where the linear sets of chromosomes are found.

O

ocher a stop codon with the sequence UGA.

octad a group of eight fungal spores contained within an ascus.

Okazaki fragments short segments of DNA that are synthesized in the lagging strand during DNA replication.

oncogene a mutant gene that promotes cancer.

oogenesis the production of egg cells.

opal a stop codon with the sequence UAA.

open complex the region of separation of two DNA strands produced by RNA polymerase during transcription.

open conformation a loosely packed chromatin structure that is capable of transcription.

open reading frame a genetic sequence that does not contain stop codons.

operator (or operator site) a sequence of nucleotides in bacterial DNA that provides a binding site for a genetic regulatory protein.

operon an arrangement in DNA where two or more structural genes are found within a regulatory unit that is under the transcriptional control of a single promoter.

ORF see *open reading frame*.

organelle a large specialized structure (often membrane-bounded) within a cell.

organizing center in plants, a region of the meristem that ensures the proper organization of the meristem and preserves the correct number of actively dividing stem cells.

origin of replication a nucleotide sequence that functions as an initiation site for the assembly of several proteins required for DNA replication.

origin of replication complex (ORC) a complex of six proteins found in eukaryotes that is necessary to initiate DNA replication.

origin of transfer the location on an F factor or within the chromosome of an Hfr strain that is the initiation site for the transfer of DNA from one bacterium to another during conjugation.

ortholog homologous genes in different species that were derived from the same ancestral gene.

outbreeding mating between genetically unrelated individuals.

ovary (1) In plants, the structure in which the ovules develop. (2) In animals, the structure that produces egg cells and female hormones.

overdominance an inheritance pattern in which a heterozygote is more vigorous than either of the corresponding homozygotes.

ovule the structure in higher plants where the female gametophyte (i.e., embryo sac) is produced.

ovum a female gamete, also known as an egg cell.

P

p an abbreviation for the short arm of a chromosome.

pair-rule gene one category of segmentation genes.

palindromic when a sequence is the same in the forward and reverse direction.

pangenesis an incorrect theory of heredity. It suggested that hereditary traits could be modified depending on the lifestyle of the individual. For example, it was believed that a person who practiced a particular skill would produce offspring that would be better at that skill.

paracentric inversion an inversion in which the centromere is found outside of the inverted region.

paralog homologous genes within a single species that constitute a gene family.

parapatric speciation (Greek, *para,* "beside"; Latin, *patria,* "homeland") a form of speciation that occurs when members of a species are only partially separated or when a species is very sedentary.

parasegments transient subdivisions that occur in the *Drosophila* embryo prior to the formation of segments.

parthenogenesis the formation of an individual from an unfertilized egg.

particulate theory of inheritance a theory proposed by Mendel. It states that traits are inherited as discrete units that remain unchanged as they are passed from parent to offspring.

paternal leakage the phenomenon that in species where maternal inheritance is generally observed, the male parent may, on rare occasions, provide mitochondria or chloroplasts to the zygote.

PCR see *polymerase chain reaction.*

pedigree analysis a genetic analysis using information contained within family trees. In this approach, the aim is to determine the type of inheritance pattern that a gene follows.

peptide bond a covalent bond formed between the carboxyl group in the last amino acid in the polypeptide chain and the amino group in the next amino acid to be added to the chain.

peptidyl site (P site) a site on the ribosome that carries a tRNA along with a polypeptide chain.

peptidyltransferase a complex that functions during translation to catalyze the formation of a peptide bond between the amino acid in the A site of the ribosome and the growing polypeptide chain.

pericentric inversion an inversion in which the centromere is located within the inverted region of the chromosome.

peripheral zone in plants, an area in the meristem that contains dividing cells that will eventually differentiate into plant structures.

petites mutant strains of yeast that form small colonies due to defects in mitochondrial function.

PFGE see *pulsed-field gel electrophoresis.*

P generation the parental generation in a genetic cross.

phage see *bacteriophage.*

pharming see *molecular pharming.*

phenotype the observable traits of an organism.

phenylketonuria (PKU) a human genetic disorder arising from a defect in phenylalanine hydroxylase.

phosphodiester linkage in a DNA or RNA strand, a linkage in which a phosphate group connects two sugar molecules together.

phylogenetic tree a diagram that describes the evolutionary relationships among different species.

physical mapping the mapping of genes or other genetic sequences using DNA cloning methods.

PKU see *phenylketonuria.*

plaque a clear zone within a bacterial lawn on a petri plate. It is due to repeated cycles of viral infection and bacterial lysis.

plasmid a general name used to describe circular pieces of DNA that exist independently of the chromosomal DNA. Some plasmids are used as vectors in cloning experiments.

pluripotent a type of stem cell that can differentiate into all or nearly all the types of cells of the adult organism.

point mutation a change in a single base pair within DNA.

polarity in genetics, the phenomenon in which a nonsense mutation in one gene will affect the translation of a downstream gene in an operon.

pollen the male gametophyte of higher plants.

polyA-binding protein a protein that binds to the $3'$ polyA tail of mRNAs and protects the mRNA from degradation.

polyadenylation the process of attachment of a string of adenine nucleotides to the $3'$ end of eukaryotic mRNAs.

polyA tail the string of adenine nucleotides at the $3'$ end of eukaryotic mRNAs.

polycistronic mRNA an mRNA transcribed from an operon that encodes two or more proteins.

polygenic inheritance refers to the transmission of traits that are governed by two or more different genes.

polymerase chain reaction (PCR) the method to amplify a DNA region involving the sequential use of oligonucleotide primers and *taq* polymerase.

polymerase switch during DNA replication, when one type of DNA polymerase (such as α) is switched for another type (such as β).

polymorphic a term used to describe a trait or gene that is found in two or more forms in a population.

polymorphism (1) the prevalence of two or more phenotypic forms in a population; (2) the phenomenon in which a gene exists in two or more alleles within a population.

polypeptide a sequence of amino acids that is the product of mRNA translation. One or more polypeptides will fold and associate with each other to form a functional protein.

polyploid an organism or cell with three or more sets of chromosomes.

polyribosome an mRNA transcript that has many bound ribosomes in the act of translation.

polysome see *polyribosome.*

polytene chromosome chromosomes that are found in certain cells, such as *Drosophila* salivary cells, in which the chromosomes have replicated many times and the copies lie side by side.

population a group of individuals of the same species that are capable of interbreeding with one another.

population genetics the field of genetics that is primarily concerned with the prevalence of genetic variation within populations.

positional cloning a cloning strategy in which a gene is cloned based on its mapped position along a chromosome.

positional information chemical substances and other environmental cues that enable a cell to deduce its position relative to other cells.

position effect a change in phenotype that occurs when the position of a gene is changed from one chromosomal site to a different location.

positive control genetic regulation by activator proteins.

positive interference the phenomenon in which a crossover occurs in one region of a chromosome and decreases the probability that another crossover will occur nearby.

posttranslational describes events that occur after translation is completed.

posttranslational covalent modification the covalent attachment of a molecule to a protein after it has been synthesized via ribosomes.

posttranslational sorting refers to protein sorting to an organelle that occurs after the protein has been completely synthesized in the cytosol.

postzygotic isolating mechanism a mechanism of reproductive isolation that prevents an offspring from being viable or fertile.

pre-initiation complex the stage in which the assembly of RNA polymerase and general transcription factors occurs at the core promoter, but the DNA has not yet started to unwind.

pre-mRNA in eukaryotes, the transcription of structural genes produces a long transcript known as pre-mRNA, which is located within the nucleus. This pre-mRNA is usually altered by splicing and other modifications before it exits the nucleus.

prezygotic isolating mechanism a mechanism for reproductive isolation that prevents the formation of a zygote.

Pribnow box the TATAAT sequence that is often found at the −10 region of bacterial promoters.

primase an enzyme that synthesizes a short RNA primer during DNA replication.

primosome a multiprotein complex composed of DNA helicase, primase, and several accessory proteins.

prion an infectious particle that causes several types of neurodegenerative diseases affecting humans and livestock. It is composed entirely of protein.

probability the chance that an event will occur in the future.

processive enzyme enzymes, such as RNA and DNA polymerase, which glide along the DNA, and do not dissociate from the template strand as they catalyze the covalent attachment of nucleotides.

product rule the probability that two or more independent events will occur is equal to the products of their individual probabilities.

proflavin a type of chemical mutagen that causes frameshift mtations.

prokaryotes (Greek, "prenucleus") another name for bacteria and archae. The term refers to the fact that their chromosomes are not contained within a separate nucleus of the cell.

prometaphase the second phase of M phase. During this phase, the nuclear membrane vesiculates, and the mitotic spindle is completely formed.

promoter a sequence within a gene that initiates (i.e., promotes) transcription.

promoter bashing the approach of making deletions in the vicinity of a promoter as a way to identify the core promoter and response elements.

proofreading the ability of DNA polymerase to remove mismatched bases from a newly made strand.

prophage phage DNA that has been integrated into the bacterial chromosome.

prophase the first phase of M phase. The chromosomes have already replicated and begin to condense. The mitotic spindle starts to form.

protease an enzyme that digests the polypeptide backbone found in proteins.

protein a functional unit composed of one or more polypeptides.

protein microarray a small silica, glass, or plastic slide that is dotted with many different proteins.

proteome the collection of all proteins that a given species can make.

proteomics the study of protein function at the genome level. It involves the study of many proteins simultaneously.

proto-oncogene a normal cellular gene that does not cause cancer, but which may incur a mutation or become incorporated into a viral genome and thereby lead to cancer.

protoplast a plant cell without a cell wall.

provirus viral DNA that has been incorporated into the chromosome of a host cell.

proximo-distal axis in animals, an axis involving limbs in which the part of the limb attached to the trunk is located proximal, while the end of the limb is located distal.

pseudoautosomal inheritance the inheritance pattern of genes that are found on both the X and Y chromosomes. Even though such genes are located physically on the sex chromosomes, their pattern of inheritance is identical to that of autosomal genes.

pseudodominance a pattern of inheritance that occurs when a single copy of a recessive allele is phenotypically expressed because the second copy of the gene has been deleted from the homologous chromosome.

pseudo-overdominance occurs when two closely linked genes exhibit heterosis. The heterosis is really due to the masking of recessive alleles.

P site see *peptidyl site.*

pulsed-field gel electrophoresis (PFGE) a method of gel electrophoresis used to separate small chromosomes or very large pieces of chromosomes.

punctuated equilibrium an evolutionary theory proposing that species exist relatively unchanged for many generations. These long periods of equilibrium are punctuated by relatively short periods during which evolution occurs at a relatively rapid rate.

Punnett square a diagrammatic method in which the gametes that two parents can produce are aligned next to a square grid as a way to predict the types of offspring the parents will produce and in what proportions.

purine a type of nitrogenous base that has a double-ring structure. Examples are adenine and guanine.

P value (e.g., in a chi square table) the probability that the deviations between observed and expected values are a matter of random chance alone.

pyrimidine a type of nitrogenous base that has a single-ring structure. Examples are cytosine, thymine, and uracil.

Q

q an abbreviation for the long arm of a chromosome.

QTL see *quantitative trait loci.*

QTL mapping the determination of the location of QTLs using mapping methods such as genetic crosses coupled with the analysis of molecular markers.

quantitative genetics the area of genetics concerned with traits that can be described in a quantitative way.

quantitative trait a trait, usually polygenic in nature, that can be described with numbers.

quantitative trait loci (QTLs) the locations on chromosomes where the genes that influence quantitative traits reside.

R

RAG1/RAG2 enzymes that recognize recombination signal sequences and make double-stranded cuts. In the case of V/J recombination in immunoglobulin genes, a cut is made at the end of one V region and the beginning of one J region.

random genetic drift see *genetic drift.*

random sampling error the deviation between the observed and expected outcomes due to chance.

realizator gene a developmental gene that plays a role in promoting the morphological characteristics of a body segment.

realized heritability a form of narrow sense heritability that is observed when selective breeding is practiced.

recessive describes a trait or gene that is masked by the presence of a dominant trait or gene.

reciprocal crosses a pair of crosses in which the traits of the two parents differ with regard to sex. For example, one cross could be a red-eyed female fly and a white-eyed male fly and the reciprocal cross would be a red-eyed male fly and a white-eyed female fly.

reciprocal translocation when two different chromosomes exchange pieces.

recombinant (1) refers to combinations of alleles or traits that are not found in the parental generation; (2) describes DNA molecules that are produced by molecular techniques in which segments of DNA are joined to each other in ways that differ from their original arrangement in their native chromosomal sites. The cloning of DNA into vectors is an example.

recombinant DNA technology the use of in vitro molecular techniques to isolate and manipulate different pieces of DNA.

recombination see *genetic recombination.*

recombinational repair DNA repair via genetic recombination.

recombination signal sequence a specific DNA sequence that is involved in site-specific recombination. Such sequences are found in immunoglobulin genes.

redundancy see *gene redundancy.*

regression analysis the approach of using correlated data to predict the outcome of one variable when given the value of a second variable.

regulatory sequence or element a sequence of DNA (or possibly RNA) that binds a regulatory protein and thereby influences gene expression. Bacterial operator sites and eukaryotic enhancers and silencers are examples.

regulatory transcription factor a protein or protein complex that binds to a regulatory element and influences the rate of transcription via RNA polymerase.

relaxosome a protein complex that recognizes the origin of transfer in F factors and other conjugative plasmids, cuts one DNA strand, and aids in the transfer of the T DNA.

release factor a protein that recognizes a stop codon and promotes translational termination and the release of the completed polypeptide.

repetitive sequences DNA sequences that are present in many copies in the genome.

replica plating a technique in which a replica of bacterial colonies is transferred to a new petri plate.

replication see *DNA replication.*

replication fork the region in which two DNA strands have separated and new strands are being synthesized.

replicative transposition a form of transposition in which the end result is that the transposable element remains at its original site, and also is found at a new site.

replisome a putative complex that contains a primosome and DNA polymerase.

repressor a regulatory protein that binds to DNA and inhibits transcription.

resolution the last stage of homologous recombination, in which the entangled DNA strands become resolved into two separate structures.

resolvase an endonuclease that makes the final cuts in DNA during the resolution phase of recombination.

response element see *regulatory sequence or element.*

restriction endonuclease (or **restriction enzyme**) an endonuclease that cleaves DNA. The restriction enzymes used in cloning experiments bind to specific base sequences and then cleave the DNA backbone at two defined locations, one in each strand.

restriction fragment length polymorphism (RFLP) genetic variation within a population in the lengths of DNA fragments that are produced when chromosomes are digested with particular restriction enzymes.

restriction mapping a technique to determine the locations of restriction endonuclease sites within a segment of DNA.

restriction point a point in the G_1 phase of the cell cycle that causes a cell to progress to cell division.

retroelement a type of transposable element that moves via an RNA intermediate.

retroposon see *retroelement.*

retrospective testing a procedure in which a researcher formulates a hypothesis and then collects observations as a way to confirm or refute the hypothesis. This approach relies on the observation and testing of already existing materials, which themselves are the result of past events.

retrotransposon see *retroelement.*

reverse genetics an experimental strategy in which researchers first identify the wild-type gene using cloning methods. The next step is to make a mutant version of the wild-type gene, introduce it into an organism, and see how the mutant gene affects the phenotype of the organism.

reverse mutation see *reversion.*

reverse transcriptase an enzyme that uses an RNA template to make a complementary strand of DNA.

reverse transcriptase PCR (RT-PCR) a modification of PCR in which the first round of replication involves the use of RNA and reverse transcriptase to make a complementary strand of DNA.

reversion   a mutation that returns a mutant allele back to the wild-type allele.

R factor   a type of plasmid found commonly in bacteria that confers resistance to a toxic substance such as an antibiotic.

RFLP   see *restriction fragment length polymorphism*.

RFLP mapping   the mapping of a gene or other genetic sequence relative to the known locations of RFLPs within a genome.

R group   the side chain of an amino acid.

rho (ρ)   a protein that is involved in transcriptional termination for certain bacterial genes.

rho-dependent termination   transcriptional termination that requires the function of the rho protein.

rho-independent termination   transcription termination that does not require the rho protein. It is also known as intrinsic termination.

ribonucleic acid (RNA)   a nucleic acid that is composed of ribonucleotides. In living cells, RNA is synthesized via the transcription of DNA.

ribose   the sugar found in RNA.

ribosomal binding site   a sequence in bacterial mRNA that is needed to bind to the ribosome and initiate translation.

ribosome   a large macromolecular structure that acts as the catalytic site for polypeptide synthesis. The ribosome allows the mRNA and tRNAs to be positioned correctly as the polypeptide is made.

ribozyme   an RNA molecule with enzymatic activity.

R loop   experimentally, a DNA loop that is formed because RNA is displacing it from its complementary DNA strand.

RNA   see *ribonucleic acid*.

RNA editing   the process in which a change occurs in the nucleotide sequence of an RNA molecule that involves additions or deletions of particular bases, or a conversion of one type of base to a different type.

RNA-induced silencing complex (RISC)   the complex that mediates RNA interference.

RNA interference   the phenomenon that double-stranded RNA targets complementary RNAs within the cell for degradation.

RNA polymerase   an enzyme that synthesizes a strand of RNA using a DNA strand as a template.

RNA primer   a short strand of RNA, made by DNA primase, that is used to elongate a strand of DNA during DNA replication.

RNase   an enzyme that cuts the sugar-phosphate backbone in RNA.

RNaseP   a bacterial enzyme that is an endonuclease and cuts precursor tRNA molecules. RNaseP is a ribozyme, which means that its catalytic ability is due to the action of RNA.

RNA splicing   the process in which pieces of RNA are removed and the remaining pieces are covalently attached to each other.

Robertsonian translocation   the structure that is produced when two telocentric chromosomes fuse at their short arms.

RT-PCR   see *reverse transcriptase PCR*.

S

satellite DNA   in a density centrifugation experiment, a peak of DNA that is separated from the majority of the chromosomal DNA. It is usually composed of highly repetitive sequences.

scaffold-attachment region (SAR)   a site in the chromosomal DNA that is anchored to the nuclear matrix or scaffold.

science   a way of knowing about our natural world. The science of genetics allows us to understand how the expression of genes produces the traits of an organism.

scientific method   a basis for conducting science. It is a process that scientists typically follow so that they may reach verifiable conclusions about the world in which they live.

segmentation gene   in animals, a gene, whose encoded product is involved in the development of body segments.

segment polarity gene   one category of segmentation genes.

segments   anatomical subdivisions that occur during the development of species such as *Drosophila*

segregate   when two things are kept in separate locations. For example, homologous chromosomes segregate into different gametes.

selectable marker   a gene that provides a selectable phenotype in a cloning experiment. Many selectable markers are genes that confer antibiotic resistance.

selection coefficient   1 minus the fitness value.

selectionists   scientists who oppose the neutral theory of evolution.

selective breeding   refers to programs and procedures designed to modify the phenotypes in economically important species of plants and animals.

self-fertilization   fertilization that involves the union of male and female gametes derived from the same parent.

selfish DNA theory   the idea that transposable elements exist because they possess characteristics that allow them to multiply within the host cell DNA and inhabit the host without offering any selective advantage.

self splicing   refers to RNA molecules that can remove their own introns without the aid of other proteins or RNA.

semiconservative replication   refers to the net result of DNA replication in which the DNA contains one original strand and one newly made strand.

semilethal alleles   lethal alleles that kill some individuals but not all.

semisterility   when an individual has a lowered fertility.

sense strand   the strand of DNA within a structural gene that has the same sequence as mRNA except that T is found in the DNA instead of U.

sequence element   in genetics, a sequence with a specialized function.

sequence-tagged site (STS)   a short segment of DNA, usually between 100 and 400 base pairs long, the base sequence of which is found to be unique within the entire genome. Sequence-tagged sites are identified by PCR.

sequencing   see *DNA sequencing*.

sequencing ladder   a series of bands on a gel that can be followed in order (e.g., from the bottom of the gel to the top of the gel) to determine the sequence of DNA.

sex chromosomes   a pair of chromosomes (e.g., X and Y in mammals) that determines sex in a species.

sex-influenced inheritance   an inheritance pattern in which an allele is dominant in one sex but recessive in the opposite sex. In humans, pattern baldness is an example of a sex-influenced trait.

sex-limited traits   traits that occur in only one of the two sexes.

sex linkage   the phenomenon that certain genes are found on one of the two types of sex chromosomes but not both.

sex pilus (pl. pili)   a structure on the surface of bacterial cells that acts as an attachment site to promote the binding of bacteria to each other. The sex pilus provides a passageway for the movement of DNA during conjugation.

sexual reproduction   the process whereby parents make gametes (i.e., sperm and egg) that fuse with each other in the process of fertilization to begin the life of a new organism.

sexual selection   natural selection that acts to promote characteristics that give individuals a greater chance of reproducing.

Shine–Dalgarno sequence   a sequence in bacterial mRNAs that functions as a ribosomal binding site.

short tandem repeat sequences (STRs)   a general name that can refer to microsatellites or minisatellites.

shuttle vector   a cloning vector that can propagate in two or more different species, such as *E. coli* and yeast.

side chain   in an amino acid, the chemical structure that is attached to the carbon atom (i.e., the α carbon) that is located between the amino group and carboxyl group.

sigma factor   a transcription factor that recognizes bacterial promoter sequences and facilitates the binding of RNA polymerase to the promoter.

silencer   a DNA sequence that functions as a regulatory element. The binding of a regulatory transcription factor to the silencer decreases the level of transcription.

silent mutation   a mutation that does not alter the amino acid sequence of the encoded polypeptide even though the nucleotide sequence has changed.

simple translocation   when one piece of a chromosome becomes attached to a different chromosome.

SINEs   in mammals, short interspersed elements that are less than 500 base pairs in length.

single-factor cross   see *monohybrid cross*.

single-strand binding protein   a protein that binds to both of the single strands of DNA during DNA replication and prevents them from re-forming a double helix.

sister chromatid exchange (SCE)   the phenomenon in which crossing over occurs between sister chromatids, thereby exchanging identical genetic material.

sister chromatids   pairs of replicated chromosomes that are attached to each other at the centromere. Sister chromatids are genetically identical to each other.

site-directed mutagenesis   a technique that enables scientists to change the sequence of cloned DNA segments.

site-specific recombination   when two different DNA segments break and rejoin with each other. This occurs during the integration of certain viruses into the host chromosome, and during the rearrangement of immunoglobulin genes.

snRNP   refers to a complex containing small nuclear RNA and a set of proteins, which are components of the spliceosome.

somatic cell   refers to any cell of the body that is not a gamete.

somatic mutation   a mutation in a somatic cell.

sorting signal   an amino acid sequence or posttranslational modification that directs a protein to the correct region of the cell.

SOS response   a response to extreme environmental stress in which bacteria replicate their DNA using DNA polymerases that are likely to make mistakes.

Southern blotting   a technique used to detect the presence of a particular genetic sequence within a mixture of many chromosomal DNA fragments.

species   see *biological species concept* and *species recognition concept*.

species recognition concept   similar to the biological species concept, but emphasizes that the sexual recognition of members of the same species acts as a positive force in evolution.

spermatids   immature sperm cells produced from spermatogenesis.

spermatogenesis   the production of sperm cells.

sperm cell   a male gamete. Sperm are small and usually travel relatively far distances to reach the female gamete.

spindle   see *mitotic spindle apparatus*.

spliceosome   a multisubunit complex that functions in the splicing of eukaryotic pre-mRNA.

splicing   see *RNA splicing*.

splicing factor   a protein that regulates the process of RNA splicing.

spontaneous mutation   a change in DNA structure that results from random abnormalities in biological processes.

spores   haploid cells that are produced by certain species such as fungi (i.e., yeast and molds).

sporophyte   the diploid generation of plants.

SR protein   a type of splicing factor.

stabilizing selection   natural selection that favors individuals with an intermediate phenotype. This results in a balanced polymorphism.

stamen   the structure found in the flower of higher plants that produces the male gametophyte (i.e., pollen).

standard deviation   a statistic that is computed as the square root of the variance.

start codon   a three-base sequence in mRNA that initiates translation. It is usually 5′-AUG-3′, and encodes methionine.

stem cell   a cell that has the capacity to divide and to differentiate into one or more specific cell types.

steroid receptor   a category of transcription factors that respond to steroid hormones. An example is the glucocorticoid receptor.

stop codon   a three-base sequence in mRNA that signals the end of translation of a polypeptide. The three stop codons are 5′-UAA-3′, 5′-UAG-3′, and 5′-UGA-3′.

strand   in DNA or RNA, nucleotides covalently linked together to form a long, linear polymer.

stripe-specific enhancer   in *Drosophila*, a regulatory region that controls the expression of a gene so that it occurs only in a particular parasegment during early embryonic development.

STRs   see *short tandem repeat sequences*.

structural gene   a gene that encodes the amino acid sequence within a particular polypeptide or protein.

STS   see *sequence-tagged site*.

subcloning   the procedure of making smaller DNA clones from a larger one.

submetacentric   describes a chromosome in which the centromere is slightly off center.

subpopulation   a segment of a population that is slightly isolated. Members of a subpopulation are more likely to breed with each other than with members that are outside of the subpopulation.

subtractive cDNA library   a cDNA library that contains cDNA inserts derived from mRNA that is expressed only under certain conditions.

subtractive hybridization   a method used to create a subtractive cDNA library.

subunit   this term may have multiple meanings. In a protein, each subunit is a single polypeptide.

sum rule   the probability that one of two or more mutually exclusive events will occur is equal to the sum of their individual probabilities.

supercoiling   see *DNA supercoiling*.

suppressor (or **suppressor mutation**)   a mutation at a second site that suppresses the phenotypic effects of another mutation.

SWI/SNF family   a group of related proteins that catalyze chromatin remodeling.

sympatric speciation (Greek, *sym*, "together"; Latin, *patria*, "homeland")   a form of speciation that occurs when members of a species diverge while occupying the same habitat within the same range.

synapsis   the event in which homologous chromosomes recognize each other and then align themselves along their entire lengths.

synaptonemal complex   a complex of proteins that promote the interconnection between homologous chromosomes during meiosis.

synteny group   a group of genes that are found in the same order on the chromosomes of different species.

T

T   an abbreviation for thymine.

tandem array (or **tandem repeat**)   a short nucleotide sequence that is repeated many times in a row.

tandem mass spectrometry   the sequential use of two mass spectrometers. As described in chapter 21, it can be used to determine the sequence of amino acids in a polypeptide.

taq polymerase   a thermostable form of DNA polymerase used in PCR experiments.

TATA box   a sequence found within eukaryotic core promoters that determines the starting site for transcription. The TATA box is recognized by a TATA-binding protein, which is a component of TFIID.

tautomeric shift   a change in chemical structure such as an alternation between the keto- and enol-forms of the bases that are found in DNA.

T DNA   a segment of DNA found within a Ti plasmid that is transferred from a bacterium to infected plant cells. The T DNA from the Ti plasmid becomes integrated into the chromosomal DNA of the plant cell by recombination.

TE   see *transposable element*.

telocentric   describes a chromosome with its centromere at one end.

telomerase   the enzyme that recognizes telomeric sequences at the ends of eukaryotic chromosomes and synthesizes additional numbers of telomeric repeat sequences.

telomeres   specialized DNA sequences found at the ends of eukaryotic, linear chromosomes.

telophase   the fifth phase of M phase. The chromosomes have reached their respective poles and decondense.

temperate phage   a bacteriophage that usually exists in the lysogenic life cycle.

template strand   a DNA strand that is used for the synthesis of a new DNA or RNA strand.

terminal deficiency   when a segment is lost from the end of a linear chromosome.

termination codon   see *stop codon*.

terminator   a sequence within a gene that signals the end of transcription.

testcross   an experimental cross between a recessive individual and an individual whose genotype the experimenter wishes to determine.

tetrad   (1) the association among four sister chromatids during meiosis; (2) a group of four fungal spores contained within an ascus.

tetraploid   having four sets of chromosomes (i.e., 4*n*).

TFIID   a type of general transcription factor in eukaryotes that is needed for RNA polymerase II function. It binds to the TATA box, and recruits RNA polymerase II to the core promoter.

thermocycler   a device that automates the timing of temperature changes in each cycle of a PCR experiment.

thymine   a pyrimidine base found in DNA. It base pairs with adenine in DNA.

thymine dimer   a mutation involving a covalent linkage between two adjacent thymine bases in a DNA strand.

Ti plasmid   a tumor-inducing plasmid found in *Agrobacterium tumefaciens*. It is responsible for inducing tumor formation after a plant has been infected.

tissue-specific gene   a gene that is highly regulated and is expressed in a particular cell type.

topoisomerase   an enzyme that alters the degree of supercoiling in DNA.

topoisomers   DNA conformations that differ with regard to supercoiling.

totipotent   a cell that possesses the genetic potential to produce an entire individual. A somatic plant cell or a fertilized egg is totipotent.

traffic signal   see *sorting signal*.

trait   any characteristic that an organism displays. Morphological traits affect the appearance of an organism. Physiological traits affect the ability of an organism to function. A third category of traits are those that affect an organism's behavior (behavioral traits).

trans-acting factor   a regulatory protein that binds to a regulatory element in the DNA and exerts a *trans* effect.

transcription   the process of synthesizing RNA from a DNA template.

transcriptional-repair coupling factor (TRCF)   a protein that recognizes when RNA polymerase is stalled over a damaged region of DNA, and recruits DNA repair enzymes to fix the damaged site.

transcription factors   a broad category of proteins that influence the ability of RNA polymerase to transcribe DNA into RNA.

transduction   a form of genetic transfer between bacterial cells in which a bacteriophage transfers bacterial DNA from one bacterium to another.

trans-effect   an effect on gene expression that occurs even though two DNA segments are not physically adjacent to each other. *Trans*-effects are mediated through diffusible genetic regulatory proteins.

transfection   when a viral vector is introduced into a host cell.

transfer RNA (tRNA)   a type of RNA used in translation that carries an amino acid. The anticodon in tRNA is complementary to a codon in the mRNA.

transformation   (1) when a plasmid vector or segment of chromosomal DNA is introduced into a host cell; (2) when a normal cell is converted into a malignant cell.

transgenic   an organism that has DNA from another organism incorporated into its genome via recombinant DNA techniques.

transition   a point mutation involving a change of a pyrimidine to another pyrimidine (e.g., C to T), or a purine to another purine (e.g., A to G).

translation   the synthesis of a polypeptide using the codon information within mRNA.

translocation   (1) when one segment of a chromosome breaks off and becomes attached to a different chromosome; (2) when a ribosome moves from one codon in an mRNA to the next codon.

translocation cross   the structure that is formed when the chromosomes of a reciprocal translocation attempt to synapse during meiosis. This structure contains two normal (nontranslocated chromosomes) and two translocated chromosomes. A total of eight chromatids are found within the cross.

transposable element (TE)   a small genetic element that can move to multiple locations within the host's chromosomal DNA.

transposase   the enzyme that catalyzes the transposition of transposable elements.

transposition   the phenomenon of transposon movement.

transposon   see *transposable element.*

transposon tagging   a technique for cloning genes in which a transposon inserts into a gene and inactivates it. The transposon-tagged gene is then cloned using a complementary transposon as a probe to identify the gene.

transversion   a point mutation in which a purine is interchanged with a pyrimidine or vice versa.

trihybrid cross   a cross in which an experimenter follows the outcome of three different traits.

trinucleotide repeat expansion (TNRE)   a type of mutation that involves an increase in the number of tandemly repeated trinucleotide sequences.

triplex DNA   a double-stranded DNA that has a third strand wound around it to form a triple-stranded structure.

triploid   an organism or cell that contains three sets of chromosomes.

trisomic   a diploid cell with one extra chromosome (i.e., $2n + 1$).

tRNA   see *transfer RNA.*

true-breeding line   a strain of a particular species that continues to produce the same trait after several generations of self-fertilization (in plants) or inbreeding.

tumor-suppressor gene   a gene that functions to inhibit cancerous growth.

two-dimensional gel electrophoresis   a technique to separate proteins that involves isoelectric focusing in the first dimension, and SDS-gel electrophoresis in the second dimension.

two-factor cross   see *dihybrid cross.*

U

U   an abbreviation for uracil.

unipotent   a type of stem cell that can differentiate into only a single type of cell.

universal   in genetics, this terms refers to the phenomenon that nearly all organisms use the same genetic code with just a few exceptions.

UPGMA method (unweighted pair group method with arithmetic mean)   a relatively simple method to construct a phylogenetic tree. This method assumes that after two species have diverged from each other, the rate of accumulation of mutations in a homologous region of DNA is approximately the same in the two species.

up promoter mutation   a mutation in a promoter that increases the rate of transcription.

up regulation   genetic regulation that leads to an increase in gene expression.

uracil   a pyrimidine base found in RNA.

UTR   an abbreviation for the untranslated region of mRNA.

U-tube   a U-shaped tube that has a filter at bottom of the U. The pore size of the filter allows the passage of small molecules (e.g., DNA molecules) from one side of the tube to the other, but restricts the passage of bacterial cells.

V

variance   the sum of the squared deviations from the mean divided by the degrees of freedom.

variants   individuals of the same species who exhibit different traits. An example is tall and dwarf pea plants.

V(D)J recombination   site-specific recombination that occurs within immnoglobulin genes.

vector   a small segment of DNA that is used as a carrier of another segment of DNA. Vectors are used in DNA cloning experiments.

vertical evolution   refers to the phenomenon that species evolve from pre-existing species by the accumulation of gene mutations, and by changes in chromosome structure and number. Vertical evolution involves genetic changes in a series of ancestors that form a lineage.

vertical gene transfer   the transfer of genetic material from parents to offspring, or from mother cell to daughter cell.

viral-like retroelements   retroelements that are evolutionarily related to known retroviruses. These transposable elements have retained the ability to move around the genome, though, in most cases, mature viral particles are not produced.

viral vector   a vector used in gene cloning that is derived from a naturally occurring virus.

virulent phage   a phage that follows the lytic cycle.

virus   a small infectious particle that contains nucleic acid as its genetic material, surrounded by a capsid of proteins. Some viruses also have an envelope consisting of a membrane embedded with spike proteins.

VNTRs   segments of DNA that are located in several places in a genome and have a variable number of tandem repeats. The pattern of VNTRs is often used in DNA fingerprinting.

W

Western blotting   a technique used to detect a specific protein among a mixture of proteins.

wild-type allele   the allele that is fairly prevalent in a natural population, generally greater than 1% of the population. For polymorphic genes, there may be more than one wild-type allele.

wobble base   the third base in an anticodon. This term suggests that the third base in the anticodon can wobble a bit to recognize more than one type of base in the mRNA.

X

X chromosomal controlling element (Xce)   a region adjacent to Xic that influences the choice of the active X chromosome during the process of X inactivation.

X inactivation   a process in which mammals equalize the expression of X-linked genes by randomly turning off one X chromosome in the somatic cells of females.

X-inactivation center (Xic)   a site on the X chromosome that appears to play a critical role in X inactivation.

X-linked genes (alleles)   genes (or alleles of genes) that are physically located within the X chromosome.

Y

YAC   see *yeast artificial chromosome.*

yeast artificial chromosome (YAC)   a cloning vector propagated in yeast that can reliably contain very large insert fragments of DNA.

Y-linked genes (alleles)   genes (or alleles of genes) that are only located on the Y chromosome.

Z

zygote   a cell formed from the union of a sperm and egg.

zygotic gene   a gene that is expressed after fertilization.

CREDITS

::

Photographs

Chapter 1
Opener: © Eye of Science/Photo Researchers, Inc.; Figure 1.2: © Roslin Institute; Figure 1.3: Advanced Cell Technology, Inc., Worcester, Massachusetts; Figure 1.4: © Biophoto Associates/Photo Researchers; Figure 1.5: © CNRI / SCIENCE PHOTO LIBRARY/Photo Researchers, Inc.; Figure 1.8: © Edmund D. Brodie III; Figure 1.9a: © Joseph Sohm; ChromoSohm Inc./CORBIS; Figure 1.9b: © Corbis/R-F Website; Figure 1.10: © March of Dimes Birth Defects Foundation

Chapter 2
Opener: © H. Reinhard/Photo Researchers, Inc.; Figure 2.1: © SPL/Photo Researchers, Inc.; Figure 2.2b: © Nigel Cattlin/Photo Researchers, Inc.

Chapter 3
Opener: © Ed Reschke; Figure 3.2 (bottom): © Leonard Lessin/Peter Arnold; Figure 3.2 (top): © Burger/Photo Researchers, Inc.; Figure 3.6a: © Biophoto Associates/Photo Researchers, Inc.; Figure 3.7 (all): © Carolina Biological Supply Company /Phototake; Figure 3.9 (all): © Ed Reschke; Figure 3.11a: © Diter von Wettstein

Chapter 4
Opener: © Robert Calentine/Visuals Unlimited; Figure 4.4a: © Zig Leszcynski/Animals, Animals; Figure 4.4b: © Jose Luis G. Grande/Photo Researchers, Inc./; Figure 4.4c: © Jane Burton/Bruce Coleman; Figure 4.4d: © Lynn M. Stone/Bruce Coleman; Figure 4.5a: © Alan & Sandy Carey/Photo Researchers; Figure 4.5b: © Hans Reinhard/Bruce Coleman; Figure 4.6: © Lynn Stone/Natural History Photography; Figure 4.7: © James A. Birchler, Professor of Biological Sciences, University of Missouri ; Figure 4.9: © Stan Flegler/Visuals Unlimited; Figure 4.12: © Bob Shanley/The Palm Beach Post; Figure 4.14a-c: Adams National Historical Park; Figure 4.14d: Courtesy of the Massachusetts Historical Society; Figure 4.16 both: © Robert Maier/Animals, Animals

Chapter 5
Opener: © Carolina Biological/Visuals Unlimited

Chapter 6
Opener: © David Scharf/Peter Arnold; Figure 6.4b: © Dr. L. Caro/Science Photo Library/Photo Researchers, Inc.; Figure 6.15a&b: © Carolina Biological Supply/Phototake

Chapter 7
Opener: © John Mendenhall, Institute for Cellular and Molecular Biology, University of Texas at Austin; Figure 7.3a: © Carolina Biological/Visuals Unlimited; Figure 7.3b: © Tim Davis /Photo Researchers, Inc.; Figure 7.6(1): © G.W. Willis, MD / Visuals Unlimited; Figure 7.6(2): Ronald G. Davidson, Harold M. Nitowsky, and Barton Childs. Demonstration of Two Populations of Cells in the Human Female Heterozygous for Glucose-6-Phosphate Dehydrogenase Variants. *PNAS* 50 (1963) f. 2, p. 484. © National Academy of Sciences, USA; Figure 7.9: © Andrzej BartkePh.D./Southern Illinois Univ. School of Medicine; Figure 7.13a: Kuroiwa, T., Kawano, S., & Hizume, M. Studies on mitochondrial structure and function in Physarum polycephalum. V. Behavior of mitochondrial nucleoids throughout mitochondrial division cycle. *J Cell Biol.* 1977 Mar;72(3):687-94. Fig. 5c; Figure 7.13b: Gibbs, SP., Mak, R., Ng, R., & Slankis, T. The chloroplast nucleoid in Ochromonas danica. II. Evidence for an increase in plastid DNA during greening. *J Cell Sci.* 1974 Dec;16(3):579-91. Fig. 1

Chapter 8
Opener: © BSIP/Phototake; Figure 8.1a (left): © Scott Camazine /Photo Researchers, Inc.; Figure 8.1a (middle): © Byron Williams,/Dept. of Molecular Biology and Genetics, Cornell University; Figure 8.1a (right): © Carlos R Carvalho/Universidade Federal de Viçosa; Figure 8.1c: © C.N.R.I./Phototake; Figure 8.4a: © Biophoto Associates /Science Source/Photo Researchers, Inc.;

Figure 8.4b: © Jeff Noonley, CdC; Figure 8.14a: © Paul J. Benke/University of Miami School of Medicine; Figure 8.14b: © Will Hart/PhotoEdit; Figure 8.20a&b: © A. B. Sheldon; Figure 8.21b: © David M. Phillips/Visuals Unlimited; Figure 8.22a: © James Steinberg/Photo Researchers, Inc.; Figure 8.22b (both): © Biophoto Associates/Science Source/Photo Researchers, Inc.; Figure 8.28: Robinson, T.J. & Harley, E.H. Absence of geographic chromosomal variation in the roan and sable antelope and the cytogenetics of a naturally occurring hybrid. *Cytogenet Cell Genet.* 1995;71(4):363-9

Chapter 9
Opener: © Ken Eward /Photo Researchers, Inc.; Figure 9.4: © Omikron/Photo Researchers, Inc.; Figure 9.12b: Pictorial Parade/Getty Images; Figure 9.13a: From "The Double Helix" by James D. Watson, 1968, Atheneum Press, NY. © Cold Spring Harbor Laboratory Archives; Figure 9.16a: © Barrington Brown/Photo Researchers, Inc.; Figure 9.16b: © Hulton|Archive by Getty Images; Figure 9.18b: © Michael Freeman/Phototake; Figure 9.24: Holbrook, S.R., Sussman, J.L., Warrant, R.W. & Kim, S.H. Crystal structure of yeast phenylalanine transfer RNA. II. Structural features and functional implications. *J Mol Biol.*, 1978 Aug 25;123(4):631-60. Fig. 2

Chapter 10
Opener: © Alfred Pasieka/Photo Researchers, Inc.; Figure 10.3: Carl Robinow & Eduard Kellengerger. The Bacterial Nucleoid Revisited. Microbiological Reviews, 1994 Jun;58(2):211-32. Fig. 4 © American Society of Microbiology; Figure 10.10b: © Simpson's Nature Photography; Figure 10.10c: © William Leonard; Figure 10.15: Markus Noll. Subunit structure of chromatin. *Nature.* 1974 Sep 20;251(5472):249-51. Fig. 2 © Macmillan Magazine Ltd.; Figure 10.16a&b: Thoma, F. & Koller, T. Influence of histone H1 on chromatin structure. *Cell* 1977 Sep;12(1) f. 2a&c, p.103. Reprinted by permission of Elsevier Science; Figure 10.17a: Photo courtesy of Dr. Barbara Hamkalo; Figure 10.18b&c: Nickerson et al. 1997. The nuclear matrix revealed by eluting chromatin from a cross-linked nucleus. *PNAS* 94: 4446-4450. Fig. 2a&b; Figure 10.19a&b: Cremer, T. & Cremer, C. Chromosome territories, nuclear architecture and gene regulation in mammalian cells. *Nature Reviews/Genetics* Vol. 2 no. 4, f. 2b&d, p. 295; Figure 10.21(1): © Dr. Gopal Murit/Visuals Unlimited; Figure 10.21(2): © Olins and Olins/Biological Photo Service; Figure 10.21(3): Photo courtesy of Dr. Barbara Hamkalo; Figure 10.21(4): Paulson, JR. & Laemmli, UK. The structure of histone-depleted metaphase chromosomes. *Cell.* 1977 Nov;12(3):817-28, f. 5; Figure 10.21(5&6), Figure 10.22a: © Peter Engelhardt/Department of Virology, Haartman Institute; Figure 10.22b: © Dr. Donald Fawcett /Visuals Unlimited

Chapter 11
Opener: © Clive Freeman, The Royal Institution/Photo Researchers, Inc.; Figure 11.3: Meselson M, Stahl, F. The Replication of DNA in Escherichia Coli. *PNAS* Vol. 44, 1958. f. 4a, p. 673; Figure 11.4: From Cold Spring Harbor Symposia of Quantitative Biology, 28, p. 43 (1963); Figure 11.8b: From Li et al. Molecular Structure of DNA polymerase I from Thermau aquaticus. (1998) *The EMBO Journal* 17, pg. 7516; Figure 11.19: Huberman, J.A., & Riggs, A. On the Mechanism of DNA Replication in Mammalian Chromosomes. *Journal of Molecular Biology* (1968) 32, Plate Ia, p. 335; Figure 11.20b: Kriegstein, H.J. & Hogness, D.S. Mechanism of DNA replication in Drosophila chromosomes: structure of replication forks and evidence for bidirectionality. *PNAS* 1974 Vol 71. f. 2 p. 137

Chapter 12
Opener: Image Courtesy of David S. Goodsell, the Scripps Research Institute; Figure 12.10a: From Seth Darst, Bacterial RNA polymerase. *Current Opinion in Structural Biology* Volume 11, Issue 2, 1 April 2001, Pages 155-162; Figure 12.10b: From Patrick Cramer, David A. Bushnell, Roger D. Kornberg. Structural Basis of Transcription: RNA Polymerase II at 2.8 Ångstrom Resolution. *Science*, Vol. 292, Issue 5523, 1863-1876, June 8, 2001. Figure 1 © AAAS; Figure 12.17: Tilghman et al. Intervening sequence of DNA identified in the structural portion of a mouse ß-glovin gene. *PNAS* 1978, Vol 75. f. 2, p. 727

Chapter 13

Opener: Image Courtesy of David S. Goodsell, the Scripps Research Institute; Figure 13.14a: © E. Kiseleva and Donald Fawcett/Visuals Unlimited; Figure 13.14b: Ramakrishnan, V. and Moore, P.B., The Ribosome in 2000: Atomic Structures at Last. *Curr. Opin. Struct. Biol.*, 11, f. 1, p. 146, 2001. © 2001 Elsevier Science; Figure 13.21: Miller, O. L. (1973) *Scientific American* Vol. 228 (3) p.35

Chapter 14

Opener: Image Courtesy of David S. Goodsell, the Scripps Research Institute; Figure 14.10B: Lewis et al. Crystal Structure of the Lactose Operon Repressor and Its Complexes with DNA and Inducer. *Science*. 1996 Mar 1; Vol. 271 Cover, © AAAS

Chapter 15

Opener: Structure of zif268 zinc fingers bound to DNA (Pavletvich and Pabo, *Science*, 252:809-817, 199 1©AAAS). Molecular graphics prepared by Song Tan, Dept of Biochemistry & Molecular Biology, Penn State using MidasPlus software (Ferrin et al, J. *Molec. Graphics* 6:13-17, 1988); Figure 15.8 (left): Courtesy of Dr. Joseph G. Gall; Figure 15.8 (right): Courtesy Dr. Oscar L. Miller

Chapter 16

Opener: © Robert Brooker; Figure 16.1(both): © Biological Photo Service; Figure 16.3a&b: Aulner et al. The AT-Hook Protein D1 Is Essential for Drosophila melanogaster Development and Is Implicated in Position-Effect Variegation. *Molecular and Cellular Biology*, February 2002, p. 1218-1232, f. 7. Vol. 22, No. 4. © 2002 American Society for Microbiology; Figure 16.5: © AP/Wide World Photos; Figure 16.20: © Dr. Kenneth Greer/Visuals Unlimited

Chapter 17

Opener: © Matt Meadows/Peter Arnold, Inc.; Figure 17.3: Perry, P. & Wolff, S. New Giemsa method for the differential staining of sister chromatids. *Nature* Vol. 251 (Sep. 13 1974) f. 2. p. 157. © Macmillan Magazines Ltd.; Figure 17.4: © Dr. John D. Cunningham/Visuals Unlimited

Chapter 18

Opener: © Argus Fotoarchiv/Peter Arnold, Inc.; Figure 18.3: Cohen et al. Construction of Biologically Functional Bacterial Plasmids In Vitro. *PNAS* 1973 Vol. 70, f. 4, p. 3243

Chapter 19

Opener: © Najlah Feanny/CORBIS SABA; Figure 19.3: © M. Greenlar/Image Works; Figure 19.4: R. L. Brinster and R. E. Hammer, School of Veterinary Medicine, University of Pennsylvania; Figure 19.6: © Yoav Levy/PhotoTake; Figure 19.7: © Jack Bostrack/Visuals Unlimited; Figure 19.14: Courtesy, Monsanto; Figure 19.15: © Richard T. Nowitz/Phototake; Figure 19.16: © Leonard Lessin/Peter Arnold, Inc.

Chapter 20

Opener: © Dr. Peter Lansdorp/Visuals Unlimited; Figure 20.4: From Ried, T., Baldini, A., Rand, T.C., and Ward, D.C. Simultaneous visualization of seven different DNA probes by in situ hybridization using combinatorial fluorescence and digital imaging microscopy. Proc. Natl. Acad. Sci. USA. 1992 Feb 15; 89(4):1388-92; Figure 20.13a: From Bruce Birren and Eric Lai. Pulsed Field Gel Electrophoresis, A practical guide. Academic Press, Inc. Figure .1, p. 109. © Academic Press, Inc.; Figure 20.13b: From Peter D. Butler and E. Richard Moxon. A physical map of the genome of Haemophilus influenzae type b. *J Gen Microbiol*. 1990 Dec; 136 (Pt 12):2333-42

Chapter 21

Opener: © Alfred Pasieka/Photo Researchers, Inc.; Figure 21.3: From Joseph L. DeRisi, Vishwanath R. Iyer, Patrick O. Brown. Exploring the Metabolic and Genetic Control of Gene Expression on a Genomic Scale. *Science* 278: p. 680-6 © AAAS; Figure 21.4b: © Medical School. University of Newcastle upon Tyne/Simon Fraser/Photo Researchers, Inc.

Chapter 22

Opener: Baylor Coll. of Med. / Peter Arnold; Figure 22.1 top: © Dr. P. Marazzi/Photo Researchers; Figure 22.1 middle: © Hiroya Minakuchi /Seapics.com; Figure 22.1 bottom: © Mitch Reardon/Photo Researchers, Inc.; Figure 22.4: © Hulton-Deutsch Collection/Corbis; Figure 22.8: From Geoffrey M. Cooper, Sharon Okenquist & Lauren Silverman, Transforming activity of DNA of chemically transformed and normal cells. *Nature* Vol. 284, 3 April 1980, pg. 420. © Macmillan Magazine Ltd.

Chapter 23

Opener: © Dr. Dennis Kunkel /Visuals Unlimited; Figure 23.2: Courtesy of E. B. Lewis, California Institute of Technology; Figure 23.7b&c, Figure 2.9(1): Christiane Nusslein-Volhard, Development, Supplement 1, 1991; Figure 23.9(2-4): Jim Langeland, Steve Paddock and Sean Carroll/University of Wisconsin - Madison; Figure 23.11a: © Juergen Berger/Photo Researchers, Inc.; Figure 23.11b: F. R. Turner, Indiana University; Page 645 (both): Horvitz, H.R. & Sulston, J. (1980) Isolation and genetic characterization of cell lineage mutants of the nematode Caenorhabditis elegans. *Genetics* 96, 435-454. Fig. 1a&b (1980); Figure 23.16a: Photodisc/Object Series/Vol. 50; Figure 23.18: © Jeremy Burgess/Photo Researchers, Inc.; Figure 23.21a-c: Elliott Meyerowitz and John Bowman 1991 Development Vol. 112:1-20; Page 660: © Stephen Small/New York University

Chapter 24

Opener: PhotoDisc/Vol. 74; Figure 24.1a: From Albert & Blakeslee, Corn and Man, *Journal of Heredity*, 1914, Vol. 5, pg. 51. By permission of Oxford University Press; Figure 24.3a left: © Photodisc Website; Figure 24.3a right: © Photodisc Website; Figure 24.10 top left: © Henry Ausloos/Animals, Animals/Earth Scenes; Figure 24.10 top right: © Roger Tidman/Corbis; Figure 24.10 bottom left: © Philip Gould/Corbis; Figure 24.10 bottom right: © David Turnley/Corbis; Figure 24.11(1): © Inga Spence/Visuals Unlimited; Figure 24.11(2): © Nigel Cattlin/Photo Researchers, Inc.; Figure 24.11(3): © Valerie Giles/Photo Researchers; Figure 24.11(4-7): © Michael P. Gadomski/Photo Researchers, Inc.

Chapter 25

Opener: © PhotoDisc Website; Figure 25.1: © COLOR-PIC/Animals, Animals/Earth Scenes; Figure 25.2(all): © Geoff Oxford; Figure 25.8b: © Corbis/R-F Website; Figure 25.12a: © O.S.F./Animals, Animals; Figure 25.13a: © Breck P. Kent/Animals, Animals/Earth Scenes; Figure 25.14: © Irene Vandermolen/Photo Researchers, Inc.

Chapter 26

Opener: © Michael & Patricia Fogden/Corbis Images; Figure 26.1 left &right: © Anthony Mercieca/Animals, Animals/Earth Scenes; Figure 26.1 middle: © Alan G. Nelson/Animals, Animals/Earth Scenes; Figure 26.3 left: © Paul & Joyce Berquist/Animals, Animals/Earth Scenes; Figure 26.3 right: © Gerald and Buff Corsi /Visuals Unlimited; Figure 26.16: From Michael Hofreiter, David Serre, Hendrik N. Poinar, Melanie Kuch & Svante Pääbo, Ancient Data (2001) *Nature Reviews Genetics*, Volume 2, p357, fig. 3. © Macmillan Magazine Ltd.

Appendix

Figure A.1a: © Michael Gabridge/Visuals Unlimited; Figure A.1b: © Fred Hossler/Visuals Unlimited; Figure A.8: © Richard Wehr/Custom Medical Stock Photo

Line Art

Chapter 1

Figure 1.1a,b: Source: U.S. Department of Energy Human Genome Program, www.ornl.gov/hgmis.

Chapter 9

Figure 9.19: From R. E. Dickerson, "The DNA Helix and How It Is Read," in *Scientific American*, December 1983, pp. 100–104. Illustrations by Irving Geis. Rights owned by Howard Hughes Medical Institute. Not to be reproduced without permission.

Chapter 12

Figure 12.10: From Darst, "Current Opinion," in *Structural Biology*, Vol. 112, p. 157. Copyright 2001 Elsevier Science. Reprinted by permission.

Chapter 14

Figure 14.10: Reprinted with permission from M. Lewis et al., *Science*, Vol 271, pp. 1247–1254, March 1, 1996. Copyright 1996 American Association for the Advancement of Science.

Chapter 20

Figure 20.1: Reprinted with permission from R. D. Fleishman, *Science*, Vol. 269, p. 28. 1995. Copyright 1995 American Association for the Advancement of Science.

Chapter 26

Figure 26.10: From Olsen and Woesl. *FASEB Journal*, Vol. 7, pp. 113–123, 1993. Permission cleared through Copyright Clearance Center.

INDEX

::